W9-BWV-960

A Guide to the Birds of Costa Rica

A GUIDE TO THE BIRDS OF COSTA RICA

by **F. GARY STILES** *and*

ALEXANDER F. SKUTCH

Plates by

DANA GARDNER

Comstock Publishing Associates

A DIVISION OF ***Cornell University Press***

ITHACA, NEW YORK

First published 1989 by Cornell University Press.
Fourth printing 1994.

Printed in the United States of America
Color plates printed by Brodock Press Inc., Utica, New York.

∞ *The paper in this book meets the minimum requirements of the American National Standard for Information Sciences—Permanence of Paper for Printed Library Materials, ANSI Z39.48-1984.*

Library of Congress Cataloging-in-Publication Data

Stiles, F. Gary.
A guide to the birds of Costa Rica.

Bibliography: p.
Includes index.
1. Birds—Costa Rica—Identification. I. Skutch, Alexander Frank, 1904- . II. Title.
QL687.C8S75 1989 598.297286 88-43444
ISBN 0-8014-2287-6 (alk. paper)
ISBN 0-8014-9600-4 (pbk. : alk. paper)

Contents

The color plates of the birds follow p. 242.

Preface **vii**

Geography and Climate **1**

Map 1. Costa Rican landforms 2

Map 2. Cities, roads, and birding localities 4

Avian Habitats in Costa Rica **21**

The Costa Rican Avifauna **34**

Bird Conservation in Costa Rica **44**

Format and Content of Family and Species Accounts **51**

An Illustrated Glossary of Anatomical Terms Used in the Text **58**

Family and Species Accounts **63**

Birding in Costa Rica **467**

Some Costa Rican Birding Localities **470**

Map 3. Birding localities in the Valle Central 471

Bibliography **478**

Index **481**

Preface

Costa Rica may be the only country in the world to have so many bird species and habitats accessible within such a small area. No larger than West Virginia, Costa Rica boasts an avifauna of more than 830 species, more than in all of North America north of Mexico. Habitats include two quite different seacoasts, highland and lowland forests, subalpine páramos, marshes, rivers, lagoons, and forested swamps as well as a wide variety of altered or man-made habitats, including coffee and cacao plantations, pastures and savannas, rice paddies, and areas of secondary growth. Within two hours' drive of San José, one can watch quetzals in highland forests, antbirds in lowland forests, or shorebirds and ibises in mangrove swamps. A magnificent system of national parks strives to protect this natural wealth against the ever-increasing pressures of deforestation and other forms of exploitation.

As human pressures on the land mount, however, the growing cost of preserving natural areas can be justified only from a knowledge and appreciation of their moral, social, and economic value. In the economic sense the rich avifauna of Costa Rica is a potential tourist attraction that has scarcely been tapped. Far more than hotels, beaches, or shops, the Costa Rican avifauna offers the informed visitor something truly beautiful and truly Costa Rican. And the members of this avifauna most uniquely Costa Rican, and most attractive to the serious observer, are the birds of the forest. Many birders will wish to visit Costa Rica precisely in order to become acquainted with these birds in their natural setting. We hope that this book will encourage such visits as well as increase visitors' knowledge of, and respect for, the country's natural heritage. We also hope that this book will introduce more Costa Ricans to the pleasures and challenges of observing the birds in their natural habitats.

We have written this guide for people who want to know where a bird lives, how it behaves, what it eats, and how it reproduces. Our treatment of these topics is necessarily brief and incomplete, but we hope we have provided a starting point for more detailed studies.

The guide has a long and checkered history; nearly seventeen years have elapsed between the germinal idea and the final product. The undertaking was begun in 1972 by Stiles, Skutch, Lloyd F. Kiff, and artist Barry MacKay. In the first few years, the only member of the group able to devote much time to it was Skutch; other duties and commitments led to the withdrawal of Kiff, MacKay, and a latecomer to the project, Susan M. Smith, who had joined us in 1973, contributing expertise on shorebirds and migrants. Dana Gardner joined the team in 1976 and became sole artist in 1979. In that year we were planning a handbook on Costa Rican birds, with a field guide to follow, but the magnitude of the task and the departure of several associates led to a major retrenchment in 1982, when we decided to devote all our efforts to the field guide.

We certainly do not claim that this guide is in any sense the "last word" on Costa Rican birds. We are all too well aware of the many gaps in our knowledge of even the basic natural history and breeding biology of a very large number of species. One of our major objectives is to draw attention to at least some of these gaps, with the hope that the frequent recurrence of such terms as "unknown," "undescribed," or "unconfirmed" will stimulate others to help fill in the blanks. Even many scientists are unaware of how little is known about the biology of the vast majority of tropical birds. Anyone who watches birds in a tropical country such as Costa Rica has a real chance to add to our knowledge. Should you encounter a hitherto undescribed behavior or nest, for example, be sure to take careful notes and then communicate your observations to an ornithological journal. Perhaps the best measure of the effectiveness of this guide will be the extent to which it spurs the collection and publication of reliable new information on the fascinating, ever-surprising birds of Costa Rica and other Neotropical countries.

Many people have helped us with this guide over the years. Colleagues, birdwatchers, and students have generously shared their information on occurrence, nesting, habits, and voices of Costa Rican birds. We are especially grateful to John Arvin, Gilbert Barrantes, Alan Brady, Rafael Campos, Michael Fogden, Frank Gill, Carlos Gómez, John W. Hardy, Carlos Herrera, David Hill, Steven Hilty, Steven Howell, Daniel Janzen, Joseph Jehl, Pedro Jordano, Louis Jost, Frank Joyce, Lloyd Kiff, Christopher Leahy, Keith Leber, Douglas J. Levey, T. James Lewis, Bette Loiselle, David McDonald, Barry MacKay, Manuel Marín, Paul Opler, Ana Pereira, Simon Perkins, Terry Pratt, George Reynard, Robert Ridgely, Julio Sánchez, Thomas Sherry, Susan Smith, Ricardo Soto, Peter Vickery, Douglas Wechsler, Bret Whitney, Thomas Will, Edwin Willis, David Wolf, and Larry Wolf. Francisco Delgado, Robert Ridgely, and Neil Smith provided helpful information on Panamanian birds; Thomas Howell and Juan Carlos Martínez furnished data for Nicaraguan birds. We had useful discussions on taxonomic and distributional questions with Eugene Eisenmann and Allan Phillips. For assistance and companionship in field and museum, we thank especially Nidia Arguedas, Rafael Campos, Isidro Chacón, Rolando Delgado, Philip DeVries, Carlos Gómez, Lloyd Kiff, Manuel Marín, Andrés Román, Loreta Rosselli, Julio Sánchez, Pamela and Edwin Skutch, Susan Smith, Erland Wimmer, and Larry Wolf. Lloyd Kiff supplied many useful references, and the Western Foundation of Vertebrate Zoology provided equipment that greatly expedited much fieldwork; initial worksheets were prepared by Joanna Barnes. Loreta Rosselli provided valuable help with the preparation of the final manuscript and index. For logistical assistance in Costa Rica, we thank the Servicio de Parques Nacionales, the Tropical Science Center, and the Organization for Tropical Studies. We acknowledge a special debt to the staff of the OTS, particularly Donald Stone and Charles Schnell, for generous help with supplies and communications on numerous occasions. For hospitality in the field, we are grateful to Amos Bien, Darryl and María Cole, William Crawford, Consuelo Rothe de Fernández, R. A. Fernández R., Rafael A. Fernández S., Walt and Elsie Fiala, Leslie Holdridge, Charles H. and Dorothea Lankester, Rafael and Cecilia Ramírez, Theodore and Lois Reynolds, Carlos Víquez, and the many Costa Ricans who over the years have allowed us to observe, camp, and work on their land. Skutch owes a special debt to Juan Schroeder F. for friendship, help, and sound advice during his early years in the Valle del General. Stiles thanks in particular Rafael Chavarría for sharing advice and local lore at Finca La Selva, during his first years in Costa Rica.

For financial support of fieldwork and writing, we thank the Frank M. Chapman Memorial Fund of the American Museum of Natural History, CONICIT, and the Vicerrectoría de Investigación, Universidad de Costa Rica. For the many courtesies rendered, Stiles is grateful to the curators and staffs of the American Museum of Natural History, the British Museum (Natural History), the Field Natural History Museum, the University of Michigan Museum of Zoology, the Museum of Vertebrate Zoology, the Dickey Collection of the University of California at Los Angeles, the U.S. National Museum, and the Western Foundation of Vertebrate Zoology.

Specimens were graciously lent to Dana Gardner by the following institutions: the Western Foundation of Vertebrate Zoology, the Universidad de Costa Rica, the Museo Nacional de Costa Rica, the University of Minnesota, the University of Michigan, Louisiana State University, the University of California at Los Angeles, and the U.S. National Museum.

Gardner is greatly indebted to Mr. and Mrs. Henry B. Guthrie, the Western Foundation of Vertebrate Zoology, and the Lida Scott Brown Fund, University of California at Los Angeles, for financial support. Gary Stiles, Lloyd Kiff, and Jon Dunn provided helpful criticism of the plates. For other help, Gardner thanks the Western Foundation of Vertebrate Zoology, Rolando Delgado M., and Frank P. Smith.

We are particularly grateful to the many friends, scientific colleagues, students, and birdwatchers who have encouraged us to persevere with the at times seemingly interminable task of completing this book; Stiles acknowledges in particular the effective pinpricks of Philip DeVries. James Clements of the Ibis Publishing Company supplied valuable encouragement during the preparation of the manuscript but was unable to follow through with plans to publish this guide. At Cornell University Press, Robb Reavill patiently answered our queries, calmed our fears, and generally kept things moving during the sometimes chaotic transfer of the project from Ibis Press. Helene Maddux shepherded the final manuscript through the production process. We appreciate the help of Nabia Bohum during the final proofreading.

Finally, we take special pleasure in acknowledging our enormous debt to Lynne Hartshorn. With the able assistance of Tony, Thyra, and Tiki, she typed and proofread the entire manuscript several times. Her careful and capable editing is directly responsible for much of the guide's internal coherence and consistency. The book's quality would have suffered seriously without her dedicated help.

F. GARY STILES
ALEXANDER F. SKUTCH
DANA GARDNER

San José, Costa Rica

A Guide to the
Birds of Costa Rica

Geography and Climate

Costa Rica's geography is dominated by two coasts and by mountains that cover more than half the land area, greatly influencing the climate. Variations in temperature and rainfall and a wide range of elevations combine to produce the rich and varied vegetation that supports the country's wealth of birds.

In this section we briefly describe Costa Rica's landforms, vegetation, and climates, in effect setting the stage for the birds. The geographical terms and regional names presented here are illustrated in Maps 1 and 2 and will be used throughout the book to describe the distributions of the birds. We recommend that people planning extensive field trips purchase more detailed maps. Excellent topographic maps of Costa Rica, on various scales, are available at the Universal, López, and Lehmann bookstores in the center of San José (gas stations in Costa Rica do not offer road maps). These maps are in Spanish, with metric units. We use the Spanish regional names in this guide to help the reader interpret the maps and ask directions. Graphs for converting metric to English units appear on the inside of the cover.

Coasts and Islands

The Caribbean coast is smooth and monotonous for most of its length, with steep sandy beaches flanked by scrub and coconut palms (Figure 1). Relief is afforded only by limestone outcrops and promontories around Puerto Limón and Cahuita and south of Puerto Viejo. The only island of any size is Isla Uvita, off Limón. There are almost no tides, and there are only three small, isolated mangrove swamps, the largest lying just north of Moín. Behind the immediate coast and separated from it by swamp forest (now being cleared for rice fields and pastures in many areas) lies the intracoastal (inland) waterway, a series of natural lagoons, sloughs, and rivers joined by artificial canals to form the major artery of coastal transportation between Moín and the Río San Juan. Waterbirds are rather sparse along this coast, with appreciable concentrations only at the river mouths, or *barras*, and in Limón harbor; the only small seabird colony (of Brown Boobies) is on a small islet next to Isla Uvita.

The Pacific coast is wholly different: the coastline itself is irregular and varied; rugged, rocky headlands alternate with broad, sandy beaches (Figure 2); mangrove swamps are common. The strong tidal fluctuations produce a broad intertidal zone that offers excellent foraging for shorebirds and other waterbirds, especially on the mudflats exposed at low tide in the two major bays, the Golfo de Nicoya and the Golfo Dulce. Scattered around the former are numerous salt-collecting ponds, or *salinas*, to which thousands of shorebirds—gulls, terns, and other birds—repair at high tide to rest or to sleep. Minor upwelling along this coast provides abundant food for many seabirds. Notable concentrations of shorebirds and herons are also found at many river mouths, especially where these are flanked by mangroves.

Along the Pacific coast are numerous islands, ranging from small stacks and pinnacles of naked rock to wooded islands with varied terrain, several square kilometers in area. The largest islands, hilltops of a sunken river basin, occur in the Golfo de Nicoya;

Maps, photos, and glossary figures by F. Gary Stiles, except for Figure 17G by L. Rosselli.

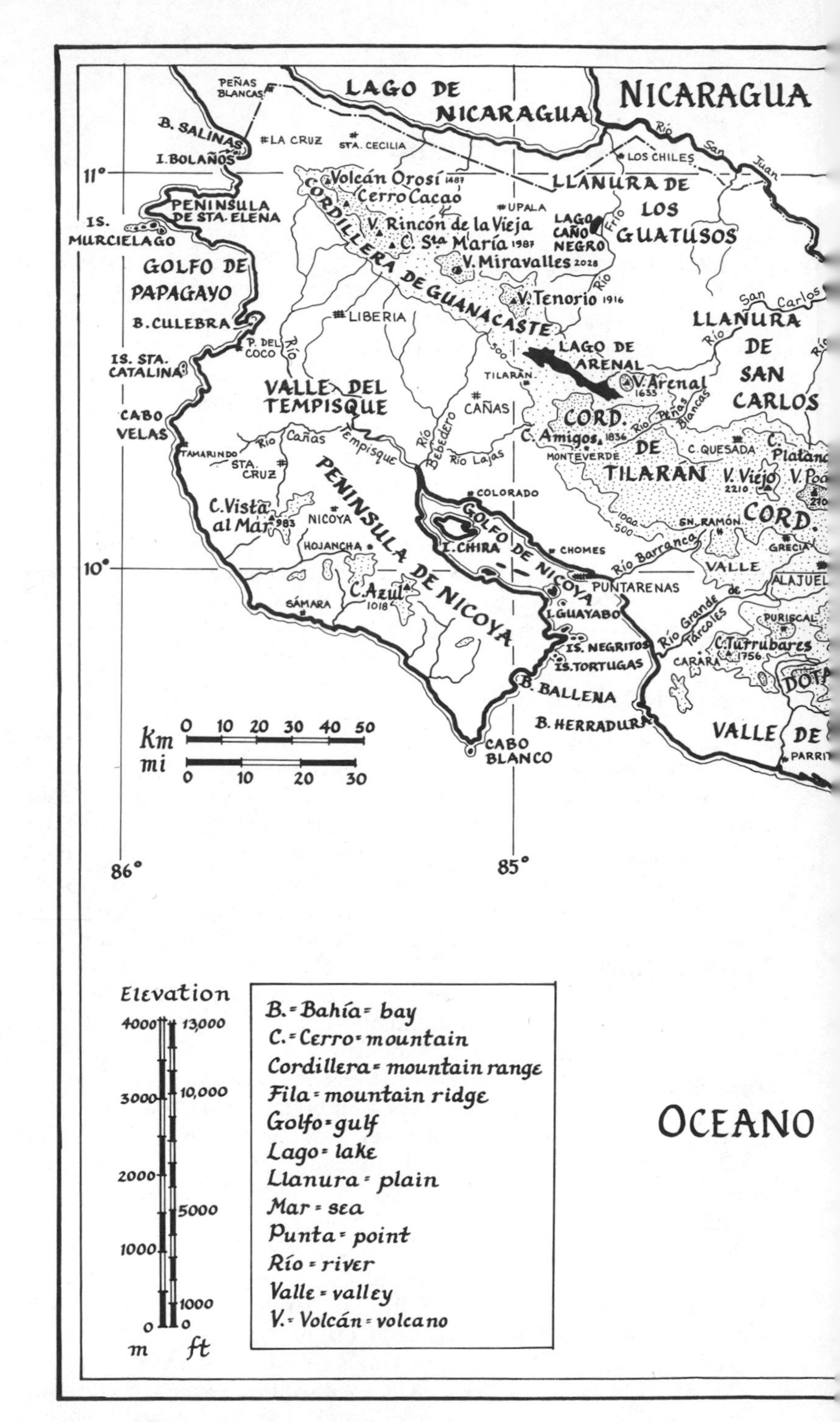

Geographical regions and major topographic features and landforms of Costa Rica.

MAP 1
83°
MAR CARIBE
Río San Juan
BARRA DEL COLORADO
LLANURA DE TORTUGUERO
TORTUGUERO
INLAND WATERWAY
Sarapiquí
PUERTO VIEJO
Río Tortuguero
Río Frío
Río Sucio
Río Jiménez
GUÁPILES
Reventazón
Pacuare
SIQUIRRES
Barva
2906
CENTRAL
V. Turrialba
3328
V. Irazu
3432
HEREDIA
CENTRAL
SAN JOSE
CARTAGO
PARAISO
TURRIALBA
Chirripó
PUERTO LIMÓN
MOIN
I. UVITA
Banano
Río Estrella
VALLE DE LA ESTRELLA
CAHUITA
PTA. CAHUITA
PTO. VIEJO
BRIBRI
Río Telire
Río Sixaola
SIXAOLA
CORDILLERA DE TALAMANCA
STA. MARÍA
C. de la Muerte
3491
3820
C. Chirripó
VALLE DE TALAMANCA
SN. ISIDRO
2000
1000
500
3280
C. Durika
C. Kamuk
3554
VALLE DEL GENERAL
BUENOS AIRES
Río General
FILA COSTEÑA
PANAMA
BAHÍA DE CORONADO
PALMAR
Río Grande de Térraba
VALLE DE COTO BRUS
Río Cotón
Río Coto Brus
C. Pando
2468
SAN VITO
Río Sierpe
CAÑAS GORDAS
I. DEL CAÑO
RINCÓN
GOLFO DULCE
GOLFITO
VILLA NEILY
84°
PENINSULA DE OSA
PTO. JIMÉNEZ
VALLE DEL COTO
PASO CANOAS
PACIFICO
PUNTA BURICA
8°

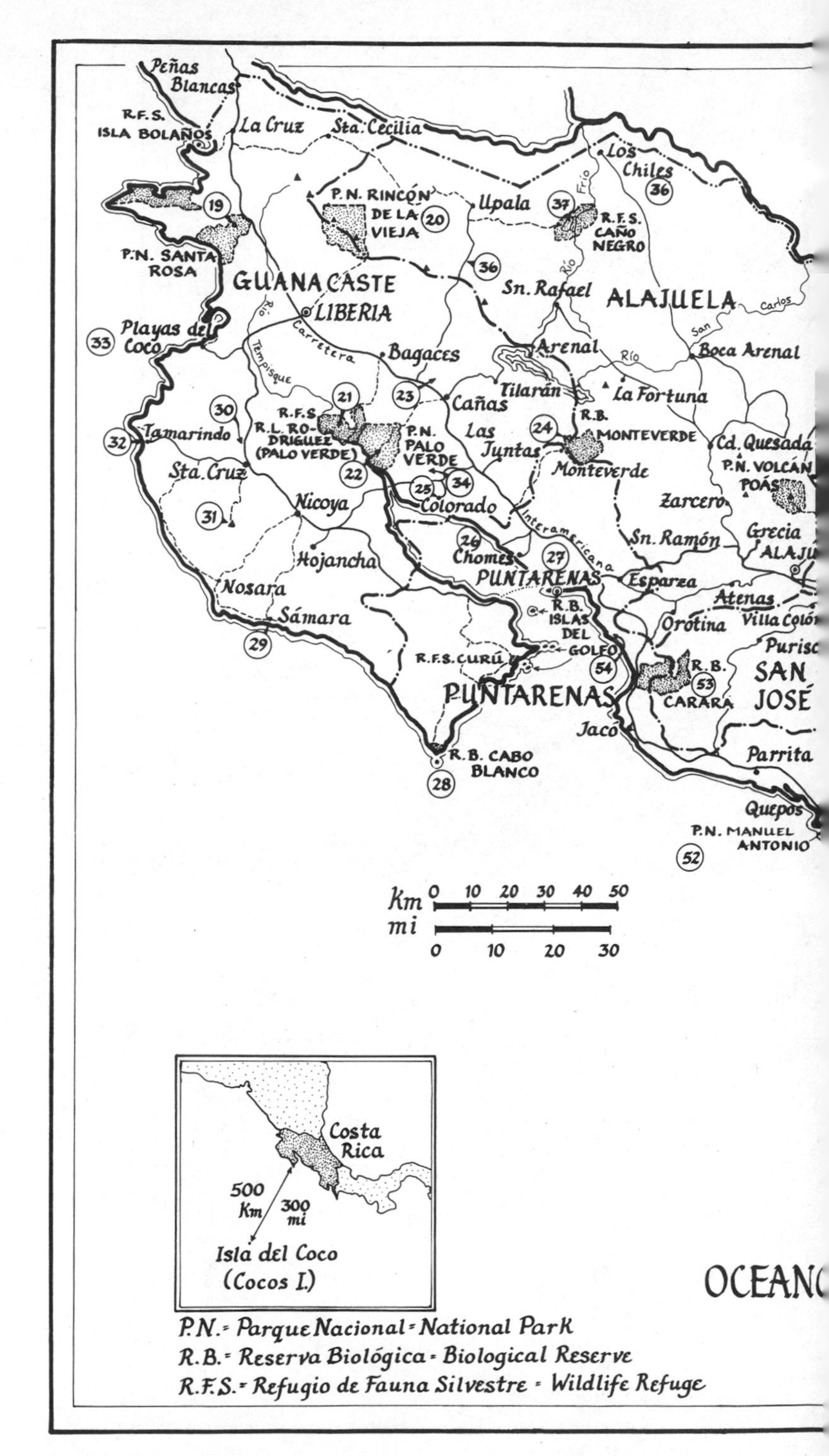

Provinces, cities and major towns, main highways, and protected areas of Costa Rica. Circled numbers refer to birding localities (see pp. 470–477 for descriptions of these localities).

MAP 2
ICARAGUA
Barra del Colorado
39
HEREDIA
LIMÓN
Tortuguero
P.N. TORTUGUERO
38
MAR
CARIBE
Sarapiquí
Puerto Viejo
R.B. LA SELVA
40
Cariari
41
Río Frío
42
46
Guácimo
Reventazón
56
Guápiles
P.N. BRAULIO CARRILLO
Siquirres
55
HEREDIA
PTO. LIMÓN
Tres Ríos
49
Moín
SAN JOSÉ
CARTAGO
Turrialba
57
Paraíso
Orosi
CARTAGO
62
Cahuita
Empalme
P.N. CAHUITA
58
50
Pto. Viejo
P.N. CHIRRIPÓ
R.B. HITOY-CERERE
Bribrí
Manzanillo
Sta. María
51
61
59
SAN JOSÉ
60
63
LIMÓN
Sixaola
P. N. LA AMISTAD
San Isidro
PANAMA
Dominical
Río General
Buenos Aires
67
69
Palmar
Río Cotón
Río Gde. de Terraba
70
San Vito
PUNTARENAS
Sabalito
68
Cañas Gordas
64
Rincón
Golfito
Villa Neily
65
66
P.N. CORCOVADO
Pto. Jiménez
Paso Canoas
Laurel
PACIFICO

Figure 1. Northern Caribbean coast, looking south from Cerro Tortuguero: inland waterway left and right of center, swamp forest in foreground. Note lack of relief.

Figure 2. Pacific coast: Playa Naranjo, Parque Nacional Santa Rosa. Broad sandy beach, river mouth, and estuary with some mangroves; rocky headlands in background.

they are the only ones with permanent fresh water and human settlement. A number of the smaller, uninhabited islands all along the coast support breeding seabirds, mostly Brown Boobies. Isla Guayabo, inside the Golfo de Nicoya, has a large colony of Brown Pelicans. Herons breed on a few small islets.

Lying in the eastern Pacific Ocean some 300mi (500km) south of Puntarenas, Isla del Coco (Cocos Island) is notable for its precipitous terrain and high rainfall (Figure 3). With only a few acres of (swampy) level ground in its entire 10mi^2 (25km^2), Cocos has never supported more than intermittent, marginal human habitation in spite of its abundant fresh water. The dense forests that cover most of the island are diverse in aspect but contain relatively few species of plants or land birds, most of which are endemic. Many seabirds breed in the treetops as well as on offshore islets.

Mountains and Valleys

The backbone of Costa Rica is a series of mountain ranges running from northwest to southeast and becoming progressively larger, higher, and geologically older as they near Panama. The northernmost range, the Cordillera de Guanacaste, consists of four volcanic massifs (Figure 4) separated by low, windswept passes. Although these massifs have eroded to varying degrees, they all preserve at least some vestiges of a conical shape, with steep slopes on their single or double peaks. Rincón de la Vieja, the only active Guanacaste volcano, has a small, active crater alongside its main peak. Except for extensive barren ash fields on Rincón de la Vieja, the upper slopes of all the Guanacaste volcanoes are covered by dense, tangled, moss-festooned cloud forest—a true elfin forest no more than 10ft (3m) high where it is exposed to strong winds. As we descend toward the Pacific, the climate of these volcanoes becomes drier. The lower slopes support a beautiful moist forest with tall, large, small-leaved trees, few epiphytes, and a very open understory of small-leaved shrubs. Farther down, this vegetation merges into the deciduous forests and savannas of lowland Guanacaste. The lower Caribbean slopes of these volcanoes support wet, lush forests that grade into tropical evergreen forests in the lowlands and in the passes, which provide a corridor that enables Caribbean flora and fauna to extend locally onto the Pacific slope. Despite considerable deforestation of the Pacific sides of the volcanoes and in the passes, the forests are mostly intact above 2500ft (750m); on the wet Caribbean slope, deforestation is slight but advancing. On the lower Pacific faces of most of these volcanoes occur old lava flows that have probably never been wooded. One may find Rock Wrens and Botteri's Sparrows at their southern geographic limit on these barren, grassy, boulder-strewn slopes.

South of a series of low hills and the deep valley of the Lago de Arenal (now much enlarged since the damming of the Río Arenal in the 1970s) lies the Cordillera de Tilarán, a broader range whose highest peaks reach 6000ft (1850m). On the Caribbean slope of this range is a lush rain forest. Windswept elfin forests and cloud forests dominate the continental divide and higher peaks (Figure 5). Continuing down the Pacific slope, one encounters progressively drier forests and pastures until the mountains drop off abruptly to the mostly deforested foothills and lowlands. To the east of this cordillera lies Volcán Arenal, a nearly symmetrical volcanic cone that exploded in 1968 after centuries of silence. This volcano and its lower neighbor, Cerro Chato with its lovely crater lake, are essentially southern outliers of the Cordillera de Guanacaste.

The Cordillera de Tilarán is itself the northern extension of the Cordillera Central, a group of four volcanic massifs that together form the northern and eastern walls of the central intermountain valleys. Much taller and older than the Guanacaste volcanoes,

Figure 3. Isla del Coco (Cocos Island). Note steep, heavily wooded terrain.

Figure 4. Volcán Orosí, the northernmost Guanacaste volcano, as seen from Hacienda Los Inocentes. Note conical form, heavily wooded slopes, clearings, and cattle ranch below.

Figure 5. Windswept elfin forest along the continental divide of the Cordillera de Tilarán above Monteverde.

these central massifs do not rise to symmetrical cones but have small cones and sometimes enormous craters on their rather flat tops. Progressively higher from west to east are four massifs that include the active volcanoes Poás and Irazú, each with extensive, desolate expanses of ash and cinders around their craters. Both Poás and Barva have cold, clear lakes in extinct craters. Fumaroles smoke in the small, elfin forest-covered cinder cone of Volcán Turrialba, the easternmost part of the Irazú massif.

All these massifs slope gently down to the intermountain valleys on the south; to the north and east they descend to the Caribbean lowlands in long, steep-sided ridges separated by deeply incised valleys that are often spectacular canyons and gorges (Figure 6). Where it has not been altered by volcanic activity, the vegetation of the upper levels of the Cordillera Central is predominantly oak forest, which on the southern slopes, toward the central valleys, has been mostly replaced by pastures for dairy cattle down to about 5500ft (1700m); much of the land down to the valley floors is in coffee plantations. Many vegetables, especially potatoes, are grown on Volcán Irazú above Cartago. Narrow fingers of forest stretch down toward the central valleys along streams, and coffee plantations are shaded by trees and banana plants, providing habitats for a variety of birds. (Unfortunately, these old-style plantations are being replaced by monocultures with little shade that use chemicals heavily for disease control, with distinctly unfavorable consequences for the birds.) Forests are much more continuous on the wetter northern and eastern slopes of this cordillera, but they are rapidly being replaced by low-grade cattle pastures, especially in the foothills.

In the geographic center of Costa Rica are two intermountain valleys separated by the low Ochomogo pass at the continental divide: the Valle Central on the Pacific slope and the much smaller Valle del Guarco on the Caribbean side, together often known as the

Figure 6. The Caribbean slope of the Cordillera Central, looking up the canyon of the Río Sucio toward Volcán Irazú from Carrillo. The flat-topped steep-sided ridges are old lava flows.

Meseta Central. The majority of Costa Ricans inhabit these valleys; the former contains the capital city of San José, the provincial capitals Alajuela and Heredia, and a number of large towns (Figure 7); the latter's major city is Cartago, at the foot of Volcán Irazú. These valleys have a pleasant climate, with elevations of mostly 3200–4500ft (1000–1400m), rolling terrain, and rich volcanic soils that are unfortunately becoming lost for agriculture because of urban sprawl. Coffee and sugarcane, with smaller areas of other crops and pasture, dominate the rural landscape of the Meseta Central; most of the little woodland that remains is second growth.

The major rivers of the Valle Central drain toward the Pacific in steep-walled wooded gorges. South and west of Ciudad Colón, Atenas, and San Ramón, the valley drops off through rugged hills to the coastal lowlands. The Valle del Guarco is flat and marshy in places; a small but distinctive avifauna, including the Sedge Wren and White-throated Flycatcher, inhabits wet scrub south and west of Cartago. Farther south and east this valley drops off abruptly to the Orosi valley, a major coffee-growing area. Here the Río Grande de Orosi flows down from the Cordillera de Talamanca to the large reservoir and hydroelectric plant at Cachí, where boating is good but waterbirds rather scarce. Below the reservoir the Río Reventazón descends in its broad canyon between rich plantations of coffee and sugarcane to the fertile valley of Turrialba and then to the lowlands at Siquirres. This valley between the Cordillera Central and the Cordillera de Talamanca, which for a century has been the main route between the Meseta Central and the Caribbean lowlands, has long been extensively deforested.

The Cordillera de Talamanca, southern Central America's highest mountain range, occupies most of southern Costa Rica. Its northern outliers, the Cerros de Escazú and the Candelaria and Puriscal ranges, form the southern wall of the Valle Central. Long

Figure 7. Panorama of the Valle Central, with Volcán Poás in the background. A mosaic of coffee plantations (dark green), pastures (light green) and sugarcane (the flat, light green area right of center).

deforested, these rugged hills are now occupied mainly by coffee farms and poor, often severely eroded cattle pastures, with mostly small patches of second-growth woodland.

The main part of the Cordillera de Talamanca consists of several large massifs rising well over 10,000ft (3200m), mostly connected by a high central ridge nearly everywhere above 8200ft (2500m), with major lateral ridges stretching to the northeast and a spur, the Dota Mountains, or Fila de Bustamante, extending toward the Pacific. The natural vegetation of the upper parts of these mountains, to timberline at about 9500ft (3000m), is a magnificent oak forest, whose understory is more or less dominated by bamboos. Above timberline lies the páramo, a chaparrallike formation of tall, gnarled shrubs in sheltered situations, a small, stiff, broomlike bamboo in open sites, and a cycadlike tree fern in the numerous swampy spots. The páramo biota descends to lower elevations where the forest has been cut or burned, as on the northernmost major massif, the Cerro de la Muerte, or "Mountain of Death," so named because many ill-prepared people "caught their death of a cold" while crossing it going to and from the Valle del General to the south, before the highway was built. The upper regions of the Talamancas were glaciated during the Pleistocene, and the highest massif, Cerro Chirripó (12,530ft, 3820m), has several fine glacial valleys with lakes formed behind old moraines (Figure 8). This massif also has the most extensive páramos, which unfortunately were burned in the last few years in fires set by man. The southernmost major massif is Volcán Barú, or Chiriquí, in Panama, the only part of the entire range with evidence of recent vulcanism.

The Caribbean slope of the Talamanca range is extremely wet. Much dense forest covers the rugged terrain, intersected by many deep, swift, rocky rivers that make travel

Figure 8. Glacial valley in the Chirripó massif, Cordillera de Talamanca; Cerro de la Muerte in background. The vegetation is páramo, with oak forest on the darker, lower ridges just visible at center.

difficult. Because of its inaccessibility, this area is the principal remaining stronghold of Costa Rica's Indian population. Along many major rivers, up to 3200ft (1000m) locally, lie extensive areas of second growth of various ages, the product of the Indians' slash-and-burn agriculture. Although extensively forested, the lower ridges of the Talamancas, where Indians hunt, support few large raptors or quadrupeds. In general, the altitudinal zonation of forest types on the Caribbean slope of the Talamancas resembles that on the Cordillera Central but is displaced slightly upward. In the southern Talamancas, many lowland birds and plants thus extend up to 3200ft (1000m) or more, in contrast to 1650–2300ft (500–700m) on the Cordillera Central. Features of special interest include Sabana Dúrika, essentially a large highland swamp at about 7000ft (2150m), which, surprisingly, has many páramo plants and birds, and several isolated lakes, such as Lago Dabagri, with a diverse aquatic biota far from other, similar areas.

On the Pacific slope of the Talamancas, lower spurs extend parallel to the main range southeast from the Cerro de la Muerte and northwest from the Chiriquí massif to form the southern coastal ranges. These low mountains, mostly below 5000ft (1500m) in elevation, support impressive rain forests on their steep Pacific slopes. Deforestation is proceeding rapidly, especially on the drier and gentler interior slopes. The San Vito–Cañas Gordas region on the Panama border, where the coast range joins the Chiriquí massif, is now a major center of coffee cultivation.

Between the coastal ranges and the Cordillera de Talamanca lies a long intermountain valley. The longer northern part, Valle del General, and the southern part, Valle de Coto Brus, were named for the major rivers draining them. Near Paso Real the Río General and Río Coto Brus join to form the Río Grande de Térraba, which flows through a narrow gap in the coastal ranges to the Pacific Ocean near Puerto Cortés. The lower General and Coto Brus valleys, between about Buenos Aires and Potrero Grande, and the slopes of the coastal ranges adjacent to the Río Grande de Térraba, are often collectively called the Térraba region.

Protected by the double rain shadow of the Talamancas and the coastal ranges, these interior valleys have a much more pronounced dry season than do the coastal lowlands. The rolling terrain has been mostly deforested, though scattered patches of beautiful evergreen tropical forest remain, as at "Los Cusingos," near El Quizarrá. In the upper reaches of the General and Coto Brus valleys, much coffee is grown; much of the central part of the Valle del General is planted to sugarcane; below, cattle pastures cover most of the land. In the Térraba region are patches of dry, open savanna with small, twisted trees and scrub. These grasslands probably date to pre-Columbian times; like most pastures, they are perpetuated by burning in the dry season. As deforestation proceeds rapidly uphill along much of the Pacific face of the Cordillera de Talamanca, the beautiful, moist-to-wet forests yield to steep, rapidly eroded pastures. In the Las Tablas area near the Panama border, however, somewhat drier conditions still support a magnificent middle-elevation forest characterized by giant cedro (*Cedrela*) and fig trees, many small-leaved trees and shrubs, and few epiphytes.

In addition to the major mountain ranges, there are smaller and more isolated mountains on the Nicoya and Osa peninsulas. Much of the Península de Nicoya has rough, hilly country, with the major massifs reaching about 3000ft (900–1000m). With heavier rainfall than in the lowlands and frequent cloud cover to mitigate the dry season, the vegetation of these peaks is somewhat lusher than that of the lowlands. Some peaks even support breeding populations of bellbirds! Unfortunately, deforestation of these mountains has been extensive. The highest hills of the much wetter Península de Osa reach 2500ft (750m) and support a beautiful cloud forest, including oak trees, a bamboo-choked understory, and a profusion of epiphytes.

Lowlands

The Caribbean coastal plain is very broad in the north and narrow in the southeast. The wettest area is the extreme northeast, toward Barra del Colorado, with its mean annual rainfall of more than 200″ (5m). To the west and south, the country becomes progressively drier, up to the foothills of the mountains. The prevalent vegetation of the Llanura de Tortuguero of the extreme northeast, as well as in the immediate vicinity of most of the Caribbean coast itself, is evergreen swamp forest, with extensive stands of the huge-leaved *Raphia* palm as well as broadleaved trees. Much of the ground in these forests is under water most or all of the year; decaying vegetation and poor drainage allow tannins to accumulate, producing the characteristic "blackwater." Such swamps are the principal habitat of a few birds, such as the rare Green-and-rufous Kingfisher.

Farther west the dry season becomes more pronounced, especially south of the Lago de Nicaragua, and forests from the Río Frío region westward are partly deciduous at this time. An extensive area of lowlands just south of the lake and along the watershed of the Río Frío (sometimes called the Llanura de los Guatusos) floods during the rainy season and dries almost completely in the dry season. The rains convert dry cattle pastures to grassy marshes, sloughs, and lagoons; the change is most dramatic at Caño Negro, where a lake 2½mi (4km) wide and more than 15ft (5m) deep forms with the rains and contracts into the bed of the Río Frío in the dry season (Figure 9). Much rice is grown on flat, partly flooded land (as at Upala), and cacao is often planted on higher ground.

Figure 9. Lago Caño Negro in the north-central lowlands at the start of the rainy season. Within a month the entire area (back to the trees in the background) will be under water; at the height of the dry season, this area is cattle pasture.

South and east of the Río Frío region lies the rolling Llanura de San Carlos, which in the last twenty years has been almost entirely converted to pasture and sugarcane. Extensive forests still remain, however, in the Poco Sol area, northward toward the Río San Juan.

The Río Sarapiquí has long been a major avenue of boat and barge traffic to the Caribbean coast, and the settlements along its banks are old. Deforestation is proceeding rapidly in the foothills of the Cordillera Central and in both directions away from the river. Eastward, there are extensive banana plantations at Río Frío (not to be confused with the Río Frío region of the north) and around Guápiles; the forests between Guápiles and the Llanura de Tortuguero have been replaced by bananas and pastures in the last fifteen years. The central Caribbean lowlands along the Guápiles-Siquirres-Limón railway, and the coastal lowlands to the south, have long been Costa Rica's main cacao-growing area. Unfortunately, in the old-style plantations, with many canopy trees left for shade, the cacao is prone to fungal disease, which has forced farmers to seek more intensive methods of cultivation (including felling of canopy trees) or to replace cacao with cattle pastures or rice fields—neither change being favorable to the avifauna.

Farther south, the two large lowland valleys, Valle de la Estrella and Valle de Talamanca, were deforested early in this century for planting bananas. The former is still a major center of banana cultivation, but most large plantations in the Valle de Talamanca were abandoned before 1950 because of disease. Today the main inhabitants of this valley are Indians who practice mostly slash-and-burn, shifting cultivation. Major banana plantations around Sixaola are now being converted to oil palm. The coastal hills between Puerto Viejo and Gandoca are still largely forested, and interesting forested swamps exist on Punta Cahuita and near Manzanillo. In general, however, virtually no forest remains on level to gently rolling terrain in the lowlands south of Limón.

On the Pacific side of Costa Rica are two major lowland areas: the dry- to moist-forested northwest, south to the mouth of the Golfo de Nicoya, and, in the extreme south of the country, the much wetter lowlands around the Golfo Dulce and on the Península de Osa. The northern and southern lowlands are connected by a narrow strip of coastal country nearly as wet as the Golfo Dulce region, except in the somewhat drier vicinity of Parrita and Quepos. Climatically, the division between the drier northwest and wetter south falls along the hills that form the southern wall of the watershed of the Río Grande de Tárcoles, from Cerro Turrubares to the Carara region.

The northern Pacific lowlands, or "Pacific northwest," are the driest major region of Costa Rica, with a severe, windy dry season lasting five to six months. The mostly rolling terrain is dissected by rivers whose valleys become increasingly steep-sided and narrow upstream toward the northern ranges and the Nicoya hills. Around the Golfo de Nicoya and in the lower basins of the Río Tempisque and Río Bebedero are flat alluvial plains, broken here and there by sharp limestone hills (Figure 10). Small alluvial plains are also scattered along the outer Pacific coast, but most of the Santa Elena and Nicoya peninsulas are rough and hilly to the ocean. The natural vegetation of most of this region is tropical deciduous forest, with most trees and shrubs leafless during the dry season. Many large trees have a characteristic umbrella shape, and the understory is dominated by tough, often spiny shrubs and vines, including some cacti. In river bottoms and other spots with a high water table, the forest is more evergreen. In moister parts of this region—for example, along the outer Península de Nicoya and south of Puntarenas—the *Scheelea* palm is abundant and the forests are partly evergreen.

In the lower Tempisque and Bebedero basins lie extensive seasonal marshes and lagoons that become largely or wholly dry toward the end of the dry season. In the region as a whole, nearly all the forest has been replaced by extensive cattle pastures, or *sabanas*, maintained by dry-season burning. Particularly in the moister areas, rice, cotton, and sugarcane are grown. More than half of the Tempisque wetlands have been drained in the last fifteen to twenty years, and a major irrigation system is planned to convert much of the remainder to rice and sugarcane fields.

In the Golfo Dulce lowlands of the southern Pacific region, the dry season, although pronounced, usually lasts only two or three months, and the forest is almost entirely evergreen. On the extensive Palmar and Coto plains, the magnificent forests have been felled, and they are rapidly diminishing in the hilly areas around the gulf and on parts of the Península de Osa. The flat lowlands are devoted mainly to rice and bananas, which in many areas are being replaced by African oil palms. The hills, which are usually covered with pasture after supporting a few crops of corn and beans, are subject to severe erosion. The only major expanse of mostly intact lowland forest in the region is the Corcovado basin (Figure 11), which in addition to truly impressive stands of trees, especially in the Llorona–San Pedrillo area, contains extensive swamps of grasses and *Raphia* palms surrounding a small lagoon. This basin will undoubtedly be the last stronghold of a number of plants, birds, and other animals endemic to southern Pacific Costa Rica and western Chiriquí, Panama.

Climate

Costa Rica has basically two seasons: the dry season, or *verano* (literally "summer"), and the rainy season, or *invierno* (literally "winter"). Much to the confusion of

Figure 10. Aerial view of the Río Tempisque basin at the end of the wet season; limestone hills of the Nicoya peninsula in the background. Note extensive seasonal marshes in foreground.

Figure 11. View of the Corcovado basin, Península de Osa; Laguna Corcovado at center. The only extensive lowland wet forest remaining in southwest Costa Rica.

northern visitors, San José is enjoying its "summer" during the Christmas holiday, and tourists on their summer vacation find Costa Rica well into its "winter." The reason is historical: the Spanish colonists came from a Mediterranean climate of hot, dry summers and wet, cool winters and related the seasons in their new land to the rainfall regime in their old home!

The timing of dry and rainy seasons varies somewhat between the Caribbean and Pacific slopes because of seasonal fluctuations in the strength of the northeasterly trade winds and the interaction of these with the major mountain chains to produce rain shadows, vortices, and eddies. On the Caribbean slope the rainy season usually begins by mid to late April and continues through mid-December in some years, late January in others. It is often interrupted by a brief, rather unpredictable dry spell, or *veranillo*, in about August or September. The wettest months are usually July and November; December is extremely wet in some years, relatively dry in others. A typical wet-season day has a few hours of sunshine in the morning and becomes cloudy by late morning or early afternoon, with rain for most of the afternoon and often continuing into the night. At unpredictable intervals between about September and February, major storms, the *temporales del Atlántico*, bring more or less continuous rain for several days. The most consistently dry months are February and March, when long stretches of sunny days, with some nocturnal or early-morning showers, alternate with occasional temporales.

The northeast trade winds, or *alisios*, are especially strong during the dry season over most of the Pacific slope, except in the south, where the high Cordillera de Talamanca produces a wind shadow–vortex effect, and the main winds come from the Pacific. Typical dry-season weather in the northwestern lowlands and Valle Central is clear and often very windy; no rain may fall for several months. The rainy season usually begins in May, with strong rains in June and a short veranillo in late June and July; the rains increase again through October and generally end by mid to late November. Especially in September and October, major storms of several days' duration, the *temporales del Pacífico*, come from the Pacific Ocean. Typical wet-season weather includes more or less sunny mornings, with afternoon rains that do not usually continue into the night. In the wetter areas of the southern Pacific region, this pattern is modified in two respects: the height of the dry season, in which little or no rain falls, is reduced to two to three months or less in most years, and the afternoon rains of the wet season tend to be longer and heavier.

In the mountains weather patterns are more variable and much affected by local topography. Often the dry season is characterized by long periods of fine, wind-driven rain or "mist" alternating with brief periods of clear, calm, sunny weather when the trade winds decrease temporarily in strength. The wettest areas in Costa Rica are the Caribbean slopes of the mountains, from the foothills well into upper elevations, where even in the dry season long spells of rainy or misty weather are not infrequent. Many of the highest peaks, especially in the Cordillera de Talamanca, are above the level of greatest rainfall; during the dry season, long spells of clear weather often occur, and many rainy-season mornings are sunny. Here the most striking aspects of the climate are the sharp and sudden changes in temperature between sun and shade and the intensity of the ultraviolet radiation in the thin, transparent air (beware of sunburn!).

Temperatures in Costa Rica vary principally with elevation rather than with time of year: the lowlands are always hot, the middle elevations cool, the high mountains cold. Daily variations in wind and cloud cover are also important: in clear weather the temperature climbs higher by day but drops lower at night. In the dry season, nighttime frosts are frequent in the mountains above 8200ft (2500m); they rarely occur in the wet

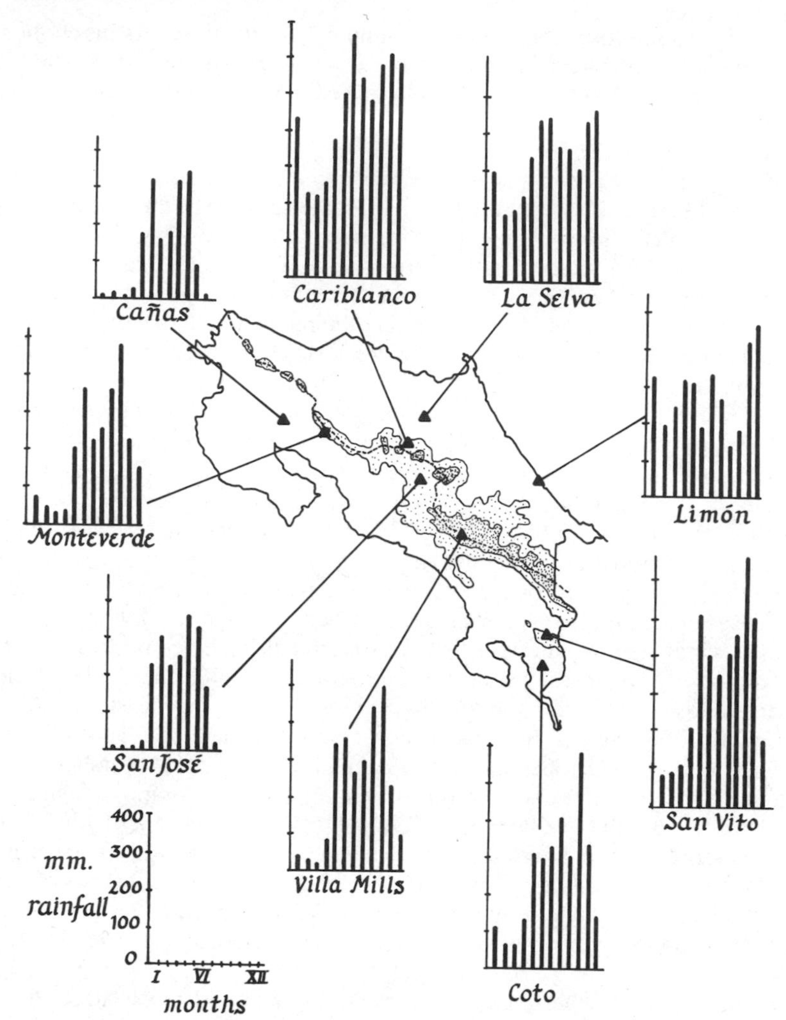

Place	Elevation	MAP[a]	MAT[a]
Cariblanco	3597ft (1090m)	223″ (5675mm)	69°F (20.4°C)
La Selva	300ft (90m)	155″ (3940mm)	76°F (24.3°C)
Puerto Limón	10ft (3m)	132″ (3350mm)	79°F (25.8°C)
San Vito	3350ft (1019m)	157″ (3988mm)	71°F (21.7°C)
Coto	100ft (30m)	118″ (3005mm)	81°F (27°C)
Villa Mills	10,170ft (3100m)	106″ (2690mm)	51°F (10.7°C)
San José	3840ft (1170m)	77″ (1951mm)	69°F (20.6°C)
Monteverde	4525ft (1380m)	96″ (2429mm)	66°F (19°C)
Cañas	145ft (45m)	70″ (1770mm)	82°F (27.8°C)

Note: The map above shows the 1000m and 2000m contours; the broken line represents the continental divide.

Source: Weather Data, Instituto Meteorológico de Costa Rica.

[a]MAP = mean annual precipitation; MAT = mean annual temperature.

Figure 12. Mean monthly rainfall of nine representative Costa Rican localities.

season. Because of differences in cloud cover, localities on the Pacific slope usually average slightly warmer than localities at the same elevation on the Caribbean slope. The only real exception to the rule of little seasonal temperature change occurs during the North Temperate winter months, when the southernmost edges of the great cold fronts penetrate the tropics and bring several days of cooler weather and strong winds locally known as the *nortes*.

These patterns are summarized briefly in the climate diagram shown in Figure 12. We have chosen a representative locality from each of the following areas: Caribbean coast (Puerto Limón), Caribbean lowlands (La Selva), wet middle elevations of the Caribbean slope (Cariblanco), the high mountain peaks (Villa Mills), the Pacific slope of the northern cordilleras (Monteverde), the Valle Central (San José), the northwestern lowlands (Cañas), the Golfo Dulce lowlands of the south (Coto), and the southern Pacific coastal range (San Vito).

Avian Habitats in Costa Rica

In this section we illustrate some of the major habitat types and characteristic plants of Costa Rica, particularly those that might be unfamiliar to visitors from the North Temperate Zone. We first present a series of forest types, for although all of Costa Rica is summarily described as the ''tropical forest biome'' in northern ecology texts and popular accounts, in reality it encompasses more kinds of forests than occur in all of temperate North America. Many Costa Rican birds inhabit only one or a few of these forest types, and the avifauna can change gradually or abruptly as one passes from one forest type to another. The most diverse avifaunas are those of the wet lowland and foothill forests: at sites like La Selva, Corcovado, and Plástico–Rara Avis (see Map 2), 200 or more species may breed, and many others occur as migrants or seasonal visitors. As one moves up the mountain slopes, the size of the avifauna decreases; in the highest oak forests of the Cordillera de Talamanca, only 40–50 species may breed, and fewer than 25 species are found above timberline. Lowland dry forests contain little more than half the number of breeding species as do wet forests, still far more than the breeding avifauna of any temperate-zone forest.

Tropical forests, especially the wetter ones, contain many more microhabitats, types of food, and places to look for food or build nests than do forests at higher latitudes. Many species of tropical birds have specialized anatomy, behavior, or both for exploiting one or more of these components. To help the reader understand the habitat requirements of most Costa Rican forest birds, we illustrate and describe some of the microhabitat types present in a tropical wet forest in Figure 15.

Finally, we treat a representative selection of nonforest and aquatic habitats. Avifaunas of the former are much smaller than those of the forests and change much less with elevation and humidity. For example, one can find most of the same 30–50 species in any area of pasture and scrub throughout the lowlands and middle elevations of Costa Rica. A partial exception is the páramo avifauna of the highest elevations, which contains a number of peculiar species, as well as many from the adjacent highland forests; many of these páramo birds have recently extended their altitudinal ranges downward as the forests have been cut. Costa Rican waterbirds are often more or less specialized for particular aquatic habitats—for example, marshes, rivers, or ponds—but occur in these habitats over a wide range of elevation and humidity conditions.

Figure 13. **Lowland Forests**

A. **Tropical Dry Forest: Palo Verde, Guanacaste; early dry season**. In the prevalent forest type of the northern Pacific lowlands, most canopy trees are leafless in the dry season, and some flower. Many large trees have short trunks and wide, umbrella-shaped crowns; smaller ones are often gnarled. The understory, which, in the photo, has not yet shed its leaves, is dominated by stiff, small-leaved, often spiny shrubs and woody lianas. Terrestrial bromeliads are rather common, but the few epiphytes are mostly cacti. Habitat of Thicket Tinamou, Banded Wren, Black-headed Trogon, and Scrub Euphonia.

B. **Evergreen Bottomland (Moist) Forest: Parque Nacional Santa Rosa**. Areas of the northwestern lowlands where the subsoil remains moist through the dry season support tall evergreen forests with large trees, including milk tree (pale trunk) and chicle (dark trunk). The middle levels are open, the understory a thicket of tall shrubs and small trees, sometimes with palms and broadleaved herbs. Many trunks bear lichens, but epiphytes are still scarce. Habitat of Long-tailed Manakin, Greenish Elaenia, Bright-rumped Attila, Spectacled Owl, Long-billed Gnatwren, and Barred Antshrike.

C. **Lowland Wet, or "Rain," Forest: Sirena, Península de Osa**. In the prevailing forest type of the Caribbean and southern Pacific lowlands, the canopy trees reach 100ft (30m) or more, with emergents such as the "ajillo" tree in the center attaining heights of 150ft (45m) (note the human figure at lower R). Tall, clean boles, often with impressive buttresses, and a wealth of lianas and epiphytes characterize such forests; many palms and broadleaved herbs occur in the understory. Habitat par excellence of antbirds, manakins, toucans, puffbirds, and other Neotropical families.

D. **Gallery Forest: Río Naranjo, Parque Nacional Santa Rosa**. Permanent or temporary streams in the dry lowlands are bordered by an evergreen forest with large ceiba, fig, or sandbox trees. Along the watercourses one finds Rose-throated Becards, Crane Hawks and Common Black Hawks, Muscovy Ducks, Social Flycatchers, and others; many dry forest birds concentrate here at the height of the dry season.

A

B

C

D

Figure 14. **Highland Forests**

A. **Subtropical Moist-Wet Forest, Elevation 3200ft (1000m): Volcán Rincón de la Vieja**. A beautiful forest characteristic of lower middle elevations on the Pacific slope, under somewhat drier conditions than higher up. Note the very clean forest floor with few palms and herbs; the understory of small-leaved woody shrubs; the large size, clean trunks, small leaves, and lack of epiphytes of the canopy trees, among which giant figs are numerous. Habitat of Rufous-and-white Wren, White-throated Robin, Ruddy Woodcreeper, and Golden-crowned Warbler.

B. **Lower Montane Rain or "Cloud" Forest, Elevation 5000ft (1500m): Monteverde, Cordillera de Tilarán**. The prevalent forest of the upper middle elevations. Note the luxuriance of moss, epiphytes, and tree ferns. Large trees reach 100ft (30m) and are so loaded with epiphytes that branchfalls and treefalls are frequent, producing an irregular canopy with many gaps. The understory is composed of soft woody shrubs, large-leaved herbs, palms, and ferns. Habitat of Resplendent Quetzal, Emerald Toucanet, Black Guan, Slate-throated Redstart, many ovenbirds and tanagers, Black-faced Solitaire, and Prong-billed Barbet.

C. **Elfin Moss Forest: summit of Volcán Orosí, Cordillera de Guanacaste**. Similar elfin forest occurs on the windswept summits of the Guanacaste volcanoes and the crest of the Cordillera de Tilarán. The canopy is wind-clipped, usually 15ft (5m) high or less, with a dense, tangled, impenetrable understory. Leaves of most trees and shrubs are thick and leathery, with stiff heavy twigs for wind resistance. Heavy cushions of moss cover nearly every surface. Highland Tinamous, Ruddy-capped Nightingale-Thrushes, and Black-and-Yellow Silky-Flycatchers are some of the birds that live here.

D. **Montane Oak Forest, Elevation 9200ft (2800m): Cerro Chirripó, Cordillera de Talamanca**. The principal forest type of high elevations on the Cordillera Central and Cordillera de Talamanca. The dominant trees are magnificent oaks reaching 125ft (40m) in height; the understory is often a dense growth of *Chusquea* bamboo. Note the abundance of moss festoons and bromeliads with red-pigmented leaves, which presumably afford protection from the intense ultraviolet radiation of these elevations. Habitat of Collared Redstart, Flame-throated Warbler, Buffy Tufted-cheek, Band-tailed Pigeon, and Mountain Robin.

B

A

C

D

Figure 15. **Forest Microhabitats: Tropical Wet, or "Rain," Forest**

A. **Forest Understory: La Selva Biological Station**. Plants of this habitat are adapted to deep shade; note the large-leaved palms and herbs. Shrubs and treelets have simple, dark leaves with pointed "drip tips" to help shed moisture. In mature forest the understory is fairly open, but when a tree falls and sunlight penetrates directly to the understory (as in the left background), dense "jungle" tangles often develop. Habitat of Great Tinamou, many antbirds and manakins, and White-whiskered Puffbird.

B. **Small Forest Stream: La Selva Biological Station**. Along such a natural "edge" the sun's rays reach the ground, favoring luxuriant growths of *Heliconia* (the large leaves in the center) and thickets and vine tangles. Habitat of Chestnut-bellied Heron, Bay Wren, Tawny-crested Tanager, and Buff-rumped Warbler. Male Long-tailed Hermits assemble in leks in the thickets; they and White-tipped Sicklebills visit the flowers of streamside *Heliconia*.

C. **Forest canopy: view into the crown of a giant *Ceiba* near Guápiles, Caribbean lowlands**. In a mature forest, most of the direct sunlight is intercepted here. Large branches are laden with bromeliads, orchids, aroids, ferns, and many other epiphytes; lianas stretch from crown to crown. Many plants and animals flourish in this light, airy world so totally different from the cool, dark understory; canopy tanagers, honeycreepers, parrots, and many other species descend to lower levels only along gaps and edges.

D. **Forest River: Río La Vaca, Golfo Dulce region**. Larger rivers that flood periodically keep land along their banks in a state of early succession. Note the low scrub on the sandbar, and the stand of "caña brava" (giant cane) and *Cecropia* tree at left. When Costa Rica was almost wholly forested, such riverbank vegetation was the main habitat for many now-widespread birds of second growth and open country, such as large flycatchers, seedeaters, and Scarlet-rumped Tanagers.

A
B
C
D

Figure 16. **Nonforest Habitats**

A. **Páramo, Elevation 10,500ft (3200m): Cerro Chirripó, Cordillera de Talamanca**. In this vegetation above timberline on Costa Rica's highest mountains the predominant plant is the stiff, broomlike dwarf bamboo *Swallenochloa*, beneath which grow bunch-grasses, mosses, and many "cushion" plants. In sheltered sites are dense patches of stiff, dense shrubs with wind-clipped tops. Habitat of Volcano Junco, Slaty Flowerpiercer, Sooty Robin, and Red-tailed Hawk.

B. **Young Second Growth along Roadside: near Palmar Sur, southern Pacific lowlands**. Such impenetrable thickets cover roadsides, overgrown pastures, and the like throughout the humid lowlands of Costa Rica. The large-leaved plants are *Heliconia* (wild plantain, or "platanillo") at the center and *Calathea lutea*, or "bijagua," at the right; the tall shrubs with white "rat-tail" inflorescences (behind the *Heliconia*) are of the genus *Piper*; the large-leaved, sparsely branching trees at the rear are *Cecropia*. Note how smaller trees are often overgrown by vines. Habitat of Great Antshrike, Slaty Spinetail, various seedeaters, Scarlet-rumped Tanager and Black-striped Sparrow.

C. **Semi-open: trail through cacao plantation, La Selva Biological Station**. Above a dense understory of cacao trees stands a broken, open canopy of planted "laurel" (*Cordia alliodora*) and remnant forest trees. Old-style coffee plantations have a similar structure: coffee understory, legume trees, such as species of *Inga* or *Erythrina*, in the overstory. Many forest birds, especially those of the canopy, range freely into such habitats, often descending far lower than in the forest itself.

D. **Savanna-Pastureland: near Taboga, Guanacaste**. Such pasture is burned annually to regenerate the grassland and prevent regrowth of woody vegetation. Note the umbrella form of the "cenízaro" (*Pithecellobium saman*) and "guanacaste" (*Enterolobium cyclocarpum*) trees, which are probably remnants of the original dry forest. In such trees many raptors nest, Laughing Falcons call, and Rufous-naped Wrens, Hoffmann's Woodpeckers, and White-throated Magpie-Jays forage.

A

B

C

D

Figure 17. **Aquatic Habitats**

A. **Open Seasonal Marsh: Palo Verde, Tempisque Basin, early dry season**. Extensive areas in this basin flood during the wet season and gradually dry during the rainless season. The trees here are ''palo verde'' (*Parkinsonia aculeata*); also visible are sedges, cattails, water hyacinth, and open water. Vast concentrations of herons, ducks, spoonbills, and other species gather in the remaining wet areas as surrounding lands dry. Such marshes are also important breeding sites for Least Bitterns, gallinules, jacanas, limpkins, grebes, rails, and other birds.

B. **Lowland River: Río Puerto Viejo near La Selva**. Dead trees stranded in the shallow water offer perches for cormorants, kingfishers, and anhingas and nest sites for Mangrove Swallows. Where vegetation hangs into the water, Sungrebes lurk. Widely fluctuating water levels cause frequent landslide scars and undercut banks, in which kingfishers burrow and Southern Rough-winged Swallows nest. The banks are lined with *Cecropia* and ''sotacaballo'' trees (*Pithecellobium longifolium*), which often support the pensile nests of Cinnamon Becards and Scarlet-rumped Caciques.

C. **Highland River: Río La Hondura, ca. 3200ft (1000m) in Parque Nacional Braulio Carrillo**. The trees along such steep, rushing, rocky rivers are often festooned with mosses. Many of the wild plantains (*Heliconia*) and broadleaved shrubs that line the banks have flowers that are pollinated by hummingbirds, such as the *Alloplectus* in the right foreground. Habitat of American Dipper, Sunbittern, Fasciated Tiger-Heron, and Torrent Tyrannulet.

D. ***Raphia* Palm Swamp: Parque Nacional Tortuguero**. Similar blackwater swamps in which tannins from decaying vegetation accumulate in the poorly drained water occur in the Río Frío district, along the Caribbean coast, on the Península de Osa, and locally elsewhere. Here the dominant plant is the *Raphia taedigera* palm (''yolillo''), with huge fronds. Note the dense growth of vines wherever broadleaved vegetation flourishes at a break in the *Raphia* canopy. Habitat of various kingfishers, herons, and other waterbirds.

A
B
C
D

Figure 17, continued.

E. **Swamp Forest: Parque Nacional Cahuita**. On poorly drained land in forested lowlands, where the ground is firmer than in the *Raphia* swamps, one finds wooded swamps with buttressed *Pterocarpus* trees and various palms, all burdened with many lianas. Habitat of Green Ibis, Rufescent Tiger-Heron, and Green-and-rufous Kingfisher.

F. **Sandy Beach at River Mouth: Río Tárcoles, Pacific coast**. Such sandbars at river mouths are often the sites where shorebirds, gulls, terns (such as these Royal Terns), and herons congregate; in contrast, one can walk along miles of unbroken sandy beach and see only an occasional heron or shorebird. Collared Plovers nest amid driftwood and flotsam above high tide, and sometimes Lesser Nighthawks breed here in loose colonies.

G. **Salt-Evaporating Ponds, or Salinas: Salina Bonilla near Colorado, on the Golfo de Nicoya**. In these shallow ponds and mudflats, and along the low earthen dikes, many shorebirds, herons, gulls, and terns concentrate to loaf and sleep at high tide or at night, then repair to the mudflats of the Golfo de Nicoya (hidden behind the mangroves in the background) to feed as the tide recedes. Other shorebirds forage mainly in the salinas themselves. Black-necked Stilts and Wilson's Plovers breed along the dikes.

H. **Mangrove Swamp at Low Tide: Rincón de Osa on the Golfo Dulce**. Similar swamps are frequent around estuaries along the Pacific coast, where the fresh water of the river mixes with tidal salt water, so that salt levels fluctuate. Characteristic trees shown here are red mangrove (*Rhizophora harrisoni*), with spreading stilt roots, and Pacific mangrove (*Pelliciera rhizophorae*), with compact, fluted buttresses. The large white flowers of the latter are the chief source of nectar for the endemic Mangrove Hummingbird. In such swamps Mangrove Vireos, Mangrove Warblers, Panama Flycatchers, and Scrub Flycatchers live, and many waterbirds rest, forage, and nest.

E F G H

The Costa Rican Avifauna

Approximately 840 species of birds have been recorded from Costa Rica and its territorial waters; the inclusion of Cocos Island and the adjacent Pacific Ocean brings the total Costa Rican bird list to more than 850 species. They include some 600 species of permanent residents that have arrived in Costa Rica from diverse sources and at different times in the geological past. A considerable number of species, especially in the highlands, evolved in southern Central America itself. Even now, as the forests that previously blocked their dispersal are destroyed, birds of open country continue to arrive and become permanent residents. In addition, more than 200 species of migrant birds reach Costa Rica each year. Most come from breeding areas in North America, but a few of the seabirds nest as far away as New Zealand, the coasts of Antarctica, or Siberia. Occasionally a few birds stray in from the north, the West Indies, the Pacific Ocean, South America, or even Europe.

Such a diverse avifauna exhibits many interesting anatomical and behavioral adaptations. Time spent watching a bird go about its daily activities will provide many fascinating insights into its uses of and adaptation to its environment—the observer can see how it feeds, breeds, and interacts with other birds of the same or different species and with the rich and varied tropical flora. In this section we discuss some aspects of the evolution, ecology, and behavior of Costa Rican birds, to help those with the interest and patience to take that ''long second look'' to interpret what they see.

Sources and Affinities of Costa Rican Birds

During most of the last 60 million years, the northern half of Central America, south to about central Nicaragua, was a tropical peninsula of the North American continent. The land that was to become southern Central America emerged as a chain of volcanic islands, somewhat east of its present position, about 50 million years ago. Through subsequent epochs, this island arc underwent vulcanism, earthquakes, and successive periods of uplift and erosion; land connections evidently existed intermittently between different islands, with the North American peninsula, and perhaps even briefly with South America. The continuous isthmus joining the North and South American continents, however, formed only within the last 3–5 million years.

Each of the early volcanic islands soon came to support an avifauna with representatives of widespread families that have demonstrated their ability to colonize remote islands, including herons, waterfowl, hawks, rails, pigeons, parrots, cuckoos, swifts, and kingfishers. Their shores were frequented by oceanic birds that could pass freely between the Atlantic and the Pacific through gaps in the archipelago, as could flightless marine animals of many kinds. Some movements of land birds between the northern and southern continents probably also occurred, with such boreal groups as the thrushes and perhaps the jays reaching South America while flycatchers and hummingbirds moved north. Some of the larger and more persistent islands became centers of evolution in their own right, notably the island that later became the modern Cordillera de Talamanca.

With the closure of the last water gaps between islands and the continents, a bridge was formed over which terrestrial plants and animals could pass between North and South America, which soon exchanged their living productions on a vast scale. Costa Rica and Panama, in particular, were enriched by contributions from the very different avifaunas of the two continents. Although interrupted by savannas during drier periods, rain forest has been continuous for much of the time from South America to Guatemala and beyond, especially on the Caribbean slope. Along this route, the peculiarly Neotropical families of largely forest-dwelling birds began to move northward. Tinamous, jacamars, puffbirds, toucans, woodcreepers, ovenbirds, antbirds, manakins, and cotingas spread over the (in a geological sense) recently formed isthmus, where today they diminish in numbers of species from south to north. In Costa Rica, these Neotropical families (except the ovenbirds) are most abundant in the lowland and foothill wet forests of the Caribbean and southern Pacific sectors. The Costa Rican avifauna shows a particularly close affinity to that of northwestern Colombia and the Pacific slope of South America as far as Ecuador because the uplift of the Andes (also geologically quite recent) greatly restricted the exchange of species between Central America and Amazonia.

The more mobile hummingbirds, flycatchers, tanagers, and icterids, probably already represented on the archipelago, became more numerous with the arrival, overland, of representatives of such South American genera as *Phaethornis, Lophornis, Heliodoxa, Attila, Serpophaga, Elaenia, Tangara, Ramphocelus, Tachyphonus*, and *Psarocolius*. The honeycreepers probably came after the land bridge had been completed. The southern continent continues to contribute to the Costa Rican avifauna, for with the destruction of forest barriers, chiefly in recent decades, such open-country birds as Yellow-headed Caracaras, Crested Bobwhites, Smooth-billed Anis, Pale-breasted Spinetails, and Red-breasted Blackbirds have been extending their ranges northward, chiefly along the Pacific side of Costa Rica. The Cattle Egret reached Costa Rica from South America, where it first appeared about a century ago in the Guianas, probably from Africa; in the last fifty years this species has spread explosively over much of the Western Hemisphere.

From the northern continent, the Costa Rican avifauna received further contributions. Some, including owls, quails, swallows, thrushes, gnatcatchers, and finches, belong to very widespread families that probably arrived, and even spread to South America, before the establishment of a land connection. Silky-flycatchers, a New World family absent from South America, most probably reached Costa Rica from the north, as did the motmots, whose center of radiation appears to have been northern Middle America. Other groups of birds, possibly of remote South American origin, including vireos, warblers, and orioles of the genus *Icterus*, became particularly rich in species in North America, from which Costa Rica probably received the ancestors of the Yellow-winged Vireo and Brown-capped Vireo, the Flame-throated Warbler, and several species of orioles. Many members of these families that breed in the North Temperate Zone return each autumn to their ancestral home in tropical America. The wrens, too, evolved many species in Mexico and the southern United States, some of which spread southward to Costa Rica.

Just as the rain forest was once continuous along the Caribbean littoral, so more arid conditions have prevailed along the Pacific coast from Mexico to the Golfo de Nicoya. The arid tropical avifauna of northwestern Costa Rica is composed largely of species that extend far into Mexico and almost certainly reached Costa Rica from the north after its territory joined that of what is now Nicaragua, if not while it was still an island. They include the White-fronted Parrot, Lesser Ground-Cuckoo, Cinnamon Hummingbird, Elegant Trogon, Turquoise-browed Motmot, Banded Wren, Spotted-breasted Oriole,

Streaked-backed Oriole, and Striped-headed Sparrow, none of which reaches southern Costa Rica. Northern Central America has continued into the present century to enrich the Costa Rican avifauna. The Inca Dove is a relatively recent arrival, and the Great-tailed Grackle, which with the destruction of the forests has been spreading widely over the country, is probably of northern stock.

The cooler climate of the Northern Hemisphere during the Pleistocene glaciations, when glaciers formed on the high summits of the Cordillera de Talamanca, made it possible for juncoes, dippers, Hairy Woodpeckers, and Rufous-collared Sparrows to reach Costa Rica from the north. The reduced temperature may also have helped a number of birds of Andean affinities, forerunners of the Buffy Tufted-cheek, Ruddy Treerunner, Red-faced Spinetail, Silvery-fronted Tapaculo, Slaty Flowerpiercer, and other species, to cross the lowlands of central Panama and become established in the mountains of western Panama and Costa Rica.

The immigrants that poured into Costa Rica did not remain unaltered in their new home but proceeded to evolve into new species, many of which remain endemic to this country and western Panama, which biologically belongs with Costa Rica. Indeed, the Costa Rica–Chiriquí highlands were probably a major center of speciation starting well before the completion of the isthmus. Now separated from other elevated regions by the lowlands of central Panama and those of southern Nicaragua, these highlands have many endemic genera, species, and subspecies. These highlands, especially the Cordillera de Talamanca, have also proved to be an isolating barrier for lowland species. The rain forests of the southern Pacific region, separated by this lofty range from similar forests on the Caribbean side, thus contain many species and subspecies distinct from, but closely allied to, Caribbean forms. Notable among the species pairs, and often considered superspecies, are the Beryl-crowned Hummingbird and Blue-chested Hummingbird, the Collared Aracari and Fiery-billed Aracari, the White-collared Manakin and Orange-collared Manakin, the Snowy Cotinga and Yellow-billed Cotinga, and the Bay Wren and Riverside Wren.

Plumage and Song

One of the chief attractions of tropical birds is the brilliance of their plumage—although many are quietly clad in browns and grays, and indeed the proportion of the breeding avifauna in tropical zones that is brilliantly colored differs little from that in temperate zones. The thoughtful birdwatcher from a northern land, where male birds are often much more colorful than females, may be surprised to find that the sexes of many of the most richly colored tropical species, including tanagers, orioles, warblers, and parrots, are alike or nearly so. Moreover, in contrast to many migrants that visit Costa Rica, these birds live in pairs and wear the same bright colors throughout the year. As far as is known, the males of only two Costa Rican land birds—the Red-legged Honeycreeper and possibly the Blue-black Grassquit—molt into a plumage like that of the females after the breeding season, when they live in flocks rather than pairs. The more brilliant birds of all kinds tend to frequent open country and the sunlit upper levels of the forest, whereas those that inhabit dimly lighted undergrowth mostly wear dark, subdued colors although often with attractively barred or streaked patterns. Many such birds have contrasting areas of buff, white, and/or black that are normally concealed or inconspicuous but can be strikingly exposed in displays.

Many Costa Rican birds, especially some of the finches, thrushes, orioles, warblers, and wrens, are superb songsters. Nevertheless, the prevailing silence during the early months of the dry season may disappoint the visitor who expects to hear profuse song in mild, sunny weather that seems to invite it. The song of many of the best songsters is confined to their breeding season, which comes later in the year. The birds that sing most constantly throughout the year include the continuously paired wrens, birds with subdued colors who forage amid dense vegetation where visibility is limited and who stay in contact with their mates by means of responsive singing. On the other hand, elegant little tanagers of the genus *Tangara*, also continuously paired, are nearly or quite songless; they live in more open places where they can see each other. Although, as in the north, the principal songsters are Oscines, many of the Suboscines have delightful voices. The dawn songs of flycatchers are mostly quaint rather than melodious, but a few are lovely. Many antbirds and others of the forest understory have simple songs of pure, whistled tones (this kind of sound carries best in a dense habitat; only in the open or high in the forest canopy do complex trills and warbles carry well for long distances). The notes of some tinamous are deeply stirring; some trogons sing melodiously; and Rufous-tailed Jacamars have charmingly elaborate songs. Although most hummingbirds have colorless, weak, or monotonous songs, some are also surprisingly accomplished vocalists, though their complex melodies often lack carrying power; in Costa Rica most hummingbird species sing, unlike most of their relatives in North America.

Although the song is usually the most distinctive vocalization of most Costa Rican birds, virtually all species have various other notes that are often just as characteristic and useful in identification. In deep forest and thickets, birds are often so difficult to see that the experienced birdwatcher depends largely upon their voices to detect and identify them.

Breeding and Molt

Most Costa Rican birds breed in monogamous pairs whose nests are well scattered even when the birds do not obviously defend territories. Because many birds remain mated throughout the year and continuously resident birds have ample time to settle down, pairs are formed and territories established inconspicuously, without the profuse singing and skirmishing that draw attention in northern lands where migratory or nomadic birds have less time to arrange these matters. Among Costa Rican land birds, only four icterids nest in colonies: two species of oropendolas whose long, woven pouches cluster in high treetops and two species of grackles who often build a number of open nests in the crown of a large tree, or in marshy vegetation. Males are less numerous than females, and pairs are not formed. Likewise, the females of all manakins and hummingbirds and several species of flycatchers, cotingas, and woodcreepers attend their nests without a mate. Among manakins, many hummingbirds, and Ochre-bellied Flycatchers, males gather in courtship assemblies, or leks, where they call and display to attract females. Males of bellbirds and some other cotingas call loudly in treetops for the same purpose, but other nonpairing males advertise their presence less conspicuously. At the other extreme, the polyandrous female jacana supplies eggs to several males, each of whom incubates them without her help. Much the same phenomenon apparently occurs in some tinamous, except that little or no pair bond is formed.

Many Costa Rican birds live in groups or small flocks consisting of a breeding pair and one or more nonbreeding birds, usually their young from previous broods, which help to feed and protect the pair's nestlings and fledglings. Such helpers are known in one or more species of toucans, puffbirds, woodpeckers, cotingas, jays, wrens, tanagers, and grosbeaks, among others. From this it is but a step to true communal nesting as practiced by anis and at least some populations of Brown Jays, among which two or more females lay their eggs in the same nest and all the parents with their helpers cooperate in rearing the young. These more elaborate breeding systems are almost always found in sedentary species with permanently defended territories.

Birds everywhere time their breeding so that their young hatch as food for them becomes more abundant. In Costa Rica, the main nesting season begins in March or April, when returning showers end the dry season of the year's early months, when shrubs and trees that bloomed in the drier weather have ripening berries, and when insect life increases. Nesting reaches its peak in April, May, or June, according to the locality, and continues on a diminishing scale until August or September. During this period the majority of insectivorous and frugivorous birds, or those with a mixed diet, raise their young. Birds with different diets often nest at other seasons. As a family, hummingbirds nest in every month, and some species breed through most of the year. Just as different species of flowers have different blooming seasons, however, different hummingbirds time their breeding to coincide with the peak blooming of the flowers they prefer.

In lowland areas with a definite dry season, many species of hummingbirds nest most freely in December or January, as the rainy season ends and as increasing sunshine, while the soil is still moist, favors the greatest profusion of flowers rich in nectar. In the highlands, many species start to nest in the middle of the rainy season, when epiphytic shrubs of the heath family (Ericaceae) begin to display spectacular clusters of red and white, orange, or purple tubular flowers that brighten the cool, wet, foggy days. Nesting among the hermit hummingbirds typically reaches its peak early in the rainy season as the wild plantains (*Heliconia*) attain their peak of bloom. Bananaquits and flower-piercers, which also drink much nectar, nest when hummingbirds do rather than with other passerines in the main nesting season. Seedeaters tend to nest later than other passerines because they wait until the grasses, which spring afresh when the rains return, have ripening seeds. In the Valle del General, Variable Seedeaters and Yellow-faced Grassquits have two breeding seasons each year, the principal one from May to August and a minor one in December and January. In this valley, White-tipped Doves, Gray-chested Doves, and Blue Ground-Doves also breed twice, from January to April and again from July to September or October.

Other birds that nest early include raptors and vultures, a few woodpeckers, and kingfishers, who must raise their young before rising rivers flood their burrows in the banks and turbid water makes fishing more difficult.

Costa Rican land birds most frequently lay two eggs in their nests; three eggs are less common and four or more rather rare. Some pigeons and cotingas lay only one. Except for hummingbirds and pigeons, which lay no more than two eggs wherever they breed, most of the widely distributed birds increase the size of their families with increasing latitude. It is likely that several factors play a role in this "latitude effect." In tropical birds, loss of nests to predators is usually more frequent than in temperate-zone birds, and small clutches may be favored because they are more easily replaced, allowing more nesting attempts. Also, small broods may attract less attention from predators. On the other hand, the greater number of hours available to temperate-zone birds for finding food may help some species feed more nestlings. Because of high losses on migration or during severe winters, temperate-zone birds usually start a breeding season with smaller populations in relation to available resources than do tropical birds. For the former, production of many young may be the best strategy to assure that some progeny will survive these difficult periods. For more sedentary tropical birds, faced with less extreme fluctuations of climate and resources, the best option may be to devote more intensive care to fewer young.

The incubation and nestling periods of Costa Rican birds tend to be considerably longer than those of their closest relatives in the North Temperate Zone, and these birds attend their fledged young for a longer time. Continuously resident tropical birds can afford to rear their families at a more leisurely pace. The longer period also facilitates the development of cooperative breeding associations, as mentioned above, in which the young of the first brood remain with their parents to help rear later broods.

With some exceptions, nesting and molting occur at different times because both make large demands upon the bird's resources. Most small land birds molt after they have raised their broods, chiefly from June to September or October. In this interval when many juveniles acquire adult colors, the birdwatcher may be puzzled by the motley patterns in evidence. After July or August, the increasing silence and secretiveness of birds molting and resting in the season of heavy rains may disappoint observers who expect much song and activity. A brief resurgence of song and even breeding may occur in a number of species, however, if conditions are still favorable once the annual molt has been completed.

In the vast majority of Costa Rican birds, the adults undergo only one complete change of plumage each year. A separate breeding plumage, acquired by changing some or all of the body feathers before the nesting season, is exceptional in nearly all groups of Costa Rican birds except the herons. Among the many species of migrants and winter residents, however, distinct winter and breeding plumages are common. Most species arrive in the fall already in dull winter dress; the more colorful breeding plumage is acquired from about February onward and may still be incomplete when the birds depart in spring. The molt to the breeding plumage nearly always involves only the body plumage and may be extensive (e.g., the Scarlet Tanager) or involve only a small number of feathers on the face or throat (some warblers). Exceptions include the Summer Tanager and Northern (Baltimore) Oriole, species in which the adult males are brightly colored year-round, and the migrant thrushes and flycatchers, which are relatively dull year-round. In all these cases only a single complete molt occurs per year, either before or after fall migration. The shorebirds are also unusual in that wing molt occurs after fall migration, but most birds arrive with the body molt to winter plumage fairly well advanced. Finally, some large birds, particularly seabirds, have very slow, protracted molts.

Most Costa Rican birds attain the adult (definitive) plumage soon after fledging, although some juvenile feathers may be retained, especially on the wings, for some time. A considerable number of species have distinctive first-year plumages—either the juvenile plumage is retained wholly or in part, or the first adult (basic) plumage differs from that of subsequent years, as in many tanagers and finches. In species of this category that are also sexually dimorphic, this first-year plumage of young males is often more or less femalelike, as in seedeaters. In these species males may sing and breed when a year old in an "immature" plumage. Relatively few Costa Rican birds require longer than a year to acquire definitive dress, and these are virtually all large species, especially among the herons and birds of prey. In this guide we describe the first and final plumages in complex sequences, but space prohibits treating each stage in detail.

Movements of the Avifauna

Although many tropical birds reside on their territories throughout the year, exceptions are numerous. Frugivorous and nectarivorous species must be more mobile than insectivorous birds to take advantage of peaks of flowering and fruiting that change with the season and the zone. In noisy flocks, macaws and other parrots fly long distances, up and down mountains or on more level courses, in search of fruiting trees. In the second half of the year, when the mountains where they nested at high and middle altitudes are drenched by chilling rains, birds of many kinds descend to warmer, if not drier, levels where their food may be more abundant, to return upward as the weather improves in the following year. Outstanding among these altitudinal migrants are Three-wattled Bellbirds, who, after breeding in the highlands, wander downward sometimes as far as the coasts. Wherever they are, the males announce their presence by stentorian calls. Resplendent Quetzals, too, descend to lower levels after nesting. Hummingbirds also appear to wander widely, disappearing from localities where they nest when drought depresses flowering. Unlike larger birds, hummingbirds tend to move downward as the dry season advances and return upward for nesting in the wet season. On the other hand, most insectivorous tropical forest birds are so strongly attached to the altitudinal zones where they breed that they are scarcely ever seen a thousand feet higher or lower.

The numerous ducks, herons, storks, ibises, and other birds that enliven the marshes of Guanacaste depart as these wetlands dry in the long, windy rainless season. Many cross the low Cordillera de Guanacaste to the Río Frío region in the northwestern Caribbean lowlands or to the Lago de Nicaragua.

Much more spectacular than the local movements of Costa Rica's resident birds is the influx of long-distance migrants from temperate and boreal North America. Among migratory land birds, warblers predominate in numbers of species and probably also in numbers of individuals, but flycatchers, swallows, thrushes, vireos, orioles, tanagers, and finches, with a few pigeons, cuckoos, and kingfishers and a number of raptors, swell the multitude. Appearing first along the Caribbean coast in August, migrants arrive in larger numbers in September and October. Many continue onward to winter in South America, but countless more remain in Costa Rica, spreading over the whole country, in both humid and semiarid regions, and extending up to the mountaintops. They share nearly every habitat and mode of foraging with the resident birds, with whom they appear to compete for food only to a minor degree and only very exceptionally fighting with them for resources. Some of the migrants, including a number of warblers and the Summer Tanager, claim individual feeding territories. While contending with others of their kind for a plot of land after their arrival in autumn, they not infrequently sing. Other migrants, more sociable, pass the months of the northern winter in small or large flocks.

Many long-distance migrants pass half or more of the year in their winter homes and are rightly regarded, not as northern birds that come south to escape winter's dearth, but as tropical birds who go north to breed—many of the migratory passerines belong to Neotropical families. The northward exodus begins in March and continues with greater intensity through April; by late May only a few laggards, or perhaps belated birds of passage from South America, remain in the country. As the time for their departure approaches, a few migrants sing a little or even much. Orchard Orioles are especially songful; and Swainson's Thrushes, passing in large numbers from their winter home in Colombia, sometimes produce more melody in Costa Rican forests than all the resident birds together. The departure of many thousands of winter visitors leaves more food for permanent residents, now starting to nest, to feed their young. In contrast to many waterbirds, no migratory land bird has been known to stay throughout the year.

The most spectacular migrations are those of the diurnal travelers. From their winter home in South America, Swainson's Hawks pass in immense flocks that often stretch from horizon to horizon, especially in spring. In the Caribbean lowlands, Turkey Vultures migrate southward and northward in multitudinous hordes. Mostly along the coast, countless Barn, Cliff, and Bank swallows stream by in loose formations most noticeable in early morning and late afternoon. Eastern Kingbirds also migrate by day in smaller flocks. Nocturnal migrants come and go more obscurely.

From South America, Costa Rica receives only a few species of migrants. Early in the year, American Swallow-tailed Kites and Plumbeous Kites, Piratic Flycatchers and Sulphur-bellied Flycatchers, and Yellow-green Vireos arrive to breed or continue farther northward, then return to the southern continent in time to avoid Costa Rica's heaviest rains. Probably because most of South America lies within the tropics, long-distance migrants are less numerous there than among the breeding birds of North America. Of those that travel northward after nesting in the South Temperate Zone, only a stray Patagonian Blue-and-White Swallow or Brown-chested Martin is likely to come as far as Costa Rica. Perhaps the most regular southern "winter" resident in Costa Rica is the Black Skimmer, for small numbers of dark-winged southern birds regularly spend from about May through September around the Golfo de Nicoya.

By early August, shorebirds of many kinds are appearing on the coasts and, to a minor degree, on inland ponds and waterways. By the end of the month, multitudes of sandpipers and plovers are foraging along the shores and other wet places, especially on the Pacific side, where such favored habitats as mudflats, mangrove swamps, rocky shores, and shallow salt-evaporating ponds (salinas) are widespread and the receding tide exposes a rich harvest of food twice daily. The Golfo de Nicoya in particular seems to be a critical fuel stop for many migrating shorebirds. Most of these shorebirds depart northward at about the same time as the land birds or a little later. In May and even into June, long, loose flocks of Black-bellied Plovers and Franklin's Gulls can be seen streaming northward along the Pacific coast. However, many yearlings that will not breed, and perhaps older individuals in poor condition, remain throughout the summer. The same is true of many gulls and terns, in which nonbreeding, summering birds are, if anything, more numerous than wintering ones.

Food and Foraging

Although the same general range of food types is available to birds in Costa Rica as in the North Temperate Zone, many of these foods are available in much greater abundance and variety year-round, so that some birds specialize in them rather than simply exploit them opportunistically. Costa Rica thus has frugivores, for example, that prefer large nutritious fruits (bellbirds), mistletoe berries (euphonias), or unripe fruits (parrots); that are specialized for obtaining fruits by reaching (toucans) or sallying (trogons); and that swallow berries whole (manakins) or mash them in the bill (many tanagers). Many Costa Rican hummingbirds have bills adapted for extracting nectar from flowers of particular lengths and curvatures, whereas others, plus the Bananaquit and the Slaty Flowerpiercer, frequently or regularly pierce their corollas. A number of species (motmots, some puffbirds, some flycatchers, etc.) are specialized "sit-and-wait" predators that take large insects, small lizards, and frogs; the Laughing Falcon eats snakes and little else. Also, at least in humid areas, there are particular places where insectivores and others seek their prey; thus we have specialized bromeliad rummagers, rolled-leaf probers, spiderweb gleaners, moss searchers, and many others.

Concentration on certain types of foods has other consequences as well. Fruits and nectar, which plants make readily available to birds who disseminate their seeds or pollinate their flowers, are in the breeding season so much easier to find than insects, which try to avoid being eaten, that many female fruit eaters and nectar drinkers can raise their broods without help from a male. Freed from domestic tasks, male hummingbirds, manakins, and others can pass their days in leks or courtship assemblies, as we have already mentioned. All Costa Rican birds in this category belong to Neotropical families; species of northern affinities regularly breed in cooperating pairs or groups, as do virtually all birds that must search for or await insect or vertebrate prey. In pollinating flowers and disseminating seeds birds play their most significant biological role in tropical forests. In the wetter regions of Costa Rica, more than half of the tree species, especially in the understory, have fruits adapted for seed dispersal by birds. In contrast, in dry regions like Guanacaste, as well as at higher latitudes, most trees have fruits dispersed by wind.

A special kind of food, rare and unfamiliar in northern lands but abundant in the tropics and frequently mentioned in our species accounts, is the aril, a soft, fleshy, nutritious tissue that partly or wholly surrounds a typically hard, indigestible seed coat. When the dry, inedible pods that contain arillate seeds dehisce to expose the bright red, yellow, or white arils, birds of many kinds flock to the feast. Migrant flycatchers, thrushes, vireos, and warblers eat many of these oil-rich arils, which help them to store fat as fuel for their long journeys.

The birdwatcher may wander through tropical woodlands, disappointed to have seen so few birds, until a mixed-species foraging flock comes into view. Then the birder may wish for extra pairs of eyes to follow the movements of a multitude of tanagers, honeycreepers, greenlets, manakins, antbirds, woodcreepers, and many other species, each seeking food in its own way. These flocks, found at all elevations, differ in composition, depending on whether they move through the canopy or the undergrowth—so that there are more tanagers, warblers, honeycreepers, and gnatcatchers in the upper levels and more antbirds, ovenbirds, woodcreepers, and manakins in the understory. The composition also varies, depending on whether the numerically dominant members take exclusively insects (e.g., antbirds), mainly fruits (some thrushes and tanagers), or a mixture of both (many tanagers, honeycreepers, etc.). Usually the flock is organized around one or more "nuclear species," more social and noisy birds that commonly forage in groups and about whom other "attendant" species gather. The attendant species are usually represented by a single individual or a pair that will not tolerate others of their kind nearby but are perfectly willing to join different species. As a flock moves through the territory of such a species, the resident individual or pair accompanies it until at the boundary it (or they) is replaced by the claimant(s) of the adjoining territory—a substitution not likely to be detected unless the birds are banded.

Although such mixed-species flocks occur in temperate-zone birds (mainly after the breeding season or during the winter), only in the tropics do they reach a high level of variety and complexity. In Costa Rica, such flocks are better developed in forest than in open country, in wetter than in drier regions, and in highlands than in lowlands. Possibly the birds gather in these mixed flocks for more productive foraging; they may play into one another's bills, as when an insect flying out to escape a woodcreeper gleaning bark is snatched up by a flycatcher. A greater advantage is probably safety from predation. At first glance, it seems unlikely that in a noisy, conspicuous flock a bird is more secure than when it is alone amid screening foliage. With many eyes watchful for approaching predators and many voices quick to sound the alarm, however, flock members may be alerted in time to save themselves. Foraging may also be enhanced indirectly; with so many watchful eyes, each individual bird can devote more of its time to foraging and less to keeping a lookout for danger.

The maximum of animation in tropical woodlands is provided by foraging army ants, usually *Eciton burchelli*, and attendant birds. As thousands of hunting ants spread widely through the ground litter and send columns up stems and trunks, small creatures hiding therein try to escape. Heedlessly exposing themselves, insects, spiders, centipedes, sow bugs, and other invertebrates, with a few frogs, lizards, or small snakes, are snatched up by watchful birds, who seldom take the ants themselves. All the activity is accompanied by a medley of calls and frequent displacements of one bird by another. The more constant, widespread ant-followers in lowlands and foothills include Bicolored Antbirds, Gray-headed Tanagers, and woodcreepers of the genus *Dendrocincla*. They are joined by many other species, including other antbirds and woodcreepers, small flycatchers, manakins, wrens, even an occasional forest-falcon, now more interested in catching insects than birds. Migrants, especially Swainson's Thrushes, hover around the outskirts of the foraging crowd. When the legion of ants moves to the forest edge or into adjoining clearings, anis, thrushes, tanagers, finches, and other birds of the open country often come to take advantage of easy foraging.

Bird Conservation in Costa Rica

In Latin America, Costa Rica is outstanding in its dedication to conservation and bird preservation. Going to great expense to compensate private owners, the country has placed more than 8 percent of its territory in national parks and equivalent reserves, which it is making a conscientious effort to protect. Outside these protected areas, however, the natural habitats and birds of Costa Rica are increasingly threatened, and the key question is whether the parks and reserves themselves will survive as pressure on the land becomes more intense. This pressure originates not only within Costa Rica but also through the country's economic dependence upon more developed nations, particularly the United States.

Threats to Costa Rica's birds are of two sorts: specific and general. The former affect only certain species and have no direct impact upon the rest of the avifauna. The two most important specific threats are the cage-bird trade and hunting. General threats affect many or all bird species over wide areas. By far the most ominous general threat is habitat destruction, but pollution from hard pesticide residues and waste is a growing menace.

In principle, hunting in Costa Rica is for sport or for subsistence, although in practice the two tend to intergrade. Open seasons and bag limits exist for most tinamous, gallinaceous birds, and ducks as well as for several species of pigeons, but the chronically shorthanded, underfunded wildlife service is able at best to achieve spotty enforcement. Species most often hunted include Crested Guans, Mourning Doves and White-winged Doves, Band-tailed Pigeons and Red-billed Pigeons, Black-bellied Whistling-Ducks, and Spotted-bellied Bobwhites, but lesser numbers of tinamous, other cracids, and ducks are also shot for sport. Almost any bird larger than a small pigeon may be hunted for food by country folk, for whom wild game is often an important source of protein; such subsistence hunting is virtually uncontrolled. Overall, hunting per se does not seem to have disastrous consequences for any Costa Rican species except the Great Curassow, Muscovy Duck, and Crested Guan. The first two, in particular, are increasingly confined to parks and reserves, where good populations often still exist. Poaching is difficult to control and even at low levels can result in extirpation of such species as guans, toucans, and wood-quails in small reserves. Only when it is combined with habitat alteration, however, does hunting characteristically affect most bird populations.

A related menace is the tendency of many rural Costa Ricans to shoot any bird of prey on sight. Usually the justification is protection of chickens or livestock. Although a few Costa Rican raptors undoubtedly kill some domestic animals, most species take other prey whenever possible; more than a few are highly beneficial, capturing many rodents, large insects, or snakes. Large, wide-ranging raptors are liable to be shot whenever they stray outside the borders of parks or reserves; Harpy Eagles and Crested Eagles, in particular, are nearing extirpation. Fortunately for most other species, the larger parks are probably big enough to preserve breeding populations if poaching can be eliminated.

The cage-bird trade is a serious problem politically as well as from the standpoint of conservation. It is an ingrained part of Costa Rican culture to have a *pajarito* or a *lora* in a cage, and some of the more valuable species, such as the Scarlet Macaw, are status

Figure 18. Two species of *Carduelis* in cages, Alajuela. The practice of trapping and selling cage birds is proving difficult to control and is having a severe impact on the populations of several species, notably macaws and parrots, some orioles, and siskins.

symbols. The prized birds whose populations have been dangerously reduced include the Scarlet Macaw (virtually extirpated except in Carara and Parque Nacional Corcovado; there are also a few in Guanacaste), the Yellow-naped Parrot, the Blue-hooded Euphonia, the Yellow-tailed Oriole, the Yellow-bellied Siskin, and the Great Green Macaw (although fractionation of habitat may be the major factor in the latter case). Other highly esteemed species, such as the Black-faced Solitaire, have enough protected habitat to maintain good populations; and there are some valued open-country birds whose ranges are expanding, such as the White-collared Seedeater. Current Costa Rican law prohibits the sale of wild birds but permits an unlimited number to be held in captivity; exportation of most species is forbidden. Enforcement of these laws is so lax and spotty that such birds as baby parakeets and parrotlets are still sold openly in some markets and at country roadsides during the breeding season. The formerly flourishing export trade of parrots, macaws, quetzals, and tanagers has been virtually halted, although some parrots are undoubtedly smuggled out through Nicaragua or Panama. The Wildlife Service attempts to exert some control by requiring bird-catchers (locally called *pajareros*) and bird owners to purchase licenses, but it estimates that no more than one-third actually do so.

Specific threats include the efforts of private landowners, municipal governments, and/or the Ministry of Agriculture to control or eradicate populations of species regarded as pests. A number of birds damage crops: whistling-ducks in young rice plantations, some parrots in cornfields and pejibaye-palm plantations, Red-winged Blackbirds and grackles in grainfields, quails among beans, and seedeaters in sorghum. Control measures have usually been clumsy and unspecific; wide-spectrum poisons in rice fields kill not only hundreds of whistling-ducks but also rails, doves, shorebirds, and many other birds. On the other hand, lacking a method to repel Great-tailed Grackles from city parks, where they often form messy and incredibly noisy communal roosts, some Guanacaste towns have simply cut down all the trees in their parks! In general, the birds that might be affected by more or less specific control measures are open-country species with large and expanding populations favored by deforestation; the problem is not how to preserve these species but how to design control methods specific to the injurious ones while avoiding widespread or long-lasting damage to the avifauna generally.

The two major general threats to the avifauna are environmental contamination and habitat destruction. The most serious contaminants are persistent residues of chlorinated hydrocarbon pesticides, especially DDT, which until recently were used in Costa Rica with no restrictions. From 200 tons of active ingredients annually before 1980, use has declined to about 40 tons—still hardly insignificant! Studies in progress on waterbirds in the Tempisque basin show that residue levels in eggs are approaching the danger point for reproductive success, and large agricultural projects planned for this area may well add to the burden. Unfortunately there is still no effective enforcement of regulations restricting use of DDT; often the farmer's only source of information is the pesticide salesman himself, who is hardly likely to recommend restraint. When DDT and other hard pesticides were banned in the United States, large American chemical companies did not stop manufacturing such profitable commodities but simply switched their marketing to south of the border. Now Costa Rican wildlife—and North American migrants—are paying the price.

A contamination threat of uncertain magnitude is the proposed transoceanic oil pipeline, if it ever goes beyond the planning stage. Fortunately, the project is for the moment mired in the local bureaucratic process and is unlikely to be implemented in the immediate future. If the degree of oil seepage in the Moín area (which includes the Caribbean coast's largest mangrove swamp) from Costa Rica's one small refinery is any indication, however, pollution will be almost inevitable if the pipeline is built.

Habitat destruction is without doubt the greatest threat to Costa Rica's avifauna as a whole. The habitats under greatest pressure are wetlands and forest, especially in the humid lowlands. Most of the once extensive marshes of the Meseta Central were drained before 1960 to make rather swampy cattle pasture, much of which has in turn been engulfed by urban sprawl. The greatest wetland area in southern Central America, the Tempisque and Bebedero river basins, has lost more than half of its marshes and lagoons to rice and sugarcane fields since 1960, and a major dike and irrigation project now planned would in the next ten to fifteen years eliminate most of the remainder outside the Palo Verde park and wildlife refuge. The White-faced Whistling-Duck may already have disappeared from Costa Rica because of this habitat destruction, and the giant Jabiru may soon follow.

Mangrove swamps are also under increasing pressure all along the Pacific coast. In addition to being destroyed to make room for salinas and shrimp-culture ponds, they are cut in order to fuel primitive stoves that evaporate water to obtain salt in the salinas and in order to make highly desirable mangrove charcoal. Among the species at risk through destruction of mangroves is the Mangrove Hummingbird, one of only three bird species endemic to Costa Rica. At this writing, no extensive mangrove swamps with good populations of Mangrove Hummingbirds are receiving any protection.

Unfortunately, even the destruction of wetlands pales beside the destruction of tropical forests. About 60% of Costa Rica's land birds, including the other two endemics, the Coppery-headed Emerald and Black-cheeked Ant-Tanager, and virtually all the species with ranges restricted to Costa Rica and western Panama, depend upon large, or at least more or less interconnected, stands of intact forest. Nearly all these species

Figure 19. Deforestation to produce low-grade cattle pasture in the Coto Brus district of the south Pacific slope. Most of the wood is simply burned in situ rather than being used.

will be at risk by the end of this century. The rate of deforestation in Costa Rica is one of the highest in the world: more than half the country's forests have disappeared since 1940, and the remainder is being lost at a rate of about 3% of the country's land area each year. The regions of most active deforestation are at present the Golfo Dulce lowlands, the lowlands and foothills of the northern Caribbean slope, and also the extreme southeast, in the Manzanillo-Gandoca districts. Most of the land between Guápiles and Tortuguero National Park, and between the Golfo Dulce and the Panamanian border, has been completely shorn of forest in the last ten or fifteen years. Within a few years, nearly all Costa Rica's remaining forest will presumably lie within the system of parks and equivalent reserves. Two questions arise. How adequate is the present park-reserve system for ensuring the survival of the country's rich avifauna if it is given adequate protection? What are the prospects for protecting this land in the foreseeable future?

It is clear from Maps 1 and 2 that Costa Rica's national parks and wildlife refuges do not conform in location to the original distribution of the country's forests. Most of the protected areas lie in the highlands, with more than half of that in the park system in the huge Parque Nacional La Amistad (190,000 hectares), which embraces most of the Cordillera de Talamanca. Costa Rica's mountain avifauna is thus much more secure than that of the lowlands. In the dry-forested northwest, much of the forest avifauna is adapted to a regimen in which most trees lose their leaves for several months each year. To some extent, these birds are better able to withstand deforestation than those of wetter forests. The relatively small areas of evergreen forest are critical seasonal resources for many species, however, and these forests are disappearing in many areas. Moreover, the deciduous forests are often threatened by wildfires that are set to burn off woody vegetation from pastureland and that, fanned by strong dry season winds, can escape into parks and reserves. Protection is most urgently needed in the mountains of the Península de Nicoya; it is hoped that the new Parque Nacional Guanacaste will result in more extensive fire control.

The greatest cause for alarm, however, is the wet lowland forests that once covered more than one-third of Costa Rica and are now reduced to less than a third of their former area, with the remainder disappearing with appalling rapidity. The two large lowland parks, Tortuguero and Corcovado, are at opposite ends of the country—and the latter is threatened by an invasion of gold seekers. At no point is a large lowland park connected by a broad wooded corridor with the highlands, to serve the needs of the many forest birds, especially fruit and nectar eaters, who migrate up and down with the changing seasons. Probably the best chance for preserving the habitat for these altitudinal migrants, which include such spectacular species as the Three-wattled Bellbird and the Bare-necked Umbrellabird, lies in preserving and broadening the forest connection between La Selva and Parque Nacional Braulio Carrillo.

Having enjoyed almost continuous habitat until recent years, birds of wet lowland forests hesitate to cross wide expanses of open country, with the consequence that populations of these birds will soon be isolated within their respective preserves. The two large lowland parks, Tortuguero and Corcovado, are probably large enough to ensure the survival of most small birds of the forest understory, but several big, wide-ranging species, such as eagles and Great Green Macaws, may be doomed. Obviously, preservation of all these birds depends upon strict protection of forest within these parks in the face of steadily increasing human pressure.

To evaluate the feasibility of such protection, it is necessary to assess the driving forces behind the rampaging deforestation, above all in the wet lowlands. Costa Rica's tradition of a stable, permissive, democratic government with its moderate socialism, including the most comprehensive social security system in Latin America, has long made the country attractive to foreign investment, which has brought increasing economic dependence upon the United States in particular by Costa Rica's eagerness for American, and more recently Japanese, manufactured goods. Through the 1960s the country's economy expanded, fueling an explosive population growth; in the 1950s, Costa Rica had the highest growth rate in the world. Costa Rica's high standard of living and economic growth were for many years supported by the opening of new lands for agriculture; improvement of existing crops compensated for the exhaustion of formerly productive lands and the urban sprawl over the fertile soils of the Valle Central. The country's supply of fresh lands to exploit is rapidly being depleted, however, and for some time newly cleared lands have been less suitable for cultivation and more quickly exhausted. The declining ability of the country's agriculture to support a growing population with a great appetite for imported luxuries is the major factor in the current economic crisis, which was triggered by the increase in oil prices from the Organization of Petroleum Exporting Countries (OPEC) and by a worldwide collapse in coffee prices.

The only way for an agrarian economy such as that of Costa Rica to increase its income and improve its balance of payments is to augment its agricultural production by devoting ever more of its land to mechanized monocultures of export-destined cash crops. As this process advances, fewer large farms are replacing the many traditional small farms, thereby decreasing the land's capacity to support the rural populace. Displaced farmers must either seek new lands, aggravating the pressure on parks and reserves, or move to the cities, exacerbating urban sprawl, pollution, and the already crushing economic burden of the social security system. In either case, the country is worse off than before!

Figure 20. Wildfire entering tropical dry forest at the height of the dry season, having escaped control when adjacent pastures were burned to control woody vegetation. Such fires progressively open up the forest, facilitating invasion by imported savanna grasses, which in turn promote even more extensive fires—a vicious cycle! Palo Verde National Wildlife Refuge,

Figure 21. The ultimate in "avian deserts," a banana plantation near Río Frío, central Atlantic lowlands.

The short-term monetary gain from these agricultural practices cannot be sustained if the land is intrinsically unsuitable for such uses. The most flagrant present example is the cattle industry: the high price of beef in the USA has accelerated the conversion of forests to low-grade pastures, where soil compaction and erosion will make the land useless in twenty to forty years. The allocation of one cow per 12–14 acres (5–6 hectares) hardly seems efficient and at the moment is economically feasible only because the American fast-food industry pays so well.

The solution to all these difficulties is complex, and it is certainly far beyond the scope of a bird guide to suggest one. We have entered into great detail to emphasize the fact that the threats faced by Costa Rica's birds form part of a much broader set of problems, many of which originate beyond Costa Rica and cannot be settled by this country on its own. The system of parks and reserves will continue to flourish only as long as the Costa Rican people and government believe that the long-term gains of conservation of natural resources, including birds, outweigh the short-term, ephemeral gains of converting the protected lands into cattle pastures. This belief would be strengthened if the parks and reserves could show short-term economic advantages as well as long-term benefits. And precisely in this area the users of this book can help.

By coming to Costa Rica to see its beautiful parks and its strikingly diverse avifauna, you can directly demonstrate your support of the costly and courageous conservation measures the country is taking in the face of discouraging odds. We also recommend staying long enough to get to know some of the less conspicuous but equally fascinating species—the flycatchers, hummingbirds, ovenbirds, and many more—as well as the more spectacular birds, such as the Resplendent Quetzal or the Three-wattled Bellbird. Be prepared, at first, for some inconvenience; Costa Ricans are extremely hospitable, but few people in either the park service or the tourist industry understand the peculiar needs and desires of modern birders. As the numbers of birdwatchers increase, however, understanding should grow. Those who wish to make direct contributions to Costa Rica's struggle to preserve its birds and natural habitats might send contributions to the Costa Rican National Parks Foundation, Apartado 236, San José 1002, Costa Rica. Costa Rica is making a large and growing sacrifice to provide you with a chance to see and enjoy this spectacular avifauna in its natural habit; your gift will help ensure that your grandchildren have this chance too. We feel that the observation of birds in a rich tropical forest is an incomparable experience, for Costa Rica's birds are truly priceless, and your contribution is urgently needed to preserve this opportunity.

Format and Content of Family and Species Accounts

Because this guide treats an avifauna of more than 800 species, the space we can reasonably devote to each is necessarily limited. Below we describe the kinds of information we include for each species and family, and how it is presented. We also explain why certain kinds of data have been excluded and where more details can be found. Most of the information that we present comes from our own observations, published and unpublished. Where we have drawn upon another source for a specific piece of information (except for facts supplied in any good ornithology text or standard checklist), the surname of the source is given in parentheses. We thus recognize the contributions of the following students of Neotropical birds: M. Alvarez del Toro, W. Beebe, G. Bello, B. Bowen, R. G. Campos, M. Capkanis, G. S. Carr, M. A. Carriker, F. Chaves, R. S. Crossin, W. Dalquest, P. Devillers, R. W. Dickerman, E. Eisenmann, P. Feinsinger, M. P. Fogden, R. ffrench, C. Gómez, R. T. Holmes, S. G. Howell, T. R. Howell, J. H. Hunt, M. L. Isler, L. F. Kiff, R. Koford, T. J. Lewis, G. Lowery, M. McCoy, D. McDonald, B. K. MacKay, M. Marín, J. T. Marshall, J. C. Martínez, E. S. Morton, H. Nanne, A. Negret, N. Newfield, P. A. Opler, S. Perkins, D. Peterson, A. R. Phillips, T. Pratt, N. S. Proctor, P. Pyle, R. S. Ridgely, S. Robinson, L. Rosselli, P. Roth, J. E. Sánchez, T. W. Sherry, P. Slud, N. G. Smith, S. M. Smith, D. W. Snow, D. Texeira, P. Vickery, T. A. Werner, A. Wetmore, N. T. Wheelwright, B. Whitney, T. Will, K. Winnett-Murray, L. L. Wolf. Our major published sources, as well as other important general references, are listed in the Bibliography. We do not consider it necessary or desirable to supply a complete listing of the literature on Costa Rican birds in this guide. The references in the Bibliography, which include several reviews of different aspects of Costa Rican ornithology, will suffice to direct the interested reader to the primary literature.

Family Accounts

Each family of Costa Rican birds is introduced by a brief account of its composition, distribution, appearance, anatomy, behavior, and reproduction. For widespread, heterogeneous families, we emphasize their Neotropical and, above all, Costa Rican species. Any feature common to all Costa Rican members of the family (e.g., all Costa Rican hummingbirds lay 2 white eggs) is mentioned here rather than repeated for each species. Because this guide does not purport to be a taxonomic work, we have adopted a rather different criterion for delimiting families than that used by, for instance, the latest checklist of the American Ornithologists' Union (AOU). Briefly, we treat as families groups that have often been given familial status in the recent past, are not certainly polyphyletic, and are sufficiently homogeneous to permit an enlightening discussion of their characters. We recognize more families than do the AOU and many authors, but we have tried not to include in any of them species that, in the light of present knowledge of avian phylogeny, are unrelated.

The arrangement of orders, families, genera, and species follows the traditional "Wetmore sequence," somewhat modified to incorporate recent changes in our understanding of the relationships of certain species and groups. At present the classification of birds, particularly at the family level, is in a state of flux. The results of new biochemical techniques in particular suggest that changes, sometimes radical ones, may be required to express adequately the evolutionary relationships of many groups of Neotropical birds. Many of these results are still unconfirmed and untested, and taxonomic changes based upon them may have to be further modified or even discarded as still other evidence accrues. For the present we feel that the Wetmore sequence is the most appropriate arrangement for this guide, since it will already be familiar to users of other Neotropical field guides and distributional lists. Within a decade or two avian systematists may have reached a new consensus on the classification of Neotropical birds; until then, we believe that adoption of broad taxonomic changes in a nontechnical work like a field guide would be premature.

Species Accounts

Names

Three or four names are given for each bird: an English name and the Latin (scientific) name; below these a recommended Spanish name, followed by, in parentheses, one or more local Spanish vernacular names when they exist. The English name is nearly always that given in the latest AOU Check-list. In a very few instances, where we have strong reasons for favoring a different name, we give this initially and include the AOU name in the "notes" section (see below). We apply the same policy to Latin names: unless both of us strongly agree on an alternative name (which in this case usually implies an alternative taxonomic treatment), we give the AOU name in the heading. In virtually all cases, our treatment is mentioned by the AOU as an alternative; we adopt no taxonomic novelties in this guide.

The recommended Spanish name is the one in widest use that seems appropriate for and specific to the bird in question insofar as we could determine what that was by a survey of the literature in this language. Only when no such name was found did we invent one, in consultation with the late Rafael L. Rodríguez, eminent linguist of the biology faculty at the Universidad de Costa Rica. We felt compelled to adopt this procedure because most Costa Rican birds have no local names, and we wished to present a Spanish name for every species. No consensus yet exists on bird names among all Spanish-speaking ornithologists. An international group based in Spain is currently working to achieve such a consensus, but its efforts are several years from fruition.

Finally, we give the local Costa Rican names for all species for which such names are known to us. As will become apparent from a cursory reading, many of these names apply to more than one species. In addition, the same species may have several names in different parts of the country, and closely related species may have very different names. Nevertheless, these names will facilitate communication between users of this guide and local people, and they will help Costa Ricans without any ornithological background to use the guide. These local names have been derived from published sources and our conversations with Costa Ricans in many parts of the country in the course of our fieldwork. Much useful information on local names has also been contributed by Julio E. Sánchez of the Museo Nacional de Costa Rica and Rafael G. Campos of Grecia.

Description

We include a more detailed description of each species than has become customary in recent field guides in order to make the book as helpful as possible to a wide variety of users, including scientists and banders as well as birders. (See Figures 22 and 23 in the Glossary.) Each description begins with the length of the bird in inches and centimeters. To standardize, nearly all lengths have been taken directly from specimens in the Zoology Museum of the Universidad de Costa Rica, prepared by Stiles and his students. Where male and female differ noticeably in the field (roughly 10–15% or more), their lengths are given separately. Because length alone often conveys an inadequate idea of the size or bulk of a bird, we have included weights for nearly all species. For these we use metric units—grams and kilograms—because the smallest English unit in common use, the ounce, is far too heavy to express conveniently the weights of small birds. Most of the weights were obtained by Stiles and his colleagues while mist-netting and collecting; for perhaps 10 percent of the species, we have relied on data from specimens in other museums or from published reports. We have tried to give average weights for birds in normal condition, excluding emaciated, fat, or very young birds, or females with eggs ready to lay. As with length, we have given a single weight for each species except when male and female differ enough for the difference to be noticeable in the field, or about 15–20%. For those unfamiliar with gram weights, we present a rough "yardstick" of weights of representative species in Table 1.

The description begins with a few brief, general indications of shape and general aspect, perhaps with one or two key field marks. Then we describe the plumages of the species, usually starting with the adult (definitive) plumage. We have tried to keep the nomenclature of plumages and colors simple. If the sexes are alike, their plumage is described under the heading **Adults**; if they differ, we subdivide into **Adult** ♂ and ♀, respectively. Unless otherwise indicated, the plumage described is the basic plumage;

Table 1. Representative weights of some Costa Rican birds

Weight[a]	Species
2	Scintillant Hummingbird (male)
5	Rufous-tailed Hummingbird, Short-tailed Pygmy-Tyrant
8	Mistletoe Tyrannulet, Tennessee Warbler
10	Slate-throated Redstart, Variable Seedeater, Blue-and-White Swallow
12	House Wren, Violet Sabrewing (male), Traill's Flycatcher
15	Red-capped Manakin, Long-tailed Tyrant, Rough-winged Swallow
20	Rufous-collared Sparrow, Common Bush-Tanager, Dusky-capped Flycatcher
25	Yellow-bellied Elaenia, Western Sandpiper
30	Scarlet-rumped Tanager, Bicolored Antbird, Swainson's Thrush
40	Tropical Kingbird, Semipalmated Plover, Hepatic Tanager
50	Buff-throated Saltator, Inca Dove, Black Tern
75	Clay-colored Robin, Barred Woodcreeper, Collared Trogon
100	Squirrel Cuckoo, Short-billed Dowitcher, White-collared Swift
150	Least Grebe, Belted Kingfisher, Crimson-fronted Parakeet
200	Resplendent Quetzal, Lineated Woodpecker, Bat Falcon, White-throated Magpie-Jay
250	Willet, Pale-billed Woodpecker, Franklin's Gull
500	Gray-headed Chachalaca, Keel-billed Toucan, American Swallow-tailed Kite
750	Spectacled Owl, Chestnut-mandibled Toucan (male)
1000	Crested Caracara, Great Tinamou, Brown Booby
1500	Osprey, Roseate Spoonbill
2500	Brown Pelican, Great Blue Heron
3000	Muscovy Duck (male)
4000	Great Curassow (male)
6500	Jabiru
7500	Harpy Eagle (female)

[a]Weights appear in grams (g). 1 ounce = approximately 28 g. 1 pound = approximately 450 g.

for species that show a seasonal plumage change (e.g., most migrants), we first describe the nonbreeding or winter (i.e., basic) plumage, then the breeding (i.e., alternate or nuptial) plumage, because the former is the usual one for most such species in Costa Rica. For immature stages, we use the heading **Immatures** if such a plumage is retained for several months or through the first year and **Young** if the plumage is worn only briefly after the young fledge. This procedure effectively sidesteps the technical problem of specifying exactly what plumage is involved: we are concerned with the bird's appearance (''feather coat''), not with plumage homologies. Often the detailed plumage sequence, particularly the timing and extent of the postjuvenile (i.e., first prebasic) molt is poorly known.

When several years are required for attainment of adult plumage, we normally describe the first-year (**Young** and/or **Immature**) and **Adult** plumages only, simply mentioning that intermediate stages exist; exceptionally, we may specify FIRST WINTER and so forth. Where several plumages are described, each succeeding description builds upon the preceding one(s); thus, we do not repeat elements of a plumage that are similar to the preceding one but merely specify what is different. Usually the last item in a description is the colors of ''soft parts'': iris (unless this is the usual dark brown or ''black,'' in which case it is omitted), unfeathered skin on the face or head, bill, and the legs and feet (for brevity, if both legs and feet are the same color, we simply say ''legs'' or ''feet,'' whichever is more conspicuous).

Habits

This section most often starts with a listing of the preferred habitats of the species as defined and described in preceding sections; the part of the habitat occupied (e.g., canopy versus understory of forest) is usually also specified. Social behavior is summarized briefly: we note whether the bird usually occurs singly, in pairs, or in pure or mixed-species flocks; whether it is territorial, a cooperative breeder, or a lek species and so forth. We also describe some of the most commonly observed foraging behaviors and some preferred foods, often including such information as the genus or family of fruits and flowers and the general type of insect. Much of this information will be more useful to scientists than to birders, but we suggest that the birder who wishes to see many fruit- and nectar-feeding species would do well to learn to recognize such plants as melastomes, *Cecropia, Heliconia*, etc. The data on foraging habits are drawn from our own field observations, from stomach contents of birds collected by Stiles and his colleagues, and from the literature, in that order. In any case, limitations of space preclude any attempt at completeness in listing the diet or foraging repertoire of any species. We also mention briefly any frequently observed display or behavioral mannerism, particularly when it might help to clinch an identification. In general, we try to present enough information in this section to give the reader at least a preliminary idea of the bird's ecology.

Voice

Here we describe the most frequently heard and diagnostic vocalizations of a species, wherever possible with information on the likely function or context—e.g., contact note, song, or aggressive sound. Often a verbal transliteration is included in italics; an exclamation point indicates very sharp, emphatic notes, and a question mark denotes an upwardly inflected, querying tone. These descriptions are in no sense an exhaustive listing of the sounds a species makes, and the reader should also realize that there is considerable geographic variation in the voices of many species, even within such a small country as Costa Rica. This section may be omitted if the bird is not known to vocalize in Costa Rica (e.g., many seabirds).

Nest

This section includes brief descriptions of the sites, forms, and materials of nests and the number, color, and pattern of eggs. In spite of many years of effort by Skutch in particular, in this section you will encounter the frustrating words "unknown" or "undescribed"; readers can help to fill gaps by publishing descriptions of any such nests and eggs they might find. If the nest and/or eggs are known from another country but not yet from Costa Rica, this fact is indicated in parentheses. We also record the approximate breeding season wherever possible. We often can estimate when a species breeds by examining the reproductive organs of collected specimens when we have never seen the nest. The reader should know, however, that breeding seasons of many tropical birds vary considerably from year to year and from place to place, and this information for most species is far from complete; a nest outside the period we mention is therefore not to be wholly unexpected. In some cases we specify the Costa Rican season—e.g., early wet season or dry season—especially in cases where rainfall rather than calendar month determines breeding. (See Figure 12.) If a bird definitely does not breed in Costa Rica (e.g., most migrants), this section is omitted.

Status

This section summarizes our knowledge of the abundance, distribution, and seasonal occurrence of each species in Costa Rica. Abundance is defined in terms of the following qualitative system. "Abundant" = many seen or heard daily (in appropriate habitat) and/or in large groups at frequent intervals; "common" = consistently recorded daily but in smaller numbers; "uncommon" = recorded regularly but usually not daily in small numbers; "rare" = recorded regularly but in very small numbers, at rather long intervals; "casual" = recorded only one or a few times in Costa Rica but normally ranges so close that an occasional occurrence is to be expected; "accidental" = recorded one or a very few times, far out of its normal range, and not likely to recur. It is obvious that this system is largely subjective and may underestimate the absolute numbers of some species that are usually silent and difficult to see (e.g., some rails and canopy hummingbirds). On the other hand, because we rely to a considerable extent on voice to detect forest birds, observers unfamiliar with their notes might at first feel that we overestimate the abundance of some species. Also, in practice the system is somewhat scaled to the particular group in question: a common hawk has a far sparser population than does a common tanager.

We describe the distribution of species in Costa Rica first in terms of the geographical sectors of the country as defined and depicted in Maps 1 and 2 and the section on geography. Sometimes we mention specific localities where a species is particularly abundant or easily seen or has a major breeding colony, as in certain waterbirds. The range of elevations usually occupied by a species is indicated in feet and meters, and the extent and timing of any altitudinal migrations are noted. The months of major migrations and periods of residence are recorded; we use the terms "winter resident," "breeding resident," and so forth to indicate a sojourn of several months during a particular time of year and "resident" without qualification when the species remains throughout the year. The terms "migrant" (spring or fall) and "visitor" imply that the species remains only briefly and indicate whether its visit is a stopover on a regular migration or more irregular, possibly a nomadic appearance.

Range

This section includes a brief statement of the species' geographic range; if this is extensive, its northern and southern limits are given, from west to east. For migrants, breeding and winter ranges are indicated. If Costa Rica itself is one limit of a species' range, this fact is mentioned without details; its distribution in the country is described in the preceding section.

Notes

This section is used principally to mention alternative English names, alternative taxonomic treatments and Latin names, and extralimital species that might reach Costa Rica and cause problems of identification. The alternative names and taxonomic arrangements given here are either those frequently used in the past but not now or those favored by one of us strongly, or both of us weakly, over the current AOU treatment. Only where both of us strongly agree on an alternative scientific name or taxonomic treatment (and only where this treatment has been mentioned as an alternative in the AOU Check-list itself) do we relegate the AOU's treatment to this section.

Virtually all the extralimital species mentioned here have occurred near Costa Rica in the recent past; in a few cases a single, unconfirmed sight record actually exists, but we think that further corroboration is required before the species is given a full account. Other species mentioned here are known from both north and south of Costa Rica and have apparently suitable habitat here (e.g., the Pearl Kite and the Wing-banded Antbird), are expanding their range toward Costa Rica (the Pearl Kite), or are open-country species that could easily do so as deforestation removes habitat barriers (e.g., the Yellowish Pipit and the Savanna Hawk). A few of the most probable of such species have been illustrated in the plates to facilitate their eventual identification if and when they arrive. Finally, we also include in this section a few species that have in the past been erroneously reported for Costa Rica (e.g., the Ruddy Duck and the Violet-bellied Hummingbird).

Information Not Included in This Guide

We do not treat subspecies; we do not wish the guide to be overly technical or too bulky, and we are too far from major ornithological collections to make or confirm many subspecific identifications. We mention geographic variation in appearance among Costa Rican birds where this is obvious in the field but do not attempt to relate it to subspecific nomenclature.

A related but more difficult decision involved the omission of precise measurements, particularly of bills, wings, tails, and eggs. In the first place, for too many species there are simply not enough specimens with complete data on gonads and skull ossification to assure an adequate sample of correctly aged and sexed birds, or the specimens are scattered widely among many museums; thus to amass such a sample would have delayed publication of this book inordinately. Moreover, it would have been necessary to determine the subspecies of every specimen, which we are not in a position to do. In his declining years Stiles may write a handbook including such information, unless, as he hopes, the AOU Check-list Committee anticipates him. For the present, mist-netters and others requiring such information are referred to Wetmore's *Birds of the Republic of Panama* (1965–1984), Blake's *Manual of Neotropical Birds* (1977), and Ridgway's *Birds of North and Middle America* (1901–1950).

We have also decided to exclude from this guide banding data from North American migrants. Stiles has been banding actively in Costa Rica since 1972, and several of his former students are now also engaged in this activity. In addition, S. M. Smith placed rings on many birds between 1974 and 1977, and a number of researchers and visitors have banded birds, especially at La Selva. We felt that the total amount of data was simply too great to be treated adequately in this guide. Moreover, the time required to gather and process data from many other banders prohibited the undertaking. Much information from Stiles's banding program, however, has found its way into the guide in the form of weight, plumage characters, and migration schedules of many species.

An Illustrated Glossary of Anatomical Terms Used in the Text

Descriptive Terms (Figure 22)

bill:

1. Upper mandible or maxilla
2. Culmen: ridge of upper mandible
3. Tomium (plural: tomia): cutting edge of bill
4. Lower mandible (often simply called mandible)
5. Commisure or rictus: corner of mouth
6. Gonys: ridge of lower mandible
7. Ramus (plural: rami): base of lower mandible

8. Nostril
9. Cere: skin surrounding nostril; may have sparse feathers
10. Rictal bristles
11. Nasal tufts
12. Lores
13. Forehead
14. Crown (may be divided into forecrown, midcrown, hindcrown); forehead plus crown called ''pileum'' (top of head)
15. Nape (adjective: nuchal): back of head
16. Malar area: extreme side of throat, below cheek
17. Cheeks: below eye to lower mandible
18. Auriculars (covering ear openings): ear-coverts
19. Hindneck
20. Side of neck
21. Back
22. Scapulars. (*Note*: back, scapulars, and, usually, lesser and middle secondary coverts are often collectively called the ''mantle'')
23. Rump
24. Upper tail-coverts
25. Rectrices or tail feathers (singular, rectrix). *Note*: numbered from central or innermost to outermost
26. Chin
27. Throat
28. Foreneck (used mainly for large or long-necked birds)
29. Chest or upper breast; the furcula is the ''wishbone''
30. Lower breast (''breast'' used without qualification includes upper and lower breast). Pectoral: adjective for ''breast''
31. Belly or abdomen
32. Sides
33. Flanks
34. Vent
35. Crissum or lower (under) tail-coverts

legs:

36. Thighs (actually feathered part of tibia)
37. Tibia (tibiotarsus)
38. Tarsus (tarsometatarsus)
39. Hallux or hind toe

40. Primaries. *Note*: numbered from innermost (proximal) to outermost (distal)
41. Secondaries. *Note*: numbered from outermost to innermost
42. Tertials or innermost secondaries (which often differ in shape or color from the other secondaries). Primaries plus secondaries (including tertials) constitute the remiges (singular: remex)
43. Primary coverts
44. Greater secondary coverts or wing-coverts
45. Middle or median secondary coverts or wing-coverts
46. Lesser secondary coverts or wing-coverts or shoulder
47. Alula or bastard wing
48. Alula covert
49. Bend of wing or wrist
50. Wing-linings (lower wing-coverts)
51. Axillars: base of underwing, ''armpit''
52. Inner web (of feather)
53. Outer web (of feather)
54. Shaft (of feather)
55. Eyelid
56. Iris (often called simply eye)

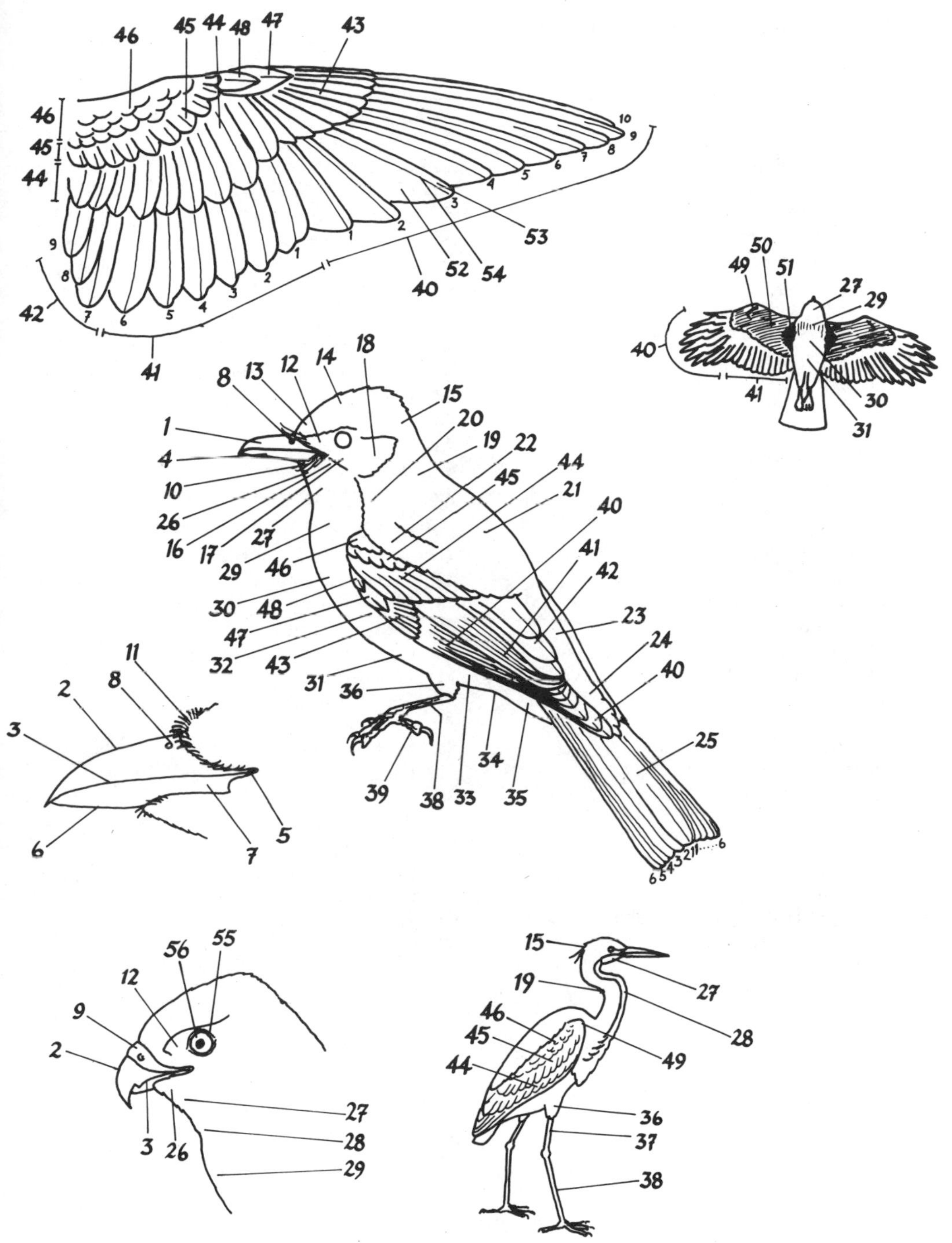
46
45
44
48
47
43
46
45
44
10
9
8
7
6
5
4
3
2
1
53
52
54
40
9
8
7
6
5
4
3
2
1
42
41
50
49
51
27
29
40
41
30
31
18
14
12
13
8
15
1
20
4
19
22
10
45
44
21
26
16
17
27
29
46
40
41
42
30
48
47
32
43
23
24
31
36
40
25
39
38
33
34
35
6
5
4
3
2
1
1
6
11
2
8
3
5
6
7
56
55
12
9
2
27
28
3
26
29
15
27
19
28
46
45
49
44
36
37
38

Patterns and Markings (Figure 23)

General

1. Cap: contrasting crown or pileum
2. Superciliary or eyebrow
3. Eye-stripe
4. Malar stripe
5. Supraloral stripe
6. Eye-spot (postocular spot)
7. Gorget: contrasting throat
8. Epaulet: contrasting shoulder-patch
9. Eye-ring
10. Ear-patch: contrasting auriculars
11. Breast-band (chest-band if across uppermost breast)
12. Half-hood
13. Spectacles
14. Hood
15. Nuchal collar (encircling hindneck)
16. Mask
17. Bib

Plumage Patterns

18. Striped
19. Coarsely or broadly streaked
20. Finely streaked
21. Mottled
22. Spotted
23. Speckled or dotted
24. Scaled
25. Coarsely scaled or scalloped
26. Chevroned
27. Finely barred
28. Barred
29. Banded

Feather Markings

30. Contrasting margin or edging
31. Submarginal marking
32. Fringe
33. Barring
34. Vermiculations
35. Shaft-streak
36. Contrasting feather-bases
37. Bases partly exposed
38. Spotted (margins)
39. Ocellated (more or less eyelike spots)

Wing Patterns

40. Contrasting leading edge
41. Speculum: contrasting patch on secondaries
42. Contrasting primary bases or ''window''
43. Carpal bar
44. Wing-stripe
45. Wing-bars
46. Contrasting margins or feather-edgings

Tail Shapes and Markings

47. Graduated tail
48. Contrasting tips to outer rectrices
49. Wedge-shaped tail
50. Contrasting corners of tail
51. Rounded tail
52. Terminal band
53. Subterminal band
54. Median band
55. Basal band
56. Truncate or square tail
57. Barred pattern
58. Notched tail
59. Contrasting outer rectrices
60. Forked tail
61. Tail-spots
62. Double-rounded tail

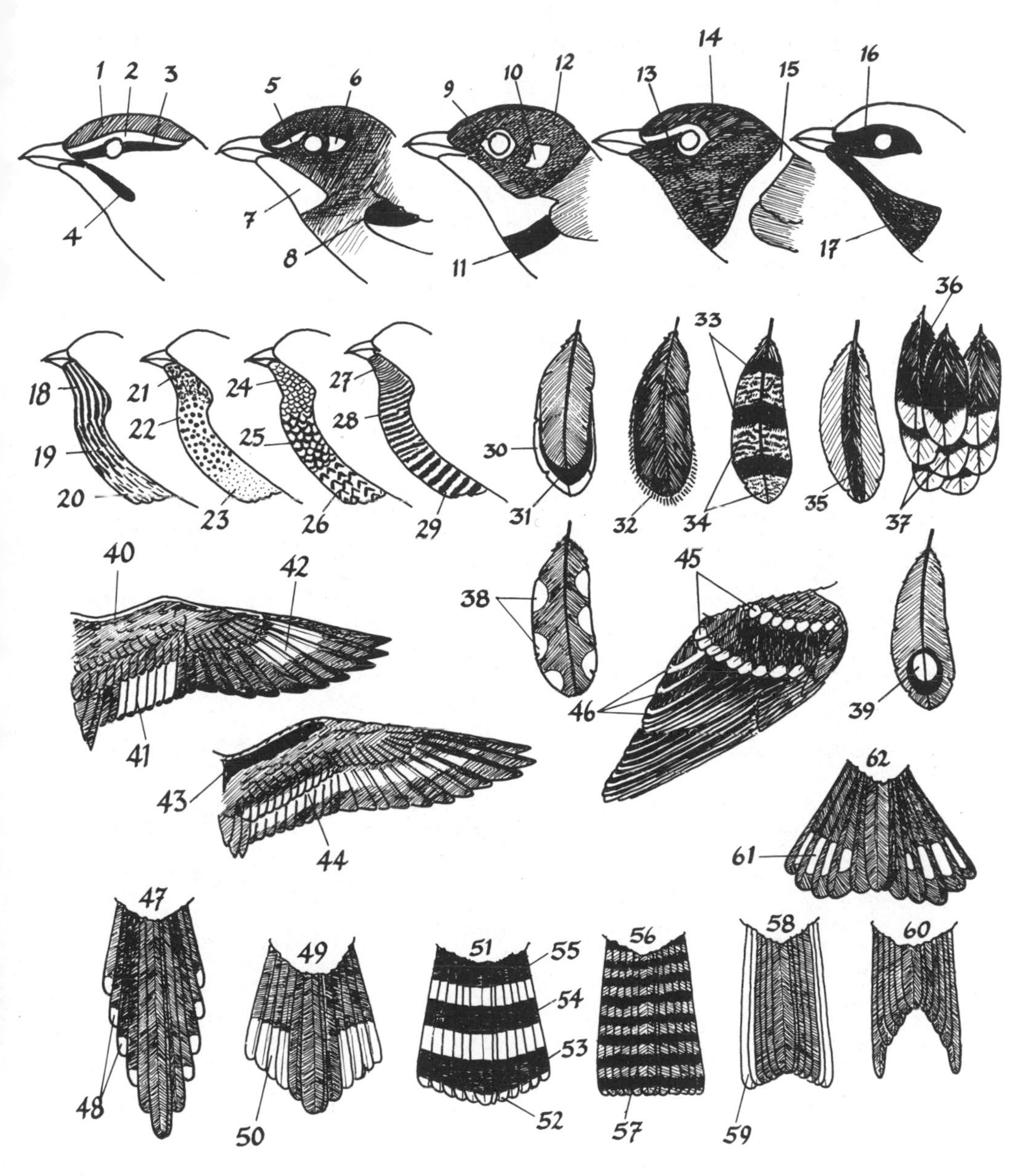
1
2
3
4
5
6
7
8
9
10
11
12
13
14
15
16
17
18
19
20
21
22
23
24
25
26
27
28
29
30
31
32
33
34
35
36
37
38
39
40
41
42
43
44
45
46
47
48
49
50
51
52
53
54
55
56
57
58
59
60
61
62

FAMILY AND SPECIES ACCOUNTS

ORDER **Tinamiformes**
FAMILY **Tinamidae:** Tinamous

The 46 species of tinamous are distributed over the American continents from eastern Mexico to the Strait of Magellan. They are stout-bodied birds with thin necks, slender bills of moderate length, very short tails, and short, often stout legs with 3 forwardly directed toes and a rudimentary, elevated hind toe. They are clad in subdued shades of brown, gray, and olive, often spotted, vermiculated, or barred; the brightest color is usually rufous or chestnut. The sexes are similar, but the females are often somewhat larger and more strongly patterned. Almost wholly terrestrial, tinamous forage over the ground in tropical and subtropical forests and thickets, in South Temperate grasslands, or amid tussocks on the high Andean puna. Their food consists largely of seeds and fruits varied with small invertebrates and, rarely, vertebrates. They try to avoid enemies by walking inconspicuously into dense vegetation but, if suddenly alarmed, may rise with a loud burst of wingbeats and hurtle away, to drop into concealing herbage at no great distance. The Great Tinamou roosts in trees, but other species sleep on the ground. Purity of tone makes the songs of some tinamous arrestingly beautiful. Tinamous' mating habits are varied, including monogamy, polygyny, and serial polyandry, but apparently in all species the male alone takes full charge of eggs and young. The 1–12 (sometimes more) eggs in a clutch, laid by 1 or more females, are highly glossy, immaculate blue, purplish, gray, chocolate, or almost black. No nest is built; the eggs are laid in a slight depression in fallen leaves and other ground litter at the base of a tree, amid fallen brush, or beneath a tussock of grass; they hatch in 19–20 days. As soon as the beautifully variegated, downy chicks are well dry, they are led away by the male parent. Chicks are defended by the male for several weeks but feed themselves during much of this time.

GREAT TINAMOU
Tinamus major Pl. 12(6)
Tinamú Grande (Gallina de Monte, Perdiz, Gongolona)

Description. 17″ (43cm); 1.1kg. Large, stout-bodied; grayish legs; very slender neck; small head. **Adults:** above dark olive-brown, barred and flecked with black; below paler, barred with black and buffy on thighs and flanks; throat whitish; crown sooty-black or (S Pacific slope) chestnut. Highland Tinamou rufescent below, especially on throat.

Habits. Forages while walking through ground cover of humid forest or adjacent tall second growth, gathering seeds, fruits, and small litter animals including insects, spiders, small lizards, and frogs; occasionally accompanies army ants; when alarmed, rises explosively with loud wingbeats and a penetrating whistle, shoots between trunks, and alights beyond view amid concealing vegetation; sleeps, and rarely rests by day, on a horizontal branch well above ground; usually solitary unless accompanied by chicks (male); very wary where hunting is frequent.

Voice. Deep, powerful, whistled notes, organlike in their velvety, swelling quality, among the most stirring sounds of tropical forest. Typically 2–4 pairs of whistles are given, the first of a pair smooth and level, the second tremulous and sliding slightly downscale. Most often heard as night descends.

Nest. Directly on ground litter at base of a trunk, between buttresses, or beside fallen branches or other shelter. Eggs 3–5, rarely more, glossy, turquoise-blue or blue-green. December–August.

Status. Common (where forest remains) resident of lowlands and foothills of entire Caribbean slope, locally to 5000ft (1500m), and S Pacific slope (N to Carara), to 5600ft (1700m) in upper Coto Brus valley; locally on Pacific face of Guanacaste volcanoes above ca. 2600ft (800m).

Range. S Mexico to W Ecuador, N Bolivia, and C Brazil.

HIGHLAND TINAMOU
Nothocercus bonapartei Pl. 12(7)
Tinamú Serrano (Gallina de Monte, Gongolona)

Description. 15″ (38cm); 850g. Fairly large; blackish pileum contrasts with cinnamon-rufous throat; legs grayish. **Adult ♂:** above olive-brown, vermiculated with brown and spotted with buff, especially posteriorly and on wings; foreneck olive-brown; breast and belly dull cinnamon to buffy-brown, obscurely vermiculated with dusky; sides and flanks olive-brown, vermiculated with black and spotted with buff. Upper mandible blackish, lower yellowish with dusky tip. ♀: similar but above and below more heavily and contrastingly marked with black. **Young:** faint white barring on head; upperparts mottled with black and rufous. Only tinamou at high elevations; more richly, contrastingly colored than Great Tinamou, which overlaps locally at middle elevations and has whitish throat.

Habits. Prowls the floor of highland forests, usually singly, keeping to dense undergrowth; wary but not especially shy; may approach motionless observer closely or walk ahead down a trail; flies rarely, when startled at close range; feeds on fallen fruits, small litter animals.

Voice. Males give a deep, throaty, resonant *huh-wowr* or *unh-heerr*, steadily and monotonously (ca. 6 calls/5sec), usually from a thicket; at a distance call sounds more whistled, only second note audible.

Nest. In a leafy depression, usually at base of tree. Eggs 2–5, blue-green, darker than those of Great Tinamou, laid by 1 or more females (?). March, June, July.

Status. Widespread but generally uncommon resident in highlands the length of the country, above 3900ft (1200m) in Cordillera de Guanacaste, mostly above 5000ft (1500m) in Cordillera Central and Cordillera de Talamanca [sporadically down to 3300ft (1000m) on Caribbean slope of former].

Range. Costa Rica to W Panama, and Andes from N Venezuela to NW Peru.

LITTLE TINAMOU

Crypturellus soui Pl. 12(17)

Tinamú Chico (Yerre, Gongolona)

Description. 9″ (23cm); 250g. Quail-sized, stout-bodied, virtually tailless. **Adults:** mostly dark brown, below paler; belly grayish-buff; crown blackish; cheeks sooty-gray; throat whitish. Iris orangish; legs and feet dull yellowish-gray. **Immatures:** above mainly cinnamon-brown, spotted with black; below grayish-brown, strongly but brokenly barred with black. Shorter-tailed than similar-sized pigeons and doves; bobwhite quail more streamlined, crested, usually in coveys; other small tinamous have red legs.

Habits. Secretive and solitary; inhabits dense second growth, weedy plantations, sugarcane fields, and forest edge in humid regions; in thickety understory of broken forest; rarely seen except when it walks across a path or other opening; flies reluctantly; eats seeds, berries, insects, occasional small frogs; also takes much grit for grinding food.

Voice. Beautifully modulated, tremulous whistles, higher-pitched than those of Great Tinamou. Full song a series of 4–8 ascending trills, each starting at a slightly higher pitch than preceding; also a longer single trill that rises, then falls.

Nest. Directly on fallen leaves amid dense, thickety concealing vegetation. Eggs 2, glossy, vinaceous-lavender, almost equally blunt on both ends. Virtually year-round, perhaps least often March–June.

Status. Locally common resident from sea level to ca. 5000ft (1500m) throughout Caribbean slope, in W and S parts of Península de Nicoya, and from about Orotina S on Pacific slope; locally along Pacific face of N cordilleras and very uncommonly in W Valle Central.

Range. S Mexico to W Ecuador, N Bolivia, and SE Brazil.

THICKET TINAMOU

Crypturellus cinnamomeus Pl. 12(18)

Tinamú Canelo (Gongolona, Perdiz)

Description. 11″ (28cm); 480g. Medium-sized, red-legged, strongly patterned; the only tinamou in most of N Pacific slope. **Adults:** upperparts heavily barred with brown or buff and black; underparts mostly cinnamon-buff, barred with black on flanks; throat white. Breast of female extensively and finely barred with black, that of male with little or no barring. Iris orangish; upper mandible blackish, lower yellowish. Similarly red-legged Slaty-breasted Tinamou much darker, absent from dry lowlands; all other tinamous and wood-quails have dull legs.

Habits. Walks deliberately on ground in forest and second growth, preferring areas with dense understory, especially spiny terrestrial bromeliads; during dry season often most numerous in evergreen gallery forest; usually seen singly; most active and vocal in early morning and late afternoon, especially during dry season; feeds on fallen fruits, seeds, occasionally small litter animals.

Voice. A resonant, hollow, whistled *tuuuuu*, often with a quaver in the middle, lasting nearly a second. In some populations (e.g., Península de Nicoya) a 2-part *tooouuuu*, the second part slightly higher in pitch.

Nest. A leafy hollow on forest floor, usually amid low vegetation and often on sloping ground. Eggs 2–7 (probably depending on number of females who lay in male's nest), lavender or purplish-brown with dull gloss. March–August.

Status. Widespread resident of N Pacific slope, usually common wherever large patches of tropical dry forest remain; from sea level to ca. 1650ft (500m) on Cordillera de Guanacaste, and throughout mountains of Península de Nicoya, to 3300ft (1000m).

Range. W and NE Mexico to Costa Rica.

Note. The very similar form *C. idoneus* of NW South America is often considered a race of *cinnamomeus*, which in turn is sometimes lumped into the Brazilian *C. noctivagus*.

SLATY-BREASTED TINAMOU

Crypturellus boucardi Pl. 12(8)

Tinamú Pizarroso

Description. 11″ (28cm); 500g. Medium-sized, dark-colored, with reddish legs. **Adult** ♂: throat whitish; head, neck, breast, and upper back dark gray; rest of upperparts dark brown, barred with black; belly tawny; flanks barred with black. Upper mandible black, lower horn-color to yellowish. ♀: similar but wing-coverts barred with buff. **Immatures:** above with buff barring more extensive and conspicuous. Little Tinamou smaller, more uniform brown with grayish legs.

Habits. Frequents dense undergrowth in wet forest and old second growth; terrestrial, usually solitary; prefers to escape danger by walking quietly away or freezing; eats fallen fruits and seeds (including palms, *Protium*, Sapotaceae), ants, and other insects, occasionally small lizards and frogs.

Voice. Call of male a deep, hollow, resonant whistle: *aaaah-ooooaaah*, first note down-slurred, second slurred up at end; at close range very ventriloquial, with a faraway quality; at a distance, resembles call of a *Leptotila* dove but with a different pattern.

Nest. In leafy hollow on forest floor, usually amid dense, low vegetation. Eggs usually 2, up to 7 (probably when more than 1 female lays in nest), dull lavender with slight gloss. March–June, perhaps through October.

Status. Locally common resident of N Caribbean lowlands, S to about Limón, up locally to ca. 2300ft (700m), reaching Pacific slope in passes between Guanacaste volcanoes.

Range. SE Mexico to Costa Rica.

Note. Sometimes *C. kerriae* of E Panama and Colombia, and *C. columbianus* of Colombia, are considered races of *C. boucardi*.

ORDER **Podicipediformes**
FAMILY **Podicipedidae:** Grebes

With 20 species, the grebe family is of nearly cosmopolitan distribution, absent only from the high Arctic, Antarctic, and some oceanic islands. Highly adapted for aquatic life, grebes have short wings, vestigial tails, legs placed far back on the body, lobed toes, and short to moderately long bills that are usually sharp although thick in the Pied-billed Grebe. Their very soft and dense plumage is mostly brownish or black above and pale below, with extensive areas of rufous or chestnut on some species. The sexes are alike or nearly so, and in some species both wear conspicuous head adornments in the breeding season. Expert divers, they subsist upon aquatic insects, crustaceans, and small vertebrates, mostly caught underwater. In the elaborate, highly vocal courtship displays of certain grebes, the sexes take identical or interchangeable roles. Their nests of coarse aquatic vegetation, with a shallow concavity in the top, usually float in shallow water, anchored to rooted plants, but may rest upon the bottom. Their white or whitish eggs, which are soon stained brownish by contact with damp vegetation, are incubated by both sexes, as far as is known, for 3–4 weeks. On leaving the nest, the parent covers the eggs with nest material. The downy, often boldly patterned young swim when a few hours old but often voyage on their parents' backs.

PIED-BILLED GREBE

Podilymbus podiceps Pl. 7(3)

Zambullidor Piquipinto (Pato de Agua, Gallardo)

Description. 13″ (33cm); 450g. Teal-sized; dark eye; thick pale bill. **Adults:** above dark dusky-brown; below mixed and mottled with dusky and white; pileum and hindneck blackish; sides of head dark brown; cheeks and neck buffy brown to grayish-brown. Breeding: throat black (often flecked with white in male); eye-ring and bill white, black ring around bill; feet black, toe webs bluish-gray. Nonbreeding: throat, eye-ring, and bill dull whitish. **Young:** boldly marked with black and white stripes and spots; buffy crown-patch; these are gradually lost as adult plumage is acquired.

Habits. Frequents freshwater ponds and lakes and smoothly flowing rivers with aquatic vegetation; dives when alarmed, reappearing at a distance, often with only its head above water; takes more fishes than Least Grebe; residents highly territorial, in pairs or family groups; wintering birds more gregarious.

Voice. A whinny followed by a loud *cow cow*

cow cow cow; a high-pitched, rattling *ke-ke-ke-ke*. Chicks utter a bright metallic *chip* and excited *weep, peep peep*.

NEST. A mass of floating vegetation with a shallow depression in the top, like that of Least Grebe but larger, often 1ft (30cm) in diameter. Eggs 5–7, white with brownish stains, covered with nest material in absence of parent. Almost throughout year.

STATUS. Resident and possibly winter resident, lowlands to middle elevations of both slopes, principally in Guanacaste, Río Frío region, and Valle Central; much local movement, depending on water levels.

RANGE. SE Alaska and N Canada to S Argentina; West Indies.

LEAST GREBE

Tachybaptus dominicus Pls. 7(4), 51(3)

Zambullidor Enano (Patillo, Pato de Agua)

DESCRIPTION. 9½″ (24cm); 150g. Our smallest swimming waterbird; appears all dark with bright yellow eye; white inner remiges conspicuous in flight; bill black, sharp-pointed. **Adults:** hindneck and upperparts blackish-brown; rest of head and neck dark grayish; underparts dusky, more or less mottled or mixed with white. Bill black with grayish to horn-color at tip; feet black. BREEDING: chin and throat black (usually flecked with white in male). NONBREEDING: throat whitish; head and neck more brownish, often tinged with cinnamon. **Young:** above blackish; head and neck striped boldly with white; below whitish; iris at first dark.

HABITS. Singly or in pairs, sometimes with 3 or 4 well-grown young, swims in open water of small, mostly vegetation-covered ponds or near margins of larger bodies of water; may occur on slow-flowing rivers, especially in dry season; dives for food and when alarmed; rarely, when closely pursued, flies low over surface; eats aquatic insects, crustaceans, small fishes and frogs.

VOICE. A sharp *beep* or *bamp* like a little tin horn; a loud, rolling *chirr* like that of White-throated Crake but drier, less bubbling. Half-grown young beg with a rapidly repeated, shrill *peep*.

NEST. A compact mass of aquatic weeds floating in still water, usually anchored to rooted plants. Eggs 3–6, glossy whitish, usually stained with brown by damp nesting material. Almost throughout year.

STATUS. Resident on both slopes from lowlands to ca. 5000ft (1525m); most numerous in Valle Central, Guanacaste, and San Vito region.

RANGE. S Texas to N Argentina; Bahamas and Greater Antilles.

NOTE. Formerly placed in the genus *Podiceps*.

EARED GREBE

Podiceps nigricollis Pl. 51(2)

Zambullidor Mediano

DESCRIPTION. 12″ (30.5cm); 300g. Larger than Least Grebe with longer, slimmer neck and bill; bill slightly upturned. **Adults** WINTER: pileum and sides of head blackish; hindneck and upperparts dark grayish-brown to dusky, except secondaries mostly white, conspicuous in flight; throat and area behind cheeks and auriculars grayish-white; foreneck, sides of breast, sides, and flanks pale grayish-brown, somewhat mottled with dusky; vent brownish; rest of underparts white. BREEDING: head, neck, upper breast black; a short rounded crest; a tuft of long, narrow golden-yellow to orangish feathers fan out from behind eye over side of head; sides and flanks mixed with chestnut to tawny. Males brighter on average. Iris scarlet, bare eye-ring orangish, bill and most of feet black, inner surfaces of toe lobes bluish. **Immatures:** resemble winter adults.

STATUS. Known in Costa Rica from one record: 2 breeding-plumaged birds seen and photographed on a glacial lake in Valle de las Morenas, at 11,500ft (3500m) on the Cerro Chirripó massif of the Cordillera de Talamanca, 22–23 April 1987 (Peterson and Capkanis). To be expected occasionally in migration or winter in Costa Rica; has recently been reported to winter rarely but possibly regularly to El Salvador (Thurber et al. 1987). All records south of Mexico have been on freshwater lakes.

RANGE. Breeds in W North America from S Canada to NW Baja California and locally to C Mexico; winters from SW USA commonly to Guatemala, rarely to El Salvador; accidental or casual in Costa Rica. Also widespread in Old World.

NOTE. Formerly called *P. caspicus*. A rufous-necked resident form from the Colombian Andes, *P. andinus* (now extinct), was sometimes included as a race of this species.

ORDER **Procellariiformes:** Albatrosses, Shearwaters, Petrels, and Allies

The most obvious feature of this order of webfooted marine birds is the tubular nostril openings, which may be located on either side of the bill or may be separate or fused atop the culmen. The bill itself is unique, being covered by heavy, horny plates separated by deep grooves, with a hooked tip. The bulging forehead of these birds contains the enlarged nasal glands, which excrete a concentrated solution of salt, thereby enabling the birds to drink seawater. Usually silent at sea, procellariiform birds are often highly vocal on the breeding grounds.

FAMILY **Procellariidae:** Shearwaters, Petrels, Fulmars, and Prions

The 55 or so species of this cosmopolitan family have long, narrow, pointed wings and mostly short legs and tails. The plumage is white, gray, brown, or black or some combination thereof; color does not vary with sex or age, but many species have dark and pale morphs. Procellariids are pelagic, coming to land only to breed, usually in colonies on oceanic islands. A few species nest on open ground or cliff ledges, but most nest in burrows or crevices that the birds enter and leave mostly at night; evidently they use their keen sense of smell to locate the correct burrow. The incubation period of the single white egg is extremely long (ca. 43–60 days or more), and males and females alternate long sessions of up to 2 weeks, during which they fast while their mate is off foraging. The downy chick is fed only once every few days on the average and grows very slowly. It is fed by regurgitation, at first on a strong-smelling oil secreted by the adult's stomach lining; this oil may also be ejected from mouth or nostrils as a defense measure.

The 8 species of procellariids recorded from Costa Rica fall into two groups: the shearwaters and the gadfly-petrels, which differ in body form and manner of flight. Specific identification at sea is often tricky; useful features include the manner of flight and colors of bill, feet, and underparts (including underwings).

BLACK-CAPPED PETREL
Pterodroma hasitata Pl. 1(15)
Petrel Gorrinegro

Description. 16″ (40cm); 300g. Bold pattern, distinctive flight, and short, heavy black bill distinguish this bird from shearwaters. Forehead, lores, and underparts white; rump and nuchal collar usually white (sometimes white reduced or lacking); cap black; upperparts dusky; underwing white with black bar across coverts, black trailing edge; legs pinkish.

Habits. Usually solitary at sea; flies swiftly, dashingly, with a series of long, swooping glides on rigid, bowed, and angled wings; at the end of each arc sweeps upward with a few quick wingbeats, then swoops away on another tack; feeds on fishes, shrimps.

Status. A single sighting off Caribbean coast, 14 August 1953 (Slud), where probably a rare but regular visitor well offshore.

Range. Breeds on mountaintops in Greater and Lesser Antilles; ranges at sea throughout the Caribbean and Atlantic Ocean.

Note. Greater Shearwater (*Puffinus gravis*) is somewhat similar in pattern but with less white on hindneck and rump and none on lores or forehead; different shape and flight; has been reported from Tortuguero on strength of a few bones found on beach (Carr). Even assuming correctness of identification, there is no guarantee that this record represents the unassisted arrival of a living bird, and we consider the occurrence of this species in Costa Rica to be hypothetical; this Atlantic Ocean species is otherwise unrecorded in E Caribbean.

DARK-RUMPED PETREL
Pterodroma phaeopygia Pl. 1(11)
Petrel Lomioscuro

Description. 17″ (43cm); 385g. A large, long-winged gadfly-petrel, the only *Pterodroma* so far recorded off the Pacific coast of Central America (others are possible). Above blackish with black cap extending down sides of face; broad forehead white; below white, with small black patch on axillars; sometimes sides of rump white. Shearwaters have very different flight, longer bills.

Habits. Ranges well offshore except near breeding areas; usually seen singly at sea; flies as described for Black-capped Petrel; eats fishes and squids; exact manner of foraging unknown but almost certainly feeds at night (as, apparently, all *Pterodroma* do).

Status. Rare but probably regular visitor well off Pacific coast, including vicinity of Cocos I. The various Costa Rica–Panama records span the period February–September.

Range. Breeds in Hawaiian and Galápagos Is.; the latter population ranges widely in E-C Pacific Ocean off Central and South America.

Note. Sometimes considered a subspecies of the Caribbean *P. hasitata*. Also likely off our area, especially Cocos I., is Juan Fernández Petrel (*P. externa*), which breeds off Chile, migrates to N Pacific: 17″ (43cm); back and upper tail-coverts paler brown than rest of upperparts; dusky cap extends less onto face; less white on forehead; no black on axillars.

PARKINSON'S PETREL

Procellaria parkinsoni Pl. 1(12)

Petrel de Parkinson

DESCRIPTION. 18″ (46cm); 675g. A large, heavily built, dark shearwater with heavy, pale, dark-tipped bill and dark feet; plumage entirely sooty blackish, but primaries below show white shafts; feet usually extend beyond tail tip in flight. Sooty Shearwater smaller, with pale underwings.

HABITS. Little known; apparently occurs singly in our waters; may be attracted to fishing boats for offal; in flight more agile and maneuverable than *Puffinus* shearwaters, with more flapping, less soaring; in calm weather may fly powerfully and directly, with steady wingbeats.

STATUS. Rare but probably regular visitor off Pacific coast, mostly well offshore; most Central American records March–April.

RANGE. Breeds on islands off New Zealand; ranges N and E to waters off Central and South America.

NOTE. Also called Black Petrel, but we prefer to avoid this name, as it was for long applied to the Black Storm-Petrel and could cause confusion. Also possible but not yet definitely recorded in our area is Flesh-footed Shearwater (*Puffinus carneipes*), which breeds off New Zealand, migrates to NE Pacific: 17″ (43cm); all dark like present species, with pinkish, dark-tipped bill, pink feet, typical shearwater (e.g., Sooty Shearwater) flight; at close range may show a whitish sheen on undersides of primaries.

PINK-FOOTED SHEARWATER

Puffinus creatopus Pl. 1(8)

Pardela Blanca Común

DESCRIPTION. 19″ (48cm); 700g. Fairly large, robust; above uniform dark gray-brown, blending into white underparts—not clean-cut; lower tail-coverts dark; underwing white with dark blotching. Bill pinkish with dark tip; feet pink. Pale-phase Wedge-tailed Shearwater has longer tail, pale scaling on back, darker bill and feet, different flight.

HABITS. Prefers cold waters and probably moves rapidly through our area; solitary or gregarious at sea; flies more slowly and heavily than Wedge-tailed Shearwater, with long, stiff-winged glides broken by easy flaps, becoming broad, sweeping arcs in strong breeze; eats more fishes than most shearwaters.

STATUS. Several definite or probable sightings in waters near Cocos I., where probably rare but regular during migrations (May–June, September–October).

RANGE. Breeds on islands off S Chile; migrates N in nonbreeding season to NE Pacific, from California to Alaska.

NOTE. Two members of Manx Shearwater group reported to range S at least occasionally to general area of Cocos I.: Townsend's (*Puffinus auricularis*)—blackish above, white below, clean-cut, with conspicuous white sides to rump; and Black-vented (*P. opisthomelas*)—dingy; dark brown upperparts blend through breast, sides, and vent to dull whitish belly. Both are rather small (ca. 14″, 35cm), with rapid flutter-and-gliding flight close to surface; both breed on islands off W Mexico.

WEDGE-TAILED SHEARWATER

Puffinus pacificus Pl. 1(10)

Pardela Colicuña

DESCRIPTION. 17″ (43cm); 400g. Medium-sized, lightly built; at sea tail appears notably long but often not obviously wedge-shaped. Above dark gray-brown, broadly scaled with paler grayish. PALE PHASE: below white, mottled with dusky-brown on sides of breast and sides and underwing—not clean-cut. DARK PHASE: very rare in our waters, below uniform blackish-brown. Bill dark gray; legs dull flesh.

HABITS. Seasonally in large flocks well off Pacific coast, though usually seen singly or in small groups close to shore; flies distinctively, lightly and gracefully rather than powerfully, with a few shallow flaps, then a buoyant glide on forward-bowed wings, sometimes veering or circling at low speed; in high winds, flies faster and more erratically; rarely follows ships; usually feeds by flying close to surface, dipping head to snatch prey, fluttering and foot-paddling to regain speed.

STATUS. Common to abundant visitor to waters between mainland and Cocos I., mostly between January and March; sporadically and in small numbers close to coast, especially near entrance to Golfo de Nicoya.

RANGE. Widespread in tropical Pacific Ocean; nearest breeding colony off W Mexico.

SOOTY SHEARWATER

Puffinus griseus Pl. 1(9)

Pardela Sombría

DESCRIPTION. 17″ (43cm); 600g. Medium-sized, rather heavy-bodied and narrow-winged, entirely sooty blackish-brown except wing-linings extensively pale gray; at close range may show paler, grayish chin. Bill relatively long and slender, blackish; feet dusky. Only dark shearwater likely in coastal waters; see Short-tailed Shearwater.

Habits. Often found in loose flocks of a few to 50 or more birds; flies swiftly, low over water, with several rapid flaps alternating with long, banking glides on straight, rigid, horizontal wings; often rests on water during calm days; forages largely by shallow dives from surface or flight; attracted to chum, may follow fishing boats.

Status. Sporadically common to abundant visitor off Pacific coast between May and October; in some years numerous in Golfo de Nicoya; not yet recorded off Caribbean coast, where occasional birds may be expected.

Range. Breeds on islands off S South America and New Zealand; during nonbreeding season makes extensive migrations into N Pacific and N Atlantic oceans.

SHORT-TAILED SHEARWATER

Puffinus tenuirostris Pl. 1(13)

Pardela Colicorta

Description. 16″ (40cm); 525g. Very similar to Sooty Shearwater and not always distinguishable; useful characters include smaller bill and rounder forehead, darker gray underwing (but sometimes a whitish patch near bend of wing); sometimes shows a vague darker cap and/or whitish chin, but overall usually seems more uniformly dark than Sooty Shearwater.

Habits. Generally similar to those of Sooty Shearwater.

Status. Accidental in our area: a single sighting under optimum conditions, ca. 40mi (60km) S of Cabo Blanco, 7 July 1978 (Stiles). Main migration route is well to W of our area, but an occasional stray is not unexpected.

Range. Breeds on islands off S Australia; migrates to N Pacific and Bering Sea during nonbreeding period.

AUDUBON'S SHEARWATER

Puffinus lherminieri Pl. 1(14)

Pardela de Audubon

Description. 12″ (30cm); 175g. Small, relatively stocky-bodied and short-winged; distinctive fluttery flight. Above blackish-brown; below white with some dark mottling at sides of breast, otherwise fairly clean-cut; crissum brownish; underwing white, leading and trailing edges dusky. Bill gray with blackish culmen and tip; feet dull flesh. Townsend's Shearwater is larger, with white sides of rump, more extensively white, clean-cut underwing, less fluttery flight.

Habits. At sea usually in small groups or singly; may approach and circle boats with fluttery, swiftlike (*Chaetura*) flight, bursts of rapid, stiff-winged flaps alternating with short glides low over water; spends much time swimming on surface; often dives for food.

Status. Uncommon visitor throughout year off Pacific coast, especially about mouth of Golfo de Nicoya; likewise rare but probably regular off Caribbean coast, which is near breeding colonies.

Range. Widespread in tropical and subtropical seas; nearest Pacific breeding colony on Galápagos Is.; in Caribbean breeds on Isla Providencia and off Bocas del Toro, Panama.

FAMILY **Hydrobatidae:** Storm-Petrels

The smallest oceanic birds, storm-petrels have relatively shorter wings and longer legs than shearwaters; the nostrils are united in a common tube atop the strongly hook-tipped bill. The 20 species have brownish-black or gray plumage, sometimes with more or less extensive areas of white on the rump and/or underparts (in some species the amount of white varies widely within or between populations, but no plumage variation is associated with sex or age). Completely pelagic outside the breeding season, storm-petrels feed on plankton and scraps gleaned from the ocean's surface. They swim buoyantly but not strongly; their legs are too weak to support them on land. Breeding is colonial, usually on islands, in nests in burrows or rock crevices. Each parent incubates the single white egg for 2–4 days continuously while the other is off feeding; incubation periods are 40–50 days. The downy chick is fed first on stomach oil, later on plankton regurgitated by adults after lengthy foraging trips.

Identification of storm-petrels at sea is often difficult. The most useful character is often the manner of flight; differences in tail shape, rump-patch, or feet are usually apparent mostly at close range.

WILSON'S STORM-PETREL

Oceanites oceanicus Pl. 2(15)

Paiño de Wilson

Description. 7″ (18cm); 35g. Medium-sized, with rather short, broad wings, short, square tail, rather long legs with yellow-webbed feet that extend beyond tail tip in flight. Plumage mostly sooty-brown, with conspicuous, square, bright white rump-patch, pale grayish diagonal band across upper wing-coverts; may show some white on vent. Leach's and Band-rumped storm-petrels have more "kinked" or angled wings, more erratic flight.

Habits. Flies steadily, directly, with almost continuous swallowlike flapping punctuated by short glides; frequently dabbles or patters with feet on surface while feeding; singly or in loose groups; often follows ships or fishing boats.

Status. Regular but sporadic seasonal visitor (April–August) off both coasts; sometimes in good numbers off outer Pacific coast but infrequent inside Golfo de Nicoya.

Range. Breeds on sub-Antarctic islands; ranges N in nonbreeding season (northern summer) to 50°N in the Atlantic and 40°N in the Pacific.

Note. Also possible in our area, especially toward Cocos I., is White-vented Storm-Petrel (*Oceanites gracilis*) (breeds Galápagos Is., recorded in Panama): smaller (6″, 15cm) than Wilson's but similar in shape; center of lower breast to vent white, like rump-patch; flight like Wilson's but wingbeats more rapid.

WHITE-FACED STORM-PETREL

Pelagodroma marina Pl. 2(11)

Paiño Pechialbo

Description. 8″ (20cm); 45g. Distinctively marked, medium-sized, with relatively broad, rounded wings, very long legs. Pale brown mantle contrasting with blackish cap and flight feathers; rump pale gray; below white; dark eye-stripe on white face.

Habits. Flies swiftly, very erratically, darting rapidly from side to side with shallow wingbeats and frequent glides; often patters over water, its very long legs conspicuous; eats marine water-striders, plankton.

Status. Probably a rare visitor to waters around Cocos I.; a single record from 60mi (100km) S of Cocos I. (late May; Beebe); more regular farther S in Galápagos Is. (July–August).

Range. Widespread in southern oceans; in Pacific breeds around Australia and New Zealand, ranging N and E to Ecuador and Cocos I.

LEAST STORM-PETREL

Oceanodroma microsoma Pl. 2(13)

Paiño Menudo

Description. 6″ (15cm); 19g. The smallest all-dark storm-petrel, with proportionately short wings and short, wedge-shaped tail that often looks rounded at sea; entirely blackish-brown except paler brown bar across upper wing-coverts; distinctly smaller than other all-dark species (Black, Leach's), with weaker, more fluttery flight.

Habits. Flies rapidly, directly to rather erratically, with quick, deep wingbeats, the effect rather swiftlike (e.g., *Chaetura*), low over water, often in troughs of waves; usually in loose groups, often in company of Black Storm-Petrels.

Status. Nonbreeding visitor and winter resident; uncommon July–November, common December–June in Golfo de Nicoya; off outer Pacific coast an uncommon fall migrant (September–October), common spring migrant (April–June).

Range. Breeds on islands around Baja California; winters S to Ecuador.

Note. Formerly placed in genus *Halocyptena*.

WEDGE-RUMPED STORM-PETREL

Oceanodroma tethys Pl. 2(12)

Paiño Danzarín

Description. 6½″ (17cm); 18g. Smallest white-rumped storm-petrel, with the weakest, most fluttery flight; distinctive, large, triangular rump-patch, which includes upper tail-coverts (tail often barely shows behind) and appears rather dull white because of dark feather-shafts; rest of plumage blackish-brown, with paler diagonal bar across upper wing-coverts.

Habits. Flies with rather fluttery wingbeats, erratically to fairly directly, low over water, often in troughs of waves; singly or in small, loose groups; does not follow ships but is often attracted to chum and offal.

Status. Peruvian-breeding race *kelsalli* a regular visitor to Pacific coastal waters; most numerous July–November, when often the commonest white-rumped storm-petrel. The nominate Galápagos race larger (8″, 20cm), with stronger flight and brighter rump; has been taken off Cocos I. along with *kelsalli*.

Range. Breeds Peru (*O. t. kelsalli*) and Galápagos Is. (*O. t. tethys*); the former ranges N to Baja California, the latter to Cocos I. and Bay of Panama.

BAND-RUMPED STORM-PETREL

Oceanodroma castro Pl. 2(14)

Paiño Rabifajeado

Description. 8″ (20cm); 40g. Medium-sized storm-petrel, with wing and tail rather intermediate in shape between Leach's and Wilson's; shearwaterlike flight often distinctive. Plumage blackish-brown with bright white band of even width narrower than that of Wilson's, across rump, making the bird appear long-tailed; paler diagonal brownish-gray band across upper wing-coverts.

Habits. At sea found in small loose groups, sometimes larger flocks; flies buoyantly, dashingly, in zigzag course, with rapid wingbeats, followed by glides with wings at or below the horizontal; when foraging, hops, runs, and dances along surface on horizontal wings; sometimes follows ships.

Status. Recorded in June off Cocos I.; a large flock in Golfo de Nicoya in May 1982, following a widespread Pacific storm (Lewis). Probably a sporadic, perhaps seasonal visitor to Pacific waters of our area, well offshore except in stormy weather. Old reports of breeding at Cocos I. appear to be without foundation.

Range. Tropical Atlantic and Pacific Oceans; in E Pacific breeds on Galápagos and Hawaiian Is.; ranges widely, occasionally to Central American coast.

Note. Often called Harcourt's or Madeira Storm-Petrel.

LEACH'S STORM-PETREL

Oceanodroma leucorhoa Pl. 2(16)

Paiño de Leach

Description. 8″ (20cm); 40g. Medium-sized, slender, long-winged, with deeply forked tail and distinctive, nighthawklike flight. Plumage mostly brownish-black, with conspicuous pale gray band across upper wing-coverts; rump varies from mostly white (more or less bisected by dusky feathers) to wholly dark; both dark- and pale-rumped birds occur in our area.

Habits. Flies swiftly, erratically, buoyantly with 1 or 2 fast, powerful flaps followed by glides on wings held well above the horizontal and noticeably kinked; sudden changes of direction impart a bounding quality; flutters less than other storm-petrels; rarely patters over surface when feeding; does not follow ships; singly or in small, loose groups in our area.

Status. Regular, uncommon to fairly common visitor well off Pacific coast but rare close to shore or in Golfo de Nicoya, where recorded mostly May–August (evidently nonbreeding birds); not yet recorded off Caribbean coast, where an occasional stray (white-rumped) would not be unexpected.

Range. Breeds on islands around N Pacific (Japan to Alaska and W Mexico) and N Atlantic; ranges to Melanesia, Peru, and Brazil. Atlantic and N Pacific populations are white-rumped; S from California in E Pacific, proportion of dark-rumped birds increases, reaching 100% in many Mexican colonies.

MARKHAM'S STORM-PETREL

Oceanodroma markhami Not Illustrated

Paiño de Markham

Description. 9½″ (24cm). Large, robust, all dark with very deeply forked tail and blue-gray cast in fresh plumage; very similar to Black Storm-Petrel and not always distinguishable in the field (flight differs somewhat, pale bar on wing-coverts longer); in hand has much shorter legs (tarsus 25mm or less; Black, 30mm or more), shafts of upper tail-coverts are white on basal half.

Habits. Pelagic; not known to approach coastal waters in our area; flies rather like a shearwater, with slow, shallow beats and frequent lifts to ca. 3ft (1m) above the water, then a long glide on nearly horizontal or slightly downturned wings, rather more leisurely than Black Storm-Petrel.

Status. Rare, probably seasonal visitor to waters around Cocos I.; 2 records, July–September.

Range. Breeds in Peru and/or N Chile (exact site unknown); disperses widely in Humboldt Current waters off W South America, N to area of Clipperton I., off W Mexico.

BLACK STORM-PETREL

Oceanodroma melania Pl. 2(17)

Paiño Negro

Description. 9″ (23cm); 55g. Large, slender, all-dark, with a deeply forked tail and very long legs. Plumage entirely brownish-black, with paler brown bar across upper wing-coverts. Best told from dark-rumped Leach's Storm-Petrel by flight; obviously larger than Least Storm-Petrel.

Habits. Flies swiftly, strongly, directly, rather like a tern (e.g., *Anous*), with a few powerful, languid, deep wingbeats alternating with glides on wings usually held well below the horizontal; usually in small, loose groups; on migration in open flocks of 50 or more that may rest on water in "rafts" during day; sometimes follows ships; readily attracted by chumming.

Status. Uncommon to locally common migrant and nonbreeding visitor off Pacific coast, with notable wintering concentrations in Golfo de Nicoya and, probably, Golfo

Dulce; most numerous October through April, but some present year-round.

Range. Breeds on islands off Baja California; winters in offshore waters from C California to N Peru.

Note. Formerly placed in genus *Loomelania*. Also possible in our area is Ashy Storm-Petrel (*O. homochroa*), recorded once off Pacific coast of Panama (Crossin); this species is smaller (8″, 20cm), chunkier, shorter-winged than Black, with pale-edged upper and lower wing-coverts; flies with very shallow wing-beats, wings normally not raised above horizontal; usually ranges S in winter to W Mexico.

ORDER **Pelecaniformes:** Pelicans, Boobies, Cormorants, and Allies

The members of this ancient and diverse order are the only birds to have all 4 toes joined by webs. Body form and proportions vary widely among the families, but all except tropicbirds lack external nostrils; only in tropicbirds are the chicks downy at hatching, and only tropicbirds lack a bare gular pouch.

FAMILY **Phaethontidae:** Tropicbirds

The 3 species of tropicbirds range widely over Earth's warmer seas. White, trimmed with black, sometimes suffused with pink or yellow, with slender, greatly elongated central tail feathers, they are outstandingly graceful. The sexes are alike. In size they are about 12–20″ (30–51cm), not including the long central tail feathers. Their stout, slightly decurved bills are red or yellow. When not breeding, they wander, alone or in usually silent pairs, far over the oceans, where they swoop low or plunge to capture small fishes and squids. Tropicbirds nest on small offshore or oceanic islands in cavities or beneath shrubs or overhanging rocks, preferably on steep cliffs overlooking water, where they can become airborne without walking or leaping; their short legs with fully webbed feet cannot support their bodies on land. In a shallow nest scrape, the female lays a single brownish egg, which she and her mate incubate, in sessions lasting 3–5 days, for a total of 41–45 days. Well covered with gray or fawn-colored down when hatched, the chicks are fed by both parents and remain on land for 11–15 weeks.

RED-BILLED TROPICBIRD

Phaethon aethereus Pl. 1(7)

Rabijunco Piquirrojo

Description. 20″ (51cm), plus 12–24″ (30–60cm) central rectrices; 700g. Mostly white; mantle barred with black; outer primaries and primary coverts black; broad black stripe from eye to nape. **Adults:** bill red; long tail streamers. **Immatures:** bill yellowish; heavier black markings on mantle (barring extends to nape) and primaries; rectrices black-tipped, none elongated.

Habits. Flies gracefully over open sea, circling and soaring with shallow, rapid wingbeats, more pigeonlike than ternlike; plunges from considerable height for fishes and squids, sometimes remaining submerged for several seconds; occasionally follows ships to catch flying fishes stirred up by them.

Voice. A sharp, piercing, whistled *keek* or *peet*, heard from flying birds at sea; also a high, screaming rattle *kee kee krrt krrt krrt*; hisses when disturbed at nest.

Status. Regular visitor in small numbers off Pacific coast, probably throughout year; breeds on islands off Mexico, Colombia, and Galápagos; unrecorded but probably regular off Caribbean coast, as small numbers breed on Swan Cay off W Bocas del Toro, Panama.

Range. Tropical and subtropical seas worldwide.

Note. Also possible off Caribbean coast is the smaller (15″, 38cm, plus tail streamers) White-tailed Tropicbird (*P. lepturus*) (breeds on islands in E Caribbean Sea): bill yellow to orange; adult has white back, black band across wing-coverts; immature has back very coarsely scaled with black rather than heavily barred; primary coverts white, no black at tips of rectrices.

FAMILY **Pelecanidae:** Pelicans

Pelicans are unmistakable: large birds with very short legs, long and ample wings, long necks, and large, straight bills with huge gular pouches. At least 1 of the 8 species occurs on every continent except Antarctica. The plumage is white, brown, or gray, usually with blackish primaries; the sexes are alike, though males average larger. The bill, pouch, iris, and facial skin change color through the nesting cycle; head and neck feathers may be molted as well. Pelicans are gregarious, breeding in colonies and often foraging in groups. They fly strongly and glide buoyantly, often in lines or in V-formation. Their staple food is fishes, captured either while swimming or by plunge-diving; species using the latter technique have a remarkable subcutaneous cushion of air sacs on the chest. The pouch is used as a scoop in foraging but not as a basket for carrying fish. Both sexes help to build the nest, incubate the 1–4 chalky white eggs, and feed the young, which hatch naked but soon acquire a coat of white down. Young beg noisily for food and are fed by regurgitation; adults are virtually voiceless. Incubation periods average ca. 4–5 weeks, fledging periods 8–9 weeks. Several years, often marked by distinctive plumages, are usually required to attain maturity.

AMERICAN WHITE PELICAN
Pelecanus erythrorhynchos Pl. 4(2)
Pelícano Blanco Americano

Description. 60″ (152cm); 7kg. Very large; white with blackish primaries and outer secondaries. **Adults:** bill, pouch, and facial skin bright yellowish; grayish smudging on nape (nonbreeding). **Immatures:** white body plumage mixed with dusky; bill and pouch grayish. Differs in color, habitat, and behavior from smaller Brown Pelican.

Habits. Frequents freshwater lagoons, open water in marshes, sometimes quiet coastal bays; thus far found in our area only singly, but farther north in groups; fishes while swimming on surface, thrusting and scooping with bill; does not dive.

Status. Accidental visitor during northern winter; one sighting at Palo Verde, 2 March 1977 (Holmes and Robinson).

Range. Breeds locally in W and C North America; winters regularly S to Guatemala, accidentally to Nicaragua and Costa Rica.

BROWN PELICAN
Pelecanus occidentalis Pl. 4(1)
Pelícano Pardo (Buchon, Pelicano, Alcatraz)

Description. 43″ (109cm); 3kg. Unmistakable; only common pelican of our area. **Adults:** body dark gray-brown; feathers of upperparts with silvery "frosting"; head and neck white; bill brownish; pouch and facial skin grayish; feet black. Breeding: neck largely black; chestnut nuchal crest; head tinged with yellowish; orbital skin reddish; bill with rosy tinge. **Immatures:** head, neck, and upperparts brown; belly white; 3 years to full adult plumage.

Habits. Frequents coastal waters, rarely venturing far out to sea; usually in small groups that fly in single file, with coordinated flaps and glides, often catching updrafts from breaking waves; forages by plunge-diving from heights of up to 30ft (10m) for shallow-swimming fishes; does not submerge; takes scraps and offal; attends fishing boats; roosts on rocks, trees, boats.

Voice. Adults usually silent, rarely give soft low grunts. Young at nest give strident piglike squeals.

Nest. A small, circular platform of sticks in the top of a shrub or tree, rarely on ground, on offshore island. Eggs 2–3, white, soon stained brownish in nest. Peak of egg-laying January–February; young fledge mostly in May.

Status. Common to abundant along Pacific coast; 4 breeding colonies known from Nicaraguan border to Golfo de Nicoya, largest on Isla Guayabo; some birds may be migrants; uncommon but widespread nonbreeding visitor year-round along Caribbean coast; very rarely wanders inland.

Range. Pacific coast from Washington to N Peru; Atlantic, Gulf and Caribbean coasts and islands from North Carolina to E Venezuela.

FAMILY **Sulidae:** Boobies

The 9 species of boobies are fairly large, streamlined marine birds with conical, pointed bills, long, pointed wings, wedge-shaped tails, short legs, and stout, often colorful feet. Their plumage varies from mostly white, with black on the flight feathers, to largely or entirely brown (especially young birds). The sexes differ at most slightly in plumage but often strikingly (at least when breeding) in colors of iris, bill, facial skin, and/or gular pouch; females average larger. Flight is strong and fast, with much gliding; all species dive from the wing to catch fishes and squids, sometimes pursuing them to great depths. Silent at sea, boobies are noisy at roosts or breeding colonies on small oceanic islands, where they gain their name by their fearlessness of man. Male and female share all nesting duties; nests vary from a slight hollow on ground to a platform of sticks in a tree; eggs are pale blue with a chalky white coating. Although some species lay up to 3 eggs, no more than 1 offspring is raised. Incubation periods are ca. 6 weeks, fledging periods 12–16 weeks. The young hatch naked but soon acquire a dense coat of down; they are fed by regurgitation. Most species need at least 2 years to acquire full adult plumage.

BLUE-FOOTED BOOBY

Sula nebouxii Pl. 1(4)

Piquero Patiazul

Description. 32″ (81cm); 1.5kg. Large, robust; at a distance upperparts appear pale at both ends, dark in middle. **Adults:** head and neck streaked with pale brown and white; tail pale brown with whitish base; white patches on rump and hindneck; below white; mantle and wings dark brown; back flecked with white. Bill bluish; iris yellow; feet bright blue. **Immatures:** head, neck, and chest more uniform gray-brown; less white on hindneck and rump; iris dark; feet and bill slaty. See young Masked Booby.

Habits. Prefers to fish in warm, clear inshore waters, usually in small groups that fly fairly high above water until fishes are spotted, then drop all together in steep diagonal dives, entering the water with a very small splash and often submerging deeply; roosts on ledges and offshore stacks.

Voice. Males give plaintive whistles; females, hoarse trumpeting quacks; heard occasionally in squabbles at roost or when pursued by frigatebirds.

Status. Irregular visitor in (usually) small numbers, at practically any time of year, to Pacific coast, especially Golfo de Nicoya, where often seen around Isla Guayabo.

Range. Warm waters along Pacific coast from Baja California to Peru, including Galápagos Is.

MASKED BOOBY

Sula dactylatra Pl. 1(3)

Piquero Blanco

Description. 34″ (86cm); 1.6kg. Large, heavy, strong-flying. **Adults:** plumage white except for blackish flight feathers; facial skin blackish; bill yellowish; feet dark gray-green. **Immatures:** head and neck dull dark brown; at least a trace of white nuchal collar; rest of upperparts uniform brown, more or less mottled with white. Young Blue-footed Booby has dark chest and white patches on hindneck and rump but not a white collar.

Habits. Pelagic, preferring warm, deep waters where flying fishes, the preferred food, abound; usually singly or in small, loose groups, sometimes associating with shearwaters. Flies strongly, steadily, with slow, powerful wingbeats and buoyant glides; occasionally accompanies ships, in shallow, swift dives catching the flying fishes stirred up off the bows; makes vertical plunges, sometimes from considerable heights, but rarely dives deeply.

Status. Uncommon visitor, probably year-round, well off both coasts, accidental close to shore.

Range. In tropical oceans worldwide. Nearest breeding colonies in Pacific off W Mexico and Galápagos Is. and in Caribbean off Venezuela.

Note. Also called White, or Blue-faced, Booby.

BROWN BOOBY

Sula leucogaster Pl. 1(1)

Piquero Moreno (Monjita, Alcatraz)

Description. 27″ (70cm); 1.1kg. Uniform dark brown upperparts and chest sharply demarcated from paler belly; wing-linings white. **Adults:** belly white; facial skin, legs, and feet bright greenish-yellow; head uniform dark brown, except in males of Pacific race, who have head largely pale gray, bill and facial skin slaty. **Immatures:** more grayish-brown overall; belly gray-brown, mottled with white, only slightly paler than chest; feet dull olive; face grayish. Young Masked Booby larger, with white chest and at least a trace of white collar; feet grayish.

Habits. Usually does not range far out to sea; most numerous about small offshore islands, sometimes feeding close to shore; in windy weather soars rapidly on stiff wings low over water like a shearwater; in calm air, flies with steady wingbeats and short glides; fishes by diagonal plunge-dives, usually from no great height; feeds singly or in small groups, but many may congregate at rich school of fishes or around fishing boats; roosts on rocks, ledges, and boats.

Voice. Raucous, guttural grunts, honks: *raak-raak-raak*, etc. Noisy in breeding and roosting aggregations, usually silent at sea.

Nest. On bare ground or on accumulation of twigs, stones, seaweed, on flat ground or ledge, on offshore island. Eggs 2, occasionally 1 or 3. Probably year-round, perhaps peak September–April.

Status. Locally common breeding resident

along both coasts; breeds on islands all along outer Pacific coast (not in Golfo de Nicoya), largest colony at Cabo Blanco; on Caribbean side breeds on islets near Isla Uvita, off Limón.

RANGE. Widespread in tropical oceans of the world.

RED-FOOTED BOOBY

Sula sula Pl. 1(2)

Piquero Patirrojo

DESCRIPTION. 28″ (72cm); 950g. Smallish, slim-winged, variably colored; pink or red on face; red feet of adults diagnostic. **Adults** PALE PHASE: white with black remiges, black spot at bend of underwing; head tinged with yellow; in Pacific (not Caribbean) tail also black. DARK PHASE: uniform grayish-brown (Pacific; 99 percent of Cocos I. colony), or gray-brown with white rump, vent, and tail (Caribbean). **Immatures:** dull brown, sometimes with trace of darker breast-band (but head and neck never darker as in Brown Booby); feet greenish.

HABITS. Pelagic when feeding, usually in small flocks; dives for fishes; chases flying fishes in air; often active at night; roosts and nests on vegetation rather than on rocks and ledges.

VOICE. A shouting, quacking *ghaaow*! when pursued by a frigatebird; around nest a deep, gurgling laugh, rapid piglike grunts, a creaking, snoring growl or hiss, or hoarse squawks.

NEST. A small platform of sticks in topmost branches of tree, usually near shore or along ridgetop. Eggs usually 2 (Cocos I.). Year-round, with probable peak in April–July.

STATUS. Abundant breeding resident on Cocos I. (numbers fluctuate seasonally); probably regular in small numbers well off both coasts of mainland but unlikely in inshore waters except perhaps after storms.

RANGE. Tropical and subtropical oceans worldwide.

FAMILY **Phalacrocoracidae:** Cormorants

The 28 species of cormorants are a practically cosmopolitan family of medium- to large-sized waterbirds with short legs, long necks and bodies, cylindrical bills with hooked tips, small gular pouches, and long, stiff tails. Most cormorants have glossy black adult plumage; a few are gray; and many southern species have white underparts. The iris, facial skin, gular pouch, and occasionally the feet may be colorful during the breeding season; the sexes are similar. Birds of coastal and inland waters, cormorants dive from the surface and pursue fishes, crustaceans, etc., underwater. Their plumage is not fully waterproof; after foraging they often stand spread-winged in the sun to dry it. Most species are gregarious, often foraging in flocks and nesting in colonies. Males and females share the chores of building the nest, incubating the eggs, and caring for the young. Nests are of sticks, seaweed, etc., placed in trees or on ledges; the 2–4 eggs are blue-green with a chalky white coating. Incubation periods are typically ca. 4 weeks. The young, naked at hatching, soon grow a coat of down; they fledge at 5–8 weeks of age. In most species, 3–4 years are required for attainment of full adult plumage.

OLIVACEOUS CORMORANT

Phalacrocorax olivaceus Pl. 4(4)

Cormorán Neotropical (Pato Chancho, Pato de Agua)

DESCRIPTION. 26″ (66cm); 1.1kg. Only cormorant in Costa Rica; head larger, neck and tail shorter than Anhinga. **Adults:** plumage mostly glossy black; head and neck browner; feathers of mantle lanceolate, dark gray with black fringes. Iris bluish-green; gular pouch yellowish-orange; bill grayish to blackish; feet black. BREEDING: head and neck glossy black with white plumes; gular pouch brighter orange, bordered by white. **Immatures:** mostly dark gray-brown, darkest on belly; head, neck, and chest buffy-brown. **Young:** gray-brown with underparts largely whitish; nestling down blackish.

HABITS. Prefers shallow, clear water at low elevations: rivers, lakes, open marshes, salt ponds, seashores; gregarious; often loafs and fishes in a group, which advances in a line, driving fishes into shallows, and dives as a unit; flies with neck outstretched, head high, body inclined, tail low, flapping steadily, rarely gliding.

VOICE. Harsh guttural croaks and grunts, most often heard at roosts or nesting areas.

NEST. A compact platform of sticks, 30–100ft (9–30m) up in tall tree. Nesting colony of ca. 400 pairs in large fig trees at N end of Lago Caño Negro. Eggs 3–4. November–March.

STATUS. Locally common in marshes of Tempisque basin and around Golfo de Nicoya; abundant in Río Frío–Caño Negro district; elsewhere widespread but generally uncommon along both coasts and larger rivers, locally to ca. 5000ft (1525m); rare, probably vagrant, in Valle Central; in recent years has become abundant (breeding?) at Lago de Arenal.

Range. Extreme SW USA to S South America.

Note. Sometimes called Neotropic Cormorant.

FAMILY **Anhingidae:** Anhingas

With 4 species, the anhinga family is widely distributed over the Eastern and Western hemispheres. These birds are easily distinguished from their closest relatives, the cormorants, by their long, sharply pointed rather than hooked bills, longer necks, and long, ample tails; their legs are very short and stout. Anhingas inhabit fresh and brackish water surrounded by trees or snags where they can perch and spread their wings to dry after an immersion, for their dark plumage rapidly becomes soaked. They spear rather than grasp the small fishes upon which they largely subsist; the triggerlike arrangement that facilitates the strong forward dart imparts to their necks a peculiar characteristic kink and is responsible for their other name, darters. In bulky nests of sticks built in trees or bushes and lined with green leaves or moss, anhingas lay 3–5 eggs that both parents incubate for ca. 4 weeks. The young, hatched naked, are fed by regurgitation by both parents. They soon sprout a coat of white or buffy down; when alarmed they drop into the water below their nest, to scramble back later if they survive. They leave the nest when about 5 weeks old and can fly 2 weeks later. At least 2 years are required for young birds to acquire full adult plumage.

ANHINGA

Anhinga anhinga Pl. 4(3)

Pato Aguja o Aninga

Description. 34″ (86cm); 1.2kg. Extremely long, thin neck; small head; long, sharp bill; long, ample wings and tail. **Adults:** body plumage glossy black; pattern of silvery spots and streaks on lanceolate feathers of mantle and wing-coverts; tail tipped with buffy-white. ♂: head and neck blackish (glossy black with white plumes when breeding). ♀: head and neck buffy-brown (buffy-white ornamental plumes when breeding). Iris red; facial skin bright blue-green when breeding, brownish at other times; feet black. **Immatures:** resemble female but are browner, especially above (little silvery pattern).

Habits. Frequents lakes, sluggish rivers, freshwater and brackish lagoons, mangroves; swims slowly and stealthily underwater, ambushing prey, which it spears with a lightning thrust of the bill; often swims with only snakelike head and neck visible; spends much time perching with wings and tail spread to dry; eats aquatic insects and other invertebrates, young caymans, small turtles, and snakes as well as fishes; often soars on thermal updrafts with neck extended and long tail spread.

Voice. Disturbed, a rapid, deep, staccato *guk-guk-guk-guk* and other guttural notes; in courtship, a whistled *eek eek eek*; mostly silent.

Nest. An untidy platform or shallow cup of sticks and dead leaves, lined with green leaves, 5–20ft (1.5–6m) above ground or water in tree, in small loose groups either alone or in larger colony of other waterbirds. Eggs 3–5, bluish-green with chalky white coating that is soon stained with brown. May–February, peak September–November.

Status. Resident, widespread but usually uncommon throughout lowlands of both slopes; abundant only near the principal breeding areas in Tempisque and Bebedero basins (Isla Pájaros, Estero Madrigal), and Río Frío–Caño Negro region.

Range. C USA to W Ecuador, N Argentina, and Uruguay.

FAMILY **Fregatidae:** Frigatebirds

The 5 species of frigatebirds are large tropical seabirds with extremely long, pointed wings, long, deeply forked tails, very small legs and feet, and rather long, hook-tipped bills. Their plumage is mostly blackish, with white on the underparts (females) and head (immatures) of most species; females are considerably larger than males. The bare red gular area can be inflated into a great red balloon by displaying males. As their body proportions suggest, frigatebirds are masters of the air. They soar effortlessly for hours and are yet capable of swift and agile maneuvers. Their small feet serve only for perching, and they require an elevated perch for takeoff. Frigatebirds feed by swooping to snatch prey, including fishes, squids, and jellyfish, from the ocean surface or by harrying boobies into disgorging their catch. Frigatebirds never voluntarily enter the ocean, as their plumage quickly becomes waterlogged. Silent at sea, they emit a variety of chatters, whistles, and croaks when nesting. The nest is a platform of sticks atop a bush or tree, built by the male with material brought by the female. Both sexes incubate the single white egg and feed the chick, which hatches naked but soon grows a coat of white down. The incubation period is long, 6–8 weeks, and the chick develops slowly, requiring 5–6 months to fledge, and remaining dependent upon its mother for at least another year. Successful pairs can thus breed only once every 2 years; young birds require several years to attain full adult plumage and probably do not breed until at least 5 years of age.

MAGNIFICENT FRIGATEBIRD

Fregata magnificens Pl. 1(6)

Rabihorcado Magno (Tijereta del Mar, Zopilote de Mar)

Description. ♂ 36″ (91cm); 1.2kg; ♀ 44″ (111cm); 1.7kg. A large frigatebird of coastal waters. **Adult ♂:** entirely blackish, at most a faint brown bar across upper wing-coverts. Bill gray; gular pouch bright pink (red in display); orbital skin blackish; feet blackish to brown. ♀: head and throat black; breast white; conspicuous pale brown upperwing bar; orbital skin violet-blue; gular pouch gray to purplish; feet reddish. **Immatures:** head and breast white; conspicuous pale upperwing bar; bill, orbital skin, and feet pale blue. White scaling on axillars, if present, diagnostic.

Habits. Usually soars high over shore or coastal waters, rarely far out to sea; congregates quickly when a school of fishes is driven to the surface; a major predator on hatchling sea turtles, which it plucks from beach or water with quick swoops; often in large concentrations about towns or fishing boats, feeding on offal; rests on rocks, vegetation of offshore islands, and rigging of boats.

Voice. High-pitched, querulous twittering and chirping notes given at nest; usually silent at sea.

Nest. A compact, hemispherical platform of twigs in topmost branches of tree or shrub, usually on steep slope of offshore island. Breeding protracted; eggs at least December–May.

Status. Present year-round on both coasts, with local concentrations, especially Golfo de Nicoya; more abundant overall on Pacific coast. Only nesting colonies in Costa Rica are at Isla Bolaños, at Bahía Salinas (ca. 200 pairs), and on small island W of Isla Guayabo in Golfo de Nicoya. The population probably includes nonbreeding visitors as well as breeding residents. Rare year-round on Cocos I., where breeding has not been reported. Occasionally a solitary bird flies many miles inland, sometimes over highlands.

Range. Tropical Atlantic (including Caribbean) and Pacific oceans.

GREAT FRIGATEBIRD

Fregata minor Pl. 1(5)

Rabihorcado Grande

Description. ♂ 35″ (89cm); 1kg. ♀ 38″ (97cm); 1.2kg. Smaller than Magnificent Frigatebird, with similar sequence of plumages; never has white scaling on axillars. **Adult ♂:** entirely blackish except for conspicuous pale brown bar across upper wing-coverts; legs reddish to brownish. ♀: pale grayish or brownish throat contrasts with black head and white breast; conspicuous whitish upperwing bar; orbital skin blue-gray; eyelids reddish; bill pale gray, gular pouch darker gray; legs red to pinkish. **Immatures:** head and breast white, the head usually more or less suffused with tawny-buff; conspicuous pale upperwing bar; feet bluish to brown. When seen with Magnificent Frigatebird (e.g., Cocos I.), this species seems more compact and agile; Magnificent, more hulking and powerful, with longer and narrower tail feathers.

Habits. Around Cocos I. this species frequently harries boobies, especially Red-footed; Magnificents tend to pick on Brown Boobies. Young birds congregate about boats circling the island, playfully chasing each other and trying to bite the tip of the radio antenna. Members of the 2 frigatebird species are often aggressive toward one another.

Voice. Soft chipping notes heard among birds interacting close to a boat; also a nasal *kack* or *ka-ack* and a rather loud whistle are given near nest.

Nest. A compact platform of sticks, on topmost branches of tree near shore or on ridgetop; usually in clusters of 10–25 pairs (Cocos I.). Possibly year-round.

Status. Abundant resident on Cocos I. (numbers vary seasonally); accidental in Pacific coastal waters, probably only following storms, as when 1 female was recorded on 31 May 1982 in Golfo de Nicoya (Stiles and Lewis).

Range. Ranges widely in tropical oceans of the world; least numerous in Atlantic.

ORDER **Ciconiiformes:** Herons, Storks, Ibises, and Allies

This order includes long-legged, long-necked wading birds placed in several families that differ in bill and foot structure and in the presence or absence of powder-down—dense patches of short, brittle, oily feathers that, when rubbed and broken, yield a powder used for cleaning and drying the plumage. Powder-down patches are usually concealed beneath thc breast feathers.

FAMILY **Ardeidae:** Herons, Egrets, and Bitterns

The 58 species of herons are characterized by long necks with modified vertebrae that permit the birds to snap their necks forward, shooting their sharp-pointed bills forth to spear or grasp small vertebrates or invertebrates, including insects. Herons normally hold their necks kinked and fly with them tucked in. Found on marshes, waterways, and shores worldwide, these birds have well-developed powder-down. The loose-textured plumage is white, gray, bluish, brownish, or purplish, mostly in simple patterns, although a few species are boldly barred, spotted, or streaked. Many species have long nuptial plumes; the colors of the bill, facial skin, and legs often change seasonally. The sexes are alike or nearly so in color, but males usually average larger. Most species have rough, guttural voices. Most herons are colonial; pairs form for the duration of a nesting attempt, and promiscuity is frequent in some species. The stick nest is often built by the female with material gathered and presented ceremoniously by the male; both sexes share in incubation and feeding the downy young. Most herons lay bluish or whitish eggs with a matte finish, but the marsh-dwelling bitterns lay brownish eggs in a nest of rushes and breed as solitary pairs. The incubation period is 16–30 days; nestling period, 35–50 days. Most high-latitude species are migratory; yearlings may not breed but remain in wintering areas through their first summer.

AMERICAN BITTERN

Botaurus lentiginosus Pl. 5(12)

Avetoro Norteño (Puncus)

Description. 25″ (64cm); 700g. Medium-sized, rather stout. **Adults:** crown chestnut; sides of head tan; throat and belly white; neck and breast buffy, broadly streaked with rusty-brown; a broad black stripe down each side of neck. Blackish remiges contrast with finely patterned brown mantle. Bill yellowish; legs greenish. **Immatures:** black neck stripe reduced or lacking. Young night-herons spotted with white above; Pinnated Bittern has barred head and neck.

Habits. Haunts freshwater marshes with tall sedges, cattails; solitary, crepuscular; walks stealthily; stands immobile to ambush prey; freezes in alarm with head and neck vertical.

Voice. When startled, a low throaty *ok-ok-ok*; in flight, a nasal *haink*.

Status. Formerly an uncommon to rare winter resident in extensive marshes of Valle Central, including upper Reventazón valley, nearly all of which have been drained; no records since early in this century.

Range. Breeds from S Alaska and C Canada to C Mexico; winters regularly to S Mexico, formerly to Costa Rica, possibly C Panama.

PINNATED BITTERN

Botaurus pinnatus Pl. 5(11)

Avetoro Neotropical (Puncus, Mirasol)

Description. 26″ (66cm); 800g. General color buff, with contrasting slaty remiges conspicuous in flight. **Adults:** throat, foreneck, and median underparts white; foreneck and breast broadly streaked with pale brown; rest of head and neck barred with black; back heavily streaked and barred with black. Iris buffy-yellow; facial skin bright yellow with brown line across lores; bill horn-color with dusky culmen; legs greenish. **Immatures:** similar, but black barring of head and neck sparser, less regular; black more blotched. Young tiger-herons are more broadly and evenly barred throughout with black; remiges barred with white.

Habits. Skulks deliberately in shallow freshwater marshes, lake borders with cattails, tall sedges, or rank growth of water hyacinths; also flooded pastures with tall grass; waits motionless for long periods to ambush prey; crouches when alarmed, lowering body and holding head vertically just high enough to see; freezes, only flushes at close range; eats fishes, frogs, rodents.

Voice. When flushed, a rough *rawk-rawk-rawk*; during breeding season male gives a deep, booming *poonk* or *poonkoo*, often at dusk or night.

Nest. A platform or shallow cup of rush stems placed in cattails. Eggs 3, brownish-olive. Breeds mostly or only in wet season.

Status. Resident locally on both slopes, principally Tempisque basin and Río Frío district, lowlands to at least 2000ft (600m), as at Turrialba.

RANGE. S Mexico to Argentina.

LEAST BITTERN

Ixobrychus exilis Pl. 6(3)

Avetorillo Pantanero (Mirasol)

DESCRIPTION. 12″ (30cm); 80g. Diminutive and richly colored. **Adults:** crown and mantle glossy black (males) or chestnut (females); white scapular stripe; large buff wing-patches; remiges slaty, tipped with buff; face and underparts mostly buff; belly white; foreneck striped with buff or rusty and white; sides with dusky shaft-streaks. Iris pale buffy-yellow; upper mandible black, lower flesh-color, tomia yellowish; legs yellow-green. **Immatures:** like female but crown and mantle paler, flecked with buff; dusky streaking below and on buffy wing-patch. No other heron or small rail has similar pattern.

HABITS. Keeps to dense stands of cattails, sedges, tall grass in freshwater marshes; usually solitary, secretive, difficult to flush; sometimes ventures out of cover to fish; retreats after capturing prey (aquatic insects, small fishes and frogs).

VOICE. When disturbed, a sharp *kwoh* or a harsh cackling; breeding male has guttural, cooing song *kookookook*; most often heard at dawn.

NEST. A compact, solidly built, shallow cup of cattail leaves, supported low over water by cattails or rushes. Eggs 3–4, occasionally 6, pale bluish. June–September.

STATUS. Rare to locally common on both slopes, principally in lowlands but occasionally to 4600ft (1400m). Breeding confirmed only in Tempisque basin (Palo Verde) and Río Frío region (Lago Caño Negro), but probably nests widely; seasonal movements related to water levels. Probably some northern birds winter October–March.

RANGE. S Canada to Peru and N Argentina; winters from S USA southward.

RUFESCENT TIGER-HERON

Tigrisoma lineatum Pl. 5(17)

Garza-Tigre de Selva (Martín Peña, Pájaro Vaco)

DESCRIPTION. 26″ (66cm); 850g. Stocky, shaggy-necked. **Adults:** head, neck, and upper breast chestnut, barred with black; brown median stripe, bordered by white, down throat and foreneck; belly grayish; mantle dusky-olive vermiculated with dull buff. Iris orangish; facial skin yellow with blackish loral stripe; upper mandible blackish, lower horn-color, shading to dusky tomia; legs dull dark olive-green. **Immatures:** head rufous; neck and upperparts rich buff to pale rufous, broadly barred or chevroned with black; center of throat, stripe down foreneck, and center of belly white. Takes at least 2 years to acquire adult plumage; subadult plumages like adult but more broadly and heavily barred.

HABITS. Haunts streams, sloughs, or swamps inside forest, rarely venturing into the open, at least by day; usually solitary; stands motionless at water's edge or on snag for long periods awaiting prey or walks very slowly in shallow water; usually rests with neck hunched, bill pointed slightly upward; when alarmed or hunting, often extends neck diagonally.

VOICE. Alarm note a hoarse *quok-quok-quok*, louder and deeper than call of night-heron; a harsh, groaning, far-carrying, long-drawn-out *kwooohh* or *kwawwh* mostly at night, probably to advertise territory.

NEST. A flimsy platform of sticks, high in a tree. Egg 1, bluish-white spotted with pale violet (in Surinam; no Costa Rican nest known).

STATUS. Uncommon to fairly rare resident of entire Caribbean slope, from sea level locally to ca. 1650ft (500m).

RANGE. E Honduras (accidental SE Mexico?) to N Argentina.

FASCIATED TIGER-HERON

Tigrisoma fasciatum Pl. 5(15)

Garza-Tigre de Río (Martín Peña, Pájaro Vaco)

DESCRIPTION. 25″ (64cm); 850g. Bill blacker, relatively shorter and thicker than in other tiger-herons. **Adults:** crown black; sides of head gray; neck blackish, finely barred with buff; mantle dusky-olive, vermiculated with dull buff; facial skin and bare sides of throat yellow; median throat feathered, white; chestnut and white median stripe down foreneck; belly dull rufous. Iris dull yellow; facial skin and base of lower mandible yellow-green, rest of bill and stripes on lores black; legs dull olive-green. **Immatures:** similar to immature Rufescent Tiger-Heron but more heavily and evenly barred with black, belly more extensively and contrastingly white; perhaps best distinguished by habitat.

HABITS. Prefers swift, rocky streams and rivers in forested hilly country; walks deliberately among stream boulders or stands patiently at water's edge awaiting prey; usually shy; flies around bend of stream or up into streamside tree when alarmed; eats fishes and large aquatic insects.

VOICE. A guttural *gwok* when alarmed.

NEST. Unknown(?).

STATUS. Locally resident on streams and rivers in foothills, mostly 800–2600ft (250–800m) along Caribbean slope from Cordillera de Guanacaste (Bijagua) to Panama; however, following introduction of trout in some highland streams, local populations now occur as high as 7900ft (2400m), as at Río Angeles N of Cerro de la Muerte.

RANGE. Costa Rica to NW Argentina and SE Brazil.

NOTE. The northern race *salmoni* has been treated as a separate species by some authors.

BARE-THROATED TIGER-HERON

Tigrisoma mexicanum Pl. 5(16)

Garza-Tigre Cuellinuda (Martín Peña, Pájaro Vaco)

DESCRIPTION. 32″ (80cm); 1.2kg. Largest tiger-heron; bare yellow throat diagnostic. **Adults:** crown black; sides of head slaty; neck finely barred with black and buff (looks olive at a distance) and with chestnut and white stripes on center of foreneck; belly rusty; back dusky, vermiculated with buff. Iris orangish; facial skin yellow-green, a pale brown stripe across lores to bill; upper mandible blackish, lower dull yellowish, tomia horn-color; legs dull olive-green. **Immatures:** crown chestnut; neck, chest, and back rich buff, all broadly barred with black; center of belly white. Iris and facial skin yellow. One or more intermediate subadult plumages with increasingly fine, more adultlike pattern.

HABITS. Prefers more open habitats and larger bodies of water than other *Tigrisoma*: riverbanks, lake margins, ponds, marshes, mangroves; in Guanacaste often numerous in gallery forest; stands motionless for long periods at water's edge, neck outstretched diagonally waiting to ambush prey, including fishes, frogs, crabs; flies heavily, with quick upstroke and slower downstroke.

VOICE. When flushed, a hoarse, raucous *howk-howk-howk*. Males repeat a hoarse, booming roar at night or dusk, especially during breeding season: *hrrrowwr! horrowr! hrrrowr!* etc.

NEST. A rather small, solidly constructed platform of sticks, usually lined with leaves, often well up in tree; usually nests solitarily. Eggs 2–3, dull white tinged with green, immaculate or indistinctly spotted with brown. Virtually throughout year, with peak in early rainy season.

STATUS. Widespread but generally uncommon resident throughout lowlands of both slopes; more common in NE lowlands (Llanura de Tortuguero, Río Frío) and Tempisque basin; extraordinarily abundant at Estero Madrigal, on Río Lajas S of Cañas; sea level to ca. 2300ft (700m), occasionally straying as high as 3600ft (1100m).

RANGE. N Mexico to NW Colombia.

NOTE. Sometimes placed in genus *Heterocnus*.

BLACK-CROWNED NIGHT-HERON

Nycticorax nycticorax Pl. 5(4)

Martinete Coroninegro (Chocuaca)

DESCRIPTION. 25″ (64cm); 800g. Stocky, bigheaded. **Adults:** crown and back glossy black; wings and tail pale gray; rest of head, neck, and underparts white; facial skin yellow-green. Iris red; bill mostly black; legs yellow. BREEDING: long white occipital plume; legs and facial skin brighter. **Immatures:** above brown, streaked with buff or white; wing-coverts and secondaries tipped with white; face, neck, and underparts whitish, streaked with brown to dusky. Iris yellow. Immature Yellow-crowned Night-Heron similar but more spotted, less streaked, with orange iris, thicker bill.

HABITS. Roosts communally by day in cattails, gallery or swamp forest, or mangroves; at night, forages singly in freshwater marshes, along streams and pond edges or on snags, waiting quietly or walking stealthily, catching fishes and frogs with quick stabs; occasionally takes rodents.

VOICE. A harsh, guttural *wok* or *kwok*, often in flight, at dusk or night; when disturbed, a deep, harsh *kwo-kwo-kwo-ko-ko*, lower and more guttural than call of Boat-billed Heron.

NEST. A shallowly concave platform of twigs that are coarser in foundation, finer in lining, in a tree overhanging water, usually well up and out toward tip of branch. Eggs 2–4, pale grayish-blue. Breeding in late rainy season, early to mid–dry season.

STATUS. Population evidently includes permanent and winter (October–March) residents; locally common in N Pacific lowlands; widespread but generally uncommon and local in Caribbean lowlands and foothills; fairly rare in Valle Central; known to breed in Tempisque basin (Isla San Pablo, Isla Pájaros) and probably Estero Madrigal, where abundant in dry season.

RANGE. S Canada to Tierra del Fuego; northern birds winter from C and S USA southward. Widespread in Old World.

YELLOW-CROWNED NIGHT-HERON

Nyctanassa violacea Pl. 5(3)

Martinete Cabecipinto

DESCRIPTION. 24″ (61cm); 625g. Big-headed, thick-billed, more slender than preceding species; pied head unmistakable. **Adults:** head black, with white to buffy forehead, crown, and postocular bar; neck and underparts gray; mantle feathers dark gray with silvery edgings. BREEDING: long white occipital plumes; cheeks, forehead more buffy. **Immatures:** above dull brown, spotted with white; below whitish, streaked with brown. Iris orange in all plumages.

HABITS. More active by day, less social, more partial to saltwater habitats than Black-crowned Night-Heron; prefers mangroves and gallery forest for roosting; forages singly along seashore, on mudflats, in salt ponds, sometimes along river and pond margins, mainly for crabs.

VOICE. Loud, quacking, dry *kwok* notes, often in accelerating series as a wild, cackling laugh; less guttural than notes of Black-crowned Night-Heron.

NEST. A sturdy platform of sticks with central depression lined with leaves, in tree overhanging water. Eggs 2–5, pale blue-green. Pairs nest singly or in small, loose groups. Breeds mainly in wet season.

STATUS. Population probably includes both permanent and winter residents (October–March or April). Most numerous in coastal areas, especially near mangroves; widespread but uncommon to rare and sporadic in lowlands of both slopes, mostly along larger rivers.

RANGE. NE USA, NW Mexico to N Peru and Brazil; winters from Texas and Florida southward.

NOTE. Sometimes placed in genus *Nycticorax*, though anatomical studies do not support this classification.

BOAT-BILLED HERON

Cochlearius cochlearius Pl. 5(2)

Pico-Cuchara (Chocuaco, Cuaca)

DESCRIPTION. 20″ (51cm); 600g. Stocky, big-headed, with large, dark eyes, huge shoe-shaped bill. **Adults:** crown, long bushy crest (usually held flat), and patch at base of hindneck black; forehead and throat white, shading to grayish-buff on face, neck and chest; rest of upperparts dull ash-gray, paler on wings and tail; wing-linings, sides, and flanks black; center of belly dull rusty. Bill and facial skin black, except dull yellow at rictus and on lower eyelids; gular pouch and legs yellow-green. BREEDING: gular pouch black. **Immatures:** much browner overall, indistinctly streaked with brown and buff on chest and belly.

HABITS. Frequents wooded riverbanks, swamp and pond margins, estuaries, mangroves; by day roosts in groups of up to 50 in trees overhanging water; at night solitary when fishing from water's edge or low snag, detecting prey by sight or perhaps touch; often forages in darker places than Black-crowned Night-Heron.

VOICE. When disturbed, a nasal, dry quacking and croaking: *kwahh-kwa-kwa-kwo-kwo-ko-ko-ko*, etc., sometimes interspersed with grunting *ook* notes.

NEST. In small colonies; a frail platform of sticks, small for size of bird, on horizontal fork overhanging water, 3–16ft (1–5m) up, rarely higher. Eggs 2, rarely 3, very pale blue with fine reddish-brown speckles, usually faint but sometimes very prominent, at thick end and occasionally scattered sparsely over entire egg. June–October.

STATUS. Resident locally in lowlands the length of both slopes; most numerous in Río Frío and Tortuguero regions, and Tempisque and Bebedero watersheds in Guanacaste; locally to 1000ft (300m).

RANGE. N Mexico to W Ecuador, Bolivia, and N Argentina.

NOTE. Formerly placed in a separate family, the Cochleariidae.

CATTLE EGRET

Bubulcus ibis Pl. 5(13)

Garcilla Bueyera (Garza del Ganado)

DESCRIPTION. 20″ (51cm); 350g. Only small white heron with a yellow bill; stockier and shorter-necked than other small egrets. **Adults:** white with yellow iris, bill, and facial skin; blackish legs; crown, back, and chest tinged with buff. BREEDING: conspicuous buffy plumes on crown, chest, and back; bill, facial skin, and legs more or less reddish. **Immatures:** no trace of buff; bill duller yellow.

HABITS. Eats mostly grasshoppers and other insects flushed by grazing livestock in pastures and savannas or by tractors in agricultural fields; sometimes groups forage by themselves in fields or open marshes; forms large communal roosts in trees overhanging water.

VOICE. Rough, dry croaking and barking notes, heard mostly as foraging birds jockey for position around grazing animals or in disputes over perches at roosts.

Nest. Usually a rather substantial shallow cup of twigs or rush stems, often lined with leaves, mostly inside mangrove or other tree or giant bamboo, often overhanging water in dense colonies. Eggs 2–4, pale blue. Mainly during rainy season.

Status. First recorded in 1954, now an abundant resident, found virtually countrywide below 6500ft (2000m), occasionally higher; numbers doubtless still increasing, because habitat continues to expand due to deforestation. For many years, almost the entire Costa Rican population withdrew from most areas to nest at Isla Pájaros (presently 10,000+ pairs) in the Tempisque basin; in the late 1970s and 1980s the population effectively outgrew this site, and new colonies were established in Golfo de Nicoya, near Caldera, in Golfo Dulce lowlands near Panama border, and at Westfalia, S of Limón. We suspect that other colonies exist or soon will be formed, especially in the Caribbean lowlands. A few northern migrants may also be present from about October through April.

Range. Breeds S Canada to S South America; northern birds migrate southward in winter. Breeding range still expanding. Widespread in Old World.

Note. Sometimes placed in genus *Ardeola*.

GREEN-BACKED HERON

Butorides striatus Pl. 6(1,2)

Garcilla Verde (Chocuaco, Martín Peña)

Description. 17″ (43cm); 210g. Small, shaggy-necked; bright yellow legs contrast with dark plumage in flight. **Adults:** cap and bushy crest black; neck maroon; throat and midventral stripe white; above steely gray-green, the wing-coverts scaled with buff; belly gray. Iris yellow; upper mandible blackish, lower yellowish; facial skin yellow with dark brown stripe from eye to bill. **Immatures:** darker, duller, more streaked than adults. **Young:** much browner, belly whitish; head, neck, and underparts broadly streaked with buffy and blackish; wing-coverts spotted with white; remiges tipped with white. Female Least Bittern smaller, with buff wing-patch.

Habits. Singly or in pairs, virtually anywhere thick vegetation adjoins or grows in shallow water: ponds, riverbanks, marshes, streams, mangroves; may feed in open, but retreats to cover when alarmed and freezes with crest raised, tail pumping; when foraging, stands quietly or walks stealthily, ambushes prey and, with quick stabs, seizes small fishes, aquatic insects, frogs.

Voice. Loud, explosive, scratchy *skow* or *kyowk* when taking flight; adult disturbed at nest gives a squalling snarl, *skowraaaaah*; a dry cackle in aggressive encounters.

Nest. A flimsy, slightly concave platform of sticks in thick vegetation over water, alone or in small, loose colonies. Eggs 2–3, occasionally up to 6, possibly laid by 2 or more females, pale blue. April–September.

Status. Locally common virtually countrywide, sea level to 6000ft (1850m); permanent resident, also winter resident, chiefly on Caribbean slope and in central highlands; migrates mostly along Caribbean coast, September–October and April–May. Also on Cocos I., where status and abundance uncertain.

Range. S Canada to Chile, Bolivia, and Paraguay; North American breeding birds winter to Colombia and Venezuela. Widespread in Old World.

Note. Costa Rican birds are members of the maroon-necked *virescens* group (Green Heron), breeding from North America to C Panama and until recently considered a separate species from the gray- or buff-necked *striatus* (Striated Heron) of W Panama to South America, and in Old World. Most *striatus* in C Panama have buff necks, now interpreted as hybrids with *virescens*, but detailed field data are lacking; similar buff-necked birds occur in *striatus* populations S to Ecuador and Brazil, beyond even the winter range of *virescens*, and are thus unlikely to be hybrids. The only Costa Rican record of *striatus* is of a buff-necked bird taken in April 1923 at Cañas, Guanacaste by A. P. Smith. Both *virescens* and *striatus* are reported from Cocos I., without breeding records. We regard species limits in *Butorides* as unresolved, and we recommend further field study, particularly in Panama.

LITTLE BLUE HERON

Egretta caerulea Pl. 5(9)

Garceta Azul

Description. 24″ (61cm); 325g. Medium-sized, rather slender and long-necked. **Adults:** body and wings dark bluish gray; head and neck dull, dark maroon; facial skin gray. Iris yellow; bill grayish with black tip; legs grayish-green. Breeding: base of bill, facial skin blue; legs black; head and neck more violaceous; lanceolate feathers on back, neck, and nape. **Immatures:** the only all-white herons with grayish, black-tipped bills, and narrow dusky tips on outer primaries.

Birds molting to adult plumage show curious calico patterns; this first dark plumage is dull, uniform gray, with little or no bluish tint, or maroon tinge on neck. See Reddish Egret.

Habits. In freshwater marshes, lagoons, rivers, estuaries, salt ponds, mudflats, and mangroves, often in loose parties or at periphery of groups of other species such as Snowy Egrets; forages singly, congregates to roost or loaf; flies with slower, deeper wingbeats than Snowy Egret, occasionally with neck extended.

Nest. A flimsy to fairly solid and bulky platform of sticks and rush stems, 6–13ft (2–4m) up in top of mangrove; eggs 2–4, pale bluish-green, June–September.

Voice. Hoarse, croaking *kraak* when startled; short, coughing *kuk* notes when in groups, as at roost.

Status. Common to abundant migrant and winter resident, fairly common nonbreeding summer resident (mostly yearlings) in lowlands of both slopes, in small numbers to 5000ft (1500m) in Valle Central. Very small numbers evidently breed during some years at I. Pájaros in lower Tempisque basin (Sánchez).

Range. Breeds from C and E USA locally through Central America to Peru and Uruguay; northern populations migratory, winter from SE USA southward.

Note. Formerly placed in genus *Florida*.

TRICOLORED HERON

Egretta tricolor Pl. 5(8)

Garceta Tricolor

Description. 26″ (66cm); 350g. Slender, with long bill and very long neck. **Adults:** head, neck, and upperparts mostly slate-gray; chestnut and white stripe down foreneck; neck and chest more purplish, streaked with chestnut; belly and wing-linings white; facial skin yellow-orange with dusky spot on lores. Iris buffy-yellow; bill horn-color shading to dusky on culmen; legs yellow. Breeding: elongated white plumes on hindneck and buffy plumes on back. **Immatures:** brownish-gray on head, neck, and mantle; otherwise like adult but without elongate plumes. **Young:** patterned like adult but neck dull reddish-brown, mantle olive-brown instead of gray.

Habits. Frequents freshwater and saltwater habitats: marshes, pond and river margins, estuaries, salt ponds, mangroves; usually seen singly, sometimes accompanying groups of other herons and egrets; feeds by standing in wait for prey or stalking stealthily; stirs with feet to flush small fishes, schools of which it sometimes pursues actively.

Voice. When alarmed, a harsh, cawing *kraaah*, more nasal and higher-pitched than call of Little Blue Heron; other deep groaning or croaking notes.

Nest. A platform of sticks with shallow central depression, often lined with leaves or grass stems. Eggs 3 (?) (1 nest with 3 young found in large mixed colony of herons and ibises); pale bluish-green. June.

Status. Widespread but generally uncommon winter resident in lowlands of both slopes, occasionally to 5000ft (1500m) in Valle Central; also rare nonbreeding summer resident (mostly yearlings); breeds in very small numbers at Isla Pájaros, perhaps elsewhere in Tempisque basin (Sánchez).

Range. SE USA, N Mexico, and West Indies to C Brazil and Peru.

Note. Sometimes placed in genus *Hydranassa*; also known as Louisiana Heron.

REDDISH EGRET

Egretta rufescens Pl. 5(7)

Garceta Rojiza

Description. 30″ (76cm); 500g. Medium-sized, long-necked; iris pale yellow. **Adults:** flesh-colored, black-tipped bill diagnostic; neck feathers lanceolate, shaggy. Dark phase: head and neck cinnamon to dull rufous; body gray. White phase: plumage entirely white. Legs blackish. **Immatures:** bill blackish, sometimes with paler, horn-colored base. Dark phase: (includes most birds seen in Costa Rica) plumage entirely dull gray, paler and more uniform than Little Blue Heron. White phase: plumage white, neck feathers shorter and broader than on adult.

Habits. Restricted to coastal areas, including salt ponds, tidal flats, and mangroves; generally seen singly but may associate loosely with other herons, notably Snowy Egrets or Little Blue Herons; walks slowly and deliberately in shallow water until it finds a school of small fishes, which it pursues with short, frantic, wing-flapping dashes to and fro, stabbing at them all the while.

Voice. Guttural croaks.

Status. Rare but probably regular winter resident along Pacific coast, especially around Golfo de Nicoya and Golfo Dulce (November–March). On Caribbean coast recorded only in fall migration (September).

Range. Breeds from Texas, Florida, Bahamas, West Indies, and NW Mexico S to Yucatán and Guatemala; S in winter to N South America.

Note. Often placed in genus *Dichromanassa*.

SNOWY EGRET

Egretta thula Pl. 5(10)

Garceta Nivosa

Description. 24″ (61cm); 375g. Plumage always entirely white; black legs with contrasting yellow feet diagnostic. **Adults:** slender black bill; contrasting yellow facial skin; lacy elongate plumes on head, breast, and back (longer and more conspicuous in breeding season). Iris yellow. **Immatures:** inner side and back of legs olive-yellow, reducing contrast with yellow feet; no elongate plumes. Immature Little Blue Heron has differently colored facial skin, bill, legs, and feet.

Habits. In marshes, lagoons, salt ponds, river mouths, or tidal flats often forages in groups, which may drive fishes with coordinated movements, or may take advantage of beating action of feeding groups of storks or spoonbills, actively fluttering up and plunging down after fishes. Along margins of lakes or large rivers where water depth drops off steeply, usually regularly spaced in individual feeding territories. Snowy Egrets usually roost communally.

Voice. In alarm, a raucous, croaking *krawwk*, deeper than note of Little Blue Heron; a sharper, more quacking *kraak* in aggressive encounters.

Nest. A platform of sticks with shallow central depression, lined with finer twigs, 6–13ft (2–4m) up in mangrove or other tree near water. Eggs 3–4, pale bluish-green. May–August.

Status. Locally common winter resident in lowlands of both slopes, sometimes to 2300ft (700m), rarely higher; uncommon to rare summer resident, mostly first-year birds. Small numbers breed in large Cattle Egret colonies in lower Tempisque basin, perhaps elsewhere.

Range. N USA to N Argentina; northern birds winter largely in Middle America.

Note. Formerly separated in genus *Leucophoyx*.

GREAT EGRET

Casmerodius albus Pl. 5(14)

Garceta Grande (Garza Real)

Description. 40″ (101cm); 950g. Largest, longest-necked all-white heron, with yellow bill and black legs in all plumages. Iris pale yellow. Cattle Egret much smaller, chunkier; other white herons differ in color of legs or bill. In flight, slow wingbeats and leisurely gliding of this species are distinctive.

Habits. Frequents marshes, estuaries, lake or river margins, tidal flats, and salt ponds; usually forages singly, although loose groups gather in marshes, but individuals, intolerant of one another when feeding, space out over available area; while foraging, stands quietly or walks slowly with kinked neck diagonal, seizing fishes and frogs with quick stabs.

Voice. A loud, hoarse, prolonged croak or several shorter croaks given in series, lower-pitched and more rattling than notes of other egrets.

Nest. A rather flimsy platform of coarse twigs with finer twigs in lining, placed atop mangrove or other tree. Eggs 2–3, blue-green. Mainly July–November in major breeding colony on Isla Pájaros, Río Tempisque; also nests March–August on small islets in Golfo de Nicoya.

Status. Widespread, locally common in lowlands, occasionally to middle elevations in Valle Central. Most Guanacaste birds are probably permanent residents, but overall the small local population is probably increased severalfold by migrants and winter residents from October to April, especially on Caribbean slope.

Range. Breeds from SE Canada and N USA to S Chile and S Argentina; northern breeders winter from S USA to N South America. Widespread in Old World.

Note. Sometimes placed in *Egretta* or *Ardea*; often called Common or American Egret.

GREAT BLUE HERON

Ardea herodias Pl. 5(6)

Garzón Azulado (Garza Ceniza, Garzón)

Description. 52″ (132cm); 2.5kg. Largest heron in Costa Rica, with slowest wingbeat. **Adults:** head white; broad black stripe at sides of crown extends to hindneck and long occipital plumes; neck brownish-gray with black-and-white midventral stripe; body and wings mostly blue-gray, with black patch on side of breast, white streaking below; thighs rufous. Iris yellow; bill yellowish; legs blackish. **Immatures:** crown gray; sides of head whitish; neck brownish; belly whitish; sides of breast streaked. Adult plumage acquired at 3 years of age.

Habits. Usually solitary; inhabits almost any shallow fresh or salt water: marshes, wet fields, margins of lagoons or sluggish streams, seashores, river mouths, mangroves; stands or walks stealthily in shallow water, stabs swiftly at fishes, frogs, rodents, crustaceans, or large insects; may establish feeding territories while wintering.

Voice. When startled, a deep, hoarse, guttural *gruuuk*.

Status. Uncommon but widespread winter resident, lowlands to middle elevations (Valle Central). A few immature or subadult birds also summer, particularly in Guanacaste, but no evidence of breeding.

Range. Breeds from S Alaska and S Canada to Chiapas and Belize, also Galápagos Is.; winters to N South America.

CHESTNUT-BELLIED HERON

Agamia agami Pl. 5(1)

Garza Pechicastaña

Description. 28″ (71cm); 550g. Very long neck and bill, relatively short legs. **Adults:** head black, with long, blue-gray occipital plumes; neck maroon, with chestnut and white midventral stripe, silvery "filigree" on lower foreneck; belly chestnut; upperparts dark glossy green. Iris orange to reddish-brown; bill blackish, shading to grayish-green at base of lower mandible; facial skin yellow-green; legs blackish. Breeding: conspicuous, elongate pale blue-gray plumes on back; facial skin reddish (males) or green (females). **Immatures:** face and neck glossy brown; mantle blackish-brown; breast buffy, streaked with dark brown, shading to whitish on belly.

Habits. Stalks along shady streams and swamps in humid forest, wading in shallow water, walking along logs and sandbars; catches small fishes in pools and riffles, frogs and lizards from banks and overhanging vegetation; usually solitary; perches in tree when alarmed.

Voice. Usually silent; may give a low *guk* in alarm; when disturbed at nest, a low, rough, rattling *kr-r-r-r-r*.

Nest. A rather flimsy, circular platform of slender twigs 3–10ft (1–3m) above water; small colony often with other herons, especially Boat-billed Heron. Eggs 2, rarely 3, bluish-green. June–September.

Status. Uncommon to rare resident in humid lowland forests of Caribbean slope and Golfo Dulce district, including Península de Osa. Only known breeding colony (ca. 15 pairs) at Westfalia, S of Limón.

Range. S Mexico to NW Ecuador, N Bolivia, and C Brazil.

Note. Also called Agami Heron.

FAMILY **Ciconiidae:** Storks

The 17 species of storks are large to enormous waders with white and/or black plumage; most have the face or the entire head and neck bare. The sexes are alike in color, but males are larger. The bill is long and massive, curved or straight. Storks lack powder-down; their toes are webbed at the base except for the hind toe, which is small and slightly elevated above the rest. Storks fly strongly, with neck extended, legs trailing behind, and slow wingbeats, and often soar to great heights. Found at temperate and tropical latitudes around the world, storks feed on a variety of animal food, including carrion. Adults are nearly voiceless but can make low vocal sounds while they select nest sites and court, and they rattle and clap their mandibles loudly. Nestlings are often very noisy. Some species are gregarious and colonial, others solitary. The nests are platforms of sticks placed on trees, ledges, or buildings; those of some species are built up year after year until they become huge. Males and females share all nesting duties, including incubating the 3–6 dull white eggs for a period of 28–36 days and feeding the chicks, which hatch naked but soon acquire a coat of whitish down. They remain in the nest for 50 to more than 100 days in the largest species.

WOOD STORK

Mycteria americana Pl. 4(6)

Cigueñón (Garzón, Guairón)

Description. 40″ (102cm); 2.5kg. Very large, with heavy bill drooped at tip. **Adults:** body plumage white; flight feathers black; head and neck bare, rough-skinned, black. Bill and legs blackish. **Immatures:** head and neck covered with grayish down; bill largely horn-color or yellowish; body feathering dull white. Adult plumage after 1 year, but head and neck are not fully bare for another year.

Habits. Frequents a variety of freshwater and saltwater habitats, often seeking areas where receding water levels concentrate fishes in shallow pools; usually gregarious, forming communal roosts and often feeding in groups that advance in a compact phalanx, with submerged, partly open bills sweeping to and fro, driving and concentrating fishes, snapping bills shut on contact; often fly and soar in groups, in open formation.

Voice. Usually silent, occasionally a soft hoarse croak or hiss when disturbed; chicks give a high-pitched, nasal *nyaa-nyaa-nyaa*.

Nest. A thin platform of sticks, sparsely lined with leafy twigs, atop mangrove or other trees by water, in dense colonies. Eggs 2–4, creamy-white with granular surface. Nests in dry season, as water levels fall.

Status. Locally common to abundant in Guanacaste (breeds Isla Pájaros, Río Tempisque) and toward end of dry season in Río Frío–Caño Negro region; often numerous around Golfo de Nicoya; elsewhere a rather rare and erratic wanderer over lowlands of both slopes and Valle Central.

Range. SE USA, NW Mexico to W Ecuador, Bolivia, and N Argentina.

JABIRU

Jabiru mycteria Pl. 4(5)

Jabirú (Galán Sin Ventura, Veterano)

Description. 53″ (135cm); 6.5kg. Huge; bill heavy, sharp-pointed, slightly upturned. **Adults:** plumage entirely white; head and neck bare, black except for red area at base of neck (brighter in male). Bill and legs black. **Immatures:** head and neck covered with sparse grayish down, rest of plumage grayish, more or less mixed with white. All egrets much smaller, heads and necks white; Wood Stork has black flight feathers.

Habits. Forages in freshwater marshes, lakes and ponds; roosts and nests in tall trees, often in wooded areas far from water; usually forages singly, walking slowly and making sudden stabs for prey, especially mud-eels, but sometimes 1 or 2 hunt at edge of group of spoonbills or Wood Storks; usually wary; flies lightly and gracefully after lumbering takeoff; often soars.

Voice. Silent except for loud bill-clapping when disturbed, especially near nest.

Nest. Solitary, a huge platform of sticks, often used year after year, well above ground in large tree in savanna or woodland. Eggs 2–4, dull white, rough-shelled. Breeds during dry season as water levels fall.

Status. Increasingly uncommon in Tempisque basin, the only Costa Rican breeding area; seasonally visits Caño Negro–Río Frío area; sporadically elsewhere in Guanacaste.

Range. S Mexico to E Peru and N Argentina.

FAMILY **Threskiornithidae:** Ibises and Spoonbills

This family of 33 species is found on all continents at tropical and temperate latitudes. Its members have shorter necks and legs than most storks or herons; some have longer tails. The plumage is white, brown, and/or black in most, but 1 species is pink, another red; the sexes are alike or nearly so, but males often average larger. In many the face, throat, or whole head is bare; all lack powder-down. Bills of ibises are slender and decurved; those of spoonbills are straight with flattened, spatulate tips. All members of the family fly strongly with extended neck but do not soar. Food includes fishes, insects, invertebrates, sometimes vegetable matter. Voices are varied; most species croak or grunt, but some whistle or toot loudly. Most inhabit marshes or shores, but a minority are birds of forest or savanna; most are gregarious and colonial. Nests of sticks, rushes, or grass are placed in trees, on the ground, or on ledges; both parents share building, incubation, and caring for the young. Ibises and spoonbills lay 2–5 eggs, white to bluish, spotted or immaculate; the chicks are downy. Incubation lasts 21–29 days, and the young fly well when 30–50 days old.

GREEN IBIS

Mesembrinibis cayennensis Pl. 4(10)

Ibis Verde (Coco Negro)

Description. 22″ (56cm); 650g. Heavy-bodied, very broad-winged, rather long-tailed; bill relatively small and slender, decurved. **Adults:** above glossy, iridescent greenish-black; below sooty-black with slight greenish gloss; small shaggy occipital crest; feathers of neck lanceolate; facial skin dark gray, more greenish above eye, pinkish on chin, and dull bluish-gray on throat. Bill pale green with horn-color along culmen, yellowish tip; legs dull pale green. **Immatures:** much duller overall, head and neck sooty with little gloss. Glossy Ibis slimmer with longer bill and legs, more pointed wings.

Habits. Frequents forested swamps, streams, pools, even muddy trails inside forest, foraging by walking about with nodding head, probing deeply into mud; may also visit open marshes beside forest; flies heavily, often above canopy, with body and neck angled upward, beak down; glides less than other ibises; usually in pairs or small groups.

Voice. Deep, mellow, rolling trills, singly or

in series, while perched or in flight: *crrrrw-crrrw-crrrw* or *co-co-co-co-correct-correct*; a more guttural *crrr-werrrw!* when perched; a sharp froglike bark when alarmed. Sometimes calls from tall snags, especially at dawn; also heard at night.

NEST. Undescribed (?).

STATUS. Resident locally throughout Caribbean lowlands, most common in Río Frío area and coastal swamps N and S of Limón.

RANGE. E Honduras to E Peru and N Argentina.

WHITE IBIS

Eudocimus albus Pl. 4(8)

Ibis Blanco (Coco)

DESCRIPTION. 25″ (63cm); 700g. Slender, decurved bill; our only ibis with much white in plumage. **Adults:** plumage entirely white except black wingtips. Iris pale blue; bare face, legs, and bill red. **Immatures:** mostly brown, with white rump and belly, mottled head and neck. Iris grayish; bill pinkish to brownish; legs olive-gray. Glossy Ibis darker, lacks white.

HABITS. Frequents various saltwater and freshwater habitats, wherever it can find soft mud in which to probe for food; gregarious, usually forages and flies in flocks; flies with a few shallow flaps, then a long glide on horizontal wings; often perches in trees; gathers in communal roosts, mostly in mangroves.

VOICE. A nasal, piglike grunting *ahnk ahnk* when disturbed; softer grunting notes, *koh-koh-koh*, when feeding or in flight.

NEST. A bulky but compact platform of twigs with a shallow central depression lined with leaves, 3–16ft (1–5m) up inside mangrove tree. Eggs usually 2, cream to greenish-white, blotched and speckled with brown. Breeds mainly in early wet season.

STATUS. Resident; center of abundance Golfo de Nicoya and Tempisque basin, breeding on Isla Pájaros in Río Tempisque; seasonally common in Río Frío region and locally farther S along Pacific coast; otherwise rare and sporadic in lowlands of both slopes.

RANGE. SE USA and NW Mexico to NW Peru and French Guiana; also Greater Antilles.

GLOSSY IBIS

Plegadis falcinellus Pl. 4(9)

Ibis Morito (Coco Negro)

DESCRIPTION. 23″ (58cm); 500g. Slender; dark; long, decurved bill. **Adults:** above and below dark, glossy chestnut-bronze (appears black at a distance); wings and tail blackish glossed with metallic green; head and neck finely streaked with dusky and whitish; facial skin gray with bluish posterior edge. Bill and feet blackish. **BREEDING:** head and neck dark glossy chestnut. Facial skin bluish-gray with bright blue border. Legs dark greenish to blackish. **Immatures:** like nonbreeding adult but body plumage duller, below gray-brown with little chestnut tinge.

HABITS. Feeds singly or in small groups, probing in soft mud and shallow water of freshwater marshes, pond margins, flooded pastures; sometimes associates with White Ibis or herons when feeding; flies strongly with a few shallow flaps and a glide on horizontal wings.

VOICE. Usually silent; when disturbed, a nasal cough or grunt.

NEST. A compact platform of dry rush stems with a shallow central depression, 6–16ft (2–5m) up in crown of mangrove tree. Eggs usually 3, dark blue-green. June–September.

STATUS. Locally common in Guanacaste; has bred in small numbers at Isla Pájaros, in Tempisque basin, since 1978. Scattered records on Caribbean slope, where probably a rare migrant; regular only in Río Frío region.

RANGE. Breeds from NE USA to Costa Rica, N Venezuela, and Greater Antilles (breeding range has expanded in recent years); winters from SE USA to N South America.

WHITE-FACED IBIS

Plegadis chihi Not Illustrated

Ibis Cariblanco

DESCRIPTION. 22″ (56cm); 475g. Very similar to Glossy Ibis in all plumages; young birds often not distinguishable. Bill generally slaty rather than blackish; facial skin often reddish and never with pale grayish or bluish border. Red iris of adults diagnostic, but immatures often have brown iris like that of Glossy Ibis. Legs and feet reddish-brown to dusky. Breeding adults have a border of white feathers around bare face.

HABITS. Similar to those of Glossy Ibis.

STATUS. One old record for Térraba valley. The species has declined somewhat in North America and probably no longer wanders as far S as Costa Rica.

RANGE. Breeds from W, C, and SE USA to C Mexico; winters S to El Salvador, formerly (?) to Costa Rica; also resident locally in South America from Colombia and Venezuela to C Chile and C Argentina.

NOTE. Has sometimes been lumped with *P. falcinellus*, but the 2 breed sympatrically in Louisiana.

ROSEATE SPOONBILL
Ajaia ajaja Pl. 4(7)
Espátula Rosada (Pato Cuchara, Garza Morena, Garza Rosada, Pato Rosado)

DESCRIPTION. 32″ (81cm); 1.4kg. Distinctive spatulate bill; only large pink bird in our area. **Adults:** head bare, greenish; neck and body white to pale pink; wings deep pink with rose-red lesser coverts; tail buffy-orange. Iris brick-red; bill greenish; legs reddish. BREEDING: facial skin suffused with orange; bill, crown and gular pouch bluish-green. **Immatures:** body pinkish-white; head covered with dull white down; wings pale pink, primaries dusky-tipped; legs and feet blackish. Three years to full adult plumage.

HABITS. Gregarious, usually feeding, roosting, and nesting in groups or flocks; frequents various freshwater or saltwater habitats—wherever shallow, open, still, or slowly-flowing water occurs; feeds by submerging bill or whole head, sweeping it from side to side while stirring up bottom mud with feet to flush small fishes, crustaceans, insects, etc., and snapping bill shut on contact. Often several forage in compact group or phalanx.

VOICE. When flying or disturbed, a nasal, rasping *keh-keh-keh-keh*; a soft grunting when feeding in groups.

NEST. A sturdy cup of sticks, lined with green vegetation, 4–16ft (1.2–5m) up inside mangrove or other tree. Eggs 2–4, dull white, blotched with different shades of brown. Nests in dense colonies, mainly in early dry season.

STATUS. Locally common to abundant, at least seasonally, in Tempisque basin, Río Frío district, and around Golfo de Nicoya; in smaller numbers elsewhere along Pacific coast; sporadic and rare over most of Caribbean slope and Valle Central. Pronounced seasonal movements reflect changes in water levels. Major (only?) breeding colony on Isla Pájaros, Río Tempisque.

RANGE. Extreme S USA and NW Mexico to C Chile and C Argentina; also Greater Antilles.

NOTE. Sometimes placed in genus *Platalea*.

ORDER **Anseriformes**
FAMILY **Anatidae:** Ducks, Geese, and Swans

The 145 species of this cosmopolitan family vary greatly in size, coloration, and proportions, but all are short-tailed and slim-necked, with 3 front toes fully webbed and the distinctive duck bill, which is more or less broad and flattened, with a nail at the tip and a comblike series of horny plates, or lamellae, along the inner side of the margin. The plumage varies widely in color and pattern, including the degree of sexual dimorphism, but it is always dense and waterproof, with an undercoat of fine grayish down. All species are at least partly aquatic, feeding on various aquatic animals and/or plants by diving, tipping up, or surface dabbling; some species browse on tender shoots or eat grain and seeds on dry land. The nests of waterfowl, placed on the ground or in cavities, are usually lined with down plucked from the parent herself. Clutch size varies from 2–16 immaculate whitish, buffy, or greenish eggs; sometimes many females lay in "dump" nests. In most ducks, the male departs after fertilization, and the female carries on nesting alone; most geese and swans and perhaps whistling-ducks form persistent, sometimes lifelong pair-bonds, and male and female share in nesting duties (although in most species the female does most or all of the incubation). The precocial young hatch with eyes open, clad in dense, usually boldly patterned down; they can run or swim and feed themselves very soon after hatching; the parents provide leadership and protection.

During the complete molt following breeding, the flight feathers are dropped nearly simultaneously in most waterfowl, which are then flightless for several weeks. In many ducks, the "eclipse" plumage acquired in this molt is soon replaced by the brighter breeding plumage, at least in males. Tree-ducks, geese, and swans have only 1 molt per year and, therefore, no seasonal change in coloration. More species of waterfowl breed in the temperate zones than in the tropics, but the high-latitude species are virtually all migratory, with many reaching tropical latitudes in winter.

BLACK-BELLIED WHISTLING-DUCK
Dendrocygna autumnalis Pl. 8(7)
Pijije Común (Piche)

DESCRIPTION. 21″ (53cm); 800g. Slender; long neck and legs; broad wings, with wide white stripe on upper surface. **Adults:** head mostly brownish; throat and eye-ring white; crown and hindneck dark brown; body mostly chestnut; belly, underwing, and tail black; crissum spotted with black and white; greater coverts and bases of primaries white, rest of remiges black; middle coverts pale grayish, lesser coverts pale brownish. Bill red; legs pink. **Immatures:** bill blackish; body mostly grayish-brown; belly mottled with black and white; wing-stripe duller but still conspicuous.

HABITS. Favors open freshwater marshes and ponds, wet pastures, shallow lagoons; gregarious, usually resting in compact flocks in or beside water by day, flying out to feed at night, often in smaller groups; swims well and can dive but usually feeds standing or walking in shallow water, taking mostly seeds but also some leaves and shoots (sometimes a pest in sprouting rice fields), mollusks, insects.

VOICE. Commonest call a shrill, whistled *wi-CHEE-chi-chee*! or some variant: *pipiCHEE chee!*, often heard overhead at night. Alarm note a piercing *yeep!* or *yip*! In flight, a down-slurred *peeuw*.

NEST. An unlined tree cavity or, probably more often, a shallow, loosely woven cup of grasses amid low vegetation on ground, often far from water. Eggs up to 15, buffy-white. Many females lay in "dump" nests, especially in cavities or nest boxes. May–October.

STATUS. Widespread resident in lowlands of both slopes, but generally uncommon and local except in Tempisque basin and Río Frío district, where abundant at some seasons; pronounced movements reflect changing water levels. Uncommon and sporadic in Valle Central; may breed in small numbers in Cartago area and in Reventazón valley.

RANGE. S Texas and NW Mexico to W Ecuador and N Argentina.

NOTE. Members of this genus are often called tree-ducks.

FULVOUS WHISTLING-DUCK

Dendrocygna bicolor Pl. 8(9)

Pijije Canelo

DESCRIPTION. 20″ (51cm); 750g. Long neck and legs, angled downward in flight; dark wings; gray to blue-gray bill diagnostic. **Adults:** mostly uniform cinnamon-buff; mantle black, broadly scaled with chestnut-rufous; white stripe down side diagnostic when swimming or perched; white V on upper tail-coverts. Legs bluish-gray. **Immatures:** like adults but the white of upper tail-coverts replaced by brown.

HABITS. In open freshwater marshes, flooded fields, pond margins; swims and dives more, perches less often in trees than other *Dendrocygna*; usually in groups or flocks; eats mostly seeds, gathered largely by tipping up and dabbling.

VOICE. A scratchy or throaty whistled *kheeeer* or *kikheeeer*.

NEST. A loosely woven, shallow bowl of grasses, on ground amid dense, low herbaceous vegetation. Eggs 10–15, white (in California; no Costa Rican record).

STATUS. Since 1975 in Tempisque basin, where the species appears to be increasing, and almost certainly breeds (e.g., at Pelón de la Bajura); up to several hundred sometimes present at Palo Verde in dry season.

RANGE. S USA to Costa Rica, definitely breeding at least S to Honduras; also greater Antilles and much of South America, Africa, and S Asia.

WHITE-FACED WHISTLING-DUCK

Dendrocygna viduata Pl. 8(8)

Pijije Cariblanco (Piche Careto)

DESCRIPTION. 17″ (43cm); 650g. Small, dark; white face diagnostic. **Adults:** upper throat white, separated from face by black band; rest of head, upper neck, and central underparts black; lower neck and chest chestnut; sides barred with black and white; mantle dark brown with rusty scaling; bill black; in flight appear all-dark. **Immatures:** dull brownish-gray with some black barring on sides; throat white; duller and smaller than Fulvous, and lack patterned wing of young Black-bellied Whistling-Ducks.

HABITS. Frequents freshwater marshes and pond margins, often in a tight little group associated with larger flocks of Black-bellied Whistling-Ducks. Like that species, it seems to forage mostly at night, resting by day in vegetation at water's edge.

VOICE. A whistled *whee-hee-heer*, the first note slurred up, the last slurred down, sometimes 2–3 times in rapid succession, the first phrase loudest; less shrill than call of Black-bellied.

NEST. A bowl of dead leaves on marshy ground. Eggs up to 9, white, tinged with cream. August–October (in Trinidad; no Costa Rican record).

STATUS. Tempisque and Bebedero basins in Guanacaste, where now extremely rare at best, due mainly to habitat destruction; presumably resident (at least formerly?).

RANGE. Costa Rica; E Panama to Bolivia and N Argentina; also Africa and Madagascar.

MUSCOVY DUCK

Cairina moschata Pl. 8(10)

Pato Real (Pato Perulero, Pato Real)

Description. ♂ 34″ (86cm), 3kg; ♀ 25″ (64cm), 1.3kg. Very large, heavy-bodied, broad-winged. **Adults:** below dusky-black; above black, strongly glossed with metallic green; conspicuous white upper and lower wing-coverts. Male not only much larger but also has conspicuous crest and extensive bare, black face with red caruncles. Bill banded with black and whitish; legs black. **Immatures:** browner, less glossy; only a small square of white on upperwing.

Habits. Inhabits forested watercourses of all sorts: streams or rivers with gallery forest, wooded swamps, mangroves; often forages in more open areas, including marshes and grainfields; eats mostly seeds, including corn and rice, and tubers, occasionally frogs, crabs, and insects; outside breeding season usually in groups or flocks; often roosts in trees; flies heavily, with head held high, body low.

Voice. A rarely heard guttural croak or quack (females only?).

Nest. In large natural cavity in tree, often high above ground, typically in gallery forest; little or no nest lining; occasionally nests on ground. Eggs 8–10, white. Mainly in wet season.

Status. Widespread but generally uncommon and local resident in lowlands of both slopes; largest concentrations in lower Tempisque basin (Palo Verde) during dry season.

Range. N Mexico to W Colombia, E Peru, and N Argentina.

AMERICAN WIGEON

Anas americana Pl. 8(6)

Pato Calvo

Description. 20″ (51cm); 700g. Medium-sized, slender, with round head; small, bluish, black-tipped bill; large patch of white on upper wing-coverts conspicuous in flight. **Adult ♂:** crown white; sides of head glossy green; neck whitish, mottled with black; mantle and sides pinkish-brown; chest purplish; belly and sides of rump white; tail and tail-coverts black. ♀: head grayish to buff, mottled with black; mantle feathers gray-brown, marked with tawny-buff; wing patch scaled with white, not solid; breast pinkish-brown; sides cinnamon.

Habits. Very alert and wary; prefers marshes, pond margins, sloughs with open shoreline; grazes on land, eating mainly leafy vegetation; does not dive; tips up to feed in shallow water; in deeper water associates with coots or Ring-necked Ducks, robbing them of the aquatic plants they bring up; flies swiftly and erratically.

Voice. Males give a whistled *wee-wheew-whew*, thinner and softer than notes of whistling-ducks; females occasionally give a low, guttural quack.

Status. Regular winter resident (October–March), usually in fairly small numbers, in Tempisque basin and occasionally elsewhere in Guanacaste; movements follow water levels; in very dry years may depart by January.

Range. Breeds from Alaska and N Canada to N USA; winters to Panama and extreme N South America; also West Indies and Hawaii.

Note. Formerly placed in genus *Mareca*. Often called Baldpate.

GREEN-WINGED TEAL

Anas crecca Pl. 8(4)

Cerceta Aliverde (Zarceta)

Description. 14″ (35cm); 325g. Very small, short-necked; dark upperwing with iridescent green speculum; underwing white. **Adult ♂:** head and neck chestnut, except glossy green side of head; most of upperparts and sides finely vermiculated with black and white; breast buffy, spotted with black, a vertical white bar on side; sides of rump buffy; belly white. ♀: mainly gray-brown, the mantle scalloped with buffy; head and neck paler with dark crown and eye-stripe and fine darker streaks; belly white. Other teals have large blue wing-patches.

Habits. Prefers freshwater marshes; should be looked for among large flocks of Blue-winged Teal and other migrant ducks.

Voice. Male gives a high, peeping *krick-et*, singly or in series; female, a weak quack.

Status. Casual to accidental winter visitor; only definite record a bird banded in Missouri, shot by hunter at Ochomogo, in 1962; also several unconfirmed reports by duck hunters of birds shot in Tempisque basin (Nanne).

Range. Breeds from Alaska and N Canada to N USA; winters regularly to C Mexico, Hawaii, and Antilles; casual to accidental as far S as Costa Rica. Widespread in Old World.

Note. New World population was formerly considered a separate species, *A. carolinensis*.

MALLARD

Anas platyrhynchos Not Illustrated

Pato Cabeciverde [Carraco (domestic forms)]

Description. 24″ (61cm); 1.1kg. Large, robust migrant duck; orange legs; dark blue speculum bordered by white. **Adult ♂:** glossy green head separated from chestnut breast by narrow white ring; body mostly pale gray; bill yellow. ♀: mottled and scaled with buffy-brown and blackish; bill black and orange. Shoveler has much larger bill, pale blue-gray wing-patch, greenish speculum.

Habits. Little known in Costa Rica; evidently most likely on freshwater ponds and marshes; feeds mostly by surface dabbling and tipping up, taking plant matter and a few small aquatic invertebrates.

Voice. A loud *quack*.

Status. Apparently once a casual to rare winter resident in Guanacaste and in Valle Central; last recorded 1950–51 by P. Slud at Turrialba (2 wintering females); no recent reports.

Range. Breeds from Alaska and N Canada to S USA; winters S to C Mexico, rarely farther at present; formerly rarely (but regularly?) to Panama. Its disappearance from Central America probably reflects changing agricultural practices and provision of abundant winter habitat in S USA and Mexico in this century. Widespread in Old World.

NORTHERN PINTAIL

Anas acuta Pl. 8(1)

Pato Rabudo

Description. 25″ (64cm); 900g. Slender, long-necked, long-tailed (especially male); bronzy speculum. **Adult ♂:** head and most of neck brown; white stripe down side of neck; foreneck, breast, and belly white; mantle and sides vermiculated with black and white; sides of rump white; tail-coverts black. ♀: crown rusty; head, neck, and breast buff, mottled and streaked with blackish; mantle and sides dark brown, scalloped with buff; belly white. Legs and feet bluish-gray.

Habits. Prefers shallow, open freshwater lagoons, marshes, and sloughs; feeds on seeds and aquatic plants by dabbling, tipping up, or diving; usually in groups or small flocks associated with wigeons or teals; flight fast, graceful.

Voice. Male gives a mellow whistled *kwee* or *kwee-hee*; female, a hoarse, muffled quack.

Status. Winter resident, usually locally and in small numbers, occasionally large flocks, principally in Tempisque basin; sporadic and local elsewhere; formerly regular on the now-drained marshes and lagoons of Valle Central. Arrives late September or October, departs January or February, depending on water levels.

Range. Breeds from Alaska and N Canada to S USA; winters to N South America, West Indies, and Hawaii. Widespread in Old World.

BLUE-WINGED TEAL

Anas discors Pl. 8(5)

Cerceta Aliazul (Pato Canadiense, Zarceta)

Description. 15″ (38cm); 400g. Large pale blue patches on upperwing plus small bill and small size distinctive. **Adult ♂:** head dark blue-gray; white facial crescent; mantle dark brown, scalloped with buff; below pinkish-buff, spotted with black. ♀: crown and eye-stripe dusky; white spot on lores; rest of head and neck buffy, finely streaked with dusky; mantle as in male; chest brown, scalloped with white; sides and belly whitish, mottled with dark brown. Bill dark gray; legs yellowish. Males, in basic (like female) plumage upon arrival in fall, acquire breeding plumage by February. See Cinnamon Teal.

Habits. Prefers freshwater marshes, ponds and sloughs, but occurs in virtually any shallow, still, or slowly flowing water in lowlands or at middle elevations: river pools, salt ponds, estuaries, flooded pastures; usually in flocks; feeds by dabbling, occasionally tipping up; eats mostly plant matter, some invertebrates. A disturbed flock flushes explosively, flies fast and erratically, in tight formation.

Voice. Male, a thin, high-pitched *tsee* or *peew*; female, a high-pitched quack. Feeding flocks emit mousy squeaking and low chattering notes.

Status. By far the most numerous migrant duck; abundant winter resident and migrant in Tempisque basin and Río Frío district; locally common or sporadic elsewhere, lowlands to middle elevations, occasionally to 9800ft (3000m) in migration; arrives September–October, remains through April or (a few) May.

Range. Breeds from Alaska and N Canada to S USA; winters to C Peru, C Argentina, and West Indies.

CINNAMON TEAL

Anas cyanoptera Pl. 8(3)

Cerceta Castaña

Description. 15″ (38cm); 400g. Size and shape of Blue-winged Teal; also has blue patch on upperwing. **Adult ♂:** mainly bright chestnut, above scalloped with black and buff. Iris red, bill black. ♀: usually indistinguishable in field from female Blue-winged, but face averages darker, with chestnut loral spot. Iris brown, bill blackish. In hand, bill is longer (usually over 41mm; female Blue-winged 40mm or less), broader at tip than at base (even width in Blue-winged).

Habits. Similar to those of Blue-winged Teal, with which it associates.

Voice. Male gives a thin whistled *peep* or *peer*; female, a somewhat more guttural quacking than female Blue-wing.

Status. Casual winter visitor in Tempisque basin; several recent sightings of 1 or 2 males with large flocks of Blue-winged Teal, November–February.

Range. Breeds in W North America from S Canada to C Mexico; winters regularly to S Mexico, casually to Costa Rica; also resident in South America to Strait of Magellan.

NORTHERN SHOVELER

Anas clypeata Pl. 8(2)

Pato Cuchara

Description. 19″ (48cm); 600g. Long spatulate bill diagnostic; only large duck with pale bluish-gray wing-coverts, conspicuous in flight; speculum green; only the very rare Mallard also has orange legs. **Adult ♂:** head dark green; chest white; belly and sides chestnut; flanks white; center of back black; tail black with outer feathers white. Bill black. ♀: scaled and spotted with brown and buff; head mostly buff with fine brown mottling; upper wing-coverts duller, grayer; bill blackish and orange.

Habits. Pairs or small groups frequent open freshwater marshes, sloughs, lagoons, occasionally salt ponds; often associate with flocks of Blue-winged Teal; feed mostly by surface dabbling, less often tipping up; filter plant matter and small animals—insects, crustaceans, mollusks—from water or soft mud.

Voice. Usually silent; male rarely gives a nasal croak; female, a low quack.

Status. Fairly common winter resident (October through March or early April) in Guanacaste; elsewhere local and in small numbers in lowlands and middle elevations.

Range. Breeds from Alaska and N Canada to C USA; winters from S USA to N South America, West Indies, and Hawaii. Widespread in Old World.

Note. Formerly placed in genus *Spatula*.

RING-NECKED DUCK

Aythya collaris Pl. 7(7)

Porrón Collarejo

Description. 16″ (41cm); 700g. Long gray wing-stripe and dark bill with pale ring toward tip diagnostic. **Adult ♂:** head, neck, breast and back glossy black (chestnut neck-ring rarely obvious); sides and flanks gray; belly and bar in front of wing white; iris yellow. ♀: mostly dull dark brown, including crown; face and foreneck gray-brown, fading to white at base of bill; eye-ring and postocular streak white; belly white; iris dark. Like other *Aythya*, floats lower in water than most ducks.

Habits. Prefers deep open water of freshwater marshes, ponds, and lakes; usually in flocks; feeds by diving for aquatic plants, insects, and mollusks; often despoiled of plants it brings to the surface by wigeons; tolerates shallower water than Lesser Scaup, remains later in dry season.

Voice. Male rarely gives a low-pitched whistle; female, a purring growl.

Status. Uncommon to locally common winter resident, especially in Guanacaste (mainly Tempisque basin); rare and local on Caribbean slope and in Valle Central; arrives late October or November, remains through January to March, depending on water levels.

Range. Breeds from E Alaska and C Canada to N and C USA; winters S to Panama.

LESSER SCAUP

Aythya affinis Pl. 7(6)

Porrón Menor

Description. 17″ (43cm); 800g. Short white wing-stripe diagnostic; iris yellow. **Adult ♂:** head black with purple gloss, notably high-crowned; chest, tail-coverts and tail black; rest of body plumage white, coarsely vermiculated with black on mantle and flanks. Bill pale bluish with black tip. ♀: mostly dull dark brown, darkest on head; mantle and sides with faint paler scaling; white patch around base of dark gray bill; belly white.

Habits. Prefers freshwater ponds, lakes, and sloughs where water is reasonably clear and 3ft (1m) or more deep; dives for aquatic plants, snails, insects, and small fishes; in small flocks, pairs, or singly; patters over water to take flight.

Voice. A weak *prrrup* in flight.

Status. Uncommon to rare winter resident; evidently less numerous than formerly and usually much outnumbered by Ring-necked Duck; regular only in Guanacaste; sporadically, in small numbers elsewhere in lowlands and middle elevations of both slopes; arrives mid to late November, leaves by February or March.

Range. Breeds from C Alaska and C Canada to NW and C USA; winters to N South America, West Indies, and Hawaii.

GREATER SCAUP

Aythya marila Pl. 7(5)

Porrón Mayor

Description. 18″ (46cm); 950g. Similar to Lesser Scaup in all plumages, best told in flight by longer white wing-stripe extending at least to middle primaries (confined to secondaries in Lesser). **Adult ♂:** head more rounded than in Lesser, with green gloss; black vermiculations on mantle finer, making back appear whiter. ♀: safely distinguished only in flight or by larger size in the hand (culmen greater than 41mm, Lesser Scaup 40mm or less).

Habits. More partial to saltwater habitats than Lesser Scaup, although habits in general similar. Any scaup along the coast should be carefully checked.

Status. Accidental; only record is sighting of 3 birds in Golfo de Papagayo during fall migration in mid-October 1974 (Stiles and S. M. Smith).

Range. Breeds in Alaska and NW Canada; winters along both coasts to N Mexico, accidental farther south. Widespread in Old World.

MASKED DUCK

Oxyura dominica Pl. 7(8)

Pato Enmascarado

Description. 13″ (33cm); 375g. Small, short-necked, chunky; long black tail usually held low, trailing in water; white patches on secondaries and greater coverts diagnostic, conspicuous in flight. **Adult ♂** Breeding: crown and face black; neck and breast chestnut; mantle scalloped with black and rusty; below cinnamon, spotted with black on sides and flanks. Bill bright blue; legs grayish. ♀: pileum and 2 broad stripes across face blackish; rest of face buffy; feathers of chest, hindneck and mantle scalloped with black and rusty; below buff, spotted or barred with black laterally; belly white. Young birds and nonbreeding males resemble female.

Habits. Putters about in quiet, shallow water amid water hyacinths and other dense, low aquatic vegetation, often in only an inch or 2 of water along mudbars; usually in pairs, occasionally in groups of 10–20; feeds by diving, taking both plant and animal food; flies swiftly low over water with rapid wingbeats; drops abruptly like a stone into vegetation; secretive but not particularly shy.

Voice. Males give a throaty *oo-oo-oo* or *kirroo-kirroo-kiroo*; females, a short, repeated hiss.

Nest. A bowl of reed stems, etc., in dense aquatic vegetation just above water level. Eggs 4–6, whitish, rough-shelled. Breeds mainly in mid- to late wet season.

Status. Uncommon and local resident in Valle Central (especially around Cartago), Guanacaste (Tempisque basin), Río Frío district, and elsewhere in lowlands and middle elevations.

Range. Texas and West Indies to S South America.

Note. Ruddy Duck (*O. jamaicensis*) was reported from Costa Rica on basis of specimen, now lost, taken in 1857 on Volcán Irazú; almost certainly the record pertains to the Masked Duck (which had not been described at the time).

ORDER **Falconiformes:** Diurnal Birds of Prey

The most obvious external feature of the members of this order is the strongly hooked bill with a naked, fleshy cere covering the base of the upper mandible around the nostrils.

FAMILY **Cathartidae:** American Vultures

The 7 extant species of American vultures are confined to the New World, where they extend from central Canada to Cape Horn, including the Falkland Islands and the Greater Antilles. In size they range from the 12kg Andean Condor, with its 10ft (3m) wingspan, down to the 1kg Lesser Yellow-headed Vulture. Except for the predominantly white adult King Vulture, they have mostly black plumage with bare heads that may be red, yellow, blackish, or fantastically variegated with brilliant colors; the sexes are alike in plumage. These vultures differ from the other Falconiformes in having perforate nostrils and weak feet with rudimentary webs; this last, along with certain other anatomical and behavioral features, has led some authors to consider them more closely related to the storks than to the hawks and falcons. Lacking a syrinx, vultures are voiceless except for low grunts and hisses. They spend much time soaring on widespread wings while searching for carrion, their principal food, which is located largely by olfaction in some species and entirely by vision in others. The King and Black vultures sometimes kill small living animals, and the Black, in particular, often varies its diet with fruit. Without building a nest, vultures lay their eggs on the floor of a cave or cavity of some sort, or on the ground in a protected place, e.g., under overhanging rocks, amid dense vegetation, or in a hollow stump. Both sexes incubate the 1–2 whitish, spotted or immaculate, rather elongate eggs for 32–58 days and regurgitate food to the downy young, which develop slowly and can fly when 10–25 weeks old. Black Vultures are abundant and important scavengers around tropical American towns and villages.

TURKEY VULTURE

Cathartes aura Pl. 13(3)

Zopilote Cabecirrojo (Zonchiche, Noneca)

DESCRIPTION. 30″ (76cm); 1.4kg. Wings and tail long and narrow, the undersurface of the flight feathers gray, contrasting with black of wing-linings and body plumage. **Adults:** head bare, red (with bluish-white band across hindneck in birds of resident race). Iris brown; bill whitish, cere red; legs flesh-yellow to whitish. **Young:** head blackish; legs grayish. Black Vulture has shorter, broader wings and short tail and is all black below except for white patch at base of primaries.

HABITS. Quarters back and forth at low to medium heights, often following odor trails upwind to carcasses hidden by vegetation; soars lightly and buoyantly with wings held above horizontal; usually alone or in 2s or 3s, never in large groups except during spectacular migratory flights; most often seen over open country but also occurs in forested regions; yields place to Black Vulture at carcasses.

VOICE. Silent; at most a low hiss or raucous grunt.

NEST. On unlined floor of a cave, crevice among rocks, or hollow stump. Eggs 2, creamy-white, heavily marked with brown. November–February.

STATUS. Common resident countrywide, though much less numerous above ca. 6500ft (2000m). Great migratory flocks pass through, principally over Caribbean lowlands, in both fall (September–October) and spring (late January–mid-May); some birds winter, principally in lowlands.

RANGE. C Canada to Tierra del Fuego; also locally in Greater Antilles.

LESSER YELLOW-HEADED VULTURE

Cathartes burrovianus Pl. 13(2)

Zopilote Cabecigualdo (Noneca)

DESCRIPTION. 25″ (64cm); 1kg. Closely resembles larger Turkey Vulture, safely distinguishable only by color of head. **Adults:** head orange to orange-yellow with blue-violet crown. Iris red; bill white. An indistinct pale brownish patch may show at the base of the primaries above, but many Turkey Vultures also approach this condition—not a reliable field mark! **Immatures:** dusky head with whitish nape.

HABITS. Frequents marshes, wet savannas, wooded margins of rivers; flies like Turkey Vulture but more often quarters very low over ground; usually seen singly; apparently prefers carcasses of fishes and reptiles; may even kill moribund fishes stranded in drying pools.

VOICE. No information (?).

NEST. In hollow in tree. Eggs 2, whitish, marked with rufous and dark chestnut (in Panama; no Costa Rican record).

STATUS. Common and almost certainly resident in Río Frío region; rare and sporadic visitor in Tempisque basin (e.g., Palo Verde), on Península de Osa, and perhaps elsewhere in Pacific lowlands.

RANGE. E and S Mexico to N Argentina.

BLACK VULTURE

Coragyps atratus Pl. 13(4)

Zopilote Negro (Gallinazo, Zoncho)

DESCRIPTION. 25″ (64cm); 1.8kg. Larger head, longer neck, more erect stance, heavier build than Turkey Vulture; tail short and square; wings broad. Plumage entirely black except for white patch at base of primaries below; head naked, wrinkled, blackish; legs whitish. Young King Vulture larger, without white in primaries, shorter tail. Black-hawks (*Buteogallus*) have banded tails, yellow ceres and legs.

HABITS. Most abundant around towns and in open country, rarer in forested regions; soars on flat wings with frequent quick, shallow flaps; often ascends to great heights to scan for food or for other vultures descending to feed; aggressive, gregarious, and opportunistic, congregates in fighting, jostling throngs at large carcasses or dumps; a major predator of hatchling sea turtles; eats fruits, including bananas and especially those of palms, e.g., oil palm and royal palm (*Scheelea*).

VOICE. Rarely hisses or grunts.

NEST. On ground beneath overhanging rocks or amid dense tangles of bracken fern, bromeliads, or other low vegetation. Eggs 2, pale grayish-green, blotched with dark brown, sometimes in a wreath. November–February or March.

STATUS. Resident countrywide from coasts to high in mountains, though uncommon to rare much above 6500ft (2000m). Part of the population may be migratory, as sustained flights in 1 direction have been recorded.

RANGE. E and S USA to C Chile and C Argentina.

KING VULTURE

Sarcoramphus papa Pl. 13(5)

Zopilote Rey (Rey Gallinazo, Rey de Zopilotes)

DESCRIPTION. 32″ (81cm); 3.5kg. Largest local vulture, in form and flight most like Black Vulture but wings even broader, tail shorter. **Adults:** plumage mostly creamy-white; mantle tinged with buff; rump and flight feathers black; neck-ruff gray; bare, wattled head gaudily variegated with orange, yellow, blue, and black. Iris white; bill dark orange. **Young:** plumage sooty-black; small neck-ruff; head blackish; bill black with orange tinge. Adult plumage after 3 years of age; subadults more or less white below, black on mantle.

HABITS. Prefers at least partly forested country, perching well up in trees and soaring mostly low above treetops, sometimes very high; may locate hidden carcasses by scent; usually solitary, at most 3 or 4 at a large carcass. When it descends to carrion, other vultures make way for it, hence the name King.

VOICE. Low, croaking noises.

NEST. In low, hollow, unlined tree stump, or in a scrape in the forest floor at the base of a spiny palm. Egg 1, unmarked creamy-white. Mainly in dry season.

STATUS. Resident countrywide, including arid and wet regions, rarely up to 4000ft (1200m). Never present in large numbers, it remains fairly common on Península de Osa, uncommon to rare in remaining wilder regions.

RANGE. Tropical Mexico to N Argentina.

FAMILY **Pandionidae:** Osprey

Although superficially resembling the larger Accipitridae (and sometimes included in that family), the Osprey has a number of anatomical peculiarities that justify ranking it as a separate family. These include its pterylosis (arrangement of feathers), biceps muscles, foot tendons, etc.; the legs and feet are also distinctive—a short, stout, heavy-scaled tarsus, a reversible outer toe, horny spicules on the sole, and long, curved claws that are all of equal length and rounded in cross section. Many of these features, particularly the foot, are specializations for grasping and holding slippery fish. The bulky stick nests of ospreys are placed on trees, ledges, or man-made structures; both sexes build the nest, which may be used and augmented year after year. Most incubation of the 3 buffy to whitish, brown-splotched eggs is by the female, who feeds the newly hatched downy chicks with fish brought by the male; both parents bring food to older nestlings and fledglings. The incubation period ranges from 30–38 days; the young fly when 50–59 days old.

OSPREY

Pandion haliaetus Pl. 17(14)

Aguila Pescadora (Gavilán Pescador)

DESCRIPTION. 23″ (58cm); 1.5kg. Large, with long wings that in flight appear bent or kinked at the wrist, next to which is a black patch below. In all plumages, above dark brown, below mostly white; head mostly white with conspicuous dark stripe from lores to nape. Iris yellowish; cere and legs grayish. **Adult ♂:** chest usually immaculate. ♀: chest usually lightly streaked with dark brown. **Immatures:** below tinged with buffy; mantle with buffy scaling.

HABITS. Requires clear, still or slowly-flowing water, salt or fresh, inhabited by surface-swimming fishes; relatively indifferent to whether adjacent land is open or forested; spots fishes while flying deliberately over water at heights of 30–100ft (9–30m), captures them by plunge-diving, then carries prey in its talons to a regular perch, commonly a high snag, and eats; occasionally takes lizards, mammals, or birds, mainly during rainy periods when silt clouds water. Individuals are usually well spaced and evidently maintain fishing territories.

VOICE. A series of sharp, thin, high-pitched *cheep* or *keeip* notes, sounding like call of much smaller bird.

STATUS. Widespread, locally uncommon to fairly common migrant and winter resident, rare but regular nonbreeding summer resident on coastal and inland waters, occasionally up to 5000ft (1500m) on both slopes; most numerous in Río Frío–Caño Negro region. Fall migration mostly September–October, spring migration March–April; migrates mainly along both coasts.

RANGE. Breeds from Alaska and C Canada to W Mexico, Belize, and the Bahamas; winters from S USA to Argentina and Chile. Widespread in Old World.

FAMILY **Accipitridae:** Hawks, Kites, and Eagles

The 205 members of this cosmopolitan family range in size from the 75g male Tiny Hawk to the 7.5kg female Harpy Eagle. Division of the group into subfamilies is based mainly on features of the skeleton, which may not reflect the appearance or behavior of the birds. Thus, although most members of the 3 groups of kites are long-winged, graceful fliers, others are quite hawklike. All members of the family have strong feet with stout, hooked claws; all are strong fliers, and many soar. The plumage is typically gray, brownish, black, and/or white, often barred, spotted, and/or streaked; the bare cere and orbital skin are often colorful. Females are larger than males but usually similar in color. Most species give loud whistles or screams as well as cackling or chattering notes. Individuals usually hunt singly, though hunting territories may be defended by pairs or by single birds. Prey is usually captured with the feet and killed and dismembered with the bill. Food includes mostly vertebrates, but some species take insects or other invertebrates, others carrion. The nest of sticks is typically built by both sexes on a tree or ledge; it is lined with green leaves. In most species only the female incubates the 1–6 whitish, usually brown-blotched eggs and feeds the downy chicks; the male hunts and delivers prey to the nest. Male and female share nesting duties more equally in a few species, and in virtually all both feed well-grown young. Incubation periods vary widely with size of the bird from 28–49 days; the young fly when 28–120 days old. Young birds usually remain with the parents for some time after fledging until they can hunt proficiently. Most species have a distinctive juvenile plumage; several years may be needed to attain adult plumage, particularly in the larger species.

GRAY-HEADED KITE
Leptodon cayanensis Pl. 17(3)
Gavilán Cabecigris

DESCRIPTION. 20″ (51cm); 440g. Medium-sized; very broad-winged; bill and feet small; tail fairly long. **Adults:** above blackish, below white; head pale gray; wing-linings black; remiges below pale gray, barred with black; tail broadly and evenly banded with pale gray and black. Iris dark; cere and legs bluish-gray. **Immatures:** variable: above blackish-brown; head and underparts vary from entirely white (except blackish crown and postocular streak) to mostly blackish with some whitish streaking on belly and wing-linings; sometimes a rusty nuchal collar. Iris yellowish; cere and feet yellow. Often such birds are best identified by shape and behavior.

HABITS. Prefers forest canopy and edge, tall second growth, or semi-open beside forest; generally sluggish and deliberate, moving inconspicuously inside thick vegetation; often soars on flat wings; eats eggs and nestlings of small birds, small frogs and lizards, and insects, especially brood of wasps and ants.

VOICE. During breeding season, birds perching on high, exposed branches give a loud barking call, a series of 15–20 *wuh* or *heh* notes, more vigorous, less grunting than call of Slaty-tailed Trogon; disturbed at nest, a screaming *aaareeee*!

NEST. A thin platform of sticks in horizontal fork, well up in forest canopy. Eggs undescribed (?). March–June.

STATUS. Uncommon and local resident in forested lowlands and foothills the length of both slopes, to ca. 2500ft (750m).

RANGE. E Mexico to W Ecuador and N Argentina.

HOOK-BILLED KITE
Chondrohierax uncinatus Pls. 14(9), 16(9)
Gavilán Piquiganchudo

DESCRIPTION. 16″ (41cm); 275g. Lightly built, with rather long tail, small weak feet; dark bill heavy, deeply hooked. **Adults:** plumage variable. Iris whitish; bare spot on lores orange, separated from greenish-yellow cere; legs yellow. GRAY PHASE: (mostly males): above slaty, below barred with gray and white. BLACK PHASE: (mostly males): mostly dull black. RUFOUS PHASE: (mostly females): face gray; crown blackish; nuchal collar rufous; mantle dark brown; below broadly barred with dark rufous and buffy-white. In all phases, tail banded with pale gray and black. **Immatures:** above blackish-brown; nuchal collar white; below white, sometimes barred with blackish; tail more narrowly and irregularly banded than in adult.

HABITS. A sedentary, sluggish denizen of wooded swamps, gallery forest, second growth or semi-open near water; hunts land snails beneath forest canopy; also takes apple snails (*Ampullaria*) from grassy or bushy swamps in wooded country; usually flies low, with a few quick flaps and a wobbly glide; occasionally soars.

VOICE. A rather musical 2- or 3-note whistle,

given from dense foliage; a harsh scream when disturbed near nest.

Nest. A small structure of sticks, in crotch or on heavy horizontal branch, well out from trunk of large forest tree, sometimes at edge, overlooking open swamp. Eggs 2, buffy-white, blotched and streaked with shades of purple, reddish-brown, and black. March–May.

Status. Widespread but generally uncommon to rare and local, in lowlands and foothills the length of both slopes, occasionally straying to Valle Central. A pair may frequent an area for several months, breed, then apparently move on.

Range. S Texas and West Indies to N Argentina.

AMERICAN SWALLOW-TAILED KITE

Elanoides forficatus Pl. 15(2)

Elanio Tijereta (Tijerilla, Gavilán Tijerilla)

Description. 23″ (58cm); 480g. Long, pointed wings; long, deeply forked tail. **Adults:** head, underparts, and wing-linings white; mantle and flight feathers black. **Young:** similar but head and chest lightly streaked with dusky; back scaled or spotted with white.

Habits. In hilly or mountainous terrain with at least scattered remnants of humid forest. This most aerial of raptors is almost always on the wing, flying gracefully and buoyantly, gliding and banking with little flapping, then swooping swiftly with a few strong wingbeats to capture prey; uses feet to pluck insects, lizards, snakes, nestling birds, occasionally fruit, from vegetation or to catch flying insects in midair; often eats in flight, holding food in 1 foot while biting off pieces.

Voice. A scratchy, shrill *kawee*, often repeated in accelerating, ascending series that ends with a squeaky descending chatter: *kawee kawee kawee kwee kwee kwee pee pee pi pi pi pi*, the whole not unlike the sudden calls of White-collared Swift but clearer, more hawklike in tone.

Nest. A rather small, shallow cup of mossy twigs, lined with Spanish moss (*Tillandsia usneoides*) or beard lichen (*Usnea*), 100–125ft (30–38m) up in tall emergent or isolated tree. Eggs 2–3, white, blotched with brown. February–May.

Status. Mainly breeding resident, a few individuals evidently present year-round, on Caribbean slope and S Pacific slopes; avoids dry Pacific NW; scarce in Valle Central; most abundant in foothills and lower mountain slopes, becoming scarce in flat coastal lowlands except when migrating; mostly 330–6000ft (100–1850m); occasionally wanders to 10,000ft (3000m). Most arrive between late December and February, depart between July and September.

Range. Breeds from SE USA to E Peru and N Argentina; most birds breeding in Central and North America pass the nonbreeding season in South America.

Note. Formerly called Swallow-tailed Kite.

BLACK-SHOULDERED KITE

Elanus caeruleus Pl. 15(1)

Elanio Coliblanco (Gavilán Bailarín)

Description. 16″ (41cm); 350g. Whiter and longer-winged than any other small hawk, with distinctive flight. **Adults:** mostly white; mantle pale gray, shading to blackish on outer primaries; black shoulders and spot beneath bend of wing. Iris red; bill black; cere and legs yellow. **Young:** below streaked with rusty; mantle strongly tinged or scaled with rusty to brownish.

Habits. Frequents open savannas, pastures, agricultural land with scattered trees; hunts by flying slowly at heights of 30–80ft (9–25m); on sighting prey (usually small rodents, lizards, large insects), hovers on slowly beating wings, drops lightly down to seize it; flies gracefully, like gull, with slightly bowed wings; often mobs larger hawks and caracaras.

Voice. Short whistled *keep* often in series; plaintive *kreeek*; other high whistled notes suggestive of a much smaller bird; mobbing or aggressive note a rasping *chaaap*!

Nest. A substantial cup of twigs, lined with finer material, usually well up in isolated tree or tall shrub. Eggs 3–5, white, heavily marked with shades of brown. January–April.

Status. Common resident of cleared areas throughout the lowlands, locally to at least 5000ft (1500m); has recently expanded range; first reported in Costa Rica in 1958; definitely breeding by early 1960s.

Range. SW USA to E Panama. Widespread in South America and in Old World.

Note. The American bird was formerly (and probably better) considered a separate species (*E. leucurus*, White-tailed Kite) from the Eurasian *E. caeruleus*.

SNAIL KITE

Rostrhamus sociabilis Pls. 14(6), 15(6)

Elanio Caracolero (Gavilán Caracolero)

Description. 17″ (43cm); 375g. Bill with long, slender hook; wings long and rounded; tail notched; tail-coverts and basal half of tail white, outer half black, with narrow white or buff terminal band. **Adult ♂:** slate-black with cere and feet red or orange. ♀: mostly blackish-brown; face buff with dark postocular stripe; underparts broadly streaked with buff. Iris red; cere and legs orange. **Immatures:** like female but buffy areas more ochraceous or rufous; iris brown; legs brownish.

Habits. Feeds almost exclusively on large snails (mostly *Pomacea* or *Ampullaria*) in freshwater marshes and sloughs. Foraging birds quarter low over marsh with slow, rather heavy flight, pluck snails from water or vegetation with feet, go to regular perch to extract snail from shell, which is dropped intact; mostly perch quietly, occasionally soar at midday.

Voice. A dry, guttural, high-pitched, nasal *kekekekek* . . . or *heh-heh-heh-heh* . . . , between the bleat of a sheep and the rattle of a large cicada.

Nest. A rather scanty, shallow cup of twigs, a few yards (m) above water, in isolated clump of trees in marsh; often 2 or more pairs nest in loose association. Eggs 3–4, buffy-white, marked heavily with brown. July–September.

Status. Locally common in Tempisque basin and Río Frío–Caño Negro region; rare and sporadic elsewhere in lowlands of both slopes. Very pronounced seasonal movements in response to changing water levels. Population has declined in recent years due to habitat destruction, especially in Tempisque basin.

Range. S Florida and SE Mexico to Ecuador and N Argentina.

DOUBLE-TOOTHED KITE

Harpagus bidentatus Pl. 16(1)

Gavilán Gorgirrayado

Description. 13″ (33cm); 180g. Accipiter-like; black median stripe on white throat; in hand or at very close range, note 2 horny "teeth" on tomium. **Adults:** above dark gray; below mostly barred with gray and white, blending to rufous barring on sides of chest (male) or solid rufous chest (female); feathers of thighs, crissum, and flanks white, fluffy (may give impression of white rump in flight); tail broadly banded with gray and black. Iris orange; cere and legs yellow. **Immatures:** above brown, flecked with buffy; below white, sparsely streaked with blackish-brown.

Habits. Usually keeps to canopy of forest or old second growth, ranging lower at openings and edges; looks for prey (lizards, large insects) from perch, captures it with a sudden swoop or pounce or runs along branches in pursuit; snatches slow-flying insects in midair; follows monkey troops or bird flocks to catch prey they flush; sometimes perches quietly on high, exposed branch; frequently soars, usually holding tail closed, wings far forward.

Voice. A lisping *tsip tsip tsip* . . . ; a high, thin, whistled *see-weeeep see-weeeet*.

Nest. A shallow saucer of sticks, usually high in tree at forest edge. Egg described as whitish, speckled with brown (not seen in Costa Rica). A nest 70ft (21m) up near San Vito de Java held 2 feathered nestlings in early May.

Status. Uncommon to fairly common resident in humid-forested lowlands and foothills, locally to ca. 5000ft (1500m) on both slopes; rare in dry-forested Pacific NW.

Range. S Mexico to W Ecuador, Bolivia, and SE Brazil.

PLUMBEOUS KITE

Ictinia plumbea Pls. 14(8), 15(4)

Elanio Plomizo

Description. 14″ (36cm); 280g. Long, pointed wings; tip of tail square or slightly notched. **Adults:** above blackish; head and underparts dark blue-gray; wings dark, showing a rufous patch in primaries from below; tail black, crossed by 2 narrow white bands. Iris red; cere and bill blackish; feet bright yellow to orange. **Immatures:** below white, heavily streaked with gray to blackish; tail with 3 broader white bands.

Habits. Soars and swoops buoyantly above canopy and along edges of broken forest, savanna groves and mangroves, less often over solid forest; plucks snakes, lizards, and large insects from vegetation; catches many insects, especially bees and termites, in flight; often migrates with American Swallow-tailed Kites.

Voice. A shrill *shirreeeer*! or *sisseeeoo*!; a *Legatus*-like *tih-wee* or *tih-deedee*.

Nest. A substantial, rather deep cup of twigs, high in tree. Egg 1, plain dull white, becoming stained with brownish as incubation proceeds (in Venezuela; no Costa Rican egg record).

Status. Breeding resident in lowlands and foothills of both slopes but relatively common only in mangroves and adjacent forests along N Pacific coast and lowlands S of Lago de Nicaragua; sometimes abundant along Caribbean coastal lowlands while migrating northward in February–March, southward in July–September.

RANGE. Breeds from S Mexico to N Argentina; Central American breeders withdraw to South America in nonbreeding season.

MISSISSIPPI KITE

Ictinia mississippiensis Pl. 15(3)

Elanio Colinegro

DESCRIPTION. 14″ (36cm); 280g. Long, pointed wings; long, black, slightly notched tail. **Adults:** head and underparts pale gray; mantle dark gray; flight feathers black except upper surface of secondaries largely white. Iris red; bill and cere black; feet dull yellowish. **Immatures:** very similar to young Plumbeous Kite but dark streaking below brown, not slaty.

HABITS. In spring and fall, migrates through Costa Rica in small to large flocks that circle in thermals to gain altitude, then glide onward with occasional flaps; sometimes maneuvers to snatch a flying insect with its feet and consumes it while continuing to travel; occasionally perches during day, unlike migrating *Buteo* hawks.

VOICE. High, thin, double whistle.

STATUS. Migrating flocks seen sporadically in both spring (late March to early May) and fall (mid-September to mid-October; later than Plumbeous Kite), in lowlands of both slopes and Valle Central. Probably most movement is through Caribbean lowlands.

RANGE. Breeds in SE USA; winters in South America.

CRANE HAWK

Geranospiza caerulescens Pl. 14(4)

Gavilán Ranero

DESCRIPTION. 19″ (48cm); 350g. Slender (but lax, fluffy plumage may impart a robust look), small-headed, very long-legged. **Adults:** slate-black with 2 white tail-bands; a curved white band across outer primaries prominent from below. Iris red; cere and bill blackish; legs orange. Birds from S Costa Rica average paler, sometimes show pale barring on posterior underparts. **Immatures:** mostly slaty-blackish; forehead, superciliaries, and throat white; cheeks streaked with white; chest and sides washed with brownish; belly, thighs, and crissum barred with whitish to dull buff. Shape, posture, and orange legs distinctive when perched; band on primaries diagnostic in flight.

HABITS. Prefers forested areas near water: gallery forest, wooded swamps; often perches at forest edge; soars over adjacent open areas, flies rather slowly and heavily; stands and walks with legs vertical, body often nearly horizontal; can bend leg backward and forward at tarsal joint, an aid in extracting its prey of lizards, frogs, insects, nestling birds, and mice from tree holes and crevices or bromeliads.

VOICE. A shrill whistled *kweeeuur*, thinner than call of Roadside Hawk; in interactions, a series of low, deep whistles *woop-woop-whooou whooou whooou*.

NEST. A rather small, shallow cup of twigs lined with finer twigs, usually in crotch well up in large tree. Eggs, usually 2(?), white. Mainly dry season to early wet season.

STATUS. Uncommon and local resident in lowlands the length of both slopes, locally to ca. 1650ft (500m); probably most numerous in Tempisque-Bebedero basins in Guanacaste and Tortuguero area of Caribbean slope.

RANGE. NW Mexico to N Argentina.

NORTHERN HARRIER

Circus cyaneus Pl. 15(7)

Aguilucho Norteño

DESCRIPTION. 21″ (53cm); 450g. Long wings; long, narrow tail; distinct facial disk; conspicuous white rump patch in all plumages. **Adult ♂:** head and upperparts pale gray; below white dotted with cinnamon; tail gray, barred with dark brown. Bill blackish with bluish base; iris, cere, and legs yellow. ♀: superciliaries, throat, and cheek-patch white; rest of head, neck, and underparts streaked with buff and dark brown; above dark brown; tail pale brown barred with blackish. Iris brownish; cere pale greenish; legs pale yellow. **Immatures:** like female but below mostly rufous with little streaking; white areas of face and rump tinged with buff.

HABITS. Quarters low over open marshes, wet savannas, and rice fields with light and buoyant flight, a few flaps and a glide, wings held in a shallow V; detects prey by sight or sound; pounces from flight; takes small mammals, birds, frogs, and insects; often perches on drier ground; in migration flies higher, with steady wingbeats.

VOICE. Usually silent.

STATUS. Presently uncommon and local winter resident in lowlands, mainly on Pacific slope; migrates largely along lower Caribbean slope; formerly regular in the now-drained marshes of Valle Central, and more numerous overall; decrease may reflect increase in habitat farther N due to deforestation; arrives early October, leaves by early May.

Range. Breeds from Alaska and N Canada to Baja California and S USA; winters to extreme N South America and West Indies. Widespread in Old World.

Note. Formerly known as Marsh Hawk.

SHARP-SHINNED HAWK

Accipiter striatus Pl. 16(3)

Gavilán Pajarero (Camaleón)

Description. ♂ 11″ (28cm), 105g; ♀ 14″ (36cm), 175g. Short, rounded wings; fairly long, square-tipped tail; iris yellow or red; cere yellow-green; legs yellow. **Adults:** above dark gray (washed with brownish in female); throat white, finely streaked with black; underparts finely barred with pale reddish-brown and white; tail gray, banded with black. **Immatures:** above brown, streaked or flecked with buffy; below whitish, streaked with brown to dull rufous, sometimes coarsely barred on flanks and sides.

Habits. Prefers scrubby second growth, hedgerows, agricultural land with scattered trees, coffee plantations, even suburban areas; usually hunts from ambush, dashing out at prey from dense cover, sometimes after a furtive, low-flying approach; captures birds up to pigeon-size and small mammals; sometimes soars to moderate heights.

Voice. Rarely heard, a thin squeal or whistle.

Status. Uncommon migrant and winter resident, mostly at middle elevations, ca. 1600–8200ft (500–2500m), occasional in lowlands; arrives by mid-October, remains until March.

Range. Breeds from Alaska and C Canada to C Mexico and Greater Antilles; winters from S Canada to Panama and West Indies.

Note. Often lumped with this species are two resident tropical forms, *A. chionogaster* (White-breasted Hawk) of N Central America and *A. erythronemius* (Rufous-thighed Hawk) of South America.

TINY HAWK

Accipiter superciliosus Pl. 16(2)

Gavilán Enano (Camaleón)

Description. ♂ 8″ (20cm), 75g; ♀ 10½″ (26.5cm), 120g. Robin-sized; iris orange; cere and legs yellow. **Adult ♂:** above dark gray; crown blackish; below white, finely barred with dusky except on throat and belly; tail banded with gray and black. ♀: similar but above washed with brownish and below with buffy. **Immatures:** below barred with buff and dark rufous; above dark brown or rufous, finely barred with black; lack the throat stripe and solid rufous on sides of chest of Hook-billed Kite. Barred Forest-Falcon has relatively longer legs and tail, dark eye; both decidedly larger.

Habits. Frequents canopy and edges of humid forest, or adjacent tall second growth; hunts mostly by waiting in concealment, darting out to attack small passerines and hummingbirds; may learn regular territorial perches of hummingbirds and try to pick them off from ambush or in rapid flight between perches; sometimes suns on open branch in morning.

Voice. Shrill *kree-ree-ree-ree* or *keer-keer-keer*.

Nest. Unknown (?). Egg of South American race bluish-white, faintly streaked and spotted with brown (?).

Status. Uncommon to rare resident (perhaps largely overlooked?) in humid lowlands and foothills, locally to 4000ft (1200m), on Caribbean slope and in Golfo Dulce region of S Pacific slope.

Range. E Nicaragua to W Ecuador and N Argentina.

COOPER'S HAWK

Accipiter cooperii Pl. 16(4)

Gavilán de Cooper (Camaleón)

Description. ♂ 16″ (41cm), 350g; ♀ 18″ (46cm), 530g. In all plumages closely resembles the Sharp-shinned Hawk in corresponding plumage but is larger, with tail rounded at tip; wings relatively shorter; head larger. **Adults:** crown usually blackish, distinctly darker than back; underparts average more finely barred than Sharp-shinned Hawk. Iris red; bill black with gray base; cere and legs yellow. **Immatures:** below more finely streaked than immature Sharp-shinned. Iris yellow.

Habits. Seems to be restricted to high elevations in Costa Rica: páramo, elfin forest, scrubby pastures, second growth; flies low, dashingly; surprises prey with sudden swift attack; probably takes birds and mammals; migrates singly, with steady, direct flight (as opposed to soaring and gliding of migrating *Buteo*).

Voice. Typically silent in migration and winter.

Status. Very rare migrant and winter resident in highlands; winters from Cordillera de Tilarán to Cordillera de Talamanca (Cerro de la Muerte) at ca. 5000–10,000ft (1500–3000m); probable migrants noted late October–early November and March, as low as 3400ft (1100m) in Valle Central.

Range. Breeds from S Canada to N Mexico;

winters commonly to Honduras, rarely but probably regularly to Costa Rica, casually to Colombia.

BICOLORED HAWK

Accipiter bicolor Pl. 16(8)

Gavilán Bicolor (Camaleón)

DESCRIPTION. ♂ 14″ (36cm), 260g; ♀ 18″ (46cm), 400g. Wings rather short, rounded; tail fairly long and rounded at tip; iris orange; cere and legs yellow. **Adults:** above dark gray, below pale gray; thighs rufous; tail blackish, broadly banded with pale gray. **Immatures:** above blackish-brown with some buffy edgings; below white to rich buff; thighs sometimes rufous; sides of head streaked with black; sometimes a vertical bar on auriculars and/or a buffy nuchal collar; iris yellow. Immatures often resemble Collared or immature Barred forest-falcons in pattern, differ in soft-part colors, shorter legs, less graduated tail.

HABITS. Hunts from lower understory well up into canopy of virgin or thinned forest, forest edge, old second growth; waits quietly in dense cover, dashes out at prey; may fly rapidly from perch to perch in canopy, scan briefly, then pursue prey flushed at close range or may fly into fruiting tree full of birds, dash off after a victim; eats mostly birds, especially thrush- to pigeon-sized, and lizards; mobs large hawks near nest.

VOICE. Members of pair call back and forth with a squalling *waaah*; a barking *keh keh keh keh*, higher-pitched, louder, less nasal than call of Slaty-tailed Trogon; young birds give a scratchy screaming *keeyaaa*.

NEST. Small and cup-shaped, of sticks lined with leaves or moss, medium to high in forest tree, near end of branch, or in mass of epiphytes. Egg white with faint rusty streaks. Breeds late dry to early wet season.

STATUS. Uncommon to rare resident in humid forested areas from sea level to ca. 5900ft (1800m) on both slopes; absent from dry-forested Pacific NW.

RANGE. SE Mexico to N Chile and N Argentina.

BLACK-CHESTED HAWK

Leucopternis princeps Pl. 17(1)

Gavilán Pechinegro

DESCRIPTION. 24″ (61cm); 1kg. Robust, broad-winged; short square tail. **Adults:** head, upperparts, and chest slate-black; rest of underparts and wing-linings white, barred finely and evenly with blackish (appears uniform gray from a distance); tail black with complete median white band, 1 or more incomplete narrow basal bands. Iris brown; cere and legs yellow. **Young:** similar, above with white scaling on wing-coverts. Contrast between black chest and pale belly diagnostic.

HABITS. Hunts mostly inside canopy or along edges of mountain forests, perching at medium heights to fairly low, making sudden stoops to ground or low vegetation for mostly slow-moving prey: snakes, frogs, large insects, crabs, occasionally mammals and birds; soars freely, often 2 or more circling high in air, calling loudly while engaging in aerial acrobatics.

VOICE. Clear, loud, far-carrying whistled notes, often melodious: *keeer* or *kleeyurrr*, sometimes repeated, accelerating into a high, clear laughing.

NEST. Unknown (?). Apparently located high in tree, in mass of epiphytes; in dry season.

STATUS. A fairly common resident of broken, humid forested terrain at middle elevations; on Caribbean slope found the length of the country, from ca. 1300–6000ft (400–1600m) in N and 3300–8200ft (1000–2500m) in S; occasionally wanders to adjacent lowlands, or up to 10,000ft (3000m) or more; on Pacific slope found well beyond continental divide only along Cordillera de Talamanca, mostly above 5000ft (1500m).

RANGE. Costa Rica to NW Ecuador.

NOTE. Also called Barred or Prince Hawk.

SEMIPLUMBEOUS HAWK

Leucopternis semiplumbea Pl. 16(10)

Gavilán Dorsiplomizo

DESCRIPTION. 15″ (38cm); 325g. Short, broad wings; fairly short tail. **Adults:** above slate-gray, below white; tail black with white subterminal band, 1 or more incomplete white bars at base. Iris yellow; cere and legs bright orange. **Young:** similar but with dark streaking on sides and throat, often blackish scaling on back; cere more yellow. Slaty-backed Forest-Falcon has longer legs and tail, dark eye, dull yellow cere and legs.

HABITS. Haunts the understory and lower canopy of forest, old second growth, cacao plantations; often at forest edge but rarely comes out into the open; does not soar above canopy; hunts by dropping onto prey (reptiles, small mammals, birds) from low perch; sometimes sits for long periods on exposed perch in canopy, sunning or preening; often permits close approach.

VOICE. A high, thin whistle on 1 pitch or slurred upward, often repeated several times in succession. In breeding season more varied whistles: a forceful *kooyalee*, an excited-sounding *kurwee wee wee wee* with upward inflection.

Nest. A platform of sticks well up in forest canopy. Eggs undescribed (?). Probable courtship and gathering of nest materials seen in early dry season.

Status. Uncommon to fairly common resident of humid-forested lowland and foothills, locally to ca. 2600ft (800m) the length of Caribbean slope.

Range. E Honduras to W Ecuador.

WHITE HAWK

Leucopternis albicollis Pl. 17(2)

Gavilán Blanco

Description. 22″ (56cm); 725g. Broad wings; rather short, ample tail. Plumage white, except primaries black, greater coverts and secondaries largely black with broad white tips (black more extensive on wing-coverts of young birds); tail with subterminal black band; wings below mostly white, the primaries tipped with black. Iris dark brown; cere gray; legs yellowish. Unmistakable if white upperparts can be seen. Soaring overhead, Black-and-White Hawk-Eagle has darker flight feathers, black mask with yellow eye, yellow cere, and longer tail, narrower wings.

Habits. Prefers forested areas in hilly terrain; hunts inside canopy or at edges and gaps; waits quietly on perch until prey exposes itself on ground, tree trunk, or branch, then glides or pounces to capture it; eats snakes, lizards, frogs, small mammals, large insects, and occasionally birds; soars frequently above canopy; flies lazily with a few flaps and a long, flat-winged glide; often perches a long while on exposed snag, sunning or preening.

Voice. A hoarse, scratchy scream *sheeeeww* or *ssshhhww*, like long-drawn-out call of Barn Owl, frequently given in flight.

Nest. An inconspicuous structure of twigs high in tree, often in mass of epiphytes. Egg pale bluish-white, marked with brown. February–May.

Status. Fairly common resident where extensive forest remains in hilly terrain (less numerous in extensive flatlands) from lowlands to lower middle elevations, reaching 4600ft (1400m) locally on Caribbean slope and S Pacific slope; absent from dry Pacific NW.

Range. S Mexico to E Peru, N Bolivia, and E Brazil.

COMMON BLACK-HAWK

Buteogallus anthracinus Pl. 13(6)

Gavilán Cangrejero

Description. 22″ (56cm); 800g. Chunky, broad-winged; short, ample tail. **Adults:** slate-black, with some paler mottling at base of primaries below (forming pale patch in flight); a single, broad white tail-band (sometimes 1 or more incomplete, narrow, partly concealed bands at base of tail). Iris brown; cere, lores, and legs yellow. **Immatures:** above dark brown, flecked with buff; head and underparts buff, heavily streaked with blackish; tail coarsely barred with white or buff and blackish.

Habits. Nearly always near water: coastal mangroves, beaches, mudflats, rivers and streams, swamps and marshes; eats mainly crabs, supplemented by reptiles and frogs, eggs and hatchlings of sea turtles, carrion; pounces on prey from low perch, where it often waits sluggishly for long periods; sometimes runs about chasing crabs on beach or mudflat; often soars, especially toward midday.

Voice. Series of 10–15 high, shrill whistles, successively louder and higher-pitched, then trailing off and speeding up: *klee klee klee KLEE KLEEE Klee kle kle ke ke ki ki* or a shorter *Klee KLEE keer ker*.

Nest. A platform of sticks, small and flimsy to large and bulky, lined with leaves, near top of small to large tree. Egg 1, rarely 2, dull white, sparsely spotted with brown. Dry season.

Status. Resident; locally common along both coasts, mostly uncommon to rare inland; from lowlands to low foothills, occasionally to 2500ft (750m).

Range. SW USA and West Indies to NW Peru and Guyana, principally in coastal areas.

Note. Primarily because of their short wings, Black-Hawks inhabiting the mangroves of the Pacific coast are sometimes considered a separate species, Mangrove Black-Hawk (*Buteogallus subtilis*). However, at least in Costa Rica, mangroves are inhabited by long-, medium-, and short-winged birds (no evident discontinuity in wing-length), and to date we can discern no diagnostic differences in color, pattern, voice, behavior, or other characters between Caribbean and Pacific coast or inland birds. Hence until detailed field studies demonstrate reproductive isolation, we prefer to consider *subtilis* as conspecific with *anthracinus*. We encourage further study of this problem.

GREAT BLACK-HAWK

Buteogallus urubitinga Pl. 13(7)

Gavilán Negro Mayor (Aguilucho, Gavilán Silbero)

DESCRIPTION. 26″ (66cm); 1.1kg. Heavy build; long legs; broad wings; short tail. **Adults:** dull black, like Common Black-Hawk, except thighs and (usually) wing-linings barred with white; upper tail-coverts largely or entirely white; tail with broad white median band, narrower white basal band that is sometimes indistinct, incomplete, or mottled. Slaty lores contrast with yellow cere. **Immatures:** very similar to immature Common Black-Hawk, but lores dark, tail much more finely barred.

HABITS. Prefers forested areas but often seen at forest edge, in semi-open, especially along rivers and streams; sometimes hunts by perching and watching in isolated trees overlooking water, as does Common Black-Hawk, but tends to take larger, more active prey: reptiles, birds (including chickens), mammals, also crabs; inside forest, flies to a perch, scans actively, moves on if no prey appears; soars less frequently than Common Black-Hawk.

VOICE. Usual call a shrill, high-pitched, thin whistle: *keeeeeeeeh*; the second half often on a higher pitch: *keeeeeeeleeeeee*! More powerful, less plaintive than whistle of Broad-winged Hawk; a shrill, harsh scream when disturbed.

NEST. A sometimes bulky platform of sticks, usually high in tree. Egg 1, whitish, spotted with reddish-chestnut and violet. One Costa Rican nest held a large nestling, nearly ready to fly, in mid-May; another held a week-old nestling in mid-February.

STATUS. Uncommon but widespread resident in lowlands to lower middle elevations, locally to 4000ft (1200m) on both slopes; more numerous than Common Black-Hawk at middle elevations or in heavily forested country, much less common in coastal areas.

RANGE. S Mexico to N Argentina.

SOLITARY EAGLE

Harpyhaliaetus solitarius Pl. 13(8)

Aguila Solitaria

DESCRIPTION. 30″ (76cm); ca. 3kg. Very large and robust; wings very broad; short tail barely extends beyond trailing edge of wings in flight. **Adults:** dark slate-gray overall; bushy nuchal crest; primaries faintly barred or mottled below with pale gray; upper tail-coverts white-tipped; tail with broad white median band, sometimes a narrow partial basal band. Iris dark; cere, lores and legs yellow. **Immatures:** resemble young black-hawks but below darker and more heavily streaked; tail irregularly barred or mottled, with wide, dark subterminal band.

HABITS. Prefers heavily forested country in hilly or mountainous terrain; often soars to great heights; calls freely in the air. Hunting behavior and food unknown.

VOICE. Usual call in flight a rapid series of short, powerful, whistled notes on same pitch: *ple ple ple ple ple ple* . . . , etc.; while perching, a long-drawn-out, powerful whistle, the second half on a lower pitch: *keeeeeerloooooo* (i.e., like Great Black-Hawk in reverse).

NEST. A bulky platform of large, coarse sticks, lined with smaller sticks and dead and fresh green leaves. Egg 1, dull white with a rough, heavily pitted surface (in NW Mexico; no Costa Rican record).

STATUS. Resident; uncommon on Caribbean slope of Cordillera Central and probably other mountain ranges, mostly 2000–5600ft (600–1700m), occasionally wandering to lowlands (La Selva); rare and local in hilly country around Golfo Dulce and on Península de Osa.

RANGE. NW Mexico to Peru and NW Argentina.

NOTE. Often placed in genus *Urubitornis*; we agree with Wetmore and others that its affinities are with *Buteogallus*.

BLACK-COLLARED HAWK

Busarellus nigricollis Pl. 17(5)

Gavilán de Ciénega (Gavilán Pescador)

DESCRIPTION. 20″ (51cm); 650g. Very broad wings; short tail that barely projects beyond trailing edge of wings in flight. Iris chestnut; cere blackish; legs dull flesh. **Adults:** plumage mostly rufous; mantle finely streaked with black; black crescent on chest sets off buffy-white head; primaries mostly black; tail chestnut with broad black subterminal band and several narrow bands toward base. **Immatures:** above brown; below dull buff; head whitish; very broad dusky-black chest patch; head and underparts streaked with blackish.

HABITS. In the vicinity of clear, shallow, stagnant or slow-moving bodies of water in flat, forested to largely open country; perches sluggishly overlooking water, on anything from fence post to tall snag; catches prey in a swoop to water surface, rarely immersing more than legs and feet (plumage easily waterlogged); takes mostly fishes, also frogs, small birds, mammals, and large insects; often soars on flat wings.

Voice. A breathy, harsh scream with an abrupt start: *BEeyurrr*; when disturbed, an insectlike buzz that slows to a sharp ticking.

Nest. A large platform of sticks 40–50ft (12–15m) up in a tree. Eggs 1–2, whitish to greenish, variously marked with brown, reddish, and gray. July–September (in Suriname; no Costa Rican record).

Status. Resident; uncommon in Río Frío region; very scarce and local in lower Tempisque-Bebedero basin, Tortuguero region, and Corcovado basin on Peninsula de Osa.

Range. C Mexico to N Argentina.

Note. Another mostly rufous hawk, the Savanna Hawk (*Buteogallus meridionalis*; Gavilán Sabanero), found from W Panama to C Argentina, will probably soon invade the Golfo Dulce lowlands, as have other open-country species following deforestation in recent years. Larger (23″, 58cm) than Black-collared Hawk, with longer tail and legs; head and underparts cinnamon-buff to pale rufous, upperparts grayish-brown; below finely barred with blackish, above with broad tawny feather-edgings; wings largely rufous; tail black with white median band and tip. Immatures resemble immature black-hawks but are more rufescent, especially on belly, crissum and wings; tail pattern like adult but median band broader and mottled with gray. Prefers open country with scattered trees; often seen on ground [see Pl. 17(4)]. Often placed in the genus *Heterospizias*.

BAY-WINGED HAWK

Parabuteo unicinctus Pl. 14(5)

Gavilán Alicastaño

Description. 21″ (53cm); 700g. Long-winged, long-tailed; upper and lower tail-coverts, base and tip of tail white; cere, lores, and legs yellow. **Adults:** mostly sooty to brownish-black; shoulders, thighs and wing-linings chestnut-rufous. **Immatures:** browner than adults, pattern similar but below broken by buffy streaking, above with rusty spotting and scaling; somewhat resemble immature black-hawks but slimmer; rufous shoulders and thighs diagnostic.

Habits. Prefers scrubby or marshy open country with scattered trees and patches of woodland; usually hunts actively, flying rapidly along edges or through open vegetation, or quartering harrierlike low over open marshes or savannas; pounces on prey from flight or dashes in pursuit like an accipiter; eats mostly birds, some mammals, lizards; hunts mostly early and late in the day; at midday sits sluggishly on elevated perch or soars high in air.

Voice. A harsh, grating, nasal scream: *jaaahr* or *nyaaah*; squalling quality reminiscent of Barn Owl.

Nest. A platform of sticks, scanty to rather substantial, usually high in large tree in savanna or at woods edge. Eggs 2, rarely 3, white, unmarked or lightly marked with brown. December–April.

Status. Locally common resident in Tempisque and Bebedero basins; uncommon elsewhere in Guanacaste and in Río Frío district. Scattered records elsewhere (Limón, San José) probably pertain to wandering individuals or migrants.

Range. SW USA to C Chile and C Argentina; northernmost populations partly migratory.

Note. Also called Harris' Hawk.

GRAY HAWK

Buteo nitidus Pl. 16(14)

Gavilán Gris (Gavilán Pollero)

Description. 16″ (41cm); 425g. **Adults:** above slate-gray (barred with pale gray in S Pacific race); upper tail-coverts white; below finely barred with gray and white; tail black with 1 white band (traces of 1–2 other narrower bands basally). Iris brown; cere and legs yellow. **Immatures:** above blackish-brown, flecked with rusty; superciliaries whitish; eye-stripe dark brown; below white, heavily streaked with dark brown; thighs lightly barred; tail pale brown, narrowly barred with black, subterminal bar widest. Immature Roadside Hawk buffier below; thighs barred with rusty; pale eyes.

Habits. Prefers broken forest, forest edge, and savanna trees, less often solid woodland, where it usually keeps to the canopy; waits quietly to ambush prey or pursues actively with swift, graceful flight; takes lizards, small mammals, large insects, some birds; often soars but not to great heights.

Voice. High, clear, whistled *pee'yurrrr* or *pee'uuurr*, less harsh than call of Roadside Hawk, less plaintive than that of Broad-winged; also a descending *keeer-keer-ker-ker-ker*.

Nest. A platform of sticks in crotch well up in tall evergreen tree. Eggs usually 2, whitish to bluish-white, plain to lightly marked with brown. December–May.

Status. Uncommon to locally common resident in NW lowlands, S to Río Grande de Tárcoles, locally to ca. 1600ft (500m) on adjacent slopes; rare and sporadic on N Caribbean slope E to Sarapiquí region; uncommon from Térraba valley S to Panama, locally to ca. 3300ft (1000m).

Range. Extreme SW USA and N Mexico to W Ecuador, N Argentina, and S Brazil.

Note. Sometimes separated in genus *Asturina*. The barred-backed populations from SW Costa Rica to South America (true *nitidus*) and the gray-backed populations to the north (*plagiatus*) are sometimes considered distinct species.

ROADSIDE HAWK

Buteo magnirostris Pl. 16(12)

Gavilán Chapulinero

Description. 15″ (38cm); 290g. A small *Buteo* with rather short, rounded wings. **Adults:** above brownish-gray; head and chest paler gray; rest of underparts buffy-white, barred with grayish on breast and pale rufous posteriorly; crissum white; primaries extensively rufous, barred and tipped with black; tail gray-brown to rufous, banded with black. Iris pale yellow; cere and legs yellow. **Immatures:** above browner; head streaked with buffy; streaked with dusky on breast; coarsely barred with dull rufous posteriorly (finely on thighs); usually some rufous in primaries; tail narrowly barred with blackish.

Habits. Prefers savannas, broken woodland, second growth, field edges; hunts by dropping from low perch onto large insects, reptiles, small mammals, rarely birds; rarely soars, and then only low and briefly; often sluggish and tame.

Voice. A petulant, angry-sounding *KREE-yurr* diagnostic; in breeding season, birds give excited-sounding nasal, barking *keh-keh-keh-keh*, perhaps in courtship.

Nest. A small platform of sticks, at medium height in tree in savanna or woodland edges, well out from trunk. Eggs usually 2, dull white, speckled and washed with brown. Dry season.

Status. Resident; common in N Pacific lowlands, uncommon in S Pacific lowlands; on Caribbean slope common only in Río Frío district but widespread and increasing with deforestation; locally to 4000ft (1200m), including (occasionally) Valle Central.

Range. E Mexico to W Ecuador and N Argentina.

BROAD-WINGED HAWK

Buteo platypterus Pl. 16(13)

Gavilán Aludo (Gavilán Pollero)

Description. 17″ (43cm); 450g. Chunky, medium-sized. **Adults:** above dark brown; below white, irregularly barred, spotted, or chevroned with reddish-brown, most heavily on breast, which may be solid reddish-brown; throat white, bordered by black malar stripes; wings below whitish with inconspicuous darker barring, remiges tipped with black; a rare melanistic phase has entirely black body plumage; tail broadly and evenly banded with black and white. Iris pale brown; cere and legs yellow. **Immatures:** above brown, with whitish streaks on head and pale spots on mantle; below white to buffy with sparse dark streaking and spotting; tail gray with narrow dark bars, the subterminal one widest.

Habits. Prefers open areas, forest edge, broken forest, semi-open; except in Guanacaste, enters closed forest infrequently; seems to hold individual winter territories, where it hunts by swooping or pouncing on prey (reptiles, large insects, small mammals, occasionally small birds) from perch at forest edge or low in tree; waits quietly on perch, or moves more actively between perches in woodland; soars frequently on sunny days; migrates in large flocks, sometimes with Swainson's Hawks or Turkey Vultures, circling in thermals to gain altitude, then gliding onward to next thermal.

Voice. A high, thin, long-drawn-out, plaintive whistle: *k'reeeeeeeeee*.

Status. Abundant migrant in fall (late September to mid-November) on Caribbean slope and in highlands; and in spring (early March–late May) on both slopes; common winter resident on both slopes from sea level to at least 6500ft (2000m). The most commonly seen hawk in Costa Rica during northern winter.

Range. Breeds in E North America; winters from S Florida and S Mexico to Peru and N Brazil.

SHORT-TAILED HAWK

Buteo brachyurus Pls. 14(1), 16(11)

Gavilán Colicorto

Description. 16″ (41cm); 480g. Rather thickset, broad-winged; tail not strikingly short; underside of wings and tail pale with faint dark barring, only tips of remiges and broad outermost tail bar obviously dark. **Adults:** above dark brown; forehead and lores white. PALE PHASE: underparts white; dark brown extends to cheeks, giving hooded look. Palest and least patterned *Buteo* below. DARK PHASE: below dark brown to blackish, contrasting with pale flight feathers; may show an indistinct pale patch at base of primaries. Iris brown; cere and legs yellow. **Immatures** PALE PHASE: below buffy with sparse dark streaking on breast. DARK PHASE: like adult but below mottled or streaked with whitish.

HABITS. In largely forested areas, also open country with scattered trees and patchy woodland; spends most of its time in the air, sometimes soaring to great heights but more often gliding and swooping with great speed and agility low over forest canopy, from which it snatches birds, lizards, and snakes.

VOICE. A shrill scream, *sheeerreeeeea*, the second part very piercing; also a softer, more nasal *keeeea* or *kleee*.

NEST. A bulky platform of sticks lined with green leaves, high in large tree. Eggs usually 2, white to bluish-white, plain to heavily marked with brown (in Florida; no Costa Rican record).

STATUS. Uncommon and local resident along foothills and lower ridges of Caribbean slope of Cordillera Central and Cordillera de Tilarán, Arenal gap, and both slopes of Cordillera de Guanacaste; possibly resident in upper Terraba valley and adjacent foothills and mountain slopes; in migration (August–October, February–April) recorded mainly in Caribbean lowlands, usually a few individuals in large flocks of other hawks; widespread but very uncommon winter resident in lowlands, mainly on Pacific slope.

RANGE. Florida and N Mexico to W Ecuador, N Argentina, and SE Brazil. Northern populations migratory.

SWAINSON'S HAWK

Buteo swainsoni Pls. 14(3), 17(7)

Gavilán de Swainson

DESCRIPTION. 21″ (53cm); 1kg. Wings long, decidedly pointed at tip. **Adults** PALE PHASE: above dark brown; forehead, lores, and throat white; brown band across chest; rest of underparts white to pale rufous; remiges gray, barred and tipped with black, decidedly darker than wing-linings. DARK PHASE: entire underparts dark brown except crissum usually buffy, barred with brown; remiges slightly paler than wing-linings; tail gray-brown, narrowly barred with black, the outermost bar widest. **Immatures:** underparts white to buffy; dark streaking on breast often forming a band; head streaked; mantle spotted and scaled with white; flight feathers as in adult; or (dark phase) below entirely dark.

HABITS. Migrates in sky-spanning flocks often numbering thousands that circle in thermals to gain altitude, then sail off on their way; descends in late afternoon to roost in trees or on open ground; rarely, if ever, feeds while migrating; when wintering, mostly seeks prey while soaring, often rather low, with wings raised in shallow V; eats insects, especially grasshoppers, and small mammals; often attends dry-season fires to capture prey flushed by advancing flames.

VOICE. Silent in migration; wintering birds give a high-pitched, plaintive scream.

STATUS. Abundant fall (late September–November) and spring (late February–early May) migrant mainly through N Caribbean slope (chiefly in spring), C and S mountains, and S Pacific slope. Sporadic and rare winter resident in lowlands of Pacific slope (Guanacaste, Térraba region).

RANGE. Breeds in W North America from E-C Alaska and NW Canada to N Mexico; winters primarily in S South America, occasionally N to Costa Rica and Florida.

WHITE-TAILED HAWK

Buteo albicaudatus Pls. 14(7), 17(6)

Gavilán Coliblanco (Gavilán Sabanero)

DESCRIPTION. 24″ (61cm); 950g. Wings and tail broad; wings slightly bent at wrist in soaring flight. **Adults:** above slate-gray; rufous shoulders; below white (rare dark phase uniform gray above and below); rump and tail white, the latter with conspicuous black subterminal band. Iris hazel; cere pale green; legs yellow. Pale phase Short-tailed Hawk smaller with inconspicuous tail-bands. **Immatures:** mostly brownish-black; shoulder usually scaled with cinnamon; tail pale gray with faint darker barring (looks all pale from distance). Bay-winged Hawk has rufous on thighs and wing-linings, mostly dark tail.

HABITS. Soars easily over open, windswept ridges, savannas, and agricultural land, often circling high in the air; may hover nearly motionless, facing into and buoyed up by the wind; attacks small mammals, birds and reptiles by sudden swoops to the ground or low vegetation; perches on trees, bushes, or ground on exposed hilltops.

VOICE. High-pitched, 2-note whistles, rather nasal in quality: *weee kerwee kerwee kerwee*, etc., or *keela keela keela*, etc.; sometimes a harsher, barking *ack kehack kehack* (Slud).

Nest. An untidy platform of sticks, sometimes lined with leaves, near top of isolated savanna tree. Eggs 2, immaculate white or lightly spotted with brown. One Costa Rican nest (near Taboga) held 2 downy young in February.

Status. Very uncommon resident of lowlands and foothills of Guanacaste, especially on approaches to Cordillera de Guanacaste; formerly locally in central highlands, but no recent records; rare but possibly resident in savannas of Térraba region.

Range. SW USA to E Peru and C Argentina.

ZONE-TAILED HAWK

Buteo albonotatus Pls. 13(1), 14(2)

Gavilán Colifajeado

Description. 21″ (53cm); 750g. Long, 2-toned wings and long tail produce a striking resemblance to Turkey Vulture in flight; sometimes only yellow cere and legs reveal its identity. **Adults:** mostly blackish, including wing-linings; remiges gray below with faint blackish barring; tail black with 3–4 pale gray bars, the outermost widest, but inconspicuous from a distance. **Immatures:** more brownish, below with paler, less distinctly barred tail and white spotting or short streaks. Other dark buteos have broader wings and tails; Common Black-Hawk chunkier with notably longer legs, shorter tail.

Habits. Prefers open country with scattered trees or thickets, especially near marshes or streams; soars with wings in shallow V, often quartering fairly low above ground, its resemblance to the innocuous Turkey Vulture often enabling it to get close to prey (birds, reptiles, small mammals), which it attacks with a sudden stoop.

Voice. Two birds in flight exchange a series of short, high screams: *kra kree-kree-kree*; disturbed near nest, a longer, lower-pitched scream: *raaaaaauu*.

Nest. A platform of sticks often lined with leafy twigs, high in tree, often in gallery woodland. Eggs 2, usually immaculate white. Dry season.

Status. Uncommon and local resident in Guanacaste, including lower slopes of Cordillera de Guanacaste; rare to uncommon resident or wanderer in lowlands and foothills elsewhere the length of both slopes, apparently primarily as migrant (April, October–November), when sometimes in company of migrating Turkey Vultures and Swainson's Hawks; probably a rare winter resident in Cartago area, perhaps elsewhere in Valle Central.

Range. SW USA to SE Brazil, Paraguay, and Bolivia; northernmost populations migratory.

RED-TAILED HAWK

Buteo jamaicensis Pl. 17(8)

Gavilán Colirrojo (Gavilán Valdivia)

Description. 23″ (58cm); 900g. Chunky, broad-winged, broad-tailed. **Adults:** above dark brown; below white, shading to cinnamon on sides, wing-linings, and thighs in resident race; chest lightly (residents) to prominently (migrants) streaked with dark brown; tail cinnamon-orange with narrow, dark subterminal band and whitish tip. Some residents are entirely cinnamon to rufous below; some migrants have entirely dark brown body plumage. **Immatures:** above spotted and streaked with white, below more heavily streaked; tail buffy, narrowly barred with dark brown (subterminal band not noticeably wider, as in all phases of Swainson's Hawk).

Habits. Soars over open or scrubby country or broken forest in highlands: páramo, pastures, volcanic slopes; hovers motionless into wind over ridgetops; captures prey by diving to ground or low vegetation; takes mostly small mammals, some lizards and birds.

Voice. A shrill, high-pitched, scratchy scream: *keeyaaahr* or *kheeeeur*, somewhat clearer, less breathy than scream of White Hawk.

Nest. A platform of sticks in tree on steep slope, with wide view (Cerro de la Muerte). Eggs of northern races white, spotted or blotched with brown (not seen in Costa Rica).

Status. Uncommon to locally common resident in highlands from ca. 3300–5000ft (1000–1500m) upward; rare in extensively forested country or at lower elevations. A few individuals sometimes seen migrating in spring and fall with large flocks of Swainson's and Broad-winged hawks; occasionally winters in Guanacaste, perhaps elsewhere in lowlands.

Range. Breeds from W Alaska and N Canada to W Panama and West Indies; northern races migratory, winter regularly to Nicaragua, occasionally to Panama.

CRESTED EAGLE

Morphnus guianensis Pl. 17(10)

Aguila Crestada

Description. 32″ (81cm); ca. 3kg. Very large; broad rounded wings; very long tail; full, pointed crest; bill and feet relatively small. **Adults:** variable; above slate-black to brownish-black, wing-coverts sometimes scaled with whitish; wings and tail below pale gray, boldly banded with black. PALE PHASE: crest black; head, neck, and chest pale grayish; rest of underparts white, more or less barred with cinnamon to pale rufous. DARK PHASE: head, neck and chest dark gray to blackish; rest of underparts white, heavily barred with blackish, or even entirely blackish. Iris brown; cere and lores blackish; feet yellow. **Immatures:** mostly white; mantle gray, more or less mottled with blackish; flight feathers below more narrowly banded with black than in adult; 3 or more years to adult plumage. Harpy Eagle more robust, especially bill, with relatively shorter tail, divided crest.

HABITS. Prefers humid lowland primary forest; also in adjacent tall second growth and semi-open; usually perches high in canopy, often on conspicuous bare branch; soars above canopy more frequently than Harpy Eagle. Hunting behavior little known, but preys on snakes, frogs, and mammals up to size of kinkajous or small monkeys, occasionally birds.

VOICE. Shrill, high-pitched whistles, sometimes 2-parted with second part higher in pitch (suggestive of Great Black-Hawk).

NEST. A bulky platform of sticks with shallow central depression, in fork of trunk of large forest tree, well up in canopy; female may deposit green leaves around edge of nest. Eggs 2, creamy-white (in Brazil; no Costa Rican record).

STATUS. Very rare resident; formerly probably widespread in humid lowlands of Caribbean and S Pacific slopes; recently reported from Sarapiquí region (La Selva) and Península de Osa (Sirena).

RANGE. N Guatemala to Bolivia, NE Argentina, and SE Brazil.

HARPY EAGLE

Harpia harpyja Pl. 17(9)

Aguila Arpía (Aguilucho)

DESCRIPTION. ♂ 38″ (96cm), 4.5kg; ♀ 42″ (107cm), 7.5kg. Huge; short, rounded wings; fairly long tail; heavy, powerful feet and bill. **Adults:** head and neck pale gray; broad, divided occipital crest; mantle, wing-linings, and broad chest-band black; below white, barred with black on thighs, flanks, and crissum; wings and tail above blackish, below pale gray, boldly banded with black (4 black tail-bands). Iris brown; bill, cere, and lores blackish; feet yellowish. **Immatures:** head and underparts white; mantle pale gray blotched with blackish; wings and tail more narrowly and irregularly banded (ca. 10 tail-bands). Several intermediate plumages; at least 4 years to full adult plumage.

HABITS. In tall wet forest on flat to hilly terrain, flies swiftly, agilely, maneuverably, like accipiter, with a few powerful flaps and a glide, in or below the canopy; rarely soars; hunts by making short flights through treetops, looking and listening for prey from perch; pursues and attacks swiftly, snatching chiefly medium-sized arboreal mammals, especially sloths and monkeys, from the side or below.

VOICE. A loud scream, *wheeeeee*; young bird gives repeated, higher-pitched version. Also croaking notes, sometimes combined with screams.

NEST. A bulky platform of sticks, lined with green leafy twigs, located above canopy level in major fork of giant emergent forest tree (usually *Ceiba*). Eggs 2, dull white (in Suriname; no Costa Rican record).

STATUS. Formerly widespread in wet-forested lowlands and mountain slopes, to at least 6500ft (2000m) on Caribbean and S Pacific slopes; now nearly extinct in Costa Rica due to hunting and deforestation. A few may survive on Península de Osa, in Cordillera de Talamanca, or in the steadily-shrinking forests in N Llanura de San Carlos near Nicaraguan border.

RANGE. SE Mexico to N Argentina and SE Brazil.

BLACK-AND-WHITE HAWK-EAGLE

Spizastur melanoleucus Pl. 17(13)

Aguilillo Blanco y Negro

DESCRIPTION. 23″ (58cm); 850g. Size and proportions of a large *Buteo* but legs fully feathered. **Adults:** head, neck, underparts and wing-linings white; lores, orbital area, and short occipital crest black; mantle blackish; remiges below pale gray, barred and tipped with black; tail below pale gray, with 4 black bands (narrow except outermost). Iris and feet yellow; cere orange. **Immatures:** similar but above browner, mixed with gray; white scaling on wing-coverts; 5 or more narrower, browner tail-bands. Distinguished from immature Gray-headed Kite by buteo-like shape, cere, and eye color; no *Buteo* has white head; immature Ornate Hawk-Eagle lacks black mask, has black barring on thighs and flanks.

HABITS. Prefers forested country but most often seen at breaks and edges; soars freely, sometimes over adjacent open lands; waits on perch for prey to pass, either pounces to ground or pursues with speed and agility in canopy; takes birds like wood-quails and toucans, mammals up to opossum-size, and large lizards.

Voice. A series of clear, ringing whistles reminiscent of call of Black-chested Hawk: *kleee kleee klee klee kle kle*, etc.

Nest. No reliable report (?).

Status. Resident; widespread but generally rare in humid-forested lowlands and foothills, locally to 3300ft (1000m), occasionally wandering to 10,000ft (3000m) on both slopes, except dry Pacific NW.

Range. S Mexico to NE Argentina and SE Brazil.

ORNATE HAWK-EAGLE

Spizaetus ornatus Pl. 17(11)

Aguilillo Penachudo (Aguilucho)

Description. 25″ (63cm); 1.2kg. Intermediate in proportions between Black-and-White and Black hawk-eagles; legs fully feathered. **Adults:** crown and long spiky crest black; rest of head, sides of neck, and sides of breast rufous, separated from white throat and chest by black stripe; rest of underparts, wing-linings, and legs white, boldly barred with black; above brownish-black; remiges below pale gray, barred with black; tail below pale gray, crossed by 4 black bands. Iris tawny-orange or yellow; cere greenish-yellow; feet yellow. **Immatures:** head and neck white with dusky streaking; short, dusky crest; above gray-brown barred with blackish; below black barring less extensive than on adult; tail with 5–11 dark bands; 2–3 years to acquire adult plumage. Immature Black Hawk-Eagle darker, more heavily streaked on head and breast.

Habits. Prefers tall wet forest in lowlands and foothills; hunts mostly inside forest, perching at medium heights to scan for prey, flying silently and swiftly between perches; captures birds up to size of guan or vulture, medium-sized mammals, large lizards and snakes, by swift pounce or dashing pursuit; soars fairly frequently though not very high, often calling loudly and engaging in aerial acrobatics.

Voice. Loud, ringing whistled cries: *huweee-whee-whi-whi-wi-wi-wi*, first note longest and slurred, second note highest, followed by descending short notes, reverse of pattern of Black Hawk-Eagle's call; also a shorter *wheeeur-whee-whi-whi-whi*. While perching, excited-sounding laughing notes; a catlike scream when disturbed.

Nest. A bulky platform of sticks ca. 40″ (1m) in diameter and 20″ (50cm) high, with sprigs of green leaves in lining and around rim, in fork of large tree, 65–100ft (20–30m) up. Egg 1, white, with faint brownish-red spots and blotches. March–June (in Guatemala); in Costa Rica a nest with 1 well-grown nestling was observed April–May 1985 in Peñas Blancas valley on Caribbean side of Cordillera de Tilarán (Fogden et al.).

Status. Uncommon to increasingly rare resident in remaining areas of extensive forest in lowlands and foothills of Caribbean and S Pacific slopes, sea level to 5000ft (1500m) locally, wandering sporadically as high as 10,000ft (3000m) or into dry forests of N Pacific slope.

Range. E Mexico to W Ecuador, NE Argentina, and SE Brazil.

BLACK HAWK-EAGLE

Spizaetus tyrannus Pls. 13(9), 17(12)

Aguilillo Negro (Aguilucho)

Description. 27″ (68cm); 1kg. Slender build; wings short, very broad and rounded, "pinched in" at base; tail very long; legs fully feathered. **Adults:** plumage mostly black; feathers of bushy occipital crest with white bases; legs, wing-linings, and tail-coverts barred with white; remiges below gray, boldly banded with black; tail black with 3 broad whitish bands. Iris yellow; cere blackish; feet yellow. **Immatures:** head and breast mostly white; dusky streaking and brown wash on crest and sides of breast; posterior underparts heavily barred with black and white; dark banding on flight feathers below narrower and finer than on adult; 2–3 years to full adult plumage.

Habits. Prefers broken forest, second-growth woodland, semi-open, and forest edge. Hunting tactics evidently similar to those of Ornate Hawk-Eagle, but on average takes smaller prey, reflecting its lighter build: birds up to size of toucans and chachalacas, including nestlings; mammals like squirrels and opossums; large lizards and snakes. Frequently soars high in air, calling loudly and melodiously.

Voice. A series of clear ringing whistles: *whick whick whick whip-WHEEEEew*, the last note longest and highest in pitch; also a single *wheeeeeu*.

Nest. A bulky platform of sticks 45ft (13.5m) up in the crown of a royal palm held 2 young in February, and in July of a later year (in Panama, N. G. Smith). Nest-building seen in

December near Carrillo, nest ca. 65ft (20m) up in large emergent forest tree on side of steep ridge.

Status. Uncommon to fairly common resident from humid lowlands to 6500ft (2000m) locally; center of abundance in foothills, 650–3300ft (200–1000m); widespread on Caribbean and S Pacific slopes; presently much more numerous than Ornate Hawk-Eagle, evidently more tolerant of habitat disturbance.

Range. C Mexico to W Ecuador, Bolivia, NE Argentina, and SE Brazil.

FAMILY **Falconidae:** Falcons and Caracaras

At first sight the 58 or so species of falcons seem a heterogeneous group, but they all share features of the skull, pectoral girdle, syrinx, leg bones, and eggshell that set them apart from the Accipitridae. The true falcons with long, pointed wings and prominently toothed bills are mostly adapted for fast, hard-hitting aerial pursuit of prey such as birds and bats. Forest-falcons have short wings, long legs and tails; they ambush and pursue prey in dense vegetation and have varied diets. The snake-eating Laughing Falcon has a long hind toe and rough, heavy tarsal scaling. The caracaras have largely bare faces, protuberant crops, and rounded wings, and mostly take rather small, slow-moving prey or carrion. Plumage is varied, but in most the female is similar in coloration to the male but much larger. Forest-falcons, Laughing Falcons, and some caracaras have loud voices and call freely, but most species give screams, whistles, or chatters mainly around the nest. Located in a cavity or in vegetation, or on a ledge or building, the nest varies from essentially nothing in most species to a substantial structure of sticks. In the true falcons the female performs nearly all of the incubation of the 2–6 brown-mottled eggs; in the caracaras a substantial proportion of incubation is by the male. Incubation takes 25–35 days. Both sexes feed the young, which remain in the nest 25–49 days, and in most species are dependent upon the parents for some weeks or months after fledging, until they can hunt for themselves.

RED-THROATED CARACARA

Daptrius americanus Pl. 14(11)

Caracara Avispera (Cacao, Come-cacao, Deslenguado)

Description. 21″ (53cm); 550g. Rather long neck, small head with bushy crown, weakly hooked bill, bare red throat and face, and long, ample tail produce a guanlike aspect. **Adults:** glossy black except for white belly, thighs, and crissum. Iris red; cere blue-gray; bill yellowish; legs orange-red. **Immatures:** facial skin duller, more orangish. No other raptor has solid black tail and wings; no Costa Rican cracid combines black plumage with red on face.

Habits. Frequents canopy and edge of humid forest, adjacent second growth, and semi-open, for much of year in groups of 2–6, which break up at start of breeding; calls freely as it flies through canopy, high over openings or in tall trees in semi-open; tears open arboreal wasps' nests, including those of fiercely stinging *Synoeca*, to extract brood; also takes adults and brood of stingless bees, and fruits (especially palms).

Voice. Extraordinarily loud and raucous. Commonest call a hoarse, trumpeting *khaaow*! or *ghaahee'ow!*, sometimes varied to *ca-ca'o, ca-ca-ca-CA'O!* At a distance suggests a macaw.

Nest. Undescribed (?).

Status. Resident; formerly widespread and fairly common in moist and wet forests of both slopes from sea level locally up to 4000ft (1200m), including parts of Valle Central. In recent years has unaccountably disappeared from most areas, even where forest remains intact; regular only in Golfo Dulce lowlands.

Range. S Mexico to W Ecuador, Bolivia, and C Brazil.

CRESTED CARACARA

Polyborus plancus Pl. 14(12)

Caracara Cargahuesos (Cargahuesos, Querque, Quebrantahuesos)

Description. 24″ (61cm); 1kg. Fairly large; rather long neck, wings, and legs. **Adults:** crown, including short crest, black; rest of head white, becoming barred with black on neck, passing into solid black of most of body and wings; large white patch in outer primaries, lightly barred with black; tail white, barred with black and with broad black terminal band. Iris brown; bare face and cere orange to red; bill whitish; legs yellow. **Immatures:** similar but much duller, browner overall; chest more or less streaked rather than barred.

Habits. Prefers open country: pastures, farmland, scrub, beaches; singly, in pairs, or small family groups following breeding, stalks erectly over ground seeking small dead animals or attacking live ones: snakes, frogs, rodents; also takes nestling birds; drives vultures from small carcasses; flies rapidly and directly with strong wingbeats and short sails; may soar in wind at no great height.

Voice. A harsh, cackling *ker-ker-ker-kerr*, etc. Toward breeding season, adults give a sharp rattle and creaking notes, often accompanying a head-tossing display.

Nest. A bulky open structure of sticks, straws, and weed stems, usually high in a palm or small-leaved tree. Eggs 2–3, rarely 4, white, clouded with shades of brown, heavily blotched with darker browns. December–March.

Status. Resident; common in dry NW up to ca. 2500ft (750m) on adjacent mountains; occasional young birds wander to W part of Valle Central; uncommon and local, but increasing, farther S along Pacific slope to Panama; on Caribbean slope uncommon in Río Frío region, an occasional visitor to San Carlos region.

Range. S USA to Tierra del Fuego and Falkland Islands.

YELLOW-HEADED CARACARA

Milvago chimachima Pl. 15(9)

Caracara Cabecigualdo

Description. 16″ (41cm); 330g. Small, lightly built, rather long-tailed; large buffy "window" in primaries. **Adults:** head, underparts, and wing-linings pale buff; a black postocular streak; back, upperwings, and underside of secondaries dark brown; tail whitish, barred with black and with broad black subterminal band. Bill and legs pale blue to greenish; cere and bare face yellow to reddish. **Immatures:** head and underparts heavily streaked with dusky; tail lacks broad subterminal band; cere bluish.

Habits. Frequents savannas, pastures, scrub, agricultural land; flies slowly, with several shallow flaps and a glide; sometimes stoops at prey from flight but more often hunts on the ground, often among cattle; makes quick dashes after mice, reptiles, and insects; also picks ticks from backs of cows; eats much carrion, especially small road kills; plunders birds' nests.

Voice. A scratchy, slightly nasal scream *keeeah* or *greeeeah*, often rapidly and excitedly repeated when 2 birds interact.

Nest. A fairly substantial platform of sticks, high in tree (1 at Parrita in crown of oil palm). Eggs 2, buffy to reddish-brown with darker splotches. December–March or April.

Status. First recorded in Costa Rica in 1973, evidently entering from Chiriquí as forest barrier in Golfo Dulce lowlands was destroyed; presently resident in lowlands and interior valleys of S Pacific slope, N at least to Orotina along coast and San Isidro inland; one seen in July 1984 on Península de Nicoya (Quebrada Honda); range evidently still expanding; breeding recorded as far N as Parrita in 1979, Tivives in 1983, and Chomes in 1988.

Range. Costa Rica to N Argentina and SE Brazil.

LAUGHING FALCON

Herpetotheres cachinnans Pl. 15(8)

Guaco

Description. 21″ (53cm); 600g. Stout-bodied, large-headed; wings short and rounded; tail long and rounded. **Adults:** head, neck and underparts white to buff; a broad black mask covering cheeks, encircling nape; upperparts dark brown; tail banded with black and pale buff. Iris dark; cere and legs yellowish. **Immatures:** head and underparts deeper buff; mantle scaled with buff; pale tail-bands narrower, more cinnamon-buff.

Habits. Prefers broken forest, edges, semiopen, second growth, savanna trees; flies slowly and steadily with quick, shallow flaps; perches for long intervals on high, exposed branches, scanning ground for snakes; drops suddenly, seizes snake in heavily-scaled talons, decapitates it with a quick bite, carries it to a perch to eat; occasionally takes lizards and rodents.

Voice. An exceptionally vocal raptor; short, nasal, laughing calls when disturbed, also a louder *wac wac* or *haw haw*. The full "guaco" call is a long, rhythmic series of loud, hollow notes with somewhat the quality of a child's shout. Often starting with a bubbly laugh, the bird commences a series of *wah* notes that gradually increase in pitch and loudness, finally breaking into a series of *wah'-co* phrases; entire sequence lasts a minute or more; sometimes 2 birds duet; often heard in evening twilight.

Nest. A large hollow high in a great tree appears to be the preferred site, but a hole in a cliff or the bulky open nest of a hawk or caracara may be occupied. Egg 1, densely mottled with shades of brown. Dry season.

Status. Resident; where not persecuted remains fairly common throughout lowlands of both slopes, rarely ascending as high as 6000ft (1850m).

Range. N Mexico to NW Peru and N Argentina.

BARRED FOREST-FALCON

Micrastur ruficollis Pl. 16(5)

Halcón de Monte Barreteado

Description. ♂ 13″ (33cm), 165g; ♀ 15″ (38cm), 200g. Small, with long graduated tail, short rounded wings, rather long legs. **Adult ♂:** above dark gray, paler gray on head and throat; below white, finely barred with blackish; tail black with 3 narrow white bars. ♀: similar but mantle dark brown. Iris pale to dark brown; facial skin, legs, and feet bright yellow. **Immatures:** above dark brown; below white to buffy; usually a pale nuchal collar; face mottled with dusky; usually a whitish bar behind cheeks; underparts plain to heavily scalloped with dusky; tail blackish with 4–5 narrow whitish bars. Young Bicolored Hawk larger, with more clean-cut face, broader whitish tail-bands, sometimes rufous thighs. Tiny Hawk smaller than adult, with different eye and cere colors, shorter tail with different pattern.

Habits. Dwells in understory of humid forest, adjacent tall second growth, and dense semiopen; flies with speed and agility through thickets; generally attacks its prey of lizards, mice, large insects, small birds (especially fledglings), and bats in a sudden dash out from dense cover; follows flocks of small birds, perhaps to pick off stragglers; attends army ant raids, but more to capture large insects than attendant birds; partly crepuscular like other forest-falcons; ascends into lower canopy to call amid concealing foliage, mostly at dawn and dusk.

Voice. Usual call a nasal, barking *kyer* or *owr* repeated many times, quite ventriloquial at close range; also a nasal, squirrel-like chuckling or cackling.

Nest. Apparently in tree cavity (?).

Status. Resident; uncommon to fairly common for a raptor, over forested portions of Caribbean slope and S Pacific slope, up to at least 6000ft (1850m); easily overlooked when present; far more often heard than seen.

Range. S Mexico to W Ecuador and N Argentina.

SLATY-BACKED FOREST-FALCON

Micrastur mirandollei Pl. 16(6)

Halcón de Monte Dorsigrís

Description. ♂ 16″ (41cm), 420g; ♀ 18″ (46cm), 550g. Relatively shorter tail and legs than other forest-falcons. **Adults:** above slate-gray; below white or buffy; face mottled with whitish; black tail crossed by 3–4 narrow white bars. Iris dark brown; cere and lores greenish-yellow; legs yellow. **Young:** above browner, scalloped with dusky on back and with dark gray on breast and sides. Semiplumbeous Hawk has shorter tail and legs, orange cere, yellow eye; Collared Forest-Falcon larger, with pale nuchal collar, blackish back.

Habits. Inhabits wet lowland forest and adjacent tall second growth and semi-open, hunting mainly in dense understory; dashes out from cover to attack prey (mostly birds and lizards), pursuing it in air or on ground; flies swiftly, agilely, silently; also runs swiftly; may call persistently from upper understory or subcanopy, especially toward dusk, sometimes before dawn.

Voice. A chanting series of 8–13 nasal, shouting *kiiih* or *kaaah* notes, the first 3–4 successively longer and lower in pitch, the next 4–6 successively higher, the last 2–3 on even pitch, loudness increasing through the series. Also reported, a higher-pitched, ventriloquial series of clear whistles that incites mobbing in small birds, which falcon then attacks (N. G. Smith).

Nest. Unknown (?).

Status. Resident in Caribbean lowlands from San Carlos and Tortuguero regions S to Panama; uncommon to rare in most areas.

Range. Costa Rica to W Colombia, Bolivia, and C Brazil.

COLLARED FOREST-FALCON

Micrastur semitorquatus Pls. 14(10), 16(7)

Halcón de Monte Collarejo

Description. ♂ 20″ (51cm), 575g; ♀ 24″ (61cm), 710g. Fairly large, slender; short rounded wings; very long rounded tail; long legs. **Adults** Pale phase: underparts and nuchal collar white to buffy; face whitish speckled with dusky; crown, auricular bar, and upperparts blackish; tail black, crossed by 4 narrow, widely spaced white bars. Dark phase: plumage mostly blackish; posterior underparts more or less barred with white to cinnamon. Iris dark brown; cere and facial skin dull greenish; legs yellow. **Immatures:** above browner, spotted with rusty. Pale phase: below with heavy dusky scalloping, breast often almost solid blackish; facial pattern and nuchal collar indistinct. Dark phase: like adult but pale barring below more extensive, more or less rufescent.

Habits. Haunts thickets in solid or broken forest, tall second growth and scrub; hunts both by ambush and by active pursuit, flying and running with amazing speed and agility through dense understory; ascends into canopy to call, mainly at dawn and dusk; eats large lizards, rodents, birds up to chachalaca-size, sometimes large insects; follows army ants to capture invertebrates stirred up by them.

VOICE. Usual call a nasal, descending *how* or *aow*, repeated several times; sometimes a structured series of 5–7 notes, becoming successively longer, louder, more slurred, the last note or 2 lower in pitch and shorter; has somewhat the quality of a small child's shout; sometimes confused with call of Laughing Falcon.

NEST. In natural cavities in large trees. Eggs undescribed (?). Dry and early wet season.

STATUS. Widely distributed resident in lowlands and foothills of both slopes, locally to 5000ft (1500m); rare over most of Caribbean slope, but more numerous in Río Frío region; more common on Pacific slope, especially dry-forested NW.

RANGE. C Mexico to NW Peru, E Bolivia, and N Argentina.

AMERICAN KESTREL

Falco sparverius Pl. 15(13)

Cernícalo Americano (Klis-klis, Camaleón)

DESCRIPTION. 11″ (28cm); 115g. A small, slender falcon with long wings and tail. Face pattern and largely rufous tail diagnostic. **Adult ♂:** back rufous, barred with black; wing-coverts blue-gray, spotted with black; tail rufous with black subterminal band and white tips; crown blue-gray with rufous center; cheeks white with vertical black malar and auricular bars; a black "eyespot" on each side of buffy nape; below white, shading to buff on breast; sides spotted with black. Iris dark brown; cere and feet yellow. ♀: above barred with rufous and black, including wing-coverts and tail; below streaked with brown; head pattern similar to male's but duller. **Immatures:** resemble adults but above more heavily barred and below streaked; cere greenish.

HABITS. Often migrates in small flocks; may soar using thermals like *Buteo* hawks; in winter, prefers open country with scattered trees, fenceposts and power lines for perching and scanning for prey; flies out lightly and gracefully, often hovers before pouncing; occasionally makes high-speed stoop in pursuit of small birds, but usual prey is grasshoppers and other insects, mice, and lizards; typically pumps tail upon alighting.

VOICE. An excited-sounding series of clear, short whistles: *klee-klee-klee* or *keu-keu-keu. . . .*

STATUS. Sporadically common fall (September–October) and spring (March–April) migrant, mostly along Pacific coast and through central highlands; widespread but generally uncommon and local winter resident on both slopes, from lowlands to middle elevations; most numerous in parts of Guanacaste (especially toward passes of the Cordillera), Río Frío area, and Valle Central; occasionally recorded between June and August, probably nonbreeders; has declined notably in recent years in Costa Rica for reasons unknown.

RANGE. Breeds from W Alaska and N Canada to NE Nicaragua, in the Bahamas and West Indies, and in South America to Tierra del Fuego. Northern populations winter to E Panama.

NOTE. The Pearl Kite (*Gampsonyx swainsonii*), resident from C Panama (possibly), N Colombia, and Trinidad to NW Peru, N Argentina, and S Brazil, has been expanding its range in Panama, and it was recently recorded in W Bocas del Toro (N. G. Smith); it could soon be found in SE Costa Rica; birds from an isolated, sedentary population in W Nicaragua may yet occur in NW Costa Rica. Kestrel-shaped but smaller [9″ (23cm); 95g], forecrown and face orange-buff, otherwise above dark, remiges and rectrices tipped with white; below mostly white with cinnamon-buff wing-linings, flanks, and thighs. Immatures above browner with pale scaling, below buffy [see Pl. 15(12)]. Prefers woodland edges and savanna trees; soars freely.

MERLIN

Falco columbarius Pl. 15(15)

Esmerejón

DESCRIPTION. ♂ 10½″ (26.5cm), 160g; ♀ 13″ (33cm), 215g. Small but powerfully built; pointed wings. **Adults:** above bluish-gray (male) or dark brown (female); below buffy, heavily streaked with dark brown; head finely streaked, with a pale eyebrow and suggestion of a dark malar stripe; tail blackish with 2–3 pale bars (gray in male, buff in female). **Immatures:** resemble females. Iris dark brown; cere and feet yellow. Lacks rufous of Kestrel; differs in shape and flight from young accipiters.

HABITS. Prefers open and semi-open country at virtually any elevation; flies powerfully and directly, with pigeonlike "rowing" wingstroke; makes dashing sallies after prey from conspicuous lookouts, or sometimes hunts strictly on the wing, soaring or hovering; sometimes consumes small prey in the air; takes mostly small birds (especially migrating swallows), also large insects; usually migrates singly, along coasts or through Valle Central.

VOICE. Virtually always silent in migration and winter.

STATUS. Rare but regular fall migrant, late September–early November; only record for spring migration mid-March; occasionally winters; recorded from Río Frío area and central highlands, rarely to 10,000ft (3000m).

RANGE. Breeds from N Alaska and Canada to extreme N USA; winters from S-C Alaska and S Canada to N South America and West Indies. Widespread in Old World.

BAT FALCON

Falco rufigularis Pl. 15(14)

Halcón Cuelliblanco

DESCRIPTION. ♂ 10″ (25.5cm), 140g; ♀ 12″ (30cm), 200g. Pointed wings; square-tipped tail; powerful build. **Adults:** head and upperparts slate-black with bluish cast; tail black, narrowly barred with pale gray; throat, sides of neck, and chest white to buffy; breast and sides black, narrowly barred with white to rufous; lower belly, thighs, and crissum chestnut-rufous. Iris dark brown; cere, eye-ring, and feet yellow. **Immatures:** above duller; chest tinged with brownish; belly and crissum paler rufous marked with black; thighs barred with buff and black. See Orange-breasted Falcon.

HABITS. Perches on high, exposed snags at forest edge, in tall trees in open, or beside river or lagoon, to scan for prey, which it captures in air after a dashing pursuit or swift, diagonal power dive from a great height; or dives and strikes at prey on outside of vegetation or on water surface; overtakes fast-flying prey like swifts, swallows, and free-tailed bats from above and behind; also seizes hummingbirds, parakeets, small waterbirds, large insects, occasionally lizards and mice; may consume small prey in the air. Flight swift with shallow, powerful wingbeats, like that of *Aratinga*; often resembles White-collared Swift in flight.

VOICE. Rapid, excited-sounding, nasal *kirkirkirkir*, etc.; a cackling *kikikik*; in aggressive interactions a higher, thinner *tsee-tsee-tsee-tsee*.

NEST. An unlined cavity, either natural or made by woodpecker, high in tree. Eggs 2–3, whitish, heavily mottled with brown. Dry season.

STATUS. Uncommon and local resident of lowlands to middle elevations, locally to 5500ft (1675m) on both slopes, mainly in humid life zones; rare in Pacific NW; evidently has decreased in recent years.

RANGE. N Mexico to W Ecuador, Bolivia, N Argentina, and SE Brazil.

APLOMADO FALCON

Falco femoralis Pl. 15(10)

Halcón Aplomado

DESCRIPTION. ♂ 15″ (38cm), 220g; ♀ 17″ (43cm), 330g. Fairly large, slender, long-tailed; striking face pattern diagnostic. **Adults:** above blackish, including postocular and malar stripes; cheeks, throat, breast, and broad stripe from above eye to nape pale buff; lower breast and sides blackish, sparsely barred with white; belly, thighs, and crissum cinnamon; tail blackish, barred with gray. Iris dark brown; cere, bare eye-ring, and feet yellow. **Immatures:** like adult but mantle browner, breast streaked with black.

HABITS. Prefers savannas, open woodland, pastures with scattered trees; waits on exposed lookout to spot prey, makes fast, dashing sallies in pursuit, returns to perch to consume its catch: mostly birds but also small mammals and large insects flushed by grass fires.

VOICE. Cackling notes, shrill whistles.

STATUS. One Costa Rican record, an immature photographed near Taboga, Guanacaste, in mid-October (Koford and Bowen). To be expected during migration and winter, especially in Guanacaste.

RANGE. Breeds from extreme SW USA (at least formerly) through Mexico, locally in Central America, and through South America to Tierra del Fuego and Falkland Is. N and S populations migratory.

ORANGE-BREASTED FALCON

Falco deiroleucus Pl. 15(11)

Halcón Pechirrufo

DESCRIPTION. ♂ 14″ (36cm), 350g; ♀ 16″ (41cm); 600g. Fairly large, powerful, large-footed. **Adults:** head and upperparts blackish-slate; chin and throat white; broad rufous chest-band; lower breast black, barred with cinnamon; belly, thighs, and crissum dark rufous; remiges below spotted or barred with buffy; tail black with 3–5 narrow whitish bands. Iris dark brown; cere, bare eye-ring, and feet bright yellow. **Young:** like adult but above scaled with tawny, streaked with black on chest and belly; thighs barred with black and buff. Bat Falcon smaller, with no rufous between white throat and black breast; Aplomado Falcon has buffy breast, bold facial stripes.

Habits. Prefers heavily forested areas in flat to hilly terrain; most often seen perching on high, exposed snag at forest edge; stoops on birds in flight, especially those of about pigeon size.

Voice. A penetrating, clear downslurred whistle, repeated several times; a sharp *ack-zeek-ackzeek.*

Status. Extremely rare, perhaps extinct in Costa Rica; formerly (?) resident on Caribbean slope, mostly in foothills to 4300ft (1300m) of Cordillera Central; no definite record for ca. 30 years; a recent sighting (December 1986) near La Selva could not be confirmed.

Range. S Mexico to S Peru and N Argentina.

PEREGRINE FALCON

Falco peregrinus Pl. 15(5)

Halcón Peregrino

Description. ♂ 16″ (41cm), 610g; ♀ 19″ (48cm), 950g. Powerful build; long pointed wings; narrow tail. **Adults:** above dark slate, blackest on head, including broad bar down cheek; below buffy to white, spotted with black on lower breast, barred with black on flanks, thighs, and wing-linings; flight feathers barred with blackish. Cere, bare eye-ring, and feet bright yellow. **Immatures:** above more brownish, scaled with buffy; below buff, heavily streaked with blackish. Larger than Merlin, which has pale eyebrow and lacks broad dark cheek-bar.

Habits. Migrates singly or in small, loose groups, sometimes soaring but more often with fast, direct, level flight; in winter prefers open wetlands or steep offshore islands, where waterbirds abound; flies remarkably swiftly, agilely, powerfully; hunts by aerial pursuit or steep power dives, striking prey in flight, recovering it in the air or from the ground; takes mostly teal, shorebirds, gulls, terns, and grackles; may harry flocks but usually attacks birds flying singly; sometimes soars during midday hours. Mostly immatures seen in Costa Rica.

Voice. Usually silent in winter or on migration; occasionally a short scream or series of shrill squeals, especially when 2 birds interact.

Status. On migration (mid-September–October, March–early May) along Caribbean coast, occasionally in Valle Central and Pacific lowlands; winters mainly around Golfo de Nicoya and in lower Tempisque basin, also on Cocos I. and probably Río Frío region, usually as widely spaced individuals (though in recent years, 2 birds have wintered regularly at Palo Verde); declined drastically in Costa Rica as North American breeding population was decimated by DDT contamination; presently increasing, although threatened by continuing heavy local use of DDT.

Range. Breeds in N and W North America (being reestablished in E) and South America; winters from S Alaska, coastal SW Canada and NE USA to Tierra del Fuego and West Indies. Widespread in Old World.

ORDER **Galliformes**: Chickenlike Birds

FAMILY **Cracidae:** Curassows, Guans, and Chachalacas

The 44 species of curassows and their relatives are large birds confined to the warmer parts of the American continents. Most live in humid forests or gallery forests in drier regions, but the chachalacas avoid closed woodland and frequent lighter vegetation. Although most abundant at low and middle altitudes, a few species ascend high into the mountains. Members of this family are clad in grays, browns, rufous, olive, white, and black, often glossed with blue or green. Their brightest colors—red, orange, yellow, blue, and violet—are displayed on their featherless parts, including bills, ceres, cheeks, dewlaps, horns, frontal knobs, and legs. Except for the curassows, the sexes are similar in coloration; in nearly all, males are larger than females. Largely arboreal, they pluck fruits and leaves from trees and vines, or descend to gather fallen fruits; a few forage mainly on the ground. The long, looped trachea, especially of the males, enables cracids to produce sounds notable for volume more than for melody. Some also make drumming or rattling sounds by vibrating their wings in the midst of a glide. In crude open nests of sticks, leaves, and rootlets, nearly always in vegetation but rarely on the ground, they lay 2–4 white or whitish eggs with rough, often pitted shells. Incubation takes 22–34 days. Soon after hatching, the downy chicks leave the nest to climb through bushes and vines, where (unlike most gallinaceous birds) they are fed from the bill by parents. The small clutch size and slow reproduction of the Cracidae makes them less able to withstand hunting than are extratropical "game" birds with larger broods. With the rapid destruction of forests where they dwell, each year these splendid birds become rarer, and several species may be on the road to extinction.

PLAIN CHACHALACA

Ortalis vetula Pl. 12(2)

Chachalaca Olivácea (Chachalaca)

Description. 22″ (56cm); 650g. Small head; long neck and tail; throat bare, bright red. **Adults:** head and neck grayish; body and wings dull olive-brown, below paler, shading to ochraceous on belly; tail blackish with green gloss and buffy-white tip. Iris brown; bill blackish; orbital skin and feet dull gray. **Young:** above greener, below buffier; barred with blackish on back, buffy on wing-coverts; rectrices pointed. Smaller Gray-headed Chachalaca has rufous in primaries.

Habits. Frequents tropical dry and moist forest, especially where interspersed with scrub and savanna; usually in groups, which may number up to 15 birds; furtive and wary, prefers to escape danger by running swiftly on ground or leaping and gliding through brushy tangles; feeds in trees or on ground, on fruit (figs, palms, Sapotaceae), seeds, leaves, flowers.

Voice. Loud, raucous *RAW-pa-haw* or *cha-cha-LAW-ka*, often by several birds in a rhythmical chorus, especially in early morning and evening, usually from well up in tree. Contact notes include peeping whistles and cackles.

Nest. A shallow saucer of twigs and plant fibers, lined with leaves, in thick vegetation. Eggs 2–4, rough-shelled, white to cream (in Texas; nest unknown in Costa Rica). Call chiefly in May–July in Costa Rica; elsewhere breeds early in wet season.

Status. Resident; common in higher mountains of Península de Nicoya, scarce and local in NW Pacific lowlands, S to head of Golfo de Nicoya (Palo Verde); very inconspicuous and easily overlooked except when calling; perhaps present but undetected elsewhere in Guanacaste.

Range. S Texas to Honduras; Costa Rica.

Note. Isolated Costa Rican population is sometimes assigned to the species *O. leucogastra* (White-bellied Chachalaca), of the Pacific lowlands of Chiapas to N. Nicaragua. However, in coloration and voice this population seems much closer to *O. vetula,* and we follow Delacour and Amadon (1973) in assigning it to this species, in spite of the considerable distributional gap.

GRAY-HEADED CHACHALACA

Ortalis cinereiceps Pl. 12(1)

Chachalaca Cabecigrís (Chachalaca, Pavita)

Description. 20″ (51cm); 500g. Differs from Plain Chachalaca by dark rufous primaries, whiter belly. **Adults:** head and upper neck dark gray, contrasting with dull olive of chest and upperparts; tail glossy greenish-black, all but central rectrices tipped with pale gray-brown. Bill and legs grayish; bare sides of throat dull red. **Young:** browner than adult, especially head; cinnamon edgings on dusky flight feathers; rectrices very pointed.

Habits. Travels in straggling flocks of 6–12 or more; prefers thickets with scattered emergent trees, often along rivers or streams; ascends to treetops at forest edge for fruit; walks gracefully along thin horizontal branches; flies like an ani, with several quick shallow flaps and a long, flat-winged glide; eats berries, the flesh of larger fruits like guava, and leaves.

Voice. Contact notes include a high, soft, peeping *white, white, white* and various cackling and purring notes. The chorus is a cacophony of raucous, *Amazona*-like squawks, *kloik, kleeuk kraahk*, and higher-pitched piping squeals, *oooeee? weeit?* Lacks the stentorian "chachalaca" calls of other chachalacas.

Nest. A broad, shallow saucer of sticks, vines, green, or dead leaves, 3–8ft (1–2.5m) up in a vine-draped shrub or tree or heap of dead branches. Eggs 3, less often 2, whitish, rough-shelled, often thickly pitted or sprinkled with tiny white flakes. February–June.

Status. Resident, often common where not severely hunted, in lowlands and foothills throughout Caribbean slope and on Pacific slope from W Valle Central and mouth of Golfo de Nicoya S to Panama; ranges to 3600ft (1100m) locally.

Range. E Honduras to NW Colombia.

CRESTED GUAN

Penelope purpurascens Pl. 12(4)

Pava Crestada (Pava)

Description. 34″ (86cm); 1.7kg. Large; bushy crest; bare red dewlap. **Adults:** plumage mostly dark olive-brown, with elongate white spots on neck and chest; glossed with bronzy-green on wing-coverts, back, and central rectrices; rest of tail and remiges blackish; lower back, rump, belly, and tail-coverts dark reddish-brown. Iris red; facial skin dark gray to dusky; bill blackish; legs dull coral-red. **Young:** body plumage vermiculated with black and scaled or flecked with ochraceous; narrow, pointed flight feathers largely rufous, vermiculated with black and tipped with buff.

Habits. Prefers forest, entering adjacent semi-open or tall second growth sparingly; in pairs or family groups forages in canopy for fruits and tender foliage, sometimes descending to ground to pick up fallen fruits and scratch for seeds, flying up when alarmed; flies at treetop height by heavy flaps and long glides; walks with agility along branches; often exposes itself gratuitously to the hunter by complaining loudly from high bough.

Voice. Most often heard is a series of loud, far-carrying, piping or "pumping" *pow* or *plee* notes; contact notes are soft, mellow whistles; dawn song a powerful "steam-whistle" call, ending in a low growl: *ku LEEErrr!* In courtship (?) produces a loud drumming with wings as it glides across an opening, by moonlight or at dawn or dusk.

Nest. A fairly bulky bowl of leafy twigs, lined with leaves, in a tree or on a stump. Eggs 2–3, dull white with rough, finely pitted shells. March–June.

Status. Resident, formerly countrywide from lowlands locally to 4000ft (1200m) on Caribbean slope and 6000ft (1850m) on Pacific slope; now absent from deforested country and becoming rare in unprotected forests.

Range. S Mexico to W Ecuador and N Venezuela.

BLACK GUAN

Chamaepetes unicolor Pl. 12(5)

Pava Negra (Pava Negra)

Description. 25″ (64cm); 950g. Stouter-bodied and shorter-tailed than chachalaca, with small head; outermost primaries narrow, sharp-pointed. **Adults:** plumage mostly glossy black; below duller, sootier. Iris red; facial skin bright blue; legs coral-red. **Immatures:** plumage duller, sootier.

Habits. Prefers dense mountain forests but if not heavily persecuted can persist in patchy forest interspersed with clearings and second growth; usually seen singly, or in pairs as nesting season approaches, in family groups afterward; walks gracefully along large branches; flies with a few flaps and a glide; takes fruits, including Lauraceae, palms, *Ardisia*, both from trees and from ground; roosts in large isolated trees.

Voice. Usually silent; startled, a low, deep *ro-rooo* or a groaning, coughing *kowr*; alarm note a ticking *tsik tsik . . .* ; during breeding season in early morning, piping calls rather like those of Crested Guan but softer, lower; a loud crackling rattle with wings as it glides between trees.

Nest. A pad or platform of leaves and twigs placed amid a mass of epiphytes, 15ft (4.5m) up in cloud forest tree. Eggs 2, white, rough-shelled. May (Bello).

Status. Resident in mountains from upper limits of forest down to ca. 3300ft (1000m), at least seasonally; common in remote areas; where hunted, scarce and wary.

Range. Costa Rica and W Panama.

GREAT CURASSOW

Crax rubra Pl. 12(3)

Pavón Grande (Pavón, Granadera)

Description. 36″ (91cm); 4kg. Very large and robust, with long tail and prominent crest of erectile, forward-curled feathers. **Adult ♂:** glossy black with white belly; yellow cere surmounted by bulbous knob. Bill dusky with pale tip; legs grayish. ♀: variable; head and crest barred with black and white; body and wings rufous; or foreparts blackish, body chestnut-rufous, wings and belly narrowly barred with black and sometimes white; tail more or less barred. Cere grayish without knob, bill yellowish. **Immature ♂:** lacks knob on cere; crest short; like darkest females in pattern but darker, more blackish overall.

Habits. Prefers forested country, where it walks over ground picking up fallen fruits, sometimes scratching for these or small animals; roosts in tree and sometimes flies to a low branch in alarm, though usually prefers to escape by running; hops up from branch to branch to take flight, a few flaps and a long glide; singly, in pairs, or in groups of up to 6.

Voice. In alarm, as when running away, a high-pitched yipping like a small dog; when disturbed, thin, high-pitched, usually descending whistles: *wheep wheet wheeeeeu*; also a low, hooting grunt. Male in breeding season gives a deep, low, resonant, ventriloquial *hummm*, repeated several times, often on different pitches.

Nest. A disproportionately small, loosely made pad of leaves and twigs with a shallow central depression, saddled in a fork or placed at the intersection of several branches and vines, 10–20ft (3–6m) up in a forest tree or thicket. Eggs 2, dull white, rough-shelled, March–May.

Status. Resident; formerly throughout lowlands of both slopes, locally to ca. 4000ft (1200m); now mostly scarce and local, good populations persisting mainly in some national parks (e.g., Corcovado, Santa Rosa, Rincón de la Vieja).

Range. S Mexico to W Ecuador.

FAMILY **Phasianidae:** Pheasants, Quails, and Allies

Most of the 165 species of this family, including the partridges and pheasants, are found from Europe to Africa and Australia. The much smaller New World division of the family (Odontophorinae) is best represented in Mexico and western United States. In all South America, only 13 species are found, 11 of them wood-quails of the genus *Odontophorus*. Although smaller and less lavishly adorned than Old World pheasants, our American species have beautiful, intricate patterns of subdued colors. Many are crested, and the sexes are similar or different in appearance. They are stout, mostly short-tailed, with short, thick bills and strong, spurless legs and feet for scratching on the ground. With short, rounded wings, they fly swiftly, but rarely far, to escape danger. They live in forests, brushland, open fields, or deserts, in small, intimate coveys, often sharing food and preening one another. They eat grains, fruits, buds, leaves, and flowers, with much smaller quantities of insects and other invertebrates. Their voices are clear and ringing, soft and low, rarely harsh; some indulge in animated duets. Bobwhites sleep on the ground, in compact groups, heads outward for escape if they are threatened. Wood-quails and wood-partridges roost in trees.

Although a few Old World pheasants build nests in trees, most members of the family nest on the ground in roughly built bowls of whatever vegetation is available. Some wood-quails make roofed nests with a side entrance. Rarely a New World quail nests on a stump or the abandoned nest of some other bird, at no great height. Clutches range from 4–20 or more eggs, the largest sets probably the product of 2 females. White to creamy or buffy, the eggs may be immaculate or marked with shades of brown. Incubation is usually by the female, but the male sometimes helps or takes full charge if his mate is lost. After an interval of 21–24 days, the downy chicks hatch and soon leave the nest to be led and guarded by both parents or the covey while they pick up their own food.

BUFFY-CROWNED WOOD-PARTRIDGE

Dendrortyx leucophrys Pl. 12(16)

Perdiz Montañera (Chirrascuá)

DESCRIPTION. 13″ (33cm); 350g. Rather chickenlike; body laterally compressed; long tail bowed upward in inverted V; short crest. **Adults:** face and throat buffy-white; head mostly brown; upper back and breast slate-gray, streaked with chestnut; rest of mantle bright brown, marked with blackish; remiges dark rufous; rump and belly dull gray-brown; tail dark brown, vermiculated with black. Iris pale gray; bill black; orbital skin and legs reddish. **Immatures:** similar face pattern but throat streaked with dusky; body browner, more or less flecked with white; wing-coverts barred with buffy. Face pattern and long tail unique among ground-dwelling birds.

HABITS. Spends most time on the ground in thickets, in forest (especially where partly logged) and second growth, including tongues of streamside woodland that extend from the mountains downward; scratches vigorously in leaf litter for seeds, fallen fruits, invertebrates; usually in small coveys; extremely wary and furtive, slinking away silently and low to the ground at least sign of danger; far more often heard than seen; roosts in groups in trees or tall shrubs.

VOICE. A loud, *whew Whit-cha, cha-wa-WHAT-cha*, or *chi-ras-KWA*, often repeated in gabbling chorus, deeper and throatier than calls of wood-quail. Alarm call in covey a soft descending tsitter or gobble.

NEST. Unknown (?). Evidently breeds at start of rainy season, ca. June–September.

STATUS. Resident locally, and often surprisingly common, in higher and wetter parts of Valle Central and adjacent slopes of Cordillera Central and Cordillera de Talamanca; also Dota region and Cerro de la Muerte massif; 3300–9200ft (1000–2800m), occasionally higher or lower. May be present along Pacific slope of Cordillera de Talamanca S to Térraba region.

RANGE. Extreme S Mexico to Costa Rica.

NOTE. Also called Buff-fronted Wood-Partridge.

SPOTTED-BELLIED BOBWHITE

Colinus leucopogon Pl. 12(13)

Codorniz Vientrimanchada (Perdiz, Codorniz)

DESCRIPTION. 9″ (23cm); 140g. Small, short-crested. **Adult ♂:** head mostly brown; eyebrow white; broad eye-stripe black; cheeks and throat white, marked with chestnut and black; upperparts grayish-brown, finely marked with black, grayish, and white; remiges dusky; breast cinnamon-brown, finely peppered with black; rest of underparts with large, round white spots edged with black. Bill black; legs dull gray. ♀: throat buffy, streaked with black; overall coloration grayer, especially breast, which is spotted with white. **Immatures:** like female but below buffy, spotted with dark brown. Only nonforest quail in its range.

Habits. Inhabits scrubby savanna, open woodland, second growth, weedy fields and pastures with high grass; during most of year in coveys of 3–15 birds, which when foraging move slowly in open formation, individuals stopping to peck or scratch for seeds, fallen fruits, insects, and grit; usually keeping in dense cover but venturing into the open in early morning and evening; sometimes a pest of bean or rice crops. When alarmed, members of a covey close ranks and run rapidly like animated bowling pins, or burst into brief, buzzy flight.

Voice. Before and during breeding season, males give a throaty, scratchy *bobWHITE* or *bob, bobWHITE*; song of hoarse, throaty phrases repeated from a perch: *pipi JEE pi JER* or *JER pi JEE*. Coveys give chuckling, chattering, peeping notes.

Nest. A cup of grass stems on ground under a thick tussock. Eggs ca. 10, white, without gloss. June–October.

Status. Locally common resident of lowlands and foothills of N Pacific slope, including Península de Nicoya; also up across Valle Central onto Caribbean slope at least as far as Paraíso; sea level to 5000ft (1500m).

Range. W Guatemala to Costa Rica.

Note. Sometimes lumped with the southern *C. cristatus*, but the northernmost race of that species differs more from *leucopogon* than do more southern forms; egg color may also differ. We believe a superspecies relationship is more appropriate.

CRESTED BOBWHITE

Colinus cristatus Pl. 12(14)

Codorniz Crestada

Description. 8″ (20cm); 125g. Differs from Spotted-bellied Bobwhite in darker upperparts and longer crest, mixed with dull buffy; rufous eyebrow and throat; pale grayish face and eye-stripe; contrasting black neck spotted with white. **Adult ♂:** breast spotted with white; posterior underparts coarsely spotted and barred with black and cinnamon-rufous. ♀: buffy rather than rufous on head; throat streaked with black.

Habits. Locally in rice and sugarcane fields, second growth and scrubby pastures; most often along roadsides in early morning, when it comes out to forage and pick up grit. Habits similar to those of Spotted-bellied, but seems to move in smaller coveys.

Voice. Notes clearer, less throaty than those of Spotted-bellied; ''bobwhite'' call sounds like *pwit pwit PWEET*; also a wheezy *WHEE-cher*, repeated 4–6 times; in coveys, chirping and peeping notes.

Nest. Unknown (?) in Central America. Eggs cream-colored, spotted and blotched with brown in South America.

Status. Locally common, presumably resident, in Golfo Dulce lowlands, from Ciudad Neily S and W to Panama; reports presumably of this species from as far N as Piedras Blancas; evidently entered Costa Rica from Chiriquí in 1970s, following deforestation; probably still expanding its range.

Range. Costa Rica to W Colombia and E Brazil.

MARBLED WOOD-QUAIL

Odontophorus gujanensis Pl. 12(9)

Codorniz Carirroja o Corcovado

Description. 10″ (25cm); 300g. Stout, compactly built with heavy bill, short tail. **Adults:** conspicuous reddish skin around dark eye; erectile crest dark brown; sides of head rusty (especially in males); neck and upper back more or less grayish; upperparts brown, vermiculated and spotted with black and buff; below brown, indistinctly barred with buff and blackish. Bill black; legs gray.

Habits. In hilly, wooded areas coveys of 5–8 walk single file over floor of forest or old second growth, keeping to deep shade, scratching in litter for fallen fruits, insects, earthworms. They avoid people by walking quietly away (but will flush if surprised at close range), and dogs by flying to low perch. Tame where not hunted, wary where persecuted.

Voice. A mellow, resonant, clearly enunciated *burst the BUBble burst the BUBble*, or *corco-VAdo* or *coco WAT-to*, by duetting pair (male gives first part, female last part), rapidly repeated many times in early morning or toward dusk, even at night. Calls include soft notes, *witty witty witty*, and a prolonged *caaa caaa caaa*.

Nest. A roofed structure with a side entrance on the ground, or a deep pocket amid fallen leaves. Eggs 4–5, white. January–August.

Status. Resident in lowlands and interior valleys of S Pacific slope, N to about Carara, from sea level to ca. 3000ft (900m). Scarce and local in most areas due to deforestation and hunting; still fairly common in forested parts of Golfo Dulce region.

RANGE. Costa Rica to E Bolivia and C Brazil.

RUFOUS-FRONTED WOOD-QUAIL

Odontophorus erythrops Pl. 12(12)

Codorniz Pechicastaña (Chirrascua, Gallinita)

DESCRIPTION. 9½″ (24cm): 280g. Unpatterned rufous underparts; crown and short crest chestnut. **Adult ♂:** face and throat black; above dark brown, heavily marked with black; wing-coverts spotted with buffy. Iris chestnut; bill black; orbital skin purplish; legs gray. ♀: similar but cheeks chestnut, throat dusky, orbital skin blackish. **Immatures:** like female but duller, spotted below with black. Only wood-quail in most of Caribbean lowlands.

HABITS. Roams the understory of wet lowland forest and adjacent dense second growth in pairs or coveys up to ca. 10 birds, foraging by scratching in leaf litter or flicking aside leaves with bill to expose fallen fruits, seeds, bulbs, tubers, insects, and other invertebrates. By calling back and forth in early morning, coveys maintain their spacing in group territories.

VOICE. Birds in covey call in gabbling chorus, individual phrases sounding like *kooLAWik kooLAWik, kooKLAWK KooKLAWK*, or *KLAWcooKLAWcoo*; soft cooing and peeping contact notes; when alarmed, a liquid rattle like that of Chestnut-backed Antbird.

NEST. A slight excavation between buttresses at foot of tree, lined with leaves and grass. Eggs 4, creamy-white.

STATUS. Formerly common and widespread in Caribbean lowlands and locally up to 2500ft (800m) in foothills of N cordilleras and Cordillera Central, to 3300ft (1000m) in Cordillera de Talamanca; with deforestation becoming increasingly scarce and local.

RANGE. E Honduras to W Ecuador.

SPOTTED WOOD-QUAIL

Odontophorus guttatus Pl. 12(11)

Codorniz Moteada o Pintada

DESCRIPTION. 10″ (25cm); 300g. Raises an orange crest (conspicuous in male, dull in female) when excited. **Adults:** crown dark brown; auricular stripe chestnut; cheeks and throat black, streaked with white; underparts olive-brown (normal phase) to rufous (rufous phase), boldly spotted with white; above dull chestnut-brown marked with black, rufous, and gray; back with white shaft-streaks. Iris orangish; bill black; legs dull gray. **Immatures:** spots below smaller, buffier, more triangular; breast and sides barred with black; cheeks and throat dusky, mottled with whitish.

HABITS. Frequents dense, thickety, often bamboo-choked understory of forest and second growth in highlands; usually in coveys of 4–10 birds that scratch vigorously in litter of forest floor for fallen fruits, seeds, tubers, invertebrates. Shy and wary, they freeze or scuttle off at any alarm; flushed at close range, they explode into buzzy flight, either to perch motionlessly on low branch or to drop abruptly back amid concealing vegetation. In early morning, several coveys often call back and forth. They roost in trees.

VOICE. Covey call consists of clear, whistled phrases: *coowit CAWit coowit COO*, or *whee to-whee to-WHIT too*, the complete phrase actually a duet; when several birds call in chorus the result is a gabbling cacophony. Contact notes are clear descending chirps and peeps; a rattling trill when alarmed.

NEST. Undescribed (?). Eggs creamy-white, sometimes spotted with brown. Probably latter part of dry season.

STATUS. Locally common resident in Cordillera Central and Cordillera de Talamanca, from 3300ft (1000m) on Pacific slope and 5000ft (1500m) on Caribbean slope up to timberline; in wet season occasionally descends to ca. 330ft (100m) in Caribbean foothills.

RANGE. S Mexico to W Panama.

BLACK-BREASTED WOOD-QUAIL

Odontophorus leucolaemus Pl. 12(10)

Codorniz Pechinegra (Gallinita de Monte, Chirrascuá)

DESCRIPTION. 9″ (23cm); 275g. Thick-set, dark-colored; sexes similar but much individual variation in amount of white on throat. **Adults:** face and most of underparts black; belly rich brown; white barring on lower breast; white throat conspicuous to virtually obsolete; some white mottling on face; above dark brown, barred and vermiculated with black, rufous, and buff. Iris brown; bill black; legs dull gray. **Immatures:** black below largely replaced by brown or chestnut; throat mottled with dusky; feathers of back edged with rusty.

HABITS. Frequents understory of cool, wet highland forest in coveys that often number 10–15 birds, who space themselves by calling back and forth in early-morning choruses, defending group territories. This species overlaps narrowly with the Rufous-fronted Wood-Quail at its lower altitudinal limits, and upward overlaps broadly with the Spotted Wood-Quail. It prefers forest interior, while Spotted occurs more in ravines, at edges and breaks.

Voice. Covey chorus a rushing gabble, individual phrases: *where-ARE-you where-ARE-you* or *kee-a WOWa kee-a WOWa* or *coweep-CHOweep*, punctuated by low hoarse chucking or clicking notes. Contact notes soft, upslurred peeps and chirps. A metallic, rattling trill when alarmed; a growling *CUH-k'krrr* in aggressive situations.

Nest. In sloping bank of small stream, a spherical burrow into leaf litter so that incubating bird fills the hollow, entrance points slightly downhill. Eggs 5, white. June (McDonald).

Status. Locally common resident of middle elevations from Cordillera de Guanacaste S into Panama; mostly on Caribbean slope but on both slopes of N cordilleras and with an isolated population in Dota region; ca. 2300–6000ft (700–1850m).

Range. Costa Rica to W Panama.

Note. Also called White-throated Wood-Quail.

TAWNY-FACED QUAIL

Rhynchortyx cinctus Pl. 12(15)

Codorniz Carirrufa

Description. 7½" (19cm); 150g. Small, stocky, thick-billed, nearly crestless. **Adult ♂:** crown and eye-stripe dark brown; rest of head rufous; throat whitish; breast slaty, shading through ochraceous-buff to white on flanks and belly; back slaty, scaled with brown; rest of upperparts brown, finely marbled and barred with buff and black. Iris chestnut; upper mandible dark gray; lower mandible and legs pale gray. ♀: head mostly dark brown; fine white and black eye-stripes; throat white; breast rich brown; posterior underparts white, boldly barred with black; above like male but browner. Much smaller than wood-quails; both sexes with unmistakable patterns.

Habits. Pairs or small coveys roam the floor of dense humid forest in foothills and adjacent lowlands; extremely inconspicuous and furtive; when alarmed, usually runs a short distance and freezes or flushes with sharp rustle of wings, to plane back to earth and hide a short distance away; forages more by pecking and digging with bill, less by scratching, than other quails; eats seeds, worms, insects.

Voice. A clear, sad, ventriloquial, dovelike *cooo* or *toot*, lasting about a second; often 2 or more birds call back and forth with coos on different pitches; imitation of call decoys birds to within a few yards. Contact note in covey a soft peeping.

Nest. Unknown (?). Eggs white, unmarked, in South America. Probably dry to early wet season.

Status. Scattered records from N Caribbean slope (S to Sarapiquí region) and adjacent Pacific slope along Cordillera de Guanacaste, between 500 and 2600ft (150–800m); almost certainly more widespread and numerous than the few records suggest but easily overlooked.

Range. E Honduras to NW Ecuador.

ORDER **Gruiformes:** Cranes, Rails, and Allies

Most members of this ancient and diverse order are more or less long-legged, either fast-running terrestrial birds or primarily wading aquatic species. The feet are incompletely or not at all webbed, even the swimming species (e.g., coots, sungrebes) having each front toe with a separate web rather than all three joined. Many species have reduced powers of flight, and flightlessness has evolved repeatedly in the order. Two main subgroups are recognized on the basis of internal anatomy: the cranes and their relatives (including the Sunbittern) and the more aquatic rails and their allies (including the Sungrebe).

FAMILY **Aramidae:** Limpkins

The Limpkin, so named for its halting gait, is the single extant species of a family that in past geologic ages was larger and more widespread. Now it is confined to the warmer parts of the New World, where it is nonmigratory. Its skeleton is cranelike, but in appearance, behavior, digestive system, and nesting it more closely resembles the rails. As it walks in a marsh, it might be mistaken for an ibis but for its more erect posture and less decurved bill. With short, broad, rounded wings, it flies for short distances with long legs and toes dangling, but on longer flights they stretch behind, in the manner of herons. Although Limpkins forage mainly on the ground in wet places, they can climb through branches and perch in trees, where they sleep. Both sexes incubate in an open nest supported amid tall marsh grasses or in branches and vine tangles near or over water, and both attend the downy, brown, precocial chicks.

LIMPKIN
Aramus guarauna Pl. 5(5)
Carao (Correa)

DESCRIPTION. 26″ (66cm); 1.1kg. Heronlike in form but neck relatively shorter, thicker; bill long, laterally compressed, slightly decurved. **Adults:** face gray-brown, head and neck finely streaked with brown and white; body and wing-coverts dark brown with lanceolate white spots; remiges dark brown glossed with greenish to purplish. Bill yellowish-horn shading to dark gray tip; legs dull greenish. **Young:** similar but less white spotting, plumage looser-textured.

HABITS. Inhabits open freshwater marshes, pond and river margins, occasionally wooded swamps; forages singly, walking about with slow, jerky gait in search of snails, mainly apple snails (*Pomacea*), which are brought to a favorite spot, their shells punctured by swift pecks, the soft parts extracted and consumed; alert, often calls loudly when observer is in sight. Flight heavy, with quick upstrokes, slower downstrokes.

VOICE. A deep, resonant squeal trailing off from loud beginning *KLEEeeo* or *krrLEEeeo*; especially at night, a harsher, twanging *kyarr* ending in a dry rattle; alarm call, a short, sharp, squealing *keew*.

NEST. A saucer of dry rushes or cattail leaves, often with a foundation of twigs, low to fairly high over water. Eggs 4–6, buffy, with brown and gray blotches. Midwet to early dry season.

STATUS. Resident, but pronounced local movements according to water levels; most abundant, at least seasonally, in Tempisque basin and Río Frío–Caño Negro region; locally elsewhere in Guanacaste and on N Caribbean slope; rare or sporadic in other lowland areas on both slopes, and to 5000ft (1500m) in Valle Central.

RANGE. SE USA, West Indies, S Mexico to W Ecuador and N Argentina.

FAMILY **Rallidae:** Rails, Coots, and Gallinules

The 132 species of this family inhabit all continents and great islands except in polar regions, and likewise many tiny islands in midocean, where numerous flightless species live (or did so until their recent extinction). Many members of the family are migratory. In length 5½–26″ (14–66cm), most have short wings and tails, narrow bodies that slip easily through dense vegetation, long legs and toes to support them on soft mud or yielding aquatic plants, and short or long, straight or decurved bills. Gallinules and coots have frontal shields. Most rails are cryptically colored in blended browns, grays, and black, frequently barred, streaked, or spotted; few are brightly clad. The sexes are nearly always alike.

Typical rails live obscurely in marshes, swamps, dense grasslands, or moist woodland, where they prefer to escape by walking instead of flying and are among the birds most difficult to know intimately, which is a pity, for they appear to be intelligent and have fascinating habits. Moorhens, gallinules, and coots are more aquatic and may frequently be seen swimming on still water; alone in the family, coots have lobately webbed feet for more effective swimming. Members of this family eat a great variety of vegetable and animal foods. Their open or roofed nests are often built by both sexes, amid dense vegetation, or floating in shallow water. Two to 16 white or buffy eggs, spotted and blotched with shades of brown and pale lilac, are incubated by both parents for periods of 18–25 days. Soon after hatching, the downy, subprecocial young leave the nest and are fed from the bill, guided and protected not only by both parents but also, in a number of species, by juveniles of an earlier brood and even by nonbreeding yearlings. They are brooded on special nests built above water or wet ground, and on such platforms adults also sleep.

SPOTTED RAIL
Pardirallus maculatus Pl. 6(4)
Rascón Moteado

DESCRIPTION. 10″ (25cm); 190g. Medium-sized, strikingly patterned. **Adults:** head and neck black, speckled with white; mantle brown and black with elongate white spots; remiges dark brown; below barred with black and white. Iris red; bill greenish-yellow with red spot at base; legs carmine-red. **Immatures:** above browner, with reduced buffy spotting; below sooty with some buffy barring.

HABITS. Skulks in cattails, tall grass, rushes or *Polygonum* thickets in freshwater marshes; furtive but not particularly shy, it may approach observer in dense vegetation but rarely ventures into open; along edges near dawn and dusk, or on dark, rainy days; often active and vocal at night; eats adult and larval insects, earthworms and other invertebrates. Pairs are apparently territorial, at least during breeding season.

Voice. A repeated groaning screech, each note introduced with a resonant pop or grunt—*g'REECH g'REECH* or *pum-KREEP pum-KREEP*—possibly territorial or aggressive; also an accelerating series of such low *pum* notes, like a distant motor starting up. Disturbed or in aggressive situations, a sharp *gek*, often repeated.

Nest. Unknown (?). Mainly in wet season.

Status. Resident (though local movements likely in response to changing water levels, especially in Guanacaste), probably widespread but local on both slopes; definitely recorded in Tempisque basin, the vicinity of Cartago and of Turrialba, and Río Frío area.

Range. C Mexico and West Indies to NW Peru, N Argentina, and S Brazil.

UNIFORM CRAKE

Amaurolimnas concolor Pl. 6(8)

Rascón Café

Description. 8″ (20cm); 95g. A small, relatively unpatterned wood-rail. **Adults:** above olive-brown; sides of head gray-brown; underparts rufous, palest on throat. Iris red; bill greenish-yellow; legs red or orange. **Immatures:** below duller and darker, more gray-brown; throat and eyebrow dull ochraceous; iris yellow.

Habits. Haunts forested swamps, streamside thickets, old wet second growth, and damp ravines, seldom leaving dense cover; walks erectly with tail cocked; pumps tail when agitated; when alarmed, scurries off with body horizontal, tail low; pecks into leaf litter and hanging debris for insects, spiders, small frogs and lizards; digs with beak in soft mud for earthworms; also eats seeds and berries.

Voice. Territorial call a series of clear whistles, loudest in the middle, then speeding up and fading away: *tooeee toooeee TOOOEEE TOOOEEE TOOEE tooee-tuee-tui*. Alarm note a sharp nasal *kek*. Low, clear whistles as contact or aggressive notes.

Nest. A loose cup of leaves and stems, in hollow of top of stump, concealed by dense vines. Eggs 4, pale buff, blotched with shades of brown and gray, mostly near large end. Wet season.

Status. Resident in wetter, forested parts of Caribbean lowlands and foothills, locally to 3300ft (1000m), and Golfo Dulce lowlands; occasional in Valle del General; uncommon and highly local but probably overlooked in many areas.

Range. S Mexico and Jamaica (formerly) to W Ecuador, N Bolivia, and S Brazil.

GRAY-NECKED WOOD-RAIL

Aramides cajanea Pl. 6(13)

Rascón Cuelligrís (Chirincoco, Pone-pone, Pomponé)

Description. 15″ (38cm); 460g. Long neck and legs; relatively short bill and tail. **Adults:** head and neck gray, washed with dark brown on hindcrown and nape; throat white; mantle brownish-olive; breast and sides cinnamon-rufous; belly, rump, tail-coverts, and tail black. Iris red; bill bright greenish-yellow; legs coral-red. **Immatures:** similar but belly sooty-black, flecked with buff.

Habits. Forages singly, walking, with short tail constantly pumping, over wet ground of streamsides, swamps, mangroves, wet second growth, or forest edge, keeping to shaded areas and thickets; warily enters plantations and dooryards; when alarmed, runs rapidly away, rarely flying; pushes aside fallen leaves and digs in soft mud with bill; eats mostly small invertebrates, frogs, seeds, berries, and palm fruits.

Voice. A prolonged duet of loud, resonant notes, melodious in distance—*chirin co chirin co co co chirin cococo*, or *cuk-cuk-cuk COOKIT COOKIT COOK co co co*, etc.—most often heard toward evening or at night; when disturbed, a deep, low "pumping" *goo-hup goo-hup goo-hup*; when alarmed, a loud, harsh cackle.

Nest. A compact, bulky mass of dead leaves and twigs with a shallow central depression, up to 10ft (3m) above ground in thicket or vine tangle. Eggs 3–5, dull white to beige, spotted and blotched with rufous and pale lilac. April–August or September.

Status. Resident and locally common countrywide from lowlands to at least 4600ft (1400m); persists in remnant streamside woods amid pastures or agricultural lands, as in Valle Central.

Range. C Mexico to N Argentina.

RUFOUS-NECKED WOOD-RAIL

Aramides axillaris Pl. 6(14)

Rascón Cuellirrufo

Description. 11″ (28cm); 275g. Smaller than Gray-necked Wood-Rail, with no gray on neck. **Adults:** head and neck rufous like breast; belly sooty-gray; throat white; a triangular gray patch on upper back contrasting with dull olive-green of back and wing-coverts; remiges mostly rufous; rump, tail, and crissum blackish. Iris red; bill greenish-yellow; legs coral-red. **Immatures:** crown dull gray-brown; sides of head pale gray; throat grayish-buff; neck and underparts grayish-brown; gray patch on back darker, more extensive.

HABITS. Little known; frequents mangrove swamps, swamp forest; most active at dawn and dusk; furtive, seldom leaving dense cover by day; eats mainly crabs.

VOICE. "An incisive, loud . . . *pik-pik-pik* or *pyok-pyok-pyok*, repeated about 8 times, often antiphonally" (ffrench).

NEST. A bowl of small twigs, lined with weed stems, dead and green leaves, 6–10ft (1.8–3m) up in tree or vine tangle. Eggs 5, like those of Gray-necked Wood-Rail (in Trinidad; no Costa Rican record).

STATUS. Uncommon and local in mangroves around Golfo de Nicoya (Lepanto, Punta Morales); also recorded from swamp forest at Bijagua, in Caribbean foothills of Cordillera de Guanacaste; possible sighting at La Selva, in Sarapiquí lowlands (Will).

RANGE. C Mexico to Ecuador and Suriname.

SORA

Porzana carolina Pls. 6(12), 51(11)

Polluela Sora o Norteña

DESCRIPTION. 8″ (20cm); 65g. Fairly small, stout-billed; contrasting black face. **Adults:** above brown, mantle with black blotches and fine white streaks; sides of head, neck, and breast slaty; face and throat black (more extensive and clean-cut in male); belly white; flanks barred with brown and white. Iris reddish; bill yellow; legs dull greenish. **Immatures:** little or no black on face; gray of head and neck replaced by buff; flanks less densely barred.

HABITS. Prefers grassy marshes, flooded fields and rice fields, reed-bordered ponds; forages singly by day; roosts communally in cattails or other thick vegetation at night; best seen in early morning and evening, when most likely to venture from thick cover; walks with tail flicking frequently, exposing white crissum; eats aquatic insects, snails, seeds.

VOICE. On winter range usually silent; occasionally clear peeping whistles, singly or in 2s or 3s, especially when alarmed.

STATUS. Migrant and winter resident, widespread but local, lowlands to ca. 5000ft (1500m) on both slopes, including Valle Central; most numerous in Tempisque basin and Río Frío area. Arrives mostly in October; remains through late February or March.

RANGE. Breeds from Alaska and N Canada to NE and SW USA; winters from S USA to Peru and Guyana, also West Indies.

YELLOW-BREASTED CRAKE

Porzana flaviventer Pl. 6(10)

Polluela Pechiamarilla

DESCRIPTION. 5″ (13cm); 25g. Tiny; bridled face; whitish underparts; broad buff dorsal stripes conspicuous in flight. **Adults:** crown and eye-stripe black; superciliaries white; above brown, with fine white streaks, black blotches; scapulars broadly edged with buff; rump dull rufous; below mostly white, tinged with buffy-yellow on sides of neck and breast. Iris reddish; bill dark greenish; legs yellowish. **Young:** similar but neck and breast with indistinct gray barring.

HABITS. In local concentrations in grassy marshes or along pond and lake margins where vegetation grows out into water; furtive, skulking, stays in thick cover most of day; rarely rises with weak, fluttering flight, yellowish legs dangling; picks small insects, snails, seeds, from water, mud, and vegetation.

VOICE. Harsh, churring, scratchy *je-je-je-jrrr*, sometimes introduced by a rolling note; a squealing *kweer*; a high, clear whistled *kleeer*.

NEST. Loosely built in a water plant. Eggs 5, sparsely spotted [in West Indies. In Costa Rica, a nest with 3 eggs in late July (Orians and Paulson)].

STATUS. Resident (but probably local movements with changing water levels) in Tempisque and Bebedero basins, Río Frío–Caño Negro district, lakes and marshes N of Upala, and probably elsewhere in lowlands of both slopes, but easily overlooked.

RANGE. S Mexico and West Indies to N Argentina and E Brazil.

BLACK RAIL

Laterallus jamaicensis Pl. 6(5)

Polluela Negra

DESCRIPTION. 5½″ (14cm); 34g. Very small, blackish; white-spotted upperparts. **Adults:** head, throat, and breast blackish (males) or dark gray (females); belly and flanks barred with blackish and white; hindneck chestnut; upperparts blackish-brown, speckled with white. Iris red; bill black; feet yellow to flesh-color. **Young:** above blacker with less white spotting; face pale gray with whitish superciliaries; throat buffy. Darker than any other small rail; chicks of larger rails lack white spotting above.

HABITS. Skulks furtively in grassy marshes and pond borders, cattails and flooded tall-grass pastures; very secretive and hard to flush; scurries away, mouselike, under vegetation; may feed at edges or perch in grass tufts in early morning; active and vocal on moonlit nights; eats small insects and other invertebrates, probably some seeds.

VOICE. A rapid *peep-peep churr*; a low clucking.

NEST. A loose cup woven of fine grass, covered by arch of grass or a woven roof, forming a globular structure with side entrance. Eggs 4–8, buffy-white, dotted with brown (in USA; no Costa Rican record).

STATUS. Resident(?), currently known only from scattered sight records: Bebedero basin (Taboga, repeatedly), Península de Osa (Rancho Quemado), Río Frío district (Medio Queso); possibly widespread but overlooked in lowlands of both slopes.

RANGE. Breeds locally in SW, C, and E USA, Mexico, Central America, Chile, Argentina, and West Indies. Winters from S USA southward.

WHITE-THROATED CRAKE

Laterallus albigularis Pl. 6(9)

Polluela Gargantiblanca (Freidora, Huevo Frito)

DESCRIPTION. 6½″ (16cm); 42g. Small; rufous neck and breast; barred flanks. **Adults:** throat white; crown olive-brown (Pacific race), or with dark gray half-hood (Caribbean race); rest of head, foreneck and breast rufous; posterior underparts barred with black and white; upperparts dark brown; sometimes with white scaling on wing-coverts. Iris red; bill pale greenish with dusky culmen; legs olive-gray. **Immatures:** crown and nape dull brown (mixed with gray in Caribbean race); throat whitish; rest of face, neck, and breast gray, flecked with cinnamon; barring of flanks dull, indistinct. No rufous in nape as in adult Gray-breasted Crake.

HABITS. Lurks amid thick, grassy vegetation that adjoins or covers shallow water in marshes, flooded pastures, margins of ponds and rivers, moving easily through and under thick herbage; difficult to flush; flies weakly with dangling legs, soon dropping back into cover; ventures into more open spots to feed at dawn and dusk or in dark, rainy weather; eats insects, spiders, seeds of grasses and sedges, algae.

VOICE. Most characteristic is a prolonged, rattling, descending churr or whinny, more bubbling than a similar call of Least Grebe. Alarm note a sharp, often metallic *chip* or *chirp*; clear, piercing, whistled calls, a burry *jeer-jeer-jeer*.

NEST. A globe with side entrance, woven of stems and leaves of grass, usually in a thick tussock, up to 4ft (1.2m). Eggs usually 2–3, whitish speckled and blotched with rufous and pale lilac. Wet season.

STATUS. Common to abundant resident throughout Caribbean lowlands, up to 5000ft (1500m), as around Cartago; on S Pacific slope, up to ca. 4000ft (1200m), as at San Vito; rarely N to Tempisque basin.

RANGE. SE Honduras to N Colombia and W Ecuador.

NOTE. Sometimes considered conspecific with the South American *L. melanophaius* but may be more closely related to *L. exilis*, with which it is widely sympatric.

GRAY-BREASTED CRAKE

Laterallus exilis Pl. 6(6)

Polluela Pechigrís

DESCRIPTION. 6″ (15cm); 33g. Small; rather dark; conspicuous rufous nape. **Adults:** crown blackish; throat whitish; rest of head, neck and breast gray; belly white; flanks barred with black and white; mantle olive-brown, more or less barred with white on wing-coverts and rump. Iris red; bill dusky with bright yellow-green spot on each side near base; legs brownish-flesh. **Immatures:** similar but little or no rufous on nape; duller barring on flanks. White-throated Crake has rufous neck and breast; Black Rail much darker with speckled mantle.

HABITS. Prefers thick, fairly tall grass in wet pastures, riverbanks and shallow marshes; slinks furtively about in a network of runways under grass; very difficult to flush; usually in pairs, probably territorial; eats small insects, earthworms, other invertebrates, and seeds.

VOICE. Commonest call an explosive, high *keek* or *peep*, often in series in which the first note is loudest, successive notes shorter and lower-pitched; a dry, ticking trill, shorter and softer than *churr* of White-throated Crake.

NEST. A globe of woven grasses, with side entrance, amid thick grass. Eggs ca. 3, creamy-white, with dark brown and gray spots and smears, heaviest toward large end (in Belize; no Costa Rican record). Probably wet season.

STATUS. Widespread but local resident in Caribbean lowlands; definitely recorded in Río Frío, Sarapiquí, and Sixaola regions; also in Golfo Dulce lowlands of Pacific slope; probably more uniformly distributed but overlooked because of its retiring habits.

RANGE. Belize to E Peru and Paraguay.

RUDDY CRAKE

Laterallus ruber Pl. 6(11)

Polluela Colorada

DESCRIPTION. 6½″ (16.5cm); 45g. Small; reddish; contrasting dark head. **Adults:** top and sides of head dark gray; throat buffy; body rufous, below paler; wings and tail dark brown. Iris red; bill black; legs greenish.

HABITS. In grassy marshes, flooded or well-drained fields and pastures with tall grass or weeds. Little known in Costa Rica; habits presumably similar to those of White-throated Crake.

VOICE. A dry, insectlike chatter of repeated notes on same pitch; a prolonged descending churr or whinny rather similar to that of White-throated Crake.

NEST. A globe woven of dry grass stems and leaves, lined with finer grass, in tussock. Eggs 3–4, whitish to buffy, spotted with brown to purplish. August–September (in Honduras; no Costa Rican record).

STATUS. Apparently a rare local resident or visitor in N Guanacaste (Miravalles, Orosí), known from two sight records (Slud).

RANGE. S and E Mexico to Costa Rica.

PAINTED-BILLED CRAKE

Neocrex erythrops Pl. 51(12)

Polluela Piquirroja

DESCRIPTION. 7½″ (19cm), 55g. Fairly small, gray-breasted rail with strikingly colored bill and legs, barred flanks. **Adults:** Pileum and hindneck brownish-slate to dark olive-brown; rest of upperparts dull olive-brown; throat white; rest of head, neck, anterior underparts slate-gray to plumbeous; flanks, thighs, abdomen, crissum, and wing linings barred with dusky and white. Iris red; bill with basal half scarlet, rest yellow-green; legs red-orange to brownish-red.

HABITS. Frequents grassy marshes, wet pastures, rice fields, drainage ditches; very furtive and difficult to observe and flush, but individuals or pairs may forage at puddles or openings beside dense vegetation at dawn and dusk.

VOICE. Loud, guttural, buzzy, froglike notes: *qur'r'r'r'rk* and *auuk*, or on series *qur'r'rk'auuk qur'r'rk'auuk* . . . (Hilty and Brown 1986).

STATUS. Known in Costa Rica from one definite sighting (Stiles and Rosselli; Río Frío, Sarapiquí lowlands; 22 August 1987). A second sighting near Hitoy Cerere in March 1985 (Pratt et al.) was this species or *N. colombianus* (see below). Recently seen and collected in Bocas del Toro, Panama, in November 1981, but not noted in April or July (N. G. Smith). Either a rare (or overlooked?) resident in Central America or a migrant.

RANGE. Widespread in E and C South America, from C Colombia (E Andes), S Venezuela, and Suriname to NW Argentina, S Brazil; several records November–February in USA suggest the species is migratory.

NOTE. Also possible in Caribbean Costa Rica is Colombian Crake (*N. colombianus*), known from Pacific slope of Colombia and N Ecuador, and with one record for C Panama. Very similar overall to *N. erythrops* but has a darker gray cap; dull buffy-brown flanks, buffy-white abdomen, crissum, and wing-linings; bill green with orange-red base, black tip; legs orange to reddish-brown. Sighting by Pratt et al. near Hitoy Cerere may have been this species, which may range to SE Costa Rica.

OCELLATED CRAKE

Micropygia schomburgkii Pl. 6(7)

Polluela Ocelada

DESCRIPTION. 6″ (15cm); 32g. Small, with white-spotted upperparts, small bill and feet. Throat white; face, sides of neck, breast, sides, and crissum rich buff; belly whitish; forehead rufous; upperparts buffy-brown, thickly spotted with white, each spot rimmed with black; remiges brown. Iris red; bill black; legs and toes salmon. Males larger than females.

HABITS. In Brazil inhabits dense grassland, is highly territorial; moves largely under vegetation along rodent runways, bursting into brief, low, fluttering flight only if suddenly startled at very close range; forages by pecking on ground for beetles, ants, grasshoppers, roaches, etc. (Negret and Texeira; in Brazil).

VOICE. A hissing trill, "like oil sizzling in a frying pan," may be an alarm call; males add 2 (or more?) *crrraaauuu* notes (in Brazil). A clear, descending rail-like whinny heard in savanna at Buenos Aires may pertain to this species.

NEST. A laterally flattened, hollow ball of dry grasses, lined with feathers; entrance on upper front (45° above horizontal); in dense vegetation 20″ (50cm) above ground. Eggs 2, dull white (in Brazil; no Costa Rican record).

STATUS. Known in Costa Rica from one specimen collected in savanna at Buenos Aires (Dickerman). Possibly more common and widespread but overlooked in Térraba region.

RANGE. Costa Rica to E Peru, Bolivia, and SE Brazil.

NOTE. Sometimes placed in genus *Coturnicops*.

COMMON GALLINULE

Gallinula chloropus Pl. 7(2)

Gallareta Frentirroja

DESCRIPTION. 13″ (33cm); 365g. Superficially ducklike; blackish, with slightly lobed toes. **Adults:** head and neck black, shading to dark slate over rest of body; mantle tinged with brownish; rump and tail black; crissum black in center, conspicuously white at sides; white stripe along side overlaps closed wing. Bill and frontal shield scarlet, bill tipped with yellow; legs greenish, except tips of tibiae vermilion. **Immatures:** throat and indistinct superciliaries white; rest of head and neck sooty-gray; upperparts browner than adult; bill dusky or yellowish; frontal shield small or absent. **Young:** natal down black; bill, bare forehead, and tips of wings reddish.

HABITS. Frequents marshes, ponds, slow streams with abundant aquatic vegetation and taller, denser cover on adjacent shores; swims buoyantly with nodding head, plucking food from water surface or reaching below, immersing head; occasionally dives; walks over ground or floating vegetation and climbs over reeds and bushes; eats seeds, leaves, roots, snails, insects, and worms; often in family groups with gray immatures helping parents to feed younger siblings; flies low, laboriously, infrequently; when alarmed, dashes over surface to cover.

VOICE. Varied loud, harsh calls; most frequently a nasal barking *nep*! or *bip*!, a deeper explosive *kup*, a low *kloc-kloc-kloc*, a strident *kr-r-r-r, kruc-kruc*, and other whinnying, chattering, shrieking notes.

NEST. A bowl of aquatic vegetation, usually with roof of bent-over cattail or sedge plants, in rooted or floating water plant or bush beside shore. Eggs 3–6, deep buff, spotted and blotched with brown and lilac. Mainly in rainy season in Guanacaste, March–June in Valle Central.

STATUS. Resident, with possibly some migrants during northern winter, countrywide from lowlands to ca. 5000ft (1500m); uncommon and local in most areas.

RANGE. S Canada to N Chile and N Argentina, including West Indies. Widespread in Old World and on many oceanic islands.

NOTE. Also called Common Moorhen, an inappropriate name for New World populations.

PURPLE GALLINULE

Porphyrula martinica Pl. 6(15)

Gallareta Morada (Calamón Morada, Gallina de Agua)

DESCRIPTION. 13″ (33cm); 235g. More colorful, less aquatic than Common Gallinule; slender, gawky build. **Adults:** head, neck and breast deep violet-blue; belly dull black; crissum white, conspicuous; back, rump, and tail dark bronzy-green; wings greenish-blue; frontal shield sky-blue. Bill scarlet with yellow tip; legs yellow. Male larger. **Immatures:** above brownish, shading to greenish-blue on wings; below buffy-white. Bill and small frontal shield dusky. **Young:** natal down black, above speckled with white; tips of wings white; bill red with black median ring and tip.

HABITS. Frequents ponds and lagoons with floating and emergent vegetation, marshes, grassy shores; swims infrequently, more often walks over floating plants, climbs through bushes and trees; eats fruits of water lilies and other aquatic plants, berries on neighboring shores, rice and other grains, aquatic insects and other invertebrates, small frogs and fishes; may take eggs or young of jacanas; found singly or in family groups containing immatures who help parents to feed and protect younger siblings.

VOICE. A variety of mostly harsh or reedy cries: a nasal, shrill *pipit pipit pipit peee-pit peee-pit*, etc., a sharp chickenlike cackle when startled, a loud shouting *gheeek!* rather like limpkin; other tooting, squawking notes.

NEST. A broad, shallow saucer or cup of dry or green grasses and other marsh vegetation, well concealed ca. 1–4ft (0.3–1.2m) above or beside water, often roofed over, and sometimes approached by long runway. Eggs 3–7, usually 4 or 5, buffy, speckled all over with chocolate and pale lilac. Breeding season prolonged at sites with permanent water; wet season in Guanacaste.

STATUS. Locally common resident countrywide in suitable habitats from sea level to at least 5000ft (1500m).

Range. E USA to N Chile and N Argentina; West Indies.

AMERICAN COOT

Fulica americana Pl. 7(1)

Focha Americana

Description. 14″ (35cm); 650g. Rather ducklike; blackish with white down-pointed bill, lobed feet. **Adults:** head and neck black, shading to dark gray on body and wings; belly mottled with whitish; secondaries tipped with white; sides of crissum white; tail black; white frontal shield tipped with maroon. Iris red; bill white, with a small dark spot near tip of each mandible; legs greenish. **Immatures:** body browner; head and neck whitish; bill dull yellowish; no frontal shield. Lacks white streak on side, as on smaller Common Gallinule.

Habits. Frequents calm, open water with plenty of algae and other aquatic vegetation, often in groups; dabbles and tips up like a duck; dives in deeper water for algae, leaves of aquatic plants, and small animals; may leave water to graze on tender grass shoots, seeds, and grain. Swimming birds constantly nod and pump the head; patter over surface to take flight.

Voice. Rather silent in winter; sometimes resonant nasal grunts and cackles, especially when disturbed.

Nest. A shallow cup or saucer woven of dried cattail leaves, low over water, supported by bent-over cattail leaves, or on floating, anchored masses of dead plants. Eggs ca. 6, buffy, thickly and finely spotted with dark brown.

Status. Primarily a migrant and winter resident, from October through April; widespread in lowlands of both slopes, up to 5000ft (1500m) in Valle Central (Ochomogo), but large concentrations only in Tempisque basin, where it has bred at least once in recent years at Palo Verde (Sánchez). Evidently a few birds may remain if lagoons do not dry up in latter part of dry season.

Range. Breeds mainly from C Canada to N Mexico, sparingly and locally to Costa Rica; also West Indies. Winters from coasts of SW Canada and USA to N Colombia; resident in the Andes from Colombia to N Chile.

FAMILY **Heliornithidae:** Sungrebes

Of the 3 species of sungrebes, also known as finfoots, the largest, measuring 24″ (60cm), inhabits Africa south of the Sahara, and the smallest, only half as long, spreads over much of tropical America, whereas that of intermediate size is confined to southeastern Asia. All have fairly long, straight bills, long necks and tails, short legs, and lobately webbed feet, like grebes and coots. Their plumage is largely brown above and white underneath. Two species have prominently striped heads and necks.

The habitat and habits of the 2 Old World species appear to differ little from those of the American species described below. Their nests and eggs, numbering 2–5 and occasionally 7, are also similar. A nest watched by Alvarez del Toro in Chiapas, Mexico, was built by both sexes. The male incubated through much of the day and the female by night. After an incubation period of only 10½–11 days, the 2 chicks hatched in a very undeveloped state, with closed eyes, pink, almost naked skin, and rudimentary feet. When disturbed, the brooding male dropped into the water and swam away with a nestling under each wing, securely ensconced in a pocket formed by a fold of skin and feathers of the side—a unique way of carrying the young to safety.

SUNGREBE

Heliornis fulica Pl. 7(9)

Pato Cantil (Perrito de Agua, Toboba)

Description. 11″ (28cm); 115g. Size of Least Grebe but seems larger because of longer pointed wings and ample tail. **Adults:** head and neck boldly striped with black and white; upperparts olive-brown; tail blackish tipped with buffy-white; below white. Iris brown; bill flesh- to horn-color, with dusky culmen; feet whitish, boldly banded with black. During breeding season female has buffy cheeks, red eyelids and bill. **Immatures:** like adults but black on neck partly replaced by dusky mottling. Young grebes have heads striped in a different pattern, short wings, and no visible tail.

Habits. Lives on streams, rivers, and sloughs, mostly in wooded country, where snags and thickets are frequent and dense vegetation hangs into water; swims on surface, pumping head like coot; rarely dives; feeds mainly by plucking insects, spiders, small frogs and lizards from overhanging vegetation; when alarmed, swims to cover or flies low and straight with shallow wingbeats; may scramble up bank into vegetation; perches on branch over water to loaf at midday or roost at night. A pair claims a stretch of stream or shoreline as its territory.

Voice. Territorial advertisement call a series of shrill, nasal barks, successively lower in pitch: *huwee! huee! huee! hueep!* Sometimes a single bark when alarmed or agitated; near nest, a sharp *kchut*.

Nest. A rather small, flimsy saucer of sticks, lined with leaves, in thick vegetation fairly low over water. Eggs 3–4, buffy white, spotted finely and uniformly with cinnamon and pale lilac. Early wet season.

Status. Locally common resident throughout Caribbean lowlands; avoids rapidly flowing water, hence not found in hilly country; also local in Corcovado basin and Coto district.

Range. C Mexico to W Ecuador, E Peru, Paraguay, and SE Brazil.

Note. Also known as American Finfoot.

FAMILY **Eurypygidae:** Sunbittern

This is one of 4 small families (the others being the trumpeters and the seriemas of South America and the Kagu of New Caledonia) that survive from a diverse and ancient stock in the Southern Hemisphere that was divided by continental drift. The single species, the Sunbittern (described below), differs from the other families both in its strongly patterned plumage and in proportions: long bill, large head, very slender neck, long and ample wings and tail. In display, often in an agonistic context, the Sunbittern spreads its wings, with the richly colored upper surface tilted forward, and fans out its raised tail to fill the gap between them, thereby forming a semicircle of plumage, in the midst of which its head stands. Its nest has rarely been found, and the best available account of its breeding is that of a pair that nested in the gardens of the Zoological Society of London more than a century ago. Both sexes built the nest and alternately incubated an egg that hatched after 27 days. Thickly covered with short down, the chick was fed by both parents with food carried in the bill; it did not leave the nest until, when it was 3 weeks old, its feathers had expanded and it could fly to the ground.

SUNBITTERN

Eurypyga helias Pl. 6(16)

Garza del Sol (Sol y Luna, Ave Canasta)

Description. 19″ (48cm); 255g. Long bill and legs, large head on very slender neck; rather heavy body held horizontally. Head black with white superciliaries and malar stripes; throat white; neck and breast brown, vermiculated with black; belly white; upperparts barred with olive-brown and blackish; upper surface of wing with striking "sun" pattern in chestnut, yellowish-buff, black, white, olive, and gray (revealed when wings spread); tail marbled with gray and white, with 2 broad bands of chestnut and black. Iris red; eyelids yellow; bill orange with black culmen; legs orange.

Habits. Prefers swift-flowing, rocky streams but sometimes on river sandbars, slow-flowing creeks, or swamp pools, generally in forested country; alone or in pairs; hops and flits from rock to rock with great agility and wades in shallows, plucking small frogs, crayfish, crabs, insects, from rocks or water; flies with 1 or 2 rather deep, heavy flaps and a long glide on flat wings.

Voice. A ringing, clear, high-pitched whistle *eeeeeeeeeeeuree*, mostly heard in early morning. Also series of shorter whistles, a ringing *Ko way*; when disturbed, a hissing, bubbling churr.

Nest. A roughly globular mass, ca. 1ft (30cm) thick, of decaying leaves, stems, and green moss with a shallowly concave top, lined with green leaves, balanced on a slender branch beside forest stream, 6–20ft (2–6m) up. Eggs 2, pinkish-buff, with scattered spots and blotches of brown and gray. March–June.

Status. Resident in diminishing numbers on Caribbean and S Pacific slopes, mostly in foothills and adjacent lowlands, ca. 300–4000ft (100–1200m), sometimes to 5000ft (1500m) in upper Coto Brus valley.

Range. S Mexico to NW Peru and Amazonian Brazil.

ORDER **Charadriiformes:** Shorebirds, Gulls, Auks, and Allies

This large and diverse order of mostly aquatic birds can be divided into 2 main groups: the shorebirds in the broad sense, with long legs, unwebbed or slightly webbed feet, and often long bills and necks; and the more aquatic types like gulls and auks, which have the 3 front toes fully webbed as well as shorter necks and bills. All share various features of the skeleton, syrinx, foot tendons, and wing feathers; all lay small clutches of rather large eggs in mostly rudimentary nests, and have more or less precocial chicks.

FAMILY **Jacanidae:** Jacanas

Distributed throughout the tropics and subtropics of both hemispheres, the 8 species of jacanas range in length from 8 to 13″ (20–33cm), not including the very long tail feathers in the nuptial attire of the Pheasant-tailed Jacana of the Far East—the only species with seasonal changes in appearance. The most striking feature of these birds of weedy ponds, marshes, margins of lakes and lagoons, and wet pastures is the extraordinary length of their toes and toenails, which enables them to distribute their weight over floating vegetation so that they do not sink and often appear to walk over water. Their medium-length bills are straight and in several species join a yellow, red, or blue frontal plate. At the bend of the wing is a knob or sharp spur, often prominently displayed when, upon alighting, jacanas hold up their wings high above their backs. Rufous, black, or dark bronze predominate in the plumage; adults of both sexes are colored alike, but females are larger than males. Immatures are different from adults and have rudimentary frontal plates. Jacanas pick up a great variety of animal and vegetable foods as they walk over water or wet ground.

At least 2 of the species, including the Northern Jacana, are polyandrous. In the breeding season, the female defends a territory that may include those of 2–4 males. Afterward, with some help from her, a male accumulates a skimpy nest of vegetable fragments on floating water plants; she gives him usually 4 buffy eggs, scribbled all over with heavy, crisscrossing blackish lines. While he proceeds to incubate them without her help, pushing his wings beneath them to separate them from the damp nest, she provides eggs for the rest of her males. If a sinking nest or rising water threatens to inundate the eggs, the male may roll or push them for yards over the lily pads to a safer spot. After an incubation period of 22–24 days, the downy chicks hatch and are led and protected by their father, who in at least some species may carry them to safety beneath a wing, legs dangling below. The mother may help to guard them.

NORTHERN JACANA

Jacana spinosa Pl. 6(18)

Jacana Centroamericana (Cirujano, Gallito de Agua, Mulita)

Description. ♂ 9″ (23cm); 95g. ♀ 10″ (25.5cm); 130g. Slender, rail-like build; long toes and claws; round wing with sharp yellow spur at wrist. **Adults:** head, neck, and breast black; body chestnut-maroon; remiges pale yellow-green tipped with blackish. Bill and frontal plate bright yellow; cere pale blue-gray, separated by line of maroon skin from frontal plate; legs greenish. **Immatures:** crown, eye-stripe and back of neck black; rest of head, neck, and underparts white; mantle olive-brown; remiges duller, yellower than adult; frontal plate small or lacking.

Habits. Inhabits ponds, marshes, flooded pastures, and riverbanks, essentially wherever aquatic vegetation, especially water lilies, water hyacinths, or grasses extend over calm or slow-flowing fresh or brackish water; flies slowly with several shallow flaps and a glide; holds wings briefly extended above back upon alightling; walks or runs jerkily over aquatic vegetation or grass, pecking and stabbing rapidly at aquatic insects, small fishes and snails, or seeds. Breeding populations often dense, resulting in frequent territorial squabbles.

Voice. Loud, harsh, scratchy, grating, most commonly a series of *jik* notes, which sometimes accelerate into a chuckle or ''jickering'' chatter; loud *jaak* or *jeek* in flight; immatures give a softer, squeaky *cheep*.

Nest. A slight accumulation of leaves and stems forming a platform of sorts on floating aquatic vegetation. Eggs 4, occasionally 3, pale brown, covered with blackish scrawls. Breeds in wet season in Guanacaste, virtually year-round in areas with permanent water.

Status. Abundant resident (with local movements according to water levels) in Guanacaste (especially Tempisque basin) and Río Frío area; locally in smaller numbers throughout lowlands of both slopes, to ca. 5000ft (1500m) in central highlands, 3900ft (1200m) on S Pacific slope.

Range. S Texas and N Mexico to W Panama; also Greater Antilles.

Note. Occasional hybridization with *J. jacana* in overlap zone in W Panama; the 2 are sometimes considered conspecific.

WATTLED JACANA

Jacana jacana Pl. 6(17)

Jacana Sureña

Description. ♂ 9″ (23cm); 95g. ♀ 10″ (25.5cm); 130g. Resembles Northern Jacana in size, shape, and yellow-green remiges. **Adults:** head, neck, body and tail entirely black, more or less glossed with purplish. Instead of a stiff, 3-pointed frontal plate has large bifid frontal wattle with a smaller wattle at either side of bill. Wattles and base of bill crimson to magenta; rest of bill yellow-orange; legs grayish. **Immatures:** resemble those of Northern Jacana but with at least a trace of wattles.

HABITS. Similar to those of Northern Jacana.

VOICE. Calls deeper, throatier, more nasal than those of Northern Jacana, less dry and squeaky. Common call *grah-grah-grah-grah*.

STATUS. A single specimen record for Costa Rica, a male in a flooded but drying rice field, acting very aggressively toward the numerous Northern Jacanas, July 1984, at Laurel on Panama border in S Pacific lowlands (Stiles).

RANGE. W Panama (sporadically to Costa Rica) to E Peru and N Argentina.

FAMILY **Haematopodidae:** Oystercatchers

Oystercatchers are found along most of the world's seacoasts, except those of tropical Africa and southern Asia; in the Old World some occur locally on inland waters. Those that breed at high latitudes are migratory. Because the family is quite uniform and most breeding populations do not overlap, the number of species of oystercatchers is much debated, with different authors recognizing between 4 and 14! All oystercatchers are 15–20″ (38–50cm) long, with plumage either wholly blackish or boldly patterned in black and white; the sexes are alike. Their long, laterally compressed bills are bright orange-red; their eyes are yellow, ringed with red; and their long, stout legs and 3 toes (the hind toe is absent) are reddish or pink and slightly webbed. They have short tails and long, pointed wings. They fly strongly, walk and run swiftly, and swim on occasion.

Walking along sandy, rocky, or gravelly beaches or sometimes over moors and farmlands, oystercatchers feed on mollusks, crustaceans, worms, and insects. With scissorlike bills, they cut the muscles that hold bivalve shells closed and then extract the soft flesh. Active by night as well as by day, they are restless and noisy, frequently repeating loud, clear whistled notes. As the breeding season approaches, 3 or more birds join in an elaborate "piping ceremony," running along with downward-pointing bills while together they utter rapid, high-pitched trills. In a shallow scrape in the ground, often decorated rather than lined with white shells, bones, straws, or other small objects, the female lays 2–4 yellowish or grayish buff eggs, heavily marked with dark brown or black. Both sexes incubate for a period of 26 or 27 days. As soon as their cryptic down is dry, the chicks leave the nest and for some days are fed by their parents. When about 5 weeks old, they can fly. After breeding, oystercatchers, particularly the migratory species, may join in large flocks.

AMERICAN OYSTERCATCHER

Haematopus palliatus Pl. 9(9)

Ostrero Americano

DESCRIPTION. 18″ (46cm); 550g. Large, heavy-bodied; long reddish bill with laterally flattened, bladelike mandibles. **Adults:** head, neck, and chest blackish; mantle and wings gray-brown except for greater coverts and all but tips of secondaries, which are white, forming conspicuous wing-stripe that extends onto primaries of northern migrants but not local breeders; underparts and upper tail-coverts white. Iris yellow; bare eye-ring and bill red to orange; legs pink. **Young:** throat gray, mottled with dusky; mantle spotted with cinnamon; bill mostly dusky-brown; iris brownish.

HABITS. Frequents flat, sandy beaches, river mouths and estuaries, especially where rocks are exposed at low tide, mudflats, rocky headlands, and salt ponds; often walks steadily for long periods along beach, feeding mainly in intertidal zone, pecking and probing mud or sand, often belly-deep in water with head submerged; eats mainly bivalve and univalve mollusks, also other invertebrates; usually in pairs on Pacific beaches.

VOICE. Clear, strong, downslurred whistle *keeew*! often repeated in flight or on ground; sharper *wreeek*, once or several times when taking flight; when alarmed, a sharp *pik*!

NEST. A scrape in sand, decorated with bits of shells or pebbles, not far above high-tide line, usually near rock or tuft of grass. Eggs 2–3, grayish-buff, spotted uniformly and thickly with dusky and gray. Late dry season through start of rains.

STATUS. Resident in small numbers on remote beaches and offshore islands the length of Pacific coast. Migrant on both coasts in fall (August–early October) and spring (April–May); uncommon winter and nonbreeding summer resident on Pacific coast, mainly in Golfo de Nicoya.

RANGE. Breeds along Atlantic-Caribbean coast from NE USA to S Mexico and from Venezuela to Argentina; and on Pacific coast from NW Mexico to C Chile; also Bahamas and West Indies. Northern populations winter to N South America.

NOTE. Sometimes considered conspecific with Old World *H. ostralegus*.

FAMILY **Recurvirostridae:** Avocets and Stilts

With about 7 species, this family is found in temperate and tropical regions over most of the world; northern populations are migratory. Avocets and stilts are 12–20″ (30–50cm) long. All have extremely long legs, those of the Black-necked Stilt longer in proportion to body size than the legs of any other birds except flamingos. Their toes are partly webbed. Their long, slender bills may be almost straight, as in stilts; strongly upcurved, as in avocets; or downcurved, as in the Asiatic Ibisbill. They have small heads, long necks, long pointed wings, and short tails. Their plumage is predominantly black and white in bold patterns, with cinnamon or rufous areas in a few species. The sexes are alike or nearly so, and seasonal changes in coloration are slight. They fly with extended necks and trailing legs, walk with long, graceful steps, and swim well.

These shorebirds frequent shallow lakes, pools, marshes, mudflats, flooded fields and pastures, in fresh, brackish, or salt water. Stilts pick insects and other small creatures from the surface of water, into which they may wade up to their bellies, or from wet ground, where they must flex their long legs to reach their food. Avocets search for small crustaceans, aquatic insects, and seeds as, advancing in flocks, they sweep their open bills from side to side in soft mud or water, sometimes immersing their heads. Social and noisy, avocets and stilts proclaim their presence by yelping or barking cries or loud, shrill whistles, but softer notes are sometimes heard.

Stilts and avocets breed more or less colonially in or beside shallow water, where sheltering vegetation may or may not be present. In dry spots or upon matted plants, little or no nest is made, but in muddy or wet spots the birds build more substantial nests of sticks, shells, or whatever. If rising water threatens to inundate their eggs, they continue to add materials until they have a high pile or until the nest floats. Four eggs are usual, but sets of 2–8 (the latter doubtless laid by more than 1 female) are found. The buffy, yellowish, or olive eggs are spotted and blotched all over with black, deep brown, and pale lilac. Both sexes incubate for periods of 22–26 days. Soon after hatching the downy young leave the nest to be brooded and guarded by the parents, who perform a variety of distraction displays in front of intruders. Sometimes many members of the colony participate in such displays.

BLACK-NECKED STILT

Himantopus mexicanus Pl. 9(12)

Cigüeñuela Cuellinegro (Soldadito)

DESCRIPTION. 15″ (38cm); 150g. Very slender; long thin neck, legs, and bill. **Adults:** top and sides of head, hindneck and upperparts black (back tinged with brownish or gray in female); forehead, lores, eyespots, entire underparts, and rump white; tail pale gray. Iris red; bill black; legs pink. **Young:** crown and nape dusky, mottled with white; some white scaling and mottling over rest of grayish-black upperparts.

HABITS. Seeks shallow salt or fresh water with a soft, muddy bottom in ponds, lagoons, salinas, sometimes tidal flats and estuaries; very social, usually in loose groups of up to 50; feeds actively in water up to belly-deep, sweeping with bill, plucking from surface of water or mud, or probing in soft mud for small aquatic insects, crustaceans, and mollusks. Flight slower, more gliding than that of other shorebirds.

VOICE. Very noisy, excitable; when disturbed or alarmed, an incessant nasal *pep* or *yip*; at night, a clear, high-pitched, repeated *peek*; in distraction display, a grating *jaahr*.

NEST. A shallow depression or scrape in low dike or mud mound, lined sparsely to rather substantially with twigs, stems, or leaves. Eggs 4, rarely 3, olive, spotted boldly with dark brown and gray. Often nests in loose colonies; upon approach of observer, many birds may give broken-wing distraction displays. April–July.

STATUS. Locally common permanent resident, principally around Golfo de Nicoya. Elsewhere migrant and winter resident (October–May) common to abundant locally in Guanacaste (especially Tempisque basin) and Río Frío region; sporadic and in small numbers elsewhere in lowlands of both slopes.

RANGE. Breeds from W and SE USA to S Chile and S Argentina; also Bahamas and West Indies; northern populations migratory, winter from S USA to N South America.

Note. Sometimes lumped with the Old World *H. himantopus*, the complex then called Common Stilt.

AMERICAN AVOCET
Recurvirostra americana Pl. 9(11)
Avoceta Americana

Description. 18″ (46cm); 350g. Large; slender; pied upperparts; slim recurved bill. **Adults:** head and neck pale gray (cinnamon in breeding season); center of back, scapulars, tips of greater coverts, secondaries, and underparts white; rest of upperparts black; rump and tail pale gray. Bill black; legs blue-gray. **Young:** similar, but head and neck suffused with tawny, mantle spotted with cinnamon-buff.

Habits. Singly or in small groups in freshwater marshes and salinas; feeds by walking in shallow water with head and neck extended, sometimes submerged, and sweeping bill from side to side over soft muddy bottom or water surface to capture aquatic insects, small crustaceans, seeds.

Voice. Sharp whistled *keep* or *keek*.

Status. Casual winter visitor to NW lowlands (Tempisque basin, vicinity of Golfo de Nicoya); 4 records, December–May.

Range. Breeds in W North America from S Canada to California, NE Mexico; winters from S USA regularly to S Mexico, rarely or casually to Costa Rica.

FAMILY **Burhinidae:** Thick-knees

Of the 9 species of thick-knees, also known as stone-curlews, 7 are widely distributed over the Eastern Hemisphere from England to South Africa and Australasia, while 2 have a more restricted range in the New World, from Mexico to Peru and Brazil, and on Hispaniola. Those that breed farthest north are migratory. In length 14–21″ (36–53cm), they are long-legged cursorial birds with thick ankle joints commonly mistaken for knees. Their 3 short, forwardly directed toes are webbed at the base; they lack a hind toe. Their rather thick bills range from short and straight to longer than the head, which is notably large, with big eyes for nocturnal vision. Their plumage, alike in the sexes, is brown, gray, or buffy, spotted and streaked. Birds of open spaces, they prefer sandy or stony terrain, savannas, bushy deserts, or riverbeds and seashores. They stand tall and walk with a stilted gait; when alarmed, they run fast, fly swiftly and low, or flatten on the ground with outstretched neck to avoid detection. By day they rest and preen in silence; in twilight they become active, to forage by night for insects, worms, mollusks, lizards, frogs, seeds and other vegetable matter. Especially when the moon shines, they are noisy, uttering a variety of piping, cackling, or croaking sounds.

Thick-knees lay their eggs on the bare ground of an unlined scrape, which some species decorate with pebbles, fragments of wood, or other small objects. Their 1–3 eggs are creamy, buffy, pale brown, or dark gray, spotted, streaked, or heavily blotched with black, brown, or pale lilac. Both sexes incubate, sitting upright, often with the off-duty partner standing or sitting nearby. After 25–27 days, the downy young hatch and soon leave the nest, led and guarded by both parents.

DOUBLE-STRIPED THICK-KNEE
Burhinus bistriatus Pl. 9(10)
Alcaraván Americano (Alcaraván)

Description. 20″ (50cm); 780g. Large; erect stance; long legs; small feet; large head with big yellow eyes; short, stout bill. **Adults:** broad black stripe on side of crown; broad white superciliaries; throat white; rest of head, neck and upperparts streaked with brown and buff; chest gray; belly white; crissum buff; lateral rectrices white, barred with brown and boldly tipped with black; remiges dark brown; pale gray bases of secondaries and white bases of inner primaries form short wing-stripe; white spot at base of outer primaries. Bill dusky, the base yellowish; legs yellowish. **Young:** similar but cinnamon spotting gives mantle a variegated appearance.

Habits. Active principally at night on savannas, pastures, stubble fields, burned areas, and openings in scrubby woodland; rests quietly, loafs and preens by day; in groups or loose flocks most of year; wary, often crouches and freezes in alarm; if approached closely, runs away, or less often lifts into straight but not rapid flight; with quick lunge or peck, picks up insects, worms, snails, scorpions, small reptiles and frogs, also some seeds and buds.

Voice. Loud, reedy, nasal trills and sputters: *pip pip prrrrripipip, prrrrip prrrrip pip pip pip pipipipip*, often long-continued—a characteristic sound of moonlit Guanacaste nights.

Nest. An unlined scrape on bare, open ground. Eggs 1 or 2, pale buffy-olive, boldly spotted with black, pale brown, and gray. January–April.

Status. Fairly common resident, at least locally, throughout Guanacaste lowlands, S sparingly to vicinity of Puntarenas.

Range. S Mexico to Costa Rica; Hispaniola; and South America from N Colombia to NE Brazil.

FAMILY **Charadriidae:** Plovers and Lapwings

With about 63 species, this family is distributed worldwide except Antarctica. In length 6–16″ (15–40cm), its members are compactly built, thick-necked, long-winged, rather short-tailed, terrestrial birds. With 3 forwardly directed toes and the hind toe much reduced or lacking, they run swiftly; they also fly well. Their short, usually rather thick bills are typically slightly swollen at the tip. Their predominantly gray and brown plumage is often boldly patterned, especially on head, neck, and breast, with black and white or, less often, chestnut or rufous. Lapwings, often prominently crested, frequently bear a sharp spur at the bend of the wing. The sexes are alike or nearly so, females usually being larger. Seasonal changes in coloration may be great, especially in the highly migratory species that breed in the far north, some of which make amazingly long flights over water. Plovers and lapwings frequent open spaces, including shores, cultivated fields, pastures, barren plains, moors, and the like, where from the ground they gather insects, spiders, mollusks, crustaceans, seeds and other vegetable matter. Their calls, heard by night as well as by day, are often melodious and far-carrying. After the breeding season, they gather in flocks.

A shallow scrape, with perhaps a few pebbles or fragments of plants placed in or around it, in any sort of open space, serves as a nest. Here the female lays 2 to (usually) 4 eggs (reported sets of 5 may mean that more than 1 female has laid in the scrape). The eggs are pyriform (i.e., the small end more or less conical) in shape and are usually some shade of buff or brown but may be olive, creamy, greenish-blue, or pinkish and are more or less heavily spotted, blotched, or scrawled with deep brown or black. During an incubation period of 24–28 days, both parents take turns on the eggs, with sometimes the male and sometimes the female doing the major share. Soon after they dry, the downy chicks leave the nest and are led by both parents while they pick up their own food. When they are threatened, their guardians try to lure the intruder away by spectacular distraction displays. When 25–40 days old, the young can fly.

LESSER GOLDEN-PLOVER

Pluvialis dominica Pl. 9(2)

Chorlito Dorado Menor

Description. 10″ (26cm); 150g. Medium-sized; dark rump; no wingstripe; pale gray underwing. **Adults** Winter: above dusky-brown, speckled with whitish to dull buff; forehead and superciliaries white; contrasting dusky ear-patch; sides of head, throat and breast whitish, mottled faintly with gray-brown, shading to white on belly. Bill black; legs dusky. Breeding: upperparts dark brown, spotted with golden-buff, separated from black face and underparts by broad white stripe from forehead to sides of breast. **Young:** similar to winter adult but above with more contrasting yellow-buff spotting; below streaked and mottled with whitish and pale grayish-brown.

Habits. On migration and in winter, single birds or small loose groups frequent open, close-cropped pastures, soccer fields, or plowed fields, especially those with wet spots or scattered puddles, often with Killdeers or Upland and Pectoral sandpipers. While migrating in spring, they mingle with Black-bellied Plovers in salinas. When foraging, they run, peer and peck, or scurry in chase of insects, beach fleas, and other invertebrates.

Voice. Clear whistled *keet*! or *pleet*! or doubled *keeleeet*! especially on taking flight.

Status. Rare fall migrant (mid-August to November); rare and local winter resident; uncommon to sporadically common spring migrant (April–May); occasional nonbreeding birds remain for summer; locally in lowlands of both slopes, to 5000ft (1500m) in central highlands; more numerous on Pacific slope.

Range. Breeds in N Canada, N Alaska and N Siberia; winters mainly in S South America, casually N to California, in Polynesia, and at S latitudes in Old World.

Note. Also called American Golden-Plover. The population breeding in Siberia and W Alaska may be a distinct species, *P. fulva*, Asiatic Golden-Plover; winters in Indo-Pacific area, accidental off Mexico; unlikely in our area.

BLACK-BELLIED PLOVER

Pluvialis squatarola Pl. 9(1)

Chorlito Gris (Avefría)

Description. 12″ (30cm); 190g. Fairly large, chunky, stout-billed; white rump and wing-stripe; black axillars in all plumages. **Adults** Winter: above dusky-brown, mixed with pale brown and whitish; forehead, eyebrow, and throat white, mottled with dusky; dusky ear-patch; rest of face, neck and breast pale gray-brown, mottled with white; belly and crissum white; tail white, barred with black. Bill black; legs grayish. Breeding: above marbled with black, pale gray and white; below black, except belly white; broad white stripe from forehead to side of breast; this plumage attained by April. **Young:** similar to winter adult but breast more heavily marked, upperparts boldly spotted with yellowish to buffy.

Habits. Prefers mudflats, salinas, sandy beaches, estuaries, and sometimes mangroves or rocky shores; during migration also found on freshwater ponds and grassy fields; may defend individual feeding territories or forage in loose groups, gathering into large flocks to loaf and sleep, often with other shorebirds, especially Willets and Knots; when foraging walks about, making sudden pecks or swift scurrying chases to catch crabs, small crustaceans, insects, marine worms.

Voice. A melodious, prolonged, mournful, slightly "breathy" whistle: *peeeuweeee* or a shorter *wheeeeur* of similar quality.

Status. On Pacific coast a common to abundant fall (August–October) and spring (March–May) migrant and winter resident, with centers of abundance in Golfo de Nicoya, Golfo Dulce, and various river mouths; fairly common nonbreeding summer resident (mostly young birds), mainly around Golfo de Nicoya; much less common on Caribbean coast, mainly at river mouths; occasional inland, mostly during fall migration.

Range. Breeds in N Alaska and N Canada; winters mainly along coasts from SW Canada and NE USA to C Chile and N Argentina. Widespread in Old World.

Note. Called Grey Plover in Old World.

SEMIPALMATED PLOVER

Charadrius semipalmatus Pl. 10(5)

Chorlitejo Semipalmado (Chorlito, Turillo)

Description. 7″ (18cm); 42g. Small, chunky; abrupt forehead; stubby bill; prominent wing-stripe. **Adults** Winter: forecrown, orbital area, stripe from bill to cheeks and breast-band (sometimes incomplete) dusky; above grayish-brown; forehead, nuchal collar, and underparts white; tail with dark brown subterminal band and white tip. Bill orange, tipped with black; eyelids and legs orange. Breeding: dusky of head and breast-band replaced by black; pattern more clean-cut, breast-band complete. **Young:** above scaled with whitish; dark areas of head and breast browner, indistinct; bill mostly black. Collared Plover has different shape of head, slender bill, no nuchal collar.

Habits. Favors mudflats, salinas, mangroves, sandy and rocky shores (especially at river mouths), occasionally visits freshwater habitats, mostly during migration; sleeps and loafs at high tide in compact flocks, often with other shorebirds; feeds singly or in loose groups on wet intertidal sand or mud, running a short distance, stopping abruptly, and pecking at small mollusks, crustaceans, and marine worms; when inland, takes insects and earthworms. Flight fast, erratic, in looser flocks than small sandpipers.

Voice. Mellow but often loud *cheweep* or *ch'weet*! Song a series of short notes accelerating into a chuckling trill reminiscent of Black-headed Trogon, often given in spring passage.

Status. Abundant migrant in fall (early August–November) and spring (late March–early May) on Pacific coast, uncommon on Caribbean coast, occasional inland, to 5000ft (1500m) in central highlands; common to abundant winter resident around Golfo de Nicoya and Golfo Dulce; locally and in smaller numbers elsewhere on both coasts, especially at river mouths. Small to moderate numbers of nonbreeders, mostly first-year birds, summer around Golfo de Nicoya.

Range. Breeds in Alaska and N Canada; winters along coasts from SE and SW USA to C Chile and S Argentina.

Notes. Sometimes lumped with the Old World *C. hiaticula*. Also possible along the Caribbean coast, especially after fall tropical storms, is Piping Plover (*Charadrius melodus*): similar in size, shape, and pattern to Semipalmated, including orange or yellow legs, eye-ring, and (mainly in summer) base of bill; but much paler above in all plumages, sandy-buff to pale grayish-buff; breast-band often incomplete. Told from similarly pale Snowy Plover by leg color, face pattern. Prefers sandy beaches; call a distinctive, penetrating *peep* or *pee-low*. Breeds C Canada to NE USA; winters along Atlantic-Gulf coasts regularly S to E Texas and Greater Antilles; has occurred in Belize.

WILSON'S PLOVER

Charadrius wilsonia Pl. 10(2)

Chorlitejo Picudo (Chorlito Gritón, Turillo)

DESCRIPTION. 8″ (20cm); 55g. Largest plover with single breast-band; large eyes, squarish head, very heavy bill. **Adults** WINTER: above grayish-brown, tinged with cinnamon above and behind eye; loral stripe, forecrown, auriculars and breast-band (narrow in center) dusky; forehead to eye, trace above and behind eye, narrow nuchal collar, underparts and narrow wing-stripe white. Bill black; legs flesh-color. BREEDING ♂: forecrown, lores, and clean-cut breast-band black. ♀: extensive cinnamon on crown, sides of head and neck, sometimes breast-band. **Young:** indistinct, incomplete brown breast-band; above with faint buffy scaling.

HABITS. On beaches, coastal mudflats and salinas, estuaries, sometimes sandbars and muddy banks of large rivers near coast; forages on wet or dry mud or sand, singly or in loose groups, with short dashes capturing small crabs, its favorite food, and other invertebrates; sleeps and loafs in compact groups, often snuggling into small depressions (footprints, etc.) in dry sand or mud, frequently with other plovers; less wary than Collared Plover.

VOICE. Usual note a sharp whistled *whit*! or *chidit*! with hard wooden quality; when alarmed, a shriller *cheet*! or *chewit*! Before and during breeding season, in aggressive encounters (males only?), a fast, gravelly song: *w'chi' diridit'-widit'*, etc.

NEST. A scrape in loose sand or earth, often on dike in salina, usually lined with a few bits of wood or shell. Eggs 2–3, pale brownish, densely spotted with dark brown and grayish, heaviest toward large end. Pairs form early in dry season; nests February–May or June.

STATUS. Permanent resident along Pacific coast, breeding commonly around Golfo de Nicoya, locally elsewhere on coast; also migrant and winter resident on both coasts, locally abundant on Pacific (Golfo de Nicoya, Golfo Dulce) side, uncommon on Caribbean, where a few are present year-round but breeding not recorded.

RANGE. Breeds from Baja California and E USA to Panama and NE Brazil, also through West Indies; on Pacific coast winters to NW Peru.

NOTE. Also (and more appropriately) called Thick-billed Plover.

KILLDEER

Charadrius vociferus Pl. 10(1)

Chorlitejo Tildío o de Dos Collares (Pijije, Tildío)

DESCRIPTION. 10″ (25cm); 95g. Fairly large, slender, long-tailed; two black breast-bands. **Adults:** crown, side of head, and mantle grayish-brown, often some rufous in wing-coverts; rump and tail cinnamon-orange; tail tipped with black and white; forehead and nuchal collar white, bordered with black; white spot above and behind eye; underparts and wing-stripe white. Eyelids red; bill black; legs flesh-color. **Young:** similar but feathers of mantle marked with dusky subterminally and fringed with buff.

HABITS. Prefers plowed fields, lawns, close-cropped savannas or pastures, muddy borders of ponds and puddles, and other inland habitats; forages in loose groups; gathers into larger flocks to sleep, although sometimes active on moonlit nights; eats insects, worms, tadpoles, isopods; alert and noisy, calling freely on sighting humans or other danger, even at considerable distance. Flight rapid but with slower wingbeats, less erratic maneuvering than smaller plovers.

VOICE. Loud, querulous, somewhat scratchy or nasal *deeeah* or *deeer*, often introduced by short notes: *didideeeah*. In alarm or excitement, a repeated *killDEEa* or *killDEE*.

NEST. A scrape in dry mud or sandy ground. Eggs 3–4, buffy, densely and boldly spotted with blackish. May–June.

STATUS. Locally common winter resident and migrant from lowlands to middle elevations on both slopes, arriving late August–September, departing April–May. Small numbers present year-round and breeding in Cartago area, probably elsewhere in central highlands. Casual in migration to 10,000ft (3000m).

RANGE. Breeds from C Alaska and N Canada to C Mexico, Bahamas, Greater Antilles, and in Costa Rica; winters from SW Canada, C and E USA to N South America; also a resident population in W South America.

SNOWY PLOVER

Charadrius alexandrinus Pl. 10(4)

Chorlitejo Patinegro (Chorlito, Turillo)

DESCRIPTION. 6″ (16cm); 40g. Small, plump, pale-backed and plain-breasted. **Adults** WINTER: above pale brownish-gray; patch on side of breast and postocular stripe darker; forehead, superciliaries, nuchal collar, wing-stripe, and entire underparts white. Bill black;

legs blackish. **Breeding:** bar on forecrown, postocular stripe, and patches on sides of breast black (male) to fuscous (female). **Young:** above paler with buffy feather-edgings; no dark patches on head or breast.

Habits. Runs about very actively on dry sand or mud above high-tide line along coast, feeding on insects, small crustaceans, and other small invertebrates; usually singly or in 2s, with other small plovers.

Voice. A rather low, mellow, whistled *pi-pee-pi*, middle note loudest, last note sometimes omitted.

Status. Very rare transient on both coasts in spring (late March) and fall (September–November); no winter records.

Range. Breeds locally along Pacific coast from Washington to Oaxaca; inland in W USA; from W Florida to Tamaulipas along Gulf of Mexico; through West Indies to N Venezuela; and on Pacific coast of South America. Northern populations winter S to Panama. Widespread in Old World.

Note. New World populations differ in size and color from those of Old World and probably represent a separate species, *C. nivosus*.

COLLARED PLOVER

Charadrius collaris Pl. 10(3)

Chorlitejo Collarejo (Turillo)

Description. 7″ (18cm); 35g. Trim, long-legged; sloping forehead; rather long slender bill; no nuchal collar. **Adults:** above grayish-brown, tinged with rufous on crown, sides of head and neck; bar on forecrown, loral streak, and breast-band black; auriculars dusky; broad forehead, spot behind eye, and underparts white; remiges darker; wing-stripe faint or absent. Bill black; legs flesh-color. **Young:** black and rufous areas on head reduced or lacking; breast-band dusky, less distinct; upperparts broadly scaled with buff. Distinguished from Wilson's Plover by slender bill, smaller size, different head shape and pattern.

Habits. Lives above high-tide line on sandy, flotsam-strewn beaches, mud and gravel bars of lowland rivers, and mudflats around drying ponds, usually singly or in pairs, sometimes loosely associated with other small plovers, occasionally in flocks; runs for short distance, stops abruptly, looks, then pecks quickly for insects, beach fleas, and other small invertebrates. Notably wary, runs swiftly, then freezes amid rocks or flotsam; flies low and straight.

Voice. Sharp, dry, whistled *prit*!, *pip*!, or *pitur*!; when alarmed, a sharp metallic *tsee*. A short, trilling song before and during breeding season.

Nest. A scrape in sand or dry soil, lined with shells, wood chips, and other bits of debris, often beside strand of beach-pea or tuft of grass. Eggs 2, pale buff, spotted boldly with dark brown. March–June.

Status. Uncommon and local but widespread resident of lowlands along both coasts, breeding mainly on beaches, gravel and mud bars near river mouths; may occur farther inland after breeding.

Range. NW and E Mexico to W Ecuador and C Argentina, mainly in coastal areas.

FAMILY **Scolopacidae:** Sandpipers and Allies

Some 85 species of this family are well distributed over the globe, except Antarctica. The greatest number of species nest in the far north, mostly in tundra, marshland, or moors, sometimes in woodland near water. However, the birds spend most of their year along the coasts or on inland wetlands over a wide range, from the northern temperate zone to high southern latitudes. Nonbreeding yearlings of most species spend their first summer well south of the nesting grounds; no species of the family breeds in Costa Rica. In size, sandpipers and their allies are 5–25″ (13–64cm), including bills that are sometimes short but mostly medium to long, slender, straight or curved, and in 1 species expanded at the tip. Legs tend to be long, with the rather long front toes slightly webbed or unwebbed, the hind toe reduced (lacking in 1 species). They have long, pointed wings, small heads, and short to medium tails.

In winter most species are brownish above, more or less mottled and streaked, and whitish to buffy below; most have distinct breeding plumages that include brighter colors (often rufous or buff) and/or more contrasting patterns. Usually this bright plumage is acquired just before or during spring migration and is mostly lost by fall migration; males are often brighter in breeding dress, and in most species are smaller than females. Most breed as territorial pairs but are highly gregarious in migration and in winter, at least for sleeping; in some species, individuals frequently hold feeding territories. Flocks in flight are often compact and beautifully coordinated, executing swift maneuvers in unison as though moved by a single impulse.

These birds subsist largely upon small invertebrates found by probing into mud or sand, often in shallow water, gleaned from the surface, or caught in the air. Their frequently melodious calls include piping and trilling and far-carrying cries. Many have spectacular aerial courtship displays, sometimes accompanied by nonvocal sounds.

Except for a few that occupy old, elevated nests of other birds, members of this family breed on the ground, in a scrape that may be bare or lined with grasses, dry leaves, or lichens and is sometimes canopied by surrounding rooted plants that the birds pull over it. The 2 to (usually) 4 buffy or olive eggs are heavily mottled. Incubation, by the male alone, the female alone, or both sexes alternately, takes 18–24 days, occasionally longer. Soon after hatching, the downy chicks leave the nest and are led by either or both parents while they pick up their own food. Exceptionally, as in snipes and woodcocks, young are fed directly by their parents until their bills grow longer.

HUDSONIAN GODWIT

Limosa haemastica Pl. 9(7)

Aguja Lomiblanca

Description. 15″ (38cm); 30g. Long, slightly upturned bill; striking black-and-white pattern on wings and tail. **Adults** Winter: above brownish-gray; crown and eye-stripe darker gray; superciliaries and underparts grayish-white, with brownish wash on foreneck and breast; rump white; tail black with white tip; wings largely black with bold white wing-stripe. Bill black with pinkish base; legs dusky. Breeding: below chestnut (male) or mixed chestnut and gray (female), barred with black. **Young:** like winter adult but below tinged with buffy, above scaled with buffy. Bill and black tail distinguish this species from Willet in winter.

Habits. Prefers lake and pond shores, inlets, and estuaries; the 2 birds seen in Costa Rica were circling over a coastal lagoon. Feeding birds probe deeply and forcefully in mud with bill, sometimes submerging head in water like a dowitcher.

Voice. A soft *ta-wit* in flight.

Status. One sight record on Pacific coast in spring migration—a pair at Chacarita, near Puntarenas, 25 May 1975 (Stiles and S. M. Smith). Scattered spring records on Pacific coast of Central America suggest a long migratory flight over E Pacific at this season.

Range. Breeds in Alaska and Canada; winters from C Chile and Paraguay to Tierra del Fuego.

MARBLED GODWIT

Limosa fedoa Pl. 9(8)

Aguja Canela

Description. 19″ (48cm); 375g. Large; long, upturned bill; mostly buffy with rusty wing-linings. **Adults:** head, neck, and underparts mostly cinnamon-buff; crown streaked with black; sides barred with black; above blackish-brown, heavily barred and spotted with cinnamon-buff to (breeding) pale rufous. Bill black with pink base; legs dull gray to blackish. **Immatures:** like adults but below with less black barring; wing-coverts edged with buff. Whimbrel browner with striped head; Long-billed Curlew more sandy-buff; both have decurved bill without pink, more horizontal stance.

Habits. Prefers mudflats for feeding, probing deeply (often inserting bill full length) for worms, crustaceans, and small clams; occasionally found on sandy beaches, especially at river mouths; forages singly or in small, loose groups; gathers in larger flocks, often with Whimbrels, for sleeping and loafing on dikes in salt ponds, mangroves.

Voice. Loud, nasal *kaWHIT*! or *ya-woik*! When disturbed, a harsh, scratchy *kurrk*; in loafing flocks, a softer *ka-wicka*.

Status. Locally common migrant (August–September and late March–April), uncommon winter resident, rare but regular nonbreeding summer resident on Pacific coast, principally around Golfo de Nicoya, smaller numbers in Golfo Dulce. Not yet recorded from Caribbean coast.

Range. Breeds mainly on Great Plains of S Canada and N USA; winters along coasts from C California and South Carolina to Colombia and N Chile.

WHIMBREL

Numenius phaeopus Pl. 9(14)

Zarapito Trinador (Cherelá, Zarceta)

Description. 17″ (43cm); 375g. Large, chunky; long, decurved bill; conspicuously striped head. **Adults:** head and neck pale buff, finely streaked with brown; broad blackish-brown stripes on each side of crown and through eye; upperparts dull brown, flecked with buffy-brown; remiges barred with buffy-brown; below pale dull buff, streaked on chest and barred on sides with brown. Bill blackish, tinged with flesh-color at base; legs dark gray.

Young: similar but more boldly spotted with buff on margins of scapulars and wing-coverts. Long-billed Curlew larger, paler, with much longer bill, no head-stripes.

Habits. Scatters singly over mudflats, edges of tidal creeks, sandy or rocky shores, occasionally flooded pastures near coast; walks with body horizontal, snatching, poking, or probing for crabs and other crustaceans, mollusks, marine worms, and insects. Gathers on salt-pond flats and dikes, in mangroves, in dense sleeping and loafing flocks, often with Willets, godwits, and Black-bellied Plovers.

Voice. Loud, harsh *pip, pip, pip* notes when alarmed; in flight or upon landing, loud, shrill laughing call of 5–8 whistled notes; in groups, especially when other birds arrive, a melodious trilling *kyurrrri-kyurrrri-kyurrri-kew*, possibly a hostile note.

Status. Locally common to abundant migrant (August–September; mid-March–early May) and winter resident on Pacific coast, especially around Golfo de Nicoya; fairly common as nonbreeding summer resident (probably first-year birds); present in much smaller numbers on Caribbean coast, mostly during migrations.

Range. Breeds in Alaska and N Canada; winters on coasts from SW and SE USA to S Chile and S Brazil. Widespread in Old World.

LONG-BILLED CURLEW

Numenius americanus Pl. 9(13)

Zarapito Piquilargo

Description. 24″ (61cm); 600g. Very large, with extremely long, slender, decurved bill. **Adults:** head, neck, and underparts rich pinkish-buff; head and neck finely streaked with black (heaviest on crown); upperparts barred and variegated with blackish-brown and buff; wing-linings cinnamon-rufous; usually shows a pale eye-ring. Bill black except flesh-colored base of lower mandible; legs gray. **Immatures:** like adults but most primaries tipped with buff; tertials more spotted (less barred) with buff.

Habits. Like Whimbrel, with which it is invariably associated in Costa Rica, usually 1 or 2 in a large sleeping or loafing group of the smaller species.

Voice. Clear, penetrating *currLEE* or *curr-LEET*! sometimes prolonged into a roll or rattle.

Status. Very rare and local but probably regular winter resident in Golfo de Nicoya area; recorded from early December to mid-April.

Range. Breeds from SW Canada to C USA; winters from S USA to Costa Rica (probably regular) and Panama (casual or accidental).

UPLAND SANDPIPER

Bartramia longicauda Pl. 11(17)

Pradero (Gansa)

Description. 12″ (30cm); 170g. Slender, long-tailed; small head and bill; large eyes. **Adults:** head and neck buffy, streaked with dark brown; usually has narrow, pale eye-ring; upperparts dark brown, broadly scaled with buff; secondaries and greater coverts barred with brown and buff; wing-linings barred with brown and white; dark primaries give wing a dark-tipped look; below pale buff, fading to white on belly and crissum, spotted or barred with black on breast and sides. Bill yellow with blackish culmen and tip; legs yellow. **Immatures:** like adult but tertials with buff marginal spots.

Habits. Frequents pastures, lawns, airstrips, plowed fields, short-grass savannas, and other open areas, usually singly or in small, scattered groups, sometimes in same field as Lesser Golden Plovers or Pectoral Sandpipers but not close to them; acts like a little thick-knee, preferring to run rather than fly if alarmed; flies like Spotted Sandpiper on stiff, downward-bowed wings; on alighting may hold wings up briefly like a jacana; eats insects.

Voice. Only call heard in Costa Rica a high-pitched, soft *kip-ip-ip-ip-ip* (up to 10 notes) when alarmed.

Status. Uncommon and local transient in both fall (late August–November) and spring (mid-March–late May), mainly in Valle Central and Pacific lowlands, much less often in Caribbean lowlands.

Range. Breeds from Alaska and SE Canada to C USA; winters in E and SE South America.

Note. Formerly called Upland Plover.

GREATER YELLOWLEGS

Tringa melanoleuca Pl. 11(2)

Patiamarillo Mayor (Pijije, Zarceta)

Description. 14″ (36cm); 160g. Fairly large, slender; long legs and neck; bill fairly long and stout, very slightly upturned. **Adults** Winter: head, neck, and chest pale grayish, finely streaked with dusky; usually a white eye-ring; loral stripe dark; mantle dark gray-brown spotted with whitish; remiges dark; no wing-stripe; belly, rump, and tail white; tail barred with dusky. Bill blackish with gray base; legs bright yellow. Breeding: similar pattern but more contrast, especially on white head and neck, which are streaked with blackish; chest, sides, sometimes belly spotted and barred with blackish. **Young:** like winter adult but above with much finer and denser pale spotting. See Lesser Yellowlegs.

HABITS. Usually singly or in small groups on mudflats, salinas, estuaries, pond margins, rain pools, sand and gravel bars of rivers and streams; walks in shallow water, often chest-deep, making abrupt dashes to snatch small fishes, tadpoles, aquatic insects, and crustaceans; wary and noisy; often with other shorebirds, especially stilts, dowitchers, Willets, sometimes Lesser Yellowlegs or Stilt Sandpipers.

VOICE. Loud, ringing *kee keer-keer-keer*! or *whee-whew-whew*! (3–4 notes) when alarmed; also a rolling, Willetlike *shree-shree-shreek*!

STATUS. Uncommon to locally fairly common migrant in fall (August or September–October) and spring (April–May); uncommon but widespread winter resident; occasional nonbreeding summer resident, lowlands to middle elevations (Valle Central) of both slopes; most numerous in salinas and lagoons around Golfo de Nicoya.

RANGE. Breeds in S Alaska and C Canada; winters mainly along coasts from N USA to Tierra del Fuego.

LESSER YELLOWLEGS

Tringa flavipes Pl. 11(3)

Patiamarillo Menor (Pijije, Zarceta)

DESCRIPTION. 11″ (27cm); 85g. Medium-sized, slender; long legs and neck; bill slender, straight, of medium length; very similar in coloration and pattern to corresponding plumage of Greater Yellowlegs but decidedly smaller, especially bill (which is blackish with at most a trace of dusky-olive at base); in breeding plumage, belly more often immaculate; voice also differs. Solitary Sandpiper smaller with dark rump, more conspicuous eye-ring; greenish legs.

HABITS. More gregarious than Greater Yellowlegs, often in small flocks, especially in coastal salinas; occupies a similar range of habitats, but seldom found along rivers; forages actively like Greater Yellowlegs, wading in water to belly-deep, picking and snatching at insects, small crustaceans; often associates with dowitchers or Stilt Sandpipers; bobs head when looking about in alarm, as do other *Tringa*.

VOICE. A sharp, somewhat nasal *tyew-tew* or *klee-hew*, accent on first syllable, softer than notes of Greater Yellowlegs; in flight, a louder, more rolling *hyeurr-hewrr*.

STATUS. Common to locally abundant fall (August–mid-October) and spring (March–early May) migrant, fairly common winter resident, very uncommon to rare nonbreeding summer resident (first-year birds) along Pacific coast and adjacent lowlands, especially around Golfo de Nicoya. In much smaller numbers on Caribbean slope and up to at least 5000ft (1500m) in central highlands.

RANGE. Breeds in Alaska, NW and C Canada; winters from S and E USA south, principally in coastal areas, to Tierra del Fuego.

SOLITARY SANDPIPER

Tringa solitaria Pl. 11(4)

Andarríos Solitario (Tigüiza)

DESCRIPTION. 8¾″ (22cm): 45g. Fairly small, slender; dark wings and rump; conspicuous eye-ring; boldly barred tail. **Adults** WINTER: head, neck, and chest pale brownish; eye-ring, throat, and belly white; upperparts olive-brown, speckled with white; lateral rectrices white, barred with black; wing-linings barred with dusky and white. Bill blackish with grayish-olive base; legs dull greenish. BREEDING: head and neck white, streaked with dark brown; upperparts darker, white spotting more conspicuous. **Immatures:** like winter adult but foreneck and chest tinged with dull buff and mottled with brown.

HABITS. Prefers freshwater ponds, marshes, streamsides, flooded pastures, and rain pools; rarely on salt ponds; avoids open mudflats, favoring areas where vegetation extends to water's edge; alone or in small loose groups, often near yellowlegs and Spotted Sandpipers; wades in shallow water, bobbing head and tilting body, sometimes pumping tail like a Spotted; forages more deliberately than yellowlegs, pecking at aquatic insects, small mollusks, crustaceans. Flight fast, buoyant, swallowlike.

VOICE. High-pitched, thin, clear whistled notes, *peet*! or, when alarmed *pee weet-weet-weet*! In flight, a shrill *piSWEE-sweet*! Voice higher pitched, more piercing than that of Spotted Sandpiper.

STATUS. Uncommon but widespread migrant in fall (August–early October) and spring (mid-March–early May); uncommon winter resident; on both slopes from lowlands regularly to 6500ft (2000m), sometimes to 10,000ft (3000m) during migrations.

RANGE. Breeds in Alaska, C and S Canada; winters from extreme S USA to Peru and Argentina.

WILLET

Catoptrophorus semipalmatus Pl. 9(6)

Pigüilo

DESCRIPTION. 15″ (38cm); 250g. Large, gray; straight, stout bill; striking wing pattern. **Adults** WINTER: above uniform brownish-gray, palest on head and neck; rump, short superciliaries, eyelids, and underparts white; tail mottled with dusky toward tip, sometimes barred with blackish; broad white wing-stripe set off by black. Bill blackish shading to dull gray at base; legs dull gray. BREEDING: head and neck whitish, heavily streaked and spotted with blackish; chest, sides, and mantle barred and spotted with black and buff. **Young:** like winter adult but above scaled with buffy to whitish.

HABITS. Frequents mudflats and salinas, mangroves, tidal channels, and, in smaller numbers, sandy or rocky shores, river mouths, and coastal lagoons; rarely found inland; singly or in small, loose groups, probes for small mollusks, pecks and snatches at crabs, other small crustaceans, small fishes, and insects; gathers in large flocks to loaf and sleep, often with other large shorebirds; stands erect, bill pointed below horizontal; bobs head like a *Tringa* when alarmed.

VOICE. Noisy; most characteristic call a loud, ringing, rolling *whree-whree-whreek*! shriller than call of Greater Yellowlegs; also a curlew-like *ku-ku-ku* . . . , and a sharp, nasal, rather gull-like *jeew*! often given by birds in resting flocks.

STATUS. Locally abundant migrant on Pacific coast in fall (August–September) and spring (late March–late May); also common to abundant in winter; uncommon to fairly common nonbreeding summer resident; largest numbers around Golfo de Nicoya. Much less numerous on Caribbean coast, no large concentrations even in migration.

RANGE. Breeds locally in NW USA, S-C Canada, and along Atlantic-Gulf coast from S Canada to S Texas, Bahamas, and West Indies; winters along coasts from USA to N Chile and N Brazil.

SPOTTED SANDPIPER

Actitis macularia Pl. 11(8)

Andarríos Maculado (Alzacolita, Piririza, Tigüiza)

DESCRIPTION. 7½″ (19cm); 40g. Fairly small; rather stout bill; short legs; dark rump; conspicuous wing-stripe. **Adults** WINTER: above dull brown, paler on sides of head and neck; brownish smudge on each side of breast; wing-coverts with dusky and buffy barring; superciliaries whitish; eye-stripe dusky; underparts white. Bill brownish with yellowish base; legs dull yellow. BREEDING: above sparsely barred with black; below boldly spotted with black. **Immatures:** like winter adult but mantle feathers with pale fringes; wing-coverts more boldly barred. Solitary Sandpiper has longer, darker legs, white-speckled mantle.

HABITS. In virtually every aquatic habitat from rain puddles to seashore; generally forages by running along water's edge, stopping and snatching at small insects, crustaceans, small fishes; usually spaced out on vigorously defended individual territories by day; gathers in small flocks to sleep; teeters continually, pumping hindquarters vigorously up and down. Flight distinctive: several quick, shallow flaps and a glide on stiff wings that are never raised above the horizontal.

VOICE. A loud, clear *peet-weet*! In alarm or aggressive interactions: *pee-weet! weet! weet!* . . . (up to 8–10 successively lower-pitched *weet* notes).

STATUS. Most widespread shorebird in Costa Rica, from both coasts up to 6000ft (1850m); generally the most numerous sandpiper well away from coast but not in large flocks or dense concentrations. Common in migration (early August–October, April–late May) and winter, very rare nonbreeding resident in summer.

RANGE. Breeds from Alaska and N Canada to S USA; winters from extreme SW Canada and SE USA to N Chile and N Argentina.

WANDERING TATTLER

Heteroscelus incanus Pl. 10(8)

Correlimos Vagamundo

DESCRIPTION. 11″ (28cm); 110g. Medium-sized, grayish; unpatterned dark rump, tail and wings. **Adults** WINTER: above uniform gray; faint paler fringes on wing-coverts; remiges more blackish; eye-stripe blackish; eyelids white; superciliaries whitish; sides of head, sides and front of neck, chest, and sides pale gray; throat and belly white. Bill blackish with yellowish base of lower mandible; legs greenish-yellow. BREEDING: sides and front of neck mottled and streaked with blackish; underparts white, barred with blackish. **Young:** like winter adult but mantle scaled with buff; wing-coverts spotted with buff; breast mottled with grayish and white.

Habits. Prefers exposed rocky promontories, rocky offshore islets, and (Cocos I.) rocky, swift streams near coast; frequently forages among rocks exposed at low tide and at stream mouths, walking deliberately and often bobbing and pumping tail like Spotted Sandpiper, picking up small mollusks, crustaceans, and other invertebrates; often associates with Ruddy Turnstones and Surfbirds on stony shores.

Voice. Very high, thin, penetrating whistles, the most piercing calls of any shorebird: *keeeea* or *keeeaweeet*; when alarmed, a shrill, laughing *keekhikikikikihihi*.

Status. Uncommon to occasionally numerous migrant along outer Pacific coast in fall (late September–early November) and spring (early March–late April); occasional individuals, mainly on offshore islands, in both winter and summer. Fairly common year-round on Cocos I.; summering birds are nonbreeding yearlings.

Range. Breeds in Alaska and NW Canada; winters from W USA to Peru along Pacific coast, also widely on islands in S and C Pacific.

RUDDY TURNSTONE

Arenaria interpres Pl. 10(6)

Vuelvepiedras Rojizo

Description. 8¾″ (22cm); 110g. Plump; short neck and legs; short, wedge-shaped, slightly upturned bill; striking pattern in flight. **Adults** Winter: head pale brownish, streaked and mottled with dusky and white; indistinct dusky stripes on cheeks and neck extending to large blackish smudge on chest; above dark brown mixed with grayish-brown; rump and base of tail white; upper tail-coverts black; white areas in scapulars and wing-coverts and broad white wing-stripe conspicuous in flight; below white. Legs orange. Breeding: head and chest with bold black-and-white pattern; upperparts rufous and black, strongly contrasting (male) or mixed with gray-brown (female). **Immatures:** like winter adult but mantle feathers fringed with buff, chest usually blacker.

Habits. Frequents rocky or sandy shores, mudflats, salinas, river mouths, tidal creeks, even adjacent wet fields, singly or in small groups; forages mostly in intertidal zone, walking busily about, overturning shells, pebbles and debris, or digging in sand for mollusks, crustaceans, fish eggs, even vegetable matter; sometimes defends small individual feeding territory but at night or high tide rests in flocks.

Voice. Alarm note, a scratchy, almost cackling *kutakuk*, or a single *kuk* or *kek*; more musical whistled notes among flock members, especially in flight; soft chattering by feeding birds.

Status. Common migrant in fall (August–October) and spring (late March–late May); locally common winter resident; uncommon nonbreeding summer resident on Pacific coast; in smaller numbers, and more local on Caribbean coast.

Range. Breeds in Alaska and N Canada; winters along both coasts from USA to Tierra del Fuego, also on islands in Pacific. Widespread in Old World.

COMMON SNIPE

Gallinago gallinago Pl. 9(5)

Becacina Común (Becada)

Description. 11″ (28cm); 105g. Plump; short legs and neck; long, heavy bill; dark wings; striped head and back. Head boldly striped with brownish-black and buffy; neck and chest buffy, mottled and streaked with dark brown; underparts white; sides and flanks barred with black; above mottled and barred with black, brown and buff, and with 2 bold, buffy stripes down each side of back; tail orangish with black subterminal bar and narrow white tip and sides. Bill dusky-brown, tinged with olive basally; legs dull greenish.

Habits. Frequents grassy marshes, pond edges, wet pastures, and rain pools; usually roosts in flocks; scatters out singly or in loose groups to feed by day; probes deeply in soft mud or gleans mud surface, grass, or shallow water for earthworms, insect larvae, small snails, and seeds; crouches and freezes in alarm, often not flushing until nearly underfoot, then bursting into low, erratic flight, finally circling around high in air to drop abruptly back to ground.

Voice. On flushing, a characteristic dry, scratchy *chahr* or *chhht*.

Status. Widespread but generally uncommon migrant and winter resident, from lowlands of both slopes well up into middle elevations, occasionally to 10,000ft (3000m) in migration. Arrives mid to late October, remains through March or April. Sizable concentrations still occur locally in Valle Central, especially around Cartago (e.g., Coris); probably wintered in thousands in these areas before most marshes were drained.

RANGE. Breeds from N Alaska and N Canada to C USA; northern populations winter from extreme SW Canada, C and NE USA to N South America; in South America resident from Colombia to Tierra del Fuego. Widespread in Old World.

NOTE. Often *Capella* is used for this genus.

SHORT-BILLED DOWITCHER

Limnodromus griseus Pls. 9(3), 11(5)

Agujeta Común

DESCRIPTION. 11″ (28cm); 95g. Medium-sized; long straight bill; white back, conspicuous in flight; no wing-stripe. **Adults** WINTER: throat and eyebrow whitish; eye-stripe dusky; above brownish-gray; center of back white; rump, tail-coverts, and tail white, barred with black; foreneck and chest pale gray mottled with dusky; belly white; spotted or irregularly barred with dusky on sides and flanks. Bill black, tinged with yellowish at base; legs dull yellow. BREEDING: above blackish-brown, marbled and scaled with cinnamon to buff; below mostly cinnamon-rufous, breast and sides spotted or barred with black; belly often white. **Immatures:** like winter adult but often tinged with buff; retain some juvenile wing-coverts and tertials, which are blackish, irregularly marked with rusty. See Long-billed Dowitcher.

HABITS. Prefers shallow salt water with soft, muddy bottom in salinas, tidal mudflats and creeks, but while migrating visits a wide variety of lowland aquatic habitats, including sandy beaches, marshes, pond margins, river sandbars, and flooded fields; feeds in compact flocks on mudflats at low tide, walking about deliberately, probing deeply, and sweeping bill side-to-side along soft, muddy bottom, taking marine worms, mollusks, crustaceans, aquatic insects, and occasionally seeds; rests in salinas; flies swiftly and erratically, in close formation, with rather deep wingbeat.

VOICE. When flushed, a rapid, abrupt *tu-tu-tu*, sometimes a single *tew*, lower-pitched than calls of yellowlegs or Long-billed Dowitcher.

STATUS. Locally abundant migrant in fall (early August–late October) and spring (late March–late May), common to abundant winter resident; rather uncommon to fairly common nonbreeding summer resident on Pacific coast, especially around Golfo de Nicoya; inland and on Caribbean slope in smaller numbers and mainly during migration, occasionally to 5000ft (1500m).

RANGE. Breeds in S Alaska, C and NE Canada; winters mainly along coasts from S USA to C Peru and C Brazil.

NOTE. Sometimes called Common Dowitcher.

LONG-BILLED DOWITCHER

Limnodromus scolopaceus Pl. 9(4)

Agujeta Silbona o Piquilarga

DESCRIPTION. 12″ (30cm); 115g. Very similar to Short-billed Dowitcher in all plumages but, sex for sex, averages larger, longer-billed, and longer-legged. In all plumages, black barring of upper tail-coverts and rectrices is heavier and more regular (on central rectrices, dark bars 2–4 times as wide as pale bars; more even width in Short-billed); in hand, note shorter distance (under 2mm) from feathers to nostril in Long-billed, 2mm or more in Short-billed (Phillips). Breeding adults have foreneck heavily spotted, breast and sides barred with blackish (more irregularly barred or spotted in Short-billed); belly always immaculate cinnamon-rufous. **Immatures:** buffy markings of tertials and wing-coverts restricted to margins.

HABITS. Prefers freshwater habitats, including pond margins, marshes, and flooded pastures, but also on smaller, more inland salinas with grassy margins; feeds largely by probing, like Short-billed, but takes more insects and earthworms.

VOICE. When flushed, a clear, penetrating whistle, *keek* or *kee*, sometimes rapidly repeated several times.

STATUS. Uncommon to rare migrant and possibly winter resident, lowlands of both slopes up to Valle Central; arrives late October–early November, departs by late April. No definite summer records, but a year-old bird banded in mid-September probably summered well to S of breeding range.

RANGE. Breeds in NE Siberia, N and W Alaska, and NW Canada; winters from S USA to Costa Rica and occasionally Panama.

SURFBIRD

Aphriza virgata Pl. 10(7)

Chorlito de Rompientes

DESCRIPTION. 10″ (25cm); 185g. Medium-sized, chunky; short bill and legs often give pigeonlike aspect; broad white wing-stripe; white rump contrasting with mostly black tail. **Adults** WINTER: head, neck, breast and upperparts mostly dull dark gray; throat, loral spot, and eye-ring white; belly white, speckled with dusky. Bill black, base of lower mandible yellow; legs and feet dull yellowish. BREEDING: head, neck, and body variegated with black and white; mantle feathers marked with rusty. **Immatures:** like winter adult but wing-coverts fringed with buffy-white. Only other shorebird with such a short bill is Ruddy Turnstone, which is smaller, browner, more patterned, with orange legs.

Habits. Usually on wave-splashed rocky promontories and on flat rock pavements below many sandy Pacific beaches as these are exposed at low tide; sporadically on mudflats and salinas; in small, compact flocks in rocky places, elsewhere singly or in small groups; walks about or clambers over rocks, pecking for small snails and other mollusks, crustaceans.

Voice. A rolling, drawn-out *churrrp*; a low *kee-weep*.

Status. Sporadically common fall migrant (mid-September to late October) along outer Pacific coast; occasional in winter, even inside Golfo de Nicoya; uncommon spring migrant (late April–May).

Range. Breeds in C Alaska; winters along Pacific coast from extreme S Alaska to Tierra del Fuego.

RED KNOT

Calidris canutus Pl. 11(19)

Correlimos Grande

Description. 10¾″ (27cm); 135g. Chunky, medium-sized; stout straight bill; rather short legs. **Adults** Winter: above dull grayish; superciliaries and lower cheeks whitish; foreneck, breast, and sides tinged with grayish, mottled with dusky; throat and belly white; rump finely barred with dusky and white (looks gray from a distance); tail grayish; rather dull wing-stripe. Bill black, base grayish-olive; legs dull greenish. Breeding: face and underparts cinnamon-rufous; male brighter. **Immatures:** like winter adults but retain juvenile wing-coverts—gray with whitish fringe set off by black.

Habits. On sandy or pebbly beaches, especially at river mouths, but in numbers only in Golfo de Nicoya, feeding on mudflats at low tide and repairing to salinas and salt-pond dikes to loaf and sleep; usually in compact flocks, often associated with Black-bellied Plovers or dowitchers; walks mincingly, pecking and snatching but usually not probing, taking small mollusks, also some crustaceans and insects.

Voice. In flight a soft, scratchy, melodious *woit-woit* or *wew-wet*; a soft, dry *k'nut*.

Status. Locally common migrant and winter resident in Golfo de Nicoya; elsewhere on Pacific coast a very uncommon migrant, mainly in fall. No inland or summer record and none for Caribbean coast. Fall migration late August to October; spring migration mainly mid-March to late April.

Range. Breeds in N Alaska and N Canada; winters from SW and E USA to Tierra del Fuego. Widespread in Old World.

SANDERLING

Calidris alba Pl. 11(11)

Playero Arenero

Description. 8″ (20cm); 55g. Fairly small, with straight, rather thick bill, conspicuous wing-stripe; blackish patch at bend of wing. **Adults** Winter: above pale brownish-gray; most of head white (dark eye more contrasting than in other small sandpipers); blackish shoulder-patch conspicuous; underparts and sides of rump white, center of rump and tail blackish, rest of tail pale gray. Bill and legs black. Breeding: head, neck, and underparts variegated with rusty and black. **Immatures:** like winter adults but wing-coverts more strongly patterned: whitish fringes, blackish shaft-streaks retained from juvenile plumage.

Habits. Most characteristic of sandy beaches but also on mudflats, repairing to upper beach or salt-pond dikes to loaf and sleep; on beaches feeds by chasing after receding waves to dig and probe in soft sand and moving slick of water, scurrying back in front of next wave; takes mainly small crustaceans (especially *Emerita*) and mollusks; usually in small flocks, but occasionally an individual defends a feeding territory along a beach.

Voice. Flight call a sharp, penetrating, often repeated *twick* or *tsik*; in feeding flocks, a soft, gabbling twitter.

Status. Common along both coasts in migration (mid-August through October and mid-March to early May) but uncommon (although widespread) in winter; no summer record.

Range. Breeds in Alaska and N Canada; winters along coasts from S Alaska and NE USA to N Chile and N Argentina.

WESTERN SANDPIPER

Calidris mauri Pl. 11(12)

Correlimos Occidental (Patudo, Becacina)

Description. 6¼″ (16cm); 25g. Small; bill relatively long with fine, slightly drooped tip. **Adults** Winter: above grayish-brown with fine dusky streaking; superciliaries, underparts, and sides of rump white; breast tinged with grayish and with sparse dusky streaking;

narrow white wing-stripe; tail pale gray. Bill and legs blackish. BREEDING: above streaked with blackish and white; crown, auriculars, and scapulars largely rufous; breast and sides spotted with blackish. **Immatures:** like winter adults but retain some buff-fringed juvenile wing-coverts. Very similar to Semipalmated Sandpiper, best distinguished in winter by bill length and voice; some adults retain a few rufous scapulars from breeding plumage well into October.

HABITS. Exploits various coastal habitats with flat or gently sloping muddy, sandy, or gravelly shores; less often inland at pond margins, rain pools, and wet fields; forages regularly in large flocks on tidal mudflats, walking and running briskly about, incessantly pecking or probing, sometimes in surface film of soft mud and water, with a rapid "sewing-machine" motion, taking small mollusks and crustaceans, aquatic insects, occasional seeds and grit; flies swiftly and erratically in compact, beautifully coordinated flocks; retires to salinas to loaf and sleep.

VOICE. A thin, reedy *kreet* or *jeep*, more plaintive and higher-pitched than note of Semipalmated Sandpiper. Various soft chattering, twittering, and warbling notes in flocks, especially at night.

STATUS. Most abundant small sandpiper on both coasts but far more numerous on Pacific; locally abundant (especially Golfo de Nicoya and Golfo Dulce) in migration (August–November; mid-March–early May) and during winter; occasional to locally common nonbreeding summer resident; locally and in small numbers inland to at least 5000ft (1500m) in central highlands, mainly during migration.

RANGE. Breeds in NE Siberia and N and W Alaska; winters from both coasts of USA to N Peru and Suriname.

SEMIPALMATED SANDPIPER

Calidris pusilla Pl. 11(13)

Correlimos Semipalmado (Patudo, Becacina)

DESCRIPTION. 6″ (15cm); 23g. Slightly smaller than Western Sandpiper with shorter, straight, more blunt-tipped bill (bill nearly always less than 22mm; Western, nearly always 22mm or more). **Adults** WINTER: plumage virtually identical to Western Sandpiper but never has rufous in scapulars in fall. Bill black; legs blackish. BREEDING: feathers of upperparts brownish-black, edged with cinnamon to buffy-brown; breast streaked with dusky-brown; lacks rufous areas of Western Sandpiper. **Immatures:** very similar in color to young Western Sandpipers when they arrive in Costa Rica.

HABITS. Habitats and behavior generally similar to those of the much more numerous Western Sandpiper. Although a few birds are often found in large flocks of Westerns, Semipalmateds tend to move mostly in compact, single-species groups or flocks. The two species sometimes mix freely when feeding on mudflats, but Semipalmateds tend to pick and snatch more for surface prey and probe less; also to wade less often than Westerns. They eat small aquatic crustaceans, insects, mollusks, and algae.

VOICE. A short, hoarse *krip* or *kiip* when flushed; a harsh *cherk* in flight; notes quite different from those of Western Sandpiper.

STATUS. Common, sometimes locally abundant migrant (mid-August to mid-November; March–early May); fairly common winter resident; occasional nonbreeding summer resident on Pacific coast, with greatest numbers around Golfo de Nicoya; much less numerous on Caribbean coast; in small numbers locally to ca. 5000ft (1500m) only during migrations.

RANGE. Breeds in N Alaska and N Canada; winters from S Mexico and S Florida to N Chile and Argentina. The population wintering and migrating in Costa Rica consists of short-billed birds from the westernmost part of the breeding range.

LEAST SANDPIPER

Calidris minutilla Pl. 11(14)

Correlimos Menudo (Menudillo) (Patudo, Becacina)

DESCRIPTION. 5½″ (14cm); 21g. Very small; brown; slender; rather short bill; legs yellowish. **Adults** WINTER: above dull brownish, streaked with blackish-brown; below white, washed with grayish-buff and finely streaked with dusky on chest; superciliaries whitish; white sides of rump; narrow wing-stripe white. Bill blackish. BREEDING: crown and hindneck streaked with brownish-black and buff; feathers of mantle brownish-black, edged broadly with buffy to bright cinnamon; breast more heavily streaked with dark brown. **Immatures:** like winter adults but retain some rufous-fringed juvenile wing-coverts.

HABITS. Prefers muddy borders of salt- or freshwater ponds, salinas, and mudflats, especially those with sparse vegetation; also on sandy beaches, river sandbars, and estuaries, usually in small groups, singly, or, from late winter onward, in 2s; seldom forms large pure flocks like other peeps; when flushed, flies low and erratically like a tiny snipe; feeds on aquatic insects, small crustaceans, mollusks, and worms by active surface gleaning, sometimes by probing in mud and shallow water.

Voice. Thin, rolling *drree* or *kreep*; distinctly more rolling than calls of other peeps.

Status. Widespread, locally common migrant (early August–October; early March–early May); uncommon to fairly common winter resident, mainly along coast, but more often inland than other peeps at this season, from lowlands to 5000ft (1500m) in central highlands; rare and local nonbreeding summer resident; much more numerous on Pacific slope in general.

Range. Breeds in Alaska and N Canada; winters mainly along coasts from NW and E USA to C Peru and E Brazil.

WHITE-RUMPED SANDPIPER

Calidris fuscicollis Pl. 11(16)

Correlimos Lomiblanco (Patudo, Becasina)

Description. 7½″ (19cm); 45g. Slightly larger, notably longer-winged than Western Sandpiper, with entirely white rump. **Adults** Winter: breast, face, and upperparts brownish-gray, faintly streaked with dusky; superciliaries, throat, and underparts white; narrow wing-stripe white. Bill black with reddish or brownish base of lower mandible; legs brownish to blackish. Breeding: above streaked with black and buffy-white; crown, auriculars and scapulars tinged with rusty; below boldly streaked with dusky or black on breast and sides. **Immatures:** like winter adults but retain buff-edged juvenile feathers in mantle and wing-coverts until spring.

Habits. Frequents sandy and muddy shores of fresh- or saltwater ponds, marshes, and salinas, also occasionally tidal mudflats and sandy beaches; usually singly or in small groups, often associated with Western or Least sandpipers; forages very methodically, often probing repeatedly in same spot, sometimes wading belly-deep; eats insects, especially aquatic larvae, and crustaceans.

Voice. A distinctive, sharp, squeaky *tzeat* or *tzeet-tzeet* or *jit-jit*, very high-pitched and mouselike, is often the best cue to the bird's presence in a large flock of peeps.

Status. Rare transient in fall (early September–mid-October) and spring (early April–late May) along both coasts and in central highlands; no winter or summer record.

Range. Breeds in N Alaska and N Canada; winters mainly in E South America, S to Tierra del Fuego.

BAIRD'S SANDPIPER

Calidris bairdii Pl. 11(15)

Correlimos de Baird (Patudo, Becasina)

Description. 7½″ (19cm); 39g. Fairly small, long-winged; dark rump; faint wing-stripe; buffy breast. **Adults** Winter: above grayish-brown, with faint dusky streaking; wing-coverts fringed with buff; rump blackish; sides buffy-white; superciliaries, throat, and belly white; breast and foreneck washed with dull buff and finely streaked with dark brown. Bill blackish with brownish base; legs black. Breeding: crown streaked with black and buff; mantle feathers blackish, broadly edged with buff; breast bright buff, streaked with dark brown. **Immatures:** like breeding adults but mantle feathers fringed with white, giving a definite scaled pattern; dark breast streaking indistinct; retains juvenile plumage, at least in part, most or all winter.

Habits. Prefers grassy margins of ponds, marshes and wet pastures; also sparsely on tidal mudflats and beaches; usually in small groups or singly, associated with other shorebirds, notably Least and Pectoral sandpipers inland or Western Sandpiper on coast; forages just above water's edge, rarely wading; alternates runs with abrupt stops to pick up prey, chiefly insects, spiders and crustaceans.

Voice. Loud rolling *drreep*, stronger than note of Least Sandpiper; also a softer, hoarser *chirrrr*.

Status. Very uncommon but probably regular migrant (September–early November; early April–early June), chiefly in the highlands but also, sparingly, along Pacific coast; almost certainly much scarcer than formerly because of widespread drainage of ponds and marshes in Valle Central during last 50 years.

Range. Breeds in Alaska, N Canada, and NW Greenland; winters from Ecuador, Paraguay, and Uruguay to Tierra del Fuego.

PECTORAL SANDPIPER

Calidris melanotos Pl. 11(7)

Correlimos Pechirrayado o Pectoral (Patudo, Becasina)

Description. ♂ 9½″ (24cm); 75g; ♀ 8″ (20cm); 48g. Male much larger than female; bigger, longer-necked than peep, with relatively stout body and small head; striped back; patterned breast abuts sharply on white belly; yellowish legs. **Adults** Winter: feathers above dark brown, edged with gray-brown;

two pale stripes down each side of back; superciliaries and throat pale; foreneck and breast dull buff, streaked with dusky; belly and sides of rump white; center of rump blackish; faint wing-stripe. Bill blackish, shading to yellowish at base. BREEDING: upperparts brighter, feathers edged with rusty; dorsal stripes broader, brighter; foreneck and breast buff, streaked with black; breast of male more mottled. **Immatures:** like winter adults but bright, buff-edged juvenile feathers retained on wing-coverts, sometimes over much of upperparts; crown feathers edged with rufous; breast more lightly streaked, sometimes almost immaculate at center.

HABITS. Prefers flooded fields, margins of ponds, and marshes; may occur in flocks on salinas and mudflats in migration, especially in spring; stalks about singly in grass, snatching abruptly at insects, its preferred food, or picking and probing in soft mud for worms, crustaceans, and insect larvae. When foraging in groups often flushes individually, to rejoin companions in compact flocks in fast, low, snipelike flight; often seen with Upland, Least, and Baird's sandpipers and Lesser Golden-Plover.

VOICE. Hoarse, rolling *jrrk*! or *dzrrt*!—deeper and creakier than call of Least or Baird's; in flight, especially in spring, adds staccato notes *dzrrt-d'd'jrrt*, also a mellow *chreet*.

STATUS. Locally common transient (late August–late November; early April–late May); no winter record; lowlands of both slopes up to 5300ft (1600m) in central highlands, also along Pacific coast; generally most numerous inland *Calidris* sandpiper. Most birds seen in fall are immatures. Occasional in summer, mainly in flooded rice fields of Tempisque basin.

RANGE. Breeds in N and W Alaska and N Canada; winters in S South America.

DUNLIN

Calidris alpina Pl. 11(10)

Correlimos Pechinegro

DESCRIPTION. 8″ (20cm); 55g. Fairly small; moderately long, heavy bill, drooped at tip; bright wing-stripe; rump white with blackish median stripe. **Adults** WINTER: above rather dark gray-brown; throat, superciliaries, most of underparts, and sides of rump white; cheeks, foreneck, and breast pale grayish, lightly streaked with dusky. Bill black; legs dark gray. BREEDING: mantle marbled with rufous and black; large black patch on belly. **Immatures:** like winter adults but mantle feathers edged with cinnamon-rufous; belly spotted with dusky; crown tinged with rusty to buffy; breast faintly streaked with brownish.

HABITS. Singly or in small groups in salinas, salt ponds, and mudflats, usually among larger flocks of Western Sandpipers; often tame; feeds by methodically probing and digging in mud and shallow water for small mollusks, crustaceans, and marine worms.

VOICE. On flushing, an abrupt *chu*; flight call a plaintive *purri* or *pyuur*.

STATUS. Very rare winter visitor and spring migrant, recorded in small numbers (up to 8 birds) on 5 occasions between late December and early May on mudflats around Golfo de Nicoya (Salina Bonilla, Chomes).

RANGE. Breeds in N and W Alaska and N Canada; winters along coasts from S Alaska and NE USA regularly to Mexico; casual to rare S to Costa Rica and Panama. Widespread in Old World.

CURLEW SANDPIPER

Calidris ferruginea Pl. 11(9)

Correlimos Zarapitín

DESCRIPTION. 7½″ (19cm); 45g. Very similar to Dunlin in winter plumage (only plumage recorded for Costa Rica) but rump entirely white, bill slimmer and somewhat more decurved, and dusky mottling on breast faint and indistinct; in hand, has longer legs (Dunlin has tarsus nearly always 27mm or less, Curlew nearly always over 27mm). BREEDING: plumage very distinctive, with head, neck, and underparts mostly chestnut. Stilt Sandpiper has longer legs and neck, and thicker, straighter bill.

HABITS. The 1 bird recorded for Costa Rica was associating with a large flock of Western and Semipalmated sandpipers on mudflats at a salina.

VOICE. Flight call a soft, whistled *chirrap*.

STATUS. Accidental; 1 record for fall migration (early November) near Colorado on Golfo de Nicoya; not otherwise recorded from Central America but may be more regular in view of the rather numerous reports from North America and West Indies.

RANGE. Breeds from N Siberia E to N Alaska; winters widely in Old World; casual to very rare along both coasts and inland in North America, S to California and West Indies; accidental in Central and South America.

STILT SANDPIPER

Calidris himantopus Pl. 11(6)

Correlimos Patilargo (Patudo, Becasina)

DESCRIPTION. 8¾″ (22cm); 55g. Long legs and neck; long, stout bill, slightly drooped at tip; rump white; tail grayish; wings dark with no wing-stripe. **Adults** WINTER: above brownish-gray, wing-coverts darker, fringed with gray; superciliaries, throat, and underparts white; foreneck and breast faintly streaked or mottled with grayish. Bill black; legs dull olive. BREEDING: crown and hindneck striped and mantle spotted with black and buffy-white; auriculars and sides of crown rusty; underparts white, barred with blackish. **Immatures:** resemble winter adults but some blackish, buff-edged juvenile feathers retained on mantle and wing-coverts through winter. Slimmer and longer-legged than dowitchers, different flight pattern and voice; grayer overall, with longer bill and darker legs than yellowlegs.

HABITS. Prefers shallow salt or fresh water with soft, muddy bottom: salinas, pond margins, marshes; often forages in compact groups that wade, frequently belly-deep, with heads and necks immersed, probing and thrusting with bill or sweeping it side-to-side, advancing slowly, stirring up mud; eats aquatic insects and larvae, small mollusks, crustaceans, and worms; often found with dowitchers and yellowlegs; the latter in particular often feed at outskirts of compact phalanxes of Stilt Sandpipers, taking advantage of prey they stir up.

VOICE. In flight a scratchy, breathy *whuur* or *whrrru*, sometimes ending in a short chatter, or a *hu-hee-hu* of similar quality; usually silent while feeding.

STATUS. Fairly common but local migrant (mid-August–late October; mid-March–late May) on Pacific slope; locally rare to common winter resident around Golfo de Nicoya; no summer record.

RANGE. Breeds in N Alaska and N Canada; winters mainly in S South America, but also regularly in Costa Rica, and sparingly N to extreme S USA.

NOTE. Formerly segregated in the genus *Micropalama*.

BUFF-BREASTED SANDPIPER

Tryngites subruficollis Pl. 11(18)

Praderito Pechianteado (Zarceta)

DESCRIPTION. 8″ (20cm); 60g. Plump body, small head, and short bill give rather pigeonlike profile; contrast between pale wing-linings and dark upperwing with at most faint wing-stripe. **Adults:** feathers of upperparts blackish, edged with buff, appearing streaked on crown, scalloped on mantle; eye-ring whitish; sides of head and neck and entire underparts buff, paler posteriorly; black spotting on sides of breast; wing-linings white; contrasting black bar on primary coverts. Bill dusky; legs yellowish. **Immatures:** like winter adults but retain some white-tipped juvenile feathers on back and wing-coverts, where dark subterminal spot becomes conspicuous with wear.

HABITS. Frequents rain pools in pastures and grassy pond margins, usually in small flocks, or 1 or 2 associate with groups of Baird's or Pectoral sandpipers; often tame; freezes in alarm, only upon very close approach rising in swift, erratic, snipelike flight; when foraging walks actively about, head bobbing, pecking and snatching at insects.

VOICE. An abrupt, harsh, dry *kwik*! or *kruk*! when flushed.

STATUS. Presently rare and sporadic in fall migration (late August–late October); no spring record, but to be expected in March and April. All records to date are from central highlands, where undoubtedly it stops less often than formerly due to widespread drainage of preferred habitats. To bé looked for in Tempisque basin and on Caribbean slope, particularly in rice-growing areas.

RANGE. Breeds in N Alaska and N Canada; winters in SE South America.

RUFF

Philomachus pugnax Pl. 11(1)

Combatiente

DESCRIPTION. ♂ 11½″ (29cm); 160g; ♀ 8½″ (21.5cm); 95g. Resembles a thickset, heavy-billed yellowlegs with buffy-edged mantle feathers, dark center of rump. Male much larger than female. **Adults** WINTER: above gray-brown, feathers with whitish to pale buff margins; superciliaries pale (usually); throat whitish; white around base of bill; foreneck and breast pale grayish-buff marked with dusky; belly and sides of rump white, tail gray-brown; remiges dusky; narrow white wing-stripe. Bill black with purplish or yellowish base; legs pink to orange-red. BREEDING ♂: unmistakable with conspicuous, variously colored ruff and head-tufts. ♀: above dark brown, the feathers conspicuously edged with golden-buff; breast bright buff, mottled with dusky. **Immatures:** like winter adults but wing-coverts darker with bright buff fringes; legs greenish to brownish.

HABITS. Does not usually associate closely with other shorebirds; walks deliberately about on shallow salt ponds and mudflats probing with bill for worms, crustaceans and mollusks.

Voice. Sometimes a low *chu-wit* or *chut-ut* when flushed.

Status. Accidental; 1 sighting each in fall (early September) and spring (late May) at Chomes, on Golfo de Nicoya. To be looked for along Pacific coast during migration periods.

Range. Breeds in N Eurasia; winters mainly in Old World tropics but casual along Pacific coast of North America S to Panama, Atlantic-Caribbean coasts, and West Indies, mainly during migrations.

FAMILY **Phalaropodidae:** Phalaropes

The 3 species of phalaropes resemble small sandpipers in most of their anatomy and are often placed in the same family. However, these most aquatic of shorebirds share a number of distinctive features: laterally compressed tarsi, lobate feet, and dense, ducklike ventral plumage that enables them to float high in the water. Besides being considerably larger than males, females, unlike most sandpipers, are much more brightly colored in the breeding season; in winter both sexes are plainly attired in white and grays.

Phalaropes nest near water, 2 species on arctic tundras around the world, 1 species on North American prairies. The former migrate and winter at sea, the latter in coastal wetlands, mostly in South America. They forage in water or on land for small fishes, insects, crustaceans, and seeds of aquatic plants. On water, they spin rapidly in small circles, evidently to stir up plankton. Social at all seasons, in winter they are highly gregarious.

During nesting, the usual roles of the sexes are reversed. After helping a male select a nest site, the female lets him prepare a scrape, in which she will deposit 3 or, usually, 4 olive-buff or creamy eggs, marked with brown or brownish-black. After her departure, he lines the scrape with fragments of surrounding vegetation, incubates the eggs for 17–20 days, and leads the precocial downy chicks. Some females are serially polyandrous, courting and depositing eggs for a second male after abandoning the first. Very few sandpipers practice polyandry, a notable exception being the Spotted Sandpiper.

RED PHALAROPE

Phalaropus fulicarius Pl. 10(11)

Falaropo Rojo

Description. 8″ (20cm); 55g. Medium-sized, with thick, rather short bill. **Adults** Winter: eye-patch blackish; dusky smudge on nape; rest of head, neck, underparts, and sides of rump white; mantle uniform gray; remiges blackish; broad white wing-stripe; center of rump and tail blackish. Bill black with some yellow at base; legs gray-brown. Breeding ♀: cap, lores and chin black; face white; neck and underparts chestnut-red; mantle streaked with dark gray and buff. ♂: paler; duller: cap streaked with buff; neck and underparts pale rufous. **Immatures:** paler gray back contrasts with dark, pale-edged wing-coverts from juvenile plumage; dark juvenile feathers often give a spotted appearance.

Habits. More pelagic than Red-necked Phalarope; in our waters usually in small numbers and well out at sea; feeds on plankton which it obtains by quick dabs at the water, often while twirling about on ocean surface.

Voice. A sharp, whistled *weet* or *keep*, higher-pitched than note of Red-necked Phalarope.

Status. Thus far known from scattered sightings of single birds or small groups, well offshore, in fall, winter, and spring (November–April); all records off Pacific coast, including vicinity of Cocos I.

Range. Breeds in N and W Alaska and extreme N Canada; winters at sea, mainly in Southern Hemisphere.

RED-NECKED PHALAROPE

Phalaropus lobatus Pl. 10(10)

Falaropo Picofino

Description. 7″ (18cm); 33g. Very fine black bill; streaked back and strong wing-stripe in all plumages. **Adults** Winter: crown, hindneck and eye-patch blackish; rest of head, neck, underparts, and sides of rump white; mantle dark gray with 2 whitish stripes down each side; remiges and center of rump blackish; legs dark blue-gray. Breeding ♀: throat and eye-spot white; rest of head and neck blackish; rich buff stripes on mantle; neck-patch rufous. ♂: paler, duller; pale rusty neck-patch extends around face; lores white. **Immatures:** like winter adults but retain buff-edged juvenile scapulars and tertials, making back appear striped.

Habits. Appears more or less regularly at

coastal lagoons and salinas during migrations, but most birds are at sea, though often within sight of shore, typically in small flocks; occasionally in enormous flocks in Golfo de Nicoya during stormy weather; rests and feeds afloat, twirling actively, picking plankton from surface; flies rapidly and directly, drops abruptly to alight on ocean.

Voice. Sharp *kit* or *whip*, somewhat like note of Sanderling.

Status. Common to abundant fall migrant (late August–early November) off Pacific coast, occurring sporadically on inshore waters; much scarcer in spring migration (April–May). A few evidently winter in Costa Rican waters; occasional inland records during fall and winter are probably of sick or injured waifs.

Range. Breeds in Alaska and N Canada; winters at sea, mainly in S Pacific and S Atlantic. Widespread in Old World.

Note. Formerly segregated in genus *Lobipes*; often called Northern Phalarope.

WILSON'S PHALAROPE

Steganopus tricolor Pl. 10(9)

Falaropo Tricolor o de Wilson

Description. ♂ 8½″ (21.5cm), 40g; ♀ 9½″ (24cm), 55g. Fairly small, slender; long neck and body; needlelike black bill; no wing-stripe. **Adults** Winter: above uniform gray, including crown, hindneck, eye-stripe; superciliaries and underparts white; rump white; tail pale grayish; remiges dusky; legs yellow. Breeding: legs black. ♀: crown and hindneck pale gray; black stripe from eye down side of neck, passing into rich chestnut as it continues onto back; a second chestnut stripe on scapulars. ♂: crown and mantle brownish-gray; eye-stripe dusky; superciliaries white, becoming buffy behind eyes; trace of chestnut on each side of base of neck. **Immatures:** like winter adults but wing coverts broadly edged with buff.

Habits. Principally in coastal salinas and lagoons, often in flocks; in much smaller numbers in freshwater ponds and puddles, marshes, and on tidal mudflats; swims in deeper water, twirling and picking rapidly at surface; often forages near stilts and dowitchers that stir up prey from bottom; wades in shallow water, picking prey from surface of mud or water or sweeping bill side-to-side over soft muddy bottom like a mini-avocet; eats aquatic insects, small crustaceans.

Voice. A distinctive nasal *herrrp* or *howrrp* like a soft honk or quack.

Status. Common to abundant in favorable habitats, especially around Golfo de Nicoya, during migrations (September–October; mid-April–late May); scattered records elsewhere in Pacific lowlands. Small numbers winter in salinas around Golfo de Nicoya, at least in some years; occasional nonbreeding birds in summer. No record for Caribbean slope.

Range. Breeds in W North America from S Canada to C USA; winters mainly from Peru and Uruguay through Chile and Argentina, occasionally N to S California and S Texas.

Note. Sometimes included in genus *Phalaropus* but seems to stand apart from the other 2 phalaropes in body proportions, behavior, and ecology.

FAMILY **Stercorariidae:** Skuas and Jaegers

Three species of jaegers and from 1–4 species of skuas comprise this family of aggressive, piratical relatives of the gulls. Its members are 15–24″ (38–61cm) in length and have strongly hooked, gull-like bills with horny ceres. Jaegers are smaller, with more pointed wings and elongated central rectrices; skuas are larger and more robust, with broader wings. The sexes are alike, but plumages are highly variable; most species have 2 or more color phases; immature plumages of some are confusingly similar, and often 3 or more years are required for attainment of adult dress.

Strong fliers, skuas and jaegers harass other birds, forcing them to drop or disgorge food, and then snatching it up in the air. They also catch lemmings and other small mammals, devour eggs and young of other birds, scavenge, and catch insects and sometimes fish. All members of the family nest on the ground on open tundras or moors—jaegers in scantily lined or bare scrapes, skuas in more substantial nests of grass and lichens. All lay 2 or, rarely, 3 brownish or greenish eggs marked with darker shades, which are incubated by both sexes for 23–30 days. Both attend the downy, semialtricial chicks. Jaegers nest in Arctic and sub-Arctic latitudes, whereas skuas nest at high latitudes in both Northern and Southern hemispheres. Breeding populations of skuas are either completely isolated or hybridize to a limited degree where their ranges overlap, so that there are numerous disagreements over species limits; the current consensus is that 3 or 4 species are involved.

SOUTH POLAR SKUA
Catharacta maccormicki Pl. 3(12)
Salteador Polar

Description. 21″ (53cm); 1.1kg. Suggests large immature gull but more massive, with larger head, hooked bill, broader wings with conspicuous white flash in primaries, wedge-shaped tail. Polymorphic, but head always similar in color to underparts; never shows rufous or cinnamon in plumage. **Adults:** head, neck, and underparts vary from pale brownish- or purplish-gray through buffy-brown to dark sooty-brown—dark birds often with slightly darker cap; all have conspicuous yellowish streaking on sides and back of neck; head and chest sometimes flecked irregularly with yellowish-white. Mantle blackish-brown, often with some pale flecking or scaling; wing-linings dusky; bases of primaries white. Bill and legs blackish. **Immatures:** like corresponding adults but pale birds show conspicuous pale nuchal collar; often suggestion of darker hood and less white in wing than adult. Bill blue with black tip; legs pale blue.

Habits. Usually alone on open sea, sometimes in small, loose groups in harbors; pursues and harries gulls, terns, boobies, forcing them to disgorge their catch; often takes offal and carrion around harbors; follows ships and attends fishing boats for scraps; unlike jaegers, alights freely on water; may catch surface-swimming fishes by snatching from flight or water surface but does not dive. Flight powerful, direct, deceptively swift.

Voice. Loud, somewhat gulllike screams.

Status. Probably sporadic visitor to both coasts, especially during Southern Hemisphere winter, but first-year birds, which often do not return to breeding areas, are possible at any season. Skuas have been sighted at Puerto Limón in August 1953 and April 1961 (Slud), Punta Guiones in December (Gómez), and in Golfo de Nicoya in May 1978, June 1979, and September 1981 (Stiles et al.). All records to date almost certainly pertain to this species, but specimen confirmation is needed, as others could occur (see below).

Range. Breeds in Antarctica; ranges into N Pacific waters as far as S Alaska and into N Atlantic off S Canada and Europe.

Note. Taxonomy and distribution of skuas still incompletely known; other forms possible in our area include Great (Northern) Skua, *C. skua* (breeds N Atlantic islands, recorded once in Belize), larger, browner than South Polar Skua, more heavily spotted with buffy, rufous, and/or cinnamon, often with distinct blackish cap; young birds darker, more rufescent than adults; Chilean Skua, *C. chilensis* (breeds S South America, ranges N in Humboldt Current, 1 possible sighting off Panama), very rusty below, distinct blackish cap, less white in wing. In addition, Antarctic or Brown Skua, *C. antarctica* (often lumped into *C. skua*) has been recorded from Lesser Antilles (recovery of bird banded in South Shetland Islands) but identification disputed; *C. maccormicki* also breeds there, and *C. antarctica* otherwise unknown N of latitude 30°S (Devillers).

POMARINE JAEGER
Stercorarius pomarinus Pl. 3(13)
Págalo Pomarino

Description. 21″ (53cm); 700g. Elongated, blunt-tipped, twisted central rectrices of adult add 6–8″ (15–20cm). Largest, most robust jaeger with stoutest bill, broadest wings with most white in primaries. **Adults** Winter: back barred or scaled with dull buffy; upper and lower tail-coverts broadly barred with whitish and dusky; indistinct dusky barring widespread on breast, neck and sides. Bill horn-color to grayish, with black tip; tarsus bluish anteriorly, rest of legs and feet black. Breeding: cap blackish; wings and upperparts dark sooty-brown; bases of all but innermost primaries white. *Pale phase:* below white, including nuchal collar; sides of head tinged with yellow; band of dusky barring across breast and sides; belly and crissum smudged or indistinctly barred with dusky. *Dark phase:* below dark, sooty-brown; yellowish wash on head sets off black cap. **Immatures:** like winter adults but lack dark cap; wing-linings barred with buffy; underparts heavily barred with dusky (dark phase entirely dark below); central rectrices short, pointed. At least 2 years to full adult plumage; several intermediate plumages.

Habits. Alone and very infrequently close to shore, more often well out to sea, especially off Pacific coast, where many associate with large flocks of Wedge-tailed Shearwaters; flight steady, powerful, but less agile than that of other jaegers; eats fishes and squids, either snatched from ocean surface or stolen from other seabirds, especially shearwaters; may approach and follow ships or fishing boats for scraps and offal; also takes carrion.

Voice. Usually silent at sea.

Status. Rare and sporadic visitor to both coasts, August through April; more common December–February off Pacific coast, when large flocks of Wedge-tailed Shearwaters occur well offshore. Most birds close to coast are immatures, which are found chiefly in Golfo de Nicoya and Golfo Dulce, in areas where gulls and terns congregate; accidental inland (Puerto Viejo de Sarapiquí, March 1986, emaciated immature).

Range. Breeds on Arctic coasts and islands; winters at sea S to Tierra del Fuego, South Africa, Australia, and New Zealand.

PARASITIC JAEGER

Stercorarius parasiticus Pl. 3(14)

Págalo Parásito

Description. 18″ (45cm); 475g. Long, pointed central rectrices of adult add 3–5″ (8–13cm). Medium-sized jaeger with narrower wings, slimmer bill than Pomarine. **Adults** Winter: feathers of back edged with dull buffy; crown and hindneck streaked; nuchal collar obsolete; upper and under tail-coverts broadly but indistinctly barred with blackish; breast largely brown, scaled with whitish. *Dark phase:* below usually all-dark. Bill and feet blackish. Breeding: cap blackish; upperparts and wings dark sooty-brown; shafts of outer 3–5 primaries and bases of most, white. *Pale phase:* nuchal collar and most of underparts white, tinged with yellowish on head and neck; often a partial or complete grayish-brown breast-band. *Dark phase:* below uniform sooty-brown; sides of head washed with yellowish; cap blackish. **Immatures:** like winter adults but upperparts more heavily barred with grayish-buff; wing-linings and crissum barred with pale gray. Base of bill often pale. Intermediate plumages exist; at least 2 years to full adult plumage.

Habits. Although principally pelagic, this is the jaeger most likely to occur inshore. Flies swiftly, dashingly, agilely, like a falcon; the most aggressive of the jaegers in persistently harrying other seabirds, forcing them to disgorge prey; chases storm-petrels, presumably to eat them rather than steal from them; occasionally perches on shore, appearing about the bulk of a Laughing Gull but more barrel-chested—and given wide berth by any Laughing Gulls present!

Voice. Usually silent at sea.

Status. Regular in small numbers during migrations (August–October and late March–April), principally off Pacific coast; rare in winter but recorded December–March in Golfo de Nicoya.

Range. Breeds on Arctic tundras; winters principally in oceans of Southern Hemisphere, sparingly N to S California, New England, and Europe.

LONG-TAILED JAEGER

Stercorarius longicaudus Pl. 3(15)

Págalo Colilargo

Description. 15″ (38cm); 300g. Long, pointed central rectrices extend another 6–10″ (15–25cm) beyond tail in adults. Smallest, slimmest, grayest jaeger; wings long and narrow, little white in primaries. Polymorphism marked in young birds, but virtually all adults are pale-phase. **Adults** Winter: feathers of mantle and cap dusky, edged with pale grayish; indistinct grayish breast-band; crissum and upper tail-coverts barred with black and white. Breeding: cap blackish; brownish-gray mantle contrasts with blackish flight feathers and underwings; shafts of outer 2 primaries white; nuchal collar and underparts white, shading to gray posteriorly; sides of head washed with yellowish. Exceedingly rare dark phase entirely dark sooty-brown with blackish cap. Bill dusky; legs pale blue, feet black. **Immatures:** like winter adults, but paler birds lack dark cap, look white-headed; many also lack dark breast-band; mantle and upperwings sooty-brown, flecked or scaled with pale buff; underwings heavily barred with dusky and buff; often considerable white on bases of primaries. Dark-phase juveniles have head and anterior underparts dark sooty-brown.

Habits. Flies buoyantly and gracefully, like tern; rarely harries other seabirds; attracted to fishing boats for scraps; also takes carrion. The one Costa Rican specimen had fresh grasshoppers as well as fish remains in the stomach, suggesting that it came inland to feed.

Voice. Usually silent at sea.

Status. One Costa Rican record, a subadult taken during spring migration (late April) in Golfo de Papagayo; probably a rare transient in our area; main migration routes far out to sea.

Range. Breeds in high Arctic; winters mainly in S Atlantic and SE Pacific oceans.

FAMILY **Laridae:** Gulls and Terns

The 80 or so species of this cosmopolitan family are about equally divided between gulls and terns. Gulls have rather stout, hooked bills, broader wings, usually rounded tails. They are omnivorous scavengers on land or water and predators on weaker birds or small animals, and sometimes rob fish-eating birds of their catch. Terns have mostly slender, sharp-pointed bills, narrower and more pointed wings, and mostly forked tails; they catch fishes and other aquatic creatures by plunge-diving; marsh-dwelling species also flycatch in the air. Terns rarely swim; gulls swim well but rarely dive.

All gulls and terns are aquatic birds with impressive powers of flight; many species make long migrations. All have short to very short legs with webbed feet and a very reduced hind toe. Their plumage is typically white with gray to blackish back and upperwings, often with contrasting black markings on the head and/or wingtips; a few are largely or wholly white, gray, or blackish. Many species have colorful bills, legs, and/or eyerings. The sexes are alike; seasonal changes in color, mainly of the head, are usual. Immatures may differ greatly from adults and require several years, and intermediate plumages, to acquire adult dress. With some notable exceptions among the terns, members of this family frequent seacoasts and inland waters rather than the high seas. They sleep on the ground or water, or sometimes on ledges; the pelagic Sooty Tern is reputed to sleep on the wing. All tend to be noisy, uttering a variety of mostly shrill and harsh notes.

Gulls and terns breed in monogamous pairs; the male feeds his mate in courtship. They nearly always nest in colonies, often numbering many thousands of pairs, mostly in open areas at least partly protected or surrounded by water. The nest may be an untidy accumulation of grasses and trash on the ground or, especially among terns, an unlined scrape in sand. Marsh dwellers make floating nests of aquatic plants. A few terns build nests in bushes and trees, and the White Tern rests its single egg on a bare branch. The eggs, in sets of 1–3, rarely 4 (the latter perhaps the product of 2 females) are gray, brown, or greenish, more or less heavily spotted and blotched with darker shades. Both sexes incubate for 20–35 days. The young, hatched downy and alert, may in a few days wander about near the nest. They are fed by both parents until they can fly when 3–8 weeks old, and often while they migrate together for great distances.

RING-BILLED GULL

Larus delawarensis Pl. 3(9)

Gaviota Piquianillada

Description. 19″ (48cm); 450g. Larger than Laughing Gull; plumage notably pale, especially in immature stages. **Adults:** head, neck, underparts, rump, tail, and trailing edges of wings white; mantle and upperwings pale gray; wingtips black, outer primaries with subterminal white spots. In winter, head and neck spotted with dusky. Iris yellow; bare eyelid red-orange; bill greenish-yellow with subterminal ring of black; legs greenish-yellow. **First winter:** head, neck, underparts, rump, and tail mainly white; dusky mottling on crown, hindneck, and rump; subterminal tail-band blackish; mantle and some wing-coverts gray-brown, mottled and scaled with grayish-buff; shoulders, greater coverts, and inner primaries extensively pale gray, remiges otherwise blackish, secondaries tipped with white. Iris brown; bill dull flesh, tipped with blackish; legs grayish-flesh. Two or 3 years to full adult plumage; bill more or less ringed by second year.

Habits. Around harbors, estuaries and river mouths, occasionally along beaches; feeds on carrion and offal but also hunts crabs and other invertebrates on mudflats and beaches. One or 2 birds often roost or loaf amid larger flocks of terns and Laughing Gulls, among which they stand out by their size and pale coloration. Flight graceful, buoyant, with faster wingbeats than Herring Gull.

Voice. Thin, high-pitched squeals; a shrill *kyow* or *keeu*!

Status. Rare but increasingly regular winter resident and visitor on both coasts, principally on Pacific; records extend from early November to late May; most refer to first- or second-year birds.

Range. Breeds in interior of North America; winters from S Canada to S Mexico and with increasing frequency to Costa Rica and Panama.

HERRING GULL

Larus argentatus Pl. 3(10)

Gaviota Argéntea

Description. 24″ (61cm); 900g. Largest Costa Rican gull, with broadest wings, heaviest bill. **Adults:** head, neck, entire underparts, rump, tail, and trailing edges of wings white (head and neck streaked with brownish in winter); mantle pale gray; wingtip black, white subterminal spots in outer primaries. Iris yellow; bare eyelids reddish; bill yellow, red spot on lower mandible; legs flesh-color. **First winter:** mainly gray-brown, streaked and spotted with grayish-buff; remiges and rectrices dusky, outer rectrices with paler spotting; bases of inner primaries extensively pale gray-brown, producing a conspicuous "window"; iris brown; bill dull flesh, tipped with blackish. Four years to full adult plumage, with several intermediate stages; always recognizable by size (twice bulk of Laughing Gull).

Habits. Chiefly a scavenger around bays and harbors; one or a few birds often loaf and roost amid large flocks of Laughing Gulls and terns on salt-pond dikes and sandpits; flies buoyantly and gracefully, but with slower, heavier wingbeats than other gulls.

Voice. Common note in winter a dry *kak-kak-kak*; also a clear laughing *kyer-kyer-kye-kekeke* . . . ; mewing, squealing notes.

Status. Uncommon to rare but increasingly regular in winter (early November through mid-May), principally on Pacific coast; mainly around bays of Golfo Dulce and Golfo de Nicoya, and harbors of Puntarenas and Limón; occasional nonbreeding summer visitor. Most records are of first- or second-year birds. Has strayed to Cocos I. (January).

Range. Holarctic; in New World, breeds widely in North America, S to N-C and E USA; winters S to Panama and West Indies.

Note. An immature gull, photographed by S. Perkins at Mata de Limon on 25 December 1986, was the size of a Herring Gull but considerably darker above, without a pale "window" on inner primaries. Two species are possible: Western Gull (*L. occidentalis*) and Lesser Black-backed Gull (*L. fuscus*); former breeds on islands off Pacific Coast from S Canada to NW Mexico, S in winter to C Mexico; latter a European species recorded with increasing frequency in winter on Atlantic slope of North America, may stray farther southward. Both are darker above than Herring Gull in all plumages, but specimens probably required to distinguish between them.

LAUGHING GULL

Larus atricilla Pl. 3(8)

Gaviota Reidora

Description. 16″ (40cm); 275g. Fairly small gull with dark upperparts, rather long, heavy bill, flat forehead. **Adults** Winter: head, underparts, rump, tail, and trailing edge of wing white, except for dusky mottling on sides and back of head; narrow, broken, white eye-ring; mantle dark slate; wingtips black. Bill dusky, tinged with red; legs blackish. Breeding: head sooty-black; bill and legs dark red. **First winter:** from adult by heavier dusky mottling on head; breast and sides mottled or smudged with dusky; wing-coverts largely dark gray-brown, edged with grayish-buff; remiges and entire wingtip dusky-black; broad subterminal tail-band blackish; bill and legs blackish. Two years to full adult plumage: second-year birds resemble adults but browner on mantle.

Habits. Frequents coastal areas, but scattered individuals or flocks occur widely on beaches, mudflats, and estuaries. Large flocks rest on salt-pond dikes or sandspits, often associated with terns and other gulls. Small numbers range far inland along large rivers, but only occasionally venture far offshore. Flies lightly and gracefully on slow and rather shallow wingbeats; banks and soars with wings fully extended; in level flight, uses steady wingbeats, few glides; feeds on carrion, offal, and scraps from fishing boats; catches fishes and shrimps in shallow water; chases crabs on mudflats; tries to rob pelicans before they can swallow their catch; hawks flying insects.

Voice. A single *kew*; several clucking *ha*s; a strident, honking laugh, *ha-ha-ha-haa-haaa*. . . .

Status. Widespread, locally common to abundant migrant, winter resident, and nonbreeding summer resident (young birds). Fall migration mainly late September–November, spring migration early April–mid-May. Most birds seen in Costa Rica are first- or second-year immatures. Largest numbers in harbors of Puntarenas and Limón; more numerous overall on Pacific coast, especially in Golfo de Nicoya; sporadic on Cocos I., chiefly young birds in fall.

Range. Breeds along Atlantic coasts from SE Canada to Texas, and through West Indies to French Guiana; also NW Mexico. Winters from Mexico and SE USA to Peru and Brazil.

FRANKLIN'S GULL

Larus pipixcan Pl. 3(5)

Gaviota de Franklin

Description. 14″ (35cm); 250g. Slightly smaller, with shorter legs than Laughing Gull; rounder head, smaller bill, more pronounced eye-ring give a more dovelike aspect; young birds and winter adults have more pronounced dusky half-hood. **Adults** Winter: mantle fairly pale gray, set off from black wingtips by conspicuous white band; primaries broadly tipped with white; head, underparts, rump, and tail white, except sides of head dusky, extending as dark mottling around nape and crown; conspicuous broken white eye-ring. Bill and legs dusky dark red. Breeding: head black; rosy bloom on underparts; bill and legs brighter red. **First winter:** resembles winter adult but wingtips broadly blackish, blending into gray mantle; wing-coverts and secondaries largely dusky, edged paler; broad black tail-band that does not reach edge of tail. Two years to full adult plumage.

Habits. Frequents harbors, river mouths, mudflats, and salt ponds; in pure flocks during migration or a few birds in larger flocks of Laughing Gulls at other seasons; more likely than Laughing Gull to occur inland, around flooded fields, marshes, and rivers; flies gracefully and buoyantly; soars more than Laughing Gull; behaves much like Laughing Gull but scavenges less; hawks more flying insects; dips to the surface to snatch fishes.

Voice. Soft clucking and mewing notes; a high, thin, piping *peee* or *pwee*; in spring migration a querulous, far-carrying *koyu*! especially when flocks pass over sitting birds or circle to land.

Status. Along Pacific coast and adjacent lowlands a fairly common fall migrant (October–November), uncommon to rare winter resident; in spring migration (April–early June) spectacular flocks fly steadily up the Golfo de Nicoya and Río Tempisque, across the Golfo de Papagayo; occasional in summer. Very rare on Caribbean coast at any season.

Range. Breeds in N Great Plains of North America; winters mainly on Pacific coast of South America, N to Guatemala or, rarely, California; casual to rare in Caribbean area.

BONAPARTE'S GULL

Larus philadelphia Pl. 3(6)

Gaviota de Bonaparte

Description. 13″ (33cm); 200g. Small, dainty, ternlike, with conspicuous white wedge on leading edge of outer wing. **Adults** Winter: head, neck, underparts, rump, and tail white, except crown tinged with gray; dusky spot on auriculars; mantle and upperwing pale gray; outer primaries and their coverts white, primaries black-tipped. Bill black; legs reddish. Breeding: head black; eyelids white; legs brighter red. **First winter:** differs from adult in dusky secondaries and wing-coverts (especially middle coverts, which form a dark diagonal band across upperwing), more extensive blackish tips and outer webs on outer primaries; black subterminal band on tail. Bill dusky with brownish or orangish base; legs dull flesh. Adult plumage attained in second year.

Habits. Mainly around harbors, estuaries, and salt ponds; flies buoyantly and boundingly, like tern; holds its slender bill pointed downward in flight, like a tern; often associates with terns while feeding or resting; forages by dipping to water surface, sometimes hovering, to snatch small fishes, crustaceans, insects.

Voice. A harsh, ternlike *keer*; soft whistled notes while feeding.

Status. Casual to rare; known in Costa Rica from 3 sightings on Pacific coast: May 1977 (2 birds in breeding plumage, Golfo Dulce, by Lewis), December 1978 and January 1980 (immatures, Chomes, by Stiles et al.).

Range. Breeds in Alaska, N and C Canada; winters along both coasts and interior to S Mexico, occasionally to Costa Rica, also Bahamas and West Indies.

Note. The Black-headed Gull (*L. ridibundus*) of Eurasia has occurred several times in the West Indies and should be looked for around Limón, especially during hurricane season. Resembles Bonaparte's Gull but larger (15″, 38cm), less ternlike; only outermost primary below white, the others dusky producing conspicuous contrast, especially from below.

GRAY GULL

Larus modestus Pl. 51(4)

Torero, Garuma

Description. 18″ (45cm); 400g. Dark, medium-sized, with conspicuous white trailing edge to wings and tail in all plumages. **Adults:** body dull gray, above darker, becoming blackish on remiges; white tips of secondaries and rectrices set off by blackish subterminal bands; head white (breeding) to dull brown with broken, pale eye-ring. Bill and legs black. **Immatures:** mostly dark grayish-brown, below paler, blacker on wings and tail; conspicuous grayish-buff feather-edgings on wing-coverts and scapulars; crissum barred with buff; secondaries tipped with pale gray; tail tipped with buffy. Two years to full adult plumage.

Habits. Frequents mainly sandy beaches; in South America often forages for small crustaceans, mollusks, and marine worms by chasing behind retreating waves like an overgrown Sanderling; also eats carrion and offal.

Voice. Mournful mewing and whistling notes.

Status. Accidental at Cocos I. (juvenile, May 1925, by Beebe); casual in Bay of Panama and to be looked for in Golfo Dulce and elsewhere along Pacific coast, especially in years of El Niño current.

Range. Breeds in deserts of N Chile; ranges along Pacific coast of South America N to Ecuador, occasionally Colombia; strays to Cocos I. and Bay of Panama.

Note. Also possible in our area, especially during northern winter, is Heermann's Gull (*Larus heermanni*): breeds NW Mexico; normally winters S along Pacific coast to Guatemala; size of Laughing Gull but in color resembles corresponding plumage of Gray Gull, except after first year with narrow white tipping on flight feathers and at least base of bill pink or red; first-year birds are uniform dark sooty-brown with paler tips to wing-coverts, yellow base to bill; juveniles have bill entirely black. One probable sighting, a juvenile among a loafing flock of Laughing Gulls on a mudflat at Chomes, 6 December 1985, by Stiles et al. [See Pl. 51(5)].

SABINE'S GULL

Xema sabini Pl. 3(7)

Gaviota de Sabine

Description. 13½″ (34cm); 200g. A small, rather short-billed gull with shallowly forked tail, conspicuous white triangle on trailing edge of wing. **Adults** Winter: head, underparts, rump, and tail white, except band of dark gray from auriculars around hindcrown; mantle and upperwing pale gray; outer primaries and their coverts black, tipped with white; secondaries, inner primaries and their coverts white. Bill black, tipped with yellowish; legs grayish. Breeding: head dark gray; bright yellow bill tip. **First winter:** pattern like winter adult, but mantle, hindneck, and upperwing brownish-gray, scaled with buffy; outer primaries and coverts fuscous-black; tail tipped with black.

Habits. In small, loose flocks or singly off Pacific coast; may occasionally rest on the outer coastal beaches, usually with other gulls or terns; flies with ternlike grace, wings continuously beating; swoops to surface, sometimes hovering briefly, to snatch small fishes and planktonic invertebrates.

Voice. Short, harsh, grating, ternlike notes and squeaky chatters, heard mainly while feeding in flock.

Status. Regular transient off Pacific coast in moderate numbers in both fall (August–November) and spring (March–June); rarely enters Golfo de Nicoya or Golfo Dulce; appears to be more numerous in spring but even then only infrequently seen on or near shore. No Caribbean record; probably casual to accidental there (1 record, central Panama).

Range. Breeds in Alaska, N Canada, Greenland and Siberia; winters in E Pacific from Panama to Chile, in Atlantic mainly N of equator; sporadic in Caribbean Sea.

Note. The genus *Xema* is sometimes merged with *Larus*. Also possible off Pacific coast, especially in years of El Niño current, is Swallow-tailed Gull (*Creagrus furcatus*): breeds Galápagos Is. and Malpelo I. off Colombia; nonbreeding birds range off W coast of South America, reported once from Panama; wings and tail like Sabine's Gull but size much larger (22″, 56cm); in nonbreeding plumages has dusky around eye and auriculars, not across hindcrown; bill long, slender; feet bright pink to flesh-color.

BLACK TERN

Chlidonias niger Pl. 2(6)

Fumarel o Charrancito Negro

Description. 9″ (23cm); 50g. Small, gray-tailed; wings and shallowly forked tail relatively shorter than on most other terns. **Adults** Winter: head and underparts white, except irregular dusky cap from eye to hindcrown and auriculars; wing-linings pale gray; mantle, upper surface of wings, rump, and tail dull slate; gray smudge at side of breast. Bill black; legs purplish-black. Breeding: head, neck, breast, and belly black; vent and crissum white. **Immatures:** differ from adult by dusky carpal bar; wing-coverts largely dusky with pale edgings; dusky in primaries; cap darker, more extensive. During first summer wings more uniform gray, but dark carpal bar retained; dark cap still more extensive, underparts mainly white.

Habits. Prefers sheltered offshore waters and bays but often ranges 6mi (10km) or more out to sea, usually not beyond sight of land; comes to shore chiefly during migrations, when flocks sometimes feed over salt ponds, flooded fields and marshes or loaf on sandbars, mudflats, and dikes; offshore, usually singly or in small, loose flocks that quickly concentrate over schools of small fishes driven to surface by larger ones below; feeds by swooping and snatching from surface; rarely dives; also eats plankton and marine water-striders; hawks insects over marshes and fields. Flight buoyant, less bounding and erratic than that of longer-winged *Sterna* terns.

Voice. Staccato, nasal notes: harsh *kik* or *keek* and a softer, more piping *peep*; a longer, scratchier *keeurrr* sometimes heard in feeding flocks.

Status. Abundant migrant (mid-September–mid-November; late April–early June) on both coasts; locally common to abundant winter resident, principally on Pacific coast (centers of abundance Golfo de Nicoya, Golfo Dulce and S Golfo de Papagayo); uncommon to locally common nonbreeding summer resident (mostly first-year birds), especially in Golfo de Nicoya.

Range. Breeds in interior of Canada and N USA; winters from S Central America to Chile and Suriname. Widespread in Old World.

Note. The genus *Chlidonias* is occasionally lumped into *Sterna*.

GULL-BILLED TERN

Sterna nilotica Pl. 2(1)

Charrán Piquinegro

Description. 15½″ (40cm); 180g. Medium-sized, rather big-headed, with short stout black bill, short notched tail, fairly broad wings. **Adults** Winter: head, neck, and underparts white, except dusky crescent before eye and dusky smudge on auriculars; upper surface of body, wings, and tail very pale gray; primaries darker but with silvery "frosting" when fresh. Legs dusky. Breeding: cap black. **Immatures:** like adults except for dusky postocular stripe; mantle spotted with brownish, remiges darker with white tips; dark subterminal smudge on tail. Bill dusky-brown; legs reddish-brown. During first summer resemble winter adults but have more extensive dark markings on head.

Habits. Prefers salt ponds, coastal lagoons, and mudflats, sometimes marshes and flooded fields inland; flies steadily and heavily for a tern, with slow wingbeats, bill only slightly pointed downward, giving a distinctly gull-like aspect; forages, usually singly, by swooping to surface, snatching up small crabs from mudflats, or small fishes, shrimps, insects, and spiders from water; sleeps and loafs singly or in small groups, sometimes with other terns, on dikes, mudflats, and sandspits.

Voice. Dry, rasping *kerreck* or *kay-dit*; various chattering calls.

Status. Along Pacific coast and adjacent lowlands, locally common fall migrant (September–early November), winter resident, and nonbreeding summer resident (mostly young birds); more numerous and widespread during spring migration (April–May), on Caribbean as well as Pacific slope; rare and sporadic on Caribbean slope at other times.

Range. Breeds locally from E and SW USA to N Mexico, also in Ecuador and Brazil to N Argentina; winters from extreme S USA and West Indies to Peru and N Argentina. Widespread in Old World.

Note. Often segregated in monotypic genus *Gelochelidon*.

CASPIAN TERN

Sterna caspia Pl. 3(1)

Pagaza Mayor o Piquirrojo

Description. 21″ (53cm); 650g. Very large, broad-winged, with heavy reddish bill and shallowly forked tail. **Adults** Winter: top of head streaked with black and white, blending into broad blackish eye-stripe; rest of head, neck, underparts, and rump white; tail grayish white; mantle and upperwing pale gray; primaries above and below blackish (silvery "frosting" above when fresh). Bill scarlet (often dusky subterminally with orange tip); legs black. Breeding: cap black. **Immatures:** mantle more brownish; some dusky feather-edgings, especially on wing-coverts; remiges more dusky. Bill brownish-red, tipped with blackish. In any nonbreeding plumage told from Royal Tern by dark forehead, primaries dark below.

Habits. Frequents estuaries, shallow tidal waters and salt ponds; often rests with flocks of other terns, but if 2–3 Caspians are present, they are usually scattered among other terns rather than together; flies heavily, like gull; plunge-dives like a tern but also settles on water to feed like a gull; occasionally robs other terns of their catch; even eats eggs of other birds (e.g., Wilson's Plover).

Voice. A deep, hoarse, croaking or growling *karrrr* or *kowk*; a distinctive, clear, sharp, far-carrying whistle *yeek*! or *yoik*!

Status. Birds in nonbreeding plumage recorded essentially year-round in small numbers around Golfo de Nicoya; no information on migrations, but birds in breeding plumage, presumably spring migrants, noted in late April and May; evidently a winter and nonbreeding summer resident locally. No record for Caribbean slope.

Range. Breeds locally from C Canada to Baja California and SE USA; winters S to extreme N South America. Widespread in Old World.

Note. Often segregated in monotypic genus *Hydroprogne*.

COMMON TERN

Sterna hirundo Pl. 2(5)

Charrán Común

Description. 14″ (36cm); 110g. Medium-sized tern with red or orange legs, slender and rather short bill, long deeply forked tail. **Adults** Winter: nape and hindcrown black, extending around to eye; midcrown mixed with black and white; rest of head, neck, underparts and most of tail white; outer rectrices edged with dusky; mantle and upperwings pale gray; primaries duskier with silvery "frosting" when fresh; primaries below extensively dusky toward tips. Bill blackish, tinged with reddish at base. BREEDING: cap uniform black; underparts pale gray; bill red with (usually) black tip. **Immatures:** from adults by buffy edgings and subterminal dusky smudges on wing-coverts, scapulars, and secondaries; well-marked dusky carpal bar; primaries more extensively dusky; tail shorter, pale gray, strongly edged with dusky. Sandwich Tern has much longer bill, shorter tail, different pattern of dusky on primaries, dark legs. See Forster's Tern.

HABITS. Most frequently in harbors, estuaries, and salt ponds along both coasts, less often off sandy or rocky shores; usually singly or in small loose groups, sometimes in large flocks during migration; flies gracefully, boundingly, with fast wingbeats, wings usually bowed or kinked; dives from flight, sometimes after hovering 10–30ft (3–9m) up, for small fishes, shrimps, and plankton; often associates with other terns while sleeping, loafing, or migrating.

VOICE. One or more high-pitched *kid* or *kip* notes; a grating *kyarr*.

STATUS. Abundant migrant in fall (late September–mid-November), sometimes also in spring (April–May); uncommon to locally common winter resident on both coasts; common nonbreeding summer resident around Golfo de Nicoya, perhaps elsewhere.

RANGE. Breeds C and E Canada, NW and E USA, Bahamas, West Indies, and islands off Venezuela; winters from S USA along both coasts and through West Indies to Peru and Argentina. Widespread in Old World.

NOTE. Two other species of medium-sized *Sterna* terns are possible in Costa Rican waters: (1) Arctic Tern (*S. paradisaea*): breeds N Canada, Alaska; winters in Antarctic waters; possible in migration off Pacific coast; similar to Common Tern in all plumages but has longer, narrower wings, longer tail; much less dusky in primaries (only narrow dark tips on outer primaries below); shorter bill, much shorter legs; darker gray below in breeding plumage, with all-red bill. (2) Roseate Tern (*S. dougallii*): breeds off E USA, locally in West Indies–Caribbean area; possible off Caribbean coast in migration or winter; black bill and brown or black legs, no dusky in primaries below; adults very pale above with extremely long forked tail; young birds with more extensive dusky on head, scaly brown mantle.

FORSTER'S TERN

Sterna forsteri Pl. 2(9)

Charrán de Forster

DESCRIPTION. 14½″ (37cm); 125g. Differs from Common Tern in heavier bill, paler primaries, slightly longer tail (extends beyond wingtips when perched; tail of Common just reaches wingtips); tail darkest on inner portion, not outer edge; distinctive head pattern in nonbreeding plumage. **Adults** WINTER: broad black stripe from eye to auriculars; nape at most washed with gray; rest of head, underparts, and outer edge of tail white; mantle, rump, and most of tail pale gray; inside of tail fork darker gray; primaries above silvery, appearing paler than rest of wing, below with less dusky than Common Tern. Bill brownish, tipped with black; legs orange-brown. BREEDING: cap black. Bill orange with dusky tip; legs reddish. **Immatures:** like adult but with pale scaling on mantle, more dusky on outer primaries, shorter tail with dark-tipped inner rectrices.

HABITS. Forages over salt ponds, freshwater marshes, lagoons, and rivers, singly or in small, loose groups, hawking flying insects like dragonflies; swoops to pluck insects or spiders from water surface or dives for fishes like a typical *Sterna*; flies with quicker, jerkier wingbeats than does Common Tern.

VOICE. A nasal, whining *zaaap* or *tza-ap*.

STATUS. Rare or casual winter resident, recorded between November and February, around Golfo de Nicoya (Chomes) and in Tempisque basin (Puerto Humo, Palo Verde).

RANGE. Breeds in interior North America (S Canada, N USA) and on coasts in S California and from NE USA to NE Mexico; winters regularly in West Indies and S to Guatemala, rarely to Costa Rica.

BRIDLED TERN

Sterna anaethetus Pl. 2(2)

Charrán Embridado

DESCRIPTION. 14″ (36cm); 100g. Medium-sized, dark-backed, with deeply forked tail; told from Sooty Tern by pale nuchal collar and narrow white forehead patch extending into short superciliary stripe. **Adults** BREEDING: cap and loral stripe black; nuchal collar pale gray; rest of upperparts dark brownish-gray; entire underparts and outer rectrices white, tinged with gray on chest and sides. Bill and feet black. NONBREEDING (rarely seen in Costa Rica): crown mostly white, streaked with black posteriorly. **Young:** like nonbreeding adult but above paler, heavily scalloped with pale buff; nuchal collar and sides of breast smudged with dusky; fairly pronounced dark carpal bar; short tail dark.

HABITS. Nests and often roosts on rocky offshore islands; forages mostly well offshore, singly or in small, loose groups; flies swiftly, gracefully, boundingly, typically 30–65ft (9–20m) above the sea; dives or swoops to pluck small fishes, crustaceans, squids, or marine water-striders from surface, which it rarely touches; sometimes rests on water, especially near breeding islands, but more often on driftwood or flotsam.

VOICE. A querulous, descending, whistled *keer* or *keew*; a scratchy, grating, catlike *meahr* (Slud, for Caribbean race).

NEST. In small colonies on steep-sided rocky, offshore islands, hiding nests under tussocks of bunch-grass. May–July. Nests not accessible in Costa Rica; in Trinidad, egg 1, creamy-white, spotted with shades of brown (ffrench).

STATUS. Breeding resident on islands off N Pacific coast, chiefly from March or April through September; rare and sporadic off this coast at other times. Known or suspected nesting areas include Is. Santa Catalina, in S Golfo de Papagayo, and islands around Península de Santa Elena. A sporadic visitor, principally August–September, off Caribbean coast (nearest breeding colonies off Belize and Venezuela).

RANGE. Breeds on tropical and subtropical islands around the world, ranging widely in adjacent seas at other times.

SOOTY TERN

Sterna fuscata Pl. 2(3)

Charrán Sombrío

DESCRIPTION. 17″ (43cm); 180g. Medium-sized; broad white forehead patch of adults extends to eye and forecrown, but not beyond. **Adults:** cap and stripe from eye to bill black; otherwise above uniform sooty-black; underparts and outer rectrices white. Bill and feet black. **Immatures:** head and upperparts dark, belly mostly white. Bridled Tern has cap darker than mantle, pale nuchal collar; noddies more uniformly dark than young Sooties, especially on belly. **Young:** head, chest, and upperparts blackish-brown, scaled with buff on mantle and wing-coverts, shading to paler grayish-brown on belly.

HABITS. Pelagic, rarely close to shore; rarely if ever alights on water but may rest on flotsam; flies buoyantly, with slow and continuous wingbeats; feeds by swooping to surface to snatch small fishes and squids; occasionally plunges to surface but avoids wetting wings; often in large flocks at sea, mixing with other seabirds over good schools of fishes.

VOICE. Shrill, squeaky or quacking *ker wacky-wack* or *ker-wekka*.

STATUS. Very rare visitor to Pacific coastal waters; no record yet for Caribbean coast, but to be looked for off either coast during and following storms. An immature bird, doubtless blown in by a storm, found on Río Sarapiquí near Puerto Viejo (April 1983). A single report of breeding on islet off Cocos I. (May 1962) not confirmed in numerous recent visits to the island.

RANGE. Breeds on tropical and subtropical islands around the world; ranges widely at sea when not nesting. Closest breeding colonies off W Mexico and Galápagos Is. in Pacific Ocean, and off Belize, Honduras and Venezuela in Caribbean.

LEAST TERN

Sterna antillarum Pl. 2(4)

Charrán Chico o Menudo

DESCRIPTION. 9″ (23cm); 45g. Very small; above much paler than Black Tern, with white, moderately forked tail. **Adults** WINTER: crown grayish; nape and stripe extending to around eye black; rest of head, neck, underparts, and tail white; above pale gray, except outer 2–3 primaries dusky. Bill blackish; feet dull yellowish. BREEDING: cap and loral stripe black; forehead and short superciliaries white. Bill yellow with black tip; feet orange-yellow. **Immatures:** mantle indistinctly spotted with dusky; distinct dusky carpal bar; crown and nape more strongly mottled with dusky; primaries more extensively dusky; bill and feet brownish.

HABITS. Prefers inshore coastal waters: bays, estuaries, salt ponds, and lagoons; avoids waves and surf; rests and loafs on sandy beaches, mudflats, and salt-pond dikes; flies more flutteringly than other terns, with rapid wingbeats; captures small fishes and invertebrates by plunge-diving; usually singly or in small, loose groups, sometimes associating with other terns; in larger flocks while migrating.

VOICE. A high, clear, thin *peelee*.

STATUS. Uncommon and local winter resident and nonbreeding summer resident; sporadically common in fall migration (late August–late October), less so in spring migration (April). Center of abundance at all seasons Golfo de Nicoya, perhaps Golfo Dulce; more widespread along Pacific coast in winter than in summer; occasional during migration on Caribbean coast.

Range. Breeds locally from coasts of SW and NE USA to Honduras, West Indies, islands off Venezuela, and interior of C USA; winters to NW and E South America.

Note. Until recently the New World populations were considered subspecies of *S. albifrons* of Old World.

ROYAL TERN

Sterna maxima Pl. 3(2)

Pagaza Real

Description. 19″ (48cm); 450g. Large, with fairly deeply forked tail; slender orange bill. **Adults** Winter: forehead and forecrown white; hindcrown and nuchal crest mostly black; rest of head, neck, underparts, rump, and tail white; tips of outer primaries below dusky; mantle, upperwings pale gray, primaries darker but with pale ''frosting'' when fresh. Bill bright orange; feet black. Breeding: cap black; crest longer; bill deeper orange. **Immatures:** mantle spotted with gray-brown; dark carpal bar contrasts with paler middle coverts; dusky smudging on remiges and tail. Bill paler and duller; feet yellowish to brownish. In first summer attains at most partial black cap with short crest. Slimmer, with longer, narrower wings, more deeply forked tail than Caspian Tern; flight ternlike on bowed wings, not gull-like.

Habits. Fishes singly or in small groups along seacoast, well out from shore beyond breakers; plunge-dives for small fishes, often from considerable height; flies more steadily, with slower wingbeats than smaller terns; may rest on flotsam at sea; often gathers in large flocks, sometimes with Sandwich Terns, to loaf or sleep on mudflats, sandspits, or salt-pond dikes; often in large, mixed tern flocks during migration.

Voice. Clear, short, shrill, high-pitched whistled notes: *keerr, kree, tsirr*, etc.; also a melodious ploverlike whistle and a grating squawk.

Status. Fairly common year-round off both coasts, sometimes locally abundant at river mouths or in Golfo de Nicoya and Golfo Dulce. Nonbreeding summering birds are mostly yearlings. Fall migrants arrive gradually, from late September into December; large flocks mainly on spring migration (April–early June). From May onward many birds in breeding plumage.

Range. Breeds from SW and E USA to NW Mexico, West Indies, and E South America to Uruguay; winters along both coasts from S USA to Peru and Argentina, and along W Africa, S Europe.

Note. Often placed in genus *Thalasseus*.

SANDWICH TERN

Sterna sandvicensis Pl. 3(4)

Pagaza Puntiamarilla

Description. 16″ (41cm); 190g. Deeply forked but fairly short tail and long, slender head and bill give rather front-heavy look in flight; long, slim wings usually bowed. **Adults** Winter: hindcrown streaked with black; short crest and sides of crown around to eyes mostly black; rest of head, neck, underparts, and tail white; above pale gray; primaries dusky with pale ''frosting'' on all but 1–2 outermost, producing dusky leading edges to wingtips. Bill black with pale yellow tip; feet black. Breeding: cap and crest black. **Immatures:** fairly distinct dusky carpal bar; brownish spotting over back and wing-coverts; remiges more extensively dusky; tail largely gray. In first summer attains partial black cap; usually retains trace of dark carpal bar.

Habits. Prefers river mouths, estuaries, harbors and lagoons but widely distributed in coastal waters, sometimes fishing well offshore; flies rapidly and agilely; plunge-dives for small fishes, often from considerable height; also takes shrimps, squids; may form large feeding flocks over rich schools of fishes near surface; typically sleeps and loafs in large groups, often with Royal Terns.

Voice. High, thin, querulous *kreee*; nasal, rasping *kaaar*.

Status. Common, sometimes locally abundant fall (September–early November) and spring (April–May) migrant; in winter common along both coasts; fairly common nonbreeding summer resident (mostly first-year birds) on Pacific coast, especially Golfo de Nicoya; less common on Caribbean coast.

Range. Breeds along coasts from E USA through West Indies (formerly off Belize) and in Eurasia from British Isles to Caspian Sea; winters on Pacific coast from Oaxaca to Peru, on Caribbean-Atlantic coast to Uruguay, and widely in Old World.

Note. Often placed in genus *Thalasseus*. The largely or entirely yellow-billed *S. eurygnatha* (breeds along coasts of E South America, winters from Venezuela and West Indies to Argentina) is often considered a race of *sandvicensis*.

ELEGANT TERN

Sterna elegans Pl. 3(3)

Pagaza Elegante

DESCRIPTION. 16″ (41cm); 235g. Fairly large; very long, slender, slightly decurved bill; crest longer, tail more deeply forked than Royal Tern. **Adults** WINTER: crown spotted with dusky; occipital crest and sides of crown extending around to eye, solid black (not mixed with white as in Royal Tern); rest of head, neck, underparts, and tail white; above pale gray; primaries darker (outer webs of 1–3 outermost dusky), silvery "frosting" when fresh; primaries below more extensively dusky than Royal Tern. BREEDING: cap and crest black. Bill bright orange, palest at tip; feet blackish. **Immatures:** mantle scaled or spotted with gray-brown; wing-coverts extensively gray-brown; remiges dusky; secondaries tipped with white; tail shorter than adult, tinged with brownish.

HABITS. At harbors, estuaries, coastal lagoons, and salt ponds, usually in small numbers and associating with Royal or Sandwich terns, especially in loafing or sleeping aggregations on mudflats or sandbars, in which the Elegants stick together in a tight group; flies more lightly and gracefully than Royal; plunge-dives for fishes and shrimps like Royal, but often from lesser heights into shallower water.

VOICE. Nasal, rasping *kazek* or *pirreet*, often repeated; also a clearer, longer *kreee-eeek*.

STATUS. Uncommon and rather local fall (September–October) and spring (April–early June) migrant and winter resident along Pacific coast; very uncommon but regular nonbreeding summer resident (first-year birds), chiefly around Golfo de Nicoya; occasional at best during migration or winter along Caribbean coast.

RANGE. Breeds from extreme S California to NW Mexico; winters mainly on Pacific coast of South America, sparingly in Central America.

NOTE. Often placed in genus *Thalasseus*.

BROWN NODDY

Anous stolidus Pl. 2(8)

Tiñosa Común

DESCRIPTION. 16″ (41cm); 160g. Dark brown with white cap and long, wedge-shaped tail that in flight appears slightly notched at tip. **Adults:** dark sooty-brown, slightly darker on flight feathers; wing-linings often noticeably paler; lores black; eyelids white; white cap blends through pale gray to dark gray-brown on nape. Bill black; feet brownish-black. **Immatures:** pale cap lacking or restricted to pale gray forehead and forecrown. Black Noddy smaller, with dark underwing and more extensive white cap in all plumages.

HABITS. Mostly offshore, in small, loose groups that quickly coalesce into sizable flocks when a school of small fishes is spotted close to the surface; snatches fish in quick swoops or while hovering but does not plunge-dive; flies swiftly on level course, with rather slow wingbeats and short glides; sometimes alights on rocky islands and promontories but more often rests on driftwood or other flotsam, the backs of sea turtles swimming at the surface, even swimming pelicans; mostly found within a few miles of shore.

VOICE. At sea, a rarely heard low, nasal, cawing *owwr*; around breeding colony, a loud, harsh, guttural *karrk* and a scolding *kwok-kwok*.

NEST. A loosely built, shallow cup of sticks, bits of shell, dry seaweed, etc., on rock ledge, ground or rarely on vegetation. Egg 1, buffy, spotted with purplish and reddish-brown. May–August or September (on Cocos I.).

STATUS. Breeds sporadically on rocky islets around Cocos I. and possibly on similar islets off Península de Santa Elena in N Guanacaste; fairly common migrant in spring (April–early June) and fall (late August–September) off Pacific coast; uncommon or local in summer (possibly breeding); rare or absent in winter. At Cocos I., uncommon to fairly common most of year, possibly disappearing September–October, returning around February. Rare off Caribbean coast.

RANGE. Breeds on tropical and subtropical islands in all major oceans; migrations poorly understood, but northern populations evidently move S after breeding.

BLACK NODDY

Anous minutus Pl. 2(7)

Tiñosa Negra

DESCRIPTION. 14″ (36cm); 100g. Smaller, darker than Brown Noddy, with more extensive white cap; tail appears more distinctly notched in flight. **Adults:** mainly dull brownish-black, including underwing; primaries blackish; lores black; white spot on lower eyelid; upper back and hindneck washed with plumbeous, blending on nape into white of cap. Bill black; feet dusky. **Immatures:** similar but nape darker, white cap terminates sharply behind.

Habits. Forages at sea, usually fairly close to breeding islands but sometimes many miles out; fishes like Brown Noddy but more often in large flocks; evidently also forages at night; flies more dashingly and erratically than Brown Noddy; often rests in large rafts on water. Before and during breeding, pairs and groups dash headlong from interior down valleys to the sea. Nesting adults and immatures at nest give bursts of rapid, vibratory nodding.

Voice. Usually silent at sea; around nesting colony, sharp, dry, nasal cackles, chatters, squeaky notes; also a plaintive, piping whistle *wheeeaeee* with rising inflection.

Nest. A substantial, shallow cup of compacted vegetable materials (seaweed, moss, dead leaves) cemented together with excrement; placed on horizontal limb of tall, open tree (especially balsa or *Cecropia* on Cocos I.), usually many pairs to a tree. Egg 1, buffy to creamy, blotched and speckled with reddish-brown and dark brown. March–September.

Status. Abundant breeding resident on Cocos I.; most or all birds evidently depart from October or November to January or February.

Range. Breeds on tropical islands of Atlantic and Pacific oceans. Most populations relatively sedentary; movements of Cocos I. population not well understood.

Note. This species is sometimes lumped with *A. tenuirostris*, the Lesser Noddy, of the tropical Indian Ocean.

WHITE TERN

Gygis alba Pl. 2(10)

Charrán Blanco (Palomita del Espíritu Santo)

Description. 12″ (30cm); 115g. Fairly small, with distinctive, awl-shaped bill and narrow black eye-ring that imparts a huge-eyed look; tail deeply forked but fairly short. **Adults:** plumage white except for eye-ring and dusky shafts to flight feathers. Bill blue at base, black tip; feet bluish-gray. **Young:** dusky spot behind eye and mottling on nape; back and wing-coverts tipped or washed with rusty.

Habits. Feeds at sea, singly or in loose flocks, usually fairly close to nesting islands but sometimes at great distances; makes U-shaped swoops, sometimes in rapid succession, to catch small fishes as they jump; does not plunge-dive, indeed rarely or never alights on water; flies rapidly and erratically, incredibly swiftly and agilely in the frequent 2- and 3-bird chases during breeding season; also flies with great agility through treetops; tame and curious, often hovering around head of observer on nesting islands.

Voice. At nest or roost, a dry, rasping *jeer-jeer-jeer-jurr-jur-jur* . . . ; fishing at sea and when hovering around a human, a shrill, piping *cheep* or *keek*.

Nest. In crotch of tree, hollow or dip in branch, or other precarious site, 15–50ft (5–15m) up; sometimes in loose colonies in favored trees. Egg 1, creamy-white, boldly blotched, spotted, or scrawled with various shades of brown. March–August or September.

Status. Abundant breeding resident at Cocos I. between February and September, absent during remainder of year; occasional or accidental off mainland Pacific coast; one sighting (July 1978) well within sight of land off Cabo Blanco (Stiles et al.).

Range. Tropical and subtropical oceans around the world.

Note. Occasionally placed in genus *Anous*; sometimes called Fairy Tern.

FAMILY **Rynchopidae:** Skimmers

Skimmers, of which 1 genus and 3 species are recognized, are widespread along the coasts of North and South America, on the great rivers of the latter, and in tropical Africa and Asia. Those that breed farthest from the tropics are migratory. In length 15–20″ (38–50cm), they are black or brownish above and white on underparts, face, and forehead. After breeding, the American and African species acquire white collars, which the Indian species wears permanently. The sexes are alike in plumage, but males are larger than females. Immatures are streaked and spotted above. The tail is short and shallowly forked, the wings long and pointed, the feet small and webbed. Skimmers have catlike, vertical pupils unique among birds, which can be narrowed in bright sunshine. Vermilion at the base in all species, the curious bill may be either black or yellow toward the end. The lower mandible, much longer than the upper, is knifelike and flexible, the sides closely striated with minute, oblique ridges. The upper mandible is exceptionally mobile and toward the end grooved to accommodate the sharp edges of the lower when the bill is closed.

Beating their wings slowly and above the horizontal to avoid wetting them, skimmers fly back and forth low over the calm water of lagoons, estuaries, and wide rivers, cutting or plowing the water with the long lower mandible of the widely open bill, which snaps shut upon making contact with a small fish, crustacean, or other aquatic creature. Although skimmers may forage at any time of day, they prefer late afternoon and night, when small fishes rise to near the surface. When not skimming over the water, the birds rest on beaches and exposed sandbars in crowded companies, sometimes numbering hundreds or thousands, all facing into the wind.

On sandy shores and sandbars, skimmers nest colonially in shallow, unlined scrapes usually at least several yards apart. Their eggs, 2–5 and sometimes as many as 7, are buff, pale blue, greenish-blue, rarely white, heavily blotched and spotted with brown, black, and pale lilac. Both sexes incubate (at least in the African species) for about a month. As soon as they hatch, the cryptically colored, downy chicks leave the nest and move over the beach, where they dig little depressions in which they lie flat and inconspicuous. Of equal length when the chick hatches, the two mandibles rapidly become unequal as the young bird becomes feathered.

BLACK SKIMMER
Rynchops niger Pl. 3(11)
Rayador Negro

Description. 19″ (48cm); ♂ 350g; ♀ 260g. Suggests a large, black-backed, long-winged tern except for its unique bill. **Adults** Winter: lores, forehead, hindneck, and underparts white; cap dusky black, mottled with white; mantle and upperwings black; secondaries and inner primaries tipped with white; wing-linings and outer rectrices white (northern birds) or dark sooty-gray (southern birds). Bill blackish, scarlet at base; feet vermilion. Breeding: cap and hindneck black like back; half or more of bill scarlet. **Immatures:** some pale-edged juvenile feathers on upperparts, in fall, mostly disappear by late winter. **Young:** feathers of upperparts with dusky bases, broad buffy edges producing scaled or mottled appearance.

Habits. Frequents areas of quiet water in rivers, estuaries, lakes, tidewaters, salt ponds, and coastal lagoons; flies deceptively fast, with slow, "tired-looking" wingbeats, the wings never dipping below the horizontal; feeding as described for the family, mostly at dawn and dusk, or at night; rests by day on mudflats or sandbars, usually with flocks of gulls and terns.

Voice. A distinctive sharp, nasal, high-pitched bark or yelp, like a small dog: *aow!* or *kowp!*, *kelp!*

Status. North American breeders are locally common fall (mid-September–late October) and spring (early April–late May) migrants and winter residents around Golfo de Nicoya; occasional and in much smaller numbers elsewhere (Río Tempisque basin, Bahía Salinas, Lago Caño Negro, Golfo Dulce); occasional nonbreeding summer visitor. Between May and October small numbers of South American breeders winter around Golfo de Nicoya and Golfo Dulce.

Range. Breeds locally in SW USA and NW Mexico, along Atlantic-Gulf coasts from NE USA to SE Mexico, and along coasts and larger rivers of much of South America. Northern birds winter from S USA south to Argentina and Chile; southern birds winter north in small numbers to Costa Rica.

ORDER **Columbiformes**
FAMILY **Columbidae:** Pigeons and Doves

The 285 species of this family are found on all continents except Antarctica, most wooded islands, and from warm tropical lowlands to snow line on Himalayan heights. The smallest are sparrow-sized, the largest as big as a curassow. The smaller ones are often called doves, the larger ones pigeons, but they overlap in size and cannot be separated by anatomy or habits. Virtually all are so similar in their compact bodies, small heads, and short necks and legs, that one who knows the domestic pigeon will have no difficulty recognizing any of them. The plumage of pigeons and doves is soft and dense; our New World species wear subdued colors in softly blended patterns (a few are boldly scaled), embellished with metallic iridescence on the neck or wing-coverts. The bill of pigeons is straight, with a fleshy cere at the base, and often rather slender and weak; hard food is broken up in the powerful, muscular gizzard with the aid of small pebbles swallowed for this purpose. The diet consists largely of seeds, nuts, and fruits, often supplemented with insects and other small invertebrates. Many pigeons forage almost wholly on the ground, walking with bobbing heads instead of hopping, but others gather berries, arillate seeds, and nuts in trees. Most build rather frail open nests in trees and shrubs, but a number nest on the ground, and a few Old World species in caves and rock crevices. Nests are usually scattered, but a few species, including the White-winged Dove in southwestern United States, and the Eared Dove of South America, breed in crowded colonies. Although 2 eggs are usual, many species lay only 1. The white or, rarely, buffy, unmarked eggs are incubated by both parents, the male sitting through much of the day, the female all the rest of the time, including the night. The young hatch helpless, thinly covered with hairlike feathers. Both parents feed them by regurgitation, at first with almost pure pigeon milk (a lipid-rich secretion of the crop lining), which after a few days is mixed with increasing amounts of solid food. Incubation periods of smaller species are 11–16 days; nestling periods 10–17 days. For larger pigeons, the periods are 17–28 days and 20–36 days, respectively.

ROCK DOVE

Columba livia Not Illustrated

Paloma Domestica (Paloma de Castilla)

Description. 12″ (30cm); 300g. The common pigeon of cities and towns; plumage highly variable, making flocks easy to recognize at a distance. **Adults:** mostly dark gray, glossed with purple and green on neck (especially males); wings gray, middle coverts and secondaries tipped with black forming 2 black bands across folded wing; rump and wing-linings white; tail with black tip. Iris orange; cere whitish; bill black; legs magenta. Often this wild-type plumage is more or less replaced or obscured by areas of white, blackish, and/or reddish-brown.

Habits. Feral birds inhabit mostly urban areas, nesting in man-made structures and foraging for seeds and scraps in streets and plazas, around markets, grain silos, and rice-processing plants, sometimes ranging into nearby croplands. Domestic birds, widely kept, are often allowed to fly free by day. Flight swift, agile, usually in flocks, which often execute strikingly coordinated aerial maneuvers. Males engage in circling courtship flights, soaring on stiff, raised wings.

Voice. A hollow, moaning cooing: *coo, cu-coo-cooorr . . .* ; a low grunt when alarmed.

Nest. A scanty cup of twigs, grass stems, etc., on ledge or girder of church steeple, bridge, warehouse, or other structure. Eggs 2. Possibly year-round.

Status. Fully domestic to completely feral in Costa Rica; feral birds mostly around larger towns and cities, especially Valle Central; flocks in rural districts are probably free-flying domestic birds.

Range. Native to Eurasia and N Africa; essentially worldwide as a commensal of man.

WHITE-CROWNED PIGEON

Columba leucocephala Pl. 51(1)

Paloma Coroniblanca

Description. 13″ (33cm); 290g. A large, dark, square-tailed pigeon with a conspicuous pale crown. **Adult ♂:** pileum shining white; nape dark maroon; hindneck metallic bronzy-green, scaled with black; rest of plumage blackish-slate, shading to dark slate-gray on rump, upper tail-coverts and belly. Iris white; orbital skin pinkish-white; cere and most of bill dark red, tip pale greenish; legs crimson. ♀: similar but crown grayish-white, body plumage tinged with brownish. **Immatures:** browner overall than female; crown grayish-brown.

Habits. Restricted to coastal forests and mangroves; arboreal, feeding on fruits and berries, occasionally snails; flight swift and direct in open, agile and twisting in vegetation; often in flock, sometimes alone.

Voice. Song a loud, deep, clear *cooo, coo-coo cooo* or *cooo-cura-cooo*.

Status. Casual or accidental along Caribbean coast. Only Costa Rican record a sighting of an adult male flying along estuary and coast at mouth of Río Tortuguero on 4 March 1987 (Vickery and Perkins). To be expected occasionally along Caribbean coast, as it often ranges from island breeding sites to coastal areas to forage; most likely from Tortuguero to Barra del Colorado, and in Sixaola region, because of proximity to breeding areas.

Range. Breeds throughout Caribbean from islands off Yucatán and Florida to NW Panama and Lesser Antilles; nearest breeding sites include Isla de Providencia, off Nicaragua, and Escudo de Veraguas, off Panama; regularly ranges to coasts of Bocas del Toro.

SCALED PIGEON

Columba speciosa Pl. 18(4)

Paloma Escamosa (Paloma Morada)

Description. 12½″ (32cm); 320g. Our only large, reddish pigeon with conspicuously scaled plumage. **Adult ♂:** crown and most of upperparts rufous- to purplish-chestnut; neck, breast, and upper back rich buff to whitish, boldly scaled with glossy greenish- to purplish-black; belly white, indistinctly scaled with purplish-brown; primaries and tail brownish-black. Bill bright red with white tip. ♀: similar but duller; mantle dull brown, neck scaling without metallic gloss. **Young:** head, neck, and breast dull brownish-gray, scaled indistinctly with dusky; wing-coverts scaled with cinnamon; bill dull reddish.

Habits. Mostly in forested regions, preferring canopy, edge, and semi-open; perches on high, exposed limbs and flies above treetops, especially when crossing open spaces; takes flight with an explosive burst of heavy wingbeats; usually seen singly, in pairs or trios; eats fruits of trees and epiphytes, including melastomes, *Hirtella*.

Voice. Song of male a deep, full, far-carrying *coooo-cu-coooo* repeated several times, or *hwooo hip-hooWOOOoh*, delivered from high, open perch.

Nest. A broad, slightly concave platform of fine twigs and branched dry inflorescences, from 150ft (45m) up in a treetop to as low as 2ft (60cm) in a thicket or amid bracken fern. Egg 1 (2 in Trinidad), white. February–August.

Status. Resident; common in lowlands S of Lago de Nicaragua, including Río Frío region; uncommon to rare over rest of Caribbean lowlands; locally common in General-Térraba–Coto Brus valleys, but absent from Golfo Dulce lowlands; sea level to 4000ft (1200m).

Range. S Mexico to W Ecuador, N Argentina, and SE Brazil.

BAND-TAILED PIGEON

Columba fasciata Pl. 18(1)

Paloma Collareja

Description. 14″ (35cm); 315g. Large, long-tailed; conspicuous yellow bill and white crescent on nape; pale gray distal third of tail separated from dark base by blackish band. **Adult ♂:** head and most of underparts dull purplish, fading to whitish on crissum; hindneck glossy iridescent green; back dark olive-gray with bronzy-green gloss; wings dark gray. Iris purplish-red, eyelids coral-red; bill, cere, and legs yellow. ♀: below grayer (less purple), less green on hindneck. **Young:** head and neck dull slate; no white crescent on nape; above with no iridescent green; below brownish with rufous suffusion or scaling.

Habits. Prefers mountainous country that is at least partly forested; usually seen in flocks of up to 30, rarely larger, that perch high in treetops and fly high over valleys, lower over ridges; flight strong, swift and direct. Favorite foods include acorns, which are plucked from the cup and swallowed whole; also fruits of *Myrica, Rapanea*; descends low for berries of *Phytolacca*.

Voice. A deep, throaty, slightly burry or rolling *c'cooo c'cooo* or *cooOOO cooOOO*, often 5–10 times in succession; also a low rattle; in flight, a nasal whinnying *nya-a-a-ah*; wings make loud slapping sound when bird takes flight and during courtship flights.

Nest. A loosely constructed, slightly concave platform of coarse sticks, usually on a horizontal branch of a tree near the trunk, 8–50ft (2.5–15m) up. Egg 1, rarely 2, white. March–May or June.

Status. Resident; common to abundant in higher parts of all major mountain ranges, down to 3000ft (900m) locally, perhaps seasonally; but large flocks of past years have been greatly reduced by shooting in many areas.

Range. Extreme SW Canada to NW Argentina.

Note. Birds from Costa Rica S were formerly often considered a distinct species, *C. albilinea*, White-naped Pigeon.

PALE-VENTED PIGEON

Columba cayennensis Pl. 18(3)

Paloma Colorada (Paloma Morada)

Description. 12″ (30cm); 250g. Only large, arboreal lowland pigeon with distinctly pale crissum. **Adult ♂:** rufous-purple mantle contrasts with mostly slate-gray head; throat whitish; nape iridescent bronzy-green; rump slaty; tail ash-gray, paler toward tip; remiges brownish-gray; below vinaceous, shading to gray on wing-linings and whitish on lower belly and crissum. Iris red; bill and cere black; legs magenta. ♀: below with less purplish, above less rufescent. **Young:** much duller; below mainly brownish-gray; above dark gray-brown; head grayish with dull chestnut scaling. Similar Red-billed Pigeon has pale bill, less contrasting pattern.

Habits. Avoids closed forest but frequents forest edge, isolated tall trees in pastures, second growth, coconut palm plantations, mangroves, and river margins; singly, in small groups or, especially following breeding, in flocks of 12 or more; eats berries of trees and shrubs, especially *Trema* and melastomes.

Voice. A resonant *cu'WOOO cu-cu WOOOO* or *c'woOOOoo-cu-c'-woOOOoo*, repeated over and over; like call of Red-billed Pigeon but usually 1 or 2 (not 3) short syllables separating longer coos; usually calls from an elevated, exposed perch.

Nest. A frail, shallow cup or concave platform of twigs and inflorescences 3–50ft (1–15m) up in shrub, small tree, or palm; sometimes a cup of grass or twigs on ground, under grass tussocks. Egg 1, white. February–June.

Status. Resident in humid lowlands of Caribbean and S Pacific, up to 2000ft (600m); especially common in coastal areas (on Pacific coast N to Tárcoles), also in Río Frío region and on Isla del Caño.

Range. S Mexico to SW Ecuador and N Argentina.

RED-BILLED PIGEON

Columba flavirostris Pl. 18(2)

Paloma Piquirroja (Paloma Morada)

Description. 12″ (30cm); 230g. Fairly large; bill ivory-white to pale pink with rose-pink base and cere. **Adults:** head, neck, and breast deep vinaceous-purple (brighter in male); back dark brownish-gray; shoulders extensively purplish-rufous; primaries dusky; rest of wing, rump, tail coverts and belly bluish-gray; tail blackish. Iris orange; bare eye-ring carmine; legs magenta. **Young:** purplish areas much duller; underparts suffused with rufous; shoulder patch darker, duller, mixed with dusky; mantle and wings more brownish.

Habits. Frequents open country with scattered trees, semi-open, agricultural areas; invades large clearings in forested areas; usually singly or in pairs, rarely larger groups perching in trees; eats berries, acorns, and buds plucked from trees or shrubs; forages on ground in open areas; locally a pest, eating sprouting seeds of corn or sorghum.

Voice. A loud, far-carrying *woooo, cuk c'c' coooo, cuk c'c' cooo*; a long-drawn-out, sonorous ascending note followed by 3 shorter notes: *cooooo cu cu coo*.

Nest. A thin, frail platform of coarse sticks, 15–80ft (4.5–25m) up on horizontal branch or in crown of a spiny palm, sometimes in a dead tree, often in a small copse, plantation, or large tree in open rather than inside dense woodland. Egg 1, white. March–August.

Status. Resident; fairly common in dry NW lowlands and Valle Central; uncommon over most of Caribbean lowlands, S to Limón; somewhat more numerous at middle elevations where clearings are extensive; occasional in S Pacific lowlands; ranges up to 7000ft (2100m).

Range. NW Mexico and S Texas to Costa Rica.

RUDDY PIGEON

Columba subvinacea Pl. 18, See (5)

Paloma Rojiza

Description. 11″ (28cm); 170g. Smallish, dark, unpatterned arboreal pigeon with dark bill; safely distinguishable from Short-billed Pigeon only by voice (though only limited overlap in elevation). **Adult ♂:** head, neck and underparts dull vinaceous; mantle dark chestnut, becoming more purplish on rump; wings and tail dark olive-brown with purplish gloss. Iris purplish-red; bare eye-ring red; bill and cere black; feet magenta. ♀: averages duller, browner overall. **Young:** head and neck brownish-gray; above dark olive-brown; feathers of wing-coverts, crown, and rump edged with dull rufous; below paler, flecked or scaled with cinnamon.

Habits. Lives chiefly in canopy of middle and high-elevation forests, also forest edge and adjacent second growth; usually singly or in pairs, occasionally small groups; perches and calls from treetops; eats fruits of trees, epiphytes, and shrubs, especially mistletoe berries; rarely descends to ground in open places for grit.

Voice. Song a clear *coo c'COO coo* or *cook for YOU too*, accent on 3rd note (rather than 2nd and 4th as in Short-billed Pigeon); also a churring growl, harsher and rougher than that of Short-billed.

Nest. A small platform of loosely stacked twigs in fork of bough near top of small tree, ca. 16ft (5m) up, in grove of trees in pasture beside forest. Egg 1, white. June–July (McCoy).

Status. Resident, widespread but usually not very numerous in mountains N to Cordillera de Tilarán. Breeds mainly from 5000ft (1500m) to timberline; often descends to 3000ft (900m) in nonbreeding season, mainly on Caribbean slope.

Range. Costa Rica to W Ecuador, Bolivia, and Amazonian Brazil.

SHORT-BILLED PIGEON

Columba nigrirostris Pl. 18(5)

Paloma Piquicorta (Dos-tontos-son)

Description. 10½″ (26.5cm); 150g. Fairly small, dark, unpatterned arboreal pigeon of lower elevations; bill short and thick, black. **Adults:** head, neck and breast dull vinaceous, shading to brown on belly; rest of upperparts dark olive-brown, glossed with purplish on lower back and rump; tail darker; primaries blackish. Female averages duller, browner.

Much less rufescent above than Ruddy Pigeon, but this difference rarely apparent in the field. Iris reddish; eye-ring and legs magenta. **Young:** head, neck and breast grayish-brown, heavily scaled or flecked with cinnamon; above dull olive-brown, scaled with dull cinnamon.

HABITS. Frequents forest canopy, edge, semiopen, and tall second growth; usually perches and calls in treetops but frequently enters neighboring low scrub or weedy clearings to forage; may alight on dirt roads or sandy stream edges to gather grit or small invertebrates; paired throughout year; eats fruits of mistletoes and *Cecropia*, berries of trees and shrubs, which it occasionally gathers from shady open ground.

VOICE. Song a far-carrying, melodious *who-COOKS-for YOU* or *cu-coo k'coo*, accented on 2nd and 4th syllables; in Spanish *dos TONtos SON*, a characteristic sound of lowland wet forests through much of the year. Also a hoarse, rolling *krooouw*, often given at midday and sometimes in aggressive situations.

NEST. A flimsy platform of twigs and dry inflorescences, 15–100ft (5–30m) up in dense vine tangle or high in tree, in tall second growth or forest. Egg 1, white. March–August.

STATUS. Common to abundant resident in humid lowlands, locally to 3600ft (1100m) on Caribbean slope and 4800ft (1450m) on S Pacific slope; absent from dry Pacific NW; uncommon in NW Caribbean lowlands S of Lago de Nicaragua.

RANGE. S Mexico to NW Colombia.

MOURNING DOVE

Zenaida macroura Pl. 18(13)

Paloma Rabuda

DESCRIPTION. 12″ (30cm); 100g. Slim, small-headed with slender bill and long pointed tail with conspicuously white-tipped outer rectrices. **Adults:** crown, nape, and hindneck slate, passing to grayish-brown over rest of upperparts; primaries dusky; wing-coverts and scapulars boldly spotted with black; face pinkish-buff, shading to pale vinaceous on foreneck and breast, buffy on belly; patch of iridescent purplish on side of neck, larger in male. Female browner overall. Bare eye-ring pale bluish; bill and cere black; legs reddish. **Young:** duller, more grayish; upperparts scaled with buffy.

HABITS. In open country with scattered trees: pastures, agricultural land, savannas, second growth; forages singly or in flocks, chiefly on ground in open areas, gathering fallen seeds of wild plants or grain in stubble fields, grit, and occasional insects; rarely plucks seeds or fruit in trees; when startled, bursts into rapid flight with whistling wings.

VOICE. A low-pitched, mournful *cooWOOoo coo, coo coo* or *coo, cooWEEu coo, coo coo*.

NEST. A shallow cup or pad of grasses and twigs, on ground under tuft of grass, or up to 4ft (1.5m) in large grass tussock. Eggs 2, white. March–July.

STATUS. Resident in E Valle Central, where especially abundant around Cartago and on lower slopes of Volcán Irazú, to at least 7500ft (2300m); occasional in Valle del General. Large numbers of northern migrants winter from Guanacaste S to Parrita area (October–March), occasionally elsewhere on both slopes.

RANGE. Breeds from S Canada to C Panama and Greater Antilles; northern populations withdraw S in winter.

NOTE. Formerly separated in genus *Zenaidura*.

WHITE-WINGED DOVE

Zenaida asiatica Pl. 18(12)

Paloma Aliblanca (Paloma Ala Blanca)

DESCRIPTION. 10¾″ (27cm); 145g. Medium-sized, brownish, with prominent white wing-band and rounded white-tipped tail. **Adults:** mostly pale brown, above darker; prominent black spot on lower cheek; posterior underparts and wing-linings gray; greater coverts white; primaries dusky; central rectrices brown, lateral ones with gray bases separated from broad white tips by median black band. Iris orange; bare eye-ring blue; bill black; legs carmine. Male has purplish suffusion on head and neck, iridescent patch of bronzy-purple on side of neck; these reduced or lacking in female. **Young:** paler and grayer, lacking purplish altogether; black cheek spot small; mantle scaled with dull buff.

HABITS. Prefers arid country with cacti, thorny scrub, and scattered trees; forages for seeds, grain, or grit on ground in open woodland and savannas; perches in trees; often nests around edges of mangroves or in "palo verde" trees in marshes; highly gregarious where abundant; may roost in large flocks in savanna trees; when alarmed, takes flight with loudly flapping wings.

VOICE. Song prolonged and complex: *guu-gu-*

g'gu guu g'gu guuu, or *coo-COO-coowoo coo-COO-coowoo coo coo cooo.* A common call is a simpler *coowook-coowoo.*

Nest. A shallow saucer of coarse sticks, 6–25ft (2–8m) up in a thorny tree, cactus, shrub, or small mangrove. Eggs 1 or 2, white with a faint buffy tinge. January–March.

Status. Fairly common to locally abundant permanent resident in dry Pacific NW, S to about Jacó, to ca. 1650ft (500m), rarely higher, on W approaches to Valle Central; occasionally strays E to San José. Large numbers of northern migrants also winter in Guanacaste (roughly November–May).

Range. SW USA to C Panama; W South America from SW Ecuador to N Chile; Bahamas and Greater Antilles.

COMMON GROUND-DOVE

Columbina passerina Pl. 18(8)

Tortolita Común (Tortolita, Palomita)

Description. 6¼″ (16cm); 40g. Small, plump, short-tailed, with dark spotting and scaling on breast. **Adult ♂:** back of head and neck gray, scaled with blackish; rest of head, neck, and underparts grayish-pink, neck and breast spotted and scaled with dusky; above grayish-brown; wing-coverts paler, pinker, boldly spotted with metallic violet; wing-linings and extensive area on remiges rufous; lateral rectrices blackish, the 1–2 outermost narrowly tipped with white. Iris reddish; bill pale red with dusky tip; legs flesh-color. ♀: browner overall, pink and gray replaced by pale brown and grayish-brown, respectively. **Young:** like adult female but above scaled with whitish and with buffy on breast; dusky markings on breast fainter.

Habits. Frequents open areas generally, especially those with bare ground: pastures, lawns, cultivated land, roadsides, openings in light woods; in pairs or small flocks; gathers seeds, berries, and occasional insects from ground; often abundant in late afternoon along unpaved country roads, evidently seeking grit; roosts and perches in trees.

Voice. Song a low, soft, clear *cooOOP* or *cowoot*, often repeated ca. once per second up to 15sec.

Nest. A slight, round pad or shallow saucer of fine grass stems, rootlets, etc., on ground beneath a sheltering shrub or tussock of grass, or in a shrub or tree up to ca. 30ft (9m), or on ledge of building. Eggs 2, rarely 3, white. January–September; nearly year-round.

Status. Common to abundant resident throughout N Pacific lowlands and Valle Central, S to entrance of Golfo de Nicoya and E to continental divide; uncommon around Cartago, and in Río Frío district; rare or absent elsewhere on Caribbean slope and on S Pacific slope.

Range. S USA to Ecuador and E Brazil; also Bermuda, Bahamas, and West Indies.

Note. Also called Scaly-breasted Ground-Dove; this and the next 2 species were formerly placed in the genus *Columbigallina.*

PLAIN-BREASTED GROUND-DOVE

Columbina minuta Pl. 18(9)

Tortolita Menuda (Tortolita, Conchita)

Description. 5½″ (14cm); 33g. Very small; differs from Common Ground-Dove by plain breast and brownish cere and bill; lacks pale head and rufous on rump of larger female Ruddy Ground-Dove. **Adult ♂:** crown and nape bluish-gray; rest of upperparts grayish-brown; wing-coverts tinged with pink and spotted with metallic violet; below pale pinkish-gray fading to pale buff on belly; lateral rectrices blackish with gray bases and narrow whitish edgings at tip; wing-linings and bases of remiges rufous. ♀: similar in pattern but head grayish-brown like back; below grayish-buff, shading to buffy-white on belly. **Young:** like adult female but scaled narrowly with buffy on mantle and cinnamon on chest.

Habits. Prefers savannas and grassy fields with scattered small trees; behavior like that of other ground-doves but rarely forms flocks.

Voice. *co way, co way* rapidly repeated; also *woo-ah woo ah, woo ah* (Wetmore).

Nest. A slight platform or shallow cup of grass or fine twigs, on ground under grass tussock or to 30ft (9m) up in shrub or small tree. Eggs 2, white. May–July in Pacific NW; also January near Parrita; possibly throughout year.

Status. Uncommon to fairly common resident locally; centers of abundance include Río Frío region, savannas of Térraba region, and deforested lowlands of Coto region E of Golfo Dulce; uncommon in Valle de Parrita, rare in N Pacific lowlands.

Range. S Mexico locally to Peru and N Paraguay.

RUDDY GROUND-DOVE

Columbina talpacoti Pl. 18(7)

Tortolita Rojiza (Tortolita o Palomita Colorada)

Description. 6½″ (16.5cm); 48g. Reddish male distinctive; female differs from Common Ground-Dove in plain breast, no red on bill, reddish rump, relatively pale head; no other small dove shows black in wing-linings. **Adult ♂:** head pale gray; forehead and throat whitish; neck, chest, back and rump purplish-chestnut; wing-coverts, distal wing-linings, and large patch on primaries rufous; bold black spots on wing-coverts; proximal wing-linings and lateral rectrices black. Iris red; bare eye-ring, bill, and cere yellowish to brownish; legs and feet flesh. ♀: duller, browner; head buff; crown and nape tinged with gray; above gray-brown; below grayish-buff, shading to dull rufous on rump, central rectrices, and crissum. **Young:** like adult female but scaled with dull rufous (males) to buff (females) on upperparts.

Habits. Prefers open areas with bare ground or sparse vegetation in pastures, agricultural fields, near dwellings, or in second growth; invades clearings in forested areas but avoids interior of woods; in pairs or flocks of 10–20, forages on ground for seeds and berries; often seen on roads, probably gathering grit; roosts gregariously amid dense foliage.

Voice. Song a soft *kitty-WOO* or *b'booOOP*, repeated up to 10 times in measured sequence, ca. 10 calls/7sec; slightly lower-pitched and less clear than song of Common Ground-Dove; also a simple, low *coo* and a longer *too-oo-wooo*.

Nest. A shallow but usually compact cup of vegetable fragments, usually 3–10ft (1–3m) up, rarely as low as 1ft (30cm), in tree, shrub, or other plant. Eggs 2, white. January–September.

Status. Resident; common to abundant in deforested lowland areas of Caribbean slope, and up to ca. 4600ft (1400m) around Cartago; also S Pacific slope, locally up to ca. 4000ft (1200m), and N along coast to Río Abangares; rare farther N in Guanacaste except locally common on Pacific slope of Península de Nicoya.

Range. N Mexico (casually, Texas) to NW Peru, N Argentina, and N Uruguay.

INCA DOVE

Columbina inca Pl. 18(11)

Tortolita Colilarga (San Juan)

Description. 8″ (20cm); 52g. Small, pale grayish, scaled with black; long white-edged tail. **Adult ♂:** forehead and underparts pale grayish-pink, shading to creamy-white on belly; above pale brownish-gray, palest on greater wing-coverts; scaled with blackish faintly on face, foreneck, and breast, more heavily elsewhere; wing-linings and bases of remiges extensively rufous; central rectrices grayish, lateral ones blackish, broadly edged and tipped with white. Iris red; bill and cere blackish; legs flesh-color. ♀: similar but below with little or no pink tinge; head and breast more heavily scaled. **Young:** like adult female but below browner, scaling less distinct; supraloral area pale; mantle feathers with buffy mottling subterminally, dark scaling more conspicuous, especially on wing-coverts.

Habits. Prefers open to lightly wooded country, scrubby second growth, lawns and dooryards; usually in pairs or small groups; largely terrestrial, gathering seeds and grit from open or sparsely vegetated ground; rests and sleeps in trees. Males call from elevated perches including electric wires and television antennas.

Voice. A rather strident *cowl-coo*!, sometimes paraphrased *No-hope*! A harsher, growling *grrr-p'hrrr* in excitement or aggressive situations.

Nest. A flimsy saucer of straws, fine twigs, and bits of weeds, 3–26ft (1–8m) up in tree (especially spiny species). Eggs 2, white. April–August, possibly to October.

Status. Presently an abundant resident throughout NW lowlands, S in diminishing numbers along coastal strip to at least Quepos; common throughout Valle Central E to Ochomogo, and uncommon to rare at least to Paraíso; has been expanding its range S in this century; first recorded in Costa Rica in 1928 in N Guanacaste; by 1937 common in S Guanacaste, by 1964 had reached Valle Central W of San José; reached Cartago and Quepos in 1970s.

Range. SW USA to Costa Rica.

Note. Often (and perhaps more appropriately) placed in genus *Scardafella*.

BLUE GROUND-DOVE

Claravis pretiosa Pl. 18(6)

Tortolita Azulada

Description. 8″ (20cm); 72g. Larger, slimmer, longer-tailed than *Columbina* ground-doves. **Adult ♂:** above bluish-gray, below paler, shading to grayish-white on face; remiges and lateral rectrices blackish; wing-coverts with bold black spots and bars. Iris red to yellowish; bare eye-ring olive-green; bill dull greenish with yellowish tip; legs flesh-color. ♀: head, neck and breast pale grayish-brown, shading to pale bluish-gray on wing-linings and posterior underparts; mantle dull brown, shading to bright chestnut on rump, tail-coverts, and central rectrices; wing-coverts spotted and barred with rufous-chestnut; lateral rectrices and remiges dusky. **Young:** like adult female but mantle scaled with cinnamon; young male grayer, wing-markings blackish.

Habits. Frequents open woodland, second growth, semi-open, edges, and clearings in wooded country, roadsides; usually singly or in pairs (the combination of a blue-gray bird and a brown bird flying together is diagnostic at any distance), occasionally in groups in good feeding areas; walks over ground, picking up seeds and small insects or grit; perches, roosts, and calls from well up in trees; males sing from high treetops.

Voice. Song a soft but resonant *coot, boop*, or *boo-oop*, repeated persistently, especially towards midday; also a short, hoarse *khoor* and croaking notes.

Nest. A frail saucer of fine twigs, tendrils, and rootlets, 3–36ft (1–11m) up in thicket, vine tangle, coffee shrub, or garden. Eggs 2, white. February–October.

Status. Resident countrywide from sea level to ca. 4000ft (1200m); most common in humid areas, especially Caribbean and S Pacific slopes; in drier NW scarce except in moist areas around Golfo de Nicoya, and along W side of Península de Nicoya.

Range. SE Mexico to W Peru and N Argentina.

MAROON-CHESTED GROUND-DOVE

Claravis mondetoura Pl. 18(10)

Tortolita Serranera

Description. 8½" (21.5cm); 95g. A large highland ground-dove with conspicuous white tail markings. **Adult ♂:** throat and face whitish, shading to bluish-slate on upperparts; wing-coverts crossed by 2 broad violet-black bars; remiges blackish; outer 3 rectrices white; foreneck and breast maroon-purple, shading through gray to white on crissum. Iris orangish; bill black; legs pink. ♀: face buffy; throat whitish; upperparts dull brown, brighter and redder on rump and central rectrices; lateral rectrices blackish, tipped with white to buffy; wing-coverts barred as in male. **Young:** resemble adult female but browner overall; wing-coverts edged with buff; wing-bars dusky and indistinct; underparts scaled or freckled with buff.

Habits. Locally and sporadically at edges and openings in highland forest and tall second growth; forages on ground, often under dense shrubbery, for seeds and fallen fruits, including those of *Phytolacca*; often appears when bamboo is fruiting, disappears when seed crop is exhausted; shy and wary, flying to branch well above ground if alarmed. Males call from medium heights or lower canopy.

Voice. A resonant *buWOOP* or *cooAH*, typically repeated 3–4 times in rapid succession, then a pause; sounds rather like a *Columbina*; a soft groaning *ghoop*, audible at close range, probably a contact note.

Nest. Undescribed (?).

Status. Probably resident in small numbers but rare and highly local; appears not to remain long in 1 locality, may be nomadic; possibly more numerous than the few records suggest; recorded at ca. 3000–10,000ft (900–3000m) the length of the country.

Range. S Mexico to W Bolivia.

WHITE-TIPPED DOVE

Leptotila verreauxi Pl. 18(14)

Paloma Coliblanca (Yuré, Coliblanca)

Description. 10¼" (26cm); 165g. Medium-sized, rather long-legged; cinnamon-rufous wing-linings and white-tipped tail conspicuous in flight; at close range note blue facial skin. **Adults:** throat whitish, shading to pinkish-buff on face, foreneck, and breast, and to white on belly and crissum; crown, nape, sides and back of neck gray-brown, glossed with iridescent purple in male (faint or lacking in female); rest of upperparts dull brown; remiges dusky; lateral rectrices blackish with broad white tips. Iris orange; bare lores and eye-ring blue; bill blackish; legs carmine. **Young:** duller, without pinkish tones or iridescence; neck, mantle, and breast scaled with dull buffy; crown mottled or streaked with dull rufous; bill reddish-brown; facial skin dull gray.

Habits. Frequents deciduous (but not evergreen) forest understory, open woodland, second growth, coffee plantations, gardens, fields, and roadsides; singly or in pairs, never in flocks, walks over ground gathering seeds, grit, occasional small insects; when startled, bursts into flight with whistling wings, flies to a low, concealed perch. Males call from dense foliage, from eye level to ca. 30ft (2–9m) up.

Voice. A rolling, growling *krrroor*, with head and pumping tail, when disturbed; song a deep, mournful *cuh-cooooaah* or *cu-coooo-ooo*, the first note soft and audible only at close range, sometimes doubled or omitted, ca. once/5–8sec. In courtship male gives a guttural, rough, broken coo: *prrr-prrr-p'p'prrr-poooo*.

Nest. A fairly substantial bowl of straws,

sticks, vines, rootlets, usually 3–10ft (1–3m) up in thicket, vine tangle, weedy field, or garden; occasionally as low as 12in (30cm) or as high as 60ft (18m). Eggs 2, less often 1, pale buffy to white. Year-round with low points ca. May–June and October–November.

Status. Resident; common to abundant in lowlands the length of Pacific slope, to ca. 2600ft (800m) along N cordilleras (higher where deforested), and to 5000ft (1500m) or more in C and S sectors; abundant throughout Valle Central and down Reventazón valley to Turrialba, but absent over rest of Caribbean slope, except for occasional, sporadic occurrences, although may yet colonize this area as deforestation proceeds.

Range. S Texas to C Argentina.

Note. Also called White-fronted Dove.

GRAY-FRONTED DOVE

Leptotila rufaxilla Pl. 18(15)

Paloma Coronigrís (Yuré)

Description. 10″ (25cm); 155g. Slightly smaller than White-tipped Dove, with gray crown and hindneck, narrower white tips to lateral rectrices, bare lores and eye-ring purplish. **Adults:** throat and forehead whitish; top and sides of head plumbeous; hindneck paler purplish-gray; rest of upperparts dull olive-brown; lateral rectrices dusky, tipped with white; face and foreneck buffy, shading through dull pinkish to white on belly (breast grayer in female); wing-linings rufous. Iris pale yellow; bill black; legs magenta. **Young:** little or no gray in crown or hindneck; feathers of mantle edged with cinnamon; foreneck and breast with indistinct darker olive scalings. Iris brownish; facial skin slaty.

Habits. Prefers understory of moist evergreen to semi-deciduous forest, old second growth, scrubby woodland, and cacao plantations; ventures into open situations adjacent to dense vegetation; forages singly or in pairs, walking over ground gathering fallen fruits, seeds, grit. Males sing from ground or perch in understory, sometimes well above eye-level.

Voice. Song lower-pitched, more guttural, shorter than that of other *Leptotila*: *grooOOoo* or *hrooow*, repeated regularly about once/3sec.

Nest. Undescribed (?). A pair with 2 recently fledged juveniles noted in May at Caño Negro.

Status. Resident in lowlands the length of Pacific slope, to 3300ft (1000m) on Península de Nicoya; also in lowlands of extreme NW Caribbean slope; common in Upala–Río Frío area and W side of Península de Nicoya; uncommon to rare elsewhere.

Range. E Mexico and Grenada to W Colombia, E Peru, S Brazil, and NE Argentina.

Note. The Central American form has often been considered a species, *L. plumbeiceps*, Gray-headed Dove, separate from *L. rufaxilla* of South America E of the Andes.

GRAY-CHESTED DOVE

Leptotila cassinii Pl. 18(16)

Paloma Pechigrís

Description. 9½″ (24cm); 165g. Tail relatively short. On Caribbean slope, the dullest, darkest *Leptotila*, with little white at tip of tail; on S Pacific slope, best told by rufous crown and nape. **Adults** Caribbean form: throat white; forehead pale gray, shading to dark grayish-brown on crown and nape; hindneck dark gray, glossed with purplish (especially in male); upperparts dark olive-brown; remiges and lateral rectrices blackish, the latter narrowly tipped with grayish-white; wing-linings cinnamon-rufous; face, foreneck, and breast purplish-gray, shading to pinkish-cinnamon (male) or buffy-brown (female) on belly; sides and flanks dark brown; center of lower belly white. Iris yellow; bare eye-ring and lores dark red; bill black; legs coral-red. Pacific form: paler overall, above more olivaceous, lateral rectrices more broadly white-tipped. **Young:** duller and browner in both forms; neck and breast scaled with cinnamon-buff; mantle feathers edged with cinnamon to rufous; nape and hindneck of Caribbean form with extensive rufous suffusion and/or rufous shaft-streaks.

Habits. In forest understory, especially near edges or small openings, but more numerous in shady old second growth, coffee and cacao plantations; in shady gardens, dooryards, and clearings mingles freely with White-tipped Doves; singly or in pairs, walks over ground gathering seeds, grit, occasional small insects; pushes aside fallen leaves with bill. Males call from low perches in thickets.

Voice. Song a long-drawn-out, mournful *cooooOOOoo* or *crrooOOoo*, loudest in middle, sometimes with rolling quality at start; longer and weaker than call of White-tipped Dove, without introductory note.

Nest. A shallowly concave platform of sticks, straws, petioles, etc., 3–16ft (1–5m) up in vine tangle or thicket, on horizontal branch or palm frond, in or near woodland. Eggs 2, white to pale buff. February–May and July–September.

Status. Common resident of Caribbean and S Pacific slopes, from sea level to ca. 2500ft (750m) on Caribbean side, and locally to 4000ft (1200m) on S Pacific slope, where found rarely N to Puntarenas and W approaches to Valle Central.

Range. S Mexico to N Colombia.

Note. The rufous-naped Pacific form is sometimes considered a separate species, *L. rufinucha*, Rufous-naped Dove; reports of this form in the Río Frío region of the Caribbean slope (Slud) evidently refer to juveniles of the Caribbean form.

PURPLISH-BACKED QUAIL-DOVE

Geotrygon lawrencii Pl. 18(21)

Paloma-Perdiz Sombría

Description. 10″ (25cm); 220g. Dark-backed, gray-breasted, with conspicuous facial pattern. **Adults:** face whitish with bold black eye- and malar stripes; rest of head, neck and breast slaty, tinged with green on neck; above dark olive-brown with center of back dark glossy purple; belly buffy; crissum white; tail dusky. Iris brown; orbital skin and legs magenta; cere dusky; bill black. **Young:** crown barred with dusky; facial pattern of adult; back blackish, faintly barred with rusty; rest of upperparts, breast and sides dusky, scalloped and barred with buff.

Habits. Prefers dense wet forest in mountainous terrain; usually singly or in pairs at breaks in dense understory; forages on ground like other quail-doves, taking seeds, fruits, worms, and insects; when alarmed, usually walks or runs quietly away, but if flushed often flies to elevated perch. Males call from well up in understory.

Voice. Song *coo-ka-krrrw* or *pum-whaa-kooow*, the third note downslurred, longer and louder than the others (the only note audible at a distance), once/3–5sec; more rolling, less resonant than calls of Buff-fronted or Chiriquí quail-doves.

Nest. A shallow cup of twigs, loosely made but fairly bulky, often upon a pad of leaves, on horizontal branch or fork, low in dense forest understory. Egg 1, buffy. June–October.

Status. Fairly common but local resident of foothill belt of Caribbean slope, ca. 1300–2600ft (400–800m) [to 3300ft (1000m) or more in SE], from Cordillera de Guanacaste S to Panama.

Range. SE Mexico; Costa Rica to E Panama.

BUFF-FRONTED QUAIL-DOVE

Geotrygon costaricensis Pl. 18(19)

Paloma-Perdiz Costarriqueña

Description. 11″ (28cm); 250g. Terrestrial; strong contrast between gray-green neck and purplish mantle; striking face-pattern. **Adults:** face and throat white with bold black eye- and malar stripes; forehead bright buff; crown, neck, and breast slate-gray, strongly tinged with green on hindneck; mantle rich purplish, shading to rufous-chestnut on wings, rump, and central rectrices; remiges and most rectrices dusky; belly dull buff. Iris brown; cere and orbital skin reddish; bill dusky with reddish base; legs carmine. **Young:** duller; forehead whitish; plumage mostly dull chestnut, barred above with dusky and below with grayish.

Habits. Similar to those of Chiriquí Quail-Dove but generally prefers higher elevations and wetter forests. Forages on ground, singly or in loosely associating pairs, working through leaf litter with bill to obtain fallen fruits, seeds, probably insects; ventures out into clearings from dense understory mainly in early morning; when flushed, flies to low limb, vibrates tail nervously.

Voice. A low, hollow *hoOOOO*, louder toward end with slightly rising inflection, often repeated regularly.

Nest. A nearly flat platform of coarse twigs, a few rootlets, and much green moss and liverworts, 13ft (4m) up on horizontal branch amid bamboo undergrowth of heavy forest. One nestling, June (only record).

Status. Resident of wet mountain forests from Cordillera de Guanacaste S along major ranges into Panama, at 3300–10,000ft (1000–3000m); center of abundance 5000–6500ft (1500–2000m) on Caribbean slope, 6500–8200ft (2000–2500m) on Pacific slope.

Range. Costa Rica to W Panama.

CHIRIQUI QUAIL-DOVE

Geotrygon chiriquensis Pl. 18(18)

Paloma-Perdiz Pechicanela

Description. 12″ (30cm); 295g. Large, reddish, terrestrial, with contrasting gray cap. **Adults:** crown and nape slaty; upperparts chestnut, becoming purplish on back; remiges dusky; face and throat buffy whitish; a bold black malar stripe; below dark rufous, fading posteriorly to cinnamon-buff; pleated effect of feathers on sides of neck. Iris dark red; bare lores slaty; bare eye-ring rose-red; cere and bill black; legs magenta. **Young:** darker, without gray cap; mantle dark brown; wing-coverts and underparts more chestnut, indistinctly barred with dull black.

Habits. Dwells in understory of mountain forests, usually singly or in pairs; feeds on ground, taking seeds, fallen fruits, some animal matter; when disturbed, walks or runs away or flushes with rattling wings, flying to a low limb; ventures into trails and clearings in early morning. During breeding season, males call incessantly from dense vegetation, 3–10ft (1–3m) above ground, sometimes from an empty nest.

Voice. A mournful, descending *whooOOoo*, loudest in the middle, lasting ca. 1sec and given at intervals of 3–5sec.

Nest. A substantial shallow cup of leaves and twigs, on end of slender tree overhanging ravine. Eggs buffy. September (Carriker).

Status. Resident at middle elevations from Cordillera de Guanacaste (at least N to Cerro Santa María) S along all major mountain ranges to Panama; mainly below Buff-fronted Quail-Dove and, on Caribbean slope, above Purplish-backed but overlaps with both; prefers drier forests, on average, than Buff-fronted. On Caribbean slope, 2000–5600ft (600–1700m), locally in SE to 6600ft (2000m); 3300–8200ft (1000–2500m) on Pacific slope.

Range. Costa Rica to W Panama.

Note. Often called Rufous-breasted Quail-Dove. Sometimes lumped with *G. linearis* (Lined Quail-Dove) of N South America; *G. albifacies* (White-faced Quail-Dove) of Mexico to N Nicaragua is also sometimes included in this complex.

OLIVE-BACKED QUAIL-DOVE

Geotrygon veraguensis Pl. 18(22)

Paloma-Perdiz Bigotiblanca

Description. 8½″ (21.5cm); 165g. A dark terrestrial dove with bold white malar stripe. **Adults:** forehead white to buffy, shading to purplish-slate on face and hindcrown; broad malar stripe silvery-white; upperparts dark olive-brown; throat whitish; neck and breast dark brownish-gray with greenish gloss; belly white; wing-linings and crissum dull cinnamon. Iris brown; orbital skin magenta; bill and cere black; legs dull coral-red. **Young:** darker and duller; upperparts scalloped with cinnamon to dull rufous; facial markings white, mixed with cinnamon; foreneck and chest mostly dull cinnamon.

Habits. In dense wet forest of lowlands and foothills; forages singly or in pairs on forest floor, usually beneath dense understory but sometimes along shady trails or in adjacent tall second growth; when alarmed, walks away quietly or flushes with a soft rattle of wings, flies a short distance, alights on ground or low perch.

Voice. A low, resonant, ventriloquial, twanging *thum* or *kuunk* like a faraway frog; in territorial disputes a sharp crack or snap produced by whacking wing against low vegetation; also a low, guttural crooning.

Nest. A shallow, often fairly bulky cup of twigs and rootlets, lined with finer twigs, saddled on liana, branch, or shrub, within a yard (m) or so of ground, often by stream or gully. Eggs usually 2, pale buff. January–July.

Status. Fairly common resident in forested portions of Caribbean lowlands, N at least to Tortuguero area, but apparently not found in drier forests S of Lago de Nicaragua; locally up to 1500ft (450m) in foothills.

Range. Costa Rica to NW Ecuador.

VIOLACEOUS QUAIL-DOVE

Geotrygon violacea Pl. 18(20)

Paloma-Perdiz Violácea

Description. 9″ (23cm); 150g. Small dove with relatively short legs, small head, unstriped face, and drooped bill. **Adult ♂:** hindneck and upper back violet, shading to rich brown on wings and chestnut on tail and wing-linings; crown, face, throat, and breast pale gray, tinged with purplish; belly white. Iris yellow-brown; bill and cere carmine; legs coral-red. ♀: duller, browner; only hindneck and upper back purplish; face and throat pale gray; neck and breast brownish, tinged with purple. Iris brown. **Young:** like female but darker, duller, without purple; above and below scalloped with cinnamon to dull rufous.

Habits. Prefers dark, heavily shaded undergrowth of forest and cacao plantations; forages for seeds and fruits on ground but evidently more arboreal than other quail-doves, often perching well into upper understory.

Voice. A rapidly repeated *oo-oo*, the first syllable higher (Slud).

Nest. Undescribed (?).

Status. Local and scarce resident in Caribbean lowlands and foothills, to at least 2000ft (600m) and on Península de Nicoya; more numerous in moist forests along Pacific face of Cordillera de Guanacaste at 2000–4000ft (600–1200m).

Range. Nicaragua to Bolivia and N Argentina.

RUDDY QUAIL-DOVE

Geotrygon montana Pl. 18(17)

Paloma-Perdiz Rojiza

Description. 9″ (23cm); 135g. Stout-bodied, short-tailed, with downwardly inclined bill; strong sexual dimorphism. **Adult ♂:** forehead rufous; rest of upperparts and broad malar stripe rufous-chestnut, glossed with metallic purple; face and throat pinkish-buff; breast vinaceous-cinnamon; belly buffy. ♀: forehead and face dull cinnamon; throat whitish; malar stripe dull brown, broad and indistinct; upperparts, sides, and flanks dull olive-brown; foreneck and breast dull cinnamon; belly dull buff. Iris orangish to brown; cere and bill dull red, tip of upper mandible brown; bare eye-ring, lores, and legs magenta (duller in female). **Young:** above olive-brown, spotted and scaled with cinnamon (more heavily in male); wing-coverts extensively rufous; below buffy-brown, broadly scaled with cinnamon (female) to nearly solid cinnamon (male) on breast.

Habits. Frequents understory of humid forest but also in adjacent old second growth, shady gardens, and cacao plantations; usually singly, walking over ground, often along a trail, picking up fallen fruits and seeds; may visit manakin leks or search under fruiting trees for seeds regurgitated by other birds; when alarmed usually freezes or flies noisily up to a low perch. Males sing from perches well up in understory, less often from ground.

Voice. A low, moaning *coooo*, confusingly similar to that of Gray-chested Dove but usually shorter, sometimes in long series, ca. 1/3sec.

Nest. A slight, loosely constructed mat of coarse twigs and petioles, brown and green leaves, 1½–10ft (0.5–3m) up on an often precarious support, such as a dead palm frond lodged among bushes or on top of a stump, broad-leaved epiphyte, horizontal branch or fork. Eggs 2, buffy to nearly white. March–August.

Status. Resident; fairly common in lowlands and foothills of S Pacific slope, to 4000ft (1200m); on Caribbean side, scarce in lowlands but more numerous in foothills, 500–3300ft (150–1000m), reaching Pacific slope of N cordilleras via low passes.

Range. S Mexico to NE Argentina; West Indies.

ORDER **Psittaciformes**
FAMILY **Psittacidae:** Parrots

With some 332 living species, parrots abound almost everywhere in the tropics except in treeless deserts; many species extend into the subtropics, a few to temperate regions. They range in length from the 3″ (8cm) pygmy parrots to 40″ (1m) macaws. Despite a variety of names, the family is quite uniform in structure and closely related to no other. The name "parrot" is used mostly for medium-sized, usually short-tailed species in the New World; macaws are big and long-tailed; parakeets are small with long or short tails; and parrotlets are small with short tails. In the family as a whole, the prevailing color is the green that makes motionless parrots so hard to detect amid green foliage, but brilliant hues are generously displayed by many species. The sexes are very different in some Old World parrots but similar or identical in all Costa Rican species.

Adapted for opening hard nuts, biting chunks out of fruit, and grinding small seeds into meal, the short, thick, hooked parrot bill combines the destructive powers of an ice pick (the sharp-pointed upper mandible), a chisel (the sharp-edged lower mandible), a file (ridged inner surface of the upper mandible), and a vise. The basic parrot diet is plant embryos and endosperm, often before the seeds mature, thus they accomplish little seed dispersal. Buds of leaves and flowers, flower parts, and nectar are often eaten; Old World lorikeets have brush-tipped tongues specialized for gathering pollen and nectar. The tongues of parrots are unusually thick and muscular, and the upper mandible is hinged at the base, giving unusual mobility. Their legs are short and strong, with 2 stout toes directed forward and 2 backward. Parrots use their bills and feet for clambering amid branches and lift food to the mouth with a foot, as few other birds do.

Parrots are monogamous, highly social birds that often fly in noisy flocks, within which pairs are often distinguishable. Their voices are mostly loud, raucous, and often astonishingly varied. Vocal learning from flock members is evidently an important part of socialization; captive birds can often learn sounds from human models, a trait that explains much of the popularity of parrots as cage birds.

A few parrots build nests, but most breed in unlined cavities that they sometimes enlarge, in trees, hollow palm trunks, or rarely cliffs and banks. The few that nest in arboreal termitaries excavate their chambers, the sexes working alternately. Often the pair will begin preparing the nest site months before they actually nest. Parrots lay 2–8 white eggs, which in most species are incubated by the female alone, fed by regurgitation by her mate; in some Old World species the male shares incubation. Incubation periods increase with the size of the birds, from 17 to ca. 35 days. The young hatch blind and naked or with sparse down. At first only the mother broods and regurgitates to them food delivered to her by her mate, but after some days both parents feed them directly. Small parrots remain in the nest 3–4 weeks, large macaws 3–4 months. Despite the enormous numbers of parrots taken from their nests for the pet trade, we know little about the reproduction of New World parrots in the wild; most of our information on their breeding comes from aviculturists.

SCARLET MACAW

Ara macao Pl. 19(1)

Guacamayo Rojo (Lapa Colorada, Lapa Roja)

Description. 33″ (84cm); 900g. Unmistakable: brilliant colors; long pointed tail and wings; large, powerful bill. **Adults:** mostly bright red; large yellow patch on wing-coverts; remiges and rectrices mostly deep blue; central rectrices red; rump and tail-coverts sky-blue. Iris yellow; facial skin pinkish-white; bill ivory-yellow and black; feet dull gray. **Young:** similar but iris brown.

Habits. Frequents solid or patchy deciduous or evergreen forests, feeding in canopy on fruits and nuts of trees (*Spondias, Hura, Eschweilera, Terminalia*, various palms); in pairs or family groups of 3–4 that sometimes unite in flocks of up to 25, and up to 50 at communal roosts in tall trees or mangroves; ranges widely daily and seasonally; flies steadily, directly, with strong, shallow wingbeats; noisy in flight, usually silent when feeding.

Voice. Loud, deep, hoarse squawks and shouts: *raaaak, rowwwka*, far louder and more resonant than voice of any *Amazona*.

Nest. In large natural cavities in tall trees or sometimes in old woodpecker holes in soft wood that it enlarges, mostly 23–80ft (7–25m) up. Eggs 1 or 2. Dry season.

Status. In 1900 abundant throughout Caribbean and Pacific lowlands; by 1950 gone from entire Caribbean slope except NW (E to Río Frío district) but still common over most of Pacific slope; at present reduced to a few pairs in Guanacaste and extreme NW Caribbean slope, with larger populations persisting in Carara and on Península de Osa (where increasingly restricted to Corcovado basin); locally to 2600ft (800m).

Range. S Mexico to Peru, Bolivia and E Brazil.

GREAT GREEN MACAW

Ara ambigua Pl. 19(2)

Guacamayo Verde Mayor (Lapa Verde)

Description. 31″ (79cm); 1.3kg. Relatively shorter-tailed and more robust than Scarlet Macaw, with much heavier bill. **Adults:** mostly yellowish-green; forehead scarlet; greater coverts greenish-blue, remiges deep blue; tail-coverts sky-blue; central rectrices brick-red tipped with greenish-blue; rest of rectrices blue above; entire tail olive-yellow below. Iris yellow; facial skin whitish, crossed by lines of red and black feathers; bill black, tipped with gray; feet dusky. **Young:** similar but duller, more olive overall; scapulars, tertials, and central rectrices edged with yellowish.

Habits. Prefers canopy of humid lowland forest, feeding largely on the fruit of the giant leguminous tree *Dipteryx panamensis*; often flies long distances to feeding trees, visiting remnant *Dipteryx* in pastures and semi-open; usually in small flocks (up to 15 birds, rarely more) that range widely over Caribbean slope, perhaps tracking the fruiting of *Dipteryx*; when feeding, silent except for occasional squawks; usually calls freely in flight.

Voice. Loud, hoarse, raucous shouts, squawks, growls: *aaaahrk, aowrk*. Voice deeper, more resonant and far-carrying than that of Scarlet Macaw.

Nest. Undescribed (?). Evidently in dry season.

Status. Widespread but increasingly scarce resident of Caribbean lowlands, appearing in numbers at a locality for a few weeks or months, then departing; presence usually coinciding with fruiting of *Dipteryx*. Extensive deforestation has isolated and possibly eliminated some seasonally important feeding areas; locally to ca. 2000ft (600m) or higher, especially along Cordillera de Guanacaste.

Range. E Honduras to NW Colombia and W Ecuador.

CRIMSON-FRONTED PARAKEET

Aratinga finschi Pl. 19(10)

Perico Frentirrojo (Cotorra)

Description. 11″ (28cm); 150g. Largest

Costa Rican parakeet; only long-tailed species with red under wing. **Adults:** mostly green, below paler and more yellowish; forehead crimson; wing-linings largely orange-red, mixed with yellow in carpal area; red spot on thigh; scattered red feathers often present on face and foreneck; all flight feathers yellowish below. Iris orange; bare eye-ring whitish; bill horn-color with grayish tip; feet dull grayish. **Young:** little or no red on forehead, none on thigh or foreneck; wing-linings more orange.

Habits. Prefers open country with scattered trees, coffee plantations, second growth, agricultural districts; sometimes a pest of corn and sorghum crops; also eats flowers and fruits of *Erythrina* and *Inga*, and fruits of *Croton, Zanthoxylum*, wild figs; highly social, flying and foraging in tightly coordinated flocks; hundreds often gather in communal roosts in palms, tall cypress or *Erythrina* trees.

Voice. Most characteristic call a dry, strident, nasal barking *kih-kih-kih-keh-keh* . . . ; higher-pitched whistled chirps and squawks, dry chattering notes.

Nest. In holes of various sorts: natural cavities, old woodpecker holes, dead palm stubs; may excavate holes in rotten stubs or masses of epiphytes, many pairs sometimes nesting close together. Eggs 2–4. Dry and early wet seasons.

Status. Common and widespread resident of low to middle elevations in deforested areas throughout Caribbean slope; has increased greatly in many areas as forests have been cut; also increasingly common on S Pacific slope as deforestation proceeds; in Guanacaste, a sporadic but increasingly regular visitor from slopes of N cordilleras; common in Valle Central, especially as postbreeding visitor.

Range. SE Nicaragua to W Panama.

OLIVE-THROATED PARAKEET

Aratinga nana Pl. 19(11)

Perico Azteco

Description. 9″ (23cm); 85g. Medium-sized, dull-colored, long-tailed parakeet with blue on wing but no contrasting markings. **Adults:** plumage mostly green; chest dull olive, shading to yellowish-olive on belly and wing-linings; many ventral feathers with dusky shafts; remiges largely blue. Iris orange; bill whitish to pale horn; cere and bare eye-ring bone-white; feet dull grayish. **Young:** similar, but iris brown.

Habits. Prefers forested and partly cutover areas; most often along rivers, forest edges, and semi-open eating favorite fruits (figs, *Inga*, some melastomes); also feeds in forest canopy on fruits of *Hura* and *Hieronyma*; usually in small flocks (up to 30 following breeding season; more may gather at good fruiting trees, especially figs). Flight fast, erratic, in tightly coordinated groups.

Voice. Dry, raucous screeches in flight, similar in quality to notes of White-crowned Parrot but higher-pitched; a sustained harsh twitter, rising at end (Slud); piercing, sometimes melodious chirps, squeakier and higher-pitched than notes of Brown-hooded Parrot.

Nest. Preferred site a hole in an arboreal termite nest, usually excavated by the birds themselves, often along riverbank or forest edge. Eggs 3–4. Dry and early wet seasons.

Status. Uncommon to locally common resident over most of Caribbean lowlands, occasionally reaching 2300ft (700m) on slopes of Cordillera Central; much less common S of Limón, where perhaps only a seasonal visitor; has decreased in many areas as forests are cut; now much less numerous overall than *A. finschi*, which evidently does better in deforested country.

Range. S and E Mexico to W Panama, mainly on Caribbean slope; also Jamaica.

Note. The Central American population is often considered a separate species from the Jamaican form, *A. astec* (Aztec Parakeet).

ORANGE-FRONTED PARAKEET

Aratinga canicularis Pl. 19(13)

Perico Frentinaranja (Catano, Periquito, Zapoyol)

Description. 9″ (22.5cm); 80g. Medium-sized, long-tailed, with blue in wings and conspicuous facial pattern. **Adults:** mostly green, below paler and yellower; breast tinged with olive; wing-linings yellowish-olive; remiges mostly blue; tail tipped with blue; forehead orange; crown dull blue. Iris yellow; bare eye-ring orange-yellow; bill whitish to pale horn; feet dull grayish. **Young:** orange of forehead less extensive (not extending to lores as in adult).

Habits. Frequents savannas with scattered trees, second growth, and forest edge as well as canopy of deciduous and evergreen forest; usually in flocks, which are larger (often 30 or more birds) and more nomadic outside breeding season, when 100 or more may gather at communal roosts; eats fruits (e.g., *Ficus, Bursera, Brosimum*), flowers (*Gliricidia, Combretum*), and seeds (*Ceiba, Inga*). Flies swiftly and directly on rapid wingbeats with brief glides.

Voice. In flight a raucous, repeated *can-can-can* (Slud); shrill screeches; notes lack the nasal quality of those of Crimson-fronted Parakeet. Also various dry, grating chirps and chatters, higher-pitched than those of White-fronted Parrot.

Nest. A cavity in an arboreal termite nest, usually excavated by the birds themselves; old woodpecker holes and natural cavities occasionally used. Eggs 3–5.Mainly in dry season.

Status. Common resident of N Pacific lowlands, locally to 3300ft (1000m) on slopes of Cordilleras de Guanacaste and Tilarán and on high mountains of Peninsula de Nicoya; regular but increasingly uncommon across Valle Central at least as far W as Grecia; formerly to San José. Population has decreased in many areas in recent years due to persecution for pet trade.

Range. W Mexico to Costa Rica.

SULFUR-WINGED PARAKEET

Pyrrhura hoffmanni Pl. 19(12)

Perico Aliazufrado

Description. 9″ (23cm); 75g. Medium-sized highland parakeet with large yellow wing-patches and long rounded tail. **Adults:** head, neck, and chest dusky-green scaled with yellowish to ochraceous-orange; ear-patch crimson; small reddish spot on chin; rest of body plumage green; outer primaries largely blue; inner remiges and greater coverts bright yellow; tail largely dull reddish. Iris grayish-brown, bare eye-ring pale yellow; bill and cere pale yellowish to horn-color; feet dull gray. **Young:** scaling of head and chest duller, less orange; much less yellow on wing.

Habits. Prefers broken, mountainous country, forested or cutover; usually in flocks of 5–15 birds; forages in canopy of forest or in smaller trees, shrubs along edges, or in second growth, gathering seeds, fruits of *Ficus, Croton, Myrtis, Miconia*; may fly down mountainside to feed, back up to roost; highly social; may allopreen in small groups, not just in pairs.

Voice. In flight, members of flock give high-pitched, grating, piercing *toweet-deet-deet-toweet* sounds, like a thin, reedy piping from a distance; while perching, a penetrating, reedy, grating *zeewheet*.

Nest. In old woodpecker hole, hollow broken stub, or other cavity at forest edge or in canopy, 25–65ft (8–20m) up. Eggs undescribed (?). Mainly dry season.

Status. Common resident of middle to upper elevations of Cordillera de Talamanca and its outlying ranges (e.g., Dota area) of S Costa Rica; regularly to 10,000ft (3000m) in dry season; particularly in wet season, descends to vicinity of Cartago at 4200ft (1300m), even to Valle del General at 2300ft (700m), evidently on daily visits.

Range. Costa Rica to W Panama.

BARRED PARAKEET

Bolborhynchus lineola Pl. 19(16)

Periquito Listado

Description. 6¼″ (16cm); 55g. Small, with short bulbous bill, fairly short graduated tail. **Adults:** mostly green, below paler, above more olive; crown and primaries more bluish; fine black barring on mantle, sides, and flanks shows only at close range; shoulders black; middle and greater coverts and rectrices tipped with black. All black markings much heavier in males than in females. Iris brown; bare eye-ring gray; cere, bill, and feet pale flesh. **Young:** paler, with black barring indistinct.

Habits. Frequents forest canopy, forest edge, second growth in mountainous country; during most of year in large, compact flocks that fly swiftly, directly, often high above ground, ranging out widely from regular roosts to forage during day, returning at sunset; during dry season often in pairs or small flocks; eats seeds, fruits, and buds of various trees and shrubs, including *Myrtis, Heliocarpus, Miconia*.

Voice. Melodious chirping, cheeping notes; resonant chirrups; scratchy, ascending *shreet* or *churree*; in chorus by flying birds, a slurred *cheer-churr*; in alarm prior to taking flight, a harsh, nasal chatter: *jur-jur-jur-jur*. Voice more mellow, less piercing and grating than that of Sulphur-winged Parakeet.

Nest. Undescribed (?). Evidently breeds in dry season, possibly early wet season (gonad data; juveniles).

Status. Fairly common to locally abundant resident of Cordillera de Talamanca; much less numerous in Cordillera Central and especially N cordilleras (where possibly a sporadic or seasonal visitor). Center of abundance ca. 6500ft (2000m) in Talamancas but at times (notably early wet season) common at 10,000ft (3000m); may descend by day to 2500ft (750m) to feed, as in Valle del General, returning uphill to roost.

Range. S Mexico to W Panama; the Andes from Venezuela to Peru.

ORANGE-CHINNED PARAKEET

Brotogeris jugularis Pl. 19(14)

Periquito Barbinaranja (Catano, Zapoyolito, Perico)

Description. 7″ (18cm); 65g. Small parakeet with conspicuous brownish shoulders, blue in wings, short pointed tail. Mostly green; below paler and yellower; more bluish on head and rump; lesser and middle coverts olive-brown; wing-linings yellow; remiges blue; orange spot on chin. Iris brown; bare eye-ring and cere white; bill pinkish to pale horn; feet flesh-color.

Habits. Prefers open country with scattered trees, semi-open, forest edge in most areas; only in dry northwest enters forest canopy regularly; flies rapidly, erratically, in coordinated flocks of up to 50 birds except during breeding, when often in pairs or small groups; eats fruits and seeds of *Ficus, Muntingia, Byrsonima, Cecropia, Ceiba*, and *Bombax*; also flowers and nectar of *Erythrina*, guava, balsa.

Voice. Noisy in flight; sometimes large flocks very noisy when feeding in large fig tree; commonly alternates dry scratchy chatters with clear, shrill, sometimes melodious chirps and *chreek* notes.

Nest. In woodpecker holes, palm stubs, holes in termitaries, natural cavities and crevices; several or many pairs may nest in large rotten tree or large termitary, each pair digging its own hole. Eggs 4–5. Dry and early wet seasons.

Status. Common to abundant resident of N Pacific lowlands, ranging locally to 4000ft (1200m) on adjacent cordilleras and in Valle Central; expanding its range rapidly in wet lowlands as forests are cut (e.g., first recorded in upper Valle del General in 1965, at La Selva in 1975), now essentially throughout lowlands of both slopes, locally to middle elevations (e.g., San Vito area).

Range. S Mexico to N Colombia and N Venezuela.

Note. Sometimes called Brown-shouldered Parakeet.

RED-FRONTED PARROTLET

Touit costaricensis Pl. 19(15)

Periquito Alirrojo

Description. 6¾″ (17cm); 80g. Chunky, with conspicuous red markings on head and wings; short, nearly square tail. **Adults:** mostly bright green, below paler, shading to yellow on chin; forehead, forecrown, lores, and orbital area bright red (more extensive in male); alula and outer wing-coverts above and below red (male), or red on upperwing replaced by blue-black (female); proximal wing-lining yellow; primaries bluish; tail mostly yellow, tipped with blue-black. Iris gray-brown; cere and bare eye-ring gray; bill yellow; feet slate-gray. **Young:** little or no red on head; iris dark brown; wing like that of adult of same sex.

Habits. Inhabits canopy and edges of mountain forest; usually seen flying in pairs or small groups (probably families) to and from feeding areas; eats fruits of trees and epiphytes, including melastomes, ericads like *Cavendishia*, and *Clusia*.

Voice. Reedy, scratchy notes, alternately slurred up or down: *dree? dree? deeah! dree? durr!* Soft, rolling *chrrik* notes while feeding.

Nest. Undescribed (?). Eggs 2–3, based on family group size of 4–5. Probably breeds during dry season.

Status. Uncommon resident of Caribbean slopes of Cordillera Central and Cordillera de Talamanca, visiting upper parts of Pacific slope locally during dry season; altitudinal movements extensive, recorded at 10,000ft (3000m) on Cerro de la Muerte in early dry season; in wet season regular at 1650–3300ft (500–1000m); occasionally to sea level, especially in SE.

Range. Costa Rica to W Panama.

Note. Sometimes lumped with *T. dilectissima* of E Panama and N South America, the enlarged species called Red-winged Parrotlet.

BROWN-HOODED PARROT

Pionopsitta haematotis Pl. 19(9)

Loro Cabecipardo (Lora)

Description. 8¼″ (21cm); 165g. Small, stocky; pale bill and eye-ring contrast with dark head; red patch on axillars. **Adults:** head and neck dull brown, darkest on face (nearly black in males); ear-patch red; chest olive; rest of body green; remiges mostly dark blue; inner webs of rectrices largely red. Iris olive to brown; bare eye-ring white; bill pale horn to ivory; cere flesh-color to orangish; feet brownish. **Young:** similar but paler, duller, without red ear-patch.

Habits. Frequents forest canopy and edges, adjacent semi-open and tall second growth, avoiding extensive open country; usually in pairs or small groups, or flocks rarely exceeding 15; flight swift and hurtling, with deep wingbeats; when feeding, silent and slow-moving, keeping more in foliage than most parrots; eats fruits and seeds of various trees and epiphytes (e.g., *Ficus, Heliocarpus, Croton, Erythrina*) as well as green leaves of certain mistletoes.

Voice. In flight, clear chirrups and rolling *keereek* notes, sometimes alternating with sharp chatters in the manner of *Brotogeris*; various gurgling, squealing, squeaking notes when perched and socializing. In general voice more mellow than those of other small parrots.

Nest. In natural cavities in forest trees. Eggs undescribed (?). Probably dry season and early wet season.

Status. Resident throughout Caribbean lowlands and in Pacific lowlands from Parrita area and Valle del General S, breeding locally to ca. 5250ft (1600m) on adjacent mountain slopes, largely withdrawing to lower elevations following breeding; fairly common in partly or entirely forested districts, scarce in extensively deforested areas.

Range. SE Mexico to W Ecuador.

WHITE-CROWNED PARROT

Pionus senilis Pl. 19(7)

Loro Coroniblanco (Chucuyo)

Description. 9½″ (24cm); 220g. Small, dark, with contrasting white cap and throat, pale bill and eye-ring, red under tail. **Adult ♂:** forehead and crown white; throat white to buffy; rest of head, neck and breast dull dark blue; belly green; remiges blue; underwings blue-green; mantle fairly dark green, shading to golden-olive on shoulders; crissum and inner webs of most rectrices red. Iris brown; orbital skin reddish; bill and cere pale yellowish; feet flesh-orange. ♀: blue stops on upper breast; lower breast scaled with blue; shoulder duller; eye-ring and feet more brownish. **Young:** feathers of cap green, edged with whitish; little or no blue on head or neck; little red on crissum.

Habits. In forest canopy but seems to prefer forest edge, semi-open, second growth, tree plantations, agricultural districts with scattered patches of woods; very social, frequently in flocks of 30–50 birds, especially following breeding season; flies rapidly, erratically with deep wingbeats; often rests on rodlike unopened frond at tip of tall palm tree; when feeding, clambers about silently and deliberately, rarely fluttering between branches; eats fruits, seeds, and nuts: ripening seeds of *Inga* and *Erythrina*, fruits of palms (including pejibaye), *Dendropanax*; sometimes a serious pest on ripening corn, pejibaye, sorghum, and fruits.

Voice. In flight dry, raucous, descending shrieks: *kreeeah* or *keeerh*, sometimes with metallic tone; various shorter notes when perched. Begging note of young a nasal grunting *aaanh-hn-aaanh-hn*.

Nest. A cavity, such as knothole or hollow branch, or (probably the preferred site) the hollow stub of a palm, unlined except for wood slivers. Eggs 3–6. January–April.

Status. Common to abundant resident of humid lowlands and foothills of Caribbean slope (where, however, progressively less numerous S from Limón) and S Pacific slope (N to Carara area), from sea level locally to 4000ft (1200m) on Caribbean slope and to 5250ft (1600m) on Pacific slope. A sporadic visitor, especially following breeding, over much of Valle Central.

Range. E Mexico to W Panama.

BLUE-HEADED PARROT

Pionus menstruus Pl. 19(8)

Loro Cabeciazul (Chucuyo)

Description. 9½″ (24cm); 220g. Small, often appearing all dark in field except for reddish spot on bill and red crissum. **Adults:** head and neck mostly bright blue, mixed with pinkish red on throat; ear-patch black; breast dark greenish with blue scaling (nearly solid blue in males); belly green; above darker green, primaries and rectrices largely blue; crissum and inner webs of rectrices rose-red. Iris dark brown; bill, cere, and eye-ring blackish with a square patch of pinkish-red on each side of upper mandible; feet brownish. **Young:** head and neck green, with blue scaling on crown and nape.

Habits. In general similar to those of White-crowned Parrot, but usually in smaller flocks (at least in Costa Rica); seems less partial to forest and more restricted to lower elevations than White-crowned in areas of overlap; often forms large communal roosts, from which small flocks range out to feed during day.

Voice. In flight, high-pitched raucous shrieks with rising inflection and distinctive metallic quality: *sheeng? shweeenk? sheweeenk?* Various shorter, rolling notes when perched.

Nest. In hollow limbs, holes in trees, palm stubs. Eggs 4 (in Trinidad; no Costa Rican record). Dry season.

Status. Fairly common resident in lowlands of Caribbean slope, S from about Limón and Siquirres; also increasingly common in Golfo Dulce district, principally on mainland side of gulf, N at least to Piedras Blancas; still rare and sporadic on Península de Osa; occasionally up to 3800ft (1170m) in San Vito region. Range expanding rapidly on Pacific slope, perhaps slightly on Caribbean slope as well; has evidently invaded Costa Rica in this century.

RANGE. Costa Rica to W Ecuador, Bolivia, and SE Brazil.

WHITE-FRONTED PARROT

Amazona albifrons Pl. 19(6)

Loro Frentiblanco (Lora)

DESCRIPTION. 10″ (25cm); 230g. Small *Amazona* with conspicuous white forehead, red on face and (males) wing-coverts, but none in secondaries. **Adults:** mostly green, scaled with black over head, breast, and upper back; forehead and forecrown white, midcrown blue; lores and orbital area red; remiges largely blue; alula and primary coverts red in male; inner webs of rectrices red basally. Iris yellow to white; bare eye-ring gray; cere and bill yellow; feet brownish. **Young:** white of forehead and forecrown more or less yellowish, less extensive than in adult (sometimes almost lacking); less red on face.

HABITS. Frequents canopy of deciduous and evergreen forest, advanced second growth, also scattered trees and forest patches in savannas and agricultural districts; usually travels in flocks, often 30–50 or more, ranging out from large communal roosts to feed; often in scattered pairs during breeding season; feeds unobtrusively in treetops, with an occasional contact squawk; eats figs, ripening seeds of *Inga* and other legumes, *Croton, Terminalia*, other seeds, nuts, blossoms, also corn, mangoes, other crops.

VOICE. A dry, grating *ka-ka-ka-ka* or *k'kak-K'Kak-K'Kack*; various squealing notes reminiscent of *Aratinga* parakeets; voice in general higher-pitched, less resonant than those of other *Amazona* species.

NEST. In natural cavities (knotholes, hollow branches, palm stubs) or old woodpecker holes, sometimes enlarged by the parrots themselves. Eggs 3–5 (according to local people). Mainly dry season.

STATUS. Common to abundant resident of lowlands and foothills of Pacific NW, from sea level to 3500ft (1100m) in mountains of Península de Nicoya and on Pacific slope of Cordillera de Guanacaste; not recorded from Valle Central or Caribbean slope.

RANGE. NW and SE Mexico to Costa Rica.

RED-LORED PARROT

Amazona autumnalis Pl. 19(4)

Loro Frentirrojo (Lora)

DESCRIPTION. 13½″ (34cm); 420g. Fairly large *Amazona* with red speculum, lores and forehead. **Adults:** mostly green, below paler; terminal half of outer rectrices pale yellow-green; usually some yellow feathers on face (more in northern than in southern populations); crown and hindneck pale blue, scaled with black; some blackish scaling on chest (especially in females); outer webs of primaries blue, of outer 4–5 secondaries red, tipped with blue-black. Iris orange; cere and base of bill yellowish-horn, shading to gray at tip; bare eye-ring pale yellowish; feet dull grayish. **Young:** iris brownish, less red on face.

HABITS. Frequents forest edge, partly cleared areas or semi-open, scattered groves in open country, less often canopy of solid forest; travels in pairs, groups, or large flocks in which pairs are evident; outside breeding season forms communal roosts, ranging out to feed in early morning, returning in late afternoon; like other parrots, after feeding often perches in bare treetops, socializing noisily; forages chiefly in crowns of tall trees, taking fruits of palms, arillate seeds of *Virola, Casearia*, and *Protium*, figs, ripening seeds of legumes, leaf buds, and some cultivated fruits including mangoes and citrus.

VOICE. Perched or flying, a wide variety of squawks, shrieks, rippling notes, shrill warbles, grunts. A diagnostic note, often given in flight, is a rather metallic *ka-link ka-link* or *k'leek k'leek*.

NEST. An unlined hole in a tall, usually dead tree, often a palm stub. Eggs 3–4. Dry season, young fledging during early rainy season.

STATUS. Resident of humid lowlands and foothills over virtually entire Caribbean slope to 2600ft (800m) locally; and S Pacific slope to 3300ft (1000m); common to abundant in partly deforested districts, uncommon in areas of dense forest; absent from dry Pacific NW except along passes of Cordillera de Guanacaste, and from Valle Central. Persecuted for cage-bird trade.

RANGE. E Mexico to W Ecuador and C Brazil.

YELLOW-NAPED PARROT

Amazona auropalliata Pl. 19(5)

Lora de Nuca Amarilla (Lora)

DESCRIPTION. 13¾″ (35cm); 480g. Only large *Amazona* on N Pacific slope; distinctive yellow nape, red speculum, mellow voice. **Adults:** mostly green, below paler; crown tinged with blue; large yellow patch on hindneck; outer webs of primaries blue, of outer 4 secondaries red, tipped with blue; broad terminal tail-band yellow-green. Iris orange; bare eye-ring and feet dull grayish; bill gray shading to blackish at tip; cere blackish. **Immatures:** lack yellow on hindneck; back and sides of neck scaled with dusky.

HABITS. Prefers deciduous and evergreen (gallery) forests, savannas with scattered trees and groves, less often tall second growth or agricultural areas; travels in pairs or small (family?) groups (evidently formerly in large flocks); flies like other large *Amazona*, steadily, directly, not swiftly, on rapid, shallow wingbeats with little or no gliding or veering; feeds high in trees, where usually silent and wary, calling only when airborne; eats fruits and seeds of *Cochlospermum, Curatella*, figs, ripening fruits of *Terminalia*, seeds of legumes, also some flowers and buds.

VOICE. Most characteristic note, often uttered repeatedly in flight, a rather deep, hoarse, rolling *karrow* or *kurra'ow*; a vast variety of squawks, toots, warbles, and other notes, delivered mostly in rather deep, mellow tones "like a person imitating a parrot" (Slud).

NEST. In such unlined natural tree cavities as hollow branches, knotholes, often in dead wood. Eggs usually 3. Dry season.

STATUS. Widespread but generally uncommon to increasingly rare resident of dry N Pacific slope, S to around Tárcoles; in most areas population has declined drastically due to persecution for cage-bird trade, where it is prized as a good "talker." Young birds are secured by climbing to nest holes or felling nest trees.

RANGE. S Mexico to Costa Rica.

NOTE. A member of the *Amazona ochrocephala* complex, sometimes considered a subspecies of *ochrocephala*, here treated as an allospecies of the *ochrocephala* superspecies.

MEALY PARROT

Amazona farinosa Pl. 19(3)

Loro Verde (Lora)

DESCRIPTION. 15″ (38cm); 600g. Large, without colorful head markings; best field marks are blackish cere, pale eye-ring. **Adults:** mostly green, below paler; feathers of top and back of head and neck tinged with dull blue; hindneck scaled with black; outer webs of primaries blue, of 3–4 outer secondaries red, tipped with blue; distal half of tail pale yellow-green. Iris orange; bare eye-ring yellowish-white; bill horn-color, shading to dark gray at tip; feet dull grayish. **Young:** iris brown, otherwise like adult.

HABITS. Prefers forested areas, generally keeping to canopy but sometimes descending, especially at gaps and edges, to middle levels or even upper understory to feed; also frequents semi-open and tall second growth near forest; travels in pairs or, more frequently, in flocks of 15–20; outside breeding season, large communal roosts may contain up to 100. As with many other parrots, flocks are noisy in flight but silent while feeding except when accompanied by raucously begging juveniles. Food includes fruits, seeds of many forest trees: palms, figs, *Brosimum*, various green legume pods (*Inga, Dussia*); arils of *Casearia* and *Virola*; some buds, flowers.

VOICE. In flight a diagnostic, raucous, repeated *k'chow chow* or *k'chepchep*; often a series of loud, resonant *kwok* notes upon alighting. Perched, a wide variety of squawks, toots, warbles, etc., somewhat lower-pitched than notes of Red-lored Parrot. Juveniles beg with nasal, grunting notes.

NEST. In natural cavities, often in dead wood, frequently enlarged by the birds themselves. Eggs usually 3. Dry season.

STATUS. Common to abundant resident of heavily forested, humid parts of Caribbean and S Pacific lowlands, especially in Golfo Dulce region; less numerous in somewhat drier General-Térraba and Río Frío regions; rarely above 1600ft (500m). Numbers decline sharply with deforestation, in contrast to those of Red-lored Parrot. Persecuted for cage-bird trade, but less severely than other large *Amazona*.

RANGE. S Mexico to W Ecuador, Bolivia, and SE Brazil.

ORDER **Cuculiformes**
FAMILY **Cuculidae:** Cuckoos

The 127 species of cuckoos inhabit tropical and temperate regions worldwide, including many remote Pacific islands; almost all species breeding at higher latitudes are migratory. Cuckoos are typically slender, long-tailed birds with laterally compressed bills decurved at the tip, and zygodactyl feet (2 toes directed forward and 2 backward); except in the ground-dwelling species, their legs are short. In size they range from a sparrow to a chachalaca. New World cuckoos are more or less plainly clad in shades of brown, gray, and buff along with black and white; their plumage may be streaked or spotted, or iridescent, with little sexual dimorphism; in the much more numerous and sometimes colorful Old World species, the sexes may differ greatly. Many

New World species have colorful facial skin. A peculiarity of cuckoos is their irregular wing molt; the primaries in particular do not grow in a regular sequence as in most other birds.

Most cuckoos are arboreal and highly insectivorous, a favorite food being caterpillars; some, especially the larger terrestrial species, also take small vertebrates, principally reptiles. Cuckoos tend to be quiet, skulking birds, although their voices, heard mostly during the breeding season, are extremely varied. Many build crude open nests of sticks and leaves; but the Old World coucals construct globular nests with a leaf-lined entrance tunnel. Many Old World species are brood parasites, laying their eggs in the nests of other birds that raise the cuckoo's chick at the expense of their own young, which are actively ejected by the young cuckoos or simply outcompeted for their parents' food offerings. In the Americas only the Striped, Pheasant, and Pavonine (South American) cuckoos are parasitic. Cuckoos' eggs are typically chalky white to plain bluish except for those of certain parasitic species, whose varied eggs often closely resemble those of their diverse hosts.

Most nonparasitic cuckoos are monogamous; males and females build the nest and share incubation (which usually begins with the first egg laid) and feeding of the consequently different-sized young. In many species the young leave the nest before they can fly and spend days clambering about in nearby vegetation. Among cuckoos, communal nesting occurs only in the anis and the Guira of the New World. These highly social species live in flocks, all members of which aid in nest-building and feeding the young. Several females often lay in the same nest, though many earlier eggs are buried in the nest or tossed out; the dominant female lays last so that most or all of her eggs are incubated. Incubation is performed mainly by the dominant pair, the female by day and the male at night.

BLACK-BILLED CUCKOO

Coccyzus erythropthalmus Pl. 21(1)

Cuclillo Piquinegro

Description. 12″ (30cm); 43g. Rather small cuckoo lacking conspicuous rufous on wings or conspicuous white in tail. **Adults:** above grayish-brown, with faint bronzy gloss; below dull white, washed with grayish-buff on throat and chest; rectrices below gray, with narrow white tips set off by blackish subterminal bars. Narrow bare eye-ring red; bill black except blue-gray base of lower mandible; legs blue-gray. **Immatures:** above brighter brown, below buffier; tail pattern blurred, indistinct; eye-ring yellowish.

Habits. Skulks silently in second growth, scrubby areas, semi-open, forest edge; perches inconspicuously, scanning surrounding vegetation deliberately; makes short sallies or running, hopping dashes to snatch large, slow-moving caterpillars, katydids. Flight headlong, hurtling, on very deep wingbeats.

Voice. Silent in Costa Rica.

Status. Known only as migrant in fall (late September–early November) and spring (April–early May); very uncommon to rare in Caribbean lowlands; very rare at best in Pacific lowlands; occasional in Valle Central, especially E part.

Range. Breeds in SE Canada, C and NE USA; winters from N Venezuela and N Colombia to Bolivia.

YELLOW-BILLED CUCKOO

Coccyzus americanus Pl. 21(2)

Cuclillo Piquigualdo

Description. 12″ (30cm); 50g. Only small cuckoo with conspicuous rufous in wings. **Adults:** above grayish-brown, auriculars darker; primaries extensively rufous; lateral rectrices black, broadly tipped with white; below dull white. Narrow bare eye-ring gray; culmen and tip of lower mandible black, rest of bill deep yellow; legs slaty. **Immatures:** bare eye-ring yellow; tail pattern blurred, indistinct.

Habits. During migration mostly in coastal scrub, second growth, hedgerows and forest edge; in winter prefers evergreen gallery forest, sometimes mangroves; keeps to thickety cover, foraging like Black-billed Cuckoo; singly or, while migrating, in small, loose, silent groups. Flight on long pointed wings seems labored, with deep wingbeats.

Voice. Silent in Costa Rica.

Status. Rare to fairly common fall migrant (mid-August–early November); generally rare and local winter resident, uncommon to rare in spring migration (late April–early June); most numerous in Pacific lowlands (principal wintering area Guanacaste, occasionally Península de Osa); regular in smaller numbers in Valle Central and Caribbean lowlands during migration. Numbers vary from year to year, probably winter only in some years.

Range. Breeds from S Canada to N Mexico and West Indies; winters from S Central America to E Peru and N Argentina.

MANGROVE CUCKOO

Coccyzus minor Pl. 21(4)

Cuclillo de Antifaz, u Orejinegro

Description. 12½″ (32cm); 65g. Larger, more robust than Yellow-billed Cuckoo, with black mask and buffy underparts. **Adults:** above grayish-brown; narrow blackish mask from bill to auriculars; lateral rectrices black, broadly tipped with white; below pale buff to deep cinnamon-buff. Bare eye-ring yellow; upper mandible black, lower rich yellow, tipped with gray; legs slaty. **Immatures:** mask indistinct; lateral rectrices browner with indistinct whitish tips; wing-coverts tinged with cinnamon and paler scaling.

Habits. Frequents gallery forest, thickets in dry forest, forest edge, second growth and scrub more than mangroves; forages deliberately, perching to scan surrounding vegetation, sometimes twisting head to grotesque angles while holding body immobile, then abruptly pursues prey by hops, runs, or sallies; eats large, slow-moving insects and *Anolis* lizards; flies with several strong flaps and a glide; wings more rounded, less "floppy" than those of Black- or Yellow-billed cuckoos.

Voice. A series of scratchy, barking *whip* notes, repeated regularly ca. 2/sec, increasing in loudness; also a single, *Myiarchus*-like *whip*! or *whik*!

Nest. A slight bowl of sticks lined with a few leaves, in tree or shrub at no great height. Eggs 2–3, bluish-green. May–September (in N Central America; no Costa Rican record).

Status. Uncommon to locally fairly common nonbreeding visitor (December–June) in lowlands and foothills of N Pacific slope, becoming rare S along coast and in Golfo Dulce district; locally to 3600ft (1100m) as in W Valle Central; on Caribbean slope regular only in extreme NW, rarely E to Río Frío region; may breed in very small numbers in lowland Guanacaste. In N Central America, a pale form (*C. m. palloris*) breeds in mangroves, a richly colored form (*C. m. continentalis*) in the interior; in Costa Rica no such division occurs: pale and dark birds occur anywhere.

Range. Breeds from N Mexico and Florida to Nicaragua, possibly to Costa Rica (or W Panama?) and through West Indies to NE South America; winters to Panama and N South America.

COCOS CUCKOO

Coccyzus ferrugineus Pl. 21(3)

Cuclillo de la Isla del Coco

Description. 13″ (33cm); 70g. Most colorful Costa Rican *Coccyzus*; only cuckoo on Cocos I. **Adults:** forehead and crown slaty, shading to grayish-brown on rest of upperparts; a narrow blackish mask from bill to auriculars; remiges mostly deep rufous; lateral rectrices black, broadly white-tipped; below uniform rich buff. Bare eye-ring yellow; upper mandible black, lower bright yellow with black tip; legs slaty. **Young:** tail pattern blurred and indistinct.

Habits. Frequents thickets, tangles, and festoons of hanging vines that occur along the island's major streams, and second growth and *Hibiscus* thickets; prefers to hop and run in dense vegetation rather than fly; forages like other *Coccyzus*, taking large insects (sphingid larvae, cicadas), and *Anolis*; least tame of the island's native landbirds. Flight consists of a few quick flaps and long glides.

Voice. A deep, dry, coughing *kcha* repeated deliberately 5–8 times, sometimes preceded by rolling or rattling notes; also a resonant, guttural *k'k'k'k''k'ru'hoo* (Slud).

Nest. Undescribed (?).

Status. Though often overlooked because of its furtive, skulking demeanor, this is undoubtedly the least common of the native landbirds of Cocos I. It is widespread in the interior forests of the island, and the total population may be larger than infrequent sightings suggest.

Range. Cocos I.

Note. Sometimes considered closely related to (or even conspecific with) *C. minor* or the South American *C. melacoryphus*.

SQUIRREL CUCKOO

Piaya cayana Pl. 21(7)

Cuco Ardilla (Bobo Chiso)

Description. 18″ (46cm); 105g. Large, slender, mostly rich rufous with very long, graduated, white-tipped tail. **Adults:** upperparts and head chestnut-rufous, becoming paler, more vinaceous on throat and upper breast; lower breast gray, becoming blackish-slate on thighs and flanks, dull black on crissum and underside of tail; all rectrices broadly tipped with white, set off above by subterminal blackish area. Iris red; bill and orbital skin yellowish-green; legs gray. **Young:** bill and orbital skin grayish; iris brown; less white at tip of tail (none on central rectrices), above not set off by black.

Habits. In pairs or singly, frequents canopy and edges of woodland, second growth, hedgerows, scattered trees in open country; leaps from branch to branch and runs along branches like a squirrel, plucking caterpillars (including those with stinging hairs and spines), katydids, cicadas, occasionally spiders and small lizards, from foliage; rarely flies far, crossing open spaces by long descending glides on flat wings, occasional bursts of quick, shallow wingbeats.

Voice. A loud, explosive metallic *kip!* or *dik!* frequently given in flight; and a loud *kip! wheeeu*; a dry, nasal, rolling *wid-d-dear* or *hic-a-roo*; particularly during breeding season, a long series of *whip* or *pwit* notes in measured cadence, ca. 3/sec; in interactions, a prolonged *churr* or rattle.

Nest. A loose foundation of long, coarse twigs supporting a cuplike mass of brown or green leaves, 2.5–40ft (0.75–12m) up in a tangle of vegetation or densely foliaged tree. Eggs 2, chalky-white. January–August.

Status. Resident countrywide, sea level to, rarely, 8000ft (2450m); common, but pairs tend to be widely scattered.

Range. NW Mexico to N Argentina.

GROOVE-BILLED ANI

Crotophaga sulcirostris Pl. 21(9)

Garrapatero Piquiestriado o Tijo (Tijo, Zopilotillo)

Description. 12″ (30cm); ♂ 80g, ♀ 70g. An ungainly-looking black bird with relatively short, rounded wings, long floppy tail, and high-ridged prominently grooved upper mandible. **Adults:** entirely black, with scaly iridescence on head and breast; wings and tail glossed with dark blue. Bill, facial skin and legs black. **Young:** similar but with smooth bills. See Smooth-billed Ani.

Habits. Frequents pastures, savannas, plantations, lawns, second growth, marshes, open woods; in loose, straggling flocks of up to 15 birds (but usually ca. 6–8), occasionally in pairs; often forages with grazing cattle, catching grasshoppers and other insects stirred up by them much more rapidly than when hunting alone; occasionally takes lizards, berries; on ground, runs swiftly, using tail as rudder for swift turns and stops; flies with several quick flaps and a long, wobbly glide on set wings; swings tail upward on landing; rests by day and sleeps at night perching on a branch in compact rows.

Voice. Most characteristic is the "tijo" call, in excitement or alarm: a querulous *tee'-ho tee'-ho tee'-ho*, often preceded by soft clucking *tuc* notes; a long series of rapid whistled *kiw* notes on same pitch, falling at end; a full, prolonged, mournful call.

Nest. A bulky cup of interlaced coarse twigs, lined with green leaves that are added daily, 2–25ft (0.6–8m) up amid dense, often thorny foliage of a tree or in a vine tangle. Eggs 3–12, rarely 15, chalky-white but soon stained with brown, and when scratched revealing a blue or blue-green shell. June–November, rarely April. Nests are built and attended by solitary pairs or 2–4 cooperating pairs.

Status. Resident countrywide from coasts to ca. 7500ft (2300m); common to abundant except in S Pacific, where Smooth-billed Ani has largely displaced it.

Range. C Texas and S Louisiana to N Chile and NW Argentina.

SMOOTH-BILLED ANI

Crotophaga ani Pl. 21(8)

Garrapatero Piquiliso (Tijo, Tinco)

Description. 14″ (35cm); ♂ 115g, ♀ 95g. Larger than Groove-billed Ani, with smooth, higher-ridged upper mandible, very different voice, more restricted distribution. **Adults:** plumage resembles that of Groove-billed Ani, but iridescence of head and neck more bronzy, of wings more violaceous. Facial skin, bill, and feet black. **Young:** duller, sootier, less iridescent overall.

Habits. Similar to those of Groove-billed Ani, in flocks in open or scrubby areas, often feeding with livestock. As in that species, birds may spread wings and tail and sun atop a shrub or fence post following a rain; occasionally follows army ants.

Voice. A querulous, whining, whistled *oooenk* or *wooyeek* or *wooiick* in alarm or excitement, often in flight.

Nest. Built by single pairs or cooperating groups, similar in site and construction to nest of Groove-billed Ani, usually 4–10ft (1.2–3m) up in thorny tree or shrub. Eggs 4–15 or more, similar to those of Groove-billed Ani, but larger. March–September.

Status. Common resident of S Pacific slope, sea level to at least 4000ft (1200m) locally, sparingly N to at least Cañas and spreading N. First recorded in Costa Rica near Panamanian border in 1931; by 1940 had reached head of

Térraba valley and by 1975, extreme S Guanacaste; has largely replaced Groove-billed Ani from Parrita area S.

Range. C Florida to W Ecuador and N Argentina, Bahamas, and West Indies.

STRIPED CUCKOO

Tapera naevia Pl. 21(5)

Cuclillo Listado (Tres Pesos)

Description. 12″ (30cm); 55g. *Coccyzus*-sized; short bushy crest, short decurved bill; alula black, conspicuous; upper tail-coverts long, half length of tail. **Adults:** crown feathers dark brown, broadly edged with rufous; superciliaries whitish; upperparts cinnamon-brown, broadly streaked with black; rectrices and remiges dusky, lateral rectrices with paler tips; below dull white, washed with brown on chest; throat and crissum brighter buff; sides of neck lightly streaked with black. Iris buff; bill horn-color, shading to black on culmen, yellow at base; legs brownish. **Young:** crown and wing-coverts spotted with buff; sides and front of neck barred with dusky.

Habits. Prefers open country with scattered trees and thickets, scrubby pastures, agricultural land, second growth; forages amid foliage like a typical cuckoo, and on ground in short vegetation, swaying from side to side, flashing the alulas like fans, perhaps to flush prey—bouts of alula-flashing alternate with sudden scurries after prey, especially grasshoppers; also flashes alulas when disturbed or frightened, or in social interactions; continuously raises and lowers crest; calls incessantly from elevated perch, in bush or tree, or on fence post.

Voice. Most common call 2 loud, clear whistles, the second a halftone higher, often repeated tirelessly, far-carrying but very ventriloquial at close range. Full song contains 5 such notes: *pee-pee-pee-PEEdee* or "tres pesos pide," usually heard in early morning; a second bird often answers with the 2-note call.

Nest. Parasitic, laying its white eggs in domed or covered nests, which it often enters just after dawn, when parents are away foraging. Recorded hosts in Costa Rica include *Synallaxis* spinetails, *Thryothorus* wrens, and *Arremonops* sparrows.

Status. Expanding its range in recent years, doubtless in response to deforestation, especially in wet lowlands of S Pacific and Caribbean slopes; presently common throughout Golfo Dulce and Térraba regions, and along coastal strip N to Tárcoles; in smaller numbers in moist lowlands around Golfo de Nicoya and in Valle Central W to about Santa Ana and Alajuela; still increasing in Valle del General; on Caribbean side spreading S from Río Frío region, now common as far S and E as Guápiles and Limón, sporadic but increasing S.

Range. S Mexico to SW Ecuador, N Argentina, and SE Brazil.

LESSER GROUND-CUCKOO

Morococcyx erythropygius Pl. 21(11)

Cuclillo Sabanero (Horera)

Description. 10″ (25cm); 63g. Usually on or near ground; deeper rufous below than Mangrove Cuckoo, with colorful facial pattern. **Adults:** above grayish-brown, strongly glossed with bronzy-green on wings and tail; crown mixed with dull buff; outer 2 rectrices tipped with dull whitish, set off by black; below cinnamon-rufous, palest on throat, darker and duller on lower belly and crissum. Bare eye-ring and lores bright yellow; bright blue skin behind eye-ring bordered by black, with white superciliaries, very narrow in front of eye, broad behind. Bill yellow with black culmen and mouth-lining; legs yellow-brown. **Young:** above duller, scaled with grayish-buff; below paler and duller, smudged with dusky; pale tips of outer rectrices indistinct.

Habits. Prefers scrubby woodland borders, second growth, savanna, understory of open woods; singly or in pairs walks or hops over ground picking up grasshoppers and other insects, or leaping to snatch them from foliage; furtive and skulking but often inquisitive, not shy; when flushed, flies a short distance, lands running; can run swiftly with head and tail low.

Voice. A series of 10 or more loud rolling or trilling notes, successively lower in pitch and more widely spaced, usually introduced by 2–3 clear, ascending whistles; one of the characteristic sounds of Guanacaste; also a clear whistle, *teeeee* on an even pitch, higher than call of Thicket Tinamou; and a rough, growling *ghaaaoow*.

Nest. A shallow bowl loosely made of dry petioles, thin sticks, and dead leaves, on ground beneath screening vegetation. Eggs 2, chalky-white. February–May.

Status. Resident on N Pacific slope, S to Río Tarcoles and in W Valle Central E to Villa Colón and Grecia; sea level to ca. 4000ft (1200m).

Range. Tropical Mexico to Costa Rica.

PHEASANT CUCKOO

Dromococcyx phasianellus Pl. 21(6)

Cuclillo Faisán

Description. 14½″ (37cm); 96g. Small-headed, crested; long ample tail; upper tail-coverts as long as tail. **Adults:** above dark brown, with dull buff feather-edges; wings and upper tail-coverts glossed with bronzy-green; rectrices grayish, with white tips set off by black; remiges and tail-coverts with small white spot at tip; white stripe behind eye; face, sides of neck, and chest buffy, streaked and spotted with dark brown; rest of underparts white. Iris yellowish; bare eye-ring yellow-green; upper mandible blackish, lower gray; legs grayish-brown. **Young:** above sooty-brown, spotted with buff on crown; wing-coverts scaled with buffy-white; rectrices and upper tail-coverts lack white at tips.

Habits. Singly, sometimes in pairs, in thickets at forest edge, in second growth or scrubby woodland, rarely far above ground; extremely furtive and secretive; takes insects like grasshoppers and cicadas, small lizards and nestlings. When surprised, it may rest on ground or a low branch, fan its great tail and flutter its wings, repeating this as it retires, in what appears to be a distraction display, though no reason for this is apparent.

Voice. Two to 4 loud, clear, successively higher-pitched whistles followed by a descending trilled or broken note, a 4–6 note call like "tres pesos" of Striped Cuckoo, but with last note a tremolo; a series of 4–5 rattling, clucking notes. Whistled calls highly ventriloquial at close range.

Nest. Parasitic; lays a whitish or pale buff egg with a wreath of rufous spots, in open or closed nest of another bird, evidently most often a flycatcher (e.g., *Rhynchocyclus, Myiozetetes*).

Status. Widespread but apparently very rare resident in lowlands the length of Pacific slope, to 2600ft (800m) locally; recorded in recent years from Parque Nacional Santa Rosa, near Esparza, and Valle del General; may be more numerous than few records indicate.

Range. S Mexico to NE Argentina.

RUFOUS-VENTED GROUND-CUCKOO

Neomorphus geoffroyi Pl. 21(10)

Cuco Hormiguero

Description. 19″ (48cm); 350g. Large; bushy-crested; long neck and tail; like a forest roadrunner. **Adults:** above dark brown, glossed with green on wings and purple on tail; crest and postocular stripe blackish; face cinnamon; below grayish-brown, shading to cinnamon on belly and dull rufous on sides, flanks and crissum; irregular black breast-band. Iris dark brown to chestnut; bill dull greenish-yellow, becoming gray at base; orbital skin slaty, becoming bright blue behind eye; legs gray. **Immatures:** much duller and darker overall, especially on neck and breast; no breast-band; bill grayish-yellow.

Habits. Haunts the floor of wet virgin forest singly, in pairs, or small (family?) groups; often forages at army-ant raids, dashing about with great speed and agility after prey flushed by ants, the long tail being used as a rudder and counterweight in sudden stops, starts, and turns; flicks aside leaves and pecks with bill; eats large insects, scorpions, centipedes, spiders, small frogs and lizards, occasionally fruit; raises and lowers crest when excited.

Voice. When feeding at army-ant swarm, a low muffled "woof" (Slud), also a loud, explosive *kchak!* evidently to frighten another bird away from a potential prey item; foraging birds maintain contact by snapping mandibles together loudly and repeatedly.

Nest. In Brazil, a broad, shallow bowl, loosely made of large twigs and lined with green leaves added daily, 8ft (2.5m) up in fork of shrub in dark undergrowth of swampy woods. Egg 1, yellowish-white (Roth). No Costa Rican record, but fledged young in July and December at La Selva and in July in SE Nicaragua (T. R. Howell).

Status. Rare resident of lowland and foothill forests of Caribbean slope, locally to 3000ft (900m) or more, barely reaching Pacific slope through passes of Cordillera de Guanacaste.

Range. NE Nicaragua to E Peru, Bolivia, and C Brazil.

ORDER **Strigiformes:** Owls

These nocturnal predators have large, forward-directed eyes set in a facial disk of radiating feathers. The disks may function like parabolic reflectors to direct sound waves to the large ear openings, which are unique among vertebrates in being differently shaped and positioned on the 2 sides, evidently for more efficient location of the sound source. Hunting by sound is also facilitated by soft filamentous processes on the plumage, which permit silent flight. Owls have strong, hooked bills and claws; 2 toes are directed forward and 2 backward, but 1 of the latter is freely reversible.

FAMILY **Tytonidae:** Barn-Owls

In addition to the nearly cosmopolitan Common Barn-Owl, this family contains 1 species on Hispaniola and 9 in the Old World from Africa through southern Asia and Indonesia to Australia. No member is known to migrate. Barn-Owls differ from typical owls in having a heart-shaped rather than roundish facial disk, longer, completely feathered legs (but featherless toes), and a serrated comb on the claw of the middle toe. In length 13–21″ (33–53cm), they are brownish above and white below, marked with gray and black. The sexes are similar in plumage, but the female is often larger than the male. The habits of the Common Barn-Owl are described below. Most of the other species prefer open grassy plains and nest on the ground; one dwells deep in Indonesian forests. In the North Temperate Zone, Common Barn-Owls lay 5–7 whitish eggs, occasionally up to 11, that are more pointed than those of typical owls. Incubation is by the female, fed by her mate, and takes about 1 month. The young, nourished by both parents, remain in the sheltered nest site for nearly 2 months.

COMMON BARN-OWL

Tyto alba Pl. 20(9)

Lechuza Ratonera (Lechuza de Campanario, Cara de Gato, Buho)

Description. 16″ (40cm); 425g. Pale-colored, with heart-shaped facial disk and long, fully feathered legs. Facial disk and underparts vary from white to rich ochraceous-buff, with fine black speckling; upperparts gray, vermiculated and spotted with white; rectrices and remiges barred with black (more heavily on buffy birds than white ones). Iris dark brown; bill white; cere pale flesh; feet pale brown.

Habits. Prefers open country with scattered trees, agricultural areas, open marshes and mudflats, as well as urban and suburban areas; usually strictly nocturnal, although sometimes active at dawn and dusk in dark weather; when hunting flies buoyantly, quartering low over ground, using sight and hearing to locate small mammals (rodents, shrews, bats), sleeping birds (especially shorebirds), and large insects; roosts by day in dense vegetation, holes in trees, deserted buildings, church steeples.

Voice. When hunting, a rasping, scratchy scream: *shaaahr* or *shwike*; when interacting with others in air, a sharp, incessant pinking; when disturbed at roost or nest, a rasping hiss.

Nest. At most a scanty accumulation of twigs, trash, etc., barely sufficient to keep eggs from rolling, in natural cavity in tree or rocks, or on ledge or platform inside building. Eggs 4–5. Nesting season apparently prolonged, perhaps year-round.

Status. Resident; widespread on both slopes from sea level to at least 6000ft (1850m) in favorable habitats, with centers of abundance in NW lowlands and Valle Central; has undoubtedly increased in recent years in wet lowlands and highlands as a result of deforestation.

Range. Essentially worldwide; in New World from S Canada to Tierra del Fuego.

FAMILY **Strigidae:** Typical Owls

Nearly everywhere except in Antarctica and on certain oceanic islands, some of the 120 species of owls are found. A few that breed at high latitudes are migratory, and those that nest in the high Arctic irrupt to the south in years of scarcity. In length 5.5–30″ (14–76cm), owls are clad in intricate patterns of grays, browns, rufous, and black that make them less conspicuous by day. Many have "horns," or "ears"—narrow lateral crests that are erected mainly during the day, especially in alarm, producing an irregular "jagged-stub" outline that may help to camouflage the birds. Most typical owls are stout-bodied and big-headed, with rounded facial disks and abundant bristly loral feathers that hide the fleshy cere at the base of the bill. Their legs are short to medium in length, their wings long and rounded, their tails short to long, unforked. Females are similar in appearance to males but are larger and heavier in most species.

Owls live in the most diverse habitats: Arctic tundra, tropical rain forest, open woodland, grasslands, deserts, and marshes. Most are arboreal, a few terrestrial. Although mainly nocturnal, some hunt in twilight or even full daylight. The prey of the larger owls includes a great variety of mammals, birds, reptiles, crustaceans, even fishes; smaller owls take many insects as well as small vertebrates. Owls' notes, sounding mysterious in the night, include hoots, whistles, shrieks, wails, tremolos, and hisses. Some sing duets. When threatened, they snap their bills loudly.

Owls rarely prepare nests for themselves. Many breed in holes in trees, among rocks, in buildings, or in almost any convenient cavity. Larger species occupy nests of crows or hawks, perhaps adding a few sticks. Some nest on the ground, and Burrowing Owls dig tunnels. Owls lay 1–14 roundish, chalky-white eggs that are incubated only or chiefly by the female, fed by her mate. After an incubation period of 26–37 days, the blind nestlings hatch well covered with whitish down. Because incubation starts with the laying of the first egg, nestmates differ greatly in age and size; when food is scarce, the oldest and largest may devour their younger siblings. While their mother broods, the father brings food for the family. When 26–42 days old, the young leave the nest. They may continue to be fed by both parents for some time, up to several months in the larger species.

TROPICAL SCREECH-OWL

Otus choliba Pl. 20(15)

Lechucita Neotropical (Estucurú o Sorococa)

DESCRIPTION. 9″ (23cm); 160g. Grayish, heavily streaked, with facial disk outlined in black, and fully feathered tarsi. **Adults** NORMAL PHASE: above grayish-brown, with black streaks and pale vermiculations; scapulars and wing-coverts boldly spotted with white; rectrices and remiges barred with cinnamon-buff; underparts white with herringbone pattern of fine dusky and rusty bars and vermiculations, heavy black shaft-streaks; legs barred with rusty. RUFOUS PHASE: (very rare): above rufous; below pale rusty; dark markings more or less reduced but black border of facial disk still diagnostic. Iris yellow; bill greenish-white; cere pale horn; feet gray-brown. **Young:** plumage fluffy, mostly buffy, finely barred with dusky; above darker.

HABITS. Prefers open woodland, streamside groves, coffee plantations, second growth, suburban areas with trees; becomes active and vocal only well after dark; perches on bare, low branches to sally or pounce on prey; in agile, fluttery flight often snatches insects in air, especially at electric lights; eats large insects, spiders, scorpions, occasionally small mammals, including bats; roosts by day in dense foliage, often pressed against tree trunk.

VOICE. Calls include several strident, screeching barks, *kwah-hwah-kwah*; a snarling note near nest; a bubbling churr or chitter when disturbed or alarmed. Song a chuckling trill followed by 2 (occasionally 1 or 3) louder toots, variously accented and spaced: *prrrr pu POO, prrrr POO poo, prrrr p'p' poo*; sometimes trill or toots alone.

NEST. A cavity in tree: knothole, hollow stub, or old woodpecker hole. Eggs 2–4. February–April.

STATUS. Resident of highlands and foothills at ca. 1300–5000ft (400–1500m) locally from Cordillera de Tilarán S to Panama, mainly on Pacific slope; the common screech-owl of Valle Central, E to Turrialba; fairly common in upper Térraba–Coto Brus area, extending N in coastal lowlands to around Orotina.

RANGE. Costa Rica to N Argentina and Paraguay.

PACIFIC SCREECH-OWL

Otus cooperi Pl. 20(11)

Lechucita Sabanera (Estucurú o Sorococa)

DESCRIPTION. 9″ (23cm); 170g. Medium-sized *Otus* with fully feathered legs and bristly toes; facial disk fairly distinct; above and below with prominent black streaking. **Adults:** facial disk pale gray, barred with dusky, bordered by white mixed with dark brown; above gray-brown, finely vermiculated with dusky and with black shaft-streaks (heaviest on crown); below white, densely and finely vermiculated with dusky and brown, with prominent black shaft-streaks heaviest on chest; rectrices and remiges barred with cinnamon-buff, thighs barred with rusty. Iris yellow; bill and cere pale greenish-gray; feet gray-brown. No rufous phase. **Young:** above grayish-buff, below whitish, becoming deep buff on chest; above and below with fine dusky barring.

HABITS. Frequents deciduous and evergreen forest, (less numerous in latter), savannas with scattered trees, second growth, and mangroves; roosts during day in thickets, vine tangles, and tree holes, but more wary and easily flushed in daylight than most owls; when hunting frequents clearings and edges, perching on low open branches, fence posts, pouncing or sallying for large insects (beetles, katydids, moths), and scorpions; often becomes active soon after dark.

VOICE. Song a series of 5–10 rather nasal, abrupt, barking notes, loudest and highest-pitched in the middle of the series; these sometimes preceded by a short, chuckling trill: *pu-pu-pu-pu-pu-pu* or *prrr pu-pu-PU-PU-PU-pu-pu*.

Nest. In knothole or old woodpecker hole. Eggs 3–4. Mainly in dry season, family groups remaining together into early wet season.

Status. Fairly common resident of lowlands and foothills of N Pacific slope, to 2600ft (800m) or more along Cordillera de Guanacaste, and 3300 (1000m) or more in the warmer, drier W Valle Central; rare or sporadic as far E as Cartago, or S of Carara, where evidently confined to mangroves.

Range. SW Mexico to Costa Rica.

Note. Sometimes considered a subspecies or "incipient species" of the *Otus asio* group; we follow the AOU in considering *cooperi* an allospecies of the *O. asio* superspecies.

VERMICULATED SCREECH-OWL

Otus guatemalae Pl. 20(12)

Lechucita Vermiculada (Estucurú o Sorococa)

Description. 8″ (20cm); 150g. Smallish, short-eared, unstreaked, with indistinct facial disk; distal tip of tarsus and toes bare. **Adults** Normal phase: above grayish-brown to buffy-brown, finely and uniformly barred and vermiculated with black; bold white spots on scapulars and wing-coverts; primaries barred with buff; facial disk and foreneck grayish-buff barred and mottled with dusky; underparts white, heavily and uniformly barred and vermiculated with black and brown; at most faint black shaft-streaks. Rufous phase: above with bright uniform rufous; below white to buffy, vermiculated with rufous and dark grayish.

Habits. Prefers wet forest in lowlands and foothills, sometimes calling or hunting at forest edge, gaps or adjacent tall second growth; usually calls from upper understory or subcanopy, chiefly from ca. 8:00 PM to midnight and again before dawn; eats mainly large insects (beetles, katydids); roosts in thickets during day; very difficult to see at night, even at relatively close range.

Voice. A common call, especially following breeding, a scratchy, sibilant *ghoor* or *khooo*; a possible aggressive note a sharp *prrrOWr* with rising inflection. Song of male a long, somewhat toadlike trill, starting softly, becoming louder, then fading and dropping slightly in pitch toward end, lasting ca. 3–5sec; female may answer with shorter, higher trill.

Nest. In old trogon nest, lined with wood chips, 16ft (5m) up in tree stub in forest. Eggs 2. March–April (Marín).

Status. Widespread, rare to locally common resident of wet lowlands of Caribbean and S Pacific slopes (Golfo Dulce district, Valle del General); ranges from sea level to 3300ft (1000m) locally. A small population recently discovered in the higher, wetter mountains of Península de Nicoya.

Range. NW and E Mexico to N Venezuela, W Ecuador, and Bolivia.

Note. Birds from N Costa Rica S are sometimes considered a separate species, *O. vermiculatus*, from *O. guatemalae*, which is then called Middle American Screech-Owl.

BARE-SHANKED SCREECH-OWL

Otus clarkii Pl. 20(10)

Lechucita Serranera (Estucurú o Sorococa)

Description. 10″ (25cm); 180g. Large, short-eared, richly colored highland screech-owl with poorly defined facial disk; distal third of tarsus bare. **Adults:** head, neck, and upperparts rich brown to dull rufous, heavily spotted, mottled, and vermiculated with black and (hindneck) buffy; scapulars and wing-coverts boldly spotted with white; flight feathers barred with cinnamon-buff; pattern of upperparts mixed with white on chest; posterior underparts mostly white with heavy dusky and rufous barring and vermiculations and heavy black shaft-streaks; thighs mostly buff. Iris yellow; bill bluish- or greenish-gray; cere and feet horn- to flesh-color. **Young:** above cinnamon-buff, speckled with white and barred with dusky; below buffy, barred with dull cinnamon.

Habits. Prefers dense mountain forest, forest edge, sometimes thinned woodland; hunts along edges, in clearings, sometimes in canopy; usually calls from well up in tree; eats large insects (beetles, orthopterans), shrews, and small rodents; often in family-sized groups, even early in breeding season (female still laying).

Voice. Song a deep whistled *hu-hu*; *HOO-HOO hoo*, third and fourth notes accented, the latter usually highest in pitch; at a distance only these 2 notes audible; sometimes varied to a 4-note *hoo HOO HOO' hooo*. Aggressive call, often given in flight, fast toots in groups of 3.

Nest. Undescribed (?). Egg-laying evidently from February to May (brood patches); fledged juveniles seen May–August.

Status. Generally uncommon resident in

mountains from Cordillera de Guanacaste (N at least to Volcán Miravalles) S to Panama, from timberline down to 3000ft (900m) in N cordilleras, or 4000ft (1200m) in Cordillera Central, Cordillera de Talamanca.

Range. Costa Rica to extreme NW Colombia.

CRESTED OWL

Lophostrix cristata Pl. 20(3)

Buho Penachudo

Description. 16″ (40cm); 400g. Medium-sized, slender with finely vermiculated plumage, rather long tail and small feet; forehead and eyebrow white, the white stripe prolonged to the very long ear-tuft, forming a striking white V when ear-tuft is erected in alarm. Dark and pale phases predominate in wet and dry areas, respectively. **Adults** Dark phase: above dark brown, spotted with white on scapulars and wing-coverts; below pale brown, becoming ochraceous posteriorly; flight feathers barred with buff; facial disk rufous, bordered with black. Pale phase: above buffy-brown, below whitish, becoming buff posteriorly; facial disk paler rufous. Iris dark orange to chestnut; bill mostly black; cere and feet dull yellowish. **Young:** head and body plumage white, fluffy; facial disk and flight feathers like adult.

Habits. Prefers forest and old second growth where thickets provide daytime roosts, especially at gaps and edges and along streams; often roosts with mate; calls from midcanopy of forest. Mode of foraging little known; apparently subsists mainly on beetles, orthopterans, roaches, caterpillars.

Voice. A low, gruff, rolling growl *k'k'k'k-krrrrr* (Eisenmann) or *k-k-KRROOOrrrr*, highest in middle. At a distance, initial short notes not audible, call sounds froglike, bubbling; at close range, strident overtones give a fierce, trumpeting quality.

Nest. Undescribed (?). Breeds in dry to early wet season (gonad data); young may accompany parents at least through September.

Status. Widespread, rare to locally fairly common resident in lowlands and foothills the length of both slopes, from sea level to at least 5000ft (1500m) around Valle Central.

Range. S Mexico to Bolivia and C Brazil.

GREAT HORNED OWL

Bubo virginianus Pl. 20(1)

Buho Grande

Description. 21″ (53cm); 1 kg. Very large, powerful, with conspicuous ear-tufts, barred underparts, and fully feathered feet. Facial disk dull cinnamon, edged with black; upperparts blackish, barred and vermiculated with gray, the feathers with tawny bases; chin and band across foreneck white; throat mixed with dusky and buff; underparts whitish to tawny, with dusky vermiculations and heavy black barring; feathers of legs and feet buffy. Iris yellow; bill and claws blackish.

Habits. Little known in Costa Rica; recorded in savanna with patchy woodland, coffee plantations and pastures with scattered dense groves and fencerows; hunts mainly over pastures, savannas, and woodland edges; roosts by day in dense tree, especially in rows of tall cypresses bordering many highland pastures. Prey includes opossums, other medium-sized mammals, and birds.

Voice. A series of deep, powerful, far-carrying hoots, typically 3–6 notes, the middle one(s) more drawn-out *hu-huuuu, huu huu huu*.

Nest. An unlined cavity of a large tree, a cave, on a protected ledge, or in abandoned nest of some large bird. Eggs 2–3. December–April (in Mexico; no Costa Rican record).

Status. Very rare resident, principally in Valle Central: recent sightings near Ochomogo, above Cartago, and near Alajuela; 1 sighting near Taboga, in S Guanacaste (Stiles).

Range. Alaska and N Canada to Tierra del Fuego.

SPECTACLED OWL

Pulsatrix perspicillata Pl. 20(8)

Buho de Anteojos (Oropopo)

Description. 19″ (48cm); 750g. Large, with powerful bill and feet, broad wings and tail, striking face pattern. **Adults:** head, neck, chest, and upperparts dark brown; rectrices and remiges barred with gray-brown; incomplete white spectacles formed by eyebrows, loral and malar bristles; a band of white across throat, bordering facial disk; underparts deep buff, more or less barred with dusky on sides. Iris yellow; bill pale horn shading to greenish-white at tip; feet grayish. **Young:** facial disk blackish, body plumage fluffy, white; wing-coverts barred with gray-brown.

Habits. In dense forest, but hunts more along edges, at clearings, in adjacent semi-open or savanna with large trees; roosts in dense vegetation, often along streams, in gallery forest; when hunting, perches on bare branches at medium height, leaning forward and scanning, striking with a swift pounce to ground or a surprisingly agile swoop to snatch prey from vegetation, taking large insects, mammals up to size of skunks and opossums, lizards, and birds (oropendolas, jays); occasionally hunts by day; frantically mobbed by jays and other birds.

VOICE. A deep, bubbling or "thrumming" *PUP-pup-pup-pup-po* or *PUM-PUM-pum-pum*, successive notes lower, weaker, faster; voice of female (?) higher-pitched; at dusk, from conspicuous perch, a hawklike scream *kerWHEEEER*! with a steam-whistle quality; juveniles give a higher, softer *keeew*!

NEST. In large natural cavity in tree. Eggs 2 (?). Dry season and early wet season; young remain with parents for up to a year after fledging.

STATUS. Widely distributed, uncommon to fairly common (for such a large raptor) resident the length of both slopes, from sea level to 5000ft (1500m) locally. More tolerant of deforestation than Crested Owl, but requires sizable wooded areas for nesting and roosting.

RANGE. S Mexico to W Ecuador, Bolivia, and N Argentina.

LEAST PYGMY-OWL

Glaucidium minutissimum Pl. 20(16)

Mochuelo Enano

DESCRIPTION. 5½″ (14cm); 60g. Slightly smaller and shorter-tailed than other pygmy-owls, with less contrasting pattern. **Adults:** head and underparts dark grayish-brown; crown speckled with white; face mottled with whitish; eyebrows and loral bristles white; black eyespots, bordered with white, on hindneck; scapulars and wing-coverts spotted with white; remiges barred with buffy-white; tail blackish with 3 white bars; below white, broadly streaked with brown to blackish on sides and flanks; sides of breast brown to cinnamon. Iris and feet bright yellow; bill and cere greenish-yellow. **Young:** crown not spotted; no eyespots on hindneck.

HABITS. Frequents canopy and edge of wet forest and adjacent tall second growth and semi-open with tall trees, such as old cacao plantations; calls most actively just before dawn, less often after dusk, rarely during day; often hunts by day in manner of other pygmy-owls, taking large insects, small lizards, and birds (e.g., tanagers, honeycreepers). Flight amid vegetation fluttery, mothlike, extremely maneuverable.

VOICE. Usual call 2–5 (typically 4) clear, unmodulated, evenly spaced toots, deeper, more resonant, faster than the clear whistles of Stripe-breasted Wren. When excited may give 10 or more toots in rapid succession, sometimes preceded by downslurred rolling note or trill.

NEST. In cavity in tree, probably most often old woodpecker hole; details unknown (?). Probably breeds in dry to early rainy seasons.

STATUS. Uncommon to locally fairly common resident of lowlands and foothills the length of Caribbean slope, from sea level to 2600ft (800m) locally. Old records from Valle Central probably represent accidental strays.

RANGE. S and W Mexico to NW Colombia, E Peru, Paraguay, and C Brazil.

ANDEAN PYGMY-OWL

Glaucidium jardinii Pl. 20(18)

Mochuelo Montañero

DESCRIPTION. 6″ (15cm); 65g. Darkest, most strongly patterned pygmy-owl; distinct rufous phase. **Adults** NORMAL PHASE: head and upperparts dark brown; below mostly white; face barred with buff, eyebrows and lores white; crown densely dotted with white; black eyespot rimmed with white on each side of hindneck; nuchal collar rufous; scapulars and wing-coverts boldly spotted with white; remiges barred with white; tail black with 4 white bars; foreneck and sides of breast rich brown, barred with buff and spotted with white; belly streaked with black. RUFOUS PHASE: head, upperparts, breast deep rufous; pale spotting buffy and sparser than on normal phase.

HABITS. Prefers canopy and edge of highland forest and adjacent semi-open, including pastures with scattered trees; hunts by flying swiftly to perch in dense foliage, peering actively about, attacking prey with a short, swift dash; if strike fails, owl usually perches rather than attempt pursuit; takes more birds than other pygmy-owls, also lizards and insects; calls mostly in early morning, late afternoon, and at night; like other pygmy-owls, switches tail from side to side when agitated.

VOICE. Usual song a long series of clear, unmodulated toots in an irregular rhythm; often notes seem to come in 2s, sometimes 3 pairs of toots only; occasionally a series of evenly spaced toots; when excited, 5 higher-pitched toots in very quick succession.

NEST. In cavity in tree, perhaps most often old woodpecker hole; 1 nest 6ft (2m) up in dead stub in pasture tree. Eggs 3. March (Marín).

STATUS. Widespread, rare to locally fairly common resident of middle to high elevations in Cordillera Central and Cordillera de Talamanca, from timberline down to ca. 3000ft (900m) on Caribbean slope and to 4000ft

(1200m) on Pacific slope; prefers wetter, more forested districts than Ferruginous Pygmy-Owl.

Range. Costa Rica to W Venezuela, Peru, and Bolivia.

Note. Has occasionally been lumped with *G. brasilianum*, but differs in voice, coloration, and ecology and is locally sympatric; others consider it closely related to the northern *G. gnoma*.

FERRUGINOUS PYGMY-OWL

Glaucidium brasilianum Pl. 20(17)

Mochuelo Común (Cuatro Ojos, Mahafierro)

Description. 6″ (15cm); 70g. Very small; partly diurnal; heavily streaked below, with eyespots on hindneck; brown and rufous phases, with intermediates. **Adults:** head and underparts grayish-brown to dull rufous; face and crown streaked finely with whitish to buff; eyebrows and loral bristles white; black patch, bordered by white on each side of hindneck; scapulars and wing-coverts spotted; remiges barred with white to cinnamon-buff; tail blackish with 5–6 white, buffy, or rufous bars; below white; sides of breast brown to rufous; breast and sides streaked with blackish to reddish-brown. Iris and legs bright yellow; bill and cere greenish-yellow; claws black. **Young:** crown more speckled than streaked; eyespots indistinct or lacking; below dark streaking paler, blurred.

Habits. In deciduous and evergreen woodland, savanna trees, semi-open, second growth, coffee plantations, suburban areas with large trees for nesting; largely crepuscular but sometimes hunts in full daylight or on dark nights, capturing large insects and small lizards, occasionally small birds, in talons after short, dashing flight from a usually concealed perch. Flight rapid and direct in open, bursts of quick wingbeats alternating with glides; amid vegetation, fluttery and agile. Regularly mobbed by small birds.

Voice. A long series of evenly spaced (ca. 2/sec), slightly upslurred or downslurred toots; when agitated, toots doubled and sequence may end with sharp bark or whinny.

Nest. In an unlined tree cavity, often an old woodpecker hole, or sometimes in a termitary, well above ground. Eggs 2–5. Dry to early wet season.

Status. Fairly common resident in NW lowlands, S to hills above Parrita, and Valle Central, regularly E to Paraíso, occasionally to Turrialba; sea level to 5000ft (1500m). No records from S Pacific slope.

Range. SW USA to N Chile, E Peru, Bolivia, and C Argentina.

BURROWING OWL

Athene cunicularia Pl. 20(14)

Lechuza Terrestre o Llanera

Description. 9″ (23cm); 150g. Fairly small, long-legged, "earless" terrestrial owl of open areas. **Adults:** head and upperparts brown; facial disk streaked and bordered with white; eyebrows white; buffy white streaks on crown; white spots on mantle and wing-coverts; rectrices and remiges barred with buffy-white; throat buffy; chest brown, spotted with buffy-white; belly buffy-white, spotted and barred with brown. Iris yellow; bill and cere horn-color; feet brownish.

Habits. Prefers savannas, pastures, and other open country; often active by day as well as night, perching on low mounds, shrubs, or fence posts; eats small rodents, insects, occasionally small birds and reptiles; when alarmed, as at approach of observer, engages in exaggerated head-bobbing.

Voice. A liquid *ka-ka-ka-ka* when alarmed; other calls during breeding season.

Status. One old specimen (20 December 1900) from SW slope of Volcán Irazú; only Panama record taken in Chiriquí at almost the same time (13 December 1900), suggesting that a single exceptional flight occurred then. A recent sight report from N Guanacaste (Fogden) requires confirmation.

Range. Breeds from SW Canada, C and SE USA to C Mexico; winters regularly to El Salvador, casual to accidental to W Panama; also resident in West Indies and, locally, in much of South America.

Note. Formerly placed in genus *Speotyto*.

MOTTLED OWL

Ciccaba virgata Pl. 20(6)

Lechuza Café (Hu de León)

Description. 14″ (35cm); 275g. Medium-sized, "earless" owl with rather long wings and tail, mottled breast, and streaked belly. **Adults:** above dark brown, barred and vermiculated with dusky and grayish-buff; scapulars spotted with buffy-white; remiges barred with pale grayish-brown; face finely streaked with whitish; eyebrows white; throat and chest dull ochraceous, mottled and vermiculated with dusky; belly buffy, heavily streaked with blackish; tail blackish, crossed by 3–4 narrow white bars. Occasional pale birds are mainly grayish-brown above, white on belly. Iris brown; cere olive-green; bill yellowish-horn; feet brownish. **Young:** mostly buffy to cinnamon, above barred with dusky; facial area whitish.

Habits. Occurs widely in forest, especially at gaps and thickety edges, also in semi-open, old second growth, coffee plantations with shade trees; roosts in low thickets by day; calls and hunts at medium height in trees, often lower, taking large insects, especially orthopterans, small mammals, especially rodents, and small snakes; not known to take birds but mobbed by small birds in daytime.

Voice. Commonest call 2–3 successively louder grunts, low and guttural, then 2 sharp downslurred hoots, the first louder: *huh-huh; WHOO', WHOO*; sometimes a single hoot, or 1–3 grunts alone; a long-drawn-out catlike screech usually heard shortly after dusk or before dawn, and various other notes.

Nest. Usually in hole in tree or palm stub; reported to use old nest of another bird. Eggs 2. February–April.

Status. Uncommon to fairly common resident from lowlands to 5000ft (1500m) locally the length of both slopes; least numerous in dry Guanacaste lowlands, where largely restricted to evergreen gallery forests.

Range. N Mexico to W Ecuador, Bolivia, and N Argentina.

Note. Also called Mottled Wood-Owl.

BLACK-AND-WHITE OWL

Ciccaba nigrolineata Pl. 20(7)

Lechuza Blanco y Negro

Description. 15″ (38cm); 350g. A dark-colored forest owl, of similar proportions to Mottled Owl, but larger. **Adults:** face and entire upperparts sooty-black; remiges barred below with whitish; tail crossed by 6–7 narrow whitish bars; facial disk bordered by white freckling; nuchal collar of white barring; entire underparts finely barred with black and white. Iris orange-brown to chestnut; bill, cere, and feet yellow-orange. **Young:** above white, closely barred with dark brown; wing-coverts black with some white barring; face whitish; below buffy-white, barred with black.

Habits. In tall forest or its immediate vicinity; in drier areas prefers evergreen or gallery forest, sometimes tall mangroves; active strictly at night; roosts in thickets by day; often hunts along edge of forest, darting out from perch to snatch prey from ground or vegetation; eats mainly large insects, especially beetles and orthopterans, small rodents, and bats, which are seized in midair; sometimes attracted to insects swarming at bright lights.

Voice. Commonest call a low grunt followed by a gruff, strident hoot: *huh, HOOoo*; these sometimes followed by 2 lower, softer, faster hoots, *huh, HOOoo hoo-hoo*, a 4-note call with pattern the reverse of that of Mottled Owl. Young birds give an ascending, breathy shriek.

Nest. Unknown (?). Dry season.

Status. Uncommon to rare and local resident of lowlands and foothills of Caribbean and N Pacific slopes, from sea level to 5000ft (1500m) locally. Not known but expected in Pacific lowlands S of Quepos.

Range. S Mexico to NW Venezuela and NW Peru.

Note. Sometimes considered a subspecies of the South American *C. huhula*.

STRIPED OWL

Asio clamator Pl. 20(2)

Buho Listado

Description. 15″ (38cm); 440g. Medium-sized; very long-horned; heavily streaked; with rather short wings and long tail. **Adults:** facial disk mostly white, bordered with black; ear-tufts black, edged with buff; entire upperparts cinnamon-buff, finely vermiculated and heavily striped with black; wings and tail barred with black; below white, heavily streaked with black, tinged with buff on belly. Iris brown to cinnamon; bill and claws black; cere grayish. **Young:** facial area cinnamon, bordered with white, black, and buff; crown buffy-white, lightly streaked with black; body plumage buff, barred with grayish.

Habits. Prefers open country with scattered thickets and woodland patches: savanna, open marshes, pastures, airstrips, sometimes rice and other agricultural fields; roosts in low thickets or on ground by day; often commences hunting at sunset, quartering low over open areas, swooping down at prey; may also watch for prey from a fence post, power line, or other exposed perch; takes mostly small mammals, large insects, some birds and reptiles.

Voice. A series of ca. 7 low, muffled hoots on same pitch (Kiff), or a single, nasal hoot lasting about 1sec, loudest and highest in middle: *hooOOOoh* or *hnnNNNnh*, hoot of female higher in pitch; also 7–8 sharp, doglike barks: *hu-how! how! how!* etc., sometimes given by mates in chorus.

Nest. A bit of flattened vegetation on ground or on flat surface at no great height [e.g., mass of epiphytes and dead leaf bases adhering to trunk of oil palm, 10ft (3m) up]. Eggs 2–4. December–March.

Status. Widespread but mostly local and uncommon resident the length of Pacific slope, from sea level to 4500ft (1400m) in Valle Central (e.g., San Pedro); centers of abundance include Guanacaste lowlands and Térraba region; on Caribbean slope recorded only from open grasslands around Cartago and rice fields near Upala; should be looked for elsewhere, as range probably expanding with deforestation.

Range. S Mexico to E Peru, N Argentina, and Uruguay.

Note. Formerly separated in genus *Rhinoptynx*.

SHORT-EARED OWL

Asio flammeus Not Illustrated

Lechuza Campestre

Description. 16″ (40cm); 350g. Medium-sized, with very short, inconspicuous ear-tufts, rather long wings and legs, streaked pattern. Facial disk buffy, streaked and edged with black; eyebrows, loral tufts, and chin white; upperparts broadly streaked with cinnamon-buff and black; rectrices and remiges rich buff, barred with blackish; below whitish to buffy, broadly streaked with dark brown on breast. Iris yellow; bill and claws blackish.

Habits. Prefers open grasslands, marshes, seashores; rests by day on ground in low vegetation; hunts from late afternoon into night, quartering back and forth low over ground like a harrier, suddenly pouncing at prey seen or heard; eats mainly rodents.

Voice. Various barking, squealing, hissing notes rarely heard outside the breeding season.

Status. Two specimens taken near San José in mid-December (1883, 1916) constitute the only Costa Rican records; accidental at best and may not occur at present. The extensive highland marshes of Valle Central that were its probable habitat have mostly been drained, and the amount of open habitat in Mexico and N Central America has vastly increased with deforestation, so that the owl need not go so far south to find abundant wintering habitat (the same may apply to the Burrowing Owl). Should be looked for in rice fields of N Guanacaste (e.g., Pelón de la Bajura).

Range. Breeds from N Alaska and N Canada to SW and E USA; winters regularly to S Mexico, perhaps Guatemala, accidentally (formerly?) to Costa Rica. Resident in West Indies and Hawaiian Is. Widespread in Old World.

UNSPOTTED SAW-WHET OWL

Aegolius ridgwayi Pl. 20(13)

Lechucita Parda

Description. 7″ (18cm); 80g. Small, big-headed, with stout feet; variable in color. **Adults:** head and upperparts uniform grayish-brown, or head and mantle darker brown; outer primaries and alula edged with white; inner secondaries and rectrices more or less spotted with white; crown sometimes finely streaked with dull whitish; eyebrows, loral tufts and chin white to deep buff; chest cinnamon to dull brown; belly white to buff. Iris yellow to tawny; cere and most of bill blackish, sides pale horn; toes flesh-color. **Young:** similar in pattern but plumage more downy; often pale streaking on chest.

Habits. Frequents forest canopy and edge, clearings and pastures with scattered tall trees, both in oak forests of high mountains and in cloud forests lower down; strictly nocturnal but evidently active soon after dusk and just before dawn. Flight fluttery, agile, with quick wingbeats; chief prey probably small rodents and shrews, supplemented by small birds and bats.

Voice. A series of 4–10 rhythmic, mellow toots on an even pitch, softer and lower than notes of pygmy-owls; when excited, a trill like that of *Otus* but much higher-pitched (Marshall).

Nest. Undescribed (?). Evidently dry and early wet seasons.

Status. Resident in Cordillera Central and Cordillera de Talamanca from timberline down locally to 8200ft (2500m); probably less rare than scattered records suggest.

Range. S Mexico to W Panama.

Note. Has sometimes been lumped with the northern *A. acadicus*. Voice reports from much lower elevations, down to 3000ft (900m), are apparently based on the amazingly owl-like call of a tree frog, *Anotheca* (Fogden).

ORDER **Caprimulgiformes:** Nightjars and Allies

All members of this order are nocturnal birds somewhat resembling owls in their soft, cryptically patterned plumage but differing in their lack of facial disks, weaker feet and bills, and huge mouths.

FAMILY **Steatornithidae:** Oilbird

This peculiar bird differs in many ways from the other members of the order Caprimulgiformes. It is the only vegetarian, apparently subsisting entirely upon the oil-rich fruits of trees like palms and Lauraceae. It is by far the most social member of the order, nesting and roosting colonially in caves and dark gorges, from which groups range nightly up to 50mi (80km) to forage, returning before dawn. The flight is strong, fast, and undulating, and immatures in particular may wander widely. Oilbirds are also adept at hovering, particularly when they pluck fruits with their strong, hook-tipped, tooth-edged bills. Their plumage is firmer, less lax and fluffy, and less densely barred and vermiculated than that of other caprimulgiform birds.

Oilbirds' nests are conical or disk-shaped masses of excrement mixed with seeds, placed on narrow ledges to which the birds unerringly find their way in total darkness by means of echolocation—a capacity apparently unique among New World birds. Both parents incubate the 2–4 white eggs for ca. 33 days and bring fruits to the young, who rapidly become fat but mature slowly. At 10 weeks of age they may weigh half again as much as their parents, but they require 3–4 months to fledge. The Oilbird's voice is loud and varied: aside from the sharp clicks used in echolocation, adults and older young emit raucous, hair-raising shrieks when intruders enter the breeding caves.

OILBIRD
Steatornis caripensis Pl. 51(8)
Guácharo

Description. 18″ (46cm); 430g. Potoo-sized, with very long wings and much heavier, stronger bill; distinctive white-spotted plumage. **Adults:** above rich brown; below and on wing-coverts dull rufous to cinnamon-brown; tail and secondaries vermiculated and narrowly barred with sooty-black; crown, middle wing-coverts, and underparts spotted with white, each spot partly edged with black; primaries dark brown, the margins spotted with white. Iris yellowish; bill and feet pale reddish-brown. **Young:** above and below much duller, darker brown; white spotting as in adult.

Status. Accidental; definitely known in Costa Rica from a partial, desiccated corpse found in late January 1986 at Villa Mills on Cerro de la Muerte, in a clearing beneath a power line that the bird might have struck (Pyle, Perkins, and S. G. Howell). One wing now in zoology museum of Universidad de Costa Rica. Not known to breed closer than Colombia; almost certainly this record pertains to a wandering individual.

Range. Breeds from NW Colombia and Trinidad to E Venezuela, Peru, and Bolivia; strays reported from E Panama and Costa Rica.

FAMILY **Nyctibiidae:** Potoos

The 5 species of potoos are confined to tropical America, from Mexico to Argentina and the islands of Hispaniola and Jamaica. In length 14–19″ (36–48cm), long-winged and long-tailed, they are clad in shades of brown, gray, buff, white, and black, intricately barred and vermiculated; the sexes look alike. In plumage they resemble the related nightjars, and like them they have huge mouths and small bills, with the tip of the upper mandible strongly decurved; they differ in lacking true rictal bristles and a comblike structure on the claw of the middle toe. Their notably large eyes are yellow or brown. Unlike nightjars, they perch upright, often on the end of a stub or post or in some other exposed situation where, if they find themselves observed, they slowly stretch upward into an elongated cryptic pose. Strictly nocturnal, they subsist upon insects and sometimes small bats that they catch on sallies from a lookout perch, in the manner of flycatchers. Heard on dark or moonlit nights, their loud calls stir the imagination. A single large, white, sparingly spotted egg, resting precariously in a shallow depression on a branch, stub, or stump, low or high, is incubated by both parents sitting upright, 1 through the day and the other through the night. After an incubation period of a full month or more, the nestling hatches well covered with short, whitish down. Fed by both parents with regurgitated insects, the young (of the Common Potoo) remains at the exposed nest site until about 50 days old, when it flies well.

GREAT POTOO
Nyctibius grandis Pl. 20(5)
Nictibio Grande (Leona, Bruja)

Description. 20″ (51cm); 600g. Large, pale-colored night bird of forest canopy; wings and tail much longer than those of any owl. **Adults:** overall pale buff, shading to white on belly; above and below lightly and irregularly barred and vermiculated with black and buffy-brown; scapulars spotted with black; irregular band of black spots across breast; wings and tail barred with black and buff, mottled with cinnamon and grayish. Iris dark brown to chestnut; bill blackish; feet pale greenish. **Young:** overall whiter; dark markings reduced.

Habits. Lives in canopy of dense lowland wet forest, sometimes ranging out to large trees in adjacent clearings or semi-open; calls most actively on moonlit nights; flies just above canopy, between large emergent trees; perches on projecting snags or open branches, from which it sallies for large flying insects, especially beetles and moths, and small bats; regularly uses high, often exposed perches in canopy or at edge as daytime roosts, perching with body vertical, head horizontal; assumes head-vertical, slit-eyed "post" posture when alarmed.

Voice. Most often a loud, far-carrying bark: *BWOW*! or *GWOK*!, given while perched or in flight; perching birds also utter a loud, guttural, snoring *GWAWWWRRRR* or *WOWWWRRRR*, evidently a territorial call.

Nest. A depression on horizontal branch or stub. Egg 1, white, spotted with dark brown to purplish-gray (in Brazil; no Costa Rican record).

Status. Resident, locally uncommon to fairly common (to judge by calling) in extensive areas of lowland wet forest the length of Caribbean slope and in Golfo Dulce area, including Península de Osa.

Range. Guatemala to E Peru, C Bolivia, and SE Brazil.

COMMON POTOO

Nyctibius griseus Pls. 20(4), 51(7)

Nictibio Común (Pájaro Estaca, Pájaro Palo)

Description. 15″ (38cm); 230g. Like a huge nightjar but distinguished by its upright stance; owls have round heads, shorter tails. **Adults:** above dull grayish-brown, streaked finely with black on crown, and mottled and vermiculated with dusky and buff; throat buffy-white with sparse dusky streaks; chest gray-brown, mottled with buff; belly paler and grayer; a band of large black spots across lower breast; erect feathers above eye suggest small "horns". Iris yellow (brilliant orange eyeshine at night); bill dusky; feet brownish. **Young:** smaller and paler than adult; nestling down dense, short, white. Great Potoo much larger, paler.

Habits. Frequents open woodland, forest edge, savanna trees; by day, most often seen in upright, elongated, cryptic posture with nearly closed eyes, perched on a stub of which it appears to be a part; when unaware of observer, fluffs plumage more and sits less upright. In twilight and at night, sallies from an exposed perch to capture large flying insects, especially beetles and moths.

Voice. A set of deep, clear, plaintive whistles, each lower in pitch than the preceding: *POO-or me, O, O, O, O* or *kloooe-kloo loo, loo, loo*, often heard on moonlit nights; this call ascribed by country people to a sloth! In Guanacaste and Valle Central, a totally different rough, squalling *kwaaah, kwa-kwa-kwa* and an abrupt, emphatic *rrrah*!

Nest. A shallow knothole at an elbow in an ascending limb or a slight depression at top of a stub, 10–60ft (3–18m) up. Egg 1, white, speckled. December–March, June.

Status. Resident countrywide from lowlands up to ca. 4100ft (1250m); generally rare but locally more common in Térraba region.

Range. Tropical Mexico to W Ecuador and N Argentina; Jamaica, Hispaniola.

Note. Quite possibly 2 species are included here, corresponding to the 2 call types: *N. jamaicensis* (Northern or Jamaican Potoo) of West Indies, Mexico and Central America S to C Costa Rica (Pacific slope) and E Honduras (Caribbean slope); and *N. griseus* (Common or Gray Potoo) from SW Costa Rica and E Nicaragua southward. There is a progression from larger, grayer, paler birds in N to smaller, buffier or browner, darker birds in S, but apparently no sharp morphological change corresponds to the abrupt change in calls. More study is needed!

FAMILY **Caprimulgidae:** Nightjars

A family of about 67 species, nightjars (or goatsuckers) occur in temperate and tropical regions worldwide except in New Zealand and most oceanic islands. Those that nest where winter is severe are migratory (1 species is known to hibernate). In length nightjars are 7–12″ (18–30cm), not including the very elongated rectrices or innermost primaries in the nuptial attire of certain tropical species. Their wings are long, their tails mostly medium to long; their legs are short, with the feet too weak to serve for more than shuffling locomotion or for normal perching. Nightjars have tiny bills but capacious mouths surrounded by bristles; their eyes are large for nocturnal vision. Their plumage is a cryptic blending of brown, buff, gray, and black, mottled, streaked, vermiculated, or barred. They often have white throats and patches of white conspicuous in flight on wings and tail. Females resemble males save that the white areas may be more buffy, less extensive, or lacking.

Nightjars inhabit woodlands, thickets and open country. By day they rest on leaf-littered ground where they are difficult to detect, or lengthwise on horizontal boughs. At twilight they become active and, from a perch or open ground with a clear view of the sky, sally to catch flying insects, often circling around to return to the same station. Nighthawks are more diurnal and aerial, often becoming active before nightfall. They fly continuously, scooping up insects in their wide mouths; on dark nights they may gather in a milling throng above bright lights that attract abundant insects. Nightjars are extremely vociferous; from on or near the ground, some repeat interminably the plaintive or loudly insistent calls that have given names to several species. Nighthawks call as they circle in the air.

With no vestige of a nest, nightjars deposit on leaf-strewn or bare ground, a rock, a sandbar, or a flat, graveled roof, 1 or 2, rarely 3, white, creamy or buffy eggs, usually spotted and blotched with darker colors. Incubated by both parents or the female alone, they hatch in 18–20 days. The chicks are well covered with soft, neutrally colored down and when a day old can hop over the ground, changing their position in response to a parent's call. They are fed by both parents, who upon the approach of a predator may make impressive gasping, hissing, ''broken wing'' distraction displays.

SHORT-TAILED NIGHTHAWK

Lurocalis semitorquatus Pl. 21(14)

Añapero Colicorto

Description. 8″ (20cm); 75g. Medium-sized, dark, with distinctive long, blunt-pointed wings and very short, square tail. **Adults:** white chevron across throat; rest of head, chest, and upperparts brownish-black, mottled with rufous; wings and tail barred with rufous; bases of scapulars and tertials white to pale gray, vermiculated with black, forming a pale band across base of wings (more conspicuous in males); belly and wing-linings cinnamon-rufous, barred with black. Bill black, shading to flesh-color on rami; feet dusky. **Young:** scapulars, tertials, chest feathers largely white, with small rufous, black-rimmed spots giving ocellated effect.

Habits. Prefers forested to partly wooded lowland areas, including old cacao plantations, semi-open, tall second growth; forages most actively at dawn and dusk, also at night, pursuing insects in flight like other nighthawks; flies erratically, like bat, with bursts of quick, shallow wingbeats and fast glides on slightly upswept wings; forages between and above canopy trees, along edges and water-courses; roosts by day lengthwise on limb of forest tree.

Voice. In flight, an upslurred *whick?* singly or in series. Breeding males give a louder *wheeyeet?* (an imitation may bring bird close overhead); in chases, a sharp *weep-weep-weep* or *weep-keeyip*.

Nest. Undescribed (?). Aerial chases and displays seen April–July; juvenile mist-netted in early September.

Status. Locally common resident in humid lowlands the length of Caribbean slope, and in Golfo Dulce and Térraba districts of S Pacific slope, sea level to 3300ft (1000m) locally, especially in Térraba area.

Range. NE Nicaragua to E Peru, N Argentina, and C Brazil.

COMMON NIGHTHAWK

Chordeiles minor Pl. 21(13)

Añapero Zumbón

Description. 9½″ (24cm); 65g. Medium-sized nightjar of open country with long, pointed wings, notched tail, conspicuous white band across primaries midway between bend and tip of wing. **Adults:** conspicuous white (male) to buffy (female) chevron on throat; head, chest and upperparts blackish, spotted with buff, most heavily on nape; scapulars edged with bright buff; secondaries and tail barred with grayish, the pale bars mottled with dusky; a white subterminal tail-band (male only); primaries blackish, the outer 5 or 6 crossed by a broad white band; posterior underparts pale buff, barred with black. Bill and feet blackish. **Young:** like adult female but throat-patch barred with dusky; above with extensive pale gray mottling.

Habits. Prefers savannas, pastures, agricultural land, open marshes, and airstrips; most often seen foraging in late afternoon or early morning, flying high in air and scooping up beetles, moths, bugs, wasps, and other flying insects; flies distinctively, erratically, with several quick flaps and a veering glide, wings always above horizontal; by day rests on ground or low perch. Large migrating flocks may soar, circling in thermals, like *Buteo* hawks; small groups or single birds may fly low using powered flight, more like swallows.

Voice. Loud, buzzy, somewhat nasal *peent* or *bzeet* in flight. Breeding males make display dives high in air, turning upward at bottom with a loud roaring sound produced by primary feathers.

Nest. Directly on ground in rolling grassland or savanna. Eggs 2, whitish, buffy, or grayish, with dark speckles. April–June.

Status. Breeds in small numbers in hilly, windswept savannas on lower Pacific slopes of Cordillera de Guanacaste (perhaps elsewhere in Guanacaste) and in Térraba district; this population evidently migrates S after breeding. Abundant fall migrant (September–early November), chiefly along Caribbean coast and adjacent lowlands, sometimes in Valle Central and Pacific lowlands; absent mid-November to late March; sporadically common spring migrant March–April over much of the country below ca. 4250ft (1300m).

Range. Breeds from N Canada to E Panama; winters throughout South America to N Argentina.

LESSER NIGHTHAWK

Chordeiles acutipennis Pl. 21(12)

Añapero Menor (Gavilán)

Description. 8¾″ (22cm); 45g. Smaller, slighter, paler than Common Nighthawk; pale band closer to wingtip than to bend of wing. **Adult ♂:** above brownish-gray, vermiculated with dusky and finely streaked with black, becoming mostly blackish on crown; tail barred widely with black, with white subterminal band; scapulars and wing-coverts boldly spotted with buff; remiges dusky, barred basally with pinkish-buff; outer 4 primaries crossed medially by broad white band; white chevron on throat; rest of foreneck and upper breast buff, mottled with black; belly pale buff, barred with blackish. ♀: similar but throat-patch and wing-band tinged with buff; no white tail-band. Bill and feet blackish. **Young:** like adult female but above paler, with little or no streaking; buff spots on wing-coverts smaller, more whitish; pale wing-band reduced or lacking; tail more narrowly barred.

Habits. Prefers open areas with scattered woods or scrub, usually near water, including savannas, open marshes, beaches, salt ponds, large rivers, and rice paddies; fairly social, often nesting and roosting in loose colonies, forming feeding aggregations at good foraging areas; in late afternoon and early morning pursues insects in flight like Common Nighthawk; after dark, rests on ground, sallying for passing insects; takes beetles, dragonflies, wasps, winged ants, craneflies; during day roosts lengthwise on low perch, occasionally on ground.

Voice. Usually silent; occasionally a low single or double *chuck* in flight; during breeding season a soft, guttural, rather toadlike trill delivered from ground; a sharp, bleating note in flight, especially in chases.

Nest. On bare sand or soil, usually in loose colonies of up to 10 pairs on open savannas or debris-strewn beach. Eggs 2, pale gray to creamy-white, covered with fine dark speckles. March–June.

Status. Fairly common breeding, perhaps permanent, resident locally on Pacific slope, especially coastal lowlands; common to abundant fall migrant (late September–early November) along Caribbean coast, sometimes along Pacific or in Valle Central; found in winter locally in Pacific lowlands, sporadically elsewhere. Composition of winter population (i.e., whether breeding birds or N migrants, and in what proportions) remains to be determined. No data on spring migration.

Range. Breeds from SW USA to Peru, Paraguay, and S Brazil; winters from N Mexico S through breeding range.

COMMON PAURAQUE

Nyctidromus albicollis Pls. 21(18), 51(10)

Tapacaminos Común (Cuyeo, Pucuyo)

Description. 11″ (28cm); 55g. Medium-sized, very long-tailed; wings long, rounded at tip; the common nightjar of country roads, with bright ruby-red eyeshine. **Adults:** above gray-brown to tawny, palest on crown, finely vermiculated and boldly streaked with black; throat chevron white; scapulars blotched with black and boldly edged with buff; wing-coverts spotted with buff; throat and chest barred and vermiculated with black and rusty; belly dull buff, barred with blackish. ♂: white band across primaries; outer 2 rectrices black, next 2 mostly white. ♀: wing-band narrower, often buffy; outer 3 rectrices indistinctly tipped with white. Bill and feet dusky. **Young:** below paler; throat-chevron buffy;

white or buffy in wings and tail like adult of same sex but less extensive and distinct.

Habits. By day, rests on shady ground in thicket, light woodland, hedgerow, coffee plantation, second growth; avoids closed forest; flushes practically from underfoot, flies lightly for a short distance, drops back to ground or, less often, alights on low perch; at night catches beetles, bugs, moths, and other insects on circling sallies from open ground, rock, or fence post.

Voice. Call a slightly burry *which* or *whip*. The full song is a burry to tremulous *who-whick who-whick whick-wick-wick-wik wik wip WHEEEEUR* or *kw-kw, kw-kw, kw-kw, kw,kw,kwah-REE-O*; from a distance only the loud last "cuyeo" note is audible; often only a burry *krrweeo* or *cuyeer* is repeated incessantly on moonlit nights during the dry season. In interactions, a perched male gives a sharp, growling *bawww*.

Nest. Directly on ground or leaves, usually in fairly open but shaded spot. Eggs 2, pale buff or pinkish-buff, spotted and blotched with shades of brown and lilac. February–April.

Status. Common to abundant resident countrywide from lowlands to ca. 5600ft (1700m) locally.

Range. S Texas to NW Peru and N Argentina.

OCELLATED POORWILL

Nyctiphrynus ocellatus Pl. 51(9)

Chotacabras Ocelado

Description. 8″ (20cm); 36g. A rather small, slender, dark nightjar; lateral rectrices white-tipped, belly spotted with white; scapulars pale grayish with ocellate black spots. **Adult ♂:** crown gray-brown laterally and posteriorly, black medially, everywhere finely vermiculated with tawny; lateral feathers slightly elongated; feathers of upperparts mostly black, vermiculated with rusty, wing-coverts with buffy to white spots at tips; scapulars gray-brown with very fine tawny vermiculations, round black spots bordered behind with buff; wings and tail blackish, barred and vermiculated with rufous tawny, all but central rectrices with narrow but conspicuous white tips; throat and facial area blackish, vermiculated with dark rufous; broad white band across lower throat; chest gray-brown with fine tawny vermiculations, forming an "apron"; posterior underparts blackish with coarser tawny vermiculations, the feathers with bold white terminal spots, passing on lower belly to indistinct pale terminal bars. ♀: similar in pattern except belly less distinctly spotted and general coloration much more rufescent. Iris dark brown, bill and feet dusky-horn, bill tipped blackish.

Habits. Frequents dense second growth several years old, dominated by small trees and vines, adjacent to forest; males sing, and sally for passing beetles and moths, from perches 6–15ft (2–5m) up.

Voice. Song of male a mellow, trilled *preeeo* repeated at intervals of ca. 5sec up to 20 times in succession.

Nest. Unknown (?). Breeds March–April (gonad data).

Status. In Costa Rica known only from a small, recently discovered breeding population near Brasilia, in the extreme NW corner of the Caribbean lowlands between Lago de Nicaragua and Cordillera de Guanacaste; may represent a recent range expansion from Nicaragua, where known from one old specimen from N Caribbean lowlands.

Range. Nicaragua and Costa Rica; NW and C Colombia to NW Ecuador. N Bolivia, SW Brazil, N Argentina, and Paraguay.

CHUCK-WILL'S-WIDOW

Caprimulgus carolinensis Pl. 21(16)

Chotacabras de Paso

Description. 12¼″ (31cm); 110g. Large, richly colored, with branched rictal bristles; relatively pointed wings. **Adult ♂:** above rich brown vermiculated with black; crown, nape, and back broadly striped with black; scapulars and wing-coverts spotted with black and buff; remiges barred with rufous and black; tail marbled and barred with black; face and underparts ochraceous, mottled with black on face and throat, barred and vermiculated with black on breast and belly; breast spotted with buff; buffy-white band across lower throat; inner webs of outer 3 rectrices broadly tipped with white. Bill and feet blackish. ♀: paler, less rufescent overall; throat-band darker buff; no white in tail.

Habits. Singly during winter in old second growth, tall hedgerows in savannas and croplands, thickets at forest edge or at gaps inside forest; rests by day on ground or low perch; forages at night by sallying up from ground or low perch in open to capture flying beetles, moths, cicadas, other medium-sized to fairly large insects.

Voice. Mostly silent; when flushed a low *chuck* (Slud).

Status. Widespread but generally uncommon to rare migrant and winter resident (October–April) from low to middle elevations the length of both slopes; inconspicuous and easily overlooked; perhaps more common than the few scattered records indicate.

Range. Breeds in SE Canada and E USA; winters from extreme S USA to Colombia and Greater Antilles.

RUFOUS NIGHTJAR

Caprimulgus rufus Pl. 21(15)

Chotacabras Rojizo

Description. 11″ (28cm); 95g. Slightly smaller, more rufescent than Chuck-will's-widow, with unbranched rictal bristles, more rounded wing. **Adult ♂:** above rich brown to dull rufous, vermiculated with black; crown, nape, and back broadly streaked with black; scapulars and wing-coverts spotted with black, edged with rich buff; remiges barred with rufous and black; tertials rich buff, heavily marbled with black; tail dull rufous, marbled and faintly barred with black; face and throat barred and mottled with rufous and black; a buffy bar across lower breast and belly mottled and vermiculated with rufous and black; belly and crissum ochraceous, barred with black; distal half of 3 outer rectrices white, edged with rusty. Bill blackish; feet grayish. ♀: paler, less rufescent overall; no white in tail. **Young:** crown largely whitish, spotted with black; scapulars extensively whitish, mottled with dusky; below, dusky barring more extensive.

Habits. Prefers forest edge, second-growth woodland, savanna thickets, mostly in hilly terrain; perches low at edges and openings to sally for flying insects; rests by day in thickety undergrowth, on ground or low perch. Males call from fairly low perches just inside woodland edge.

Voice. Song a resonant, somewhat burry *chuck, wick-wick-WEE-o*, the first syllable audible only at close range (Ridgely). At height of breeding season, males may sing virtually all night long.

Nest. At most a few dry or green leaves laid on ground in sheltered spot. Eggs 1–2, creamy, blotched with gray and pale lilac. January–May (in Panama and Trinidad; no Costa Rican record).

Status. Rare and local resident of lowlands and foothills, to 3300ft (1000m) locally, on both slopes of S Costa Rica, N to hills above Parrita on Pacific slope, Valle de la Estrella on Caribbean; perhaps most numerous in Térraba region.

Range. Costa Rica to N Argentina and S Brazil.

WHIP-POOR-WILL

Caprimulgus vociferus Pl. 21(17)

Chotacabras Gritón o Ruidoso

Description. 9″ (23cm); 50g. Fairly small; grayer and more streaked above than other Costa Rican nightjars. **Adult ♂:** above grayish, finely vermiculated with dusky and rather heavily streaked with black; scapulars blotched with black, edged with buff; wing-coverts browner, mottled with buff and dusky; remiges blackish, barred with ochraceous-buff and mottled with grayish; face and throat heavily barred and mottled with tawny and black; white chevron on throat; chest mottled with grayish and black and spotted with buff; belly dull buff, barred with blackish. Lateral 3 rectrices broadly tipped with white. Bill black; feet grayish. ♀: overall paler, buffier; lateral rectrices tipped with buff.

Habits. Strictly nocturnal; rests during day on ground or lengthwise on low branch, in thicket at forest edge, in hedgerow or gallery forest; when flushed, flies silently for a few yards, often alighting on a higher perch; forages by sallying from low perch in open, like other *Caprimulgus*.

Voice. Silent in Costa Rica.

Status. Casual to very rare winter resident (mid-November–late March) the length of Pacific slope, from sea level to ca. 4000ft (1200m) in Valle Central.

Range. Breeds from SE Canada to SE USA, and from SW USA to Honduras; winters from SE USA and N Mexico to W Panama.

Note. Populations of E and W North America differ in voice, egg color, and slightly in plumage; may be different species. Costa Rican birds represent eastern form.

DUSKY NIGHTJAR

Caprimulgus saturatus Pl. 21(19)

Chotacabras Sombrío

Description. 9″ (23cm); 55g. Very dark, relatively unpatterned arboreal nightjar of high mountains. **Adult ♂:** head, chest, and upperparts sooty-black, spotted with rufous; narrow throat-band buffy-white; chest spotted with pale buff; scapulars and tertials extensively buff marbled with black; remiges blackish, barred with cinnamon-rufous; tail chestnut, barred and mottled with black; 3 outer rectrices broadly white-tipped; belly cinnamon-buff, barred with black. Bill black; feet grayish. ♀: paler, rufous more extensive; throat-band reduced or absent; no white in tail. **Young:** above rufous, mantle barred with black; scapulars and wing-coverts more buffy; breast dull chestnut, densely barred with black; belly clear grayish-buff; tail like adult female.

Habits. Prefers edges and openings in canopy of mountain forests, including the páramo-forest transition, highland pastures with scattered tall trees, and gaps; perches on open branches and snags (rarely on ground) to sing (males) and to sally for flying insects, especially beetles and moths; strictly nocturnal, starting to call or forage after dusk, ceasing before dawn; males may sing throughout clear moonlit nights during breeding season; in foggy or rainy weather, only a song or 2 after dusk and before dawn.

Voice. Song: *chup-wheer-purrWHEEEW*, the *chup* low-pitched, audible only at close range, the *wheer-purr* with a rolling quality, the final *wheeew* loudest, with a burry tone; also *chuck, wheer-puRREEE*; sometimes a scratchy *wheer* in flight.

Nest. Directly on leaf litter, beneath herbage on sloping ground at forest edge. Egg 1, dull chalk-white. February– or March–April; may commence active singing as early as December.

Status. Locally common resident, widely distributed above 6500ft (2000m), from Cordillera Central S through Cordillera de Talamanca, including Dota area, and above 5000ft (1500m) in Cordillera de Tilarán.

Range. Costa Rica to W Panama.

WHITE-TAILED NIGHTJAR

Caprimulgus cayennensis Pl. 21(20)

Chotacabras Coliblanco

Description. 8¼″ (21cm); 40g. Small, slender, with prominent nuchal collar; male has long, distinctively marked tail. **Adult ♂:** above grayish, vermiculated with dusky and buff and finely streaked with black (heavily on crown); pale rufous nuchal collar; scapulars blotched with black and edged with buff; wing-coverts spotted and remiges barred with buff; a broad white band across bases of outer 4 primaries; central rectrices grayish-buff, barred and marbled with black; inner webs of remaining rectrices mostly white, crossed by diagonal black band; throat buffy-white; breast cinnamon-buff, barred and vermiculated with dusky; belly pale buff; sides and flanks barred with black. Bill black; feet grayish. ♀: lacks white in tail; wing-band cinnamon. **Young:** like female, but prominent buffy tips on wing-coverts and scapulars; crown blackish, spotted with buff; remiges tipped and rectrices edged with buff.

Habits. Prefers open, windswept savannas, pastures, low scrub, large clearings such as airstrips; roosts on ground, often under low shrub or tussock, rarely perching on branches; forages by sallying up from open ground for flying insects, including beetles, bugs, moths, damselflies, and hymenopterans; occasionally forages aerially at dawn over scrubby vegetation (ffrench).

Voice. A soft, staccato ticking by flushed bird; high *see-see* notes sporadically in flight (ffrench). Song a high, thin *pt-cheeeeeeeee*, the second note a long-drawn-out, rising whistle that falls slightly at end.

Nest. Directly on ground, often near or under bush or grass tussock. Eggs 2, pinkish-buff, scrawled and blotched with reddish or purplish-brown. February–June (in Trinidad; in Costa Rica fledglings seen in May).

Status. Resident, widely but locally distributed on Pacific slope from N Guanacaste S, including Térraba and Golfo Dulce regions, from sea level to ca. 2600ft (800m); sporadically over S two-thirds of Caribbean slope, where status unknown.

Range. Costa Rica to N Brazil; also Lesser Antilles (Martinique).

ORDER **Apodiformes:** Swifts and Hummingbirds

Members of this order share the ability to rotate their wings from the base so as to derive power from the upstroke as well as the downstroke, the wing remaining fully extended; all have very short legs and cannot walk or hop. However, in other respects swifts and hummingbirds are so different that the anatomical features they share might result from convergence; in any case, each group has long evolved separately.

FAMILY **Apodidae:** Swifts

A family of about 75 species, swifts occur in temperate and tropical regions worldwide except in New Zealand and many oceanic islands; those breeding at higher latitudes migrate to the tropics for the winter. These most aerial of birds measure 3½–10″ (9–25cm) in length, and have compact, streamlined bodies, long pointed wings, and tails that are mostly short and blunt, but long and deeply forked in a few species. Swifts never perch, walk, or intentionally alight on the ground (from which they might not be able to take wing); their strong-clawed feet and very short legs are adapted only for clinging to vertical surfaces. Clinging ability is improved in some species by the ability to direct all 4 toes forward or by the additional support provided by spine-tipped tails.

Although often confused with the unrelated swallows, swifts are easily distinguished by their shallow, rapid wingstrokes and frequent glides on rigid, swept-back wings. Wholly insectivorous, they catch all their food as, in loose flocks, they range widely and erratically, usually high in the air. Swifts' bills are tiny, but with their extraordinarily wide gapes they need no rictal bristles to aid them in flycatching. They are clad somberly in black, brown, and gray, often glossed with bluish or greenish, sometimes with white or rufous rumps, flanks, or collars; the sexes are nearly always alike. Swifts' voices vary from soft twitterings to sharp, loud, shrill grating shrieks; as far as is known, they do not sing. They usually sleep gregariously, generally in protected spots like rock crevices or caves, chimneys, or hollow trees; some species are reputed to sleep while cruising high in the air.

Swifts are monogamous. Copulation often occurs in flight, male and female coupling in the air and separating after tumbling earthward for several seconds. Their nests, built by both sexes, are diverse, but most appear to be glued together and to their support by saliva abundantly secreted from glands that enlarge greatly as the breeding season approaches. Some Far Eastern cave swiftlets make nests of pure saliva (highly esteemed by Oriental gourmets) or of saliva mixed with bark, lichens, or feathers. Some swifts fasten brackets of twigs, plucked with their feet in flight, inside a chimney or hollow tree; others build cuplike nests of plant fibers and feathers in crevices in cliffs, inside buildings, or in old swallow nests. Swallow-tailed swifts hang long sleeves of compacted plant down and feathers beneath rock outcrops, stout boughs, or eaves; a projecting bracket inside the tube supports the eggs. The Old World Palm Swift uses saliva to fasten a spoon-shaped bracket of fibers and feathers to the underside of a hanging palm frond and to attach its eggs thereto; it incubates them clinging upright. Only in the American genus *Cypseloides* and a few closely related genera is saliva apparently not used. These birds place their conical or disk-shaped nests of moss, liverworts, and mud on rock faces or in niches near water, often behind waterfalls; often attachment is provided by rooting of the living plants that form the nest (Marín and Stiles).

Swifts' white, elongate eggs, in sets of 1–5 or 6, are incubated by both parents for 16–28 days. The nestlings, hatched naked, soon grow a dense coat of down; they are attended by both parents and fly from the nest when 25–65 days old; in unfavorable weather their residence in the nest may be prolonged.

BLACK SWIFT

Cypseloides niger Pl. 22(2)

Vencejo Negro

Description. 7″ (18cm); 35g. Fairly large, with a rather long tail, notched distinctly in male and slightly in female. **Adults:** sooty-black, above glossed with bluish, below slightly paler and grayer; whitish scaling on forehead and, especially in females, on belly and flanks. Bill and feet black. **Immatures:** white scaling extensive over entire head and body except upper back, throat, and sides of head. Second-year birds are browner, with little pattern; adult plumage acquired at end of second year.

Habits. Roosts and nests in groups on ledges or in crevices of mountain cliffs, usually near or behind waterfalls; ranges daily down to valleys and foothills to forage, generally in small flocks and often in company of other mountain-dwelling swifts; flies swiftly and directly when traveling; when foraging alternates bursts of quick, powerful wingbeats with rapid swooping glides punctuated by abrupt changes in direction; generally flies and forages higher in air than *Chaetura* swifts; often forages with other swifts at leading edges of rainstorms. Food includes especially hymenopterans (winged ants, bees, wasps), flies, and beetles.

Voice. Generally silent; during courtship chases, a high-pitched twittering or chattering.

Nest. A bulky cup or disk of moss, in niche or cave of wet rock face, often near waterfall and supported by vegetation. Egg 1 (in North America; nest not known in Costa Rica). Birds in breeding condition May–July (Kiff).

Status. Breeding resident and migrant, April–October. Flocks apparently migrating over Valle Central in April–early May and September–early October; specimen of northern race taken mid-April on Península de Osa; perhaps migrates mostly or entirely on Pacific slope. Almost certainly breeds locally in Cordillera Central and Cordillera de Talamanca;

as yet unrecorded on Caribbean slope but probably occurs, at least in mountains.

Range. Breeds from S Alaska to Costa Rica and Antilles; winters from Mexico and Greater Antilles to N South America.

Note. Often previously separated in genus *Nephoecetes*.

WHITE-CHINNED SWIFT

Cypseloides cryptus Pl. 22(4)

Vencejo Sombrío

Description. 6″ (15cm); 36g. Fairly large, robust, with broad head, large eyes, short, broad wings, and short, truncate tail. **Adults:** sooty-black; below slightly paler and browner; small white spot on chin, at base of bill; sides of forehead scaled with grayish. Bill and feet black. **Immatures:** similar, but with more or less extensive white scaling on belly, flanks, and crissum, but no definite spot on chin; forehead paler brown. A browner, relatively unpatterned second-year plumage, with faint pale scaling on abdomen and less white on chin.

Habits. Roosts and nests, evidently individually or in loose groups, in steep-sided gorges or on cliffs where water cascades down in curtains or veils, or sometimes behind large waterfalls; accompanies traveling and foraging flocks of other swifts; flies more heavily and directly than Chestnut-collared Swifts, with rapid, rather batlike wingbeats; eats mainly flying ants.

Voice. Sharp chips and more melodious chirping notes; also explosive, staccato clicking notes.

Nest. A compact mass of mud and plant matter, chiefly mosses and liverworts with slight depression on top, lined with green grass or fine leaves, attached to vertical rock face under ledge or outcrop from which water continuously drips, keeping nest continuously wet. Egg 1. May–August (Marín and Stiles).

Status. Uncommon to rare breeding, perhaps permanent, resident in mountains the length of the country. Definite breeding sites include Zapote de Upala, Cordillera de Guanacaste; gorge of Río Tiribí, above Tres Ríos; and waterfall on Río Sardinal, Rara Avis. Doubtless forages with other swifts at low to middle elevations on both slopes.

Range. Locally from Belize to N and W South America.

SPOT-FRONTED SWIFT

Cypseloides cherriei Pl. 22(5)

Vencejo de Cherrie

Description. 5½″ (14cm); 23g. Medium-sized, slender, with notably broad head, large eyes, and striking facial markings (hardly visible in the field); tail fairly long, truncate; rectrices pointed. **Adults:** above dull black, below slightly browner; bold white spots in front of and behind eye; often a small white area on chin. Bill and feet black. **Immatures:** white spots behind eye and on chin smaller or absent; pronounced white scaling on belly and flanks. May have a browner, more uniform second-year plumage.

Habits. Roosts and nests in rocky gorges of mountain streams; flies somewhat more heavily and directly, less erratically than Chestnut-collared Swift, with which it sometimes associates; eats mainly flying ants. The small, seemingly all-dark, square-tailed *Cypseloides* swifts seen singly or in pairs from time to time in mountains foraging high over forested gorges in clear weather or lower down and with other swifts around rainstorms may be this species.

Voice. A rarely-heard high, thin chipping.

Nest. A rounded or conical mass of mud and plant matter, chiefly mosses, liverworts, and filmy ferns, flattened to slightly concave on top, attached to vertical rock face (usually under slight overhang) or in niche in shaded gorge, above mountain stream or behind falling water. Egg 1. May–July (Marín and Stiles).

Status. Uncommon to rare breeding, perhaps permanent resident; virtually nothing known of seasonal movements and may be more numerous than the few records indicate; definitely recorded only from Pacific slope of Cordillera Central (breeding, upper Río Tiribí above Tres Ríos) and Cordillera de Talamanca (Helechales) but almost certainly more widespread, at least while foraging.

Range. Costa Rica, Colombia, N Venezuela.

CHESTNUT-COLLARED SWIFT

Cypseloides rutilus Pl. 22(3)

Vencejo Cuellicastaño

Description. 5½″ (14cm); 22g. Only slightly larger than *Chaetura* swifts but with much longer, slimmer, more tapered wings, longer truncate to slightly notched tail; chestnut-rufous collar of males and some females diagnostic. **Adults:** above glossy sooty-black; below somewhat paler and browner; collar of males narrowest and sharpest on nape, broadest and dullest on chest, sometimes includes most of throat. In females collar absent, incomplete or occasionally as extensive as in males. Bill and feet black. **Young:** collar lacking but feathers of crown, face, nape, and sides of throat sooty-black, edged with dull, dark rufous; body plumage tinged with brown.

Habits. Nests and roosts in rocky ravines and gorges of mountain streams, ranging out daily in small groups or flocks to forage, often in foothills or valleys, using open areas as preferred flyways; typically flies higher and faster than *Chaetura* swifts, with bursts of stiff-winged flaps interspersed with fast, swerving glides on wings usually held below the horizontal; often joins other swifts feeding at leading edge of rainstorms, where updrafts bring many insects into the air; foods include ants and termites, beetles, bugs, wasps.

Voice. Sharp, scratchy, dry sputtering notes and high-pitched chatters; voice in general drier, more buzzy and metallic than that of *Chaetura* swifts; most noisy in courtship chases high over open valleys and ridges, before and early in breeding season.

Nest. A compact conical mass of plant matter, mostly green moss, with some mud, attached to vertical rock face, usually protected by overhang, over mountain stream, in gorge or ravine or sometimes under bridge. Eggs 2. April–August or September.

Status. Locally common resident in mountains, especially Cordillera Central and Cordillera de Talamanca; breeds mainly 5000–8000ft (1500–2450m); forages regularly down to 1000ft (300m), occasionally to sea level, as at Carara; less numerous, possibly only a sporadic visitor, in lower N cordilleras.

Range. NW and SE Mexico to E Peru and Bolivia.

Note. Formerly placed in genus *Chaetura*.

WHITE-COLLARED SWIFT

Streptoprocne zonaris Pl. 22(1)

Vencejón Collarejo (Golondrón)

Description. 8¾″ (22cm); 98g. Very large, heavily built, with conspicuous white collar; tail slightly notched, appearing square or slightly rounded when spread. **Adults:** plumage mostly black, glossed with bluish on back and chest; white collar narrowest and clearest on hindneck, broader and duller on breast because of dusky feather-bases. Bill and feet black. **Young:** overall duller, more sooty black; white collar much reduced to virtually absent; belly feathers narrowly edged with grayish.

Habits. Roosts and nests in small groups to flocks of 50 or more, generally in wet crevices or caves in mountains, especially near waterfalls; ranges out to forage over virtually the entire country, returning at dusk or even later. Flight appears leisurely but is deceptively fast. Traveling birds fly directly, alternating glides with bursts of powerful flaps; when foraging they circle, climb, dive, and veer while seeming hardly to flap at all, capturing flying insects, especially beetles, flies, wasps, bees, winged ants and termites, often in front of rainstorms in company of other swifts, especially species of *Cypseloides*.

Voice. Often very noisy, especially during breeding season; various loud hissing or grating screeches, chattering and buzzing notes: *chee'yah! cheent!* or *kit-jeeah!* Sometimes a large flock, circling high in air, suddenly bursts into a chorus of screeching. Loud, piercing, grating shrieks during headlong courtship chases, which may attain speeds exceeding 60mph (100kph), often low over the ground.

Nest. A shallow saucer or platform of mud, moss and insect chitin, on ledge in cave, usually near or behind waterfall; rarely lays directly on ledge. Eggs 2. March–July.

Status. Common and widespread resident, definitely nesting in Cordillera Central and Cordillera de Talamanca, perhaps elsewhere; outside breeding season, smaller groups roost in scattered sites in mountains throughout; foraging birds seen virtually countrywide but most frequently over wet lowlands and middle elevations.

Range. C Mexico to Peru, N Argentina, and SE Brazil; also locally in West Indies.

CHIMNEY SWIFT

Chaetura pelagica Pl. 22(8)

Vencejo de Paso

Description. 5″ (13cm); 21g. Large, robust *Chaetura* with less contrast between throat, rump, and body plumage than in any resident species. Crown, hindneck and mantle dull blackish, shading to dark grayish-brown on rump, tail-coverts, tail, and most of underparts; throat and chest slightly paler, more grayish. Bill and feet black.

Habits. Migratory behavior resembles that of swallows: single birds or loose groups fly steadily, at heights of a few feet to 330ft (100m) or more. Along coast large numbers often pass in long straggling lines, sometimes accompanying swallows; occasionally 1 turns aside briefly to pursue some insect. In stormy weather thousands may gather into feeding flocks over open areas, sometimes in company of other swifts. Food includes hymenopterans, dipterans, and other flying insects.

Voice. A high-pitched twittering or chippering, somewhat louder and sharper than notes of Gray-rumped Swift.

Status. Sporadically common to abundant migrant in fall (early October–early November) on Caribbean slope, especially close to the coast but occasionally to 3600ft (1100m) over mountain slopes; spring migration mid-March–late April, along Caribbean coast.

Range. Breeds from C and S Canada to SE USA; winters from W Peru to Bolivia and C Brazil.

VAUX'S SWIFT

Chaetura vauxi Pl. 22(9)

Vencejo Común o Grisáceo

Description. 4¼″ (11cm); 18g. The common small dark swift of the highlands; noticeably (but not conspicuously) pale rump and throat. Like other *Chaetura*, has a short, spine-tipped tail. **Adults:** mostly glossy black; rump, upper tail-coverts, and breast dull, dark grayish-brown; throat pale gray; feathers of rump and tail-coverts narrowly fringed with white when fresh. Bill and feet black. **Young:** feathers of throat with very extensive dusky bases; broader fringes on rump feathers; above, duller black.

Habits. In pairs, small groups, sometimes flocks of 20 or more, above varied terrain from pastures and cities to forest; forages over wide areas, high above ground in sunny weather, lower in rain; often joins other swifts feeding at edges of rainstorms; when foraging flies with bursts of stiff-winged flaps alternating with fast glides on horizontal wings; takes beetles, moths, flying ants and termites, bugs, flies, wasps. At start of breeding season birds sail rapidly about in pairs or trios, wings held upward in V, calling loudly.

Voice. More varied than those of other species of *Chaetura*: a thin, sharp chipping; high-pitched, rippling chatters; sibilant squeaking and buzzy notes.

Nest. A shallow half-cup of matchstick-sized twigs cemented together with saliva, attached to vertical surface in hollow tree, dark attic, or other dark cranny, in a small colony or singly. Eggs 3. March or April to July; courtship chases seen as early as late January.

Status. Common resident of highland areas, mostly between 2300–6600ft (700–2000m), ranging occasionally up to 10,000ft (3000m) or down into lowlands, the length of both slopes, including higher hills of Península de Nicoya; a breeding population at Bahía Ballena on Península de Nicoya.

Range. Breeds from SE Alaska to C California, E and S Mexico to E Panama and N Venezuela; winters from C Mexico S through breeding range.

Note. Populations from S Mexico southward may represent a species, *C. richmondi* (Dusky-backed Swift), distinct from northern *C. vauxi*.

BAND-RUMPED SWIFT

Chaetura spinicauda Pl. 22(11)

Vencejo de Rabadilla Clara

Description. 4¼″ (11cm); 18g. A blackish, slim-winged *Chaetura* with sharply contrasting pale rump-band and distinctly pale throat. Mostly sooty-black, glossed with blue; rump pale grayish, the feathers with narrow white edgings when fresh; upper tail-coverts dull black; sides of head and neck dusky; throat and chest pale grayish, the feathers with dusky bases that show through with wear, giving mottled effect. Bill and feet black. See Gray-rumped Swift.

Habits. Flies over forest, semi-open, pastures, and agricultural land in pairs, small groups, or flocks of up to 50 after breeding season; often associates with other species of swifts in mixed flocks where feeding conditions are good, as in the updrafts ahead of rainstorms; sometimes plucks prey from foliage of canopy, even hovering briefly to do so; eats beetles, wasps, bees, winged ants and termites, mayflies, other flying insects.

Voice. High-pitched squeaky notes, soft twitterings, lower-pitched chatters, less sharp and loud than the notes of Vaux's Swift.

Nest. In a tree hole, 30ft (9m) up. Eggs undescribed (?). February–June (in Trinidad; no Costa Rican record).

Status. Common to abundant resident of lowlands and foothills of S Pacific slope from Quepos and upper Valle del General S, especially numerous in Golfo Dulce area; ranges locally up to 4000ft (1200m) in humid areas (e.g., San Vito), but in drier General-Térraba region keeps more to valleys, below 3000ft (900m).

Range. Costa Rica to W Colombia and C Brazil.

GRAY-RUMPED SWIFT

Chaetura cinereiventris Pl. 22(10)

Vencejo Lomigrís

Description. 4″ (10cm); 17g. Small, slender, blackish *Chaetura* with distinctly pale rump and throat. **Adults:** mostly glossy black; rump and upper tail-coverts dull gray, the feathers with narrow whitish fringes when fresh; sides of head and neck dark slate; throat pale gray, with broad dusky feather bases that show through with wear, giving mottled effect. Bill and feet black. **Young:** overall duller and browner, especially on throat. Band-rumped Swift of Pacific slope has dark tail-coverts (thus shows a dark area behind the pale rump).

HABITS. In loose flocks of 20 or so, flies over forest, open habitats, or waterways, hovering to glean insects from canopy foliage as well as foraging in usual aerial manner; forages higher toward midday or in clear weather; may join other swifts at edges of rainstorms or other good foraging sites; eats flying ants, bees, wasps, beetles, flies. In dry weather flocks swirl low over rivers and ponds, skimming water to drink and bathe.

VOICE. Soft, high-pitched rippling twitters; a light chittering.

NEST. Of sticks, in chimney, up to 4 eggs (in Brazil; no Costa Rican nest seen, though several birds seen entering large hollow tree in Sarapiquí lowlands). March–July.

STATUS. Common to abundant resident in wet lowlands the length of Caribbean slope, from sea level to ca. 2000ft (600m) in foothills and valleys.

RANGE. E Nicaragua to NW Panama; Colombia and Venezuela to W Ecuador, NE Argentina, and SE Brazil. Also Grenada, Lesser Antilles.

NOTE. Sometimes lumped with the West Indian *C. martinica*.

LESSER SWALLOW-TAILED SWIFT

Panyptila cayennensis Pl. 22(7)

Vencejo Tijereta Menor (Macuá)

DESCRIPTION. 5″ (13cm); 18g. Strikingly patterned, graceful, with long, deeply forked tail, extremely pointed lateral rectrices and outer primary. Mostly glossy blue-black; throat and upper breast white, this continuing as narrow collar around hindneck; white spot on lores; white tuft on each side of rump; secondaries and inner primaries edged with white when fresh. Bill and feet black.

HABITS. Flies singly or in pairs, occasionally small groups, over forest, pastures, cropland, towns, and broad rivers, often with or above flocks of *Chaetura*, usually high above ground in clear weather, often foraging low with other swifts around rainstorms. Flight slower, more erratic and swallowlike than that of other swifts while foraging; bursts of rapid, fluttering wingbeats alternating with curving, buoyant glides; tail usually closed, spikelike, but sometimes, for abrupt changes in direction, opened so widely as to appear rounded. In courtship chases, flight is extremely rapid and twisting. Food includes flying ants and termites, small beetles and bugs, flies, wasps.

VOICE. Soft chattering notes in flight; at nest, a light *chee-chee-chee* (ffrench).

NEST. A sleevelike structure to 3ft (1m) long, of plant downs, feathers, and other soft materials, cemented together with saliva, fastened to underside of a branch or to trunk of a forest tree, or to some man-made structure; entrance at bottom of the often free-hanging tube; a shelf near top holds eggs and nestlings. Eggs 2–3. January–June; both parents sleep in nest before, during, and after nesting.

STATUS. Resident in lowlands, valleys, and foothills of Caribbean slope up to 2600ft (800m), and on Pacific slope from Parrita S, up to at least 3300ft (1000m) on slopes around General, Térraba, and Coto Brus valleys.

RANGE. S Mexico to E Peru and C Brazil.

NOTE. The nest, the bird, or any part of either are still regarded by some country folk as a powerful aphrodisiac (when appropriately prepared by the local witch, or *brujo*).

GREAT SWALLOW-TAILED SWIFT

Panyptila sanctihieronymi Pl. 22(6)

Vencejo Tijereta Mayor

DESCRIPTION. 8″ (20cm); 50g. Essentially a large version of *P. cayennensis*; tail relatively longer; outer 2 rectrices and outer primary sharply pointed. Mostly glossy blue-black; throat, upper breast, and narrow collar around hindneck white; spot above and in front of eye, patch on flank, tips of secondaries and inner primaries white. Bill and feet black.

HABITS. All birds seen in Costa Rica have been flying high overhead; flight more direct and powerful, less fluttery than that of *P. cayennensis*.

VOICE. Unknown (?) in Costa Rica.

STATUS. Known in Costa Rica from 5 sightings (March, April, September, November) at La Selva, in Sarapiquí lowlands, and above Valle del General. The dates suggest that these might have been migrants, but the species has not been recorded farther S.

RANGE. S Mexico to N Honduras, casual or accidental in Nicaragua and Costa Rica.

FAMILY **Trochilidae:** Hummingbirds

The 330 species of hummingbirds comprise the second largest, and undoubtedly the best-known, strictly New World avian family. Although found from Alaska to Tierra del Fuego, hummingbirds are far more numerous in the tropics; species breeding at high latitudes winter in warmer regions. Hummingbirds are notable for their small size, 2¼–8″ (5.7–20cm), iridescent colors, manner of flight, and pugnacity, all of which relate to their extreme specialization for nectar feeding. However, all hummingbirds also take small insects and/or spiders, which represent their main source of proteins and other nutrients; nectar is essentially a highly concentrated, conspicuous, and predictable source of energy.

Hummingbird flight involves rapid wingbeats and a unique rotation of the entire wing at the shoulder joint; changing the angle of the wingbeat permits flight in any direction including backward or hovering in place—both essential for extracting nectar from long floral tubes. Among birds their small size gives hummingbirds the highest metabolic rates and energy needs relative to their weights. An important energy-saving device is the ability to enter torpor on cold nights, regulating their body temperature 30–50°F (17–28°C) below active levels—essentially a nightly hibernation. Given adequate nectar supplies and torpor, some hummingbirds can prosper on perpetually cold and wet tropical mountains or nearly up to snow line on bleak Andean heights.

Hummingbirds are important pollinators, especially in tropical highlands where cool temperatures limit insect and bat activity. Flowers specialized for hummingbird pollination are tubular and rich in nectar and usually some shade of red, pink, or orange (though hummingbirds visit flowers of all colors). Floral tubes often correspond closely to the length and curvature of the bills of their chief pollinators, which vary from long and decurved in many hermits, to short and straight, even slightly recurved in a few. Most hummingbirds are very aggressive and intolerant at flowers; individuals of some species regularly hold feeding territories from which they eject all other hummers regardless of species or sex. Less pugnacious, the hermits and females of many species use regular foraging routes to visit widely scattered flowers; small hummingbirds, in particular, visit many small, insect-pollinated flowers.

In all carefully studied species, the female carries out the entire nesting effort without direct male aid. In a few species a male may defend a female's nest or flowers, but usually male mating territories are entirely separate from female nesting areas. Males of some hummingbird species gather in courtship assemblies or leks, in which each male holds a small mating territory; some hold mating territories at flowers, whereas others adopt mating stations away from either flowers or other males. Males of most tropical hummingbirds advertise their availability to females by their mostly weak, squeaky, and monotonous songs; a few engage in elaborate aerial displays. In most species males are more colorful than the often confusingly similar females, and may sport bright gorgets or crowns, long tails, frills, or other adornments that probably enhance territory defense as well as facilitating rapid species recognition. Both sexes are dull-colored in most hermits and bright in species where females as well as males hold territories at flowers. Produced mostly by the microscopic structure of the feathers rather than by pigments, the iridescent colors of hummingbirds change with the angle of viewing and appear black in poor light (which often makes identification difficult).

Hummingbirds' nests are usually neat, compact cups of soft, downy material, decorated on the outside with mosses and/or lichens, and held together and attached to their support, usually the twig of a bush or tree, with cobweb. Hermits attach their nests beneath the tip of a palm frond or a strip of some great leaf, which forms a roof over them. Hummingbirds virtually always lay 2 elongate, papery-shelled, white eggs that hatch in 15–19 days, rarely longer. Fed by their mother with regurgitated nectar and insects, the young remain in the nest for 20–26 days, occasionally more, and fly well when they leave.

In Costa Rica, all hummingbirds are called *gurrión*.

WHITE-TIPPED SICKLEBILL

Eutoxeres aquila Pl. 23(8)

Pico de Hoz

DESCRIPTION. 5″ (13cm); 11g. Very large, robust hermit with strongly decurved bill, streaked underparts and long, rounded, white-tipped tail. **Adults:** above dark bronzy-green; feathers of rump and upper tail-coverts fringed with buffy-white; rectrices pointed, white-tipped; throat and breast blackish, broadly streaked with white to buffy. Bill black except basal half of lower mandible bright yellow; feet flesh-horn. **Young:** broader buffy fringes over entire upperparts; rectrices more acutely pointed; ventral streaking more buffy.

HABITS. Ranges widely in forest understory, adjacent second growth, and along forest edges, visiting scattered flowers (especially those of *Heliconia* with pendant inflorescences and strongly decurved corollas, also *Centropogon granulosus*); flies swiftly and powerfully over distances of 1mi (1.6km) or more; invariably perches to feed, often clinging acrobatically to inflorescence with strong, heavy feet. Males form small courtship assemblies in *Heliconia* thickets.

VOICE. High, thin, sharp piercing *tsit*ting notes. Song a long series of variably squeaky notes in regular rhythm.

NEST. A loose, coarsely made cup of fine rootlets, fungal rhizomorphs and other fine plant fibers, with short "tail" of debris, attached by spiderweb to underside of strip of *Heliconia* leaf or broad dicot leaf, often over stream, 3–12ft (1–4m) up. January–May.

STATUS. Widespread but usually uncommon in wet-forested foothills and adjacent lowlands, N at least to Volcán Santa María on Caribbean slope, to Parrita on Pacific slope. Center of abundance ca. 1000–2300ft (300–700m) on Caribbean slope, or up to 3300ft (1000m) on Pacific; locally up to 4000ft (1200m).

RANGE. Costa Rica to W Ecuador and NE Peru.

BRONZY HERMIT

Glaucis aenea Pl. 23(4)

Ermitaño Bronceado

DESCRIPTION. 4″ (10.5cm); 5.3g. Medium-sized hermit with moderately decurved bill, rounded tail with rufous base separated from white tip by black subterminal band. **Adults:** above bronzy-green, including central rectrices; upper tail-coverts fringed with pale buff; crown more brownish; dusky mask bordered by buffy malar stripe and postocular spot; below pale cinnamon-rufous, brightest on chest and sides. Upper mandible black, lower dull yellowish with dusky tip; feet flesh-orange. **Young:** throat and breast dull bronze, feathers fringed with buff; buffy fringes on most of upperparts, including sides of crown; belly dull cinnamon.

HABITS. Favors thickety second growth and streamsides, swampy areas with tall grass or herbaceous growth and forest edge; visits scattered or clumped flowers, especially those of *Heliconia* and banana; gleans insects and spiders in dense foliage. Males call and display separately, usually in thickets along streams or edges; they guard female's nest but do not help to build, incubate, or feed young.

VOICE. Call a thin *tseew* or *tseet*, often given in series; a *tsit*tering trill in aggressive interactions. Male's song (?) alternates high, thin whistles with liquid sputters and warbles.

NEST. A well-built to moderately flimsy cup of plant fibers, especially fungal filaments and cobweb, usually with a messy "tail" of debris, attached to underside of strip of *Heliconia* or banana leaf, 1½–20ft (0.5–6m) up. January–August.

STATUS. Widespread, locally common resident of wet lowlands of Caribbean and S Pacific slopes, to 1000ft (300m) or sporadically to 2500ft (750m) in General and Coto Brus valleys.

RANGE. E Nicaragua to W Panama; Pacific lowlands of Colombia and Ecuador.

NOTE. Sometimes lumped with the larger *G. hirsuta*, Rufous-breasted Hermit, of Panama and South America.

BAND-TAILED BARBTHROAT

Threnetes ruckeri Pl. 23(5)

Ermitaño Barbudo

DESCRIPTION. 4¼″ (11cm); 5.8g. Medium-sized hermit with sharp contrast between dark throat and bright chest; tail rounded, white at base and tip; rectrices bluntly pointed. **Adults:** above bronzy-green; upper tail-coverts fringed with buff; lores and auriculars dusky; postocular spot and malar stripe dusky; throat dusky; chest orange-rufous; rest of underparts grayish-buff. Upper mandible black (males have a yellow streak near nostril); lower mandible yellow, tipped with black; feet flesh-yellow. **Young:** feathers of upperparts fringed with cinnamon-buff; chest feathers dull bronzy, fringed with buff, little or no rufous; secondaries narrowly edged with buff at tips.

HABITS. Prefers forest understory, edges, and old second growth; visits mostly scattered flowers, especially those of *Costus*, *Heliconia*, and banana; often slits or pierces long-tubed flowers, especially those of *Calathea*, to reach nectar; gleans insects and spiders in thickets, largely from under and upper sides of leaves. Males sing, alone or in small leks, from regular perches in dense undergrowth.

VOICE. Call a high, thin *tseep* or *tsee-tseep*. Caribbean males sing a high, thin syncopated *didiDIT dew dew*, often repeated; songs of Pacific birds longer, more complex, including trills and warbles.

NEST. A more or less loosely made cup of fine rootlets, fungal rhizomorphs, other fine plant fibers and cobweb, usually with a short "tail" of debris, attached to underside of strip of *Heliconia* or banana leaf, 6–13ft (2–4m) up. February–May (Caribbean slope); June–September (Pacific).

STATUS. Common to locally abundant resident of wet-forested lowlands of Caribbean slope, to ca. 2000ft (600m), and S Pacific slope to 2600ft (800m); some sporadic postbreeding movements, especially of young birds, to 4000ft (1200m).

Range. SE Guatemala to W Ecuador and W Venezuela.

LONG-TAILED HERMIT

Phaethornis superciliosus Pl. 23(2)

Ermitaño Colilargo

Description. 6″ (15cm); 6g. Brownish lowland hermit with conspicuous facial stripes and long, attenuate white-tipped central rectrices, very long decurved bill. **Adults:** above bronzy-green, the feathers fringed with buffy, most broadly on upper tail-coverts; head dark brown, with pale buffy facial stripes; underparts pale grayish-brown to buff; lateral rectrices largely black, tipped with buffy-white. Upper mandible black, lower dull orange tipped with dusky; feet flesh-horn. **Young:** similar but feathers of crown, nape, throat and upperparts with broader tawny-buff fringes.

Habits. Dwells in forest understory, especially along edges and streams, and in adjacent tall second growth where its favorite foodplants (*Heliconia, Costus, Aphelandra, Passiflora vitifolia*) are most abundant; visits scattered flowers along regular foraging beats that are often ½mi (1km) or more long; gleans spiders and insects from foliage, dead branches, and spiderwebs. Males form singing-courtship assemblies of up to 25 birds in thickets, especially along streams.

Voice. Flight call a high, squeaky *sweep* or *tseeip*; in aggressive interactions a descending, accelerating series of 3–5 piercing whistles; song of male a single squeaky, buzzy or grating note (varies between leks, or males on a single lek) repeated monotonously, ca. 1/sec.

Nest. A neat, thick-walled cup of plant fibers and cobweb, with messy-looking "tail" of dead leaves and other debris, attached to underside of tip of leaf of understory palm, strip of *Heliconia* or banana leaf, 4–15ft (1.2–4.5m) up. Mainly January–August in Caribbean lowlands; May–September, rarely January, in Pacific.

Status. Common to abundant resident of lowlands the length of Caribbean slope and on S Pacific slope, sporadically and locally as high as 3300ft (1000m); rare and local, mainly in evergreen gallery forest, in N Pacific lowlands.

Range. S Mexico to Bolivia and C Brazil.

GREEN HERMIT

Phaethornis guy Pl. 23(3)

Ermitaño Verde

Description. 6″ (15cm); 6g. Female more strongly patterned, male dark; white-tipped central rectrices long in female, shorter and pointed in male; very long, decurved bill. **Adult ♂:** above and below mostly dark iridescent green, becoming deep blue on buff-edged tail-coverts; lateral rectrices mostly black, edged with whitish at tip; facial stripes ochraceous, short and narrow, sometimes obsolete; lower belly dark gray. ♀: head dusky-green with broad ochraceous-buff facial stripes; underparts mostly gray. Upper mandible black, lower red with dusky tip; mouth-lining red; feet dusky-flesh. **Young:** like female but head and upperparts with ochraceous-buff fringes; males have relatively shorter tail. Males require 2 years to attain adult plumage.

Habits. Visits flowers, including *Heliconia, Costus*, and *Razisea*, along regular foraging routes in understory and edges of forest or adjacent tall second growth; ascends to subcanopy for flowers of *Columnea*; gleans spiders from webs and vegetation. Up to 20 males form a lek in dense forest understory of mountain ridges or swales.

Voice. Flight call *chreek* or *syurk*; singing males repeat monotonously a single note *chrrrk, twurrp*; squeals and popping sounds in chases and displays on lek. Notes in general squeaky, more metallic and lower-pitched than those of Long-tailed Hermit.

Nest. A compact cup of plant down, treefern scales, and spiderweb, with untidy "tail" of debris, attached to underside of tip of leaflet of understory palm, or sometimes a strip of *Heliconia* leaf. February–September.

Status. Common to locally abundant resident of wet mountain forests, from 1650–2600ft (500–800m) to ca. 6500ft (2000m) the length of both slopes. Some movement to lower elevations following breeding, especially by young birds.

Range. Costa Rica to W Colombia and SE Peru.

Note. Sometimes called Guy's Hermit.

LITTLE HERMIT

Phaethornis longuemareus Pl. 23(1)

Ermitaño Enano

Description. 3½″ (9cm); 2.6g. Very small hermit with bold facial stripes, slightly decurved bill, bright underparts, and long, buff-tipped tail. **Adults:** above bronzy-green; upper tail-coverts broadly fringed with dull rufous; tail largely dusky, the central rectrices tipped with whitish, others with buff; face blackish with pale buff stripes; below pale rufous to cinnamon-buff. Tail of female averages longer, more pointed. Bill black except basal half of lower mandible bright yellow; feet yellowish-flesh. **Young:** above with extensive rusty fringes.

HABITS. Prefers dense understory of forest, forest edge, second growth, gallery woodland, semi-open, and gardens; visits many small, often insect-pollinated flowers; expert at piercing corollas of larger flowers to obtain nectar; has foraging route of a few hundred yards; gleans tiny insects and spiders from webs and vegetation. Up to 25 males form a lek in a thicket, perching usually within 10–20in (25–50cm) of ground.

VOICE. Flight call a high, thin, upslurred *seeik* or *pseet*; aggressive note a descending series of thin, sharp *seep* notes. Song of male a high-pitched phrase of thin, squeaky notes and trills, repeated several to many times; ventriloquial.

NEST. A compact cup of plant down, fine fibers, and spider web, with short "tail" of debris, attached to underside of tip of leaf of understory palm, small dicot leaf, or strip of banana leaf, 1½–12ft (0.5–3.5m) up. Breeding prolonged; peaks April–July, November–December.

STATUS. Common to abundant resident throughout lowlands, and foothills of Caribbean slope and S Pacific slope, including interior valleys, to 5000ft (1500m) locally; uncommon to rare and local in evergreen gallery forests of N Pacific lowlands.

RANGE. S Mexico to W Ecuador, E Peru, and N Brazil.

GREEN-FRONTED LANCEBILL

Doryfera ludoviciae Pl. 23(11)

Pico de Lanza Frentiverde

DESCRIPTION. 4½″ (11.5cm); 5.7g. Dark highland hummingbird with very long, slender, slightly upturned bill and short, rounded tail. **Adult ♂:** mostly iridescent dusky-green; forehead glittering blue-green; crown and nape dark bronzy; small white postocular spot; upper tail-coverts deep blue; tail black, tipped with gray. ♀: paler overall, below more grayish; little or no bright feathering on forehead; pileum more purplish; tail tipped with paler gray. Bill and feet black. **Young:** resemble adult female but pileum duller, more bronzy.

HABITS. Dwells in canopy, at edges, and along streams of mountain forests; feeds mainly at long-tubed, pendant flowers of epiphytic Ericaceae (*Satyria, Cavendishia, Psammisia*), mistletoes (*Psittacanthus*); at large clumps males may be territorial and sing. Females nest and fly more along streams and ravines; young birds, in particular, may descend to shrub level to feed at short-tubed flowers at clearings and edges.

VOICE. Sharp, dry, crackling, sputtering, or snapping notes while feeding; thin, piercing squeaks in flight; song of male a dry, sputtering warble.

NEST. A bulky cup of treefern scales, mosses, and cobweb, usually attached to rootlet or twig under rock overhang in dark, wet ravine, occasionally under bridge or eaves. August–January.

STATUS. Widespread but generally uncommon resident of wet mountain forests, mainly on Caribbean slope, from Cordillera de Tilarán S to Panama, chiefly 2500–7500ft (750–2300m); sometimes, perhaps seasonally, to 8500ft (2600m); on Pacific slope rarely below 5000ft (1500m).

RANGE. Costa Rica to W Venezuela, Peru, and Bolivia.

SCALY-BREASTED HUMMINGBIRD

Phaeochroa cuvierii Pl. 23(10)

Colibrí Pechiescamado

DESCRIPTION. 4¾″ (12cm); ♂ 9.5g, ♀ 8g. Large, dull-colored, with white corners of tail; shaft of outermost primary thickened and flattened, especially in adult males. **Adults:** mostly bronzy-green, dullest on crown; feathers of throat, breast, and sides broadly fringed with dull buff; belly grayish-buff; corners of tail dull white, set off by diagonal blackish band. Bill black, except pink basal half of lower mandible; feet black. **Young:** outer primary with little (males) or no (females) thickening of shaft; above buffy feather-edgings prominent, especially on crown, nape and upper tail-coverts.

HABITS. Prefers forest edge, tall second growth, semi-open, clearings with scattered trees, and mangroves; very aggressive; sometimes territorial at flowers, notably those of terrestrial bromeliads (*Bromelia*) in Guanacaste, and trees like *Genipa, Inga*, and *Erythrina*; in mangroves prefers flowers of the Pacific mangrove, *Pelliciera*. Males form loose singing and courtship assemblies of up to 8 birds in canopy of second growth or low mangroves.

VOICE. Aggressive note a piercing *cheet*. Song of males very loud, persistent, variable: phrases of 4–8 notes, series of sharp chips or squeaks interspersed with short trills and sputters.

NEST. A well-built cup of soft plant down,

sometimes including fungal rhizomorphs, heavily decorated on outside with mosses and lichens, often lichens in lining, the whole held together with cobweb; typically saddled on horizontal branch of small tree 6–25ft (2–8m) up. May–January.

Status. Locally common resident in lowlands and lower foothills up to 4000ft (1200m) the length of Pacific slope; in Guanacaste breeds mostly in mangroves, occurs more widely during wet season; a common resident in Río Frío area; elsewhere on Caribbean slope, a sporadic visitor, mostly in dry season.

Range. Belize to N Colombia.

Note. Currently segregated in genus *Phaeochroa* but morphologically and behaviorally seems quite typical of *Campylopterus*.

VIOLET SABREWING

Campylopterus hemileucurus Pl. 23(9)

Ala de Sable Violáceo

Description. 6″ (15cm); ♂ 11.5g, ♀ 9.5g. Very large; ample tail with prominent white corners; shafts of outer 2 primaries greatly thickened and flattened in adult males, less so in females and young; bill decurved, especially in females. **Adult ♂:** head, underparts, and upper back deep violet; wing-coverts and posterior upperparts dark green; tail blackish, distal half of outer 3 rectrices white. ♀: above entirely dark green; below gray with violet gorget; tail as in male. Bill and feet black. **Young:** above and below with prominent dull buffy feather-edgings; no violet; rectrices narrower than in adult; primary shafts with little (males) or no (females) thickening; males, below dusky bluish-green.

Habits. Prefers understory and edges of mountain forests, especially streamsides and ravines; also patches of woods in disturbed areas, old second growth, banana plantations; rarely territorial at flowers; less aggressive and dominant than its size suggests. Favorite flowers include *Heliconia*, bananas, sometimes understory shrubs like *Cephaelis*. Up to 10 males form a lek, singing from heights of 7–13ft (2–4m) in saplings in forest understory or edge.

Voice. At flowers, both sexes give sharp, penetrating, explosive twitters or dry chatters. Song of male a long series of evenly spaced but variable notes: *cheep tsew cheep tik-tik tsew cheep* . . . , high-pitched, piercing and ventriloquial.

Nest. A bulky, well-constructed cup, mainly of green moss, lined with fine fibers and plant down, largely held together with spider web, placed on low, slender horizontal branch of small tree or bamboo, typically overhanging ravine or stream, 3–20ft (1–6m) above ground or water. May or June–October.

Status. Locally common resident of mountains the length of the country, on both slopes; breeds principally 5000–8000ft (1500–2400m), or down to 3300ft (1000m) in N ranges; often descends to 1300ft (400m) outside breeding season.

Range. S Mexico to W Panama.

WHITE-NECKED JACOBIN

Florisuga mellivora Pl. 23(17)

Jacobino Nuquiblanco

Description. 4¾″ (12cm); 7g. Fairly large, with relatively short, thick bill; males strikingly patterned; females have notably scaly throat. **Adult ♂:** head, neck, and throat deep blue; breast, rest of upperparts, and central rectrices bright green; nape and belly white; lateral rectrices white, narrowly tipped and edged with black. ♀: above bright green; throat dull bluish-green, heavily scaled with dull white; belly dull white; tail green, broadly tipped with dark blue, outer rectrix with white edge and tip. Many females have more or less malelike plumage, sometimes with buff facial stripes; a few are distinguishable from males only by slightly smaller size and longer bills. Bill and feet black. **Young:** plumage much bronzier than adults; often buff facial stripes.

Habits. Frequents mostly forest canopy or treetops in semi-open; often lower at edges, clearings, and in second growth; females nest in dense understory. Both sexes visit many flowers of trees (*Inga, Vochysia, Erythrina, Symphonia*) and epiphytes (*Norantea, Columnea*), also *Heliconia*; males, in particular, are aggressive, but rarely territorial at flowers. Both sexes often hawk flies for long periods, hovering and darting high in air.

Voice. High, thin, staccato *tsit*ting notes; males sing from regular perches in treetops, a long series of high, thin *tseep* notes.

Nest. A soft, felted cup of pale-colored plant down, on upper surface of leaf of understory palm, where sheltered from above by another leaf, 3–10ft (1–3m) up. Incubating females float off nest in a mothlike distraction display. January–June.

Status. Seasonally common breeding resident in lowlands and foothills of Caribbean and S Pacific slopes; most breeding below ca. 1600ft (500m) on Caribbean slope or 2500ft (750m) on Pacific slope; becomes very rare between September and December in many areas, but nature and extent of seasonal movements poorly understood.

Range. S Mexico to W Ecuador, Bolivia, and Amazonian Brazil.

BROWN VIOLET-EAR

Colibri delphinae Pl. 23(6)

Colibrí Orejivioláceo Pardo

Description. 4½″ (11.5cm); 6.5g. Fairly large, dull-colored; conspicuous pale malar stripe; rather short, stout bill; ample tail. **Adults:** mostly dull grayish-brown; back glossed with green; tail-coverts dusky, broadly edged with cinnamon; tail bronze-green, with narrow ochraceous tip set off by greenish-black subterminal band; violet patch from below eye to auriculars; lores and malar stripe buffy-white; throat-patch glittering green, blue at lower edge; breast feathers with dusky centers. Bill black; feet dusky. **Young:** entire upperparts broadly fringed with cinnamon-buff; little or no violet on face.

Habits. Dwells mainly in canopy of forest, tall second growth and semi-open; descends to shrub level at edges and clearings; feeds mainly at small, short-tubed, mostly insect-pollinated flowers of trees (*Inga, Calliandra, Symphonia*), and epiphytes (*Clusia*, Marcgraviaceae), where dominant to smaller hummers but rarely territorial; often flycatches high in air at openings. Breeding males form loose singing assemblies in lower canopy along forested ridges or forest edge.

Voice. Call a short dry *chip* or *chichip* or a more sibilant, slurred *tsip*. Song a series of 5–10 emphatic, squeaky *tseeup* or *ksip* notes of ca. 2/sec, followed by a variable pause, then repeated.

Nest. A small cup of plant down "saddled on a small twig of a low bush at about 4ft (1.2m) from the ground under bamboos" (ffrench, in Trinidad; no Costa Rican record). November–April or May.

Status. Uncommon and local breeding resident at lower middle elevations, ca. 1300–3600ft (400–1100m) on Caribbean slope and to ca. 5200ft (1600m) on Pacific slope, in all major mountain ranges, also highlands of Península de Osa. May wander widely into adjacent lowlands outside breeding season.

Range. Guatemala to Bolivia, E Brazil, and Trinidad.

GREEN VIOLET-EAR

Colibri thalassinus Pl. 23(7)

Colibrí Orejivioláceo Verde

Description. 4″ (10.5cm); 5g. Medium-sized; almost wholly green with conspicuous tail-band. Bill slender, slightly decurved to straight. **Adult ♂:** mostly bright, fairly pale green; throat and breast glittering green, breast tinged with blue; broad patch of violet from below eye across auriculars; broad subterminal band of blue-black on tail. ♀: glittering green restricted to throat; breast duller, bronzier. Bill and feet black. **Young:** crown, hindneck, and rump fringed with dull rufous; below lacks glittery green, the feathers narrowly edged with buff.

Habits. Prefers open, brushy highland areas with scattered trees, including pastures, hedgerows, second growth; also along highland streams, at large gaps in forest, and forest edge; visits flowers of many herbs, shrubs, epiphytes, and trees; sometimes territorial but displaced by more aggressive species (*Panterpe, Lampornis*). During much of year males sing interminably from high perches overlooking open spaces with flowers, often several in same area but usually not in well-defined leks.

Voice. Call a sharp, dry, rather low-pitched *chut* or *zut*; sometimes many such notes run together into a chatter. Song a vigorous but unmelodious 2- to 4-note phrase—*CHEEP chut-chut, chip CHEEP chut*, or *CHEET-chup*—repeated rapidly for minutes on end through much of day.

Nest. A fairly substantial cup of treefern scales, down, dry grass blades, mosses, bound with cobweb and decorated with bits of moss and lichen, placed on downward-drooping twig, rootlet, or bamboo stem at forest edge, streamside, or overhanging roadbank, 3–10ft (1–3m) up. October–March.

Status. Common to abundant resident in highlands S from Cordillera de Tilarán; more numerous on Pacific slope, where deforestation is greater; breeds at ca. 5200–10,000ft (1600m–3000m) in Cordillera Central and Cordillera de Talamanca, down to ca. 4750ft (1450m) in Cordillera de Tilarán; wanders down to ca. 3300ft (1000m), outside breeding season.

Range. C Mexico to W Panama; Andes from N Venezuela to Bolivia.

GREEN-BREASTED MANGO

Anthracothorax prevostii Pl. 23(13)

Manguito Pechiverde

Description. 4¾″ (12cm); 7.5g. Large savanna hummingbird with broad purple tail. **Adult ♂:** bronzy-green above; central rectrices purplish-bronze, lateral rectrices magenta, edged with black; center of throat and breast black, shading laterally through blue to bright green on sides of neck, sides and belly; crissum purplish. ♀: sides of neck and sides green; rest of underparts white, with broad black central stripe; outer 3 rectrices purplish with blue-black subterminal band, white tips. Occasional females have malelike plumage. Bill and feet black. **Young:** like adult female but rufous feathering between green of sides and white of underparts; above with rusty feather-edgings.

HABITS. Prefers savannas, coffee plantations and other semi-open areas with scattered tall trees; also edges of gallery forest, mangroves, and tall second growth; visits flowers of trees, especially legumes (*Inga, Caesalpinia, Erythrina*), and lianas; rarely territorial; highly insectivorous, often flycatching high in air. Males use scattered tall trees as song perches, also sing in flight between them.

VOICE. Usual call a liquid *tsup* or *tseep*; aggressive note *pzzt!* Male's song a buzzy, "zingy" *kazick-kazee*, usually rapidly repeated 3–4 times.

NEST. A shallow cup of pale-colored plant down, at most sparsely decorated on outside with bits of lichen or bark, usually on outer twig of leafless or sparsely foliaged tree high above ground. December–April or May.

STATUS. Locally common resident on Caribbean slope W from Río Frío region, S of Lago de Nicaragua; also on N Pacific slope, chiefly in Tempisque and Bebedero watersheds, and around Golfo de Nicoya S to Tárcoles; uncommon to rare elsewhere in Guanacaste and in Valle Central, E at least to Turrialba. Evidently spreading on N Caribbean slope following deforestation; recent records S and E to La Selva and Guápiles.

RANGE. NE Mexico to NW Peru; islands in W Caribbean.

NOTE. The Panama race *veraguensis*, with little or no black on throat of male, sometimes considered a separate species, may eventually be found on S Pacific slope.

VIOLET-HEADED HUMMINGBIRD

Klais guimeti Pl. 25(11)

Colibrí Cabeciazul

DESCRIPTION. 3″ (7.5cm); 2.8g. Small, with prominent square white postocular spot, pale-tipped tail. **Adult ♂:** head, including throat, bright violet-blue; rest of upperparts bronzy-green; tail more bluish, becoming black subterminally; lateral rectrices tipped with pale gray; rest of underparts ash-gray, spotted with green laterally. Bill black; feet dusky. ♀: similar but only crown bright greenish-blue; throat pale gray; lateral rectrices more broadly tipped with dull white. **Young:** like female but with little or no blue on crown.

HABITS. Frequents canopy and edge of humid forest, also tall second growth, open country, and gardens with flowering trees and shrubs; prefers small flowers including *Warscewiczia, Stachytarpheta, Taetsia*, and *Hampea*, at which males may set up territories if larger, more dominant hummingbirds are absent. Males form leks around a forest gap or along edge of tall second growth, each male singing daylong from 1–3 defended perches on high, slender, open twigs.

VOICE. While foraging, high-pitched, sharp, dry staccato chips, often run into a shrill sputtering in interactions. Song of male consists of high-pitched, sharp, squeaky to tinkling notes in short phrases: *che tewink tewink tewink* or *ki chu si-chew chu si-chew*.

NEST. A tiny, thick-walled cup, largely of green moss, lined with seed down, often on a pendulous twig or vine above a shaded stream, less often in deep forest far from water, 3–15ft (1–4.5m) up. January–April or May.

STATUS. Fairly common resident of foothills and adjacent lowlands of Caribbean slope, chiefly 160–3300ft (50–1000m), and on S Pacific slope, above 1000ft (300m).

RANGE. N Honduras to NW Bolivia and extreme W Brazil.

WHITE-CRESTED COQUETTE

Lophornis adorabilis Pl. 25(5)

Coqueta Crestiblanca

DESCRIPTION. 2¾″ (7cm); 2.7g. Very small; tail and underparts largely rufous; white rump-band; male has ornate crest and cheek-tufts. **Adult ♂:** forehead and lores coppery-bronze; erect white crest on forecrown; rest of upperparts bronzy-green; lower rump (behind rump-band) and upper tail-coverts purplish-bronze; rectrices chestnut-rufous, edged with bronze; throat, auriculars, and long wispy cheek-tufts glittering green; foreneck white; rest of underparts cinnamon-rufous, flecked with green on sides of breast. Bill red with black tip; feet dusky. ♀: duller, without crest and cheek-tuft; face and forecrown dusky-bronze; throat and chest white, throat speckled with bronzy-green; black subterminal tail-band; upper mandible all black. **Young:** like adult female but tail pattern duller, less distinct; young male has throat more heavily flecked with green.

Habits. Wanders widely, appearing when the small flowers of *Inga, Vochysia, Lonchocarpus, Stachytarpheta*, are abundant, then disappearing for long intervals; in forest keeps high in canopy, coming down to shrub level at edges, and in second growth; hovers slowly through foliage gleaning small spiders and insects. In courtship, hovering male oscillates from side to side in short arcs, in front of perched or hovering female. Flight when foraging very quiet, steady; hovers with tail cocked.

Voice. Soft liquid *tseep*ing when feeding.

Nest. A minute downy cup, well encrusted with lichens, 16–60ft (5–18m) up in clearing or at forest edge, often rather exposed. December–February.

Status. Resident, chiefly on S Pacific slope, N to Cordillera Central, occasionally crossing to Caribbean slope in C Costa Rica; 1000–4000ft (300–1220m).

Range. Costa Rica and W Panama.

Note. Also called Adorable Coquette; sometimes included in the genus *Paphosia*. Paucity of records of males in full regalia between April and August suggests that they go into eclipse after breeding season, but more observations are needed.

BLACK-CRESTED COQUETTE

Lophornis helenae Pl. 25(6)

Coqueta Crestinegra

Description. 2¾″ (7cm); 2.8g. Very small; white rump-band; spotted underparts; male crested. **Adult ♂:** crown and gorget glittering green; wispy, black, hairlike crest; cheeks striped with buff and black; lower throat black, the lateral feathers elongated and pointed; above bronzy-green; lower rump (behind rump-band) blackish; lateral rectrices rufous, edged with black; breast-band bronzy; belly white, spotted with bronze. Bill red with black tip; feet blackish. ♀: lacks crest, black or green on head and throat; blackish mask; throat buffy-white, spotted with bronze; lateral rectrices with subterminal black and green band. **Young:** resemble adult female.

Habits. Mostly high in forest canopy, coming lower, sometimes to shrub level, at edges, gaps, semi-open, and brushy hedgerows (especially those with flowering *Stachytarpheta*); visits mostly small, typically insect-pollinated flowers on mass-flowering trees (e.g., *Hymenolobium, Cordia, Dipteryx, Vochysia*), and epiphytes (*Clusia, Norantea*); also gleans tiny arthropods from twigs and foliage; subordinate to larger hummingbirds; usually not aggressive but poaches in territories of other species. Flight very quiet and steady; white rump-band imparts resemblance to hovering sphinx moth.

Voice. Nearly always silent; occasionally a soft, dry chipping.

Nest. A small cup at end of twig of tree at forest edge; ca. 25ft (8m) up. Late March (Will).

Status. Very uncommon (or often overlooked) resident of Caribbean slope, chiefly 1000–4000ft (300–1200m) from Cordillera de Guanacaste S to Río Reventazón drainage; occasional, probably as a seasonal visitor, on Pacific slope in upper Valle Central; descends irregularly, but perhaps not infrequently, to Caribbean lowlands; may even breed down to ca. 330ft (100m).

Range. S Mexico to Costa Rica.

RUFOUS-CRESTED COQUETTE

Lophornis delattrei Pl. 25(4)

Coqueta Crestirrojiza

Description. 2¾″ (7cm); 2.8g. Very small; white or buffy rump-band; dull bronzy underparts; male has bizarre crest. **Adult ♂:** crown and long shaggy crest rufous; rest of upperparts bronzy-green; lower rump (behind rump-band) and upper tail-coverts purplish-bronze; rectrices mostly chestnut-rufous, edged with bronze; throat glittering green, subtended by a "necklace" of white, pointed feathers; rest of underparts dark bronzy-green mixed with dull rufous; crissum cinnamon-rufous. Bill dull coral-red, tipped with dusky; feet dusky. ♀: face, forecrown dull cinnamon-rufous without crest; throat pale rufous or buffy, speckled with dusky-bronze posteriorly; below duller, duskier than male; subterminal blackish tail-band. **Young:** like adult female but throat grayish, flecked with white.

Habits. Unknown (?) in Costa Rica; evidently wanders widely seeking abundant small flowers, as do other coquettes.

Status. Known in Costa Rica from 4 specimens taken in October of different years (1892–1906) at San Pedro, E of San José in Valle Central at ca. 3900ft (1200m); nearest record is W Bocas del Toro, Panama; to be looked for on S Caribbean slope.

Range. Costa Rica to N Colombia and E Bolivia.

GREEN THORNTAIL

Discosura conversii Pl. 25(1)

Colicerda Verde

Description. ♂ 4″ (10cm), ♀ 3″ (7.5cm); 3g.

Small, dark, with conspicuous white rump-band; male has long, wirelike tail; female has bold white mustache. **Adult ♂:** above dark green; below glittering green with center of breast bright blue; thigh-tufts and patch on flanks white; upper tail-coverts black, tipped with green; rectrices blue-black with white shafts, the outer 3 pairs long, wirelike. ♀: broad malar stripe white; below mostly blackish, spotted with green, and a green breast-band; tail forked, the outer rectrices pointed, white-tipped. Bill and feet black.

HABITS. In treetops of forest canopy and edges; visits mostly small, numerous, insect-pollinated flowers of trees (e.g., *Warscewiczia, Inga, Pithecellobium*), epiphytes (*Clusia*, Marcgraviaceae), also flowering shrubs like *Stachytarpheta* along edges and hedgerows; hawks tiny flies and wasps. Hovering flight very quiet, steady, with tail held cocked up at nearly a right angle. Breeding males perch on high bare twigs, sometimes give dive displays.

VOICE. Usually silent; a soft, squeaky chipping in interactions.

NEST. Undescribed (?). Probably breeds November–April.

STATUS. Uncommon resident of lower middle elevations of Caribbean slope, ca. 2300–4600ft (700–1400m) on Cordillera Central and Cordillera de Talamanca; in early wet season (June–August) descends to 200ft (60m) in foothills and adjacent lowlands where it mainly visits *Warscewiczia*.

RANGE. Costa Rica to W Ecuador.

NOTE. Has usually been separated in genus *Popelairia*.

FORK-TAILED EMERALD

Chlorostilbon canivetii Pls. 24(13), 25(10)

Esmeralda Rabihorcada

DESCRIPTION. 3¼″ (8cm); 2.6g. Small; male bright green with deeply-forked tail; female gray below with bold white postocular stripe. **Adult ♂:** above bronzy-green, brightest on forehead; below glittering emerald-green; white thigh-tufts; rectrices blue-black, tipped with bronzy. ♀: below pale gray; tail double-rounded, distally blue-black, 2 outer rectrices white-tipped. Upper mandible black (sometimes red basally in males), lower red with black tip or all black; feet black. **Young:** like adult female but with buffy feather-edges on face, nape, and rump; below tinged with buff; tail of young male blue-black, shallowly forked, outer 2 rectrices gray-tipped.

HABITS. Prefers scrubby savannas, second growth, edges of disturbed forest, cultivated areas and gardens; subordinate to and chased by most other hummingbirds; visits many small, short-tubed, largely insect-pollinated flowers of shrubs, herbs, and vines ignored by other species; very nervous while foraging, pumping tail and darting rapidly. Males sing persistently from low perches in brush along edges, give a shallow dive display with a dry sputter at bottom during aggressive (and courtship?) interactions.

VOICE. Note a dry, scratchy *chut* or *chit*, sometimes run together into a soft, staccato chatter. Male has a high, thin, monotonous song of repeated simple phrases like *tsippy-tsee tsee* or *tseee tseeeree* (Slud).

NEST. A neat little cup of pale-colored plant down and fibers, distinctively decorated on outside with chips and strips of bark, usually 3–10ft (1–3m) up in shrub or herbaceous plant. November–March or April.

STATUS. Widespread but generally uncommon in lowlands, valleys and adjacent mountains the length of Pacific slope, to at least 2600ft (800m) along Cordillera de Guanacaste and to 5000ft (1500m) around Valle Central and General-Térraba region; extends onto Caribbean slope along Cordillera de Guanacaste and down Reventazón valley at least to Turrialba; more abundant locally in parts of Valle Central, on slopes of Cordilleras de Guanacaste and Talamanca.

RANGE. NW and C Mexico to C Panama.

NOTE. Birds from S Costa Rica to C Panama are sometimes considered a separate species, *C. assimilis*, Garden Emerald, distinguished from the northern *canivetii* by less forked tail, duller crown and solid black bill in males. Because birds from the Pacific coastal strip and foothills S of the Golfo de Nicoya are more or less intermediate, we treat *assimilis* as a race of *canivetii*. We consider *canivetii* (including *assimilis*) to be an allospecies of the *C. mellisugus* superspecies.

CROWNED WOODNYMPH

Thalurania colombica Pls. 24(4), 25(15)

Ninfa Violeta y Verde

DESCRIPTION. ♂ 4″ (10cm), 4.5g; ♀ 3½″ (9cm); 4g. Male very dark, with deeply forked tail; female has double-rounded tail, pale throat and chest contrasting with dark belly. **Adult ♂:** crown, upper back, shoulders, and belly deep violet; rest of upperparts dark green; throat and chest glittering green; tail blue-black. ♀: above bright green, more bluish posteriorly; distal half of tail blue-black, outer 3 rectrices white-tipped; throat and chest pale gray; belly dull, dusky green.

Bill black; feet dusky. **Young ♂:** little or no violet or bright green; above dusky-green; below dusky-bronze; tail shorter than adult. ♀: face, nape and rump fringed with buff.

Habits. Prefers wet forests of lowlands and foothills, including adjacent edges, tall second growth and semi-open. During breeding season males frequent forest canopy, visiting flowers of epiphytes (bromeliads, ericads, *Columnea*); females stay mainly in understory, where they visit flowering shrubs like *Besleria*; both sexes visit flowers like *Cephaelis* at gaps. After breeding, many birds concentrate at flowers of *Heliconia*, where males defend territories. Both sexes glean insects and spiders from foliage; males do more aerial flycatching.

Voice. A high, dry, fast *kip* or *kyip*, often reiterated in an excited chatter. Aggressive note a dry, scratchy *chut-t-t-t*. Song (?) of squeaky *tseep* or *ksit* notes, monotonously repeated.

Nest. A compact cup of treefern scales, plant down, and cobweb, decorated outside with lichens and moss, on horizontal twig under overhanging leaf, 3–16ft (1–5m) up in forest. February–May or June.

Status. Very common resident of wet lowlands, valleys and foothills from sea level to ca. 2500ft (750m) throughout Caribbean slope and on Pacific slope N to Quepos-Parrita area; after breeding may move up adjacent mountain slopes to 4000ft (1200m).

Range. C Mexico to N Colombia and W Venezuela.

Note. Sometimes lumped with the dull-crowned *T. furcata*, Common or Fork-tailed Woodnymph, of Amazonia. Also reported from Costa Rica on basis of 2 old, probably mislabeled specimens, is Violet-bellied Hummingbird (*Damophila julie*), otherwise known from South America E only to C Panama. Male resembles male woodnymph but lacks purple on back and crown and has rounded tail; female whiter below, without dark belly, tail more rounded and with gray-tipped outer rectrices; both sexes have pink base of lower mandible.

FIERY-THROATED HUMMINGBIRD

Panterpe insignis Pl. 24(12)

Colibrí Garganta de Fuego

Description. 4¼″ (11cm); 5.7g. Usually appears shiny dark green with blue tail, small white postocular spot, slender bill; brilliant throat and crown visible mainly from above. **Adults:** crown royal-blue; sides of head and hindneck black; back bright green; upper tail-coverts bluish; tail dark blue; throat glittering rosy copper-orange; center of breast violet-blue; belly green to bluish. Bill black except rose-pink basal half of lower mandible; feet dusky. **Young:** similar but with rusty feather-edgings on face and nape.

Habits. Mostly in forest canopy, but descends freely to shrub level at edges, openings, and in second growth; favors flowers of epiphytes, especially ericads (*Cavendishia, Macleania*), bromeliads, vines (*Tropaeolum, Bomarea*), and shrubs (*Centropogon*); often pierces long-tubed flowers, or uses holes made by *Diglossa* or bumblebees; active, noisy, aggressive; frequently defends rich patches of flowers. Breeding males allow females to share their flowers.

Voice. High-pitched, liquid or sharp and penetrating twitters, chatters, and chipping notes; males appear to lack a true song; a buzzy note in courtship.

Nest. A bulky cup of treefern scales, plant down, and cobweb, heavily decorated outside with mosses and lichens, near tip of drooping bamboo stem or on rootlet dangling beneath overhanging bank, 6–13ft (2–4m) up. August–January.

Status. Common to abundant resident of high mountains from Cordillera de Guanacaste (Volcán Miravalles) S to Panama; upward from 4600ft (1400m) on N cordilleras, and from 6600ft (2000m) on Cordilleras Central and de Talamanca. After breeding some descend as low as 2500ft (750m).

Range. Costa Rica to W Panama.

SAPPHIRE-THROATED HUMMINGBIRD

Lepidopyga coeruleogularis Pls. 24(8), 25(13)

Colibrí Garganta de Zafiro

Description. 3½″ (9cm); 3.7g. Fairly small, with rather deeply forked tail, no facial stripes. **Adult ♂:** above green, more bronzy on tail-coverts and central rectrices; lateral rectrices blue-black; throat and chest violet-blue; rest of underparts green to blue-green; crissum feathers broadly edged with white; thigh-tufts white. Bill black except basal half or more of lower mandible red; feet black. ♀: lateral rectrices tipped with grayish-white; below white, speckled with green laterally, becoming solid green on sides. **Young:** like female but below grayer; above with narrow cinnamon-buff feather-edgings.

Habits. Prefers open, scrubby vegetation, especially near the coast.

STATUS. Recorded from Costa Rica on the basis of an old, possibly mislabeled specimen; one unconfirmed sighting of a male near Potrero Grande, May 1962 (R. Ryan, N. Boyajian); the species is otherwise known W to W Chiriquí (Puerto Armuelles) very close to Costa Rican border. Regardless of whether one rejects these old records, the species is to be expected on S Pacific slope, especially in view of the number of open-country species that have already invaded this area from adjacent Panama following the cutting of intervening forest.

RANGE. Costa Rica (?) or W Panama to N Colombia.

BLUE-THROATED GOLDENTAIL

Hylocharis eliciae Pl. 24(9)

Colibri Colidorado

DESCRIPTION. 3½″ (9cm); 3.7g. Tail bright golden-bronze; swollen red base of bill. **Adult ♂:** above bronzy-green, shading to purplish-bronze on upper tail-coverts; throat and chest deep blue to purplish-blue; lower breast green; sides and flanks bronzy-green; belly grayish-buff. Bill red with black tip; feet dusky. ♀: underparts mostly dull buffy, passing to bronzy-green on sides and flanks; throat and breast spotted with blue; outer 2–3 rectrices tipped with cinnamon; culmen blackish. **Young:** below with very little blue, veiled by broad grayish-buff fringes; tail duller bronze.

HABITS. In humid areas frequents open woodland, semi-open, and tall second growth, but in dry regions prefers evergreen gallery forest; in both, freely enters open areas to visit flowers of shrubs like *Stachytarpheta* or *Hamelia*, herbs like *Heliconia*; rarely territorial at flowers. Up to 10 males gather in a lek, in late wet and dry seasons, each singing throughout the day on 1 or a few defended perches on horizontal twigs in upper understory or lower canopy.

VOICE. Call a high, buzzy *tzip* or *tzet*; aggressive note a sharp, liquid, descending twitter. Song of male highly variable between (and occasionally within) leks; typically a phrase of 5–8 notes, the first a piercing *tseee*, followed by a series of single or double notes, or short trills; a male usually gives 1–3 such phrases, pauses, then repeats: *seee; sa se sa se sasese; tseet twosip twosip twosip*; or *zeee wrrrr zewet zewet zewet*, etc.

NEST. A cup with strongly incurved rim, of fine fibrous or downy materials, with a few bits of gray lichen on the outside, 7–20ft (2–6m) up amid second growth, in gallery forest understory, or in a garden. December–April.

STATUS. Resident in lowlands and valleys the length of both slopes, to 2500ft (750m) in Valle del General and to 3000ft (950m) in W Valle Central; rare over most of Caribbean slope but common in Río Frío region; more numerous on Pacific slope but local in dry-forested NW.

RANGE. S Mexico to E Panama.

WHITE-BELLIED EMERALD

Amazilia candida Pl. 24(1)

Amazilia Pechiblanca

DESCRIPTION. 3¼″ (8.5cm); 4g. Bicolored, above greenish-bronze, tinged with purplish on forehead, crown, and upper tail-coverts; tail dull bronze, outer rectrices with slightly paler tips set off by blackish subterminal band; below white, speckled with bronze-green on sides of throat, breast, and sides. Upper mandible black, lower pink with black tip; feet dusky. Female Mangrove Hummingbird has blackish tail, green upper tail-coverts, more green spotting on breast.

HABITS. Little known in Costa Rica; farther north prefers forest edge, semi-open, second growth; feeds largely at shrubs (e.g., *Cephaelis, Hamelia*) and herbs (*Heliconia*). Males may form loose courtship assemblies, singing from low perches in dense vegetation.

VOICE. Song of male a monotonous *chiperty chink kachink kachink*, repeated persistently. Calls include high, light *tsip* and chittering notes.

STATUS. Casual stray or at best very rare resident in Costa Rica, at least formerly; no recent record. Old specimen records scattered widely: Reventazón valley, around San José, Térraba valley, and Puerto Jiménez on Peninsula de Osa.

RANGE. C Mexico to Costa Rica.

BERYL-CROWNED HUMMINGBIRD

Amazilia decora Pl. 24(18)

Amazilia Corona de Berilo

DESCRIPTION. 3½″ (9cm); 4.5g. Medium-sized, dark, rather nondescript with fairly long bill. **Adult ♂:** crown glittering blue-green; upperparts, central rectrices, lower breast, and sides bronzy-green; upper tail-coverts purplish; lateral rectrices blue-black with broad grayish tips; chin dark green; lower throat and chest violet-blue; belly buffy-gray. ♀: crown dull; throat and chest pale gray, heavily spotted with green and (on lower throat) blue; tail tipped with paler gray. Bill black except pink basal half of lower mandible; feet dusky. **Young:** little or no bright green or blue; crown, throat, and breast dull bronzy-green, feathers broadly edged with pale buff.

Habits. Prefers open woodland, coffee plantations, second growth, gardens with scattered trees; in forested areas at edges, along streams and riverbanks and in clearings; visits flowers like *Inga, Hamelia*, other shrubs and trees, epiphytes like *Satyria*, also *Heliconia*; males sometimes defend feeding territories and form leks of up to 12 birds in second growth, trees at edge of woods, or shady gardens, usually away from flowers, singing from perches 8–20ft (2–6m) up, through most of year except at height of dry season.

Voice. Call a dry, metallic *tzip* or *tzit*. Song of male monotonous, persistent, squeaky: *seek chiseek chiseek chiseek* . . . or *tsweet tswe we we we*. . . . A low rippling trill in interactions.

Nest. A deep, neat cup of plant down and bast fibers, cobweb, sometimes treefern scales, sparsely decorated with bits of lichen, 6–14ft (2–4m) up on horizontal twig of tree or shrub. December–January, May–July or August.

Status. Irregularly common resident of lowlands, foothills, and valleys of S Pacific slope, N to about Carara, from sea level locally to ca. 4000ft (1200m) (e.g., Coto Brus valley, San Vito); in many areas varies in abundance, sometimes disappearing and reappearing unpredictably.

Range. Costa Rica to W Panama.

Note. Sometimes considered conspecific with *A. amabilis* of Caribbean slope. Usually called Charming Hummingbird.

BLUE-CHESTED HUMMINGBIRD

Amazilia amabilis Pl. 24(14)

Amazilia Pechiazul

Description. 3¼″ (8.5cm); 4.2g. Medium-small, rather nondescript; male with shorter tail than male woodnymph, green and violet reversed on crown and breast; female appears speckled below. **Adult ♂:** crown and malar area glittering green; upperparts bronzy-green; central rectrices purplish-bronze; lateral rectrices blue-black with faint dark gray tips; chin dark green; throat and chest blue-violet; belly grayish. Bill black except pink basal half of lower mandible; feet dusky. ♀: crown and face bronzy-green; throat and chest dull white, heavily speckled with green anteriorly and blue medially; lateral rectrices more distinctly gray-tipped. **Young:** resemble female but with buffy feather-edgings, below duller and bronzier.

Habits. Prefers open woodland, tall second growth, banks of streams and rivers, semi-open, large treefall gaps; visits flowers of trees (*Inga, Bravaisia, Hamelia*), various shrubs and herbs (*Heliconia*). Males sometimes territorial at large clumps of nectar-poor flowers of *Warscewiczia* trees; at richer flowers dominated and chased by woodnymphs, plumeleteers, etc.; form loose singing assemblies of 2–8, perching on bare twigs 6–25ft (2–8m) up and 30–60ft (9–18m) apart along streams, edges, and clearings, usually away from flowers.

Voice. Metallic *tsink* or *tsit* notes; a little descending rippling trill in interactions; male song a louder, squeaky, long-continued *tsip tsew tsew tseek tsew tseek*. . . .

Nest. A compact cup of plant down, bast fibers, sometimes a few fungal rhizomorphs, sparsely decorated with bits of lichen and moss, ca. 6ft (2m) up in small tree at forest edge. February–May.

Status. Uncommon to locally common resident in lowlands the length of Caribbean slope, except drier NW sector S of Lago de Nicaragua, mostly below 1000ft (300m), occasionally to 1600ft (500m) in foothills, perhaps mostly as postbreeding wanderer.

Range. NE Nicaragua to W Ecuador.

MANGROVE HUMMINGBIRD

Amazilia boucardi Pl. 24(2)

Amazilia Manglera

Description. 3¾″ (9.5cm); 4.5g. Only white-bellied hummingbird likely in its restricted range and habitat; female differs from rare White-bellied Emerald in blackish tail and green-spangled chest. **Adult ♂:** above uniform bronzy-green, including central rectrices; throat and chest bright bluish-green, becoming bronzy on lower breast and sides; belly dull white; tail notched, blackish lateral rectrices tipped with dark gray. ♀: below white, speckled with green laterally and across chest; lateral rectrices tipped with pale gray. Bill black except pinkish base of lower mandible; feet dusky.

Habits. Restricted to mangroves where its favorite flower, Pacific mangrove (*Pelliciera rhizophorae*), is abundant; also visits flowers of other trees (e.g., *Lonchocarpus*), vines, and epiphytes in and adjoining mangroves; males aggressive but usually not territorial at flowers. During breeding season males give a shallow dive display.

Voice. A pebbly *djt*; a rapid descending twitter (Slud).

Nest. A relatively small cup of balsa floss,

other pale-colored plant down, and cobweb, decorated with bits of lichen; on mangrove twigs, usually overhanging water, 3–13ft (1–4m) up. October–February.

Status. Locally common resident of mangroves along Pacific coast from Golfo de Nicoya S to Golfo Dulce (Puerto Jiménez).

Range. Costa Rica.

STEELY-VENTED HUMMINGBIRD

Amazilia saucerrottei Pl. 24(15)

Amazilia Culiazul

Description. 3½″ (9cm); 4.5g. Medium-sized, all-green hummingbird with bronzy rump and notched blue-black tail. **Adult ♂:** above bronzy-green, shading to bronze on wing-coverts and lower back, purplish-bronze on rump; upper tail-coverts and tail dark steel-blue to blue-black; below entirely glittering dark green, with white thigh-tufts and blue crissum. ♀: lower breast and belly duller green; crissum feathers edged with grayish-buff; outer rectrices purplish at tip. Upper mandible black, lower rose-pink with black tip; feet black. **Young:** below dull, dark bronze-green.

Habits. Prefers second growth, scrubby savanna with scattered trees, coffee plantations, and gardens; regularly at openings and edges of evergreen gallery forest, especially during dry season; visits flowers of many kinds of trees (*Inga, Pithecellobium, Tabebuia, Genipa*), shrubs (*Hamelia, Stachytarpheta*), vines, epiphytes, herbs (*Lobelia*). Both sexes are aggressive, often territorial at flowers. Males sing singly, usually from open perch overlooking defended flowers.

Voice. High, sharp *tsip* or *chit* notes, sometimes in sputtering series, softer and lighter than notes of Rufous-tailed Hummingbird. Song of male a buzzy, squeaky *bzz WEEP-wup*.

Nest. A compact cup of pale-colored plant down and cobweb, usually heavily decorated on outside with lichens, on outer twig of small tree 6–23ft (2–7m) up. December–April.

Status. Common to abundant resident of N half of Pacific slope (extending E to Río Frío region on Caribbean slope), S to hills above Parrita and Dota region, across Valle Central, down Reventazón valley on Caribbean slope at least as far as Turrialba; from lowlands up adjacent mountain slopes to 4000ft (1200m) in N cordilleras, and to 5900ft (1800m) S of Valle Central, as at El Copey de Dota.

Range. W Nicaragua to Costa Rica; also Colombia and NW Venezuela.

Note. The Central American population differs strikingly in voice and behavior from true *saucerottei* of South America and probably represents a distinct species, *A. sophiae*, Blue-vented Hummingbird.

INDIGO-CAPPED HUMMINGBIRD

Amazilia cyanifrons Not Illustrated

Amazilia Gorriazul

Description. 3¾″ (9.5cm). Essentially like Steely-vented Hummingbird but larger, with blue crown. **Adult ♀:** (male unknown) forecrown bright greenish-blue, passing to green on hindcrown; rest of upperparts rather dark bronzy-green, becoming more bronzy on wing-coverts and rump, purplish-bronze on upper tail-coverts; tail blue-black; below bright green, center of belly grayish; crissum purplish-blue with grayish feather-edgings.

Habits. Unknown, except that the single female was collected at "a pretty high point" on the Pacific face of Volcán Miravalles on 10 September 1895, by C. F. Underwood. This species has not been encountered during recent visits to this volcano.

Status. At best an extremely rare and local resident of Cordillera de Guanacaste. Differs too much in bill length and color to be a stray from the South American population of *A. cyanifrons*, or an aberrant Steely-vented Hummingbird, nor can its characters easily be accounted for by hybridization between any 2 local species. Perhaps the foremost ornithological mystery of Costa Rica.

Range. Costa Rica; the Magdalena Valley of Colombia.

Note. The lone Costa Rican specimen has sometimes been considered to represent a separate species *A. alfaroana*, Alfaro's Hummingbird.

BLUE-TAILED HUMMINGBIRD

Amazilia cyanura Pl. 24(16)

Amazilia Coliazul

Description. 3½″ (9cm); 4.5g. Virtually identical to Steely-vented Hummingbird except secondaries rufous, with dusky tips; bases of inner primaries also rufous; tail slightly more violet-blue; broader buffy edgings on crissum feathers.

Habits. Prefers open woodland, edges, shade trees in coffee plantations, and young second growth; especially visits *Inga* flowers.

Status. Accidental, known from a specimen taken October 1904 near San José and a sighting at La Selva in May 1958, in Sarapiquí lowlands (Slud). To be looked for in N Guanacaste and lowlands S of Lago de Nicaragua.

Range. S Mexico to S Nicaragua on Pacific slope; accidental in Costa Rica.

SNOWY-BELLIED HUMMINGBIRD

Amazilia edward Pl. 24(3)

Amazilia Vientriblanca

Description. 3½″ (9cm); 4.6g. Only the male Mangrove Hummingbird also has green breast sharply contrasting with white belly; no overlap in range. Above bronzy-green, shading to bronze on lower back, purplish-bronze on rump and upper tail-coverts; tail purplish-black with narrow bronzy tips (wider in females) on lateral rectrices; wing-coverts bright copper-bronze; throat and breast bright green; crissum purplish. Bill black except pinkish basal half or more of lower mandible; feet blackish.

Habits. Prefers brushy savanna, scrubby woodlands, coffee plantations, roadside trees, gardens, and forest edge; visits flowers of shrubs (e.g., *Palicourea, Stachytarpheta*) and trees (*Calliandra, Inga, Vochysia*); occasionally territorial. Males sing solitarily from inconspicuous perches in trees or shrubs, usually not near flowers. In most areas fluctuates markedly in numbers, presumably reflecting local shifts in flowering.

Voice. Call a light *tip* or *tsip*; song a soft *be beebee, d'beebee* or *tseer tir tir* (Slud).

Nest. A compact cup of plant down, heavily decorated on outside with lichens, sometimes mosses, 5–30ft (1.5–9m) up in shrub or small tree. Late October–January.

Status. Locally common in foothills and lower mountain slopes at 1000–5300ft (300–1600m) in and around Térraba, General, and Coto Brus valleys of S Pacific slope, and into Panama via the San Vito–Sabalito area; evidently absent from wetter Golfo Dulce district.

Range. Costa Rica to E Panama.

Note. Also known as Snowy-breasted Hummingbird.

CINNAMON HUMMINGBIRD

Amazilia rutila Pl. 24(11)

Amazilia Canela

Description. 3¾″ (9.5cm); 4.8g. Medium-sized; only Costa Rican hummingbird with both uniform cinnamon underparts and rufous tail. **Adults:** above bronzy-green; upper tail-coverts edged with rufous; rectrices chestnut-rufous, edged with bronze at tip; entire underparts uniform cinnamon-rufous (averaging paler in females), palest on throat. Bill of male red with black tip, of female with upper mandible mostly black, only base and culmen red; feet dusky. **Young:** rufous feather-edging on face, crown, and rump; below often paler; upper mandible black.

Habits. Prefers deciduous forest, brushy savanna and edges, second growth; usually avoids closed gallery forest; visits flowering trees (*Genipa, Tabebuia*), shrubs (*Helicteres*), and epiphytes (*Combretum*). Both sexes are aggressive, sometimes territorial at flowers. Males sing singly or in small, loose groups from dense brush along streams and woodland edge.

Voice. Call a buzzy, scratchy *tzip*, drier than note of Rufous-tailed Hummingbird. Song a short phrase of 3–6 high, thin, whistled *tsee* notes, lower and faster toward end.

Nest. A tidy cup of seed down and treefern scales, covered with gray lichens, all bound with cobweb, 3–16ft (1–5m) up in shrub or tree at forest edge or in understory. December–May or June.

Status. Common resident of lowlands and foothills of N Pacific slope, S to vicinity of Orotina and Tárcoles, up to 1650ft (500m) or more on adjacent mountain slopes and to 3300ft (1000m) in W part of Valle Central, where regular E to Grecia, sporadic as far as San José; accidental on Caribbean slope (Reventazón valley, La Selva).

Range. W Mexico to Costa Rica.

RUFOUS-TAILED HUMMINGBIRD

Amazilia tzacatl Pl. 24(10)

Amazilia Rabirrufa

Description. 4″ (10cm); 5.2g. Only medium-sized, green-breasted hummingbird with rufous tail. **Adult ♂:** above bronzy-green; upper and lower tail-coverts and tail deep rufous; throat and chest glittering green; lower breast and sides bronzy; belly grayish-buff; rectrices edged with purplish-bronze at tip. Bill mostly red with black tomia and tip. ♀: below glittering green less extensive; rectrices edged more broadly with bronze. Upper mandible black, tinged with red basally; feet dusky. **Young:** rufous feather-edgings on face and crown; bronzy edges of rectrices do not extend to tips of feathers; upper mandible black.

Habits. Over most of its range prefers non-forest habitats: open scrub, second growth, thickety edges, semi-open, coffee plantations, gardens, dooryards, entering humid forest along breaks and gaps to forage; in dry

Pacific NW, chiefly along rivers in gallery forest; feeds at a wide variety of flowers (*Hamelia, Heliconia, Stachytarpheta, Cephaelis*, and banana); highly aggressive and often territorial at rich clumps of flowers. Males sing mostly from dawn to sunrise, from scattered perches near flowers or in small, loose assemblies where flowers are few or scattered.

Voice. Call a low-pitched *chup* or *chut*, sometimes in sputtering series; aggressive notes include 1 or several shrill *see* notes, rising and accelerating; in chases a metallic *tsunk*. Male's song a lisping, colorless phrase of 3–5 thin, whistled notes, in deliberate rhythm: *tse we ts' we* or *tse tseu wip tsik tsew*; phrase usually repeated several times, then a pause.

Nest. A compact cup of plant fibers and down, bits of dead leaves, cobweb, decorated rather heavily on outside with bits of lichen, sometimes moss, 3–20ft (1–6m) up on slender, horizontal, fairly open twig. May–January on Pacific slope, October–June on Caribbean slope—practically throughout year except during annual molt.

Status. Most abundant and widespread Costa Rican hummingbird from sea level on both slopes to 6000ft (1850m) locally, wherever forest has been removed, and along banks of large rivers in forested country; least common, most local in dry Pacific NW.

Range. NE Mexico to W Venezuela and W Ecuador.

STRIPED-TAILED HUMMINGBIRD

Eupherusa eximia Pls. 24(17), 25(14)

Colibrí Colirrayado

Description. 3¾″ (9.5cm); 4.3g. Medium-sized; conspicuous rufous secondaries; outer 2 rectrices largely white with dark outer margins, giving spread tail a striped look from above. **Adult ♂:** above bronzy-green; tail mostly dusky-bronze, outer rectrices tipped with blackish; secondaries bright rufous, tipped with black; below glittering green, duller on belly. ♀: below pale gray, speckled with bronze-green laterally; outer rectrix almost entirely white. Bill black; feet dusky-flesh. **Young:** face, upperparts, sides, and flanks with buffy feather-edgings.

Habits. Prefers cool, wet forest, dwelling mainly in canopy but frequently descending to shrub level in gaps, along edges, or in adjacent semi-open or second growth; only nesting females found regularly in heavily shaded understory; visits flowers of trees (*Inga*), shrubs (especially Acanthaceae, Rubiaceae), and epiphytes (Ericaceae, Gesneriaceae), sometimes piercing long corollas; males, in particular, very aggressive and sometimes territorial at flowers. Males sing regularly at gaps and edges, perching at medium heights.

Voice. Call a sharp, piercing *peet* or *bzeet*, often repeated rapidly. Song of male long-continued: 1–3 squeaky, sometimes metallic notes, then a low, dry, insectlike trill, then 1–3 more squeaks. Aggressive note a sharp buzz, often accompanied by spreading tail.

Nest. A neat cup of pale-colored plant down and fibers, some treefern scales, decorated on outside with mosses and lichens, usually including many bits of a bright red lichen (no other Costa Rican hummer uses these); 3–10ft (1–3m) up in dark understory, often near stream. September or October–March or April.

Status. Resident of middle elevations the length of the country; in most areas local and patchily distributed; breeds mostly from 2600ft (800m) in N cordilleras, and from 4000ft (1200m) in higher mountains, up to ca. 6500ft (2000m) in Cordillera de Talamanca. After breeding, some may descend to 1000ft (300m) in foothills, or ascend to ca. 8000ft (2450m), rarely higher.

Range. E Mexico to W Panama.

BLACK-BELLIED HUMMINGBIRD

Eupherusa nigriventris Pls. 24(21), 25(12)

Colibrí Pechinegro

Description. 3¼″ (8cm); 3.4g. Fairly small; rufous in secondaries; outer 3 rectrices white. **Adult ♂:** forecrown, face, and underparts velvety black; thigh-tufts and crissum white; above bronzy-green, duskier on central rectrices; secondaries rufous, tipped with black. ♀: below pale gray, spotted with bronze-green laterally. Bill black; feet dusky-flesh. **Young:** face, crown, hindneck, and rump with dusky feather-edgings; in males, black of adult replaced by dark sooty-bronze.

Habits. Lives in cool, wet forests of lower middle elevations; males in particular found mostly in canopy, feeding at flowers of various trees (e.g., *Pithecellobium, Inga*) and especially epiphytes (Ericaceae, *Columnea, Elleanthus, Norantea*); females frequently descend to shrub level along edges and at gaps, visiting flowers of *Cephaelis, Witheringia*, and *Besleria*. Males may defend canopy flowers not claimed by the larger, dominant *Lampornis hemileucus*; often sing from tall shrubs at edges of gaps.

Voice. A sharp, high-pitched *tseep* or *peet*; a high *tsit*tering in chases. Male's song a high, thin, sputtering warble.

Nest. A small cup chiefly of treefern scales and spiderweb, sparingly decorated with bits of mosses and lichens, in forest understory or at edge, 6–13ft (2–4m) up in shrub, often where protected from above by large overhanging leaves. October–March, sometimes as early as August.

Status. Uncommon to locally abundant resident of Caribbean slope of Cordillera Central and Cordillera de Talamanca, breeding from ca. 3000–6500ft (900–2000m); after nesting may range downslope to 2000ft (600m), rarely lower.

Range. Costa Rica to W Panama.

WHITE-TAILED EMERALD

Elvira chionura Pls. 24(19), 25(16)

Esmeralda Coliblanca

Description. 3″ (7.5cm); 3.2g. Small; flashing white tail boldly marked with black. **Adult ♂:** above bronzy-green, shading to dark coppery-bronze on upper tail-coverts and central rectrices; lateral 3 rectrices white, tipped with black; below mostly glittering green; belly and crissum white. Bill black, except pale pink basal half of lower mandible; feet dark gray. **♀:** central rectrices more bronzy, lateral rectrices white with black subterminal band; throat and breast dull white medially, speckled with bronze-green laterally, passing to bronzy-green on sides, with pale gray feather margins. **Young:** face, crown, and rump with pale grayish-buff feather edgings; tail pattern less distinct; male below duller green than adult, the feathers with dull grayish-buff fringes.

Habits. Frequents humid mountain forests, semi-open, shaded gardens; males more in canopy, females more in understory; visits flowers of epiphytes (Ericaceae, *Clusia*), trees (*Symphonia, Inga, Quararibea*), shrubs (*Palicourea, Stachytarpheta*), also banana. Small groups (2–4) of males gather in the morning at gaps or edges, perching 16–25ft (5–8m) up and singing from regular perches or in flight during circuits of the clearing, and engaging in frequent chases.

Voice. High-pitched, buzzy notes in chases; soft scratchy chipping notes while foraging. Male's song a prolonged, thin, scratchy twitter mixed with buzzy or gurgly notes, rising and falling in pitch.

Nest. Undescribed (?). Probably June–November.

Status. Locally common resident on Pacific slope of Cordillera de Talamanca, N to Dota region, and in coastal ranges of S Pacific slope at 3300–5600ft (1000–1700m), wandering down to 2500ft (750m) or up to 6500ft (2000m).

Range. Costa Rica to W Panama.

COPPERY-HEADED EMERALD

Elvira cupreiceps Pls. 24(20), 25(17)

Esmeralda de Coronilla Cobriza

Description. 3″ (7.5cm); 3.2g. Small; short, distinctly decurved bill; outer 3 pairs of rectrices white, marked with blackish near tip. **Adult ♂:** crown, upper tail-coverts, and central rectrices coppery-bronze, otherwise above bronze-green; below glittering green; center of belly dark purplish in northern populations; thighs and crissum white; white lateral rectrices narrowly tipped with dusky. **♀:** below dull white, speckled with green laterally; irregular black subterminal band on outer rectrices. Bill black except pink basal half of lower mandible; feet dusky-flesh. **Young ♂:** below dull bronzy-green; crown only slightly tinged with coppery. **♀:** face, head, nape, and rump with cinnamon feather-edgings.

Habits. In and beside wet highland forest; males frequently high up in canopy, both sexes at all levels at edges, gaps, and semi-open, only females regularly in dark understory; visits mostly small blossoms of trees (*Pithecellobium, Quararibea, Guarea*), epiphytes (Ericaceae, *Clusia*), and shrubs (especially scattered *Besleria* in understory, by females). Small groups of males perch at middle heights in trees around a gap or in scattered pasture trees, singing and chasing each other.

Voice. A high, thin, liquid *quip* or *quit* or rapid high sputtering in chases. Males sing a high, thin, twittering and warbling song, while perched or in flight, often continuously for 10–20sec.

Nest. A neat little cup of plant down and treefern scales, outside heavily decorated with mosses and a few bits of lichen, 3–10ft (1–3m) up in shrub in forest understory, at edge, or by roadside. October–March.

Status. Fairly common resident of Caribbean slope from Cordillera de Guanacaste to N side of Reventazón valley, breeding between ca. 2300–5000ft (700–1500m), and ranging as low as 1000ft (300m) afterward; reaches Pacific slope of N cordilleras above ca. 4000ft (1200m).

Range. Costa Rica.

SNOWCAP

Microchera albocoronata Pl. 25(8)

Copete de Nieve

Description. 2½″ (6.5cm); 2.5g. Tiny, short-billed; short, rounded tail. Male unmistakable, female whiter below than most small sympatric species. **Adult ♂:** crown shining white; rest of head and body deep wine-purple; central rectrices purplish-bronze, lateral rectrices white, indistinctly tipped with dusky. ♀: above bronzy-green; below dull white; postocular spot white; auriculars dusky; lateral rectrices dull white with dusky subterminal band. Bill and feet black. **Young:** like adult female but duller, below grayer; back feathers edged with grayish-buff to rusty. Upper tail-coverts and central rectrices of male more copper-bronze; purple feathers first appear below along midline, producing striking dark medial stripe.

Habits. Prefers canopy and edges of wet forest, also adjacent semi-open and second growth; feeds mostly on small flowers of trees (*Warscewiczia, Inga, Pithecellobium*), vines (*Gurania*), shrubs (*Hamelia*), and epiphytes (*Norantea, Cavendishia*). Males may defend feeding territories against other snowcaps, but are usually displaced by larger hummingbirds. During breeding season small, loose groups of males gather at forest edge, each singing from one or more regular perches on exposed twigs 13–25ft (4–8m) up, frequently chasing and fighting. During postbreeding movement to lowlands, usually found at flowering *Warscewiczia* trees.

Voice. A soft, high-pitched, dry *tsip*; buzzy notes and chitters in aggressive interactions. Male's song a soft, sputtering, warbling medley: *tsitsup tsitsup tsitsup tsew ttttt-tsew* or *tsip-tsee tsippy tsippy tsippy tsip-tick tsew*.

Nest. A tiny cup of treefern scales, plant down, and cobweb, lightly decorated with green moss and a few bits of green lichen around the rim, 5–10ft (1.7–3m) up on lower twig of tree, or vine hanging from larger tree. January–May.

Status. Breeds locally the length of Caribbean slope, chiefly 1000–2600ft (300–800m), locally to 3300ft (1000m); after breeding most descend to adjacent lowlands; occasionally wanders to ca. 4600ft (1400m); most common along Cordillera Central; uncommon to rare along Cordillera de Talamanca; reaches Pacific slope at passes through Cordillera de Guanacaste.

Range. Honduras to W Panama.

RED-FOOTED PLUMELETEER

Chalybura urochrysia Pl. 23(12)

Colibrí Patirrojo

Description. 4¼″ (11cm); ♂ 7g, ♀ 6g. A large, dark, noisy hummingbird with conspicuous red or pink feet and a long, broad tail. **Adult ♂:** above dark bronzy-green, shading to purplish-bronze on upper tail-coverts; tail purplish-black, slightly notched; throat and breast glittering dark green; belly dusky-bronze; crissum sooty-blackish. ♀: above like male; below dull ash-gray, speckled with green laterally; tail more rounded, bronzier than that of male, outer 2–3 rectrices tipped with dull gray. Bill black except dull pink basal half of lower mandible; feet bright pink to red. **Young:** crown, nape, and rump with buffy to cinnamon fringes.

Habits. In forested regions but prefers edges, stream banks, gaps, and shady second growth; very aggressive and dominant at flowers, males often holding territories at large clumps of *Cephaelis* or *Heliconia*, in particular; females invade territories of other species; nevertheless often shy and nervous toward observer, easier to hear than to see.

Voice. Loud metallic chipping and chattering notes; when excited, a high shrill sputter. Male has a soft, scratchy, warbled song *ter-pleeleeleelee ter-pleeleeleelee ter-pleeleelee-lee ter-pleee*.

Nest. A deep, compact cup of kapok floss and other fine, pale plant fibers and spiderweb, heavily decorated on the outside with green moss and a few bits of lichen, 1½–5ft (0.5–1.5m) up in small shrubs near stream or trail in forest. February–May, in some years probably as early as December.

Status. Widespread but usually uncommon resident of lowlands and foothills up to 2300ft (700m) throughout Caribbean slope, except drier lowlands S of Lago de Nicaragua.

Range. E Honduras to NW Ecuador.

Note. Also known as Bronze-tailed Plumeleteer, but the Costa Rican form *C. u. melanorrhoa*, once considered a separate species, does not have a bronzy tail. This form hybridizes extensively with truly "bronze-tailed" *C. u. urochrysia* in extreme W Panama.

WHITE-BELLIED MOUNTAIN-GEM

Lampornis hemileucus Pl. 24(5)

Colibrí Montañés Vientriblanco

Description. 4″ (10.5cm); ♂ 6.2g, ♀ 5g. Combination of white postocular stripe and white underparts diagnostic; tail distinctly bronzy. **Adult ♂:** crown, face, and chin glittering green; otherwise above bronzy-green; upper tail-coverts and tail bronzy; a dusky subterminal band on lateral rectrices; center of throat violet; otherwise below white, spotted with green laterally. ♀: lacks glittering green on head and violet on throat; lateral rectrices with grayish tips. Bill black; feet dusky-flesh. **Young:** head, back, and sides with rusty fringes; chin and throat of male dull bronzy.

Habits. Prefers canopy and edges of wet forest, ranging down to shrub level in open understory and at gaps; visits a wide variety of flowers, especially epiphytic Ericaceae (*Cavendishia, Thibaudia*), of which males vigorously defend large clumps during breeding season by persistent song and swift chases through the canopy. Males, in particular, are very aggressive at flowers; they often dominate larger species like *Heliodoxa* when both visit the same plants.

Voice. Call a high-pitched *deet* or *deedeedeet*; various distinctively high-pitched, liquid twitters and trills in interactions; in chases during breeding season, males emit a continuous squeaky sputtering. Male's song is a medley of squeaks, dry or liquid trills and sputters.

Nest. Undescribed (?). August or September–March.

Status. Locally common resident of lower middle elevations of Caribbean slope, ca. 2300–4600ft (700–1400m), from Cordillera de Tilarán and Volcán Arenal S to Panama.

Range. Costa Rica to W Panama.

Note. This species and the next 2 (as well as the *L. viridipallens-sybillae* complex of N Central America) could well be separated from *Lampornis* as the genus *Oreopyra*.

PURPLE-THROATED MOUNTAIN-GEM

Lampornis (castaneoventris) calolaema

Pl. 24(7)

Colibrí Montañés Gorgimorado

Description. 4″ (10.5cm); ♂ 6g; ♀ 4.8g. Medium-sized; conspicuous white postocular stripe over dusky auriculars. **Adult ♂:** crown glittering pale blue-green; rest of upperparts bright bronzy-green; gorget purple; chest glittering green to dusky green; belly gray; notched tail blue-black. ♀: lacks bright crown and gorget; entire underparts cinnamon-rufous; double-rounded tail green with blue-black subterminal band and white tips on lateral rectrices. Bill black; feet dusky. **Young ♂:** crown, gorget and chest dull dusky-green; first purple feathers on throat with extensive pale bases. ♀: sides of crown and auriculars with rusty feather-edges.

Habits. Prefers forested areas in steeply sloping, broken terrain; mainly in canopy but moves down to shrub level freely at edges and breaks, or out into adjoining second growth or semi-open. Males, in particular, very aggressive, dominant, often territorial at flowers; visit and often defend flowers of epiphytic Ericaceae (*Satyria, Cavendishia*) and Gesneriaceae (*Columnea*), or understory shrubs (*Cephaelis*); often sing while on flower-centered territories.

Voice. Usual call a sharp, penetrating, buzzy *zeet* or *zeep*; higher-pitched, scratchy, chattering notes in interactions. Male's song high, thin and dry, a complex medley of sputtering and warbling notes.

Nest. A compact, deep cup of pale-colored plant downs, fibers, and treefern scales or fine, dead leafy liverworts, decorated on outside with bits of moss and lichen; 2–12ft (0.7–3.5m) up in understory shrub, small tree, or vine, usually at edge or opening. October–March or April, occasionally June.

Status. Common to locally abundant resident of N cordilleras above 2600ft (800m) on Caribbean slope and above 3300ft (1000m) on Pacific; at upper middle elevations, ca. 3900–8200ft (1200–2500m), on Cordillera Central, La Carpintera, and extreme N end of Cordillera de Talamanca; also (formerly?) common on Pacific side of Dota heights and upper parts of Pacific coastal ranges, S to Panama; absent from most of Cordillera de Talamanca. Descends to 1000ft (300m) locally after breeding.

Range. S Nicaragua to W Panama.

Note. The taxonomy of the *L. castaneoventris* Mountain-gem complex is poorly understood. Three distinct, semiisolated groups evidently hybridize to a limited extent: purple-throated, blue-tailed *calolaema*; white-throated, gray-tailed *cinereicauda*; and white-throated, blue-tailed *castaneoventris* (restricted to Panama). The latest AOU Check-List lumps *cinereicauda* into *castaneoventris* while maintaining *calolaema* as a separate species, although there is probably less evidence of hybridization between the former 2 than between either of them and *calolaema*. We believe that all 3 should be given the same taxonomic treatment, whether as allospecies of a superspecies or as subspecies of a single species, *L. castaneoventris* (Variable Mountain-gem). Without further data (difficult to obtain because of extensive deforestation in most of the critical contact zones), choice between these alternative treatments is arbitrary.

GRAY-TAILED MOUNTAIN-GEM
Lampornis (castaneoventris) cinereicauda Pl. 24(6)
Colibrí Montañés Coligrís

Description. 4″ (10.5cm); ♂ 6.2g, ♀ 5g. **Adults:** very similar to the respective sex of Purple-throated Mountain-gem, differing as follows: ♂: gorget white, usually with a few scattered purple or bluish feathers around edge; chest glittering pale green; tail ash-gray, darker toward tip. ♀: tail mostly dull bronzy-green to dusky-green, distinctly duller than upper tail-coverts. **Young ♂:** crown and gorget dull dusky-green; first white feathers in gorget often tinged with buff. ♀: sides of crown and auriculars with rusty feather-edgings.

Habits. Generally similar to those of Purple-throated Mountain-gem, which it replaces in Cordillera de Talamanca; when the 2 occur together, as at N end of Talamancas or Dota area, the Gray-tailed occupies higher elevations. Preferred flowers include epiphytic Ericaceae (*Cavendishia, Satyria*), bromeliads, shrubs like *Centropogon* and *Alloplectus* at edges and gaps. At higher elevations this species is subordinate at flowers to the more aggressive Fiery-throated Hummingbird and occurs more in open understory.

Voice. Contact-spacing note a high, squeaky *pick* or *pipipick*; also a sharp, thin, buzzy, rolling *zeet* or *zeep*. In interactions, high-pitched, penetrating, grating chatters. Males have a sputtery, bubbly song.

Nest. A deep, compact cup of fine brown plant fibers and treefern scales, heavily decorated outside with moss and bits of lichen, in understory shrub, 3–10ft (1–3m) up, often at forest edge or by trail. October–April.

Status. Fairly common resident of oak forests of Cordillera de Talamanca S about to Panama border (beyond which it is abruptly replaced by blue-tailed *L. castaneoventris*), from 6000ft (1850m) up to timberline; most numerous around 8000ft (2400m); descends to ca. 5000ft (1500m) after breeding.

Range. Costa Rica.

Note. See Purple-throated Mountain-gem.

GREEN-CROWNED BRILLIANT
Heliodoxa jacula Pl. 23(15)
Brillante Frentiverde

Description. ♂ 5″ (13cm), 9.5g; ♀ 4¾″ (12cm), 8g. Large; rather stout; bill straight; feet strong. **Adult ♂:** forehead, crown, lores, throat, and breast glittering green; rest of upperparts and belly bright bronzy-green; long, deeply forked tail blue-black; white spot behind eye; foreneck with metallic violet spot; crissum and thighs white. Bill black; feet dusky. ♀: white spot behind eye and white stripe below eye; shallowly forked tail black with outer feathers narrowly white-tipped; below white, heavily spotted with bright green, which becomes solid on sides. **Young:** resemble adults of same sex but duller, below bronzier (and on crowwn of males); throat extensively cinnamon or buff.

Habits. In wet highland forest, from middle understory well up into canopy, also at gaps, edges, and adjacent tall second growth; feeds largely at candelabralike inflorescences of large epiphytic *Marcgravia* vines, male intermittently defending large plants; female also visits understory flowers (e.g., *Drymonia conchocalyx*); during early rainy season often visits flowers of *Heliconia* along streams, riverbanks or in second growth. Flight fast, darting; nearly always perches to feed, clinging to inflorescence with feet.

Voice. Usual call a loud, squeaky *kyew* or *tyew*, often rapidly repeated when agitated; loud sputtering notes and squeaks in chases.

Nest. Undescribed (?). July or August to about January.

Status. Fairly common resident at ca. 2300–6600ft (700–2000m) from Cordillera de Guanacaste S to Panama, mainly on Caribbean slope but locally on Pacific slope on N cordilleras, and especially on Cordillera de Talamanca and coastal ranges of S Pacific slope; outside breeding season descends occasionally to 330ft (100m).

Range. Costa Rica to W Ecuador.

MAGNIFICENT HUMMINGBIRD
Eugenes fulgens Pl. 23(16)
Colibrí Magnífico

Description. 5″ (13cm); ♂ 10g, ♀ 8.5g. Largest and longest-billed hummingbird of the high mountains. **Adult ♂:** crown violet; gorget bright bluish-green; sides and back of head black; postocular spot white; rest of upperparts and chest dark bronzy-green; belly grayish; tail dark bronze, strongly notched. ♀: above bronzy-green; tail more bronzy, shading to blackish subterminally, outer 3–4 rectrices tipped with grayish-white; postocular stripe white; auriculars dusky; below dull gray, spotted laterally with bronzy, mottled with dusky on throat. Bill and feet black.

Young: resemble adult female but below darker, browner (especially throat), scaled with pale buff; above with buffy fringes, most prominent on head and neck.

HABITS. Lives in oak forests of high elevations, especially at edges and breaks in canopy or in adjacent second growth; during breeding season males defend territories where giant thistle (*Cirsium*) is numerous; they also ascend to canopy to visit flowers of epiphytes (Ericaceae) and vines (*Passiflora*). Females visit long-flowered species of *Centropogon*, usually several small bushes scattered along foraging route; both sexes visit other shrubs at gaps and edges, especially *Fuchsia* and *Cestrum*.

VOICE. Usual call a rather guttural, rolling *nrrt* or *drrrk*. In aggressive encounters emits a rapid stream of high, clear, liquid chips run together. Males have a soft, low-pitched song of burbling, scratchy, buzzy notes, given in courtship interactions or while perched on territory.

NEST. A bulky cup of soft plant down, fine fibers, and rootlets, the outside heavily decorated with moss and bits of lichen, near tip of descending bamboo stem ca. 10ft (3m) up in forest understory or at edge of small gap. November–March.

STATUS. Fairly common resident of higher parts of Cordillera Central and Cordillera de Talamanca, from timberline down to ca. 6600ft (2000m) but most abundant above 8000ft (2500m); sporadically, perhaps seasonal visitor, down to 6000ft (1850m).

RANGE. SW USA to W Panama.

NOTE. The large form *E. spectabilis* of Costa Rica and Panama is sometimes considered a species separate from the smaller, blacker-breasted *E. fulgens*, Rivoli's Hummingbird, of SW USA to Nicaragua.

PURPLE-CROWNED FAIRY

Heliothryx barroti Pl. 23(14)

Colibrí Picopunzón

DESCRIPTION. 4½″ (11.5cm); 5.6g. Immaculate white underparts; tail long (especially in female), graduated, conspicuously white-sided; bill tapers to needlelike point. **Adult** ♂: forehead and crown flashing violet; velvety black mask terminating posteriorly with a metallic violet spot, bordered below by glittering green malar stripe; upperparts bright metallic green; 2 central pairs of rectrices blue-black, outer 3 white. Bill black; feet dusky. ♀: lacks violet and glittering green on head; tail longer, outer 3 rectrices with a black band basally. **Young:** head and upper back with rufous fringes; throat spotted or smudged with dusky; male lacks violet or glittering green on head.

HABITS. Frequents canopy and edges of forest, descending to lower levels in semi-open, tall second growth, and shaded gardens; flight exceptionally light and graceful; gleans small insects from outer foliage of trees, or hawks them in the air; extracts nectar from flowers by piercing their bases, including flowers larger than itself or tough, waxy blooms like *Heliconia*; not territorial but quite aggressive, able to withstand attacks of most territorial hummingbirds.

VOICE. A high, thin, staccato *tsit*ting, often with metallic or tinkling quality; when excited, runs many *tsit* notes together into a high chittering.

NEST. A cup with incurved rim, often tapering below to cone, largely of plant down with little or no green moss or lichen, 20–65ft (6–20m) up near tip of slender branch, often over stream or at forest edge. October–March, sometimes through July (especially Pacific slope).

STATUS. Uncommon to fairly common resident over most of Caribbean slope, from lowlands to ca. 4250ft (1300m), and on S Pacific slope N to Carara, up to 5500ft (1675m); rare in drier lowlands S of Lago de Nicaragua, absent from dry NW.

RANGE. SE Mexico to SW Ecuador.

NOTE. Sometimes lumped with *H. aurita*, Black-eared Fairy, of E South America.

PLAIN-CAPPED STARTHROAT

Heliomaster constantii Pl. 23(19)

Colibrí Pochotero

DESCRIPTION. 4¾″ (12cm); 7.5g. Fairly large, rather dull, with relatively long, straight, stout bill and short tail, distinctive white stripe down center of rump. **Adult** ♂: above dark bronzy-green; crown duskier; rectrices greenish-bronze, with distal third black, outer 3 white-tipped; bold white postocular and malar stripes; chin and face dusky; gorget iridescent rose-red to orange-red; below mostly dull gray, flecked with bronze-green laterally; center of lower breast, belly, and tuft on each side of rump white. ♀: dusky of chin extends to middle of gorget, the feathers of which are edged with dull buff; white tips of lateral rectrices more extensive. Bill black; feet dusky. **Young:** entire gorget dusky; gorget, chest, sides, and upperparts broadly edged with dull buff; rump stripe tinged with buff.

Habits. Frequents canopy and edge of dry and evergreen gallery forest, also coffee plantations, second growth, dry scrub, and isolated savanna trees; visits mainly flowers of *Erythrina* with long corollas (which males sometimes defend), *Ceiba, Bombacopsis, Tabebuia*, and other trees, and occasionally understory *Heliconia*; frequently flycatches and gleans high in canopy.

Voice. A high-pitched, fairly soft and melodious slurred *tseep* or *cheek*; in chases, a high, piercing twittering.

Nest. A shallow cup of pale-colored plant down, variously decorated outside with bits of bark and gray lichen, near tip of branch, usually in exposed position, well up in tree at woods edge or in savanna. October–January.

Status. Uncommon to fairly common resident in lowlands and hills of N Pacific slope, to 2600ft (800m) along Cordillera de Guanacaste and 3300ft (1000m) on Península de Nicoya; decreasing abundance E across Valle Central regularly to San José, rarely to Cartago; rare in General-Térraba region.

Range. NW Mexico to Costa Rica.

LONG-BILLED STARTHROAT

Heliomaster longirostris Pl. 23(18)

Colibrí Piquilargo

Description. 4½″ (11.5cm); 7g. Fairly large; long, straight bill; rather short tail; broad white streak down center of rump. **Adult ♂:** forehead and crown glittering metallic blue to blue-green; rest of upperparts rather dark bronze-green; rectrices black distally, lateral 2–3 pairs white-tipped; small postocular spot and broad malar stripe white; chin blackish; gorget dark metallic purple; breast gray, flecked with bronze-green laterally, becoming solid bronze-green on sides; center of lower breast and belly dull white; white tufts on each side of rump. Bill and feet black. ♀: crown dusky-bronze with little or no blue; white malar stripes broader and gorget narrower, its anterior half dusky. **Young:** gorget entirely dusky; gorget, chest, sides, head, and upperparts with broad buffy fringes.

Habits. Frequents forest canopy and edge, tall second growth, shady gardens and pastures, coffee and banana plantations in humid regions; favors flowers of *Erythrina* with long corolla tubes; defends large trees with abundant flowers against all other hummingbirds, or flies between scattered small trees over a wide area; also visits flowers of banana and *Heliconia*, where it may be subordinate to other species; catches small insects in the air; sleeps on high, fully exposed perches.

Voice. When foraging, a high, liquid, rather weak *tsip tsip* or *tseep*; squeaky twitters in aggressive encounters.

Nest. A broad, shallow bowl of mosses, liverworts, and plant down with a few gray lichens on outside, fastened by cobweb to upper side of dead branch, 15–35ft (4.5–10.5m) up, where fully exposed but appearing like an excrescence on the branch. December–February.

Status. Fairly common resident of General-Térraba–Coto Brus area of S Pacific slope, from lowlands up to ca. 4000ft (1220m); uncommon in Golfo Dulce region and rare virtually throughout Caribbean lowlands, occasionally up to 4600ft (1400m), as in Reventazón drainage; very rare, perhaps only a wanderer, in Valle Central and around Golfo de Nicoya (Slud).

Range. S Mexico to NW Peru, Bolivia, C Brazil, and Guianas.

MAGENTA-THROATED WOODSTAR

Calliphlox bryantae Pl. 25(2)

Estrellita Gorgimorada

Description. ♂ 3½″ (9cm), ♀ 3″ (7.5cm); 3.5g. Larger, longer-billed than *Selasphorus* species; conspicuous white or buffy patch on each side of rump; male has long forked tail. **Adult ♂:** above bronze-green; central rectrices short, black-tipped; lateral rectrices long, black with rufous shafts and edges; postocular spot white; gorget metallic purple; conspicuous white collar across foreneck; chest and sides green; belly rufous. Bill and feet black. ♀: above like male; auriculars dusky; throat buffy; white collar less distinct; green of breast and sides mixed with rufous; tail double-rounded, the outer 3 rectrices with rufous bases, black subterminal area, buffy tips. **Young:** like adult female but below paler; above with buffy fringes, especially on head and hindneck; lateral rectrices with relatively less (male) or more (female) pale tipping.

Habits. Prefers forest edge and clearings, scrubby pastures, second growth, and semiopen; visits flowering herbs (*Lobelia*), shrubs and trees (*Inga, Quararibea*). Where flowers are abundant, both sexes are very aggressive, sometimes territorial, although usually supplanted or ignored by larger hummingbirds. Males give song and dive displays when alone on feeding territories and also in loose groups, possibly leks, away from flowers. Feeding birds hold tail cocked and closed; males open tail in aggressive displays (Feinsinger).

VOICE. A dry *cht*; when excited, a liquid twittering; in encounters, a hard, dry, rolling twitter: *chrrrrt*; sometimes in chases, a low, whistled *teww*. Song of male a low, rapid, sputtering, gurgling medley.

NEST. Undescribed (?).

STATUS. Resident locally from Cordillera de Guanacaste (Santa María) S along Pacific slope, including upper reaches of Valle Central, Dota area, and Cordillera de Talamanca, to Panama; reaches Caribbean slope in mountain passes of N cordilleras, also upper Reventazón valley, from 2300ft (700m) (especially in Cordillera de Guanacaste) to 6000ft (1850m) in Talamancas. Pronounced local seasonal movements related to shifts in flowering.

RANGE. Costa Rica to W Panama.

NOTE. Often placed in genus *Philodice*.

RUBY-THROATED HUMMINGBIRD

Archilochus colubris Pl. 25(9)

Colibrí Garganta de Rubí

DESCRIPTION. 3¼″ (8cm); ♂ 2.7g, ♀ 3g. Whiter below than other small hummers of its range; inner primaries narrow, bluntly pointed (female) to sharply angled (male) at tips. **Adult ♂:** above bronzy-green, including central rectrices; tail deeply forked, lateral rectrices black, pointed; postocular spot white; gorget ruby-red; white collar across foreneck; rest of underparts grayish-white, spotted with green laterally. ♀: cheeks and auriculars dusky; underparts entirely grayish-white, spotted with green laterally; tail rounded, the lateral rectrices bluntly pointed with green to grayish bases, black subterminal band, broad white tips. Bill and feet black. **Young:** like adult female but above with buffy feather-edges, especially on rump and nape; lateral rectrices with bases more dusky, tips more rounded (especially females); males usually show some red feathers on throat and less white in tail.

HABITS. Prefers brushy second growth and deciduous forest, occasionally gallery forest; also flowering hedges, shade trees in coffee plantations and rural dooryards; visits flowers of a variety of herbs, shrubs, vines, and trees, especially those producing large numbers of small, often insect-pollinated blossoms, or scattered flowers. Generally subordinate to most local species, it can rarely hold feeding territories but may poach from territories of larger species.

VOICE. Call, a liquid downslurred *tyew* or *tew*; in interactions, a scratchy chattering.

STATUS. Uncommon to locally common winter resident in N Pacific lowlands; rare to occasional across Valle Central and in S Pacific lowlands; very uncommon but regular in Los Chiles–Río Frío district, but on rest of Caribbean slope a very rare vagrant, mostly during fall migration, mid-October–November. Spring departure late March to mid-April.

RANGE. Breeds in E North America from S Canada to Florida and Texas; winters from NW Mexico and S Texas regularly to Costa Rica, casually to W Panama.

SCINTILLANT HUMMINGBIRD

Selasphorus scintilla Pl. 25(7)

Chispita Gorginaranja (Chispitas, Colibrí Mosca)

DESCRIPTION. 2½″ (6.5cm); ♂ 2g, ♀ 2.3g. Tiny; very rufescent below; white collar on foreneck. **Adult ♂:** above bronzy-green; upper tail-coverts edged with rufous; tail broadly striped with black and rufous; gorget brilliant orange-red, elongated at sides; front and sides of neck white; breast and belly mostly cinnamon, spangled with green, shading to whitish down midline; crissum rufous. Bill and feet black. ♀: similar but throat buff, speckled with dusky; central rectrices largely green, lateral rectrices with a broad black subterminal band. **Young:** resemble female but above with rusty fringes, especially on crown, nape; tail of young males with much less green in central rectrices.

HABITS. Frequents brushy forest edge, scrubby pastures, hedgerows, young second growth, coffee plantations, occasionally rural dooryards and suburban gardens; visits a variety of mostly small, often insect-pollinated flowers: *Salvia, Lantana, Hyptis, Rubus*; usually a furtive poacher at territories of larger, dominant hummingbirds (*Amazilia, Lampornis, Colibri*) at *Stachytarpheta* hedges. During breeding season males occupy territorial posts on tall bushes or trees overlooking open areas, often with *Salvia* flowers; chase or give conspicuous dive displays to repel or intimidate intruders.

VOICE. A soft, high, liquid chipping or *tsipping*; a sharper chatter in interactions; solitary birds are almost always silent. Males do not sing; in flight their attenuated outer primaries produce a high, thin insectlike trilling.

NEST. A tiny cup of pale-colored plant floss from thistles and other composites, grass heads, treefern scales, etc., decorated on outside with mosses and lichens, sometimes lined with small feathers, 3–13ft (1–4m) up on outside of large shrub or grass tussock beside an open space. September–February.

Status. Widely distributed, scarce to locally common resident at ca. 3000–7000ft (900–2100m) on Pacific slope from Cordillera de Tilarán (Monteverde, where possibly only a vagrant) S along slopes of Cordillera Central overlooking Valle Central, and along Cordillera de Talamanca to Panama; extends on to Caribbean slope in Cartago-Paraíso area and upper Reventazón valley; strays occasionally to Caribbean slope of Cordillera Central (e.g., Montaña Azul, Virgen del Socorro). After breeding some may ascend to 8000ft (2450m) as on Volcán Poás.

Range. Costa Rica and W Panama.

VOLCANO HUMMINGBIRD

Selasphorus flammula Pl. 25(3)

Chispita Volcanera (Chispitas, Colibrí Mosca)

Description. 3″ (7.5cm); ♂ 2.5g, ♀ 2.8g. The common small hummer of high elevations. Varies geographically: brightness and redness of males' gorgets, amount of buffy below and black in tail increase from S to N. **Adult ♂:** above bronzy-green; central rectrices mostly green to mostly black, lateral rectrices mostly black, edged and tipped with rufous; gorget rose-red, mauve-purple, grayish-purple, or purplish-green; below mostly white, including collar across foreneck; sides of breast more or less suffused with buffy or pale cinnamon and spotted with green. Bill and feet black. ♀: throat whitish, speckled with dusky-bronze; lateral rectrices with bases more or less rufous, broad black subterminal band, the outer 3 broadly tipped with white to buff. **Young:** resemble adult female but above with prominent buffy feather-edgings, lateral rectrices with green or dusky extending to base; females (not males) have small buffy area at tip of central 2 pairs of rectrices.

Habits. Prefers open, brushy areas: páramo, scrubby second growth produced by volcanic eruptions, landslides, or human disturbance, also gaps and edges of stunted elfin forest; visits a wide variety of mostly small, often insect-pollinated flowers: *Fuchsia, Castilleja, Salvia, Vaccinium, Rubus, Miconia*; also extracts nectar from long-tubed flowers via holes pierced by bumblebees or flowerpiercers. Breeding males defend conspicuous perches and surrounding open areas, which usually have flowers nearby; sing and give towering dive displays. Females, sometimes males, visit scattered flowers or poach furtively from territories of larger, dominant hummingbirds, especially Fiery-throated Hummingbirds.

Voice. Usually silent, occasionally soft chips while foraging; in interactions, a high, thin chatter or twitter. Males on territory give a high, thin, whistled song: *teeeeeuu*. In chases males alternate songs and chatters.

Nest. A compact little cup of pale-colored plant down (*Cirsium* and *Senecio*) and cobweb, heavily decorated on outside with mosses and lichens, 3–16ft (1–5m) up in shrub or tree, usually on outermost branch, or on rootlet dangling from overhanging bank (which nearly always faces S or E). August–February.

Status. Common to abundant resident from 6000ft (1850m) in Cordillera Central and 6600ft (2000m) in Cordillera de Talamanca up to highest peaks; outside breeding season descends as low as 4000ft (1200m); also recorded from Cerros de Escazú.

Range. Costa Rica to W Panama.

Note. The red-gorgeted, black-tailed, buff-vented form from N Cordillera Central (Viejo, Poas, and Barva volcanoes) was formerly considered a separate species, *S. simoni*, Cerise-throated Hummingbird, but details of plumage and displays show that it is a race of *S. flammula*.

ORDER **Trogoniformes**
FAMILY **Trogonidae:** Trogons

The 40 species of trogons are widely distributed over the warmer regions of the Old World and the New. They are best represented in America, where they occur in practically all wooded areas of the continental tropics and even beyond them to southern Arizona, while Cuba and Hispaniola each has an endemic species. Arboreal birds of medium size, they have compact bodies, short necks, short stout bills, and short legs with the first and second toes directed backward (rather than the first and fourth, as in other zygodactyl or "yoke-toed" birds). Their plumage is notably soft, dense, and dry-textured. Among the most colorful birds of tropical America, the males have glittering green, blue, or violet upper plumage and chests. The posterior underparts are contrasting yellow, orange, or red. Females are plainer, brown, gray, or slaty where the males are green, but with bellies colored nearly as brightly as those of their mates. Outstanding in its magnificence is the male Resplendent Quetzal, whose long, slender tail-coverts extend far beyond his tail in an elegant train.

Trogons are sedate birds who perch upright, with long tails directed almost straight downward. Their flight tends to be undulatory, with intermittent bursts of rapid wingbeats. Their food is chiefly arthropods and fruits, both of which they pluck from stem or foliage at the end of a sudden upward or outward sally, without alighting. They vary their diet with an occasional lizard, frog, or snail. Their calls or songs are simple but often melodious. They nest in holes, which both sexes carve in decaying trunks, occupied termitaries, large arboreal wasps' nests, rarely arboreal ants' nests, or compact root masses of epiphytes. The cavity may be a shallow niche that leaves much of the sitting bird exposed or a well-enclosed chamber entered through an ascending tube. On the unlined bottom they lay 2 or 3 white to pale blue eggs, which the female incubates by night and the male for much of the day, often sitting continuously for many hours. The young, hatched naked, are fed and brooded by both parents. Incubation requires 17–19 days; nestling period is 14–ca. 30 days.

RESPLENDENT QUETZAL

Pharomachrus mocinno Pl. 26(1)

Quetzal (Quetzal)

DESCRIPTION. 14″ (36cm), plus up to 25″ (64cm) for male's streamers; 210g. Male unmistakable even without streamers because of laterally compressed, helmetlike crest that extends forward to cover base of bill; female and young crestless but with distinctively barred tail, gray breast. **Adult ♂:** mostly glittering green; elongated wing-coverts extend to sides of breast; 4 central upper tail-coverts elongated into slender, flexible streamers; remiges and central rectrices black; lateral rectrices white; lower breast maroon, shading to bright crimson on belly. Bill yellow; feet olive-gray. ♀: much duller; wing- and tail-coverts only slightly elongated; head dull bronze-green; breast gray; belly paler crimson; lateral rectrices coarsely barred with black and white; upper mandible black. **Immatures:** like adult female but above more bronzy, tail more coarsely barred. **Young:** above dark sooty-brown; scapulars, wing-coverts, and tertials with extensive buffy bases and edges; below buff, fading to white on belly, indistinctly barred or smudged with dusky.

HABITS. Prefers damp, epiphyte-laden mountain forests, frequenting canopy and edges; usually solitary or in pairs, though several may gather at a fruiting tree; may travel in small loose flocks following breeding; sallies to pluck fruits or snatch insects, small frogs and lizards, or snails; favorite fruits include various Lauraceae, also *Symplocos*.

VOICE. Most often heard is a sharp, cackling *perwicka*, upon taking flight or in agitation; male's song of deep, smooth, slurred notes in simple patterns: *keow kowee keow k'loo keow k'loo keeloo . . .*, often strikingly melodious; also whining notes, chiefly in nesting season; in steeply rising display flight, male calls *very good, very good*.

NEST. A deep, woodpeckerlike, unlined cavity with a single entrance, 14–90ft (4.3–27m) up in decaying trunk in forest or nearby clearing. Eggs 2, pale blue. March–June (often 2 broods).

STATUS. Fairly common resident of forested or partly forested mountains from Cordillera de Tilarán S to Panama; upward from 4000ft (1200m) in Cordillera de Tilarán, and from 5000ft (1500m) to over 10,000ft (3000m) in Cordilleras Central and de Talamanca; following breeding makes limited altitudinal movements; persists in largely deforested areas if remnant woods contain good feeding and nesting trees.

RANGE. S Mexico to W Panama.

NOTE. Pronounced *ket-saal'*.

SLATY-TAILED TROGON

Trogon massena Pl. 26(2)

Trogón Coliplomizo

DESCRIPTION. 12″ (30cm); 145g. Large, red-bellied; tail of adults below uniform blackish-gray. **Adult ♂:** face and throat dull black; rest of head, neck, chest, and upperparts glossy green; wing-coverts white vermiculated with black (appearing pale gray in field); remiges dull black; tail above dark bronzy-green; posterior underparts red. Iris brown to dark orange; bill and bare eye-ring orange-red; feet brownish. ♀: head, neck, and upperparts dark slate-gray, breast paler gray; belly and crissum red; most of upper mandible black. **Im-**

matures: like respective adults but lateral rectrices more pointed and narrowly tipped and barred with dull white; secondaries edged with buff; male has upper breast gray, and wing-coverts much darker than adult.

Habits. Prefers canopy and middle levels of humid lowland forest, where often difficult to detect, but comes lower at edges and adjacent semi-open or second growth; mostly solitary, except during breeding season; sallies for various fruits, including *Coussarea, Hamelia, Guatteria*, small palms; also snatches caterpillars, katydids, other insects, and small lizards from foliage; may follow monkey troop or cacique flock, catching insects flushed by them.

Voice. Usual call a nasal, barking *uh, wuk,* or *ow*, repeated many times steadily on same pitch; lower in pitch and less melodious than notes of most trogons; also, especially in agitation, low throaty notes rapidly repeated, sometimes accelerating into a prolonged woody rattle.

Nest. An unlined chamber entered through an ascending tunnel, carved by both sexes in an occupied arboreal termitary or decaying trunk, 8–50ft (2.5–15m) up. Eggs 3, white to bluish-white. March–June.

Status. Common resident of lowlands and foothills of Caribbean slope (except rare in drier lowlands W of Upala) and S Pacific slope, N to Carara, rarely to Río Barranca; also reaches Pacific slope through passes in Cordillera de Guanacaste; locally to 4000ft (1200m).

Range. SE Mexico to NW Ecuador.

LATTICE-TAILED TROGON

Trogon clathratus Pl. 26(3)

Trogón Ojiblanco

Description. 12″ (30cm); 130g. Large, red-bellied; best told from Slaty-tailed Trogon by voice, eye and bill color. **Adult ♂:** face and throat dull black; rest of head, neck, chest, and upperparts iridescent green, shading to dark bluish-green on tail; underside of tail blackish, finely and sparsely barred with grayish-white; wing-coverts finely vermiculated with black and white, appearing gray; lower breast and belly rose-red. Iris white; no eye-ring; bill yellow; feet brownish. ♀: head, chest, and upperparts slate-gray, washed with sooty on face and crown; lower breast olive-brown; belly pale rose; wing-coverts blackish, faintly vermiculated with pale gray. Culmen broadly black, rest of bill yellow. **Immatures:** much like respective adults, but rectrices more pointed, more coarsely and irregularly barred; chest of male largely brown.

Habits. Dwells in lower canopy and middle levels of wet foothill forest, sometimes coming to edge or into adjacent semi-open, or upper understory to feed; eats various Lauraceae and other fruits, large insects, occasionally small frogs and lizards; usually solitary except during breeding season.

Voice. A rapid series of ca. 15 loud, resonant, clucking *Kwa* notes that rises in pitch and loudness to a crescendo in the middle, then becomes faster, lower, and softer; clearer, less nasal, more patterned than calls of Slaty-tailed Trogon. Softer, more bubbling calls at nest.

Nest. A cavity in rotting stub or snag, occasionally an arboreal termitary, 16–25ft (5–8m) up. Eggs undescribed (?). February–May.

Status. Uncommon to fairly common resident of lower mountain slopes and foothills barely entering adjacent lowlands, of Caribbean slope from Panama N at least to Volcán Miravalles; breeds 300–3600ft (90–1100m), descending from higher parts of its range in mid to late rainy season.

Range. Costa Rica to W Panama.

BAIRD'S TROGON

Trogon bairdii Pl. 26(6)

Trogón Vientribermejo

Description. 11″ (28cm); 95g. Fairly large; posterior underparts deep orange; distinctive tail pattern. **Adult ♂:** face and throat dull black; rest of head, chest, and upperparts deep blue to violet-blue, shading to greenish-blue on back and tail; wing-coverts and remiges black; underside of tail white; lower breast, belly, and crissum deep orange to vermilion. Bare eye-ring pale blue; bill silvery bluish-gray; feet dusky. ♀: mostly dark slate-gray, blacker on face and throat; wing-coverts and underside of tail black, narrowly barred with white; belly and crissum deep orange. Upper mandible mostly black, lower dark gray. **Immatures:** similar to adult female but rectrices more pointed, below more irregularly barred; little (male) or no (female) orange on belly.

Habits. Frequents canopy of tall rain forest, sometimes descending to upper understory or venturing into adjacent tall second growth or semi-open to visit fruiting trees and shrubs; usually solitary or in pairs, sometimes family groups following breeding; when excited, male rapidly spreads and folds tail, flashing white; plucks fruits or insects from foliage on an outward dart, without alighting.

VOICE. A prolonged, rolling series of barking notes, first level, then accelerating and falling abruptly, sometimes terminating with several widely spaced notes on a lower pitch; also an accelerated series of soft notes; clearer and more melodious than call of Slaty-tailed Trogon, which does not change pitch; a liquid low twitter or ripple between members of pair during courtship and nest-building; a sharp cackle when agitated.

NEST. A rounded, unlined chamber entered via an ascending tunnel, 6–50ft (2–15m) up in a large decaying trunk in forest or adjacent clearing. Eggs 2–3, white. April–August.

STATUS. Fairly common resident where forest remains, in lowlands and foothills of S Pacific slope, locally to 4000ft (1200m); regularly N to Carara, occasionally as far as Orotina.

RANGE. Costa Rica and adjacent SW Panama.

NOTE. Also called Vermilion-breasted Trogon. Sometimes considered a race of the White-tailed Trogon, *T. viridis*, of Panama and South America, which has a yellow belly; the female's tail is mostly white below with a little bit of barring at the base; the usual call does not drop strongly in pitch. This species has recently been reported from W Bocas del Toro, Panama (N. G. Smith), and may appear in the Sixaola region of Costa Rica.

BLACK-HEADED TROGON

Trogon melanocephalus Pl. 26(10)

Trogón Cabecinegro (Viuda Amarilla)

DESCRIPTION. 10½″ (27cm); 90g. The only yellow-bellied Costa Rican trogon with an unbarred tail. **Adult ♂:** head, chest, and wings dull black; back iridescent green; rump, and upper tail-coverts deep blue; tail above bluish-green with black tip; belly and crissum deep yellow, fading to whitish adjacent to dark chest; tail below black, the outer 3 rectrices with broad, square white tips. Bare eye-ring pale blue; bill pale blue-gray; feet dusky. ♀: mostly blackish-slate, darkest on throat and face; wings blackish; below like male except white tips to rectrices narrower; most of upper mandible black. **Immatures:** resemble adults of same sex but duller; white tips to rectrices more extensive on outer web, which often shows some irregular white barring medially; secondaries edged with white spots. **Young:** mostly grayish-brown; wing-coverts spotted with buff; posterior underparts whitish, blotched with dusky.

HABITS. Frequents deciduous woodland and evergreen gallery forest, sometimes tall second growth, alone, or in small parties of both sexes when competing for mates; several may gather at fruiting trees; plucks insects and fruits in fluttering, hovering sallies; takes arillate seeds like *Trichilia*, berries, drupes like *Spondias*.

VOICE. A rapid, rattling series of clear, barking notes that accelerates into a chuckling trill that falls in pitch; harsher, more nasal, less resonant than call of Baird's Trogon; also low whining notes while carving nest cavity; a more steadily repeated, barking *chu chu chu. . . .*

NEST. An unlined chamber carved into the heart of an occupied termitary 5–25ft (1.5–8m) up in tree, on fence post, or other low support. Eggs 3, white. March–July.

STATUS. Common resident of lowlands and hills of N Pacific slope, S to vicinity of Orotina and Tárcoles, and across lowlands S of Lago de Nicaragua to Río Frío district; strays S to head of Golfo Dulce, and E to Grecia along river bottoms in W Valle Central.

RANGE. E Mexico to Costa Rica.

NOTE. Sometimes considered a race of Citreoline Trogon, *Trogon citreolus*, of Mexico.

ELEGANT TROGON

Trogon elegans Pl. 26(7)

Trogón Elegante

DESCRIPTION. 11″ (28cm); 78g. Slender, long-tailed; only red-bellied trogon of northern Pacific lowlands. **Adult ♂:** face and throat dull black; rest of head, chest, and upperparts metallic green; wing-coverts finely vermiculated with black and white (appearing gray); remiges blackish, primaries edged with white; white band separates green chest from bright red posterior underparts; tail above bronzy, tipped with black; lateral rectrices finely barred with black and white, broadly tipped with white. Bare eye-ring orange; bill yellow; feet pale horn. ♀: brown where male is green; white bar on auriculars; white eye-ring interrupted below eye; central rectrices cinnamon-rufous; breast-band pale buff; central underparts white; only sides, flanks and crissum red; tail below like male but inner webs of lateral rectrices largely uniform black. **Immatures:** like respective adults but duller, with rectrices longer, slimmer, lateral ones more coarsely barred, less distinctly white-tipped; male has center of belly white, faint auricular bar, central rectrices smudged with rufous toward tip. **Young:** mostly dull brown; posterior underparts mottled and barred with brown and white; wing-coverts spotted with white.

HABITS. Usually in taller and denser dry forest more than Black-headed Trogon, especially on hillsides and in ravines and gullies, also evergreen forest; usually keeps well up in canopy, coming lower to nest or feed; sallies to snatch berries, arillate seeds, insects, caterpillars, from foliage.

VOICE. A hoarse, throaty, downslurred *kahr, kowr*, or *khowur*, repeated ca. 5–10 times, then a pause before next series; begging young give a longer, squalling *kowwwrr*.

NEST. An unlined (rarely lined?), woodpecker-like cavity in a low rotting trunk 6–20ft (2–6m) up; may actually occupy, perhaps enlarge, old woodpecker hole. Eggs 3 (to 4?), white or bluish-white. April–July (often 2 broods).

STATUS. Uncommon to locally common resident of N Pacific lowlands and foothills, from Nicaraguan border S to around Bagaces, the head of Golfo de Nicoya, and the base of Península de Nicoya; to ca. 2500ft (750m) on slopes of Cordillera de Guanacaste.

RANGE. SE Arizona to Costa Rica.

NOTE. The northernmost race of this species has been called Coppery-tailed Trogon.

COLLARED TROGON

Trogon collaris Pl. 26(5)

Trogón Collarejo (Viuda Roja, Quetzal Macho)

DESCRIPTION. 10″ (25cm); 70g. Medium-sized, with conspicuous white breast-band and red belly; no overlap with Elegant Trogon. **Adult ♂:** forehead, face, and throat black; rest of head, neck, chest, and upperparts metallic green, including black-tipped central rectrices; wings black, the coverts narrowly barred and vermiculated with white; lateral rectrices black, narrowly barred with white, the terminal bar widest. No conspicuous bare eye-ring; bill yellow; feet grayish to pale brown. ♀: olive-brown where male is green; tail above dark rufous, tipped with black; wing-coverts olive-brown, finely vermiculated with black; interrupted white eye-ring; sides of breast more or less brownish; belly duller red; lateral rectrices gray, finely freckled with white, and with black subterminal bars, white tips; culmen black, rest of bill pale yellow. **Immatures:** resemble respective adults but duller, rectrices narrower, more heavily and coarsely marked with white. **Young:** mostly rich brown, mixed with buff on breast, shading to deep buff on belly; crissum rufous; wing-coverts spotted with buff.

HABITS. Prefers humid mountain forests, frequenting the lower canopy and upper understory, coming lower along sides and in adjacent tall second growth; usually singly or in pairs, sometimes in small groups both before and after breeding season; somewhat more insectivorous than most trogons, taking beetles, homopterans, caterpillars (including spiny types), crickets; often lurks on outskirts of mixed-species foraging flocks, probably to catch prey startled out of concealment by other birds; also takes many small fruits.

VOICE. Usual call 2–4 clear, mellow, descending notes, *cow cao* or *cow cao-cao*, the first note slightly lower in pitch; occasionally several accelerating notes follow the first *cow*, the call ending in a chuckling trill; when perturbed, a prolonged *chur-r-r-r* while raising and lowering tail.

NEST. A shallow, unlined niche that leaves most of the sitting bird exposed or a more enclosed cavity 4–16ft (1.2–5m) up in a usually slender decaying stub. Eggs 2, white. January–April.

STATUS. Common resident of middle and upper elevations on Cordillera Central and Cordillera de Talamanca; breeds upward from 2300ft (700m) on Caribbean slope, and 3000ft (900m) on Pacific slope, to at least 7500ft (2300m); wanders occasionally up to at least 9200ft (2800m) and, following breeding, down to 500ft (150m) on Caribbean slope and to 2000ft (600m) on Pacific.

RANGE. Tropical Mexico to NW Ecuador and N Bolivia.

NOTE. Also called Bar-tailed Trogon.

ORANGE-BELLIED TROGON

Trogon aurantiiventris Pl. 26(4)

Trogón Vientrianaranjado

DESCRIPTION. 10″ (25cm); 70g. Similar to corresponding sexes of Collared Trogon but with orange instead of red posterior underparts.

HABITS. Similar to those of Collared Trogon; fruits include *Hasseltia*, *Rubus*, and species of Ericaceae, Myrtaceae, Rubiaceae, Lauraceae, and Symplocaceae (Wheelwright et al.).

VOICE. Identical to that of Collared Trogon.

NEST. An unlined cavity in rotten stub, 6–16ft (2–5m) up. Eggs 2, white. March–May.

STATUS. Uncommon to fairly common resident of middle and upper levels of Cordillera de Tilarán and Cordillera de Guanacaste, above ca. 2500ft (750m); rare and local at ca. 3000–6000ft (900–1850m) on Cordillera Central, Dota area, and Cordillera de Talamanca.

RANGE. Costa Rica and W Panama.

NOTE. Although the AOU Check-List treats the Orange-bellied Trogon as a full species, it has long been questioned whether it is but a partially localized color phase of Collared Trogon. No other partly sympatric trogon species are so similar in voice, plumage, and ecology; moreover, in some areas (e.g., Caribbean side of Volcán Miravalles, San Lorenzo area between Cordillera Central and Cordillera de Tilarán), many birds have darker red-orange bellies approaching those of *collaris*, whereas occasional *collaris* in many areas have paler, more orangish bellies. Although we strongly suspect that *aurantiiventris* is simply a color phase, pending an adequate study of the living birds, we here follow the current practice and treat it as a species.

BLACK-THROATED TROGON

Trogon rufus Pl. 26(9)

Trogón Cabeciverde

DESCRIPTION. 9″ (23cm); 57g. A small, yellow-bellied trogon with green head and chest, barred tail. **Adult ♂:** face and throat black; rest of head, neck, chest, and upperparts bright metallic green, the central rectrices blue-green with black tips; wing-coverts finely vermiculated with black and white, appearing gray; posterior underparts rich yellow; lateral rectrices evenly barred with black and white, with broader white tips. Bare eye-ring pale blue; bill greenish-yellow; feet grayish. ♀: head, neck, chest, and upperparts olive-brown, the wing-coverts finely vermiculated with black; central rectrices rufous, tipped with black; lateral rectrices like those of male; below duller yellow than male, sides of breast washed with brown. Bare eye-ring blue-gray; culmen black. **Immatures:** like adults but duller, lateral rectrices more pointed, more coarsely barred; young male has throat, breast, and wing-coverts largely brown, central rectrices tipped indistinctly with dull rufous.

HABITS. Prefers deep shade of lower and mid-levels of humid forest and tall second growth, sometimes cacao plantations or other semi-open with dark, dense understory; flies up, snatches fruit or insects (beetles, crickets, caterpillars), from vegetation; alone or in pairs.

VOICE. When cautious or alarmed, a churring or rattling *krrrr* or a nasal *nyurrrrrp*, while raising and lowering tail; song, 2–5 clear, mellow, descending whistles in a deliberate series, the later notes lower in pitch (reverse of song of Collared Trogon).

NEST. A shallow, unlined niche that leaves most of sitting bird exposed, 2½–20ft (0.75–6m) up in slender decaying stub. Eggs 2, white. February–June.

STATUS. Common resident of Caribbean slope (except drier lowlands S of Lago de Nicaragua) and S Pacific slope, N to Carara, and in smaller numbers to near Puntarenas; has been recorded, undoubtedly as a stray, from near head of Golfo de Nicoya; reaches Pacific slope of Cordillera de Guanacaste at lower middle elevations via low passes; lowlands locally to 3300ft (1000m).

RANGE. Honduras to W Ecuador and NE Argentina.

VIOLACEOUS TROGON

Trogon violaceus Pl. 26(8)

Trogón Violáceo

DESCRIPTION. 9″ (23cm); 56g. Small, yellow-bellied, tail barred; differs from preceding species in colors of head and breast, voice. **Adult ♂:** head dull black, shading to metallic violet-blue on neck and chest; rest of upperparts metallic bluish-green; central rectrices black-tipped; wing-coverts finely vermiculated with black and white, appearing gray; posterior underparts bright yellow, fading to whitish at dark chest; lateral rectrices barred evenly with black and white, with broader white tips. Bare eye-ring yellow; bill pale gray; feet dark gray. ♀: head, neck, chest, and upperparts dark gray, darkest on face; incomplete white eye-ring; wing-coverts blackish, faintly and finely barred with white; below duller yellow than male, sides of breast washed with gray; inner webs of lateral rectrices largely black; upper mandible mostly blackish. **Immatures:** resemble adult female (except that male has green back and central rectrices) but lateral rectrices more pointed, more coarsely barred.

HABITS. In dry and moist regions, prefers canopy, especially evergreen gallery forest; in humid areas frequents forest edge, semi-open, clearings with scattered tall trees, and tall second growth, at or above medium heights, rarely low; sallies, hovers to pluck fruits and insects from vegetation; may perch near wasp nest, seize wasps on the wing; alone or in pairs.

VOICE. A long series of clear, downslurred

whistled notes on same pitch: *kwer kwer kwer kwer . . .* , or *kew kew kew . . .* ; also a disturbance note similar to but more resonant than that of Black-throated Trogon, a nasal *nyrrrrp* while raising and lowering tail.

NEST. In a chamber carved in a high gray, top-shaped arboreal nest of *Parachartergus* wasps, less often in a blackish, stalactitelike, arboreal nest of *Azteca* ants, in a termitary, in a decaying trunk, or in a dense mass of roots of a large fern or other epiphyte. Eggs 2–3, white. March–June.

STATUS. Resident countrywide from lowlands up to ca. 2700ft (830m) on Caribbean slope, 4000ft (1200m) on Pacific side, including W end of Valle Central; uncommon and local in dry NW, common in more humid areas.

RANGE. Tropical Mexico to W Ecuador, N Bolivia, and Amazonian Brazil.

NOTE. Also called Gartered Trogon.

ORDER **Coraciiformes:** Kingfishers, Motmots, and Allies

Members of this order present considerable diversity of body forms, beaks and habits, but all are characterized by syndactylous feet, with the 2 outermost of the 3 forwardly directed toes joined for much of their length; only the first toe is directed backward.

FAMILY **Alcedinidae:** Kingfishers

Most of the 90 species of kingfishers live in the tropics of the Eastern Hemisphere, from West Africa to Polynesia, with a few extending into the temperate zones. The New World has only 6 species, all of which occur in Costa Rica. Though found from Alaska to Tierra del Fuego, New World kingfishers are also best represented in the tropics. In size kingfishers are 4–18″ (10–46cm). They have short necks, large heads, bills that are usually long, stout, straight, and sharp-pointed, and short legs. Many Old World kingfishers are brilliantly colored or boldly patterned in black and white; some have long, racket-tipped tails, recalling the New World motmots. Most are forest dwellers that eat insects and other invertebrates and sometimes small vertebrates. The New World species are more quietly clad in dark green or blue-gray, with white and/or chestnut underparts; all are primarily fishers, although they may also take insects and crustaceans. Many Old World species nest in cavities in termitaria or holes in trees, but all New World species dig burrows into earthen banks, usually along streams. Both sexes, working alternately, excavate the burrows, where, in an enlarged, unlined terminal chamber, the 3–8 white eggs are laid. Both parents incubate, taking long turns of up to 24 hours in the larger species. After an incubation period of 19–26 days the young hatch, blind and naked. They are attended by both parents, who do not practice nest sanitation, so that the nest chamber becomes littered with fish bones and other wastes. The young fledge when 25–38 days old and remain dependent upon the parents for much of their food for up to several weeks thereafter.

RINGED KINGFISHER

Ceryle torquata Pl. 27(1)

Martín Pescador Collarejo

DESCRIPTION. 16″ (41cm); 290g. Very large; conspicuous bushy crest; long, stout bill. **Adult ♂:** above blue-gray, the crown and crest finely streaked with black; wing and tail feathers black, spotted and barred with white, and broadly edged with blue-gray; throat, broad collar encircling neck, spots before and behind eye, and crissum white; rest of underparts deep rufous. Bill blackish, shading to dark horn on base and sides of lower mandible; feet brownish. ♀: similar, but chest slate-blue, separated by a narrow band of white from rufous belly. **Immatures:** resemble adult female, but slaty feathers of chest edged and suffused with rufous; white breast-band more distinct on female; belly paler rufous; upperparts finely streaked with black.

HABITS. Frequents deep, smoothly flowing lowland streams, shores of lakes and lagoons of fresh or salt water, estuaries, and broad tidal channels; plunges for fishes from overhanging branch, snag, or electric wire, in general choosing more elevated perches than other kingfishers; flies high with regular wingbeats between bodies of water; solitary or in pairs.

VOICE. A harsh, loud, hard *kleck*, uttered in flight; a loud, rapid, mechanical rattle, especially when disturbed, heavier and more explosive than call of Belted Kingfisher; members of pair on a perch alternate low chatters while raising and lowering tail.

NEST. A burrow in a bank, ca. 6–8ft (2–2.5m)

long, 6″ wide by 4″ high (15×10cm), with unlined chamber at end. Eggs 3–5, white, glossy. Dry season when streams are low.

STATUS. Fairly common resident throughout the lowlands of both slopes, occasionally up to ca. 3000ft (900m).

RANGE. S Texas to Tierra del Fuego; Lesser Antilles.

NOTE. Has sometimes been placed in the genus *Megaceryle*.

BELTED KINGFISHER

Ceryle alcyon Pl. 27(2)

Martín Pescador Norteño

DESCRIPTION. 12″ (30.5cm); 150g. Resembles Ringed Kingfisher but much smaller, with mostly white underparts. **Adult ♂:** upperparts, including bushy crest, and broad breast-band blue-gray; remiges and rectrices black, edged with blue-gray; secondaries and rectrices barred with white; a white patch at base of outer primaries, conspicuous in flight; underparts, collar encircling neck, and spot in front of eye white; flanks blue-gray. Bill black, except base of mandible gray; feet blackish. ♀: similar but with a deep rufous band across lower breast. **Immatures:** resemble respective adults but with the blue-gray breast-band mixed or suffused with cinnamon-rufous.

HABITS. Frequents coastlines, shores of rivers, lakes, estuaries; establishes individual territory upon arriving in fall, to the accompaniment of much chattering and chasing; sometimes displaced by the larger Ringed Kingfisher; plunges for fishes from an elevated perch or after hovering 16–20ft (5–6m) above water.

VOICE. A loud, dry, prolonged rattle "like a fishing reel," in interactions or when flying about in territory.

STATUS. Fairly common migrant, especially along Caribbean coast; locally common winter resident, especially in Río Frío district and in coastal areas; also on Cocos I.: arrives by mid-September; departs by late April, occasionally early May.

RANGE. Breeds from C Alaska and N Canada to N Mexico; winters from C USA to N South America, Bermuda, West Indies, and Galápagos Is.

AMAZON KINGFISHER

Chloroceryle amazona Pl. 27(4)

Martín Pescador Amazónico

DESCRIPTION. 11½″ (29cm); 110g. Largest green-backed kingfisher, with notably long, heavy bill, smooth crest. **Adult ♂:** above dark metallic green, extending to sides of breast; inner webs of rectrices and remiges spotted with white; white spot below eye; breast deep rufous; rest of underparts and narrow nuchal collar white. Bill and feet black. ♀: similar, but lacks rufous, with a more or less incomplete green breast-band; sides of breast spotted with green. **Immatures:** resemble adult female but wing-coverts spotted with white; breast of male tinged with chestnut. See Green Kingfisher.

HABITS. Prefers fast- or slow-flowing rivers and large creeks, also channels in mangroves, estuaries, and lake shores; solitary or in pairs; plunges for fishes from perch or after hovering above water.

VOICE. A harsh, sharp *klek* or *zeck*; a woody, staccato *chrit* alone or in sputtering series, becoming a sharp, dry rattle or chatter that may accelerate practically into a drone when perturbed; during courtship or early in breeding season, song is a fast series of whickering whistled notes, rising then falling in pitch: *see see see . . . sue sue sue sue . . .* or *joy joy joy joy. . . .*

NEST. An unlined burrow in a stream bank, 4–5ft (1.2–1.5m) long, ca. 4″ wide by 3″ high (10×7.5cm). Eggs 3–4, glossy white. January–March.

STATUS. Common resident, especially along rivers, from both coasts locally to ca. 3000ft (900m).

RANGE. Tropical Mexico to N Argentina.

GREEN KINGFISHER

Chloroceryle americana Pl. 27(5)

Martin Pescador Verde

DESCRIPTION. 7″ (18cm); 37g. Smaller than Amazon Kingfisher, with white spotting on wings, conspicuous flash of white in tail. **Adult ♂:** above dark metallic green, outer webs of remiges spotted with white; inner webs of rectrices extensively white; breast deep rufous; remaining underparts and narrow nuchal collar white; lower breast spotted with green. Bill black except pale gray along gonys; feet dark gray. ♀: similar but lacks rufous; a green band across chest, and a second, more spotted, sometimes incomplete band across lower breast. **Immatures:** resemble respective adults but breast markings reduced; rufous of male paler, reduced to narrow band; green bands of female often incomplete, the posterior one reduced to a few spots on sides.

HABITS. Frequents small woodland streams,

marshes, and rain pools, as well as borders of wider waterways; alone or in pairs; plunges from low twigs, emergent rocks, or fence wires for small fishes, sometimes aquatic insects.

VOICE. A dry, rasping *dzeew*, often combined with a dry, snapping sputter when excited: *dzeew dzeew kuk-kuk-kuk dzeew*; a subdued *tick tick tick*, much quieter than calls of larger kingfishers.

NEST. A burrow 2–3ft (60–90cm) long in a stream bank, usually screened by vines or exposed roots. Eggs 3–6, glossy white. February–April.

STATUS. Widespread, locally common resident from lowlands of both slopes up locally to 4000ft (1200m), occasionally higher.

RANGE. C Texas to N Chile and C Argentina.

GREEN-AND-RUFOUS KINGFISHER

Chloroceryle inda Pl. 27(3)

Martín Pescador Vientrirrufo

DESCRIPTION. 8½″ (21.5cm); 60g. The only New World kingfisher lacking white on belly or crissum. **Adult ♂:** above dark metallic green, with fine whitish to rufous spots on wings and tail; cheeks and throat rich buff, shading to deep rufous over rest of underparts; narrow rufous streak over lores; narrow nuchal collar rufous. Bill black, except gonys flesh-color; feet grayish. ♀: similar but with a dark green chest-band, speckled with white. **Immatures:** wing-coverts, wings, and tail spotted more densely with white to buff; green breast-band heavily spotted and suffused with rufous, especially in male.

HABITS. Prefers forest swamps, less often small forest streams, keeping to deep shade where often difficult to see; plunges from low twig or vine for small fishes, aquatic insects; solitary or in pairs.

VOICE. A rolling, hard, sharp *drrrt* with an antbirdlike quality; a dry, staccato *dit-dit-dit* or (Ridgely) a crackling *trit-trit-trit*; song of shrill, thin, high-pitched notes *week week week* . . . or *wick wick wick wick* . . . , higher and thinner than corresponding notes of Amazon Kingfisher.

NEST. Undescribed (?). Probably dry season (molt and gonad data).

STATUS. Uncommon to rare and local resident of Caribbean lowlands, especially coastal swamps (e.g., Tortuguero, Cahuita, Manzanillo); reported from Río Frío region and Sarapiquí lowlands (La Selva).

RANGE. SE Nicaragua to W Ecuador, C Bolivia, and C Brazil.

AMERICAN PYGMY KINGFISHER

Chloroceryle aenea Pl. 27(6)

Martín Pescador Enano

DESCRIPTION. 5″ (13cm); 18g. Very small; rufous underparts with contrasting white belly. **Adult ♂:** above dark metallic green; secondaries sparsely spotted with buff; lateral rectrices with white bases, inner webs spotted with white; cheeks and throat deep orange-buff, shading to rich dark rufous on breast, sides, and flanks; center of belly and crissum white. Bill black, except gonys yellowish; feet dusky. ♀: similar, but broad breast-band dark green. **Immatures:** resemble the respective adults but below much paler and duller; wing-coverts speckled with rufous; male has traces of green breast-band.

HABITS. Frequents small woodland streams, pools, puddles, and small channels in mangroves, usually staying in shade; plunges for small fishes and aquatic insects from low perch; catches damselflies and other insects in the air on fast, buzzy sallies; solitary or in pairs, often quite tame; incessantly pumps tail, sometimes simultaneously bobbing head.

VOICE. A soft, dry, staccato ticking, sometimes accelerating into a rattle or buzz; song may be series of more musical chirps.

NEST. A burrow about 1ft (30cm) long, in an earthen bank by a stream or path or in mass of clay heaved up by an uprooted tree, not always near water. Eggs 3–4, white. March–April, possibly through June.

STATUS. Uncommon but widely distributed resident of lowlands of both slopes, occasionally to 2000ft (600m).

RANGE. Tropical Mexico to W Ecuador, C Bolivia, and C Brazil.

NOTE. Usually known as the Pygmy Kingfisher.

FAMILY **Momotidae:** Motmots

The 9 species of motmots are small to middle-sized arboreal birds confined to tropical America, where they are best represented in Mexico and Central America. Their bills, rather short, broad, and slightly downcurved, have more or less serrated tomia. They are beautifully clad in blended shades of green, blue, and rufous; the sexes are alike or nearly so. On most (all Costa Rican species except the Tody Motmot), the 2 long, central tail feathers, completely vaned when they first grow out, become racket-tipped by the loss of the loosely attached barbs above the terminal disk or spatule, leaving a length of naked shaft—the birds appear not to trim their tails deliberately. Motmots frequent a variety of habitats from high rain forest and arid scrub to shady plantations and suburban gardens. Solitary or in pairs, they perch long in the shade, often swinging their tails pendulumlike from side to side, and dart out or down to capture insects, other invertebrates, or small reptiles. Most vary their diet with fruits. Their notes range from dull and wooden to ghostly hoots and delightfully melodious undulations. They nest in unlined burrows that both sexes dig in roadside or stream banks or in the side of an obscure pit or mammal's den. Some occupy niches in masonry or the sides of wells in limestone country. They lay 2–5 glossy round white eggs, which both sexes incubate for 18–22 days, sitting for many hours without interruption. Both feed the naked newly hatched young. When 24–31 days old, they are well feathered and fly from the burrow.

TODY MOTMOT

Hylomanes momotula Pl. 27(12)

Momoto Enano

DESCRIPTION. 6½″ (16.5cm); 32g. Very small; tail without racquets; dark with white belly and mustache. **Adults:** forehead and forecrown olive-green, shading to dull dark rufous over rest of crown and hindneck; rest of upperparts dark dull green; wings and tail more olive; a short blue-green streak above eye; area around eye and from eye to auriculars black; lores, nasal tufts, and line below black mask buffy white; cheeks, throat, and breast olive-green, tinged or mixed with brownish, with a broad dull white stripe on each side of throat (sometimes reduced in male); lower breast tinged with bluish, shading to dull white on belly. Upper mandible black, lower horn-color, shading to black on tomia and tip; feet pale olive. **Young:** much duller; forehead, crown, and hindneck mostly gray-brown; throat dull olive, streaked with buff, with little or no white; breast and sides gray-brown; belly pale buff; little blue over eye.

HABITS. Frequents shady understory of moist evergreen forest, especially along dark ravines; solitary or in pairs; perches quietly at about eye level, sallying out to pluck insects, spiders, or snails from vegetation; catches butterflies (including *Morpho*) and dragonflies in flight.

VOICE. A soft, high sputtering like a small squirrel, evidently when alarmed; also "a far-carrying, resonant *kwa-kwa-kwa-kwa* . . . reminiscent of Prong-billed Barbet" (Ridgely).

NEST. Undescribed (?). A juvenile just out of nest in late May.

STATUS. Very uncommon and local resident of Pacific slope of Cordillera de Guanacaste, at ca. 1600–3300ft (500–1000m), barely reaching Caribbean slope through passes, as at Bijagua.

RANGE. S Mexico to W Colombia.

BROAD-BILLED MOTMOT

Electron platyrhynchum Pl. 27(9)

Momoto Piquiancho

DESCRIPTION. 12″ (30.5cm); 60g. Small; appears very big-headed in the field; broad, flat bill with keeled culmen. **Adults:** chin blue-green; narrow black mask extending to auriculars; black spot on chest; rest of head, neck and breast cinnamon-rufous, darkest on crown; belly bluish-green; above fairly dark green; primaries and racket-tipped tail bluish-green. Bill black; feet dusky. **Young:** darker and duller, especially below, where rufous largely replaced by dark olive-green; crown washed with olive; short blue-green streak above eye; bill with whitish tip.

HABITS. Prefers humid forest and older second growth, shady semi-open; usually perches and calls from lower canopy to upper understory, but comes lower to feed on fruits of *Heliconia*; primarily insectivorous, sallying to pluck cicadas, other insects, spiders, centipedes, small frogs and lizards from vegetation, sometimes from ground; eats many of the giant, fiercely stinging ant, *Paraponera*; catches butterflies and dragonflies in flight; alone or in pairs.

VOICE. Usual call a deep, hoarse, croaking, ventriloquial *cwaa* or *kawk*, often doubled, with something of the quality of a train whistle, most often heard at dawn; sometimes, especially during breeding season, 1 bird will give longer, higher-pitched *kawk kawk* . . . , while a second gives a lower, faster *kwak-kwak-kwak*. . . . Fledglings utter soft, mellow, almost soprano notes.

NEST. A burrow 33–39″ (0.8–1m) long in a vertical bank, the entrance screened by vegetation, with an enlarged, unlined terminal chamber. Eggs 2–3. February–April or May.

STATUS. Fairly common resident of Caribbean lowlands and foothills the length of the country, except drier areas S of Lago de Nicaragua; to 3300ft (1000m) in N and 5000ft (1500m) in

Talamancas; reaches Pacific slope of Cordillera de Guanacaste.

Range. N Honduras to W Ecuador, C Bolivia, and C Brazil.

KEEL-BILLED MOTMOT

Electron carinatum Pl. 27(10)

Momoto Pico Quilla

Description. 12½″ (32cm); 65g. Resembles Broad-billed Motmot in size and shape but mostly green, only forehead rufous; bright blue superciliaries longer than young Broad-billed. Forehead deep rufous, abruptly separated from dark green of crown; narrow black mask extending to auriculars; chin blue-green; large breast-spot black; rest of head and body green, above darker and more olive; lower breast and belly tinged with cinnamon; remiges and rectrices bluish-green; racket-tipped tail. Bill black with tip horn-color; feet dusky.

Habits. Prefers tall humid forest, especially near gullies and streams, in hilly country. In general, behavior resembles that of Broad-billed Motmot, with which it is sympatric in Costa Rica. Solitary or in pairs, perches and forages mostly at low to medium heights.

Voice. A loud, far-carrying *cut cut cadack*, resembling the cackling of a hen (Lowery and Dalquest, in Mexico); apparently louder and sharper than call of Broad-billed Motmot (Fodgen).

Nest. Undescribed (?).

Status. Rare resident of Caribbean foothills of Cordillera de Guanacaste and Cordillera de Tilarán, between 1000 and 3000ft (300–900m). In Peñas Blancas Valley of Cordillera de Tilarán, at S extreme of its distribution, an individual of this species apparently mated with a Broad-billed Motmot; no young have been produced (Fodgen). Reports of the present species farther S require confirmation because of possible confusion with young Broad-billed Motmots.

Range. SE Mexico to Costa Rica.

TURQUOISE-BROWED MOTMOT

Eumomota superciliosa Pl. 27(11)

Momoto Cejiceleste (Pájaro Bobo)

Description. 13½″ (34cm); 65g. Fairly small; striking pattern; large tail racquets on very long naked shafts. **Adults:** narrow black mask extending to auriculars; broad superciliaries and narrow streak below front of mask pale turquoise; center of throat black, bordered laterally by turquoise stripe; center of back and area behind eye rufous; belly pale cinnamon-rufous; rest of head, neck, and body rich olive-green, darkest on crown; remiges and rectrices including racquets pale greenish-blue, boldly tipped with black. Bill and feet black. **Young:** crown dusky blue-green with paler tips; turquoise brow very short and narrow; green of body more bluish, less olive; little or no rufous on back or black and turquoise on throat; belly paler and duller; rectrices, including racquets, narrower and duller.

Habits. Prefers deciduous woodland and evergreen gallery forest, also frequenting savanna trees and low scrubby thickets, especially in wet season; sometimes at edges of mangrove swamps; perches quietly for long intervals, swinging tail pendulumlike from side to side, then darting out suddenly to seize a beetle or other insect, spider, or small lizard or snake from the ground; catches butterflies, bees, or dragonflies in flight; beats prey loudly against perch before swallowing it; flight swift and undulatory, rarely long-continued; alone or in pairs or family groups.

Voice. A rough, deep, hoarse *kowk* or *cawak*, often repeated; sometimes a longer *kup-kup-kup-kup-KOWK KOWK*; when agitated or alarmed, *howk-huk* or *howk-a-huk*.

Nest. A burrow 2–8ft (0.6–2.5m) long in earthen bank, ca. 3″ (8cm) in diameter, usually beside road or stream, with unlined terminal chamber. Eggs 3–5. May–June. Reported to nest colonially in Yucatán but does not appear to do so in Costa Rica.

Status. Common resident of lowlands and hills of N Pacific slope, locally to ca. 2600ft (800m) on slopes of Cordillera de Guanacaste; regularly S to about Tárcoles, with 1 record for Parrita area; rare and sporadic in Valle Central.

Range. SE Mexico to Costa Rica.

RUFOUS MOTMOT

Baryphthengus martii Pl. 27(7)

Momoto Canelo Mayor (Pájaro Bobo)

Description. 18″ (46cm); 195g. Very large, robust; long slim tail with rather small racquets; bill decurved, strongly serrate. **Adults:** broad mask black; small black spot on chest; rest of head, neck and breast cinnamon-rufous, shading to deeper rufous on lower breast; back, wing-coverts, and sides deep green; belly and crissum greenish-blue; rectrices and remiges dark blue. Iris dark red; bill black; feet blackish. **Young:** similar to adults but paler and duller, with no black on chest, mask duller black and smaller; lack tail racquets.

Habits. Frequents tall humid forest, shady older second growth, and semi-open, usually perching and calling from well up in canopy, but descending freely to understory to forage; eats many fruits, especially those of palms and *Heliconia*; captures insects and spiders, small frogs and lizards on sallies to ground or vegetation, or catches fishes and crabs like a kingfisher; forages with army ants; solitary or in pairs, but more often in groups than other motmots, especially at dawn.

Voice. Hollow, resonant, owl-like hooting notes, often 2–3 loud notes followed or answered by a rapid, rolling series of several more on a lower pitch: *hoop hoop huhuhuhuhu*, often heard as a chorus at dawn; in agitation or aggressive situations, a loud, harsh, dry chatter, like that of a large squirrel.

Nest. A long, often winding burrow 12–16ft (4–5m) long, in a stream bank, or in the side of a pit or mammal's den; eggs not seen. March–June.

Status. Common resident of lowlands and foothills of Caribbean slope, locally to 3000ft (900m) in N and 4600ft (1400m) in S, near Panama.

Range. NE Honduras to W Ecuador, NE Argentina, and S Brazil.

Note. Often considered a race of the South American Rufous-capped Motmot (*B. ruficapillus*), which lacks racquet-tipped tail feathers, below is less rufous, and has a different voice (Whitney).

BLUE-CROWNED MOTMOT

Momotus momota Pl. 27(8)

Momoto Común (Pájaro Bobo)

Description. 15½″ (39cm); 120g. Large; crown and mask bordered with blue; slightly decurved, strongly serrate bill. **Adults:** center of crown and broad mask that tapers to a point on auriculars black; crown encircled by blue that is paler and more greenish on forehead, more violet behind eye; mask and small black breast-spot narrowly edged with blue; throat bluish-green; rest of upperparts green; remiges and rectrices greenish-blue; the breast and upper belly vary from green with faint olive tinge to ochraceous-olive; lower belly and crissum more bluish-green. Iris red; bill black; feet blackish. **Young:** resemble adults but entire crown suffused with greenish-blue, the borders indistinct; black mask smaller, more sooty; no breast-spot; lack tail racquets; iris dark brown.

Habits. Very adaptable, frequenting rain forest, drier woodland, wooded ravines, shady second growth, hedgerows, shady gardens, semi-open, and coffee plantations. Singly or in pairs, perches in shade at no great height, swinging long tail slowly from side to side; sallies to ground to capture large spiders, earthworms, insects, small lizards and snakes, sometimes hopping in pursuit; also plucks insects and fruits from foliage; accompanies army ants; visits feeders for bananas and bread.

Voice. A soft, resonant *hoop-hoop* or *coot coot*, heard most often at dawn, from which motmots take their name; a low, hollow *whoo-whoo* hoot, especially near burrow; in aggressive interactions a hoarse, dry, coughing bark, sometimes in series like the chatter of a large squirrel; a long series of hoots that accelerates into a gobble when alarmed.

Nest. A long, often winding burrow, 5–14ft (1.5–4m) long and ca. 4″ (10cm) in diameter, with unlined terminal chamber, in exposed bank of road or stream or in the side of a pit or hollow in ground, well concealed. Eggs 3–4. March–May.

Status. Resident throughout lowlands and middle elevations of Pacific slope, to 5000ft (1500m) in N and 7000ft (2150m) in S; common except in dry NW, where local and largely restricted to gallery forest; common on Caribbean slope from Cartago area to around Turrialba and uncommon to rare at lower middle elevations on N slope of Cordillera Central; absent from most of Caribbean lowlands, but fairly common in drier areas S of Lago de Nicaragua.

Range. NE Mexico to NW Peru and N Argentina.

Note. Also called Blue-diademed Motmot, perhaps a more appropriate name for most of the species, as only the northernmost race in Mexico has the entire crown blue.

ORDER **Piciformes:** Woodpeckers, Jacamars, Toucans, Puffbirds, and Allies

The main external character of this varied order of arboreal, hole-nesting birds is the zygodactyl foot, with the second and third toes directed forward, the first and fourth backward; occasional species in several families have lost the fourth toe.

PLATES

The birds depicted on these plates are adults unless otherwise noted. Species that occur only at certain elevations or in restricted geographic areas of Costa Rica are also noted. All birds on a plate are drawn to the same scale, except on Plates 51 and 52 and on those plates where a line divides the plate into two halves drawn to different scales; also, on some plates birds in flight are drawn to a smaller scale than those perched.

Plate 1: Large Seabirds

1. **Brown Booby** (*Sula leucogaster*), p. 76. (*a*) Hood brown; feet and gular pouch yellow. (*b*) ♂ (Pacific race): face pale gray; bill grayish; gular pouch dark. (*c*) Immature: brown; trace of hood; bill and feet duller.
2. **Red-footed Booby** (*Sula sula*), p. 77. Pink on face; feet red. (*a*) Pale phase: white with black remiges (Pacific race with black tails). (*b*) Dark phase (Pacific race): uniform brown (Caribbean race with white tails). (*c*) Immature: brown with trace of dark breast-band; feet greenish.
3. **Masked Booby** (*Sula dactylatra*), p. 76. (*a*) Large; white; more black on wings than 2a; face dark. (*b*) Immature: head and throat brown; nuchal collar white.
4. **Blue-footed Booby** (*Sula nebouxii*), p. 76. (*a*) Mantle dark brown; head and tail pale brown; white on rump, upper back; feet blue. (*b*) Immature: darker, duller, above less white; feet slaty.
5. **Great Frigatebird** (*Fregata minor*), p. 79. More compact, outer rectrices shorter than 6. (*a*) ♀: chin and throat grayish; eye-ring reddish. (*b*) Immature: head tawny-buff. (*c*) ♂: pale upperwing bar more pronounced than 6c. Cocos I.
6. **Magnificent Frigatebird** (*Fregata magnificens*), p. 79. (*a*) ♀: entire head black; eye-ring blue to purplish. (*b*) Immature: head white; bill blue. (*c*) ♂: upperwing bar brown. Larger, gawkier than 5.
7. **Red-billed Tropicbird** (*Phaethon aethereus*), p. 74. (*a*) Long tail; bill red; barred mantle; eye-stripe black. (*b*) Immature: bill yellowish; short tail black-tipped.
8. **Pink-footed Shearwater** (*Puffinus creatopus*), p. 70. Above dark, below white (not clean-cut); bill and feet pink. Cocos I.
9. **Sooty Shearwater** (*Puffinus griseus*), p. 70. Dark; wing-linings pale; long bill; often pale chin. See 13.
10. **Wedge-tailed Shearwater** (*Puffinus pacificus*), p. 70. Conspicuously long tail. (*a*) Pale phase: white below, moderately clean-cut. (*b*) Dark phase. In both, pale scaling on back.
11. **Dark-rumped Petrel** (*Pterodroma phaeopygia),* p. 69. (*a*) Above: uniformly dark; black cap with "ear flaps." (*b*) Below: note underwing, black patch on axillars.
12. **Parkinson's Petrel** (*Procellaria parkinsoni*), p. 70. Large, robust; heavy pale bill with dark tip; short tail; below pale primary shafts.
13. **Short-tailed Shearwater** (*Puffinus tenuirostris*), p. 71. Dark, including wing-linings (but sometimes pale sheen); small bill; often shows darker cap, paler throat.
14. **Audubon's Shearwater** (*Puffinus lherminieri*), p. 71. Very small, clean-cut; distinctive fluttery flight.
15. **Black-capped Petrel** (*Pterodroma hasitata*), p. 69. (*a*) Above: rump and nuchal collar white (sometimes reduced). (*b*) Below: note underwing pattern, dark bar at side of breast.

1a
1b
2a
3b
3a
2b
4a
1c
4b
2c
5a
6a
6c
5b
6b
5c
7a
7b
11b
8
9
11a
10b
10a
12
15a
13
14
15b

Plate 2: Small Terns and Storm-Petrels

1. **Gull-billed Tern** (*Sterna nilotica*), p. 159. Wing very pale; short tail; stout bill black; small dark patch behind eye (black cap in breeding plumage).
2. **Bridled Tern** (*Sterna anaethetus*), p. 160. (*a*) Breeding: above dark; narrow white forehead and short eyebrow; nuchal collar whitish. (*b*) Nonbreeding: crown white, hindcrown mottled. (*c*) Young: back scaly; shorter tail; sides of breast smudged with dusky.
3. **Sooty Tern** (*Sterna fuscata*), p. 161. (*a*) Above uniform black; broad forehead white; no eyebrow. (*b*) Immature: underparts largely blackish.
4. **Least Tern** (*Sterna antillarum*), p. 161. Very small. (*a*) Breeding: bridled head; above very white; leading edge of primaries dark; bill yellow with black tip. (*b*) Immature: pale crown; dark carpal bar. Nonbreeding adult similar, without carpal bar.
5. **Common Tern** (*Sterna hirundo*), p. 159. (*a*) Nonbreeding: from eye around nape black; wingtips below extensively dark; outer web of outer rectrices dark. (*b*) Immature: dark carpal bar; shorter tail. (*c*) Breeding: cap black; bill red.
6. **Black Tern** (*Chlidonias niger*), p. 158. (*a*) Immature: irregular dark nape-patch; dark smudge on sides of breast; dark carpal bar (lacking in winter adult); above dark, including short tail. (*b*) Breeding: head and breast black.
7. **Black Noddy** (*Anous minutus*), p. 163. Very dark; sharply defined white cap (less extensive in immature); tail often appears notched in flight. Cocos I.
8. **Brown Noddy** (*Anous stolidus*), p. 163. Larger, browner than 7; white cap blends into dark nape. Immature lacks white cap.
9. **Forster's Tern** (*Sterna forsteri*), p. 160. Silvery upperwing; black stripe from eye to auriculars (not nape); dark on inside of tail fork.
10. **White Tern** (*Gygis alba*), p. 164. Entirely white; awl-shaped black bill; appears big-eyed.
11. **White-faced Storm-Petrel** (*Pelagodroma marina*), p. 72. White face pattern and brown mantle contrast with black flight feathers, gray rump.
12. **Wedge-rumped Storm-Petrel** (*Oceanodroma tethys*), p. 72. Tail barely shows behind white tail-coverts. Very small Peruvian race along coast; larger Galápagos race at Cocos I.
13. **Least Storm-Petrel** (*Oceanodroma microsoma*), p. 72. Very small; all dark; wedge-shaped tail.
14. **Band-rumped Storm-Petrel** (*Oceanodroma castro*), p. 72. Even-width white band across rump; tail longer than 15, forked.
15. **Wilson's Storm-Petrel** (*Oceanites oceanicus*), p. 71. Square white rump-patch extends to crissum; short square tail; long legs; wings broad, straight.
16. **Leach's Storm-Petrel** (*Oceanodroma leucorhoa*), p. 73. Kinked wings; nighthawklike flight. (*a*) Dark-rumped form. (*b*) White-rumped form: dark center of rump. Intermediates: rump partly white or gray.
17. **Black Storm-Petrel** (*Oceanodroma melania*), p. 73. Very large; dark; tail deeply forked; in hand note long legs; more ternlike flight than 16.

1
2a
3a
3b
2c
2b
5b
5c
5a
4a
4b
9
7
6b
6a
10
13
12
8
11
14
15
16a
16b
17

Plate 3: Large Terns, Gulls, and Jaegers

1. **Caspian Tern** (*Sterna caspia*), p. 159. Heavy bill reddish; crown and forehead streaked; short tail; wingtips below dark.
2. **Royal Tern** (*Sterna maxima*), p. 162. (*a*) Long bill orange; forehead and forecrown white; black of hindcrown does not reach eye. (*b*) Breeding: cap black.
3. **Elegant Tern** (*Sterna elegans*), p. 163. Very long, slender bill orange; longer tail than 2; black of hindcrown reaches eye.
4. **Sandwich Tern** (*Sterna sandvicensis*), p. 162. (*a*) Long, slender bill black with yellow tip; relatively short tail. (*b*) Breeding: cap black.
5. **Franklin's Gull** (*Larus pipixcan*), p. 156. Rounder head, smaller bill than 8. (*a*) Breeding: hood black; black wingtip framed in white. (*b*) First winter: half-hood dusky; broad broken white eye-ring; incomplete black tail-band.
6. **Bonaparte's Gull** (*Larus philadelphia*), p. 157. (*a*) Breeding: small; hood black; white wedge in forewing. (*b*) First winter: head white; auricular spot black; less white in wing.
7. **Sabine's Gull** (*Xema sabini*), p. 158. Black wedge in forewing; white triangle on rear of wing.
8. **Laughing Gull** (*Larus atricilla*), p. 156. (*a*) Breeding: hood black; narrow broken white eye-ring; dark mantle blends into black wingtips. (*b*) First winter: compared with 5b, less distinct half-hood, tail-band reaches edge of tail. (*c*) Nonbreeding: indistinct half-hood; above darker; narrower eye-rings; more sloping forehead than 5. [See also Pl. 51(6).]
9. **Ring-billed Gull** (*Larus delawarensis*), p. 155. (*a*) First winter: larger; above paler than 8 with rounder head; bill pinkish with black tip; legs paler. (*b*) Ringed bill; pale mantle.
10. **Herring Gull** (*Larus argentatus*), p. 155. Much larger than 9. (*a*) Bill yellow with red spot; legs pink. (*b*) First winter: browner than 9; pinkish bill ringed or tipped with black.
11. **Black Skimmer** (*Rynchops niger*), p. 165. Like large tern; above black; rather long neck; bill with scarlet base, long lower mandible.
12. **South Polar Skua** (*Catharacta maccormicki*), p. 153. Like large, very robust and powerful dark gull with white flash in primaries. Body plumage variable in color.
13. **Pomarine Jaeger** (*Stercorarius pomarinus*), p. 153. (*a*) Pale phase: elongated central rectrices blunt, twisted; large, robust, distinct breast-band. Dark phase mostly or entirely dark brown. (*b*) Immature: larger, more white in wing, more distinctly barred rump than 14b.
14. **Parasitic Jaeger** (*Stercorarius parasiticus*), p. 154. (*a*) Pale phase: long-pointed central rectrices; often no breast-band; less white in wing than 13, more than 15. Dark phase dark brown, from 13 by tail. (*b*) Immature: from larger 13 by indistinctly barred rump.
15. **Long-tailed Jaeger** (*Stercorarius longicaudus*), p. 154. Above grayer than 14; long tail-streamers; little white in wing. Dark phase extremely rare. Immature like 14 but smaller; in hand note blue legs, black feet.

1
2a
3
4a
6a
6b
7
5a
5b
8a
8b
9a
2b
4b
8c
9b
10b
11
10a
13a
14a
12
13b
14b
15

Plate 4: Large Waterbirds

1. **Brown Pelican** (*Pelecanus occidentalis*), p. 75. Upperparts silvery-brown. (*a*) Breeding: note dark neck, red-tinged bill. (*b*) Nonbreeding: neck white; bill brownish. (*c*) Immature: browner, especially head and neck.
2. **American White Pelican** (*Pelecanus erythrorhynchos*), p. 75. Large; white with black primaries; bill yellowish. Very rare visitor.
3. **Anhinga** (*Anhinga anhinga*), p. 78. Long, sharp-pointed bill. (*a, b*) ♀: head and neck buffy. (*b*) Flying: broad wings; long tail; outstretched neck. (*c*) ♂: head and neck black. (*d*) Swimming: only neck and head above water.
4. **Olivaceous Cormorant** (*Phalacrocorax olivaceus*), p. 77. (*a*) Nonbreeding: head and neck brownish. In flight head and neck angled up, tail low. (*b*) Breeding: head and neck glossy black with white plumes. (*c*) Young: below whitish; above grayer. (*d*) Swimming: low in water.
5. **Jabiru** (*Jabiru mycteria*), p. 88. (*a*) Flying: remiges white. (*b*) Huge; plumage all white; red on lower neck; thick, sharp, slightly upturned bill.
6. **Wood Stork** (*Mycteria americana*), p. 87. (*a*) Flying: flight feathers black; outstretched neck. (*b*) Bare, blackish head and neck. (*c*) Immature: whitish down on head and neck; bill yellowish.
7. **Roseate Spoonbill** (*Ajaia ajaja*), p. 90. Only pink waterbird; note spatulate bill. (*a*) Flying: remiges pink. (*b*) Note rose-red shoulder, bare head greenish. (*c*) Immature: much paler; whitish down on head and neck.
8. **White Ibis** (*Eudocimus albus*), p. 89. (*a*) Flying: wingtips black; outstretched neck. (*b*) Bare face and decurved bill red. (*c*) Immature: wings brown; back and neck with white patches; belly white.
9. **Glossy Ibis** (*Plegadis falcinellus*), p. 89. (*a*) Flying: very dark; slender; outstretched neck; in nonbreeding plumage head and neck finely streaked. (*b*) Breeding: more slender, longer legs, neck, and bill than 10, more chestnut.
10. **Green Ibis** (*Mesembrinibis cayennensis*), p. 88. (*a*) Flying: broad, rounded wings, longish tail; flies with body held low; heavy-looking. (*b*) Thickset; green gloss; shaggy crest.

4a
4b
4c
3b
1a
3a
3c
2
3d
1b
1c
4d

6a
10a
9a
5a
8a
7a
5b
10b
9b
6b
7c
8c
6c
8b
7b

Plate 5: Herons and Limpkin

1. **Chestnut-bellied Heron** (*Agamia agami*), p. 87. Very long bill; long neck; silvery "filigree" on foreneck; belly chestnut. Immatures much browner.
2. **Boat-billed Heron** (*Cochlearius cochlearius*), p. 83. Huge shoe-shaped bill; large eyes. (*a*) Cap, crest, sides, and wing-linings black. (*b*) Immature: duller, browner; below faintly streaked.
3. **Yellow-crowned Night-Heron** (*Nyctanassa violacea*), p. 83. (*a*) Body gray; pied head. (*b*) Immature: bill thicker than 4b with upperparts more spotted, below less streaked; iris orange.
4. **Black-crowned Night-Heron** (*Nycticorax nycticorax*), p. 82. (*a*) Crown and back black; neck and underparts white; wings pale gray. (*b*) Immature: more streaked than 3b; iris yellow.
5. **Limpkin** (*Aramus guarauna*), p. 124. Brown, spotted with white; suggests young night-heron but different shape.
6. **Great Blue Heron** (*Ardea herodias*), p. 86. Very large; long neck; upperparts slaty.
7. **Reddish Egret** (*Egretta rufescens*), p. 85. Immature, dark phase: larger, more uniform, duller gray than adult 9; iris yellow; base of bill horn-color or brownish.
8. **Tricolored Heron** (*Egretta tricolor*), p. 85. Above dark gray; neck and chest purplish; neck stripes, belly, and wing-linings white.
9. **Little Blue Heron** (*Egretta caerulea*), p. 84. (*a*) Dark bluish-gray; head and neck purplish (except in second-winter birds). (*b*) Immature: white; legs greenish; base of bill gray; tips of primaries dusky.
10. **Snowy Egret** (*Egretta thula*), p. 86. (*a*) Breeding: note ornate plumes; bill and legs black; face and feet yellow. (*b*) Immature: inside of legs yellowish; face duller; no long plumes.
11. **Pinnated Bittern** (*Botaurus pinnatus*), p. 80. Buffy; neck barred with black; remiges slaty.
12. **American Bittern** (*Botaurus lentiginosus*), p. 80. Above browner than 11, neck streaked; black stripe on side of neck; remiges blackish.
13. **Cattle Egret** (*Bubulcus ibis*), p. 83. (*a*) Breeding: bill and legs reddish; buffy plumes on crown, back, chest. (*b*) Immature: bill yellow; legs blackish.
14. **Great Egret** (*Casmerodius albus*), p. 86. Large; white; long neck; bill yellow; legs blackish.
15. **Fasciated Tiger-Heron** (*Tigrisoma fasciatum*), p. 81. (*a*) Very dark; thick bill black. (*b*) Immature: averages most tawny, with most extensively white belly of any immature tiger-heron. Rocky rivers.
16. **Bare-throated Tiger-Heron** (*Tigrisoma mexicanum*), p. 82. Large; bare throat yellow. (*a*) Finely barred neck; looks olive at a distance; sides of head gray. (*b*) Immature: broadly barred with rich buff and black.
17. **Rufescent Tiger-Heron** (*Tigrisoma lineatum*), p. 81. (*a*) Neck chestnut-maroon, barred with black. (*b*) Immature: averages most rufescent of any immature tiger-heron. Wooded swamps.

1
2a
2b
3b
4b
3a
4a
5
6
7
8
9a
9b
10b
10a
11
12
13a
13b
14
16a
16b
15b
15a
17b
17a

Plate 6: Marsh and Stream Birds

1. **Green-backed (Striated) Heron** (*Butorides striatus*), p. 84. Like 2a but neck gray or buff.
2. **Green-backed (Green) Heron** (*Butorides s. virescens*), p. 84. (*a*) Back greenish; neck maroon. (*b*) Young: below heavily streaked; above dusky.
3. **Least Bittern** (*Ixobrychus exilis*), p. 81. (*a*) ♂: back black with white stripe; neck and wing-patch rich buff. (*b*) ♀: browner; neck more streaked. (*c*) Flying: wing-patches buffy.
4. **Spotted Rail** (*Pardirallus maculatus*), p. 124. Above spotted with white; below barred; bill yellow with red spot at base.
5. **Black Rail** (*Laterallus jamaicensis*), p. 126. Small; dark; mantle spotted; nape chestnut.
6. **Gray-breasted Crake** (*Laterallus exilis*), p. 127. Small; breast gray; back brown; nape rufous.
7. **Ocellated Crake** (*Micropygia schomburgkii*), p. 128. Small; buffy; above boldly spotted. Savannas, Terraba region.
8. **Uniform Crake** (*Amaurolimnas concolor*), p. 125. Mostly rufous; bill yellow-green; iris red; legs reddish.
9. **White-throated Crake** (*Laterallus albigularis*), p. 127. Back brown; neck and breast rufous; barred flanks. (*a*) Pacific race. (*b*) Caribbean race: hood gray.
10. **Yellow-breasted Crake** (*Porzana flaviventer*), p. 126. Very small; bridled face; below buffy-white; above with buff stripes.
11. **Ruddy Crake** (*Laterallus ruber*), p. 128. Below rufous; hood gray.
12. **Sora** (*Porzana carolina*), p. 126. (*a*) Neck and breast gray; face black; bill yellow. (*b*) Immature: browner, less clean-cut. [See also Pl. 51(11).]
13. **Gray-necked Wood-Rail** (*Aramides cajanea*), p. 125. Neck gray; mantle olive; breast rufous; belly black; bill greenish-yellow; legs red.
14. **Rufous-necked Wood-Rail** (*Aramides axillaris*), p. 125. Smaller than 13 with neck rufous like breast; gray patch on back. Pacific mangroves.
15. **Purple Gallinule** (*Porphyrula martinica*), p. 129. (*a*) Purple and green; frontal shield sky-blue; red bill with yellow tip; legs yellow. (*b*) Immature: buffy; wings bluish.
16. **Sunbittern** (*Eurypyga helias*), p. 131. (*a*) Striped head; slim neck; above barred; white spots on wing-coverts; bill and legs orange. (*b*) Flying: striking wing pattern.
17. **Wattled Jacana** (*Jacana jacana*), p. 132. Like 18 but plumage all black; facial wattles red; note long toes. S Pacific slope, rare.
18. **Northern Jacana** (*Jacana spinosa*), p. 132. (*a*) Head and neck black; body purplish-chestnut; bill and frontal shield yellow. (*b*) Flying: remiges yellow-green. (*c*) Immature: above brown; below white; crown and eye-stripe black.

1
2a
2b
3a
3b
3c
4
5
6
7
8
9a
9b
10
11
12a
12b
13
14
15a
15b
16a
16b
17
18a
18b
18c

Plate 7: Swimming and Diving Waterbirds

1. **American Coot** (*Fulica americana*), p. 130. Head and neck black; body dark gray; bill and frontal shield white; sides of crissum white.
2. **Common Gallinule** (*Gallinula chloropus*), p. 129. (*a*) Bill red and yellow; lateral stripe and sides of crissum white. (*b*) Immature: duller, paler.
3. **Pied-billed Grebe** (*Podilymbus podiceps*), p. 67. (*a*) Breeding: thick ringed bill white; throat black. (*b*) Immature: brownish; thick pale bill.
4. **Least Grebe** (*Tachybaptus dominicus*), p. 68. (*a*) Breeding: neck and head slaty; throat black. (*b*) Nonbreeding and immature: browner; throat whitish; eye yellow. (*c*) Young: striped head and neck. [See also Pl. 51(3).]
5. **Greater Scaup** (*Aythya marila*), p. 95. Very like 6, but in flight note long white wing-stripe. Pacific coast, accidental.
6. **Lesser Scaup** (*Aythya affinis*), p. 94. (*a*) ♂: sides white; head and chest dark; bill blue. (*b*) ♀: brown; sharply white at base of bill. (*c*) Flying: short white wing-stripe.
7. **Ring-necked Duck** (*Aythya collaris*), p. 94. Ringed bill. (*a*) ♂: back black; sides gray with white bar at front. (*b*) ♀: eye-ring white; face whitish. (*c*) Flying: wing-stripe gray.
8. **Masked Duck** (*Oxyura dominica*), p. 95. (*a*) ♂, breeding: largely chestnut; face black; bill blue. (*b*) ♀: 2 dark stripes across buffy face. (*c*) Flying: speculum white.
9. **Sungrebe** (*Heliornis fulica*), p. 130. (*a*) ♂, immature: striped head and neck; ample tail trails in water. (*b*) ♀, breeding: cheeks buffy; bill red; note banded feet. (*c*) Flying: broad wings.

1
2a
2b
3a
4a
4b
3b
4c
5
6b
6a
6c
7b
7a
7c
8b
8c
8a
9b
9a
9c

Plate 8: Ducks

1. **Northern Pintail** (*Anas acuta*), p. 93. (*a*) ♂: head brown; neck-stripe white; long tail. (*b*) ♀: slender; rather buffy; pointed tail. (*c*) Flying: speculum bronzy; slender; pointed tail.
2. **Northern Shoveler** (*Anas clypeata*), p. 94. Long spatulate bill. (*a*) ♂: head dark green; chest white; sides chestnut. (*b*) ♀: bill marked with orange; coarsely scaled. (*c*) Flying: large wing-patch blue (♂) or gray-blue (♀).
3. **Cinnamon Teal** (*Anas cyanoptera*), p. 93. ♂: chestnut with blue wing-patch; bill black; iris red. ♀ usually indistinguishable from 5b.
4. **Green-winged Teal** (*Anas crecca*), p. 92. (*a*) ♂: head chestnut and green; spotted breast; white bar on side. (*b*) ♀, flying: very small; speculum green; no wing-patch.
5. **Blue-winged Teal** (*Anas discors*), p. 93. Most abundant migrant duck. (*a*) ♂: white crescent on dark gray head. (*b*) ♀: small; above with scaly pattern; blackish bill with pale spot near base. (*c*) Flying: large wing-patch pale blue.
6. **American Wigeon** (*Anas americana*), p. 92. (*a*) ♂: crown white; sides of head green; body tinged with purplish. (*b*) ♀: head grayish; sides purplish-buff. (*c*) ♂, flying: large wing-patch white. (*d*) ♀, flying: smaller wing-patch broken, grayish-white.
7. **Black-bellied Whistling-Duck** (*Dendrocygna autumnalis*), p. 90. (*a*) Bill and feet reddish; neck and breast rusty; belly black; whitish on wing. (*b*) Immature: duller; belly gray; bill blackish. (*c*) Flying: broad wing-stripe white.
8. **White-faced Whistling-Duck** (*Dendrocygna viduata*), p. 91. (*a*) Dark; face white; back barred. (*b*) Immature: dull; grayish; breast rusty; no wing-stripe. (*c*) Flying: wings and tail dark, unpatterned. NW, rare.
9. **Fulvous Whistling-Duck** (*Dendrocygna bicolor*), p. 91. (*a*) Buffy; side stripe white; back barred with chestnut. (*b*) Flying: wings dark; white V on upper tail-coverts.
10. **Muscovy Duck** (*Cairina moschata*), p. 92. Wing-coverts white. (*a*) ♂: glossy black; crest, caruncles. (*b*) ♀, flying: slight crest; smaller than male. (*c*) Immature, flying: brownish-black; small wing-patch white.

2c
6c
5c
1c
6d
4b
3
1b
5b
1a
5a
4a
6b
2a
2b
6a
7c
7a
7b
8a
8b
8c
9b
9a
10a
10c
10b

Plate 9: Shorebirds, 1: Large Species

1. **Black-bellied Plover** (*Pluvialis squatarola*), p. 136. (*a*) Breeding: black underparts and marbled back separated by white stripe; belly white. (*b*) Winter: square head; large eyes; stout bill. (*c*) Flying: axillars black; rump white; above with white wing-stripe.
2. **Lesser Golden-Plover** (*Pluvialis dominica*), p. 136. (*a*) Breeding: like 1 but above spotted with golden-buff, belly black. (*b*) Winter: like 1 but buffier overall, superciliaries broader, bill finer. (*c*) Flying: underwing uniform gray.
3. **Short-billed Dowitcher** (*Limnodromus griseus*), p. 145. (*a*) Winter: long bill; rather short legs; rump and tail barred; superciliaries white. (*b*) Breeding: below reddish; above with cinnamon feather-edgings. (*c*) Flying: white stripe up center of back. [See also Pl. 11(5).]
4. **Long-billed Dowitcher** (*Limnodromus scolopaceus*), p. 145. Winter: very like 3a but black barring of rump and tail heavier; breast more uniform (less spotted or mottled) in all plumages. See text.
5. **Common Snipe** (*Gallinago gallinago*), p. 144. (*a*) Long bill; short legs and neck; striped head and back. (*b*) Flying: dark wings; striped back; orange tail with banded tip.
6. **Willet** (*Catoptrophorus semipalmatus*), p. 143. Large; rump white; straight, heavy bill. (*a*) Breeding: barred and spotted pattern. (*b*) Winter: above uniform grayish; below whitish. (*c*) Flying: striking wing-stripe; tail grayish.
7. **Hudsonian Godwit** (*Limosa haemastica*), p. 140. (*a*) Breeding: below chestnut; upturned bill. (*b*) Flying: white rump and narrow wing-stripe contrast with black tail and wings.
8. **Marbled Godwit** (*Limosa fedoa*), p. 140. Upturned bill with pinkish base. (*a*) Warm buffy to cinnamon; erect stance. (*b*) Flying: wing-linings rufous.
9. **American Oystercatcher** (*Haematopus palliatus*), p. 133. Bill red or orange; iris yellow; legs pink; bold pattern.
10. **Double-striped Thick-knee** (*Burhinus bistriatus*), p. 135. (*a*) Large; erect stance; superciliaries white; black stripe on side of crown; large eyes yellow. (*b*) Flying: broken wing-stripe; conspicuous tail pattern. N Pacific lowlands.
11. **American Avocet** (*Recurvirostra americana*), p. 135. Long, thin recurved bill; pied upperparts. In breeding plumage head and neck cinnamon.
12. **Black-necked Stilt** (*Himantopus mexicanus*), p. 134. Black and white. (*a*) Very long legs and neck; slim straight bill. (*b*) Flying: wings and back black; rump white.
13. **Long-billed Curlew** (*Numenius americanus*), p. 141. (*a*) Larger, buffier, much longer-billed than 14; head not striped. (*b*) Flying: wing-linings cinnamon.
14. **Whimbrel** (*Numenius phaeopus*), p. 140. (*a*) Large; brownish; striped head; fairly long decurved bill. (*b*) Flying: relatively unpatterned—no wing-stripe or contrasting wing-linings.

1a
2a
3a
4
5a
1b
2b
3b
5b
3c
1c
2c
7b
9
6c
8b
6a
7a
8a
6b
14b
10b
13b
12b
10a
13a
11
12a
14a

Plate 10: Shorebirds, 2: Small Plovers, Phalaropes, and Rocky Shore Species

1. **Killdeer** (*Charadrius vociferus*), p. 138. (*a*) Fairly large; 2 breast-bands. (*b*) Flying: strong wing-stripe; orange-buff rump; tail with banded tip.
2. **Wilson's Plover** (*Charadrius wilsonia*), p. 138. Thick bill black; square head; legs flesh-color. (*a*) ♂, breeding: complete breast-band black. (*b*) Winter; indistinct, often broken brown breast-band; no black in face; breeding ♀ similar but with cinnamon on face and sides of neck. (*c*) Flying: large head and bill; narrow wing-stripe.
3. **Collared Plover** (*Charadrius collaris*), p. 139. (*a*) Long legs flesh-color; slender bill black; sloping forehead; breast-band black. (*b*) Flying: faint wing-stripe.
4. **Snowy Plover** (Charadrius alexandrinus), p. 138. Above pale; bill and legs black; no breast-band or conspicuous facial pattern.
5. **Semipalmated Plover** (*Charadrius semipalmatus*), p. 137. Stubby orange bill with black tip; legs orange. (*a*) Breeding: facial markings and breast-band black. (*b*) Winter: these markings dusky or dark brown. (*c*) Flying: bold wing-stripe.
6. **Ruddy Turnstone** (*Arenaria interpres*), p. 144. Short wedge-shaped bill; short legs orange. (*a*) Breeding: harlequin head-pattern; upperparts rufous. (*b*) Winter: dusky smudge on chest; above gray-brown. (*c*) Flying: bold wing pattern.
7. **Surfbird** (*Aphriza virgata*), p. 145. Rather pigeonlike, with short bill. (*a*) Winter: above and breast dark gray; legs yellow. (*b*) Flying: bold wing-stripe and rump white; tail black with white tip.
8. **Wandering Tattler** (*Heteroscelus incanus*), p. 143. (*a*) Breeding: below uniformly barred. (*b*) Winter: above dark gray; chest grayish; superciliaries white; legs yellow-green. (*c*) Flying: dark, unpatterned.
9. **Wilson's Phalarope** (*Steganopus tricolor*), p. 152. (*a*) ♀, breeding: crown pearl-gray; striking black and chestnut stripes; male much duller, smaller. (*b*) Winter: above gray, including crown; below white; needle bill; eye-stripe faint. (*c*) Winter, flying: wings dark; rump white.
10. **Red-necked Phalarope** (*Phalaropus lobatus*), p. 151. Winter. (*a*) Swimming: back with pale stripes; nape and eye-patch black; forecrown white; needle bill. (*b*) Flying: back and wings striped.
11. **Red Phalarope** (*Phalaropus fulicarius*), p. 151. Winter. (*a*) Swimming: from 10 by unmarked gray back, thicker bill with (usually) some yellow at base. (*b*) Flying: back unmarked; wing-stripe.

1a
2b
4
2a
3a
5b
5a
1b
5c
2c
8c
3b
7b
6c
8a
6a
7a
8b
6b
9a
9c
10b
11b
9b
10a
11a

Plate 11: Shorebirds, 3: Sandpipers

1. **Ruff** (*Philomachus pugnax*), p. 150. (*a*) Winter: like a thickset yellowlegs but buffier; above scaled with buff; legs orangish. (*b*) Flying: narrow white U-shape on upper tail-coverts, sometimes interrupted in center.
2. **Greater Yellowlegs** (*Tringa melanoleuca*), p. 141. Long neck; long legs yellow; bill rather thick, slightly upturned, longer than head.
3. **Lesser Yellowlegs** (*Tringa flavipes*), p. 142. (*a*) Smaller than 2 with slim, straight, shorter bill. (*b*) Flying: wings dark; rump white; barred tail.
4. **Solitary Sandpiper** (*Tringa solitaria*), p. 142. (*a*) Smaller than 3 with greenish legs, darker upperparts; eye-ring white; slimmer than 8 with longer neck and legs. (*b*) Flying: wings and rump dark; barred tail.
5. **Short-billed Dowitcher** (*Limnodromus griseus*), p. 145. Winter: long straight bill; short legs yellowish; above gray with barred tail; rump and center of back white. [See also Pl. 9(3).]
6. **Stilt Sandpiper** (*Calidris himantopus*), p. 149. (*a*) Winter: longer, slimmer neck than 5 with longer blackish legs; long bill slightly drooped at tip. (*b*) Breeding: ear-patch rusty; below barred. (*c*) Flying: rump white; tail gray; wings dark.
7. **Pectoral Sandpiper** (*Calidris melanotos*), p. 148. Thickset; buffy; streaked breast abuts sharply on white belly; legs yellow; short bill; pale streaks on back. In flight rump appears mostly dark.
8. **Spotted Sandpiper** (*Actitis macularia*), p. 143. Teeters. (*a*) Winter: dark smudge at sides of breast; legs yellowish. (*b*) Breeding: below spotted. (*c*) Flying: strong wing-stripe; barred tail.
9. **Curlew Sandpiper** (*Calidris ferruginea*), p. 149. Winter: like 10 but slimmer, more decurved bill, below paler; rump white. Accidental or casual.
10. **Dunlin** (*Calidris alpina*), p. 149. Winter: larger than 12, above darker; bill drooped at tip. Very rare.
11. **Sanderling** (*Calidris alba*), p. 146. (*a*) Winter: very pale, especially head; contrasting dark shoulder. (*b*) Flying: conspicuous wing-stripe; center of rump dark.
12. **Western Sandpiper** (*Calidris mauri*), p. 146. (*a*) Winter: most common peep; above grayish; long drooped black bill; legs black. (*b*) Breeding: much rufous in face and back. (*c*) Flying: fairly distinct wing-stripe; dark center of rump.
13. **Semipalmated Sandpiper** (*Calidris pusilla*), p. 147. Winter: very like 12 but bill shorter, blunt-tipped, straight. In breeding plumage, brighter brown but lacks rufous of 12.
14. **Least Sandpiper** (*Calidris minutilla*), p. 147. Winter: smaller, browner than 12 or 13; legs yellowish.
15. **Baird's Sandpiper** (*Calidris bairdii*), p. 148. (*a*) Immature: breast buffy; back more or less scaled; legs black; long wings extend beyond tail. (*b*) Flying: rump mostly blackish.
16. **White-rumped Sandpiper** (*Calidris fuscicollis*), p. 148. Rump white. (*a*) Winter: above grayer, whiter on breast than 15, with long wings. (*b*) Flying: faint wing-stripe.
17. **Upland Sandpiper** (*Bartramia longicauda*), p. 141. Erect stance; small head with large eyes; long neck; in flight shows dark primaries, no wing-stripe.
18. **Buff-breasted Sandpiper** (*Tryngites subruficollis*), p. 150. Underparts buffy; upperparts scaled; bill short; in flight, underwing white with dark wrist-mark.
19. **Red Knot** (*Calidris canutus*), p. 146. (*a*) Winter: large; thickset; grayish; stout straight bill. (*b*) Breeding: below rufous; eye-ring white.

1a
2
3a
4a
1b
4b
8c
3b
6c
15b
16b
12c
5
6a
6b
7
11b
9
8b
10
11a
8a
12a
12b
13
14
17
15a
16a
18
19a
19b

Plate 12: Chachalacas, Guans, Quails, and Tinamous

1. **Gray-headed Chachalaca** (*Ortalis cinereiceps*), p. 118. (*a*) Slender; olive with gray head; bare throat red; tail-tip buffy-white; rufous in primaries. (*b*) Flying.
2. **Plain Chachalaca** (*Ortalis vetula*), p. 118. From 1 by lack of rufous in primaries, more ochraceous belly, whiter tail-tip. Guanacaste only (esp. Península de Nicoya).
3. **Great Curassow** (*Crax rubra*), p. 119. Very large; curly erectile crest. (*a*) ♂: black; bill-knob yellow. (*b*) ♀: variable; blackish-brown to rufescent; wings and tail more or less barred.
4. **Crested Guan** (*Penelope purpurascens*), p. 118. Large; brown, below spotted with white; bare dewlap red; bushy crest.
5. **Black Guan** (*Chamaepetes unicolor*), p. 119. Black with blue face; not prominently crested. Mountains.
6. **Great Tinamou** (*Tinamus major*), p. 65. Large; tailless; olive-brown; head more grayish; flanks buffy, barred with black; legs gray.
7. **Highland Tinamou** (*Nothocercus bonapartei*), p. 65. More rufescent overall than larger 6, with contrasting slaty head. Highlands.
8. **Slaty-breasted Tinamou** (*Crypturellus boucardi*), p. 67. From larger 6 and 7 by red legs, dark gray breast. (*a, b*) ♂: above uniform brown. ♀ has buffy barring on wings. N Caribbean slope.
9. **Marbled Wood-Quail** (*Odontophorus gujanensis*), p. 121. Facial skin reddish; inconspicuous marbled pattern. S Pacific slope.
10. **Black-breasted Wood-Quail** (*Odontophorus leucolaemus*), p. 122. Below black, flecked with white; throat white (sometimes inconspicuous).
11. **Spotted Wood-Quail** (*Odontophorus guttatus*), p. 122. Below varies from brown to rufous, spotted with white. (*a*) Brown phase, ♂: throat black, streaked with white; prominent orange crest erected in alarm. (*b*) Rufous phase, ♀: less contrast between forehead and crest.
12. **Rufous-fronted Wood-Quail** (*Odontophorus erythrops*), p. 122. ♂: crown and underparts rufous; face and throat black (much less extensive in female). Caribbean slope.
13. **Spotted-bellied Bobwhite** (*Colinus leucopogon*), p. 120. Below with prominent white spotting. (*a*) ♂: breast brownish, little spotting. (*b*) ♀: breast heavily spotted. Pacific NW, Valle Central.
14. **Crested Bobwhite** (*Colinus cristatus*), p. 121. ♂: face largely rufous; underparts spotted with cinnamon-rufous. Golfo Dulce area only.
15. **Tawny-faced Quail** (*Rhynchortyx cinctus*), p. 123. Smaller than wood-quails. (*a*) ♂: face rufous; breast gray. (*b*) ♀: throat and eye-stripe white; barred belly.
16. **Buffy-crowned Wood-Partridge** (*Dendrortyx leucophrys*), p. 120. Long pleated tail; conspicuous white facial pattern; orbital skin red; breast slaty.
17. **Little Tinamou** (*Crypturellus soui*), p. 66. Very small; uniformly dark; throat white; legs dull.
18. **Thicket Tinamou** (*Crypturellus cinnamomeus*), p. 66. Most rufescent and heavily barred tinamou; legs red; breast of ♂ unmarked, of ♀ barred. NW.

1b
2
4
5
1a
3b
3a
6
7
8a
13a
9
10
14
13b
12
15b
15a
11a
11b
17
16
8b
18

Plate 13: Vultures and Large Black Raptors

1. **Zone-tailed Hawk** (*Buteo albonotatus*), p. 109. Flying: very like 2 and 3; barring on wings and tail inconspicuous from a distance; note feathered head, yellow cere. [See also Pl. 14(2).]
2. **Lesser Yellow-headed Vulture** (*Cathartes burrovianus*), p. 96. (*a*) Slightly smaller, browner than 3; note head color. (*b*) Flying: safely told from 3 only by head color; above may show pale shafts of outer primaries.
3. **Turkey Vulture** (*Cathartes aura*), p. 96. Long wings and tail; flight feathers gray; soars with wings in shallow V. (*a*) Migrant: head red; note horizontal stance. (*b*) Resident: note whitish band on nape. Young have blackish heads, dull nuchal band.
4. **Black Vulture** (*Coragyps atratus*), p. 96. (*a*) Head black; short tail; erect stance. (*b*) Flying: broad wings held flat while soaring; white patch in primaries.
5. **King Vulture** (*Sarcoramphus papa*), p. 97. (*a, b*) Harlequin head; body creamy-white; flight feathers black; broad wings; short tail. (*c, d*) Young: from smaller 4 by all-black primaries, pale eyes.
6. **Common Black-Hawk** (*Buteogallus anthracinus*), p. 104. (*a*) Chunky; cere and lores yellow; thighs blackish. (*b*) Flying: single white tail-band; rump dark; below white at base of primaries. (*c*) Immature: buffy; below heavily streaked with black; coarsely barred tail.
7. **Great Black-Hawk** (*Buteogallus urubitinga*), p. 105. (*a*) Larger, grayer than 6a with dark lores, barred thighs. (*b*) Flying: white rump and incomplete basal tail-band. (*c*) Immature: like 6c but more finely barred tail.
8. **Solitary Eagle** (*Harpyhaliaetus solitarius*), p. 105. (*a*) Much larger, grayer than 7a with dark thighs, shaggy short crest. (*b*) Flying: short tail, single broad white tail-band; very broad wings; white scaling on upper tail-coverts. (*c*) Immature: like 7c but tail and breast usually darker.
9. **Black Hawk-Eagle** (*Spizaetus tyrannus*), p. 111. (*a*) Slimmer than 6, 7, 8 with long many-banded tail, yellow eyes; legs barred, fully feathered. (*b*) Flying: broad, banded wings "pinched in" at base; long tail. [See also Pl. 17(12).]

1
2b
3b
5d
2a
4b
5b
4a
3a
5a
5c
6b
7b
6a
7a
8b
6c
7c
8a
9b
8c
9a

Plate 14: Black Raptors

1. **Short-tailed Hawk** (*Buteo brachyurus*), p. 107. (*a*) Dark phase: brownish-black; indistinct tail bars; usually white on face. (*b*) Flying: pale, indistinctly barred flight feathers; only terminal tail-bar distinct. [See also Pl. 16(11).]
2. **Zone-tailed Hawk** (*Buteo albonotatus*), p. 109. (*a*) Blacker than 1 with longer, more widely banded tail. (*b*) Immature: tail narrowly banded; below with white spots or short streaks. [See also Pl. 13(1).]
3. **Swainson's Hawk** (*Buteo swainsoni*), p. 108. Dark phase: browner than 1 or 2; wings and tail longer, narrower, darker than 1; usually shows pale crissum. [See also Pl. 17(7).]
4. **Crane Hawk** (*Geranospiza caerulescens*), p. 101. (*a*) Cere and bill blackish; iris red; long legs orange. (*b*) Immature: grayer; white on face; below with buffy-white barring. (*c*) Flying: broad wings with white crescent in primaries; banded tail.
5. **Bay-winged Hawk** (*Parabuteo unicinctus*), p. 106. (*a*) Shoulders and thighs chestnut; belly, rump, and base of tail white. (*b*) Immature: pattern obscured by streaks and spots. (*c*) Flying: wing-linings chestnut; long tail with white base.
6. **Snail Kite** (*Rostrhamus sociabilis*), p. 99. (*a*) ♂: slender hooked bill; cere and legs red or orange; belly and base of tail white. (*b*) Flying: long slender wings; slightly notched tail. [See also Pl. 15(6).]
7. **White-tailed Hawk** (*Buteo albicaudatus*), p. 108. Immature: very dark (below often nearly uniform dark brown); shoulder cinnamon; from 5b by broader wings, more uniformly pale tail; thighs not rufous. [See also Pl. 17(6).]
8. **Plumbeous Kite** (*Ictinia plumbea*), p. 100. (*a*) Dark gray, head paler; cere dark; iris red; shorter legs, longer wings than 4. (*b*) Flying: tail square to notched; long pointed wings with rufous patch in primaries. [See also Pl. 15(4).]
9. **Hook-billed Kite** (*Chondrohierax uncinatus*), p. 98. Black phase: heavy hooked bill; iris white; bare spot on lores orange; cere greenish; long tail. [See also Pl. 16(9).]
10. **Collared Forest-Falcon** (*Micrastur semitorquatus*), p. 114. Dark phase: slender; short rounded wings; long rounded tail with narrow white bars; belly and thighs with cinnamon barring. [See also Pl. 16(7).]
11. **Red-throated Caracara** (*Daptrius americanus*), p. 112. Guanlike; face red; cere bluish; bill yellow; tail black; belly white.
12. **Crested Caracara** (*Polyborus plancus*), p. 112. (*a*) Neck and face white; crown black; crissum, rump, base of tail white; bare face reddish. Immature browner. (*b*) Flying: white "window" in primaries. Looks black in the middle, white on the ends.

1b
3
4c
5c
1a
2a
4a
5a
4b
2b
5b
7
6b
9
8b
8a
6a
12b
10
11
12a

Plate 15: Kites and Falcons

1. **Black-shouldered Kite** (*Elanus caeruleus*), p. 99. (*a*) White with pale gray mantle; shoulders black; iris red. (*b*) Young: streaking and scaling rusty. (*c*) Flying: long, pointed, kinked wings with black wrist mark; long tail white.
2. **American Swallow-tailed Kite** (*Elanoides forficatus*), p. 99. Below white with black flight feathers; long deeply forked tail.
3. **Mississippi Kite** (*Ictinia mississippiensis*), p. 101. (*a*) Flying: long pointed wings; slightly notched tail black; body gray; paler head. (*b*) Immature: below with rusty streaking; feet yellowish.
4. **Plumbeous Kite** (*Ictinia plumbea*), p. 100. Immature: like 3b but below with dusky streaking; feet yellow-orange. [See also Pl. 14(8).]
5. **Peregrine Falcon** (*Falco peregrinus*), p. 117. (*a*) Flying: cap and cheek-bar black; throat and chest whitish; barred pattern. (*b*) Immature, soaring (to smaller scale): below streaked heavily; wings appear longer, more pointed than *Accipiter;* tail longer than *Buteo.*
6. **Snail Kite** (*Rostrhamus sociabilis*), p. 99. ♀: below buffy, heavily streaked; sides dark; superciliaries pale; slender hooked bill; tail notched with white base and tail-coverts. [See also Pl. 14(6).]
7. **Northern Harrier** (*Circus cyaneus*), p. 101. Conspicuous white rump-patch; facial disk. (*a*) ♀: below heavily streaked. (*b*) ♀, flying: long narrow wings and tail. (*c*) ♂: pale grayish; darker primaries; below with cinnamon specks.
8. **Laughing Falcon** (*Herpetotheres cachinnans*), p. 113. "Bandit mask" on square, buffy head; long banded tail; short broad wings.
9. **Yellow-headed Caracara** (*Milvago chimachima*), p. 113. (*a*) Head and underparts buffy; postocular streak black; bill bluish; cere and lores orangish. (*b*) Immature: streaked; cere blue-gray. (*c*) Flying: pale tail with dark subterminal band; pale patch in primaries, dark secondaries.
10. **Aplomado Falcon** (*Falco femoralis*), p. 116. Bold facial pattern; barred breast; belly cinnamon.
11. **Orange-breasted Falcon** (*Falco deiroleucus*), p. 116. Like huge 14 but with rufous across chest; much darker than 10.
12. **Pearl Kite** (*Gampsonyx swainsonii*), p. 115 (in American Kestrel, Note). Very small; face buffy; thighs cinnamon. Not yet recorded in Costa Rica; most likely in SE.
13. **American Kestrel** (*Falco sparverius*), p. 115. (*a*) ♂: conspicuous facial bars; back and tail reddish; wings blue-gray. (*b*) ♀: above more barred and below streaked; wings brown.
14. **Bat Falcon** (*Falco rufigularis*), p. 116. (*a*) Throat and sides of neck whitish; breast barred with black and white; thighs and belly chestnut. (*b*) Flying: dark; pointed, barred wings; throat white (often looks collared).
15. **Merlin** (*Falco columbarius*), p. 115. Back slaty; head and underparts streaked with dark brown; trace of cheek-bar (much less than larger 5b).

1c
2
5b
5a
3a
1a
3b
6
7a
1b
4
7b
14b
9c
7c
8
9b
10
11
9a
12
13a
13b
14a
15

Plate 16: Hawks, Kites, and Forest-Falcons

1. **Double-toothed Kite** (*Harpagus bidentatus*), p. 100. (*a*) Sides of chest (♂) or entire chest (♀) rufous; throat stripe black. (*b*) Immature: below white with fine dark streaks. (*c*) Flying: rather narrow wings held far forward, tail closed; tail-coverts white, fluffy.
2. **Tiny Hawk** (*Accipiter superciliosus*), p. 102. (*a*) Below with fine dark barring; iris orange; cere yellow. (*b*) Immature: below more rufous (some are rufous above).
3. **Sharp-shinned Hawk** (*Accipiter striatus*), p. 102. (*a*) Below with fine reddish barring; above slaty. (*b*) Immature, flying: below streaked; long tail square-tipped; rather short, broad wings.
4. **Cooper's Hawk** (*Accipiter cooperii*), p. 102. (*a*) Larger than 3, more capped. (*b*) Immature, flying: tail more rounded, more broadly white-tipped than 3; streaking below less extensive.
5. **Barred Forest-Falcon** (*Micrastur ruficollis*), p. 114. (*a*) ♂: below with fine blackish barring; mantle gray. (*b*) ♀: mantle brown. (*c*) Immature: below buffy to white; coarser barring often restricted to sides; nuchal collar.
6. **Slaty-backed Forest-Falcon** (*Micrastur mirandollei*), p. 114. Above slaty; below white to buffy; cere and legs yellow; eye dark.
7. **Collared Forest-Falcon** (*Micrastur semitorquatus*), p. 114. Large; short wings; long rounded tail; above blackish. (*a*) Pale phase: collar; cere greenish; iris dark; below white or buff. (*b*) Immature: below heavily barred and scaled. [See also Pl. 14(10).]
8. **Bicolored Hawk** (*Accipiter bicolor*), p. 103. (*a*) Below pale gray; thighs rufous; iris red; cere yellow. (*b, c*) Immature: variable; below white to buffy, with or without rufous thighs.
9. **Hook-billed Kite** (*Chondrohierax uncinatus*), p. 98. Bill heavy, hooked; iris white; bare spot on lores orange. (*a*) Gray phase (mostly ♂♂). (*b*) Rufous phase (mostly ♀♀). (*c*) Immature: below white; nuchal collar (usually); faint barring or none. [See also Pl. 14(9).]
10. **Semiplumbeous Hawk** (*Leucopternis semiplumbea*), p. 103. Chunky; short tail; note orange legs and cere, yellow eye. Caribbean slope.
11. **Short-tailed Hawk** (*Buteo brachyurus*), p. 107. (*a*) Pale phase: dark hood; forehead and throat white; faint tail bars. (*b*) Immature: head and underparts buffy; sparse dark streaking. [See also Pl. 14(1).]
12. **Roadside Hawk** (*Buteo magnirostris*), p. 107. (*a*) Head and chest gray; rusty barring on belly; iris yellow; rufous in remiges. (*b*) Immature: from 13b and 14b by rufous-barred thighs, buffy cast.
13. **Broad-winged Hawk** (*Buteo platypterus*), p. 107. (*a*) Breast rufous to brown; belly scaled (variable); iris pale brown. (*b*) Immature: below streaked, usually less heavily than 14b; thighs barred, spotted or scaled. (*c*) Flying: evenly banded tail.
14. **Gray Hawk** (*Buteo nitidus*), p. 106. (*a*) Above pearly-gray; below with fine gray barring. (*b*) Immature: below with heavy dark streaking, thighs lightly barred. (*c*) Flying: single distinct white tail-band with others indistinct.

1c
13c
14c
3b
4b
1b
1a
2b
2a
3a
5c
5b
5a
6
4a
7a
7b
8a
8b
8c
9a
9b
9c
10
11b
11a
12a
12b
13b
13a
14b
14a

Plate 17: Large Raptors

1. **Black-chested Hawk** (*Leucopternis princeps*), p. 103. (*a*) Finely barred underparts contrast with dark head and chest. (*b*) Flying: very broad wings; short tail.
2. **White Hawk** (*Leucopternis albicollis*), p. 104. (*a*) White with black markings on wings and tail. (*b*) Flying: broad rounded wings; short tail.
3. **Gray-headed Kite** (*Leptodon cayanensis*), p. 98. (*a*) Head gray; small bill; above blackish; below white. (*b*) Flying: broad rounded wings with black wing-linings; long rounded tail. (*c*, *d*) Immature: highly variable, some mostly black; distinguish by shape and behavior.
4. **Savanna Hawk** (*Buteogallus meridionalis*), p. 106 (in Black-collared Hawk, Note). Mostly rufous; long-legged. To be looked for on S Pacific slope.
5. **Black-collared Hawk** (*Busarellus nigricollis*), p. 105. (*a*) Mostly rufous with white head; black crescent on chest. (*b*) Immature, flying: streaked chest buffy; note broad rounded wings, very short tail.
6. **White-tailed Hawk** (*Buteo albicaudatus*), p. 108. (*a*) Above slaty; below white; shoulder rufous; tail white with black subterminal band. (*b*) Flying: dark cheeks give half-hood effect; long broad wings and tail; conspicuous tail-band. [See also Pl. 14(7).]
7. **Swainson's Hawk** (*Buteo swainsoni*), p. 108. (*a*) Pale phase: throat white; chest-band brown; belly white to cinnamon, sometimes heavily spotted or scaled with dusky. (*b*) Flying: pale wing-linings contrast with darker flight feathers; only subterminal tail-bar conspicuous. [See also Pl. 14(3).]
8. **Red-tailed Hawk** (*Buteo jamaicensis*), p. 109. (*a*) Resident race: belly cinnamon; breast nearly unmarked white; tail rusty. (*b*) Migrant, flying: tail rusty; chest streaked; belly white; broad wings. (*c*) Immature: below streaked, whiter than young 7 with less of a chest-band; pale remiges in flight.
9. **Harpy Eagle** (*Harpia harpyja*), p. 110. Huge, robust. (*a*) Divided crest; chest-band black; very heavy feet. (*b*) Immature, flying: below very pale; youngest without trace of chest-band; broad wings and longish tail.
10. **Crested Eagle** (*Morphnus guianensis*), p. 109. Very large; slimmer than 9. (*a*) Pale phase: from 9 by tawny barring below, no chest-band, pointed crest. (*b*) Dark phase: dark chest; below with heavy dark barring. (*c*) Immature, flying: very pale; note longer, slimmer wings and tail than 9b.
11. **Ornate Hawk-Eagle** (*Spizaetus ornatus*), p. 111. (*a*) Long spiky crest; sides of neck and breast rufous; underparts barred. (*b*) Immature, flying: very pale; broad wings; long tail.
12. **Black Hawk-Eagle** (*Spizaetus tyrannus*), p. 111. Immature, flying: much darker, more streaked than 11b; wings broader, more pinched in at base. [See also Pl. 13(9).]
13. **Black-and-white Hawk-Eagle** (*Spizastur melanoleucus*), p. 110. (*a*) Head and underparts white; iris yellow; mask and short crest black. (*b*) Flying: shape like *Buteo,* tail shorter and wings less broad and narrow-based than 11b.
14. **Osprey** (*Pandion haliaetus*), p. 97. Long, narrow, kinked wings, dark spot at wrist; head white with dark eye-stripe.

1b
5b
6b
7b
2b
3b
1a
2a
8b
3a
3c
3d
4
5a
6a
7a
8a
8c
10c
12
14
9b
11b
13b
9a
10a
11a
10b
13a

Plate 18: Pigeons and Doves

1. **Band-tailed Pigeon** (*Columba fasciata*), p. 167. Large; white crescent on nape; banded tail; bill yellow. Highlands.
2. **Red-billed Pigeon** (*Columba flavirostris*), p. 168. Purplish with grayer belly and tail; bill whitish; cere rose-pink.
3. **Pale-vented Pigeon** (*Columba cayennensis*), p. 167. Gray head, white belly and crissum, and fairly pale gray tail contrast with ruddy-purple body; bill black.
4. **Scaled Pigeon** (Columba speciosa), p. 166. Conspicuously scaled neck and underparts; bill red with pale tip.
5. **Short-billed Pigeon** (*Columba nigrirostris*), p. 168. Uniform purplish-brown; small bill black. **Ruddy Pigeon** (*C. subvinacea*) (p. 168) very similar, slightly more reddish, especially on wing-coverts (distinguish in field best by voice; at higher elevations).
6. **Blue Ground-Dove** (*Claravis pretiosa*), p. 171. (*a*) ♂: bluish-gray; dark wing-spots. (*b*) ♀: browner, longer-tailed than 7, 8, or 9; note chestnut wing-pattern.
7. **Ruddy Ground-Dove** (*Columbina talpacoti*), p. 170. (*a*) ♂: reddish with pale gray head. (*b*) ♀, flying: duller, browner with pale head; from below note black wing-linings.
8. **Common Ground-Dove** (*Columbina passerina*), p. 170. (*a*) ♂: face and neck pinkish; crown gray; patterned breast. (*b*) ♀, flying: browner; patterned breast; note rufous in primaries, blackish tail.
9. **Plain-breasted Ground-Dove** (*Columbina minuta*), p. 170. (*a*) ♂: like 8a in color but smaller, unmarked breast. (*b*) ♀: from larger 8b by unmarked breast.
10. **Maroon-chested Ground-Dove** (*Claravis mondetoura*), p. 172. (*a*) ♂: slaty with maroon breast; face and belly pale; white in tail. (*b*) ♀: brown with pale face; conspicuous wing-bars; white in tail. Highlands, rare.
11. **Inca Dove** (*Columbina inca*), p. 171. (*a*) Pale grayish with scaly pattern; long tail. (*b*) Flying: rufous in primaries; white sides to long tail.
12. **White-winged Dove** (*Zenaida asiatica*), p. 169. (*a*) Brownish with white in wing; spot below eye black. (*b*) Flying: rounded tail with white corners; wing-stripe white. N Pacific slope.
13. **Mourning Dove** (*Zenaida macroura*), p. 169. Long pointed tail. (*a*) Brownish; black spots on wings. (*b*) Flying: lateral rectrices tipped with white.
14. **White-tipped Dove** (*Leptotila verreauxi*), p. 172. (*a*) Below grayish-buff; eye-ring blue; iris orange. (*b*) Flying: rounded tail with white-tipped lateral rectrices; wing-linings, inner webs of primaries rufous.
15. **Gray-fronted Dove** (*Leptotila rufaxilla*), p. 173. Like 14 but crown lead-gray, cheeks more buffy, orbital skin purplish, iris pale yellow.
16. **Gray-chested Dove** (*Leptotila cassinii*), p. 173. (*a*) Pacific form: very like 15 except for rufous nape. (*b*) Caribbean form: darker, grayer; lateral rectrices tipped with grayish-white.
17. **Ruddy Quail-Dove** (*Geotrygon montana*), p. 176. (*a*) ♂: above ruddy-purplish; striped face; underparts buffy. (*b*) ♀: browner; face buffy; note head shape.
18. **Chiriquí Quail-Dove** (*Geotrygon chiriquensis*), p. 174. Rufous breast contrasts with slaty cap; malar stripe black. Highlands.
19. **Buff-fronted Quail-Dove** (*Geotrygon costaricensis*), p. 174. Breast and head gray, contrasting with purplish mantle; forehead buffy; black malar stripe on white face. Highlands.
20. **Violaceous Quail-Dove** (*Geotrygon violacea*), p. 175. Above violet (♂) or ruddy-brown (♀); below whitish; no malar stripe; drooped bill.
21. **Purplish-backed Quail-Dove** (*Geotrygon lawrencii*), p. 174. Like 19 but duller, darker; above dark brown with purple gloss. Caribbean slope.
22. **Olive-backed Quail-Dove** (*Geotrygon veraguensis*), p. 175. Olive-brown with silvery-white malar stripe; throat and forehead pale. Caribbean slope.

1
2
3
4
5
6a
6b
7b
8b
9b
7a
8a
9a
10a
10b
11a
11b
13b
12b
12a
13a
14b
14a
15
16b
17b
17a
16a
18
19
20
21
22

Plate 19: Parrots

1. **Scarlet Macaw** (*Ara macao*), p. 177. Red, blue, and yellow; long tail. Unmistakable.
2. **Great Green Macaw** (*Ara ambigua*), p. 177. Green with red forehead and tail; rump and remiges blue. Caribbean slope.
3. **Mealy Parrot** (*Amazona farinosa*), p. 183. Large; cere blackish; eye-ring pale; nape bluish.
4. **Red-lored Parrot** (*Amazona autumnalis*), p. 182. (*a*) Fairly large; forehead and lores red, cere and eye-ring paler; crown blue. (*b*) Flying: red speculum (3 is similar); central rectrices not pale-tipped.
5. **Yellow-naped Parrot** (*Amazona auropalliata*), p. 182. (*a*) Fairly large; yellow patch on nape; cere blackish; eye-ring pale. (*b*) Flying: red speculum; tail pale-tipped. N Pacific slope.
6. **White-fronted Parrot** (*Amazona albifrons*), p. 182. (*a*) ♂: forehead white; crown blue; extensive red on face, forewing (restricted or lacking in ♀). (*b*) Young: forehead yellowish (sometimes almost lacking). (*c*) ♂, flying: red on forewing, secondaries blue. N Pacific slope.
7. **White-crowned Parrot** (*Pionus senilis*), p. 181. (*a*) Head and breast dull blue; crown and throat white; shoulder bronzy. (*b*) Flying: wings blue; red under tail.
8. **Blue-headed Parrot** (*Pionus menstruus*), p. 181. Head and breast brighter blue than 7; bill black with reddish spot; red under tail and blue wings conspicuous in flight, as in 7.
9. **Brown-hooded Parrot** (*Pionopsitta haematotis*), p. 180. (*a*) Hood dark brown; ear-patch red; bill and eye-ring pale. (*b*) Flying: red patch under wing.
10. **Crimson-fronted Parakeet** (*Aratinga finschi*), p. 177. (*a*) Largest parakeet; forehead crimson; eye-ring whitish. (*b*) Flying: reddish and yellow on underwing.
11. **Olive-throated Parakeet** (*Aratinga nana*), p. 178. (*a*) Eye-ring pale; head, throat, and chest dull, no red. (*b*) Flying: primaries blackish below; little color or pattern. Caribbean slope.
12. **Sulfur-winged Parakeet** (*Pyrrhura hoffmanni*), p. 179. (*a*) Long tail reddish; ear-patch red; scaly pattern on head; eye-ring pale. (*b*) Flying: yellow in wings. Cordillera de Talamanca.
13. **Orange-fronted Parakeet** (*Aratinga canicularis*), p. 178. Forehead orange; crown blue; eye-ring yellow; blue in primaries; long pointed tail. N Pacific slope.
14. **Orange-chinned Parakeet** (*Brotogeris jugularis*), p. 180. (*a*) Shoulders brownish; orange spot on chin. (*b*) Flying: short pointed tail; wing-linings yellow.
15. **Red-fronted Parrotlet** (*Touit costaricensis*), p. 180. (*a*) ♂: face and wing-patch red (less extensive in ♀); short, square tail yellow with black tip. (*b*) Flying: red and yellow underwing.
16. **Barred Parakeet** (*Bolborhynchus lineola*), p. 179. (*a*) Green with black shoulder (especially ♂); fine black barring. (*b*) Flying: looks all-green; short pointed tail. Highlands.

1
2
3
4a
4b
5a
5b
6a
6b
6c
7a
7b
8
9a
9b
10a
10b
11a
11b
12a
12b
13
14a
14b
15a
15b
16a
16b

Plate 20: Owls and Potoos

1. **Great Horned Owl** (*Bubo virginianus*), p. 192. Very large, ''horned''; underparts barred; face rusty. Very rare.
2. **Striped Owl** (*Asio clamator*), p. 195. Long ''horns''; buffy below with heavy black streaking.
3. **Crested Owl** (*Lophostrix cristata*), p. 192. (*a*) Pale phase, alarm: conspicuous ''horns'' erected, (*b*) Dark phase: fine vermiculations below with no bars or streaks; dark breast.
4. **Common Potoo** (*Nyctibius griseus*), p. 198. Perches more or less erectly; brownish-gray with black spots on breast; shoulders black; scapulars pale; may include 2 species (see text). [See also Pl. 51(7).]
5. **Great Potoo** (*Nyctibius grandis*), p. 197. Larger, much paler than 4.
6. **Mottled Owl** (*Ciccaba virgata*), p. 194. (*a*) Eyes dark; breast barred and mottled; belly boldly streaked. (*b*) Young: buffy; facial disk pale.
7. **Black-and-white Owl** (*Ciccaba nigrolineata*), p. 195. Facial disk, head, and upperparts blackish; collar and underparts barred; bill yellow.
8. **Spectacled Owl** (*Pulsatrix perspicillata*), p. 192. (*a*) Large; robust; white spectacles, yellow eyes, and pale bill contrast with dark brown head and upperparts; belly buffy. (*b*) Young: white with dark facial disk.
9. **Common Barn-Owl** (*Tyto alba*), p. 189. Heart-shaped facial disk; long legs; head and underparts white to deep buff; variable dark speckling.
10. **Bare-shanked Screech-Owl** (*Otus clarkii*), p. 191. Dark; richly colored; coarsely vermiculated; inconspicuous ''horns''; lower tarsi bare. Highlands.
11. **Pacific Screech-Owl** (*Otus cooperi*), p. 190. (*a*) Gray-brown; below with black streaking; facial disk lightly outlined; fairly conspicuous ''horns.'' (*b*) Young: fluffy; indistinctly barred; no ''horns.'' N Pacific slope.
12. **Vermiculated Screech-Owl** (*Otus guatemalae*), p. 191. Below with little streaking, mostly vermiculations; inconspicuous ''horns.'' (*a*) Normal phase. (*b*) Rufous phase.
13. **Unspotted Saw-whet Owl** (*Aegolius ridgwayi*), p. 196. Small; brown; plumage variable, often with paler eyebrows, but below never streaked or barred. Highlands.
14. **Burrowing Owl** (*Athene cunicularia*), p. 194. Long-legged; below barred; above spotted. Terrestrial, accidental.
15. **Tropical Screech-Owl** (*Otus choliba*), p. 190. Below boldly streaked; facial disk boldly outlined in black; fairly conspicuous ''horns.''
16. **Least Pygmy-Owl** (*Glaucidium minutissimum*), p. 193. Very small; crown speckled; breast cinnamon-brown. Caribbean slope.
17. **Ferruginous Pygmy-Owl** (*Glaucidium brasilianum*), p. 194. Crown streaked; tail more narrowly barred than 16. (*a*) Brown phase. (*b*) Rufous phase (some birds intermediate). Note ''false eyes'' on nape. Pacific slope.
18. **Andean Pygmy-Owl** (*Glaucidium jardinii*), p. 193. (*a*) Normal phase: dark, rich brown; white speckling on crown, breast, and back. (*b*) Rufous phase: nearly unspotted. Highlands.

1
2
3a
3b
4
5
8a
8b
7
6a
6b
9
10
11b
11a
13
12b
12a
18b
18a
15
14
17b
16
17a

Plate 21: Cuckoos and Nightjars

1. **Black-billed Cuckoo** (*Coccyzus erythropthalmus*), p. 184. Bill black and gray; eye-ring red (adult) or yellowish (immature); rectrices gray with narrow black and white tips.
2. **Yellow-billed Cuckoo** (*Coccyzus americanus*), p. 184. Bill largely yellow; primaries rufous; rectrices black with broad white tips.
3. **Cocos Cuckoo** (*Coccyzus ferrugineus*), p. 185. Size of 4 but more richly colored. Only cuckoo on Cocos I.
4. **Mangrove Cuckoo** (*Coccyzus minor*), p. 185. Underparts pale to deep buffy; mask black; bill and tail like smaller 2.
5. **Striped Cuckoo** (*Tapera naevia*), p. 187. Bushy crest; white facial stripes; above striped; chest not streaked.
6. **Pheasant Cuckoo** (*Dromococcyx phasianellus*), p. 188. Crested; conspicuous eye-stripe; broad rectrices narrowly white-tipped; chest streaked. Pacific slope, rare.
7. **Squirrel Cuckoo** (*Piaya cayana*), p. 185. Large; rufous with gray belly; white-tipped rectrices.
8. **Smooth-billed Ani** (*Crotophaga ani*), p. 186. Larger than 9 with smoother, higher-ridged bill. S Pacific slope, but range expanding.
9. **Groove-billed Ani** (*Crotophaga sulcirostris*), p. 186. Bill grooved, high-ridged; long tail; short rounded wings.
10. **Rufous-vented Ground-Cuckoo** (*Neomorphus geoffroyi*), p. 188. Very large; crested; bill yellowish; breast-band black.
11. **Lesser Ground-Cuckoo** (*Morococcyx erythropygius*), p. 187. Blue facial skin set off by black and white; below rufous. N Pacific slope.
12. **Lesser Nighthawk** (*Chordeiles acutipennis*), p. 200. (*a*) ♂: below barred; above pale; tail-band white (lacking in ♀). (*b*) ♂, flying: white band crosses primaries toward tip (♀ has buffy band).
13. **Common Nighthawk** (*Chordeiles minor*), p. 199. Flying: slightly larger than 12, but white band (both sexes) crosses primaries midway.
14. **Short-tailed Nighthawk** (*Lurocalis semitorquatus*), p. 199. Very dark; very short tail; no white in wings; batlike in flight.
15. **Rufous Nightjar** (*Caprimulgus rufus*), p. 202. (*a*) Large; overall reddish. (*b*) In hand note unbranched rictal bristles. (*c*) Tail of male: distal half of outer 3 rectrices mostly white, edged with rusty; female has lateral rectrices narrowly tipped with cinnamon-rufous.
16. **Chuck-will's-widow** (*Caprimulgus carolinensis*), p. 201. Similar to 15 but slightly larger and less reddish. (*a*) In hand note branched rictal bristles. (*b*) Tail of ♂: outer 3 rectrices with distal half of inner web white; ♀ has these rectrices broadly tipped with deep buff.
17. **Whip-poor-will** (*Caprimulgus vociferus*), p. 202. Rather grayish; finely patterned; no white on wing; white (♂) or buff (♀) corners of tail.
18. **Common Pauraque** (*Nyctidromus albicollis*), p. 200. (*a*) Tail long; scapulars boldly patterned; belly finely barred; white band on primaries. (*b*) ♂: tail-stripes white. (*c*) ♀: lateral rectrices with narrow white tips; wing-band narrower, often buffy. [See also Pl. 51(10).]
19. **Dusky Nightjar** (*Caprimulgus saturatus*), p. 202. Very dark; corners of tail white (♂) or buff (♀). High mountains.
20. **White-tailed Nightjar** (*Caprimulgus cayennensis*), p. 203. (*a*) ♂: rather pale; prominent nuchal collar; breast-spots pale; long notched tail; wing-band white. (*b*) Spread tail of ♂: largely white. (*c*) ♀, flying: wing-band cinnamon; dark tail barred.

1
2
3
4
5
6
7
8
9
10
11
12a
12b
13
14
15a
15b
15c
16a
16b
17
18a
18b
18c
19
20a
20b
20c

Plate 22: Swifts and Swallows

1. **White-collared Swift** (*Streptoprocne zonaris*), p. 206. Very large; complete white collar (faint in young birds).
2. **Black Swift** (*Cypseloides niger*), p. 204. Large; tail rather long, notched (especially in ♂); forehead usually pale.
3. **Chestnut-collared Swift** (*Cypseloides rutilus*), p. 205. Collar chestnut (usually reduced or absent on ♀); tail with slight notch or truncate; more tapered wings than *Chaetura*.
4. **White-chinned Swift** (*Cypseloides cryptus*), p. 205. Size of 2 but tail short and truncate, wings short; in hand note white chin (pale brown in immature).
5. **Spot-fronted Swift** (*Cypseloides cherriei*), p. 205. Size of 3 but lacks collar; at short range note spots before and behind eye.
6. **Great Swallow-tailed Swift** (*Panyptila sanctihieronymi*), p. 208. Pattern of 7 but much larger. Rare.
7. **Lesser Swallow-tailed Swift** (*Panyptila cayennensis*), p. 208. Throat and flank-spots white; long tail (usually held closed, spikelike).
8. **Chimney Swift** (*Chaetura pelagica*), p. 206. Large robust *Chaetura;* uniformly dark with throat slightly paler.
9. **Vaux's Swift** (*Chaetura vauxi*), p. 207. Smaller, slimmer than 8; throat and rump slightly paler than body. (*a*) Above. (*b*) Below.
10. **Gray-rumped Swift** (*Chaetura cinereiventris*), p. 207. Small; rump and upper tail-coverts pale gray; throat whitish. Caribbean slope.
11. **Band-rumped Swift** (*Chaetura spinicauda*), p. 207. Small; pale band across rump; upper tail-coverts dark. S Pacific slope.
12. **Barn Swallow** (*Hirundo rustica*), p. 343. (*a*) Long forked tail; below rusty with dark breast-band. (*b*) Immature: below paler, tail shorter, breast-band indistinct.
13. **Cliff Swallow** (*Hirundo pyrrhonota*), p. 343. (*a*) Forehead pale; chest dark; belly white. (*b*) Flying: rump buffy; rather short tail. Immatures: pattern of head and chest indistinct. [See also Pl. 52(1).]
14. **Brown-chested Martin** (*Phaeoprogne tapera*), p. 342. Pattern of 17 but much larger. Rare southern migrant.
15. **Gray-breasted Martin** (*Progne chalybea*), p. 342. (*a*) Below dingy grayish-brown; belly white; forehead dark. (*b*) Flying: above dark blue (♂) to mostly dusky (♀).
16. **Purple Martin** (*Progne subis*), p. 342. (*a*) Young ♀: like 15 but paler forehead, below more heavily marked. (*b*) Flying: note grizzled nuchal collar. (*c*) ♂: all dark.
17. **Bank Swallow** (*Riparia riparia*), p. 345. (*a*) Small; above brown; below white; breast-band. (*b*) Flying: grayer above than 18, 19.
18. **Northern Rough-winged Swallow** (*Stelgidopteryx serripennis*), p. 344. (*a*) Brownish wash across chest; throat at most faintly tinged with buff; crissum mostly white. (*b*) Flying: rump dark.
19. **Southern Rough-winged Swallow** (*Stelgidopteryx ruficollis*), p. 344. (*a*) Throat tawny-buff to cinnamon; belly usually tinged with yellow; crissum feathers usually black-tipped. (*b*) Flying: rump usually pale.
20. **Blue-and-white Swallow** (*Notiochelidon cyanoleuca*), p. 345. (*a*) Small; short-tailed; below white; crissum black. (*b*) Young: below tinged with brownish to buffy; above dusky. (*c*) Flying: from 21–23 by black crissum, shorter and broader wings.
21. **Tree Swallow** (*Tachycineta bicolor*), p. 346. Larger than 20, above greener, white under tail.
22. **Violet-green Swallow** (*Tachycineta thalassina*), p. 347. Like 21 but above with violet, white on flanks and behind eyes.
23. **Mangrove Swallow** (*Tachycineta albilinea*), p. 346. White line over eye; rump white; above steely-green.

1
2
3
4
5
6
7
8
9a
9b
10
11
12a
12b
13a
13b
14
15a
15b
16a
16b
16c
17a
17b
18a
18b
19a
19b
20a
20b
20c
21
22
23

Plate 23: Hummingbirds, 1: Hermits and Large Species

1. **Little Hermit** (*Phaethornis longuemareus*), p. 211. Very small; below cinnamon; long central rectrices buffy-white.
2. **Long-tailed Hermit** (*Phaethornis superciliosus*), p. 211. Below grayish-buff; facial stripes buffy; long central rectrices with white tips.
3. **Green Hermit** (*Phaethornis guy*), p. 211. (*a*) ♂: glossy dark green; rump tinged with bluish; facial stripes faint; central rectrices with short, spiky white tips. (*b*) ♀: below gray; facial stripes ochraceous; central rectrices like 2.
4. **Bronzy Hermit** (*Glaucis aenea*), p. 210. Below uniform cinnamon; rounded tail with rufous base, black subterminal band, and white tip.
5. **Band-tailed Barbthroat** (*Threnetes ruckeri*), p. 210. Dusky throat contrasts with rufous chest; tail with white base, rectrices more pointed than 4.
6. **Brown Violet-ear** (*Colibri delphinae*), p. 214. Brown; rather hermitlike facial stripe but short straight bill; rump cinnamon; tip of tail ochraceous.
7. **Green Violet-ear** (*Colibri thalassinus*), p. 214. Rather pale green; tail-band dark blue; ear-patch violet. Highlands.
8. **White-tipped Sicklebill** (*Eutoxeres aquila*), p. 209. Very large; underparts streaked; strongly decurved bill.
9. **Violet Sabrewing** (*Campylopterus hemileucurus*), p. 213. Very large. (*a*) ♂: body deep violet; back dark green; broad tail with striking square white corners. (*b*) ♀: below gray with violet gorget. Middle elevations.
10. **Scaly-breasted Hummingbird** (*Phaeochroa cuvierii*), p. 212. Large; dingy; below scaly-green; tail with dull white corners set off by blackish.
11. **Green-fronted Lancebill** (*Doryfera ludoviciae*), p. 212. ♂: bill long, thin, slightly up-turned; forehead glittering blue-green (lacking in ♀); short, rounded bluish tail. Middle elevations.
12. **Red-footed Plumeleteer** (*Chalybura urochrysia*), p. 225. (*a*) ♂: below glittering dark green; tail purplish-black; feet red. (*b*) ♀: below dull gray; feet bright pink; tail tipped with dull gray. Caribbean slope.
13. **Green-breasted Mango** (*Anthracothorax prevostii*), p. 214. (*a*) ♂: below glittering green, shading to blue-black medially; broad tail purple. (*b*) ♀: below white with broad dark central stripe. Occasional ♀'s resemble ♂'s.
14. **Purple-crowned Fairy** (*Heliothryx barroti*), p. 228. (*a*) ♂: below white; long tail with white sides; crown violet; mask dark. (*b*) ♀: tail longer, no violet in crown.
15. **Green-crowned Brilliant** (*Heliodoxa jacula*), p. 227. (*a*) ♂: large; forehead and underparts glittering green; throat with purple "stickpin"; white postocular spot; tail forked. (*b*) ♀: below white, spangled with green; white stripe below eye; tail notched.
16. **Magnificent Hummingbird** (*Eugenes fulgens*), p. 227. Large; long straight bill. (*a*) ♂: throat green; crown violet; breast dark bronzy; prominent white postocular spot. (*b*) ♀: below gray; tail tipped with white; postocular stripe white. Highlands.
17. **White-necked Jacobin** (*Florisuga mellivora*), p. 213. (*a*) ♂: hood blue; nape and belly white; tail mostly white. (*b*) ♀: below with scaly pattern; outer rectrices edged and tipped with white. Many ♀♀ more or less like ♂.
18. **Long-billed Starthroat** (*Heliomaster longirostris*), p. 229. ♂: crown bright blue-green (duller in ♀); gorget purple (smaller in ♀); facial stripes white; white stripe down rump; long straight bill.
19. **Plain-capped Starthroat** (*Heliomaster constantii*), p. 228. ♂: gorget red (anterior half black in ♀); crown unmarked; facial stripes; long straight bill; rather short tail; white stripe down rump. N Pacific slope.

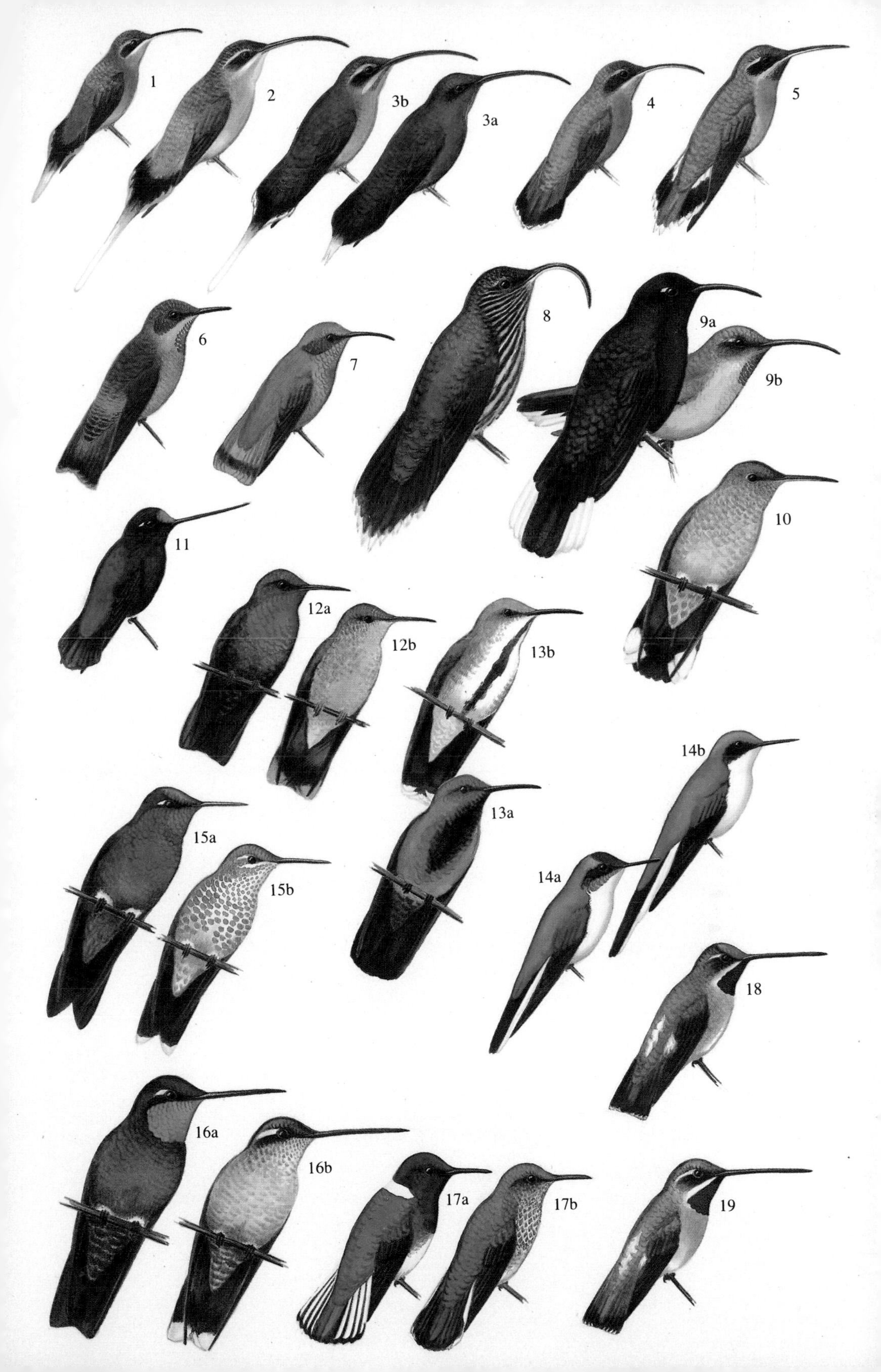
1
2
3b
3a
4
5
6
7
8
9a
9b
10
11
12a
12b
13b
14b
13a
15a
15b
14a
18
16a
16b
17a
17b
19

Plate 24: Hummingbirds, 2: Medium-sized Species and Dark Males

1. **White-bellied Emerald** (*Amazilia candida*), p. 219. Below white with some green spotting on sides of breast; tail greenish-bronze with dark subterminal band. Rare.
2. **Mangrove Hummingbird** (*Amazilia boucardi*), p. 220. (*a*) ♂: throat and breast bright bluish-green; belly whitish; tail blackish, rather long. (*b*) ♀: like 1 but more extensively spotted with green on sides; tail blacker. Pacific mangroves.
3. **Snowy-bellied Hummingbird** (*Amazilia edward*), p. 222. Throat and breast glittering green; belly white; rump, tail-, and wing-coverts bronzy. General–Térraba–Coto Brus region.
4. **Crowned Woodnymph** (*Thalurania colombica*), p. 217. ♂: dark; slender; long forked tail; throat and chest glittering green; crown and belly violet. [♀ on Pl. 25(15).]
5. **White-bellied Mountain-gem** (*Lampornis hemileucus*), p. 225. ♂: broad white postocular stripe combined with white underparts diagnostic; ♀ lacks violet gorget. Caribbean foothills.
6. **Gray-tailed Mountain-gem** [*Lampornis* (*castaneoventris*) *cinereicauda*], p. 227. ♂: crown glittering pale blue-green; throat and postocular stripe white; tail gray. ♀ like 7b but tail duller. Cordillera de Talamanca.
7. **Purple-throated Mountain-gem** [*Lampornis* (*castaneoventris*) *calolaema*], p. 226. (*a*) ♂: like 6 but gorget purple, tail blue-black. (*b*) ♀: below mostly pale rufous; postocular stripe white.
8. **Sapphire-throated Hummingbird** (*Lepidopyga coeruleogularis*), p. 218. ♂: throat and chest violet-blue; deeply forked tail. Possible along Panama border on Pacific slope. [♀ on Pl. 25(13).]
9. **Blue-throated Goldentail** (*Hylocharis eliciae*), p. 219. (*a*) ♂: throat and chest blue; belly buffy; tail golden; bill mostly red, with broad base. (*b*) ♀: below mostly buffy, spotted with blue; tail duller, tipped with buffy.
10. **Rufous-tailed Hummingbird** (*Amazilia tzacatl*), p. 222. Throat and breast glittering green; belly grayish; tail-coverts and tail rufous. Bill of ♂ mostly red, of ♀ and immature, mostly black.
11. **Cinnamon Hummingbird** (*Amazilia rutila*), p. 222. Below entirely cinnamon-rufous; tail rufous. Bill of ♂ mostly red, of ♀ and immature, mostly black. N Pacific slope.
12. **Fiery-throated Hummingbird** (*Panterpe insignis*), p. 218. In good light, blue crown, copper-orange throat and breast diagnostic; otherwise looks dark, glossy green with blue-black tail; postocular spot white. Highlands.
13. **Fork-tailed Emerald** (*Chlorostilbon canivetii*), p. 217. ♂: small; deeply forked tail; below glittering green; rump and tail-coverts green. [♀ on Pl. 25(10).]
14. **Blue-chested Hummingbird** (*Amazilia amabilis*), p. 220. (*a*) ♂: crown and malar area glittering green; chest blue-violet; tail dark. (*b*) ♀: below whitish, spotted with green and (on chest) blue. Caribbean slope.
15. **Steely-vented Hummingbird** (*Amazilia saucerrottei*), p. 221. Below rather dark glittering green; tail blue-black; rump purplish-bronze. Mainly N Pacific slope.
16. **Blue-tailed Hummingbird** (*Amazilia cyanura*), p. 221. Very like 15, best told by rufous in secondaries; tail more violet-blue.
17. **Striped-tailed Hummingbird** (*Eupherusa eximia*), p. 223. ♂: throat and breast glittering green; belly whitish; secondaries rufous; white in tail. [♀ on Pl. 25(14).]
18. **Beryl-crowned Hummingbird** (*Amazilia decora*), p. 219. (*a*) ♂: crown glittering blue-green; chest dark violet-blue; belly grayish; rather long bill. (*b*) ♀: below pale gray; breast spotted with green and blue; tail dark, tipped with pale grayish. S Pacific slope.
19. **White-tailed Emerald** (*Elvira chionura*), p. 224. ♂: small; throat and breast glittering green; belly white; tail mostly white with black terminal band. S Pacific slope. [♀ on Pl. 25(16).]
20. **Coppery-headed Emerald** (*Elvira cupreiceps*), p. 224. ♂: small; crown, upper tail-coverts and central rectrices coppery-bronze; tail mostly white; decurved bill. Mostly Caribbean slope. [♀ on Pl. 25(17).]
21. **Black-bellied Hummingbird** (*Eupherusa nigriventris*), p. 223. ♂: face and underparts velvety black; crissum and most of tail white; rufous in secondaries. Caribbean slope. [♀ on Pl. 25(12).]

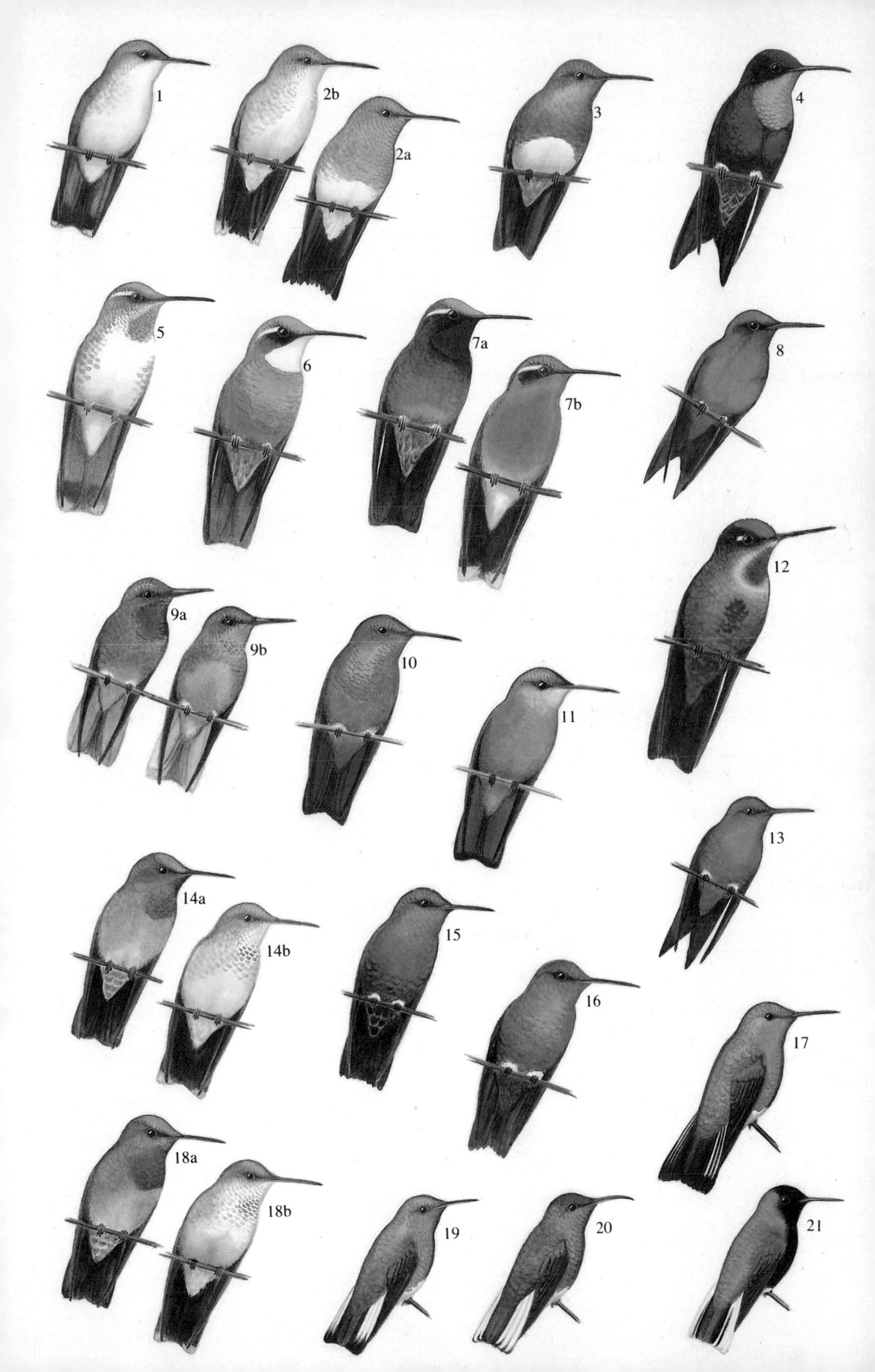
1
2b
2a
3
4
5
6
7a
7b
8
12
9a
9b
10
11
13
14a
14b
15
16
17
18a
18b
19
20
21

Plate 25: Hummingbirds, 3: Small Species, Females

1. **Green Thorntail** (*Discosura conversii*), p. 216. Rump-band white. (*a*) ♂: dark green; wirelike tail. (*b*) ♀: mustache and patch on lower side white. Caribbean slope.
2. **Magenta-throated Woodstar** (*Calliphlox bryantae*), p. 229. (*a*) ♂: gorget purple; collar white; buffy-white patch on sides of rump; long forked tail. (*b*) ♀: note rump patches, rufous underparts; larger than 3 or 7, with longer bill.
3. **Volcano Hummingbird** (*Selasphorus flammula*), p. 231. Below mostly white; ♂♂ with green sides of breast. (*a*) Talamanca ♂: gorget purplish-gray or purplish-green. (*b*) Poas-Barva ♂: gorget rose-red. (*c*) Irazú-Turrialba ♂: gorget dull purple. (*d*) ♀: speckled throat; outer 3 rectrices tipped with white or buffy.
4. **Rufous-crested Coquette** (*Lophornis delattrei*), p. 216. Rump-band white. (*a*) ♂: bushy crest rufous; green gorget with "necklace." (*b*) ♀: face rufous; breast green. Casual to accidental.
5. **White-crested Coquette** (*Lophornis adorabilis*), p. 215. Rump-band white. (*a*) ♂: crest white; gorget and cheek-tufts green; below rufous. (*b*) ♀: throat and chest white; belly rufous. S Pacific slope.
6. **Black-crested Coquette** (*Lophornis helenae*), p. 216. Rump-band white. (*a*) ♂: long, wispy black crest; gorget green; belly with bronze spots. (*b*) ♀: black mask; spotted belly. Caribbean slope.
7. **Scintillant Hummingbird** (*Selasphorus scintilla*), p. 230. (*a*) ♂: gorget orange-red; collar white; rufous on tail and sides of breast. (*b*) ♀: smaller, more rufescent than 3d.
8. **Snowcap** (*Microchera albocoronata*), p. 225. (*a*) ♂: deep purple; crown shining white; white on tail. (*b*) ♀: below whiter than other tiny hummers; short bill; short tail largely whitish. Caribbean slope.
9. **Ruby-throated Hummingbird** (*Archilochus colubris*), p. 230. (*a*) ♂: gorget red; deeply forked tail black. (*b*) ♀: most like 10 but below whiter; postocular spot white.
10. **Fork-tailed Emerald** (*Chlorostilbon canivetii*), p. 217. ♀: below grayer than 9b, with white postocular stripe over dusky auriculars; tail white-tipped, double-rounded. [♂ on Pl. 24(13).]
11. **Violet-headed Hummingbird** (*Klais guimeti*), p. 215. Large, square white postocular spot. (*a*) ♂: head and chest violet-blue. (*b*) ♀: larger, below grayer than 8b.
12. **Black-bellied Hummingbird** (*Eupherusa nigriventris*), p. 223. ♀: below gray; tail mostly white; rufous in secondaries. Caribbean slope. [♂ on Pl. 24(21).]
13. **Sapphire-throated Hummingbird** (*Lepidopyga coeruleogularis*), p. 218. ♀: below white; rather deeply forked tail. [♂ on Pl. 24(8).]
14. **Striped-tailed Hummingbird** (*Eupherusa eximia*), p. 223. ♀: below pale gray; tail largely white; rufous on secondaries; larger than 12, more green on sides of throat and breast. [♂ on Pl. 24(17).]
15. **Crowned Woodnymph** (*Thalurania colombica*), p. 217. ♀: pale gray throat and chest contrast with dark belly. [♂ on Pl. 24(4).]
16. **White-tailed Emerald** (*Elvira chionura*), p. 224. ♀: below white, sides spotted with green; tail white with bold subterminal black band. [♂ on Pl. 24(19).]
17. **Coppery-headed Emerald** (*Elvira cupreiceps*), p. 224. ♀: decurved bill; whiter below than 12, no rufous in secondaries; broken subterminal black band on white tail. [♂ on Pl. 24(20).]

1b
1a
2a
2b
3a
3b
3c
3d
4a
4b
5a
5b
6a
6b
7a
7b
8a
8b
9a
9b
10
11a
11b
12
13
14
15
16
17

Plate 26: Trogons and Jacamars

1. **Resplendent Quetzal** (*Pharomachrus mocinno*), p. 232. Much larger than other trogons. (*a*) ♂: crest unmistakable; tail white; long tail-coverts before and during breeding season. (*b*) ♀: barred tail; no other trogon has green chest, mostly gray underparts. Highlands.
2. **Slaty-tailed Trogon** (*Trogon massena*), p. 232. Large; belly red; no breast-band; tail dark gray. (*a*) ♂: above green; shoulder pale; eye-ring and bill reddish. (*b*) ♀: above gray.
3. **Lattice-tailed Trogon** (*Trogon clathratus*), p. 233. (*a*) ♂: like 2 but tail faintly barred; iris white; bill yellow. (*b*) ♀: note brownish breast. Caribbean slope.
4. **Orange-bellied Trogon** (*Trogon aurantiiventris*), p. 235. ♂. Except for orange to orange-red bellies, sexes identical to corresponding sex of 5. Mainly N cordilleras.
5. **Collared Trogon** (*Trogon collaris*), p. 235. Small; red-bellied; breast-band white. (*a*) ♂: above green; barred tail; lacks conspicuous eye-ring. (*b*) ♀: above brown; tail below gray with dark bar setting off white tips to rectrices; note eye-ring. Highlands, except N cordilleras.
6. **Baird's Trogon** (*Trogon bairdii*), p. 233. (*a*) ♂: dark blue; belly deep orange; tail white below. (*b*) ♀: less red below than 2b, with barred tail; bill and eye-ring bluish. S Pacific slope.
7. **Elegant Trogon** (*Trogon elegans*), p. 234. (*a*) ♂: like 5a but eye-ring orange; broader white tips to rectrices. (*b*) ♀: white bar on auriculars; sides of breast brown; tail like ♂. Guanacaste only.
8. **Violaceous Trogon** (*Trogon violaceus*), p. 236. Belly yellow; barred tail. (*a*) ♂: head and breast mostly violet-blue; eye-ring yellow. (*b*) ♀: head dark gray.
9. **Black-throated Trogon** (*Trogon rufus*), p. 236. (*a*) ♂: like 8a but head and breast green; eye-ring bluish; bill yellow. (*b*) ♀: mostly brown; tail above rufous, below barred.
10. **Black-headed Trogon** (*Trogon melanocephalus*), p. 234. (*a*) ♂: larger than 8a without barred tail or pale shoulder; eye-ring bluish. (*b*) ♀: duller; above grayer than ♂; narrower white tips to rectrices.
11. **Great Jacamar** (*Jacamerops aurea*), p. 243. Large; stout bill; head bluish; back coppery-purplish; tail below blackish. (*a*) ♂: lower throat white. (*b*) ♀: no white on throat. Caribbean slope.
12. **Rufous-tailed Jacamar** (*Galbula ruficauda*), p. 243. Long, slim bill; metallic green breast-band; underside of tail rufous. (*a*) ♂: throat white. (*b*) ♀: throat buffy.

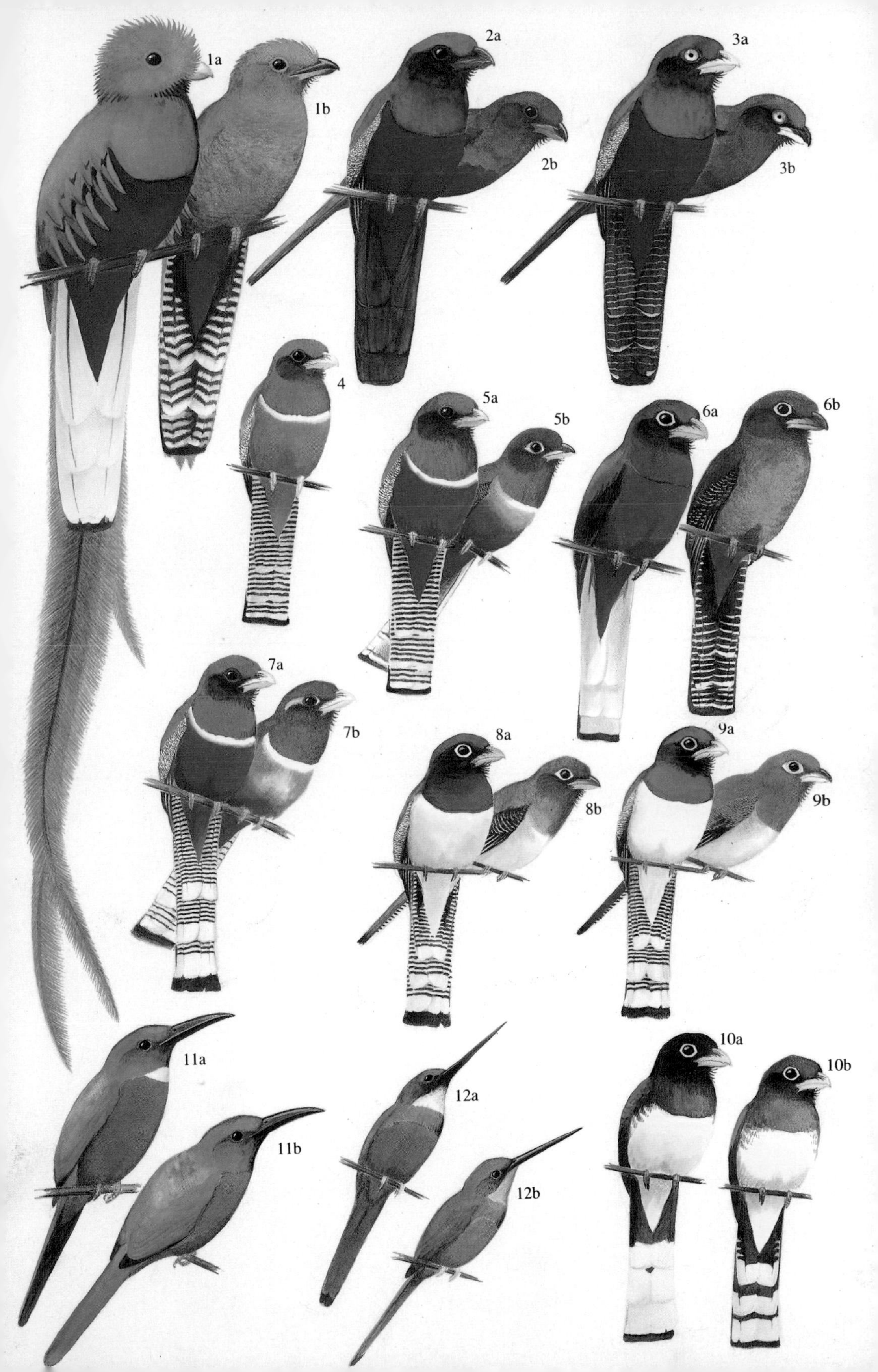
1a
1b
2a
2b
3a
3b
4
5a
5b
6a
6b
7a
7b
8a
8b
9a
9b
10a
10b
11a
11b
12a
12b

Plate 27: Motmots, Kingfishers, Toucans, and Large Woodpeckers

1. **Ringed Kingfisher** (*Ceryle torquata*), p. 237. Very large; bushy crest; collar white. (*a*) ♂: breast and belly rufous. (*b*) ♀: chest-band slate-blue.
2. **Belted Kingfisher** (*Ceryle alcyon*), p. 238. Smaller than 1; below mostly white; in flight shows white wing-patch. (*a*) ♂: breast-band blue. (*b*) ♀: breast-bands blue and rufous.
3. **Green-and-rufous Kingfisher** (*Chloroceryle inda*), p. 239. Above dark green; below rufous, paler on throat. (*a*) ♂: breast unmarked. (*b*) ♀: chest-band green, flecked with white. Caribbean lowlands.
4. **Amazon Kingfisher** (*Chloroceryle amazona*), p. 238. Above dark green; below white; very little white in wings or tail. (*a*) ♂: breast rufous. (*b*) ♀: breast and sides spotted with green.
5. **Green Kingfisher** (*Chloroceryle americana*), p. 238. Smaller than 4 with relatively smaller bill and crest, more white in wings and tail. (*a*) ♂: breast rufous. (*b*) ♀: 2 indistinct green breast-bands.
6. **American Pygmy Kingfisher** (*Chloroceryle aenea*), p. 239. Much smaller than 3, with white belly. (*a*) ♂: no breast-band. (*b*) ♀: breast-band green.
7. **Rufous Motmot** (*Baryphthengus martii*), p. 241. Large; long-tailed; mostly rufous; broad mask, small chest-spot. Caribbean slope.
8. **Blue-crowned Motmot** (*Momotus momota*), p. 242. Center of crown and mask black; blue diadem; throat bluish-green; underparts vary from green to tawny.
9. **Broad-billed Motmot** (*Electron platyrhynchum*), p. 240. Much smaller than 7 with smaller mask, larger chest-spot; chin blue; belly green; bill broad and flat. Caribbean slope.
10. **Keel-billed Motmot** (*Electron carinatum*), p. 241. Size and shape of 9 but only rufous on forehead; from larger 8 by green crown, only blue over eye. N Caribbean slope.
11. **Turquoise-browed Motmot** (*Eumomota superciliosa*), p. 241. Very large "long-stalked" racquets; broad eyebrows pale turquoise; mask and center of throat black; back rufous; belly cinnamon-rufous. Northwest.
12. **Tody Motmot** (*Hylomanes momotula*), p. 240. Very small; lacks racquets; note indistinct streaking below, blue over eye, rufous crown and nape.
13. **Lineated Woodpecker** (*Dryocopus lineatus*), p. 257. Large; crested; note white stripe up neck, continuing across face to bill. (*a*) ♂: crest and malar stripe red. (*b*) ♀: forehead and malar stripes black.
14. **Pale-billed Woodpecker** (*Campephilus guatemalensis*), p. 258. Larger, stouter than 13 with less pointed crest, no facial stripes. (*a*) ♂: head red. (*b*) ♀: front of crest black.
15. **Collared Aracari** (*Pteroglossus torquatus*), p. 248. Chest black; yellow below with black breast-spot, red and black band across belly; bill yellowish and black, tomia serrated.
16. **Fiery-billed Aracari** (*Pteroglossus frantzii*), p. 249. Bill brilliant orange; red band across belly; larger breast spot than 15. Replaces 15 on S Pacific slope.
17. **Emerald Toucanet** (*Aulacorhynchus prasinus*), p. 248. Small; mostly green with blue throat; chestnut on tail and crissum; bill yellow and black. Highlands.
18. **Keel-billed Toucan** (*Ramphastos sulfuratus*), p. 250. Black with yellow-green face, yellow bib; crissum red; upper tail-coverts white; rainbow bill.
19. **Chestnut-mandibled Toucan** (*Ramphastos swainsonii*), p. 250. From smaller 18 by yellow and maroon bill.
20. **Yellow-eared Toucanet** (*Selenidera spectabilis*), p. 249. Only toucan with throat, breast, and belly black; bicolored bill. (*a*) ♂: head black; ear-tufts yellow. (*b*) ♀: crown and nape chestnut; no ear-tufts. Caribbean slope.

1a
1b
2b
2a
7
8
9
4a
3a
3b
4b
11
10
5a
5b
6a
6b
12
15
13b
13a
16
18
17
14b
14a
19
20a
20b

Plate 28: Barbets, Puffbirds, and Woodpeckers

1. **Prong-billed Barbet** (*Semnornis frantzii*), p. 247. Chunky. ♂: orange-tawny; thick bill grayish; ♀ lacks long black nape feathers. Highlands.
2. **Red-headed Barbet** (*Eubucco bourcierii*), p. 246. Thick bill yellow; streaked sides. (*a*) ♂: head red. (*b*) ♀: auriculars bluish; black around eye; rest of head and neck orangish.
3. **White-necked Puffbird** (*Bucco macrorhynchos*), p. 244. Large; black and white; very large head; heavy bill.
4. **White-fronted Nunbird** (*Monasa morphoeus*), p. 246. Slaty, with bristly white feathers surrounding base of orange-red bill. Caribbean slope.
5. **Lanceolated Monklet** (*Micromonacha lanceolata*), p. 245. Small; above brown, below white with bold black streaks.
6. **White-whiskered Puffbird** (*Malacoptila panamensis*), p. 245. Big-headed, with white mustache and bristly feathers at base of bill; iris red. (*a*) ♂: more cinnamon. (*b*) ♀: browner, more heavily streaked.
7. **Pied Puffbird** (*Bucco tectus*), p. 244. Much smaller than 3; note white eyebrow, white in scapulars and tail. Usually in groups. Caribbean lowlands.
8. **Chestnut-colored Woodpecker** (*Celeus castaneus*), p. 257. ♂: head tawny-buff with ragged crest; body deep chestnut, marked obscurely with black; ♀ lacks crimson malar area. Caribbean slope.
9. **Cinnamon Woodpecker** (*Celeus loricatus*), p. 257. ♂: head not paler than back; below paler than 8, black markings very conspicuous; short crest; ♀ lacks red throat. Caribbean slope.
10. **Rufous-winged Woodpecker** (*Piculus simplex*), p. 256. Mostly greenish; underparts barred with yellowish; primaries rufous. (*a*) ♂: crown and malar stripe red; slight crest. (*b*) ♀: red on nape only.
11. **Golden-olive Woodpecker** (*Piculus rubiginosus*), p. 256. Above olive; below barred; face mostly whitish; crown slaty. (*a*) ♂: nape, borders of crown, and malar stripe crimson. (*b*) ♀: only nape crimson.
12. **Red-rumped Woodpecker** (*Veniliornis kirkii*), p. 255. ♂: small; above olive with red rump; yellow on hindneck; below barred; streaky red crown lacking in ♀.
13. **Smoky-brown Woodpecker** (*Veniliornis fumigatus*), p. 255. Small; brownish; tail black. (*a*) ♂: crown mottled with red. (*b*) ♀: crown blackish.
14. **Black-cheeked Woodpecker** (*Melanerpes pucherani*), p. 252. Above mostly black; back barred; rump white; sides of head and neck black; white spot behind eye. ♂ with crown and nape red; ♀ with center of crown black. Caribbean slope.
15. **Acorn Woodpecker** (*Melanerpes formicivorus*), p. 252. Harlequin face pattern; rump and wing-patch white; streaked breast. ♂: pileum entirely red; ♀ forecrown black. Highlands.
16. **Hoffmann's Woodpecker** (*Melanerpes hoffmannii*), p. 253. Back barred; belly yellow. (*a*) ♂: nape yellow; crown red. (*b*) ♀: lacks red crown. Mainly N Pacific slope, spreading.
17. **Red-crowned Woodpecker** (*Melanerpes rubricapillus*), p. 253. Similar to 16, but nape and belly red. S Pacific slope.
18. **Golden-naped Woodpecker** (*Melanerpes chrysauchen*), p. 252. Like 14 but nape yellow, center of back white, with barring only on wings. (*a*) ♂: crown red. (*b*) ♀: lacks red on head. S Pacific slope only.
19. **Hairy Woodpecker** (*Picoides villosus*), p. 254. ♂: fairly small; below brownish; white stripes on face and down center of back. ♀ lacks red patch on nape. Highlands.
20. **Yellow-bellied Sapsucker** (*Sphyrapicus varius*), p. 254. Belly yellowish, long wing-stripe white; barred wings; back mottled. (*a*) ♂: face striped; crown and throat red. (*b*) Immature: back and breast more brownish, mottled; facial pattern indistinct.

1
2a
2b
3
4
5
6b
6a
7
8
9
10a
10b
11b
11a
12
13a
13b
14
15
16b
16a
17
18b
18a
19
20a
20b

Plate 29: Woodcreepers, Furnariids, and Piculet

1. **Olivaceous Piculet** (*Picumnus olivaceus*), p. 251. Small, chunky; short, striped tail; note dotted nape; ♂ crown orange-flecked.
2. **Streaked Xenops** (*Xenops rutilans*), p. 275. Like 3 but streaked below, superciliaries whitish. Highlands.
3. **Plain Xenops** (*Xenops minutus*), p. 274. Small; below tawny-brown; conspicuous malar stripe silvery; wing and tail stripes. Lower elevations than 2.
4. **Ruddy Treerunner** (*Margarornis rubiginosus*), p. 269. Bright rufous; throat and broad superciliaries whitish; long, weakly ''spine-tipped'' tail. Highlands.
5. **Spotted Barbtail** (*Premnoplex brunnescens*), p. 268. Below with buffy spots neatly ringed with black; above rich brown; tail weakly ''spine-tipped.'' Middle elevations.
6. **Wedge-billed Woodcreeper** (*Glyphorhynchus spirurus*), p. 261. Small; looks rather square-headed; bill wedge-shaped; narrow superciliaries and breast-streaks buffy; in flight note buffy wing-stripe.
7. **Olivaceous Woodcreeper** (*Sittasomus griseicapillus*), p. 261. Small; slender; olive-gray body contrasts with chestnut wings and tail; rufous in primaries.
8. **Streaked-headed Woodcreeper** (*Lepidocolaptes souleyetii*), p. 265. Fairly small; pale, slender, slightly decurved bill; broad buffy breast-streaks edged with black; streaked pileum.
9. **Spotted-crowned Woodcreeper** (*Lepidocolaptes affinis*), p. 265. Larger than 8 but similar pattern; note spotted pileum. Higher elevations than 8.
10. **Long-tailed Woodcreeper** (*Deconychura longicauda*), p. 260. Fairly small, slender; buffy throat trails off into spots, not streaks, on breast; long tail-spines.
11. **Plain-brown Woodcreeper** (*Dendrocincla fuliginosa*), p. 259. No streaking or spotting; throat and face look grayish, with trace of dark malar stripe; note rather pale iris.
12. **Tawny-winged Woodcreeper** (*Dendrocincla anabatina*), p. 260. More pattern than 11; note dull buffy throat and fine superciliaries, rufous in wing; head often appears ragged, unkempt. Mainly S Pacific slope.
13. **Ruddy Woodcreeper** (*Dendrocincla homochroa*), p. 260. Entirely rufescent, throat paler; note dull gray lores, unkempt appearance of head.
14. **Brown-billed Scythebill** (*Campylorhamphus pusillus*), p. 266. Unmistakable long, slender, decurved bill; above and below with buffy streaking.
15. **Black-banded Woodcreeper** (*Dendrocolaptes picumnus*), p. 263. Large; only woodcreeper to combine buffy streaking on chest and black barring on belly; stout straight bill. Middle elevations.
16. **Black-striped Woodcreeper** (*Xiphorhynchus lachrymosus*), p. 264. Most strikingly patterned woodcreeper; back and breast feathers pale buff, edged with black; wings rufous; straight, strong bill.
17. **Buff-throated Woodcreeper** (*Xiphorhynchus guttatus*), p. 263. Dark, straight, fairly long and stout bill; malar streak dark; broad streaks on breast buffy, on back and head streaks narrow.
18. **Strong-billed Woodcreeper** (*Xiphocolaptes promeropirhynchus*), p. 262. Very large, with very stout bill; face rather pale and finely streaked; indistinct dark cheek-stripe. Middle elevations, rare.
19. **Barred Woodcreeper** (*Dendrocolaptes certhia*), p. 262. Large; no other woodcreeper has body entirely barred; stout dark bill.
20. **Spotted Woodcreeper** (*Xiphorhynchus erythropygius*), p. 264. Body plumage rather olive, spotted with yellowish; broad indistinct pale eye-ring.
21. **Ivory-billed Woodcreeper** (*Xiphorhynchus flavigaster*), p. 264. Like 17 but slightly larger, with pale bill, below with more extensive and contrasting buffy streaking. Guanacaste only.

1
2
3
4
5
6
7
8
9
10
11
12
13
14
15
16
17
18
19
20
21

Plate 30: Furnariids and Terrestrial Antbirds

1. **Buffy Tuftedcheek** (*Pseudocolaptes lawrencii*), p. 269. Buffy of throat extends back over lower cheeks; tail orange-rufous. Highlands.
2. **Spectacled Foliage-gleaner** (*Anabacerthia variegaticeps*), p. 271. Spectacles buffy-orange; throat buffy, obscurely scaled posteriorly; below with no streaks; notched tail rufous. Middle elevations.
3. **Buff-throated Foliage-gleaner** (*Automolus ochrolaemus*), p. 272. Conspicuous buffy throat and spectacles. Lowlands.
4. **Striped Foliage-gleaner** (*Hyloctistes subulatus*), p. 270. Throat buffy; underparts broadly but indistinctly streaked with buff; head finely streaked.
5. **Ruddy Foliage-gleaner** (*Automolus rubiginosus*), p. 272. Large; dark; unpatterned; throat and chest ochraceous. Coto Brus area.
6. **Buff-fronted Foliage-gleaner** (*Philydor rufus*), p. 271. Tawny-buff face contrasts with dark eye-stripe, grayish crown; tail bright rufous, looks notched. Middle elevations.
7. **Streaked-breasted Treehunter** (*Thripadectes rufobrunneus*), p. 271. Large; stout black bill; throat and breast with ochraceous streaks. Middle elevations.
8. **Lineated Foliage-gleaner** (*Syndactyla subalaris*), p. 270. (*a*) Bill rather short, sharp-pointed; throat buffy; fine breast-streaks. (*b*) Young: superciliaries and below largely orange-rufous. Middle elevations.
9. **Scaly-throated Leaftosser** (*Sclerurus guatemalensis*), p. 274. Dark, especially on wings and tail; throat pale, scaled with blackish posteriorly. Lowlands.
10. **Gray-throated Leaftosser** (*Sclerurus albigularis*), p. 273. Gray throat contrasts with dark rufous breast; tail black; rump chestnut. Middle elevations.
11. **Tawny-throated Leaftosser** (*Sclerurus mexicanus*), p. 273. Dark brown with rufous throat and chest, paling to buffy on chin. Middle elevations.
12. **Red-faced Spinetail** (*Cranioleuca erythrops*), p. 268. Slender; mostly brownish-olive; half-hood, wings, and long tail rufous. Young with orange-rufous face, dark eye-stripe and crown.
13. **Ochre-breasted Antpitta** (*Grallaricula flavirostris*), p. 289. Small; tailless; rather like a miniature 14 with a yellow bill; malar stripe black. Middle elevations.
14. **Fulvous-bellied Antpitta** (*Hylopezus fulviventris*), p. 289. Tailless; pileum dark slate; back olive; below mostly white; breast ochraceous, indistinctly scaled or streaked with black. Caribbean slope.
15. **Black-faced Antthrush** (*Formicarius analis*), p. 286. Long-legged; tail usually cocked; face black with rufous behind; underparts grayish.
16. **Rufous-breasted Antthrush** (*Formicarius rufipectus*), p. 287. Like 15 but top and back of head, underparts chestnut-rufous. Middle elevations.
17. **Black-headed Antthrush** (*Formicarius nigricapillus*), p. 286. Very dark; head and underparts black; orbital skin conspicuous. Caribbean foothills.
18. **Spectacled Antpitta** (*Hylopezus perspicillatus*), p. 288. Conspicuous buffy spectacles; below white; breast boldly streaked with black.
19. **Black-crowned Antpitta** (*Pittasoma michleri*), p. 287. Very large; below heavily scaled with black. (*a*) ♂: head mostly black. (*b*) ♀: face and throat largely rufous. Caribbean foothills.
20. **Scaled Antpitta** (*Grallaria guatimalensis*), p. 288. Large; below rufous; malar stripe pale; above broadly scaled with black. Middle elevations.
21. **Wing-banded Antbird** (*Myrmornis torquata*), p. 286 (in Black-faced Antthrush, Note). ♂: face, throat, and breast black; belly gray; bold buffy wing-bars. ♀ similar but throat and breast rufous. (No definite Costa Rican record.)

1
2
3
4
5
6
7
8a
8b
9
10
11
12
13
14
15
16
17
18
19a
19b
20
21

Plate 31: Antbirds

1. **Fasciated Antshrike** (*Cymbilaimus lineatus*), p. 276. (*a*) ♂: crown black; iris red, finely barred with black and white. (*b*) ♀: crown rufous; barred with black and buff. Caribbean slope.
2. **Great Antshrike** (*Taraba major*), p. 276. Large; eyes red; heavy hooked bill; below white. (*a*) ♂: above black. (*b*) ♀: above rufous.
3. **Barred Antshrike** (*Thamnophilus doliatus*), p. 276. (*a*) ♂: smaller, more crested, more broadly barred than 1; eyes whitish. (*b*) ♀: above rufous; below buffy; striped face and crest.
4. **Russet Antshrike** (*Thamnistes anabatinus*), p. 278. Above brown; wings brighter; note pale superciliaries, dark eye-stripe (sometimes indistinct); below olive-buff, breast darker.
5. **Slaty Antshrike** (*Thamnophilus punctatus*), p. 277. (*a*) ♂: below slaty; above black; wings boldly marked with white. (*b*) ♀: brown and buff; wings marked with buffy-white. Caribbean slope.
6. **Black-hooded Antshrike** (*Thamnophilus bridgesi*), p. 277. (*a*) ♂: mostly black; belly slaty; dotted white wing-bars. (*b*) ♀: below grayish-olive, streaked with whitish. S Pacific slope.
7. **Ocellated Antbird** (*Phaenostictus mcleannani*), p. 285. Appears brown above and rufous below, heavily spotted with black; throat black; conspicuous blue facial skin. Caribbean slope.
8. **Dusky Antbird** (*Cercomacra tyrannina*), p. 282. (*a*) ♂: below slaty; above blackish-slate; smaller, slimmer bill, much less white on wing and tail than 5. (*b*) ♀: above olive-brown; below rufous. No colorful orbital skin.
9. **Bicolored Antbird** (*Gymnopithys leucaspis*), p. 284. Below white with dark sides; orbital skin blue.
10. **Immaculate Antbird** (*Myrmeciza immaculata*), p. 284. Conspicuous blue orbital skin becomes white behind eyes; plumage dark, unmarked. (*a*) ♂: black. (*b*) ♀: dark brown. Middle elevations.
11. **Chestnut-backed Antbird** (*Myrmeciza exsul*), p. 283. Orbital skin blue; head blackish; body chestnut-brown. (*a*) ♂: black extends to upper belly. (*b*) ♀: black less extensive.
12. **Spotted Antbird** (*Hylophylax naevioides*), p. 285. (*a*) ♂: brightly patterned; note wing-bars, band of spots across breast; throat black. (*b*) ♀: much duller, at most trace of spots; broad wing-bars rusty.
13. **Bare-crowned Antbird** (*Gymnocichla nudiceps*), p. 283. (*a*) ♂: black with white wing-bars; face and forecrown bare, bright blue. (*b*) ♀: head, underparts, and wing-bars rufous; orbital skin blue; bill silvery.
14. **Dull-mantled Antbird** (*Myrmeciza laemosticta*), p. 284. ♂: foreparts mostly blackish; iris red; dotted wing-bars. ♀ similar, throat spotted with white. Caribbean foothills.

1a
1b
2a
2b
3a
3b
4
5a
5b
6a
6b
7
8a
8b
9
10a
10b
11a
11b
12a
12b
13a
13b
14

Plate 32: Antwrens, Spinetails, and Other Small Understory Birds

1. **Dotted-winged Antwren** (*Microrhopias quixensis*), p. 281. Dotted shoulders; single broad wing-bar white; graduated tail tipped with white; slender bill. (*a*) ♂: velvety black. (*b*) ♀: above dark slate; below rufous.
2. **White-flanked Antwren** (*Myrmotherula axillaris*), p. 280. (*a*) ♂: blackish; dotted wing-bars; flanks white. (*b*) ♀: above brownish; head gray; eye-ring and below whitish. Caribbean slope.
3. **Slaty Antwren** (*Myrmotherula schisticolor*), p. 281. (*a*) ♂: dark slate; throat and breast black; narrow wing-bars. (*b*) ♀: brown; face and breast tinged with ochraceous. Middle elevations.
4. **Tawny-crowned Greenlet** (*Hylophilus ochraceiceps*), p. 380. Crown ochraceous; above olive-brown; breast dull buffy-yellow; iris pale.
5. **Checker-throated Antwren** (*Myrmotherula fulviventris*), p. 280. Brownish; below paler; wing-bars buffy; iris pale; heavy bill. (*a*) ♂: throat black, spotted with white. (*b*) throat pale buffy. Caribbean slope.
6. **Plain Antvireo** (*Dysithamnus mentalis*), p. 278. (*a*) ♂: head slaty; back olive; narrow wing-bars; underparts without streaks. (*b*) ♀: above browner; crown rufous; eye-ring white; no wing-bars.
7. **Streaked-crowned Antvireo** (*Dysithamnus striaticeps*), p. 279. (*a*) ♂: head gray, streaked with black; wing-bars white; breast streaked with black. (*b*) ♀: crown rufous, streaked with black; eye-ring and wing-bars whitish; streaked breast. N Caribbean lowlands.
8. **Spotted-crowned Antvireo** (*Dysithamnus puncticeps*), p. 279. (*a*) ♂. (*b*) ♀. Differ from corresponding sex of 7 by more spotted crown, breast streaking finer and less contrasting. S Caribbean lowlands.
9. **Slaty Spinetail** (*Synallaxis brachyura*), p. 267. (*a*) Dark gray; crown and wings rufous; long wispy tail. (*b*) Young: browner, with mottled throat, shorter tail.
10. **Rufous-rumped Antwren** (*Terenura callinota*), p. 281. Warblerlike; wing-bars yellow; rump rufous; eye-stripe dusky. (*a*) ♂: cap black. (*b*) ♀: cap olive, like back. Caribbean slope.
11. **Pale-breasted Spinetail** (*Synallaxis albescens*), p. 267. (*a*) Above brownish; below whitish; cap and wings rufous; face gray; long wispy tail. (*b*) Young: little or no rufous in cap or on wings; below buffy. S Pacific slope.
12. **Silvery-fronted Tapaculo** (*Scytalopus argentifrons*), p. 290. (*a*) ♂: blackish; eyebrow pale; rufous barring posteriorly. (*b*) ♀: paler, browner; throat dark gray, Highlands.
13. **Zeledonia** (*Zeledonia coronata*), p. 404. Very dark; short tail; long legs; crown orange. Highlands.
14. **Tawny-faced Gnatwren** (*Microbates cinereiventris*), p. 372. Face tawny; throat white; belly gray; bill broader than 15, without white in tail. Caribbean slope.
15. **Long-billed Gnatwren** (*Ramphocaenus melanurus*), p. 371. Head much less patterned than 14; bill very long and slender; tail long, graduated, white-tipped.

1b
1a
2b
2a
3b
3a
4
5b
5a
6b
6a
7b
7a
8b
8a
9b
9a
10b
10a
11b
11a
12a
12b
13
14
15

Plate 33: Manakins and Becards

1. **White-collared Manakin** (*Manacus candei*), p. 303. (*a*) ♂: pileum and wings black; broad collar white; belly yellow. (*b*) ♀: olive-green with yellow belly; legs orange. Caribbean slope.
2. **Orange-collared Manakin** (*Manacus aurantiacus*), p. 302. ♂: like larger 1 but white of collar replaced by orange. ♀ resembles 1b (no overlap). S Pacific slope.
3. **Gray-headed Manakin** (*Piprites griseiceps*), p. 303. Note white eye-ring on gray head, rather long tail, yellow underparts; usually tertials have inner webs largely pale yellow, forming stripe on side of back; flycatcherlike but with manakin bill. Caribbean slope.
4. **Long-tailed Manakin** (*Chiroxiphia linearis*), p. 301. (*a*) ♂: black with blue back; crest red; long central rectrices. (*b*) Immature ♂ (several plumage stages): intermediate. (*c*) ♀: olive-green, below paler; legs orange; central rectrices somewhat elongated. N Pacific slope.
5. **Lance-tailed Manakin** (*Chiroxiphia lanceolata*), p. 301. Resemble corresponding sex of 4 (no overlap); ♂ more greenish-black with shorter central rectrices. Coto Brus area.
6. **Red-capped Manakin** (*Pipra mentalis*), p. 299. (*a*) ♂: head red; eyes white; thighs yellow. (*b*) ♀: from other ♀ manakins by short tarsi, iris sometimes pale; below pale olive to yellowish.
7. **Blue-crowned Manakin** (*Pipra coronata*), p. 300. (*a*) ♂: black with blue crown. (*b*) ♀: deeper green than other ♀ manakins.
8. **White-crowned Manakin** (*Pipra pipra*), p. 300. (*a*) ♂: velvety black; pileum white. (*b*) ♀: head grayish; iris reddish; throat grayish; breast pale olive. Caribbean slope.
9. **White-ruffed Manakin** (*Corapipo leucorrhoa*), p. 302. (*a*) ♂: glossy blue-black; gorget white. (*b*) ♀: throat pale gray; breast olive; iris and legs dark.
10. **Thrushlike Manakin** (*Schiffornis turdinus*), p. 304. Olive-brown; wings and tail browner; may show grayish feathering around eye. Cotingalike.
11. **Cinnamon Becard** (*Pachyramphus cinnamomeus*), p. 291. Above rufous; below paler; throat and superciliaries palest. Smaller, with more contrasting pattern than Rufous Mourner or Rufous Piha.
12. **Rose-throated Becard** (*Pachyramphus aglaiae*), p. 293. (*a*) ♂ (Pacific race): slaty; cap black; bushy crest; little or no rose in throat. ♂ of Caribbean race entirely blackish with paler belly. (*b*) ♀: above rufous; below buffy; cap gray.
13. **White-winged Becard** (*Pachyramphus polychopterus*), p. 292. (*a*) ♂: gray and black; wings and scapulars boldly edged with white. (*b*) ♀: note white spectacles, buffy wing-bars, buffy tips to rectrices.
14. **Barred Becard** (*Pachyramphus versicolor*), p. 291. Note finely barred underparts, eye-ring. (*a*) ♂: above black; much white in wing. (*b*) ♀: above olive with gray cap; rufous in wing. Highlands.
15. **Black-and-white Becard** (*Pachyramphus albogriseus*), p. 292. (*a*) ♂: from 13a by white stripe over lores, gray back. (*b*) ♀: from 13b by rufous cap, black line through eye bordering cap; below more yellowish.

1a
2
4a
5
1b
4b
3
4c
6a
6b
8a
8b
7a
7b
9a
10
9b
11
14b
12b
13b
14a
12a
13a
15b
15a

Plate 34: Tityras, Cotingas, Sharpbill, and Mourners

1. **Masked Tityra** (*Tityra semifasciata*), p. 293. Facial skin and base of bill red; primaries and tail-band black. (*a*) ♂: above grayish-white; face black. (*b*) ♀: above brownish; hood darker.
2. **Black-crowned Tityra** (*Tityra inquisitor*), p. 294. No bare facial area; smaller than 1; cap black. (*a*) ♂: above pale gray. (*b*) ♀: face rufous; back brownish.
3. **Yellow-billed Cotinga** (*Carpodectes antoniae*), p. 297. Bill yellow; otherwise like corresponding sex of 4. S Pacific slope.
4. **Snowy Cotinga** (*Carpodectes nitidus*), p. 296. Bill black and gray. (*a*) ♂: shining white; crown tinged with blue-gray. (*b*) ♀: above gray; wings with bold white margins. Caribbean slope.
5. **Turquoise Cotinga** (*Cotinga ridgwayi*), p. 296. Resembles corresponding sex of 6 but smaller; ♂ has black eye-ring, spots on back, and shoulders. S Pacific slope.
6. **Lovely Cotinga** (*Cotinga amabilis*), p. 295. (*a*) ♂: brilliant blue; throat and belly purple. (*b*) ♀: below with distinctive speckled pattern; above scaled; note rather dovelike head shape. Caribbean slope.
7. **Sharpbill** (*Oxyruncus cristatus*), p. 299. Above olive; below spotted; note sharp-pointed bill. Crest (nearly always concealed) red in ♂, smaller and orange in ♀.
8. **Speckled Mourner** (*Laniocera rufescens*), p. 295. Rufous; breast faintly barred; wings darker with rufous wing-bars; yellow tufts at sides of breast usually concealed. Cotingalike head shape.
9. **Rufous Mourner** (*Rhytipterna holerythra*), p. 314. Rufous; below slightly paler; bill pale at base; kingbird-sized.
10. **Rufous Piha** (*Lipaugus unirufus*), p. 295. Similar to 9 but larger, stouter; robin-sized; different voice.
11. **Purple-throated Fruitcrow** (*Querula purpurata*), p. 297. Black; rather short, square tail; bill silvery. (*a*) ♂: throat purple. (*b*) ♀ and immature: all black. Caribbean slope.
12. **Three-wattled Bellbird** (*Procnias tricarunculata*), p. 298. (*a*) ♂: body chestnut; head white; 3 wormlike wattles. (*b*) ♀: below yellow, heavily streaked with dark olive. Immature ♂'s resemble ♀'s but usually show distinct wattles.
13. **Bare-necked Umbrellabird** (*Cephalopterus glabricollis*), p. 297. Very large; black; heavy bill. (*a*) ♂: umbrellalike crest; inflatable bare red throat (largely concealed when not inflated). (*b*) ♀: short crest gives flat-headed look. (*c*) Flying: very broad wings, front-heavy silhouette. Caribbean slope.

2a
3
1a
2b
1b
4a
4b
7
5
8
9
6a
10
6b
11a
11b
12b
13c
13a
13b
12a

Plate 35: Flycatchers, 1: Medium and Large Species

1. **Tropical Kingbird** (*Tyrannus melancholicus*), p. 307. Note notched, all-dark tail, dark mask, olive wash on chest, thick bill.
2. **Western Kingbird** (*Tyrannus verticalis*), p. 308. Outer rectrices white; breast and back paler, bill smaller than 1; mask less pronounced.
3. **Eastern Kingbird** (*Tyrannus tyrannus*), p. 307. Above blackish; below white; white tail-band.
4. **Fork-tailed Flycatcher** (*Tyrannus savana*), p. 306. Black head contrasts with pale back; tail very long, deeply forked, all black.
5. **Scissor-tailed Flycatcher** (*Tyrannus forficatus*), p. 306. From 4 by pale head, orangish at sides of breast and beneath wing, white in rectrices.
6. **Bright-rumped Attila** (*Attila spadiceus*), p. 310. Note yellow rump, streaked breast, slightly upturned bill, reddish eye (brown in young), erect stance.
7. **Gray Kingbird** (*Tyrannus dominicensis*), p. 308. Above paler than 3; thick bill and notched tail recall 1. Rare migrant.
8. **Piratic Flycatcher** (*Legatus leucophaius*), p. 308. Above brownish; indistinctly streaked breast; belly pale yellow; conspicuous white superciliaries; smaller and duller than 10, 11.
9. **Golden-bellied Flycatcher** (*Myiodynastes hemichrysus*), p. 312. From all other kiskadee-type flycatchers by dark malar stripe. Middle elevations.
10. **Sulphur-bellied Flycatcher** (*Myiodynastes luteiventris*), p. 310. Heavily streaked; rump and tail rufous; below brightest yellow on belly.
11. **Streaked Flycatcher** (*Myiodynastes maculatus*), p. 311. From 10 by weaker malar stripe, thicker bill with lower mandible mostly pale, yellow brightest on breast; different voice. Pacific slope.
12. **Boat-billed Flycatcher** (*Megarhynchus pitangua*), p. 309. Heavy bill (culmen noticeably convex); back olive; little rufous in wings or tail; orange in crown usually concealed.
13. **Great Kiskadee** (*Pitangus sulphuratus*), p. 313. Above browner than 12, with slimmer bill, much rufous in wings and tail; yellow in crown frequently visible; very different voice.
14. **Social Flycatcher** (*Myiozetetes similis*), p. 313. Coloration like 12 but much smaller, with short, rather broad bill; voice distinctive.
15. **Gray-capped Flycatcher** (*Myiozetetes granadensis*), p. 312. Similar to 14 in size and shape, and often found with it, but head paler gray, superciliaries shorter and less distinct; different voice.
16. **White-ringed Flycatcher** (*Coryphotriccus albovittatus*), p. 309. From 14 by longer bill, white superciliaries extending broadly around crown, yellow crown-patch often visible; different voice. Caribbean slope.
17. **Great Crested Flycatcher** (*Myiarchus crinitus*), p. 316. Darker, more brightly colored than 18 with bill pale at base; note rufous in wings and tail.
18. **Brown-crested Flycatcher** (*Myiarchus tyrannulus*), p. 315. Paler than 17, above more brownish; in hand, mouth-lining flesh-color. Note black bill. N Pacific slope.
19. **Ash-throated Flycatcher** (*Myiarchus cinerascens*), p. 316. Most like 18 but below paler, base of mandible pale; dusky tips to inner webs of rectrices. Very rare migrant.
20. **Panama Flycatcher** (*Myiarchus panamensis*), p. 314. Most like 18 but lacks rufous in wings and tail. Pacific mangroves.
21. **Dusky-capped Flycatcher** (*Myiarchus tuberculifer*), p. 316. Below as brightly colored as 17 but much smaller; note dark cap.
22. **Nutting's Flycatcher** (*Myiarchus nuttingi*), p. 315. Virtually identical to 18 in coloration but smaller, with different voice; in hand, note orange mouth-lining. Northwest.
23. **Royal Flycatcher** (*Onychorhynchus coronatus*), p. 326. Distinctive orangish tail and rump, dotted wing-bars, long flat bill; crest usually closed, giving ''hammer-headed'' look. (*a*) ♂: crest spread. (*b*) ♀: crest closed; crest smaller, paler.

1
2
3
4
5
6
7
8
9
10
11
12
13
14
15
16
17
18
19
20
21
22
23b
23a

Plate 36: Flycatchers, 2: Small Species, Salliers

1. **Torrent Tyrannulet** (*Serpophaga cinerea*), p. 333. Pale with contrasting dark cap, wings, tail; wags tail. Highland streams.
2. **Dark Pewee** (*Contopus lugubris*), p. 319. Crested; very dark with paler throat and belly; faint wing-bar. Middle elevations.
3. **Ochraceous Pewee** (*Contopus ochraceus*), p. 319. Size and shape of 2; breast and wing-bars tinged with ochre. Highlands.
4. **Olive-sided Flycatcher** (*Contopus borealis*), p. 317. Dark sides contrast with white of central underparts; stouter than 2 or 3.
5. **Black Phoebe** (*Sayornis nigricans*), p. 305. Blackish with white belly. Streams.
6. **Long-tailed Tyrant** (*Colonia colonus*), p. 305. Blackish; white band surrounding gray crown; white down back; long central rectrices. Caribbean slope.
7. **Western Wood-Pewee** (*Contopus sordidulus*), p. 317. Very like 8 and not always distinguishable except by scratchy or burry call; averages darker, especially on breast, crissum, lower mandible, anterior wing-bar.
8. **Eastern Wood-Pewee** (*Contopus virens*), p. 318. Above averages paler, greener than 7 with brighter wing-bars; mostly clear, whistled calls. See text.
9. **Tropical Pewee** (*Contopus cinereus*), p. 318. Smaller than 7 or 8; perches lower; more contrasting plumage, distinctly yellowish belly; pale lores.
10. **Tawny-chested Flycatcher** (*Aphanotriccus capitalis*), p. 324. Size and shape of *Empidonax*, but brightly colored; breast and wing-bars ochraceous; white throat and spectacles contrast with gray head. N Caribbean slope.
11. **Tufted Flycatcher** (*Mitrephanes phaeocercus*), p. 323. Small; crested; below bright ochraceous; pale wing-bars. Highlands.
12. **Acadian Flycatcher** (*Empidonax virescens*), p. 320. Above olive; throat white; belly and eye-ring yellow; wing-bars tinged with tawny.
13. **White-throated Flycatcher** (*Empidonax albigularis*), p. 322. Above and on chest brown; throat white; dull eye-ring; wing-bars tinged with brownish. Very local.
14. **Alder Flycatcher** (*Empidonax alnorum*), p. 321, and **Willow Flycatcher** (*Empidonax traillii*), p. 321. Throat white; eye-ring dull; belly tinged with yellow. See text.
15. **Least Flycatcher** (*Empidonax minimus*), p. 322. Small; above grayish, especially on head; eye-ring and wing-bars bright white. Rare migrant.
16. **Black-capped Flycatcher** (*Empidonax atriceps*), p. 323. Very dark; eye-ring incomplete, broad behind eye. High elevations.
17. **Bran-colored Flycatcher** (*Myiophobus fasciatus*), p. 326. Like a brown *Empidonax* with streaked breast; concealed yellow crown-patch. S Pacific slope.
18. **Cocos Flycatcher** (*Nesotriccus ridgwayi*), p. 324. Like a dingy *Empidonax* but with longer bill; superciliaries pale. Endemic to Cocos I.
19. **Yellowish Flycatcher** (*Empidonax flavescens*), p. 322. Below yellow, tinged with olive; eye-ring extends into triangular postocular spot; wing-bars yellowish. Middle elevations.
20. **Yellow-bellied Flycatcher** (*Empidonax flaviventris*), p. 320. Below yellow, washed with olive on chest and throat; distinct eye-ring; wing-bars tinged with yellow.
21. **Sulphur-rumped Flycatcher** (*Myiobius sulphureipygius*), p. 325. Rump pale yellow; wings and tail blackish; breast ochraceous. Redstartlike behavior.
22. **Black-tailed Flycatcher** (*Myiobius atricaudus*), p. 325. Smaller body than 21, with paler breast, back more olive. S Pacific slope.
23. **Ruddy-tailed Flycatcher** (*Terenotriccus erythrurus*), p. 324. Small; below buffy; rufous on wings and tail; large eyes; bushy crown grayish.
24. **Olive-striped Flycatcher** (*Mionectes olivaceus*), p. 340. Throat and breast olive, finely streaked with paler olive to yellowish; conspicuous pale postocular spot.
25. **Ochre-bellied Flycatcher** (*Mionectes oleagineus*), p. 340. Olive-green with ochraceous belly; very dull, unpatterned.
26. **Slaty-capped Flycatcher** (*Leptopogon superciliaris*), p. 339. Head mostly slaty; face mottled with whitish; dark ear-patch; wing-bars buffy. Middle elevations.
27. **Sepia-capped Flycatcher** (*Leptopogon amaurocephalus*), p. 340. Like 26 but head browner, chest somewhat greener. Caribbean lowlands.

1
2
3
4
5
6
7
8
9
10
11
12
13
14
15
16
17
18
19
20
21
22
23
24
25
26
27

Plate 37: Flycatchers, 3: Elaenias, Tyrannulets, Flatbills, and Others

1. **White-throated Spadebill** (*Platyrinchus mystaceus*), p. 327. Tail short; broad, flat bill; bold face pattern; above brownish; breast tinged with ochraceous; crown-patch yellow. Middle elevations.
2. **Stub-tailed Spadebill** (*Platyrinchus cancrominus*), p. 327. Resembles 1 but smaller, paler breast; little or no yellow in crown. Guanacaste.
3. **Golden-crowned Spadebill** (*Platyrinchus coronatus*), p. 328. Crown cinnamon-rufous; much greener than 1 or 2. Wet lowlands.
4. **Scale-crested Pygmy-Tyrant** (*Lophotriccus pileatus*), p. 331. Short, broad crest rufous-tipped, usually held flat; iris pale; breast lightly streaked.
5. **Black-capped Pygmy-Tyrant** (*Myiornis atricapillus*), p. 332. Tiny; stubby tail; cap blackish; spectacles white.
6. **Northern Bentbill** (*Oncostoma cinereigulare*), p. 331. Note peculiar downcurved bill, pale iris; breast grayish, faintly streaked.
7. **Common Tody-Flycatcher** (*Todirostrum cinereum*), p. 330. Above blackish; iris pale; tail rather long, graduated, white-tipped; below yellow.
8. **Black-headed Tody-Flycatcher** (*Todirostrum nigriceps*), p. 329. Cap black; throat white; iris dark; back yellowish-olive; tail dark.
9. **Slate-headed Tody-Flycatcher** (*Todirostrum sylvia*), p. 330. Slaty cap contrasts with white loral stripe and narrow broken eye-ring; thick bill; breast gray.
10. **Mistletoe Tyrannulet** (*Zimmerius vilissimus*), p. 337. Conspicuous yellow margins on wing; head gray with paler superciliaries; iris pale; below whitish, faintly streaked; short bill.
11. **Yellow Tyrannulet** (*Capsiempis flaveola*), p. 333. Below yellow; above olive; conspicuous pale superciliaries and wing-bars; rather vireolike pattern.
12. **Rufous-browed Tyrannulet** (*Phylloscartes superciliaris*), p. 332. Bridled face; white spot at base of bill; superciliaries rufous; warblerlike. Caribbean slope.
13. **Brown-capped Tyrannulet** (*Ornithion brunneicapillum*), p. 338. Cap brown; conspicuous white superciliaries and spot below eye; short bill and tail; below yellow; breast washed with olive.
14. **Yellow-bellied Tyrannulet** (*Ornithion semiflavum*), p. 339. From 13 by gray cap, breast clearer yellow; little overlap.
15. **Yellow-crowned Tyrannulet** (*Tyrannulus elatus*), p. 337. Size and shape of 10 but note conspicuous wing-bars and much stronger head pattern; partly concealed yellow crown-patch.
16. **Yellow-olive Flycatcher** (*Tolmomyias sulphurescens*), p. 328. Head gray with white spectacles and pale iris; margins to wing feathers greenish-yellow; flat bill.
17. **Yellow-margined Flycatcher** (*Tolmomyias assimilis*), p. 329. From 16 by dark iris, bright yellow margins to wings; also by voice, habitat. Caribbean slope.
18. **Zeledon's Tyrannulet** (*Phyllomyias zeledoni*), p. 338. Somewhat resembles 10 but note wing-bars, yellow underparts faintly streaked with olive; face has grizzled look; iris reddish. Middle elevations.
19. **Northern Beardless-Tyrannulet** (*Camptostoma imberbe*), p. 336. Small; dingy; short bushy crest; dull wing-bars; superciliaries pale, indistinct. N and NW.
20. **Southern Beardless-Tyrannulet** (*Camptostoma obsoletum*), p. 336. Wing-bars and underparts brighter than 19; little overlap. Often cocks tail. S Pacific slope.
21. **Greenish Elaenia** (*Myiopagis viridicata*), p. 335. Above olive-green, head duller; indistinct pale superciliaries; crown-patch yellow (usually concealed); faint wing-bars; breast olive; belly yellow.
22. **Scrub Flycatcher** (*Sublegatus modestus*), p. 336. Short bill and short bushy crest suggest small elaenia; breast grayish; belly and wing-bars yellow. Pacific mangroves.
23. **Eye-ringed Flatbill** (*Rhynchocyclus brevirostris*), p. 329. Big-headed; flat bill; conspicuous white eye-ring; breast dull olive with faint paler streaks; tail usually appears notched.
24. **Mountain Elaenia** (*Elaenia frantzii*), p. 334. Greenish; head bushy but not really crested; small bill; note broad pale edgings to tertials. Highlands.
25. **Lesser Elaenia** (*Elaenia chiriquensis*), p. 334. Smaller, browner than 24; note pale patch in secondary edgings; concealed crown-patch white. Pacific slope.
26. **Yellow-bellied Elaenia** (*Elaenia flavogaster*), p. 333. Largest, most crested elaenia; belly yellow; note wing-bars, pale edges of tertials; crown-patch white, often conspicuous.

1
2
3
4
5
6
7
8
9
10
11
12
13
14
15
16
17
18
19
20
21
22
23
24
25
26

Plate 38: Wrens and Nightingale-Thrushes

1. **Sedge Wren** (*Cistothorus platensis*), p. 351. Small; buffy; streaked back; rump cinnamon. Vicinity of Cartago.
2. **Rufous-naped Wren** (*Campylorhynchus rufinucha*), p. 352. Large; below white; cap and eye-stripe black; nape and back rufous; above barred. N Pacific slope.
3. **Banded-backed Wren** (*Campylorhynchus zonatus*), p. 351. Large; above boldly barred; breast spotted; belly cinnamon.
4. **Rock Wren** (*Salpinctes obsoletus*), p. 360. Overall mottled appearance; black and buff bands at tip of tail. N Guanacaste, local.
5. **Striped-breasted Wren** (*Thryothorus thoracicus*), p. 354. Face, throat, and breast striped with dusky and white. Caribbean slope.
6. **Banded Wren** (*Thryothorus pleurostictus*), p. 355. Sides and crissum boldy barred with black; above bright brown. N Pacific slope.
7. **Spotted-breasted Wren** (*Thryothorus maculipectus*), p. 356. Breast densely spotted. N of Cordillera de Guanacaste.
8. **Rufous-breasted Wren** (*Thryothorus rutilus*), p. 357. Throat and cheeks black, spotted with white; breast rufous. Mainly S Pacific slope.
9. **Rufous-and-white Wren** (*Thryothorus rufalbus*), p. 353. Above deep rufous; below white; darker and duller on sides, flanks; crissum barred; malar stripe.
10. **Riverside Wren** (*Thryothorus semibadius*), p. 354. Underparts finely barred. S Pacific slope.
11. **Black-bellied Wren** (*Thryothorus fasciatoventris*), p. 356. Broad bib white; face and underparts black; flanks barred. S Pacific slope.
12. **Bay Wren** (*Thryothorus nigricapillus*), p. 354. Body rich chestnut; throat white; head black, marked with white. Caribbean slope.
13. **Black-throated Wren** (*Thryothorus atrogularis*), p. 356. Very dark, antbirdlike; face and throat black; sparse white streaks on auriculars, superciliaries. Caribbean slope.
14. **Gray-breasted Wood-Wren** (*Henicorhina leucophrys*), p. 359. Tail short; breast gray; superciliaries and facial streaks white. Highlands.
15. **White-breasted Wood-Wren** (*Henicorhina leucosticta*), p. 359. From 14 by white breast; from *Thryothorus* by short tail. Lower elevations than 14.
16. **Timberline Wren** (*Thryorchilus browni*), p. 358. Broad superciliaries white; white in wing. High elevations.
17. **Plain Wren** (*Thryothorus modestus*), p. 352. Conspicuous white superciliaries; above dull; tail brighter; below unpatterned; belly buffy. Caribbean birds (Canebrake Wren) much grayer, notably larger.
18. **House Wren** (*Troglodytes aedon*), p. 357. Small; no white markings; superciliaries and underparts buffy.
19. **Ochraceous Wren** (*Troglodytes ochraceus*), p. 358. From 18 by brighter, more ochraceous superciliaries, throat; forest. Highlands.
20. **Whistling Wren** (*Microcerculus luscinia*), p. 361. Long bill and legs; stubby tail; throat pale; breast gray (scaled in young). Terrestrial. S half of country.
21. **Nightingale Wren** (*Microcerculus philomela*), p. 360. Much darker than 20, with extensive dark scaling. N half of country, excluding lowland Guanacaste.
22. **Song Wren** (*Cyphorhinus phaeocephalus*), p. 361. Throat and breast rufous; orbital skin pale; peculiarly shaped bill. Caribbean slope.
23. **Slaty-backed Nightingale-Thrush** (*Catharus fuscater*), p. 368. Very dark; belly pale; note pale iris; bill, eye-ring, and legs orange. Middle elevations.
24. **Black-billed Nightingale-Thrush** (*Catharus gracilirostris*), p. 370. Bill black; no eye-ring; olive breast contrasts with pale throat, gray belly. High elevations.
25. **Black-headed Nightingale-Thrush** (*Catharus mexicanus*), p. 368. Head black; back and breast dark; belly pale; note orange bill, eye-ring, legs; iris dark.
26. **Orange-billed Nightingale-Thrush** (*Catharus aurantiirostris*), p. 369. Above brown; paler and brighter than 25; breast gray; bill, eye-ring, and legs orange. Birds of S Pacific slope have gray heads.
27. **Ruddy-capped Nightingale-Thrush** (*Catharus frantzii*), p. 369. Above brown, brightest and reddest on crown; lower mandible orange; no eye-ring; legs brownish. Highlands.

1
2
3
4
5
6
7
8
9
10
11
12
13
14
15
16
17
18
19
20
21
22
23
24
25
26
27

Plate 39: Thrushes, Jays, Silky-flycatchers, Catbird, and Dipper

1. **Gray-cheeked Thrush** (*Catharus minimus*), p. 367. Cheeks gray, faintly streaked; eye-ring faint or absent; breast heavily spotted with black.
2. **Swainson's Thrush** (*Catharus ustulatus*), p. 366. Spectacles buffy; cheeks mottled, not streaked; breast heavily spotted.
3. **Veery** (*Catharus fuscescens*), p. 367. Above brighter brown, breast less spotted than 1 or 2; no eye-ring.
4. **Wood Thrush** (*Hylocichla mustelina*), p. 366. Larger, more boldly spotted than 1 or 2; head brighter, more rufescent than rest of upperparts.
5. **Pale-vented Robin** (*Turdus obsoletus*), p. 364. Above rich brown, below paler; belly and crissum white; bill black. Caribbean slope.
6. **Mountain Robin** (*Turdus plebejus*), p. 365. Duller, grayer than 8; bill black; iris dark. Highlands.
7. **Sooty Robin** (*Turdus nigrescens*), p. 365. Large; ♂ blackish, ♀ dark sooty-brown; iris pale; eye-ring, bill, and legs orange. High elevations.
8. **Clay-colored Robin** (*Turdus grayi*), p. 363. Dull brown; throat striped; bill yellowish; iris reddish-brown.
9. **White-throated Robin** (*Turdus assimilis*), p. 363. Throat white, the anterior part streaked with black; eye-ring yellow; bill and feet yellowish.
10. **Black-and-yellow Silky-flycatcher** (*Phainoptila melanoxantha*), p. 374. Chunky, often looks small-headed. (*a*) ♂: head and upperparts black; rump and flanks yellow. (*b*) ♀: face gray; back olive; yellow only on flanks. Highlands.
11. **American Dipper** (*Cinclus mexicanus*), p. 350. Gray; wings and tail darker; long pale legs. Nods, bobs, and continually pumps tail. Mountain streams.
12. **Gray Catbird** (*Dumetella carolinensis*), p. 362. Slender; long-tailed; dark gray with cap, wings, and tail blackish; crissum rufous.
13. **Black-faced Solitaire** (*Myadestes melanops*), p. 365. Slate-gray; face black; bill, eye-ring, and legs orange; pale margins on remiges.
14. **Cedar Waxwing** (*Bombycilla cedrorum*), p. 373. Soft brown, blending to yellow belly; note mask, crest, and yellow tail-band.
15. **Long-tailed Silky-flycatcher** (*Ptilogonys caudatus*), p. 373. Long pointed tail with white "windows," pointed crest. ♂: yellow head and crissum contrast with blue-gray body; ♀ duller, more olive. Highlands.
16. **Silvery-throated Jay** (*Cyanolyca argentigula*), p. 349. Dark purplish-black; throat and brows silvery to purplish-white. High elevations.
17. **Azure-hooded Jay** (*Cyanolyca cucullata*), p. 349. Dark blue; hood sky-blue, bordered with white on brows. Middle elevations.
18. **White-throated Magpie-Jay** (*Calocitta formosa*), p. 347. Above blue; long tail white-tipped; long curled crest. N Pacific slope.
19. **Brown Jay** (*Cyanocorax morio*), p. 348. Large; mostly dark brown; belly and tips of outer rectrices white; young with yellow eye-rings, bill more or less yellow.
20. **Black-chested Jay** (*Cyanocorax affinis*), p. 348. Head and breast black; facial markings pale blue; iris yellow; above dark purplish; tail broadly white-tipped.

1
4
6
7
2
5
8
3
10b
10a
9
11
14
15
13
12
16
17
18
19
20

Plate 40: Vireos, Warblers, and Bananaquit

1. **Green Shrike-Vireo** (*Vireolanius pulchellus*), p. 375. Above bright green, below paler; throat yellow; blue on nape; heavy hooked bill.
2. **Rufous-browed Peppershrike** (*Cyclarhis gujanensis*), p. 375. Heavy pale bill; superciliaries rufous; iris orange; breast yellow.
3. **Red-eyed Vireo** (*Vireo olivaceus*), p. 378. Olive back contrasts with gray crown, bordered with black; iris red. Immatures have brown eyes, and flanks and crissum washed with yellow.
4. **Black-whiskered Vireo** (*Vireo altiloquus*), p. 379. Above browner than 3, with dark malar streak. Very rare migrant.
5. **Yellow-green Vireo** (*Vireo flavoviridis*), p. 378. Below much more yellow than 3; crown not bordered with black. Iris brick-red in adult, brown in young.
6. **Yellow-throated Vireo** (*Vireo flavifrons*), p. 377. Spectacles yellow; broad wing-bars; yellow throat and breast contrast with white belly.
7. **Lesser Greenlet** (*Hylophilus decurtatus*), p. 381. Head pale gray; spectacles white; small, rather short-tailed.
8. **Scrub Greenlet** (*Hylophilus flavipes*), p. 380. Iris pale, bill pinkish; above olive; below mostly yellow. S Pacific slope.
9. **Solitary Vireo** (*Vireo solitarius*), p. 377. Broad spectacles white; head blue-gray; back olive; broad wing-bars. Rare migrant.
10. **Yellow-winged Vireo** (*Vireo carmioli*), p. 376. Wing-bars yellow; spectacles yellow, broken behind eye. Highlands.
11. **Mangrove Vireo** (*Vireo pallens*), p. 376. Above grayish; below white; loral stripe yellow; eyes pale; narrow wing-bars. Pacific mangroves.
12. **White-eyed Vireo** (*Vireo griseus*), p. 376. Eyes white; spectacles yellow; broad wing-bars; gray of neck contrasts with olive of crown and back. Rare migrant.
13. **Warbling Vireo** (*Vireo gilvus*), p. 379. Head grayish; superciliaries white; below whitish; note pale lores. Very rare migrant.
14. **Brown-capped Vireo** (*Vireo leucophrys*), p. 379. Cap brownish; superciliaries white; posterior underparts pale yellow. Highlands.
15. **Philadelphia Vireo** (*Vireo philadelphicus*), p. 379. Pileum grayish; superciliaries white; breast varies from pale to fairly bright yellow.
16. **Rufous-capped Warbler** (*Basileuterus rufifrons*), p. 403. Crown and cheeks rufous; superciliary and spot below eye white; nape grayish; below yellow.
17. **Black-cheeked Warbler** (*Basileuterus melanogenys*), p. 403. Crown rufous; superciliaries white; cheeks black. Highlands.
18. **Golden-crowned Warbler** (*Basileuterus culicivorus*), p. 402. Above grayish-olive; broad black stripe on sides of crown with orangish central stripe; below yellow. Lower elevations than 20.
19. **Orange-crowned Warbler** (*Vermivora celata*), p. 385. Resembles 22 but darker, duller, with faint streaks below. Very rare migrant.
20. **Three-striped Warbler** (*Basileuterus tristriatus*), p. 402. Bold black pattern on dull buffy head; center of crown orangish; below pale dull yellowish. Middle elevations.
21. **Worm-eating Warbler** (*Helmitheros vermivorus*), p. 384. Black stripes on bright buffy head; long sharp-pointed bill.
22. **Tennessee Warbler** (*Vermivora peregrina*), p. 385. Small, sharp-pointed bill; superciliaries narrow. (*a*) ♂, breeding: superciliaries and underparts white; head gray. (*b*) Immature ♀: head olive; breast and superciliaries yellowish. Other plumages intermediate.
23. **Buff-rumped Warbler** (*Phaeothlypis fulvicauda*), p. 404. Rump and base of tail bright buff. Bobs and wags tail. Streams.
24. **Bananaquit** (*Coereba flaveola*), p. 382. Bill sharp-pointed, decurved; superciliaries and wing-spot white; throat gray; below yellow. Immatures much duller, more olive.

1
2
3
4
7
6
5
8
9
10
11
12
13
14
15
16
19
17
18
22b
20
21
22a
23
24

Plate 41: Gnatcatchers and Warblers

1. **White-lored Gnatcatcher** (*Polioptila albiloris*), p. 370. Tail long, conspicuously white-edged. (*a*) ♂, breeding: cap black. (*b*) ♂, nonbreeding (approx. Sept.–Jan.): white line through lores, over eye. (*c*) ♀: like nonbreeding ♂ but cap gray. N Pacific slope.
2. **Tropical Gnatcatcher** (*Polioptila plumbea*), p. 371. Below grayer than 1, less white on wing, much broader superciliaries. (*a*) ♂: crown black. (*b*) ♀: crown gray.
3. **Tropical Parula** (*Parula pitiayumi*), p. 387. ♂: above dark grayish-blue; back olive; single wing-bar; bicolored bill. ♀ with less pronounced black mask, less orange on breast. Middle elevations.
4. **Northern Parula** (*Parula americana*), p. 386. From 3 by partial eye-ring, 2 wing-bars. (*a*) ♂: note chestnut breast-band. (*b*) ♀: above more olive, breast-band reduced or absent. Very rare migrant.
5. **Golden-winged Warbler** (*Vermivora chrysoptera*), p. 384. Note yellow in wing and crown, gray upperparts. (*a*) ♂: throat and eye-stripe black. (*b*) ♀: throat and eye-stripe dark gray. Young resemble adults but have white chins.
6. **Blue-winged Warbler** (*Vermivora pinus*), p. 384. Below golden-yellow; wings bluish-slate with 2 wing-bars. (*a*) ♂: forecrown yellow; eye-stripe black. (*b*) Immature ♀: crown olive; eye-stripe dusky. Hybrids between 5 and 6: see text.
7. **Flame-throated Warbler** (*Parula gutturalis*), p. 386. (*a*) Throat and chest intense orange; above gray with black triangle on back. Black mask more extensive in male. (*b*) Young: duller, more brownish; trace of wing-bars and eye-stripe. High elevations.
8. **Blackburnian Warbler** (*Dendroica fusca*), p. 392. Note pale stripes on back, dark cheeks outlined with yellow or orange. (*a*) ♂, breeding: orange and black; broadly streaked below. (*b*) Immature ♀: olive and yellow; above and below with narrow streaks. Other plumages intermediate.
9. **Bay-breasted Warbler** (*Dendroica castanea*), p. 393. (*a*) ♂, breeding: crown, throat, and sides chestnut; mask black; neck-patch buffy. (*b*) Immature ♀: above olive; back lightly streaked; sides, flanks, and crissum tinged with buff. Other plumages intermediate; flanks usually tinged with chestnut.
10. **American Redstart** (*Setophaga ruticilla*), p. 401. (*a*) ♂: above black; belly white; bright orange on wings, tail, and sides of breast. (*b*) ♀: wing and tail pattern like ♂ but dusky and yellow; spectacles white; upperparts grayish. Immature ♂ similar.
11. **Blackpoll Warbler** (*Dendroica striata*), p. 393. (*a*) ♂, breeding: cap black; cheeks white; streaks down sides black. (*b*) Immature ♀: from 9 by at least trace of streaking on sides, little or no buffy tinge on flanks and crissum. Breast often tinged with yellow. Rare migrant.
12. **Yellow-throated Warbler** (*Dendroica dominica*), p. 392. Throat yellow; black and white face pattern; black streaks on sides; back gray.
13. **Black-and-white Warbler** (*Mniotilta varia*), p. 383. Boldly streaked with black and white. (*a*) ♂: throat and cheeks black. (*b*) ♀: cheeks gray; throat white.
14. **Pine Warbler** (*Dendroica pinus*), p. 394. From 8, 9, 11 by lack of streaks on back. (*a*) ♂: below bright yellow; sides streaked with olive. (*b*) Immature ♀: above brownish; dull wing-bars; below with buffy tinge, little or no yellow. Very rare migrant.

1c
2b
3
1b
2a
4b
1a
4a
5a
6a
5b
6b
7a
7b
8a
9a
8b
10a
9b
10b
11a
12
11b
13a
14a
13b
14b

Plate 42: Warblers with Yellow Underparts and No Wing-Bars

1. **Prothonotary Warbler** (*Protonotaria citrea*), p. 383. Wings gray, back olive; white in tail; rather large bill. (*a*) ♂: head orange-yellow like breast. (*b*) ♀: head mostly olive.
2. **Yellow Warbler** (*Dendroica petechia*), p. 387. (*a*) ♂: mostly yellow; above yellowish-olive; below with chestnut streaking. (*b*) Immature ♀: much more olive, below often only tinged with yellow; yellow flash in tail; no dark markings on head. Other plumages intermediate. (*c*) ♂, Cocos I.: cap rufous.
3. **Mangrove Warbler** (*Dendroica petechia erithachorides*), p. 388. (*a*) ♂: head chestnut. (*b*) ♀: larger than ♀ 2, usually rufous tinge on head, especially on crown. Young grayer than 2b. Mangroves.
4. **Wilson's Warbler** (*Wilsonia pusilla*), p. 400. Small; bright yellow on face and underparts. (*a*) ♂: cap black. (*b*) ♀: olive crown (sometimes some black) contrasts with yellow face.
5. **Hooded Warbler** (*Wilsonia citrina*), p. 399. Face and forehead yellow; white in tail. (*a*) ♂: distinctive black hood. (*b*) Immature ♀: from 4b by yellow forehead, tail. Adult ♀ usually has traces of black hood.
6. **Collared Redstart** (*Myioborus torquatus*), p. 402. Face yellow; crown rufous; upperparts and breast-band blackish; white-edged tail. High elevations. Often spreads, flashes tail.
7. **Slate-throated Redstart** (*Myioborus miniatus*), p. 401. Head, chest, upperparts blackish; crown dark rufous; white-edged tail; below bright yellow; behavior like 6. Middle elevations.
8. **Canada Warbler** (*Wilsonia canadensis*), p. 400. Above gray; eye-ring white; white belly and crissum contrast with yellow of rest of underparts. (*a*) ♂: necklace and crown-spots black; distinct yellow stripe over black lores. (*b*) Immature ♀: trace of necklace; crown brownish, lores indistinct. Other plumages intermediate.
9. **Yellow-breasted Chat** (*Icteria virens*), p. 399. Large; spectacles white; face blackish; throat and breast bright yellow; belly white.
10. **Common Yellowthroat** (*Geothlypis trichas*), p. 397. From other yellowthroats by buffy-white belly, small size. (*a*) ♂: black mask bordered with whitish. (*b*) ♀: eye-ring buffy to white; crown tinged with rufous; breast duller yellow. Immature ♂ has trace of mask.
11. **Olive-crowned Yellowthroat** (*Geothlypis semiflava*), p. 398. Above olive; below yellow; bill black. (*a*) ♂: extensive black mask without contrasting border. (*b*) ♀: no contrasting markings. Caribbean slope.
12. **Masked Yellowthroat** (*Geothlypis aequinoctialis*), p. 398. ♂: like 11a but black mask bordered with gray. ♀ most like 11b (no overlap). San Vito area only.
13. **Gray-crowned Yellowthroat** (*Geothlypis poliocephala*), p. 399. Thick bicolored bill; broken eye-ring. (*a*) ♂: small mask black; crown gray. (*b*) ♀: lores blackish; less gray on crown; belly tinged with buff.
14. **Mourning Warbler** (*Oporornis philadelphia*), p. 396. (*a*) ♂: hood gray, marked with black on throat. (*b*) ♀: hood grayish; throat buffy to white. (*c*) Immature: head olive with little trace of hood; throat yellow (chest faintly marked with black in males). Eye-ring variable in extent in females and young, absent in males. [See also Pl. 52(5).]
15. **Kentucky Warbler** (*Oporornis formosus*), p. 396. Broad spectacles yellow. (*a*) ♂: cheeks black; crown black with gray scaling. (*b*) Immature ♀: face and crown olive. Other plumages intermediate.
16. **MacGillivray's Warbler** (*Oporornis tolmiei*), p. 397. ♂: like 14a but conspicuous broken eye-ring. Other plumages like 14, tell in hand by measurements (see text).

1a
1b
2a
3a
2c
2b
3b
4a
4b
5b
5a
6
8b
8a
7
10b
10a
9
11b
11a
13a
12
14a
13b
14b
15a
16
14c
15b

Plate 43: Warblers, Chiefly with Wing-Bars and Streaks

1. **Black-throated Blue Warbler** (*Dendroica caerulescens*), p. 389. (*a*) ♂: face, throat, and sides black; wing-patch white; upperparts slate-blue. (*b*) ♀: above brownish; below buffy-yellowish; eye-spot and narrow superciliaries whitish; smaller wing-patch. Rare migrant.
2. **Cerulean Warbler** (*Dendroica cerulea*), p. 391. (*a*) ♂: above blue; cheeks dark; breast-band (often faint in fall). (*b*) Immature ♀: below dingy; above olive, tinged with bluish on crown; faint streaks on sides. Adult ♀ brighter, crown more bluish.
3. **Palm Warbler** (*Dendroica palmarum*), p. 394. Above brownish; rump greenish; wing-bars dull; crissum yellow. Wags tail. (*a*) ♂, breeding: breast yellow; chestnut in crown. (*b*) Immature ♀: below mostly whitish; crown dull; streaked sides. Other plumages intermediate.
4. **Chestnut-sided Warbler** (*Dendroica pensylvanica*), p. 392. (*a*) ♂, breeding: crown yellow; facial stripes black; sides chestnut. (*b*) ♀, winter: eye-ring white; cheeks gray; crown and upperparts olive; wing-bars yellowish; above with faint streaks or none; winter ♂ similar, usually has trace of chestnut on sides.
5. **Prairie Warbler** (*Dendroica discolor*), p. 394. Small; back olive. Wags tail. (*a*) ♂: face and underparts yellow and black; back streaked with chestnut. (*b*) Immature ♀: gray and white facial pattern; breast yellow; sides lightly streaked; rather dull wing-bars. Rare migrant.
6. **Cape May Warbler** (*Dendroica tigrina*), p. 389. Below heavily streaked; rump yellow to olive-green; back darker. (*a*) ♂, breeding: cheeks chestnut; below yellow, streaked with black. (*b*) Immature ♀: below with olive wash; rump greenish, no yellow. Other plumages intermediate.
7. **Hermit Warbler** (*Dendroica occidentalis*), p. 391. Face yellow, little or no olive on cheeks. (*a*) ♂ breeding: most of head yellow; throat black. (*b*) Immature ♀: throat whitish; crown olive; back tinged with grayish, streaked. Other plumages intermediate. Very rare migrant.
8. **Townsend's Warbler** (*Dendroica townsendi*), p. 390. Dark cheek-patch sharply defined; breast extensively yellow. (*a*) ♂, breeding: crown, cheeks, and throat black. (*b*) Immature ♀: cheeks olive; throat yellow. Other plumages intermediate. Rare.
9. **Black-throated Green Warbler** (*Dendroica virens*), p. 390. Face yellow; cheeks streaked or mottled with olive; breast pale yellow or white. (*a*) ♂, breeding: crown olive; throat black. (*b*) Immature ♀: cheek-patch indistinct; faint streaks on sides. Other plumages intermediate.
10. **Magnolia Warbler** (*Dendroica magnolia*), p. 388. Rump yellow; distinctive tail pattern. (*a*) ♂, breeding: above mostly black; below with black chest-patch and streaks. (*b*) Immature ♀: indistinct whitish chest-band; below with faint streaks; head gray; back dull olive; eye-ring white.
11. **Yellow-rumped (Myrtle) Warbler** (*Dendroica coronata*), p. 389. Rump and sides of breast yellow; above and sides streaked with black. (*a*) ♂, breeding: above gray; yellow in crown; breast mostly black; throat white. (*b*) Immature ♀: above brown, no yellow in crown; below streaking less distinct. (*c*) "Audubon's Warbler" (*D. c. auduboni*), ♂, breeding: very rare western race told in all plumages by bright or dull yellow throat.
12. **Ovenbird** (*Seiurus aurocapillus*), p. 395. Center of crown orangish, bordered with black; eye-ring; below with black streaks.
13. **Louisiana Waterthrush** (*Seiurus motacilla*), p. 395. From 14 by broader, whiter superciliaries, larger bill, buff-tinged flanks and crissum. Mostly highland streams.
14. **Northern Waterthrush** (*Seiurus noveboracensis*), p. 395. Superciliaries and underparts usually tinged uniformly with buff or yellowish. Mostly lowland streams, mangroves.

1a
1b
2a
2b
3a
3b
4a
4b
5a
5b
6a
6b
7a
7b
8a
8b
9a
9b
10a
10b
11a
11b
11c
12
13
14

Plate 44: Icterids

1. **Spotted-breasted Oriole** (*Icterus pectoralis*), p. 411. Sides of breast spotted with black; no wing-bars. N Pacific slope.
2. **Streaked-backed Oriole** (*Icterus pustulatus*), p. 413. Back orange to olive, streaked with black; much white on wing. N Pacific slope.
3. **Yellow-tailed Oriole** (*Icterus mesomelas*), p. 411. Large; yellow and black; outer rectrices yellow. Caribbean slope.
4. **Yellow-headed Blackbird** (*Xanthocephalus xanthocephalus*), p. 414. (*a*) ♂: black; head yellow; wing-patch white. (*b*) ♀: browner; face and chest ochre-yellow; throat paler; lower breast streaked. Casual.
5. **Black-cowled Oriole** (*Icterus dominicensis*), p. 410. No wing-bars. (*a*) Mostly black; shoulder, rump, and belly yellow. (*b*) Immatures: black on face and throat; wings and tail dusky; back olive. Caribbean slope.
6. **Orchard Oriole** (*Icterus spurius*), p. 410. (*a*) ♂: chestnut and black; 1 wing-bar. (*b*) ♀: below yellow; 2 white wing-bars. (*c*) Immature ♂: like ♀ but face black.
7. **Northern (Baltimore) Oriole** (*Icterus g. galbula*), p. 412. (*a*) ♂: only orange oriole with black head, orange in tail. (*b*) ♀: below orange; above olive-brown; more or less blotched or spotted with black. (*c*) Immature: above brownish-olive; below ochraceous-orange (♂ brighter), no black.
8. **Montezuma Oropendola** (*Psarocolius montezuma*), p. 406. Very large; chestnut with yellow tail; facial skin bluish and pink; bill orange-tipped.
9. **Chestnut-headed Oropendola** (*Psarocolius wagleri*), p. 405. Large; body black; head chestnut; bill ivory; eyes pale.
10. **Giant Cowbird** (*Scaphidura oryzivora*), p. 407. (*a*) Large, size of 9; iris red; rather small head; ♂ with neck-ruff (iris and bill of young birds yellow). (*b*) Flying: note upswept wingtips.
11. **Scarlet-rumped Cacique** (*Cacicus uropygialis*), p. 406. Red rump often hard to see; long, pointed bill ivory; eyes pale blue. Canopy.
12. **Yellow-billed Cacique** (*Amblycercus holosericeus*), p. 407. Black; iris yellow; bill ivory-yellow. Thickets.
13. **Red-breasted Blackbird** (*Sturnella militaris*), p. 414. (*a*) ♂: black with red breast. (*b*) ♀: streaked, brownish; breast suffused with pinkish. S Pacific slope.
14. **Red-winged Blackbird** (*Agelaius phoeniceus*), p. 413. (*a*) ♂: black; shoulder red, edged with buff. (*b*) ♀: brownish; below paler, heavily streaked. N and NW only.
15. **Bronzed Cowbird** (*Molothrus aeneus*), p. 408. ♂: glossy black; neck-ruff; eyes red. ♀ duller, without conspicuous neck-ruff. [See also Pl. 52(7).]
16. **Great-tailed Grackle** (*Quiscalus mexicanus*), p. 409. (*a*) ♂: large; glossy purplish-black; long graduated tail; iris yellow. (*b*) ♀: above dusky; below brownish; superciliaries and iris pale.
17. **Nicaraguan Grackle** (*Quiscalus nicaraguensis*), p. 409. (*a*) ♂: much smaller than 16a. (*b*) ♀: below much paler than 16b. Río Frío region only.

1
2
3
4a
4b
5a
5b
6a
6b
6c
7a
7b
7c
8
9
10a
10b
11
12
13a
13b
14a
14b
15
16a
16b
17a
17b

Plate 45: Euphonias and Olive-Green Tanagers

1. **Yellow-crowned Euphonia** (*Euphonia luteicapilla*), p. 419. (*a*) ♂: throat dark; entire crown yellow. (*b*) ♀: above olive; below yellow, unpatterned.
2. **Spotted-crowned Euphonia** (*Euphonia imitans*), p. 421. (*a*) ♂: throat dark; crown obscurely spotted, extending behind eyes. (*b*) ♀: forehead and belly rufous. S Pacific slope.
3. **Olive-backed Euphonia** (*Euphonia gouldi*), p. 421. (*a*) ♂: back glossy olive-green; forehead yellow; belly rufous, surrounded by yellow. (*b*) ♀: forehead and belly rufous, more yellow below than 2b. Caribbean slope.
4. **Scrub Euphonia** (*Euphonia affinis*), p. 419. (*a*) ♂: throat dark; forehead yellow. (*b*) ♀: forehead yellowish; crown and nape gray; breast washed with olive. NW only.
5. **Yellow-throated Euphonia** (*Euphonia hirundinacea*), p. 420. Thick bill. (*a*) ♂: throat and forehead yellow. (*b*) ♀: below yellowish with belly extensively whitish.
6. **Tawny-capped Euphonia** (*Euphonia anneae*), p. 418. (*a*) ♂: cap orange-cinnamon; throat dark. (*b*) ♀: forehead rufous; below gray. Middle elevations.
7. **White-vented Euphonia** (*Euphonia minuta*), p. 418. (*a*) ♂: small; throat dark; belly white. (*b*) ♀: throat whitish; breast yellowish; belly white.
8. **Thick-billed Euphonia** (*Euphonia laniirostris*), p. 420. Thick bill. (*a*) ♂: throat and entire crown yellow. (*b*) ♀: below bright olive-yellow. S Pacific slope.
9. **Blue-hooded Euphonia** (*Euphonia elegantissima*), p. 417. (*a*) ♂: crown and nape sky-blue; throat dark; belly and forehead rufous. (*b*) ♀: olive; below paler; hood blue; forehead rufous; throat cinnamon-buff.
10. **Golden-browed Chlorophonia** (*Chlorophonia callophrys*), p. 416. (*a*) ♂: brilliant green; crown and nape blue; eyebrow and belly yellow. (*b*) ♀: duller; only forehead yellow. Highlands.
11. **Blue-and-gold Tanager** (*Buthraupis arcaei*), p. 428. Like a huge euphonia with dark sides and flanks; note red eyes. Caribbean slope.
12. **Ashy-throated Bush-Tanager** (*Chlorospingus canigularis*), p. 440. Sides of head grayish-olive, unpatterned; grayish-white throat and belly contrast with yellow-olive chest and sides; note dark malar stripe. Caribbean foothills.
13. **Common Bush-Tanager** (*Chlorospingus ophthalmicus*), p. 439. Head dark brownish; conspicuous white postocular spot. Middle elevations.
14. **Sooty-capped Bush-Tanager** (*Chlorospingus pileatus*), p. 440. Broad broken white superciliaries on blackish head. High elevations.
15. **Sulphur-rumped Tanager** (*Heterospingus rubrifrons*), p. 437. Slaty with darker wings; rump yellow; white tufts at sides of breast. S Caribbean slope.
16. **Olive Tanager** (*Chlorothraupis carmioli*), p. 433. Olive-green, paler below; throat brighter; no contrasting markings; bill black. Caribbean slope.
17. **Gray-headed Tanager** (*Eucometis penicillata*), p. 437. Hood gray; slightly crested; above olive; below yellow. Pacific slope.
18. **Dusky-faced Tanager** (*Mitrospingus cassinii*), p. 438. Very dark, dingy; rather long, slender pale bill and iris suggest icterid. Caribbean slope.
19. **Palm Tanager** (*Thraupis palmarum*), p. 429. Dull olive (purplish tints in male) with contrasting black wings.

1a
1b
2a
2b
3a
3b
4a
4b
5a
5b
6a
6b
7a
7b
8a
8b
9a
9b
10a
10b
11
12
13
14
15
16
17
18
19

Plate 46: Honeycreepers and Tanagers

1. **Shining Honeycreeper** (*Cyanerpes lucidus*), p. 426. (*a*) ♂: throat black; back blue; legs yellow; slim, decurved bill. (*b*) ♀: above green; breast streaked with bluish.
2. **Red-legged Honeycreeper** (*Cyanerpes cyaneus*), p. 426. (*a*) ♂: throat blue; back black; legs red; crown turquoise; wing-linings yellow. (*b*) ♀: green; below indistinctly streaked; legs reddish-brown.
3. **Blue Dacnis** (*Dacnis cayana*), p. 427. Bill sharp-pointed. (*a*) ♂: bright blue; black in wings, on throat and back. (*b*) ♀: fairly bright green with blue head.
4. **Scarlet-thighed Dacnis** (*Dacnis venusta*), p. 427. (*a*) ♂: above blue; below black; thighs red. (*b*) ♀: above green, with blue tints; below pale dull buff; thighs cinnamon.
5. **Emerald Tanager** (*Tangara florida*), p. 422. ♂: bright pale green; crown and rump yellow; square ear-patch black. ♀ duller, lacks yellow crown. Caribbean foothills.
6. **Silver-throated Tanager** (*Tangara icterocephala*), p. 422. ♂ bright golden-yellow; malar stripe black; throat whitish; back streaked. ♀ duller, more greenish.
7. **Green Honeycreeper** (*Chlorophanes spiza*), p. 425. Bill black and yellow, decurved; note red iris. (*a*) ♂: bluish-green; face and crown black. (*b*) ♀: grass-green, below paler.
8. **Speckled Tanager** (*Tangara guttata*), p. 422. Above green; below heavily spotted; face yellow; lores black.
9. **Bay-headed Tanager** (*Tangara gyrola*), p. 424. ♂: head chestnut; above green; below bright blue; golden tints on nape and shoulder. ♀ paler and duller.
10. **Rufous-winged Tanager** (*Tangara lavinia*), p. 424. (*a*) ♂: head and wings chestnut-rufous; body bright green; back tinted with golden; center of belly blue. (*b*) ♀: paler, duller, lacks rufous head. Caribbean foothills.
11. **Plain-colored Tanager** (*Tangara inornata*), p. 424. Dingy gray; below paler; shoulder blue (often concealed); belly buff. Caribbean slope.
12. **Spangled-cheeked Tanager** (*Tangara dowii*), p. 425. Mostly black; belly cinnamon; pale blue-green spotting on cheeks; scaling on chest; rump pale blue-green; dark blue in wings. Highlands.
13. **Golden-hooded Tanager** (*Tangara larvata*), p. 423. (*a*) Hood golden; mask black, edged with blue; body largely black; belly white. (*b*) Young: much duller; black and blue reduced, head more greenish.
14. **Black-and-yellow Tanager** (*Chrysothlypis chrysomelas*), p. 439. Small; slim-billed; warblerlike. (*a*) ♂: bright yellow and black. (*b*) ♀: from warblers by whitish breast-tufts, black bill; belly whitish. Caribbean foothills.
15. **Blue-gray Tanager** (*Thraupis episcopus*), p. 428. ♂: pale bluish with brighter blue wings and tail. ♀, young: grayer.
16. **White-lined Tanager** (*Tachyphonus rufus*), p. 436. Bicolored bill. (*a*) ♂: black; white in wing-linings and scapulars shows in flight. (*b*) ♀: rufous; below paler.
17. **Tawny-crested Tanager** (*Tachyphonus delattrii*), p. 437. (*a*) ♂: black; rounded crest tawny-orange. (*b*) ♀: uniform dark brown. Caribbean slope.
18. **White-shouldered Tanager** (*Tachyphonus luctuosus*), p. 436. (*a*) ♂: black; shoulder white (larger in Pacific race, which also has concealed orange crown-patch). (*b*) ♀: above olive; below yellow (Pacific race: top and sides of head gray).

1b
2b
3b
4b
1a
2a
3a
4a
5
6
7a
7b
8
9
10a
10b
11
12
13a
13b
14a
14b
15
16a
16b
17a
17b
18a
18b

Plate 47: Larger Red or Yellow Tanagers

1. **White-throated Shrike-Tanager** (*Lanio leucothorax*), p. 435. Heavy hooked bill; throat whitish. (*a*) ♂: yellow; head and wings black (and rump in S Pacific race). (*b*) ♀: above brown; below yellowish; chest and throat buffy-brown.
2. **Western Tanager** (*Piranga ludoviciana*), p. 432. (*a*) ♂, breeding: yellow and black; note wing pattern, orange head (faint in winter). (*b*) ♀: back olive; below yellow; wing-bars. N Pacific slope.
3. **Crimson-collared Tanager** (*Phlogothraupis sanguinolenta*), p. 430. Black with thick whitish bill; crown, collar, rump, and belly deep red (replaced by dull rufous in young). Caribbean slope.
4. **Scarlet-rumped Tanager** (*Ramphocelus passerinii*), p. 429. (*a*) ♂: velvety black; rump brilliant scarlet; bill silvery, inflated lower mandible. (*b*) ♀ (Pacific race): olive; orange on rump and breast. (*c*) ♀ (Caribbean race): olive; head tinged with gray; below paler; no orange.
5. **Summer Tanager** (*Piranga rubra*), p. 430. Note thick pale bill. (*a*) ♂: entirely rose-red. (*b*) ♀: above yellow-olive; below ochraceous-yellow with orangish tinge; wing-linings yellow.
6. **Hepatic Tanager** (*Piranga flava*), p. 431. Note thick blackish bill. (*a*) ♂: rich brick-red; cheeks darker. (*b*) ♀: above olive; below yellowish (sometimes tinged with ochraceous); dark cheeks.
7. **White-winged Tanager** (*Piranga leucoptera*), p. 432. Bold white wing-bars. (*a*) ♂: bright red; black wings and small mask. (*b*) ♀: mostly olive; wings blackish; no overlap with 2.
8. **Scarlet Tanager** (*Piranga olivacea*), p. 431. (*a*) ♂, breeding: scarlet; wings and tail black; bill pale. (*b*) ♀: above more olive; below with less orange than 5b, wings dusky; wing-linings white. Migrant only.
9. **Flame-colored Tanager** (*Piranga bidentata*), p. 433. Streaked back and wing-bars. (*a*) ♂: orange-red. (*b*) ♀: yellowish-olive.
10. **Red-crowned Ant-Tanager** (*Habia rubica*), p. 434. (*a*) ♂: dusky-red; crown bright red. (*b*) ♀: olive; below paler; crown-patch yellow. Pacific slope.
11. **Black-cheeked Ant-Tanager** (*Habia atrimaxillaris*), p. 435. ♂: throat and crown-patch salmon; face black; upperparts dark gray. ♀ duller. Golfo Dulce area.
12. **Red-throated Ant-Tanager** (*Habia fuscicauda*), p. 434. (*a*) ♂: darker, duller than 10a with bright red, contrasting throat; crown-patch small. (*b*) ♀: olive; below paler; contrasting yellowish throat; dull, concealed crown-patch. Caribbean slope.
13. **Rosy Thrush-Tanager** (*Rhodinocichla rosea*), p. 438. (*a*) ♂: blackish; throat and breast rose-red; superciliaries red over lores, then white. (*b*) ♀: cinnamon-rufous where male is red. Mainly General-Térraba region.

1a
1b
2a
2b
3
4a
4b
4c
5a
5b
6a
6b
7a
7b
8a
8b
9a
9b
10a
10b
11
12a
12b
13a
13b

Plate 48: Grosbeaks, Buntings, and Larger Finches

1. **Black-headed Saltator** (*Saltator atriceps*), p. 441. Large; throat white; head largely black; back yellowish-olive.
2. **Buff-throated Saltator** (*Saltator maximus*), p. 442. Throat buffy, bordered with black; head gray; superciliaries white; back bright olive, less yellowish than 1.
3. **Grayish Saltator** (*Saltator coerulescens*), p. 442. Above more grayish than 1 or 2; throat and superciliaries white; malar stripes black.
4. **Streaked Saltator** (*Saltator albicollis*), p. 443. Below white, heavily streaked with olive; tip of bill often yellow. Mainly S Pacific slope.
5. **Black-faced Grosbeak** (*Caryothraustes poliogaster*), p. 443. Face black; head and breast mustard-yellow; heavy bill pale at base. Caribbean slope.
6. **Slate-colored Grosbeak** (*Pitylus grossus*), p. 444. ♂: dark gray; bill red; throat white. ♀ paler and duller with orange bill. Caribbean slope.
7. **Black-thighed Grosbeak** (*Pheucticus tibialis*), p. 444. Large; below yellow, with dull ochre wash; wings, back, and thighs black; wing-patch white. Highlands.
8. **Rose-breasted Grosbeak** (*Pheucticus ludovicianus*), p. 445. (*a*) ♂, breeding: black; breast rose; belly and wing markings white; in fall this pattern largely obscured by brown streaking. (*b*) ♀: heavy bill pale; head stripes; above and below with dark brown streaking; wing-bars.
9. **Black-headed Grosbeak** (*Pheucticus melanocephalus*), p. 445. (*a*) ♂: head black; body cinnamon-rufous; wings black and white. (*b*) ♀: from 8b by buffy underparts, lightly streaked only on sides. Casual to accidental.
10. **Blue-black Grosbeak** (*Cyanocompsa cyanoides*), p. 446. Very heavy bill black. (*a*) ♂: dull blue-black; eyebrows paler. (*b*) ♀: rich dark brown; below tinged with rufous.
11. **Blue Grosbeak** (*Guiraca caerulea*), p. 445. (*a*) ♂: blue with rufous wing-bars; bicolored bill. (*b*) ♀: brown; below paler; wing-bars buffy. Much larger than 12b. Flicks tail.
12. **Indigo Bunting** (*Passerina cyanea*), p. 447. (*a*) ♂: breeding: deep blue; in fall mixed with brown. (*b*) ♀: brownish; buffy breast; below brown streaking more or less distinct; blue tint on shoulders; conical bicolored bill; wing-bars indistinct, buffy. [See also Pl. 50(6).]
13. **Painted Bunting** (*Passerina ciris*), p. 447. (*a*) ♂: colorful, unmistakable. (*b*) ♀: no other small finch olive and yellow.
14. **Black-headed Brush-Finch** (*Atlapetes atricapillus*), p. 457. Throat and central underparts white; sides and crown-stripes gray; head otherwise black. S Pacific slope.
15. **Yellow-throated Brush-Finch** (*Atlapetes gutturalis*), p. 456. Above blackish; crown-stripe white; below whitish with yellow throat.
16. **Chestnut-capped Brush-Finch** (*Atlapetes brunneinucha*), p. 456. Conspicuous white throat, forehead spots; chest-band and sides of head black; cap chestnut, bordered laterally by yellowish.
17. **Orange-billed Sparrow** (*Arremon aurantiirostris*), p. 458. Bill orange; head striped; throat white; breast-band black; shoulder-patch yellow, more or less concealed.
18. **Prevost's Ground-Sparrow** (*Melozone biarcuatum*), p. 459. Bold black and white facial pattern on rufous head; large black breast-spot.
19. **White-eared Ground-Sparrow** (*Melozone leucotis*), p. 460. White and yellow pattern on black head; throat and breast black.
20. **Large-footed Finch** (*Pezopetes capitalis*), p. 455. Stout; head and throat mostly black, striped with gray on crown; body olive; below paler. High elevations.
21. **Sooty-faced Finch** (*Lysurus crassirostris*), p. 457. Mostly dark olive; belly yellow; cap rufous-chestnut; face and throat blackish; malar stripe white. Caribbean slope.
22. **Yellow-thighed Finch** (*Pselliophorus tibialis*), p. 455. Dark gray; blacker on head, wings, tail; conspicuous yellow thighs. Highlands.

1
2
3
4
5
8b
9b
6
7
8a
9a
10a
10b
11a
11b
12b
12a
13b
13a
14
15
16
17
18
19
20
21
22

Plate 49: Seedeaters, Flowerpiercer, and Finches

1. **Ruddy-breasted Seedeater** (*Sporophila minuta*), p. 450. Small. (*a*) ♂: above grayish; below and on rump rufous; wing-spot white. (*b*) ♀: buffy-brown; wing feathers with buffy margins.
2. **White-collared Seedeater** (*Sporophila torqueola*), p. 448. (*a*) ♂: complete white collar, wing-bars; back sometimes blacker than shown; belly white to buff. (*b*) ♀: only female seedeater with wing-bars.
3. **Variable Seedeater** (*Sporophila aurita*), p. 449. (*a*) ♂, Caribbean race: black; wing-linings, center of belly, and spot on primaries white. (*b*) ♂, Pacific race: black more extensive than 2a; wing-spot; no buffy tinge. (*c*) ♀, Pacific race: above olive-brown; belly whitish. Caribbean ♀ browner, especially below.
4. **Yellow-bellied Seedeater** (*Sporophila nigricollis*), p. 450. (*a*) ♂: head black; below yellowish; back dark olive. (*b*) ♀: more olive than 3c, below tinged with yellowish.
5. **Slate-colored Seedeater** (*Sporophila schistacea*), p. 448. (*a*) ♂: slaty; belly, wing-patch and mustache white (evidently only in old birds); bill yellow. (*b*) ♀: brownish, paling to whitish on belly.
6. **Yellow-faced Grassquit** (*Tiaris olivacea*), p. 447. (*a*) ♂: yellow throat and superciliaries contrast with blackish face and breast. (*b*) ♀: olive green; paler throat and superciliaries.
7. **Blue-black Grassquit** (*Volatinia jacarina*), p. 452. (*a*) ♂: glossy blue-black; in flight, white flash at base of wing. (*b*) ♀: brownish; below streaked.
8. **Cocos Finch** (*Pinaroloxias inornata*), p. 452. Chunky; pointed, decurved bill. (*a*) ♂: entirely black. (*b*) ♀: below heavily streaked. Young similar but with yellow bills. Cocos I. only.
9. **Slaty Flowerpiercer** (*Diglossa plumbea*), p. 454. Bill with abrupt terminal hook, base of lower mandible flesh-color. (*a*) ♂: gray. (*b*) ♀: brownish; below more or less faintly streaked. Highlands.
10. **Thick-billed Seed-Finch** (*Oryzoborus funereus*), p. 451. (*a*) ♂: larger than 3a, with much heavier bill. (*b*) ♀: rich brown; below tinged with chestnut.
11. **Blue Seedeater** (*Amaurospiza concolor*), p. 450. (*a*) ♂: dull blue-black. (*b*) ♀: dark rich chestnut-brown, unpatterned.
12. **Peg-billed Finch** (*Acanthidops bairdii*), p. 454. Note peculiar, bicolored bill; tail looks notched. (*a*) ♂: slate-gray; below paler. (*b*) ♀: browner; below faintly streaked; wing-bars buffy. Highlands.
13. **Pink-billed Seed-Finch** (*Oryzoborus nuttingi*), p. 451. (*a*) ♂: black; huge bill pale. (*b*) ♀: rich brown, tinged with chestnut; bill mostly dark; 10b much smaller. Caribbean slope.
14. **Slaty Finch** (*Haplospiza rustica*), p. 453. (*a*) ♂: slate-gray; bill conical, less thick than 11. (*b*) ♀: olive-brown; below paler; streaked on breast. Highlands.

1a
1b
2a
2b
3a
3b
3c
4a
4b
5a
5b
6a
6b
7a
7b
8a
8b
9a
9b
10a
10b
11a
11b
12a
12b
13a
13b
14a
14b

Plate 50: Siskins, Sparrows, Meadowlark, and Others

1. **Yellow-bellied Siskin** (*Carduelis xanthogastra*), p. 465. (*a*) ♂: black with yellow belly; yellow in wings and tail. (*b*) ♀: above olive; below yellow; yellow in wings. Highlands.
2. **Lesser Goldfinch** (*Carduelis psaltria*), p. 465. Below yellow; white in wing. (*a*) ♂: above black; bicolored bill. (*b*) ♀: above olive.
3. **Grassland Yellow-Finch** (*Sicalis luteola*), p. 453. Largely yellow; above olive, streaked with brown. Rare; recorded in Guanacaste; to be expected on S Pacific slope.
4. **Rusty Sparrow** (*Aimophila rufescens*), p. 461. Large; cap rufous; malar stripe black. NW, local.
5. **Botteri's Sparrow** (*Aimophila botterii*), p. 462. Smaller, paler than 4, with short faint malar streak, buffy breast. NW, very local.
6. **Indigo Bunting** (*Passerina cyanea*), p. 447. Immature ♀: more streaked, less buffy below than adult [See Pl. 48(12)], trace of buffy wing-bars; blue faint or absent on shoulder.
7. **Striped-headed Sparrow** (*Aimophila ruficauda*), p. 461. Head striped with black and white; breast smudged with gray; tail dull cinnamon. NW.
8. **Grasshopper Sparrow** (*Ammodramus savannarum*), p. 460. Breast buffy; crown dark with pale central stripe; lores yellow; looks flat-headed, with short tail.
9. **Chipping Sparrow** (*Spizella passerina*), p. 463. Rather slim, long-tailed; rump and underparts gray; eye-stripe black; superciliaries white; crown rufous. Accidental migrant.
10. **Wedge-tailed Grass-Finch** (*Emberizoides herbicola*), p. 464. Sparrowlike, with long pointed tail; note bicolored bill. Térraba region.
11. **Volcano Junco** (*Junco vulcani*) p. 462. Iris yellow; bill pinkish; above brown and streaked; below gray. High elevations.
12. **Lincoln's Sparrow** (*Melospiza lincolnii*), p. 463. Sides of head grayish; crown striped with bright brown; note pale eye-ring, buffy breast-band with dark streaking; smaller, slimmer than 13b. Rare migrant.
13. **Rufous-collared Sparrow** (*Zonotrichia capensis*), p. 463. (*a*) Striped head with short crest; nape rufous; "bow tie" black. (*b*) Young: head browner; below streaked; little or no rufous.
14. **Black-striped Sparrow** (*Arremonops conirostris*), p. 459. Head gray, striped with black; back olive; shoulder-patch yellow, more or less concealed.
15. **Olive Sparrow** (*Arremonops rufivirgatus*), p. 458. Head grayish-buff, striped with dark brown; concealed yellow shoulder-patch. Smaller, buffier than 14. N Pacific slope.
16. **Eastern Meadowlark** (*Sturnella magna*), p. 414. Above streaked; below yellow with black V on breast; in flight, white sides to short tail conspicuous.
17. **House Sparrow** (*Passer domesticus*), p. 466. (*a*) ♂: bib black; cheeks white; crown gray; nape chestnut. (*b*) ♀: short and stout; below grayish; head dull brown with pale superciliaries. Cities and towns.
18. **Bobolink** (*Dolichonyx oryzivorus*), p. 415. (*a*) ♂, breeding: below black; nape buff; scapulars white. (*b*) ♀, winter: rich buff, below tinged with yellow; head striped; sides and back streaked. Rare migrant.
19. **Dickcissel** (*Spiza americana*), p. 415. (*a*) ♂, breeding: suggests a miniature 16 but note thick bill, gray on head, rufous shoulder. (*b*) Immature ♀: like 17b but below tinged with yellow and sparsely streaked; shoulders tinged with rufous. Other plumages intermediate.

1a
1b
2a
2b
3
4
6
7
5
9
8
10
12
11
13a
13b
14
17a
17b
15
16
18b
19a
18a
19b

Plate 51: Accidental, Hypothetical, and Recently Added Species

Note. For comparison and to facilitate identification, we show most of these species with a species well known in Costa Rica.

1. **White-crowned Pigeon** (*Columba leucocephala*), p. 166. Large; dark; white crown diagnostic (duller but still evident in immatures). Caribbean coast.
2. **Eared Grebe** (*Podiceps nigricollis*), p. 68. Larger than 3; iris red. (*a*) Breeding: very dark; tuft of yellow feathers behind eye. (*b*) Nonbreeding: blackish above, grayish below; throat to behind auriculars white. Inland ponds and lakes: once, Cerro Chirripó.
3. **Least Grebe** (*Tachybaptus dominicus*), p. 68. Smaller, more brownish than 2; iris yellow. [See also Pl. 7(4).]
4. **Gray Gull** (*Larus modestus*), p. 157. Second year, nonbreeding. Adult grayer with white head in breeding plumage. Juvenile browner. In all plumages shows broad white terminal tail-band, white on trailing edge of wing, black bill; slightly larger than 5 and 6. Accidental, Cocos I. (especially El Niño years?).
5. **Heermann's Gull** (*Larus heermanni*), p. 157 (in Gray Gull, Note). Size of 6; very dark. (*a*) Nonbreeding: head white, speckled with brownish; white tips on rectrices, secondaries; bill red with black tip. Head pure white in breeding plumage. (*b*) Immature: uniform dark brownish-gray; base of bill reddish. (*c*) Juvenile: blackish-brown; amount of pink or yellow at base of bill varies, sometimes lacking. Several sightings, Golfo de Nicoya (Chomes).
6. **Laughing Gull** (*Larus atricilla*), p. 156. Nonbreeding: size of 5 but white below. [See also Pl. 3(8).]
7. **Common Potoo** (*Nyctibius griseus*), p. 198. Hunting pose: appears huge-eyed; perches erectly but head not held vertically. [See also Pl. 20(4).]
8. **Oilbird** (*Steatornis caripensis*), p. 197. Larger than 7 with much stronger bill; reddish-brown, speckled with white; nocturnal. Accidental: one record.
9. **Ocellated Poorwill** (*Nyctiphrynus ocellatus*), p. 201. ♂: smaller, much darker than 10; note white-tipped lateral rectrices, white-spotted belly. ♀ similar but more rufescent overall. Extreme NW Caribbean slope.
10. **Common Pauraque** (*Nyctidromus albicollis*), p. 200. ♀: browner than 9, different tail-pattern, scapulars conspicuously scalloped, not spotted or "ocellated." [See also Pl. 21(18).]
11. **Sora** (*Porzana carolina*), p. 126. From 12 by bill and leg colors; amount of black on face varies. [See also Pl. 6(12).]
12. **Painted-billed Crake** (*Neocrex erythrops*), p. 128. Note gray foreparts, barred flanks; bright red-and-yellow bill diagnostic.

1
2a
2b
3
4
5a
5b
5c
6
7
8
9
10
11
12

Plate 52: Hypothetical Species and Recent Additions, continued

1. **Cliff Swallow** (*Hirundo pyrrhonota*), p. 343. Note pale rump; adults (but not all immatures) have pale foreheads, black on chest. [See also Pl. 22(13).]
2. **Cave Swallow** (*Hirundo fulva*), p. 343 (in Cliff Swallow, Note). Very like 1 but forehead and rump darker, throat and chest paler. Possible during migrations, especially along Caribbean coast.
3. **Nashville Warbler** (*Vermivora ruficapilla*), p. 385. Smaller than 5, with finer bill, conspicuous eye-ring, and median throat yellow. (*a*) ♂: head gray, rufous in crown (inconspicuous). (*b*) ♀: paler, duller; belly largely white. Recent sightings on both slopes.
4. **Connecticut Warbler** (*Oporornis agilis*), p. 397. Larger, heavier than 5, with complete white eye-ring, relatively short tail with long lower tail-coverts. (*a*) ♂: complete gray hood, throat often paler, chest darker. (*b*) ♀: hood browner than 5, more complete across breast. See also MacGillivray's Warbler, p. 397. Several recent sightings.
5. **Mourning Warbler** (*Oporornis philadelphia*), p. 396. ♀: hood grayer than 4b, usually incomplete across breast; eye-ring variable but virtually never broad and complete as in 4b. [See also Pl. 42(14).]
6. **Melodious Blackbird** (*Dives dives*), p. 408. Only Costa Rican icterid with entirely black plumage *and* dark eye; from smaller 7 by longer bill and tail. Recently recorded from N Pacific slope; may be invading.
7. **Bronzed Cowbird** (*Molothrus aeneus*), p. 408. Note red eye, neck-ruff (lacking in 6). [See also Pl. 44(15).]

FAMILY Galbulidae: Jacamars

The 15 species of jacamars are confined to wooded regions of continental tropical America (including Trinidad), chiefly at low altitudes. Most species have glittering metallic plumage, much like that of hummingbirds. They have short legs and long, pointed, usually very slender bills, which reach past the fragile wings of butterflies and dragonflies to grasp their bodies firmly and to keep their fluttering wings, as well as the stings of bees and wasps, away from the bird's face while it knocks its prey against a perch. Jacamars appear to be wholly insectivorous, often capturing large brilliant butterflies, such as *Morpho* and swallowtails, many glittering beetles and bees, and other insects. Among nonpasserines, some jacamars are outstanding for their elaborate songs.

Jacamars nest in burrows that both sexes dig with their bills in earthen banks, in wooded slopes, in the wall of clay uplifted when a great tree is uprooted by wind, or in termitaries. The unlined chamber at the inner end is soon covered with glittering chitinous parts of insects regurgitated by the parents, upon which lie 2–4 glossy white eggs. Male and female share incubation, the latter sitting through the night; the incubation period is 19–21 days. In contrast to the naked hatchlings of other piciform birds, those of jacamars have copious long whitish down. The nestlings soon become loquacious, repeating weak versions of the adults' trills and calls while they wait for the food that both parents bring to them. By the time they fledge at about 3 weeks of age, the nest hole is befouled with excrement and regurgitated hard parts of insects.

RUFOUS-TAILED JACAMAR

Galbula ruficauda Pl. 26(12)

Jacamar Rabirrufo (Gorrión de Montaña)

DESCRIPTION. 9″ (23cm); 27g. Slender, long-tailed, with short, rounded wings and a long, slender, sharp-pointed bill, which with its iridescent plumage gives the bird the aspect of an overgrown hummingbird. **Adult ♂:** upperparts, including wings, tail, face, and broad breast-band, glittering iridescent green; primaries black; usually some black on chin; throat white; posterior underparts and outer 3 pairs of rectrices rufous. Bill black; feet yellowish-brown. ♀: similar but throat buffy, posterior underparts usually paler. **Immatures:** resemble respective adults but green of upperparts darker, duller, bronzier; breast-band broader and less clean-cut; for some time after fledging, bills noticeably shorter than those of adults.

HABITS. Frequents shady forest edge, especially along streams, open understory of woodland and semi-open, tall second growth; especially partial to cacao plantations; rests on horizontal perch with jauntily uptilted bill, looking actively from side to side for flying insects, which it seizes on a swift, buzzy sally; takes many glittering beetles and bees, butterflies, and dragonflies; may perch above rotting fruits on forest floor to catch butterflies attracted thereto; solitary or in pairs.

VOICE. Usual call a loud, penetrating whistle that cuts off abruptly: *wheeert!* or *peeeur!*; sometimes a clear *preeet* repeated rapidly 3–10 times; song a prolonged, high-pitched, ascending trill, which may be clear and melodious or dry and buzzy.

NEST. A burrow 11–20″ (28–51cm) long in bank, or in earth among root mass of uprooted tree; or a short tunnel to a chamber excavated in a termitary. Eggs 2–4. March–June.

STATUS. Fairly common resident of lowlands and foothills of Caribbean slope and S Pacific slope, N regularly to Carara and rarely to head of Golfo de Nicoya; regularly to 2500ft (750m), rarely to 4000ft (1200m).

RANGE. S Mexico to W Ecuador and NE Argentina.

NOTE. The Central American form is sometimes considered a separate species, *G. melanogenia* (Black-chinned Jacamar), although hybridization with the true *ruficauda* occurs in N Colombia, and in Costa Rica the amount of black on the chin is highly variable.

GREAT JACAMAR

Jacamerops aurea Pl. 26(11)

Jacamar Grande

DESCRIPTION. 12″ (30cm); 63g. Larger and more robust than Rufous-tailed Jacamar, with stouter, shorter, slightly decurved, sharp-pointed bill. **Adult ♂:** chin, sides of face, and entire upperparts glittering metallic green, shading to bluish on chin, forehead and tail, and to coppery-purplish on back; primaries black; lower throat white; rest of underparts deep rufous; underside of tail blue-black. Bill black; feet dark horn. ♀: similar but no white on throat. **Immatures:** duller, bronzier, and above less iridescent, with little or no coppery on back; below paler and duller; rectrices narrowly tipped with dull bronze.

Habits. Frequents midlevels and canopy of humid forest and shady neighboring semi-open and tall second growth, coming lower at edges and gaps; sallies to seize butterflies, dragonflies, beetles, and other insects in midair or to pluck small lizards, insects, or spiders from vegetation; beats larger prey against perch before swallowing it; usually in pairs, sometimes solitary.

Voice. Members of a pair call back and forth with a variety of soft buzzing, grating, and mewing notes; song a loud, clear, penetrating high whistle *keeeyeeeeeew*! that sounds more like the call of a hawk.

Nest. A chamber in an arboreal termitary 9–50ft (3–15m) up. Eggs undescribed (?). March–May or June.

Status. Rare to very uncommon resident of Caribbean lowlands and foothills, up to ca. 1600ft (500m), N to Llanura de San Carlos and Llanura de Tortuguero.

Range. Costa Rica to W Ecuador, N Bolivia, Amazonian Brazil, and Guianas.

FAMILY **Bucconidae:** Puffbirds

The 30 species of puffbirds are confined to continental tropical America, chiefly in warm lowlands. Large heads, abundant lax plumage (beneath which their short legs may be all but invisible) and short tails make many of them appear stout and "puffy," whence the name. Their bills, of short or medium length, may be notably stout and hooked or more slender, decurved, and tapering. Lacking bright spectral colors, they are clad in shades of brown, rufous, and gray, often heavily barred or streaked; some are boldly patterned in black and white, or are almost wholly black, with vivid bills. They are among the birds often called *bobo* (stupid) because of their energy-saving manner of foraging: while they perch in apparent lethargy, their eyes are keen to detect insects and other small invertebrates that move within their range of vision, perhaps 60ft (18m) distant. With a rapid dart, a puffbird seizes its prey and carries it back to its perch, against which it may beat it loudly before swallowing it. Puffbirds vary their diet with small frogs and lizards, but most take little or no fruit. Their voices range from thin and weak to stentorian; some join in ringing choruses. Puffbirds nest in holes that both sexes carve in occupied termitaries, in leaf-lined burrows in the forest floor, and occasionally in other cavities. Two or 3 white, glossy eggs are incubated by both parents, and both attend the naked young, in nunbirds assisted by helpers. Incubation periods in this family are unknown; the few recorded nestling periods range from 20–30 days.

WHITE-NECKED PUFFBIRD

Bucco macrorhynchos Pl. 28(3)

Buco Collarejo

Description. 9½″ (24cm); 105g. Stout, big-headed, boldly patterned, with large, heavy, hooked bill. **Adults:** forehead and superciliary region white; crown, nape, and broad stripe from bill through eye to hindcrown black; throat, upper breast, cheeks, and collar around hindneck white; broad breast-band, remaining upperparts, wings and tail black, the mantle feathers with narrow white fringes when fresh; sides and flanks sooty-black, scaled with white; remaining underparts white. Iris red; bill and feet black. **Young:** forehead, throat, and belly washed with grayish-buff; throat narrowly scaled with dusky; breast-band narrow; upperparts duller black, the feathers with soft grayish fringes; iris brown.

Habits. Frequents forest canopy and edge, semi-open, and clearings and second growth with scattered tall trees; perches for long periods on open high branch or snag, immobile except for head movements, then makes sudden, long, direct sally to seize a large insect or small lizard from foliage; carries prey back, beats it against perch, then swallows it; flight direct and fast on rapidly beating wings; solitary or in pairs.

Voice. Usually silent; during breeding season a long bubbling trill, at a constant pitch or rising slightly, then falling; occasionally a sad, descending, ventriloquial *peeeur*.

Nest. In a cavity that both sexes excavate in a large arboreal termitary. Eggs unknown (?). March–May.

Status. Widely distributed but generally uncommon resident of lowlands of both slopes, chiefly below 2000ft (600m).

Range. S Mexico to W Ecuador and NE Argentina.

Note. Formerly placed in genus *Notharchus*.

PIED PUFFBIRD

Bucco tectus Pl. 28(7)

Buco Pinto

Description. 6″ (15cm); 27g. Patterned like a miniature White-necked Puffbird but much more active; upper mandible terminates in 2 small prongs, lower in a hook that fits between them (like Prong-billed Barbet in reverse). **Adults:** above mostly black; lores and postocular streak dull white; forehead speckled with white; a white patch on scapulars; rectrices tipped with white, the outer ones with a median white bar on inner webs; below mostly white with a black breast-band that does not connect with black of upperparts; flanks sooty-black; indistinctly scaled with white. Iris dark brown; bill and feet black. **Young:** above brownish-black, spotted with buffy-white on wing-coverts; scapular tufts, throat and chest tinged with buff; breast-band narrower; forehead speckled with buff.

HABITS. Frequents canopy and edge of humid forest, adjacent clearings, and second growth with scattered tall trees, semi-open; during most of year travels in small groups of ca. 6; usually in pairs during breeding season; individuals maintain contact by loud calls; perches in treetops looking actively about; frequently pumps tail; makes darting sallies to seize insects from air or foliage; eats dragonflies, bees, wasps, beetles.

VOICE. Loud, clear, thin, high-pitched whistles: *pweee pweee pweee* or *weereee weereee wreee weeea*, in various patterns, the notes in quality not unlike those of Broad-winged Hawk but much shorter, repeated.

NEST. A cavity in a termitary high in a tree, attended by both parents. Eggs undescribed (?). May.

STATUS. Resident, but evidently with pronounced local movements, in lowlands of Caribbean slope N to Llanura de Tortuguero (La Suerte); uncommon to increasingly rare as forests are cut; sea level to 1000ft (300m).

RANGE. Costa Rica to E Peru and Amazonian Brazil.

NOTE. Formerly placed in genus *Notharchus* (which is now united with *Bucco*).

WHITE-WHISKERED PUFFBIRD

Malacoptila panamensis Pl. 28(6)

Buco Barbón

DESCRIPTION. 7″ (18cm); 42g. Stout, brownish, streaked below; prominent facial bristles and white malar tufts. **Adult ♂:** upperparts bright olive-brown; crown, mantle, and wing-coverts finely spotted with cinnamon-brown to cinnamon-buff; tail cinnamon-brown; remiges dusky, edged with cinnamon; face and forehead finely streaked with cinnamon-buff; short whitish tufts on forehead; throat buffy-white; chest tawny-brown to cinnamon, shading to cinnamon-buff on belly and whitish on crissum; underparts with indistinct darker streaking. Iris red; upper mandible blackish; lower yellowish-horn with dusky tip; feet yellowish-brown. ♀: similar in pattern to male but more contrast; upperparts grayish-brown, spotted and streaked with pale buff; throat buff; posterior underparts white, indistinctly streaked with dark brown. **Young:** resemble adult female (males somewhat brighter), upperparts more barred or scaled, less spotted with dull buff; throat scaled with brown; malar tufts small; belly more narrowly streaked.

HABITS. Perches quietly in understory to middle levels of forest, shady semi-open (especially cacao plantations), or old second growth; sallies out or down to snatch insects, spiders, small frogs or lizards from ground or vegetation, carrying them back and beating them against perch before swallowing; alone or in pairs; sometimes attends mixed flocks of small birds, or foraging swarms of army ants; switches tail from side to side when excited.

VOICE. High, thin, sibilant whistles, seemingly from a much smaller bird: *tseeeeeip; pseeeeeeer; seeeeeet?* Generally silent but complains interminably when nest or fledglings seem to be in peril.

NEST. A burrow 6–22″ (15–56cm) long and ca. 2½″ (6cm) in diameter, level or descending at a slight angle into gently sloping or nearly level ground or into a little hillock on forest floor, with a short entrance tunnel or collar of twigs and dead leaves extending out an additional 2–3″ (5–8cm); enlarged terminal chamber lined with dead leaves. Eggs 2, rarely 3. March–June.

STATUS. Fairly common resident of lowlands and lower foothills of Caribbean slope (except drier areas of extreme NW) and S Pacific slope N to head of Golfo de Nicoya, but very rare N of Carara; reaches Pacific side of Cordillera de Guanacaste via low passes; to 3000ft (900m) locally on Caribbean slope, 4000ft (1200m) on Pacific slope.

RANGE. SE Mexico to W Ecuador.

NOTE. Also called White-whiskered Softwing.

LANCEOLATED MONKLET

Micromonacha lanceolata Pl. 28(5)

Monjito Rayado

DESCRIPTION. 5″ (13cm); 19g. Suggests a diminutive, strongly marked White-whiskered Puffbird. **Adults:** forehead, lores, eye-ring, and well-developed facial bristles white, separated by narrow black line from rich brown of crown and upperparts; all but central feathers of short, dull brown tail with black subterminal band, pale brown tip; below mostly white, boldly streaked with black; flanks and crissum tawny-buff. Bill black, legs and feet olive-gray. **Immatures:** upperparts lightly scaled with buff; black streaks below edged with buff.

HABITS. Little known; evidently prefers wet forests of lower middle elevations; alone, often as attendant of mixed-species flocks of small birds in middle and upper levels of forest; sallies for insects from exposed perch; takes berries (Gómez).

VOICE. Usually silent; rarely 1–5 high, thin, plaintive whistled notes with rising inflection, each slightly higher in pitch than the last.

NEST. Undescribed (?).

STATUS. A rare resident on Caribbean slope of Cordillera Central and Cordillera de Talamanca, at ca. 1300–4500ft (400–1350m); recent sightings at Virgen del Socorro (by Río Sarapiquí), Tapantí, Carrillo, and near Colonia Palmareña, near S end of Cordillera de Tilarán.

RANGE. Costa Rica to E Peru and W Brazil.

WHITE-FRONTED NUNBIRD

Monasa morphoeus Pl. 28(4)

Monja Frentiblanca (Julío)

DESCRIPTION. 11½″ (29cm); 105g. Slimmer, longer-tailed than other puffbirds, with tapering, slightly decurved, bright orange-red bill. **Adults:** forehead, lores, and chin with short, stiff, bristly, erect white feathers; rest of head dull black, shading to dark slate-gray over rest of body, palest on belly; wings and tail more blackish. Iris brown; feet dark gray. **Young:** facial tufts cinnamon-buff; plumage in general tinged with brownish; throat, chest, and wing-coverts edged with dull brown; bill paler orange, marked basally with black.

HABITS. Straggles through canopy of tall rain forest and out into adjoining shady semi-open and clearings with scattered trees, in loose flocks of 4–8, rarely more; snatches insects, spiders, small frogs and lizards from foliage or bark by long, swift darts from lookout perch; catches slow-flying insects in midair; often accompanies flocks of caciques or other vigorously foraging birds (e.g., oropendolas), or traveling troops of monkeys, catching prey flushed by these.

VOICE. Surprisingly varied; common call an unclear descending whistle followed by a short rippling trill: *peeeur-r-r-r-r*. Also a loud, mournful *how how how*; various undulating or rippling trills, churrs, rattles. Especially during breeding season, 3–10 individuals, perching a few inches apart on a horizontal limb or vine at midheight of forest, with heads thrown back, join in a chorus of loud, continuous gobbling, barking notes, sometimes for up to 20min.

NEST. A burrow sloping downward in level or sloping forest floor, 32–55″ (80–140cm) long and ca. 3″ (8cm) in diameter, the mouth surrounded by a collar of dead leaves and twigs, the enlarged terminal chamber lined with leaves; attended by 3–6 adults. Eggs 2–3. March–May.

STATUS. Common resident (where forest still remains) of lowlands and foothills of Caribbean slope, locally up to 2300ft (700m), especially in SE; rare or absent in drier lowlands S of Lago de Nicaragua.

RANGE. E Honduras to N Bolivia and SE Brazil.

FAMILY **Capitonidae:** Barbets

Most of the 74 species of barbets inhabit the Old World tropics, especially Africa, the home of several terrestrial species of the savannas as well as the more typical arboreal forms. In length 3½–12″ (9–30cm), barbets are stocky birds with short tails and wings, short legs, strong feet, and stout to very thick bills, from the base of which spring tufts of bristles responsible for the family name (from the French *barbe*, beard). Many are brilliantly colored in patterns of red, orange, yellow, green, or blue; some are plainly attired. The sexes may differ greatly in coloration or may be almost the same. Many have strong voices that they use freely, often dueting with their mates or joining in choruses. They consume mostly fruits, supplemented by insects, including, in some Old World species, many termites. All of the 13 species that inhabit the New World, chiefly northern South America, are arboreal. They carve woodpeckerlike holes in dead trees, the 2 sexes sharing the labor. On the unlined floor barbets lay 2–5 white eggs, which both sexes incubate for 13–15 days. The young, hatched naked, are fed by both parents, assisted, in several of the African species, by 1 or more helpers. The nestling period is quite long, 4–7 weeks in some species. After flying, the young are led back to sleep with their parents in the nest cavity. Later, larger numbers may lodge together in a hole in a tree.

RED-HEADED BARBET

Eubucco bourcierii Pl. 28(2)

Barbudo Cabecirrojo

DESCRIPTION. 6″ (15cm); 35g. Chunky, big-headed; both sexes strikingly colored; conspicuous, stout yellow bill. **Adult ♂:** lores, forehead, and chin black; rest of head, throat, and chest deep red, shading abruptly to orange on breast and to yellow, heavily streaked with dull green on posterior underparts; upperparts, wings, and tail dull green, on sides of neck separated from red head by a vertical bar of bluish-white. Iris brick-red; bill greenish-yellow; feet olive-green. ♀: lores, forehead, and chin black; throat pale green; forecrown and sides of neck deep orange, this continuing as a band across upper breast; rest of crown dark ochraceous-olive, suffused with orange; cheeks and short streak over eye pale blue; lower breast pale olive-green. **Immatures:** like corresponding adults but red or orange on crown and chest duller and less extensive.

HABITS. Prefers wet forest and adjoining thickets and semi-open; forages high or low, often in dense vine tangles, frequently hanging inverted; probes rolled dead leaves, sometimes holding them beneath a foot and biting them to expel hidden insects and spiders; also gleans insects from foliage and moss on branches; takes some fruit, including berries of *Myrica* ("arrayán") and Ericaceae, guava, bananas at feeder; solitary or in pairs, often as attendants in mixed-species flocks.

VOICE. Usually silent; squirrel-like sputters and chatters in territorial disputes; a low, cicadalike rattle; song of male a prolonged, resonant, ventriloquial, somewhat toadlike trill *krrrrrrrrrr*, delivered from high in tree.

NEST. In cavity 2–8ft (0.6–2.5m) up in fence post or slim rotting stub. Eggs 2–3. March–June.

STATUS. Resident at middle elevations on both slopes of Cordillera de Talamanca, mainly on Caribbean slope of Cordillera Central, and along upper slopes of Cordillera de Tilarán; ca. 1300–5300ft (400–1600m) on Caribbean slope, and 3000–6000ft (900–1850m) on Pacific; between October and February descends to lower elevations: 1000ft (300m) on Caribbean slope, and 2400ft (750m) on Pacific.

RANGE. Costa Rica to NE Peru.

NOTE. Sometimes placed in the genus *Capito*.

PRONG-BILLED BARBET

Semnornis frantzii Pl. 28(1)

Barbudo Cocora (Cocora)

DESCRIPTION. 6¾″ (17cm); 62g. Chunky, short-tailed, brownish; bill short and very thick, the upper mandible terminating in a small hook, the lower in 2 little prongs that fit around it. **Adults:** lores, chin, and narrow area around eye sooty-black; forehead and forecrown orange-tawny; rest of upperparts brownish-olive, darkest on crown where suffused with orange-tawny, shading to dark olive-green on wings, rump, tail-coverts, and tail; cheeks, throat, and chest ochraceous-olive; a patch of bluish-gray on each side of breast; posterior underparts olive-green, with golden gloss on belly; male bears a tuft of stiff, glossy, elongate black feathers on nape. Iris red; bill silvery-gray except tip and culmen black; feet olive-gray. **Immatures:** paler and duller than adults; little or no bright area on forehead. Iris dark brown; bill dusky, shading to horn-color basally; orbital skin pale dull yellowish (this area feathered in adults).

HABITS. Prefers cool, wet, moss-festooned mountain forest and adjacent old second growth and clearings with large trees; monogamous and strongly territorial during nesting season; at other times in flocks of up to a dozen or more; up to 16, possibly more than 1 flock, may roost together in a hole in a tree; eats mainly fruits of trees and epiphytes but descends to fruiting shrubs along edges; swallows fruits whole or squeezes them to swallow the juice, discarding the husk; also takes some insects and flower petals; squeezes nectar from tubular flowers; holds large items beneath foot to tear them apart; flight direct, with rapid, buzzy wingbeats.

VOICE. A resonant, far-carrying, rather throaty *cwa-cwa-cwa-cwa* . . . duet by mated pair or by larger groups in chorus; low, dry, rattling notes when moving in groups; a nasal, squalling *kwaaah* in agonistic encounters.

NEST. An unlined, woodpeckerlike cavity 11–60ft (3.5–18m) up in a dead, but not excessively rotten, stub or snag. Eggs 4–5. March–May.

STATUS. Fairly common resident of middle elevations of both slopes, from Cordillera de Tilarán S; ca. 2500–8000ft (750–2450m) on Caribbean slope, occasionally down to 1650ft (500m); and 5000–8000ft (1500–2450m) on Pacific.

RANGE. Costa Rica and W Panama.

FAMILY **Ramphastidae:** Toucans

Forty-two species of toucans inhabit rain forests and lighter woodlands of continental tropical America, chiefly at low altitudes; only a few species live high in the mountains. In length they are 12–26″ (30–66cm), including their disproportionately huge bills, which are surprisingly light, hollow structures, crisscrossed internally by many thin supporting rods. Their plumage is varied, often largely black, relieved by red, yellow, and white; or olive-brown and blue; or chiefly green. Their bills are often vividly tinted. Except in toucanets of the genus *Selenidera*, the sexes are alike, but males' bills tend to be longer—a difference readily noticed in some pairs in the field. Their calls, seldom musical, include croaks, barks, rattles, or high, sharp notes. In small, loose flocks, toucans straggle through the treetops, rarely descending. Largely frugivorous, they vary their diet with insects and other small invertebrates, small lizards and snakes, birds' eggs and nestlings. Smaller species commonly nest in holes carved by woodpeckers, which they may enlarge, but the big *Ramphastos* toucans prefer cavities resulting from decay in living trees. Here, on the unlined floor that is often covered with regurgitated seeds, toucans lay 2–4 white eggs. Both sexes incubate, sitting very restlessly for such large birds. The young, hatched naked after 16 or more days' incubation, are attended by both parents during a nestling period of 6 weeks or more. After fledging, young aracaris sleep with their parents in the nest cavity. Aracaris are the only toucans known to lodge socially in tree holes throughout the year.

EMERALD TOUCANET

Aulacorhynchus prasinus Pl. 27(17)

Tucancillo Verde (Curré)

Description. 11½″ (29cm); 180g. Smallest and shortest-billed Costa Rican toucan, the only one with green body. **Adults:** crown and hindneck olive-green, shading to grass-green over rest of upperparts; rectrices bluish distally, tipped with chestnut; lores, area below eye, and entire throat dark blue; rest of underparts pale yellowish-green; crissum chestnut. Orbital skin bluish-slate; sides of upper mandible bright yellow; base of culmen, broad stripe along tomium, and entire lower mandible black; a dark red patch around nostrils; a narrow yellow (upper mandible) and white (on lower) line encircling base of bill; legs bright olive-green. **Young:** throat paler, more grayish-blue; underparts paler and duller; dark areas of bill dusky-horn; no pale line encircling base.

Habits. In straggling flocks of ca. 5–10 birds, in upper levels of montane forests, forest edge, and adjacent second growth, semi-open, and clearings with scattered trees; restless and excitable, often cocking tail or stretching neck to peer about; flight direct on rapid, buzzy wingbeats, ending with a short sail to the perch; eats many fruits, also insects, an occasional small lizard, and eggs and nestlings of other birds.

Voice. Usual call a loud, far-carrying, dry *rrip rrip rrip rrip* or *curré curré curré* reminiscent of the sound of a crosscut saw, often continued for minutes on end; among members of flock, a soft *eeaaah*; in aggressive encounters or alarm, a sharp, staccato, creaking rattle; various croaking or barking notes.

Nest. In hole carved by woodpeckers or other unlined cavity, 7–90ft (2–27m) up, in forest or adjacent clearing. Eggs 3–4. March–July (usually 2 broods).

Status. Common to abundant resident of middle elevations the length of the country, upward from ca. 2600ft (800m) on Caribbean slope and 3300ft (1000m) on Pacific, commonly to 8000ft (2450m) and in much smaller numbers to 10,000ft (3000m); occasionally wanders much lower after breeding.

Range. Tropical Mexico to N Venezuela and E Peru.

Note. The blue-throated form of Costa Rica and W Panama is sometimes considered a species (Blue-throated Toucanet, *A. caeruleogularis*) distinct from the otherwise white-throated Emerald Toucanet.

COLLARED ARACARI

Pteroglossus torquatus Pl. 27(15)

Tucancillo Collarejo (Cusingo, Tití, Felix)

Description. 16″ (41cm); 230g. Slender, medium-sized, with long graduated tail, short rounded wings, large bill. **Adults:** head, neck, and chest glossy black; a narrow rufous nuchal collar; upperparts glossy dark olive-green; remiges and rectrices dark olive; rump and upper tail-coverts red; underparts mostly bright yellow, suffused with red; a round black spot on upper breast; broad black band across upper belly, more or less mixed with red; thighs chestnut. Iris bright yellow with black spot behind pupil; facial skin black, shading to dull red behind eye; most of upper mandible dull pale yellowish, often maroon near base; stripe along culmen, tip, coarse serrations along tomium, and entire lower mandible black; a narrow white line completely encircles base of bill; legs bright olive-green. **Young:** much duller: head and chest sooty-black; back dark brownish-olive; rump duller red; below pale yellow with indistinct breast-spot and band; bill lacks basal white line and prominent serrations; pattern indistinct.

Habits. Straggles through middle and upper levels of forest, semi-open, old second growth in flocks of 6–15, coming lower along edges, in gaps, or young second growth; flight level, on rapidly beating wings, short glide to perch; springy, jaylike hops along branches; eats mainly fleshy or arillate fruits (aroids, *Protium*, palms, *Cecropia*), some insects, small lizards, eggs and nestlings of small birds; up to 6 sleep in hole in tree, folding tails over backs.

VOICE. Call a loud, high, sharp *seek*, *pseek*, *pisseek*, or *pink*, emphatic and penetrating; aggressive note a rasping *grahhrr*; in alarm, a high squealing *eeeyeeek* (as when attacked by forest-falcon); a rapidly repeated sharp *pitit* in excitement.

NEST. In old woodpecker hole or natural cavity in tree, 20–100ft (6–30m) up. Eggs 3. January–May. Up to 5 adults may attend a nest.

STATUS. Resident, common to abundant in lowlands and foothills throughout Caribbean slope (except where extensively deforested), locally up to 3300ft (1000m), reaching 3900ft (1200m) in SE; on Pacific slope generally uncommon to rare from Nicaraguan border S to Las Juntas; more numerous on Península de Nicoya and lower slopes of Cordillera de Guanacaste; rare wanderer in Valle Central.

RANGE. S Mexico to W Ecuador.

FIERY-BILLED ARACARI

Pteroglossus frantzii Pl. 27(16)

Tucancillo Piquianaranjado (Cusingo)

DESCRIPTION. 17″ (43cm); 250g. Very similar to its close relative, the Collared Aracari, which it replaces on S Pacific slope, differing mainly in its strikingly colored bill. **Adults:** plumage like that of Collared Aracari, except that a line of deep chestnut separates the black of the head and dark green of the back; black breast-spot larger and more extensive laterally; band across belly broader, entirely red except for a more or less broken black anterior border; facial skin mostly red, only lores black. Bill very different: upper mandible mostly deep orange to vermilion, shading to bright yellow-green at base, serrations of tomium very small, white; basal half of culmen and entire lower mandible black; white line encircles base of bill; iris and feet like Collared Aracari. **Young:** much duller; head sooty-black; back darker and duskier; below paler yellow; breast-spot small and indistinct; breast-band mixed with black and rufous; no white line around base of bill; iris dull orangish.

HABITS. Generally resembles Collared Aracari. Roams through upper levels of humid forest and adjacent clearings in straggling bands of 10 or less, seeking fruit, insects, and nestling birds; sometimes descends almost to ground to eat berries; up to 5 sleep together in old woodpecker hole, with rest of band in another hole nearby.

VOICE. Call a high, sharp, metallic, squeaky *pink, pity, kaseek* or *kissick*, like call of Collared Aracari but more often 2-noted; a harsh croak in aggressive situations; a weak, reedy rattle (Slud).

NEST. In old woodpecker hole 20–100ft (6–30m) up, unlined but bottom covered with regurgitated seeds. Eggs 2. January–April. Three or more adults may attend a nest.

STATUS. Resident of S Pacific slope up to 5000ft (1500m) locally, N to about Orotina and hills above Atenas (Cuesta de Aguacate); common except at N extreme of range, or where deforestation extensive.

RANGE. Costa Rica and W Panama.

NOTE. Sometimes considered a subspecies of Collared Aracari.

YELLOW-EARED TOUCANET

Selenidera spectabilis Pl. 27(20)

Tucancillo Orejiamarillo

DESCRIPTION. 14½″ (36cm); 220g. The only sexually dimorphic Costa Rican toucan, and the only one with solid black breast and belly. **Adult ♂:** head and most of underparts glossy black, except long yellow ear-tufts; large tufts of yellow-orange on flanks; crissum red; thighs chestnut; upperparts olive-green; primaries blackish; tail bluish-slate. Iris deep red; facial skin bright yellow-green, shading to turquoise above eye and to yellow-orange on malar area; upper mandible mostly pale, dull greenish-yellow, a broad stripe along tomia and entire lower mandible dusky-horn, shading to blackish near tip; inside of bill vermilion; legs blue-gray. ♀: similar except forehead, crown, and hindneck deep chestnut; no ear-tufts. **Young:** duller, below more sooty-black; belly tinged with olive; little or no orange on flanks or (male) yellow on head; crissum paler red; crown and nape of female dark sooty-brown.

HABITS. Prefers canopy of tall wet forest, descending lower in adjacent old second growth or shady semi-open; during most of year in pairs or small groups of 3 or 4, sometimes alone; following breeding sometimes in groups of 4–8; eats fleshy and arillate fruits (*Hampea, Protium, Dendropanax, Guatteria*, aroids), occasional large insects or small lizards, possibly nestling birds.

Voice. Calling male throws head back to 1 side, flashing vermilion bill lining, and flops tail up and to other side with each croak, as though to disassemble himself; in aggressive encounters, a hoarse sputter; song a hoarse, low croak, lower and rougher than that of Keel-billed Toucan and with an initial click: *K'Krrruk K'Krrruk K'Krrruk. . . .*

Nest. Unknown (?).

Status. Resident in foothills of Caribbean slope the length of the country, reaching Pacific slope locally in Cordillera de Guanacaste; breeds mainly 1000–4000ft (300–1200m). In mid to late wet season, from September through January, occasionally into March, many descend to adjacent lowlands.

Range. SE Honduras to NW Colombia.

KEEL-BILLED TOUCAN

Ramphastos sulfuratus Pl. 27(18)

Tucan Pico Iris (Curré Negro)

Description. ♂ 18½″ (47cm), 500g; ♀ 17″ (44cm), 380g. Large; black with yellow bib; very colorful bill. **Adults:** mostly black, suffused with maroon on hindneck and upper back and with olive on lower back and belly; upper tail-coverts white; crissum red; lower face, sides and front of neck bright yellow, bordered posteriorly with red. Bill mostly pale green to yellow-green with maroon tip, orange on side of upper mandible, greenish-blue along side and gonys of lower mandible, and dusky bars along tomia; black line encircles base of bill; iris olive-green; facial skin bright yellow, becoming chartreuse around eye; legs bright blue. **Young:** sooty-black, above lacking maroon suffusion; red of lower border of bib and crissum paler, duller, less extensive; bill pattern duller, less distinct. Bill notably short for some time after fledging.

Habits. Troops through upper levels of forest and adjacent tall second growth and semi-open in small, loose flocks of up to 6, sometimes visiting remnant tall pasture trees, coming low along edges for berries; flies with strong undulations, closing wings briefly following a burst of rapid, shallow beats, then spreading them and gliding, then more wingbeats; feeds on fruits, including large-seeded, arillate types (*Virola*, *Protium*), small-seeded berries of many kinds, and *Cecropia* catkins; also insects and an occasional small lizard or snake; throws food from bill tip to throat by an upward toss of the head.

Voice. A harsh and monotonous croaking *crrrik crrrik crrrik*, with a resonant wooden or mechanical quality, like sound of winding an old clock; in chorus, reminiscent of a pondful of frogs; while calling swings head and great bill widely in all directions; aggressive note a short, sharp, hard, dry rattle, castanetlike.

Nest. A deep tree hollow or cavity resulting from decay, 9–90ft (2.7–27m) up, often in living tree, the bottom covered with regurgitated seeds. Eggs 3–4. January–May.

Status. Common resident of wooded regions throughout Caribbean lowlands, to 4000ft (1200m) locally; uncommon to rare over N half of Pacific slope, including Valle Central, becoming more common on lower slopes of Cordillera de Guanacaste, hills of Península de Nicoya, and those S and W of Valle Central; very rare to accidental vagrant S of Parrita-Quepos area.

Range. Tropical Mexico to N Colombia, NW Venezuela.

Note. Also (and more appropriately) called Rainbow-billed Toucan.

CHESTNUT-MANDIBLED TOUCAN

Ramphastos swainsonii Pl. 27(19)

Tucan de Swainson (Quioro; Dios-te-dé; Gran Curré Negro)

Description. ♂ 22″ (56cm), 750g; ♀ 20½″ (52cm), 580g. Very large, black with yellow bib, bicolored bill. **Adults:** mostly black with strong maroon suffusion on crown, hindneck, upper back, breast; bib deeper yellow than that of Keel-billed Toucan, bordered by a narrow line of white, followed by a broader line of red; upper tail-coverts white, crissum red. Iris olive; facial skin chartreuse, shading to yellow behind and below eye; bill above bright yellow and below dull dark maroon, this shading to black along the dividing line that runs diagonally along the side of the upper mandible; legs bright blue. **Young:** sooty-black, above tinged with dull chestnut-maroon above; red of bib and crissum paler and less extensive; yellow of bib duller; lower bill duskier; young birds do not attain full-sized bills for several months after fledging.

Habits. Generally resemble those of Keel-billed Toucan, which it dominates at fruiting trees; with undulatory flight, small flocks straggle through canopy of forest and semi-open, out into clearings with scattered tall trees, seeking mainly fruits, supplemented with insects, occasional snakes and lizards, and nestling birds; holds large item beneath foot while tearing it apart; bathes in water-filled hollows high in trees; feeds mate.

Voice. A shrill, yelping, resonant *Dios te dé, te dé*, or *keeuREEK kirick kirick*, or *yo-YIP* *a-yip, a-yip*, often repeated incessantly. Toward evening, flocks gather in emergent trees or on tall dead snags and call in chorus, often answered by another group some distance away. Aggressive note a loud, mechanical-sounding rattle, deeper than that of Keel-billed Toucan.

NEST. A cavity in living tree resulting from decay, or occasionally an old woodpecker hole in a large dead tree. Eggs unknown (?). January–June.

STATUS. Resident of wooded lowlands and foothills on Caribbean slope, reaching 4000ft (1200m) locally; generally common, except in drier lowlands S of Lago de Nicaragua; abundant resident of S Pacific slope from lowlands to 6000ft (1850m) along Cordillera de Talamanca, extending N to Carara.

RANGE. E Honduras to N Colombia.

NOTE. Sometimes considered a race of the South American *R. ambiguus*. Also called Swainson's Toucan.

FAMILY **Picidae:** Woodpeckers

The woodpecker family, with 210 species, is found worldwide except in the polar regions, Australia, New Zealand, Madagascar, and remote oceanic islands. In size, woodpeckers range from piculets only 3½″ (9cm) long to the 22″ (56cm) Imperial Woodpecker. Although a few are dull-colored, many are handsomely attired in red, yellow, green, or black and white; some have high crests. The sexes of even the most brilliant tend to be alike except for small but conspicuous differences on the head. With toes that spread widely forward and backward to clamp them to upright trunks, acute tail feathers to prop them up, and chisel-like bills, woodpeckers are well adapted for foraging on trunks and branches. Despite this specialization, they are versatile birds: many are expert flycatchers; they vary their diet with fruits and nectar; some consume great numbers of ants; a few drill holes in bark and drink the sap that fills them; others forage largely on the ground. Their flight is strongly undulatory. Their voices are varied; a rolling *churr* is frequent, and some have melodious calls. Nearly all beat tattoos on resonant wood. As far as is known, all sleep in holes that they carve in trees, most of them singly, the more social species in pairs or family groups. A few open-country species dig holes in earthen banks. All woodpeckers nest in unlined cavities, where tropical species usually lay 2–4 glossy white eggs, while species of high latitudes not infrequently have twice as many. Both sexes incubate for 11–18 days, with the male, usually the more domestic member of the pair, occupying the nest at night. The nestlings, hatched naked, remain in the hole for 20–35 days. They are fed by both parents, directly from the bill in many species, by regurgitation in those that consume great quantities of ants. The most social woodpeckers lead their fledglings back to the nest hole, where the whole family continues to lodge.

OLIVACEOUS PICULET

Picumnus olivaceus Pl. 29(1)

Carpenterito Oliváceo (Telegrafista)

DESCRIPTION. 3½″ (9cm); 11g. By far the smallest Central American woodpecker, without rigid projecting tips to rectrices. **Adult ♂:** pileum black, the crown with fine, short orange streaks; nape minutely dotted with white; rest of upperparts brownish-olive, the secondaries edged with olive-green; tail black with outer edge and median stripe pale buffy-yellow; throat whitish, scaled with dusky laterally; chest pale olive-brown; belly dull yellowish, indistinctly streaked with brownish. ♀: entire pileum finely dotted with white. Bill black, lower mandible grayish at base; feet gray. **Young:** overall, duller and browner; pileum dusky, indistinctly flecked with grayish-brown.

HABITS. Frequents lighter woodland, forest edge, tall scrub, shady plantations and gardens, singly or in family groups, usually below medium heights; without using tail for support, climbs over thin branches, twigs, and vines, into which it busily pecks, removing ants, termites, and their brood; hitches crosswise along branches with tail cocked downward like a furnariid, or perches upright like a songbird.

VOICE. A fine, rapid twitter or trill, soft and clear or shrill and insectlike; a sharp, sibilant *sst* or *pst*, sometimes repeated; does not drum but taps very rapidly while foraging.

NEST. A shallow cavity with round doorway barely 1″ (2.5cm) in diameter, carved by both sexes in rotting fence post or slender dead trunk of soft wood, 3–30ft (0.9–9m) up. Eggs 2–3. January–May. Fledglings are led back to the nest, where the whole family continues to sleep.

STATUS. Uncommon to fairly common resident in General-Térraba region and N along coastal strip to hills above Parrita, but not in more humid Golfo Dulce region; reaches 4600ft (1400m) on slopes of Cordillera de Talamanca and in Dota region; uncommon in drier Caribbean lowlands S of Lago de Nicaragua, W from Río Frío region to around Santa Cecilia.

Range. Guatemala to W Ecuador.

ACORN WOODPECKER

Melanerpes formicivorus Pl. 28(15)

Carpintero Careto

Description. 8¼″ (21cm); 85g. Black and white with bold harlequin face pattern. **Adult ♂:** crown and nape red; black area around base of bill, broadest on chin; forehead white, throat pale yellow, these joined by a white line down the lores; white rump and patch on primaries; rest of head and upperparts, wings, tail glossy blue-black; breast and sides heavily streaked with black and white; belly yellowish-white. Iris white or yellow; bill black; feet grayish. ♀: similar but forecrown black. **Young:** resemble adult male but crown and nape much duller; lower hindneck grayish; breast more gray, black streaks less heavy and distinct; iris brown or gray.

Habits. Frequents highland forest and clearings with scattered dead trees, pastures with remnant forest trees; sociable and very active, living in groups of ca. 3–6, all of whom may sleep in same high hole or 2 adjacent holes; seldom found far from oaks; stores acorns whole or piecemeal in crevices of bark or wood, or amid epiphytes, or occasionally in neatly fitted, specially made holes; an expert flycatcher, sallying from snags or fence posts.

Voice. Nasal whinnying, laughing, or rattling calls: *rack-up, rack-up* or *whaaka-whaaka*, varied to a more rolling *r-r-r-rack-up*, a more strident *rraaa-har-har, reep-hi-hur-hurr*, or *rub a dub dub*. Loquacious, members of group frequently calling to each other, sometimes in chorus.

Nest. A hole excavated 20–70ft (6–21m) up in dead, usually well decayed tree; contents not examined in Costa Rica. April–August. Up to 5 adults may incubate and feed nestlings.

Status. Resident at middle and upper elevations of Cordillera Central and Cordillera de Talamanca, from timberline down to 5000ft (1500m), straying to 3000ft (900m); center of abundance above 6000ft (1850m). Numbers vary locally and from year to year according to abundance and fruiting of oaks, which in any locality produce good acorn crops at irregular intervals. Rare resident or possibly sporadic visitor on Cordillera de Tilarán (Slud).

Range. W USA to Colombia.

GOLDEN-NAPED WOODPECKER

Melanerpes chrysauchen Pl. 28(18)

Carpintero Nuquidorado

Description. 7″ (18cm); 60g. Golden nape and broad median white stripe down back. **Adult ♂:** forehead and nape golden-yellow; crown red; rump, upper tail-coverts, center of back, and spotting on tertials white, otherwise upperparts black, including sides of head; throat and breast buffy-olive; center of belly orange-red; sides, flanks, and crissum barred with black and yellowish. Bill black; feet gray. ♀: similar but no red on head; forecrown black, rest of crown and nape dull yellowish. **Young:** resemble parents of same sex but duller, yellow of forehead often narrower.

Habits. Prefers canopy and middle levels of humid forest, but extends increasingly into semi-open and scattered trees as forest shrinks; eats wood-boring beetles and their larvae, other insects; a skilled flycatcher; takes much fruit: *Cecropia*, figs, arillate seeds, bananas. Young remain with parents until start of following breeding season; whole family of ca. 3–6 move as a loose group, sleep together in a high hole.

Voice. A resonant *churr*; a short, loud, rattling or laughing trill, 3–5 times in rapid succession on same pitch; several short rattles; both sexes beat rapid tattoos.

Nest. A hole 16–100ft (5–30m) up, usually in massive dead trunk. Eggs 3–4. March–June; rarely 2 broods. Unlike most woodpeckers, female as well as male sleeps in nest during breeding.

Status. Resident in remaining forests of S Pacific slope from lowlands to 5000ft (1500m) locally, N along coast to around Carara; becoming scarce and local as forests are cut.

Range. Costa Rica and W Panama.

Note. Formerly classified in genus *Tripsurus*, which was lumped into *Centurus*, in its turn merged with *Melanerpes*.

BLACK-CHEEKED WOODPECKER

Melanerpes pucherani Pl. 28(14)

Carpintero Carinegro

Description. 7¼″ (18.5cm); 63g. No other Caribbean woodpecker has red below. **Adult ♂:** forehead golden-yellow; crown and nape red; sides of head black; a white spot above and behind eye; rump white; rest of upperparts black, barred with white on the back and spotted on the remiges; below buffy-olive, paler and grayer on throat; sides, flanks, and crissum barred with black and yellowish; center of belly red. Bill black, feet dull olive. ♀: similar but crown mostly black. **Young:** resemble adult of same sex but duller; white markings above reduced and indistinct; belly with little red, mottled with black.

HABITS. Frequents upper and middle levels of humid forest, semi-open, scattered tall trees in clearings, and old second growth; gleans insects from trunks, branches, and lianas; pecks and probes cracks and crevices in bark, bases of epiphytes, and dead wood for termites, beetles, grubs, and caterpillars; pecks into *Cordia* twig nodes for ants; hawks flying insects; eats much fruit, including berries, arils, and *Cecropia* catkins; drinks nectar from large flowers of balsa and kapok trees; usually in pairs that may sleep in the same hole or singly.

VOICE. Most often heard is a series of ca. 4 short, rattling trills on the same pitch; a longer, smoother rattle; a loud, full-toned *krrrr*; both sexes beat tattoos.

NEST. A hole 20–100ft (6–30m) up in a dead trunk or branch. Eggs 2–4. March–June.

STATUS. Common resident of lowlands and foothills throughout Caribbean slope, to 2300ft (700m) in N and 3000ft (900m) in SE, extending locally onto Pacific slope of Cordillera de Guanacaste at lower middle elevations; less numerous from Río Frío region W; numbers decline with extensive deforestation.

RANGE. SE Mexico to W Ecuador.

NOTE. Formerly classified in genus *Tripsurus* or *Centurus*.

HOFFMANN'S WOODPECKER

Melanerpes hoffmannii Pl. 28(16)

Carpintero de Hoffmann

DESCRIPTION. 7″ (18cm); 68g. Above boldly barred; head and underparts mostly pale brownish. **Adult ♂:** nasal tufts yellow; forehead white; most of crown red; nape golden-yellow; rest of head, including hindcrown, and most of underparts grayish-buff, the flanks and crissum irregularly barred with black; center of belly yellow; rump white; distal primaries and most of tail black; rest of upperparts evenly barred with black and white. Orbital skin dull brownish-gray; bill black; feet grayish. ♀: similar but crown mostly whitish; yellow on nape reduced to a band at base of hindneck. **Young:** like adult of same sex but duller, above barring less distinct, head and underparts darker and more olive, mottled with black on belly and flanks.

HABITS. Prefers deciduous forest, light woodland, second growth, shade trees in coffee plantations and gardens, scattered trees in pastures, hedgerows; avoids dense wet forest but may invade when forest has been cut; pecks into decaying wood for ants, beetles, and their larvae, often coming down below eye level on fence posts and stumps; probes and flakes dead bark; eats many fruits, including figs, *Cecropia* catkins, arils; drinks nectar of balsa, African tulip tree, and other large flowers; quick to mob pygmy-owls; paired throughout year but usually sleeps singly in holes, rarely 2 together.

VOICE. A hard, metallic rattle, in repeated short bursts; a prolonged, nasal rattle or sputter, somewhat wavering in pitch; in excitement a querulous, grating, *woick-a woick-a woick-a* . . . or *wicka wicka wicka* . . . , often accompanied by a vigorous nodding of the entire body.

NEST. A hole 5–30ft (1.5–9m) up in a dead trunk or branch, sometimes only 3ft (1m) up in fence post. Eggs 2–3. February–June; often 2 broods.

STATUS. Common to abundant resident of N Pacific slope, S to about Tárcoles, including Valle Central and adjacent slopes up locally to 7000ft (2150m); on Caribbean slope E from Ochomogo down Reventazón valley regularly to Turrialba; from passes in Cordillera de Guanacaste E to Río Frío region, which it has apparently reached in the last 30 years. Scattered but increasingly numerous records elsewhere on Caribbean slope suggest the species is expanding its range in this region as forests are cut.

RANGE. S Honduras to Costa Rica.

NOTE. Sometimes classified in the genus *Centurus* and often considered a race of the more northern Golden-fronted Woodpecker (*M. aurifrons*); very closely related to *M. rubricapillus*, Red-crowned Woodpecker, with which it apparently hybridizes extensively between Tárcoles and Parrita. Throughout this region most birds have the nape and belly some shade of orange.

RED-CROWNED WOODPECKER

Melanerpes rubricapillus Pl. 28(17)

Carpintero Nuquirrojo

DESCRIPTION. 6¾″ (17cm); 55g. Only woodpecker on S Pacific slope with barred back.

Adult ♂: forehead white; nasal tufts yellow; crown and nape red; rump white; rest of upperparts and most of wings and tail evenly barred with black and white; sides of head and underparts grayish-buff to pale buffy-olive, palest on throat; center of belly red; flanks and crissum barred irregularly with black. Orbital skin dull grayish-brown; bill black; feet grayish. ♀: similar but crown whitish, only lower nape pale red. **Young:** like adult of same sex but duller, more olive on head and underparts, above less distinctly barred; young female has lower nape yellowish, not red; belly mottled.

Habits. Frequents edges and more open parts of forest, gallery woodland, mangroves, semi-open, second growth; occurs high in wooded habitats, often much lower in more open ones; pecks into decaying wood for ants, beetles, grubs; also gleans small insects and spiders from trunks and branches, even foliage; eats many fruits, including bananas at feeders; visits balsa flowers for nectar; paired throughout year but sleeps singly in hole.

Voice. A drawled rattle or churr, *krr-r-r-r*, often prolonged; various other calls resemble those of Hoffmann's Woodpecker; both sexes beat tattoos.

Nest. A hole 11–75ft (3.4–23m) up, usually in slender dead tree or branch. Eggs 2. February–June; 2 broods.

Status. Common to abundant resident on S Pacific slope, N to Parrita and San Isidro, from sea level to ca. 5300ft (1600m); when forest is cleared this species often increases, while Golden-naped Woodpecker declines in numbers. N of the Quepos-Parrita area most individuals appear more or less intermediate between this species and Hoffmann's Woodpecker.

Range. Costa Rica to N South America.

Note. Often classified in genus *Centurus*.

YELLOW-BELLIED SAPSUCKER

Sphyrapicus varius Pl. 28(20)

Carpintero Bebedor

Description. 7¼″ (18.5cm); 50g. Medium-sized; white stripe on folded wing; upperparts variegated with black, white, and brownish. **Adult ♂:** forehead and crown red, narrowly bordered with black; nasal tufts and stripe across cheeks to side of neck white; anterior throat red, broadly framed in black; upperparts black, irregularly barred and blotched with yellowish-white and buff on sides of back; outer middle and greater coverts white, forming broad longitudinal stripe; remiges and rectrices spotted with white; underparts pale yellow, washed with brownish laterally; sides and flanks with irregular streaks or V-shaped marks of black. Bill blackish; feet grayish-green. ♀: wing-stripe less extensive; throat whitish; crown varies from red like male to entirely black; streaking below heavier. **Immatures:** throat whitish, mottled with red in male; crown dusky, spotted with buff; head pattern obscured by dusky mottling; back mixed with brown; breast grayish-buff, scalloped with dusky; wing-stripe less extensive.

Habits. Frequents montane forests and larger trees in pastures, clearings, semi-open, even suburban areas; drills horizontal rows of shallow holes for sap, favoring willow and poplar trees; forages like a typical woodpecker, excavating in dead or dying wood for beetles and their larvae, flaking off bark to expose ants and their brood; solitary; adult males rarely seen in Costa Rica.

Voice. Usually silent in Costa Rica; occasionally a rather nasal, catlike *naaah*.

Status. Very uncommon migrant and winter resident between about late October and late March, chiefly in highlands at 3000–10,000ft (900–3000m) the length of the country; an occasional bird may also turn up in lowlands of either slope, mostly during migration. Only migratory woodpecker in Costa Rica.

Range. Breeds from NW Canada to S USA; winters to C Panama and West Indies.

HAIRY WOODPECKER

Picoides villosus Pl. 28(19)

Carpintero Serranero o Velloso

Description. 6¾″ (17cm); 42g. Fairly small; below brown; above black with striped face and broad whitish stripe down back. **Adult ♂:** red band across nape; crown, cheeks, malar stripe, and most of upperparts black; superciliaries and broad median stripe on back white, more or less tinged with brown; lores, cheek stripe, underparts, and 3 outer rectrices dull brown; sides of breast streaked with black; remiges sparsely spotted with white. Bill black, becoming gray on gonys; feet dusky-olive. ♀: similar but nuchal band white, confluent with superciliaries. **Young ♂:** duller, above more sooty; crown dusky, many feathers loosely tipped with red, giving streaked effect; paler red on nape. ♀: similar but no red on nape and sometimes little or none on crown.

Habits. Frequents wet, epiphyte-burdened montane forests and neighboring clearings and semi-open; forages mostly in middle and upper levels of forest but comes much lower, sometimes to ground in second growth and clearings; pecks industriously into dead and dying trees, old stalks of giant thistle, stems of soft shrubs like *Senecio*; eats beetles and their grubs, crickets, flies, spiders; sometimes accompanies mixed-species flocks; sleeps singly in holes that are usually carved by males.

Voice. Note a sharp, hard *pick* or *bip* or a fuller, stronger *keick* or *beep*; a sharp, hard, descending and accelerating laugh; a high, clear, rapid *bic-bic-bic-bic-bic*; both sexes beat short, rolling tattoos.

Nest. A hole 10–60ft (3–18m) up. Eggs 3. February–April.

Status. Common resident of highland forests from Cordillera de Tilarán S into Panama, mostly from 5000ft (1500m) up to timberline but occasionally down to 4000ft (1200m) on Caribbean slope; abundance centered in oak forests of higher elevations.

Range. Alaska and N Canada to W Panama and Bahamas.

Note. Formerly placed in genus *Dendrocopos*, which is now united with *Picoides* by the AOU.

SMOKY-BROWN WOODPECKER

Veniliornis fumigatus Pl. 28(13)

Carpintero Pardo

Description. 6″ (15cm); 34g. Small, brownish, virtually unpatterned, no crest; our plainest woodpecker. **Adult ♂:** crown and nape slaty, the feathers broadly tipped with red giving a mottled effect; forehead, lores, and cheeks dull olive-brown, passing to brighter brownish-olive, tinged with ochraceous to reddish over back and wing-coverts; rump and upper tail-coverts dull brown; remiges and rectrices blackish-brown; wing-linings and base of remiges barred with dusky and buffy-white; throat dark grayish; breast dark olive-brown, becoming paler and brighter over posterior underparts. Bill black except base of mandible pale gray; feet gray. ♀: similar but crown blackish-brown, the feathers broadly edged with olive-brown. **Young:** resemble adult of same sex but duller, above without reddish tinge; throat more brownish; young male has red of crown paler, duller, less extensive.

Habits. In forest in mountainous terrain, usually forages high in broken canopy or lower at edges, often in vine tangles or on slender branches; in lowlands prefers forest edge, as along riverbanks, and second growth, where it often occurs down to eye level, pecking industriously into thin stems and vines; eats mainly small wood-boring beetles and their larvae; paired throughout year but sleeps singly in hole; in family group following breeding.

Voice. A hard, rolling, gravelly *keer-keer-keer-keer* or *krrr-krrr-krrr*; a metallic *wick* or *pwik*, softer and more slurred than note of Hairy Woodpecker; in interactions, a squeaky *wick-a wick-a* or *tsewink tsewink*; beats prolonged, rapid tattoos.

Nest. A hole 5–25ft (1.5–8m) up in fence post or trunk. Eggs 4. February–May.

Status. Widespread resident of lowlands and middle elevations of Caribbean slope, locally up to 6000ft (1850m); on Pacific slope in General-Terraba region at 1300–6000ft (400–1850m), including San Vito region, but absent from wetter Golfo Dulce lowlands.

Range. Tropical Mexico to W Peru and NW Argentina.

RED-RUMPED WOODPECKER

Veniliornis kirkii Pl. 28(12)

Carpintero Lomirrojo

Description. 6″ (15cm); 30g. Small, brownish, below barred; red rump unique among Costa Rican woodpeckers. **Adult ♂:** nasal tufts and face dusky-olive; crown and hindneck slaty, the feathers tipped with red producing a streaky effect, bordered behind eye by a line of golden-yellow streaks that extends around lower hindneck and behind auriculars; upperparts mostly olive-green, glossed with yellow to orange on back and wing-coverts; lower rump and upper tail-coverts orange-red; remiges and rectrices dusky, edged with olive green; wing-linings and bases of remiges barred with buffy-white and dusky; chin grayish, passing to dusky-olive on rest of underparts, which are barred narrowly anteriorly with buff, more broadly posteriorly with dull whitish. Bill dusky, base of mandible pale gray; feet dull gray. ♀: crown and hindneck blackish-brown, the feathers tipped with paler brown, not red. **Young:** resemble adult of same sex but above duller, no yellow or orange gloss; yellow markings of head and nape greenish, more diffuse; crown of male more diffusely streaked with paler red; below with barring broader and less distinct; outer rectrices barred with dull brown.

Habits. In canopy of lowland wet forest, coming lower in adjacent second growth, mangroves, light woods; forages on trunks and limbs, often amid thick foliage; pecks steadily to expose larvae and adults of small wood-boring beetles; usually seen singly or in pairs.

Voice. A weak, squeaky *k'wink,k'wink* or *kee-yik kee-yik*, repeated 2–4 times; a nasal *keeer*, softer than note of Rufous-winged Woodpecker; beats a rapid, often prolonged tattoo.

Nest. A hole in a tree, 10–25ft (3–8m) up. Eggs 3 (in Trinidad; no Costa Rican record); fledged young seen June–July.

Status. Uncommon resident of Sixaola region, near Panama border on Caribbean slope, and Golfo Dulce region (N to Río Terraba watershed) on S Pacific slope.

Range. Costa Rica to W Ecuador and Trinidad and Tobago.

GOLDEN-OLIVE WOODPECKER

Piculus rubiginosus Pl. 28(11)

Carpintero Verde Dorado

Description. 8″ (20cm); 75g. Gray crown contrasts with buffy cheeks; above bright olive, below barred. **Adult ♂:** forehead and crown slaty, bordered by broad crimson superciliaries and band across nape; sides of head pale buff; broad malar stripe crimson; back and wing-coverts bright ochraceous-olive with yellow gloss, the feathers often tinged with reddish distally; remiges and rectrices more blackish; wing-linings yellowish; throat and chin buffy-yellow, finely mottled and streaked with dusky; breast barred with buffy-yellow and dark olive; posterior underparts, rump, and upper tail-coverts greenish-yellow, indistinctly barred with olive. Iris dark red; bill blackish; feet gray. ♀: similar but only nape crimson. **Young:** resemble respective adult; duller, above less golden; below dark markings heavier, duskier; red of nuchal band paler, duller, extending as streaks onto hindcrown; superciliaries and malar stripes of male duller; female malar stripe dusky.

Habits. In canopy and edges of forest, adjacent semi-open and second growth, trees bordering fields and coffee plantations and in shady gardens, in mountainous terrain; usually forages high, often amid dense foliage or clusters of epiphytes or on mossy branches; descends to low stubs and fence posts along hedgerows and edges; eats ants, termites, beetles and their larvae; usually seen singly; sleeps singly in hole throughout year.

Voice. A loud, clear, somewhat jaylike *deeeeh* or *keeeep*; a prolonged, gravelly roll or trill, higher-pitched and more rapid than that of a *Melanerpes* and more constant in pitch; a dry *churr*; in interactions, a liquid *woick-woick-woick. . . .*

Nest. A hole excavated in a stub or tree 4–60ft (1.2–18m) up. Eggs 2–4. February–May.

Status. Resident of middle elevations the length of the country, at ca. 2500–6500ft (750–2000m) on Caribbean slope and 3500–7000ft (1050–2150m) on Pacific slope; most abundant in Cordillera Central and Cordillera de Talamanca, including Valle del Guarco and mountain slopes overlooking Valle Central and General-Térraba region.

Range. NE Mexico to NW Peru and NW Argentina.

RUFOUS-WINGED WOODPECKER

Piculus simplex Pl. 28(10)

Carpintero Alirrufo

Description. 7″ (18cm); 55g. Greenish, with short red nuchal crest, irregularly barred belly, large rufous wing-patch. **Adult ♂:** forehead, crown, hindneck, and broad malar stripe red; cheeks, throat, chest, and upperparts olive-green, brightest on back; remiges mostly cinnamon-rufous, barred with olive, tipped with dull black; tail dull black; chest spotted with buffy-yellow; belly yellowish-buff, barred with dusky. Iris pale blue-gray to yellowish; bill black with pale gray base; feet olive. ♀: similar, only hindneck (including crest) red. **Young:** resemble adults of same sex (except young male lacks red on forehead and malar region), but throat, foreneck, and chest duskier, indistinctly mottled with buffy-olive; belly duller and more irregularly barred.

Habits. Frequents canopy and edge of humid forest, sometimes large trees in adjacent semi-open; singly or in pairs forages mostly high in trees, pecking into mossy and rotting branches, lianas, dead wood caught in vine tangles; taps incessantly and irregularly; eats ants, beetles and their larvae; occasionally accompanies mixed-species flocks.

Voice. A loud, sharp, nasal *deeeah*, higher-pitched than note of Golden-olive Woodpecker and clearer, less nasal than that of Slate-colored Grosbeak; a loud, emphatic series of jaylike, downslurred, slightly nasal notes: *heew heew heew heew . . .* ; beats prolonged tattoos.

Nest. A hole excavated in recently dead tree or rotten stub, 8–16ft (2.5–5m) up. Eggs 2–4. February–May.

Status. Uncommon to fairly common resident of lowlands and foothills throughout Caribbean slope, locally up to 2500ft (750m); on

S Pacific slope, N to Parrita and probably to Carara, locally up to 3000ft (900m).

Range. N Honduras to extreme W Panama.

Note. Sometimes considered a subspecies of *P. leucolaemus* of E Panama and N South America, but differs substantially in both facial pattern and vocalizations; we consider it a distinct species.

CHESTNUT-COLORED WOODPECKER

Celeus castaneus Pl. 28(8)

Carpintero Castaño

Description. 9″ (23cm); 100g. Dark-colored, with paler head and long, ragged, swept-back crest. **Adult ♂:** head and neck ochraceous-buff, becoming golden-buff on crest; broad malar stripe crimson; below deep chestnut, chevroned with black; above slightly paler rufous-chestnut, with bars and chevrons of black; outer primaries and rectrices black distally; wing-linings, sides, and flanks golden-buff (the latter concealed when bird is perched). ♀: similar but lacks crimson on face. Iris chestnut; bill ivory-yellow, tinged with blue-green at base; feet dark olive to grayish. **Young:** average duller, darker; below with black markings sparser and more irregular; malar area indistinctly mottled with dusky.

Habits. Prefers dense forest, forest edge, and trees with dense foliage and epiphytes in semi-open, mainly in canopy and subcanopy, down to middle levels along edges; eats mainly ants and termites, pecking into termite tunnels, *Cecropia* trunks; also pries off bark flakes,eating what is uncovered; sometimes takes fruit.

Voice. A sibilant, thick, descending whistle: *peew* or *kheeu*, often followed by 2–4 sharp nasal notes: *kheeu, wet-wet-wet*; when excited, a series of sharp double notes, *wik-kew wik-kew wik-kew*; drums a rapid tatoo lasting ca. 1½sec, shorter and softer than that of Black-cheeked Woodpecker.

Nest. In hole in soft, living or recently dead tree, excavated by both sexes, 13–70ft (4–21m) up. Eggs 3–4 (in Belize; not seen in Costa Rica). February–May.

Status. Uncommon resident throughout Caribbean lowlands, locally to 2500ft (750m) in foothills.

Range. S Mexico to NW Panama.

CINNAMON WOODPECKER

Celeus loricatus Pl. 28(9)

Carpintero Canelo

Description. 8¼″ (21cm); 83g. Resembles Chestnut-colored Woodpecker but has shorter, more compact crest; underparts paler, with black markings heavier and more conspicuous. **Adult ♂:** cheeks, chin, and throat crimson, center of throat spotted with black; rest of head, neck, and chest rufous-cinnamon, shading to chestnut-rufous over rest of upperparts, which are marked with spots and short bars of black; breast and belly cinnamon-buff, the entire underparts boldly scalloped or chevroned with black; wing-linings rufous; primaries and tail boldly barred with cinnamon-buff and black. ♀: head rufous-cinnamon, lacking red; underparts darker, deeper cinnamon. Iris brick-red; bill mostly greenish-ivory, culmen horn-color; feet brownish-gray. **Young:** resemble adult female but throat and cheeks mottled with dusky; below with black markings sparser and more irregular.

Habits. Frequents humid forest and nearby semi-open, old second growth, and clearings; mainly in canopy in forest, coming much lower in open areas; pecks into swollen nodes of laurel (*Cordia alliodora*) twigs and *Cecropia* trunks for ants and into tunnels for termites.

Voice. A ringing, penetrating series of 3–5 notes, accelerating and descending: *peee-peew-peu-pu*; when agitated, a sharp, descending, rolling parrotlike chatter.

Nest. A hole excavated by both sexes in soft wood of living or recently dead tree, 20–30ft (6–9m) up. Eggs undescribed (?). March–May.

Status. Uncommon to fairly common resident of lowlands and foothills of Caribbean slope, from sea level to 2500ft (750m) locally, except rare to absent in drier lowlands S of Lago de Nicaragua.

Range. E Nicaragua to NW Ecuador.

LINEATED WOODPECKER

Dryocopus lineatus Pl. 27(13)

Carpintero Lineado

Description. 13″ (33cm); 197g. Slimmer than Pale-billed Woodpecker, with more pointed crest, striped face, below with indistinct pattern, back stripes widely separated. **Adult ♂:** crown and crest red; malar stripe darker red; sides of head sooty-black; chin and throat streaked with dusky and white; a white stripe from bill across cheeks, down sides of neck; outer webs of outer scapulars white, forming a stripe on each side of back; rest of neck, breast, and upperparts dull black; wings and tail brownish-black; wing-linings pale, dull yellow; posterior underparts dull grayish-buff to buffy-white with smudgy bars, spots, and narrow shaft-streaks of sooty-black. Iris pale yellow; feet grayish. Bill yellowish-ivory in N Costa Rica, blackish in S, and more or less intermediate in the central areas of both slopes. ♀: similar but forehead and malar stripe black. **Young:** resemble adult of same sex but duller and browner; crest shorter, paler red.

HABITS. At forest edge and in gallery forest in drier areas, but seldom enters closed wet forest, preferring semi-open, shaded gardens, trees in pastures, second growth; flies with strong undulations; forages on trunks and large limbs, prying up bark, pecking deeply into rotting wood for beetles and their larvae; takes many ants and their brood, often pecking into trunks and branches of *Cecropia* and devouring ants as they emerge; takes fruits (*Heliconia*); paired throughout year but sleeps singly in hole.

VOICE. A loud, ringing, far-carrying *wicwicwicwicwic* . . . or *weep weep weep weep* . . . , mates often calling back and forth; a loud, sharp *pik*, sometimes followed by a rolling note, *pik-urrrr-r-r*, sometimes varied to *ch'whirrr* or *k'rroo*. Both sexes drum with a few slow taps, followed by an accelerating series.

NEST. In a tree hole 10–100ft (3–30m) up (rarely lower). Eggs 3–4. January–May.

STATUS. Uncommon to fairly common resident throughout lowlands of both slopes, locally up to ca. 3600ft (1100m); most numerous where new clearings interrupt forest.

RANGE. N Mexico to NW Peru and N Argentina.

PALE-BILLED WOODPECKER

Campephilus guatemalensis Pl. 27(14)

Carpintero Picoplata (Dos Golpes, Carpintero Chiricano)

DESCRIPTION. 14½″ (37cm); 255g. Large, with bushy crest; more robust than Lineated Woodpecker; no facial stripes; broad yellowish-white stripes down back form a V. **Adult ♂:** head red; yellowish-white stripes down each side of neck almost converge on lower back; rest of neck, breast, upperparts black; remiges and rectrices more brownish; wing-linings yellow; posterior underparts greenish-yellow, heavily barred with blackish. Iris yellow; bill ivory, tinged with bluish-gray at base; orbital skin and feet grayish-brown. ♀: similar but throat and front of crest black. **Young:** resemble adult female but duller, below with less pattern; sides of head dull black, faintly tinged with red.

HABITS. Frequents middle levels to fairly high in forest, often coming lower at edges and gaps or in adjacent semi-open or old second growth; sometimes enters clearings with scattered trees; flies with strong undulations; digs deeply into decaying trunks, removing large flakes and splinters, eating mostly wood-boring beetles and their larvae; takes some fruit; paired year-round but sleeps singly in hole.

VOICE. Loud, staggering nasal rattles and sputters; a loud bleating note; breeding birds utter whining or moaning sounds and a low *keeu keeu keeu keeu*. Drum diagnostic: two powerful taps in quick succession.

NEST. A deep cavity 16–50ft (5–15m) up in a large trunk. Eggs 2. August–December.

STATUS. Resident in lowlands of both coasts, up to 3300ft (1000m) in N, and to 5000ft (1500m) on S Pacific slope; common in wholly or partly forested areas but has declined markedly in extensively deforested regions.

RANGE. N Mexico to W Panama.

NOTE. Also called Flint-billed Woodpecker.

ORDER **Passeriformes:** Passerines (Perching Birds)

About 60% of all avian species belong to this huge order, which includes most of the small birds familiar to people around the world. Passerines range in size from little bigger than the average hummingbird, to the 30″ (76cm) raven, as large as a big hawk. The most obvious external feature of the order is the "perching" foot, with the opposing hind toe as long as the 3 forward-directed ones and on the same level, with a notably large and strong claw; but a number of structural features of the wing, palate, and foot, and a unique type of sperm, also set the passerines apart as a natural group. The order is usually divided into 2 major groups: the suborder Passeres, or "oscines," the true songbirds with a complex syrinx, including some 80 percent of the order; and one or more suborders of "suboscines" with a simpler syrinx but a more advanced middle ear apparatus. The suboscines are almost entirely a New World group and are represented in Costa Rica by the families listed from woodcreepers through flycatchers; the families from swallows through Old World sparrows are oscines.

The origin and affinities of the passerines are poorly known because of their scanty fossil record. There is no doubt that the Passeriformes have undergone the most extensive and rapid evolutionary radiation of any bird group in the modern era. Passerines have come to occupy a wide variety of ecological niches with little concomitant change in structure, making their classification at the family level extremely difficult and controversial. Very few passerine families are clear-cut, without intermediate forms linking them to other families; often species only distantly related come to resemble each other closely through their occupation of similar ecological niches.

FAMILY **Dendrocolaptidae:** Woodcreepers

With about 50 species, this family is confined to wooded regions of the American continents and closely adjacent islands, from Mexico to northern Argentina, mostly at low and middle altitudes. Its members are about 5–14″ (12.5–36cm) in length and are clad in shades of brown, chestnut, rufous, and buff, sometimes with touches of white or black. Most are variously streaked, spotted, or banded. The sexes look alike. Largely, if not wholly insectivorous, woodcreepers are highly specialized for foraging on the trunks and branches of trees. The stiff shafts of their tail feathers project beyond the vanes and are usually downcurved, the better to engage irregularities in the bark and provide support while they climb with sharp-nailed toes. Their bills range from short and wedgelike to moderately long and stout, or very long, slender, and strongly decurved. Although most of their food is obtained by gleaning from bark and probing into fissures, some occasionally dig into decaying wood with vigorous, woodpeckerlike blows, thereby justifying their older name of woodhewers. Their songs tend to be repetitions of a single note, which is often loud and clear; a few have surprisingly complex vocalizations. Adults sleep singly in crannies in trees or hollow palm stubs, which may be open to the sky.

Some woodcreepers are monogamous, the sexes taking equal shares in building, incubating, and rearing the young, whereas others form no lasting pair bonds and the female alone attends the nest. Woodcreepers nest in holes in trees or, less often, termitaries, sometimes in old woodpeckers' holes, although a less conspicuous natural cavity is preferred. They appear never to carve their own nest chambers but may try by pecking to broaden a too-narrow doorway. They carry in flakes of bark, leaves, mosses, or lichens to make a crude nest upon which the female lays 2, less often 3, white eggs. Incubation periods range from ca. 17–21 days. The young, hatched with sparse down, remain in the nest for 18–24 days and usually closely resemble their parents in plumage.

The difficulty of seeing ventral surfaces of woodcreepers climbing trunks in dimly lighted forests often hinders identification, but voices and bills provide helpful clues.

PLAIN-BROWN WOODCREEPER

Dendrocincla fuliginosa Pl. 29(11)

Trepador Pardo

Description. 8¼″ (21cm); 42g. Without prominent streaking; bill moderately long, nearly straight; dark malar stripe on grayish-looking face. **Adults:** above brown, duller on pileum, becoming rufous-chestnut on tail-coverts and tail and dull rufous on remiges; sides of head and throat grayish-olive to dull olive-brown with inconspicuous buffy shaft-streaks; a broad but indistinct dusky malar stripe; below olive-brown, shading to dull rufous-chestnut on belly and crissum. Iris grayish to brown; bill black; feet dark gray. **Young:** similar to adults but face and throat tinged with buff; below slightly paler.

Habits. Prefers low and middle levels of humid forest, tall second growth, and semi-open; regularly accompanies army ants, clinging vertically to trunks or less often perching crosswise on branch or horizontal vine above the swarm, flying out or dropping to ground to seize fleeing insects and spiders; hammers loudly on bark or wood to uncover insects. Although several may gather at army ant swarm, usually solitary, with individual territories; the sexes associate only briefly for courtship and fertilization.

Voice. Usual call a sharp *psick* or *pseek* with sibilant, sometimes hoarse quality; a laughing call, and a prolonged low rattle delivered with closed bill; song a prolonged *churr* ending in distinct, not unmelodious notes.

Nest. In a hollow stub of palm or stout bamboo or knothole or other cavity in tree, 3–30ft (1–9m) up; if deep, cavity is filled with much green moss before a nest is built of leaves, fibers, and seed down. Eggs 2. May–August.

Status. Fairly common resident of Caribbean

lowlands and foothills, extending through low passes to Pacific slope of Cordillera de Guanacaste; less common in drier lowlands S of Lago de Nicaragua; sea level to 2500ft (750m).

Range. SE Honduras to W Ecuador, Bolivia, NE Argentina, and Guianas.

TAWNY-WINGED WOODCREEPER

Dendrocincla anabatina Pl. 29(12)

Trepador Alirrubio

Description. 7½″ (19cm); 40g. Brownish, unstreaked, with dull buffy throat and contrasting tawny-rufous remiges; straight bill and tail rather short; nuchal feathers often have a shaggy, unkempt look. **Adults:** head and body mostly olive-brown, brightest on wing-coverts and paler, more cinnamon on belly; tail-coverts and tail rufous-chestnut; blackish inner webs of primary coverts contrast with tawny-rufous remiges; primaries tipped with dusky; a narrow buffy stripe from above eye back over auriculars that have narrow buffy shaft-streaks; throat dull buff (appearing paler, more conspicuous when fluffed out), trailing off to narrow shaft-streaks on chest. Iris grayish to yellow-brown; bill black with gray on gonys; feet dark grayish. **Young:** resemble adults but throat duller, with faint darker scaling; postocular streak broader and less distinct.

Habits. In humid forests, older second growth, semi-open, and mangroves; climbs up trunks, usually to no great height, using tail as prop; habitually forages with army ants, clinging above them on trunk, and catches fleeing insects, spiders, small lizards on long aerial sallies; usually solitary and notably pugnacious, driving others of its kind from ant swarm; sleeps alone in tree hole, often evicting woodpeckers and using their dormitories; female associates with male only briefly before laying; nervous and excitable, often flicking wings, fluffing crown and throat feathers.

Voice. Call a sharp, querulous *tyew, deew* or *deyeew*; song a high, sharp trill or rattling *churr*, often prolonged; a high-pitched, excited-sounding *whehehehehehehe*.

Nest. In a hollow stump, palm or bamboo stub, or woodpecker hole, 5–20ft (1.5–6m) up; deeper cavity is filled with much green moss before nest of fibrous or papery bark, rootlets, and whitish lichens is built. Eggs 2. March–July.

Status. Common resident throughout lowlands and foothills of S Pacific slope, N to around Carara and rarely to Puntarenas area; sea level to 4000ft (1200m), occasionally to 5000ft (1500m) along slopes of Cordillera de Talamanca.

Range. SE Mexico to W Panama.

RUDDY WOODCREEPER

Dendrocincla homochroa Pl. 29(13)

Trepador Rojizo

Description. 8″ (20cm); 44g. Nearly uniform rufescent with grayish lores; bill straight; tail rather short; feathers of nape often have a scruffy, unkempt look. **Adults:** mostly rich rufous-brown, palest on belly, shading to cinnamon-rufous on throat and to rufous-chestnut on crown, wings, tail-coverts, and tail; lores dull gray. Iris reddish-brown; bill dusky-brown shading to horn-color on tomia; feet pale grayish. **Young:** virtually identical, slightly more rufous on throat and belly.

Habits. Frequents humid and semi-arid woodlands and adjacent semi-open and trees in clearings; regularly forages with army ants, clinging upright to trunks or saplings, flying out or dropping to ground to seize insects and spiders fleeing the ants; usually solitary, but 2–3 may gather at ant swarm; when excited or interacting, birds fluff their throats and crowns, chase each other; on S Pacific slope, often at same ant swarm as Tawny-winged Woodcreeper, which chases it.

Voice. Call note a sharp, squeaky *kink* or *quink*; a descending, scratchy, nasal, querulous *deeeeah*, often with a quaver or roll at the end: *squeeirrr*! or a prolonged *churr*ing.

Nest. In hollow palm trunk or cavity in tree, 2–16ft (0.6–5m) up; if cavity is deep, bird may nearly fill it with moss or leaves before constructing a nest of bark and fibers. Eggs 2–3. April–June.

Status. Uncommon resident of lowlands the length of Pacific slope, becoming more numerous on lower slopes of N cordilleras and hills of Península de Nicoya, to ca. 4500ft (1350m), and Cordillera de Talamanca, to 5000ft (1500m); rare and sporadic in W end of Valle Central and in lower middle elevations of Caribbean slope at 2000–3300ft (600–1000m), occasionally down to 300ft (90m); uncommon in lowlands S of Lago de Nicaragua and E to Río Frío region.

Range. S Mexico to N Colombia and N Venezuela.

LONG-TAILED WOODCREEPER

Deconychura longicauda Pl. 29(10)

Trepador Delgado

DESCRIPTION. 7¼″ (18.5cm); 24g. Slender; long tail spines; spotted-breasted; bill thin, straight, of moderate length; nuchal feathers often look ragged as in *Dendrocincla* species. **Adults:** above rich brown, crown with narrow buffy streaks and blackish scaling; remiges, tail-coverts, and tail rufous-chestnut; primary coverts and alula dusky; narrow superciliaries and eye-ring buff; sides of head streaked with dull buff and dusky; chin and throat dull buff; breast dark olive-brown with large buffy spots; rest of underparts paler olive-brown with fine, sparse, buffy streaking; wing-linings cinnamon-buff. Iris hazel; upper mandible blackish, lower gray; legs plumbeous. **Young:** similar but throat with indistinct brownish scaling, breast spots larger and less distinct.

HABITS. Frequents lower and middle levels of humid forest; quiet and retiring; usually solitary, sometimes in pairs; often joins flocks of antwrens, antvireos, tanagers (e.g., Olive Tanager) moving through upper understory or subcanopy; forages mostly on trunks and large branches.

VOICE. A long series of soft but resonant *chip* or *chrit* notes that starts slowly, speeds up into a slow trill, then slows into a jerky rhythm; less shrill and more patterned than laugh of Tawny-winged Woodcreeper.

NEST. In a deep hollow in top of a dead trunk ca. 30ft (9m) high, thickly lined with dry leaves. Eggs 2. April (only Costa Rican record).

STATUS. Very uncommon resident of lower middle elevations of Caribbean foothills the length of the country, at 1300–3600ft (400–1100m); uncommon in forested portions of Golfo Dulce lowlands, N to around Palmar; and to 4200ft (1300m) on mountains overlooking Coto Brus valley.

RANGE. SE Honduras to E Peru, C Brazil, and Guianas.

NOTE. Central American races are sometimes separated as a distinct species, *D. typica*, Cherrie's Woodcreeper.

OLIVACEOUS WOODCREEPER

Sittasomus griseicapillus Pl. 29(7)

Trepadorcito Aceitunado

DESCRIPTION. 6″ (15cm); 14g. Small, grayish, unstreaked; bill short and slender. **Adults:** head, neck, and underparts grayish-olive, darker and more olive on crown and chest; back and wing-coverts olive-brown; remiges, rectrices, rump, and tail-coverts chestnut-rufous; bases of inner webs of inner primaries and secondaries buff, forming conspicuous stripe when wing is spread. Upper mandible dusky, lower gray; feet grayish. **Young:** below slightly paler, above brighter, especially on rump; wing-coverts edged with rufous.

HABITS. Frequents dense forest and adjacent shady old second growth in humid regions and more open woodland and gallery forest in drier lowlands; climbs actively up trunks and branches at all heights, plucking small insects and spiders from bark surface and sallying in pursuit of prey flushed from hiding; eats beetles, homopterans, small wasps, ants, caterpillars, and eggs and pupae of various insects; usually solitary.

VOICE. A very sharp, fine, rapid trill with a sibilant, rattling quality: *preeeeew*, higher and thinner than that of Streaked-headed Woodcreeper, more attenuated than that of Plain Xenops.

NEST. In a tall, slender, hollow palm trunk, the wide doorway opening skyward ca. 40ft (12m) up, probably lined with small dead leaves and leaf fragments. Eggs undescribed (?). Incubation in April.

STATUS. Uncommon and local resident of NW lowlands, becoming more common on Pacific slopes of N cordilleras above ca. 2000ft (600m), continuing along slopes of Cordilleras Central and de Talamanca, up to ca. 5600ft (1700m) into Panama; on Caribbean slope locally the length of the country between ca. 1600 and 5000ft (500–1500m).

RANGE. C Mexico to NW Peru, N Argentina, and Guianas.

WEDGE-BILLED WOODCREEPER

Glyphorhynchus spirurus Pl. 29(6)

Trepadorcito Pico de Cuña

DESCRIPTION. 6″ (15cm); 16.5g. Small, with distinctive short, wedge-shaped bill. **Adults:** top and sides of head dark brown; forehead, lores, and cheeks indistinctly mottled with buff; auriculars finely streaked with buff; a buffy stripe from above eye back over auriculars; back and wing-coverts rich brown; rump, upper tail-coverts, and tail chestnut-rufous; throat rich buff, indistinctly scaled with dusky; breast olive-brown with short, lanceolate buffy streaks that posteriorly become finer and sparser; belly paler, grayer; remiges tawny-rufous, tipped with dusky; bases of inner webs of inner primaries and secondaries buffy, forming conspicuous stripe when wing is spread. Bill black, basal half of lower mandible gray; feet grayish. **Young:** similar but buffy streaking on breast less distinct, dark scaling on throat heavier, crown feathers with sooty borders.

Habits. Frequents humid forests and neighboring semi-open, old second growth, thinned woodland; climbs slowly up trunks at all heights, using tail for support, with head in continuous rapid motion, flicking off tiny bits of bark and evidently picking off prey too small to be detected with binoculars; shows a strong preference for certain kinds of trunks with very fine, flaky bark (e.g., *Inga coruscans* at La Selva); sometimes descends a short way tail-first; takes very small insects and spiders, or larger items when feeding young; usually alone.

Voice. Call a sneezing *schip* or *sfik*, sometimes repeated many times; song a fine, rapid, rather warblerlike trill that tapers off sharply; young birds out of nest give a breathy, scratchy descending series of 5–6 notes, the first loudest: *keekekekiki*.

Nest. A pad or shallow cup of fine rootlets or other dark fibrous materials, in a natural cavity with a narrow entrance up to 20ft (6m) high but usually much lower; in decaying stump, in space between buttresses, closed space in deeply furrowed trunk; the nest itself sometimes at or even below ground level. Eggs 2. March–June.

Status. Common to abundant resident of Caribbean lowlands and foothills, to 3600ft (1100m), occasionally to 4500ft (1400m), rarely higher, extending onto Pacific slope of Cordillera de Guanacaste locally; also common in lowlands and foothills of S Pacific slope, N to about Carara, and locally up to ca. 5000ft (1500m).

Range. S Mexico to N Bolivia, C Brazil, and Guianas.

STRONG-BILLED WOODCREEPER

Xiphocolaptes promeropirhynchus Pl. 29(18)

Trepador Gigante

Description. 12″ (30cm); 140g. Very large, with long stout bill and buffy facial stripes. **Adults:** pileum blackish-brown; back and wing-coverts rich brown; crown and upper back finely streaked with buff; rump dark rufous; remiges and tail deep rufous-chestnut; face mixed with dark brown and buff, with fairly distinct broad, buffy postocular and malar stripes; chin and center of throat buffy-white; rest of underparts olive-brown; breast finely streaked with buff; belly barred with black. Iris chestnut; bill olive-horn; feet olive. **Young:** more patterned; throat spotted and barred with buff and black; chest with broad buffy streaks that are bordered with fine black spots; wing-coverts with ochraceous-buff tips set off by dusky.

Habits. Little known. Frequents dense wet forest, singly or in pairs, foraging from near ground up to middle levels on trunks and thick branches; usually silent and surprisingly inconspicuous, seldom associating with mixed-species flocks.

Voice. A powerful *cooo-WEEEW*, the first note thin, rising and querulous, the second loud, snapping; sometimes varied to *coo-WHEEE-WEEEW*, somewhat resembling call of Spotted Woodcreeper in pattern but deeper and stronger.

Nest. Undescribed (?).

Status. Rare resident at lower middle elevations, ca. 1600–5600ft (500–1700m), along N and E slopes of Cordillera Central, and around headwaters of Río Reventazón on N end of Cordillera de Talamanca (perhaps continuing along Caribbean slope to Panama), and the Pacific face of this cordillera on slopes overlooking Coto Brus and Río Cotón drainages.

Range. C Mexico to Guyana, Peru and Bolivia.

BARRED WOODCREEPER

Dendrocolaptes certhia Pl. 29(19)

Trepador Barreteado

Description. 11″ (28cm); 73g. Large; long, stout, dark bill; entire head and body uniformly barred. **Adults:** crown dull rufous; back and lesser wing-coverts olive-brown; sides of head and neck, most of underparts cinnamon-brown; sides of breast darker and more olive; all these areas uniformly barred with black, the barring below more undulating; greater coverts, rump, upper tail-coverts, and tail rufous-chestnut. Upper mandible black, lower dusky with tomia horn-color; feet olive to grayish. **Young:** similar but barring often reduced on throat and finer, less distinct on belly; crown more ochraceous.

Habits. Frequents forest, old second growth, thinned woodland, semi-open, and trees in neighboring clearings; usually forages on trunks and thick branches at lower and middle levels; often attends army ant swarms, perching low and sometimes chasing smaller woodcreepers, forcing them higher; drops to ground for fleeing insects, spiders, small frogs and lizards; sometimes scales off bark, probes into crevices and epiphytes for beetles, bugs, crickets; solitary or in pairs; rather lethargic.

VOICE. A variety of querulous, whistled calls, loud and clear: *toooit-toooit-toooit-tooo* or a longer, almost gobbling *tewy-tewy-tewy-toooey-toooey-toooey-toooit-tew-tew-tew* that rises then falls, with an urgent, excited-sounding quality; calls very actively at dawn, sometimes at dusk.

NEST. In a hollow palm stub or natural cavity, 5–20ft (1.5–6m) up, lined with bits of bark and leaves. Eggs 2. May–July.

STATUS. Common resident throughout Caribbean lowlands, up to 2000ft (600m) in N, and 3000ft (900m) in SE; uncommon and local in NW lowlands, but more numerous on Península de Nicoya and slopes of N cordilleras, locally to 4200ft (1300m); more common in lowlands and foothills from Carara S to Panama, including interior valleys, and to 4000ft (1200m) on neighboring slopes.

RANGE. S Mexico to Bolivia, N Brazil, and Guianas.

BLACK-BANDED WOODCREEPER

Dendrocolaptes picumnus Pl. 29(15)

Trepador Vientribarreteado

DESCRIPTION. 10¾" (27cm); 65g. Unique among Costa Rican woodcreepers in having throat and chest streaked, belly barred; bill stout, rather short. **Adults:** pileum and hindneck olive-brown, scaled with black and finely streaked with buff; back richer brown with fine buffy shaft-lines; rump, upper tail-coverts, remiges, and rectrices rufous-chestnut; sides of head dark brown; indistinct buffy postocular streak and partial eye-ring; feathers of throat buffy with black dots and brown edgings laterally, giving the effect of broad, indistinct streaking; chest olive-brown, more narrowly and distinctly streaked with buff; posterior underparts cinnamon-brown, finely and evenly barred with black. Bill blackish shading to pale horn on tomia and tip; feet gray. **Young:** similar but throat more scaled, less streaked; head pattern indistinct, pileum more blackish; black barring on belly indistinct.

HABITS. In wet forests of middle elevations; forages on thick trunks and branches and dead stubs at all heights but mostly fairly high; usually solitary, silent, rather sluggish and inconspicuous; eats wood-boring beetles, caterpillars, crickets.

VOICE. A soft, nasal *wrenh* with upward inflection, by members of pair (Ridgely).

NEST. In hole in tree. Eggs 2 (in Colombia; no Costa Rican record). Nest-building recorded in April; woodcreepers of this genus evidently pair for nesting.

STATUS. Very uncommon to rare (at least seldom seen) at middle elevations on Caribbean slope of Cordillera Central, in Dota region, and on Cordillera de Talamanca, at ca. 3000–6500ft (900–2000m).

RANGE. S Mexico to Bolivia, NW Argentina, Brazil, and Guianas.

BUFF-THROATED WOODCREEPER

Xiphorhynchus guttatus Pl. 29(17)

Trepador Gorgianteado

DESCRIPTION. 8½" (21.5cm); 48g. Fairly large; brown, buffy throat and breast streaking; bill long, stout, straight. **Adults:** crown and hindneck blackish-brown, thickly spotted with buff; back and wing-coverts olive-brown, the upper back with buffy, black-margined streaks; rump, upper tail-coverts, tail, and tertials rufous-chestnut; remaining remiges tawny-rufous to russet; sides of head dark brown, indistinctly streaked with buff; a broader, buffy postocular streak over auriculars; throat bright buff; malar streak dusky; lower throat darker buff, scaled with dusky, passing into broad, buffy, dusky-margined streaks on breast; posterior underparts cinnamon-brown with narrow, indistinct buffy streaks. Upper mandible black, lower pale grayish to horn-color, shading to dusky tip; feet slaty. **Young:** similar but throat faintly scaled with dusky; breast streaks broader, more broken by dark tips to feathers; less buffy spotting on crown.

HABITS. Frequents humid forests, especially at edges and breaks, also semi-open, old second growth, mangroves; emerges into adjacent shady clearings to forage and nest; creeps up trunks from near ground well into canopy and outward or downward along branches, poking and probing into crevices, tufts of moss or epiphytes, and knotholes; flakes off bits of dead bark; takes beetles, roaches, crickets, spiders, beetle and lepidopteran larvae, earwigs, and occasional small lizards or frogs; does not pair, usually solitary but may join mixed flocks; rarely forages with army ants.

VOICE. Call a loud, clear, rolling *cheer*, often repeated several times; in alarm or uncertainty, a drier *chu* or *choe*. Song, a series of ca. 7–20 upward-slurred *wick* or *doy* notes, clear and querulous, rises, holds for most of the sequence, then falls in pitch; usually given from concealing foliage high in tree.

NEST. In a hollow palm stub or large bamboo, a

cranny among roots of strangling fig or nook in a building near forest, the entrance 3–21ft (1–6.5m) up. A deep cavity is filled with hundreds of pieces of stiff bark and wood fragments, upon which the eggs lie, often far below entrance. Eggs 2. March–July.

Status. Resident, common from lowlands to ca. 2100ft (650m) on Caribbean slope and to 3000ft (900m) on S Pacific slope; rare and local in N Pacific lowlands, mainly in moist gallery forest.

Range. E Guatemala to N Bolivia, Brazil, and Guianas.

IVORY-BILLED WOODCREEPER

Xiphorhynchus flavigaster Pl. 29(21)

Trepador Piquiclaro

Description. 9½″ (24cm); 60g. Fairly large, heavily streaked; bill long, strong, straight, notably pale. **Adults:** head blackish-brown, streaked with buff, broadly on pileum and narrowly on face; back and wing-coverts olive-brown with broad, black-margined, buffy streaks; upper tail-coverts, tail, and remiges chestnut-rufous; throat bright buff; narrow malar streak black; foreneck bright buff scaled with black, passing to broad, buffy, black-margined streaks on breast; rest of underparts dull brown, with sparser, dull buffy streaking. Bill silvery-horn except for blackish base of upper mandible; feet dull olive. **Young:** similar but throat feathers with sooty fringes; streaking on breast and back duller, less distinct; bill brownish. Larger than Buff-throated Woodcreeper with more contrasting pattern, paler bill, different voice.

Habits. Prefers tall evergreen forest with few epiphytes, less often deciduous forest, forest edge, and semi-open; forages mostly on limbs well up in canopy, less frequently on trunks; may accompany mixed flocks in canopy but more often seen alone or in pairs; eats large insects like crickets, cockroaches, and beetles.

Voice. Call note a sharp squeak; a short 4- or 5-note laugh. Song a rapid series of 15–20 high, thin notes, falling in pitch, slowing, and becoming louder toward end, sounding like a long-drawn-out whinny.

Nest. Unknown (?).

Status. Resident in extreme NW; locally common at 1300–3000ft (400–900m) along Pacific slope of Cordillera de Guanacaste and in higher hills of Península de Nicoya; in lowlands uncommon and local S to Santa Rosa, rarely to Palo Verde; uncommon and local on Caribbean slope along Nicaraguan border, E to Los Chiles.

Range. NW and E Mexico to Costa Rica.

BLACK-STRIPED WOODCREEPER

Xiphorhynchus lachrymosus Pl. 29(16)

Trepador Pinto (Relinchero)

Description. 9½″ (24cm); 58g. Most strikingly patterned woodcreeper; boldly patterned head and body contrast with rufous wings and tail. **Adults:** feathers of pileum, orbital area, hindneck, and back with broad blackish borders, U-shaped to teardrop-shaped, with pale buffy centers, producing spotted pattern on head, streaking on back; greater coverts, remiges, rump, upper tail-coverts, and tail rufous; throat pale buff, scaled with black posteriorly; black edges on pale buffy feathers become broader on breast to produce a streaked pattern that blurs and is indistinct on posterior underparts. Upper mandible dark horn, lower silvery; feet olive-gray. **Young:** pattern less distinct, especially on back and belly; black areas duller and browner; throat more heavily and indistinctly scaled with sooty-black.

Habits. Frequents middle and upper levels of wet forest and adjacent semi-open, sometimes coming lower in shady clearings or along edges; climbs jerkily up trunks and large branches, pecking and flaking off bits of dead bark, probing into moss and lichens, examining termite and wasp nests; sometimes makes fluttering sallies after fleeing prey; takes beetles, crickets, bugs, cicadas, ants, spiders, moths, occasionally small lizards; solitary or in pairs; often joins mixed flocks in canopy and subcanopy.

Voice. Call a loud, vigorous *whee here-here* or *wee tir-tir-tir*; a loud, clear *doweeet*, more emphatic and less querulous than notes of Barred Woodcreeper; song a high-pitched, long-drawn-out, descending whinny; also a loud, rolling, descending *cheer* or *chirrrw* resembling that of Buff-throated Woodcreeper.

Nest. In a palm stub or tree hole, with entrance 2–20ft (0.6–6m) up, lined with wood chips, bits of bark. Eggs 2. March–June.

Status. Fairly common resident of humid lowlands throughout Caribbean slope and on S Pacific slope, including General–Térraba–Coto Brus valleys, N in small numbers to Carara; locally up to 3300ft (1000m) on Caribbean slope and to 4000ft (1200m) on S Pacific slope.

Range. E Nicaragua to NW Ecuador.

SPOTTED WOODCREEPER

Xiphorhynchus erythropygius Pl. 29(20)

Trepador Manchado

Description. 9″ (23cm); 50g. More olive than other woodcreepers, with distinctive spotted pattern, buffy eye-ring; bill long, stout, and straight. **Adults:** pileum dusky-olive, the forehead and forecrown with narrow buffy streaks; back and wing-coverts olive-brown with fine, buffy shaft-streaks on upper back; rump, upper tail-coverts, tail, and tertials rufous-chestnut; remaining remiges russet to tawny-rufous; sides of head dark olive, indistinctly spotted and streaked with buff; a broad, indistinct, buffy eye-ring; throat yellowish-buff, spotted with olive; rest of underparts olive, spotted with yellowish-buff. Culmen blackish, rest of bill silvery-horn; feet slaty. **Young:** overall more brownish; spotting below less distinct, especially on throat and belly.

Habits. In humid forest and adjacent shady semi-open and old second growth, chiefly in mountainous terrain; climbs vertical trunks but more often creeps along branches, especially their undersides, sometimes working head-downward out toward the tips; pokes and probes into moss tufts and crannies in masses of epiphytes; dislodges leaves and trash with sideways flick of the bill; eats various insects including roaches, beetles, crickets, and earwigs, also spiders and occasional small frogs or salamanders, which it beats vigorously against perch; singly or in pairs, regularly joins mixed-species flocks moving through upper understory or canopy.

Voice. A high, thin, eerie set of 2–3 long-drawn-out, descending, clear whistles, successively lower in pitch, *piiiiiiiiiiiir piiiiiiiiiir piiiiiiiiiiir*; a high, sharp, rolling *djeer* or *keerw*, weaker and shorter than call of Buff-throated.

Nest. In palm stub or other cavity or crevice, entrance 6–30ft (2–9m) up. Eggs 2. March–June.

Status. Fairly common resident of middle elevations on both slopes, descending uncommonly to humid lowlands along Caribbean and S Pacific slopes, but not in dry-forested NW; center of abundance 2300–4800ft (700–1450m) on both slopes, sometimes up to 5600ft (1700m); very uncommon below ca. 300ft (100m) in lowlands, especially well away from mountains.

Range. S Mexico to W Ecuador.

STREAKED-HEADED WOODCREEPER

Lepidocolaptes souleyetii Pl. 29(8)

Trepador Cabecirrayado

Description. 7½″ (19cm); 28g. Slender, brownish, extensively streaked; bill slender, slightly decurved, moderately long. **Adults:** pileum and hindneck dark brown, thickly streaked with buff; back and wing-coverts olive-brown, the upper back with fine, black-margined buffy streaks; rump, tail-coverts, tail, remiges chestnut-rufous; sides of head and neck streaked with dark brown and pale buff; chin and throat pale buff; rest of underparts dull brown, densely marked with pale buffy, black-margined streaks that are broadest on breast. Upper mandible brownish, lower pale horn; feet olive-gray. **Young:** throat duller with sooty fringes; streaking above and below denser but less distinct, edged with sooty; bill darker.

Habits. Prefers light woodland, shady plantations, clearings with scattered trees, gardens, forest edge, but occasionally ventures into dense rain forest; also in gallery forest and mangroves; climbs up trunks and outward along branches, gleaning, prying off flakes of loose bark, and lifting tufts of moss to dislodge beetles, roaches, bugs, ants, spiders, eggs, and pupae; takes moths, wasps, bees; occasionally pecks into decaying wood; paired throughout year; does not join mixed flocks.

Voice. A simple, descending, clear trill, soft and melodious, with which mates answer each other; when alarmed or perturbed, with twitching wings a harsher, more rattling, shorter trill; call a plaintive *pyuu*.

Nest. In an inconspicuous cranny in dead or living tree, occasionally a woodpecker's hole lined with flakes of hard or corky bark, 15–80ft (4.5–24m) up. Eggs 2. March–June.

Status. Resident throughout lowlands of both slopes, locally to 5000ft (1500m), though increasingly uncommon above 3000ft (900m); rare in W Valle Central; most common in humid areas, uncommon in dry NW, where still the most numerous woodcreeper.

Range. S Mexico to NW Peru, N Brazil, and Guyana.

SPOTTED-CROWNED WOODCREEPER

Lepidocolaptes affinis Pl. 29(9)

Trepador Cabecipunteado

Description. 8½″ (21.5cm); 35g. Bill pale, slender, decurved; crown spotted; below heavily streaked. **Adults:** pileum and nape olive-brown, each feather with blackish border and buffy central spot; back and wing-coverts olive-brown, the upper back often with fine, buffy shaft-streaks; rump, tail-coverts, remiges, and rectrices chestnut-rufous; sides of head indistinctly streaked with buff and blackish; chin and throat buff; rest of underparts olive-brown, densely marked with broad, black-edged buffy streaks, this pattern extending as a narrow collar around base of hindneck. Bill pale silvery-horn, with dusky patch at base of culmen; feet olive-gray. **Young:** similar but throat duller, streaking below blurred by heavier, less distinct dark margins; crown spots smaller, less distinct; bill darker. Only woodcreeper at high elevations.

HABITS. In mossy, epiphyte-burdened montane forest and adjacent clearings with scattered trees, tall second growth; forages as do other woodcreepers, creeping up trunks and large branches, with its slender bill extracting beetles, crickets, larvae, from bark, mosses, and lichens; paired throughout year; joins flocks of bush-tanagers and other small birds moving through middle and upper levels of forest.

VOICE. Call a plaintive, squeaky *deeik*; song a thin, reedy to nasal note followed by a rattling trill: *deeeeeeah, hihihihihi*; sometimes 2 reedy whistles without trill, *deeee-deeeeih*.

NEST. In cavities or fissures in tree trunks or in old hole carved by woodpecker or barbet, the floor covered with small flakes of hard, thin bark, 2–26ft (0.6–8m) up. Eggs 2. March–June.

STATUS. Uncommon to fairly common resident of higher parts of all mountain ranges, upward from ca. 3300ft (1000m) in N cordilleras; in Cordilleras Central and de Talamanca, from timberline down to 5000ft (1500m) on Pacific slope and 4000ft (1200m), rarely 2500ft (750m), on Caribbean.

RANGE. C Mexico to N Bolivia.

BROWN-BILLED SCYTHEBILL

Campylorhamphus pusillus Pl. 29(14)

Trepador Pico de Hoz

DESCRIPTION. 9″ (23cm); 40g. Medium-sized; prominently streaked; distinctive, very long, slender, decurved bill. **Adults:** pileum and hindneck blackish-brown, streaked with buff; back and wing-coverts rich brown, with fine, buffy shaft-streaks; rump and upper tail-coverts dark rufous; wings and tail chestnut; sides of head streaked with blackish and buff; chin and throat deep buff, with broad, dusky-brown streaks; remaining underparts olive-brown, paler posteriorly; foreneck and breast with narrow, buffy streaking that fades out on belly. Bill brownish-horn with dusky base of upper mandible; feet olive-green. **Young:** similar but streaking above and below darker buff, broader and less distinct; bill darker.

HABITS. Prefers dense wet forest, especially in hilly terrain; climbs trunks and branches at all heights, even dropping to fallen logs; uses long bill to probe for beetles, earwigs, ants, spiders, into tufts of moss, in deep crevices in furrowed trunks, between thick lianas twined together, into clustered palm fruits, in bromeliad tanks, and among bases of palm fronds; eats egg cases of roaches, larvae; usually seen singly, often accompanying mixed-species flocks in understory or lower canopy.

VOICE. Elaborate and varied series of fine, clear, slurred single or double whistles, and rapid trills; sometimes seems to sing 2 songs simultaneously: a fine, clear, ascending trill and at the same time a loud, clear *tewe tewew twee tewe tewe we we we we* or some variant thereof; sometimes begins with 1–3 clear whistles like those of Spotted Woodcreeper, then continues with descending, accelerating medley of trills and whistles; especially vocal at dawn and dusk.

NEST. Unknown (?). Recently fledged juveniles seen May–July. Evidently pairs for breeding.

STATUS. Uncommon resident of lower middle elevations, ca. 1000–5000ft (300–1500m) on Caribbean slope of Cordilleras de Tilarán, Central, and de Talamanca; on S Pacific slope found in General–Térraba–Coto Brus valleys, to 5600ft (1700m) on Cordillera de Talamanca and coastal ranges, and sparingly throughout lowlands of Golfo Dulce region, especially in hilly areas.

RANGE. Costa Rica to W Ecuador and Guyana.

FAMILY **Furnariidae:** Ovenbirds

A family of 214 species, ovenbirds spread widely over the whole of South America, from the lowlands to the high, cold páramos and punas; but northward none has extended beyond Mexico, and none has reached the West Indies. Of all neotropical avian families, this is the most heterogeneous in external form, habits, and nidification, resulting in a diverse array of names: spinetail, castlebuilder, treerunner, barbtail, tuftedcheek, foliage-gleaner, xenops, and leaftosser, to mention only a few. From warbler- to thrush-sized, ovenbirds are clad chiefly in shades of brown and gray, from bright rufous and chestnut to blackish, often streaked or spotted or with patches of white or black. Some have notably long tails, which may be tipped with spines. The sexes, almost always alike, tend to remain throughout the year in sedentary pairs or family groups, which often join mixed-species flocks. Their food consists almost wholly of insects and other small invertebrates, which they gather in the most diverse ways: gleaning from foliage, creeping over trunks, pecking into decaying stems and vines, tossing aside fallen leaves on the ground. Their notes are often dry and repetitious, but a few, notably the leaftossers, sing delightfully. Duets and responsive singing or calling are frequent.

No other New World family exhibits a greater diversity of nest architecture. The family receives its name from the massive domed structures of hardened clay or mud, miniatures of the baking ovens (*hornos*) widespread in Latin America, built by several species of ovenbirds (*horneros*) in South America. Virtually all members of this family either build covered nests or breed in cavities, whether in holes in trees, burrows in the ground, fissures in cliffs, closed structures made by other birds, or crannies in man's constructions. The many covered nests are exceedingly diverse in form and materials, fastened amid marsh vegetation or hung high in trees. Often the nests are massive structures of interlaced sticks, veritable avian castles entered through long tunnels; sometimes they contain a number of separate rooms. Ovenbirds lay 2–4 unmarked eggs, rarely more, which are most often white but may be tinted with blue, green or pale buff. Nearly always the sexes take almost equal shares in building, incubating and rearing the young. Incubation periods are 14–21 days and nestling periods 14–26 days, rarely longer.

PALE-BREASTED SPINETAIL

Synallaxis albescens Pl. 32(11)

Arquitecto Güitío

Description. 5½″ (14cm); 13g. Slender, rather wrenlike, with a long, pointed, graduated, wispy tail. **Adults:** forehead and sides of head dull gray; crown, nape, and wing-coverts rufous; rest of upperparts, including wings and tail, grayish-brown; lores and chin whitish; lower throat and foreneck gray, grizzled with white; breast gray, shading to dull white on belly; flanks brownish. Iris yellowish-brown to buff; upper mandible black, lower gray with dusky tip; legs pale gray to flesh-color. **Young:** above dull brown, with a smaller, duller patch of cinnamon on shoulder; remiges edged with russet; cheeks and throat buff; chest, sides, flanks buffy-brown; belly buffy-white. Slaty Spinetail is larger, much darker.

Habits. Lurks in weedy pastures, abandoned fields with low scrubby growth, savannas with trees scattered amid coarse grass; hops and clambers deliberately through dense low vegetation, gleaning beetles, crickets, caterpillars, other small insects and spiders; rises to top of grass tussock or shrub, or into low tree to sing, often on open twig, tail partly spread; flies low but directly, dropping abruptly into cover.

Voice. Call a staccato *bip*, often rapidly repeated, and followed by a prolonged, rattling *churrr*; song a harsh, grating, emphatic, dry *bet-chu* or *guit-teéo*, singly or repeated incessantly, 35–40/min, up to half an hour.

Nest. A bulky, globular structure of straws and fine twigs, covered by a thick thatch, entered through a long, horizontal, tunnel-like extension, and lined with pieces of downy leaves, usually mixed with bits of snake skin, 2–5ft (0.6–1.5m) up in grass tussock or vine-draped shrub. Eggs 2, less often 3, white. December–July.

Status. Resident, common in savannas and agricultural areas in Golfo Dulce, General, and Térraba regions of S Pacific slope, from lowlands to 4000ft (1200m); spreading with deforestation, has colonized upper Valle del General and San Vito–Coto Brus region within last 20 years.

Range. Costa Rica to N Argentina, SE Brazil, and Guianas.

Note. Also called Pale-breasted Castle-builder.

SLATY SPINETAIL

Synallaxis brachyura Pl. 32(9)

Arquitecto Plomizo

Description. 6″ (15cm); 18.5g. Dark, rather wrenlike; wings short and rounded; tail long, graduated, pointed, and wispy; legs long and strong. **Adults:** forehead and sides of head sooty-gray; crown, nape, wing-coverts, and bases of primaries rich rufous; back dark brownish-olive; rest of wings and tail dark sooty-olive; chin and throat slaty, grizzled with white, the feathers of lower throat with black bases, forming a darker patch; breast dark slate; belly paler; flanks dusky-olive. Iris reddish-brown; bill black; legs blue-gray to blackish. **Young:** above uniform dull sooty-olive; wing-coverts and bases of primaries edged with dull dark rufous; lores and chin dull yellowish; underparts pale grayish-olive, belly tinged with buffy to yellowish.

Habits. Prefers young second-growth thickets

of overgrown pastures, roadsides, riverbanks; skulks like a wren through tangled vegetation, deliberately poking and gleaning for beetles, bugs, caterpillars, grubs, eggs and adults of spiders; sings from high in dense vine tangles; drops to near ground in alarm; flies reluctantly with dangling legs and tail, a buzzy flutter; cocks tail and flicks wings when agitated; paired year-round.

Voice. A hard, heavy, low-pitched *churrr* or *chu-chu-chrrrr* with a wooden grating quality that accelerates and falls in pitch and intensity, often responsively by members of pair foraging out of sight; call a sharp, dry *chip* or *kyip*.

Nest. A bulky, globular structure of interlaced twigs, ca. 14″ high by 17″ long (36×43cm), the chamber covered by thick thatch, entered through a long, horizontal passageway, and lined with a pad of fragmented downy leaves bound together by cobweb; scraps of snakeskin often stuck in the walls here and there; 1½–15ft (0.4–4.5m) up in shrub or vine-draped tree. Eggs 2–3, white to pale greenish. January–September.

Status. Common resident throughout lowlands of Caribbean slope, up to 5000ft (1500m) around Cartago; and on S Pacific slope N to San Isidro and along coast to around Carara; to 4000ft (1250m) locally.

Range. N Honduras to W Ecuador and E-C Brazil.

Note. Also called Slaty Castlebuilder.

RED-FACED SPINETAIL

Cranioleuca erythrops Pl. 30(12)

Colaespina Carirroja

Description. 6″ (15cm); 16g. Slender, creeperlike, with a long, pointed, graduated tail; bill short, sharp, slender. **Adults:** forehead, crown, cheeks, wing-coverts, and edges of primaries bright rufous; rest of upperparts and wings brownish-olive; tail dull rufous; chin gray, grizzled with white; rest of underparts pale brownish-olive. Iris reddish-brown; culmen dusky, rest of bill pale brownish-horn; legs brownish to olive. **Immature:** crown and auriculars like back; lores and incomplete eye-ring rufous; stripe from above eye back over auriculars ochraceous; cheeks, throat, and chest strongly tinged with ochraceous to dull rufous; wing-coverts edged with dull rufous.

Habits. Prefers upper understory and lower canopy of wet forests of middle elevations, coming lower at forest edge and adjacent old second growth and semi-open; very active, creeping and climbing along branches, gleaning and rummaging in mosses, lichens, and curled dead leaves lodged in vine tangles; clings to twigs in all attitudes, often upside down; eats many beetles, crickets, roaches, caterpillars, spiders, also *Cecropia* protein corpuscles; singly or in pairs, regularly joins mixed-species flocks.

Voice. Contact note a short, rolling, high-pitched *prrreep*; a squeaky *chew-seep*; song an ascending, rapid series of querulous, squeaky, lisping *sfi* or *tsu* notes that may turn into a high, rather swiftlike twitter.

Nest. A very bulky, more or less bell-shaped mass of green moss and bits of thin, dry herbaceous vines, often with dead leaves incorporated, that hangs from end of drooping leafy branch of tree or vine 16–40ft (5–12m) up; inconspicuous entrance near bottom leads to nest chamber by a narrow, ascending tunnel; chamber lined with fine straw and leaves. Eggs not seen; 1 nest held 2 newly-hatched young. March–June.

Status. Fairly common resident of middle elevations the length of both slopes, mainly from ca. 7500ft (2300m) down to 2300ft (700m) on Caribbean slope and 4000ft (1200m) on Pacific, locally somewhat higher or lower.

Range. Costa Rica to Colombia and W Ecuador.

SPOTTED BARBTAIL

Premnoplex brunnescens Pl. 29(5)

Subepalo Moteado

Description. 5½″ (14cm); 17g. Small, rather plump; dark brown, below heavily spotted; shafts of rectrices project as soft spines. **Adults:** pileum dark olive-brown, streaked with buff on forehead, scaled with blackish on crown and nape; back rich brown, passing to dark rufous on rump, both with faint darker scalloping; wings and tail dark chestnut-brown; sides of head dusky; lores, superciliaries, eye-rings, and fine streaking on auriculars ochraceous-buff; throat ochraceous-buff, scaled with blackish posteriorly; rest of underparts dark brown, thickly spotted with elongate or teardrop-shaped buffy spots. Upper mandible blackish, lower flesh-color with dusky tip; legs dark brown. **Young:** similar but sooty scaling on throat heavier, buffy spots below rounder and less distinct; buffy shaft-streaks on back.

Habits. Prefers wet, moss-bedecked mountain forests and adjacent old second growth and clearings with trees and stumps covered with vines and epiphytes; climbs up vertical or inclined trunks and branches from near ground to upper understory, rarely higher, often moving sideways or clinging inverted to undersides of limbs, searching through moss, crevices in bark, epiphytes; uses its soft-tipped tail for a prop less often than do woodcreepers; eats small insects and spiders, including larvae and eggs; rarely perches crosswise on a twig; single adults regularly join mixed flocks led by antwrens or *Basileuterus* warblers; pumps tail when excited.

VOICE. Call a sharp, thin, piercing, high-pitched *tseep* or *peek*; song a rapid series of such notes that form a dry rattle or trill.

NEST. A massive, more or less globular structure of green mosses or liverworts, bound together with fine, dark-colored rootlets, attached to rough or pitted surface of upright or fallen trunk or vertical rocky cliff, or in a crotch, often near tip of drooping branch; entrance at bottom, from which tubular passage leads straight up to chamber, where a shallow depression in a thick mass of mosses and liverworts holds the eggs. Eggs 2, white. March–June. Adults sleep singly in similar but often slightly built nests.

STATUS. Common resident of middle elevations the length of both slopes, from ca. 2000ft (600m) on Caribbean slope and 3300ft (1000m) on Pacific, up to ca. 8000ft (2450m).

RANGE. Costa Rica to N Venezuela and C Peru.

RUDDY TREERUNNER

Margarornis rubiginosus Pl. 29(4)

Subepalo Rojizo

DESCRIPTION. 6¼″ (16cm); 18g. Bright reddish-brown; suggests a short-billed little woodcreeper but tail ends in projecting spines that are soft, not rigid. **Adults:** crown and auriculars chestnut-rufous; back, wing-coverts, rump, and tail bright rufous; primaries edged with cinnamon-rufous; cheeks, sides of neck, nuchal collar tawny-ochraceous; stripe from eye back over auriculars whitish to ochraceous-buff; throat whitish; rest of underparts tawny-ochraceous, shading to cinnamon-rufous on flanks and crissum; breast with fine buffy-white spots rimmed with black. Bill flesh-color except base of culmen dusky; legs dull flesh. **Young:** virtually identical, throat feathers with faint sooty fringes.

HABITS. Frequents all levels of montane forests, up to middle of canopy, as well as adjacent edges, thickets, and clearings with scattered trees; climbs vertical trunks, using its spine-tipped but flexible tail for support less than do woodcreepers; creeps along mossy branches, clings to twigs and bamboo stalks, often upside-down; probes and rummages in tufts of moss and lichens, dead leaves lodged in vines or epiphytes, for small insects, spiders, their eggs and larvae; rarely descends to ground; alone or in pairs, frequently joins mixed flocks or bands of *Chlorospingus* tanagers.

VOICE. A variety of high, thin, sharp, twittering notes, including a fine, descending trill and sharp *tsit* notes.

NEST. Unknown (?). Fledged juveniles seen May–July (Villa Mills).

STATUS. Common resident of high mountains, up to timberline, the length of the country, upward from ca. 4000ft (1200m) in Cordillera de Guanacaste, 5000ft (1500m) in Cordillera de Tilarán, and 6000ft (1850m) in the Cordilleras Central and de Talamanca, extending somewhat lower on Caribbean than Pacific slope.

RANGE. Costa Rica and W Panama.

BUFFY TUFTEDCHEEK

Pseudocolaptes lawrencii Pl. 30(1)

Trepamusgo Cachetón

DESCRIPTION. 8″ (20cm); 48g. Bold buffy cheek-patches; conspicuous orange-rufous tail appears notched. **Adults:** pileum and hindneck brown, streaked with buff and dusky; back, shoulders and tertials russet, faintly scalloped with dusky, shading to cinnamon-rufous on rump and upper tail-coverts; wings blackish with ochraceous-buff wing-bars; sides of head dusky with a narrow, buffy postocular streak and eye-ring; a tuft of rich buffy elongated feathers on cheek and side of neck; throat pale buff with faint sooty scaling posteriorly; breast with smudgy buffy spotting and dusky scaling, becoming dull buffy-olive with indistinct paler smudging on belly; sides, flanks, and crissum russet to dull cinnamon-rufous. Bill black, except tomia and gonys yellowish; legs olive-green. **Young:** lacks buffy streaking on pileum; upper back more heavily scalloped with sooty; flanks brighter, more orange-rufous; below sooty scaling heavier.

HABITS. In wet, epiphyte-laden mountain forests and adjacent clearings with scattered trees; creeps actively along mossy trunks and branches and through vine tangles, gleaning and probing; rummages actively in bromeliads, tossing out leaves and trash in all directions to uncover beetles, roaches, spiders, even salamanders; takes moths, caterpillars, grubs; alone or in pairs, regularly joins mixed flocks moving through upper understory and lower canopy.

Voice. Call a sharp, ringing, metallic *pwik* or *pweent*; song a liquid, gurgling trill or twitter that ascends, then slows and descends at the end, often preceded by 1–2 sharp clear notes: *peek peek prrrrrrreeeeee*.

Nest. In old woodpecker hole in rotting stub, 16–30ft (5–9m) up, with thick lining of loose brown treefern scales, with which the solitary parent covers the egg upon leaving nest. Egg 1, white. January–April.

Status. Rare resident in Cordillera de Tilarán above 5200ft (1600m); uncommon to fairly common in Cordilleras Central and de Talamanca from 6500ft (2000m) up to timberline; rarely strays down to 4000ft (1200m) on Caribbean slope.

Range. Costa Rica and W Panama; also Andes of Colombia and Ecuador.

Note. Sometimes considered a race of the South American *P. boissonneautii*.

STRIPED FOLIAGE-GLEANER

Hyloctistes subulatus Pl. 30(4)

Trepamusgo Rayado

Description. 7″ (18cm); 33g. Slender, with rather long, straight, sharp bill; bright buffy throat; rest of head and body with blurred buffy streaking. **Adults:** sides and top of head, hindneck, and back olive-brown, the feathers with blackish edges and buffy shaft-streaks; wing-coverts and remiges largely tawny-rufous; rump and tail-coverts dark rufous; tail chestnut-rufous; throat buffy, trailing off into indistinct streaking on breast and upper belly; rest of underparts tawny-olive; wing linings orange-rufous. Upper mandible dusky, lower dark horn; legs olive. **Young:** throat with faint dusky scaling; below with buffy irregular streaking.

Habits. Prefers primary forest, sometimes venturing into adjacent tall second growth or semi-open; singly or in pairs, nearly always accompanies mixed flocks of antwrens, greenlets, etc.; rummages and gleans actively in dense tangles from upper understory to lower canopy, poking into bromeliads and accumulations of trash lodged in vegetation, tossing debris about to expose insects, spiders, small frogs and lizards. When excited, as in aggressive interactions, puffs out throat into a bright yellowish-buff ball; often depresses tail below body when foraging.

Voice. Often noisy: call a loud, harsh, rasping *chook* or *churk*, sometimes repeated; a loud, dry chatter *zeck-zeck-zeck*. . . . Song a series of 8–10 or more loud, heavy, wooden notes: *kick kick kick*.

Nest. Undescribed (?).

Status. Uncommon resident of most of Caribbean lowlands, except drier area S of Lago de Nicaragua, and in Golfo Dulce region, N along coast to around Dominical; up to ca. 2000ft (600m) in foothills of both slopes but absent from General–Térraba–Coto Brus valleys.

Range. E Nicaragua to W Ecuador, SE Peru and Amazonian Brazil.

Note. Also called Striped Woodhaunter; sometimes included in the genus *Philydor*.

LINEATED FOLIAGE-GLEANER

Syndactyla subalaris Pl. 30(8)

Trepamusgo Lineado

Description. 7½″ (19cm); 37g. Bill wedge-shaped at tip; small buffy throat-patch and below with fine streaking; young below with extensive orange-rufous. **Adults:** above olive-brown, shading to tawny-rufous on wings and chestnut-rufous on upper tail-coverts and tail; crown and upper back with fine, buffy shaft-streaks; sides of head olive-brown, streaked with buff; a nuchal collar of buffy streaks margined with black; chin and upper throat buff; rest of underparts brownish-olive, with sharp, fine buffy streaks that fade out on belly. Bill black with gonys pale horn; legs olive. **Young:** postocular stripe, nuchal streaks, throat, and breast orange-rufous, speckled with black on chin.

Habits. Prefers thickets and vine tangles in understory of primary and old secondary forest, mostly at treefalls, along narrow streams and other small gaps in canopy; forages mainly by rummaging and probing into clumps of moss and epiphytes, accumulations of trash lodged in vines, branches, and treefern crowns, taking beetles, crickets, roaches, other insects, spiders, small frogs and lizards; alone or in pairs, nearly always accompanying mixed-species flocks.

Voice. Call a dry, harsh *tzuk*, like the sound of hitting brush with a machete; song a series of 6–10 sharp, dry, scratchy *djit* or *kik* notes that usually speeds up in the middle, slows again at the end.

Nest. Undescribed (?).

Status. Fairly common resident of middle elevations of both slopes the length of the country, from ca. 7500ft (2300m) down to

2000ft (600m) on Caribbean slope and 3300ft (1000m) on Pacific.

Range. Costa Rica to W Ecuador, E Peru, and W Venezuela.

Note. Sometimes placed in genus *Philydor*.

SPECTACLED FOLIAGE-GLEANER

Anabacerthia variegaticeps Pl. 30(2)

Trepamusgo de Anteojos

Description. 6¼″ (16cm); 24g. Spectacles orangish; below no streaks; tail usually appears notched. **Adults:** crown and face dark olive-brown with fine, buffy shaft-streaks; eye-ring and postocular stripe cinnamon-orange; rest of upperparts dark ochraceous-olive, tinged with rufous on wing-coverts; tail dark rufous; chin pale buff; lower throat olive-buff, faintly scaled with dusky; underparts tawny-olive; wing-linings orange-rufous. Upper mandible dark gray, lower paler; legs olive. **Young:** similar, but dusky scaling on throat and chest more pronounced.

Habits. Prefers hanging vine tangles, dead branches, and dense foliage from upper understory well up into canopy in primary and tall secondary forest; alone or in pairs, regularly accompanies mixed-species flocks; rummages and gleans actively, hopping and clambering up vines, thin branches, and the undersides of treefern fronds, poking into leafy tangles and tufts of moss, sometimes fluttering after dislodged prey; takes beetles, homopterans, katydids, moths, other insects, and spiders; often flicks wings nervously.

Voice. Usual call an explosive, sibilant *squick* or *squeek*; also a scratchy, squeaky *kweeeah* and a scratchy rattle. Song a long series of squeaky, emphatic *skew* notes in a regular cadence, sometimes rising or falling in pitch.

Nest. Undescribed (?).

Status. Resident of middle elevations the length of Caribbean slope and on Pacific slope of Cordillera de Talamanca, including Dota area; reaches Pacific slope in higher parts of N cordilleras; at ca. 2600–6000ft (800–1850m) on Caribbean slope, though not below ca. 4000ft (1200m) in SE; on Pacific slope somewhat higher, to 6600ft (2000m) above Coto Brus valley.

Range. S Mexico to W Panama; W Colombia and Ecuador.

Note. Sometimes considered conspecific with the South American *A. striaticollis* and sometimes placed in the genus *Philydor*; has often been called Scaly-throated Foliage-gleaner.

BUFF-FRONTED FOLIAGE-GLEANER

Philydor rufus Pl. 30(6)

Trepamusgo Rojizo

Description. 7½″ (19cm); 34g. Face and underparts extensively ochraceous; no streaking; tail appears notched. **Adults:** forehead, superciliaries, orbital area, and throat cinnamon-orange to ochraceous, becoming paler and duller on breast, shading to ochraceous-tawny on belly and flanks; crown, spot on lores and auriculars grayish-olive, shading to pale olive-brown or tawny-olive on back; wings and tail rufous; wing-linings cinnamon-orange. Upper mandible blackish, lower gray; legs olive. **Young:** similar but slightly duller; crown grayer; face and throat more buffy, less ochraceous.

Habits. Prefers primary and old secondary forest, usually from upper understory well up into canopy but sometimes descending much lower along edges and breaks; gleans actively on limbs and in foliage, often hopping along a nearly horizontal limb, swinging body from side to side, out to the tip where it rummages in the leaves, often hanging upside down on the swaying twig; takes mainly cockroaches, katydids, beetles, and caterpillars, and spiders and their eggs; alone or in pairs, usually accompanies mixed-species flocks.

Voice. Call a sharp, dry, metallic *chik* or *tzik*, more resonant than somewhat similar note of White-throated Spadebill; song a squeaky, woodpeckerlike *woika-woika-woika*. . . .

Nest. In a burrow in bank or hole in masonry wall or tree. Eggs 2, white (in South America; no Costa Rican record).

Status. Generally uncommon to rare resident of middle elevations at 2600–7500ft (800–2300m) on Caribbean slope from Cordillera de Tilarán S, and between 4000 and 8000ft (1200–2500m) in Dota area and Cordillera de Talamanca on Pacific slope.

Range. Costa Rica and W Panama; also from N Venezuela to W Ecuador, NE Argentina, and S Brazil.

STREAKED-BREASTED TREEHUNTER

Thripadectes rufobrunneus Pl. 30(7)

Trepamusgo Cuellirojizo

Description. 8½″ (21.5cm); 54g. Large, robust foliage-gleaner with stout black bill, distinctive ochraceous throat and breast streaking. **Adults:** pileum dark olive-brown, broadly and indistinctly scaled with blackish; back and shoulders rich brown, shading to dark chestnut on remiges and rump, and dark rufous-chestnut on tail-coverts and tail; sides of head dusky, streaked finely with ochraceous-buff; throat and sides of neck ochraceous, trailing off into ochraceous streaking that is heavy on breast, fading out on belly; remaining underparts tawny-olive to olive-brown, darkest on flanks. Legs dull olive. **Young:** similar but throat and breast streaking paler, more buffy, the streaking continuing onto belly.

HABITS. In cool, wet, epiphyte-laden mountain forests and adjacent tall, dense second growth, especially in deep, dark ravines; actively gleans and rummages in dense undergrowth, vine tangles, accumulations of debris lodged in vegetation, bromeliads and other epiphytes, for beetles, roaches, katydids, caterpillars, and other insects, spiders, frogs, salamanders, and small lizards; usually solitary but often joins mixed flocks of antwrens, warblers, bush-tanagers.

VOICE. Call a loud, grating *zeck*; song a burry *chi-wáwr, chi-wówr* with the rolling quality of call of Boat-billed Flycatcher; members of mated pair call back and forth with a hard, grating, ascending chatter or twitter.

NEST. In a burrow in a steep bank, about 2ft (60cm) long, a broad, shallow mat or saucer of dark fibrous rootlets. Eggs 2, white. February–August.

STATUS. Uncommon resident of upper Cordillera de Guanacaste and Cordillera de Tilarán, more common on Caribbean slope of Cordillera Central and along both slopes of Cordillera de Talamanca, including Dota area; from 2300ft (700m) on Caribbean slope and 4000ft (1200m) or more on Pacific, up to 8200ft (2500m), rarely to 10,000ft (3000m).

RANGE. Costa Rica and W Panama.

BUFF-THROATED FOLIAGE-GLEANER

Automolus ochrolaemus Pl. 30(3)

Hojarrasquero Gorgianteado

DESCRIPTION. 7¼″ (18.5cm); 42g. Bright buffy throat and spectacles, plain underparts distinctive; larger, stouter, more a thicket-haunter than Spectacled Foliage-gleaner. **Adults:** upperparts dark olive-brown, darkest on pileum, brightening to dark chestnut on remiges and to chestnut-rufous on tail-coverts and tail; sides of head dark brown with sparse buffy streaks on auriculars; lores, eye-ring, and streak from above eye back over auriculars rich buff; throat rich buff, with some dusky scaling on foreneck; remaining underparts buffy-brown medially, dark brown on sides and flanks; sometimes a suggestion of indistinct buffy streaks on chest. Culmen blackish, shading to yellowish-horn on tomia and lower mandible; legs olive. **Young:** crown tinged with chestnut; facial area tinged with rufous; spectacles faint; feathers of lower throat and chest with broader olive to sooty fringes, producing a mottled or scaled effect.

HABITS. Travels singly or in pairs through lower levels of wet forest and adjacent old, shady second growth and cacao plantations, often with mixed flocks; rummages for insects, spiders, small frogs and lizards in curled dead leaves and accumulations of trash above ground in thickets and vine tangles; often hangs inverted; sidles up thin stems, creeps along branches; occasionally drops to ground and flicks aside leaves with bill; often depresses tail; puffs out throat in excitement.

VOICE. Call a nasal, burry *rack*. On Pacific slope, song a loud, harsh rattle lasting 2–5sec and repeated 10–30 times/min at dawn in breeding season; or lower and more prolonged, sometimes for 45sec without interruption. In Caribbean lowlands, song is sharper, higher, more rapid and whinnying, usually falling and slowing toward end, with peculiar nasal twang or metallic ring, at dawn and sometimes at dusk; beats time with tail.

NEST. At end of burrow 18–30″ (46–76cm) long, in earthen bank beside stream or trail, a broad, shallow bowl or pad of curving rachises of compound leaves. Eggs 2–3, white. February–May.

STATUS. Common resident in lowlands and foothills of Caribbean and S Pacific slopes up to 4000ft (1200m) locally; barely reaches N Pacific slope through low passes in Cordillera de Guanacaste.

RANGE. S Mexico to NW Ecuador, N Bolivia, and Amazonian Brazil.

NOTE. Also called Buff-throated Automolus.

RUDDY FOLIAGE-GLEANER

Automolus rubiginosus Pl. 30(5)

Hojarrasquero Rojizo

DESCRIPTION. 8″ (20cm); 52g. Robust, strong-billed, dark, with bright ochraceous throat, no prominent streaking. **Adults:** pileum and hindneck chestnut, back and wing-coverts dark olive-brown, passing to chestnut-brown on remiges and deep chestnut on tail-coverts and tail; sides of head dark brown with fine, pale shaft-streaks on auriculars; cheeks rufous; chin, throat, and chest bright ochraceous to cinnamon-rufous, passing to tawny-olive on posterior underparts; flanks and thighs dark brown; pale shaft-streaks may show on chest. Upper mandible black, lower dusky-horn; legs grayish-olive. **Young:** throat paler, especially anteriorly; chest with indistinct buffy streaking; indistinct dark scalloping over much of underparts.

Habits. Prefers understory thickets in forest, along edges, and adjacent shady second growth, especially in ravines; rummages and gleans actively among dead leaves and trash caught up in branches and vines; sidles up vertical stems and vines; pecks at decaying branches; sometimes descends to ground to push aside litter with bill; occasionally forages with army ants; takes cockroaches, katydids, beetles, spiders, small frogs and lizards; paired year-round.

Voice. Call a nasal *ta-whoip* or *ka-whick*; a sharp, descending, nasal *nyeeaah* or *knaaayr knaayr*, between a whine and a scream; a staccato, nasal chatter.

Nest. A shallowly concave mat of fine rootlets, lined chiefly with soft, yellowish-brown plant fibers, at end of burrow ca. 6ft (2m) long in bank of dark ravine. Eggs 2, white. May (in Mexico; no Costa Rican record).

Status. Locally common resident around Coto Brus valley on Pacific slope near Panama border, 3600–4600ft (1100–1400m); reported from Cordillera de Tilarán above Monteverde at ca. 5000ft (1500m), but confirmation required.

Range. C Mexico discontinuously to W Ecuador, NW Bolivia, and Guianas.

Note. Also called Ruddy Automolus.

TAWNY-THROATED LEAFTOSSER

Sclerurus mexicanus Pl. 30(11)

Tirahojas Pechirrufo

Description. 6″ (15cm); 28g. Stout-bodied, small-headed, with short, rounded tail and long, slender bill. **Adults:** above dark brown, darkest on crown; rump and upper tail-coverts dark rufous-chestnut; tail blackish; sides of head blackish-brown; chin buffy, throat cinnamon-rufous, shading through dark rufous on chest to dark olive-brown on rest of underparts. Upper mandible black, shading to dusky tip; lower flesh-color, tip dark horn; legs blackish. **Young:** duller, above more sooty; throat paler, with fine dusky scaling on lower throat and breast.

Habits. Hops over ground in wet mountain forests, tossing leaves aside with bill or pulling them toward itself to expose beetles, ants, roaches, other insects, and spiders; pecks into decaying wood for grubs; solitary or in pairs; when alarmed, gives a double wing-flick.

Voice. Call a sharp, squeaky *zick* or *tseek*; song a series of 4–5 thin, slurred, progressively lower-pitched and shorter whistles: *pseeer-pseeer-pseer-psee-pse*.

Nest. In chamber at end of burrow 20″ (50cm) long and 4″ (10cm) in diameter, a loose cup of dry twigs. Eggs 2, dull white. December–April.

Status. Uncommon resident of middle elevations the length of both slopes, between ca. 2300 and 5000ft (700–1500m) on Caribbean slope and 3300–6000ft (1000–1850m) on Pacific, except absent from Valle Central and facing slopes.

Range. SE Mexico to Peru, N Bolivia, and E Brazil.

Note. Members of the genus *Sclerurus* are sometimes called by the misleading name leafscraper.

GRAY-THROATED LEAFTOSSER

Sclerurus albigularis Pl. 30(10)

Tirahojas Gargantigrís

Description. 6¾″ (17cm); 38g. Largest leaftosser; distinctive contrast between gray throat and rufous breast. **Adults:** pileum dark brown, faintly scalloped with blackish; back and wing-coverts dark chestnut-brown, shading to dark brown on remiges and chestnut on rump and upper tail coverts; tail blackish; sides of head dark grayish-olive, shading to gray on throat; breast dark rufous; belly sooty, washed with dark olive; sides, flanks, and crissum dark chestnut-brown. Upper mandible black, lower horn-color with blackish tip; legs brownish-black. **Young:** similar but breast duller and darker; throat darker, scaled with dusky.

Habits. Hops over ground in wet mountain forests with head erect and tail low, stopping to flick leaves rapidly aside and probe muddy ground with bill; often seems to prefer trails or other open spots on forest floor for foraging; eats beetles, spiders, roaches, other invertebrates, tiny frogs; when disturbed, flies with a squeak to low perch, sits with depressed tail, flicking wings nervously; often sympatric with Tawny-throated Leaftosser.

Voice. Call a squeaky *swick*, softer and less sharp than note of Tawny-throated; song (in Trinidad; ffrench) a phrase of 5 syllables, the first note soft and low, the next 2 louder and higher, the last 2 weaker, shorter, and sharper, the whole repeated over and over.

Nest. A few midribs of leaves on floor of nest chamber at end of curving burrow 12–18″ (30–46cm) long in vertical bank, often above a trail. Eggs 2, white (in Trinidad; no Costa Rican record).

Status. Uncommon resident of wet middle elevations, primarily on Caribbean slope at ca. 2000–5000ft (600–1500m), in Cordilleras de Guanacaste, de Tilarán, and Central; reaches Pacific slope in upper parts of N cordilleras; not yet recorded from Cordillera de Talamanca.

Range. Costa Rica to N Bolivia, Guianas, Trinidad, and Tobago.

SCALY-THROATED LEAFTOSSER

Sclerurus guatemalensis Pl. 30(9)

Tirahojas Barbiescamado

Description. 6½″ (16.5cm); 35g. Only leaftosser with patterned throat, without chestnut rump. **Adults:** above dark brown, more olive-brown on wings; tail blackish; sides of head more olive-brown with buffy shaft-streaks; throat white, scaled with blackish; chest feathers cinnamon-buff to tawny-olive with dusky edges, fine buffy shaft-streaks; posterior underparts dark olive-brown, tinged with sooty on belly. Bill black except base of lower mandible pinkish; legs dusky. **Young:** throat buffy, heavily and indistinctly scaled or streaked with dusky; chest more ochraceous; below rich brown scaled indistinctly with blackish.

Habits. Prefers mature wet forest with sparse undergrowth; hops over ground, squats with flexed legs, tail spread and braced on ground, rapidly flicking leaves right and left with bill to expose small invertebrates; when disturbed, often rises to low perch with a sharp squeak, then darts away through the undergrowth; sometimes clings upright to tree trunk near ground, using tail as a prop, but does not climb; continually gives nervous wing-flicks while foraging.

Voice. Call a sharp, sometimes metallic *pick* or *zick*; song, delivered by both sexes, a rapid series of 10–15 sharp, strong, clear whistles, the first 4–5 rising in pitch, the last ones gradually descending; or a longer series lasting 5–6sec that rises progressively in pitch, and may be repeated with scarcely a pause for minutes on end.

Nest. A loose, shallow bowl of rachises of compound leaves, in chamber at end of curving burrow 21–32″ (53–81cm) long; in vertical bank beside stream or trail, or in wall-like mass of clay heaved up by roots of fallen tree. Eggs 2, white. May, October, December.

Status. Uncommon resident of wet lowlands and foothills, to 2600ft (800m) throughout Caribbean slope, barely reaching Pacific slope through low passes in N cordilleras; and on S Pacific slope, to 3300ft (1000m) locally.

Range. S Mexico to Colombia.

PLAIN XENOPS

Xenops minutus Pl. 29(3)

Xenops Común

Description. 4¾″ (12cm); 12g. Small; body unstreaked, with short, sharp, slightly upturned bill, crescent-shaped silvery malar stripe; wings and tail boldly patterned. **Adults:** above olive-brown, with fine, buffy shaft-streaks on crown; tail-coverts, central and 2 outermost rectrices cinnamon-rufous, rest blackish; secondaries cinnamon-rufous at base and tip, black medially; primaries and primary-coverts with medial portion cinnamon-rufous, rest blackish; bases of inner webs of secondaries and inner primaries ochraceous-buff, producing conspicuous stripe on spread wing; face olive-brown with buffy streak from above eye over auriculars, buffy mottling on cheeks; underparts pale brownish-olive. Upper mandible black, lower grayish; legs gray. **Young:** resemble adults but throat darker buff, with heavier dark smudging.

Habits. In lower and middle levels and along edges and breaks in wet forest, in taller second-growth woods, shady plantations, and clearings with scattered trees; hops and hitches sideways along slender hanging vines without using tail for support or clings, titlike, in all attitudes to slender decaying twigs, vines, and large petioles, pecking into center for eggs, larvae, and mature insects like earwigs, small katydids, ants, and termites; also probes dead rolled leaves; solitary or in pairs, regularly in mixed-species flocks including antwrens, antvireos; adults sleep singly in cavities in trees.

Voice. Call a high, sharp, piercing *peet* or *cheet*; song a fine, sharp, rapid trill resembling that of Olivaceous Piculet, repeated for many minutes at dawn in the breeding season, and often preceded by 1–2 call notes.

Nest. A shallow hole in a decaying trunk or limb, with neat, round entrance 1″ (2.5cm) in diameter, carved by piculets or by the birds themselves, well lined with shredded bast fibers, 5–30ft (1.5–9m) up. Eggs 2, white. January–May.

Status. Uncommon to locally common resident of humid lowlands and foothills on both slopes; rare and local in dry NW, where confined to moist evergreen gallery forest; sea

level to 3600ft (1100m) on Caribbean slope and to 5000ft (1500m) on Pacific.
Range. S Mexico to W Ecuador, NE Argentina, and C Brazil.

STREAKED XENOPS
Xenops rutilans Pl. 29(2)
Xenops Rayado
Description. 5″ (13cm); 13g. Similar to Plain Xenops but below broadly streaked; white malar mark less conspicuous. Pileum, hindneck, and back olive-brown, finely streaked with buffy; rump, upper tail-coverts, and most of tail cinnamon-rufous, the inner webs of rectrices 2–4 largely black; wings with cinnamon-rufous edgings, the bases of inner webs of inner primaries and secondaries, also wing-linings ochraceous-buff; sides of head dusky, with narrow whitish superciliaries and fine whitish streaking on auriculars; throat dull white; rest of underparts pale grayish-brown, tinged with rufous posteriorly, and streaked with dull white, most broadly on chest. Upper mandible black, lower flesh-color; legs gray.
Habits. Forages like Plain Xenops of lower elevations but is more confined to forest, where it more often frequents high canopy but is also found lower at edges and in second growth; regularly joins mixed-species flocks; in zone of overlap (especially in SW) often in same flock as Plain Xenops, and foraging in the same manner.
Voice. Call a high, thin, sibilant *tsip* or *zis*, sometimes repeated several times rapidly on same pitch; song (in Trinidad, ffrench) a rapid series of shrill twittering notes rising to a climax and then diminishing, sometimes with a final distinct note.
Nest. A natural cavity in a tree, lined with a few petioles and roots, 8–15ft (2.5–4.5m) up. Eggs 2, white (in Trinidad; no Costa Rican record).
Status. Rare resident on Caribbean slope of Cordillera Central, more common along both slopes of Cordillera de Talamanca, including Fila Cruces, from ca. 2600ft (800m) on Caribbean slope and 4000ft (1200m) on Pacific up to ca. 8000ft (2450m).
Range. Costa Rica to N Argentina and Guianas.

FAMILY **Formicariidae:** Antbirds

About 250 species of antbirds inhabit the tropical American mainland and nearby islands but are absent from the Antilles. Birds of warm forests and thickets, they are most abundant in the vast forests of Amazonia and the Guianas; only a few live at cool heights. In the forest, most stay low; some forage at midheights. Although many antbirds join mixed flocks, they do not congregate in large groups of their own species; when not nesting they live in pairs, family groups, or singly. In size, antbirds are about 3–14″ (7.5–36cm). Finding no vernacular designations for them, naturalists have given them names of more familiar birds of other lands. The smallest, known as antwrens, flit through trees and shrubbery much like warblers. Next in size are antvireos, who search through foliage more deliberately. Many middle-sized species are called simply antbirds. Some middle-sized to large species, with a little hook at the tip of a strong bill, are known as antshrikes. Antthrushes are fairly large and forage on the ground, over which they walk like small rails. Plump, short-tailed, long-legged antpittas hop over the ground.

Shade dwellers, antbirds are clad in black, gray, olive, buff, rufous, and brown, often with large areas of white; some are conspicuously barred or streaked. In many, the dark feathers of the back, shoulders, crown, or leading edges of the wings have white bases that form conspicuous patches when the feathers are spread in moments of excitement. The eyes are often offset by brightly colored, naked skin. Although not brilliant songsters, many antbirds have attractive utterances that usually consist of the repetition of similar notes, forming a trill, a roll, or a series of clear whistles.

The family receives its name from the few species that regularly follow army ants, catching small fugitives but not the ants themselves. Antbirds are almost wholly insectivorous, but some of the larger species also take small frogs, lizards, snakes, and fruits. The usually cuplike nest is attached by its rim to a horizontal fork of a tree or shrub; a few nest in holes in trees. They lay 2 eggs, rarely 3. Male and female share almost equally in nest building, incubating, and caring for the young. Most hatchlings are naked, but those of hole-nesting antthrushes are densely covered with dark down. Incubation periods are 14–20 days, and nestling periods 9–18 days, both longest in cavity nesters.

FASCIATED ANTSHRIKE

Cymbilaimus lineatus Pl. 31(1)

Batará Lineado

Description. 7″ (18cm); 41g. Heavy-bodied and bushy-crowned, with a very stout, sharply hooked bill; plumage finely barred above and below. **Adult ♂:** above black, very narrowly barred with white everywhere except crown and hindneck; below finely and evenly barred with black and white. ♀: crown chestnut-rufous; forehead and rest of upperparts black, thickly barred with buff; below paler buff, especially on throat, barred heavily with black on sides of breast, more lightly on throat and center of breast and belly. Iris red; upper mandible black, lower mandible and legs pale blue-gray. **Young:** resemble adult female but crown barred with black; markings on wings and tail whitish; below paler, throat and breast often whitish, and black barring fainter and less distinct.

Habits. Prefers thickets and vine tangles at low to middle heights around gaps in forest, along streams and edges, and in tall second growth; sluggish and furtive, sits for long periods peering about, then makes sudden sally or hops rapidly and heavily to snatch crickets, katydids, beetles, bugs, caterpillars, or other insects, or spiders and small lizards from vegetation; almost always hidden in vegetation; flies rapidly but not strongly with quick, buzzy wingbeats between thickets; paired throughout year.

Voice. Song clear, ventriloquial, upward-slurred whistles, the first 3–4 successively higher in pitch, the last 1–2 sometimes lower; also a sputtering series of hard, heavy, dry, metallic notes, sometimes speeding up into a chatter or interspersed with clear, descending whistles, and often followed by a little whine.

Nest. Cup attached by rim in a horizontal fork, of dark fibers so loosely woven that light passes through, 7–25ft (2–7.5m) up amid foliage. Eggs 2, creamy-white, spotted with chocolate-brown and pale lilac. April–June.

Status. Fairly common resident throughout Caribbean lowlands, except uncommon to rare in drier areas just S of Lago de Nicaragua; to 2000ft (600m) along Cordillera de Guanacaste and to 4000ft (1200m) in Cordilleras Central and de Talamanca.

Range. E Honduras to N Bolivia and Amazonian Brazil.

GREAT ANTSHRIKE

Taraba major Pl. 31(2)

Batará Grande

Description. 8″ (20cm); 75g. Large, stout, bicolored, with bushy crest, vivid red eyes, and heavy black bill abruptly hooked at tip. **Adult ♂:** above black, including sides of head and neck; narrow wing-bars white; a concealed patch of white in middle of back; below white except flanks sooty, lower belly and crissum black. ♀: pattern like male, but rufous-chestnut replaces black; wing-bars faint or absent; lores and orbital area blackish; underparts often tinged with buff. Legs bluish-gray. **Immatures:** resemble adults but sides of breast finely barred or scaled with sooty; reduced white patch on back. **Young ♂:** above sooty-black, indistinctly barred with dull rufous; below narrowly barred with blackish, the breast and sides tinged with grayish-buff, vent and crissum barred with buff. ♀: above broadly and indistinctly barred with blackish; below buff, barred indistinctly with blackish on lower throat, breast, and sides. Iris brown.

Habits. Lurks in impenetrable young second-growth thickets with rank grasses, vines and tangled shrubbery, stands of great-leaved herbs (*Heliconia, Calathea*), bamboo thickets and canebrakes along streams; usually stays low but may rise above densest growth to sing, especially in early morning; alarmed, slinks silently away in dense cover, rarely flying; loudly rustles dry curled leaves, seeking adult insects and larvae; gleans insects, spiders, small lizards from foliage; usually in pairs.

Voice. A deep, throaty rattle, *shrrrrw*; a prolonged, deep *churr*. Song a dry, nasal trill or roll that accelerates in "bouncing-ball" pattern and ends with scratchy, squalling, nasal whine: *tok tok too to to to t' t' t' trrr-waaanh.*

Nest. A bulky cup suspended by rim; frame of coarse dry vines supports middle layer of strips of monocot leaves, lined with thinner vines; 5–8ft (1.5–2.5m) up in thicket, often near edge. Eggs 2, dull white with blackish and chocolate blotches, spots and crisscross streaks of purplish-brown and pale lilac. April–July.

Status. Common resident of wet lowlands and foothills, locally to 3300ft (1000m), on Caribbean slope and S Pacific slope, N to Carara; rare and local N to Puntarenas area.

Range. SE Mexico to NW Peru, NE Argentina, and E Brazil.

BARRED ANTSHRIKE

Thamnophilus doliatus Pl. 31(3)

Batará Barreteado

Description. 6¼″ (16cm); 28g. Smaller, with hooked bill less heavy than Fasciated

Antshrike; iris pale yellow, bushy crest. **Adult** ♂**:** crown feathers black with white bases; rest of upperparts black, coarsely barred with white; sides of head and throat indistinctly streaked with black and white; rest of underparts broadly, evenly barred with black and white. ♀: crown rufous-chestnut; rest of upperparts rufous; sides of head and nuchal collar indistinctly streaked with black and white; below buff, paler on throat and belly; sides of throat and foreneck sparsely flecked with black; breast with faint sooty scaling or spotting. Upper mandible blackish, lower bluish-gray; legs plumbeous. **Young** ♂**:** above and below barred with buff and sooty-black, less distinctly than adult. ♀: like adult but above and below more or less barred with sooty-black.

Habits. In humid areas prefers low, second-growth thickets and hedgerows; in drier areas inhabits understory of evergreen gallery forest, and (in wet season) deciduous forest; hops and flits heavily as it gleans beetles, ants, bugs, grasshoppers, caterpillars, other insects, and spiders in thick vegetation; usually in pairs.

Voice. Soft, slurred, whistling notes; a snarling nasal *naah* or *chaar* in alarm or agitation. Song a loud, wooden or nasal rattling trill that descends and accelerates, usually ending in an abrupt, barking, upwardly inflected *wank*; male and female often sing responsively, female's voice higher-pitched.

Nest. A deep, thin-walled cup fastened by rim in horizontal fork; of fine vines, tendrils, fungal rhizomorphs, often a few tufts of green moss on outside; 3–30ft (1–9m) up in thicket. Eggs 2, rarely 3, white, with scratches, blotches, and speckles of chocolate or purplish-brown, over whole surface or concentrated at thicker end. January–June.

Status. Resident throughout lowlands of both slopes, up to ca. 4600ft (1400m) in Valle Central, including Cartago area; uncommon and local in humid regions; common in dry NW and lowlands S of Lago de Nicaragua.

Range. NE Mexico to NW Peru and NE Argentina.

BLACK-HOODED ANTSHRIKE

Thamnophilus bridgesi Pl. 31(6)

Batará Negruzco

Description. 6½″ (16.5cm); 27g. Very dark, rather long-tailed antshrike of southern Pacific slope. **Adult** ♂**:** entire upperparts, head, and chest black, shading to blackish-slate on rest of underparts; wing-coverts and outer 3 rectrices with small white spot at tip; wing-linings and concealed patch at center of back white. ♀: head and neck black; top and sides of head finely streaked with white; rest of upperparts dark sooty-brown; wing-coverts and tail black; wing-coverts and outer 2–3 rectrices with white spot at tip; rest of underparts dark grayish-olive; throat, breast, and crissum broadly streaked with dull white. Bill black; legs dark gray. **Immature** ♂**:** like adult but duller, foreneck and breast narrowly streaked with white. ♀: below browner than adult.

Habits. Prefers thickets in and along edges of forest, in second growth of all ages, sometimes mangroves; forages by gleaning insects and spiders from leaves and twigs, moving with abrupt, heavy hops and flits; usually in pairs; sometimes notably tame and curious.

Voice. A loud series of 6–10 nasal barks all on same pitch. Song a low, rather wooden, accelerating trill, usually all on 1 pitch, ending with a downward-inflected bark: *cow cow cow co co k'k'k'COW*; a far-carrying *cack cack cack*. As in other antbirds, the tail beats time to the notes.

Nest. A roomy cup attached by rim in horizontal fork, the thin fabric of fine, dark fibers decorated with a few tufts of green moss, 2–12ft (0.6–3.6m) up in thicket at edge or gap in forest. Eggs 2, dull white, wreathed with bright brown and pale lilac blotches and spots. February–August.

Status. Common resident of S Pacific lowlands and foothills to 3600ft (1100m), N to Carara area; uncommon to rare and local farther N, to Volcán Tenorio.

Range. Costa Rica and W Panama.

SLATY ANTSHRIKE

Thamnophilus punctatus Pl. 31(5)

Batará Plomizo

Description. 5½″ (14cm); 24g. Relatively small and heavy-billed; boldly marked wings and tail. **Adult** ♂**:** above black, sides of the crown and back with slaty fringes; scapulars and tertials edged with white; wing-coverts and rectrices tipped with white; a concealed white patch in center of back; rest of plumage dark slate. ♀: above olive-brown, tinged with rufous on crown; wings and tail blackish-brown, the wing-coverts and rectrices tipped with pale buff; below paler olive-brown, smudged with gray on throat. Upper mandible black, lower gray with black tip; legs gray. **Young** ♂**:** above dull grayish-brown with paler fringes; below pale olive-brown, shading to pale gray on throat and belly, with pale buffy fringes; wing-coverts with buffy spot at tip. ♀: similar but above more cinnamon-brown, below paler and more buffy.

Habits. Frequents thickets in forest understory and edges, old second growth, and semi-open; partial to shady cacao plantations; from a low perch scans deliberately for prey, attacks with a series of swift, heavy hops or a short, fluttering sally; takes surprisingly large insects and spiders, including walking sticks, katydids, and cicadas, also small lizards; beats and mandibulates large prey before swallowing; usually in pairs that sometimes join mixed flocks of small birds; occasionally forages with army ants; in aggressive interactions erects back feathers revealing white.

Voice. A low, accelerating wooden roll or barking trill that ends with an abrupt, loud bark with upward inflection: *ri-ri-ri-r-r-r-r-r-r wank*; alternatively, a barking note with downward inflection may precede an ascending roll; contact notes include softer versions of the roll and a series of fine, slurred whistles.

Nest. A cup of dark fibers suspended by rim in horizontal fork, so thin that eggs are often visible through bottom; outside more or less decorated with green moss, 3–12ft (1–3.6m) up. Eggs 2, whitish, heavily spotted and blotched with shades of brown, especially on large end. January–September.

Status. Common to abundant resident throughout Caribbean lowlands, to 2300ft (700m) in N and 3300ft (1000m) in SE.

Range. Guatemala to W Ecuador, N Bolivia, and SE Brazil.

RUSSET ANTSHRIKE

Thamnistes anabatinus Pl. 31(4)

Batará Café

Description. 5½″ (14cm); 21g. Small arboreal antshrike whose colors suggest an ovenbird, but with heavy hooked bill, vireo-like behavior. **Adults:** above tawny-brown to olive-brown, shading to tawny-rufous on wings and rufous on upper tail-coverts and tail; broad eye-stripe and indistinct malar stripe olive-brown; superciliaries, cheeks and throat olive-buff, shading to brighter olive on breast and grayish-olive on belly and flanks; male has concealed patch of cinnamon-orange on center of back. Iris brick-red; upper mandible black, lower pale gray; legs olive-green. **Immatures:** similar to adults but throat paler, more buffy; wing-coverts with broad rufous fringes.

Habits. Travels through upper understory to lower canopy of forest, forest edge, also in adjacent tall second growth or semi-open, usually accompanying mixed flocks of tanagers, greenlets, warblers, etc.; alone or in pairs; gleans and probes foliage at tips of branches, hopping deliberately and scanning, sometimes hanging from twigs, inspecting leaf bases and pulling apart leaf clusters and curled leaves to obtain a wide variety of adult and larval insects (beetles, bugs, katydids, cicadas), and spiders.

Voice. A variety of squeaky whistled notes, especially an upslurred *sweek* and a doubled *sweesik*; song a rapid sequence of plaintive notes that gets louder, then softer: *cheep cheep CHEEP CHEEP cheep*.

Nest. A deep cup fastened in horizontal fork by rim, of pieces of dead leaves bound together by fine, dark fungal rhizomorphs, 23–50ft (7–15m) up at forest edge or in semi-open. Eggs 2, dull white, speckled with brown. April–June.

Status. Uncommon to fairly common resident of Caribbean foothills and adjacent lowlands, becoming very uncommon in coastal districts and in drier lowlands S of Lago de Nicaragua; up to 3000ft (900m) in NW, and 5000ft (1500m) in SE; on S Pacific slope common N to about Quepos, and very uncommon N to Carara; sea level to 5000ft (1500m) in upper Coto Brus valley.

Range. S Mexico to N Bolivia.

PLAIN ANTVIREO

Dysithamnus mentalis Pl. 32(6)

Batarito Cabecigrís

Description. 4½″ (11.5cm); 14.5g. Small, stout, short-tailed, big-headed, with rather heavy, slightly hooked bill. **Adult ♂:** top and sides of head dark slate, a stripe above auriculars slightly paler; rest of upperparts dark grayish-olive; 2 narrow white wing-bars; concealed white patch on shoulder; throat pale gray; breast, sides, and flanks grayish-olive; belly and crissum pale yellow; wing-linings white. ♀: pileum tawny-rufous; rest of upperparts olive-brown; sides of head washed with grayish, with narrow white eye-ring; throat pale gray, tinged with olive; breast, sides, flanks pale brownish-olive; belly pale yellow. Upper mandible black, lower mandible and legs pale gray. **Immatures:** resemble adults but below paler, throat whitish and chest yellow like belly; remiges of male browner; female has wing-coverts edged with tawny, eye-ring buff.

Habits. Frequents understory of mature forest or adjacent tall second growth; in pairs, family groups, or less often alone; sometimes joins mixed-species flocks or is joined by a small furnariid or woodcreeper; hunts busily by quick hops and short sallies, pausing now and then to scan surrounding vegetation; gleans beetles, moths, ants, spiders, katydids, from foliage and twigs.

Voice. Contact note a soft, clear, slurred *tew-tew*; a protracted, mournful *cher cher cher* . . . when disturbed. Song a series of clear whistled notes that rises, then falls and accelerates into a roll with a wooden quality; also gives a melodious, rippling series of 4–6 successively higher clear notes.

Nest. A deep, thin-walled cup, suspended by rim in horizontal fork, of dark fibrous rootlets and fungal filaments, usually decorated on outside with moss, 2–7ft (0.6–2m) up, in forest. Eggs 2, dull white, flecked and blotched with purplish-brown, most heavily on large end. February–July.

Status. Common resident of humid middle elevations of both slopes of N cordilleras and Cordillera de Talamanca, Caribbean slope of Cordillera Central, chiefly 2300–5000ft (700–1500m), up locally to 8200ft (2500m), mainly on Cordillera de Talamanca and Dota region; uncommon in General and Coto Brus valleys and in Golfo Dulce region; strays to lowlands on Caribbean slope rarely, perhaps seasonally.

Range. S Mexico to NW Peru and NE Argentina.

STREAKED-CROWNED ANTVIREO

Dysithamnus striaticeps Pl. 32(7)

Batarito Pechirrayado

Description. 4¼″ (11cm); 17g. Chubby body, notably short tail and heavy bill hooked at tip, below heavily streaked. **Adult ♂:** top and sides of head slate-gray, pileum broadly streaked with black; upperparts grayish-olive; 2 narrow, often broken wing-bars; lesser wing-coverts black, spotted with white; sides of neck and breast slaty; flanks olive-brown; rest of underparts white, tinged with buff on lower belly; throat and breast heavily streaked with blackish. ♀: pileum rufous, lightly streaked with black; rest of upperparts olive-brown; pale wing-bars and shoulder-spots buffy; cheeks mottled with buffy and dusky; a narrow white eye-ring; throat whitish; rest of underparts buffy, palest on belly and washed with olive on sides and flanks; throat and breast streaked with dusky. Iris grayish; upper mandible black, lower mandible and legs gray. **Immature ♂:** resembles adult female but above grayer, below whiter.

Habits. Pairs or, less often, single birds travel with mixed-species flocks of antwrens, greenlets, etc., through understory of mature forest and adjacent tall second growth; snatches insects and spiders from foliage by sudden hops and sallies, usually after scanning about deliberately from perch.

Voice. Contact notes soft, slurred, clear, single or double whistles; complaint a low *pur-r-r* interspersed with slight, mournful, whistled notes. Song a series of rich whistled notes that rises, grows louder, then falls, accelerating the while and ending in a roll, longer and more complex than song of Dusky Antbird.

Nest. A thin cup of dark fibers attached by rim in slender horizontal fork, outside lightly decorated with green moss, 5–14ft (1.5–4m) up. Eggs 2, whitish, blotched and flecked with chocolate, very heavily on large end. March–June.

Status. Common resident of N Caribbean lowlands and foothills, becoming scarce at its southern limit S of Limón; locally up to 2600ft (800m) on Cordillera de Guanacaste, barely overlapping its Pacific slope; and to 2300ft (700m) on Cordillera Central.

Range. E Honduras to Costa Rica.

SPOTTED-CROWNED ANTVIREO

Dysithamnus puncticeps Pl. 32(8)

Batarito Cabecipunteado

Description. 4½″ (11.5cm); 17g. Told from closely similar Streaked-crowned Antvireo, which it replaces in extreme SE Costa Rica, by spotted crown and more lightly streaked underparts. **Adult ♂:** pileum slate-gray and black, spotted with white; rest of upperparts slate-gray, tinged with olive on rump; narrow white wing-bars; sides of head slaty, cheeks barred or mottled with white; below white, finely streaked with dusky on breast; flanks brownish; lower belly buff. ♀: pileum tawny-rufous, with indistinct brighter spots and fine black streaks; rest of upperparts grayish-olive; narrow buffy wing-bars; face brownish, mottled with dusky and whitish on cheeks; a narrow whitish eye-ring; throat whitish; breast buffy, finely streaked with dusky; belly whitish; sides and flanks brownish mixed with buff. Iris gray; upper mandible black, lower mandible and legs bluish-gray.

Habits. Prefers understory of humid forest, traveling in pairs or family groups, sometimes joining mixed-species flocks; foraging habits resemble those of Streaked-crowned Antvireo but reported to walk over ground sometimes.

Voice. A short, rolling, descending *prrrr*; faint chirps. Song a series of clear whistled notes, the first 3–4 ascending, then several on same pitch, then an accelerating, descending series ending in a roll or trill, resembles slightly less melodious song of Streaked-crowned.

Nest. A cup suspended in horizontal fork by rim, ca. 7ft (2m) up in undergrowth. Eggs 2. April–July (in Panama; no Costa Rican record).

Status. Common resident of lowlands and foothills to 2600ft (800m) in Sixaola region near Panama border on Caribbean slope, sparingly N to around Cahuita (Fila Carbón).

Range. Costa Rica to W Ecuador.

CHECKER-THROATED ANTWREN

Myrmotherula fulviventris Pl. 32(5)

Hormiguerito Café

Description. 4″ (10cm); 10.5g. Dull brownish with buffy wing-bars and notably heavy bill, iris often pale; male has checkered throat. **Adult ♂:** above olive-brown, forehead and face paler and somewhat mottled with dusky; upper tail-coverts and tail brighter, more cinnamon-brown; wing-coverts blackish, broadly tipped with buff; below paler, more buffy, with grayish wash across breast; throat black, heavily spotted with white. ♀: similar but throat buff, obscurely marked with black; center of breast strongly tinged with buff. Iris pale yellow to reddish-brown; upper mandible blackish, lower gray; legs slaty. **Young:** brighter, above more cinnamon-brown; wing-bars broad but less distinct; throat sooty-gray; rest of underparts bright cinnamon-brown, foreneck and breast faintly streaked with buffy-white.

Habits. In understory of mature wet forest and adjacent tall, shady second growth; usually in pairs or small groups that are nuclei of mixed flocks; actively gleans foliage, often hanging acrobatically; frequently ransacks curled dead leaves and vine tangles, and investigates leaf-bases of epiphytic aroids; eats eggs, larvae, and adults of various insects and spiders. At territorial boundaries (as when a flock crosses), 2 males perch ca. 1ft (30cm) apart, lower their heads, fluff their plumage, and sway back and forth in unison but in opposite directions, singing loudly.

Voice. Contact note in flock a high, thin *cheep*, softer and less emphatic than alarm note; alarm note a high, sharp *peeesk* or *tseeet*. Song a long series of high, loud, emphatic notes: *tseek-tseek-tseek-tseek. . . .*

Nest. A deep pouch, 6″ (15cm), with oblique opening at top, of fine fungal filaments and rootlets supporting a thick layer of dead leaves, lined with fine fibers, suspended from terminal fork of thin, drooping twig, 16″–6ft (0.4–1.8m) up in forest undergrowth. Eggs 2, white, with fine spots and scrawls of reddish-brown and pale lilac, mostly in wreath around large end. March–August.

Status. Resident throughout Caribbean lowlands, common except in drier area S of Lago de Nicaragua, up to ca. 2300ft (700m) in foothills.

Range. NE Honduras to W Ecuador.

Note. Also called Fulvous-bellied Antwren.

WHITE-FLANKED ANTWREN

Myrmotherula axillaris Pl. 32(2)

Hormiguerito Flanquiblanco

Description. 3½″ (9cm); 8.5g. Small, with slender bill slightly hooked at tip; flank feathers notably long, pale, and silky. **Adult ♂:** above slate-black; wing-coverts and rectrices blacker and tipped with white spots; scapulars edged with white (usually concealed); below velvety black; flank plumes white. ♀: top and sides of head gray, shading to olive-brown over rest of upperparts; wing-coverts and rectrices tipped with buffy-olive; narrow eye-ring whitish; auriculars with fine white streaks; throat white; rest of underparts buffy, palest on flank plumes; sides of breast washed with gray-brown. Bill black (or lower mandible gray in female); legs plumbeous. **Immature ♂:** below slate-gray, flank plumes mixed with brownish and white; throat grizzled with white; wings as in young. **Young:** above cinnamon-brown; below paler and brighter, shading to whitish on throat and belly; wing-coverts tipped with small olive-buff spots; throat more or less barred with gray in male.

Habits. Usually in pairs, travels with mixed-species flocks of other antwrens, antvireos, greenlets, etc., through middle and upper understory of mature wet forest and adjacent tall shady second growth; warblerlike, actively gleans insects and spiders amid foliage; as flock crosses territorial boundary, members of neighboring pairs display at same sex, calling and turning from side to side, exposing flank plumes as they raise 1 wing at a time.

Voice. Usual call a whining, nasal *eer-uur* or *eer-eruuur*; low churring and twittering notes, including a sweet *cheeup cheeup*. Song, by both sexes in encounters, a descending series of clear, ascending, whistled *pwee* notes that may end in a little roll.

Nest. A deep cup, composed largely of dead leaves and leaf skeletons, held together, attached at rim, and lined with black fungal rhizomorphs, 8″–13ft (0.2–4m) up in forest.

Eggs 2, white, speckled with reddish-brown, chiefly in wreath at large end. March–July.

Status. Fairly common resident of Caribbean lowlands, except rare in drier areas S of Lago de Nicaragua, up to 2300ft (700m) in N, 3000ft (900m) in SE.

Range. NE Honduras to W Ecuador, N Bolivia, and SE Brazil.

SLATY ANTWREN

Myrmotherula schisticolor Pl. 32(3)

Hormiguerito Pizarroso

Description. 4″ (10cm); 9.5g. Rather slender and long-tailed; white on wing-coverts of male less extensive than in other antwrens; female lacks pale spots on wing-coverts, below brighter brown than other female antwrens. **Adult ♂:** dark slate; sides of head, throat, breast, and upper belly black; wing-coverts blackish, narrowly tipped with white; concealed white patch on shoulder. ♀: above olive-brown, wings and tail tinged with cinnamon; below buffy-brown, paler and more buffy on throat, and with ochraceous tinge on breast and belly. Upper mandible black, lower black in male, flesh-color in female; legs gray. **Immature ♂:** resembles adult female but duller, darker, more grayish; wing-coverts faintly tipped with olive-buff. Males sometimes breed before acquiring full black shield on throat.

Habits. Travels through understory of wet forest and adjacent tall second growth, usually in pairs, accompanying mixed-species flocks of antvireos, ovenbirds, woodcreepers, and *Basileuterus* warblers; in southwest often with Red-crowned Ant-Tanagers; actively gleans small insects and spiders from foliage, sometimes from undersides of branches that it bends over to inspect; probes rolled leaves and tufts of moss, often clinging acrobatically.

Voice. Usual call a squeaky, whining *cheeur, skeé yew* or *cheer-cheeur*, less nasal than similar note of White-flanked; a low, ascending *churr*; alarm note a sharp squeak. Song, rarely heard, a soft, low *t'weet t'weet t'weet t'weet weet weet weet weet.*

Nest. A rather deep cup of black fungal rhizomorphs, so thin that eggs are often visible through the fabric, suspended by rim in a slender horizontal fork of sapling, 3–7ft (0.9–2m) up in forest. Eggs 2, white or cream, blotched, speckled, or scrawled with reddish or purplish-brown, in wreath or all over. March–July.

Status. Fairly common resident of wet middle elevations the length of both slopes (thus excluding Valle Central and facing slopes), from ca. 2300ft (700m) in N and 3300ft (1000m) in SE to 5500ft (1700m) on Caribbean slope; on Pacific slope, only above ca. 3300ft (1000m) in NW, but down to sea level in Térraba and Golfo Dulce regions and up locally to 6500ft (2000m) on Cordillera de Talamanca.

Range. S Mexico to W Ecuador and E Peru.

RUFOUS-RUMPED ANTWREN

Terenura callinota Pl. 32(10)

Hormiguerito Lomirrufo

Description. 4½″ (11.5cm); 7g. Slender bill, long tail; suggests a small flycatcher or warbler in color and shape; rufous rump often concealed beneath wings. **Adult ♂:** pileum and narrow postocular stripe black; superciliaries white; back mostly olive-green, with concealed area of rufous and black; wings and tail blackish, edged with olive-green; 2 bold yellow wing-bars; a concealed yellow patch on shoulder; side of head, throat, and breast pale gray; belly pale yellow, tinged with olive. ♀: similar but pileum and eye-stripe olive-brown; no yellow in shoulder or rufous and black in back. Upper mandible black, lower pale gray; legs blue-gray.

Habits. Prefers middle levels of wet, mossy forests; an active, nervous gleaner of small beetles, caterpillars, crickets, other insects, and spiders, from foliage at tips of twigs; may cling acrobatically below twig but does not hover or sally; often flicks wings nervously; usually in pairs, which often accompany mixed-species flocks of warblers, tanagers, ovenbirds, etc.

Voice. Usually silent; occasionally an energetic series of squeaky notes, *schi-schi-schi*, sometimes interspersed with sibilant sputters.

Nest. Undescribed (?). Breeding indicated in March by female carrying nest material (moss).

Status. Locally uncommon to fairly common resident of lower middle elevations, 2500–3600ft (750–1100m), on Caribbean slope, from S end of Cordillera de Tilarán (Jamaical area) to Cordillera Central and probably locally S along Cordillera de Talamanca, at progressively higher elevations.

Range. Costa Rica to W Ecuador, E Peru, and Guyana.

DOTTED-WINGED ANTWREN

Microrhopias quixensis Pl. 32(1)

Hormiguerito Alipunteado

Description. 4¼″ (11cm); 8.5g. Warblerlike; slender bill; long, graduated, white-tipped tail; a broad white wing-bar below dotted shoulders; concealed white patches on shoulder and in center of back. **Adult ♂:** mostly velvety black; sides and flanks dark slate; middle and lesser coverts with white dot at tip, greater coverts broadly white-tipped; all but central rectrices with white spot at tip. ♀: top and sides of head and hindneck dark slate, shading to black on lower back and rump; below entirely rufous-chestnut, paler posteriorly. Bill black; legs blackish. **Immatures:** like adults but retain brown-edged remiges. **Young ♂:** above and below dark sooty-brown, shading to dull cinnamon-brown on belly; wing-coverts dusky, the greater coverts broadly and middle coverts very narrowly and indistinctly tipped with white; rectrices more pointed. ♀: like male but below much more extensively cinnamon-brown to dull rufous.

Habits. In pairs or family-size group that travels alone or with other antwrens in mixed-species flocks through understory of wet forest, or more often at gaps and along edges, in tall second growth and dense semi-open, including cacao plantations; forages mainly in thickets and vine tangles, deliberately gleaning insects and spiders from outer foliage, often in more exposed position than other antwrens; rarely follows army ants. At territorial boundaries rivals of same sex display at each other, erecting back feathers to expose white, swaying back and forth, and singing.

Voice. Usual call a mellow whistled *pyeer* or a sharper *peep*, sometimes combined: *peep peep a pyeer*; soft twitters and chirps; a plaintive *tweo* when distressed. Song a series of 5–8 progressively higher-pitched whistled notes leading up to a little falling trill or rattle: *chee chee chee chee chee che che chr'r'r'r*.

Nest. A pouch or deep cup attached by rim to slender twigs amid clustering foliage, of partly decayed leaves held together and lined by fine dark fibers, 3½–40ft (1–12m) up. Eggs 2, white, spotted all over with brown, more heavily blotched on thick end. January–August.

Status. Common resident throughout lowlands of Caribbean and S Pacific slopes (N sparingly to around Carara), up locally to 3300ft (1000m); reaches Pacific slope in NW through low passes.

Range. SE Mexico to W Ecuador, N Bolivia, and C Brazil.

Note. Also called Velvety Antwren.

DUSKY ANTBIRD

Cercomacra tyrannina Pl. 31(8)

Hormiguero Negruzco

Description. 5¾" (14.5cm); 18g. Medium-sized, with rather slender bill, long tail, and no skin exposed on face; like an oversized antwren. **Adult ♂:** above blackish-slate; below dark slate, tinged with olive on flanks; shoulders and center of back with concealed white areas; wing-coverts black, edged with white forming narrow wing-bars; outer 2–3 rectrices narrowly edged with whitish at tip. ♀: above olive-brown, tinged with cinnamon on wings and tail; wing-coverts sometimes edged narrowly with buff; below tawny-ochraceous to dull rufous, darkest on breast, tinged with olive on sides and flanks. Bill black, legs gray. **Immatures:** like adult female but darker, especially males, which often have some gray feathering on head and breast. **Young:** above and below dull brown; belly paler and more tawny to cinnamon-brown; upperparts, throat, and breast with narrow indistinct sooty fringes.

Habits. Prefers tall, impenetrable thickets at openings in forest, along edge, or beside forest in second growth and semi-open; in understory of evergreen gallery forest; paired throughout year; hops and creeps deliberately while scanning for prey; captures beetles, wasps, bugs, homopterans, caterpillars, and spiders after a quick flit or dash along perch, or a brief hover; does not associate with other birds.

Voice. Contact note *kick* or *kick-ew* with a somewhat nasal quality; alarm note a harsh, abrupt, descending rattle. Song a series of whistled notes, starting distinctly and accelerating to a rattling trill, all on same pitch, sung responsively by members of pair, female's song higher in pitch and softer.

Nest. A deep, pensile pouch with oblique opening at top and thick walls of dead leaves and fibers, slung below fork of slender drooping branch or vine, 2–10ft (0.6–3m) up. Eggs 2, dull white, spotted all over with reddish-brown, most heavily on large end. February–September.

Status. Common resident in humid lowlands and foothills of both slopes, to ca. 3300ft (1000m) on Caribbean slope and 4000ft (1200m) on Pacific; uncommon and local in lowlands N of Carara to Tempisque basin and on Península de Nicoya; reaches N Pacific slope through passes along Cordillera de Guanacaste.

Range. SE Mexico to W Ecuador and Amazonian Brazil.
Note. Also called Tyrannine Antbird.

BARE-CROWNED ANTBIRD

Gymnocichla nudiceps Pl. 31(13)
Hormiguero Calvo

Description. 6½″ (16.5cm); 34g. Medium-sized, stout-bodied, with rather broad, rounded tail, conspicuous blue skin on face, and silvery-gray bill. **Adult ♂:** plumage black except for 2 white wing-bars; lesser coverts scaled with white; outer primary edged with white; concealed patch of white in middle of back; wing-linings white; entire forehead, forecrown, and sides of head bare, bright blue, tinged with greenish on forehead and purplish below eye. ♀: only lores and orbital area bare; pileum and hindneck dull rufous-chestnut; rest of upperparts dark brown; 2 rufous wing-bars; entire underparts deep rufous; flanks washed with dull brown. Iris reddish; legs blue-gray. **Immature ♂:** dull black; crown feathered; wings dark sooty-brown, no wing-bars. ♀: like adult but without wing-bars (both sexes retain juvenile wing-coverts). **Young:** dark sooty-brown without wing-bars.

Habits. In thickets along forest edge and in second growth, especially in swampy areas where large-leaved herbs (*Heliconia, Calathea*) abound; usually in pairs; often forages with army ants; at other times deliberately gleans cockroaches, beetles, bugs, orthopterans, other insects, and spiders from stems and foliage, usually near ground; pounds tail when disturbed.

Voice. Usual call a shrill, downslurred *cheeah*; song a series of up to a dozen loud, sharp *cheer* notes, all on same pitch and accelerating slightly at end; contact note a softer, cheeping *tseeuw*, often repeated with rising and falling cadence: *tsee tseeuw tsuu*; in alarm, a rattling chatter.

Nest. Unknown (?).

Status. Uncommon resident in most of Caribbean lowlands, becoming very common from Río Frío region W; fairly common to uncommon in lowlands and interior valleys of S Pacific slope, N sparingly to around Carara; to 2300ft (700m) on Caribbean slope, 4000ft (1200m) on Pacific.

Range. Belize to N Colombia.

CHESTNUT-BACKED ANTBIRD

Myrmeciza exsul Pl. 31(11)
Hormiguero Dorsicastaño

Description. 5½″ (14cm); 28g. Heavy-bodied, short-tailed, with rather heavy bill and conspicuous pale blue orbital skin. **Adult ♂:** head, neck, and breast slate-black, shading to dark slate on belly; flanks, vent, and crissum dark brown; upperparts, wings, and tail chestnut; alula edged with white, sometimes white dots at tips of wing-coverts. ♀: head and neck blackish-slate, tinged with brownish, the throat often paler, shading on foreneck to dark chestnut brown (Caribbean slope) or chestnut-rufous (Pacific) over rest of underparts. Bill black, legs blackish. **Immature ♂:** below paler, more slaty; both sexes retain juvenile alula, and below often with a few sooty feathers. **Young:** mostly dull blackish, the feathers with grayish bases giving smudgy effect; belly and flanks tinged with brown; alula narrowly edged or tipped with dull buff.

Habits. Lurks in dark undergrowth of wet forest, especially in thickets along streams or edges or at overgrown treefall gaps; in tall, rank second growth, canebrakes; hops and creeps in low vegetation, rummaging amid dead leaves and trash caught in vine tangles or leaf bases of small palms; hops over ground, occasionally flicking aside leaves with bill; eats insects and spiders, occasionally small lizards and frogs; infrequently forages with army ants or with mixed-species flocks; pounds tail and fluffs body, sleeks head, droops wings in territorial disputes; paired throughout year.

Voice. Call a harsh, grating, sometimes nasal *waaa* or *naaar*; scold or alarm note *wittit wittit wittit* or a sputtering, rather metallic chatter. Song a clear, whistled *peeet peeew* or *peeet peeet peeew*, that of female higher-pitched; easily imitated, birds often respond by approach and tail-pounding.

Nest. A compact or bulky and untidy cup of lengths of vine and assorted vegetation, often including large, projecting dead leaves, sparsely lined with fungal filaments and fibrous rootlets, usually concealed amid low vegetation 10–16″ (25–40cm) up. Eggs 2, dull white, blotched, speckled, and streaked with purplish or rufous-brown, very heavily on thick end. April–August.

Status. Very common resident of lowlands and foothills of Caribbean and S Pacific slopes, locally to 3000ft (900m); rare and local on Pacific slope N of Carara, to about

Cañas and Taboga, where restricted to moist evergreen forest.

Range. E Nicaragua to W Ecuador.

DULL-MANTLED ANTBIRD

Myrmeciza laemosticta Pl. 31(14)

Hormiguero Alimaculado

Description. 5½″ (14cm); 25g. Dark, largely terrestrial, with dotted wing-bars and red iris; no bare orbital area. **Adult ♂:** chin and throat black; rest of head, neck, breast slate-black; posterior underparts dark brown; upper back dull olive-brown, shading to dull chestnut on tail-coverts and tail; middle and lesser coverts black, with white dots at tips; greater coverts brighter brown, with buffy to white dots at tips; white line along bend of wing; concealed white patch in back. ♀: similar but paler; throat spotted with white; middle and greater coverts brown with buffy tips; little or no white in back. Upper mandible black, lower dark gray; legs plumbeous. **Young:** iris brown; head washed with brown; no black below; throat faintly mottled with buff.

Habits. Prefers thickets along forest streams, especially in dark ravines, in wet foothills; hops deliberately and unobtrusively over ground and through dense, low tangles, poking at leaf litter, sometimes turning leaves, gleaning beetles, roaches, crickets, insect larvae, spiders, and sowbugs from ground or low foliage; frequently pumps tail, more emphatically when excited; in aggressive interactions, raises back feathers to expose white patch.

Voice. Calls include a rolling, grating descending *dzhrw* or *jeew*, sometimes repeated, and a harsh, explosive *sput*. Song a rather lilting series of 3–4 clear, high whistles, followed by 3–5 others on a different pitch, usually lower: *tsee tsee tsee, tyew-tyew-tyew-tyew*.

Nest. Undescribed (?). Eggs (in Ecuador) reported to be pinkish, varyingly marked with reddish-brown and dark purple spots and lines.

Status. Locally common within a narrow altitudinal belt between 1000 and 2500ft (300–750m) [to 3300ft (1000m) in SE] along Caribbean foothills the length of the country.

Range. Costa Rica to W Venezuela and NW Ecuador.

IMMACULATE ANTBIRD

Myrmeciza immaculata Pl. 31(10)

Hormiguero Inmaculado

Description. 7¾″ (19.5cm); 40g. Fairly large, dark, unmarked; orbital skin conspicuous; tail rather long. **Adult ♂:** plumage entirely black, except white line across bend of wing at wrist, conspicuous in flight or displays. ♀: above dark chestnut-brown; below slightly paler; forehead, sides of head, chin, and tail blackish; white on wrist like male. Iris chestnut; skin of lores and around eye bright blue, becoming white behind eye; upper mandible black, lower pale gray; legs black. **Young:** uniform dull brownish-black; orbital skin dull blue; no white at wrist.

Habits. Roams through understory of wet forest in hilly country; rummages and gleans in low vegetation, leaf litter, and accumulations of trash caught in vines, shrubs, and small palms, for beetles, earwigs, orthopterans, bugs, roaches, other insects, spiders, millipedes, scorpions, small frogs and lizards; regularly follows army ants but does not usually join mixed-species flocks; wags or pounds long tail incessantly.

Voice. When disturbed, a rolling sputter or chatter like that of Chestnut-backed; alarm note a loud, jaylike *queep*; also a sharp *kee-whit*; song a series of loud, emphatic, clear whistles, regularly spaced: *tee teer teer teer teer teer teer teer*.

Nest. Unknown (?). Fledged young in June–July (Parque Nacional Braulio Carrillo).

Status. Fairly common resident of lower middle elevations the length of the country (N at least to Volcán Miravalles) on Caribbean slope at ca. 1000–5600ft (300–1700m); in Dota region and locally along Cordillera de Talamanca on Pacific slope, chiefly above 3000ft (900m).

Range. N Honduras to W Ecuador and E Colombia.

BICOLORED ANTBIRD

Gymnopithys leucaspis Pl. 31(9)

Hormiguero Bicolor

Description. 5¾″ (14.5cm); 30g. Medium-sized with stout body, short tail; below white, conspicuous facial skin. **Adults:** entire upperparts uniform chestnut-brown; cheeks and auriculars blackish; sides, flanks, and crissum duller brown; rest of underparts white. Skin of lores and orbital area pale blue. Iris chestnut; upper mandible black, lower mandible and legs pale blue-gray. **Young:** above like adults but duller; middle and greater wing-coverts edged with rufous, forming indistinct wing-bars; throat dark sooty-gray; breast brown; belly grayish with indistinct paler smudging; orbital skin grayish-blue.

Habits. In forest understory, nearly always

foraging with army ants, which it follows into adjacent shady second growth and semi-open; clings to thin upright stems near ground, to which it descends to capture fugitive insects and spiders, not eating the ants themselves; rarely follows a person, seizing insects he stirs up; paired throughout year; several pairs often exploit same ant swarm, with dominance relations between them related to distance from home territory.

VOICE. Alarmed or disturbed, a nasal *per-r-r-r-r* or *cheeurrr* or a low *churr*; song a series of clear, thin, high-pitched, rather querulous whistles that speeds up and rises in pitch and either ends on a high note or becomes lower and slower; sometimes terminates with 1 or more harsh, nasal drawls. Very vocal; a birdwatcher often locates army ants at a considerable distance by this antbird's songs and calls.

NEST. A thin mat of rootlets and fibers resting upon leaf fragments, in hollow center of small stump of palm or other tree, opening upward and usually less than 3ft (1m) up. Eggs 2, whitish or cream, spotted and longitudinally streaked with reddish-brown and deep red. March–January but chiefly April–September.

STATUS. Common resident of humid lowlands and foothills throughout Caribbean slope, and on Pacific slope N to about Carara and in low passes of N cordilleras; up to 5000ft (1500m) on Caribbean slope, to 5600ft (1700m) on S Pacific slope.

RANGE. N Honduras to W Ecuador, NE Peru, and NW Brazil.

NOTE. Birds of Central America and W of Andes in South America are sometimes considered a separate species, *G. bicolor*.

SPOTTED ANTBIRD

Hylophylax naevioides Pl. 31(12)

Hormiguero Moteado

DESCRIPTION. 4½″ (11.5cm); 18g. Small, plump, short-tailed; strikingly patterned, with broad wing-bars. **Adult ♂:** top and sides of head slate, tinged with brownish on pileum and hindneck; back chestnut-rufous with concealed white patch; rump to base of tail duller, more olive-brown; middle of tail black, tip cinnamon; wings blackish, wing-coverts and secondaries tipped with cinnamon-rufous; throat black; breast and belly white, breast crossed by a band of large black spots; sides, flanks, and crissum grayish. Bill black; legs pale gray. ♀: head mostly olive-brown; back duller rufous; throat dull white; breast crossed by a smudgy band of dull gray mixed with buff; flanks and sides more extensively brownish-gray. Lower mandible gray. **Young:** resemble adult female but head mostly sooty-gray; below mostly brownish, breast-band indistinct; no white in back, the feathers with extensive dull bases, sooty fringes.

HABITS. In understory of humid forest, where it frequently forages with army ants, following them into adjacent tall second growth and shady semi-open; more often seen away from ants than Bicolored, which dominates it at ants; clings to thin vertical stems near ground, rarely rising above eye level or alighting on ground except to seize insects or spiders stirred up by ants; active; often notably fearless of man; in pairs throughout year.

VOICE. A sharp downslurred rattle less shrill than that of a *Myrmeciza*; disturbed, a sharp *skeeip* or *psip*, sometimes repeated steadily up to 10 times; a low *churr*, a soft *peep*. Song a thin, rather soft, burry *peede weede weede weede weede weede* or *peede peede peede sip sip sip* and a high-pitched trill.

NEST. A deep, thin-walled cup of fine dark fibers, often with bits of moss, dead leaves, or twiglets decorating outside; attached by rim to slender twigs or fork 1–3ft (0.3–1m) up in forest undergrowth. Eggs 2, whitish, heavily blotched and speckled all over with chocolate-brown, the larger marks often more longitudinal. April–July.

STATUS. Common resident of lowlands and foothills throughout Caribbean slope, locally up to 2600ft (800m) in N and 3300ft (1000m) in SE; reaches Pacific side of Cordillera de Guanacaste, where especially abundant between 2000 and 2500ft (600–750m).

RANGE. E Honduras to W Ecuador.

OCELLATED ANTBIRD

Phaenostictus mcleannani Pl. 31(7)

Hormiguero Ocelado

DESCRIPTION. 8″ (20cm); 55g. Largest and least numerous of the regular army-ant followers of Caribbean slope; long-tailed, with large, bare cobalt-blue area on face, black eyelashes. **Adults:** pileum grayish-brown; chin, throat, and upper breast black; a collar of rufous-chestnut around neck, broadest on chest; back and wing-coverts grayish-brown, this color almost obscured by large black spots, each bordered posteriorly with cinnamon-buff; tertials edged with buff at tips; rump and upper tail-coverts brown; tail black; lower breast rufous, shading to cinnamon-brown posteriorly; breast and belly heavily marked with large black spots; crissum black with broad cinnamon fringes. Iris chestnut; bill black, sometimes with pale tip; legs pale flesh. **Young:** resemble adults but pileum indistinctly barred with grayish and sooty black; above pattern blurred by sooty fringes to most feathers; throat and breast sooty-black, fading to sooty-brown posteriorly; breast irregularly spotted with dull rufous.

HABITS. In forest undergrowth, following army ants with Spotted and Bicolored antbirds, occasionally into old second growth and shady semi-open; perches on erect stems and branches, often at eye level, makes dashing sallies to ground or vegetation to snatch fleeing large insects, spiders, scorpions, other invertebrates, and small vertebrates; dominates smaller antbirds; in pairs or small groups of up to 6; habitually pumps tail up and down.

VOICE. Alarmed or disturbed, a penetrating, rolling *cheeer*, sharper than note of Bicolored; hoarse, low scratchy notes while foraging. Song resembles that of Bicolored Antbird but shorter, more mellow, less penetrating: a series of clear whistled notes that rises, speeding up only slightly, then falls and slows, some 10–12 notes in all; sometimes abbreviated to ca. 6–7 rising notes and a final descending one.

NEST. Undescribed (?).

STATUS. Uncommon resident of lowlands and foothills of Caribbean slope, reaching Pacific slope locally through low passes in N cordilleras; sea level to 4000ft (1200m).

RANGE. E Honduras to NW Ecuador.

NOTE. Also called Ocellated Antthrush.

BLACK-FACED ANTTHRUSH

Formicarius analis Pl. 30(15)

Gallito Hormiguero Carinegro

DESCRIPTION. 6¾″ (17cm); 60g. Stout, long-legged; resembles small rail as it walks over ground with short tail cocked. **Adults:** above dark brown, more chestnut on upper tail-coverts; area behind eye to side of neck dark rufous; chin, throat, cheeks, and lores black, a white spot on lores; chest and flanks dark olive, shading to dull gray on lower breast and sides, and to dull whitish on center of belly; crissum tawny-brown. A crescent of bluish-white skin behind and below eye; bill black; legs brownish. **Young:** forecrown brown, lightly spotted with black; cheeks, chin, and throat dull white to buffy, with sooty-brown feather tips (Pacific slope), or sooty-black, smudged with olive (Caribbean); underparts more olive, the feathers with dusky bases, producing smudged effect.

HABITS. Walks sedately over floor of mature forest or tall second growth, tilting forward with each step, flicking aside leaves with bill, picking up insects, snails, spiders, other invertebrates, rarely a small snake, lizard, frog, or fallen fruit; sometimes forages on outskirts of army-ant swarm; rarely perches on branch 10–16ft (3–5m) up, resting or preening; paired and territorial throughout year but forages singly, from time to time stopping, throwing back head, and singing.

VOICE. Song typically 3 full, mellow whistled notes, the first higher and accented: *keep two two*, but the number of 2s may vary from 1–10; female song higher in pitch and softer; readily imitated: bird often responds by close approach, pumping tail with each song. Alarm note, also given upon going to roost, a sharp *kip* or *twip*.

NEST. In slender hollow trunk, stump, or taproot opening upward, to side, or both, 20″–11ft (0.5–3.5m) up; a mat of slender petioles and flower stalks resting upon a pad of coarse dead leaves; often far below entrance. Eggs 2, white but soon stained and speckled all over with brown. March–September; sometimes 3 broods.

STATUS. Common resident of lowlands and foothills of Caribbean slope, to 1600ft (500m) in N and 4000ft (1200m) in SE; and on S Pacific slope (N to Carara) up to 5000ft (1500m); reaches N Pacific slope locally through low passes.

RANGE. SE Mexico to N Bolivia and E Brazil.

NOTE. Also possible on Caribbean slope is Wing-banded Antbird (*Myrmornis torquata*); known from SE Nicaragua and Panama and possibly sighted once on Fila Carbón in SE; like a small (5½″, 14cm) antthrush but with 3 buffy wing-bars, blue orbital skin; above brown, cheeks and throat black, belly dark gray [see Pl. 30(21)].

BLACK-HEADED ANTTHRUSH

Formicarius nigricapillus Pl. 30(17)

Gallito Hormiguero Cabecinegro

DESCRIPTION. 7″ (18cm); 70g. Darkest antthrush, with most conspicuous orbital skin. **Adults:** head, throat, and breast black (extending farther back in male), shading to dark gray posteriorly (tinged with brown in female); flanks pale brownish-olive; wing-linings largely buff; above rich dark brown; upper and under tail-coverts chestnut; tail blackish. Bill black; orbital skin bluish-white; legs dusky. **Young:** mantle fringed with sooty; head and chest duller, more sooty-black; chest fringed with dull olive; lower breast and sides with faint, dusky streaks; bill tip white.

Habits. Walks over floor of primary forest, sometimes entering adjacent tall second growth, peering into crevices and tangles, under fallen branches, sometimes turning up leaves to look beneath; eats small litter animals including crickets, roaches, earwigs, sowbugs, spiders, tiny reptiles and frogs. Usually holds tail cocked, may pump it when agitated; normal stance erect but stretches to fullest height while singing; territorial year-round.

Voice. A sharp *chirp* or *chwip* in alarm and before roosting in evening; a little nasal whinny in aggressive interactions. Song a rapid, pulsating series of ca. 20 deep, resonant, whistled notes, the first 2–3 slower, more staccato, the next 6–8 rising in pitch, the last 10–12 on the same pitch, with the final 2 notes slower, the entire series lasting 4–5sec.

Nest. In hollow palm stub 2m tall, ca. 1m below epiphyte-covered entrance, a shallow, bulky cup of leaf petioles with a few dead leaves (Marín). Eggs 2, dull white, soon stained with brown. April–May.

Status. Locally common within a narrow altitudinal belt of very wet foothill forest the length of Caribbean slope, 1300–2600ft (400–800m) in N and C Costa Rica, and 2000–4000ft (600–1200m) in SE; replaced below this belt by Black-faced Antthrush, above it by Rufous-breasted Antthrush.

Range. Costa Rica to W Ecuador.

RUFOUS-BREASTED ANTTHRUSH

Formicarius rufipectus Pl. 30(16)

Gallito Hormiguero Pechicastaño

Description. 7½″ (19cm); 75g. Terrestrial, heavy-bodied, with cocked tail; rufous breast diagnostic. **Adults:** face and throat black; rest of head dark chestnut; back, rump, and wings dark brown; upper and lower tail-coverts chestnut; tail blackish; wing-linings buff; breast deep rufous; belly dull cinnamon-rufous; sides and flanks dusky. Rufous of breast more extensive, belly brighter in male. Bill black; orbital skin dull bluish-white; legs dusky. **Young:** feathers of face and throat sooty to brownish, edged with chestnut; lower breast and belly extensively grayish; above more olive-brown.

Habits. Frequents the floor of primary forest in mountainous country; habits appear similar to those of other species of *Formicarius*, which it replaces abruptly at higher elevations.

Voice. Song a pair of clear whistles, *toot-toot*, the second on the same pitch or very slightly higher than the first; song of female higher in pitch. A loud *chrip* in alarm or before roosting; a dry, snapping *snuk* when disturbed.

Nest. Undescribed (?).

Status. Uncommon to fairly common resident of middle elevations on Caribbean slope, from Cordillera de Tilarán S to Panama, ca. 2800–5000ft (850–1500m) in N and C Costa Rica, 4000–6000ft (1200–1850m) in SE.

Range. Costa Rica to W Ecuador and E Peru.

BLACK-CROWNED ANTPITTA

Pittasoma michleri Pl. 30(19)

Tororoi Pechiescamoso

Description. 7½″ (19cm); 110g. Very large, heavy-billed, long-legged, short-tailed; below boldly scaled. **Adult ♂:** most of head black, the face flecked with rufous, the auriculars mostly rufous-chestnut; back olive-brown, heavily streaked with black, becoming rich brown on rump, tail, and wings; wing-coverts and tertials spotted with buff; throat black, flecked with rufous and white; rest of underparts mostly white, heavily scaled with black, almost uniform black on chest; belly tinged with buff; crissum brown; flanks olive-brown. ♀: black markings on underparts less heavy, especially on face, throat, and breast, which are often strongly ochraceous. **Young:** crown sooty; face dull brown mixed with rufous; back indistinctly barred and scaled with ochraceous and black; wing-coverts broadly tipped with ochraceous; below scaling duller, sootier, less distinct.

Habits. Frequents understory of primary forest and tall second growth, hopping on the ground in great, springy bounds; rummages in leaf litter, sometimes digging in soft earth; regularly attends army-ant raids, where it dominates all other antbirds; hops swiftly in pursuit of prey flushed by the ants; takes crickets, roaches, earwigs, other insects, whip-scorpions, scorpions, spiders, small reptiles and frogs; sings from horizontal branch low in understory; paired throughout year but usually forages singly.

Voice. In alarm or excitement, a harsh, sharp, guttural, coughing *kuk kuk kuk* (often 10 or more notes) like a huge squirrel. Song a very long, rapid series of short, clear, "pinking" whistles, powerful and mellow, lasting a minute or more, starting very rapidly and gradually decelerating throughout.

Nest. A large, thin-walled cup of fine, dark

rootlets, placed upon a bed of weathered leaves lodged in the crown of an understory palm, 3ft (1m) above ground. Eggs 2, pinkish buff, with scattered purplish blotches and cap of dense brown spots. April (in Panama; no Costa Rican record).

Status. Resident locally in wet foothills of Caribbean slope from Cordillera de Tilarán S to Panama, ca. 1000–2600ft (300–800m), up to 3300ft (1000m) in SE.

Range. Costa Rica to NW Colombia.

SCALED ANTPITTA

Grallaria guatimalensis Pl. 30(20)

Tororoi Dorsiescamado

Description. 7″ (18cm); 98g. Large, plump, long-legged, nearly tailless; below rufous, above scaled; bill heavy, rather short; big, dark eyes. **Adults:** forehead olive-brown; crown and hindneck slate-gray, back and rump olive, all broadly scaled with black; wings and tail cinnamon-brown, primaries with paler and brighter edgings that contrast with dusky coverts; lores, malar stripe, and crescent behind eye whitish; sides of throat dusky; center of throat and foreneck tawny-buff, bordered posteriorly with dusky; rest of underparts bright tawny or rufous, slightly paler on belly. Upper mandible dusky, lower horn-color with dusky tip; legs gray. **Young:** crown, throat, and breast dusky with broad teardrop-shaped streaks of buff; belly dull rufous; back dark brown scaled with blackish; wing-coverts broadly edged with rufous.

Habits. Hops over ground with long, springy bounds in wet mountain forests; flicks leaves aside with bill; digs and probes in soft mud for earthworms, large insects, other invertebrates, and small frogs; when alarmed, rises to perch as much as 10ft (3m) up, glances around, then dives into undergrowth and disappears; usually solitary; wanders widely after breeding season, especially juveniles.

Voice. Song a low-pitched, resonant series of notes that starts as a trill, rises in pitch and slows to distinct, pulsating notes, stops abruptly, heard only at dawn; a low piglike grunt or croak in alarm.

Nest. A crude saucer of large dead leaves, vines, sticks, and moss, lined with fine rootlets and moss, 2–7ft (0.6–2m) up on stump, mossy log, or liana by tree trunk, in thick vegetation. Eggs 2, greenish-blue. May–July.

Status. Very uncommon resident of middle elevations the length of Caribbean slope ca. 2600–5400ft (800–1650m); on Pacific slope of N cordilleras, Cerros de Escazú (at least formerly) and perhaps locally on Cordillera de Talamanca.

Range. C Mexico to W Ecuador and N Bolivia.

SPECTACLED ANTPITTA

Hylopezus perspicillatus Pl. 30(18)

Tororoi Pechilistado

Description. 5″ (12.5cm); 48g. Plump, big-eyed, long-legged, nearly tailless; conspicuous buffy spectacles and streaked breast; terrestrial. **Adults:** pileum, hindneck, and mantle slaty, shading to grayish-olive on rump and scapulars; mantle finely streaked with buff; wing-coverts brown, broadly tipped with buff; primaries edged with cinnamon-buff, contrasting with blackish primary coverts; outer web of alula buff; broad eye-ring and lores buff, separated by narrow black line; cheeks and patch on auriculars dull buff; blackish malar streak; chin, throat, breast, and belly white, breast heavily streaked with black and, laterally, with buff; flanks and crissum ochraceous-buff. Upper mandible black, lower flesh-color with dusky tip; legs pale gray. **Young:** upperparts mostly chestnut, the wing-coverts tipped with buff and edged with rufous; below chestnut-rufous, paler and mixed with buffy on belly.

Habits. Hops over ground in humid forest, sometimes running for short distances like a thrush (*Turdus*); flicks leaves aside with bill; pausing, rhythmically puffs out and contracts breast feathers while jerkily half-spreading and closing wings, perhaps to flush prey; eats insects, spiders, and other invertebrates; when alarmed, flies up to low perch to look around, then plunges into undergrowth and vanishes; forages solitarily but may be paired throughout year.

Voice. A series of ca. 6–9 mellow whistles with increasing, then decreasing tempo: *cow—cow-cow cow cow cow-cow—cow* or *pee pee peepeepeepee pew pew pew*, more complex and hollow-toned than Black-faced Antthrush's longer songs, which do not vary in tempo; song of female higher in pitch; a soft, clear *cowee cowee cowee*, possibly a contact note; alarm note an explosive *teew*; also loud rattle, dying away at end.

Nest. A bulky, crudely built, often precariously supported shallow bowl of coarse sticks and dead leaves, sparsely lined with rootlets, tendrils, and fine petioles, 2–5ft (0.6–1.5m) up in forest shrub. Sometimes a little lining is added to neater nest abandoned by some other bird. Eggs 2, broad and blunt, grayish, heavily and coarsely blotched with dark brown, which may be nearly solid on large end. April–August.

STATUS. Uncommon to fairly common resident in lowlands and foothills of Caribbean and S Pacific slopes (N at least to Quepos area), up to 4000ft (1200m).

RANGE. NE Honduras to W Ecuador.

NOTE. Also called Streak-chested Antpitta. This and following species sometimes placed in the genus *Grallaria*.

FULVOUS-BELLIED ANTPITTA

Hylopezus fulviventris Pl. 30(14)

Tororoi Pechicanelo

DESCRIPTION. 5″ (13cm); 44g. Lacks both conspicuous eye-ring and heavy streaking below; underparts largely rich ochraceous. Pileum dark slate with faint sooty scaling, shading through grayish-olive on back to dull olive on rump and upper tail-coverts; primaries edged with tawny at base, contrasting with blackish coverts; broad, dull buffy eye-ring; face mixed with buff and blackish; throat and center of lower breast and belly white; rest of underparts rich ochraceous-buff to pale rufous; breast and sides indistinctly streaked with black. Upper mandible black, lower grayish with blackish tip; legs dull flesh.

HABITS. Prefers dense, low, tangled thickets of shrubs and vines along forest streams, in treefalls, at forest edge, or impenetrable young second growth in small abandoned clearings; forages on ground and logs for insects and spiders; sings from horizontal branches up to 5ft (1.5m) above ground; fluffs plumage when it calls; apparently in territorial pairs throughout year.

VOICE. Song a rapid, ascending series of ca. 10–13 clear whistles, the first double-toned, the second with a slight catch, the last 3–4 on same pitch, the entire series becoming louder and slightly faster toward end. When alarmed or excited, a hard, metallic, rolling *trrrrr*.

NEST. Undescribed (?).

STATUS. Widespread, locally common resident of lowlands and foothills the length of Caribbean slope, from sea level to ca. 2000ft (600m) in most areas, and 3000ft (900m) in SE. Seldom seen, its abundance is indicated by the often-heard song.

RANGE. NE Honduras to Ecuador.

OCHRE-BREASTED ANTPITTA

Grallaricula flavirostris Pl. 30(13)

Tororoi Piquigualdo

DESCRIPTION. 4″ (10cm); 18g. Small, nearly tailless, long-legged, terrestrial; conspicuous pale bill; big round eyes set off by buffy eye-rings. **Adults:** face and upperparts olive-brown, washed with gray on nape; eye-ring and line on side of forehead rich buff, separated by black crescent on lores; throat rich buff, with a black streak down each side; narrow, often concealed white crescent on lower throat; broad breast-band ochraceous-buff; rest of underparts white; breast and sides lightly streaked with black; flanks washed with olive. Bill yellow-orange with dusky base; legs horn-color. **Young:** similar but face pattern and black streaking below faint and indistinct.

HABITS. In understory of primary forest and adjacent tall second growth, preferring thickets in treefall gaps; gleans beetles, crickets, other insects, and spiders while hopping along logs and branches or on ground; usually sings 3–10ft (1–3m) up in thicket.

VOICE. Song a high, rattling trill that drops, then rises in pitch, lasting 1–2sec; reminiscent of trill of Long-billed Gnatwren but notes drier, more staccato; alarm note a scratchy *bzeet*; a short, rattling chatter when excited.

NEST. Undescribed (?).

STATUS. Resident; scarce to fairly common in wet lower middle elevations at ca. 2500–4300ft (750–1300m) on Caribbean slope of Cordillera Central and probably Cordilleras de Tilarán and Talamanca, where not yet recorded; and up to 6000ft (1850m) in Dota region, along Pacific face of Talamancas, and on Fila Cruces above San Vito.

RANGE. Costa Rica to W Ecuador and N Bolivia.

FAMILY **Rhinocryptidae:** Tapaculos

Of the 27 species in this exclusively New World family, 1 occurs in Costa Rica and western Panama. All the rest occur in South America, chiefly in the highlands and the South Temperate Zone. Tapaculos are 4¼–10″ (11–25cm). Their soft, loose plumage is mainly gray, brown, chestnut, and black, with touches of white, often barred, spotted or scaly. The sexes are usually alike. At the base of their short, rather thick bills are nostrils covered with a movable flap. With stout legs and feet, tapaculos hop over the ground or run swiftly across open spaces between sheltering bushes, holding their tails erect like miniature roosters, earning for 1 species the name *gallito*, or "little cock." With short, rounded wings, they fly weakly and reluctantly. Their food is chiefly insects and other small invertebrates, along with some seeds, which they find by scratching, chickenlike, on the ground. Secretive inhabitants of dark undergrowth in dense forests or of scrub, they make their presence known by the persistent use of loud and sometimes melodious voices, difficult to trace to their sources. Yet they are curious and peer out from concealment at the person who does not obviously watch them.

Nests of these shy, elusive birds are exceedingly difficult to find. In a bank green with selaginellas and ferns high in the Andes, a nest of the Unicolored Tapaculo fitted neatly into a globular niche that the birds had lined all around with capillary, branching, leafless, black stems of liverworts and mosses. In Chile, some species build nests of grasses and rootlets in long tunnels that they dig in banks and sides of ravines; others occupy abandoned burrows of small mammals in rocky places; still others prefer natural cavities in trunks, often high above the ground; or they hide their nests behind loose bark of old trees or root masses in banks. In thorny shrubs, a few tapaculos build, of grasses or twigs, closed structures with a side entrance. Tapaculos lay 2–4 immaculate white eggs that are large for the birds' size. Both sexes incubate and feed the young, but otherwise little is known of their breeding biology.

SILVERY-FRONTED TAPACULO

Scytalopus argentifrons Pl. 32(12)

Tapaculo Frentiplateado

Description. 4¼" (11cm); 17g. A plump, dark, often noisy little bird of highland thickets; tail usually carried partly cocked. **Adult ♂:** mostly sooty-black, paler on lower breast and belly; forehead and short superciliaries silvery-gray; wings and tail brownish-black; flanks and crissum barred with tawny-rufous, sometimes tertials also. ♀: above dark, rich brown, scaled with black; feathers of back with black centers; face, throat, and chest dark sooty-gray, washed with olive-brown; belly blackish, broadly edged with tawny-buff. Bill black; legs fuscous; anterior claws white. **Young:** like adult female but feathers of throat dull grayish-buff to tawny with blackish centers; rest of underparts sooty-black, heavily scaled with buffy-brown.

Habits. Skulks in dense undergrowth of highland forests and adjacent second growth, especially thickets and bamboo tangles along streams and ravines, so inconspicuous and secretive that it would be considered very rare were it not for its loud, distinctive calls; hops and creeps about, often on ground, peering into all sorts of nooks and crannies for larvae, pupae, and adult insects (beetles, earwigs, crickets), spiders and their egg cases, and sowbugs; usually seen singly but sedentary and possibly paired throughout year; curious, often responding to squeaking but virtually never leaving concealment.

Voice. Usual call or scold a series of ca. 3–7 sharp, hard, regularly-spaced notes, *kew-kew-kew-kew-kew*, with a resonant, wooden quality. What may be the song is a much longer, more nasal and rattling chatter, *cht-cht-cht-cht*. . . .

Nest. Undescribed (?).

Status. Fairly common resident of middle and upper elevations, from Cordillera de Guanacaste (N at least to Volcán Miravalles) to Panama, from timberline down to ca. 5000ft (1500m) on Pacific slope, and locally down to 3300ft (1000m) on Caribbean slope.

Range. Costa Rica and Panama.

FAMILY **Tityridae:** Tityras and Becards

Long considered cotingas, the tityras and becards have recently been transferred to the Tyrannidae by some authorities because of similarities in the anatomy of skull and syrinx. Because in some other features these birds resemble cotingas more than flycatchers, this shift is controversial; until their relationships are better established, we prefer to treat the tityras and becards as a separate family. One feature peculiar to these birds is the greatly reduced, more or less spinelike ninth primary of the adult males.

Of the 20 species in this family, 19 are spread over the warmer parts of the American continents and closely adjacent islands from extreme southern USA to northern Argentina; one is isolated on Jamaica. The 3 species of tityras are stout and largely white, trimmed with black; 2 have red skin around the eyes. The 17 species of becards are smaller, big-headed, and clad in black, gray, rufous, chestnut, olive, yellowish, and white in various combinations, with a patch of rose or pink on the throats of males of 2 species. Becards are largely insectivorous; tityras take larger insects as well as more fruits; all pluck their food from trees, frequently without alighting. A few have melodious voices. Tityras nest in holes in trees, into which they carry many dead leaves, twigs, and other bits of plants. Becards build bulky, covered nests with a doorway in the side or bottom, either swinging at the end of a drooping branch or supported in a fork. In the family as a whole, eggs number from 2 to, rarely, 6, with a grayish, brownish or brownish-white ground marked with darker shades. The female builds the nest, sometimes with more or less help from her mate; as far as known, she does all the incubation, but her mate regularly helps to feed the young. Incubation periods are 18–21 days, nestling periods 20–30 days.

BARRED BECARD

Pachyramphus versicolor Pl. 33(14)

Cabezón Ondeado

Description. 4¾″ (12cm); 14g. Smallest becard and only one barred below; conspicuous eye-ring. **Adult ♂:** pileum, hindneck, and back glossy black; scapulars, lesser and middle wing-coverts largely white, edged with black; greater coverts and remiges black, edged with white; rump, upper tail-coverts, and tail slate, lateral rectrices edged with white at tip; sides of head and neck, sides of throat greenish-yellow, shading to white over rest of underparts; face, underparts except center of belly finely barred with blackish. ♀: pileum and hindneck slate-gray; rest of upperparts olive-green; wing-coverts mostly rufous; remiges dusky, edged with buff on tertials and rufous on most of secondaries; eye-ring yellow; sides of head, neck, and breast yellow-olive, rest of underparts greenish yellow; throat, breast, and sides narrowly barred with dusky. Upper mandible black, lower bluish-gray; legs gray. **Immature ♂:** much duller, more greenish than adult; pileum and hindneck dark sooty-gray; rest of upperparts dull olive-green mixed with sooty-black; white of scapulars and wings replaced by greenish-yellow; below mostly dull greenish-yellow, indistinctly barred with dusky except on belly; eye-ring yellowish.

Habits. In middle and upper levels of highland forest, coming lower along edges and in nearby semi-open; singly or in pairs, often accompanies mixed-species flocks; in family groups following breeding season; more active and animated than other becards; sallies to snatch insects from vegetation, often striking against foliage or hovering briefly; also takes small fruits (e.g., *Trema, Urera*).

Voice. An incessant, soft, high-pitched *weet weet weet weet* . . . or a more excited-sounding *tseep tseep tseep tseep* . . . ; a high, weak trill and twittering notes.

Nest. A roughly globular structure ca. 1ft (30cm) in diameter with entrance at side of bottom, saddled in fork of slender branch near top of tree, 50–75ft (15–23m) up, of green moss, slender dead vines, and leaves, chiefly of bamboo and lined with same; typically in tree apart from closed forest. Eggs 2, brownish-white, speckled with dark brown, chiefly at large end (Marín). April–June.

Status. Uncommon resident in upper parts of Cordilleras Central and de Talamanca along both slopes and in Dota region, mostly at 5000–8200ft (1500–2500m), but during first half of year occasionally to 10,000ft (3000m); very uncommon to rare above 5000ft (1500m) in Cordillera de Tilarán.

Range. Costa Rica to NW Ecuador and N Bolivia.

CINNAMON BECARD

Pachyramphus cinnamomeus Pl. 33(11)

Cabezón Canelo

Description. 5½″ (14cm); 22g. Both sexes rufous, unlike most becards; adult males with short, sharp-pointed 9th primary. **Adults:** upperparts uniform rufous to rufous-tawny; primary coverts darker, more dusky; below cinnamon-buff to tawny-buff, tinged with rufous across breast; short, narrow superciliaries buffy; lores dusky. Upper mandible blackish, lower gray; legs dark gray. **Immatures:** like adults but paler overall; upperparts brighter rufous; belly paler buff; 9th primary of males not reduced.

Habits. In edges and breaks, especially along rivers and streams, mangroves, open woodland, and semi-open; sometimes ranges up into forest canopy or out into nearby low second growth; eats insects, including beetles, katydids, homopterans, ants, caterpillars, also spiders, and berries, all snatched from foliage in flight.

Voice. A variety of high, thin, plaintive, slurred whistles; song of ca. 6 slight, plaintive notes, ascending in pitch: *dee dee dee dee dee de*; female's song shorter and weaker; other notes descending: *deeeuu dew dew, dew dew*, etc.

Nest. A bulky, roughly globular structure with inner chamber reached by ascending tunnel from lower side, of green moss, long brown fibers, dry bamboo leaves, rootlets, and vegetable down, hanging 8–50ft (2.5–15m) up from tip of drooping slender branch, often near active wasps' nest. Eggs 3–4, brownish-white, mottled and streaked with olive-brown, chiefly at thicker end. March–July.

Status. Common resident of Caribbean lowlands, up to 2500ft (750m) in foothills; on Pacific slope uncommon and local in or beside mangroves from Panama border N to Golfo de Nicoya.

Range. SE Mexico to NW Ecuador and NW Venezuela.

WHITE-WINGED BECARD

Pachyramphus polychopterus Pl. 33(13)

Cabezón Aliblanco

Description. 5¾″ (14.5cm); 21g. Rather long, graduated tail; wings and tail conspicuously marked with white (male) or cinnamon-buff (female). **Adult ♂:** pileum glossy blue-black; back, wings, and tail black; scapulars, wing-coverts, and secondaries broadly edged with white; broad white tips on all but central rectrices; sides of head, entire underparts, rump, and upper tail-coverts dark slate-gray, paler on belly. Bill silver-gray with black tip; legs plumbeous. ♀: above greenish-olive; wings and tail blackish; wing-coverts edged with cinnamon-buff, secondaries with buff; all but central rectrices tipped broadly with cinnamon-buff; below pale yellow, washed with olive on breast and sides and grayish on throat; broken white eye-ring; whitish line over lores. Upper mandible blackish, lower blue-gray. **Immature ♂:** like adult but retains juvenile remiges and some yellow on belly. **Young:** like adult female but above browner; feathers of crown with dusky tips; wing and tail markings less clean-cut; below paler yellow (male with grayish wash over breast, face, and throat).

Habits. In humid areas prefers light woodland, semi-open, shady plantations, and old second growth; mainly in evergreen gallery forest in dry northwest; also in mangroves; singly or in pairs, forages in middle or upper parts of trees, sallying to snatch caterpillars, beetles, leafhoppers, spiders, from foliage; eats many small berries; associates loosely with mixed-species flocks.

Voice. Song a series of 6–9 clear, mellow, downslurred notes, the first slightly higher in pitch, the last 2–3 slower: *chew, chewchew-chewchew, chew, chew, chew*; sometimes song rises in pitch and becomes weaker; female gives weaker version, also soft, liquid warbles at nest.

Nest. Roughly globular with side entrance, 7–8″ (18–20cm) in diameter, of shredded bast fibers, threadlike inflorescences, Spanish moss, feathers, etc., lined with grass, bamboo leaves, and broad strips of monocot leaves, in fork of slender branch, 14–125ft (4–38m) up in isolated tree, usually well screened by foliage. Eggs 3–4, pale gray, speckled all over with brown, most heavily in a wreath. April–August.

Status. Resident in lowlands of both slopes, up to 3300ft (1000m) on Caribbean slope and 4000ft (1200m) on Pacific; uncommon and local in dry NW; common elsewhere.

Range. Guatemala to N Argentina and Guianas.

BLACK-AND-WHITE BECARD

Pachyramphus albogriseus Pl. 33(15)

Cabezón Cejiblanco

Description. 5½″ (14cm); 20g. Resembles White-winged Becard but male has white over lores, gray back; female has rufous cap, bold white eyebrow. **Adult ♂:** pileum glossy black; wings black, wing-coverts and secondaries broadly edged with white; back, rump, and base of tail dark gray, subterminal area of rectrices black, tips white; sides of head, nuchal collar, and entire underparts pale gray; lores blackish; broad white supraloral stripes meet across forehead. Bill blue-gray with black tip; legs blue-gray. ♀: pileum dull rufous, edged with black; white supraloral stripe extends back as superciliary; narrow black stripe through eye to hindcrown; above dark olive-green; wings and tail dusky, with buffy edgings on wing-coverts and secondaries; tail tipped with buff; nuchal collar, cheeks, breast, and sides pale yellowish-olive; throat whitish; belly pale yellow. Bill blackish with gray tomia. **Young:** like female but duller; pileum feathers edged with sooty-black; white facial stripe indistinct; cheeks grayish; no nuchal collar; wing markings cinnamon-buff, less distinct.

Habits. Prefers upper understory and canopy of wet forest, sometimes edges and adjacent tall second growth; usually in pairs, often accompanying mixed-species flocks of tanagers, furnariids, etc.; perches sluggishly, scanning nearby foliage, often cocking or twisting head to odd angles, then making a rapid, fluttering sally, often hovering to snatch fruit or insect.

Voice. A variety of high, thin whistled notes, in tone like those of Cinnamon Becard but less mournful: *swee-sweeet* or *weeea-weeeeur*, as members of pair often call back and forth; song of male *chew-chewy ch'chewy ch'chewy*.

NEST. A globular mass of dead leaves, moss, bits of vine, etc., with entrance at bottom, built into vertical fork near tip of trunk or branch of tree in subcanopy or at forest edge, 23–65ft (7–20m) up. Eggs undescribed (?). March–April.

STATUS. Uncommon resident of wet midelevation forest, ca. 2600–6000ft (800–1850m), from Cordillera de Tilarán S to Panama, chiefly on Caribbean slope but locally on Pacific slope [Dota region, formerly (?) Valle Central, possibly Cordillera de Talamanca]; descends to lower elevations, especially on Pacific slope, during second half of year.

RANGE. Costa Rica to NW Peru and N Venezuela.

ROSE-THROATED BECARD

Pachyramphus aglaiae Pl. 33(12)

Cabezón Plomizo

DESCRIPTION. 6½" (16.5cm); 33g. Large, heavy-billed; males much darker in Costa Rica than in northern races, with rosy throat-patch faint or lacking. **Adult ♂:** pileum glossy black; rest of upperparts, including wings, tail, and sides of head, blackish-slate; below slate-gray, darkest on breast, with at most a faint tinge of rose on throat (Pacific race); or with upperparts, throat, and breast slate-black, shading to paler gray on belly (Caribbean race). Upper mandible black, lower blue-gray; legs plumbeous. **♀:** pileum dark slate-gray to blackish; back tawny-rufous, shading to brighter rufous on wings and tail; underparts buffy, tinged with ochraceous on breast, cheeks and sides of neck. Bill black. **Immature ♂:** like adult female but variable gray on head and body gives dingy aspect. **Young:** like adult female but above duller with trace of ochraceous nuchal collar; below paler, throat whitish.

HABITS. Frequents canopy of deciduous woods and evergreen gallery forest, scrubby second growth; singly or in pairs, roams about, often with other small birds; catches insects, especially caterpillars, by sallying and striking foliage; sometimes hawks flying insects; takes many fruits, including arillate seeds (e.g., *Stemmadenia*).

VOICE. A variety of thin, high-pitched, strident or squeaky whistles, often with rising inflection: *wheeeiii, p'eeeeeii*, etc.; dawn song a long-continued *wheeuu-whyeeeuur, wheeuu-whyeeeuur*. . . .

NEST. A very bulky, globular to bell-shaped structure with entrance at or near bottom, up to 30" (76cm) long, of dry vines, grass stems, soft twigs, leaves, often with bits of *Ceiba* (kapok) floss or *Inga* flowers woven in, decorated with bits of green vegetation, 13–70ft (4–21m) up, at tip of slender drooping branch, often over water. Eggs 3–4, brownish-white, speckled with dull brown, mostly in wreath. April–June.

STATUS. Uncommon to fairly common resident in lowlands of N Pacific slope, including Península de Nicoya and S to about Orotina, up to ca. 1000ft (300m). A rare visitor, probably a migrant, between about November and March in Caribbean lowlands; a few recent records for San Vito area.

RANGE. SE Arizona and S Texas to Costa Rica, possibly extreme W Panama.

NOTE. Formerly placed in the genus *Platypsaris*.

MASKED TITYRA

Tityra semifasciata Pl. 34(1)

Tityra Carirroja (Pájaro Chancho)

DESCRIPTION. 8¼" (21cm); 88g. Stout, big-headed, with banded tail; base of heavy bill and skin of lores and orbital area red. **Adult ♂:** forehead, behind eye, cheeks, and chin black, framing red skin; rest of upperparts, including middle and lesser wing-coverts, tertials, and base of tail pale gray; distal half of tail black with broad, white terminal band; rest of wings black; rest of head and underparts white. Iris reddish-brown to brick-red; tip of bill blackish; legs plumbeous. **♀:** top and sides of head and chin dark sooty-brown; rest of upperparts grayish-brown, becoming pale gray on lesser and middle coverts and tertials; rest of wing dull black; tail dusky with white terminal band; throat and foreneck white; rest of underparts pale gray. **Immature ♂:** like adult female but above somewhat paler and grayer. **Young:** like adult female but above paler, more brownish, the feathers with dull cinnamon fringes and dusky shaft-streaks.

HABITS. Roams through canopy of forest and crowns of taller trees along edges, in semi-open, or in clearings or savannas, in pairs or small, straggling flocks; usually sallies to pluck fruit, large insects, or an occasional small lizard from foliage, sometimes hovering briefly; or hops heavily along branches to snatch from perch; eats figs, many arils (*Casearia, Trichilia*), lauraceous fruits.

VOICE. Most often a dry, nasal, grunting *reek-*

reek or *reek-rack*, the second note higher; or a longer series of such notes on 2 pitches; various dry, insectlike notes.

NEST. In woodpecker hole or other cavity in tree, partly filled with loose litter of leaf fragments, bits of twigs and inflorescences, beneath which eggs are hidden during female's absences and nestlings lie until about 2 weeks old, usually 40–100ft (12–30m) up, rarely 11ft (3.4m). Eggs 2, dark buff, heavily marbled with brown, especially on thicker end. March–July; 2 broods.

STATUS. Common resident of lowlands and middle elevations the length of both slopes, in humid and dry regions, to 5000ft (1500m) on Caribbean slope and 6000ft (1850m) on Pacific, rarely to 7500ft (2300m).

RANGE. N Mexico to W Ecuador, E Peru, and Amazonian Brazil.

BLACK-CROWNED TITYRA

Tityra inquisitor Pl. 34(2)

Tityra Coroninegra

DESCRIPTION. 7¼″ (18.5cm); 50g. Smaller than Masked Tityra, without bare face; bill very thick. **Adult ♂:** pileum and lores black; nape white; rest of upperparts pale gray, including lesser and middle wing-coverts, tertials, and basal half of tail; rest of tail black with narrow white terminal band; rest of wings black, except inner webs of primaries white basally, forming conspicuous patch in flight; cheeks white, shading to grayish-white on underparts. Upper mandible black, lower silvery-gray; legs dark gray. ♀: forehead buff to dull rufous; crown dull black; sides of head chestnut; nape mottled with grayish, white and brown; back and rump grayish-brown; lesser and middle coverts, tertials, and base of tail dull ash-gray; rest of wings and tail blackish, tail narrowly tipped with white; throat white; rest of underparts pale gray. Upper mandible dusky, lower grayish. **Immature ♂:** like adult female but forehead white, mantle paler and grayer. ♀: pileum duller, mixed with dull gray. **Young:** like adult female but pileum mixed with chestnut and black; nape white, mottled with buff and dusky; feathers of back with darker centers and pale fringes, producing smudgy effect; wing-coverts and tertials tinged with buff and mottled with dusky; underparts tinged with buff.

HABITS. Wanders in pairs or family groups through forest treetops and scattered trees in nearby clearings; scans foliage deliberately until it spots a caterpillar or large insect, then darts up to snatch it while on the wing; eats many small fruits; often lower in trees than Masked Tityra, but the 2 regularly occupy the same areas, even nest in different holes in the same tree.

VOICE. Low, weak, dry or nasal notes, softer than those of Masked Tityra, often in 2s: *chet-chut, chaa-cherp*; a peculiar thin *corre corre*.

NEST. In woodpecker hole or similar cavity in tree, 40–100+ft (12–30+m) up, partly filled with loose litter of leaf fragments, flower stalks, etc. Eggs undescribed (?); once 3 nestlings. March–June.

STATUS. Uncommon resident in lowlands of both slopes, rarely up to 4000ft (1220m).

RANGE. C Mexico to W Ecuador and NE Argentina.

NOTE. Sometimes separated in monotypic genus *Erator*.

FAMILY **Cotingidae:** Cotingas

Despite its reduction in recent classifications from about 90 to 65 species, the Cotingidae remains an extraordinarily diverse family, strictly confined to wooded regions of the mainland Neotropics. Ranging in size from 3 to 18″ (7.5–46cm), cotingas include some of the smallest as well as largest of passerine birds. Males may be wholly white, wholly black, or largely red, orange, green, and yellow, lustrous blue and purple, rufous, or gray. Some have crests, curious wattles, or areas of colorful skin. Females of the more ornate species are more plainly attired. Although a few cotingas are exclusively frugivorous, most appear to prefer a mixed diet of fruit and insects, both usually gathered high in trees. Although some are renowned for their far-carrying voices, others are strangely silent, even when trying to attract a female. Many do not pair; the males, high in trees or rarely on the ground, often in loose or closely spaced groups, call and display to draw females, who build their nests, incubate their eggs, and rear their young unassisted. Nests may be fairly bulky open structures loosely constructed of twigs, neater cups, brackets of mud, and pieces of plants attached to vertical rock faces, or mats barely large enough to hold a single egg and nestling. Many species lay only 1 egg, some 2 or rarely 3, which are gray, grayish-brown, buffy, tan, or olive, with darker markings. Contrasting with the many solitary females, a few cotingas are assisted in rearing the young by attentive mates, and the pair may even have nest helpers. Cotingas' incubation periods of 17–28 days and nestling periods of 21–44 days are exceptionally long for passerine birds.

RUFOUS PIHA

Lipaugus unirufus Pl. 34(10)

Piha Rojiza (Guardabosque)

Description. 9″ (23cm); 75g. Nearly uniform rufous; size and shape of *Turdus* but with bigger head and thicker bill; larger and stouter than Rufous Mourner. Crown rufous-brown with fine, paler shaft-streaks; back and rump duller rufous-brown; wings and tail brighter rufous; throat pale rufous; breast rufous-brown to cinnamon-brown; belly paler cinnamon-brown. Bill horn-color, darker on culmen and palest at base; legs gray to olive.

Habits. Frequents canopy and middle levels of wet forest and adjacent shady semi-open, occasionally visiting clearings with scattered trees; does not pair; usually in small, loose groups but not compact flocks; does not associate more than casually with mixed-species flocks; perches for long periods, hunched forward and looking all about, then makes a sudden sally to snatch fruit, insect, or spider from foliage, often hovering briefly; notably fond of palm and lauraceous fruits and large caterpillars; rarely descends to ground to seize insect or scorpion; often would go undetected except for its loud voice, but usually calls at long and unpredictable intervals.

Voice. A variety of loud, clear, explosive whistles: a shrill, assertive *peeer* or *wheeeo*; a longer *whee-er-wet* or *wheeoweet*; a softer *wheer-weet* or *pee-ha*; a heavy, metallic, rolling *chrrrg*, often in series; a longer, more musical trill; often calls in response to sudden loud noise: a falling branch, handclap, or axe blow (hence its Spanish name, *guardabosque*, forest guard).

Nest. A slight saucer of coiled tendrils, barely sufficient to hold the egg, visible through bottom, supported by thin horizontal twigs or saddled on a slender branch 17–35ft (5–10m) up in forest; built and attended by female alone. Egg 1, grayish, heavily mottled with brown, almost solidly on large end. March–August.

Status. Fairly common resident in lowlands of Caribbean and S Pacific slopes, in foothills to 2000ft (600m) in N and C areas, and to 4000ft (1200m) in S parts of both slopes.

Range. S Mexico to W Ecuador.

SPECKLED MOURNER

Laniocera rufescens Pl. 34(8)

Plañidera Moteada

Description. 8″ (20cm); 48g. In head shape resembles *Cotinga* or *Carpodectes*, not the relatively unpatterned Rufous Mourner or Rufous Piha. **Adults:** head and body rufous, darkest on crown, palest on chin; narrow dusky scaling or barring over entire upperparts and breast, fading out on belly; wing-coverts dusky, broadly tipped with rufous, forming spotted wing-bars; a usually concealed patch of yellow or orange-tawny on sides of breast. Bill blackish, except grayish base of lower mandible; legs gray. **Young:** above and below more heavily barred; face and lower throat washed with grayish; sparse black spotting on breast; rufous spots on wing-coverts set off by black.

Habits. Frequents upper understory of primary forest, often by streams, edges, or wooded swamps; flies rapidly and directly to perch, then holds body still while twisting head to odd angles, scanning surrounding vegetation; by sudden, fluttering sallies plucks insects, their larvae and pupae, small lizards, or fruits from foliage; singly may accompany mixed-species flocks of antwrens, furnariids, tanagers; apparently does not form pairs; males sing from regular perches 10–33ft (3–10m) up, in small, loose groups that are probably leks, individuals spaced at intervals of 160–330ft (50–100m).

Voice. Usually silent. Song, given with neck outstretched and bill wide open, a plaintive, clear whistled *ueeYEE eeeYEEE eeeYEEE eeeYEEE* (3–5 or more phrases, the first shorter and lower in pitch), or *peetiyeet* (Slud), weaker than song of Thrushlike Manakin but similar in tone.

Nest. Unknown (?).

Status. Rare, local resident of wet lowlands and foothills, up to 2300ft (700m), the length of Caribbean slope, and in Golfo Dulce region of Pacific slope, N to Quepos area.

Range. S Mexico to NW Ecuador.

Note. In some recent treatments placed in the Tyrannidae; its true relationships are obscure, but its voice and behavior seem more akin to those of certain cotingas or *Schiffornis*; we tentatively leave it in the Cotingidae.

LOVELY COTINGA

Cotinga amabilis Pl. 34(6)

Cotinga Linda

Description. 7½″ (19cm); 72g. High, dove-like forehead, broad bill; male has outer 4 primaries narrow, 7th primary very short. **Adult ♂:** mostly brilliant blue; chin, throat, upper chest, and center of belly deep purple, interrupted by blue band across lower breast; wing and tail feathers black with greenish-blue margins; tail-coverts nearly as long as tail. ♀: above dark grayish-brown, the feathers broadly edged with whitish, producing scaled effect; below pale buff to whitish with faint dusky spots on throat, larger and more distinct spots on breast; wings dark brown, the coverts with broad, buffy borders, the remiges narrowly edged with whitish. Bill and legs blackish. **Young:** like female but secondaries and wing-coverts edged with cinnamon-buff, those of pileum with pale brown; pale edgings on feathers of back, broader, appear spotted rather than scaled.

HABITS. Stays high in canopy of wet forest or nearby shady semi-open or shady clearings; plucks fruit, especially of Lauraceae, from crowns of tall trees in fluttering sallies; takes insects and small lizards; does not form lasting pairs; several may gather at fruiting trees.

VOICE. Sound between rapid tinkle and rattle, heard whenever male flies, may be made by wings rather than voice; no other notes recorded for male. Female utters loud, agonized shrieks when defending young; when concerned for them, repeats a low, clear *ic ic ic*. . . .

NEST. Probably a shallow bowl; only known nest, hidden among epiphytes in high treetop, held at least 1 nestling in May (Montaña Azul); attended only by female.

STATUS. Rare and local (has probably decreased in recent years because of deforestation) resident of foothills and adjacent lowlands on Caribbean slope from Cordillera de Guanacaste (Volcán Miravalles) S to near Panama border; most records between 1000 and 5500ft (300–1700m), but to sea level in SE.

RANGE. S Mexico to Costa Rica, possibly extreme NW Panama.

TURQUOISE COTINGA

Cotinga ridgwayi Pl. 34(5)

Cotinga Turquesa

DESCRIPTION. 7″ (17.5cm); 55g. Smaller than Lovely Cotinga; outer 2 primaries of male short, the 9th greatly attenuated; tail-coverts not elongated. **Adult ♂:** mostly intense glossy blue; chin, throat, upper breast, and center of belly deep violet, interrupted by narrow band of blue across lower breast; feathers of upperparts and wing-coverts with broad black bases that show through blue as obscure spotting; eye-ring black; remiges and rectrices black, edged with greenish-blue. ♀: above dusky, spotted with white on crown and hindneck, scaled with buff elsewhere; often a blue gloss on rump; below buffy, obscurely and finely spotted with blackish on throat, conspicuously on breast. Upper mandible blackish, lower gray to horn-color with blackish tip; legs dark gray. **Young:** like adult female but wing-coverts edged with cinnamon rather than buff.

HABITS. Wanders widely through treetops in or near wet forest, singly or in loose flocks of 6 or less; rests for long intervals on high, exposed perch; eats fruits of trees, including figs, *Cecropia*, and *Citharexylum*, and of the parasitic mistletoe *Psittacanthus*; descends low to gather berries of pokeweed *Phytolacca* in clearings; evidently does not form pairs; only female attends nest.

VOICE. Flying males produce a low, soft twitter, probably with wings; distress call of female a raucous shriek.

NEST. A slight, shallow cup of coiled tendrils and long, wiry fungal strands, saddled on horizontal limb, the middle of 3 proceeding from fork, 3ft (1m) from trunk, and 30ft (9m) up in isolated tree near forest. Eggs 2, buffy, speckled all over with brown, most heavily on thicker end. March.

STATUS. Resident in lowlands and foothills of Pacific slope from Carara S to Panama, from sea level to 6000ft (1850m) along lower slopes of Cordillera de Talamanca; at any 1 locality appears irregularly and may be absent for long intervals.

RANGE. Costa Rica and W Panama.

SNOWY COTINGA

Carpodectes nitidus Pl. 34(4)

Cotinga Nivosa

DESCRIPTION. 9″ (22.5cm); 105g. Medium-sized, stout-bodied, with broad wings and short tail, broad bill, and high, dovelike forehead. **Adult ♂:** above very pale bluish-gray, darkest on crown, remiges, and rectrices; below white; appears entirely shining white in treetop. ♀: top and sides of head slaty; rest of upperparts duller ash-gray; wings and tail dusky, the secondaries and wing-coverts broadly edged with white; broad eye-ring white; throat and breast pale gray, shading to white on posterior underparts. Culmen black, rest of bill blue-gray; legs blackish. **Immature ♂:** like adult female but above paler gray; sides of head and underparts white; breast tinged with bluish-gray. **Young:** like adult female but above whitish, indistinctly scaled with dusky.

HABITS. Frequents canopy of wet forest and semi-open, coming lower in adjacent tall second growth, at gaps, or along streams; plucks fruits while perching or after a sally and brief hover; prefers fruits of Lauraceae, mistletoes, and figs; pairs or small groups often rest quietly in treetops, screened by foliage in open; flight rapid and direct.

VOICE. Rarely heard, a dry, scratchy *chih* or *chee*, sometimes repeated rapidly 2–8 times, sounding like an oriole's chatter but slower.

NEST. Unknown (?); only known nestling in March.

STATUS. Resident throughout Caribbean lowlands and foothills up to ca. 2500ft (750m); evidently wanders widely; ranges from common to rare in any locality.

RANGE. N Honduras to W Panama.

YELLOW-BILLED COTINGA

Carpodectes antoniae Pl. 34(3)

Cotinga Piquiamarillo

DESCRIPTION. 8½″ (21.5cm); 98g. Resembles Snowy Cotinga but no overlap in range; bill largely yellow. **Adult ♂:** crown very pale bluish-gray; back, rump, and tail white, faintly tinged with gray; wings and underparts white. Bill bright yellow with black stripe on culmen; legs blackish. ♀: forehead and crown dark ash-gray; rest of upperparts somewhat paler; wings and tail blackish, the wing-coverts and secondaries broadly edged with white; sides of head dull slate; broad eye-ring white; throat and breast pale gray, shading to white posteriorly. Base of bill bright yellow, shading to blackish on culmen and tip.

HABITS. Stays high in mangroves and trees of neighboring forests but also visits tall treetops in clearings near woodland; eats fruits of Lauraceae, mistletoes, melastomes; often in loose groups, especially in mangroves, where birds fly about actively. In March and April, a solitary male in foothills displayed by swooping from 1 dead branch to another in the same high treetop, also sidled along branches, always silently.

VOICE. A dovelike or trogonlike *cah* or *cow*, ending in a throaty scrape (Slud).

NEST. Unknown (?).

STATUS. Resident on S Pacific slope, from mouth of Río Tarcoles S; locally common, especially in extensive mangroves, but evidently wanders widely, to 2500ft (750m) in foothills.

RANGE. Costa Rica and W Panama.

NOTE. Sometimes considered a subspecies of Snowy Cotinga, *C. nitidus*; sometimes called Antonia's Cotinga.

PURPLE-THROATED FRUITCROW

Querula purpurata Pl. 34(11)

Quérula Gorgimorada

DESCRIPTION. ♂ 11″ (28cm), 115g; ♀ 10¼″ (26cm), 100g. Fairly large, black, with broad, strong bill, and long, broad wings. **Adult ♂:** above glossy black, including wings and tail; below somewhat duller; chin black; a broad, glossy dark purple expansible gorget. ♀: plumage entirely black, less glossy above than male. Bill silver-gray (with black tip in female); legs black. **Immatures:** dull black like adult female. **Young:** dull brownish-black, barred faintly with deeper black; wings and tail dull black.

HABITS. Roams through upper levels of wet forest in closely integrated groups of 3–8 individuals, sometimes associated with White-fronted Nunbirds and oropendolas; comes lower in semi-open and along edges for fruits of *Heliconia* and *Hamelia*, which it usually takes while perching; in canopy, usually plucks fruits and sometimes insects in a swooping sally; often shakes tail upon alighting or when calling; suns itself with wings spread like a vulture; entire group roosts on a horizontal branch, in close contact.

VOICE. A variety of mellow, full-bodied, whistled calls, most commonly a wavering *kweeowee* or *queruwee*, often repeated 2–3 times rapidly, the first time higher in pitch; a rising *woooeep*; especially when flock is traveling; various dry, scratchy notes: *hwwwwk* like a man hawking to spit, or *hwak-hwak* like a thirsty duck.

NEST. A loosely built, flimsy saucer of twigs and dry panicles, through which egg is often visible, 20–50ft (6–15m) up in tree, often at edge of forest. Egg 1, "dark olive thickly covered with blackish-brown markings" (in Guyana, Snow). May–August. Several adults may feed the nestling.

STATUS. Fairly common resident in lowlands and foothills the length of Caribbean slope from sea level to ca. 1500ft (450m), or to 2000ft (600m) in SE.

RANGE. Costa Rica (and almost certainly SE Nicaragua) to W Ecuador, N Bolivia, and Amazonian Brazil.

BARE-NECKED UMBRELLABIRD

Cephalopterus glabricollis Pl. 34(13)

Pájaro-sombrilla Cuellinudo (Pájaro Danta)

DESCRIPTION. ♂ 16″ (41cm), 450g; ♀ 14″ (36cm), 320g. Large, chunky, black, with heavy, broad bill; big head (made to seem much larger by spectacular umbrella crest, especially male); long, broad wings; short tail. **Adult ♂:** entirely black, above strongly glossed with blue, producing scaly effect on mantle; below duller, duskier black, feather shafts paler. Skin of throat bright orange-red, that of chest vermilion, forming an inflatable sac that becomes scarlet in full display. ♀: above duller black, with less bluish gloss; shorter crest makes head appear flat, forehead bulbous; small orangish bare patch on each side of foreneck. Upper mandible black, lower gray with black at base; legs dark gray. **Young:** resemble adult female but plumage grayer; crest shorter and fluffier; smaller and paler bare patch on neck.

HABITS. Frequents upper understory to mid-canopy of primary forest, sometimes visiting fruiting trees in adjacent tall second growth; eats fruits of palms, Lauraceae, Annonaceae, and some large insects, including orthopterans and caterpillars, small lizards and frogs, plucked from vegetation on noisy sallies, or gleaned with heavy, jaylike hops along branch; beats prey vigorously against perch before swallowing; usually seen singly in nonbreeding season but often migrates in small groups, which prior to breeding may include displaying males. On breeding grounds, males perch and display in loose groups or ''exploded leks'' in subcanopy.

VOICE. At fruiting trees, contact notes include low coughs, grunts, or throaty, guttural chuckles; begging note of young, a soft whine. In display, male calls *HOOM hik-ratch t'hoom K'*. Leaning slightly forward, with bill nearly closed, foreneck sac inflated like a huge ripe tomato, he emits the loud, far-carrying, initial *HOOM*, like a heavy mallet striking an oil drum; then, throwing his head back and opening his bill, he utters the dry, hacking *hik-ratch*; with deflating sac, he emits the second, softer *hoom*; while regaining his upright posture he gives the final dry, staccato *k'*.

NEST. Unknown (?). Males display March–May or June, the probable breeding season.

STATUS. Uncommon and local; spends most of year in foothills and adjacent lowlands of Caribbean slope (N at least to Volcán Miravalles), females chiefly below 650ft (200m), males 330–1600ft (100–500m); migrates upslope to breed locally between 2600 and 6500ft (800–2000m).

RANGE. Costa Rica and W Panama.

NOTE. Sometimes considered conspecific with *C. ornatus* of South America.

THREE-WATTLED BELLBIRD

Procnias tricarunculata Pl. 34(12)

Campanero Tricarunculado (Pájaro Campano, Rin-ran)

DESCRIPTION. ♂ 12″ (30cm), 220g; ♀ 10″ (25cm), 145g. Large, with heavy body, broad flattened bill. **Adult ♂:** entire head, neck, and chest immaculate white; rest of plumage chestnut-rufous, below darkest; skin of lores and orbital area dull gray, with short, scattered, bristly black feathers; 3 long, wormlike wattles dull black. Bill black with silver line along tomia; mouth lining black; legs dark gray. ♀: above olive-green with narrow yellow eye-ring; fine yellow streaking on top and sides of head; below bright yellow, striped heavily with dark olive-green except on lower belly and crissum; bill black, except base of mandible dusky-horn. **Immatures:** like adult female but below streaking much more blurred and indistinct; male larger, with wattles of varying length.

HABITS. Frequents middle and upper levels of mountain forest during breeding season; at other times lower in tall second growth, visits tall trees in semi-open; plucks fruits of Lauraceae and other trees, usually while perched but sometimes with a short, heavy sally; males sing and display from high perches, often exposed above canopy and nearly invisible from below, though final phase of courtship on open perch in subcanopy; does not pair, nest evidently built and attended only by female; outside breeding season males call sporadically from dense canopy vegetation.

VOICE. Song of male varies with locality; includes loud, far-carrying notes, mostly wooden or metallic, without resonance, but in some regions somewhat more bell-like; dry staccato notes, high penetrating whistles, rolling or nasal notes. Examples: *BONK seee k'k' see k'berk; bonk see k'k' berk see see see BONK* (Monteverde, Braulio Carrillo); in Coto Brus valley a simpler *krik BRENK . . . BRENK jreet jrrrt*, etc. The loud notes (BONK, BRENK or BUCK) are audible for ⅓mi (0.5km) or more. While he emits these notes, male gapes widely, exposing cavernous black mouth; at conclusion of song, he flies out a short distance, turns sharply, and returns to perch, where he briefly spreads his tail and retracts his neck; when female visits display perch, the 2 exchange places in a simple ritual.

NEST. Undescribed (?). Breeding season (as judged by gonad data of specimens and displays) mainly March–June but much variation between years and localities.

STATUS. Breeds mainly 4000–7500ft (1200–2300m) in Cordillera de Tilarán, Caribbean slope of Cordillera Central, both slopes on Cordillera de Talamanca, and as low as 3000ft (900m) on Cordillera de Guanacaste and highest mountains of Península de Nicoya; ranges sporadically to 10,000ft (3000m) or more following breeding; during second half of year descends to foothills and lowlands of both slopes; at some foothill localities (e.g., Valle del General) has been heard in every month and rarely is present year-round.

RANGE. E Honduras to W Panama.

SHARPBILL

Oxyruncus cristatus Pl. 34(7)

Picoagudo

DESCRIPTION. 6½″ (16.5cm); 42g. Stout body; rather small head; sharp-pointed bill; short, strong legs; below spotted but totally unthrushlike. **Adult ♂:** crown black, mixed with olive; concealed scarlet erectile crest; hindneck mottled with dusky; rest of upperparts olive-green, with yellowish wing-bars; face and underparts pale yellowish, brightest on belly; face and throat scaled, breast and sides spotted with black; outer primary with serrate leading edge. Iris brick-red; upper mandible dusky, lower grayish; legs gray. ♀: crown more olive; crest smaller, orange; outer primary not serrate. Iris orange. **Young:** no orange in crown; black markings less heavy; wing-bars brighter and broader.

HABITS. Dwells in canopy of wet forest, usually keeping high in trees; of stolid demeanor, sits quietly and sluggishly, then moves abruptly; flies horizontally with bursts of fast, buzzy flaps and hurtling glides; hops rapidly through foliage, often clinging or hanging acrobatically; uses sharp-pointed bill to pry and poke into tufts of moss and rolled or appressed leaves for insects and spiders, or to pry open dehiscing pods with arillate seeds; takes berries; solitary, but may accompany mixed-species flocks. During breeding season, 3–5 males form small leks high in canopy, singing from perches 330–1000ft (100–300m) apart.

VOICE. Song a long-drawn-out, wiry, descending trill, *eeeeeuuuurrrrr*; otherwise silent.

NEST. A shallow cup of leaf petioles, with cardboardlike outer coating of lichens, mosses and cobweb, saddled on small twig in crown of tall canopy tree (in Brazil; no Costa Rican record). Eggs undescribed (?). In Costa Rica, breeds March–June.

STATUS. Uncommon and local resident at 2300–4600ft (700–1400m) from Cordillera de Guanacaste (Volcán Miravalles) S along Caribbean slope to Panama; also (formerly?) in Dota region. Some descend after breeding to ca. 1300ft (400m).

RANGE. Costa Rica discontinuously to E Peru, SE Brazil, and Paraguay.

NOTE. Recent biochemical, behavioral and anatomical evidence indicates that the Sharpbill, long placed in its own monotypic family, is a member of the Cotingidae.

FAMILY **Pipridae:** Manakins

The 60 species of manakins are confined to wooded regions of tropical continental America, including Trinidad and Tobago, chiefly in warm lowlands. Most are small with stout bodies, usually short tails, and short, broad bills. Adult males are nearly always much more ornate than females. Their prevailingly black plumage is relieved by contrasting areas of intense red, orange, yellow, blue or white, on the head, encircling the neck, on back or breast. Some have crests or elongated tails. Females are much plainer, often unadorned olive or grayish. Largely frugivorous, manakins pluck berries on a short flight, without alighting, and they snatch insects from foliage in the same manner. A few have melodious calls; others compensate for the limitations of their voices by making with their wings whirring sounds, a snap like that made by breaking a dry twig, or volleys of such snaps. Manakins are not known to pair. Males of many species gather in courtship assemblies or leks, where, with vocal or wing sounds and the most varied antics, they attract females. Usually each male displays at his own private station, in a tree or bush, less often over a bare patch of ground that he has cleared of debris, in hearing if not in sight of his competitors. In a few species, 2 or more males join in coordinated displays on the same perch. With no help from a mate, the female builds her nest, incubates her 2 eggs for 18–21 days, and rears her young, who remain in the nest for 13–20 days.

RED-CAPPED MANAKIN

Pipra mentalis Pl. 33(6)

Saltarín Cabecirrojo

DESCRIPTION. 4″ (10cm); 16g. Notably short legs and puffy thigh feathers; bill and legs appear pale. **Adult ♂:** top, sides, and back of head brilliant scarlet; thighs bright yellow; chin and wing-linings pale yellow; rest of plumage black. Iris white; upper mandible pale horn, lower mandible and legs flesh-color. ♀: above and on chest dull olive-green; throat and belly paler; thighs more yellowish;

occasionally a few red feathers on head. Iris dull brown, rarely white; upper mandible dark horn, lower flesh-color; legs pale brownish-flesh. **Immature ♂:** like adult female but iris white, usually some red on head, sometimes a few black feathers on body. **Young:** like adult female.

Habits. Frequents lower and middle levels of mature wet forest, adjacent tall second growth and shady clearings, often visiting fruiting trees at gaps and edges; plucks small fruits of melastomes, aroids, Rubiaceae, *Guatteria*, on the wing with a rapid, buzzy sally; solitary except in courtship assembly, where several males perform on thin horizontal branches well above ground and 10–100ft (3–30m) apart. Their displays include backward slides with legs stretched up and yellow thighs exposed; pivoting through 180 degrees, and looping flights. In forests where Blue-crowned Manakins occur, stays higher, in upper understory and subcanopy.

Voice. Most often a short, explosive *psip*, often preceded by a high, thin *p'tsweeeeee*; full call in display *psit psit psit p'tsweeee psip*, final note sharply emphatic; also several *psit* notes followed by a buzz; a series of snaps made by male's wings, sounding like a typewriter, or snapping whirrs also made by wings. Females nearly always silent.

Nest. A shallow cup fastened by rim in horizontal fork, of fine, usually brown fibers, with fragments of dead leaf attached to bottom, 5–35ft (1.5–10.5m) up in forest. Eggs 2, dark grayish-buff mottled with brown, especially in wreath. March–July.

Status. Common resident of lowlands of Caribbean and S Pacific slopes (N at least to Manuel Antonio), locally up to 3500ft (1050m) in foothills, reaching Pacific slope of Cordillera de Guanacaste via low passes.

Range. SE Mexico to NW Ecuador.

Note. Also called Yellow-thighed Manakin.

BLUE-CROWNED MANAKIN

Pipra coronata Pl. 33(7)

Saltarín Coroniceleste

Description. 3½″ (9cm); 12g. Small; abrupt forehead and flat crown conspicuous in adult male. **Adult ♂:** velvety black except crown bright blue. Iris reddish-brown; upper mandible black, lower blue-gray; legs black. ♀: forehead dull yellowish; rest of upperparts dull green, not olive like other female manakins; chin whitish; throat grayish mixed with green; breast and sides dull green; belly dull olive-yellow. Iris dark brown; legs dark gray. **Immature ♂:** like adult female, above somewhat grayer; often tinged with blue on rump; a few blue feathers in crown. **Young:** like adult female but above duller, more grayish, below grayish-buff.

Habits. Frequents understory of humid forest and taller second growth; plucks berries and snatches insects from foliage by a darting upward sally, dropping back to a lower perch; solitary except in males' courtship assemblies, where each performs on horizontal twigs 3–40ft (1–12m) up, within hearing of several others. Relatively simple display consists of rapid, darting flights between horizontal branches of small trees, accompanied by song but no wing sounds, males displacing each other on perches.

Voice. A soft, clear, somewhat rattling trill, *prrrreew* or *chk-rrreew*; males on courtship assembly sing *pi pipipipi chu-WAAK*, the last notes harsh, nasal, froglike; also a sharp *k'wek k'wek*, unaccompanied by other notes.

Nest. A tiny, shallow cup attached by rim in fork of slender horizontal twig, of fine, pale-colored fibers, covered below with dry, papery leaf fragments and sometimes green moss, bound with cobweb, 20″–7ft (0.5–2m) up in forest undergrowth. Eggs 2, dull white to pale gray, heavily mottled with brown or rufous-brown, chiefly in wreath. February–June.

Status. Common resident of S Pacific slope, N regularly to Quepos, rarely to Carara, from lowlands to 4500ft (1350m) locally; in Sixaola region of extreme SE Caribbean lowlands, to 4000ft (1200m).

Range. Costa Rica to NW Ecuador, N Bolivia, and C Brazil.

Note. Sometimes called Velvety Manakin.

WHITE-CROWNED MANAKIN

Pipra pipra Pl. 33(8)

Saltarín Coroniblanco

Description. 4″ (10cm); 14g. Male has long, flattened, white crown-patch that can be erected into a low crest; note female's brick-red eye, grayish crown. **Adult ♂:** velvety black, except pileum and hindneck white. Iris brick-red; upper mandible black, lower silver-gray; legs dark gray. ♀: above olive-green; top and sides of head and hindneck strongly tinged with slate-gray; throat grayish-olive; breast brighter olive; belly paler olive-yellow; flanks dull olive-green. Bill grayish-horn, culmen blackish; legs grayish. **Immature ♂:** like adult female, head often more strongly washed with gray. **Young:** like adult female but below duller and darker, with little yellow tinge on belly; iris brown.

Habits. Prefers dense understory of humid foothill forest and adjacent tall second growth; makes sudden, buzzy sallies up to snatch fruits of melastomes, *Cephaelis, Psychotria, Phytolacca*; flicks wings nervously while perched; solitary except for courtship assemblies of males, which are smaller (ca. 2–4 birds) and looser than those of other manakins. Each male perches on 1 of several bare horizontal branches 10–40ft (3–12m) up within a radius of ca. 160ft (50m) and just within earshot of others; flies between perches with butterflylike, slow, deep wingbeats.

Voice. A scratchy to buzzy, sibilant *jeea'eeeh* or *cheeahi*, often with a nasal, almost sneering tone; male on song perch sings a louder version of this call, preceded by 1 or more popping notes, with each of which he pumps his head and body up and down: *p'p'p' cheeea'aeeh*.

Nest. Unknown (?).

Status. Resident at lower middle elevations ca. 2600–5000ft (800–1500m) of Cordillera Central and Cordillera de Talamanca.

Range. Costa Rica to NE Peru and SE Brazil.

LONG-TAILED MANAKIN

Chiroxiphia linearis Pl. 33(4)

Saltarín Toledo (Toledo)

Description. 4½″ (11.5cm); elongated central rectrices add 4–6″ (10–15cm) to adult males and ca. 1″ (2–3cm) to females; 19g. Abrupt forehead, elongated central rectrices, and orange legs distinctive; only manakin in most of its range. **Adult ♂:** mostly black; crown, including bifid crest on hindcrown, glossy crimson; back sky-blue; long, very narrow central rectrices. ♀: above and on breast olive-green; throat paler, washed with grayish; posterior underparts pale olive, becoming whitish on lower belly and crissum; rarely some red in crown. **Immature ♂:** requires 3–4 years to attain full adult dress; by 1 year has red crown, dusky face, more or less elongated central rectrices; by 2 years has rest of head blackish, some blue on back and black below. **Young:** like adult female but somewhat paler on belly.

Habits. Frequents dry or humid forest with abundant undergrowth, especially common in gallery forest in northwest, also borders of mangrove swamps, tall second growth; sallies to snatch fruits, especially of dark-leaved understory tree *Ardisia revoluta*; pairs or trios of males perform courtship display in dense, shady tangles. From stations ca. 1ft (30cm) apart 2 males alternately make fluttering leaps straight upward, to fall again into the same spots, the leaps becoming lower as the tempo of the dance increases. In "cartwheel" dance, usually given with female present, each male in turn flutters up and backward to descend to the spot from which his dance partner has meanwhile shuffled forward. Only 1 male of each pair or trio, he with fullest adult dress, if differences exist, copulates with female.

Voice. Very varied, including a short, sharp *weet* or *pwit*; a nasal, whining *waaah*; a clear, resonant *heer-ho*. Two males on or near display perch call in unison, a clear, ringing *to-lay-do*; while dancing, males repeat a nasal, catlike *miaow-raow*; dance ends with a piercing *pweet*.

Nest. A shallow cup fastened by rim in a horizontal fork, of fungal rhizomorphs, mosses, leaf blades, petioles, bast fibers, grasses, and cobweb, with dry leaves hanging from the outside, 2–7ft (0.6–2m) up, often over dry streambed or open space. Eggs 2, less often 1, buffy, lightly or heavily spotted with chocolate-brown. April–July.

Status. Common, locally abundant resident from lowlands to ca. 5000ft (1500m) on N Pacific slope, S to Carara and Dota region, including Valle Central; to Caribbean slope from Ochomogo to (formerly?) Juan Viñas and locally along Cordillera de Guanacaste.

Range. S Mexico to Costa Rica.

LANCE-TAILED MANAKIN

Chiroxiphia lanceolata Pl. 33(5)

Saltarín Coludo (Toledo)

Description. 5¼″ (13.5cm); pointed central rectrices extending an additional ½″ (1cm) or more in adult males, less in females; 19g. Resembles Long-tailed Manakin but central rectrices much shorter, males duller black. **Adult ♂:** mostly black, below with strong dull gray-green tinge; feathers of crown glossy crimson, elongated into short crest (not bifid); back pale blue. Iris chestnut; bill black; legs bright orange. ♀: above and on breast olive-green, shading to pale grayish-olive on throat and to pale yellowish to grayish-white on belly. Bill dusky. **Immature ♂:** like Long-tailed, requires 3–4 years to attain adult plumage by (probably) similar sequence; in younger males red of crown is paler, more scarlet, than in adult.

Habits. Frequents middle and upper understory of moist forest, especially where partly thinned, tall second-growth scrub in more open country; forages mainly by sallying and hovering to pluck fruit from shrubs and small trees. During breeding season pairs of males engage in synchronized display on traditional perches, as does Long-tailed Manakin.

Voice. Usual call a clear, mellow *peew* or *keer*; also a sharp *kip*, a nasal, descending *waaanh*, 1 or several whistled *tyooi* notes, and a slow, liquid trill like a telephone ringing; advertising song, given in unison by 2 males, a clear *to-wit-do*, rather more broken than the *to-lay-do* (*toledo*) of Long-tailed Manakin.

Nest. A small, shallow cup of grass and leaf fibers, slung between the 2 branches of a low, horizontal, forked twig. Eggs usually 2, buffy with reddish-brown spotting, mainly in a wreath around large end. August–September (in Panama; no Costa Rican record).

Status. Center of abundance around head of relatively dry Coto Brus valley, from about Las Alturas through La Unión, around to Sabanilla, ca. 3300–5000ft (1000–1500m), in scattered local concentrations; presumably resident. An old report from Golfito but no recent one.

Range. Costa Rica to N Venezuela.

WHITE-RUFFED MANAKIN

Corapipo leucorrhoa Pl. 33(9)

Saltarín Gorgiblanco

Description. 4″ (10cm); 12.5g. Small, with rather long tail; male has outermost primary very short and narrow; female told from other species by gray throat, dark legs. **Adult ♂:** mostly glossy blue-black; crissum white; erectile white gorget covering chin, throat and sides of neck (in S Pacific race, white less extensive on center of throat). Upper mandible black, lower silver-gray; legs dark gray to blackish. ♀: above olive-green; throat pale gray, more or less washed with olive; breast, sides and flanks olive-green; belly pale greenish-yellow; lower mandible dull gray with dusky tip. **Immature ♂:** like adult female but throat often paler gray, sometimes with varying amount of white. During second year, males are mostly olive-green, with black mask and largely white throat (Rosselli). **Young:** like adult female but somewhat darker, more olive on throat and belly.

Habits. Frequents lower and middle levels of wet forests and nearby shady clearings and tall second growth; on breeding grounds often somewhat gregarious, foraging in loose parties of both sexes; more solitary in lowlands, where 1–2 birds may briefly join mixed-species flock with tanagers as nucleus; plucks berries and insects on upward or outward sally. Display ground a mossy fallen log in forest, shared by 3–4 adult males who, 1 at a time, descend to it obliquely by a slow, undulating flight with tail raised, to alight with gorget fluffed and immediately give a little fluttering jump as though the log were electrified, then fly off as another male alights.

Voice. A high, thin, rolling *prreeet*, a softer, thin, sharp *seee*, and a sharp squeal. As male alights on log he makes a dull snapping *flap* or *fut* with wings, followed immediately by a vocal *chee waa* or *chee-rup*, the *chee* rather strident, the short second note given during the hop.

Nest. A shallow cup or hammock slung between thin arms of horizontal fork, of long, thin, brown filaments or blackish fungal rhizomorphs with leaf skeletons attached to bottom, 16–23ft (5–7m) up in forest. Eggs 2, whitish, heavily marked with brown. April–June.

Status. Common resident the length of the Caribbean slope, breeding mainly at 1300–3000ft (400–900m), higher in SE; and on S Pacific slope, breeding mainly between 3500 and 5000ft (1100–1500m) on the Cordillera de Talamanca and coastal ranges, and above 2000ft (600m) on the higher hills of the Península de Osa; descends to the foothills and adjacent lowlands between about July and December or January, and sporadically at other seasons.

Range. E Honduras to NW Venezuela.

Note. Populations of Central America and W South America often classified as a distinct species, *C. altera*, from the more eastern *leucorrhoa* (White-bibbed Manakin).

ORANGE-COLLARED MANAKIN

Manacus aurantiacus Pl. 33(2)

Saltarín Cuellinaranja (Hombrecillo)

Description. 4″ (10cm); 15.5g. Adult males have outer 5 primaries very narrow for distal half; other remiges with thickened, bowed shafts, so that wings produce a rustling or ticking sound in flight. **Adult ♂:** pileum, wings, and broad band across midback black; cheeks, throat, breast, and broad collar around hindneck extending to shoulders and upper back glossy, bright orange; lower back, rump and tail olive-green; posterior underparts golden-yellow, tinged with olive. Bill blackish; legs orange. ♀: above olive-green; rump and upper tail-coverts paler and more yellowish; throat yellowish-olive; breast, sides, and flanks olive-green; belly bright olive-yellow. **Immature ♂:** like adult female but throat and breast tinged with golden. **Young:** like adult female but legs yellowish to flesh-orange.

Habits. Usually low in more open parts of rain forest and margins, taller second growth, shady plantations, gardens; sallies for berries and insects; solitary except in courtship assembly, where each male removes all litter from small, roundish patch of ground, above which he leaps back and forth between thin, upright stems, snapping loudly with each jump. When female comes they jump together, crossing each other above bare court. Two or 3 males often display together as they travel through woods.

Voice. A clear *cheeu*; when disturbed, *chee-yu* or thin, tense *chee*. Loud single snaps, like sound of breaking dry twig, volleys of snaps and whirrs, are made by male's wings, which rustle audibly in flight.

Nest. A shallow cup slung between arms of thin, horizontal forked twig, usually of pale bast fibers without leaves or mosses attached to bottom, 2–8ft (0.6–2.5m) up, rarely higher, in forest, coffee plantation, shady garden, or above stream. Eggs 2, pale gray or blue-gray, heavily mottled with brown, the marks on side often longitudinal. March–June, rarely September.

Status. Locally common resident from Carara S on Pacific slope, from lowlands to 3600ft (1100m).

Range. Costa Rica and W Panama.

Note. Sometimes classified as a race of Golden-collared Manakin, *M. vitellinus*, of Panama and Colombia.

WHITE-COLLARED MANAKIN

Manacus candei Pl. 33(1)

Saltarín Cuelliblanco (Bailarín)

Description. 4¼″ (11cm); 18.5g. Wings of adult male modified as in Orange-collared Manakin; no other manakin in its range combines yellow belly and orange legs. **Adult ♂:** pileum, wings, broad band across midback, and tail black; sides of head, throat, breast and broad collar around neck, extending to shoulders and upper back, white; rump and upper tail-coverts olive-green; posterior underparts bright yellow. Bill black. ♀: upperparts, head, and breast olive-green, slightly paler on throat and lower rump; sides and flanks yellowish-olive; belly bright yellow, more or less tinged with olive. **Immature ♂:** like adult female but sometimes with throat paler, more grayish. **Young:** like adult female but duller, belly pale olive with little yellow, legs pinkish-orange.

Habits. Prefers tangled margins of humid forest, banks of woodland streams, dense, tall second growth, old cacao plantations; usually stays in low thickets; sallies to pluck fruits as do other manakins; often moves in loose groups, the males periodically snapping their wings and displaying, often protruding feathers of throat beyond bill to form a beard. Courtship displays much like those of Orange-collared Manakin but clears much larger courts, sometimes over 4ft (1.2m) across, amid vertical saplings.

Voice. Call a rolling *preew* or *wheeyu*; males sometimes call *prrr-weeu*, the first note rolling, the second clear; with wings they make rustling or whiffling sounds in flight as well as single loud snaps or rolling series of same; in undulatory flight to court, a peculiar low *brrrt*.

Nest. A shallow cup suspended from arms of horizontal fork, of black fungal rhizomorphs and fine brown fibers, often lined with filamentous brown *Myriocarpa* inflorescences that may dangle below for up to 4ft (1.2m); outside usually decorated with green moss; 3–10ft (1–3m) up. Eggs 2, whitish, speckled with brown and with broad wreath of brown streaks. April–August.

Status. Common resident of lowlands and foothills of Caribbean slope, barely reaching Pacific slope of Cordillera de Guanacaste through low passes; sea level to 2300ft (700m).

Range. SE Mexico to extreme W Panama.

GRAY-HEADED MANAKIN

Piprites griseiceps Pl. 33(3)

Saltarín Cabecigrís

Description. 4¾″ (12cm); 16g. Rather long-tailed for a manakin; flycatcherlike coloration; conspicuous white eye-ring gives a big-eyed, staring look. **Adults:** head mostly slate-gray, mixed with olive on crown and nape; rest of upperparts olive-green; center of throat yellow, shading to olive-green on sides of throat, across breast, and on sides and flanks; belly yellow, inner webs of tertials largely pale yellow. Upper mandible dark gray, lower pale gray; legs plumbeous. **Young:** similar but head olive, without gray.

Habits. Frequents upper understory and middle levels of forest and adjacent tall second growth, descending to shrub level at gaps and edges; usually seen singly, often accompanying understory mixed-species flocks of small antbirds, greenlets, etc.; sits quietly, peering about; sallies abruptly to pluck fruits or seize beetles, small grasshoppers, katydids, or other insects from vegetation. Males sing from dense foliage in upper understory or subcanopy, usually in loose groups, each barely within hearing of others.

Voice. Call a soft, liquid, rolling *purrr* or *wurrr*, not nasal like note of bentbills; soft *chips*, flicking tail with each note. Song an elaborate, structured medley of staccato and rolling notes: *pik pík prrrity pee-kit prrrity peer* or *wip-píp pitpit-purrurr pip-purrr*.

Nest. Unknown (?).

Status. Uncommon and local resident of Caribbean lowlands and foothills, most numerous 330–2000ft (100–600m), up to 2500ft (750m) locally; principally from Limón and Siquirres N, but recently encountered in Sixaola area (Suretka) near Panama border; possibly ranges continuously along base of Cordillera de Talamanca.

Range. E Guatemala to Costa Rica.

THRUSHLIKE MANAKIN

Schiffornis turdinus Pl. 33(10)

Tordo-saltarín

Description. 6¼″ (16cm); 35g. Larger, with longer tail and bill than other manakins; male not brightly colored. **Adults:** above deep olive-brown, wings and tail brighter, more russet; throat and chest brighter olive, often with ochraceous tinge (especially in males), throat sometimes paler; posterior underparts grayish-olive to dull olive-brown. Bill blackish with tomia horn-color, base of lower mandible gray; legs plumbeous. **Young:** like adults but duller; feathers of head with sooty-black fringes; breast with sooty tinge.

Habits. Moves deliberately through lower levels of mature forest, rarely entering tall, shady second growth beside forest, often clinging to vertical stems; more insectivorous than other manakins, eating caterpillars, beetles, katydids, and cicadas as well as berries and arillate seeds, all snatched from vegetation on short sallies; solitary, not forming courtship assemblies or pairing. Males sing from widely spaced perches, may be territorial.

Voice. Song of male 2 or (usually) 3 high, clear, beautifully modulated whistles, the first slurred downward (may be very long-drawn-out, short, or absent), the second slurred up, the third short: *peeeeeuuuu puuuweeet-pu*, or *keew kyoooooweet-ku*; or sliding up the scale in 3 parts, the first longest and the second shortest, the entire song lasting 1–2sec, delivered up to 7 times/min; a low, descending rattle and a short, rippling *prrrw*.

Nest. A bulky cup of large dead leaves and leaf skeletons, thickly lined with blackish fungal rhizomorphs, 3–5ft (1–1.5m) up in spiny palm, vine tangle, or small epiphyte, supported from below. Eggs 2, glossy, pale buff, spotted and blotched with black, dark brown, and pale lilac, chiefly in wreath. February–August, chiefly May.

Status. Resident the length of Caribbean slope and on S Pacific slope, N to Carara; uncommon to fairly common in foothills at 1000–2500ft (300–750m) on most of Caribbean slope, and up to 3300ft (1000m) in SE; very uncommon to rare in lowlands; on Pacific slope fairly common at 1650–4000ft (500–1200m), to 5500ft (1700m) above Coto Brus valley; uncommon in wet lowlands.

Range. SE Mexico to W Ecuador, N Bolivia, and SE Brazil.

Note. This species differs in numerous ways from other manakins (anatomy, coloration, diet, voice, nest, etc.) and is probably more closely allied to the cotingas or even the flycatchers; has also been called Thrushlike or Brown Mourner. The birds from Mexico to E Panama may be a separate species (*S. veraepacis*, Brown Manakin) from true *S. turdinus*, of E Panama and South America.

FAMILY **Tyrannidae:** American or Tyrant Flycatchers

With about 384 species, this is the largest avian family confined to the Western Hemisphere, where it spreads widely over the continents and islands, from Alaska to Tierra del Fuego. Its members occur in dense lowland rain forests to treeless páramo and puna of the high Andes, from marshes to deserts, from grasslands to mangrove swamps. The smallest flycatchers, 2½″ (6.5cm) long, are among the smallest of birds; the largest, 12″ (30cm) long, are jay-sized (several long-tailed species measure longer but have smaller bodies). They tend to be plainly attired in shades of olive, gray, and brown, making some of the smaller species very difficult to distinguish in the field (or even in the hand!). Many species display much yellow; a few are largely black or white; and very few are brilliantly colored. Widespread among flycatchers are crown-patches of red, vermilion, yellow, or white, usually concealed by darker feathers except when spread in moments of excitement. With rare exceptions, the sexes are too similar to be readily distinguished in the field, and the young differ little from adults. Flycatchers rarely sing except at dawn and less often in evening twilight, when their prolonged performances may be harsh, quaint, or arrestingly beautiful. Their calls, loud or soft, are exceedingly diverse and often help in identifying species with similar plumages. Species that breed at higher latitudes are highly migratory; some, such as the kingbirds, travel conspicuously by day.

Flycatchers' bills tend to be broad and flat, adapted for catching insects, which some pursue on spectacular aerial sallies, but perhaps more, especially of the smaller species, pluck from leaf or stem as they dart through vegetation. Many add berries and arillate seeds to their fare; a few are mainly frugivorous. Some of the larger species capture small lizards and frogs, or even small mammals. Even fishes are eaten by a few, including the Great Kiskadee.

Of the relatively few flycatchers that have been carefully studied, most are monogamous. Males not infrequently help to build nests but scarcely ever incubate and as a rule feed and defend the young. Males of a few species take no share in nesting; some gather in courtship assemblies and call to attract females, much as do manakins and hummingbirds, without, however, the elegant plumage and elaborate displays of these 2 families.

Flycatchers' nests are as diverse as their habitats and habits. Many species build cuplike nests in trees or shrubs, or rarely, on the ground. A few plaster nests of mud on protected vertical surfaces. Some place their nests in holes in trees, in buildings, birdhouses, or clefts in cliffs. In the tropics, many build covered nests, which may be roofed structures with a side entrance, supported from below, or a great variety of more elaborate pendant nests. Species difficult to differentiate by appearance often build very different nests. Many tropical flycatchers lay only 2 eggs, those of higher latitudes up to 6. The eggs may be plain white or variously pigmented. Incubation periods are 12–23 days, nestling periods, 14–28 days, both being longer among tropical species.

BLACK PHOEBE

Sayornis nigricans Pl. 36(5)

Mosquero de Agua

Description. 6″ (15.5cm); 21g. Medium-sized, blackish; rarely found far from water. **Adults:** head, neck, and most of underparts dull black to slate-black; center of belly white, flanks and crissum sooty-gray; rest of upperparts blackish-slate; 2 gray wing-bars; secondaries and tertials edged with grayish-white; tail blackish, with outer webs of outer rectrices edged with white. Bill and legs blackish. **Young:** plumage mainly dingy brownish-black; belly and wing-bars dull cinnamon-brown to buffy-brown; feathers of back and rump with pale brownish fringes.

Habits. Singly or in pairs, along boulder-strewn mountain streams, at coffee *beneficios* where much water is used, above streets in streamside towns, and in similar situations; pumps tail up and down; captures dragonflies and other insects in the air, often darting up from a boulder in midstream, or picks them from bare mud or the cement floor used for drying coffee, dropping down from fence post, wire, or other lookout; enters porches to glean insects caught in spiderwebs; roosts on service wires on porch or beneath eaves, or in recesses among streamside rocks.

Voice. Call a simple *chip*, often sharp and brisk; song, usually delivered at dawn but also in full daylight if bird is excited, a rather dry *fe-be*, repeated many times while the bird perches or circles in the air, holding his body almost upright.

Nest. A thick-walled cup of dried mud or clay, strengthened by fragments of plants and lined with fine grass, rootlets, and a few feathers, attached to vertical surface beneath bridge or overhanging rock. Eggs 2–3, plain white or more or less spotted with reddish-brown. March–May.

Status. Resident on both slopes, chiefly 2000–6000ft (600–1850m); descends to lower altitudes after nesting.

Range. SW USA to NW Argentina.

LONG-TAILED TYRANT

Colonia colonus Pl. 36(6)

Mosquero Coludo

Description. 5″ (13cm), not including long, narrow central rectrices that add 4–4¾″ (10–12cm) to males, 2–3½″ (5–9cm) to females; 15g. Rather short, broad bill; black plumage with pale cap and back diagnostic. **Adult ♂:** mostly dull black, grayer on belly; crown sooty-gray, bordered by broad white band extending from forehead past eye, over auriculars, to sides of hindneck; an irregular stripe of grayish-white down center of back. Bill and legs blackish. ♀: similar but paler and grayer on posterior underparts; back stripe averages grayer, crown darker. **Young:** paler than adults; dark sooty-gray with paler gray belly; faint whitish stripe bordering crown; central rectrices broader, barely projecting beyond rest of tail.

Habits. Nests in and forages from dead or dying trees or snags, riddled with woodpecker holes, that project well clear of other vegetation in humid forested areas: recent clearings, semi-open, abandoned plantations, along waterways, beside treefall gaps in forest, or on snags in upper canopy; hawks flying insects, especially stingless bees, often on long sallies with intricate maneuvers; usually in pairs that perch not far apart on exposed twigs or branches.

Voice. Mates maintain contact by frequently repeated high, thin, *peee* or *sweet* with upward inflection, often 2–3 times in rapid succession; song of male includes longer phrases in a deeper, more plaintive voice: *weeee we we we* or *bee wee we wee*, the longer notes with a slight quaver; a sharp, sibilant *chip* in interactions.

Nest. A thick mat of rachises of compound leaves, in old woodpecker's hole or other cavity, 25–100ft (8–30m) up in tree or palm trunk. Eggs white (in South America; not seen in Costa Rica). March–July.

Status. Common resident of Caribbean lowlands, to ca. 2000ft (600m), rarely higher, in foothills.

Range. NE Honduras to W Ecuador, NE Argentina, and Guianas.

SCISSOR-TAILED FLYCATCHER

Tyrannus forficatus Pl. 35(5)

Tijereta Rosada (Tijerillo)

Description. 7¾″ (19.5cm) to tips of central rectrices; long outer rectrices add up to 6″ (15cm) to length of males, 3½″ (9cm) to females; 40g. Outer primary greatly attenuated (1–3mm wide) for terminal 20–30mm in males, 10–15mm in females. **Adult ♂:** pileum and hindneck pale gray, shading to duller ash-gray, tinged with pink, on back and rump; wings blackish, the coverts and secondaries broadly edged with whitish; upper tail-coverts and tail black, except outer 2 rectrices white or pinkish-white, broadly tipped with black; below white, tinged with gray on breast; sides vermilion; flanks, crissum, and wing-linings salmon-pink to pale vermilion; concealed vermilion crown-patch. Bill blackish; legs black. ♀: similar but duller, pink areas paler; crown-patch reduced or absent. **Immatures:** like adult female but above browner, below less pink; outer primary with little or no attenuation at tip.

Habits. Frequents savannas, fields, and open country in general; scattered individuals perch on telephone wires, fences, or exposed twigs on savanna trees between sallies to catch flying insects; larger numbers gather in trees that offer berries or arillate seeds, facing into wind with long tails streaming, then suddenly fly out all together, sometimes returning to same tree; in the evening hundreds converge from surrounding country to roost in trees in marshes, mangroves, and towns, flying swiftly, directly and high above the ground.

Voice. Usual call a dry *kip* like that of Gray-capped Flycatcher; in interactions around roost, a dry sputter or twitter; at close range, a dry rustling, probably made by emarginated outer primary, can be heard as bird flies.

Status. Common, locally abundant winter resident in dry NW, less numerous in Valle Central and adjacent slopes to 7500ft (2300m); rare and sporadic on S Pacific slope; arrives mid-October to early November, departs by mid-April.

Range. Breeds in S-C USA and NE Mexico; winters from S USA to W Panama.

Note. Formerly called *Muscivora forficata*.

FORK-TAILED FLYCATCHER

Tyrannus savana Pl. 35(4)

Tijereta Sabanera (Tijerilla)

Description. 6¾″ (17cm) to end of central rectrices; greatly elongated, ribbonlike outer rectrices add up to 8″ (20cm) to length of males, 5½″ (14cm) to females; 28g. Outer 3 primaries deeply incised, the narrow outer segment of primaries 9 and 10 ca. 15mm long in males, 10mm in females. **Adults:** top and sides of head black, with large, concealed yellow crown-patch; back, rump and edgings of wing-coverts and secondaries pale gray, wings otherwise blackish; tail black, bases of outer rectrices edged with white; below wholly white. Bill and legs black. **Immatures:** like adults but outer primaries less incised, the narrow outer segments broader. **Young:** duller, browner; feathers of rump, upper tail-coverts, and wing-coverts edged with cinnamon; no bright crown-patch; outer primaries not incised.

Habits. Prefers savannas and other open, grassy places, where it often perches 3ft (1m) or less up on shrub, stalk, or fence wire, its white breast gleaming from afar; sallies upward to catch flying insects, or drops down to snatch them from herbage or ground; sometimes associates closely with wintering Yellow-rumped Warblers; eats berries and fruits of royal palm, plucking them in flight or while clinging to the fruit cluster; flies swiftly, with long tail streamers rippling. Males have a slow, butterflylike courtship flight. Scores or hundreds congregate from surrounding fields to roost in crowded savanna trees, sometimes in fruit trees near houses.

VOICE. Call a low, weak, somewhat creaking *jek* or *jiit*, sometimes rapidly repeated; a lower, more bleating *ek-ek-ek-ek* . . . ; a rapid clicking; a dry rattle made with wings, especially in male's courtship flight.

NEST. A shallow cup of vegetable fibers and leaves lined with seed down, 3–35ft (1–10m) up in tree or shrub. Eggs 2–3, glossy white, spotted with chocolate and lilac, mainly in a wreath. March–June.

STATUS. Common resident of Térraba region and locally elsewhere on S Pacific slope, mostly 330–1700ft (100–520m); may nest sporadically in Valle Central (no recent reports); sporadically and in small numbers, perhaps as migrant, in Guanacaste and up to 8000ft (2400m) in central highlands, mainly during dry season. Movements of this wide-ranging bird need clarification.

RANGE. Breeds in scattered localities from SE Mexico to C Argentina and Uruguay; at least more northern and southern populations migratory, wintering irregularly from SE Mexico to N Argentina and S Brazil.

NOTE. Formerly known as *Muscivora tyrannus*.

EASTERN KINGBIRD

Tyrannus tyrannus Pl. 35(3)

Tirano Norteño

DESCRIPTION. 7½″ (19cm); 40g. Blackish and white, with conspicuously white-tipped tail; outer 2 primaries abruptly emarginated for terminal 10mm in males, more shallowly and for outer 5–8mm in females. **Adults:** top and sides of head dull black, with concealed vermilion crown-patch (smaller in female); back and rump blackish-slate; tail black, with a broad white terminal band; wing-coverts and secondaries edged with pale gray to white; below white, breast strongly tinged with gray. Bill and legs black. **Immatures:** above paler and duller, more brownish; lack crown-patch; outer primaries not emarginated; often with patchy look in fall due to incomplete post-juvenile molt prior to migration.

HABITS. Migrates by day in loose flocks of a few individuals to hundreds, flying very high on clear days and lower, often at treetop height, in early morning and evening or during cloudy weather. At such times groups often rest briefly in trees and hawk insects or eat berries or arillate seeds; large numbers congregate in areas of scrub along Caribbean coast to eat berries of *Cordia curassavica*. As night approaches, they gather to roost in isolated clumps of trees or amid tall, coarse grass, often with other small birds. On both spring and fall migrations, same roost may be occupied for a month by kingbirds, perhaps a succession of different individuals.

VOICE. A high, thin, scratchy *keer* or dry chattering notes, heard mainly about roosts; nearly always silent while migrating.

STATUS. Passage migrant along both slopes in fall (late August or early September to late October) and spring (late March to mid-May); often common in Pacific lowlands, up to 5500ft (1700m) in central highlands and inland in Caribbean lowlands, but extraordinarily abundant along Caribbean coast.

RANGE. Breeds from N-C and SE Canada to S-C and SE USA; winters from Colombia to N Chile and N Argentina.

TROPICAL KINGBIRD

Tyrannus melancholicus Pl. 35(1)

Tirano Tropical (Pecho Amarillo)

DESCRIPTION. 8¼″ (21cm); 40g. Large, active, yellow-bellied flycatcher with dark, notched tail; outer 4 (female) or 5 (male) primaries distinctly emarginate at tip, the narrow tip of outermost ca. 10mm long and 2–3mm wide in males, shorter and broader in females. **Adults:** top and sides of head gray; concealed vermilion crown-patch (smaller in female); back and rump grayish-olive; wings blackish, coverts edged with gray and secondaries with whitish; upper tail-coverts and tail black; lores to auriculars dusky; throat grayish-white, shading to pale gray on foreneck, yellow-olive on chest; posterior underparts bright yellow. Bill and feet black. **Immatures:** like adults but little or no vermilion in crown or emargination of primaries. **Young:** above browner, below much paler; feathers of pileum fringed with buff; back tinged with buff; feathers of wings and tail edged with cinnamon to yellowish-buff; throat white; breast grayish-buff to buffy-olive; posterior underparts pale yellow.

HABITS. Frequents open places of all sorts, from savannas, agricultural lands, and banks of large rivers to towns and cities—wherever it can find exposed, elevated perches and scattered trees for nesting; captures wasps, bees, dragonflies, and butterflies on long, often intricate aerial sallies; often waits near flowers or above mud puddles to dive at butterflies; eats many berries, plucked while perching or on the wing; often pursues and buffets large predatory birds; outside breeding season gathers in large communal roosts with Fork-tailed or Scissor-tailed flycatchers.

Voice. Sharp, high-pitched, scratchy twitters and chatters; often members of pair call back and forth with vibrating wings; dawn song of male a medley of short, clear notes and thin, ascending trills: *pit prrrrr pit pit prrrrr* . . . ; at close range a sharp crackle, evidently made with wings, is audible from flying birds.

Nest. A broad, shallow, untidy cup of thin dead vines, grasses, weed stems, and tendrils, lined with finer materials and sometimes horsehair, often so slight that eggs can be glimpsed through bottom, 6–40ft (1.8–12m) up in shrub or tree in exposed situation, sometimes low above water. Eggs 2–3, whitish or pale buff, blotched with reddish-brown or pale brown and lilac. March–July.

Status. Abundant resident countrywide from lowlands to 6000ft (1850m), uncommon to 8000ft (2400m).

Range. SE Arizona to C Argentina and Guianas.

GRAY KINGBIRD

Tyrannus dominicensis Pl. 35(7)

Tirano Gris

Description. 8½″ (21.5cm); 44g. Large; white underparts suggest Eastern Kingbird; thick bill and notched tail, as well as pattern of emargination on primaries, like those of Tropical Kingbird. **Adults:** upperparts gray, crown with narrow, dusky shaft-streaks and concealed patch of vermilion (smaller in female); upper tail-coverts, tail, and wings blackish; wing-coverts, remiges, and rectrices edged narrowly with whitish; lores to auriculars dusky; underparts white, washed with gray across breast. Bill and feet black. **Immatures:** similar but with no crown-patch or emargination of primaries.

Habits. In Costa Rica recorded as isolated individuals migrating with Eastern Kingbirds along Caribbean coast; behaves much like that species. Resident Tropical Kingbirds respond aggressively to Gray Kingbirds but ignore Easterns.

Voice. Usually silent; once, in interaction with Tropical Kingbird, a rolling *preer*, less thin and scratchy than call of Tropical.

Status. Casual or very rare passage migrant (early September to early October; mid to late March) in immediate vicinity of Caribbean coast; as yet no winter records; 1 sighting of a single bird, undoubtedly a stray, near Puerto Jiménez on Golfo Dulce in May (Lewis).

Range. Breeds in SE USA, Antilles, and N South America; winters from Greater Antilles to C Panama, Venezuela, and Guianas.

WESTERN KINGBIRD

Tyrannus verticalis Pl. 35(2)

Tirano Occidental

Description. 8″ (20cm); 40g. Resembles Tropical Kingbird but below paler, smaller bill; tail square-tipped and white-edged; tips of outer 3 (female) or 4 (male) outer primaries attenuate, the outermost very narrow for terminal 15–20mm in males, 10mm in females. **Adults:** pileum and hindneck ash-gray, with concealed vermilion crown-patch (smaller in females); back and rump grayish-olive; wings blackish, the coverts and secondaries edged with grayish-white; rump and upper tail-coverts black, the outer rectrices edged with white; lores and auriculars dusky, shading to pale ash-gray on breast, and canary-yellow on posterior underparts. Bill and legs black. **Immatures:** like adults but without crown-patch; outer primary slightly (male) or not (female) emarginate.

Habits. Prefers open country with scattered trees that serve as lookout perches for aerial flycatching, often by intricate maneuvers; varies diet with berries; often associates with Scissor-tailed or Fork-tailed flycatchers while wintering in Costa Rica, joining communal roosts of these species.

Voice. Shrill metallic chatters, heard mainly at roost.

Status. Sporadic winter resident, chiefly on Pacific slope, S to Térraba region, between about November and April, from sea level to 3300ft (1000m) in W Valle Central; may be fairly common at a site in 1 year, rare or absent the next.

Range. Breeds from SW Canada to NW Mexico; winters from S Mexico to Costa Rica.

PIRATIC FLYCATCHER

Legatus leucophaius Pl. 35(8)

Mosquero Pirata

Description. 6″ (15cm); 26g. Medium-sized, noisy, brownish, with rather short bill and streaked breast. **Adults:** crown sooty-brown with large, concealed yellow patch; white superciliaries encircling crown, speckled with dusky on forehead, tinged with brown on nape; rest of upperparts dull brownish-olive; wings and tail more dusky, the wing-coverts and secondaries edged with dull yellowish, the upper tail-coverts with cinnamon; a broad dusky stripe from lores to auriculars; lower cheeks and throat white, shading to pale yellow on posterior underparts; brownish malar stripe; breast and sides heavily but indistinctly streaked with dull dark brown. Bill black, legs dusky. **Young:** similar but without crown-patch; superciliaries buff; wing-coverts and rectrices edged with cinnamon; below paler, with less streaking.

HABITS. Frequents open country with scattered trees, tall second growth, forest edge, gallery forest, and semi-open; hawks flying insects, including dragonflies, from exposed treetop perch; eats many small berries and green catkins of *Cecropia*; usually in pairs.

VOICE. A loud, clear, shrill, ringing *whiye'eee* or *pee-e-e-e*, often followed by 3–5 shorter lower notes: *tititi* or *pee-de-de-de*; alarm call a short, deep, rather mellow *dee*, rapidly repeated.

NEST. Captures closed nests of other birds by persistently harassing them and throwing out eggs or less often nestlings; prefers roofed nests of *Myiozetetes* flycatchers, pensile nests of *Tolmomyias* and *Rhynchocyclus*, bulky structures of becards, swinging pouches of orioles and oropendolas; carries in few or many small dead leaves to form a loose litter on bottom of stolen structure. Eggs 2–3, brownish-gray, suffused and blotched with brown, especially on thicker end. March–June.

STATUS. Breeding resident, arriving in S Costa Rica in late January or early February, not until 2–4 weeks later in N areas; common on both slopes from lowlands to 5000ft (1500m), infrequently to 6000ft (1850m); departs by late September or early October.

RANGE. Breeds from SE Mexico to NW Ecuador, N Argentina, and S Brazil; Mexican and Central American populations withdraw to South America after breeding.

WHITE-RINGED FLYCATCHER

Coryphotriccus albovittatus Pl. 35(16)

Mosquero Cabecianillado

DESCRIPTION. 6¼″ (16cm); 24g. Size of Social Flycatcher but with longer bill, crown ringed by white, yellow crown-patch often visible; similarly patterned Great Kiskadee much larger, with rufous in wings and tail. **Adults:** crown black with golden-yellow central patch; white superciliaries narrow across forehead, join broadly across nape; back and rump dark olive; wings and tail blackish, the tertials conspicuously edged with yellowish-white; broad stripe from lores through eyes and cheeks to side of neck black; throat white; rest of underparts bright yellow. Bill and legs black. **Young:** similar but above browner; crown barred with rusty, without yellow patch; wing-coverts and rump feathers edged with cinnamon; tertials edged with buffy-yellow.

HABITS. Roams in groups of 2–5 in trees that border waterways or other edges, into tall trees in semi-open, or through upper canopy of forest; perches on exposed twigs on outside of tree crowns, scanning foliage below, then sallies out and down to pluck insects and spiders from upper surfaces of leaves or inflorescences; gathers berries in flight; frequently snaps bill while flying, especially if another bird is nearby; occasionally accompanies mixed-species flocks including tanagers and Black-faced Grosbeaks.

VOICE. Usual call a clear, slightly nasal descending whistle followed by a rattling trill that rises in pitch and may slow down at the end: *wheeereeeeee-e-e-e* or *wheeeurrrrr-rreek*; a prolonged, higher-pitched, querulous, almost-clear trill that slows down and ends with several disjointed notes.

NEST. In old woodpecker hole or niche among bromeliad cluster 30–50ft (10–15m) up; contents not examined in Costa Rica. In South America, a different race builds a cup-shaped nest of grasses in old woodpecker hole or abandoned cacique nest. Eggs 2, cream-color, streaked and blotched with chocolate-brown, solidly in wreath. March–June.

STATUS. Locally common resident of Caribbean lowlands, occasionally to 2000ft (600m) in foothills; uncommon in drier lowlands E of Río Frío.

RANGE. E Honduras to NW Ecuador, N Brazil, and Guianas.

NOTE. Sometimes placed in the genus *Conopias*; has been called *Conopias parva*.

BOAT-BILLED FLYCATCHER

Megarhynchus pitangua Pl. 35(12)

Mosquerón Picudo

DESCRIPTION. 9″ (23cm); 70g. Very large; bill very heavy, the culmen noticeably convex; more olive on back, less rufous in wing than similar-sized Great Kiskadee. **Adults:** crown and sides of head dull slate-black, with concealed crown-patch yellow to orange; white superciliaries extend to sides of forehead, nearly meet on nape; rest of upperparts dull olive; wings and tail more dusky, remiges and rectrices narrowly edged with cinnamon; throat white, rest of underparts bright yellow. Bill and legs blackish, mouth lining whitish. **Young:** above darker and browner; feathers of entire upperparts, including crown, wings, and tail, edged with cinnamon to dull rufous; no crown-patch; superciliaries tinged with yellow.

Habits. Frequents canopy and edge of humid or dry forest, semi-open, savanna with scattered tall trees, tall second growth, and shady gardens; usually in pairs or family groups of 3–5 that wander restlessly; locates insects while perching, then flies up to pluck them from leaf or bark without alighting; beats larger prey loudly against perch before swallowing; especially fond of cicadas; rarely hawks flying insects; sallies to snatch figs, berries, and arillate seeds; chases toucans and other predators.

Voice. Usual call a nasal, rolling or stuttering *neeeeeah*; a clear, rolling *cheeur*; a loud, rapidly repeated *choip choip*. Dawn song consists of repetitions of a loud, clear *cheer*, punctuated by a slurred *bo-oy*, or a repeated *cheer chirree*.

Nest. A shallow bowl of twigs, dry vines, roots and rhizomes of epiphytes gathered from trees rather than ground, 20–100ft (6–30m) up on branch of tree usually standing alone. Eggs 2–3, whitish, everywhere speckled with brown and pale lilac, densely on thicker end. March–July.

Status. Common resident countrywide, lowlands to 6000ft (1850m), rarely 6500ft (2000m); descends from higher altitudes after breeding season.

Range. Tropical Mexico to NW Peru, N Argentina, and S Brazil.

BRIGHT-RUMPED ATTILA

Attila spadiceus Pl. 35(6)

Atila Lomiamarilla

Description. 7″ (18cm); 40g. Fairly large, big-headed, with long, stout bill, strongly hooked at tip that often appears slightly upturned; highly variable in color. **Adults:** top and sides of head and hindneck olive-green to brown, variably streaked with black; forehead and/or cheeks mixed with yellow in some; back chestnut-brown to brownish-olive; rump and upper tail-coverts bright yellow to ochraceous-buff; tail dark brown to cinnamon-rufous; wings dark brown, wing-coverts edged with brown, olive, or cinnamon-buff; throat grayish-white, yellow, or olive, streaked with dusky; breast yellow, olive or gray, with olive, dusky, gray, or brown streaking, or none; belly white, pale buff, or pale yellow, flanks, thighs, and crissum whitish, yellow, or buff. Iris brick-red, brighter in male; bill dark brown, base of lower mandible (sometimes both mandibles) horn- to flesh-color; legs gray. **Young:** consistently differ from adults in brown eye, narrow cinnamon fringes to crown feathers, tertials usually broadly edged with cinnamon.

Habits. Frequents wet and dry forests and nearby shady clearings and gardens, semi-open, tall second growth; active and noisy, moves about restlessly, from high in trees to ground; constantly wags tail up and down through wide arc; scans all about, then makes short sally to vegetation or ground for insects, spiders, small frogs or lizards; or captures prey while hopping over ground; takes many berries and arillate seeds; sometimes forages with army ants; usually solitary while foraging, even in breeding season, when a strong pair bond is formed.

Voice. Varied and freely used by both sexes: a loud, ringing, assertive *beat-it, beat-it, beat-it naow*; a less emphatic, clear *ooo weery weery weery weery weery woo*; a briefer *we her her*; sharp, rattling notes in flight; a prolonged, low, soft trilling while inspecting nest site or (female) incubating.

Nest. A bulky cup of fibrous rootlets, rachises of compound leaves, and pieces of green fern frond or moss, lined with similar but finer materials in niche or nook amid epiphytes, between buttresses or stump sprouts, or in bank of road or stream, 2½–10ft (0.8–3m) up, often well outside forest. Eggs 3–4, dull white to pinkish or pale buff, spotted and blotched with brown, cinnamon-rufous, and pale lilac, especially in a wreath. March–July.

Status. Resident countrywide from lowlands to at least 6000ft (1850m); common where at least patchy forests remain.

Range. NW Mexico to W Ecuador, Bolivia, and SE Brazil.

Note. Sometimes called Polymorphic Attila; formerly the genus *Attila* was placed in the Cotingidae; in plumages, nesting habits and voice seems most closely related to *Myiodynastes*.

SULPHUR-BELLIED FLYCATCHER

Myiodynastes luteiventris Pl. 35(10)

Mosquero Vientriazufrado (Pecho Amarillo)

Description. 8″ (20cm); 45g. Large, with thick bill and prominent dark streaking, rufous tail; has darker bill, chin, and malar area than Streaked Flycatcher, with belly at least as yellow as breast. **Adults:** upperparts buffy-brown, broadly but indistinctly streaked with blackish; a large, concealed golden-yellow crown-patch; wings blackish with cinnamon edgings on shoulders; middle and greater coverts and secondaries broadly edged with pale buff; above with whitish stripes; below with blackish lores and auriculars; throat white, chin, and malar area heavily streaked to almost solid dusky, center of throat lightly streaked; rest of underparts pale yellow, heavily streaked with blackish on breast and sides. Bill blackish, base of lower mandible flesh-color; legs dusky. **Young:** similar but feathers of upperparts brighter, with cinnamon-buff bases and edges; no yellow in crown, but feathers with cinnamon-orange bases; wing-coverts edged with rufous.

Habits. During breeding season prefers dry forest and borders of wetter forest, semi-open, and open country with scattered tall trees; noisy and quarrelsome; usually keeps high in trees, foraging for insects with spectacular sallies high in the air or shorter darts amid vegetation; eats many berries and arillate seeds (e.g., *Bursera, Trichilia*), usually plucked while flying past; in pairs and, after May or June, usually in family groups.

Voice. Calls are high, thin, sibilant whistles, forceful or petulant: *weel-yum, squeeeah, pyeeeeuh*, etc., often preceded by staccato notes *p'p'p'peeee-ya*; dawn song a soft, liquid, melodious *tre-le-re-re* or *cheree-chirit*, tirelessly repeated for many minutes.

Nest. In old woodpecker hole or natural cavity in tree, 10–90ft (3–27m) up, which female fills to near doorway with coarse twigs to support neat, shallow cup of petioles, dry inflorescences, leaf rachises, and finer twigs, with no soft lining. Eggs 2–3, white, mottled all over with cherry-red and faintly with pale lilac. April–June.

Status. Sporadically common passage migrant the length of both slopes from lowlands to upper middle elevations; uncommon to fairly common breeding resident of N half of both slopes, S to Valle Central and Reventazón drainage, from lowlands to 6500ft (2000m) on Pacific slope but not below 2000ft (600m) on Caribbean. Arrives between early March and early April; migrants from N begin to appear by early August; depart by mid-October.

Range. Breeds from SE Arizona to Costa Rica; winters in South America E of Andes in Peru and Bolivia.

STREAKED FLYCATCHER

Myiodynastes maculatus Pl. 35(11)

Mosquero Listado

Description. 8″ (20cm); 45g. Resembles Sulphur-bellied Flycatcher but bill heavier and paler; throat and chin white; below yellow (if any) most intense on breast. **Adults:** upperparts buffy-brown, brighter and more cinnamon on pileum, heavily and indistinctly streaked with blackish; a large concealed golden-yellow crown-patch; wings blackish, remiges and coverts edged with bright cinnamon, tertials with pale yellow; upper tail-coverts and rectrices rufous, with black stripes along shafts; superciliaries pale yellowish; broad stripes from lores to auriculars brownish-black; cheeks pale yellowish; malar area streaked with dusky; chin and throat white, lightly streaked with blackish; breast pale yellow to yellowish-white; belly white; flanks and crissum pale yellow; breast, sides, sometimes upper belly heavily streaked with blackish. Upper mandible black, lower pale horn to flesh-color with dusky tip; legs dark gray. **Young:** crown rufous, heavily streaked with black, without yellow; wing-coverts edged with rufous; below little or no yellow.

Habits. Frequents edges and openings in forest, semi-open, mangroves, open country with scattered trees; in dry northwest prefers evergreen gallery forest; solitary or in pairs, sometimes loose flocks in migration; hawks flying insects, especially wasps and flying ants, from high perches, but more often makes short sallies and hovers in dense vegetation to snatch cicadas, other insects, and small lizards; eats many berries and arillate seeds.

Voice. A sharp, dry *dik* or *chek*; a dry, nasal *chuk-yi chuk-yi* or *tsu-ka' tsu-ka'*, often repeated, especially when excited. At dawn and dusk persistently repeats a soft, clear, liquid *kawee-teedly-wink* or *whee-cheerily-chee*.

Nest. In hole in tree, amid leaf-bases of palm tree, in a bird box, rarely on ledge or niche under eaves of house, 18–75ft (5.5–23m) up. If the cavity is deep, female fills it with coarse materials to near the entrance, then builds a shallow nest of dry petioles and rachises of compound leaves and flower stalks, lined with finer rachises. Eggs 2–3, glossy whitish, heavily marked with cherry-red and pale lilac. March–May.

Status. Resident and passage migrant on Pacific slope; rare vagrant on most of Caribbean slope, but may breed regularly in lowlands S of Lago de Nicaragua; breeds fairly commonly throughout NW lowlands, infrequently to 3500ft (1100m) along facing mountain slopes and in W Valle Central; uncommon on S Pacific slope, from lowlands to 5000ft (1500m) in Coto Brus valley; definitely less common in Guanacaste during nonbreeding season, but it is uncertain whether the breeding population departs wholly or in part, to be replaced by northern migrants; N migration February–March, S in late July to early September.

Range. Breeds from tropical Mexico to C Peru, N Argentina, and S Brazil; populations at extremes of range are migratory; winters from Costa Rica to Bolivia and S Brazil.

GOLDEN-BELLIED FLYCATCHER

Myiodynastes hemichrysus Pl. 35(9)

Mosquero Vientridorado (Pecho Amarillo)

Description. 8″ (20cm); 41g. Differs from all other kiskadee-type flycatchers by its dark malar stripe. **Adults:** crown dark gray, streaked finely with black, and with concealed patch of bright yellow; rest of upperparts dark greenish-olive; wings and tail dusky; tertials edged with pale yellow, other remiges and wing-coverts narrowly edged with cinnamon; broad blackish stripe from lores to auriculars, bordered by white superciliaries and cheek stripes; broad malar stripe dusky; chin white; rest of underparts bright yellow. Bill and feet black. **Young:** crown and back much browner; no crown-patch; wing-coverts edged with rufous; below paler yellow.

Habits. Frequents edges and canopy of wet, epiphyte-burdened mountain forest, especially along rivers and streams or at treefall gaps; perches at medium heights to high in trees; hawks flying insects from exposed perch but more often snatches insects and fruits from vegetation in short upward sallies; paired throughout year, in family groups for several months following breeding season.

Voice. Usual calls resemble those of Sulphur-bellied Flycatcher but less thin, more raucous and squealing: *syup, seeeik*, or *seek-a-skeeeir*, etc. When alone, often a plaintive *pee-ah* or *peeeir*, resembling call of Social Flycatcher; when disturbed, a repeated heavy chipping; dawn song a clear, melodious *tree-le-loo*, repeated many times.

Nest. A shallow, solidly but coarsely made cup of fine rootlets and moss, often camouflaged with moss, in cranny amid clump of epiphytes on high branch, old woodpecker hole, or in niche among ferns or other plants on vertical cliff or bank, 20–100ft (6–30m) up. Eggs 3, pinkish-buff blotched faintly with gray and heavily with pale and dark reddish-brown, especially in wreath around thick end. March–May.

Status. Fairly common resident of Caribbean slope at 2300–6000ft (700–1850m) from Cordillera de Guanacaste (N at least to Volcán Miravalles) to Panama; locally on Pacific slope in higher parts of N cordilleras; to 7500ft (2300m) locally along Cordillera de Talamanca.

Range. Costa Rica and W Panama.

Note. Occasionally considered a race of Golden-crowned Flycatcher, *M. chrysocephalus*, of South America.

GRAY-CAPPED FLYCATCHER

Myiozetetes granadensis Pl. 35(15)

Mosquero Cabecigrís (Pecho Amarillo)

Description. 6½″ (16.5cm); 30g. Medium-sized, with rather short bill; resembles Social Flycatcher but has white forehead, lacks broad superciliaries. **Adult ♂:** forehead and short, indistinct superciliaries white; crown gray, usually forming short, bushy crest with usually concealed central patch of vermilion; rest of upperparts olive-green; wings and tail dusky; wing-coverts edged with olive, remiges and rectrices narrowly edged with cinnamon; sides of head dusky; throat white; rest of underparts bright yellow. Iris pale brown; bill and legs black. ♀: similar but crown smooth, with bright crown-patch reduced or lacking. **Young:** crown grayish-olive, without vermilion; rectrices and wing-coverts broadly edged with tawny or cinnamon.

Habits. Frequents agricultural land with scattered trees, second growth, semi-open, and forest edge, especially along rivers and streams; may move through forest canopy between clearings or stream courses; in pairs or family groups, sometimes in roving flocks of up to a dozen following breeding season; snatches flying insects on often spectacular sallies; takes many berries and arillate seeds.

Voice. Usual call, often repeated incessantly, a dry, sharp *bip* or *wic*; in interactions a variety of harsh, strident, rather angry-sounding staccato and burry notes: *kurr keer ch'beer, k'keer keer jeer k'beer*; dawn song a loud and hoarse, repeated *kip kip kip-k'beer* or *hic, bit of a cold*.

Nest. Bulky, roofed structure with a wide side entrance, of straws and weed stems, like that of Social Flycatcher and often in same small tree, 3–60ft (1–18m) up, mostly 5–25ft (1.5–8m). Eggs 2 or 3, rarely 4, dull white, speckled and blotched with brown, chocolate, and pale lilac, most heavily in wreath around larger end. February–June, rarely August.

Status. Common to abundant resident in lowlands of Caribbean slope and S Pacific slope, locally to 5500ft (1650m) along Pacific slope of Cordillera de Talamanca and, rarely, in central highlands.

Range. E Honduras to NW Peru, N Bolivia, and W Brazil.

SOCIAL FLYCATCHER

Myiozetetes similis Pl. 35(14)

Mosquero Cejiblanco (Pecho Amarillo)

Description. 6¼″ (16cm); 27g. Medium-sized, with rather short bill; above greener, with more grayish head markings than other kiskadee-type flycatchers. **Adults:** pileum gray, mixed with whitish on forehead; a large, concealed orange-red crown-patch; broad white superciliaries that converge but do not meet on nape; rest of upperparts olive-green to dull olive; wings and tail dusky, the wing-coverts edged with olive to dark grayish; remiges and rectrices narrowly edged with yellowish; sides of head dark sooty-gray; throat white; rest of underparts bright yellow. Bill and legs black. **Young:** similar but feathers of wings and tail broadly edged with cinnamon; crown-patch lacking or very small and paler orange; sides of head paler gray.

Habits. Frequents pastures, agricultural land, clearings with scattered trees, shady gardens, and banks of rivers and ponds; catches insects on long aerial sallies or shorter darts to snatch them from vegetation; takes many berries, arillate seeds, and other fruits, plucked while perching or in flight; often forages over ground; enters shallow water to catch tadpoles; paired throughout year, family groups not persisting long after breeding season.

Voice. Calls include a harsh, sharp *teeer* or *peeeeur*, sometimes softened to a plaintive *pe-ah* or *chee*; a rolling, scolding *wheer*, often repeated several times, occasionally alternating with short notes; a chipping *wit*; and a series of *chu* notes. Most calls are squeakier and thinner on S Pacific slope. Dawn song on S Pacific slope a clearly enunciated *chips-a-cheery*, elsewhere a less clear *k'cheery k'cheery chip k'cheery* or some garbled variant.

Nest. A bulky roofed structure with wide side entrance, of straws and weed stems, 6–50ft (2–15m) up in tree or shrub that is often thorny, frequently beside wasps' nest or that of another flycatcher, often near or over water, sometimes on manmade structure. Eggs 2–4, white or creamy, sometimes pinkish, speckled and blotched with shades of brown and pale lilac, mostly in cap or wreath at large end. February–June.

Status. Common resident countrywide from lowlands to 5600ft (1700m), occasionally higher; at highest elevations chiefly during nesting season.

Range. NW Mexico to NW Peru, NE Argentina, and S Brazil.

Note. Also called Vermilion-crowned Flycatcher. Sometimes the birds from C and E Costa Rica northward are considered a separate species, *M. texensis* or Social Flycatcher, from *M. similis* of SW Costa Rica to South America, called Vermilion-crowned Flycatcher.

GREAT KISKADEE

Pitangus sulphuratus Pl. 35(13)

Bienteveo Grande (Cristofué, Pecho Amarillo)

Description. 9″ (23cm); 68g. Very large, brightly colored; much rufous in wings and tail; bill stout but not swollen, culmen straight. **Adults:** crown and sides of head black; a large, partly concealed yellow crown-patch, broad white superciliaries that encircle crown; rest of upperparts olive-brown; wing-coverts edged with rufous; remiges with bases and edges extensively rufous, tips dusky; rectrices dusky with inner webs extensively rufous; throat white; rest of underparts bright yellow. Bill and legs black. **Young:** rufous edges on wings and tail more extensive; no yellow in crown; below slightly paler.

Habits. Prefers open country with scattered trees and shrubs, light woods, savannas, and suburban gardens, often near water; boisterous and noisy, mobs large predatory birds but sometimes plunders nests of smaller birds; catches large insects, small lizards, snakes, and frogs, even mice, earthworms, and spiders by dropping from low perch, sometimes hopping about on ground; plunges into shallow water for small fishes, tadpoles, and insects; takes fruit from perch or on the wing; in pairs or family groups; mated pairs often give wing-shivering, crest-raising displays.

Voice. A vigorous, strident *KIS ka DEE*, *KICK a DEER*, or *SPEAK to MEE*, sometimes varied to *KIK KIK KIK a DEER* or *SPEAK to ME, to ME, to ME*, or *KIK-weer*; soft churring notes about the nest; a shrill *eeek* while mobbing a predator.

Nest. A bulky, often untidy, roofed structure with wide side entrance, of straws and weed stems, often including Spanish moss, string, bits of rags, paper and other trash, sometimes entire nests of small birds like seedeaters; on solid horizontal support of tree or manmade structure, or in bull's-horn acacia or cactus, 4–40ft (1.2–12m) up. Eggs 3, less often 4, pale buffy or creamy-white, spotted with brown. February–June, occasionally October.

Status. Common to abundant resident from lowlands to 5000ft (1500m), occasionally higher, except in S Pacific, from which it was absent until within the last 20 years and is still uncommon in many areas; reached upper Valle del General in late 1970s. Evidently was prevented from occupying this region in the past by forest barrier; increased greatly on Caribbean slope with deforestation.

Range. NW Mexico and S Texas to C Argentina.

RUFOUS MOURNER

Rhytipterna holerythra Pl. 34(9)

Plañidera Rojiza

Description. 8″ (20cm); 40g. Entirely rufous, with bill pale at base; smaller and slimmer than Rufous Piha, with relatively longer tail. **Adults:** above cinnamon-brown to russet, brighter and more cinnamon-rufous on tail; remiges darker brown, broadly edged with cinnamon-rufous, contrasting with dusky-tipped primary coverts; underparts bright cinnamon to ochraceous-tawny, darker and often with brownish wash on breast. Upper mandible blackish, tinged with horn at base, lower flesh-color to pale horn with dusky tip; legs gray. **Young:** similar but brighter, above more rufescent and paler, below more cinnamon-buff.

Habits. Frequents middle and upper levels of humid forest and adjacent semi-open and old second growth, shady plantations, and thinned woods; sits for considerable time on twig in vegetation, scanning all around, sometimes twisting head to odd angles, then makes sudden sally to snatch large, slow-moving caterpillars, katydids, and walking sticks from foliage; plucks many berries and arillate seeds in flight; often accompanies mixed-species flocks; alone, in pairs or family groups.

Voice. Usual call a drawled, weary-sounding *right here* or *waay teeer*, first note slurred up and second down. Song a more vigorous *wee hi hr weeur-weeur-weeur. . . .*

Nest. Unknown (?). Breeds March–June (gonad data and observations of fledglings).

Status. Common resident in lowlands and foothills of Caribbean slope and S Pacific slope, up to 4000ft (1200m); reaches Pacific slope of Cordillera de Guanacaste through low passes but absent from NW lowlands and Valle Central.

Range. SE Mexico to NW Ecuador.

Note. Recent anatomical studies show that the genus *Rhytipterna*, long included in the Cotingidae, belongs in the Tyrannidae and is closely related to *Myiarchus*.

The Genus *Myiarchus*: These flycatchers are distinguished by their unpatterned, bushy-crested heads, gray throats and breasts, and rather long, stout bills. Most show rufous in wings and tails.

PANAMA FLYCATCHER

Myiarchus panamensis Pl. 35(20)

Copetón Colipardo

Description. 7½″ (19cm); 32g. Very plain; lacks extensive rufous in wings and tail. **Adults:** pileum dusky-olive; sides of head dull grayish; rest of upperparts grayish-olive; wings and tail dusky, wing-coverts and secondaries edged with gray, rectrices with indistinct pale grayish tips; throat and breast pale gray, rest of underparts pale yellow. Bill black except base of lower mandible sometimes pinkish; mouth lining orange; legs black. **Young:** similar but wing and tail feathers edged with bright cinnamon.

Habits. Confined to mangrove swamps in Costa Rica; hawks flying insects from open perches and sallies to snatch insects from vegetation; hops and runs along ground or mangrove roots like a *Turdus*, sallying up at flying or resting insects; eats many berries. Mainly in subcanopy and interior of tall mangroves, while the related Dusky-capped Flycatcher frequents the upper canopy and, in NW, the Brown-crested Flycatcher occurs mainly along mangrove edge.

Voice. Call a scratchy *gwee heer* or *kuwiheer*; in alarm, a more emphatic and grating *gwiheek*; dawn song *kweea-hurr, kwiheer, kweeahurr* repeated in various combinations, the *hurr* with a burry, rolling quality.

Nest. In hole in tree, 13–40ft (4–12m) up (contents not examined in Costa Rica), lined with a bulky mass of rootlets, vines, leaf fragments, plant down, animal hairs and snakeskin. Eggs 2–3, pale greenish-white or cream, striped and spotted with dark brown and reddish-brown (in Panama). April–May.

Status. Locally common resident in mangrove swamps along Pacific coast from Golfo de Nicoya to Panama.

Range. Costa Rica to N Colombia and NW Venezuela.

Note. Previously considered a race of the widespread South American *M. ferox*, Short-crested Flycatcher.

BROWN-CRESTED FLYCATCHER

Myiarchus tyrannulus Pl. 35(18)

Copetón Crestipardo

Description. 7½″ (19cm); 34g. Large; short, bushy crest; conspicuous rufous in wings and tail; resembles Great Crested Flycatcher but slightly smaller, above browner and below paler, with all-black bill. See Nutting's Flycatcher. **Adults:** above grayish-brown; wings dusky, wing-coverts and secondaries edged with whitish, primaries with broad, rufous edges and rufous bases; central rectrices dusky, the others with dusky outer and rufous inner webs; throat and breast pale gray; belly pale yellow. Bill black; mouth-lining flesh-color; legs blackish. **Young:** similar but rufous on wings more extensive; wing-coverts edged with cinnamon.

Habits. Prefers open country with scattered trees, scrub, edges of woodland, and mangrove swamps; mostly stays from upper understory to midcanopy levels; makes short sallies to snatch insects from foliage or pluck berries or arillate seeds; hawks flying insects; quick to mob pygmy-owls; especially when disturbed, puffs out throat and bobs head like a lizard; alone or in pairs.

Voice. Characteristic call an incessantly repeated *wit* or *whitp*, usually with scratchy or burry quality; a rolling *wheeh* or *here* or a repeated *come here*. Dawn song a medley of sharp, short notes and rolling trills, burry or clear and rather melodious: *which-wirrick; wik-wheer, which-wi wi-wirreeer*, etc.

Nest. In hollow fence post or stub open at top, old woodpecker hole, or similar cavity, the entrance 3–20ft (1–6m) up, the nest as much as 3ft (1m) below entrance; a shallow cup of hair, feathers, grass, moss, bark, etc., usually with scraps of snakeskin. Eggs 2–4, creamy-white, heavily marked with blotches, spots, and streaks of brown, purplish and lavender. March–May or June.

Status. Common to abundant resident of lowlands and foothills of Pacific slope from Nicaragua S to around Orotina, occasional as far as Punta Leona; to 3000ft (900m) on slopes of N cordilleras, and E to Santa Cecilia on Caribbean slope.

Range. SW USA to Costa Rica, and from Colombia and Guianas to N Argentina.

Note. The considerably larger and more olive race breeding from SW USA to Honduras may occur in Costa Rica in winter, S to Valle del General; specimen confirmation required. Also called Wied's Crested Flycatcher.

NUTTING'S FLYCATCHER

Myiarchus nuttingi Pl. 35(22)

Copetón de Nutting

Description. 6¾″ (17cm); 24g. Nearly identical in color to Brown-crested Flycatcher but smaller, with crown often noticeably darker than back; in hand, note orange mouth-lining. **Adults:** above grayish-brown, pileum darker and sometimes faintly tinged with cinnamon; wings dusky, wing-coverts and secondaries broadly edged with buffy-white, primaries with cinnamon-rufous; much less rufous on inner webs of primaries than in Brown-crested; central rectrices dusky, the others with inner web rufous, outer dusky; throat pale gray, rest of underparts pale yellow. Bill and feet black. **Young:** below paler; wing-coverts edged with cinnamon.

Habits. Frequents interior and edges of deciduous and evergreen forest and second growth, usually from understory to medium heights, mostly lower than Brown-crested Flycatcher; sallies and hovers within foliage to snatch insects and berries; less often hawks flying insects; alone or in pairs, in family groups after nesting.

Voice. Usual call *chik* or *pik*, higher, thinner, and clearer than note of Brown-crested; a high, clear *wheeer*, heard occasionally at midday, higher and less plaintive than similar call of Dusky-capped Flycatcher. Dawn song a clear whistled *pi-wheeer*, interspersed with scratchy notes and a trilling *preeer*, higher and thinner than that of Brown-crested.

Nest. In cavity in fence post, 1–4ft (0.3–1.2m) high, a cup of fur, feathers, and catkins, with shreds of reptile skin. Eggs 3–5, creamy-white, heavily blotched and streaked with reddish-brown, purple and black. March–May (in Oaxaca, Mexico; no Costa Rican record).

Status. Uncommon to fairly common resident of lowlands and foothills of dry NW, S to around Cañas; to 4000ft (1200m) on slopes of Cordillera de Guanacaste.

Range. NW Mexico to Costa Rica.

Note. Also known as Pale-throated Flycatcher.

GREAT CRESTED FLYCATCHER

Myiarchus crinitus Pl. 35(17)

Copetón Viajero

Description. 8″ (20cm); 35g. Largest and most brightly colored *Myiarchus*, much greener on back than Brown-crested, with pale base of lower mandible; mouth-lining orange. **Adults:** above olive to greenish-olive; wings dusky, wing-coverts edged with pale gray to buffy, tertials with yellowish to white; primaries edged on inner and outer webs with cinnamon-rufous; central rectrices dusky, the others with inner webs rufous, outer dusky, more or less edged with cinnamon-rufous; throat and breast gray, sides of breast washed with olive; rest of underparts yellow. Bill blackish, except base of lower mandible pale horn; legs black. **Immatures:** very similar, but wing-coverts usually with cinnamon edgings; rufous edgings broader on wings and tail. Both adults and immatures arrive in fresh plumage in fall.

Habits. Frequents upper levels of forest, thinned woodland, semi-open, taller second growth, and clearings with scattered trees; usually solitary, may hold individual territories in winter; hawks flying insects or plucks them from foliage on short sallies; eats many berries; rarely raises crest in winter home.

Voice. A loud whistled *wheeep* or *wheeeik* with rising inflection, often repeated several times in succession.

Status. Common passage migrant on both slopes, from lowlands to ca. 6000ft (1850m); common winter resident on Pacific slope, uncommon on Caribbean slope, from lowlands to 4500ft (1400m), occasionally higher; arrives in late September or early October, departs by mid to late April, occasionally early May.

Range. Breeds in S-C and SE Canada and C and E USA; winters from S Florida and Cuba to N South America.

ASH-THROATED FLYCATCHER

Myiarchus cinerascens Pl. 35(19)

Copetón Garganticeniza

Description. 7½″ (19cm); 27g. Very similar to Brown-crested Flycatcher, but below paler with smaller bill, different tail pattern and voice. Above grayish-brown, slightly darker on head; lores gray; wings dusky, coverts and secondaries edged with pale grayish, primaries with rufous; lateral rectrices with outer webs dusky, inner webs rufous with dusky at tips (never any dusky on inner webs of Brown-crested's rectrices except along shaft); throat and breast grayish-white; belly yellowish-white. Bill black; legs blackish; mouth-lining flesh-color.

Habits. Unknown in Costa Rica; farther N in Central America inhabits tropical dry forest, savanna, and dry scrub.

Voice. Call a sharp *pip*, often with metallic quality; a rolling *kip-wheer*! often repeated; clearer, less hoarse than notes of Brown-crested Flycatcher.

Status. Definitely known from a specimen collected 9 March 1934 in foothills N of Ciudad Quesada, at 800ft (250m) at S edge of San Carlos lowlands. To be expected occasionally in Río Frío region and in N Guanacaste; any pale *Myiarchus* on Caribbean slope should be closely examined; in Guanacaste worn Brown-cresteds can be nearly as pale below.

Range. Breeds from NW USA to C Mexico; winters from SW USA regularly to Honduras, casually (?) to Costa Rica.

DUSKY-CAPPED FLYCATCHER

Myiarchus tuberculifer Pl. 35(21)

Copetón Crestioscuro

Description. 6½″ (16.5cm); 20g. Small *Myiarchus* with contrasting dark cap; brightly colored but little rufous in adults' remiges or tail. **Adults:** pileum dusky to blackish; rest of upperparts olive; wings and tail dusky, wing-coverts paler toward tip and edged with rufous; remiges and rectrices with narrow cinnamon-rufous edgings; sides of head dark olive; throat and breast gray; rest of underparts fairly bright yellow. Bill and legs black. **Young:** above browner, below much paler than adults; wing-coverts, remiges, and rectrices with broad rufous edgings.

Habits. Frequents edges and more open parts of forest, forest canopy, light woodlands, shady plantations, clearings with scattered trees, old second growth, and mangroves; usually solitary after nesting season but may be paired by December; catches insects on aerial sallies or plucks them from foliage; descends to ground for grasshoppers and spiders; eats many berries and arillate seeds.

Voice. A plaintive, clear whistle, *weeeeer* or *wheeeu*, higher-pitched in middle, often one of the only avian sounds during the heat of midday; in interactions a buzzy trill that rises and falls or a gravelly chatter. At dawn and dusk and often on cloudy afternoons sings a medley of a sharp, whistled *whit*, a long-drawn-out, plaintive *wheeeu*, a staccato *whedu*, and a harsh trill, all delivered in no fixed order and continued for minutes on end, especially in breeding season.

NEST. In a cavity in a fence post or tree, sometimes an old woodpecker hole, 3–100ft (1–30m) up, of soft fibers, seed down, hair, feathers, sometimes moss or snakeskin. Eggs 3, dull white or buffy, heavily blotched and speckled with chocolate in a wreath on thicker end, with longitudinal streaks or dots along sides. April–June.

STATUS. Common resident countrywide from lowlands to 4000ft (1200m) and in smaller numbers to 6000ft (1850m).

RANGE. Extreme SW USA to NW Peru, N Argentina, and SE Brazil.

OLIVE-SIDED FLYCATCHER

Contopus borealis Pl. 36(4)

Pibí Boreal

DESCRIPTION. 6¾″ (17cm); 32g. A large, thickset pewee with heavy bill, rather short tail, and below with distinctive "open jacket" pattern. **Adults:** above dark grayish-olive, more blackish on pileum; wings and tail dusky, wing-coverts edged with gray-brown and secondaries with whitish; sides of head and neck, sides, and flanks olive-gray, the lores and narrow eye-ring speckled with white; median underparts white with faint yellow tinge; a tuft of silky white feathers on lower side nearly always concealed. Upper mandible black, lower dull orangish with dusky tip; legs black. **Immatures:** like adults but pale edgings on wing feathers usually broader and brighter.

HABITS. Almost anywhere in migration, in exposed snags and open branches; in winter mostly around edges and clearings, or broken canopy of highland forest and semi-open; makes long sallies after flying insects of many sorts, often returning to the same conspicuous perch; consistently solitary; usually territorial.

VOICE. Usual call a clear *pik*, often repeated, resembling calls of other large pewees; song, rarely heard in fall and winter but more often toward spring, a high-pitched, vigorous *quick-FREE-beer* or *whip-WEE-wheer*.

STATUS. Uncommon to fairly common fall (late August to late October) and spring (mid-March to early June) migrant countrywide from lowlands to 8000ft (2500m), rarely to 10,000ft (3000m); often very abundant along Caribbean coast in September and early October; rare winter resident on both slopes, chiefly in foothills and mountains at 2000–7500ft (600–2300m).

RANGE. Breeds from C Alaska and E Canada to extreme NW Mexico and E USA; winters mainly in Andes from Colombia and Venezuela to SE Peru, in small numbers in Central America and S Mexico.

NOTE. Formerly classified in the monotypic genus *Nuttallornis*.

WESTERN WOOD-PEWEE

Contopus sordidulus Pl. 36(7)

Pibí Occidental

DESCRIPTION. 5½″ (14cm); 13.5g. Very similar to Eastern Wood-Pewee and not always distinguishable except by voice, but averages darker and grayer overall. **Adults:** above dusky-olive to olive-gray, crown darker and duskier; wings and tail dusky, with grayish wing-bars, the anterior one often very dull; secondaries edged with grayish-white; throat grayish-white; sides of neck, breast, sides, and flanks pale grayish-olive, this sometimes continuous across breast; belly white with little or no yellow tinge; feathers of crissum and wing-linings with extensive grayish bases, whitish tips. Upper mandible black, lower black except basal third or less dull orangish; legs black. **Immatures:** at least posterior wing-bar and edgings to secondaries brighter, more whitish; anterior wing-bar usually tinged with dull gray-brown; at least distal third (and usually over half) of bill dark. In the hand, many Easterns and Westerns can be separated by the relatively longer tails of the former, specifically the distance between the tip of the longest upper tail-covert and the tip of the central rectrix: if the wing measures 82mm or more (males), only in Easterns does the length of the exposed part of the central retrix exceed 30mm, only in Westerns is it less than 29mm; if the wing measures 79mm or less (females), only in Easterns is this length more than 27mm, only in Westerns under 26mm (overlap at intermediate values in each case).

HABITS. Similar to those of Eastern Wood-Pewee during migration but tends to be more numerous in highlands, Eastern in lowlands; apparently paired birds sometimes seen June–July at forest edges and pasture hedgerows in mountains.

VOICE. A burry *beeeiih* or *pheeer*, upwardly inflected or descending, harsher than notes of Eastern Wood-Pewee.

STATUS. Common to abundant transient on both slopes in spring (late March–late May, perhaps into early June) and fall (late July or early August to mid-November), mostly above 2300ft (700m), occasionally to 10,000ft (3000m); often outnumbers Eastern Wood-Pewee in highlands. A possible rare breeding resident at 5000–8500ft (1500–2600m), but confirmation of nesting lacking; very rare to casual winter resident, mostly above 4000ft (1200m) in central highlands.

RANGE. Breeds from E-C Alaska and W Canada to highlands of Honduras, possibly to Costa Rica; winters from Colombia and Venezuela to Peru and Bolivia, casually to Costa Rica.

NOTE. Formerly known as *Myiochanes richardsonii*.

EASTERN WOOD-PEWEE

Contopus virens Pl. 36(8)

Pibí Oriental

DESCRIPTION. 5½″ (14cm); 14g. A fairly small flycatcher told from *Empidonax* by its lack of a conspicuous eye-ring, fly-hawking habits, longer and more pointed wings, and short legs; larger and duller than Tropical Pewee. See Western Wood-Pewee. **Adults:** upperparts grayish-olive, darker and duskier on crown; wings and tail dusky, with 2 grayish-white wing-bars, secondaries with whitish margins; eye-ring faint, narrow; throat and center of breast dull whitish; belly white, tinged with yellow; sides of neck and breast, and sides pale grayish-olive (occasionally continuous across center of breast); crissum whitish, the longest feathers with concealed grayish bases; wing-linings pale grayish-olive, the feathers with broad yellowish-white margins. Upper mandible black, lower orangish with narrow dusky tip; legs black. **Immatures:** wing-bars and edges of secondaries brighter, more contrasting, white to warm buffy-white on anterior wing-bar; lesser wing-coverts and upper tail-coverts edged with buffy to white; up to distal half of lower mandible dark. Fall adults of Eastern and Western wood-pewees are in worn plumage; immatures are fresh.

HABITS. In all sorts of situations on migration, wherever it finds trees and shrubs, from edges, clearings, and canopy of tall forest to low scrub and at all heights in trees; from exposed perch sallies out to catch flying insects, especially bees, wasps, beetles, and flies, often returning to same lookout; does not take fruit; solitary, though in a large migratory wave many individuals may be in view, each maintaining a foraging space ca. 30ft (9m) in diameter.

VOICE. A high, clear whistled *pee-a-wee* or a shorter, ascending *pee-we* or *puueee*, sometimes scratchy but not thick and burry like note of Western Wood-Pewee.

STATUS. Abundant transient in fall (mid-August to mid-November) and spring (early March to mid-May) from lowlands to ca. 5000ft (1500m) and in much smaller numbers to 8000ft (2450m), rarely higher; rare winter resident, mainly below 4000ft (1200m).

RANGE. Breeds in S Canada and E-C USA; winters from Colombia and Venezuela to Peru and W Brazil, casually N to Costa Rica.

NOTE. Formerly placed in genus *Myiochanes*.

TROPICAL PEWEE

Contopus cinereus Pl. 36(9)

Pibí Tropical

DESCRIPTION. 5″ (13cm); 12.5g. Smallest and most brightly colored pewee, with dark cap, yellowish belly, and pale lores. **Adults:** pileum dark brown to dusky-olive; rest of upperparts dark olive; wings and tail dusky with 2 pale brownish to grayish-white wing-bars; secondaries with grayish-white margins; lores mixed with whitish; cheeks, sides of neck, breast, and sides grayish-brown to grayish-olive; throat white, often tinged with yellow; posterior underparts pale yellow; wing-linings dusky with yellowish-buff margins. Upper mandible blackish, lower orangish with dusky tip; legs black. **Young:** above browner, with conspicuous buffy to whitish fringes; lower breast and belly whiter; wing-bars and pale edgings on secondaries bright buff to yellowish.

HABITS. Frequents light woods, woodland edges, hedgerows, scrubby pastures, shady gardens and dooryards, open country with scattered trees and mangroves; usually perches fairly low, often on fence wire or low branch; sallies out less far than wood-pewees to catch flying insects, including small bees and wasps, beetles, moths, and flies, and less likely to return to same perch; habitually vibrates tail upon landing; usually solitary.

VOICE. A sharp, scratchy to metallic *peet* or *fweet*, often repeated incessantly, mainly in breeding season; a little rattling trill, *chee* or *preer*. Dawn song a repeated, high, sharp *weet weet weet*, punctuated by a lower, more musical *we-ye*, sometimes prolonged into a little warble.

NEST. A broad, shallow, compact cup of gray lichens, tendrils, fibers, seed down, bits of weed stems, and green moss, firmly bound with much spiderweb, 7–40ft (2–12m) up in shrub or tree. Eggs 2–3, dull white with a wreath of bright brown and pale lilac blotches around thicker end and a few small spots elsewhere. March–June.

STATUS. Resident; common in Caribbean lowlands and foothills to 2300ft (700m), and in diminishing numbers up Reventazón drainage to ca. 4300ft (1300m) around Cartago; uncommon to fairly common in Valle Central and on S Pacific slope to 4000ft (1200m), to 5000ft (1500m) in Coto Brus area; rare and local, mostly in mangroves, in dry NW lowlands.

RANGE. S Mexico to N Argentina, Paraguay, and S Brazil.

DARK PEWEE

Contopus lugubris Pl. 36(2)

Pibí Sombrío

DESCRIPTION. 6½″ (16.5cm); 23g. Large, dark, prominently crested pewee, without conspicuous pale throat or wing-bars. **Adults:** above dark sooty-olive to sooty-gray, darkest on pileum; wings and tail blackish, the wing-coverts edged with dark gray, secondaries with gray margins; below slightly paler, throat tinged with paler gray, shading to dark grayish-olive on flanks and to pale yellowish to buffy-white on center of belly and crissum. Upper mandible black, lower orange; legs black. **Young:** above browner, rump and upper tail-coverts with rufous fringes; wing-coverts broadly margined with dull rufous; throat and breast dull brownish-gray, belly buffy-white.

HABITS. Frequents wet montane forests, usually at edges and openings, or in adjoining semi-open or clearings with scattered trees; sallies for flying insects from usually high, exposed perches, often at very top of tree and usually returns to same lookout, often vibrating tail upon landing; solitary after nesting season; mobs Emerald Toucanets, especially while nesting.

VOICE. A loud, staccato *wic* or *whip*, repeated incessantly; a loud, clear to throaty *weer*, lower in pitch and heard less often; dawn song, repeated over and over from treetop in nesting season, *fred-rick-fear*, the *rick* ascending in pitch, the *fear* lower.

NEST. A broad, shallow cup with massive walls of green mosses, liverworts, and lichens, bound together with spiderweb, lined with fibrous rootlets and coarse vegetable fibers, often saddled over descending branch, 16–60ft (5–18m) up. Eggs undescribed (?). March–June.

STATUS. Resident between ca. 4000 and 7000ft (1200–2150m), from Cordillera de Tilarán S; mainly on Caribbean slope of Cordillera Central, on both slopes of Cordillera de Talamanca.

RANGE. Costa Rica and W Panama.

NOTE. Sometimes considered a race of the South American *C. fumigatus*.

OCHRACEOUS PEWEE

Contopus ochraceus Pl. 36(3)

Pibí Ocráceo

DESCRIPTION. 6½″ (16.5cm); 23g. Large, crested pewee with notched tail, below with distinct ochraceous tinge. **Adults:** above olive-green, darker and browner on crown; wings blackish with 2 buffy-olive to ochraceous-olive wingbars; anterior underparts ochraceous-olive, palest on throat, shading abruptly to pale yellow on belly and crissum; paler and duller in worn plumage, but always at least some ochraceous tinge on chest. Upper mandible black, lower orange; legs black. **Young:** similar but wing-bars brighter, more ochraceous; lesser coverts edged with ochraceous-buff.

HABITS. At breaks in canopy of montane oak forest, along streams, at treefall gaps and edges, and in tall, broken second growth near forest; from projecting stubs and branches with a clear view makes long sallies to hawk bees, flies, butterflies, moths, beetles, and other flying insects, vibrating its tail upon returning to perch. In coloration, crest and behavior, suggests an overgrown Tufted Flycatcher.

VOICE. Call a sharp, piercing *pwit*, often repeated 2–3 times, higher and shriller than that of Black-capped Flycatcher; song a high, thin, piercing *peeeeyit* or *peeeeyeeet*, the first syllable accented.

NEST. A bulky cup, covered outside with moss, saddled on slender, horizontal limb high in canopy at forest edge. Eggs undescribed (?). March.

STATUS. Uncommon to rare and local in Cordillera Central (Irazú-Turrialba massif) and Cordillera de Talamanca ca. 7200–10,000ft (2200–3000m).

RANGE. Costa Rica to W Panama.

THE GENUS *Empidonax*: In the hand, many species can be identified by measurements of the wingtip, with the wing folded in its normal position against the body (see the figure below). The 2 most useful measurements are "formula A," the difference in length between the 5th and 10th primaries (negative if

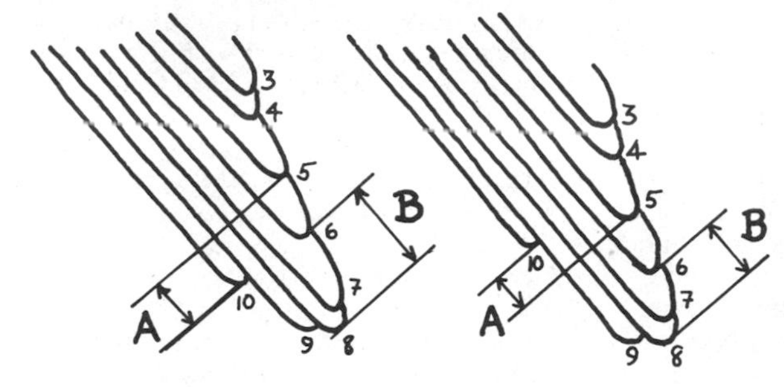

Figure 24. Empidonax wing formulas. Left: pointed wing (formula A negative, formula B large). Right: rounded wing (formula A positive, formula B small).

primary 10 is longer than primary 5), and "formula B," the difference between the longest primary (usually number 8) and primary 6. Formula A is always *positive* (5 longer than 10) in all resident species of *Empidonax* and usually in Least; it may be positive or negative in Yellow-bellied and Willow; and is nearly always *negative* in Alder and Acadian flycatchers. Only in these latter two is formula A often *more* negative than −3. Formula B is usually less than 5 in the resident species, Least, and Yellow-bellied; between 2 and 6 in Willow, 3 and 8 in Alder, and 5 and 10 in Acadian. As might be expected from their close relationship, overlap is greatest in these measurements between Alder and Willow flycatchers. In the hand, most (but not all) individuals of these sibling species can be separated by "Stein's Formula": 7.95 + 0.15 (Formula B − Formula A) = predicted bill length (measured from anterior edge of nostril to tip). Willows usually have bills *longer*, Alders *shorter*, than is "predicted" by this formula (Phillips et al. 1966). (All measurements in millimeters.)

YELLOW-BELLIED FLYCATCHER

Empidonax flaviventris Pl. 36(20)

Mosquerito Vientriamarillo

Description. 4¾″ (12cm); 11g. Green-backed, with olive-yellow throat; eye-ring and wing-bars yellowish; wing relatively short and rounded; outer web of 6th primary usually emarginate. **Adults:** above olive-green; eye-ring, wing-bars, and edgings of tertials pale yellow, often fading to white by spring; throat yellow, washed with olive to grayish; chest yellowish-olive; belly yellow, washed with olive on flanks. Upper mandible black, lower yellowish to pale orange; mouth-lining orange; legs dark gray. Many adults are molting outer primaries as they arrive in fall. **Immatures:** similar but usually with pale edges or fringes on tips of secondaries, primary coverts and rectrices.

Habits. On migration occurs in various habitats, from low scrub to forest; in winter prefers understory of primary or secondary forest, scrubby woodland, and shady clearings, where individuals hold winter territories; forages for insects and sometimes fruits by sally-gleaning, sometimes hovering, mostly inside vegetation, usually alighting on a different perch after each sally.

Voice. A sharp *pweet* or *tooweet*, apparently to mark its territory; also a harsh, squeaky *chew* or *cheep*. Especially early in their stay they are quite vocal, calling loudly as they patrol their territories in early morning and evening.

Status. Common fall migrant (late August through October) on both slopes, chiefly in lowlands but regularly to at least 5000ft (1500m); locally common winter resident throughout lowlands of both slopes, somewhat more numerous on Pacific, to 4000ft (1200m) in central highlands; sporadically common spring migrant in lowlands of both slopes (March to mid-May, occasionally later).

Range. Breeds from N Canada to extreme N-C and NE USA; winters from S and E Mexico to E Panama.

ACADIAN FLYCATCHER

Empidonax virescens Pl. 36(12)

Mosquerito Verdoso

Description. 5¼″ (13.5cm); 12g. Rather large *Empidonax*; above very green, with buffy-white to tawny wing-bars, long pointed wing; dull mouth-lining; gray legs. **Adults:** above olive-green; bold yellowish-white eye-ring; secondaries and tertials edged with whitish to buff; throat white (rarely pale yellow); chest yellowish-olive; lower breast white; belly, sides, and flanks pale yellow (brighter than Alder or Willow). Upper mandible black, lower whitish; mouth-lining flesh-color to yellowish. **Immatures:** similar; rectrices usually narrowly fringed with buff at tips; primary coverts and secondaries narrowly edged with buff. Both adults and immatures arrive in fresh plumage in fall.

HABITS. On migration occurs from open scrub and young second growth to primary and secondary forest; in winter prefers thickets and gaps in forest understory and edge, where solitary and apparently territorial; forages by snatching insects and a few fruits from vegetation in flight, alighting on a different perch each time; occasionally joins mixed flocks.

VOICE. A sibilant *pseep* or *wheest*, sometimes quite loud and penetrating; a softer, shorter *psip*.

STATUS. Common migrant in Caribbean lowlands in both fall (mid-September–late November) and spring (early March–mid-May); in fall uncommon but regular along Pacific slope of Cordillera de Guanacaste and Valle Central; sporadic elsewhere on Pacific slope. Winters uncommonly in Caribbean lowlands, locally to ca. 1600ft (500m) in foothills, and at least occasionally in lowlands and valleys of S Pacific slope.

RANGE. Breeds in extreme SE Canada and E USA; winters from E Nicaragua to N South America.

ALDER FLYCATCHER

Empidonax alnorum Pl. 36(14)

Mosquerito de Charral

DESCRIPTION. 5″ (13cm); 12g. Virtually indistinguishable in field from Willow Flycatcher, although voice may help. **Adults:** above dull grayish-olive, usually tinged with greenish on crown and sides of neck; eye-ring narrow, whitish; wing-bars white, anterior one often tinged with buff; secondaries sharply edged with white; throat white; chest washed with brownish-olive to dull olive; lower breast white; sides, flanks, and belly pale yellow. Upper mandible black, lower dull orangish to horn-color; mouth-lining orange; legs black. **Immatures:** above brighter olive; wing-bars broader, buffier (or often pale yellow); primary coverts tipped with olive; often fairly broad white edging on secondaries. In fall, adults are in worn plumage, young in fresh.

HABITS. Mostly in old fields, low scrub, hedgerows, and semi-open, usually near ground; seldom enters tall, shady second growth or woodland, where it may ascend to lower canopy; migrates in definite waves, often in company of Willow Flycatcher, passing through rapidly, seldom tarrying more than a day or 2 in 1 locality; forages by short sallies to pluck insects or berries from vegetation; sometimes hawks flying insects; habitually flicks tail upward.

VOICE. Usual call a piping *peep* or *pip*; song a hurried, scratchy *wee-bee'o* or *wiz-birw'*, with accent on second syllable, heard occasionally during spring migration.

STATUS. Abundant fall migrant (late August–early November) in Caribbean lowlands and across Valle Central, in decreasing numbers in Pacific lowlands; occasionally as high as 6500ft (2000m); more in lowlands of both slopes, less in highlands during spring migration (March–late May). In fall, adults pass through first, young mostly come later. No definite winter records. Outnumbers Willow Flycatcher by ca. 4:1 during fall migration.

RANGE. Breeds from C Alaska and N Canada to NE USA; winters from Colombia and Venezuela to N Argentina.

NOTE. Formerly considered conspecific with Willow Flycatcher, the 2 species collectively called Traill's Flycatcher.

WILLOW FLYCATCHER

Empidonax traillii Pl. 36(14)

Mosquerito de Traill

DESCRIPTION. 5″ (13cm); 12g. Virtually identical to Alder Flycatcher but has relatively longer bill, more rounded wing, and duller and browner coloration. All these differences are subtle, and most individuals are not safely distinguished in the field, except by voice. **Adults:** above dull olive-gray, usually tinged with brown; wing-bars and edgings of tertials white to grayish-white, anterior wing-bar sometimes tinged with brownish; eye-ring very dull whitish, often incomplete; throat white; chest washed with brownish-olive; belly whitish to pale yellow. Upper mandible black, lower dull orangish to horn-color; legs black. **Immatures:** above brighter olive, usually with brownish cast; wing-bars brighter, tinged with yellowish to dull buff; below brighter yellow than adult but usually paler than Alder.

HABITS. Often migrates with Alder Flycatcher and behaves similarly but perhaps less likely to enter woodland in fall. So far as known, only Willow is present in winter, when it prefers brushy savanna edges and second growth.

VOICE. Usual call a sharp *whit* or *whip*, softer and drier than call of Alder Flycatcher; song a burry, explosive *fitz-bew* or *pitch'ew*, with accent on first syllable, occasionally heard during spring migration.

Status. Fairly common fall and spring migrant (mid-August–late October, peak late September; and mid-March–late May) from lowlands to middle elevations of both slopes; uncommon winter resident, mostly in Pacific lowlands; rare at middle elevations (Valle Central) and on Caribbean slope.

Range. Breeds from S Canada to SW and NE USA; winters from S Mexico to NW Colombia.

Note. Willow and Alder flycatchers were formerly considered a single species called Traill's Flycatcher; because they are so similar, more information is required to clarify their winter distributions.

WHITE-THROATED FLYCATCHER

Empidonax albigularis Pl. 36(13)

Mosquerito Gargantiblanco

Description. 4¾″ (12cm); 12g. Small, very brown *Empidonax* with short, rounded wings, very dull eye-ring. **Adults:** above brownish-olive; eye-ring narrow, indistinct, buffy; wing-bars dull buff to pale brown; throat white; chest washed with brownish; belly yellowish-buff; wing-linings bright buff, usually crissum also. Upper mandible black, lower flesh-orange; mouth-lining orange; legs blackish. **Young:** above duller, more sooty-brown, with more contrasting, bright buffy to cinnamon wing-bars.

Habits. Prefers brushy marshes, wet, scrubby second growth, hedgerows, and swampy pasture edges; usually stays within 3–6ft (1–2m) of ground; active, sallying to snatch insects from vegetation or hawk them in flight; eats beetles, small wasps, nymphs of mantids and grasshoppers, damselflies, homopterans, and flies; also relatively small amounts of fruit; habitually vibrates or shivers wings and tail upon landing.

Voice. A variety of scratchy, burry notes: *wrick* or *rrrip* with rising inflection, or a more broken *pit-a-rip*; song a longer, louder *wheet* or *whirrt*, sometimes doubled: *whit-whirrt*.

Nest. A neat cup of dry grass blades and plant fibers, 3–6ft (1–2m) up in shrub overlooking lower vegetation. Eggs usually 2, buffy-white with pale and dark brown spots, mostly near large end. April–June or July.

Status. Locally common resident in its patchy habitat in Valle del Guarco, roughly in an arc around S side of Cartago, extending from Paraíso to Coris. Scattered records elsewhere, especially at lower to middle elevations on Caribbean slope, may be of migrants from N Central America. Unrecorded but likely in Cañas Gordas district, on Pacific slope near Panama, where it has been found breeding in neighboring Chiriquí.

Range. Breeds locally from N Mexico to W Panama; winters from C Mexico to C Panama.

LEAST FLYCATCHER

Empidonax minimus Pl. 36(15)

Mosquerito Chebec

Description. 4½″ (11.5cm); 10g. Smallest, grayest migrant *Empidonax*, with relatively small bill and notched tail, bright white eye-ring and wing-bars; outer web of 6th primary emarginate. Above grayish-olive; crown and nape gray with little or no greenish tinge; prominent white eye-ring, wing-bars, and edgings of tertials; throat white; chest washed with dull olive; lower breast white; belly, sides, and flanks tinged with yellow. Upper mandible black, lower grayish to flesh-color; legs blackish.

Habits. In brushy clearings and forest borders, thinned woodland, and tall second growth; generally keeps fairly low, from shrub level to upper understory; forages by sallying and hover-gleaning inside vegetation, like other migrant *Empidonax*; in winter solitary and sedentary, perhaps territorial.

Voice. A sharp *whit* or *pwit*, stronger than note of Willow Flycatcher; song, occasionally heard in spring, a sharp dry *che-BEK*.

Status. Very rare migrant and winter resident (early October through at least late April) in lowlands of both slopes; at least on migration, to 4000ft (1200m) or more in central highlands.

Range. Breeds from W and SE Canada to NW and E USA; winters from N Mexico regularly to Nicaragua, rarely to Costa Rica, casually to C Panama.

YELLOWISH FLYCATCHER

Empidonax flavescens Pl. 36(19)

Mosquerito Amarillento

Description. 5″ (12.5cm); 12g. Often appears slightly crested; eye-ring broader behind eye and more prominent than in any migrant *Empidonax*; below yellower, with darker wing-bars, than Yellow-bellied Flycatcher. **Adults:** above greenish-olive; wings blackish, wing-bars buffy-olive to ochraceous-buff; eye-ring pale yellow; belly yellow, tinged with olive; breast strongly tinged with ochraceous. Upper mandible black, lower flesh-orange; legs gray. **Young:** above bright brownish-olive; wing-bars and margins of secondaries broader and brighter, from buffy-yellow to cinnamon-orange; rectrices indistinctly tipped with buff; below paler yellow, chest washed with brownish, sides tinged with buff, belly almost white.

Habits. In cool, damp mountain forests, especially in openings and at edges, in shady pastures, and amid second growth; catches insects on short aerial sallies from low to moderately high perches, plucks them from foliage or bark of trees while hovering or drops briefly to ground for insects and spiders; eats some berries; solitary after breeding season; periodically flicks tail; usually does not associate with mixed flocks.

Voice. A thin, high-pitched, rather colorless or wiry *seeee* or *tseeep*, sometimes shortened to a stronger *tsick*, in series. At dawn in nesting season, from a high perch sings *seee seee chit* ca. 20 times/min.

Nest. A deep cup of green mosses and liverworts, interlaced with fibrous rootlets and lined with vegetable fibers, horsehair, papery bark, grasses, 6–15ft (2–4.5m) up in shallow niche in massive trunk, earthen bank, or cliff, often screened by drooping ferns and grasses, or saddled in vertical fork. Eggs 2–3, dull white, speckled and blotched with pale brownish-cinnamon or rusty-brown, densely on broader end. March–June.

Status. Common resident in mountains the length of the country, between ca. 2600ft (800m), rarely 2000ft (600m), and 7000ft (2150m) on Caribbean slope, and 4000–8000ft (1200–2450m) on Pacific slope.

Range. SE Mexico to W Panama.

BLACK-CAPPED FLYCATCHER

Empidonax atriceps Pl. 36(16)

Mosquerito Cabecinegro

Description. 4½″ (11.5cm); 9g. Dark, short-crested head contrasts with white eye-ring that is very broad behind eye and interrupted above it. **Adults:** pileum and hindneck sooty-black; rest of upperparts brownish-olive; wings and tail blackish, 2 wing-bars and edges of secondaries paler brown; outer rectrices with pale outer margins; cheeks dark olive-brown; sides of throat and neck, entire breast, pale brownish-olive; throat buffy-gray; belly buffy-yellow. Upper mandible black, lower orangish with dusky tip; legs blackish. **Young:** similar but head browner; wing-bars and edges of secondaries paler and more buffy; belly paler.

Habits. Frequents upper canopy of montane oak forests, coming much lower at edges and breaks, in second growth and pastures with scattered trees, and at edges of páramo; often behaves like pewee; sallies from exposed perch to catch flying insects, especially flies, beetles, and moths, often returning to same lookout; vibrates tail on landing or flicks it up quickly, lowers it slowly; often very tame; solitary after breeding season.

Voice. Call a simple, whistled *chip* or *whit*; in breeding season, a loud *keep-keer*; in interactions, an excited-sounding *keep-keep-keep-keep*.

Nest. A cup of fine grasses and moss, lined with fine vegetable fibers, horsehairs, small feathers, and beard lichen, in pastures, attached to grass pendant from top of low vertical bank or saddled in upright fork 6–40ft (2–12m) up in shrub or tree. Eggs 2, white or creamy, unmarked. March–May.

Status. Common resident of higher parts of Cordillera Central, Cordillera de Talamanca, and Dota region, mostly from 8000ft (2450m) up to 11,000ft (3300m) or more; locally resident down to 6900ft (2100m), as at Sabana Dúrika; during height of rainy season may appear as low as 6000ft (1850m).

Range. Costa Rica and W Panama.

TUFTED FLYCATCHER

Mitrephanes phaeocercus Pl. 36(11)

Mosquerito Moñudo

Description. 4¾″ (12cm); 8.5g. Small, with pointed crest, distinctive ochraceous underparts; behaves like a little pewee. **Adults:** above greenish-olive, including crest; tail and wings blackish, 2 wing-bars and edgings to secondaries pale olive to buffy, tertials with whitish margins; throat and chest bright ochraceous to cinnamon, shading through cinnamon-buff to pale or buffy-yellow on belly and crissum. Upper mandible black, lower orange-yellow; legs black. **Young:** pileum sooty-brown, rest of upperparts dark brownish-olive; feathers of entire upperparts with cinnamon-buff fringes; wing-bars bright cinnamon-orange; breast pale cinnamon-rufous; throat cinnamon-buff; belly pale buff.

Habits. In heavy montane forest clearings and edges, and broken canopy; ventures into clearings with scattered trees and tall second growth; from exposed low or high perches sallies to catch flying insects, often by intricate aerial maneuvers, often returning to same perch and usually vibrating tail upon landing; briskly active; in pairs throughout year.

Voice. A high-pitched, rapid series of liquid, rather confiding notes: *weet weet weet* . . . or *pui-pui-pui-pui-pi-pi*, often upon returning to perch; a thin *seer* or *peew* or, especially in breeding season, *peew-peew* or *peep-wit wit wit*; quaint, prolonged dawn song a very rapid series of high, thin notes: *bip-bip-bip-dididi-up-bip-bip-bibibiseer*.

NEST. A shallow saucer of green moss, liverworts, and gray-green lichens, with middle layer of fine, dark rootlets and lining of narrow, branching, foliaceous lichens, 13–90ft (4–27m) up on dangling vine or branch or slender upright or horizontal branch, usually well concealed amid mosses, ferns, bromeliads, or other epiphytes. Eggs 2, dull white, with a wreath of brownish blotches around thicker end. April–June.

STATUS. Resident on both slopes at ca. 2300–10,000ft (700–3000m), rarely down to 1500ft (450m); most abundant 4000–7000ft (1200–2150m).

RANGE. NW Mexico to NW Ecuador and E Bolivia.

RUDDY-TAILED FLYCATCHER

Terenotriccus erythrurus Pl. 36(23)

Mosquerito Colirrufo

DESCRIPTION. 3½″ (9cm); 7g. Very small, with large eyes, rather long rufous tail, and prominent rictal bristles. **Adults:** head, back, and shoulders grayish-olive, tinged with ochraceous on forehead; eye-ring, rump, upper tail-coverts, tail, broad edgings of wing-coverts and remiges, cinnamon-rufous; throat buff; breast cinnamon-rufous to ochraceous; belly paler and more buffy. Upper mandible black, lower pale flesh with dusky tip; legs pale brown to pale orangish. **Young:** above similar but brighter, more ochraceous; edges of wing-coverts rufous; tail darker rufous, tinged with dusky at tip; breast darker, tinged with olive.

HABITS. In lower and middle levels of wet forest and taller second growth; plucks small insects from foliage in darts from twig to twig, engaging in intricate aerial pursuits of fleeing insects, especially homopterans; usually solitary, sometimes joins mixed flocks; rapidly twitches wings above back, both together; sometimes vibrates them rapidly, making weak whirring sound; apparently does not pair, nest built and attended by female alone.

VOICE. A spirited but not loud *seeoo see* or *peeer peet*, the first note lisping or rolling, the second sharp and emphatic; repeats this phrase over and over at dawn; in interactions an urgent-sounding *pee peew peew peew* and variants.

NEST. Elongated, pyriform, with side entrance protected by visor, of dark-colored fibers and fragments of dead leaves, suspended from slender drooping twig or dangling vine, 5–15ft (1.5–4.5m) up in forest undergrowth. Eggs 2, white, blotched with chocolate in a wreath around thicker end and sparingly elsewhere. March–May.

STATUS. Common resident in wet lowlands and valleys of Caribbean slope and S Pacific slope, from sea level to 3300ft (1000m), to 4000ft (1200m) or more along Cordillera de Talamanca in Coto Brus region.

RANGE. SE Mexico to N Bolivia, C Brazil, and Guianas.

TAWNY-CHESTED FLYCATCHER

Aphanotriccus capitalis Pl. 36(10)

Mosquerito Pechileonado

DESCRIPTION. 4¾″ (12cm); 11g. Resembles *Empidonax* in size, shape, and pattern, but much more colorful; lacks yellow rump of *Myiobius*. Head rather dark gray, somewhat tinged with olive in female; broken whitish spectacles; back olive-green, tinged with ochraceous; wings dusky, bright ochraceous-buff wing-bars and margins of secondaries; throat buffy-white; breast ochraceous, shading to fairly bright yellow on belly. Upper mandible black, lower flesh-color with black tip; legs gray.

HABITS. Prefers dense vegetation along forest streams and edges and in old second growth near forest; singly or in pairs, following a regular route, forages in dense, low vegetation, chiefly by sudden upward sallies to snatch insects from undersides of leaves and twigs, especially beetles, bugs, and ants; may spread tail and close it nervously upon landing, but does not pump it.

VOICE. Usually a rapid phrase, last note loudest and burry, *chee chee spt't cheew* or *chit it-it chee'yew*; sometimes a longer, more *elaborate* phrase, *choot choot choot ch-ch-ch-chttttree'ih* (Slud), reminiscent of song of Gray-headed Manakin.

NEST. Undescribed (?).

STATUS. Local, uncommon resident of N Caribbean slope S to about Turrialba and Limón; lowlands to ca. 2600ft (800m), locally to 3300ft (1000m) on NE slope of Cordillera de Guanacaste.

RANGE. E Nicaragua to Costa Rica.

COCOS FLYCATCHER

Nesotriccus ridgwayi Pl. 36(18)

Mosquerito de la Isla del Coco

DESCRIPTION. 5″ (13cm); 11g. Only flycatcher on Cocos I.; long bill; rather long tail;

considerable variation in color. **Adults:** above grayish-olive to dark brownish-olive; tail and wings darker, with dull buffy to yellowish-buff wing-bars; usually faint, paler superciliaries; below pale grayish-buff, buffy-white or pale yellowish, washed with brownish to olive across breast. Upper mandible black, lower pale horn with dusky tip; legs dark gray. **Young:** browner than adults; superciliaries and wing-bars tawny to cinnamon.

HABITS. In forest, tall *Hibiscus* scrub and *Annona* swamp, regularly visiting second growth; forages at all heights, even picking prey from ground, but spends most time in upper understory and lower canopy; frequently hunts in crown of tree ferns; hawks flying insects; sallies, hovers, or runs along branch to snatch prey from vegetation; capable of fast, agile, aerial pursuit of fleeing prey inside foliage; takes a variety of insects, chiefly planthoppers, also fruits; paired and territorial throughout year.

VOICE. Usual call a dry, descending, accelerated tsitter, almost a trill, sometimes interspersed with staccato notes; members of pair often sing rapidly back and forth.

NEST. A small cup of plant fibers, on thin twig amid outermost foliage, usually near top of tree. Egg usually 1, creamy-white (Sherry). January–April or May.

STATUS. Common resident throughout wooded areas on Cocos I., from sea level to highest hills.

RANGE. Endemic to Cocos I.

SULPHUR-RUMPED FLYCATCHER

Myiobius sulphureipygius Pl. 36(21)

Mosquerito Lomiamarillo

DESCRIPTION. 4¾″ (12cm); 12g. Lively, redstartlike behavior; large eyes; prominent rictal bristles; rather long, ample tail. **Adult ♂:** upperparts mostly dark olive; a large, erectile, mostly concealed, yellow crown-patch; rump pale yellow; upper tail-coverts, tail, and wings black, the wing-coverts and remiges edged with dark olive; throat, eye-ring, and streaking on auriculars yellowish to buffy-white; an indistinct spot or bar of dusky below eye; breast and sides ochraceous-tawny; belly pale yellow. Upper mandible gray to black, lower whitish to flesh-color with dusky tip; legs grayish. ♀: similar but little or no yellow in crown. **Young:** above browner; no yellow in crown; below paler, the feathers with sooty fringes that give smudgy look to breast and sides.

HABITS. Hops and flits through lower and middle levels of wet forest, older second growth, and gallery forest, often with mixed flocks of small birds, sometimes with army-ant followers, but hardly ever with another of its kind; fans tail and lowers wings, displaying rump; often flits both wings sideways; plucks insects from foliage in short sallies or pursues them in the air with intricate maneuvers; takes many beetles, bees, homopterans, and flies; often catches prey flushed by other birds in mixed flock; not known to eat fruit. Does not pair; nest built and attended by female alone.

VOICE. A sharp, dry *psit* or *spik*, frequently while foraging; song a lilting, clear *tseuu tseuu tseuu tseer tseer* with variants, by male only (?).

NEST. A pyriform structure of fine, brownish vegetable strands, with an apron completely covering doorway in side of roundish chamber, so that bird enters by upward flight; pendant from slender, drooping twig or hanging vine, often at woodland edge or above stream or path, 4–35ft (1.2–10.5m) up. Eggs 2, white or pinkish, finely speckled all over with chocolate. March–June.

STATUS. Widely distributed, fairly common resident of lowlands of Caribbean and S Pacific slopes (N to about Carara), up to ca. 2600ft (800m) in N and C sectors, and 4000ft (1200m) in SE and Térraba and Coto Brus areas.

RANGE. S Mexico to W Ecuador.

NOTE. Sometimes considered a subspecies of *M. barbatus* of Amazonia.

BLACK-TAILED FLYCATCHER

Myiobius atricaudus Pl. 36(22)

Mosquerito Colinegro

DESCRIPTION. 4¾″ (12cm); 10g. Closely resembles Sulphur-rumped Flycatcher but paler and duller below, with smaller body and longer tail. **Adults:** above dark olive-green; partly concealed yellow crown-patch (smaller in female); a narrow, dull, buffy eye-ring; rump pale yellow; upper tail-coverts and tail black; wings blackish with faint olive edgings; throat pale dull yellow; breast pale brownish to buffy-brown, shading to pale yellow on belly. Upper mandible black, lower dark flesh; legs gray. **Immatures:** like adults but without yellow in crown.

HABITS. In light woods, scrubby second growth, borders of mangroves, and forest edge; enters closed forest only in evergreen gallery woodland where Sulphur-rumped Flycatcher of similar behavior is absent.

VOICE. Sharp, rather wiry *tsit* or *wit*, weaker

than note of Sulphur-rumped; song a "simple, rather sweet *cheer-cheer-cheer*" (Slud).

Nest. Similar to that of Sulphur-rumped and also attended only by female, hangs 16″–10ft (0.4–3m) up, often near or above a stream or lake shore. Eggs 2, pinkish or faintly buffy-orange, with slightly darker wreath around thicker end. April (in Costa Rica) to July (in C Panama).

Status. Uncommon resident from lowlands to ca. 3000ft (900m) on S Pacific slope, rare and local N to lower Tempisque basin.

Range. Costa Rica to NW Peru, S Brazil, and S Venezuela.

BRAN-COLORED FLYCATCHER

Myiophobus fasciatus Pl. 36(17)

Mosquerito Pechirrayado

Description. 4¾″ (12cm); 9.5g. Like a small, brownish *Empidonax* with streaked breast; a conspicuous whitish patch formed by edgings of secondaries. **Adults:** above cinnamon-brown, darker on pileum; a concealed yellow crown-patch (large in male, small in female); wings and tail dusky; 2 broad, buffy wing-bars; margins of secondaries dark basally, then abruptly buffy-white; lores and indistinct eye-ring pale buff; throat and breast pale buffy-yellow, shading to pale yellow on belly; breast and sides streaked with grayish-brown (more heavily in male). Upper mandible black, lower flesh-color; legs black. **Young:** above brighter, more rufescent; wing-bars rufous; no crown-patch.

Habits. Frequents low, scrubby second growth, overgrown pastures, and brushy savannas; avoids woodland; flits inconspicuously between low perches, snatching insects from foliage or pursuing them in flight; eats beetles, bugs, ants, small wasps, and flies, also berries; pumps or vibrates tail while perched; alone or in pairs.

Voice. A rather mellow, whistled note, repeated rapidly: *whee he he he he* or *wee, wee, wee*; a slight, low, clear trill; a harsh, weak croak. Dawn song (heard in Ecuador) a low, soft, but far-carrying *chite* repeated ca. 1/sec for many minutes.

Nest. A vireolike cup attached by rim to arms of a fork, 4–12ft (1.2–3.6m) up in shrub or small tree, of grasses, fine stems, vines, rootlets, often with green moss on outside, all bound with spiderweb. Eggs 2, dull white or pale buff, speckled and blotched with reddish-brown in wreath around thicker end, sparingly elsewhere. April–June.

Status. Locally abundant but spottily distributed resident from ca. 1000–4000ft (900–1200m) in General–Térraba–Coto Brus region, apparently not in wetter Golfo Dulce lowlands of S Pacific slope.

Range. Costa Rica to N Chile, C Argentina, and E Brazil.

ROYAL FLYCATCHER

Onychorhynchus coronatus Pl. 35(23)

Mosquero Real

Description. 6¾″ (17cm); 21g. Slender, long-tailed, with long, flat bill and a long, fan-shaped crest that is nearly always folded, giving a "hammer-headed" silhouette. **Adults:** upperparts olive-brown, except normally concealed crest that is orange-red in male, pale orange in female, each feather tipped with a black spot margined with blue-violet; middle and greater wing-coverts with buffy spot at tip; tertials with a buffy margin at tip; lower rump and upper tail-coverts cinnamon-buff; tail tawny-orange, shading to brownish at tip; throat buffy-white; rest of underparts cinnamon-buff to buffy-yellow; breast smudged and obscurely barred with olive-brown. Iris pale to medium brown; upper mandible blackish, lower horn-color to yellowish with dusky tip; legs orangish. **Young:** crest poorly developed; back feathers with blackish subterminal bar, buffy fringe; larger buffy tips on tertials; breast more conspicuously barred.

Habits. Frequents dense forest, tall second growth, semi-open, and gallery forest, mostly along shaded wet or dry streambeds and ravines, riverbanks, even including shaded waterways between open fields; perches very erect, the tail held vertically downward; captures dragonflies, butterflies, homopterans, and other insects in flight or darting sallies from branch to branch beneath canopy; beats larger prey against perch to remove wings; solitary except in pairs when nesting. In the hand gives a spectacular "cobra" display with crest fully extended in a fan, head vibrating and swaying from side to side. Undisturbed birds often spread their crests widely while preening, especially in the rain. A male occasionally gives the full head-swaying display while approaching his mate or repelling an intruding bird from their nest.

Voice. A loud, mellow, hollow-sounding whistle, usually 2-syllabled: *keeeyup* or *keeeyew*, with quality of jacamar or Great Crested Flycatcher; song of male a long series of higher, sharper notes with most peculiar intonation.

Nest. A uniquely slender, elongated, pendant,

brownish mass of plant fibers, rootlets, small living epiphytes, dead leaves, and a little green moss, 2–6ft (0.6–1.8m) long, with a shallow, open-fronted niche near the middle that holds the eggs; suspended from drooping twig or hanging vine 8–20ft (2.4–6m) above a shaded stream where sheltered from wind. Eggs 2, deep reddish-brown, becoming paler on the narrower end. March–May or June.

Status. Uncommon to fairly common resident on Pacific slope throughout lowlands up to ca. 2500ft (750m), and in NW Caribbean lowlands E to Río Frío region; very uncommon to rare and local elsewhere throughout Caribbean lowlands below ca. 1300ft (400m); after breeding, may ascend to 3000ft (900m) on Pacific slope.

Range. S Mexico to NW Peru, N Bolivia, and SE Brazil.

Note. Central American birds are perhaps best regarded as a species, *O. mexicanus*, separate from *O. coronatus*, Amazonian Royal Flycatcher, of South America; here we follow the AOU Check-List.

WHITE-THROATED SPADEBILL

Platyrinchus mystaceus Pl. 37(1)

Piquichato Gargantiblanco

Description. 3¾″ (9.5cm); 11g. Very small, big-headed and stub-tailed, with a very broad, flat bill. **Adults:** above brownish-olive, darker on pileum and sides of head; a concealed, erectile crest of bright yellow on hindcrown and nape (reduced or lacking in female); eye-ring, postocular streak, and bar on auriculars buff; loral spot, chin, and throat white; breast, sides, and flanks bright tawny-brown, shading to pale buffy-yellow on rest of underparts. Bill black, except tip and tomia of lower mandible pale horn; legs pale gray. **Young:** above brighter, more rufescent, no yellow in crown; throat and breast pale grayish-brown, shading to buffy-white on belly.

Habits. Frequents understory of cool, wet mountain forests and adjacent tall, shady second growth; perches quietly, scanning surrounding foliage, then sallies up to pluck small insects from undersides of leaves and twigs; flies low and rapidly; when excited, male sings with yellow crown feathers erected into fanlike crest; solitary except during nesting season.

Voice. Call a sharp, surprisingly loud *pick* or *chip*, not unlike call of Hairy Woodpecker; song of male a rattling, ascending trill ending in an abrupt squeak, sometimes preceded by a rippling, descending trill.

Nest. A neat cone, open at top, plastered all over with thin, papery pieces of decaying leaves that drape below bottom, and a fragment of snakeskin; thin middle layer of pale-colored vegetable fibers; contrasting lining of black fungal fibers; 3–10ft (1–3m) up in upright fork of shrub. Eggs 2, white, with slight yellowish tinge and faint rufous wreath. March–May.

Status. Common resident of wet middle elevations the length of both slopes between ca. 2300 and 7000ft (700–2150m).

Range. Costa Rica to W Ecuador, NE Argentina, and S Brazil.

STUB-TAILED SPADEBILL

Platyrinchus cancrominus Pl. 37(2)

Piquichato Norteño

Description. 3½″ (9cm); 10g. Very similar to White-throated Spadebill, but smaller and paler, with much less yellow in crown. **Adults:** above olive, tinged with ochraceous; crown and sides of head darker, browner; wings blackish, with ochraceous-olive edgings; spectacles, postocular streak, and auricular spot pale yellowish; throat white; breast dull ochraceous to buffy-brown; belly pale yellow. Upper mandible black, lower mandible and legs pale horn. Male has small concealed patch of yellow in center of crown. **Young:** above browner, feathers with smudgy dark fringes; face pattern faint and indistinct; fairly distinct fulvous wing-bars; throat and breast pale grayish-buff, shading to whitish on belly; no yellow on crown.

Habits. Prefers understory of dry to moist, deciduous to semievergreen and gallery forest; forages low, perching near or on ground, scanning, then sallying upward to pluck mostly concealed insects from undersides of leaves or twigs; sometimes searches for insects in ground litter, without turning or moving leaves; takes many homopterans, ants, beetles, and spiders; alone or in pairs, which call incessantly back and forth.

Voice. A 2- or 3-note descending ripple or twitter: *kiku* or *kikiku*.

Nest. A thin-walled cup, fitting firmly into vertical fork of understory shrub; of fine grasses and bark strips mixed with bits of leaves, lined with black fungal rhizomorphs. Eggs 2, dark buff, with reddish-brown dots and blotches around large end. May.

Status. Uncommon, local resident of N Pacific lowlands, S to Parrita, and up to 4300ft (1300m) along slopes of Cordillera de Guanacaste, where it overlaps with White-throated

Spadebill; uncommon in drier lowlands S of Lago de Nicaragua, between Los Chiles and Santa Cecilia.

RANGE. S Mexico to Costa Rica.

NOTE. Formerly considered a race of *P. mystaceus*, White-throated Spadebill, but differs in voice, coloration and size; the 2 are locally sympatric without known hybridization.

GOLDEN-CROWNED SPADEBILL

Platyrinchus coronatus Pl. 37(3)

Piquichato Coronirrufo

DESCRIPTION. 3¼″ (8.5cm); 9g. Tiny, with big head, stubby tail, broad flat bill, and prominent rictal bristles; above greener, below yellower than other spadebills. **Adults:** crown cinnamon-rufous to russet, mixed with olive and bordered laterally with black; a concealed, yellow crown-patch on male, a smaller, more orange patch on female; rest of upperparts dark olive-green; wings and tail dusky with paler olive edgings; lores, eye-ring, postocular streak, and auriculars pale olive-yellow; a black stripe or bar below eye, another behind auriculars; below pale yellow strongly tinged with olive on breast and sides. Upper mandible black, lower whitish; legs gray. **Young:** forehead and crown brownish-olive; rest of upperparts grayish-olive; wing-coverts and remiges edged with ochraceous-olive; pale facial markings whitish to buffy; below whitish to very pale yellow.

HABITS. Frequents dark understory of wet forest and adjacent tall second growth; perches quietly, peering all about, then abruptly sallies up to pluck beetles, caterpillars, small grasshoppers, other insects, and spiders from underside of leaves or twigs, drops down to a new perch; often upon alighting or calling flicks both wings open and shut; rarely accompanies army ants; in pairs throughout year.

VOICE. A long-drawn-out, weak, high-pitched, rippling trill that falls, then rises, easily mistaken for sound of an insect; a soft, cricketlike chirp, often repeated 2–3 times. A twanging sound when flying between perches may be made by wings.

NEST. A neat little cup of bast fibers, treefern scales, fungal rhizomorphs, bound with spiderweb and decorated on outside with green moss and bits of spider egg cases, lined with fine fibers, sometimes with a messy tail of debris, 3–13ft (1–4m) up in fork of upright stem. Eggs 2, short and blunt, dull white or creamy-buff, spotted and blotched with rufous, brown, and pale lilac. April–June.

STATUS. Common resident of wet lowlands, up to 2300ft (700m) on Caribbean slope and 4000ft (1200m) on S Pacific slope; to Pacific slope along Cordillera de Guanacaste via low passes.

RANGE. Honduras to W Ecuador, Amazonian Brazil, and Guianas.

YELLOW-OLIVE FLYCATCHER

Tolmomyias sulphurescens Pl. 37(16)

Piquiplano Azufrado

DESCRIPTION. 5″ (13cm); 14.5g. Small, rather long-tailed, with pale iris and spectacles, and notably broad, flat bill. **Adults:** crown and hindneck gray, tinged with olive; rest of upperparts bright olive-green; wings and tail dusky, the feathers with bright yellowish-olive edges; narrow spectacles grayish-white; cheeks and throat pale gray, shading to pale yellowish-olive on chest and sides, and sulphur-yellow on rest of underparts. Iris whitish to pale yellow; upper mandible blackish, lower whitish; legs gray. **Young:** below paler, more whitish; pileum much more olive; a broken white eye-ring (not spectacles). Iris dark brown or olive (requires several months to become as pale as in adults).

HABITS. In dry areas often in forest canopy, also tall second growth; in humid regions only in open woods, taller second growth, shady gardens, pasture trees, and forest edge; inside crown of tree, sallies up to snatch beetles, ants and bugs from undersides of leaves and twigs, drops down to a new perch; takes berries; solitary except when nesting.

VOICE. A high, thin, sibilant to slightly scratchy *pssssst* or *bzeeeek*, sometimes rapidly repeated, becoming sharper and more insistent with each repetition.

NEST. Pendant, retort-shaped, of matted, fine blackish rootlets and fungal rhizomorphs, suspended from slender, usually exposed twig or dangling vine, often at woodland edge, at roadside, or in isolated tree, often in bull's-horn acacia or near wasps' nest, 5–35ft (1.5–11m) up. Eggs 2–3, creamy-white, sometimes tinged with rufous, speckled with brown or cinnamon, chiefly on thicker end. April–June.

STATUS. Common to abundant resident of lowlands and foothills of Pacific slope, to 3000ft (900m) in N and 4500ft (1350m) in S; uncommon to common in Caribbean lowlands, locally up to 4600ft (1400m) in central highlands.

RANGE. S Mexico to NW Peru, N Argentina, and S Brazil.

NOTE. Also called Yellow-olive Flatbill.

YELLOW-MARGINED FLYCATCHER

Tolmomyias assimilis Pl. 37(17)

Piquiplano Aliamarillo

Description. 5″ (13cm); 15g. Big-headed, with broad, flat bill; differs from Yellow-olive Flycatcher in dark iris, much more conspicuous margins of wing feathers, and voice. **Adults:** crown, nape and sides of head dark gray, more or less washed with olive; spectacles dull white; rest of upperparts olive-green; wings dusky with conspicuous yellow margins of remiges and greater coverts; throat and breast pale gray, often washed or faintly streaked with yellowish, shading to yellow on belly. Iris olive to dark brown; upper mandible black, lower pale horn with darker tip; legs gray. **Young:** margins of wing feathers broader, less distinct, tinged with ochraceous; little gray on head.

Habits. From medium heights to high in canopy of wet forest, sometimes coming lower along edges or breaks, or in adjacent tall second growth; usually in pairs or small groups, sometimes singly, but nearly always accompanying mixed flocks of greenlets, warblers, honeycreepers, and tanagers; forages mostly by upward sallies to snatch insects from underside of leaves; occasionally pursues insects dislodged or flushed; eats mostly beetles, also many bugs, worker ants, and leafhoppers, and relatively small amounts of fruits; often holds tail partly cocked upward.

Voice. A series of 3–5 high-pitched, sibilant, very emphatic whistles, *tsew-tseep tseep tseep*.

Nest. A retort-shaped structure of fine, black plant fibers, chiefly fungal rhizomorphs, similar to that of Yellow-olive Flycatcher but usually situated much higher, often 30–70ft (9–21m) up and usually near wasps' nest. Eggs not seen. April–June.

Status. Locally common resident of Caribbean lowlands (except drier NW sector S of Lago de Nicaragua), from sea level to ca. 2000ft (600m) along Cordillera Central, locally to 3300ft (1000m) in Cordillera de Talamanca in SE.

Range. Costa Rica to NW Ecuador, N Bolivia, and E Brazil.

EYE-RINGED FLATBILL

Rhynchocyclus brevirostris Pl. 37(23)

Piquiplano de Anteojos

Description. 6″ (15cm); 23g. Stout, big-headed, with a notably broad, flat, massive bill; tail usually appears distinctly notched. **Adults:** above olive-green; wings and tail dusky with paler olive-green edgings; conspicuous white eye-rings; lores and cheeks grayish; a paler grayish patch on auriculars, bordered by a dusky patch posteriorly; throat and breast dull olive-green, with faint yellowish or pale grayish shaft-streaks; lower breast and sides paler, faintly streaked with yellowish; belly pale yellow. Males have the barbs of the outer primaries stiffened and comblike. Upper mandible blackish, lower pale horn; legs gray. **Young:** very similar but belly brighter yellow, breast virtually unstreaked; sides more coarsely streaked; edgings of wings more yellowish; little or no dusky on auriculars.

Habits. Frequents upper understory and middle levels of humid forest and adjacent tall second growth and riverside trees; rather sluggish, perching quietly and turning head from side to side, then making sudden upward sally to snatch beetle, bug, caterpillar, or other insect from underside of leaf or twig; eats berries and arillate seeds; occasionally forages with army ants; often accompanies mixed flocks, especially those led by tanagers (*Chlorothraupis, Chlorospingus*) and Black-faced grosbeaks; solitary, does not form pairs.

Voice. A high-pitched, scratchy, harsh whistle, *seeep* or *sweeeip*; a longer *weeep weeep wip-wip-wip* with a sputtering quality.

Nest. A retort-shaped structure, with rounded chamber entered through downwardly directed spout, 12–18″ (30–45cm) long, suspended from a slender, drooping twig or vine, 5–40ft (1.5–12m) up, often over open space or small stream inside forest; of rootlets and coarse fibers, often incorporating large dead leaves and live leaves of supporting twig, bits of moss and debris. Eggs 2, pale reddish-brown or pinkish-buff, mottled or faintly spotted with dark red or chestnut, especially in wreath around thicker end. March–June. At all seasons adults sleep alone in similar but more slightly built nests.

Status. Uncommon to fairly common resident on Caribbean slope, S Pacific slope (N to about Carara), and locally on upper Pacific slope of N cordilleras; lowlands to 5600ft (1700m), to 6900ft (2100m) in upper parts of Coto Brus valley.

Range. S Mexico to NW Ecuador.

BLACK-HEADED TODY-FLYCATCHER

Todirostrum nigriceps Pl. 37(8)

Espatulilla Cabecinegra

Description. 3¼″ (8cm); 6.3g. Tiny, with long, relatively heavy, blunt-tipped bill;

strong contrast between black cap and white throat. **Adults:** pileum and sides of head black; upperparts otherwise olive-green, paler and more yellowish on upper back; tail and wings blackish, 2 pale yellow wing-bars, remiges edged with pale yellow; rest of underparts bright yellow. Bill black; legs blue-gray to blackish. **Young:** head duller black; throat tinged with yellow; breast and belly paler yellow; yellow wing markings tinged with olive.

HABITS. Frequents canopy of wet forests, forest edge, tall second growth, and semi-open, especially tall *Cordia* trees shading cacao plantations or pastures; alone or in pairs; often forages in flowering trees (e.g., *Inga*), making sudden, swift sallies to snatch flies, small wasps, and other insects from foliage or flowers; takes ants from extrafloral nectaries and *Cordia* twig nodes; sidles along twig while wagging tail from side to side.

VOICE. A measured series of (usually) 5–8 high, sharp chips, the first somewhat lower, the latter ones speeding up slightly: *tzip tsip tsip tsip tsip tsip-tsip-tsip*; less often a single low *chip*, more resonant than that of Common Tody-Flycatcher. Incubating female utters a slight, soft trill.

NEST. A short, pendant, pyriform pouch with side entrance protected by visor, of matted, pale-colored plant fibers and seed down, 20–50ft (6–15m) up in tree at edge or in open, often beside wasps' nest. Eggs not seen; in Colombia, egg white with a few yellowish-brown spots on thicker end. March–May.

STATUS. Locally common in Caribbean lowlands, to 2300ft (700m), occasionally straying as high as 5000ft (1500m); barely reaches Pacific slope along Cordillera de Guanacaste through low passes.

RANGE. Costa Rica to W Ecuador and NW Venezuela.

COMMON TODY-FLYCATCHER

Todirostrum cinereum Pl. 37(7)

Espatulilla Común

DESCRIPTION. 3¾" (9.5cm); 6.5g. Very small, big-headed, with long, straight, flattened, blunt-tipped bill, and rather long, graduated tail. **Adults:** forehead, forecrown, lores, and upper cheeks black, shading to dark sooty-gray on nape and dark olive-green on rest of upperparts; wings blackish with pale yellow edgings; rectrices black, tipped with white, most broadly on short outer ones; underparts entirely yellow. Iris pale yellow, suffused with red above pupil; bill black with narrow tip and base of lower mandible whitish; legs bluish-gray. **Young:** similar, but pileum and cheeks dark sooty-gray; yellow wing markings tinged with buff; below paler yellow; iris dark (requires several months to become pale).

HABITS. Frequents dooryard shrubbery, shady plantations, open groves, second growth, scrub, and borders of gallery forest and mangroves; avoids interior of dense forest but enters canopy along edges; from eye level to high treetops, captures small beetles, flies, parasitic wasps, and other insects by swiftly darting against foliage; often wags tail from side to side while hopping sideways along branches; paired throughout year.

VOICE. Mates answer each other with resonant cricketlike trills; a cricketlike *tick* or *chip* often repeated in an irregular rhythm or in measured time; at dawn in nesting season, a slight, sharp *tic* delivered at rates of up to 110/min for many minutes. Rivals are pursued with clicking mandibles.

NEST. An elongated, pensile structure with visor-shaded side entrance and loose tail dangling below rounded chamber, of most varied vegetable materials, bound with spiderweb, and well lined with feathers, seed down, etc., 3 to, rarely, 100ft (1–30m) up, at end of slender twig or hanging vine. Eggs 2–3, immaculate white or rarely with a few fine brownish spots. February–July. Nests are not used as dormitories; adults and fledged young roost in trees, the latter in close contact.

STATUS. Common to abundant resident countrywide except in dense forests, lowlands to 3500ft (1150m), locally up to 5000ft (1500m).

RANGE. Mexico to NW Peru, E Bolivia, and S Brazil.

NOTE. Also called Black-fronted Tody-Flycatcher.

SLATE-HEADED TODY-FLYCATCHER

Todirostrum sylvia Pl. 37(9)

Espatulilla Cabecigrís

DESCRIPTION. 3¾" (9.5cm); 7.5g. Very small; big-headed, with thick blunt-tipped bill, shorter and less flattened than that of Common Tody-Flycatcher. **Adults:** pileum and hindneck slate-gray, darkest anteriorly; otherwise above olive-green; wings and tail blackish, the wing-coverts with yellow margins and wing-bars, the remiges edged with pale olive-yellow, the outer rectrices edged with whitish; a white line from bill to above eye; lower eyelid white; cheeks pale gray; throat and breast pale gray, indistinctly streaked with white; center of belly white; sides, flanks, and crissum pale yellow, tinged with olive on sides. Iris reddish-brown, tawny, or purplish; bill blackish with pale tip; legs pale gray. **Young:** pileum and hindneck grayish-olive; wing-bars tinged with buff; below paler, without streaking on breast; iris darker, browner.

HABITS. In humid regions prefers dense, young second growth; understory of dry woods and evergreen gallery forest; inside dense foliage, hops about, making fast, buzzy sallies, striking against leaves as it snatches leafhoppers, beetles, flies, small wasps, or spiders; continually flicks both wings partly open and shut; perches less erectly than bentbill; paired throughout year.

VOICE. A rather low, throaty, resonant *tic* or *kwip* or *tuc*; a short, descending, trilled *tu-u-u* or *trrrr*, less nasal and buzzy than call of bentbill, often preceded by a sharp note, *tue-trrrr*.

NEST. An elongated, pensile structure with side entrance, of grasses, plant epidermis, and fibers, swinging free in small, clear space amid thicket, usually under heavy canopy of foliage, 6–10ft (1.8–3m) up. Eggs 2, white, with chocolate spots in a wreath and sparingly scattered elsewhere. April–June.

STATUS. Resident countrywide from lowlands to ca. 3150ft (950m); most abundant on S Pacific slope, uncommon and local in NW and over most of Caribbean slope.

RANGE. S Mexico to Colombia, N Brazil, and Guianas.

NORTHERN BENTBILL

Oncostoma cinereigulare Pl. 37(6)

Piquitorcido Norteño

DESCRIPTION. 3¾″ (9.5cm); 7.3g. Very small, with faintly streaked breast and unique, downcurved bill. **Adults:** pileum dull olive-green with dusky streaking; crown tinged with slate (more prominently in males); rest of upperparts dull olive-green; wings and tail dusky with brighter olive to yellow-olive edgings; sides of head grayish; lores whitish; throat and breast pale gray, indistinctly streaked with whitish; lower breast tinged with olive-green, streaked with yellow, becoming plain yellow on belly; sides and flanks tinged with olive. Iris flesh-color to buffy; bill blackish with whitish tomia; legs flesh-color to pale gray. **Young:** iris brown; below paler and less streaked; wing-coverts tipped with buffy-yellow.

HABITS. In humid regions prefers fairly tall second-growth thickets, scrubby semi-open, and forest edge; in drier areas in understory of lighter woods, tall deciduous, and open evergreen forest; after scanning deliberately from a perch, abruptly flies obliquely upward to pluck beetles, leafhoppers, small grasshoppers, other insects, and spiders from undersides of leaves; usually forages below ca. 15ft (5m); does not pair, is solitary even in breeding season.

VOICE. Call a dry, throaty, sneering *naaaahrr* or *gr-r-r-r-r*, sometimes preceded by a short *kip* or *pip*; other harsh, short notes; a descending, almost melodious trill.

NEST. A nearly globular, pensile structure with visor-shielded side entrance, of pale-colored fibers with green moss on outside, lined with seed down; attached to slender twig 1–2ft (30–60cm) up on border of thicket. Eggs 1–2, whitish, with wreath of very fine, pale brown spots. March–June.

STATUS. Common resident from lowlands up to ca. 2500ft (750m) on Caribbean slope and 4000ft (1200m) on Pacific; least numerous in dry NW lowland.

RANGE. S Mexico to W Panama; NW Colombia.

NOTE. The Southern Bentbill, *O. olivaceum*, of C Panama to N Colombia, is sometimes considered a race of the present species, which is then called simply Bentbill.

SCALE-CRESTED PYGMY-TYRANT

Lophotriccus pileatus Pl. 37(4)

Mosquerito de Yelmo

DESCRIPTION. 3½″ (9cm); 8g. Very small, plump, below streaked; elongated crown feathers forming a helmetlike crest that usually lies flat, appearing big-headed and neckless. **Adult ♂:** feathers of pileum black, with broad rufous edgings that make flattened crest appear almost solid rufous; otherwise above olive-green; wings and tail dusky with olive-green edgings and 2 yellowish wing-bars; eye-ring cinnamon-rufous; sides of head tawny-olive; throat whitish, shading to pale yellow on belly; breast washed with olive, throat and breast streaked with dull gray. ♀: similar but crest less developed, sides of head cinnamon-rufous. Iris buff to pinkish-white; bill black, except base of lower mandible pale flesh; legs pale gray to horn-color. **Young:** crown tawny to cinnamon, without crest; broad eye-ring yellowish; wing-bars tinged with ochraceous; below whitish, tinged with buff on breast, with little yellow on belly or streaking on throat.

Habits. Frequents middle and upper understory of wet, mossy mountain forests, especially at gaps and edges with a dense growth of tall shrubs and small trees; forages by sallying upward, sometimes hovering briefly, to pluck beetles, other small insects and spiders from foliage or branches; solitary, does not pair; males call persistently in same small area between about January and June.

Voice. Noisy males constantly repeat a series of sharp, wooden to metallic, very loud and emphatic notes: *chug-jig jig jig jig jig*, all on same pitch or rising and accelerating; may start or end with a descending, rolling *cheeu-rig*; a softer, more melodious double whistle; both sexes give a sharp, rolling, cricketlike *preet*.

Nest. A pensile, pouch-shaped structure with visor-shielded side entrance, of grass stems, leaf petioles and other plant fibers, lined with plant down and decorated on outside with strands of green moss, with a short or long tail of bast fibers hanging below, suspended from twig or tip of treefern frond 5–12ft (1.5–3.5m) up. Eggs 2, creamy-white, unmarked. March–May.

Status. Common resident of foothills and lower mountain slopes from ca. 5500ft (1700m) down locally to 1000ft (300m) on Caribbean slope and 2500ft (750m) on most of Pacific slope; in hills of Península de Osa above 330ft (100m).

Range. Costa Rica to W Ecuador and E Peru; also SW Brazil.

BLACK-CAPPED PYGMY-TYRANT

Myiornis atricapillus Pl. 37(5)

Mosquerito Colicorto

Description. 2½″ (6.5cm); 5.2g. Tiny; blackish cap; white spectacles; very short tail. **Adults:** pileum black (duller, more sooty in female); nape and sides of head dark slate; rest of upperparts olive-green; tail and wings dusky with olive-green edgings, more yellowish on secondaries; chin, throat, and center of breast white, shading to pale yellow on belly; sides of neck and breast pale gray. Bill black; legs flesh-color to orangish. **Young:** pileum dark sooty-brown; upperparts more brownish; edgings of wing-coverts and underparts tinged with buff.

Habits. Frequents forest canopy, coming much lower at edges and gaps, in tall second growth, or semi-open; darts up to pluck small beetles, caterpillars, homopterans, other insects, and spiders from undersides of leaves, sometimes hovering; alone or in pairs, or family groups following breeding.

Voice. Call a reedy, sharp *tseep* or *keep* with rising inflection, readily confused with notes of insects or small frogs; a variety of soft, whistled, and trilling calls among members of family group.

Nest. Pensile, ca. 6″ (15cm) long, with round side doorway, covered with green mosses and liverworts; bottom of rounded chamber thickly padded with fine, pale bast fibers; attached to slender twig or petiole 4–22ft (1.2–7m) up in forest or nearby opening. Eggs 2, white with pale brown blotches forming wreath and a few scattered elsewhere. March–May.

Status. Fairly common resident throughout Caribbean lowlands, except drier region S of Lago de Nicaragua, up to 2000ft (600m) or more locally.

Range. Costa Rica to W Ecuador.

Note. Sometimes placed in genus *Perissotriccus*; by some considered conspecific with the South American *M. ecaudatus* and called Short-tailed Pygmy-Tyrant.

RUFOUS-BROWED TYRANNULET

Phylloscartes superciliaris Pl. 37(12)

Mosquerito Cejirrufo

Description. 4¼″ (11cm); 7g. Active, warblerlike; striking bridled face; white nasal tufts; superciliaries often indistinct. **Adults:** pileum gray; superciliaries rufous, blending into forehead; sides of head, throat and breast grayish-white; auriculars, nasal tufts brighter white; a blackish line from below eye around auriculars to hindcrown; rest of upperparts olive-green; narrow yellowish wing-bars; breast pale gray; belly pale yellow. Bill blackish; legs gray. **Young:** pileum mostly olive to brownish; little or no rufous in superciliaries; bridle broader, paler; chest smudged with olive; wing-bars broader.

Habits. Frequents subcanopy and canopy of wet forest, sometimes coming much lower along edges and at gaps; spritely, often pertly cocking tail, or flicking 1 wing at a time; hops bouncily and sallies in outermost foliage of crown of tree, often darting up to snatch beetles and other insects from undersides of leaves; takes small fruits (e.g., *Miconia, Trema*); usually in pairs or family groups, often with mixed flocks of tanagers, honeycreepers, warblers, etc.

Voice. Call a vigorous, arresting *swick* or *squeet*; a breathy, emphatic *pisseet* or a sharp *swee-swee-swee*.

Nest. Undescribed (?). Evidently breeds March–May or June (gonad data; presence of fledglings).

Status. Resident at ca. 2300–4000ft (700–

1200m) from Cordillera de Guanacaste (N at least to Cerro Santa María) to Cordillera Central; not yet reported from Cordillera de Talamanca but almost certainly present, probably at progressively higher elevations S; locally common in Cordillera Central, uncommon northward.

Range. Costa Rica to Colombia and NW Venezuela.

YELLOW TYRANNULET

Capsiempis flaveola Pl. 37(11)

Mosquerito Amarillo

Description. 4″ (10.5cm); 8g. Very small, slender, with long tail; superciliaries, wing-bars, and rather small bill suggest a warbler or vireo. **Adults:** above olive-green; wings dusky with yellow-olive edgings and 2 yellowish wing-bars; rectrices dusky with olive-green edgings; nasal tufts and superciliaries pale yellow, set off by dusky-olive eye-stripe; sides of head paler olive; entire underparts bright yellow, palest on throat and washed with olive-green on breast and sides. Bill black except base of lower mandible pale gray to flesh-color; legs gray. **Young:** upperparts brownish-olive with tawny wash over crown and hindneck; wing-bars tawny-buff; superciliaries tinged with buff; below much paler yellow, washed with buffy-brown on breast.

Habits. Wanders restlessly through thickets of shrubs and small trees in dense young second growth, overgrown pastures, coffee plantations, and along riverbanks; catches beetles, flies, homopterans, and other insects on short sallies across open spaces or plucking them from foliage; eats many small berries; in loquacious pairs or family groups throughout year.

Voice. Members of pairs or groups continually converse with a medley of rather soft, mellow *wick* or *pwik* notes with an upward inflection, punctuated by little chattering, rolling, or sputtering notes: *wick wick wick wick . . . tu wick wick-week brrr wick wick,* etc.

Nest. A substantial cup of grass blades, pale-colored fibers, shreds of plant epidermis, etc., with some green moss on outside, 7–20ft (2–6m) up in shrub, tree, or maize plant. Eggs 2, white, unmarked or very lightly speckled with reddish-brown. November–August.

Status. Fairly common resident throughout Caribbean lowlands to 2000ft (600m), and S from about Carara on Pacific slope, from lowlands to ca. 4000ft (1200m).

Range. Nicaragua to NE Argentina and SE Brazil.

Note. A few authors place this species in the genus *Phylloscartes*.

TORRENT TYRANNULET

Serpophaga cinerea Pl. 36(1)

Mosquerito Guardarríos

Description. 4″ (10cm); 8g. Small, perky, tail-wagging flycatcher of rocky highland streams; small bill, distinctive pattern. **Adult ♂:** pileum, sides of head, and nape black, with a concealed white crown-patch; back, shoulders, and rump pale gray; upper tail-coverts dark gray; tail and wings blackish, with 2 grayish-white wing-bars; below very pale gray, shading to white on throat and center of belly. Bill and legs black. ♀: similar but pileum duller and grayer; crown-patch reduced or lacking. **Young:** pileum and face dusky to sooty-brown; upperparts washed with buffy-brown; wing-bars broader, tinged with yellowish; breast fringed with brown.

Habits. Confined to rushing mountain streams and rivers, where mated pairs hold linear territories throughout year; in very rainy weather, when streams flood, may forage along roadsides; stands on emergent rocks, pumping tail up and down; catches damselflies, stoneflies, flies, and other insects on intricate aerial sallies, often low over water; or snatches insects from vegetation along banks or wet boulders, often where only momentarily exposed between surges that they barely escape; restlessly active.

Voice. A sharp, rapidly repeated, piercing *chip* or *tseep*, audible above din of rushing water, uttered in unison by mates coming together on rock; male (?) at dawn repeats a similar note more slowly, over a longer interval.

Nest. A substantial cup of fine roots and fibers, lined with soft materials, usually including many downy feathers, nearly always well covered outside with green moss, on branch 2–13ft (0.6–4m) above water or rocky shore. Eggs 2, pale buff or dull whitish, unmarked. January–June.

Status. Common resident of wet middle elevations from Cordillera de Tilarán S on both slopes at ca. 800–6000ft (250–1850m) on Caribbean, and 2000–6500ft (600–2000m) on Pacific.

Range. Costa Rica to N Bolivia and NW Venezuela.

YELLOW-BELLIED ELAENIA

Elaenia flavogaster Pl. 37(26)

Elainia Copetona (Copetoncillo, Bobillo)

Description. 6″ (15cm); 25g. Medium-sized, rather small-headed and small-billed, with a prominent bushy crest. **Adults:** upperparts olive to greenish-olive; a central white crown-patch exposed when crest is erected; wings and tail dusky; 2 dull buffy wing-bars; remiges edged with yellowish, with broader, paler edgings on tertials; narrow eye-ring white; sides of head grayish-brown; throat and breast pale gray, shading to pale yellow posteriorly; breast washed with olive. Bill blackish, except base of lower mandible horn- to flesh-color; legs black. **Young:** above browner; crest shorter, without white center; wingbars and edgings of tertials brighter, pale yellowish; below paler.

Habits. Frequents scrubby second growth, gardens, plantations, and other open areas with scattered trees, usually at medium heights; catches insects on short aerial sallies; eats many small fruits and arillate seeds, gleaned from perch or plucked on short sallies; active and animated, in pairs throughout year, never flocking.

Voice. Call a drawled, harsh, scratchy and sibilant *wheer* or *beeer*; also *wheer-chup* or *well-chip*, the first syllable prolonged, the second often clipped; mates join in jumbled duets of harsh, scratchy notes; dawn song a harsh, assertive *we do* or *chebeer jur-jeer*, rapidly repeated for many minutes.

Nest. A neat, shallow cup of fine vegetable materials bound together with much spiderweb, decorated externally with green or gray lichens or green moss, well lined with downy feathers; 4–60ft (1.2–18m) up in a fork or saddled over horizontal limb; varies from substantial to scanty. Eggs 2, short, broad, and blunt, whitish to pale buff, spotted and blotched with brown and pale lilac, chiefly in a wreath. March–July, rarely August; 2 broods.

Status. Resident countrywide, lowlands to 6000ft (1850m); generally common, least so in NW lowlands; has increased greatly in many areas with deforestation.

Range. S Mexico and Lesser Antilles to NW Peru, N Argentina, and S Brazil.

LESSER ELAENIA

Elaenia chiriquensis Pl. 37(25)

Elainia Sabanera

Description. 5¼″ (13.5cm); 17.5g. Small, brownish elaenia with short, inconspicuous crest; abruptly pale edgings of middle secondaries form distinctive patch on folded wing. **Adults:** upperparts grayish-olive; a concealed white crown-patch; wings and tail dusky; 2 dull whitish wing-bars; tertials edged with pale grayish, middle secondaries with yellowish-white except at bases; lores and narrow eye ring dull whitish; throat, breast, and sides pale brownish-gray, palest on throat and tinged with olive on breast; center of breast and belly pale yellow to cream-color. Bill blackish, with basal half of lower mandible pinkish (more extensively pale than Yellow-bellied Elaenia); legs blackish. **Young:** similar but above browner, without crown-patch; wing-bars dark buff; underparts paler.

Habits. Frequents weedy fields, scrubby savanna, brushy marshes, plantations, and gardens; catches insects on short aerial sallies and eats many small berries and arillate seeds, occasionally bananas at feeders; gleans insects and spiders from foliage; usually perches screened by foliage, often quite low; pairs very strongly territorial in breeding season, attacking and routing even other flycatcher species that approach their nest.

Voice. Call a clear, querulous *wheew* or *whoeeu*, frequently repeated, sometimes varied to a burry *wheer*; a thick *tse'-be*; dawn song of breeding male a dry, lusterless *a we d' de de*, tirelessly repeated. In evening twilight he flies almost straight up 100ft (30m) or more, repeating similar notes, then dives into low vegetation. Pairs do not duet.

Nest. A cup usually much less compact than that of Yellow-bellied Elaenia, composed of grass, weed stems, bits of decaying leaves, etc., bound with spiderweb, often scantily lined with feathers (fewer than in Yellow-bellied Elaenia's nest), usually with green moss but no lichens on outside, 15″–35ft (0.4–11m), mostly 5–10ft (1.5–3m) up in tree or shrub. Eggs 2, exceptionally 1, dull white with wreath of pale brown or rufous-brown spots, sometimes a few scattered elsewhere. April–June; 2 broods.

Status. Common to locally abundant in General-Terraba–Coto Brus valleys and adjacent slopes to ca. 5000ft (1500m); an isolated colony between Cartago and Paraíso in central highlands, and another at base of Volcán Miravalles in Guanacaste (*fide* Slud); permanent or perhaps only breeding resident, since rarely seen between about September and early January, suggesting emigration, probably to Panama or South America.

Range. Costa Rica to NW Ecuador, C Bolivia, and SE Brazil.

MOUNTAIN ELAENIA

Elaenia frantzii Pl. 37(24)

Elainia Montañera (Tontillo, Bobillo)

Description. 6″ (15cm); 20g. Slender greenish elaenia, fluffy-crowned but not really crested; most of inner webs of tertials whitish. **Adults:** above dull olive to brownish-olive; little or no concealed white in crown; wings and tail dusky, with 2 whitish wing-bars, the anterior one often dull, olive-tinged; remiges edged narrowly with yellowish; sides of head olive; narrow eye-ring and line over lores dull yellowish or whitish; throat, breast, and sides pale yellowish-olive, tinged with grayish-white on chin and throat; belly pale dull yellow. Bill blackish, basal half or more of lower mandible flesh-color; legs blackish. **Young:** above browner; wing-bars brighter, more yellowish; yellow of underparts much paler, more whitish; throat and breast with little yellow tinge.

Habits. In canopy of mountain forests, but prefers edges and clearings, second growth, and pastures with scattered trees and shrubs; usually keeps to thick foliage, making short sallies to pluck insects and spiders from leaves, twigs, or bark, even from ground or pasture grass; catches insects in air; eats many berries and arillate seeds, plucked in short, fluttering sallies or from perch; usually solitary except when nesting.

Voice. A long-drawn-out, down-slurred *peeeer* or *peeeah* in burry tone, or a clearer whistled *peer* or *pee-er*, longer and less plaintive than call of Mistletoe Tyrannulet; dawn song of male given from treetops for many minutes during breeding season, a rather buzzy *d'weet d'weet* or *ch'weet ch'weet*, sometimes with interspersed, gurgling, short syllables, *ch'weet ch'weet ch'weeza-chur ch'weet*.

Nest. A compact cup, with green mosses or liverworts, often mixed with lichens, covering inner layer of fibrous rootlets, decaying fibrous leaf-sheaths, and black fungal rhizomorphs; downy feathers in lining; 6–50ft (1.8–15m) up in shrub, tree, or tall bamboo. Eggs 2, dull white to pale buffy, spotted and blotched with pale cinnamon, chiefly in wreath. April–June.

Status. Breeds commonly throughout highlands from ca. 4000ft (1200m) in N cordilleras and 6000ft (1850m) in Cordillera Central and Cordillera de Talamanca, infrequently lower; on highest mountains of Península de Nicoya; withdraws from higher elevations between August or September and late January, down to 3000ft (900m), but is generally uncommon, suggesting some emigration from Costa Rica.

Range. Guatemala to Colombia and W Venezuela.

GREENISH ELAENIA

Myiopagis viridicata Pl. 37(21)

Elainia Verdosa

Description. 5″ (13cm); 13g. Like a small, brightly colored elaenia with a rather long, slender bill, and without conspicuous wing-bars. **Adults:** above olive-green, more brownish on crown; concealed bright yellow crown-patch (reduced in female); wings and tail dusky with olive-green edgings, brightest on secondaries; indistinct pale grayish superciliaries; white eye-ring interrupted before and behind eye; auricular feathers with pale shafts; throat and breast pale gray, breast tinged with olive, especially laterally; rest of underparts pale yellow; anterior underparts often appear faintly streaky. Bill black except base of lower mandible flesh-horn; legs blackish. **Young:** above much browner, crown with little or no yellow; wing-coverts tipped with buffy-brown; breast and throat tinged with buff; belly paler yellow.

Habits. Prefers groves of medium-sized trees or light second-growth woods, especially near streams; edges and more open parts of evergreen forest in dry regions; forages within crowns of trees, plucking beetles, caterpillars, small grasshoppers, and other insects from foliage on short sallies, usually after a period of scanning from a perch; eats berries and arillate seeds; usually alone, occasionally with other birds.

Voice. A high, thin, rolling or sibilant *speeeer* or *seeeees*, more rolling, less scratchy than note of Yellow-olive Flycatcher. High, thin, prolonged dawn song varies from bird to bird: *see-e seer seer see-e*, or *peer-weer peer-weer peer-weer*, or *peer weedyum peer weedyum*.

Nest. A shallow cup attached by rim in a forked twig or saddled in leaf axil, of tendrils, rootlets, rachises of compound leaves, and spiderweb, so thin that eggs are visible through bottom, 20–35ft (6–10m) up in tree. Eggs 2, white, streaked and blotched with chocolate and lilac. May–June.

Status. Fairly common to very uncommon resident throughout Pacific slope, least numerous in Golfo Dulce region, from lowlands to ca. 3600ft (1100m), locally to 5000ft (1500m); common in lowlands of NW Caribbean slope from Río Frío W; rare in upper Reventazón drainage.

Range. Tropical Mexico to W Ecuador, N Argentina, and E Brazil.

SCRUB FLYCATCHER

Sublegatus modestus Pl. 37(22)

Mosquero Gorgigrís

DESCRIPTION. 5½" (14cm); 13.5g. Like a small, brownish-backed elaenia with a darker gray throat and breast and a short, bushy crest. **Adults:** above grayish-brown; wings dusky with dingy grayish-buff wing-bars; supraloral stripe faint, grayish-white; eye-ring narrow, indistinct; throat and breast pale gray, shading to pale dull yellow on belly and crissum. Bill black, base of lower mandible flesh-color; legs dusky. **Young:** above browner, scaled with dull white; throat and breast indistinctly scaled or smudged with brownish; wing-bars and margins more whitish.

HABITS. Restricted to mangroves and their immediate vicinity, preferring dense young mangroves 15–30ft (5–9m) high; in tall mangroves found mainly at gaps and edges, and in adjacent second growth and scrubby woodland; usually perches fairly low, sometimes on mangrove root; sallies and hovers to snatch ants, beetles, caterpillars, and other insects from vegetation, or hawks insects in flight; takes fruits; looks about alertly between sallies; often pumps tail once or twice on landing; in general, acts rather like an *Empidonax*; usually solitary.

VOICE. A clear whistled *duweep*, a more scratchy *dhweer*; especially in breeding season, a more urgent-sounding *weep* or *weep-a-weep*.

NEST. A small cup of plant fibers, ca. 10–20ft (3–6m) up in mangrove or other small tree. Eggs 2, creamy-white, spotted finely with chestnut-brown to lavender, mostly toward large end. April–June or July.

STATUS. Locally common resident around Golfo de Nicoya; not recorded anywhere along outer coast and rare in mangroves of Golfo Dulce (Golfito, Puerto Jiménez).

RANGE. Costa Rica to Uruguay and N Argentina.

NOTE. Birds from Costa Rica S to Amazonian Brazil are often considered a separate species *S. arenarum*, from the more southern *S. modestus* (Short-billed Flycatcher).

SOUTHERN BEARDLESS-TYRANNULET

Camptostoma obsoletum Pl. 37(20)

Mosquerito Silbador

DESCRIPTION. 3¾" (9.5cm); 7.5g. Very small, active, with small bill, bushy crown, and 2 prominent wing-bars. **Adults:** pileum dull brownish-olive, shading to dull olive-green over rest of upperparts; wings dusky with pale yellowish or white wing-bars and edges to secondaries; tail dusky with narrow whitish tip; indistinct grayish-white superciliaries; narrow eye-ring white; throat dull white; rest of underparts pale yellow, slightly tinged with olive on breast. Bill and legs black. **Young:** above browner, rump and upper tail-coverts buffy-brown; wing-bars dull cinnamon-buff; below pale buffy-white. More brightly colored, with much more prominent wing-bars, than Northern Beardless-Tyrannulet, with which it overlaps between the lower Tempisque basin and Puntarenas.

HABITS. In open woods, second growth, scrub, savannas with scattered trees, shady gardens and plantations, rarely in forest; often cocks tail or flags it up and down; raises crown feathers in a low, loose crest; forages like a vireo or warbler, hopping and flitting through treetops or shrubbery, plucking ants, beetles, scale insects, other small insects, and spiders from leaves and twigs, sometimes while hovering; rarely engages in aerial flycatching; eats many berries, especially of mistletoes; alone or in pairs.

VOICE. A series of 2–10 clear, thin, high-pitched whistles in falling cadence, becoming progressively softer and shorter: *wheeeew wheew wheeew wheew weew weu*, etc. Dawn song of male, repeated for many minutes from a high, exposed perch, a penetrating, far-carrying *te be be be*.

NEST. Roughly globular with side entrance, of dry liverworts, mosses, lichens, fine leaf rachises, and spiderweb, lined with seed down, tucked into a hanging mass of liverworts, a richly branched inflorescence, a large, curled dead leaf, or cluster of prickly "achiote" pods, 16–90ft (5–27m) up. Eggs 2, white, finely spotted with cinnamon, chiefly in a wreath. December–February or March.

STATUS. Fairly common resident of S Pacific slope, from lowlands to ca. 2500ft (750m); decreases in abundance N of Carara; rare and local, in patches of evergreen forest, in lower Tempisque basin.

RANGE. Costa Rica to C Peru, N Argentina, and S Brazil.

NORTHERN BEARDLESS-TYRANNULET

Camptostoma imberbe Pl. 37(19)

Mosquerito Chillón

DESCRIPTION. 3¾" (9.5cm); 7.5g. Very small, short-billed, dull-colored, with dull wing-bars and short, bushy crest. **Adults:** above and on breast dull, grayish-olive; throat and superciliaries paler; eye-ring narrow, whitish; wing-bars dull whitish to grayish-buff; tertials edged with white; belly pale yellowish to dull white. Upper mandible black, lower flesh-color, shading to horn-color at tip; legs gray. **Young:** above browner; feathers of back edged with olive-brown; below tinged with buff; wing-bars broader, ochraceous-buff.

HABITS. In deciduous forest, semi-open, tall scrub, and edges of savanna; avoids tall evergreen forest, enters gallery forest only along roads or breaks; forages from undergrowth well into canopy, making short upward sallies to snatch insects or occasionally small berries from vegetation; habitually flicks tail through a shallow arc and often gives nervous double wing-flicks between sallies; usually alone or in pairs.

VOICE. A piercing, jacamarlike whistle, *fleeet* or *peeeeuk*; a descending laughing ripple; song 4–5 loud, piercing, thin whistles: *peee-peeew-peeeew-peeeuu*, middle notes loudest, dropping slightly in pitch toward end, more forceful than notes of Scrub Euphonia.

NEST. A small, globular structure with side entrance, in twigs at tip of branch, in clump of epiphytes, on palm stem, 3–30ft (1–9m) up. Eggs usually 2, white, finely speckled with reddish-brown. March–June.

STATUS. Resident in N Pacific lowlands S to Puntarenas, to ca. 2600ft (800m) on adjacent slopes of Cordillera de Guanacaste; common S to about Cañas-Taboga area, scarcer farther S; a common resident in lowlands S of Lago de Nicaragua, from Río Frío area W to Santa Cecilia-Orosí.

RANGE. Breeds from extreme SW USA and N Mexico to Costa Rica; N populations migratory, winter from N Mexico southward.

NOTE. Previously sometimes considered conspecific with Southern Beardless-Tyrannulet, but sympatric without interbreeding in W-C Costa Rica.

MISTLETOE TYRANNULET

Zimmerius vilissimus Pl. 37(10)

Mosquerito Cejigris

DESCRIPTION. 3¾″ (9.5cm); 8.5g. Very small; short bill; rather long legs and tail; pale iris; conspicuous yellow wing margins; no wing-bars. **Adults:** pileum dull gray; rest of upperparts dull olive-green; wings blackish, coverts and secondaries with yellow edgings; tail dusky; short superciliaries pale gray; lores dusky; throat and breast grayish-white with indistinct pale gray streaks; center of belly white; sides and flanks dull yellowish-olive. Iris yellowish, buff or grayish-brown; bill black, with brownish base of lower mandible; legs blackish. **Young:** crown olive with little grayish tinge; iris darker; yellow margins of wing-coverts broader but less clean-cut, often tinged with olive; superciliaries tinged with yellow; underparts tinged with buff.

HABITS. Found virtually anywhere in humid regions where mistletoes abound, from canopy of tall forest to second growth, scattered trees in pastures and clearings, shady gardens and plantations; usually stays fairly high in trees; eats mainly mistletoe berries but also takes other small berries, beetles, small insects, and spiders, all of which it plucks from foliage in short flits and sallies; lively and animated, often standing with legs seemingly extended to maximum extent and with tail cocked; alone or in pairs.

VOICE. Call a loud, clear, plaintive *peeer* or *peeeu*, often repeated, sometimes varied to *peeyup*; dawn song a rather melancholy-sounding *yer-de-dee, yer-de-dee, pe-pe-pe* or *deeu deeu dee tee-a-weedy*, sometimes followed by a faint rattle or trill.

NEST. Roughly globular with side entrance, of mosses, pale-colored plant fibers, and downs, stiff rootlets and cobweb, set amid tufts of moss hanging beneath a branch; inside a large, curled dead leaf; amid roots, rhizomes, or branches of a small, dangling epiphyte such as an orchid; inside a Yellow-olive Flycatcher's nest or attached to its bottom; 6–50ft (2–15m) up. Eggs 2, dull white, speckled or blotched with cinnamon or rufous. February–July, rarely August; 2 broods.

STATUS. Common resident countrywide from lowlands to 10,000ft (3000m), except absent from N Guanacaste and rare in rest of N Pacific lowlands and Valle Central and relatively uncommon in high mountains.

RANGE. S Mexico to N Venezuela.

NOTE. Formerly placed in the genus *Tyranniscus*; usually called Paltry Tyrannulet.

YELLOW-CROWNED TYRANNULET

Tyrannulus elatus Pl. 37(15)

Mosquerito Coroniamarillo

DESCRIPTION. 4″ (10cm); 8g. Resembles Mistletoe Tyrannulet but has black crown and yellow crown-patch, underparts and wing-bars. **Adults:** pileum black with partly concealed bright yellow crown patch (more extensive in male); dull white eye-ring interrupted by blackish eye-stripe; sides of head and hindneck gray; rest of upperparts olive-green; wings and tail dusky; 2 wing-bars and edges of secondaries pale yellowish; cheeks and throat pale gray; chest yellowish-olive; posterior underparts pale yellow. Bill and legs black. **Young:** head, chest, and back sooty-brown, darkest on crown, with indistinct grayish to buffy scaling; no yellow on crown; belly dull buffy-yellow; dull buffy wing-bars.

Habits. Prefers scrubby second growth, hedgerows, pastures with scattered trees and groves; usually in pairs, foraging well up in trees by short flits and sallies in foliage. Staple food probably mistletoe berries, but various other small fruits and some insects, especially small beetles, are eaten.

Voice. A clear, whistled *whee-peer* or *waay-TEEER*; contact note between members of pair a short, dry buzz.

Nest. A shallow cup of fine twigs and fungal filaments, some of which hang below, where several leaf petioles sprout from branch (Eisenmann). Eggs 2, cream-colored, unmarked (in South America). January–August (in Panama; no Costa Rican record).

Status. Uncommon to locally common resident in Golfo Dulce lowlands, N at least to Piedras Blancas but not yet recorded from Península de Osa; evidently has invaded from Panama in recent years following removal of forest.

Range. Costa Rica to W Ecuador, N Bolivia, and Amazonian Brazil.

ZELEDON'S TYRANNULET

Phyllomyias zeledoni Pl. 37(18)

Mosquerito Frentiblanco

Description. 4½″ (11.5cm); 11g. Gray head and small bill recall Mistletoe Tyrannulet, but note conspicuous wing-bars, bicolored bill, and brighter yellow belly. **Adults:** head slate-gray; white line mixed with gray from forehead over eye; sides of head mottled and streaked with gray and white; rest of upperparts olive-green; yellow wing-bars and edgings to secondaries; throat white, faintly streaked with pale gray; breast olive, indistinctly streaked with yellowish. Iris reddish-brown; upper mandible black, lower flesh-orange; legs dark gray. **Young:** above browner; little or no gray on head; throat tinged with yellow; belly tinged with buff; rectrices with narrow buffy tips.

Habits. Frequents forest canopy, mostly above midheight but descending to upper understory at edges and gaps; active, warbler-like, foraging in outer foliage, sallying to pluck insects from upper and under surfaces of leaves, flicking its wings between sallies, working through tree with short flits and dashes, occasional longer flights; usually travels in small groups or pairs that sometimes associate with mixed flocks; does not cock tail (unlike Mistletoe Tyrannulet).

Voice. Call a sharp, sibilant *tsip* or *tseeet*; high, thin hissing whistles and trills in interactions; song a deliberate series of 3–6 piercing, emphatic *tseet* notes, ca. 3 per 2sec.

Nest. Undescribed (?).

Status. Uncommon to locally rare resident at ca. 3000–6000ft (900–1850m) on Caribbean slope of Cordillera de Tilarán, Cordillera Central and Cordillera de Talamanca, also Dota region and, locally, on Pacific slope of Cordillera de Talamanca in S.

Range. Costa Rica to W Panama.

Note. Formerly placed in genus *Acrochordopus*; sometimes lumped with the larger, green-headed *P. burmeisteri* (Rough-legged Tyrannulet) of S South America; or *P. leucogonys* of N South America is merged in *P. zeledoni*, in which case the English name is White-fronted Tyrannulet.

BROWN-CAPPED TYRANNULET

Ornithion brunneicapillum Pl. 37(13)

Mosquerito Gorricafé

Description. 3¼″ (8cm); 7g. Tiny, with short bill and tail, conspicuous superciliaries and cap. **Adults:** pileum sooty-brown; broad white superciliaries meeting across forehead; dusky areas before and behind eye separated by white spot below eye; rest of upperparts olive-green; wings and tail dusky, with olive edgings; underparts yellow, clouded, and indistinctly streaked with yellow-olive on throat and breast. Bill and legs black. **Young:** crown paler, more olive; upperparts tinged with brown; white markings of face narrower, less distinct; wing-coverts indistinctly tipped with olive-buff; below paler yellow.

Habits. Frequents canopy of tall wet forest, coming much lower along edges and at gaps, in adjacent second growth or semi-open; usually in pairs, accompanying mixed flocks of greenlets, gnatcatchers, honeycreepers, etc.; actively rummages and gleans like a tit or antwren, poking into rolled dead leaves, often hanging acrobatically; eats beetles, bugs, small caterpillars, other small insects, and spiders, small berries, including mistletoes.

Voice. Loud, arresting, high, clear, whistled notes: 1 or 2 slurred upward, followed by 4–6 successively lower and faster, *peek peek peee pee-pee-pee-pi-pi*; sometimes a longer final note rises in pitch.

NEST. A shallow, untidy cup of fine brownish twigs or leaf petioles mixed with bark shreds, 40ft (12m) up in tree in semi-open (in Panama; no Costa Rican record). Eggs undescribed (?). June–August (from sightings of fledglings and gonad data).

STATUS. Fairly common in Caribbean lowlands, N at least to Tortuguero, but not known in drier lowlands W of Río Frío; to 3000ft (900m) locally on slopes of Cordillera Central and Cordillera de Talamanca.

RANGE. Costa Rica to W Ecuador and N Venezuela.

NOTE. Sometimes placed in genus *Microtriccus*; by some considered a race of the Yellow-bellied Tyrannulet, *O. semiflavum*, but both occur at Lago Caño Negro and perhaps elsewhere in Río Frío region.

YELLOW-BELLIED TYRANNULET

Ornithion semiflavum Pl. 37(14)

Mosquerito Cejiblanco

DESCRIPTION. 3¼″ (8cm); 7g. Very similar to Brown-capped Tyrannulet and best distinguished in field by voice; cap gray not brown, tail slightly longer; underparts clearer yellow. **Adults:** pileum slate-gray; white superciliaries meeting on forehead; lores dusky; a small white suborbital spot; sides of head and rest of upperparts olive-green; wings and tail dusky with olive edgings; underparts yellow, tinged with olive on sides of breast. Bill black, legs dark gray. **Young:** crown and upperparts tinged with brownish; below paler yellow, throat and breast with faint streaks or smudges of olive-gray.

HABITS. Frequents forest edge, semi-open, and tall second growth, less often forest canopy; singly or in pairs, accompanies mixed flocks, but is found more often by itself than is Brown-capped Tyrannulet; like that species gleans actively for small insects and berries in foliage but does not usually sally or hover.

VOICE. A high, thin *pyeeer*, longer and higher-pitched than call of Mistletoe Tyrannulet; a series of 3–5 such whistles that are successively lower in pitch and faster, ending with an upslurred note, *pyeeer peeer peer-pew-puweet*, thus lacking the initial rising notes of Brown-capped Tyrannulet, and less vigorous and full-throated in general.

NEST. Undescribed (?).

STATUS. Uncommon resident of the coastal lowlands and Golfo Dulce sector of S Pacific slope, N to about Carara; from Río Frío region W to Caribbean slope of Cordillera de Guanacaste, up to 2000ft (600m) locally in both regions.

RANGE. S Mexico to Costa Rica (and almost certainly adjacent SW Panama).

SLATY-CAPPED FLYCATCHER

Leptopogon superciliaris Pl. 36(26)

Mosquerito Orejinegro

DESCRIPTION. 5″ (12.5cm); 12g. Small; rather long tail; narrow bill; buffy wing-bars; mottled face with crescent-shaped blackish ear-patch. **Adults:** pileum slate-gray, shading to olive-green over rest of upperparts; wings dusky with olive-yellow edgings, brighter and more yellow on tertials, and 2 broad yellowish-buff wing-bars; tail dusky with olive edgings; lores, cheeks, and orbital area finely mottled and streaked with gray and white; posterior part of auriculars dull blackish; throat pale grayish-olive, passing into yellowish-olive on breast and sides, and yellow on belly. Bill blackish except base of lower mandible pale horn; legs gray. **Young:** pileum more olive, with little gray; face pattern less distinct; wing-bars more orange, less buff; underparts, especially throat, paler.

HABITS. Prefers upper understory and lower canopy of wet forest and forest edge in foothills and lower mountains; often the only individual of its kind in a mixed flock; stops briefly on slender, exposed perch to scan, then sallies forth to capture insects in air or from foliage; takes many berries; restless, rarely delays long in 1 spot; often nervously flicks 1 wing at a time open and shut.

VOICE. Call a sharp, sibilant, emphatic *wsst*, sometimes sharpened to *swick*; this often followed by a hard little rattle: *wsst-tsrrrr* or a mocking *peet-yer*; also *hit chu' hit chu' hit chu'*, like a little sneeze, and *hit cheee*, the second syllable thin and prolonged.

NEST. Globular or pyriform, with side entrance shielded by visor, composed chiefly of thin, dark brown rootlets, with paler fibers, the globular chamber lined all around with finely shredded, pale-colored bast fibers and tufts of silky seed down, attached at top to splinter beneath a log, 4ft (1.2m) above rapids in deep ravine. A Costa Rican nest held 1 nestling in early March. In Trinidad, 2 white eggs. February–July.

STATUS. Uncommon to fairly common resident of Caribbean slope from Cordillera de Tilarán S, Dota region, and Pacific slope of

Cordillera de Talamanca, including coastal ranges, 2000–5200ft (600–1600m).

Range. Costa Rica and Trinidad to W Ecuador and N Bolivia.

SEPIA-CAPPED FLYCATCHER

Leptopogon amaurocephalus Pl. 36(27)

Mosquerito Cabecipardo

Description. 4¾″ (12cm); 10g. Resembles Slaty-capped Flycatcher but crown brown, dark ear-patch less distinct, wing-bars more ochraceous. **Adults:** pileum and hindneck dark brown; lores, cheeks, and behind eyes mottled with dusky and buff; a patch of dark brown on posterior auriculars; back and rump olive-green; upper tail-coverts and tail russet-brown; wings dusky, with buffy-olive edgings and 2 buffy-ochraceous bars; chin and throat grayish-olive; chest and sides deeper olive; belly pale yellow. Iris pale brownish-yellow; bill blackish, except base of lower mandible pale horn; legs dusky. **Young:** similar to adults but crown darker, more sooty; wing-bars brighter, more ochraceous; belly paler yellow.

Habits. Frequents understory of humid forest, shady cacao plantations, and old second growth; less active than Slaty-capped Flycatcher, perching for longer periods looking about, making short sallies to catch insects in the air or pluck them from foliage; eats grasshoppers, small butterflies, spiders, and berries; sometimes accompanies mixed flocks; nervously flicks wings 1 at a time, like Slaty-capped Flycatcher.

Voice. A loud, sharp, harsh *pitch'er*, often repeated; a sharp, metallic chipping and chatter. In Panama, a low *pree-ee-ee-ee* delivered with trembling wings by a breeding male (Wetmore).

Nest. Globular or elliptical with side entrance, of green and dry mosses, very fine roots, plant stems, and grasses, lined and decorated on exterior with seed down, swinging from slender rootlets or other strands beneath rock or log, in deep ravine, above stream or beside forest trail. Eggs 2–3, white. April–May (in Mexico; no Costa Rican record).

Status. Rare and local resident in foothills and adjacent lowlands of N Caribbean slope, S to Cordillera Central, and to Pacific side of N cordilleras; very locally in Térraba–Coto Brus region of S Pacific slope; lowlands to 4300ft (1300m).

Range. S Mexico to C Bolivia, N Argentina, and Paraguay.

OLIVE-STRIPED FLYCATCHER

Mionectes olivaceus Pl. 36(24)

Mosquerito Ojimanchado

Description. 5″ (13cm); 14.5g. Small, dull-colored, with yellowish postocular spot and finely streaked throat and breast; bill cylindrical; rictal bristles much reduced. **Adults:** above olive-green; wings and tail dusky with olive-green edgings and faint paler wing-bars; throat and breast dull olive-green, finely streaked with yellowish; belly pale yellow; flanks olive. Bill black except base and gape orangish; mouth lining white; legs grayish. Male has inner web of 9th primary deeply emarginated. **Young:** streaking on throat and breast faint, grayish; belly duller and darker, often tinged with ochraceous; postocular spot faint; wing-bars darker, more ochraceous; 9th primary of male not emarginated.

Habits. Prefers middle levels of wet forest; descends to shrub level at edges or in adjacent second growth; usually alone but may accompany mixed flocks of antwrens, bush-tanagers, etc.; perches alertly, sometimes rapidly flicking 1 wing at a time open and shut; forages mostly by hover-gleaning sallies; takes more insects than closely related Ochre-bellied Flycatcher, but at least half its diet consists of berries and arils, especially *Trema, Clusia*, and *Casearea*. During breeding season, males appear to establish scattered individual singing territories in upper understory.

Voice. Usually silent. Song of male a thin, sibilant trill that rises and falls interminably with a cicadalike effect: *pseeeuuueeeeuuu-eeeeuuueee* . . . , with bill continuously wide open, head swaying forward and backward.

Nest. A slender, pyriform structure with a side entrance, externally of green moss, the round central chamber lined with soft, pale brown fibers; suspended from free-hanging vine or aerial root of aroid, often beside a vertical cliff, bank or base of large tree 3–6ft (1–2m) up. Eggs 2–3, white. April–July.

Status. Common breeding resident of middle elevations the length of both slopes, 2600–6500ft (800–2000m) on Caribbean, and 3300–7000ft (1000–2150m) on Pacific. Following breeding, most birds move downslope, many reaching 160ft (50m) in adjacent lowlands, few remaining much above 4000ft (1200m).

Range. Costa Rica to NW Ecuador and C Peru.

OCHRE-BELLIED FLYCATCHER

Mionectes oleagineus Pl. 36(25)

Mosquerito Aceitunado

Description. 5″ (12.5cm); 13g. Small, nearly unpatterned, plainly colored but with distinctive ochraceous belly; bill rather slender, cylindrical; rictal bristles much reduced; adult males have outer 3 primaries, especially the outermost, narrowed and pointed at tip. **Adults:** above greenish-olive; wings and tail dusky with olive edgings; sides of head, chin, and throat more grayish, breast greenish-olive, posterior underparts ochraceous-buff, tinged with olive on sides. Bill blackish, shading to horn-color or flesh-orange at base of lower mandible; many adults (only males?) have yellow or orange mouth corners; inside of bill white; posterior mouth-lining black; legs gray. **Young:** similar but wing-coverts and tertials narrowly tipped and edged with ochraceous; belly paler, more buffy; outer primaries of males with little or no emargination.

Habits. Prefers lower levels of humid forest, tall second growth, semi-open, and clearings with scattered trees near woodland; forages by short sallies and hovers to pluck small insects, berries, and arillate seeds; sometimes pauses briefly, as on inflorescence of *Heliconia* or aroid, to gather fruits, which comprise by far the greater part of its diet; sometimes accompanies mixed flocks; persistently twitches 1 wing at a time up over back; usually solitary except that, between about February and September, up to 6 males gather in loose courtship assemblies, each one perching ca. 6–30ft (2–9m) up in shaded understory and 50–160ft (15–50m) or more from his fellows, singing through much of day; does not pair.

Voice. Usually silent. Song of male weak, unmelodious, in 2 parts, the first a series of widely spaced notes or phrases, the second a measured series of louder, wooden chips, sometimes preceded by staccato notes; from any distance only this part is audible: *whip wit whip wit wit chipchip chip chip chip chip*, or *pik ch'wik pik ch'wik ch'wik K-WIT-K'WIT-K'WIT-K'-WIT*.

Nest. An elongated pyriform structure, 12″ or more long by ca. 3½″ wide (30×9cm), with a side entrance, thickly covered with green moss and generously lined with bast fibers, attached to a slender, dangling epiphytic root, hanging vine, or twig, often close beside a mossy cliff, vertical bank, or trunk, 2–12ft (0.6–3.6m) above ground or water. Eggs 2, more often 3, white. March–July.

Status. Common to abundant resident of lowlands and foothills of Caribbean slope and Pacific slope S from Carara; to 3000ft (900m) on Caribbean slope and 4000ft (1200m) on Pacific, rarely higher; uncommon and local in dry NW lowlands, mainly in evergreen gallery forest; more numerous in hills and on W side of Península de Nicoya.

Range. S Mexico to W Ecuador, Bolivia, and Amazonian Brazil.

Note. Formerly placed in the genus *Pipromorpha* and sometimes then called Oleaginous Pipromorpha.

FAMILY **Hirundinidae:** Swallows

Numbering about 80 species, swallows and martins occur worldwide except in polar regions, New Zealand, and some smaller oceanic islands. Those that nest where winters are severe migrate long distances, traveling by day and catching insects as they go, gathering at night into populous roosts. In size 4–9″ (10–23cm), they have long, pointed wings, moderate to long and deeply forked tails, short flat bills, and short legs with which they can only shuffle along. Some are plainly attired in grays and browns; others display blue, green, violet, or red, seldom of the brightest shades. Sexual and seasonal differences in plumage are usually slight or lacking. Their songs, sometimes sweetly varied, commonly lack range and force; their calls are slight, shrill notes or weak twitters. As befits their larger size, martins have deeper voices. Swallows subsist almost wholly upon insects, which they catch in the air as they circle tirelessly over the countryside. A few vary their diet with berries. To rest, they perch on high, exposed twigs or wires, each maintaining a minimum distance from its nearest neighbors. At night, some sleep in sheltered places in buildings or in empty nests; others roost socially in reed beds, sugarcane fields, or trees.

The nest, built by both sexes or the female alone, may be loosely made of vegetable materials in a tree cavity, birdhouse, burrow in the ground, or one of man's constructions or maybe a massive structure of mud or clay, cuplike or well enclosed, plastered in a sheltered spot on the vertical side of a cliff or house. The 3–7 eggs are white, either immaculate or spotted with red, brown, or lilac. Incubation is performed by both sexes or the female alone, for 13–16 days, in a few species for 18 or 19 days. Fed by both parents, the young remain in the nest for 18–28 days and occasionally longer, not leaving until they can fly well.

PURPLE MARTIN

Progne subis Pl. 22(16)

Martín Purpúrea (Golondrón)

Description. 6¾″ (17cm); 45g. Large; all-dark male unmistakable; female differs from smaller Gray-breasted Martin by pale feather-tips on forehead, neck, and nuchal collar. **Adult ♂:** head and body glossy dark purplish-blue; wings and tail blackish with bluish gloss. ♀: above duller, only nape and scapulars solid blue, dusky feather-bases exposed elsewhere; feathers of forehead, nuchal collar, sides of neck and throat, and chest dusky-brown with more or less conspicuously pale tips; lower breast and belly whitish to grayish-buff with dusky shaft-streaks and feather-bases. Bill and legs black. **Immatures:** identical to female adult. **Young:** like adult female but duller and more uniform; only nape and scapulars strongly glossed with bluish (less purplish than adult female); pale scaling on neck less pronounced.

Habits. Migratory behavior as described for other swallows, but tends to fly higher (except during foul weather) and more often in single-species flocks. Occasionally a migrating flock descends at dusk, especially in bad weather, to roost on telephone wires or trees. Occasional inland strays may associate with flocks of Gray-breasted Martins but are perhaps more often seen singly, usually over open areas.

Voice. Usually silent while migrating; around perches and roosts, gives loud churring notes, much like those of Gray-breasted Martin.

Status. Irregularly common to abundant migrant along Caribbean coast in both fall (August–mid-October) and spring (late January–April); rarely and sporadically inland, including Pacific slope, especially in fall (September–October); a fall vagrant on Cocos I.

Range. Breeds from S Canada to C Mexico; winters in South America E of Andes, S to Bolivia and SE Brazil.

Note. Also possible in our area during southern winter is Southern Martin (*P. elegans*), which breeds in S South America and winters in N South America; recorded in S Florida and C Panama. Male indistinguishable in field from *P. subis*; in hand, tail more deeply forked; female darker below (little or no white on belly) than female or young Purple or Gray-breasted martins. Most likely April–September; in particular, any all-dark martin in June–July could be this species.

GRAY-BREASTED MARTIN

Progne chalybea Pl. 22(15)

Martín Pechigrís

Description. 6½″ (16.5cm); 40g. Largest and heaviest resident swallow with a large, broad bill, notched tail, and brownish throat and tail. **Adult ♂:** head dusky, crown and nape glossed with dark steel-blue; rest of upperparts glossy, dark steel-blue; wings and tail black with bluish gloss; chin and throat grayish-buff, with dusky feather-bases; sides of neck and breast dusky-brown, fringed with pale grayish; lower breast and sides paler grayish; rest of underparts white, with scattered fine, dark shaft-streaks. Bill and feet black. ♀: similar but above mostly dusky, blue reduced to scattered glossy feather tips, especially on the crown, back, and upper tail-coverts; posterior underparts dingier, more grayish-white. **Immature ♂:** like adult female but below browner; pale fringes on breast broader and more buffy; belly dingier, more brownish-gray, with more dusky shaft-streaks.

Habits. Gregarious; frequents open places, especially around buildings, bridges, and new clearings with scattered dead trees; flies strongly and directly, glides for considerable distances on outstretched wings; forages and often perches higher than other swallows found with it; catches wasps, bees, dragonflies, beetles, mayflies, flies, and other insects in flight, often on a long downward swoop from a high perch; outside breeding season hundreds gather to roost on wires or girders.

Voice. Loud, gravelly, rolling, churring notes that sound musical at a distance, like *churrr, cheeerp*, or *chreet*; song a full, throaty medley of churring warbles and trills.

Nest. A shallow, loosely built cup of dry grass, weed stems, dead leaves, twiglets, in a cranny in building, bridge, birdhouse, or other man-made structure; in wilder areas, high in an old woodpecker hole or other cavity in a dead tree in a clearing, or in a cranny amid rocks. Eggs 2–5, white. March–May.

Status. Locally common to abundant resident countrywide, up to ca. 5600ft (1700m); absent from many breeding areas from about September to January, but probably its movements are within Costa Rica rather than long-distance migrations.

Range. N Mexico to NW Peru, N Argentina, and S Brazil; migratory at both extremes of range.

BROWN-CHESTED MARTIN

Phaeoprogne tapera Pl. 22(14)

Martín de Ríos

Description. 7″ (18cm). Color, pattern, and

size suggest an overgrown, washed-out Bank Swallow. **Adults:** above dull brown, rectrices and remiges blackish; below white, with broad breast-band and sides grayish-brown; an irregular sprinkling of black spots, partly veiled by white feather tips, along midline of breast to upper belly. Bill and feet black. **Young:** similar but above and below duller, grayer; breast-band broader and less distinct; below with little or no black spotting; less contrast overall.

STATUS. One to 3 birds have been noted on at least 6 occasions, between June and mid-September of different years, in vicinity of San José, flying over open areas with Blue-and-White Swallows and Gray-breasted Martins; probably best considered a casual migrant and visitor during the Southern Hemisphere winter: to be looked for especially on S Pacific slope.

RANGE. Breeds from N Colombia and Guianas to N Argentina and S Brazil; in SW Ecuador and W Peru; southern breeding populations migrate northward, winter from Bolivia and S Brazil N regularly to C Panama, casually to Costa Rica.

NOTE. Sometimes included in genus *Progne*.

CLIFF SWALLOW

Hirundo pyrrhonota Pls. 22(13), 52(1)

Golondrina Risquera

DESCRIPTION. 5¼" (13.5cm); 20g. Buffy rump, streaked back and dark throat-patch diagnostic. **Adults:** forehead white to rufous; crown and back glossy dark blue, back sparsely streaked with whitish; rump cinnamon-buff; wings and tail blackish; face and throat mostly chestnut; a large black patch on lower throat; sides of neck and nuchal collar brownish-gray; chest grayish-brown tinged or suffused with cinnamon; lower breast and belly white. Bill and feet black. **Young:** overall paler, duller; above largely sooty-gray; throat and face grayish-buff, only cheeks tinged with chestnut; throat-patch sooty-black, indistinct; rump paler buff; forehead and chin mottled with white.

HABITS. Migrates with Barn and Bank swallows, usually second or third in order of abundance. An occasional wintering bird forages and roosts with flocks of Barn Swallows. During migration, mixed-species flocks of swallows often tarry for a few minutes to several days in good feeding areas. Then and in winter they mostly feed low over ground or water on small flying insects, and between foraging bouts perch in groups on electric wires or barbed wire fences, each bird maintaining its distance from others.

VOICE. A low *chur* or *chirt*; in alarm or excitement, a sharper *keer*.

STATUS. Very common fall (late August or early September–late October) and spring (early March–late May) migrant along both coasts; less common inland, but large numbers pass through Valle Central in fall, crossing continental divide at Ochomogo from W to E, and in smaller numbers from E to W in spring; winters in very small numbers but fairly regularly, especially in Guanacaste.

RANGE. Breeds from Alaska and N Canada to S Mexico; winters chiefly in S South America, occasionally N at least to Costa Rica.

NOTE. Often segregated in genus *Petrochelidon*. There is an old, unconfirmed report of the closely related Cave Swallow (*H. fulva*) for Costa Rica; very similar to Cliff Swallow, differing chiefly in chestnut forehead, buffy throat and chest without dark patch; breeds from SW USA to S Mexico and Greater Antilles, also in Ecuador and Peru; northern populations migratory but winter range unknown (though presumably migrates through Central America, thus expected in Costa Rica). No recent records, but to be looked for especially along Caribbean coast in fall [see Pl. 52(2)].

BARN SWALLOW

Hirundo rustica Pl. 22(12)

Golondrina Tijereta

DESCRIPTION. 5½–6¾" (14–17cm); 17g. Slender, long-winged, with deeply forked tail (longer in males than in females, in adults than in young). **Adults:** above glossy dark blue; forehead, throat, and chest dark rufous, separated from pale cinnamon posterior underparts by incomplete dark breast-band; all but central rectrices with white spots on inner web. Females average paler below than males. Bill and feet black. **Immatures:** above duller, mixed with dusky on back; head mostly dusky; forehead blotched with buffy-white; throat and chest dull cinnamon; belly buffy-white; tail shorter, outer rectrices more blunt.

HABITS. Migrates in spectacular flights along both coasts, often with smaller numbers of other swallows, flying low over ground or water in early morning, evening and in bad weather, usually higher during sunny midday. Individuals are spaced as they stream by, low over hilltops, high over valleys; often one veers aside in brief pursuit of insect, then resumes its journey. In winter prefers open areas including lawns, pastures, rice fields, and marshes; gathers in large communal roosts, from which smaller groups disperse daily to forage.

Voice. A dry *krit* or *krip*, sometimes doubled; sharper, more metallic notes in squabbles at perches.

Status. Abundant migrant in fall (early to mid-August through October) and spring (early March–late May or early June) along both coasts and, more sporadically and in lesser numbers, throughout lowlands of both slopes and up through Valle Central; uncommon to rare at higher elevations but occasionally reaches 10,000ft (3000m). Widespread, uncommon to locally abundant winter resident on both slopes from lowlands to ca. 5000ft (1500m). A few seen in July may be early fall migrants.

Range. Breeds from Alaska and N Canada to C Mexico; winters regularly from Costa Rica and West Indies to Tierra del Fuego, casually N to S USA. Widespread in Old World.

SOUTHERN ROUGH-WINGED SWALLOW

Stelgidopteryx ruficollis Pl. 22(19)

Golondrina Alirrasposa Sureña

Description. 4¾″ (12cm); 15g. Fairly small, brownish swallow with shallowly notched tail, cinnamon throat, and usually pale rump; adult males have barbs of outer web of outer primary stiffened and recurved, producing rough, filelike effect. **Adults:** above grayish-brown, darkest on pileum; rump white to grayish, usually distinctly paler than back; upper tail-coverts, tail, and wings brownish-black, the tertials edged with grayish or white; throat tawny-buff to bright cinnamon; breast and sides paler grayish-brown; posterior underparts white, belly usually tinged with yellow, and the longest lower tail-coverts usually tipped with black. Bill and legs black. **Young:** cinnamon of throat spreads over chest, sides, and flanks; feathers of back wing-coverts and tertials tipped or edged with buff.

Habits. Prefers open areas, especially near streams or roads with vertical earthen banks for nesting; tirelessly circles in loose flocks, low over the ground to fairly high, catching flying ants, beetles, flies, wasps, and other insects; rests on dead snags or electric wires; sometimes roosts gregariously in fields of sugarcane.

Voice. Call a rather liquid, gurgling *jroit* or *troit*, or a sharp, loud *jrrt*, stronger and less dry than call of Northern Rough-winged Swallow; a more complex *chreet chreet trit-trit-troit*, in chases and interactions; song of soft, clear notes with rarely a brief, musical trill.

Nest. A shallow bowl of grasses and dead leaves, in an abandoned burrow of motmots or kingfishers or other narrow tunnel in an exposed roadside or streamside bank, usually within sight of the entrance. Rarely may despoil a jacamar of its burrow, even ejecting the latter's eggs. Eggs 4, rarely 5, immaculate white. April—May, rarely June.

Status. Locally common to abundant resident throughout Caribbean lowlands, locally to 3300ft (1000m), rarely higher; and in lowlands and foothills of S Pacific slope, to 6000ft (1850m) in Coto Brus region; in post-breeding dispersal, appears sporadically at higher elevations in N Pacific lowlands between about August and February.

Range. E Honduras to NW Peru and C Argentina.

NORTHERN ROUGH-WINGED SWALLOW

Stelgidopteryx serripennis Pl. 22(18)

Golondrina Alirrasposa Norteña

Description. 5″ (12.5cm); 15.5g. Compared with Southern Rough-winged Swallow, has longer wing, slightly longer tail, and smaller bill; in the field, best distinguished by pale throat (at most tinged with buff, never cinnamon), and dark rump; under tail-coverts usually much less heavily marked with dusky. **Adults:** above grayish-brown, with darker tail and wings; tertials narrowly edged with grayish-white when fresh; chest, sides, and flanks pale grayish-brown; throat paler, sometimes tinged with buff (especially in residents); rest of underparts dull white, occasionally faintly tinged with yellow; longest lower tail-coverts immaculate or more or less dusky at tip. Bill and feet black. Males have rough outer primary like Southern Rough-winged. **Young:** feathers of upperparts with broad, indistinct cinnamon-brown edgings; wing-coverts with broad cinnamon-buff tips; tertials broadly edged with dull cinnamon-buff; throat and chest dull cinnamon to buffy-brown.

Habits. Similar to those of Southern Rough-winged; breeding birds occur mostly at higher elevations but some overlap; migrants sometimes journey with Barn or other swallows but also form pure flocks, especially in spring.

Voice. Generally drier, rougher than that of Southern Rough-winged. Usual call a harsh *brrrt* or a more rolling *jeet*, sometimes doubled; a gurgling *zhrit*; in interactions, a buzzy *jee-jee-jee* or *jrrr-jrrr-jrrr-jrrr*.

Nest. A well-made, shallow bowl of grasses, dry leaves, rootlets, etc., in a burrow in an exposed bank, often dug by motmots or kingfishers. Eggs 4, rarely 5, white, unmarked. April–May.

Status. Common breeding resident of N cordilleras and central highlands, at ca. 1000–6000ft (300–1850m); some dispersal into lowlands during nonbreeding season; uncommon to fairly common migrant in fall (early September–late October) and spring (early March–early May), especially along both coasts; between September and about February, locally common in lowlands of both slopes, up to ca. 5000ft (1500m) in N and C highlands. Because it is generally impossible to distinguish Costa Rican breeders from northern migrants in the field, the precise distribution of each group during N winter and local nonbreeding season is poorly known.

Range. Breeds from SE Alaska and SE Canada to Costa Rica; winters from S USA to Panama.

Note. Usually considered conspecific with *S. ruficollis* in the past and sometimes still so classified on the basis of the superficially intermediate coloration (but not measurements or juvenile plumage) of the form of *ruficollis* breeding on Pacific slope of Costa Rica and adjacent Panama. In Costa Rica, Northern and Southern Rough-winged swallows overlap and often breed in close association, at ca. 1000–3300ft (300–1000m) on Caribbean slope, apparently with little or no hybridization. If all forms are considered conspecific, the English name is Rough-winged Swallow.

BLUE-AND-WHITE SWALLOW

Notiochelidon cyanoleuca Pl. 22(20)

Golondrina Azul y Blanco

Description. 4¼″ (11cm); 10g. Small, with shorter tail, relatively shorter and broader wings than migrant swallows. **Adults:** upperparts entirely glossy dark blue to violet-blue; wings and tail blackish, glossed with blue; below white, tinged with warm buff in fresh plumage (females average duller white, with stronger buffy tinge), except flanks, crissum, and wing-linings blue-black. Bill black, feet blackish. **Young:** upperparts dusky with some bluish gloss; throat and chest clouded with grayish-brown to dull buff; flanks and crissum sooty. Migrants from southern South America (race *patagonica*) differ from residents in having black crissum mixed with white.

Habits. Frequents towns, agricultural areas with scattered trees, openings in montane forests, and upland bodies of water; forages in loose, circling flocks, catching small insects including wasps, flies, homopterans, beetles; flight slower, more fluttering and maneuverable than that of longer-winged migrant or rough-winged swallows; perches on overhead wires or thin bare branches; mated throughout year, each pair sleeping in its own niche or cranny.

Voice. Song a thin, weak, prolonged trill, sliding upward toward end, often beginning with a gurgling sputter, *dzzzhreeeeee*, frequently uttered in flight; a crackling buzz. Full, plaintive monosyllables and low, harsh notes express alarm or irritation.

Nest. A shallow bowl of straws and weed stems, lined with downy feathers, in a hole in a tree or bank, cranny in a building, crevice in masonry, or any snug niche. Eggs 2–4, white. March–June; 2 broods. After fledging, young continue to return to the nest space to sleep with their parents until about 2months old.

Status. Resident countrywide ca. 1300–10,000ft (400–3000m) or more in high mountains; most abundant at middle and upper elevations, rare in Guanacaste. Between May and September large resident population may be augmented by a few migrants from South America.

Range. Nicaragua to Tierra del Fuego; southernmost populations migrate N during the southern winter, regularly to C Panama, casually to S Mexico.

Note. Sometimes placed in monotypic genus *Pygochelidon*; some authorities lump *Notiochelidon* with the South American genus *Atticora*.

BANK SWALLOW

Riparia riparia Pl. 22(17)

Golondrina Ribereña

Description. 4½″ (11.5cm); 12.5g. Smaller, grayer above than rough-winged swallows, with white throat, distinct dark breast-band. **Adults:** above grayish-brown, darker on wings and tail; lores dusky; indistinct whitish supraloral stripe; underparts white with a band of dark grayish-brown across chest; some dark brown spotting down midbreast, largely obscured by white feather tips. Bill and feet black. **Young:** similar but breast-band broader, paler, less distinct; feathers of upperparts and breast-band with pale fringes, especially on rump and upper tail-coverts; little or no dark spotting on breast. In fall, adults are in worn plumage, young are in fresh.

HABITS. Migrates with Barn and Cliff swallows, often flying lower than them; may tarry for a few minutes to several days where food is abundant, as over newly harvested fields, freshly mown grass, pastures, ponds, or marshes. In winter found mainly about marshes and wet pastures.

VOICE. A rarely-heard, dry *chrrt*; a soft chatter.

STATUS. Abundant migrant in fall (late August or early September to early November) from lowlands of both slopes, especially along coasts, to 5000ft (1500m) in Valle Central; casual to very rare winter resident, principally in lowlands of Tempisque basin and Río Frío; abundant spring migrant along both coasts (early March–mid-May); rare inland.

RANGE. Breeds from Alaska and N Canada to S USA; winters mainly from E Panama to Peru and N Argentina, casually in Central America. Widespread in Old World, where known as Sand Martin.

TREE SWALLOW

Tachycineta bicolor Pl. 22(21)

Golondrina Bicolor

DESCRIPTION. 5½″ (14cm); 20g. Larger, with longer wings and tail than Blue-and-White Swallow; tail deeply notched; lower tail-coverts white. **Adult ♂:** above glossy dark greenish-blue, down to level of eye; wings and tail blackish; below entirely white. **♀:** above duller, feathers with extensive dusky bases; breast often washed with grayish. First-year females mostly dusky above. Bill black, legs dusky-flesh. **Young:** above mostly grayish-brown, with strong greenish gloss; dusky wash across breast and down sides.

HABITS. Forages in pastures or savannas near ponds, marshes, or salinas, sometimes over open water; usually in small numbers among larger flocks of migrant swallows, both during fall migration and in winter; occasionally in larger pure flocks, especially during spring migration; associates with resident Blue-and-White Swallows in highlands or Mangrove Swallows in lowlands.

VOICE. Soft, staccato *chick* or *crick* in foraging flocks; high, thin, hissing whistles during aggressive interactions at perches.

STATUS. Irregularly rare to locally common migrant (early September–late October; March–April) and winter resident, from lowlands of both slopes to ca. 5000ft (1500m) in central highlands. Numbers vary widely; rare to occasional most years but with irruptions in which flocks of 100 or more appear in favorable localities (e.g., Coris, near Cartago; Palo Verde).

RANGE. Breeds from Alaska and N Canada to S USA; winters from S USA to Costa Rica, sporadically to Panama.

NOTE. Sometimes placed in genus *Iridoprocne*.

MANGROVE SWALLOW

Tachycineta albilinea Pl. 22(23)

Golondrina Lomiblanca

DESCRIPTION. 5″ (13cm); 14g. Rump whiter, contrasting more with upperparts than in Southern Rough-winged or Cliff swallows, both of which have darker throats and chests; larger Tree Swallow lacks white rump. **Adults:** upperparts, including wing-coverts, mostly dark steel-green (becoming more bluish with wear), duller in females; rectrices and remiges blackish, tertials with white margins; narrow white supraloral stripe; below white, tinged with gray on breast and sides. Bill and feet black. **Young:** above dingy grayish, with faint green gloss; rump and underparts white, with dusky-brown wash across breast.

HABITS. Prefers wide expanses of fairly still, open water of lakes, rivers, sloughs, estuaries, salinas, sometimes marshes and wet pastures; social outside breeding season, congregating in flocks of up to 50 where snags, wires, or other projecting objects provide numerous perches low over water; skims close to water surface to catch flies, winged ants, wasps, homopterans, and other small flying insects; occasionally forages over fields, rarely above canopy of forest beside water. Breeding pairs defend nest site and foraging area, often a stretch of river.

VOICE. Note a scratchy to rolling *jeet* or *jrrt*.

NEST. A skimpy to bulky, untidy cup of grass, horsehair, or rootlets, in a cranny or hole in a snag, bridge, or building, over or beside water, often less than 6ft (2m) up. Eggs 3–5, white. March–April.

STATUS. Uncommon to locally abundant resident of both slopes, chiefly in lowlands but up to at least 3300ft (1000m) in favorable spots (e.g., Lago Dabagri).

RANGE. N Mexico to E Panama; coastal N Peru.

NOTE. Sometimes placed in genus *Iridoprocne*.

VIOLET-GREEN SWALLOW

Tachycineta thalassina Pl. 22(22)

Golondrina Verde Violácea

Description. 5″ (12.5cm); 14g. Much more white on sides of head than Tree or Mangrove swallows, which lack purplish or bronzy tones above; rump with white sides and dark center. **Adult ♂:** crown and nape bronzy-green to purplish-bronze; sometimes purplish nuchal collar; back and wing-coverts pale green, tinged with bronze; center of rump and upper tail-coverts mixed with violet and dark green; remiges and tail blackish; below entirely white, extending over auriculars to above eye; lores dusky. ♀: much duller, above more dusky; chest tinged with gray. Bill black; feet dusky-flesh. **Young:** above dull grayish with slight bronzy gloss; throat and chest pale grayish; chest tinged or smudged with dusky-brown.

Habits. Occurs singly or in flocks; often associating with other swallows, especially the similar appearing Tree, Mangrove, and Blue-and-White swallows.

Voice. A thin, soft *chit* or *chwit.*

Status. Casual to rare, sporadic migrant and winter visitor, recorded from mid-November to late March in N Pacific lowlands (Taboga, Colorado, Bebedero) and at several mountain sites (Monteverde, Cerro de la Muerte) and once on Caribbean coast (Matina).

Range. Breeds in W North America from Alaska S to C Mexico; winters from SW USA and NW Mexico regularly to Honduras, casually to W Panama.

FAMILY **Corvidae:** Jays, Crows, and Allies

Jays, crows, ravens, magpies, and their relatives constitute a nearly cosmopolitan family of about 102 species. In size they are 8–28″ (20–71cm); ravens are the largest passerines. Their bills are mostly short and strong. Their coloration, alike in male and female, varies from mostly or entirely black in crows and ravens to blue, purple, green, yellow, and white in many of the jays, which are often crested. Their voices, loud and freely used, tend to be harsh rather than melodious; such singing as is done by the American jays is usually a sotto voce medley that is often pleasing but audible only at close range. Members of this family are often considered the most intelligent of birds; certainly they are among the most versatile in their feeding habits, for in addition to a wide variety of animal foods, from carrion to eggs and nestlings of other birds, they consume many fruits, nuts, and grains. Their nests, usually in trees but sometimes in holes or rock crevices, or on the ground, are mostly open, consisting of a foundation of coarse sticks that supports a cup of rootlets or other finer materials. Magpies build covered nests. Clutches range from 2–7 eggs, which are usually blue, green, or gray, with darker markings. Except among nutcrackers, they are normally incubated only by the female, who takes long sessions and is fed by her mate, sometimes by helpers as well. After an incubation period of 16–21 days, the young hatch naked or with sparse down. They remain in the nest for 20–24 days in the smaller species, up to 5 or 6 weeks for the raven. They are nourished with food brought partly in the throat pouch and partly in the bill, not only by both parents but also, in a number of New World jays, by few or many helpers, all of whom live yearlong in closely knit extended families.

WHITE-THROATED MAGPIE-JAY

Calocitta formosa Pl. 39(18)

Urraca Copetona (Urraca, Piapia Azul)

Description. 18″ (46cm); 205g. A large blue-and-white jay with a long, graduated tail and an extravagant crest of long, forwardly curved feathers. **Adults:** forehead and crest feathers vary from black to mostly white with black tips; back, rump, and upper tail-coverts grayish-blue, shading to brighter sky-blue on crown and wings and to cobalt-blue on tail; outer 4 rectrices with broad white tips; border of crown black, continuing, as a crescent behind auriculars, to side of neck; a curving black breast-band, usually narrow but rarely covering most of breast; face and underparts otherwise white; malar area sometimes tinged with blue. Bill and legs black. **Young:** head and underparts fluffy; crown extensively white; black on head reduced to crescent behind auriculars; breast-band faint; crest short; above duller, more grayish than adult.

Habits. Travels in noisy, straggling flocks of 5–10 through thorny scrub, savanna trees, groves near houses and along watercourses, deciduous and gallery forest; hunts through foliage, hanging tangles, leaf bases of bananas, and ground litter for large insects like caterpillars, katydids, roaches, beetle grubs, small frogs and lizards; eats many fruits, including cultivated varieties and maize; takes eggs and nestlings of other birds; sips nectar from large blossoms of balsa; loudly mobs Spectacled Owls and other predators, including man; flies with several strong flaps and a flat-winged glide; on ground progresses by springy hops.

Voice. Extremely varied, including a grating, quizzical, upslurred *reek*? a harsh, guttural *raah* or *reeah* when scolding, and metallic squeals. Other harsh notes alternate with mellow, liquid calls, often *weep weep weep*. While resting inconspicuously, these jays sing a medley of low gurgles, trills, and sputters. Hunger call of nesting female a loud, plaintive *pee-ah*.

Nest. A bulky pile of coarse sticks supporting a neatly finished cup of wiry roots and fibers, 20–100ft (6–30m) up in a tree. Eggs 3–4, gray, finely flecked all over with brown. February–July. Breeds cooperatively, with several adults attending a nest.

Status. Common resident of N Pacific slope, including Península de Nicoya, S to near San Mateo and Orotina, occasionally to Carara. Lowlands to 2600ft (800m), occasionally to 4000ft (1200m) along N cordilleras but not in Valle Central; E to about Brasilia on extreme NW Caribbean slope.

Range. C Mexico to Costa Rica.

Note. If the Black-throated Magpie-Jay (*C. colliei*) of NW Mexico is considered a race of *C. formosa*, the English name of the species is Magpie Jay.

BROWN JAY

Cyanocorax morio Pl. 39(19)

Urraca Parda (Piapia)

Description. 15½″ (39cm); 235g. Large, with graduated white-tipped tail; no crest but frontal feathers stiff and erect, the anterior ones curving forward over base of bill; a small, inflatable sac in furcular (chest) region. **Adults:** head, chest, and entire upperparts dark brown, darkest on face; lower breast grayish-brown, shading to creamy-white on belly and crissum; all but central pair of rectrices with broad white tips. Bill and bare eye-ring black; legs dusky. **Young:** similar but bill and bare eye-ring bright yellow; legs brownish-yellow. The yellow parts darken gradually and in highly variable patterns over several years, facilitating recognition of individuals.

Habits. Straggling, noisy flocks of 6–10 wander through open woodland, banana and coffee plantations, forest edge, open country with scattered trees, and second growth, foraging from ground to treetops, seeking insects, spiders, small lizards and frogs by rummaging and gleaning in foliage or ground litter, flaking off dead bark, probing rotten wood or rolled dead leaves held beneath a foot, making short sallies to catch dragonflies and other flying insects, eating many wild and cultivated fruits, including *Cecropia* spikes and maize, drinking nectar of banana and balsa flowers; occasionally pillaging nests of smaller birds. Flies boundingly, deep wingbeats alternating with long glides; on the ground, advances with long, springy hops.

Voice. Usual call a strident, often repeated *chaa* or *pyaaaah* that varies in pitch and intonation according to the situation. At close range, a popping sound, made by inflating the furcular sac, is audible at the start of each call: *p'yaaaah*; contact note at close range *p'ouw* or *p'uw*. Females on nest utter a plaintive, whining *pee-ah*.

Nest. A bulky foundation of coarse sticks, often with a middle layer of flexible vines, supporting an inner cup of long, fibrous roots pulled from the ground. Eggs 2–7 (the larger sets apparently laid by 2 or more females), pale, chalky blue-gray, thickly speckled with brown, especially on large end. March–June. Breeds cooperatively, up to 10 individuals attending a single brood.

Status. Common resident of deforested parts of N cordilleras, central highlands, and Caribbean lowlands, from sea level to 8200ft (2500m) on volcanoes; still expanding range with deforestation of Caribbean slope; increasingly common in lowland Guanacaste and S in coastal lowlands and inland valleys of Pacific slope; reached Parrita area before 1970; a straggler recorded in Valle del General in 1974.

Range. S Texas to W Panama, casually to C Panama.

Note. Often classified in genus *Psilorhinus* (and perhaps better so, as the furcular sac is unique); the form with white-tipped tail, found from C Mexico southward, was formerly classified as a species *P. mexicanus* or White-tipped Brown-Jay, separate from the northern, brown-tailed *P. morio*.

BLACK-CHESTED JAY

Cyanocorax affinis Pl. 39(20)

Urraca Pechinegra

Description. 14″ (36cm); 205g. Large; striking pale eye and facial spots; feathers of frontal area stiffened and erect, forming a tuft from over eye to nostrils; tail conspicuously white-tipped. **Adults:** pileum, sides of head, throat, and breast black; blue spots above and below eye; a short blue malar stripe; nape purplish, shading to brownish-purple on back and rump; wings and tail dark violet-blue, all rectrices broadly tipped with yellowish-white; posterior underparts pale creamy-yellow. Iris pale yellow; bill and feet black. **Young:** head and neck dull black, blue facial markings absent or much reduced; frontal feathers less stiff and erect; upperparts duller, browner; breast grayish-brown; belly buffy-yellowish; iris brownish-yellow.

Habits. Moves quietly through middle and upper levels of gallery woodland, forest, shaded cacao plantations, tall second growth, and adjacent banana plantations and clearings; forages from ground to high in trees for insects, small lizards and frogs, and fruits of many sorts; occasionally attends army-ant raids; shy and retiring, usually remaining in dense vegetation; mobs predatory birds and snakes.

Voice. Usual call a harsh, loud *chuck* or *kyuck-kyuck* (Eisenmann); in alarm, a rattling, nasal *chep-chep-chep*; a variety of other squeaky and metallic rattling sounds; gives a "whisper song," a medley of subdued notes.

Nest. A bowl of fine twigs supported by a bulky foundation of coarse sticks, in an upright fork of a tree or near the end of a branch. Eggs 3–5, pale buff or brownish-white, heavily spotted and blotched with olive-brown and tinged with gray. April–May (in Colombia; no Costa Rican record).

Status. Uncommon and local resident of lowlands and foothills of SE Caribbean slope, from Panama border N to near valley of Río Estrella, up to at least 4000ft (1200m).

Range. Costa Rica to N Colombia and NW Venezuela.

AZURE-HOODED JAY

Cyanolyca cucullata Pl. 39(17)

Urraca de Toca Celeste (Piapia de Montana)

Description. 11½″ (29cm); 88g. Medium-sized, dark blue with contrasting sky-blue hood; no crest but feathers of forehead stiffened, plushlike. **Adults:** forehead, sides of head and neck, throat, breast, and upper back black; rest of body, wings and tail dark blue; forecrown silvery-white, shading to pale blue on rest of crown, nape, and hindneck. Iris dark red; bill and legs black. **Young:** black areas more sooty, blue areas darker and duller, including hood; no white on forecrown.

Habits. Moves in pairs or groups of 3–5 through upper understory and subcanopy, and along edges of wet mountain forests; noisy when traveling but forages quietly, deliberately poking and probing into moss and epiphytes, gleaning from foliage and crevices in bark, holding rolled leaves or large prey against perch with feet and pulling them apart with bill; eats beetles, katydids, homopterans, larvae, spiders, small vertebrates, berries of *Miconia* and Ericaceae, and other fruits.

Voice. Most often heard is a loud, explosive, grating *jeet-jeet* or *djeeek-djeeek-djeeek*, often when flock is moving and sometimes interspersed with softer whistles and rasping *djak* notes. Scold a dry, raucous *raaaah*; contact note a soft, clear, upslurred *woyt*.

Nest. A cup 2″ (5cm) deep of interlaced thin fibrils and twigs, lacking a soft lining, on a coarsely constructed foundation of larger, longer twigs, 16–23ft (5–7m) up in small tree at forest edge. Each of 2 nests held 2 nestlings and was attended by a pair without helpers. Eggs not seen. April–June (Winnett-Murray et al.).

Status. Widely distributed but generally uncommon at ca. 2600–7000ft (800–2100m) along Caribbean slope from Volcán Arenal and Cordillera de Tilarán S to Panama; Pacific slope in higher parts of Cordillera de Tilarán, NE of San Ramón and in Dota region.

Range. E Mexico to W Panama.

SILVERY-THROATED JAY

Cyanolyca argentigula Pl. 39(16)

Urraca Gorgiplateada

Description. 10½″ (26.5cm); 65g. Only jay of high mountains; dark with conspicuous whitish throat; no crest but frontal feathers stiffened, erect, plushlike. **Adults:** most of head, neck, breast, and upper back black, shading to dark violet-blue on rest of body; wings and tail brighter violet-blue; throat, foreneck, and a narrow band along sides and front of crown, glossy white (Cordillera Central) or purplish-white (Cordillera de Talamanca). Bill and legs black. **Young:** crown, hindneck, back, and rump dusky-purplish; sides and front of crown bordered indistinctly with purplish-blue; face, sides of neck, and chest sooty-black; throat and foreneck dull white; rest of underparts dark dull purplish-blue.

Habits. Flocks of 4–10 range over large territories in oak forests of high mountains, and after breeding often join in loose aggregations of up to 30 birds who roost communally. With frequent calling, flocks move in straggling fashion through canopy and subcanopy, coming lower at edges and nearby second growth, foraging quietly and deliberately, working through each tree, rummaging in foliage, bromeliads, and tufts of mosses and lichens on large branches, eating insects, small frogs, lizards, salamanders, berries of *Macleania* and *Miconia*, and other fruits.

Voice. Most frequent call a harsh, nasal, somewhat scratchy *jew-jeah-jeah* or a single *nyaaaah*; scold a harsher *zhraaak*; upon taking flight, 2–4 sharper calls: *nyat nyat nyat*. Young birds give a higher, faster *nyaah nyaah*.

Nest. Undescribed (?). Breeding roughly March–June.

Status. Thinly distributed in high mountains, mostly at 6600–10,500ft (2000–3200m); occasionally, perhaps seasonally, down to ca. 4300ft (1300m), from Irazú-Turrialba massif S along Cordillera de Talamanca to Panama.

Range. Costa Rica to W Panama.

FAMILY **Cinclidae:** Dippers

Unique among passerines in its aquatic habits, this small family contains only 5 species, 1 confined to North and Central America, 2 to South America, and 2 to the Old World. Chunky birds 6–8″ (15–20cm) long, they have short bills, wings, and tails and long legs. Their dense plumage is brown or gray, marked with white in 2 species; the sexes are alike. Dippers live along swiftly flowing streams in mountainous or hilly country. To catch aquatic invertebrates and tiny fishes, they plunge boldly into rushing water, swimming with their wings more than their webless feet or walking on the bottom. On land they bob up and down, like many other small waterbirds. Their songs, like their call notes, are loud enough to be heard above the clamor of mountain torrents. On a ledge or in a niche above a stream, sometimes behind a waterfall, dippers build, largely of green moss, a domed nest with a side entrance, lined with grass and leaves. Here the female lays 3–7 white eggs, which she alone incubates for 15–17 days. Her mate helps to feed the young, who leave the nest when 20–25 days old.

AMERICAN DIPPER

Cinclus mexicanus Pl. 39(11)

Mirlo Acuático Plomizo

Description. 6½″ (16.5cm); 46g. Plump-bodied, short-tailed, with rather small head and slender bill and long, pale legs. **Adults:** above slate-gray; below paler, shading to dusky-gray on pileum and sides of head and pale brownish-gray on chin and throat; an interrupted white eye-ring; wings and tail blackish with gray edgings; in fresh plumage, greater wing-coverts narrowly edged with white at tip. Bill blackish; legs pale, dull flesh. **Immatures:** above paler gray, below whitish; narrow whitish wing-bars; secondaries edged with whitish.

Habits. Lives along rushing mountain streams, often in dark, gloomy gorges; stands upon projecting boulders, bobbing up and down and pumping tail; plucks prey from wet rock surface, immerses head and neck or submerges entire body in pools to seize backswimmers, water bugs, dragonfly and caddisfly larvae, other aquatic insects, and occasionally tiny fishes. Pairs defend linear territories along streams during most or all of year.

Voice. A sharp, metallic *jik*, often repeated, or a metallic, rasping *jee-jee-jee-jee*, etc. Song a piercing *peee peee pi-jur pi-jur p'jeet*, with variations, the final double notes harsh and grating; sometimes a prolonged and elaborate medley.

Nest. Usually a bulky roofed structure with side entrance, mostly of green moss and a few rootlets, placed on ledge or niche 3–13ft (1–4m) above stream; rarely a simple cup of mosses and rootlets placed in a deep niche or under an overhang that provides the roof. Eggs 2–4, white. Late February–May.

Status. Locally common resident the length of Caribbean slope and on both slopes of Cordillera de Talamanca at ca. 2600–8200ft (800–2500m), occasionally down to 2000ft (600m) after breeding; on Pacific slope of Cordillera Central only above ca. 5500ft (1700m).

Range. N Alaska to W Panama.

FAMILY **Troglodytidae:** Wrens

All 60 species of wrens inhabit the Western Hemisphere, from Alaska to Tierra del Fuego and the Falkland Islands and in the Antilles; 1 species has spread widely over Eurasia and northern Africa. Slender-billed, 4–9″ (10–23cm) long, they are clad in shades of brown, rufous, chestnut, buff, gray, black, and white, often prominently barred, streaked or spotted; the sexes look alike.

Almost wholly insectivorous, they seek their food in forests (where mostly they remain low), thickets, open groves, grasslands, marshes, or deserts. Many of the tropical species live yearlong in pairs, the two members remaining in contact by voice while they skulk amid dense vegetation. Many are superb songsters who sing duets, often antiphonally, the mates skillfully articulating their alternating phrases into a single song; churrs and scolding notes are freely given. With the exception of the Black-capped Donacobius of South America, recently transferred to this family, all wrens, as far as is known, make either well-enclosed nests amid vegetation or open nests in cavities of some sort. Indefatigable builders, wrens of both sexes construct roofed nests with a side entrance that are often bulky, not only for eggs, but also as dormitories to be used by single individuals or whole families together. Those that breed in holes sleep in almost any nook or cranny. Tropical wrens lay 2–5 eggs, northern species 4–8 or more. Wrens' eggs are white or, rarely, blue or brownish, immaculate or more often speckled or blotched with brown or pale lilac. Only females incubate, sometimes fed by their mates. After a period of 13–19 days, the young hatch naked or with sparse down. They remain in the nest for 14–19 days and are fed by both parents, in a few species assisted by helpers. After the young fledge, the parents or their auxiliaries usually lead them back to the nest or to some other snug shelter at nightfall.

SEDGE WREN

Cistothorus platensis Pl. 38(1)

Soterrey Sabanero

Description. 4¼″ (11cm); 9.5g. Small, slender, rather long tailed; streaked back; cinnamon rump conspicuous in flight. **Adults:** crown and back black, narrowly streaked with pale buffy to white; hindneck and scapulars buffy-brown; rump cinnamon-brown; throat and belly white; breast buffy with faint dark flecking; flanks cinnamon-buff. Upper mandible black, lower mandible and legs flesh-color. **Young:** crown and back dull brown with faint paler streaking; barring on wings and tail irregular and indistinct; breast with little buff.

Habits. Skulks low in flooded pastures with tall grass, wet bushy meadows, and open marshes, usually keeping out of sight but singing or scolding from a more exposed perch in a tussock or bush, especially in response to squeaking; pert, active, irascible; often cocks tail or wags it side-to-side; forages by gleaning and probing amid grasses, leaves, and twigs for beetles and other small insects; flies buzzily, with tail held low; usually dives into cover after a short exposure.

Voice. A dry, insectlike, series of *chut* notes sometimes accelerating into a scratchy chatter. Songs include a great variety of dry, sputtery trills, ticking or buzzy notes, and high, thin whistles; typically a phrase is rapidly repeated several times, then bird shifts to a different one. Scold a buzzy, nasal *wrrrrrrt*.

Nest. A hollow ball of grass blades with a side entrance, low, often only a few inches (cm) up, in a dense tussock of grass or rushes. Eggs 4, white. April–September.

Status. Locally abundant resident in Valle del Guarco S and E of Cartago, and around Ochomogo on continental divide; occasionally strays W to beyond San José; to be sought in San Vito–Cañas Gordas area (recorded in W Chiriquí).

Range. Breeds from S Canada to Tierra del Fuego; high-latitude populations migratory, tropical ones sedentary.

Note. Populations of North and Central America are sometimes regarded as a species, *C. stellaris*, separate from *C. platensis* of South America. Also called Short-billed Marsh-Wren.

BANDED-BACKED WREN

Campylorhynchus zonatus Pl. 38(3)

Soterrey Matraquero

Description. 6½″ (16.5cm); 29g. Fairly large wren, above boldly barred and below spotted. **Adults:** pileum brownish-gray, spotted with black; hindneck black, streaked with white; rest of upperparts, wings, and tail broadly and heavily barred with black and buffy-white to tawny; superciliaries narrow, white; sides of head mottled and streaked with dusky and whitish; chin, throat, and breast white, heavily spotted with black; posterior underparts cinnamon to ochraceous-tawny, with some black spotting on sides and thighs. Iris reddish-brown; upper mandible dark gray, lower flesh-color; legs horn-color. **Immatures:** crown dull blackish; nape tawny-buff, smudged with dusky; rest of upperparts dull cinnamon, irregularly and indistinctly barred with blackish; superciliaries and face more buffy; throat dull white, shading through buff on breast to cinnamon on belly, irregularly mottled or scaled with dusky.

Habits. At gaps and edges in wet forest but more frequently in thinned forest, semi-open, riverside groves, trees and shrubbery around houses, and second growth; usually in extended family groups of 4–12; forages high in forest, often lower elsewhere, creeping over branches, sometimes with head or back downward, probing into crevices and bromeliads, prying up flakes of bark, pulling off lichens, gleaning in foliage, occasionally descending to rummage in ground litter for insects and spiders.

VOICE. Dry, harsh, rasping; song of varied phrases, a medley of slurred, rolling notes, dry chatters, and staccato sputters; members of pair often join in animated, cacophonous duets; members of group call back and forth with rasping chatters and dry *zek*s.

NEST. A very bulky globular structure with a wide side entrance, of the most varied materials, 6–100ft (2–30m) up in trees or shrubs, often hidden in a clump of bromeliads. Eggs 3–5, white, unmarked or speckled faintly with brown. March–June. Breeds cooperatively; fledglings are led to sleep in breeding nest or similar structure; families remain together, lodging in a succession of nests, throughout the year.

STATUS. Common resident on Caribbean slope from lowlands to 5600ft (1700m).

RANGE. C Mexico to NW Ecuador.

RUFOUS-NAPED WREN

Campylorhynchus rufinucha Pl. 38(2)

Soterrey Nuquirrufo (Chico Piojo, Salta Piñuela)

DESCRIPTION. 6¾″ (17cm); 36g. Large; above boldly patterned; below white; broad white-tipped tail; rather long, decurved bill. **Adults:** pileum and eye-stripe black; broad superciliaries white; hindneck and upper back rufous; lower back and rump cinnamon-brown, obscurely barred with buffy-white and black, the rump more streaked; wings and central rectrices broadly barred with black and grayish-white to pale brownish-gray; remaining rectrices mostly black, with broad white tips somewhat spotted with black; underparts white, tinged with buff posteriorly. Iris rufous; bill black except base of lower mandible gray; legs bluish-gray. **Immatures:** crown and eye-stripe sooty-black; superciliaries and underparts more buffy; rufous of hindneck and back duller, indistinctly spotted and barred with black and buff; pale barring on wings broader and more buffy.

HABITS. Active, noisy pairs or family groups of 3–5 frequent deciduous and gallery forest, open woodland, second growth, low scrub, savanna trees, and groves around houses; forage for beetles, crickets, larvae, roaches, other insects, and spiders, by gleaning from foliage, probing and poking into crevices in bark, epiphytes, buildings; often bold and inquisitive.

VOICE. Contact between members of pair or group maintained by short, harsh, rasping notes. Mates often join in intricate duets, the song consisting of rich, mellow, throaty whistles alternating with dry chatters and gurgles in repeated phrases of 3–8 notes.

NEST. A roughly globular structure with a side entrance shielded by a visor, of straws, rootlets, and vegetable fibers, copiously lined with seed down, 5–25ft (1.5–8m) up in usually thorny tree, shrub, or cactus, especially bull's-horn acacia. Eggs 3–5, white, heavily blotched and speckled with brown, gray and blackish, mostly on large end. April–June. After breeding, families sleep together in nests like those used for breeding; family groups mostly break up before next breeding season.

STATUS. Common resident of lowlands of N Pacific slope, S to Carara and Punta Leona; sea level to 2600ft (800m), rarely higher on slopes of N cordilleras and in W Valle Central to near Grecia; on extreme NW Caribbean slope, E to around Brasilia.

RANGE. C Mexico to Costa Rica.

PLAIN WREN

Thryothorus modestus Pl. 38(17)

Soterrey Chinchirigüí (Chinchirigüí)

DESCRIPTION. Two populations in Costa Rica, differing in size and color, but both characterized by white superciliaries, nearly unmarked upperparts, mostly whitish underparts, plain crissum. (1) **"Plain Wren":** 5″ (13cm); 18g. ADULTS: above dull brown, brightening to cinnamon-brown on rump, wings, and tail; remiges faintly and narrowly barred with dusky; tail more broadly barred with black; eye-stripe brown; cheeks and auriculars white, indistinctly streaked with dusky; below white, tinged with buffy-brown on breast; sides, flanks, lower belly, and crissum cinnamon-buff. Iris reddish-brown; upper mandible black, lower pale gray with dark tip; legs gray. IMMATURES: above slightly duller and grayer, especially on pileum; superciliaries narrow and dull; throat and breast more grayish, tinged with brown on breast; belly dull buff; sides, flanks and crissum buffy-brown. Iris grayish-brown. (2) **"Canebrake**

Wren": 5½" (14cm); 23g. ADULTS: above grayish-brown, little or not at all brighter on rump, wings, and tail; tail more heavily barred with black; below much grayer, sides of breast gray; flanks, lower belly, and crissum pale grayish-brown. IMMATURES: slightly brighter, above more olive; below with buffy tinge, especially on belly, flanks, and crissum; dark barring of tail finer, more broken.

HABITS. Lurks in weedy fields, low, dense second growth, overgrown pastures and gardens, and brushy hedgerows; avoids closed woodland; gleans for insects and spiders in dense tangles, probing into hanging debris and rolled leaves; restless and active, rarely emerging into open; paired and territorial throughout year.

VOICE. Contact note an incessant, dry *chut*, often rapidly repeated; when disturbed, a dry *churr* and rasping, scolding notes; a dry *chi-wurp*. Two distinct song types: a clear, rich, ringing *chinchiri-gwee* given antiphonally, the male singing *chinchiri* and the female contributing a perfectly synchronized *gwee*; and a shorter, high-pitched, piercing 2- or 3-note phrase; either version repeated several times in rapid succession. Songs and calls of "Canebrake Wrens" are generally lower-pitched and more throaty; this form also gives series of rich or harsh, slurred whistles.

NEST. **"Plain Wren":** a roughly globular structure with visor-shielded side entrance, its thick walls composed of grasses and weed stems, softly lined with seed down, 2–10ft (0.6–3m) up in bush, grass tussock, or vine tangle. Eggs 2, rarely 3, white, unmarked. January–September. **"Canebrake Wren":** undescribed (?). Adults of both forms sleep in flimsier, more pouchlike nests; juveniles may lodge in old nests of other species, often those of Bananaquits.

STATUS. **"Plain Wren":** resident the length of Pacific slope from sea level to 6500ft (2000m) locally, reaching Caribbean slope via low passes in N cordilleras and across Valle Central down to at least Turrialba in Reventazón drainage; common except in lowland Guanacaste, where confined to wetter sites. **"Canebrake Wren":** common in coastal lowlands the length of Caribbean slope. With deforestation "Canebrake Wrens" are expanding their range inland, where they reached La Selva in the early 1970s.

RANGE. S Mexico to C Panama.

NOTE. The "Canebrake Wren" is sometimes considered a separate species, *T. zeledoni*; as the 2 forms may now be coming into contact, a study to determine their status is needed.

RUFOUS-AND-WHITE WREN

Thryothorus rufalbus Pl. 38(9)

Soterrey Rufo y Blanco

DESCRIPTION. 5½" (14cm); 25g. Differs from Plain Wren in having brighter upperparts, more heavily streaked cheeks, and barred crissum; from Banded Wren in unmarked, dingy sides and flanks; voice and habitats distinctive. **Adults:** above deep rufous, duller and browner on pileum; wings and tail barred with black; superciliaries white; brown stripe through eye; cheeks boldly streaked with black; throat and median underparts white; sides and flanks grayish-brown; crissum barred with black and white. Iris reddish-brown; upper mandible black, lower mandible and legs pale gray. **Young:** above darker and duller; crown scaled with dull black; breast tinged with buffy to brownish and mottled with blackish; crissum barred with brown and black; iris brown.

HABITS. In dry northwest, occurs in evergreen gallery forest, moist forests on adjacent mountain slopes, and locally in mangroves; in wetter areas, mainly in open, scrubby woodland; nearly always in pairs; forages rather deliberately, creeping along branches and logs or over the ground, hopping through thickets, probing and gleaning for spiders, cockroaches, sowbugs, beetles, crickets, and other invertebrates; flicks tail from side to side when excited.

VOICE. Calls include a hard, staccato chatter and a rapid, dry sputter; scold a sharp, scratchy *chet*, delivered singly or in rapid-fire series. The beautiful, leisurely, low-pitched, flutelike song consists of slow trills, clear whistles, and occasional sharp notes, in complex phrases delivered singly; pairs often sing antiphonally, the male giving the main phrase, his mate adding several final notes.

NEST. Retort-shaped, of grass stems and rootlets or largely of black fungal rhizomorphs, 6–10ft (2–3m) up in shrub, tree, or spiny palm. Eggs 3–4, greenish-blue, unmarked. April–August.

STATUS. Locally common resident of N Pacific slope, including at least W Valle Central, from sea level to 3600ft (1100m); abundant in mountains of Península de Nicoya, wherever forest remains; farther S, increasingly local, known mainly from lower Térraba valley. An old record for "Angostura" probably refers not to the locality of this name near Turrialba but to one above Esparza, where this wren occurs.

Range. Extreme SW Mexico to N Colombia and NW Venezuela.

STRIPED-BREASTED WREN

Thryothorus thoracicus Pl. 38(5)

Soterrey Pechirrayado

Description. 4½″ (11.5cm); 17g. Fairly small, short-tailed; throat and breast striped. **Adults:** above brown, duller and grayer on pileum, brighter and more russet on rump and upper tail-coverts; wings and tail heavily barred with grayish-brown and black; superciliaries white; sides of head and neck streaked with black and white; throat and breast striped with blackish and white; upper belly grayish-white mottled with black; sides and flanks dull brown; crissum buffy-brown barred with black. Iris reddish-brown; upper mandible black, lower pale gray; legs dark gray. **Young:** above brighter, more russet, crown scaled with black; throat and chest grayish-brown, irregularly striped with white, each stripe bordered with dusky; posterior underparts russet-brown to dull cinnamon. Iris dull brown.

Habits. Haunts thickets at woodland edge, in more open parts of forest, along streamside, in dense plantations of cacao and coffee, mostly below 20ft (6m); hunts insects and spiders by poking into moss tufts, hanging trash, rolled leaves, and gleaning from branches and foliage, often hanging and reaching acrobatically; paired throughout year.

Voice. Call a rolling *cherk* or *chrrik*; scold a series of rather soft, guttural chatters and sputters. Two distinct songs: the first includes varied, short melodious phrases of 3–5 loud, mellow, full-voiced whistles, repeated by the male just rapidly enough for his mate to slip a softer, often more involved phrase between his notes: *who's to SEE/me, little me/who's to SEE/me, little me*; the second a series of clear whistles on same pitch, delivered chiefly at dawn, at a uniform rate of 3/2sec, easily imitated and suggesting a pygmy-owl rather than a wren. Juveniles sing sweet, rambling songs, quite different from those of adults.

Nest. Roughly globular, with a rounded chamber entered through an antechamber with doorway facing downward, of fibrous materials and leaf skeletons, covered on outside with green moss, lined with finer fibers; saddled over thin, horizontal branch 5–20ft (1.5–6m) up in shrub, cacao tree, or vine tangle, often at edge. Eggs 2–3, bluish-white. March–July. Similar nests serve as dormitories.

Status. Common resident of lowlands and foothills the length of Caribbean slope, reaching Pacific slope locally via low passes in N cordilleras; sea level to 3300ft (1000m), rarely higher.

Range. Nicaragua to C Panama.

BAY WREN

Thryothorus nigricapillus Pl. 38(12)

Soterrey Castaño

Description. 5¾″ (14.5cm); 27g. Medium-sized, richly colored, with striking head pattern. **Adults:** top and sides of head black with contrasting white spots on lores, upper and lower eyelids, and auriculars; a white malar stripe enclosed by black; throat white; rest of plumage rich rufous-chestnut, below paler; wings, tail and crissum heavily barred with black. Iris chestnut; upper mandible black, lower pale gray with black tip; legs dusky. **Young:** similar but duller; head more sooty, white markings indistinct; indistinct dark barring on scapulars and flanks; often with dusky scaling across chest.

Habits. Prefers thickets along streams or lagoons in generally forested country, sometimes dense second growth at forest edge; breeds in pairs, but young apparently remain with parents for several months; forages by gleaning foliage, twigs, and branches, poking into hanging trash and vine tangles for beetles, roaches, crickets, caterpillars, other insects, and spiders.

Voice. Calls include dry, harsh, grating chatters and sputters like ant-tanager; a harsh, rolling *prrrrk*, and a sharp, metallic *chrink*. Song distinctive for its forceful, explosive delivery and rapid repetition of rich-toned, clear, slurred whistles, trills, and warbles; sometimes antiphonal, female inserting short notes among phrases of male and often initiating the duet.

Nest. Retort-shaped with short entrance tube, of weed stems, rootlets and strips of *Heliconia* leaves, sparsely decorated with green moss and lined with leaf skeletons, 6–16ft (2–5m) up in crotch of upright or nodding branch. Eggs 2–3, white, speckled with reddish-brown, especially around large end. March–August.

Status. Common and widespread resident the length of Caribbean slope, from sea level to ca. 2300ft (700m) in N, 3300ft (1000m) in S; less numerous in drier lowlands just S of Lago de Nicaragua.

RANGE. E Nicaragua to W Ecuador.

NOTE. Our form, *T. n. castaneus*, is sometimes considered a species separate from the more southern, boldly barred *T. nigricapillus* of E Panama and South America.

RIVERSIDE WREN

Thryothorus semibadius Pl. 38(10)

Soterrey Pechibarreteado

DESCRIPTION. 5″ (13cm); 21g. Fairly small, with distinctive, finely barred underparts. **Adults:** above rufous-chestnut, the wings and tail finely barred with black and cinnamon-rufous; narrow white superciliaries bordered above and below with black; eye-ring white; sides of head and neck streaked with black and white; underparts mostly white, tinged with buff on belly, becoming buffy-brown on flanks and crissum, everywhere except upper throat narrowly and evenly barred with black. Iris reddish-brown; upper mandible black, lower pale gray; legs dark gray. **Young:** similar but above duller; crown finely scaled with blackish; below with black barring somewhat blurry; iris duller brown.

HABITS. Prefers dense vegetation fringing broader streams or swampy openings in forest, but, especially in rainy weather, often ventures into thickets at edge of forest, brushy banks, or shaded gardens away from water; hunts insects and spiders by gleaning in dense foliage, ransacking vine tangles, and probing curled dead leaves; in pairs or families of 3 or 4.

VOICE. Male and female maintain contact with sharp *weet-poo* or *peet-t'churr* and answer each other with softer, tinkling notes sometimes in flight. A harsh, rasping *churrr* and a staccato dry sputter when annoyed. Songs extremely varied, usually consisting of repeated loud, ringing 2- or 3-note phrases: *victory, victory, victory, checker, checker, checker*, etc.; sings duets, the composite phrases of 5–7 notes: *victory/we-do-it/victory/we-do-it*.

NEST. A bulky, globular structure saddled over a horizontal twig, the wide vestibule balancing the well-enclosed egg chamber, the former entered through a wide doorway facing downward and inward, of fibrous materials and more or less green moss. Eggs 2, white with fine, faint speckles of pale brown. December–August. Similar nests are occupied at night at all seasons, usually by a single bird, sometimes as many as 3, probably a parent with young.

STATUS. Common resident on S Pacific slope, N to about Carara; lowlands to 4000ft (1200m).

RANGE. Costa Rica to W Panama.

NOTE. Has sometimes been considered a subspecies of the Bay Wren, *T. nigricapillus*.

BANDED WREN

Thryothorus pleurostictus Pl. 38(6)

Soterrey de Costillas Barreteadas

DESCRIPTION. 5¼″ (13.5cm); 20g. Diagnostic white underparts barred laterally with black. **Adults:** above cinnamon-brown, wings narrowly and tail broadly barred with black; superciliaries white; broad eye-stripe brown; cheeks white, indistinctly streaked with black; underparts white, tinged with buffy-brown on flanks; sides, flanks, and crissum heavily barred with black. Iris rufous; bill black, except base of mandible gray; legs pale gray. **Young:** above somewhat duller brown; below dull white, with breast, sides, and flanks faintly mottled and smudged with dusky, the flanks tinged with brown; iris dull brown.

HABITS. Frequents openings in deciduous or evergreen woodland, open woods, scrubby, thorny second growth, brushy ravines, sometimes mangrove swamps; in pairs or family groups, searches for insects and spiders in undergrowth, sometimes on ground; pushes up fallen leaves with bill; probes curled dead leaves and crevices in bark and roots, and in leaves of small spiny palms; wags tail from side to side when excited.

VOICE. When disturbed, a light, staccato ticking or sputter; a nasal, hard, rattling *cherrrt*; often alternates a metallic roll with a hard rattle: *kert rrruk kert rrruk*. A superb songster; songs consist of complex, varied phrases containing loud, clear whistles and melodious trills; usually repeats a given phrase several times, then shifts to another.

NEST. A retort-shaped structure with long entrance tube directed downward, saddled 3–8ft (1–2.5m) up in crotch of small tree, often a bull's-horn acacia or spiny palm, sometimes beside wasps' nest, compactly built of weed stems, fibrous rootlets, and the like, and lined with fine grasses or sometimes entirely of fine, yellow grass inflorescences. Eggs 3–4, immaculate white or pale greenish-blue. May–August.

STATUS. Abundant resident in dry NW lowlands, S to about Orotina, up locally to 2600ft

(800m) on slopes of N cordilleras and in W Valle Central, E occasionally to Grecia.

Range. C Mexico to Costa Rica.

BLACK-THROATED WREN

Thryothorus atrogularis Pl. 38(13)

Soterrey Gorginegro

Description. 6″ (15cm); 24g. Antbirdlike, black face and chest with little white on head or barring on wings and tail. **Adults:** sides of head, throat, and breast black; superciliaries faint white; auriculars with sparse white streaks; above uniform chestnut-brown, below slightly paler; wings dusky with chestnut-brown edgings; tail black with sparse, sometimes nearly obsolete, grayish-brown barring; crissum black, barred with buffy brown. Iris rufous; bill black except base of lower mandible blue-gray; legs blackish. **Young:** above and below dull olive-brown, becoming dusky on face; faint buffy streaking on auriculars.

Habits. In dense second growth at forest edge, gaps, abandoned clearings, or along streams, generally preferring taller vegetation than Plain Wren, less tied to streams than Bay Wren; usually in pairs; forages in vine tangles and thickets like other *Thryothorus*, taking insects, spiders, and other small arthropods.

Voice. Calls include a fast, nasal to wooden, rattling *praaaaaht* and a guttural, rolling *beewrrr* or *bweeurrrr*. Song of rich, clear, often slurred whistles and warbles usually ending in a vigorous trill. After repeating 1 phrase several times, the bird shifts to another; sometimes sings antiphonally, the female adding notes to end of phrase. Song slower, mellower, less varied than that of Bay Wren.

Nest. Undescribed (?). Probably breeds April or May–August.

Status. Locally common resident of lowlands and foothills, up to ca. 3600ft (1100m), the length of Caribbean slope, reaching Pacific slope around bases of Guanacaste volcanoes.

Range. E Nicaragua to NW Panama.

Note. Sometimes *T. spadix*, Sooty-headed Wren, of E Panama and Colombia, is lumped with this species.

BLACK-BELLIED WREN

Thryothorus fasciatoventris Pl. 38(11)

Soterrey Vientrinegro

Description. 6″ (15cm); 30g. Large, rather long-tailed *Thryothorus* with boldly contrasting white bib and black belly. **Adults:** above dark rufous-chestnut, duller on forehead; wings faintly and tail heavily barred with black; sides of head sooty-black; superciliaries faintly grayish-white; throat, foreneck, and chest white; rest of underparts dull black, barred with pale gray to brown on lower belly, flanks, and crissum. Iris chestnut; upper mandible black, lower pale gray; legs black. **Young:** above somewhat duller; throat and chest more grayish; rest of underparts dull brown with faint blackish barring; iris dull brown.

Habits. Haunts thickets of *Heliconia* and *Calathea* near streams, scarcely venturing into adjacent tangled second growth or forest understory; deliberately gleans insects and spiders from large-leaved plants, probing inflorescences and leaf bases, sometimes all but disappearing into curled, hanging dead leaves and rummaging noisily; in pairs throughout year.

Voice. Scold a low, rasping chatter or sputter. Song of complex phrases of rich, clear, mellow, slurred whistles and shorter notes, less vigorous and explosive than that of Bay Wren and without trills: *go HERE go tiddle-o, goo WAAY to here go ch'wo*, singly or repeated 2–3 times.

Nest. A globular structure with a side entrance shielded by a visor, loosely woven wholly of dry strips of *Heliconia* leaves, 3–6ft (1–2m) up in *Heliconia* thicket, usually where several leaf petioles cross. Eggs unknown (?). Probably May–July.

Status. Locally common resident in Golfo Dulce region and lower Térraba valley of S Pacific slope, N in smaller numbers along coastal lowlands to around Carara; sea level to 1650ft (500m).

Range. Costa Rica to W Colombia.

SPOTTED-BREASTED WREN

Thryothorus maculipectus Pl. 38(7)

Soterrey Pechimoteado

Description. 5″ (13cm); 17g. Densely spotted breast and belly diagnostic. **Adults:** pileum and nape chestnut-brown; rest of upperparts bright brown; tail paler and heavily barred with black; narrow superciliaries white and postocular stripe brown; sides of head streaked with black and white; below white, heavily spotted with black except center of throat and chin; flanks buffy-brown; crissum barred with black and white. Iris rufous; bill silver-gray with black culmen; legs slaty. **Young:** similar but with smaller, grayer, indistinct spots on brown-tinged breast and belly.

Habits. Skulks in thickets along rivers, second growth, dense cacao and citrus plantations, canebrakes, and forest edges; prefers taller, thicker vegetation than Plain Wren, drier or scrubbier sites than Bay Wren; like other *Thryothorus*, forages by gleaning, probing in vine tangles, hanging debris, rolled leaves, and dense foliage for insects and spiders; usually in pairs.

Voice. Calls include a sharp *churr*, a drawled, ascending, chattering *reeeeeeik* when disturbed and clicking notes when concerned for young. Song of male a melodious phrase of 5–7 clear, ringing whistles, sometimes including a short trill, often several times repeated; often sings antiphonally, female interpolating a shorter, 2–4 note phrase between those of male.

Nest. A globular structure with a side entrance and little or no entrance tunnel, compactly made of grass stems and leaves, lined with fine fibers, seed down, and feathers; in crotch of shrub or thorny tree 3–20ft (1–6m) up in thicket. Eggs 3–4, white, with heavy reddish-brown blotches and streaks. April–June or July. Similar nests serve as dormitories.

Status. Locally fairly common resident of lowlands S of Lago de Nicaragua, from Río Frío district (Los Chiles; Playuclas) W at least to Upala and Canalete, sea level to at most 650ft (200m).

Range. E Mexico to Costa Rica.

Note. Sometimes considered a race of *T. rutilus*, Rufous-breasted Wren; sometimes the Speckled Wren, *T. sclateri*, of Ecuador and Peru, is included in *maculipectus*, or all 3 are lumped under the name *T. rutilus*, Speckled Wren.

RUFOUS-BREASTED WREN

Thryothorus rutilus Pl. 38(8)

Soterrey Carimoteado

Description. 5″ (12.5cm); 16g. A rather small, arboreal *Thryothorus* with distinctive spotted face and throat contrasting with plain rufous breast. **Adults:** pileum russet, narrowly bordered laterally by black; rest of upperparts olive-brown; tail grayish-brown, heavily barred with black; narrow superciliaries white; sides of head and throat black, thickly speckled with white; breast cinnamon-rufous, becoming paler and duller posteriorly, the flanks tinged with olive-brown; crissum barred with buff and black. Iris rufous; upper mandible dark gray, lower pale gray with dark tip; legs bluish-gray. **Young:** similar but duller, especially on crown and posterior underparts; breast paler, more cinnamon; spots on throat and face larger, less well defined.

Habits. Frequents thickets and vine tangles, bamboo thickets at forest edge, borders of coffee plantations, tall second growth, and gardens in humid regions; searches in dense vegetation up to 30ft (9m) for beetles, crickets, grasshoppers, roaches, bugs, caterpillars, ants, other insects, and spiders; sleeps singly in nests made by self or stolen from other wrens or Bananaquits; paired throughout year.

Voice. Calls include a prolonged, questioning *reep* or *ch'reeep*, and a sharp *churr* when annoyed. Song of male a phrase of 5–7 clear, ringing whistles, repeated deliberately; sings antiphonally, phrase of male followed immediately by 3–4 notes of his mate, the 2 songs so well blended that they seem 1; juveniles sing softly warbled, rambling songs.

Nest. A bulky, globular structure with a round side entrance, of dry bamboo leaves, grass blades, dicotyledonous leaves, straws, etc., softly lined with seed down, 4″–50ft (0.1–15m) up amid low herbage or high vine tangles. Eggs 2–4, usually 3, whitish, more or less densely speckled with cinnamon or rufous. March–July, rarely January.

Status. Common resident in foothills and valleys of Térraba–General–Coto Brus region, mainly 1000–4000ft (300–1200m); rare or absent in wetter lowlands of Golfo Dulce or coastal regions, but isolated local populations N on Pacific slope, mainly in foothills, to base of Volcán Tenorio; recorded from lowlands at head of Golfo de Nicoya.

Range. Costa Rica and Trinidad-Tobago to Colombia and Venezuela.

HOUSE WREN

Troglodytes aedon Pl. 38(18)

Soterrey Cucarachero (Soterrey, Zoterré, Cucarachero)

Description. 4″ (10cm); 12g. Small, brown, obscurely patterned, the common wren around human dwellings. **Adults:** above brown, barred with black on wings and tail, and with faint, fine darker barring on back, at least in fresh plumage; superciliaries faintly buffy; sides of head brown, mottled and streaked with paler brown; throat and belly whitish to pale buff, breast dark dull buff; sides and flanks buffy-brown. Upper mandible blackish, lower grayish-horn to dull flesh; legs horn-color. **Young:** darker, usually more grayish; often darker mottling or scaling on breast; sometimes throat and breast rufescent; barring on tail more broken.

Habits. Natural habitat probably low second growth and beached snags and stumps along large rivers; now found mainly about human habitations and in man-made habitats;

searches incessantly for insects and spiders in weedy fields, low thickets, dooryards, plantations, piles of brush or stones, crevices in bark, exposed tree roots, and buildings, which it often enters; lively, animated, irascible; paired throughout year; sleeps at all seasons in crannies in buildings, holes in trees, niches in banks, bunches of bananas, adults always singly, or the female with her fledged young.

Voice. Calls include a variety of staccato, rattling, burbling, or chattering notes, which become harshly grating when nest is molested. Male has large repertory of rapid, cheerful songs that include sputtering notes, clear or scratchy slurred whistles, and bubbling trills, more prolonged and varied than songs in its North American relatives; often female answers her mate's song with twitters or slight, clear or wheezy trills.

Nest. In nooks and crannies in the most varied situations, often in buildings or a niche in a bank. If the space is large, it is nearly filled with coarse twigs, which support a cup of finer materials, lined with feathers. Eggs 3–5, usually 4, whitish, densely marked with brown. December–September; 3–4 broods.

Status. Resident countrywide from lowlands to 9000ft (2750m); mostly common, except in extensively forested areas and in dry NW lowlands.

Range. S Canada to Tierra del Fuego; northern populations winter S to S Mexico.

Note. Birds from S Mexico southwards are often classified as a separate species, *T. musculus* (Southern House-Wren), from the northern *T. aedon* (Northern House-Wren).

OCHRACEOUS WREN

Troglodytes ochraceus Pl. 38(19)

Soterrey Ocroso

Description. 3¾″ (9.5cm); 9.5g. Very small, resembling House Wren but brighter, especially on face, with broad ochraceous-buff superciliaries. **Adults:** upperparts and broad postocular stripe tawny-brown, paler and tinged with ochraceous on forehead and forecrown; wings and short tail barred with black; superciliaries, cheeks, and sides of neck ochraceous-buff, paling to rich buff on throat and breast and whitish on belly; sides and flanks ochraceous-tawny; crissum barred with buff and black. Bill black with tomia and base of lower mandible horn-color; legs brownish. **Young:** similar but superciliaries paler buff, underparts with extensive dusky mottling and scaling.

Habits. Frequents cool, wet, moss- and epiphyte-laden mountain forests and edges, and adjacent tall second growth and pastures with scattered old trees; in pairs year-round and in family groups of 3–4 for several months following breeding, often accompanying mixed-species flocks through midlevels of forest; creeps up or down vertical trunks and along large branches, seeking insects and spiders amid moss, matted roots, and clustered stems of epiphytes, or in tangles of vines; sometimes descends to rummage and glean on epiphyte-laden stumps.

Voice. Call a rolling, thin, high-pitched *peeew* or *preeer*; a low, weak, dragging *churr*; song a varied medley of high, thin, slurred whistles and liquid trills in a subdued tone.

Nest. Hidden in a dangling mass of epiphytes that has broken away from a bough and is suspended by its still-adhering roots, swinging in the breeze, 16–50ft (5–15m) up. No details of nest and eggs. April–May. Similar sites are occupied for sleeping.

Status. Fairly common resident of mountains on both slopes, from Cordillera de Tilarán S, mainly from ca. 8000ft (2450m) down to 3000ft (900m) on Caribbean slope and 4000ft (1200m) on Pacific; rarely to 10,000ft (3000m) in Cordillera de Talamanca.

Range. Costa Rica to E Panama.

Note. Sometimes considered conspecific with the Mountain Wren, *T. solstitialis*, of N South America, but differs strikingly in voice and behavior, has much shorter tail, and seems to us to be perfectly distinct; some authors include the similarly distinct *T. rufociliatus* of N Central America in the same species; we strongly favor full species status for all 3.

TIMBERLINE WREN

Thryorchilus browni Pl. 38(16)

Soterrey del Bambú

Description. 4″ (10cm); 14g. Fairly small, stub-tailed, with broad superciliaries and patch on folded wing. **Adults:** above bright tawny-brown, wings and tail narrowly barred with black, and outer 5–6 primaries broadly edged with white; broad superciliaries and eye-rings white; broad postocular stripe brown; sides of neck streaked with black and white; below white, more or less spotted or smudged with dusky, especially on sides; flanks buffy-brown. Upper mandible blackish, lower flesh-color; legs pale brownish. **Young:** superciliaries, throat, and chest tinged with olive-buff; nape, sides of neck, and chest indistinctly streaked with dusky.

HABITS. Prefers thickets, especially bamboos, in páramo and subpáramo, second growth, or at edges and breaks in oak forest; active, inquisitive, often cocks and pumps tail; gleans in foliage tangles, probes rolled leaves and crevices; creeps along branches, sometimes hops on ground; flutters to pluck prey from underside of leaf or twig; takes small insects, caterpillars, spiders; probably paired year-round.

VOICE. Scold a sharp, harsh, *shook* with rising inflection; a fast, ascending series of rolling, rasping *bee* notes; a high, thin, scratchy whistled *bheeee* or *dzeee* when disturbed. Song a fairly complex, variable phrase of rapid warbles and short notes often ending in a high, thin, somewhat scratchy whistle; often rapidly repeated several or many times, giving effect of a squeaky swing.

NEST. A hollow ball with a side entrance, woven entirely of bamboo leaves, with sparse lining of fine fibers, 3–10ft (1–3m) up in shrub or bamboo. Eggs 2, white, finely speckled all over with pale brown. April–June.

STATUS. Common resident at 9200–12,000ft (2800–3600m), mostly above timberline, but with an isolated colony in Sabana Dúrika at ca. 7200ft (2200m) on Caribbean slope; from Irazú-Turrialba massif of Cordillera Central S along Cordillera de Talamanca to Panama.

RANGE. Costa Rica and W Panama.

NOTE. Sometimes placed in genus *Troglodytes*, but in song, behavior, and nest seems also close to *Henicorhina*; for the present we maintain it in its own monotypic genus.

WHITE-BREASTED WOOD-WREN

Henicorhina leucosticta Pl. 38(15)

Soterrey de Selva Pechiblanco

DESCRIPTION. 4″ (10cm); 16g. Fairly small, with very short tail; white throat and breast, streaked cheeks diagnostic. **Adults:** pileum varies from dull brown with black border to mostly black; rest of upperparts dull chestnut, barred with black on wings and tail; superciliaries white; sides of head and neck streaked with black and white; throat, breast, and upper belly white; sides gray; posterior underparts cinnamon to tawny-brown. Bill black, legs dusky. **Young:** underparts mostly dull brownish-gray, shading to whitish on throat; flanks more olive-brown; dark head markings sooty-blackish, white streaks indistinct; above darker and duller. Young Gray-breasted Wood-Wrens have streaked throats.

HABITS. Frequents lower understory of wet forests and adjacent tall, shady second growth; searches for insects, sowbugs, spiders, centipedes and other invertebrates among tangles of fallen branches and vines, debris caught in leaf bases of small understory palms, ferns, and herbage on or near ground; often bold and inquisitive but adept at remaining concealed; in pairs or family groups.

VOICE. Usual call a distinctive, sharp, somewhat burry *cheek, bzeet*, or *teleet*; scold a low, sputtering *churrrr chut-ut-ur*; an explosive *tuck*, a flutelike whistle. Both sexes sing simple, varied phrases of 2–5 loud, clear, melodious, whistled notes: *cheer, oweet-oweet; cheery weather; cheero cheero cheer*. Usually a phrase is repeated 1 to several times, then, after an often lengthy pause, the bird continues with same or different phrase.

NEST. A compact, roofed structure with side entrance shielded by visor, of fibrous rootlets, leaf skeletons, stems of mosses, and liverworts, with more or less green moss on outside, lined with fine materials and often downy feathers, on ground or rarely 2ft (60cm) up amid dense concealing herbage or fallen branches. Eggs 2, immaculate white or rarely lightly speckled. February–May. Dormitory nests, occupied by single individuals or parents with fledglings, are of flimsier construction and much higher, usually 2–10ft (0.6–3m) up in slender saplings or vine tangles.

STATUS. Common resident of lowlands and lower foothills of Caribbean slope, up to 3000ft (900m) in N and 4000ft (1200m) in SE; reaches Pacific slope of N cordilleras but absent from dry lowlands. On S Pacific slope, rare in lowlands but common in foothills, mainly at 1000–6000ft (300–1850m), rarely higher, and N to Carara.

RANGE. C Mexico to NE Peru and Suriname.

NOTE. Also called Lowland Wood-Wren.

GRAY-BREASTED WOOD-WREN

Henicorhina leucophrys Pl. 38(14)

Soterrey de Selva Pechigrís

DESCRIPTION. 4¼″ (11cm); 18g. Differs from White-breasted Wood-Wren of lower elevations in its gray breast and striped throat in all plumages. **Adults:** pileum dull dark brown to blackish; rest of upperparts dark chestnut-brown, barred with black on wings and tail; superciliaries white and postocular stripe blackish; sides of head and neck dull black, streaked with white; chin and throat white, striped with blackish; breast and upper belly slate-gray, darker on sides; flanks and crissum rich tawny to russet-brown. Bill black; legs dusky. **Young:** similar but dark markings of head more sooty, white streaking less distinct; above paler, and on flanks more olivaceous.

Habits. Frequents lower understory of highland forest, including bamboo thickets, bush-choked ravines, and adjacent dense second growth; in pairs or family groups; active, incessantly and busily searching for small invertebrates in low herbage and ground litter, fallen limbs or trees, piles of fallen epiphytes or other debris on ground or caught in low vegetation; furtive, adept at disappearing into foliage or beneath hanging banks, but sometimes inquisitive and irascible.

Voice. Calls include dry, staccato notes and sputters; harsh, rasping churrs when disturbed. Song a tumbling, flowing, often highly melodious medley of clear, rich whistles and warbles seeming to lack a set pattern; Unlike White-breasted Wood-Wren, male and female join in lively duets, especially in nesting season.

Nest. A globular structure with a vestibule or antechamber entered through a downward-facing doorway, of fine black fibrous rootlets covered with more or less green moss, often among vegetation overhanging a roadside bank, path, or edge of a ravine, at no great height. Eggs 2, white. March–June. Male and female sleep together in similar nest.

Status. Common resident of mountains throughout, from timberline down to 2600ft (800m) on Caribbean slope and 3600ft (1100m) on Pacific.

Range. C Mexico to W Ecuador and N Bolivia.

ROCK WREN

Salpinctes obsoletus Pl. 38(4)

Soterrey Roquero

Description. 5½″ (14cm); 22g. Heavily mottled, rather long-tailed; bill long, slightly decurved. **Adults:** above dull grayish-brown, obscurely mottled with dusky, wings barred with dusky; rump and upper tail-coverts dull buff, barred with black; tail mostly brownish-gray, barred with black, with black subterminal band and dull buffy tip; superciliaries paler; postocular stripe dark; cheeks dull buff, smudged with dusky; below dull white, washed with buff on breast and flanks, everywhere densely spotted and irregularly barred with blackish; crissum broadly barred with black. Upper mandible gray, lower paler gray; legs dusky. **Young:** above more or less barred with dusky; below with spotting blurred and indistinct; grayish-buff spotting on wing-coverts.

Habits. In widely dispersed pairs inhabits grassy, windswept, boulder-strewn slopes, open rockslides, rocky canyons, and arroyos with sparse vegetation; perches on boulders or low vegetation to sing or scold; often depresses tail, pumps body up and down; forages by hopping on ground, creeping over boulders, poking into crevices and grass tussocks in search of insects and spiders. Flight low, fast, whirring, usually for short distances.

Voice. Scold a nasal *jeee-jijijijit*; also a dry *jer-weet*. Song composed of phrases of 1–2 short, metallic trills or rattling warbles, often with burry quality, repeated rapidly 3–6 times: *chree chree chree cher-weeza cher-weeza cher-weeza cher-weeza*.

Nest. A shallow cup of grass and plant fibers, lined with finer materials and feathers, in cavity or short burrow in or under rocks. Eggs 4, white, speckled with reddish-brown. May–June.

Status. Locally common resident along Pacific slope of Cordillera de Guanacaste, S to Volcán Miravalles, 1600–5200ft (500–1600m).

Range. SW Canada and W USA to Costa Rica.

NIGHTINGALE WREN

Microcerculus philomela Pl. 38(21)

Soterrey Ruiseñor

Description. 4″ (10cm); 18g. Small; very dark; stubby tail; terrestrial; above and below scaled with black. **Adults:** above rich, dark brown, scalloped with dull black; wing-coverts with a subterminal, bright brown bar or buffy dot; throat and breast dull gray, heavily scaled with blackish (breast often nearly solid black); posterior underparts dark brown, barred or scalloped with black. Bill black, gonys sometimes pale brownish; legs black. **Immatures:** like adults but below more contrasting pale scaling. **Young:** above, paler brown subterminal marks set off by dark fringes, increasing contrast; below dusky with more or less conspicuous pale gray scaling, rarely fine buffy streaks; throat paler.

Habits. Prefers dense forest, especially in ravines and foothills; walks over forest floor, incessantly pumping its tail as it peers and pokes into leaf litter, fallen branches, and debris for insects, spiders, centipedes, sowbugs, and other invertebrates; usually alone; males sing from ground or low vegetation.

Voice. When agitated, a soft, sharp *tchut* or

chek or a more sibilant *schip*. Song amazing, consisting of a long series of clear whistles on different pitches, delivered about 2/sec for up to 30sec, sounding like a slightly tone-deaf person whistling a tune; at close range a short opening motif of rapid, ascending notes. Juveniles beg with a soft, sad *peew* or *peek*.

Nest. Undescribed (?). Breeds from May or June through at least September.

Status. Locally common resident from lowlands to ca. 4000ft (1200m), occasionally to 4600ft (1400m) on N Caribbean slope, S to hills N of San Ramón (La Balsa) and N side of Volcán Turrialba, near Guácimo; reaches Pacific slope around bases of Guanacaste volcanoes; a possibly isolated population on hills west of Tortuguero.

Range. Extreme SE Mexico to Costa Rica.

Note. Often considered a subspecies (into which *M. luscinia* is lumped) of the South American *M. marginatus*; the entire complex formerly called Nightingale Wren. However, *philomela* is quite distinct in coloration and voice. See Whistling Wren.

WHISTLING WREN

Microcerculus luscinia Pl. 38(20)

Soterrey Silbador

Description. 4¼″ (10.5cm); 19g. Resembles Nightingale Wren but paler, less heavily marked. **Adults:** above rich chestnut-brown, scaled with blackish on crown; wing-coverts with a rusty subterminal bar; throat whitish; breast dull gray, often with brownish flecking and paler scalloping; belly and flanks rich brown with faint or no dark barring. Bill black except base of lower mandible brownish; legs blackish. **Immatures:** like adults but below retaining some scaling; breast and upper belly more brownish. **Young:** throat scaled with dusky, breast with pale gray and dark brown, belly with dusky; upperparts scaled or barred with blackish.

Habits. Haunts the floor and lower understory of forest or old second growth, walking about pumping hindquarters like a miniature dipper; males may mount stumps, logs, or buttresses (but never twigs) to sing. Food and foraging habits like those of Nightingale Wren but may also attend army ants.

Voice. Call a sharp *tchip* or *tchik*. Song remarkable: a brief opening motif of short, fast, ascending notes is followed by a long series of high, piercing, whistled notes, each lasting 1sec or more, that become progressively more widely spaced (up to 10sec between final notes) and lower-pitched; entire song may take 2–4min. Whistles of latter part of song may be doubled.

Nest. Undescribed (?). Breeds roughly May–September.

Status. Fairly common resident of wet lowlands and foothills up to ca. 3300ft (1000m) on Caribbean slope and 5600ft (1700m) on S Pacific slope, N to watersheds of Río Reventazón and Río Barranca, respectively.

Range. Costa Rica to E Panama.

Note. Possibly a subspecies of the South American *M. marginatus* complex, called Scaly-breasted Wren by the AOU although only N Andean races have scaly breasts as adults, Amazonian birds being white-breasted; if *M. luscinia* is included, Whistling Wren would be the best name for the entire complex.

SONG WREN

Cyphorhinus phaeocephalus Pl. 38(22)

Soterrey Canoro

Description. 4¾″ (12cm); 25g. Plump; short-tailed; peculiar high-ridged culmen; skin around eye imparts an antbirdlike aspect. **Adults:** above dark brown, wings and tail barred with black; sides of head and neck, throat and breast rufous-chestnut; belly dull gray; sides and flanks olive-brown. Orbital skin pale gray to bluish-gray; bill black; legs blackish. **Young:** upperparts extensively barred or scaled with dull black; rufous of face and neck bordered behind by a whitish or buffy line; belly grayish-olive.

Habits. Travels through understory of wet forest and adjacent tall second growth in compact family groups of 3–6; seeks insects, spiders, and other invertebrates by hopping over ground, lifting fallen leaves with bill, and letting them fall back in place if nothing edible is exposed; rummages and probes in fallen debris on ground or caught in leaf bases of small understory palms and shrubs; accompanies mixed flocks and forages with army ants, often with Spotted Antbirds.

Voice. Common call a low-pitched, guttural, incessantly repeated *cutta* or *cuttakut* that at a distance sounds like croaking of frogs or burbling of falling water. Song consists of clear, well-spaced whistled notes. A single bird gives these clear notes while uttering a seemingly continuous stream of low guttural notes, or 2 birds (possibly a mated pair) contribute, often producing strikingly musical sequences.

Nest. A retort-shaped structure 2–8ft (0.6–2.5m) up in the crotch of a slender, upright sapling, with the chamber on one side and the downwardly directed entrance tube on the other side of the fork, of coarse vegetable materials and lined with leaf skeletons. Eggs 2, white with a wreath of pale brown speckles. January–May, rarely through October. Often a number of these nests in various stages of repair are found in the same small area; families of 3–5 sleep in them.

Status. Common resident of Caribbean slope, from lowlands up to ca. 3500ft (1050m); reaches Pacific slope of Cordillera de Guanacaste via low passes.

Range. NE Honduras to W Ecuador.

Note. Sometimes lumped with the somewhat differently patterned Musician Wren, *C. aradus*, of South America E of the Andes, which has a still more complex song.

FAMILY **Mimidae:** Mockingbirds, Thrashers, and Catbirds

From southern Canada to southern South America, 31 species of this exclusively New World family are found, but they occur in greatest numbers and diversity in the southwestern United States, Mexico, and the West Indies. Only those that breed farthest north are migratory, and only 1 reaches Costa Rica. In length 8–12″ (20–30cm), they are short-winged, long-tailed birds, with rather slender, short to moderately long bills that range from nearly straight to strongly decurved. Most are gray or brown above, with paler underparts that may be plain, streaked, spotted, or scaly, but a few are blue or glossy black. The sexes are alike and wear the same colors throughout the year. Most forage on or near the ground in open woodland, brush, gardens, or semidesert, where, with their bills, they toss the litter aside or dig into the soil for small invertebrates. Fruits fill out their diet. Their songs are richly varied; some are famous mimics. The nest, sometimes built by both sexes but probably more often by the female alone, is a bulky cup or bowl, of twigs, rootlets, dead leaves, and the like, usually low in a bush or cactus, rarely on the ground or at moderate height. The 2–5 or, rarely, 6 eggs are blue-green, bright blue, buff, or whitish, immaculate or marked with brown. The female usually incubates them alone for a period of 12 or 13 days. Fed by both parents, the young remain in the nest for 11–15 days. Thrashers and mockingbirds vigorously defend their nests against humans and other intruders.

GRAY CATBIRD

Dumetella carolinensis Pl. 39(12)

Pájaro-gato Gris

Description. 8″ (20cm); 35g. Slender, with longer tail than any wren or thrush. **Adults:** pileum black; tail and wings blackish; rest of upperparts dark slate, underparts paler; crissum chestnut. Iris dark brown; bill and feet black. **Immatures:** similar; in hand usually show a few shorter, more brownish, juvenile wing-coverts; iris often dark purplish.

Habits. Prefers dense second growth, hedgerows, and forest edge; secretive and skulking but may venture out of cover at dawn and dusk; hops on ground with tail cocked, wings flicking; gleans insects from foliage, moving deliberately, often twisting head to scan, stretching neck to reach prey; eats many fruits, including berries of *Phytolacca* and *Cordia curassavica* and arillate seeds. Sometimes in loose flocks on migration; scattered, sedentary individuals in winter.

Voice. Most distinctive is a nasal, rather harsh, catlike mewing; alarm note a dry, staccato, explosive *chut* or a dry sputtering.

Status. Common but unobtrusive fall migrant (occasionally early September but not common before mid-October through mid-November) in Caribbean lowlands, mostly below 1600ft (500m) but in small numbers up to 5000ft (1500m); occasional in fall and spring in Valle Central and on S Pacific slope. In winter, uncommon to locally common in Caribbean lowlands, rare in Golfo Dulce lowlands; remains through late April or early May. Spring passage scarcely noticeable and only along Caribbean coast.

Range. Breeds in S Canada, E and C USA; winters from S USA to C Panama and West Indies.

Note. Tropical Mockingbird, *Mimus gilvus*, is similar in size and shape but paler on head and body (whitish below), with white in wings and tail; recorded sporadically, especially in S Pacific lowlands but almost certainly as escaped cage birds (S Mexico to El Salvador, Lesser Antilles, South America; introduced and established in C Panama, but wild birds unlikely in Costa Rica).

FAMILY **Turdidae:** Thrushes, Robins, Solitaires, and Allies

With about 300 species spread widely over the world, except New Zealand and some oceanic islands, the thrush family contains some of the most familiar birds. Many are renowned songsters notable for complex or continuously flowing melodies. Most diverse in the Old World, thrushes reached the Neotropics via North America; only the genera *Catharus* and *Turdus* have many species south of Mexico. New World thrushes are mostly plainly attired in shades of brown, gray, rufous, black, and white. Many have bright bills, legs, and eye-rings, but some Old World species are more colorful. The sexes tend to be similar, but juveniles of most species (and adults of some) have spotted plumage. In size 5–13″ (12.5–33cm), most thrushes have rather slender bills and "booted" tarsi with no obvious scaling. Species breeding at higher latitudes are migratory, and many tropical species make regular altitudinal or local migrations. In so large a family, foraging techniques and diets vary; many eat much fruit, whereas others are mainly insectivorous; many forage on the ground for earthworms and other invertebrates. Nests, usually built by the female with little help from her mate, are cuplike, of vegetable materials, in some species reinforced by a layer of hardened mud, situated amid the foliage of trees or shrubs, in holes in trees, birdhouses, crevices in rocks, on the ground, or in burrows underground. Members of this family lay 2–6 or, rarely, more eggs, which are blue, blue-green, or otherwise tinted, immaculate or heavily marked with shades of brown. Incubated chiefly or wholly by the female, they hatch in 12–14 or, rarely, 15 days. Blind and sparsely downy at birth, nestlings are fed by both parents but brooded only by their mother. At ages 12–16 days, the young leave the nest.

In some recent classifications, this family has been treated as a subfamily of the Muscicapidae (Old World Flycatchers and relatives). However, new biochemical evidence casts doubt upon this arrangement, and we prefer to maintain the Turdidae as a full family.

WHITE-THROATED ROBIN

Turdus assimilis Pl. 39(9)

Mirlo Gorgiblanco (Yigüirro Collarejo)

DESCRIPTION. 8¾″ (22cm); 72g. Yellow bill and eye-ring, white crescent on foreneck, and boldly striped throat diagnostic. **Adults:** top and sides of head blackish, tinged with olive; rest of upperparts dark grayish-olive; tail blackish; chin and throat white, streaked with black; foreneck white; breast, sides, and flanks pale grayish-olive to olive-brown, shading to white on belly and crissum. Bare eye-ring bright yellow; bill dull yellow, shading to dusky or horn-color at base of culmen; legs brownish yellow. **Immatures:** like adults but retain some juvenile wing-coverts. **Young:** above dark olive with buffy shaft-streaks on pileum and back; wing-coverts with buffy spots or margin at tips; throat, foreneck, and breast buff, heavily spotted with black; belly whitish, less spotted; bill dusky.

HABITS. Most often in moist and wet forests but enters neighboring bushy clearings, stands of tall wild cane, and riverside trees to nest, and visits fruiting trees in hedgerows, savannas, and second growth, especially outside the breeding season; forages for insects, earthworms, and other invertebrates on branches and on ground, where it may brush aside litter with bill to expose prey; eats many fruits; often forms flocks outside breeding season.

VOICE. Call a peculiar, often nasal or froglike *kek, nrrk* or *krrt*; alarm note deep, harsh, and guttural; in alarm or when going to roost in evening, a full, mellow, mournful whistled *peeyuu* or *whuueeet*. A superb songster with a large repertory of contrasting phrases, frequently delivered in doublets, the rich, melodious, robinlike caroling phrases often interspersed with thin, rather chaffy chatters and trills.

NEST. A bulky cup with a thick middle layer of mud mixed with vegetable fragments and a fibrous lining, usually well covered outside with green moss, 3.5–25ft (1–8m) up, often in a small, vine-draped tree. Eggs 2–3, dull white or pale blue, heavily mottled all over with reddish-brown. March–early June.

STATUS. Breeds in foothills and lower mountain slopes, commonly on Pacific slope and rarely on Caribbean side, mainly at 2600–6000ft (800–1850m), higher in Coto Brus region; formerly in upper Valle Central but now a rare straggler at best; following breeding descends to lower elevations; recorded between August and December at Golfito, Rincón, and other sites in Golfo Dulce region.

RANGE. N Mexico to W Ecuador.

NOTE. Sometimes considered conspecific with *T. albicollis*, White-necked Robin or Thrush, of N and E South America; also called White-throated Thrush.

CLAY-COLORED ROBIN

Turdus grayi Pl. 39(8)

Mirlo Pardo (Yigüirro)

DESCRIPTION. 9¼″ (23.5cm); 76g. Told from all other brownish *Turdus* by yellowish bill and reddish-brown iris. **Adults:** above, across chest, and on flanks olive-brown, shading to buffy-brown on belly and crissum; throat pale buff, streaked with olive-brown. Bill greenish-yellow, brighter and more yellow on tomia; legs brownish-gray. **Immatures:** like adults but retaining some juvenile wing-coverts. **Young:** above browner with smudgy darker scaling on back, buffy spots or fringes on tips of wing-coverts; below cinnamon-buff to ochraceous-buff, spotted with dusky on breast and sides; iris dull brown.

Habits. Frequents cultivated districts of all sorts, suburban lawns and gardens, pastures with scattered trees, second growth; eats earthworms, slugs, larval and adult insects, an occasional lizard, and fruits of many kinds; forages much on ground, pushing litter aside with bill; readily attracted to feeders; accompanies army ants; quite aggressive in immediate vicinity of nest, but evidently does not defend large territories; vigorously mobs Brown Jays; outside breeding season penetrates well into heavy forest, canebrakes, and dense second growth to forage but nests in more open surroundings; on alighting, habitually spreads and closes tail while flipping it up and down.

Voice. Call a throaty *tock*, variously inflected; a hoarse, sometimes nasal barking *pup pup pup pup* or *toc toc toc*, etc.; a low rattle; a querulous *keeyoooo* or *jerereee* when concerned for nest or young, also when going to roost; a thin, sibilant, lisping note. Song a long-continued caroling of varied phrases, mostly rich and melodious, containing slurred whistles, warbles, short trills, and now and then dry or piercing notes, with much variation among different males; heard rarely except before and during breeding, the song is said by local people to "call the rains" and has earned the *yigüirro* its status as Costa Rica's national bird.

Nest. A massive, broad bowl composed of coarse vegetable materials, with much mud in the middle layer, lined with coarse rootlets, bast fibers, rachises of compound leaves, 3–100ft (1–30m) up on a firm support well screened by foliage, sometimes in open shed, on windowsill, or in tree cavity. Eggs 2 or 3, rarely 4, pale to bright blue, thickly mottled all over with rufous and pale lilac. March–July; sometimes 2 broods.

Status. Resident countrywide, lowlands to 8000ft (2450m); generally abundant but less numerous in dry NW and at upper elevations.

Range. NE Mexico (casual in Texas) to N Colombia.

Note. Also called Garden, or Gray's, Thrush.

PALE-VENTED ROBIN

Turdus obsoletus Pl. 39(5)

Mirlo Vientriblanco (Yigüirro de Montaña)

Description. 9″ (23.5cm); 80g. No other large thrush has contrasting white lower belly and crissum. **Adults:** above brown, somewhat brighter and more russet on wings and tail; throat pale grayish-brown with inconspicuous darker streaking; breast and sides paler, more olive-brown, flanks darker; wing-linings ochraceous-buff. Bill black; legs grayish-brown. **Immatures:** like adults but retain some juvenile wing-coverts. **Young:** above somewhat darker brown with faint, paler shaft-streaks and darker fringes; wing-coverts with spots or fringes of buff at tips; throat pale grayish-brown; breast and sides buffy-brown, heavily spotted with darker brown; belly buffy-white with paler spotting.

Habits. Frequents middle and upper levels of humid forests and adjacent tall second growth; visits pasture trees and lower second growth especially outside breeding season; mainly arboreal but sometimes forages for invertebrates in ground litter; eats mainly fruits of palms and Lauraceae, arillate seeds, and berries of many sorts; pairs for breeding, but at other seasons forms wandering flocks of 10–30, moving between fruiting trees.

Voice. Calls include a thin, dry, twittering *bzeeek* and a throaty *wuk*; in alarm or when going to roost, a series of 2–4 querulous whistles, successively higher in pitch, *woeep-woeep-woeep-woeep*? Song a long-continued caroling, rich and somewhat throaty in tone, resembling that of Clay-colored Robin but somewhat more rapidly delivered and less varied, lacking trilled phrases.

Nest. A large, bulky bowl of fibers with some mud incorporated into the foundation, the exterior of green moss, the lining of fine, dark rootlets, 16–60ft (5–18m) up on branch, usually in or beside a clump of epiphytes, occasionally in crown of tree fern. Eggs 2–3, pale blue-green, spotted and blotched with pale reddish-brown, especially at large end. April–May.

Status. Common breeding resident at 2500–4000ft (750–1200m), occasionally to 5250ft (1600m) on Caribbean slope, N at least to Cerro Santa María; after breeding descends lower in foothills and into adjacent lowlands, regularly to 330ft (100m), occasionally to sea level.

Range. Costa Rica to W Ecuador.

Note. Often called Pale-vented Thrush. Sometimes classified as a subspecies of *T. fumigatus* of E South America.

MOUNTAIN ROBIN

Turdus plebejus Pl. 39(6)

Mirlo Montañero (Yigüirro de Montaña)

DESCRIPTION. 9½″ (24cm); 86g. Grayer than Clay-colored Robin, with black bill; grayer and more uniform than Pale-vented Robin. **Adults:** above dark, dull olive-brown, including wings and tail; below paler grayish-brown, palest on belly; throat paler, with inconspicuous darker streaking; wing-linings brighter buffy-brown; crissum darker, feathers broadly fringed with buff. Bill black; legs blackish. **Immatures:** like adults but with some juvenile wing-coverts retained. **Young:** above more olive, wing-coverts with terminal spots or fringes of buff; throat buffy-brown; breast heavily mottled or spotted with dark brown and buff; belly grayish-buff, less spotted with dark brown.

HABITS. Frequents tall, epiphyte-burdened mountain forests and adjacent clearings, pastures with scattered, moss-draped trees; outside breeding season, flocks also visit open groves of small, berry-laden trees and low second growth; eats arillate seeds and berries of many sorts; forages on large branches and on ground in forest for insects and other invertebrates; in cloudy and rainy weather, hops over ground in mountain pastures.

VOICE. In flight, a high, thin *seee* or *peeent*; a frequently uttered *whip whip* or *pip pip pip*, lighter than similar call of Clay-colored Robin; a deeper *tock tock tock* when fearful for its nest; a low, questioning *toc*. Song an almost endless, mechanical succession of weak, unmelodious notes with little variation in pitch, most unthrushlike: *chir chip chip cher chip chip cher cher tsup chip . . .* or *chirk cheep churry chirk chirk chip cher. . . .*

NEST. A roomy cup of green moss, with a middle layer of dry bamboo leaves and a thin lining of black rootlets, well concealed amid epiphytes 10–40ft (3–12m) up in isolated tree near forest or forest canopy. Eggs 2–3, blue-green. March–June.

STATUS. Breeds from timberline down locally to 4300ft (1300m) in mountains throughout; following breeding withdraws from higher elevations, down to 3000ft (900m) at height of wet season, when usually in flocks; returns to high elevations January–February.

RANGE. S Mexico to W Panama.

SOOTY ROBIN

Turdus nigrescens Pl. 39(7)

Mirlo Negruzco (Escarchado, Escarchero)

DESCRIPTION. 10″ (25.5cm); 96g. Large, dark, with contrasting orangish bill, legs, and eye-ring, and pale eye. **Adult ♂:** uniform brownish-black, with lores, orbital area, wings, and tail sooty-black. Iris pale gray; bill, bare eye-ring, and legs bright orange. ♀: similar but paler and browner overall, sometimes with faint darker streaking on throat. Bill, eye-ring, and legs yellowish-orange. **Immatures:** like adult female but retain some juvenile wing-coverts. **Young:** above sooty-brown, with blackish fringes and fine, buffy shaft-streaks; wing-coverts spotted or margined with buff at tips; below buff, streaked with blackish on throat and very heavily spotted with sooty-black elsewhere; flanks and crissum blackish, streaked with buff. Eye-ring, bill, and legs brownish-yellow.

HABITS. Frequents open areas at high elevations including páramo, bushy scrub around volcano cones, open bogs, pastures, and low second growth; enters oak forest mainly along edges and openings; forages for insects and spiders mainly on open ground, hopping about, poking into grass tussocks and cushions of low herbs, turning fallen leaves with bill; eats many berries, especially Ericaceae, *Solanum*, and melastomes, blackberries, and arillate seeds; pairs for nesting but usually seen singly or in loose groups at good foraging sites.

VOICE. Call a distinctive, low, grating, harsh *grrrrk* or *grrek*, often repeated; a thin, rolling *prrreee*. Song consists of short, gurgling, buzzy or squeaky phrases, each repeated 3–6 times, then after a pause of several seconds, a new phrase is repeated: *chuweek chuweek chuweek, seechrrrzit seechrrrzit seechrrzit, tseeur tseeur tseeur tseeur*.

NEST. A very bulky cup of twigs, rootlets, lichens, and mosses with a tightly woven lining of soft, fine grasses, 7–26ft (2–8m) up in tall shrub or tree. Eggs 2, immaculate greenish-blue. March–May.

STATUS. Common to abundant resident of high elevations of Cordillera Central and Cordillera de Talamanca, above ca. 8200ft (2500m), occasionally down to 7000ft (2150m) mainly after the breeding season.

RANGE. Costa Rica and W Panama.

BLACK-FACED SOLITAIRE

Myadestes melanops Pl. 39(13)

Solitario Carinegro (Jilguero)

DESCRIPTION. 6¾″ (17cm); 33g. Short, broad orange bill contrasts with black face; legs shorter than those of other thrushes. **Adults:** above mostly slate-gray, below paler; forehead, face, and chin black; wings and tail black with slaty edgings, except in a black patch on edges of inner secondaries. Bill and legs bright orange. **Young:** above sooty-gray to grayish-olive, spotted with buff, each spot set off by sooty fringe; wing-coverts tipped with buff; below dull gray, heavily and indistinctly spotted with buff, becoming whitish, mottled with sooty and buffy on belly. Culmen dusky, rest of bill and legs paler orange.

Habits. Lurks in dense understory of shrubs and bamboos of wet mountain forests, frequently ascending well into canopy to forage or sing; persists in wooded, brush-choked ravines amid pastures and cultivation; eats many fruits, in quest of which it may venture forth into trees in pastures or semi-open, or young second growth; plucks berries and arillate seeds from perch or sometimes with a hovering sally; pairs for breeding; may form loose flocks in nonbreeding season, when many move to lower elevations.

Voice. Call an ascending, nasal *ghank* or a liquid *quirt*; alarm note a drier, more buzzy *shweee*. Song consists of high, thin, clear whistles, drawn-out and leisurely, with fluty transitions and liquid undulations, occasional metallic notes with overtones, sounding exquisitely beautiful in its natural surroundings but often harsh and jarring in urban surroundings, for the *jilguero* is prized as a cage bird for its song.

Nest. A scanty to bulky cup of green mosses and foliaceous liverworts, lined with dark fibrous rootlets and stems of mosses and liverworts, in a niche in a mossy bank, in a crevice or dark fissure in a trunk, or in a large tuft or ball of moss in a vertical fork, 4–16ft (1–5m) up. Eggs 2–3, white to pale pinkish, spotted with bright rufous-brown, especially on large end, sometimes in a wreath. April–June.

Status. Breeds from 3000ft to at least 9000ft (900–2750m) the length of both slopes; following breeding many move lower, regularly to 1500ft (450m), occasionally to 330ft (100m), at least on Caribbean slope. Common in more inaccessible or protected areas, but persecution for the cage-bird trade has greatly lowered its numbers in many regions.

Range. Costa Rica and W Panama.

Note. Sometimes considered a race of the Andean Solitaire, *M. ralloides*. A few authors have recently placed the genus *Myadestes* in the Bombycillidae.

WOOD THRUSH

Hylocichla mustelina Pl. 39(4)

Zorzal del Bosque

Description. 7¼″ (18.5cm); 48g. White underparts more boldly and extensively spotted than other migrant thrushes; eye-ring white; head rufescent. **Adults:** pileum cinnamon-rufous to russet, shading to cinnamon-brown on back and wings, and olive-brown on rump, tail-coverts, and tail; sides of head streaked with buffy-white and dusky; below white, tinged with buff on breast, heavily spotted with black on sides of throat, breast, and sides. Upper mandible black, lower pale horn with dusky tip; legs pale flesh. **Immatures:** like adults but some juvenile wing-coverts with buffy spot or fringe at tip.

Habits. Frequents forest undergrowth, low moist thickets, banana plantations, old cacao plantations; hops on ground and logs, seeking insects, spiders, earthworms, and other invertebrates in litter, sometimes turning leaves with bill; eats much fruit, sometimes visiting low second growth for berries of *Phytolacca*; usually alone and possibly territorial in winter, sometimes in loose groups during migration.

Voice. Calls include a sharp *pit-pit-pit* and a more liquid *quirt-quirt-quirt*, often heard in response to disturbance or at dusk; a nasal, rattling *trrrrrr, trrrr*. Song, sometimes given at dawn prior to spring departure, consists of phrases of 3–5 fluty notes, often ending in a trill.

Status. Uncommon to fairly common fall migrant (late September–mid-November), uncommon but widely distributed winter resident from lowlands of both slopes (rare in dry NW and Valle del General) to 4600ft (1400m) on Caribbean slope and 5600ft (1700m) on S Pacific slope, occasionally higher; spring migration March–April, scarcely noticeable except along Caribbean coast.

Range. Breeds in S Canada, E and C USA; winters from S Texas to NE Colombia.

SWAINSON'S THRUSH

Catharus ustulatus Pl. 39(2)

Zorzal de Swainson

Description. 6¼″ (16cm); 28g. Distinguished from other migrant *Catharus* by its conspicuous buffy spectacles; Wood Thrush larger, above much brighter and below more extensively spotted. **Adults:** above olive-brown to grayish-olive, including wings and tail; sides of head brown with fine, buffy shaft-streaks on auriculars, buffy mottling on cheeks; throat and breast buff, with a blackish malar stripe and heavy black spotting on lower throat and breast; posterior underparts white, tinged with olive-brown on sides and flanks; upper belly more or less spotted with olive-brown. Bill black except base of lower mandible pale horn to flesh-color; legs pale brownish-flesh. **Immatures:** like adults but some greater wing-coverts tipped or fringed with pale buff, or with buffy shaft-streaks.

HABITS. Prefers lower levels of humid forests and thickets, semi-open, and old second growth, but in migration apt to occur almost anywhere, including urban areas and open country; eats many fruits and arillate seeds, relatively few insects and invertebrates, for which it may forage on outskirts of army ant swarms; solitary or, during migratory waves, in loose flocks that seem to fill the woods.

VOICE. Call a sharp, clear, upwardly inflected *pwick* or *quirt*; during migration a piercing *kweet* or *fweep*, often heard from birds passing overhead at night. Sings freely on spring migration, a flutelike series of rather hurried-sounding, upward-spiraling trills, *wreeya-wreeeya-weeeya-wee*; a large migratory wave may contribute more song than residents.

STATUS. Abundant fall migrant (mid-September–November) in Caribbean lowlands and central highlands, locally to ca. 8000ft (2450m); in Pacific lowlands uncommon to rare in fall, and not before October; in spring (April–late May) an abundant migrant along foothills and in adjacent lowlands of both slopes, including central highlands, occasionally to ca. 9000ft (2750m), and relatively rare in coastal lowlands; rare winter resident of both slopes from sea level to ca. 5000ft (1500m).

RANGE. Breeds in Alaska, Canada, N and W USA; winters mainly in South America E of Andes, in small numbers to N Mexico.

NOTE. Formerly classified in genus *Hylocichla*; also called Olive-backed or Russet-backed Thrush. Occasional individuals have narrow, inconspicuous spectacles.

GRAY-CHEEKED THRUSH

Catharus minimus Pl. 39(1)

Zorzal Carigrís

DESCRIPTION. 6½″ (16.5cm); 31g. Very similar to Swainson's Thrush but with at most a faint, narrow eye-ring; sides of head more conspicuously streaked. **Adults:** above olive or grayish-olive, including wings and tail; sides of head gray, with whitish shaft-streaks; eye-ring narrow, pale, dull buff, or lacking; throat buffy-white, with a blackish malar streak; breast pale buff, heavily spotted with black; posterior underparts white; flanks washed with olive-brown; sides and sometimes upper belly spotted with grayish-olive. Bill black, except base of lower mandible flesh-color to pale horn; legs brownish-flesh. **Immatures:** like adults but some wing-coverts edged or spotted with buff at tips.

HABITS. Favors streamside thickets, open woodland, taller second growth, old cacao plantations, but in migration may turn up almost anywhere, often in small numbers among larger waves of Swainson's Thrushes of similar habits.

VOICE. Call a scratchy, thin *pseer* or *peew*; in migration gives a piercing *pweep* very similar to the Swainson's "migration call."

STATUS. Very uncommon fall migrant (early October–mid-November), mainly on Caribbean slope and to ca. 5000ft (1500m) in central highlands; casual winter resident in lowlands of both slopes; apparently unrecorded in spring; its date of departure unknown.

RANGE. Breeds in Alaska, N and E Canada, and extreme NE USA winters mainly in South America E of Andes, casually to Costa Rica.

NOTE. Formerly placed in genus *Hylocichla*.

VEERY

Catharus fuscescens Pl. 39(3)

Zorzal Dorsirrojizo

DESCRIPTION. 6½″ (16.5cm); 32g. Brighter brown above than Swainson's or Gray-cheeked thrushes, with less heavily spotted breast, no eye-ring. **Adults:** above tawny-brown to dull rufous-brown; sides of head brown with buffy shaft-streaks; center of throat and chin whitish; sides buffy, streaked with brown; chest buff, spotted with dusky-brown; belly white, sometimes with faint grayish-olive spotting posteriorly; flanks and sides washed with olive. Bill blackish except base of lower mandible pale horn; legs pale horn. **Immatures:** similar, but some wing-coverts edged with buff at tip.

HABITS. In second growth and forest, often in company of other migrant *Catharus*; forages on ground, often near water, hopping and gleaning in leaf litter; takes many fruits (*Cordia, Urera, Miconia*), during migration.

VOICE. Call a sharp *pwit* or *pwik*.

STATUS. Uncommon to sporadically common fall transient (late September–late October) along Caribbean coast; rare inland but fairly regular in small numbers in Valle Central; rare spring transient (March–April) mainly along Caribbean coast; no winter record.

RANGE. Breeds in S Canada, E and C USA; winters from N Colombia to Guyana and C Brazil.

BLACK-HEADED NIGHTINGALE-THRUSH
Catharus mexicanus Pl. 38(25)
Zorzal Cabecinegro

DESCRIPTION. 6¼″ (16cm); 30g. Darker above than Orange-billed Nightingale-Thrush, especially on head; Slaty-backed Nightingale-Thrush has pale eye, darker breast. **Adult ♂:** pileum and hindneck black; rest of upperparts dark brownish-olive; sides of head dark gray, shading to dull white, and on throat indistinctly smudged with gray; breast dull gray, tinged with olive laterally, shading to white, sometimes tinged with pale yellow on belly; sides and flanks gray, tinged with olive. Narrow eye-ring and bill bright orange, dusky at base of culmen; legs paler orange. ♀: similar but pileum brownish-black; culmen extensively dusky. **Young:** above dark sooty-brown, faintly streaked with dull buff; lesser and middle coverts with buffy spot at tip; greater coverts with paler fringes; throat, sides, and flanks buff, with sooty smudging and scaling; breast dull buff, heavily scaled with sooty-brown; belly white.

HABITS. Dwells in dark understory of primary forest, foraging on or near ground, progressing by quick, springy hops and sudden stops, often pecking through leaf litter or moss without scratching; takes beetles, caterpillars, and other insects; eats much fruit (*Psychotria, Cephaelis*). Males sing 3–10ft (1–3m) up in shrubs, mostly at dawn and in afternoon and evening.

VOICE. Call a querulous, ascending *ghwwww?* or *dzeeeet?*, with rather buzzy quality; a hard, dry rattle when agitated. Song composed of thin, flutelike phrases of 3–8 notes, including clear whistles and trilled, buzzy, or harsh notes; a phrase often repeated several times in succession, as in Orange-billed Nightingale-Thrush, but song much more melodious. Song of Slaty-backed is simpler, without trilled phrases.

NEST. A substantial cup of mosses and rootlets, lined with leaf skeletons, decorated outside with green moss, 3–10ft (1–3m) up in fork of understory shrub. Eggs 2, pinkish-white, speckled with reddish-brown. March–July.

STATUS. Common resident of foothill belt at 1000–2600ft (300–800m), locally to 3300ft (1000m) the length of Caribbean slope; on Pacific slope of N cordilleras between 2300–4300ft (700–1300m), locally to 5000ft (1500m) in somewhat drier forests than on Caribbean slope; evidently formerly widespread in Valle Central.

RANGE. E and S Mexico to W Panama.

SLATY-BACKED NIGHTINGALE-THRUSH
Catharus fuscater Pl. 38(23)
Zorzal Sombrío (Arremendado, Jilguerillo)

DESCRIPTION. 6¾″ (17cm); 35g. Darkest nightingale-thrush, with diagnostic pale iris. **Adults:** above dark gray, shading to slate-black on head, wings, and tail; throat, breast, sides, flanks, and crissum slate-gray; center of lower breast and belly white. Iris pale gray; eye-ring and bill orange except culmen tinged with dusky (more strongly in female); legs paler orange. **Immatures:** like adults but with brownish tinge on wings, and sometimes on throat and breast. **Young:** above sooty-black, tinged with brown anteriorly and on wings, throat dull buff, smudged with blackish; foreneck and breast sooty-black, indistinctly streaked and spotted with buffy-brown; sides and flanks dark sooty-gray; belly paler gray, washed with buffy-brown anteriorly; most of bill blackish.

HABITS. Mostly on or near ground in wet mountain forests, ascending to eye level or higher for fruits or, occasionally, to sing; hops bouncily, stops and stands pertly, nervously flicking wings; forages by gleaning in leaf litter or low vegetation for insects, spiders, and sowbugs; takes many berries of Rubiaceae, Gesneriaceae, etc. Interspecifically territorial against other forest nightingale-thrushes where their altitudinal ranges overlap.

VOICE. Usual call a nasal, mewing, ascending *maaaaah?*; a clear, high, whistled *poeeeee* in alarm or agitation. Song of clear, flutelike whistles with phrases of 2 and 3 notes often alternating: *peee leee, peee-o-lay* or *one two, three-to-two* (Slud); resembles song of solitaire but simpler, lower in pitch, notes shorter, no harsh notes intermixed.

NEST. Substantial cup of rootlets and plant fibers, lined with black *Marasmius* rhizomorphs, heavily decorated on outside with green moss, 20″–10ft (0.5–3m) up in fork of understory shrub or small tree. Eggs 2, blue-green, speckled with dark brown. April–August.

STATUS. Common resident 2600–7500ft (800–2300m) on Caribbean slope, from Panama N at least to Cerro Santa María, and locally on Pacific slope at somewhat higher elevations, mainly on Cordillera de Tilarán and Dota region.

Range. Costa Rica to NW Venezuela, N Colombia, and Bolivia.

RUDDY-CAPPED NIGHTINGALE-THRUSH

Catharus frantzii Pl. 38(27)

Zorzal Gorrirojizo

Description. 6¼″ (16cm); 28g. Similar to Orange-billed Nightingale-Thrush but lacks orange eye-ring; bill bicolored; crown usually distinctly brighter than back. **Adults:** pileum and hindneck bright russet, shading to olive-brown, rest of upperparts somewhat tinged with russet; sides of head olive-gray; throat grayish-white, indistinctly streaked or smudged with grayish; breast, sides, and flanks gray, washed with olive, shading to white on center of belly and crissum. Upper mandible black, lower orange; legs brownish. **Young:** upperparts russet-olive; face sooty-gray, shading to paler gray on center of throat; breast, sides, and flanks olive, smudged with darker olive-brown, passing to white on belly; upper belly indistinctly spotted or smudged with olive-brown; legs pale flesh.

Habits. Lurks in undergrowth of wet mountain forests and neighboring thickets, including ravines choked with bamboo and tall second growth in otherwise open areas; searches for insects and other invertebrates on or near the ground, sometimes hopping out into roads or pastures with tall adjoining vegetation; varies diet with berries; solitary or in pairs.

Voice. Call a whistled, ascending *whoeeet* or, in alarm or agitation, a slightly burry *correeee*. Song beautiful, varied, but subdued, consisting of a wide variety of phrases that include flutelike trills, clear whistles, caroling warbles; successive phrases are different, each male seeming to possess an extensive repertory; usually in evening or in dim, misty weather, from a low, inconspicuous perch.

Nest. A bulky cup of green mosses and liverworts mixed with weed stems, pieces of vines and straw, lined with fine fibrous rootlets and leaf skeletons, 3–13ft (1–4m) up in forest undergrowth, tangled thicket, or shrubs. Eggs 2, pale grayish or greenish-blue, mottled all over with brown or cinnamon, most heavily on thick end. April–June.

Status. Fairly common resident in mountains the length of both slopes, above ca. 4500ft (1350m) in N cordilleras, and ca. 5000–8200ft (1500–2500m) on Cordilleras Central and de Talamanca.

Range. C Mexico to W Panama.

ORANGE-BILLED NIGHTINGALE-THRUSH

Catharus aurantiirostris Pl. 38(26)

Zorzal Piquianaranjado (Jilguerillo, Inglesito)

Description. 6¼″ (16cm); 27g. Resembles Ruddy-capped Nightingale-Thrush but has orange eye-ring and upper mandible; above brightest on wings and tail, not crown. **Adults:** upperparts bright brownish-olive, shading to russet on wings, dull russet on tail; sides of head grayish-olive (or, on S Pacific slope, pileum and hindneck slate-gray, sides of head paler gray); throat whitish to pale gray; breast, sides and flanks gray, more or less tinged with olive; belly and crissum white. Bare eye-ring and most of bill bright orange; base (males) or most (females) of culmen dusky; legs paler orange. **Young:** above duller olive-brown, indistinctly spotted or streaked with buff; face sooty; throat dull olive-buff to whitish, smudged with dusky; breast dull sooty-olive, indistinctly spotted with buff; sides buffy-white with smudgy dark scaling; belly dull white; bill mostly blackish.

Habits. Frequents thickety second growth, forest edge, plantations and gardens in humid regions, and understory of forest; forages mostly on ground for insects, spiders, and other invertebrates, ascends to medium heights for berries and arillate seeds; upon alighting often cocks and slowly lowers tail 1 or more times, often with a flick of the wings; alone or in pairs; usually shy and retiring; aggressive toward migrant *Catharus*.

Voice. Call a sharp *mew*, reminiscent of Gray Catbird but sometimes prolonged into a nasal chatter. Song consists of short phrases of 3–6 notes that include chips, squeaks, warbles, and trills but too high-pitched, thin, hurried, and jumbled to be melodious; typically repeats a phrase many times before changing to another.

Nest. A bulky, thick-walled cup of coarse, dry materials mixed with green moss, lined with fine tendrils, rootlets, grass stems, or inflorescences, 3½″–10ft (9cm–3m) up in thicket, coffee bush, maize plant, fruit tree, or hedge, never far from dense cover. Eggs 2, rarely 3, blue or pale blue, speckled and blotched all over with brown and pale lilac. March–August.

Status. Common resident on Pacific slope of Cordilleras Central and de Tilarán, including Valle Central, at ca. 2000–7500ft (600–2300m), down Reventazón drainage at least to Turrialba; recorded from Carrillo, possibly a stray; an isolated population above 2000ft (600m) in hills of Península de Nicoya; in General–Térraba–Coto Brus region of S Pacific slope at ca. 1300–5000ft (400–1500m).

Range. S Mexico and Trinidad to Colombia and Venezuela.

Note. The gray-headed races from SW Costa Rica southward have sometimes been considered a separate species, *C. griseiceps*.

BLACK-BILLED NIGHTINGALE-THRUSH

Catharus gracilirostris Pl. 38(24)

Zorzal Piquinegro (Cuitiento)

Description. 5¾″ (14.5cm); 21g. Smaller than other nightingale-thrushes, at higher elevations; the only one without orange on bill. **Adults:** forecrown and face slate-gray, shading to bright olive-brown on rest of upperparts; below pale gray, fading to white on center of belly, with a broad band of olive across breast; flanks olive-brown; crissum paler. Bill black; legs brownish. **Young:** above duller, more olive; wing-coverts with pale fringes; forecrown, face, and throat sooty-gray, tinged with olive; rest of underparts duller and darker than adult; breast and sometimes upper belly with sooty smudging.

Habits. Frequents cold, wet highland oak forests, second growth, pastures, clearings, and patches of denser shrubbery in páramo; forages on ground or on large horizontal or inclined branches to well up in canopy; progresses by springy hops, sudden stops, or short dashes after insects or spiders; probes tufts of mosses and lichens on limbs; turns fallen leaves with bill; eats many berries; often perches on stump or low branch; quickly raises tail, often with a flick of the wings, then lowers it more slowly; alone or in pairs.

Voice. Call a high, thin, descending *pseeeew* or a penetrating, ascending *seeet*; a short, nasal chatter. Song of 1–3 clear whistles, then a tinkling, sometimes jumbled trill that may rise and fall as it fades away, often sounding blurred or "fuzzy"; successive phrases different and usually separated by a pause of several seconds.

Nest. A bulky cup of green moss and rootlets with a finely woven lining of grass stems and fine rootlets, 3–16ft (1–5m) up in dense shrub or small tree. Eggs 2, greenish-blue, speckled with reddish-brown or deep chestnut, most heavily on thicker end. March–June.

Status. Common to abundant resident on Cordillera Central and Cordillera de Talamanca from ca. 7000ft (2150m) on Caribbean slope and 8200ft (2500m) on Pacific, up to timberline, and uncommonly in páramo up to at least 11,500ft (3500m).

Range. Costa Rica and W Panama.

FAMILY **Sylviidae:** Gnatcatchers, Gnatwrens, and Old World Warblers

Ornithologists do not concur on the classification of gnatcatchers and gnatwrens, although most agree that they are closely related to the Old World warblers. This family of nearly 300 species is spread widely over the Eastern Hemisphere, with only 15 species of gnatcatchers, gnatwrens, and kinglets in the West. Rarely as long as 7″ (18cm), members of this family are slender-billed birds that flit through woodlands, thickets, savannas, and marshes, restlessly seeking small insects and spiders. Most are plainly attired in olives, browns, grays, black, and white, with rarely more than a touch of bright color. The sexes may be alike or different. Conforming to this general pattern, our few tropical American species are very small, active birds that glean insects amid foliage, gnatcatchers mostly in the trees, gnatwrens at lower levels of woodlands. Their songs are clear, slow trills. In contrast to the great variety of covered, pouchlike, pendant, or leaf-enclosed as well as open nests built by Old World warblers, the tropical American species build cuplike nests, high in trees or near the ground. Both sexes construct the nest and incubate the 2–3 white, spotted eggs, which hatch in 13–17 days, longest in the gnatwrens. Both parents feed and brood the young, which hatch naked and leave the nest when 12–15 days old.

WHITE-LORED GNATCATCHER

Polioptila albiloris Pl. 41(1)

Perlita Cabecinegra

Description. 4¼″ (11cm); 7g. Small, slender, with long bicolored tail. **Adult ♂:** cap glossy black, extending to lores and eye during breeding; between about September and January, lores and narrow line over eye white; rest of upperparts pearly blue-gray; remiges black, tertials edged with white; central rectrices black, outer ones white; below white, washed with pale blue-gray across chest. ♀: cap slightly darker gray than back, head pattern as on nonbreeding male. Bill black except base of lower mandible blue-gray; legs plumbeous. **Young:** like adult female but upperparts tinged with brownish.

HABITS. Prefers open, scrubby woodland and second growth; along edges but not inside taller deciduous or evergreen forest; usually in pairs, hopping and flitting in foliage, actively gleaning beetles, bugs, small moths, caterpillars, leafhoppers, other insects, and spiders, or pursuing them with short, fluttering sallies; nervous, excitable, continually wagging and cocking long tail.

VOICE. Usual call a scratchy, dry or sibilant *seeah* or *pseek*; scold a longer, drier *mew* than that of Tropical Gnatcatcher; song a sputtering series of scratchy notes and chips.

NEST. A neat, dainty cup of fine fibers and downs, bound with spiderweb and decorated on outside with bits of lichen and bark, 3–10ft (1–3m) up in small tree or shrub, often bull's-horn acacia. Eggs 2–3, white, with reddish speckling, especially toward large end. April–June or July.

STATUS. Resident in NW lowlands, locally to 2500ft (750m); common from Nicaraguan border S to lower Tempisque basin, increasingly rare S on Península de Nicoya and to mouth of Golfo de Nicoya; toward S limit of range virtually restricted to mangroves and adjacent scrub and second growth.

RANGE. S Mexico to Costa Rica.

NOTE. Formerly sometimes considered conspecific with Tropical Gnatcatcher (*P. plumbea*).

TROPICAL GNATCATCHER

Polioptila plumbea Pl. 41(2)

Perlita Tropical

DESCRIPTION. 4″ (10cm); 6.5g. Small, with long, slender bill and conspicuously white-edged tail; told from White-lored Gnatcatcher by its white lores and broad white superciliaries. **Adult ♂:** pileum, postocular stripe, and sides of neck black; back and rump bluish-gray; remiges black, edged with gray, tertials edged with whitish; central rectrices black, outer ones white; lores, superciliaries and cheeks white, shading to grayish-white on most of underparts; center of belly white. ♀: similar except black on head replaced by dark gray. Bill black, except base of lower mandible whitish; legs plumbeous. **Immature ♂:** like adult female but superciliaries shorter. **Young:** above gray with some brown feather-edgings; superciliaries indistinct; cheeks washed with brown; often an interrupted white eye-ring; remiges with duller gray edges; below whiter than adult.

HABITS. Flits airily about the canopy of forest and tall second growth, descending freely to shrub level at edges; visits trees in clearings or hedgerows, but avoids dark undergrowth; gleans spiders and their eggs, caterpillars, beetles, leafhoppers, bugs, and other insects from outer twigs and foliage, often hovering; paired throughout year; often accompanies mixed-species flocks of greenlets, honeycreepers, etc. in forest canopy; habitually cocks and flits tail jauntily.

VOICE. Calls include a thin, buzzy *zeet-zeet* and a fine, thin, nasal *chaaa*; scolds with a series of scratchy whines or mews, *seeur seeur seeur*. Song a simple trill of high, thin whistles that accelerates and falls in pitch toward end, *swee see see si si si si si si siu su su su*; may become slower toward end; higher, thinner and more prolonged than song of gnatwrens.

NEST. A dainty cup resembling that of a hummingbird, of soft vegetable fibers, with bits of lichen, moss, or liverwort fastened to the outside by spiderweb, softly lined with fine fibers and seed down, usually saddled on a branch or in a fork 6–25ft (2–8m) up. Eggs 2 or 3, white, finely speckled with brown. March–June.

STATUS. Common resident of lowlands of both slopes, except scarce or absent in N Guanacaste and on most of Península de Nicoya; sea level to 5000ft (1500m); rare in Valle Central and on adjacent slopes.

RANGE. S Mexico to W Peru and Amazonian Brazil.

LONG-BILLED GNATWREN

Ramphocaenus melanurus Pl. 32(15)

Soterillo Picudo

DESCRIPTION. 4¾″ (12cm); 10g. Extremely long, slender bill with small terminal hook; graduated, white-tipped tail; striped throat; rusty face. **Adults:** pileum and hindneck ochraceous olive-brown; back and wings dark grayish-olive; tail blackish-brown, the outer 3–4 pairs of rectrices white-tipped; sides of head and neck bright cinnamon; throat white, striped with black; rest of underparts buff, fading to nearly white on center of belly. Iris pale brown; bill dark horn with pinkish tomia; legs gray. **Immatures:** like adults but retain juvenile wings; above more uniform brown, below paler buff, throat streaked less heavily. **Young:** pileum and nape buffy-brown; rest of upperparts dull brown, with much less contrast than in adult; face duller, more tawny; throat buff, smudged with dusky and white; breast grayish-brown, fading to whitish on belly; wings with olive-brown edgings.

Habits. Haunts dense, tangled vegetation along edges and breaks in forest, in tall second growth, or overgrown plantations, usually low but sometimes rising to 30ft (9m) in vine tangles; hops, flits, and climbs wrenlike as it searches for small insects and spiders, snatching fast-moving prey with tip of its long bill; wags its long tail loosely up and down; paired throughout year; does not associate with mixed flocks.

Voice. Calls include surprisingly loud, clear, staccato whistles; a sharp, wrenlike ticking when perturbed, and a low, dry *churr* that may be broken into a rather deep, chattering *chrchrchrchr* . . . in agonistic interactions. Song a prolonged trill, either all on same pitch or rising, then falling, varying in tone from clear and melodious to dry and harsh.

Nest. A deep, compact cup of dry grasses, leaves, and fibrous bark, sometimes covered by green moss, lined with fine fibers, 6–13″ (15–33cm) up amid low, dense vegetation. Eggs 2, white, finely spotted with cinnamon. April–June.

Status. Common resident countrywide from lowlands to 4000ft (1200m); in dry NW, somewhat more local, favoring moist sites with evergreen vegetation.

Range. S Mexico to NE Peru, C and SE Brazil.

Note. Middle American populations, S to Colombia, were formerly classified as a separate species, *R. rufiventris*, from South American *R. melanurus*, Straight-billed Gnatwren.

TAWNY-FACED GNATWREN
Microbates cinereiventris Pl. 32(14)
Soterillo Caricafé

Description. 4″ (10cm); 11g. Bill long, but shorter and broader than that of Long-billed Gnatwren; tail short; white throat and tawny face. **Adults:** upperparts, including wings and tail, dark brown, tinged with olive on back; sides of head and neck bright tawny or ochraceous, separated by a broad black streak from white of chin and throat; foreneck streaked with black and white; breast and belly dark slate, shading to dark olive on sides and flanks. Upper mandible black, lower pale horn with dusky tip; legs gray. **Young:** above darker, more sooty-brown; sides of head dull buffy-brown; throat pale grayish-olive, shading to dark dull olive-brown on breast, sides, and flanks; center of belly dark grayish-olive.

Habits. Moves restlessly through dark undergrowth of humid forest, hopping among low stems, over logs and ground, gleaning and probing in ground litter and foliage for ants, other insects, and spiders; in pairs or small family groups, often accompanying mixed flocks of antwrens, antvireos, etc.

Voice. Calls include a nasal, burry, angry-sounding *nyaaar* or *meeyaar*; a harsh dry chatter like that of a wren or ant-tanager; in interactions or when agitated, a high, penetrating, whistled *peeee*. Song a deliberate series of lisping, scratchy, slurred notes mixed with clear whistles and harsh chatters: *cher-chwa seee chew cher ch' ch' ch' ch' tweet tsur tsur cher*. . . .

Nest. A cup of green moss, papery bark, petioles and bits of dead leaves, held together by vegetable fibers, lined with leaf skeletons and fine fibers, attached to vertical branches of a shrub, 20″ (50cm) up in forest. Eggs 2, white, spotted all over with reddish and dark brown. April–May.

Status. Fairly common resident on Caribbean slope, from lowlands to 3000ft (900m) in N, 4000ft (1200m) in SE.

Range. SE Nicaragua to W Ecuador and SE Peru.

Note. Also called Half-collared Gnatwren.

FAMILY **Bombycillidae:** Waxwings

This family is usually limited to the single genus and 3 species of waxwings but is sometimes expanded to include the 4 species of silky-flycatchers and the Hypocolius of southwestern Asia, making a family of 8. Waxwings breed in North Temperate and subarctic regions. In length 6–8″ (15–20cm), they are crested birds clad in softly blended browns and grays, with tails bordered terminally with yellow or (in the Japanese Waxwing) red. They bear, at the tips of their secondary feathers, corpuscles that resemble drops of red sealing wax. The sexes are alike. With short, rather thick bills, they eat great quantities of fruits and catch insects in the air. They lack advertising song but warble softly while courting and nesting. In a cuplike nest built by both sexes in a tree or shrub, the female lays 3–6 pale blue or blue-gray eggs, spotted with black, and incubates them alone while she is fed generously by her mate. After a period of 12–13 days the eggs hatch, and the male helps to feed the nestlings, who remain in the nest for about 16 days. After the breeding season, waxwings gather in flocks that rest in treetops and wander irregularly in search of food. In some years the Cedar Waxwing comes as far south as Costa Rica and Panama.

CEDAR WAXWING
Bombycilla cedrorum Pl. 39(14)
Ampelis Americano

DESCRIPTION. 6½″ (16.5cm); 32g. Sleek, long-winged, rather short-tailed, with a pointed, swept-back crest, a short, broad bill, and short legs. **Adults:** head and breast buffy-brown, shading to pale yellow on belly, grayish-brown on back, and slate-gray on wings, rump and tail; tail tipped with bright yellow; crissum white; narrow mask black, outlined above by white and separated from blackish chin by sharp white malar streak. Bill and feet black. **Immatures:** black mask small, mostly confined to lores; below streaked with dull white and olive-brown; remnants of this plumage sometimes visible as late as January. In flight the pointed wings and short, yellow-tipped tail are distinctive.

HABITS. Nearly always found in compact nomadic flocks of 5–20 in the highlands, visiting a variety of fruiting trees, shrubs and epiphytes, mostly in semi-open, second growth, and forest edge; typically perches in bare treetops, often making kingbirdlike fly-catching sallies. Flocks may perch quietly for some time, then abruptly, with a burst of calling, fly rapidly and directly to another tree.

VOICE. A sibilant, almost hissing, high-pitched *seeeee*, not loud but penetrating; a chorus of these notes from a flock passing overhead is immediately recognizable.

STATUS. Sporadic winter visitor (December through April, in some years not departing before mid-May) chiefly at 1650–6500ft (500–2000m), in central highlands, or along N cordilleras; occasionally singly or in small groups in NW lowlands; in numbers only every 3 or 4 years, particularly from Valle Central S.

RANGE. Breeds from S Alaska, N and E Canada to S USA; winters from S Canada to C Panama, casually to N South America.

FAMILY **Ptilogonatidae:** Silky-flycatchers

Confined to Central America, Mexico and the southwestern United States, this small family has only 3 genera with 4 species, 2 of which live in Costa Rica. About the size of thrushes, these birds have short, rather broad bills, relatively short wings, and short legs, with plumage of gray, brown, olive, and black, some with touches of yellow and white. All but the Black-and-Yellow Silky-flycatcher are crested. The sexes differ in coloration, greatly in the northern Phainopepla, less in the Central American species. Their food consists of insects, usually caught in the air, and many berries. Although they call freely, they are poorly endowed with song. Their cup-shaped nests in trees and shrubs, often high, are built by both sexes of the Long-tailed Silky-flycatcher but chiefly or exclusively by the male Phainopepla. Their 2 or 3 eggs (rarely 4 in the Phainopepla) are gray or grayish-white, heavily marked with brown or lilac. Both sexes of the Phainopepla incubate, but only the female Long-tailed Silky-flycatcher does so. Hatched with abundant white down, the young are fed by both parents and remain in the nest for 18–25 days—surprisingly long for an open-nesting passerine bird.

LONG-TAILED SILKY-FLYCATCHER
Ptilogonys caudatus Pl. 39(15)
Capulinero Colilargo (Timbre, Pitorreal)

DESCRIPTION. ♂ 9½″ (24cm); ♀ 8¼″ (21cm); 37g. Slender, with pointed crest and graduated tail, the central rectrices slender and projecting, especially in male; bill and legs short. **Adult ♂:** forehead and forecrown pale gray; eye-ring bright yellow; rest of head, neck, and chest olive-yellow; breast, upper belly, back, rump, upper tail-coverts, and wing-coverts bluish-gray; remiges black, edged with gray; tail black, the outer 4 rectrices with large white spots on inner webs; flanks olive-green, center of belly paler; crissum yellow. ♀: forehead and forecrown darker gray; rest of head and body olive green, brightest on flanks; center of belly whitish; tail shorter and duller black. Bill and legs black. **Immatures:** like adults but with shorter central rectrices, white markings on tail indistinct. **Young:** above brownish-olive with paler fringes; crest rudimentary; wing-coverts and secondaries edged with reddish-brown to bright olive; below ochraceous-olive, shading to buffy-yellow on belly, brighter yellow on crissum; eye-ring whitish; tail short.

Habits. Pairs for breeding but during most of year ranges widely in straggling flocks through forested and partly cutover areas of high mountains; flight high and undulating; perches on topmost exposed twigs of tall trees; sallies to capture flying insects, often in prolonged, intricate aerial maneuvers; plucks many berries while perched, especially those of mistletoes and *Solanum*, for which it often descends to shrub level in clearings and second growth.

Voice. Perched or in flight, a repeated *chechip, chechip* or *chididit*, the tone varying from wooden and cricketlike to clear and belllike; in flight, a more prolonged, crackling or rattling *che-e-e-e-e-e*; seems to have no true song, but a low, lisping whisper song is rarely heard; juveniles give a high *cheeep*.

Nest. A neat cup or broad, thick, shallow bowl, composed mostly or entirely of well compacted, gray-green beard lichen (*Usnea*), with no special lining; often in loose colonies; 6–60ft (2–18m) up in trees or shrubs, often in pasture, and often amid concealing growth of beard lichen. Eggs 2, pale gray, blotched and spotted with dark brown and pale lilac. March–June.

Status. Common resident in Cordilleras Central and de Talamanca, mainly from timberline down to ca. 6000ft (1850m); outside breeding season occasionally wanders down to 4000ft (1200m).

Range. Costa Rica and W Panama.

BLACK-AND-YELLOW SILKY-FLYCATCHER

Phainoptila melanoxantha Pl. 39(10)

Capulinero Negro y Amarillo

Description. 8¼″ (21cm); 56g. Crestless, rather thrushlike; heavy-bodied, with broad, rounded tail and fairly short legs. **Adult ♂:** head, back, wings, and tail black; rump bright yellow; breast olive-green; sides and flanks bright yellow and olive-green in varying proportions; center of belly slate-gray. ♀: pileum black; hindneck dark gray, shading to sooty-gray on face and throat, and to dark olive-green on breast and remaining upperparts, including rump; posterior underparts like male. Iris dark red; bill black; feet blackish. **Young:** like adult female but duller; pileum sooty-black; no gray on hindneck; breast faintly streaked with dusky; little or no yellow on flanks.

Habits. Frequents middle levels of highland forest and adjacent second growth and edges, venturing into clearings to feed; eats mainly berries of trees, epiphytes, and shrubs, including *Drimys, Phytolacca*, Araliaceae, Ericaceae, mistletoes; rather sluggish, often spending long periods stuffing itself with fruits; may briefly accompany bush-tanager flocks but soon lags behind, eating fruits; plucks insects from foliage and catches them in the air, usually not returning to same perch; alone, in pairs, or small, loose flocks after breeding.

Voice. High, sharp, rather soft *tsit*ting and *tseep*ing notes and twitters, suggesting a much smaller bird; tempo and volume of notes increase before birds fly; appears not to sing.

Nest. A large, compact cup of green moss mixed with a few thin stems and fern fronds, lined with fine rootlets and stems, 5–13ft (1.5–4m) up in a dense shrub or sapling. Eggs 2, grayish-white, densely spotted everywhere with pale gray, brownish-purple and dark brown (Kiff). April–May.

Status. Uncommon to fairly common resident of mountains throughout country, upward from 4000ft (1200m) in N cordilleras and from 6000ft (1850m) in Cordillera Central and Cordillera de Talamanca; may wander lower outside breeding season.

Range. Costa Rica and W Panama.

FAMILY **Vireonidae:** Vireos, Greenlets, Shrike-Vireos, and Peppershrikes

The vireo family includes 43 species of arboreal birds with more or less hook-tipped bills, 4–7″ (10–18cm) in length, confined to the continents and islands of the Western Hemisphere. Typical vireos are plainly attired in olive or gray above, with grayish, whitish, or yellow underparts, with or without eye-rings or wing-bars. Most species breed in North America and the West Indies, and most are highly migratory; even some that nest within the tropics perform long migrations. Greenlets are small, sedentary, tropical vireos with greenish, olive, yellow, and gray plumage, often with yellow eyes. The larger and somewhat more colorful shrike-vireos and peppershrikes, sometimes placed in their own families (Vireolaniidae and Cyclarhidae, respectively), are also mainly tropical, with stouter, more strongly hooked bills. In the family as a whole, the sexes are at most only slightly different.

Members of this family inhabit tropical lowland, tropical montane and temperate-zone forests and scrub. With deliberate movements that contrast with the sprightlier flitting of wood-warblers, they search through foliage for insects and spiders, sometimes hanging inverted. They hold large or tough items, such as cocoons, beneath a foot while they tear them apart with the bill. Many also eat considerable quantities of berries and arillate seeds. On the whole, they are persistent rather than brilliant songsters, the males repeating the same simple phrase over and over in a leisurely fashion, some even singing loudly while they sit in the nest. As far as known, all members of the family build similar nests: cups or pouches attached by the rim in a forked horizontal twig, with the bottom unsupported. They differ chiefly in materials and thickness, some being so thin that the eggs are visible through the bottom. Tropical species lay 2 or 3 eggs, northern species 3 or 4, rarely 5. The eggs are white, pinkish-white or creamy, nearly always spotted with brown or lilac. Males may or may not help to build, but in a number of species they take turns at incubation, and nearly always they help to feed the young, which hatch naked or with sparse down. Incubation periods are from 12–16 days, nestling periods 10–15 days.

RUFOUS-BROWED PEPPERSHRIKE
Cyclarhis gujanensis Pl. 40(2)
Vireón Cejirrufo

Description. 5½″ (14cm); 31g. Stout, big-headed, with very heavy bill and conspicuous pale orange iris. **Adults:** crown dull brown, bordered by broad cinnamon-rufous superciliaries that meet on forehead; chin, cheeks, sides and back of neck gray; rest of upperparts olive-green; below mostly yellow, fading to white on belly and crissum. Bill pale silvery-gray; legs flesh-color. **Immatures:** like adults but belly more extensively white, breast often tinged with olive. **Young:** crown more grayish; superciliaries shorter and much paler; wing-coverts with narrow ochraceous fringes; below mostly dull white, shading to pale yellow on sides and flanks; iris dull brown.

Habits. Prefers forest edge, broken forest, semi-open, trees in pastures, hedgerows, gardens, and second growth; frequents evergreen gallery forest and mangroves; forages in outer foliage at all heights, from tall treetops to shrubbery; moves with slow deliberation, ending periods of scanning foliage with swift, heavy hops to seize prey, often reaching far out or hanging upside-down; eats insects and spiders; often holds caterpillars or cocoons beneath a foot while tearing them apart; paired throughout year but often forages alone.

Voice. Call a variable series of 3–8 loud, slurred whistles, inflected upward or downward, first or second note loudest, the others successively lower, heard at long intervals. Juveniles utter a loud, repeated *wick*. Song a short, melodious, grosbeaklike warble of 5–7 notes, heard through most of year, variously transliterated as *we don't believe it, we've been wishing to meet you, we're here to greet you*, etc.

Nest. A hemispheric cup, fastened by its rim in a terminal fork, often so thin that the eggs are visible through the bottom. One was made of richly branched gray lichens and a little green moss bound together with much spider-web and lined with coarse, wiry materials; 7–35ft (2–10m) up. Eggs 2–3, pinkish-white, speckled with bright brown. March–July.

Status. Common resident of Cordilleras de Tilarán, Central, and de Talamanca and of Dota region, also Valle Central E to near Turrialba in Reventazón drainage; mainly 2300–8000ft (700–2450m); locally in Río Frío region and in dry NW; most numerous on Península de Nicoya and around Golfo de Nicoya, S to Parrita in mangroves.

Range. NE Mexico to C Argentina and S Brazil.

GREEN SHRIKE-VIREO
Vireolanius pulchellus Pl. 40(1)
Vireón Esmeraldino

Description. 5½″ (14cm); 30g. Chunky, short-tailed, big-headed, with heavy, hooked bill. **Adult ♂:** forehead, nape, and orbital area bright blue; crown and cheeks green, mixed or tinged with blue; rest of upperparts bright green, tinged with olive on wings and tail; throat bright yellow, paler and often whitish on chin; rest of underparts yellowish-green, clouded with olive on flanks, shading to yellow on lower belly and crissum. ♀: similar but crown and face less blue, chin always yellow. On S Pacific slope, both sexes have only nape blue, rest of head bright green. Upper mandible black, lower silvery; legs gray. **Young:** above olive-green; indistinct yellowish stripes above and below eye; wing-coverts tipped with yellowish; below dull yellow.

Habits. Frequents forest canopy and tall trees in semi-open, coming lower along forest edges and gaps; solitary or in pairs; territorial most or all of year; forages deliberately, scanning foliage, then hopping to capture prey, sometimes clinging or hanging to work through rolled leaves; takes relatively large insects, including many caterpillars, also berries and arillate seeds. An individual or pair may accompany a mixed flock of tanagers, greenlets, and honeycreepers through canopy.

Voice. Call a dry, scratchy *raah*; in aggressive interactions, a sharp, scratchy twitter or chatter. Song, often the best clue to the bird's presence, a clear, whistled *peeya-peeya-peeya*, sometimes increased to 4 notes or with the second syllable dropped, *pee-pee-pee* or *peer-peer-peer-peer*.

Nest. A mossy, vireolike cup, 45ft (14m) up. Eggs undescribed (?). May (in Panama; no Costa Rican record).

Status. Locally common resident of humid lowlands and foothills the length of Caribbean slope, and from Carara area S on Pacific slope, from sea level to 3300ft (1000m), higher in S on both slopes; rare below 1000ft (300m) in SE, and in drier Río Frío district S of Lago de Nicaragua.

Range. S Mexico to E Panama.

Note. Sometimes placed in genus *Smaragdolanius*.

WHITE-EYED VIREO

Vireo griseus Pl. 40(12)

Vireo Ojiblanco

Description. 5″ (13cm); 12g. A small vireo best recognized by bright yellow spectacles and white iris, throat, and wing-bars; much more brightly colored than Mangrove Vireo. **Adults:** above olive-green, strongly tinged with gray on nape and sides of neck; 2 yellowish-white wing-bars; broad yellow stripe over blackish lores joins narrow, interrupted yellow eye-ring; below dull white, tinged with buff on sides of throat and breast; sides and flanks pale yellow, mixed with olive. Upper mandible black, lower gray; legs dark gray. **Young:** similar, but iris brownish.

Habits. Both birds found in Costa Rica were in areas of dense young second growth dominated by the shrub *Cordia curassavica,* whose red berries they were eating; both were alone and silent.

Voice. Not heard in Costa Rica. Calls include a harsh mewing note and a short, sharp ticking.

Status. Casual or rare fall migrant, known from two birds mist-netted along the S Caribbean coast (Puerto Vargas, 4 November 1978; Rio Vizcaya, 19 October 1985). Possibly regular in small numbers, especially during periods of tropical storms in the West Indies.

Range. Breeds from extreme SE Canada to C Mexico and Florida; winters from SE USA to Greater Antilles and N Nicaragua, casually to W Panama.

MANGROVE VIREO

Vireo pallens Pl. 40(11)

Vireo de Manglar

Description. 4½″ (11.5cm); 11g. Small, dull-colored, with broad yellowish-white loral stripe continuing narrowly halfway around eye; narrow wing-bars. **Adults:** above dull grayish-olive, brightest on forehead; wings dusky with 2 white wing-bars, remiges edged with dull yellowish; lores and cheeks dull grayish-buff; below dull white, tinged with buffy-yellow on breast, sides, and flanks. Iris pale brown to yellow; upper mandible grayish, lower flesh-color with tip horn-color; legs grayish. **Young:** above grayish-brown; below pale grayish-buff; belly whitish; facial pattern indistinct.

Habits. Exclusively in mangroves, where it seems to prefer dense, young red mangroves (*Rhizophora*); usually keeps fairly low, hopping actively through foliage and roots, pausing briefly to scan for spiders, small beetles, leafhoppers, and other prey, which it captures with a quick dash; takes some fruit; rarely hovers, does not sally; usually solitary or in pairs.

Voice. Two birds together keep up a squeaky sputtering; song is a harsh, loud, grating *jeeweeweewee* or a burry *chewy-chewy-chewy-chewy* somewhat like the call of the Yellow-throated Vireo but higher-pitched, more slurred.

Nest. Cup-shaped; in mangroves. Eggs 2–3. April–June (in Belize; no Costa Rican nest seen, but juveniles in July–August).

Status. Locally common resident in mangroves of N Pacific coast, from Bahía Salinas around Península de Nicoya to Golfo de Nicoya, S at least to Tivives.

Range. NW Mexico to Costa Rica on Pacific coast; Yucatán Peninsula to SE Nicaragua on Caribbean.

Note. Sometimes considered conspecific with White-eyed Vireo, *V. griseus*.

YELLOW-WINGED VIREO

Vireo carmioli Pl. 40(10)

Vireo Aliamarillo

Description. 4½″ (11.5cm); 13g. Small, below yellowish, with conspicuous yellow wing-bars and broad, interrupted spectacles. **Adults:** above greenish-olive; wings blackish, with broad, pale yellow wing-bars and edgings on tertials; primaries edged with olive-yellow; broad yellowish-white superciliaries, confluent with broad, interrupted whitish eye-ring; lores dusky; chin whitish; rest of underparts pale yellow, washed with olive on throat, breast, and sides. Upper mandible blackish, lower pale gray; legs gray. **Young:** above more brownish; wing-bars tinged with ochraceous-olive; superciliaries buffy-white; throat whitish; rest of underparts very pale yellow, tinged with buff on breast.

Habits. Frequents canopy of mountain forests and trees in adjacent clearings, occasionally descending to undergrowth or venturing into neighboring second growth; gleans insects and spiders in foliage with deliberate movements; takes berries; often joins mixed flocks, or accompanies Flame-throated Warblers.

Voice. Calls include a nasal *net* and an oriolelike *chwick*. Song consists of short, abrupt phrases of 2–3 notes separated by longer pauses, in a high, falsetto tone, including distinctive, slurred, burry notes: *viree chichui chuyee; viree viree cheeyu; viree witchum viree witchum*.

Nest. A small cup fastened by its rim in the fork of a twig, of green liverworts and mosses, lichens, pieces of green leaves, and silken egg cases of spiders or insects, producing a mosaic of many colors, 9–65ft (3–20m) up in tree or shrub. Eggs 2, white, with dark spots on thicker end. March–June. Both sexes build, incubate, and feed young.

Status. Common resident of upper parts of Cordillera Central, Dota region and Cordillera de Talamanca, from ca. 6500ft (2000m) to timberline; after breeding many descend to lower elevations, locally down to 5000ft (1500m) at height of rainy season.

Range. Costa Rica and W Panama.

YELLOW-THROATED VIREO

Vireo flavifrons Pl. 40(6)

Vireo Pechiamarillo

Description. 5½″ (14cm); 18g. Warblerlike in color and pattern but with bigger head, heavier hooked bill. Above bright olive-green; scapulars and rump slate-gray; 2 broad white wing-bars, secondaries broadly edged with white; spectacles, throat, and breast yellow; belly white; flanks tinged with gray. Upper mandible dark gray, lower blue-gray with darker tip; legs dark gray.

Habits. Prefers canopy, edge of forest, tall second growth, trees in semi-open, shaded gardens, and coffee plantations; in migration in more open areas and low scrub; singly in winter, probably territorial; associates freely with mixed flocks of greenlets, gnatcatchers, tanagers, and warblers in canopy; slowly and deliberately gleans foliage, usually high in trees, frequently pausing to scan surrounding vegetation, taking relatively large insects, rarely fruit; often cocks tail.

Voice. Call a descending phrase of 3–5 harsh, grating notes, very distinctive and often the best clue to the bird's presence; after arrival in fall and before departure in spring, males sometimes sing with short phrases like those of Yellow-green Vireo, but intervals between phrases much longer, notes burry.

Status. Widespread, common migrant and winter resident, arriving as early as mid-September but more regularly in October, remaining through late April; found from sea level to at least 5500ft (1700m) the length of both slopes, locally to 6600ft (2000m) in migration.

Range. Breeds in SE Canada, C and E USA; winters from extreme S USA and E Mexico to Colombia, Venezuela, and West Indies.

SOLITARY VIREO

Vireo solitarius Pl. 40(9)

Vireo Solitario

Description. 5½″ (14cm); 17g. Large; conspicuous white spectacles on gray head; broad wing-bars. **Adults:** above olive-green; crown and sides of head slaty; wing-bars and most of underparts white; sides and flanks mixed with olive-green and yellow. Upper mandible black, lower mandible and legs blue-gray. **Immatures:** very similar, but crown usually tinged with olive, primary coverts edged with brownish.

Habits. Prefers forest edge and semi-open, usually remaining high; seen singly, often accompanying mixed flocks of warblers, tanagers, and other vireos; eats fruits and insects in about equal proportions.

Voice. Loud nasal notes occasionally; not known to sing in Costa Rica.

Status. Very rare or casual winter visitor (December–late March) on Pacific lowlands (Palo Verde) and N and C highlands (Monteverde, Cerro de la Muerte).

Range. Breeds from S Canada to Honduras;

winters from S USA to Costa Rica, W Panama, and Cuba.

RED-EYED VIREO

Vireo olivaceus Pl. 40(3)

Vireo Ojirrojo

DESCRIPTION. 5½″ (14cm); 17g. No wing-bars; much stronger face-pattern than Yellow-green Vireo, with much less yellow below. **Adults:** above olive-green; pileum slate-gray; white superciliaries bordered by blackish stripes along sides of crown and through eyes; below dull white; cheeks, sides of breast, and flanks washed with olive; crissum tinged with pale yellow. Iris brick-red; upper mandible dark gray, lower pale gray; legs bluish-gray. **Immatures:** iris brown to chestnut; yellow tinge on crissum much stronger, extending to flanks.

HABITS. In trees or tall shrubbery almost everywhere, from forest canopy and edge to second growth, suburban gardens, and city parks; numbers fluctuate daily as waves of migrants pass through, usually in loose groups that may join mixed flocks of resident species or other migrants; gleans foliage for insects and spiders, sometimes clinging acrobatically; takes many berries (*Trema*, *Cordia*) and arillate seeds (*Clusia, Lacistema*).

VOICE. Usually silent; in fall or spring may give a nasal, whining *nyaa* in alarm or agitation; occasionally, when northbound in spring, voices a song like that of Yellow-green Vireo, but phrases longer, delivery less rapid.

STATUS. Abundant migrant in fall (late August–early November) and spring (early April–late May), with occasional stragglers trailing up to a month behind on both N and S journeys; principally in lowlands but often common to 4300ft (1300m) in Valle Central; small numbers sporadically reach at least 7000ft (2150m) in fall.

RANGE. Breeds from S Alaska and NW Canada to NW and E USA; winters in W Amazonia.

YELLOW-GREEN VIREO

Vireo flavoviridis Pl. 40(5)

Vireo Cabecigrís (Cazadora, Fraile)

DESCRIPTION. 5½″ (14cm); 18.5g. A fairly large vireo, no wing-bars; below yellower and with less distinct face pattern than Red-eyed Vireo. **Adults:** pileum slate-gray, indistinctly bordered with dusky; superciliaries pale gray; eye-stripe dusky; cheeks pale olive; rest of upperparts olive-green, brightest on edges of wings and tail; below dull white, the sides of breast, sides, and flanks olive-yellow; crissum bright yellow. Iris brick-red; upper mandible grayish-horn, lower pale gray; legs blue-gray. **Immatures:** like adults but wing coverts duller, below yellow less extensive; iris brownish. **Young:** pileum pale grayish-brown; back and shoulders buffy-brown; wing-coverts dull olive with yellowish-olive fringes; superciliaries white, but narrow and poorly defined; below white, tinged with yellow on sides, flanks, and crissum. Iris dark brown; bill and legs pale horn.

HABITS. Frequents crowns of trees in clearings, gardens, light woods, savannas, and canopy and middle levels of deciduous and evergreen forest; flits and hops in outer foliage, pausing frequently to scan for insects, especially caterpillars and beetles, which it gleans from leaves and twigs; also eats many berries, including those of mistletoes, *Trema, Cordia*, and melastomes, and arillate seeds, notably of *Clusia* and *Stemmadenia*. Males sing freely upon arrival; singing declines somewhat until the rains approach, when nesting commences and song reaches a peak; pairs break up August–September.

VOICE. Calls include a little rattle or churr, and a whining, nasal, burry *nyaaah* expressing alarm or anxiety. Song consists of short, 1–3 note phrases, repeated tirelessly at rates of 50–55/min during peak of singing: *viree veer viree fe'e vireo chirip viree. . . .*

NEST. A cup ca. 2½″ (6.5cm) broad and high, attached by its rim to the arms of a stout forked twig, of the most varied materials, including grass blades, fragments of broad leaves, strips of papery epidermis and bark, lined with slender, curved rachises, 5–40ft (1.5–12m) up in a shrub or tree, usually below 12ft (3.5m). Eggs 3, less often 2, white, with a wreath of brown or chocolate spots. March–July. Only female builds and incubates; male helps to feed young.

STATUS. Common to abundant breeding resident from lowlands to at least 5000ft (1500m) the length of Pacific slope, least numerous in wetter, more forested districts; on Caribbean slope uncommon except in drier Río Frío region. Arrives in S Costa Rica in late January or early February, not until mid-February to early March in Valle Central; departs by mid-October, occasionally early November.

RANGE. Breeds from S Texas to C Panama; passes nonbreeding season in W Amazon basin.

NOTE. Often considered a race of the Red-eyed Vireo, *V. olivaceus*.

BLACK-WHISKERED VIREO

Vireo altiloquus Pl. 40(4)

Vireo Bigotudo

DESCRIPTION. 5½" (14cm); 17g. Closely resembles Red-eyed Vireo but above more brownish, especially on crown, which is only faintly edged with dusky; best mark is distinct dusky malar streak separating white throat from olive-washed cheeks. Iris brick-red in adult, reddish-brown in young birds.

HABITS. The one bird recorded in Costa Rica was in low dense scrub dominated by *Cordia curassavica* and *Piper* shrubs beside a coconut plantation on the Caribbean coast; it gleaned insects deliberately in low vegetation and took *Cordia* berries.

VOICE. Not heard in Costa Rica.

STATUS. Accidental or casual fall migrant; one record (Puerto Vargas, 16 September 1978). Possibly more regular than the limited data suggest; several records in C Panama between late August and January (Ridgely).

RANGE. Breeds S Florida, West Indies; winters from Greater Antilles to N South America.

PHILADELPHIA VIREO

Vireo philadelphicus Pl. 40(15)

Vireo Amarillento

DESCRIPTION. 4½" (11.5cm); 11.5g. A small vireo without wing-bars, somewhat resembling Tennessee Warbler (with which it often associates) in color but with relatively bigger, rounder head, thicker bill, broader facial stripes, yellow crissum. **Adults:** pileum gray, tinged with olive; rest of upperparts grayish olive-green; superciliary and area below eye white, separated by dusky eye-stripe; cheeks washed with olive; underparts vary from mostly fairly bright yellow, paler on throat and belly, to mostly whitish, distinctly yellow only on breast and crissum. Upper mandible dusky-horn, lower pale horn; legs gray. **Immatures:** like adults but sometimes a few shorter, duller, pale-tipped greater wing-coverts retained from juvenile plumage.

HABITS. Prefers light woodland and forest edge, coffee plantations, hedgerows, gardens, and taller second growth; regularly in forest only in dry NW but may visit forest canopy elsewhere during migration; in winter solitary or in small, loose groups, often joining mixed flocks of other small birds or groups of Tennessee Warblers; gleans small insects, especially beetles, from foliage; takes small berries and arillate seeds rich in oil.

VOICE. Usually quiet and unobtrusive; occasionally utters a soft descending, twittering *cheeur*, especially when agitated.

STATUS. Fairly common fall migrant and winter resident on Pacific slope, less numerous on Caribbean slope, from lowlands to 5500ft 1650m), rarely to 7000ft (2130m) in migration; arrives in late October, departs unobtrusively by late April, rarely early May.

RANGE. Breeds in Canada and extreme N USA; winters from Yucatán Peninsula and Guatemala to C Panama, rarely N Colombia.

WARBLING VIREO

Vireo gilvus Pl. 40(13)

Vireo Canoro

DESCRIPTION. 4¾" (12cm); 12g. Resembles Philadelphia Vireo but duller, paler; facial stripes less distinct, pale lores; no yellow on breast. Pileum grayish-brown, shading to dull grayish-olive over rest of upperparts; superciliaries white; eye-stripe dusky, obscured on lores by white feather tips; underparts dull white, tinged with buffy laterally and on cheeks; flanks and crissum sometimes tinged with pale yellow. Upper mandible dark gray, lower pale gray; legs dark gray.

HABITS. In winter seems to prefer light woodland and savanna groves; usually keeps well up in trees, often associating with canopy flocks of warblers, other vireos, and greenlets; gleans insects in foliage, sometimes hovering; also takes fruit. On migration may associate with other vireos or occur in semi-open and second growth.

VOICE. Mewing notes, less harsh and grating than those of Red-eyed Vireo.

STATUS. Apparently a casual to rare winter visitor and migrant; at least 6 sightings to date, late December–late April, scattered over N half of both slopes, from lowlands to ca. 2600ft (800m). Specimen confirmation desirable.

RANGE. Breeds from S Alaska and N Canada to C Mexico; winters from N Mexico to Nicaragua, casually to Costa Rica.

BROWN-CAPPED VIREO

Vireo leucophrys Pl. 40(14)

Vireo Montañero

DESCRIPTION. 4¾" (12cm); 12g. Fairly small; above brownish; whitish throat contrasts with yellow underparts; no wing-bars. **Adults:** pileum dull brown; rest of upperparts dull olive, greener on rump and edges of flight feathers; superciliaries white; eye-stripe brownish; cheeks brownish-white; throat and chest dull white; rest of underparts pale yellow, washed with olive on flanks. Upper mandible blackish, lower pale grayish with dusky tip; legs gray. **Young:** above buffy-brown, darkest on crown, tinged with ochraceous on scapulars; wing-coverts grayish-brown with yellowish fringes; superciliaries indistinct or interrupted; below paler than adult.

HABITS. Frequents canopy and edges of mountain forest, tall second growth, trees in clearings and pastures; actively gleans foliage for insects, sometimes hanging acrobatically to probe rolled leaves; takes many berries; often joins mixed flocks of warblers and bush-tanagers; solitary or in pairs, or in family groups immediately after breeding.

VOICE. Call a sharp sibilant *pisss* or *twist*, sometimes repeated rapidly. Song a full, rich warbling that alternately rises and falls; resembles songs of certain finches or yellowthroats (e.g., Olive-crowned) but phrases usually shorter, often with slightly harsh or burry quality.

NEST. Unknown (?). Probably breeds March–June.

STATUS. Fairly common resident of mountains from Cordillera de Tilarán S to Panama, mostly at 5000–8000ft (1500–2450m), down to 4000ft (1200m) locally outside breeding season.

RANGE. E-C Mexico to NW Bolivia.

NOTE. Sometimes considered conspecific with the Warbling Vireo, *V. gilvus*.

SCRUB GREENLET

Hylophilus flavipes Pl. 40(8)

Verdillo Matorralero

DESCRIPTION. 4½″ (11.5cm); 13g. Small, chunky; bill rather heavy, sharp-pointed, conspicuously pale; iris pale yellow. **Adults:** above olive-green, including edges of flight feathers; face and chin pale grayish, shading to olive-yellow on chest and fairly bright yellow over rest of underparts. Bill and legs flesh-color. **Young:** pileum buffy-brown; upperparts brownish-olive, tinged with ochraceous; superciliaries whitish, indistinct; face tinged with buff; throat whitish; rest of underparts pale yellow, tinged with buff on breast. Iris brown; bill dusky, shading to pale horn at base of lower mandible.

HABITS. In dense, low scrub, oil palm plantations, trees in pastures and clearings, forest edge, and tall second growth, especially dense stands of bamboo on abandoned banana plantations; searches deliberately for insects amid foliage, often clinging upside-down to examine rolled leaves or search undersides of leaves and twigs; probes curled dead leaves held beneath foot; eats berries and arillate seeds, including those of *Clusia* and *Bocconia*; solitary or in pairs.

VOICE. Calls include harsh, nasal, buzzy, or scratchy notes. Song a simple phrase of slurred or double notes, *cheree cheree cheree* or *chi-cheer chi-cheer chi-cheer chi-cheer*, rapidly repeated 10 or more times in a full, mellow, far-carrying voice, or a single note repeated many times.

NEST. A hemispheric cup of fine, pale-colored fibers, sometimes with sprays of green moss on outside, sometimes so thin that eggs are visible through bottom, suspended by its rim in slender horizontal fork of tree, shrub, or tall bamboo, 13–33ft (4–10m) up. Eggs 2–3, white, finely spotted with chocolate. April–July, once December.

STATUS. Resident, locally common on Pacific slope from Carara S to Panama, rarely N to Caldera; lowlands to 3000ft (900m).

RANGE. Costa Rica to Venezuela and Tobago.

TAWNY-CROWNED GREENLET

Hylophilus ochraceiceps Pl. 32(4)

Verdillo Leonado

DESCRIPTION. 4¼″ (11cm); 11g. Small, chunky, suggesting an antwren, with which it is often found; iris pale gray. **Adults:** nasal area golden-yellow, shading to bright ochraceous-tawny on forehead and crown, dull olive on hindneck, and brownish-olive on back; rump and upper tail-coverts more greenish; tail and edges of primaries russet-brown; sides of head grayish-olive, shading to pale grayish on throat, with faint paler mottling; breast dull buffy-yellow, shading to pale yellow, tinged with olive, on rest of underparts. Bill dusky-horn, base of lower mandible paler horn; legs flesh-color to pale horn. **Young:** above paler, crown more buffy-brown with no yellow; below paler, more strongly tinged with buff. Iris darker gray; bill and legs dusky.

HABITS. Prefers lower levels of humid forest with abundant undergrowth; active and restless, usually in pairs or family groups, accompanying mixed flocks of antwrens, antvireos, gnatwrens, ant-tanagers, flycatchers, and other small birds; gleans grasshoppers, roaches, moths, caterpillars, ants, other insects, and spiders from foliage, holding larger items beneath a foot while it dismembers them; occasionally takes small berries.

VOICE. Call an incessantly repeated, loud, rather harsh, nasal *doy doy doy* or *chewy chewy*; rarely a subdued trill. Song a high, clear, penetrating whistle, longer than note of Striped-breasted Wren: *eeeeeeeee*, either level or upwardly inflected and quite ventriloquial.

NEST. A sturdy cup attached by its rim with spiderweb to the arms of a horizontal fork, of bast fibers and seed down, covered with green moss, 3–20ft (1–6m) up in a sapling or small tree. Eggs 2 (color not seen). March–April.

STATUS. Common resident of Caribbean slope and of S Pacific slope N to Carara, reaching Pacific side of N cordilleras through low passes; lowlands to 4000ft (1200m).

RANGE. S Mexico to W Ecuador, C Bolivia, and C Brazil.

LESSER GREENLET

Hylophilus decurtatus Pl. 40(7)

Verdillo Menudo

DESCRIPTION. 4″ (10cm); 9g. A small, rather short-tailed vireo, with white spectacles that contrast only slightly with pale gray head. **Adults:** head gray, paler on face; line over lores and eye-ring white; rest of upperparts, including wings and tail, olive-green; below dull white, tinged with yellowish on breast; sides and flanks washed with olive-green; crissum olive-yellow. Upper mandible dark gray, lower pale gray to horn-color; legs pale gray. **Young:** pileum pale brownish-olive, shading to dull olive-green on back, brighter on rump; sides of head dull buff, spectacles narrow and indistinct; below white, tinged with buffy on throat and breast.

HABITS. Hunts restlessly through forest canopy and edge, crowns of trees in semi-open or clearings, tall second growth, sometimes descending to forage in shrubbery in open areas, searching amid green foliage and prying into curled dead leaves, often hanging with head and back downward; eats insects and spiders, often varied with small berries and arillate seeds; in pairs or small groups, often accompanying mixed flocks of gnatcatchers, honeycreepers, warblers, and other small birds.

VOICE. When perturbed, 2–4 harsh, downslurred, successively lower, nasal, whining notes: *neeah-neeah-neeah-neeah* or a single, petulant *sheew*. Song consists of short phrases of 3–8 warbled, whistled notes separated by pauses, with the last note almost always downslurred, suggesting that of Yellow-green Vireo but sweeter and more melodious, the phrases longer and the delivery less hurried: *chi chi' cher cher cher cher chiri cher, wicheet wich chi cher*. Often a given phrase is repeated many times. Females answer mates with a shorter version.

NEST. A deep cup, almost a pouch, attached by rim to diverging branches of a drooping, leafy bough, mainly of whole small dead leaves and strips of larger ones, held together by a few fibers and much spiderweb, sparsely lined with fine, pale fibers, 14–45ft (10–14m) up. Eggs 2, white, spotted and blotched with pale brown. February–May.

STATUS. Common to abundant resident of lowlands and foothills of both slopes, from sea level to 3000ft (900m) in N and 4000ft (1200m) in S; least common in extensively deforested areas.

RANGE. NE Mexico to W Ecuador.

NOTE. Gray-crowned birds from C Panama northward were formerly considered a species separate from green-crowned populations to the south (*H. minor*) and called Gray-headed Greenlet, but the 2 hybridize extensively in C Panama.

FAMILY **Coerebidae:** Bananaquit

With the transfer of the honeycreepers and conebills to the Thraupidae and the flowerpiercers to the Emberizidae, the Bananaquit remains alone. In some recent classifications, it has been reduced to a subfamily of the Parulidae. However, it differs from wood-warblers (and indeed, from all other nine-primaried oscines) in many ways, and it is probably best kept in a separate family of which it is the only species.

The more than 30 recognized races throughout its wide range testify to the plasticity and adaptability of the Bananaquit. Its slightly decurved, sharp-pointed bill is ideal for probing small flowers or piercing large ones for nectar or fruits for juice; its rather short legs and stout feet permit it to hang and cling at all angles to forage. Its nesting habits are unique, for both sexes, alone or together, build both dormitory and breeding nests, with males the more active builders. The breeding season is long and apparently timed to take advantage of abundant flowering; several broods per year may be raised. The female alone incubates the 2–3 eggs for 12–13 days. The young are fed by regurgitation by both parents, and nestlings normally remain in the nest for 17–19 days, a very long period for such a small bird. After the young fledge they often sleep in nests of wrens, flycatchers or other birds or in the open until they can build their own dormitories; the female continues to sleep alone in the breeding nest or in some other nest.

BANANAQUIT
Coereba flaveola Pl. 40(24)
Reinita Mielera (Pinchaflor, Santa Marta)

Description. 3¼″ (9cm); 9.5g. Small, chunky, with sharp-pointed, decurved bill, short tail and legs. **Adults:** pileum and broad stripe through eye dull black, separated by a long white superciliary; remaining upperparts dark olive-gray, except rump and upper tail-coverts olive-yellow; outer rectrices narrowly tipped with dull white; primary coverts black; edges of 4–5 outer primaries white at base, forming conspicuous patch on folded wing; throat gray; rest of underparts bright yellow, tinged with olive on flanks, and paling to whitish on crissum. Bill black; legs slate-color. **Immatures:** above dull sooty-olive, darkest on pileum; wing-patch very small; superciliaries yellowish-olive; throat, breast, and sides dull olive-yellow; belly brighter; flanks and crissum tinged with dull buff.

Habits. Alone or in pairs frequents canopy and edge of humid forest, open woods, plantations, second growth, dooryards, gardens, and hedgerows with abundant flowers; probes small flowers and pierces the base of long, tubular corollas for nectar, often while clinging head downward; gleans small insects and spiders from foliage, or creeps over trunks and limbs searching for them; pierces berries to suck out juice; often eats protein corpuscles of *Cecropia*; in highlands often joins mixed flocks of tanagers, warblers, honeycreepers, etc., in forest canopy; visits feeders to drink juice of oranges or nibble daintily at bananas.

Voice. Call a high, piercing, sharp *tsip* or *tseep*, sometimes repeated in a sputtering twitter in excitement; a sharp, metallic chipping. Song high, thin, scratchy, insectlike: *tsee-tsee-tsee-tsee-tzzeew*, the last note trilled, or *tzee zheeew-zheeew* with 2, sometimes more, thin trills; sings persistently most of year.

Nest. A compact globe with a round doorway facing obliquely downward, of the most varied vegetable materials, often with much green moss, softly lined with seed down, fine fibers, or feathers. Persistent builders, at all seasons Bananaquits make dormitories that resemble breeding nests and sleep in them singly. Nests of both kinds are mostly 5–10ft (1.5–3m), rarely 50ft (15m), up in shrubs, trees, and vine tangles. Eggs 2, dull white, spotted with brown. Breeds nearly throughout year, except March–May on S Pacific slope, July–September in Caribbean lowlands (varies locally with flowering); 2 or more broods.

Status. Common to abundant resident throughout Caribbean slope, and on Pacific slope from about Carara S, from sea level to 4000ft (1200m), locally to 5000ft (1500m).

Range. S Mexico and West Indies to NE Argentina and S Brazil.

FAMILY **Parulidae:** Warblers

A fairly homogeneous family of about 110 species, the warblers (in the Old World called wood-warblers) are essentially confined to the continents and islands of the New World, from the Arctic Circle to central Argentina; nearly all species breeding in North America migrate to the tropics for the nonbreeding season. Most species resident in the tropics live in the mountains. Small, mostly slender-billed, dainty birds, warblers range in size from 4″ to, rarely, 7″ (10–18cm). Yellow is the prevalent bright color in their exceedingly varied plumage, but orange, red, chestnut, and blue adorn many species, often in intricate patterns. Among the migratory species the sexes usually differ in color and often have distinct breeding and nonbreeding "winter" plumage; both sexes of sedentary tropical species wear the same bright or dull colors throughout the year.

Most warblers inhabit forests, light woods, shady plantations, thickets, and gardens; a few prefer treeless grasslands, marshes, or streams. Most species are primarily insectivorous, gleaning their prey from foliage while perched or hovering or pursuing it with aerial sallies; some vary their diets with berries, arillate seeds, and nectar. Warblers are nearly always monogamous, and strongly territorial in the breeding season or year-round in tropical forms. Some migrants claim individual territories in their winter home, whereas others flock; residents and migrants associate with mixed flocks of resident birds, especially in the highlands. The songs of warblers are mostly simple and rather unmelodious, but a few songs are superb; most species sing freely, at least while breeding. Most northern warblers build cup nests, high in trees, in bushes, or on the ground; but 1 northern species and many tropical species construct roofed nests with a side entrance, on a bank or the ground, or they tuck their open nests amid sheltering mosses or larger epiphytes. Generally the female builds alone, but males of a few species help substantially. In the tropics, sets of 2 or 3 eggs are usual, whereas at higher latitudes 3–5 are common. Warblers' eggs are white or tinted with green, blue, or pink, more or less heavily marked with brown, chestnut, lilac, or black, rarely immaculate. As far as is known, only females incubate, sometimes fed by their mates. Incubation periods of northern species are 11–12 or, exceptionally, 14 days; of tropical species, 14–17 days. Young hatch with sparse down, are fed by both parents, and remain in the nest for 8–15 days, again longer in tropical than in northern species. Immatures usually differ in coloration from their parents.

BLACK-AND-WHITE WARBLER

Mniotilta varia Pl. 41(13)

Reinita Trepadora

Description. 5″ (13cm); 11g. Only small, boldly streaked, trunk- and branch-creeping bird. **Adult ♂:** head boldly striped with black and white; back black, streaked with white; wings black with 2 white wing-bars; cheeks and throat black; rest of underparts white, heavily streaked with black on breast, sides, and crissum; alula covert black with white tip. ♀: similar but cheeks gray, throat and breast white; sides streaked with dull blackish; flanks and (often) face tinged with buff. Upper mandible blackish, lower horn-color, gonys dusky; legs brownish. **Immatures:** resemble adult female; alula covert dull black with subterminal white spots. Male has white sides streaked with black; female's sides washed with buff or brownish and streaked with dusky.

Habits. In almost any tall scrub during migration, but in winter prefers good-sized trees in forest, semi-open, old second growth, parks, and gardens; forages by creeping up and down trunks, along large branches, gleaning insects, their eggs and pupae, spiders, etc., from bark, usually well above ground; does not use tail as a prop; rarely hovers to pluck prey from underside of limb. Evidently establishes individual territories in winter, but individuals often accompany mixed foraging flocks of warblers, greenlets, honeycreepers, and other small birds in canopy.

Voice. A thin, weak *tsit* or *tseep*, given freely but hard for humans to hear. Upon arrival in fall, many birds sing for a brief period, probably while establishing territories; song a high, very thin *see, sir see, sir see, sir*.

Status. Sporadically common fall migrant (chiefly late August–October) from lowlands to at least 8200ft (2500m) on both slopes; in winter widespread but uncommon resident from sea level to ca. 6600ft (2000m), but center of abundance ca. 1500–5000ft (450–1500m) and more numerous on Pacific slope, notably along Cordillera de Guanacaste and Cordillera de Talamanca. Departs by late March to mid-April.

Range. Breeds from NW and E Canada to SE USA; winters from extreme S USA and West Indies to N South America.

PROTHONOTARY WARBLER

Protonotaria citrea Pl. 42(1)

Reinita Cabecidorada

Description. 5″ (13cm); 12.5g. Combination of unmarked gray wings, olive back, rich yellow face or entire head diagnostic; bill large but fine-pointed. **Adult ♂:** entire head and upper breast orange-yellow, at most narrow olive fringes on crown feathers; belly rich yellow; crissum white; back olive; wings and central rectrices blue-gray, outer rectrices with outer webs and tips black, inner webs mostly white. Bill dark gray, base of lower mandible flesh-color; legs gray. ♀: similar but pileum olive, face tinged with olive; crown and nape feathers with yellow bases; breast paler; belly extensively white; less white in tail. **Immature ♂:** like adult but crown feathers heavily edged with olive, partly obscuring yellow; edges of white areas in outer rectrices blurred. ♀: like adult but little or no yellow in forehead and crown, face yellow-olive.

Habits. Prefers thickets adjoining rivers, streams, ponds, and lagoons, also mangroves; mostly stays within 20ft (6m) of ground; singly, in groups, or often in male-female pairs, singles and pairs sometimes associating with mixed flocks; forages deliberately, using long bill to probe bark crevices, rolled leaves and tangles, taking mostly insects and spiders, occasionally fruits or nectar.

Voice. A soft, sibilant *psit* or *tssp*; when agitated a sharper, more insectlike *tseeip* or *kseek*.

Status. Locally common fall migrant (late August–October) from lowlands to ca. 5000ft (1500m), especially near both coasts; in winter fairly common in mangroves along Pacific coast, uncommon in riparian areas from lowlands locally to ca. 4300ft (1300m) in Valle Central and in Río Frío district; rare elsewhere on Caribbean slope. Departs by early to mid-March.

Range. Breeds in E USA and extreme SE Canada; winters from S Mexico to N South America.

WORM-EATING WARBLER

Helmitheros vermivorus Pl. 40(21)

Reinita Gusanera

Description. 5″ (13cm); 13g. Sharp contrast of black stripes on bright buffy head diagnostic. **Adults:** upperparts uniform dull olive; head, throat, and breast bright buff, head with 4 clean-cut black stripes; chin and belly whitish; flanks washed with olive. Bill horn-color with dusky culmen; legs brownish-flesh. **Immatures:** like adults but tertials fringed with rusty.

Habits. In winter frequents thickets and forest undergrowth, preferring evergreen forest in dry northwest; probably territorial, but individuals frequently accompany mixed flocks with other warblers, antwrens, etc., sometimes rising to middle levels of forest; probes and searches rolled dead leaves, gleans dense foliage, vine tangles, sometimes ground litter, for insects and spiders; pecks into rotten twigs for termites.

Voice. Call a rolling *zeet* or *zreet*, sometimes rapidly repeated several times; also a *chip*. Song not heard in Costa Rica.

Status. Uncommon fall migrant (early September, rarely late August, through October), mainly on Caribbean slope; in winter uncommon and local over both slopes from lowlands to about 5000ft (1500m), rarely higher; departs by mid to late April.

Range. Breeds in E and C USA; winters from S Mexico and the Bahamas to C Panama.

GOLDEN-WINGED WARBLER

Vermivora chrysoptera Pl. 41(5)

Reinita Alidorada

Description. 4½″ (11.5cm); 8.5g. Bright yellow bars or patch on wing diagnostic; above gray; throat dark; bill very slender, sharp-pointed. **Adult ♂:** pileum bright yellow; otherwise above slate-gray; tertials tinged with olive; middle and greater wing-coverts mostly bright yellow; inner webs of outer 3 rectrices largely white; chin, throat and sides of head black, set off by white superciliaries and malar stripes; rest of underparts white, washed with gray laterally. ♀: pattern similar but black on head replaced by gray; yellow of pileum largely obscured by olive feather tips; above washed with olive. Upper mandible black, lower dusky-horn; legs dusky. **Immatures:** resemble respective adults but chin largely pale gray to whitish.

Habits. In winter prefers wooded areas, especially forest canopy and at edges or gaps, or in adjacent tall second growth or semi-open; in migration may occur in low scrub. Individuals apparently establish winter territories, within which they often accompany mixed flocks of antwrens, greenlets, and other warblers, gleaning small insects and spiders at all heights, specializing in gaping and probing into rolled dying or dead leaves in thickety tangles for caterpillars, spiders, and crickets.

Voice. A short, rather soft *chip* or *tsup*.

Status. Fairly common fall migrant (early September–late October); uncommon but widespread winter resident of Caribbean lowlands and middle elevations of both slopes, to 8200ft (2500m) on migration, 7000ft (2150m) in winter; down to 2300ft (700m) on Pacific slope but virtually absent from lowlands. Most numerous at 2300–4600ft (700–1400m) on Caribbean slope; in Valle Central, rare in winter but more common during migration. Departs mid to late April.

Range. Breeds in SE Canada and NE USA; winters from S Mexico and Guatemala to Colombia and Venezuela.

BLUE-WINGED WARBLER

Vermivora pinus Pl. 41(6)

Reinita Aliazul

Description. 4½″ (11.5cm); 8.5g. Small; mostly yellow head and underparts; dark eye-stripe; wing-bars. **Adult ♂:** head yellow, except feathers of hindcrown edged with olive; eye-stripe black; back olive; tail and wings gray with 2 white wing-bars; rump yellowish; underparts yellow except crissum white. ♀: eye-stripe dusky; crown yellowish-olive; cheeks washed with olive. Upper mandible blackish, lower gray; legs dark gray. **Immature ♂:** like adult female, except eye-stripe often black. ♀: eye-stripe dusky-olive; crown olive like back (little or no yellow on feather bases).

Habits. Much like those of Golden-winged Warbler but seems more partial to second growth, semi-open, and hedgerows; pries open rolled leaves by gaping, probes them for insect larvae and spiders; hovers under leaves or hangs acrobatically under leaves and twigs to snatch prey.

Voice. A squeaky *swik* or *sirk* (Slud).

Status. Very uncommon to rare migrant and winter resident, early to mid-September through mid to late April; most numerous in Caribbean lowlands, especially near coast, but scattered birds winter on both slopes, lowlands to ca. 5000ft (1500m).

Range. Breeds in extreme SE Canada and E USA; winters from S Mexico to C Panama.

Note. Hybridizes frequently with Golden-winged Warbler; the commonest hybrid phenotype is "Brewster's Warbler," which resembles a Golden-wing but with face pattern of Blue-wing: black eye-stripe but no black on cheeks or throat. About 5 sightings of "Brewster's Warbler" in Costa Rica, mostly from middle elevations of both slopes (Guayabo, San Vito, Monteverde), also La Selva.

TENNESSEE WARBLER

Vermivora peregrina Pl. 40(22)

Reinita Verdilla

Description. 4½" (11.5cm); 8.5g. Small, plainly clad, with fine, sharp-pointed bill. Above mostly olive-green, the greater coverts with more yellowish tips forming a faint wing-bar in some; narrow pale superciliaries and dark eye-stripe; below white, except throat and breast more or less yellowish; small white spots on inner webs of outer 1–2 rectrices. **Adult** ♂ Winter: pileum olive-gray; throat, breast, and superciliaries white, strongly tinged with yellowish or buff. Breeding: pileum gray, superciliaries and throat white. ♀: pileum olive, usually duller than back; throat, breast, face, and superciliaries yellow, strongly tinged with olive. Bill dark gray, tomia and gonys horn-color; legs dark gray. **Immature** ♂**:** like adult female, but pileum often brighter, below sometimes whiter. ♀: pileum as bright as back; superciliaries, throat, and breast brighter yellow, with little olive tinge. Philadelphia Vireo has thicker bill, bigger head, broader superciliary and eye-stripe. In Guanacaste, face often stained orangish by *Combretum* pollen during dry season.

Habits. During winter prefers semi-open, second growth, coffee plantations, gardens; regular in canopy and at edges of forest; in migration may occur almost anywhere; social, usually in small flocks that move about within a definite area, often in company of greenlets, honeycreepers, and other warblers; actively gleans in foliage; probes rolled leaves for insects and spiders; visits flowers of many sorts for nectar (sometimes individuals defend flowers for short periods); pierces berries to suck juice; eats many small berries, especially those of mistletoe, also protein corpuscles of *Cecropia*; attends feeders for bananas.

Voice. High-pitched *tsit* and *tseep* notes: one can often locate large blooming trees (*Erythrina, Eucalyptus*) by the "tseeping" swarm of Tennessees visiting the flowers! Territorial birds give a sharper *chip*.

Status. During fall migration common to abundant over nearly the entire country, less often at high elevations; usually arrives mid to late September, with large waves from mid-October into November; in winter common to abundant from sea level to ca. 7500ft (2300m) on both slopes, generally more numerous at middle elevations than in lowlands and on Pacific than Caribbean slope; very abundant in Valle Central; late in dry season, commonly to at least 10,000ft (3000m) in montane forests when oaks flower and put forth new leaves; departs in April or early May.

Range. Breeds in Alaska, much of Canada, and extreme N USA, winters from S Mexico to Colombia and Venezuela.

ORANGE-CROWNED WARBLER

Vermivora celata Pl. 40(19)

Reinita Olivada

Description. 4¼" (11cm); 8g. Small, fine-billed; above dingy olive-green; below yellowish with indistinct olive streaking on throat, breast and sides; pale eye-ring interrupted by dusky eye-stripe. **Adult** ♂**:** orange-rufous crown-patch partly concealed; back and crown washed with gray; eye-ring yellowish. ♀: little or no rufous in crown; eye-ring whitish; below often tinged with brown, dark streaking heavier than in male. Upper mandible black, lower gray, tomia whitish; legs dusky. **Immatures:** like adult female but above and below with more brownish, throat often whitish. ♂: often some rufous in crown. ♀: lacks rufous in crown; crissum often distinctly olive rather than yellow. Most like Tennessee Warbler but easily distinguished by darker streaked underparts, indistinct superciliaries.

Habits. Like Tennessee Warbler, freely takes both nectar and fruits as well as insects; most likely to be seen singly, perhaps with Tennessee Warblers.

Voice. Call a sharp *kip* or *tsip*.

Status. Accidental fall migrant; only Costa Rican report a banded bird recovered in Limón (26 September 1974), when the first waves of Tennessee Warblers normally reach Caribbean slope of Costa Rica.

Range. Breeds from W Alaska to S California and SE Canada; winters from S USA to Guatemala and Belize.

NASHVILLE WARBLER

Vermivora ruficapilla Pl. 52(3)

Reinita Cachetigrís

Description. 4¼" (11cm); 7.5g. Gray half-hood suggests *Oporornis,* but much smaller with finer bill; pale throat and complete white eye-ring in all plumages. **Adult ♂:** slate-gray half-hood extends to sides of throat; a partly concealed chestnut crown-patch; rest of upperparts olive-green, brightest on rump; center of throat, breast, sides, and crissum bright yellow; center of belly white, flanks washed with olive-green. ♀: similar but pileum tinged brownish and with little or no chestnut; back washed with gray, sides and flanks with olive. Upper mandible blackish, lower flesh-gray, tomia paler; legs dusky. **Immature ♂:** like adult female but often more chestnut in crown; chin and sometimes upper throat whitish. ♀: similar but duller; pileum and often face strongly washed with brown, passing to gray at sides of throat; center of throat mostly buffy-white, sides and flanks washed with brownish-olive.

Habits. In Costa Rica evidently prefers dense shrubbery along forest edges and roadsides; only solitary individuals seen to date. Farther N in Central America, where regular in winter, resembles Tennessee Warbler in occurring mostly in small roving flocks, visiting flowers freely, and taking small berries and arillate seeds as well as gleaning actively for small insects.

Voice. Calls include a high, thin, dry *tsip* and a louder, more metallic *chip*.

Status. Probably a casual winter visitor, presently known from sightings at La Pacifica, 14 December 1970 (Opler), and Virgen del Socorro, 8 March 1987 (Newfield). Like several other migrants outside their normal ranges, can evidently turn up at low to middle elevations on either slope.

Range. Breeds from SW Canada to C California, and S-C and SE Canada to NE USA; winters from NW Mexico and S Texas regularly to El Salvador (Thurber et al. 1987), rarely or casually to Costa Rica and W Panama.

FLAME-THROATED WARBLER

Parula gutturalis Pl. 41(7)

Reinita Garganta de Fuego

Description. 4¾″ (12cm); 11g. Sharp-pointed bill; contrast between orange-yellow to vermilion throat and chest and whitish belly diagnostic; no wing-bars. **Adults:** above slate-gray, with a large black triangle on back; lores, eye-ring, and malar area black (extending to auriculars in male, usually not in female); posterior underparts centrally white, grayer on sides and flanks. Bill black, except base of lower mandible yellow; legs brownish. In some pairs, male's throat deeper orange than female's. **Immatures:** like adults but throat and breast paler; may retain some juvenile wing-coverts through first year. **Young:** above brownish-gray, with little or no black; wing-coverts indistinctly tipped with dull buff; throat and breast dull buff, posterior underparts grayish-white.

Habits. Frequents canopy and edge of mountain forests and scattered trees in adjoining clearings, sometimes descending into low, bushy growth on sides of ravines; very active, gleans caterpillars, insects, and spiders from foliage at tips of branches, sometimes hanging acrobatically; probes into tufts of lichens or moss on twigs; takes small berries, especially of mistletoes; territorial while breeding, adults often accompanying Yellow-winged Vireos, bush-tanagers, or migrant warblers; juveniles often flock; in wet season flocks of 30 or more of all ages troop through forest canopy.

Voice. Call a sharp, high-pitched *chit* or *chip*. Song a weak, dry, insectlike buzz with rising inflection, preceded by several short notes or a little trill: *pee pipipipi shwaaa*?

Nest. A bulky cup of green mosses and liverworts, lined with fibrous materials, on a low bank or amid epiphytes to 70ft (21m) up in a tree; roofless but always well shielded above by epiphytes, especially bromeliads, or by dense grass and moss on a bank. Eggs 2, dull white, immaculate or everywhere faintly and finely sprinkled with pale brown. March–May.

Status. Common resident of Cordillera Central (except rare N of Volcán Barva) and Cordillera de Talamanca, from 7000ft (2150m) to timberline, locally down to 6000ft (1850m) on Caribbean slope; in latter part of wet season sometimes appears as low as 4600ft (1400m).

Range. Costa Rica and W Panama.

NORTHERN PARULA

Parula americana Pl. 41(4)

Parula Norteña

Description. 4¼″ (11cm); 8g. Resembles Tropical Parula but with 2 wing-bars, interrupted eye-ring; no orange tinge on throat, but usually a dark breast-band. **Adult ♂:** above bluish-gray, with olive-green triangle on back; eye-ring and wing-bars white; lores black; chin, throat, and breast yellow with black and chestnut breast-band; belly white; flanks washed with grayish-buff. ♀: upperparts washed with olive; no black on lores; little or no chestnut in reduced breast-band. Upper mandible black, lower yellow; legs yellowish. **Immature ♂:** breast-band largely obscured by pale yellow fringes. ♀: breast-band faint or absent, replaced by dingy white; flanks and belly often very buffy.

Habits. At woodland edges, in semi-open and tall second growth, usually from middle heights to high in canopy; gleans actively amid foliage at tips of twigs, often hovering to snatch small insects and spiders; seen singly but often with mixed flocks of small birds.

Status. Very rare migrant and winter resident, recorded between late October and early April ca. 1300–4000ft (400–1200m) on both slopes (Villa Quesada, Turrialba, San Pedro).

Range. Breeds in SE Canada and E USA; winters from C and E Mexico to Costa Rica, also West Indies.

TROPICAL PARULA

Parula pitiayumi Pl. 41(3)

Parula Tropical

Description. 4″ (10cm); 7g. Very small, with slim, conspicuously bicolored bill and single wing-bar, rarely traces of a second. **Adult ♂:** above dark grayish-blue, with large triangle of olive-green on back; lores, orbital area, and cheeks black; greater wing-coverts broadly tipped with white, sometimes 1–2 of outer middle coverts also; outer 3 rectrices with white spots on inner web; below bright yellow, suffused with orange-tawny on lower throat and breast, passing to white on crissum. ♀: similar but only lores and orbital area black, orange suffusion on breast much fainter. Upper mandible black, lower bright yellow; legs yellowish-brown. **Young:** above much duller gray, with little or no olive in back or black in face; below uniform pale yellow with little or no orange on breast; wing-bar dull and indistinct.

Habits. Prefers canopy and edge of tall, wet highland forest and taller trees in adjacent semi-open, clearings, and second growth; very active and restless; gleans insects from outer foliage and twigs, often hovering to snatch prey from undersides of leaves or flitting against vegetation; takes small berries and protein corpuscles of *Cecropia*; alone or in pairs, sometimes with flocks of migrant warblers and other small birds.

Voice. Call a soft, high *pit* or *sip*, becoming sharper, louder, and often rapidly repeated in excitement or agitation. Song a rapid series of high, thin notes followed by buzzy trills on different pitches: *tsew tsew tsip tsip tsrrrrrr ts'sreeeeee*, with variations, tirelessly repeated.

Nest. Tucked into a thick cushion of green moss, 30ft (9m) up in small tree in clearing near forest; 3 nestlings in early June (only Costa Rican record). Elsewhere builds roofed nests, eggs 2 to 4, white, spotted and blotched with chestnut.

Status. Fairly common resident of Caribbean slope at 2000–6000ft (600–1850m), occasionally down to 1000ft (300m); and above 3000ft (900m) on Pacific side, from Cordillera de Guanacaste S, excluding Valle Central.

Range. N Mexico to NW Peru, N Argentina, and E Brazil.

Note. Also called Olive-backed Warbler.

YELLOW WARBLER

Dendroica petechia Pl. 42(2)

Reinita Amarilla

Description. 4½″ (11.5cm); 9g. Only warbler with yellow tail flash; no contrasting facial markings or wing-bars. **Adults:** above yellowish-olive; below yellow; remiges, rectrices, and all coverts edged with greenish-yellow; inner webs of all rectrices mostly yellow. ♂: forehead, face, and underparts bright yellow; breast, sides, and flanks streaked with chestnut, finely (occasionally not at all) in fall, boldly in spring. ♀: forehead like back in color; face washed with olive; below usually paler yellow, with chestnut streaking sparse or absent. Bill dark gray, tomia whitish; legs yellowish-olive. **Immatures:** primary coverts edged with brownish. ♂: like adult female, but below more often streaked. ♀: dullest, most greenish on face and underparts, often pale yellowish-olive, rather than yellow.

Habits. Frequents second growth, brushy pastures and hedgerows, agricultural lands, semi-open, mangroves, gardens, and dooryards; solitary and territorial in winter; forages actively, gleans small insects in foliage, sometimes hovering or engaging in brief aerial pursuits; in migration may join flocks of other migrant warblers, but rarely joins mixed flocks in winter.

Voice. Call a sharp *chip* or *chit*, louder and stronger than similar note of Chestnut-sided Warbler; song a whistled *sweet-sweet-sweet. I'm-so-sweet*; rarely given in subdued form while recent arrivals settle territorial boundaries and before the spring departure.

Status. Abundant fall migrant (September–

October, with small numbers arriving by mid-August); common winter resident from lowlands of both slopes to 5000ft (1500m), sometimes to 7000ft (2150m); departs by early to mid-May, with major migratory waves noted mainly along Caribbean coast.

Range. Breeds from Alaska and C Canada to C Mexico; winters from S USA and West Indies to Peru, Bolivia, and Amazonian Brazil; resident in West Indies.

MANGROVE WARBLER

Dendroica petechia erithachorides Pl. 42(3)

Reinita de Manglar

Description. 5″ (12.5cm); 11.5g. Similar to Yellow Warbler but larger, nearly always with some rufous about the head, except in dingy, grayish young. **Adult ♂:** above mostly olive-green; below golden-yellow; entire head and throat rufous-chestnut; breast and sides lightly streaked with chestnut; inner webs of rectrices mostly yellow. ♀: only crown rufous-chestnut; face suffused with rufous; breast sparsely streaked. Upper mandible dark horn, lower paler, tomia whitish; legs yellow-brown. **Immatures:** below paler yellow than adults, often with some grayish patches on neck and sides of breast; no rufous in crown; male usually with patchy rufous about face and throat; juvenile primary coverts retained. **Young:** dingy; above grayish, tinged with olive, including edges to primary coverts; below grayish-white, sometimes tinged with pale yellow or buff.

Habits. Confined to mangroves and immediate vicinity, in permanent, sedentary pairs that vigorously exclude other Mangrove Warblers from their territories but sometimes tolerate or ignore migrant Yellows, at other times expel them as well. Foraging behavior generally similar to that of Yellow Warbler but more deliberate, with more gleaning and less hovering, mainly in foliage of mangroves, at any height, also in tangled aerial roots, sometimes even on ground.

Voice. Call a loud, strong *chip*, heavier than call of Yellow Warbler; song a short, rapid, loud, whistled phrase: *see weecha weecha wheet* with variations, usually ending on a high, upslurred note.

Status. Common resident of mangroves along Pacific coast from Golfo de Nicoya S, but apparently absent from outer Península de Nicoya to N; on Caribbean, only from Moín to Matina.

Range. Along Pacific coast from NW Mexico to Peru; along Gulf-Caribbean coast from E Mexico to Venezuela.

Note. The distinctive Cocos I. race [*aureola*; see Pl. 42(2c); also resident on Galapagos Is.] is larger, 5¼″ (13.5cm), 12.5g, with chestnut cap, heavy breast streaking in male, little or no chestnut in female; common in coastal scrub and *Hibiscus*. The Mangrove Warbler complex is generally considered to be conspecific with the Yellow Warbler; we treat the 2 separately here for convenience and to emphasize their ecological and behavioral distinctness in Costa Rica.

MAGNOLIA WARBLER

Dendroica magnolia Pl. 43(10)

Reinita Colifajeada

Description. 4¼″ (11cm); 8.5g. In all plumages the blackish tail with broad white band across middle of all but the central rectrices diagnostic; 2 white wing-bars; rump and underparts yellow. **Adult ♂** Winter: head gray; eye-ring white; back olive-green, streaked with black; upper tail-coverts black; sides heavily streaked with black. Breeding: pileum blue-gray, separated from black cheeks by white stripe; back black; below heavy black streaking converges to a solid patch on chest. ♀ Winter: like male but chest dull, pale yellow; back and sides lightly streaked with dusky; upper tail-coverts edged with gray. Bill blackish; legs dusky-brown. **Immatures:** above browner, with less contrast between head and back; male with dusky streaking on sides and back; female nearly or quite unstreaked; both with yellowish-white band across chest; upper tail-coverts broadly edged with olive-gray.

Habits. In winter frequents open groves, thickets, and gallery woodland; usually alone, possibly territorial, but often joins mixed flocks of other warblers, greenlets, gnatcatchers, and honeycreepers; forages for small insects and spiders by gleaning upper surfaces of leaves, sometimes fluttering down from higher perch or hovering, or pursuing fleeing prey in air.

Voice. Call a distinctive *tlep* or *t'zek*; not heard singing in Costa Rica.

Status. Very uncommon migrant and winter resident in lowlands of both slopes, locally to 5000ft (1500m) in central highlands; in winter most numerous in Guanacaste and Río Frío district, in migration along Caribbean coast; present from mid-September to mid-April.

Range. Breeds from N Canada to NE USA;

winters from C Mexico and Bahamas to C Panama and West Indies.

CAPE MAY WARBLER

Dendroica tigrina Pl. 43(6)

Reinita Tigrina

Description. 4¾″ (12cm); 11g. Below with heavy streaking in all plumages; sides of neck pale, with dark cheeks; yellowish rump; rather dull wing-bars. **Adult ♂** Winter: above grayish-olive, streaked with black, very heavily on crown; cheeks olive, mixed with chestnut; face, throat, and sides of neck yellow; rest of underparts whitish; breast and sides streaked with black. Breeding: crown black; cheeks chestnut; sides of neck yellow; underparts yellow, heavily streaked with black. ♀ Winter: like male but above streaked more lightly with dusky and below with dusky-olive; no chestnut and little or no yellow on face and neck. Breeding: duller than male, with no chestnut on face; crown mostly olive. Bill blackish; legs dusky-brown. **Immature ♂:** like adult but greener and above less heavily marked with black, the crown feathers olive-gray with concealed black centers; below with streaking less distinct. ♀: like adult but crown tinged with brown; below with streaking less distinct; little or no yellow on breast.

Habits. Mainly in open groves, garden trees, semi-open areas; gleans and hovers to snatch small insects and spiders in foliage or clings to mossy branches to extract larvae or other small invertebrates, carrying prey to perch to eat it; takes small berries; fond of nectar, joining throngs of Tennessee Warblers at flowering *Eucalyptus* and *Erythrina* trees and dominating them in aggressive encounters; solitary.

Voice. Call a high *sit* or *sip* like that of Tennessee Warbler; song, heard once in May, several high, thin, rather sweet whistled notes.

Status. Apparently a rare winter resident, recorded from both coasts and up to 4250ft (1300m); late November to mid-May.

Range. Breeds in Canada and N USA; winters from C Florida through West Indies, casually from Yucatán to Panama.

BLACK-THROATED BLUE WARBLER

Dendroica caerulescens Pl. 43(1)

Reinita Azul y Negro

Description. 4¾″ (12cm); 10g. Best field mark is a white patch at base of primaries, smaller and duller in female. **Adult ♂:** above dark grayish-blue; face, throat, and sides black; rest of underparts white. ♀: above olive-brown, tinged with gray; narrow superciliaries and spot below eye whitish; below dull buffy-yellow, tinged with olive on breast. Bill blackish; legs dusky to brownish. **Immature ♂:** like adult but above tinged with olive; chin mixed with white. ♀: like adult but above without gray tinge, below duller and more buffy; superciliaries and eye-spot duller, tinged with buff; whitish spot on primaries very small and dull.

Habits. Prefers forest canopy, edges, and semi-open in hilly country; actively gleans foliage, flits and sallies for small insects; visits flowers (e.g., *Norantea*); single birds often accompany mixed flocks.

Voice. A soft, dry ticking.

Status. Casual to rare migrant and winter resident; 6 sight records (January–March) at 2000–4300ft (600–1300m) on Caribbean slope of Cordillera Central (Carrillo, La Hondura, Cariblanco) and in Coto Brus region (San Vito).

Range. Breeds in SE Canada and E USA; winters principally in West Indies and N South America, occasionally in Central America.

YELLOW-RUMPED (MYRTLE) WARBLER

Dendroica coronata Pl. 43(11)

Reinita Lomiamarilla

Description. 5″ (13cm); 12.5g. A large warbler with a bright yellow rump, yellow on sides of breast. **Adult ♂** Winter: above mostly grayish-brown, streaked with black; forehead and lower back gray; distinct yellow crown-patch; 2 white wing-bars; below white, washed with brown and streaked with dusky on breast and sides; bright yellow patch on each side of breast. Breeding: above bluish-slate, streaked with black; cheeks and forehead black; superciliaries, eyelids, and throat white; below white, with large black patch on breast; sides streaked with black. ♀ Winter: like male but above browner, lacking gray on forehead and back; crown-patch partly concealed; anterior wing-bar brownish; upper tail-coverts grayish-brown with black centers. Breeding: as in fall but brighter, above and below with more clean-cut and contrasting pattern. Bill black; legs dark brown. **Immature ♂:** like adult female but upper tail-coverts black, edged with gray; anterior wing-bar usually white. ♀: crown-patch faint, concealed; throat tinged with buff; small yellow patches on sides of breast tinged with buff; upper tail-coverts mostly brownish; wing-bars dull; above with faint streaking.

Habits. Frequents pastures, savannas, roadsides, low scrub, and other open places; hops over ground picking up small insects and spiders or plucking them from grass; flies up to trees or shrubs when alarmed or to rest; usually in groups or small flocks, sometimes associating with Fork-tailed Flycatchers in Térraba district or with Palm Warblers along Caribbean coast during migration.

Voice. Call a low *chup* or a hoarse *chep*; a slight *tsip*.

Status. Uncommon to sporadically common fall migrant along Caribbean coast, often appearing by mid-September but not regular before mid-October; irregular and local winter resident countrywide from lowlands to at least 5000ft (1500m) in central highlands; departs late March.

Range. Breeds from Alaska and N Canada to Guatemala and NE USA; winters S to E Panama and West Indies.

Note. Western forms of this species, formerly considered a separate species and called Audubon's Warbler *D. auduboni* [see Pl. 43(11c)], winter regularly to Honduras and have been recorded several times in Costa Rica; they differ from the above description mainly in having the throat distinctly yellowish in all plumages.

TOWNSEND'S WARBLER

Dendroica townsendi Pl. 43(8)

Reinita de Townsend

Description. 4¾" (12cm); 9g. Resembles Black-throated Green Warbler but below with more yellow; cheek-patch darker and more distinct; streaked back. **Adult** ♂ Winter: feathers of crown and cheek-patch black, edged with olive; sides of head and neck bright yellow; back olive-green, streaked with black; remiges, upper tail-coverts, and rectrices blackish, edged with gray; 2 white wing-bars; inner webs of outer 3 rectrices largely white; throat black, partly veiled by yellow feather-edgings; breast and sides bright yellow, streaked with black; belly white. Breeding: crown, cheek-patch and throat solid black. ♀ Winter: similar but crown and cheeks olive; facial markings, throat, and breast dull yellow; sides washed with olive, narrowly streaked with dusky. Breeding: crown and cheeks black with olive edgings; throat mottled with black. Bill black; legs blackish. **Immatures:** like adult female, but male has bases of cheek and throat feathers black, sides more heavily streaked with black; female duller, with very little streaking; belly tinged with buff.

Habits. In disturbed forest, edges, second growth, and clearings; actively gleans small insects and caterpillars in foliage at all heights, occasionally hovering to pluck them from undersides of leaves; hawks flying insects; singly in Costa Rica, sometimes with mixed flocks of warblers, notably Black-throated Greens, and other small birds.

Voice. A soft, high-pitched *tsip*.

Status. A rare but regular migrant and winter resident (late September–early April) in highlands at 4500–10,000ft (1350–3000m); in recent years, wintering birds seen consistently at Monteverde and at Villa Mills on Cordillera de Talamanca.

Range. Breeds from Alaska to NW USA; winters from California to Costa Rica, casually to W Panama.

BLACK-THROATED GREEN WARBLER

Dendroica virens Pl. 43(9)

Reinita Cariamarilla

Description. 4¾" (12cm); 9g. Sides of head yellow, enclosing indistinct olive cheek-patch; above olive-green, with at most a trace of dark streaking; 2 white wing-bars. **Adult** ♂ Winter: little olive on cheek; throat and chest black, sides heavily streaked with black, the black markings partly veiled by pale yellowish fringes; posterior underparts white; wings edged with olive-gray; inner webs of outer 3 rectrices mostly white. Breeding: throat and chest solid black; lower breast often tinged with yellow. ♀ Winter: like male but above with no streaking; throat feathers yellowish-white with concealed black bases; some black on sides of breast; sides streaked with dusky. Breeding: throat and sometimes breast mottled with black. Bill black; legs dusky or brownish. **Immatures:** like adult female but alula and primary coverts dusky, edged with brownish-gray. ♂: black feather bases on throat and chest more extensive, only partly concealed; sides narrowly streaked with black. ♀: little or no black on throat, sides very narrowly streaked with dusky, face tinged with olive.

Habits. Prefers forest canopy and edges, pasture trees, and semi-open, but sometimes forages low in scrubby second growth; actively gleans small insects and spiders from upper and lower surfaces of leaves, often pirouetting and flitting on outside of foliage; sallies for flying insects; seen alone, in groups of up to 15, or with mixed flocks of small canopy insectivores. Groups and flocks troop through a wide area; lone birds often spend long periods working foliage of a single tree.

Voice. High-pitched, weak *tsit*, sometimes in series.

Status. Common migrant and winter resident (early to mid-October through mid-April) chiefly in highlands though in Caribbean lowlands during migration; winters mainly at 3300–10,000ft (1000–3000m), most abundantly at 5000–8000ft (1500–2450m) in Cordilleras Central and de Talamanca, lower in N cordilleras.

Range. Breeds across Canada and in E USA; winters from extreme S USA to Greater Antilles and C Panama.

HERMIT WARBLER

Dendroica occidentalis Pl. 43(7)

Reinita Cabecigualda

Description. 4¾″ (12cm); 9g. Resembles Black-throated Green Warbler but cheek-patch reduced to dusky or olive mottling; yellow of face usually extends to forehead; above grayish-olive, streaked with black. **Adult ♂** Winter: forehead and most of crown yellow like face; throat feathers black, edged with whitish; rest of underparts white, lightly streaked with black on sides. Breeding: head mostly yellow; hindneck and throat black. ♀ Winter: forehead yellow; crown olive, streaked with black; throat yellowish, smudged with black; below tinged with buffy; sides lightly streaked with dusky. Breeding: head mostly yellow, with dark mottling on crown and throat. Bill black; legs dusky. **Immatures:** like adult female but male whiter below with more black on throat; female very buffy below, also on wing-bars; little or no yellow on forehead or black on throat; above more olive (but still much grayer than Black-throated Green, which never has streaked crown).

Habits. Little known in Costa Rica; seen singly, mostly in hedgerows or at forest edge; forages in foliage of conifers, especially the widespread, introduced Guatemalan Cypress (*Cupressus lusitanica*).

Status. Casual winter resident at middle elevations, mostly on Pacific slope (December–March); several sightings at Monteverde, one on Volcán Irazú, one from near San Vito; also once on Caribbean slope near Colonia Palmareña.

Range. Breeds in W USA; winters from C California regularly to Nicaragua, casually to Costa Rica; accidental in W Panama.

CERULEAN WARBLER

Dendroica cerulea Pl. 43(2)

Reinita Cerúlea

Description. 4¼″ (11cm); 8g. Blue adult male distinctive, with white wing-bars and no yellow below; female and young greenish above with at least a tinge of blue and wing-bars tinged with yellow. **Adult ♂** Winter: above grayish-blue; below white; back and sides streaked with black; traces of a black breast-band; cheeks blackish; feathers of sides of crown with black bases; rectrices and remiges edged with bluish. Breeding: above brighter blue; distinct black breast-band; postocular stripe white. ♀ Winter: above olive-green, tinged with bluish on crown, rump, and upper tail-coverts; superciliaries yellowish; cheeks dusky-olive; below dull yellow, shading to white on chin and crissum; sides washed with greenish and streaked with dusky-olive; primary coverts edged with gray. Breeding: similar but brighter, above more bluish, more sharply streaked on sides. Upper mandible black, lower gray; legs grayish-brown. **Immature ♂:** like adult female but primary coverts edged with brownish; back faintly streaked with black; rump more bluish; feathers of sides of crown with black bases; belly white; sides streaked with blackish. ♀: above duller, with little or no bluish tinge; below more greenish, with little streaking.

Habits. Frequents forest canopy, gaps and edges, semi-open, usually high in trees, actively gleaning foliage and sallying for small insects from the tips of twigs; in fall usually in loose groups or waves that remain in a locality a day or 2.

Voice. Note a sibilant *ksip* or *tsip*.

Status. Rare to sporadically common fall migrant (late August or early September to mid-October) in Caribbean lowlands and foothills from sea level to 1600ft (500m), occasionally to 5000ft (1500m), with small numbers reaching Pacific slope in N and C highlands; abundance varies widely from day to day and between years. No winter records; in spring regular only along Caribbean coast in April, with occasional records inland, including Caribbean and Pacific lowlands.

Range. Breeds in extreme SE Canada and E

USA; winters from Colombia and Venezuela to Bolivia.

BLACKBURNIAN WARBLER

Dendroica fusca Pl. 41(8)

Reinita Gorginaranja

DESCRIPTION. 4½″ (11.5cm); 8.5g. In all plumages has dark cheek-patch framed by yellow or orange superciliary and side of neck, 2 pale stripes on each side of back and 2 white wing-bars. **Adult ♂** WINTER: center of forecrown, facial markings, throat, and breast deep yellow to orange, fading to white on lower belly and crissum; sides broadly streaked with black; feathers of cheeks and upperparts black, edged with buff; inner webs of outer 4–5 rectrices mostly white. BREEDING: cheeks and upperparts black, without buffy edgings; crown-patch, facial markings, throat, and breast deep orange. ♀ WINTER: crown-patch yellow, partly concealed; above brownish-olive, streaked with black; facial markings, throat, and breast pale yellow to yellow-orange; sides streaked with dusky. BREEDING: like male but black feathers broadly edged with brownish-olive, orange areas paler or yellowish. Upper mandible black, lower horn-color with dusky tip; legs grayish or dusky. **Immature ♂:** like adult but above with brown feather-edgings much heavier; black streaking on sides usually blurred. ♀: duller and paler than adult with crown-patch concealed or absent, sides indistinctly streaked with dull olive.

HABITS. Prefers forest canopy and edge, semi-open, tall second growth; during migrations also in low scrub, hedgerows, and gardens; forages in outer foliage of trees and shrubs, actively snatching small insects and spiders from leaves while perched or hovering or pursuing them in flight; usually solitary in winter, but often joins mixed flocks of tanagers, honeycreepers, warblers, and other small birds.

VOICE. Usual call a high, light *tsik*; not known to sing in Costa Rica.

STATUS. Common to abundant fall migrant (late August–late October), principally at middle elevations but locally from lowlands to high mountains on both slopes. Common winter resident at 3000–5000ft (900–1500m) and in smaller numbers down to 1500ft (450m), up to 8200ft (2500m); more numerous on Caribbean slope; spring migration throughout April, mainly along Caribbean coast.

RANGE. Breeds from C and E Canada to SE USA; winters from Costa Rica to Peru and Bolivia.

YELLOW-THROATED WARBLER

Dendroica dominica Pl. 41(12)

Reinita Gorgiamarilla

DESCRIPTION. 5¼″ (13.5cm); 10g. Gray and white, with yellow throat, black facial stripes and streaks on sides. **Adults:** throat and chest bright yellow; crown black (more extensive in male), rest of upperparts slate-gray; 2 broad, white wing-bars; superciliaries, sides of neck, and posterior underparts white; stripes through eye and down cheek black, the latter confluent with heavy black streaks on sides. Upper mandible black, lower gray; legs blackish. **Immatures:** similar but flanks, tertials, primary coverts, and often back washed with brownish, especially in female.

HABITS. Prefers semi-open, old second growth, thinned woodland, sometimes suburban gardens; a rather sluggish, slow-moving gleaner, hopping deliberately along twigs and narrow branches, usually well up in shrubs or trees; generally alone.

VOICE. A rarely heard, soft *chip*; usually silent.

STATUS. Rare migrant and winter visitor (mid-September to mid-March) with scattered records from both coasts to ca. 4400ft (1350m) in central highlands.

RANGE. Breeds in E USA and Bahamas; winters from SE USA to S Central America and West Indies.

CHESTNUT-SIDED WARBLER

Dendroica pensylvanica Pl. 43(4)

Reinita de Costillas Castañas

DESCRIPTION. 4½″ (11.5cm); 9g. In all plumages has 2 yellowish wing-bars; in fall and winter note white eye-ring, gray face contrasting with bright yellowish-green pileum. **Adult ♂** WINTER: above bright olive-green with partly concealed black streaking on back and usually on crown; wings and tail blackish with greenish to gray margins; outer 2–3 rectrices with inner webs extensively white; below grayish-white, shading to pale gray at sides of neck and breast; usually a broad chestnut stripe down each side. BREEDING: crown bright yellow; rest of upperparts greenish-yellow, heavily streaked with black; face and underparts white; a broad black stripe bordering crown, another extending from lores down cheeks to join a broad chestnut stripe along side. ♀ WINTER: like male but above with no streaking and at most a trace of chestnut on side. BREEDING: resembles male but duller, especially on crown; black on face and chestnut on sides less extensive. Upper mandible blackish, lower horn-color with dusky tip; legs dark gray. **Immatures:** resemble adult female but males often show some streaking above and chestnut on side.

HABITS. In middle and upper levels of forest, freely descending to shrub level at edges and gaps, in second growth, semi-open, and shady gardens; hops rapidly through foliage, gleaning small insects, caterpillars, and spiders and pursuing fleeing prey in the air; also takes berries; solitary and territorial in winter, often joining mixed flocks.

VOICE. Call a sharp *chip* or *chirp*, softer than note of Yellow Warbler; song a whistled *please-please-please-to-meet-cha*, heard frequently prior to spring departure in some years.

STATUS. Common to abundant fall migrant (mid-September, rarely late August, to early November) countrywide from lowlands to 6000ft (1850m), occasionally higher; common winter resident, except uncommon in dry NW, up to ca. 5000ft (1500m) on both slopes; spring migration mainly through Caribbean lowlands (early April–mid-May).

RANGE. Breeds in S Canada and N and E USA; winters from S Mexico to E Panama, casually to Trinidad and Venezuela.

BAY-BREASTED WARBLER

Dendroica castanea Pl. 41(9)

Reinita Castaña

DESCRIPTION. 5″ (13cm); 11.5g. Fairly large; streaked back; 2 white wing-bars; below nearly always tinged with chestnut or buffy, with no dark streaking. **Adult** ♂ WINTER: above olive-green, streaked with black; usually some chestnut in crown; throat whitish; rest of underparts buffy-white, washed with greenish-yellow laterally; flanks, sides, and often breast strongly tinged with chestnut; crissum buff. BREEDING: crown, throat, breast, and sides chestnut; sides of head black; a buffy patch on side of neck. ♀ WINTER: like male but no chestnut in crown, only flanks tinged with chestnut; above with black streaking reduced, especially on rump. BREEDING: much duller than male, below with much less chestnut. Upper mandible black, lower gray; legs dusky. **Immatures:** like adult female but male above usually more heavily streaked, below often with more chestnut; female with little or no streaking above and no chestnut below; chest and sides washed with buffy-yellow. See Blackpoll Warbler.

HABITS. In winter prefers forest edge and semi-open, sometimes forest canopy or second growth; virtually anywhere in fall, from interior of forest to pastures; deliberately gleans twigs and foliage, peering about, then hopping or flitting after prey, rarely hovering or sallying, taking largely slow-moving insects, including many larvae; eats much fruit and occasionally sips nectar. Seen singly in winter, often with mixed flocks including other warblers; in definite flocks or waves during migration.

VOICE. A soft chipping; a high *cheep* or *zeet*.

STATUS. Common to abundant fall migrant (late September–mid-November) in Caribbean lowlands and foothills, decreasing in abundance upward but regular to at least 6000ft (1850m), and across Valle Central; rare in Pacific lowlands; in winter uncommon to rare but regular resident in lowlands and locally to ca. 4000ft (1200m) on both slopes. Spring migration April to early May, mostly along Caribbean coast.

RANGE. Breeds from N Canada to NE USA; winters from Costa Rica to Colombia and Venezuela.

BLACKPOLL WARBLER

Dendroica striata Pl. 41(11)

Reinita Rayada

DESCRIPTION. 5″ (13cm); 12g. Very similar to Bay-breasted Warbler in fall and winter but lacks any trace of chestnut, rarely a faint buffy tinge below; sides of breast with at least faint darker streaking; in hand, outer web of primary 6 never emarginated like those of 7 and 8; nearly always emarginated in Bay-breasted. **Adult** ♂ WINTER: upperparts olive-green, shading to grayish on upper tail-coverts, and streaked with black; 2 white wing-bars; below dull white, suffused with pale yellow or greenish-yellow anteriorly; sides of throat and breast usually streaked with black. BREEDING: pileum black; upperparts olive-gray; cheeks and underparts white; malar stripe black; sides and upperparts streaked with black. ♀ WINTER: resembles male but black streaking above well marked only on back, faint elsewhere; anterior underparts usually pale greenish-yellow; sides of breast, sides, and sometimes flanks streaked with dull olive. BREEDING: above grayer, more heavily streaked with black, including crown; below whiter, streaked with black on sides. Upper mandible blackish, lower horn-color; legs dusky to flesh-horn. **Immatures:** resemble adult female, but male above more heavily streaked, below with less yellowish; female below with little streaking, sometimes crissum and belly faintly tinged with buff.

HABITS. In Costa Rica prefers edges and second growth, foraging primarily by gleaning;

usually alone, but may accompany mixed flocks.

VOICE. Not heard in Costa Rica.

STATUS. Casual to very rare fall migrant and winter resident, recorded at least 10 times from lowlands to lower middle elevations the length of both slopes. Evidently occasional individuals stray across Caribbean to Costa Rica during fall migration (late October–November), wander about irregularly, may remain at least to January.

RANGE. Breeds from Alaska and N Canada to extreme NE USA; winters mainly E of Andes in South America, casually in Central America.

PINE WARBLER

Dendroica pinus Pl. 41(14)

Reinita de Pinos

DESCRIPTION. 5″ (13cm); 12g. Rather large, sluggish; 2 wing-bars, often dull grayish or brownish rather than white; throat and breast more or less yellowish with olive streaking laterally. **Adult ♂:** above olive-green; faint yellowish superciliaries and suffusion on neck behind olive cheeks; wing-bars white or pale gray; throat and breast yellow; posterior underparts white; sides streaked broadly with olive. ♀: similar but upperparts, flanks, and wing-bars usually with brownish tinge; yellow of breast paler, often tinged with olive; sides less heavily streaked. Bill and legs blackish. **Immatures:** like adult female but male below less brownish; yellow of breast usually ends abruptly rather than fading to white of belly; female browner overall, entire underparts usually washed with buff, with little or no yellow.

HABITS. Frequents semi-open and old second-growth, moving deliberately and gleaning methodically for long periods in a small area, then abruptly flying to a distant spot; forages mostly well up in trees, often on branches and twigs just within the foliage.

VOICE. The 2 birds seen in Costa Rica were silent.

STATUS. Accidental fall visitor; 10 sightings of at least 2 individuals 22 August–10 September 1976, near San Pedro de Poás at 4900ft (1500m), and Ciudad Universitaria at 4000ft (1200m), in upper Valle Central (Stiles and S. M. Smith).

RANGE. Breeds in SE Canada, E USA, Bahamas, and Hispaniola; winters in SE USA and West Indies; accidental in Costa Rica.

PRAIRIE WARBLER

Dendroica discolor Pl. 43(5)

Reinita Galana

DESCRIPTION. 4½″ (11.5cm); 8g. Small, with yellow underparts, streaked sides; pale superciliaries and large spot below eye, dark malar area and cheeks. **Adult ♂:** above bright olive-green with back obscurely streaked with chestnut; 2 dull white to yellowish wing-bars; superciliaries and spot below eye yellow, the latter framed by black stripes; below bright yellow, sides streaked with black. ♀: similar but facial stripes dusky-olive, little or no chestnut in back; wing-bars grayish. Upper mandible blackish, lower dark horn; legs dusky. **Immatures:** superciliary and spot below eye whitish, rest of face grayish; wing-bars brownish or lacking; male has sides streaked with dusky, female has fainter dusky-olive streaking.

HABITS. Little known in Costa Rica; evidently prefers scrubby woodland edges and second growth, open groves; sometimes with flocks of other migrant warblers, greenlets, etc., actively gleaning foliage from shrub height to well up in trees, occasionally on ground, taking small insects and spiders; incessantly wags tail up and down.

VOICE. Call a soft, dry *chit*.

STATUS. Casual fall migrant and winter visitor, recorded from N and S Pacific lowlands, Caribbean lowlands and central highlands (late August–February); apparently an occasional bird strays in from Caribbean, perhaps driven by a tropical storm during fall migration.

RANGE. Breeds in extreme SE Canada and E USA; winters from extreme S USA through West Indies, straying to Central America.

PALM WARBLER

Dendroica palmarum Pl. 43(3)

Reinita Coronicastaña

DESCRIPTION. 5″ (12.5cm); 10.5g. Brownish, terrestrial, with streaked breast, greenish rump and yellow crissum. **Adult ♂** WINTER: above dull brown, streaked with black; crown usually spotted or suffused with chestnut; rump greenish-yellow; faint buffy-brown wing-bars; superciliaries buff; usually a whitish spot below eye; below pale yellow, tinged with buff and streaked with dusky on breast, sides, and flanks. BREEDING: cap bright chestnut; cheeks mixed with chestnut; superciliaries, throat and chest brighter yellow. ♀ WINTER: similar but with little chestnut in crown; underparts mostly buffy-white, only lower breast and crissum distinctly yellow. BREEDING: like male but chestnut and yellow of head duller, belly at most tinged with yellow. Bill blackish, except base of lower mandible pale horn; legs dusky. **Immatures:** resemble adult female but male often has more chestnut in crown, female buffy below with only a faint yellow tinge.

Habits. Prefers open areas near water, including wet pastures, flooded fields with short grass, roadside puddles, lawns, and salina dikes; forages on ground for small insects and spiders, retreating to bushes or trees when alarmed; also takes berries; incessantly wags tail; associates with Yellow-rumped Warblers in migration, and with flocks of seedeaters in winter.

Voice. Rarely heard, a slight *tsp*.

Status. Uncommon to rare migrant in fall (mid-October to November, rarely late September) and spring (late February–early April), mainly along Caribbean coast; casual to very rare winter resident recorded from lowlands to ca. 2600ft (800m) the length of both slopes.

Range. Breeds from N Canada to NE USA; winters from SE USA to West Indies, Yucatán, and rarely or casually S Central America.

OVENBIRD

Seiurus aurocapillus Pl. 43(12)

Reinita Hornera

Description. 5½″ (14cm); 18g. Large, chunky, terrestrial, tail-pumping warbler with eye-ring, striped crown, and streaked breast. **Adults:** feathers of center of crown and hindneck orange, tipped with olive-green; broad black stripe on each side of crown; rest of upperparts and cheeks olive-green; broad white eye-ring; below white, tinged with olive on flanks; malar stripe black; breast, sides, and flanks heavily streaked with black. Upper mandible dusky, lower horn-color; legs flesh-color. **Immatures:** olive tips of crown feathers broader, often completely obscuring orange; tertials fringed with tawny to dull rufous; sometimes outer greater coverts tipped with buff.

Habits. Prefers shady understory of forest with well-developed shrub layer, second growth and tall scrub; on migration in shaded gardens; walks steadily about on ground, picking up insects, spiders, and other small invertebrates; bobs head and constantly flips tail up, depresses it slowly; flicks partly cocked tail when alarmed or agitated; usually solitary and territorial in winter.

Voice. Note a sharp, hard *chut* or *tsuk*, often repeated rapidly when bird is agitated.

Status. Fairly common fall migrant (mid-September to late October), winter resident, and spring migrant (late March–early May), from lowlands to 5000ft (1500m) on both slopes. Obvious migratory waves only along Caribbean coast, where occasional birds may arrive as early as first week of September.

Range. Breeds in C and E Canada and E USA; winters from NW Mexico and SE USA to West Indies and N South America.

NORTHERN WATERTHRUSH

Seiurus noveboracensis Pl. 43(14)

Reinita Acuática Norteña (Menea Cola, Tordo de Agua)

Description. 5¼″ (13.5cm); 15g. Large, terrestrial, tail-pumping warbler with buffy-white superciliaries, heavy streaking below. **Adults:** above dark grayish-brown, including stripe through eye; cheeks mottled with buffy and brown; below varies from white, with faint yellow tinge on belly, to entirely pale yellow; throat more or less speckled with blackish; breast, sides, and flanks heavily streaked with blackish; usually pale spots at tips of inner webs of outer 1–2 rectrices (larger in males). Upper mandible blackish, lower horn-color; legs pale horn. **Immatures:** very similar but tertials fringed with dull buff; rectrices lack pale spots.

Habits. Prefers shady margins of streams, rivers, and ponds, wooded swamps and mangroves; in migration sometimes in open second growth or at wet spots in trails or roads; usually solitary, defending individual territories in winter (though either more social or territories very small in mangroves); walks over wet or muddy ground or stands on rocks or snags, constantly bobbing head and teetering hindquarters, picking up insects, crustaceans and other small invertebrates from ground or water's edge.

Voice. A distinctive loud metallic *tsink* or *chink*.

Status. Common migrant and winter resident (mid to late August through mid-May) from lowlands to 5000ft (1500m), rarely higher, on both slopes; often extremely abundant in September along Caribbean coast; in winter, most numerous in Caribbean lowlands and mangroves along Pacific coast.

Range. Breeds from Alaska and N Canada to N USA; winters from extreme S USA and N Mexico to Ecuador and Peru.

LOUISIANA WATERTHRUSH

Seiurus motacilla Pl. 43(13)

Reinita Acuatica Piquigrande (Menea Cola, Tordo de Agua)

Description. 5½″ (14cm); 18g. Very similar to Northern Waterthrush but bill larger, superciliaries white (sometimes tinged with brownish over lores), throat usually spotless, rich buff tinge on flanks and crissum. **Adults:** above dark grayish-brown, tinged with olive on rump and tail-coverts; broad eye-stripe dark brown, blending into mottled cheeks; malar stripe dark brown; below mostly white; breast, sides, and flanks heavily streaked with dark grayish-brown; tips of outer webs of 1–2 outer rectrices narrowly edged with white (more pronounced in males). Upper mandible blackish, lower horn-color; legs flesh-color. **Immatures:** very similar but tertials fringed with deep buff to dull tawny, little or no white in tail.

Habits. Prefers clear, swift-flowing, rocky streams in at least partly forested country, avoiding both extensive openings and still water (where Northern Waterthrush may be found). Habits and mannerisms generally resemble those of Northern; individuals stake out stretches of streams as foraging territories in winter.

Voice. A loud, metallic *tsink* or *tseek*, similar to that of Northern Waterthrush but sometimes higher-pitched, more penetrating.

Status. Uncommon to locally common migrant and winter resident (early to mid-August to mid-April), chiefly in foothills and highlands, at 1000–6600ft (300–2000m), occasionally to 8500ft (2600m); sporadically in lowlands, mostly on Caribbean slope during migration, and occasionally down to 330ft (100m) in winter.

Range. Breeds in SE Canada and E USA; winters from S Florida and N Mexico to West Indies, Colombia, and Venezuela.

KENTUCKY WARBLER

Oporornis formosus Pl. 42(15)

Reinita Cachetinegra

Description. 5″ (13cm); 13g. In all plumages has broad, incomplete yellow spectacles, bright yellow underparts, olive-green upperparts. **Adult ♂:** crown feathers black with gray tips, broadest on hindcrown; hindneck washed with gray; a yellow supraloral stripe continues around eye to below it; lores, cheeks, and broad stripe down each side of neck black; sides and flanks washed with olive-green. ♀: similar but feathers of crown usually with broader, brownish-gray tips; black feathers of cheeks usually tipped with olive; black on sides of neck usually less extensive. Upper mandible black, lower dark horn; legs flesh-color. **Immatures:** like adult female; male usually with black stripe on side of neck longer but narrow; female with crown and face dusky-olive.

Habits. Prefers shady understory of wet forest and tall second growth, thickets at forest edge or in gaps, in migration also in low scrub; hops over ground and logs, runs and flits along twigs in low shrubbery, gleaning ants, beetles, caterpillars, and other insects from undersides of leaves; habitually holds tail high, flicks it nervously; raises crown feathers when agitated; maintains individual winter territories.

Voice. A sharp *tchip*, *chewp*, or *chup*, rather variable in pitch, louder and heavier than note of Chestnut-sided Warbler.

Status. Widespread, uncommon to fairly common migrant and winter resident (early to mid-September through late April) from lowlands regularly to 4000ft (1200m) on both slopes, locally to 6000ft (1850m), especially on S Pacific slope; least numerous in lowland Guanacaste, where mostly restricted to wettest gallery forests.

Range. Breeds in E USA; winters from S Mexico to N Colombia and N Venezuela.

MOURNING WARBLER

Oporornis philadelphia Pls. 42(14), 52(5)

Reinita Enlutada

Description. 4¾″ (12cm); 11.5g. Above olive-green and below yellow with pale legs and bill, and always at least a trace of gray or brownish hood. **Adult ♂:** entire head, neck, and chest gray, mottled with black on throat; feathers of chest with black bases that become more conspicuous as gray fringes wear off. ♀: similar but hood grayish-brown, more grayish on face; throat whitish, throat and chest usually strongly tinged with buff; often a narrow, usually incomplete whitish eye-ring. Culmen blackish, rest of bill pale horn to flesh-color; legs pale flesh. **Immatures:** resemble adult female but throat yellow, usually tinged with buff; sides of lower throat and chest extensively smudged with olive to brownish; young males usually with black feather bases on chest that may show as dull mottling; yellowish eye-ring may be partial, complete or absent. See MacGillivray's Warbler.

Habits. Skulks in low thickets in young second growth and overgrown, weedy pastures; gleans small insects and spiders in low growth or from ground litter; eats protein corpuscles of small *Cecropia* trees; solitary and territorial in winter.

Voice. Call, often best clue to bird's presence, a sharp, hard, dry *chit* or *chet*, given repeatedly when bird is agitated. Song a beautiful, prolonged warble, occasionally heard in Costa Rica for weeks before spring departure (Skutch).

Status. Common fall migrant (mid-September through early November; a few arrive late August in some years) and winter resident in lowlands of both slopes, to 4600ft (1400m) in central highlands; spring departure late April to mid-May.

Range. Breeds in Canada and NE USA; winters from S Nicaragua to E Ecuador and S Venezuela.

MACGILLIVRAY'S WARBLER

Oporornis tolmiei Pl. 42(16)

Reinita de Tupidero

Description. 4¾″ (12cm); 11.5g. Very similar in all plumages to Mourning Warbler and not always distinguishable. Adults of both sexes have conspicuous broken white eye-ring, interrupted fore and aft; young have less striking interrupted yellowish-white eye-ring similar to, but usually more clean-cut than, eye-rings of many young Mournings. With bird in hand, note relatively longer tail of MacGillivray's; difference between wing and tail lengths (wing minus tail) of MacGillivray's is nearly always 9mm or less; of Mourning, 11mm or more.

Habits. Prefers second growth, scrubby old fields, hedgerows; habits very similar to those of Mourning Warbler but more restricted to highlands in winter.

Voice. Call a dry, thin *chit* or *tsik*, usually lighter than that of Mourning Warbler.

Status. Uncommon migrant and winter resident (mid-September to mid-May); in migration from lowlands to at least 6600ft (2000m), rare in Pacific lowlands; in winter mainly ca. 3300–6600ft (1000–2000m); most numerous in N cordilleras and central highlands.

Range. Breeds from S Alaska to W USA; winters from NW Mexico to W Panama.

CONNECTICUT WARBLER

Oporornis agilis Pl. 52(4)

Reinita Ojianillada

Description. 5¼″ (13cm); 14g. Plumages similar to those of smaller Mourning Warbler but always has complete white eye-ring; lower tail-coverts long, reaching nearly to tip of tail; legs and feet larger. **Adult ♂:** olive-green above, yellow below; complete slate-gray hood without black on chest; throat often paler gray. ♀: hood brownish-gray with well-defined posterior border below forming a distinct chest-band (lacking or only vaguely suggested in Mourning); throat pale buffy. Upper mandible horn-color to dusky, lower pinkish; legs flesh-color. **Immatures:** resemble adult female but hood and throat browner, eye-ring tinged with buff; chest-band more olivaceous in males, browner in females.

Habits. In Costa Rica prefers second growth and brushy forest edges, especially in moist spots; keeps low, often descending to ground over which it walks rather than hops; rather lethargic, slow-moving, solitary, sometimes staying for long periods in same small area.

Voice. Call a loud, metallic *pik* or *peek*.

Status. Casual or very rare migrant and winter visitor, known from 3 sightings (October–March) in Pacific lowlands and valleys: Puerto Jiménez (Lewis), La Pacífica (Campos), and San Isidro del General (Proctor). Usually migrates through West Indies, evidently straying rarely, perhaps regularly to Central America; probably can occur at low to middle elevations on either slope.

Range. Breeds across S-C Canada, extreme N-C USA; winters from NE Colombia to Amazonian Brazil.

COMMON YELLOWTHROAT

Geothlypis trichas Pl. 42(10)

Antifacito Norteño

Description. 4½″ (11.5cm); 9.5g. Smaller than local yellowthroats; browner than Masked or Olive-crowned; females and immatures have forehead tinged with rufous, never gray as in Gray-crowned; buffy-white belly distinguishes it from all *Oporornis*. **Adult ♂:** above brownish-olive, shading to olive-green on rump and tail; black mask covering forehead and sides of head, bordered behind by grayish-white; throat, breast, and crissum bright yellow; sides and flanks bright buffy-brown. ♀: lacks mask; crown tinged with rufous; interrupted buffy eye-ring; yellow less extensive on breast. Upper mandible black, lower mandible and legs pale horn to flesh-color. **Immatures:** like adult female but males have feathers on side of head with black bases suggesting a mask; females with little yellow below, often only a tinge on throat or chest.

Habits. Prefers reedy marshes and dense, scrubby vegetation in wet areas or adjoining open water; secretive, usually keeping low and hidden in vegetation, foraging by deliberately gleaning beetles, homopterans, etc., from leaves and stems; usually solitary in winter, often in loose groups or waves during spring and fall migrations.

Voice. Note a dry, husky *chet* or *dzht*.

Status. Uncommon to locally common migrant (mid-October–November and April–early May) from lowlands to middle elevations on both slopes; widespread winter resident in appropriate habitat, but common locally only in highlands (around Cartago, Cañas Gordas); very uncommon to rare in lowlands (e.g., Tempisque basin, Río Frío region).

Range. Breeds from SE Alaska and NE Canada to S Mexico and SE USA; winters from S USA to West Indies and E Panama, casually in Colombia.

MASKED YELLOWTHROAT

Geothlypis aequinoctialis Pl. 42(12)

Antifacito Sureño

Description. 5¼″ (13.5cm); 17g. Size of Gray-crowned Yellowthroat but male has more extensive black mask; female has pale gray cheeks, no black on lores, no buff on flanks; bill less heavy. **Adult ♂:** broad black mask from forehead to cheeks, bordered behind by gray; rest of upperparts olive-green; underparts bright yellow, washed with olive on sides and flanks. ♀: no mask; indistinct, pale superciliaries above pale gray cheeks; otherwise like male. Bill blackish except tomia and gonys pale horn; legs horn- to flesh-color.

Habits. Frequents grassy marshes, rush and cattail beds, and brushy flooded pastures, also well-drained fields when scattered trees and shrubs rise above lush herbage; shares its habitat with wintering Common Yellowthroat, whereas Gray-crowned occupies adjacent drier sites; skulks in dense vegetation, feeding on insects; usually in pairs.

Voice. Call a sharp, fine *stit* or *chip*; scold a nasal *jurrr* or *churr*; song a prolonged warble of repeated 3–4 note phrases that becomes progressively higher, weaker and faster, ending in a garbled flourish, 1 version sounding like *tuwichywer tuwichywer tuwichywer tuweecha tuweecha tuweecha chee-che-che chit*. Such a song lasts ca. 5sec; several may be run together with hardly a pause; delivered from a low perch or as the male rises steeply to a height of 13–20ft (4–6m).

Nest. A bulky cup, mostly of broad grass blades, lined with brown fibers, 18″ (46cm) up in a dense tussock of grass. Eggs 2, white, spotted with pale lilac and brown, mostly on thick end. May (1 record).

Status. Locally common resident in San Vito–Coto Brus region near Panama border at 3000–4000ft (900–1200m).

Range. Costa Rica to N Argentina and Uruguay.

Note. The form inhabiting Costa Rica and adjacent W Panama is separated by the length of Panama from other populations and is often considered a separate species, *G. chiriquensis*, Chiriquí Yellowthroat.

OLIVE-CROWNED YELLOWTHROAT

Geothlypis semiflava Pl. 42(11)

Antifacito Coroniolivo

Description. 5″ (12.5cm); 15.5g. Differs from other yellowthroats in lacking gray or rufous in face or crown and white or buffy on underparts. **Adult ♂:** black mask very extensive, including forecrown and extending to sides of neck, not separated by gray or white from olive-green of rest of upperparts; below bright yellow, strongly washed with olive-green on sides and flanks; crissum and sometimes breast tinged with olive-green. ♀: lacks mask; short stripe from bill over eye yellow; lores and cheeks olive-green; olive more extensive on sides and flanks. Bill blackish with whitish tomia; legs dull orange to yellowish. **Young:** above browner, more olive; below paler, duller yellow, washed with buff on lower throat and breast and with buffy-olive on sides and flanks.

Habits. Frequents wet pastures with tall grasses and brush, dense second growth and cattails in marshy sites, tall grass adjoining rivers and ponds, generally preferring wetter sites with taller vegetation than Gray-crowned Yellowthroat; gleans small insects and spiders while skulking in dense foliage; usually in pairs; often cocks tail or flicks it from side to side when agitated.

Voice. Call a sharp *chip* or *kip*; a short, dry, descending chatter when alarmed or agitated; song a prolonged, rich-toned warble of 2 and 3-note phrases, with a rolling finale: *wichaychu wichaychu wichaychu witchetychay-chu witchety chay-chu witchety chay-chu witchety-chrrrr*, delivered from top of a grass stalk or shrub, or in flight, sometimes as male rises 25ft (7.5m) up, then glides down into the herbage.

Nest. A cup of grass stems and leaves, lined with finer grass, 10–30″ (25–76cm) up in large, dense grass tussock. Eggs 2, white, sparsely sprinkled with fine black and dark brown spots, sometimes in a wreath at thick end. April–June.

Status. Common resident throughout Caribbean lowlands, locally to 3300ft (1000m), rarely higher in central highlands (e.g., Tapantí); reaches Pacific slope only W of Lago de Arenal.

Range. NE Honduras to W Ecuador.

GRAY-CROWNED YELLOWTHROAT

Geothlypis poliocephala Pl. 42(13)

Antifacito Coronigrís

Description. 5¼″ (13.5cm); 15.5g. Differs from other yellowthroats in its less extensive black mask, thicker bill, and longer, more graduated tail. **Adult ♂:** lores, orbital area, and narrow band across forehead black; pileum and auriculars slate-gray; rest of upperparts greenish-olive, brighter and greener on wings and tail; small white spots on upper and lower eyelids; cheeks and sides of neck olive-green; below bright yellow, paler on belly; flanks washed with buffy-olive. ♀: lores blackish; narrow band of slate-gray across forehead, over eyes and lores; pileum grayish-brown; sides of head, back, and rump brownish-olive; wings and tail greener; below with yellow less extensive than or male, belly mostly buffy-white; lower breast and crissum tinged with buff; sides and flanks buffy-brown. Culmen black, rest of bill and legs pale horn to flesh-color. **Immatures:** resemble adult female, but buffy on sides and flanks more extensive, extending across breast; males have more extensive yellow below and dusky and gray on face; females have little or no gray.

Habits. Frequents pastures and savannas with some tall grass and brush, low second growth, bracken fern, and sugarcane; in pairs or alone; gleans beetles, leafhoppers, caterpillars and other larvae, other insects, and spiders in dense, low growth; occasionally sallies for flying insects; takes some berries; often cocks or wags its long tail; perches and sings from tops of tall shrubs or small trees but dives into thickets when alarmed.

Voice. Its varied calls include a buzzy *dee-deet* or *tew-tezeet*, the last note higher; a rapid set of 3 rich-toned, slightly burry notes in falling cadence, *jeejeejew* or *cheeyuwur*, sometimes expanded into a descending series of 5 or more rolling *chewy* or *chereep* notes; a sharp chip in alarm. Two distinct songs: one a rich, finchlike warble of repeated phrases, similar to that of Olive-crowned Yellowthroat but shorter and not ending in a burry note; the other a fast series of clear whistled *weep* notes that accelerates into a descending trill.

Nest. A cup of dry grass blades and dead leaves, lined with fine grasses or other fibers, including horsehair, 6–20″ (15–50cm) up in a large grass tussock. Eggs 2, white, lightly spotted with black and dark brown, mostly near large end. May–July.

Status. Common resident of lowlands the length of both slopes, to 5000ft (1500m) or more in central highlands; has increased greatly on Caribbean slope and in Golfo Dulce region in last 20 years as large areas have been converted from forest to pasture.

Range. N Mexico to W Panama.

Note. Sometimes placed in the monotypic genus *Chamaethlypis* and called Ground Chat.

YELLOW-BREASTED CHAT

Icteria virens Pl. 42(9)

Reinita Grande

Description. 7″ (18cm); 25g. Very large, robust, heavy-billed warbler with white spectacles and yellow breast. **Adults:** above olive-green, tinged with gray on crown; lores and below eye deep black (male) to sooty-black (female); cheeks dark gray; stripe over lores, broken eye-ring and malar stripe white; throat and breast bright yellow; belly white; crissum tinged with buff; flanks washed with brownish-olive. Bill grayish-horn with blackish culmen; legs gray. **Immatures:** above averaging browner; supraloral stripes tinged with olive; lores dusky; olive wash on sides of breast; stronger buffy tinge on flanks and crissum. Face averages duller, more olive in females.

Habits. Skulks near ground in thickets of young second growth and woodland edges; sedentary, solitary, possibly territorial during winter; deliberately gleans foliage for insects and spiders, also notably fond of fruit (*Phytolacca*, melastomes, *Cordia*).

Voice. Usual call a low-pitched, grating, heavy *chut* or *chup*; a heavy, metallic "chucking."

Status. Widespread but generally uncommon migrant and winter resident (mid to late September through late April) in lowlands and foothills of both slopes, locally to ca. 4500ft (1400m) in Valle Central but not recorded from dry NW lowlands.

Range. Breeds from S Canada through USA to C Mexico; winters from NW Mexico and SE USA to W Panama.

HOODED WARBLER

Wilsonia citrina Pl. 42(5)

Reinita Encapuchada

Description. 5″ (12.5cm); 10.5g. Hooded males unmistakable; females are the only warblers that combine olive-green upperparts, yellow underparts, and a white flash in tail. **Adult ♂:** forehead and face bright yellow, framed by black hood extending to upper back, sides of neck, and chest; above otherwise olive-green, below yellow, washed with olive on sides and flanks; inner webs of outer 3 rectrices mostly white. ♀: yellow forehead and face surrounded by olive, the male's hood suggested to varying extent by black smudging on crown, sides of head, neck, and chest. Upper mandible black, lower grayish-horn; legs flesh-color. **Immature ♂:** like adult but feathers of hood with olive edgings. ♀: like adult but little or no black; usually an olive eye-stripe and wash on auriculars.

Habits. Frequents low, moist thickets, forest edge, second-growth woods, and hedgerows; usually stays low, sallying or hovering to pluck insects or spiders from foliage; frequently opens and closes tail, flashing white spots; holds individual winter territories but often accompanies mixed flocks therein.

Voice. A weak *chip* or a higher, sharp *cheep*; when agitated a louder, stronger *tchip*, resembling call of Chestnut-sided Warbler.

Status. Very uncommon to rare migrant and winter resident (late September through late April or early May) in Caribbean lowlands, Valle Central, General–Coto Brus region, and locally elsewhere up to 4000ft (1200m).

Range. Breeds in extreme SE Canada, E USA; winters from C and E Mexico to Panama.

WILSON'S WARBLER

Wilsonia pusilla Pl. 42(4)

Reinita Gorrinegra

Description. 4¼″ (11cm); 7g. Very small; in all plumages identified by contrast of bright yellow lores, superciliaries and forehead with olive-green or black crown; no wing-bars or tail-spots. **Adult ♂:** crown glossy black; rest of upperparts olive-green; forehead, lores, superciliaries, and throat golden-yellow; rest of underparts bright yellow; cheeks, sides, and flanks washed with olive-green. ♀: similar but black cap small, limited to forecrown, or absent; throat and face paler yellow. Upper mandible blackish, lower horn-color; legs flesh-orange. **Immatures:** resemble adults but underparts and face more extensively washed with olive; male's cap partly obscured by olive fringes.

Habits. Frequents canopy, openings and edges of forest, second growth, coffee plantations, brushy fields, dooryard trees and shrubbery; active, constantly flicking wings and tail; hunts small insects and spiders by gleaning, sometimes hanging upside-down, hovering, or pursuing them in flight with short, fluttering sallies; often in small groups on migration; in winter usually solitary and territorial but often accompanies mixed flocks or associates with other species, such as pairs of Rufous-capped Warblers, in drier areas.

Voice. Call an emphatic, rather low-pitched and nasal *chet* or *chip*; song a rapid series of whistled *cheep* notes ascending in pitch, sometimes heard just before spring departure.

Status. Common to abundant migrant and winter resident (mid-September through mid-May), mostly in highlands; in migration often common in Caribbean lowlands, rare and sporadic in Pacific lowlands; winters mainly above 3000ft (900m), becoming progressively more abundant up to timberline and the only migrant found regularly in páramo. As dry season intensifies, those below 3000ft (900m) appear to migrate upward, especially on Pacific slope.

Range. Breeds from Alaska and N Canada to SW and NE USA; winters from S USA to C Panama.

CANADA WARBLER

Wilsonia canadensis Pl. 42(8)

Reinita Pechirrayada

Description. 4¾″ (12cm); 10.5g. Only gray-backed warbler without wing-bars or crown-stripes; yellow underparts with contrasting white crissum; always at least a trace of necklace. **Adult ♂:** forecrown black with gray fringes; yellow supraloral stripe joining white or yellowish eye-ring; lores and below eye black; sides of head, rest of upperparts gray; a line of sharp black streaks across breast. ♀: feathers of crown with black bases, broad grayish-brown tips; lores dusky, no black on face; upperparts usually tinged with brown; breast streaks blurry, dusky or olive with little or no black. Upper mandible dark horn, lower paler; legs flesh-color to orangish. **Immatures:** like adult female but male usually has black spots mixed with dusky breast streaking; female above browner with no black in crown, supraloral stripe and streaking on breast often dull and indistinct.

Habits. Frequents thickets in taller second growth and forest, coffee plantations, semi-open, and hedgerows; actively gleans small insects from leaves and twigs, sometimes hovering; often cocks tail; in loose groups or waves on migration, often with other warblers; may join antwren flocks in forest or forage with army ants; sometimes alone.

VOICE. A rather quiet, dry *chut*, sometimes a low sputtering.

STATUS. Common to abundant fall migrant (early September–early November, occasionally later) from lowlands to 6000ft (1850m), rarely higher, on both slopes; uncommon spring transient (April–early May), chiefly along Caribbean coast, sporadically elsewhere; very rare winter resident in lowlands and middle elevations on both slopes.

RANGE. Breeds from C Canada to E USA; winters from Colombia and Venezuela to E Peru and N Brazil, casually in Central America.

AMERICAN REDSTART

Setophaga ruticilla Pl. 41(10)

Candelita Norteña (Raya Roja)

DESCRIPTION. 4¾″ (12cm); 8.5g. Bill broad; well-developed rictal bristles; black T on yellow or orange tail; yellow or orange patches on wings and sides of breast. **Adult ♂:** head, chest, sides, and upperparts black; basal two-thirds of all but central rectrices and bases of most remiges orange; sides of breast and wing-linings orange; rest of underparts white, sometimes suffused with orange. ♀: throat and spectacles dull white; head gray; rest of upperparts dark grayish-olive; dusky and yellow replace black and orange of male on wings and tail, except less yellow in primaries; breast-patch and wing-linings bright yellow; rest of underparts dull white, often tinged with buff. Upper mandible black, lower mandible and legs gray. **Immature ♂:** like adult female but above browner, below more buffy; breast-patches often distinctly orange; by spring may show a few black feathers on head, but does not acquire definitive plumage until after first breeding season. ♀: like male but little yellow on breast or wing.

HABITS. Frequents middle and upper levels of forest, light woods, semi-open, and tall second growth, much lower at gaps and edges; sometimes in low scrub while migrating; captures insects on aerial sallies, maneuvering with great agility amid branches; sometimes gleans or hovers; often droops wings and fans tail; solitary, probably territorial in winter, but may accompany mixed flocks.

VOICE. A sharp *tsip* or *chit*, softer than call of Yellow Warbler but similar in quality.

STATUS. Uncommon, occasionally common fall migrant (mid-August, rarely late July, through late October) and uncommon and local winter resident, from lowlands of both slopes to ca. 5000ft (1500m), rarely higher in migration; uncommon spring migrant (April–early May), mainly in Caribbean lowlands.

RANGE. Breeds from SE Alaska and E Canada to N Florida; winters from extreme S USA to West Indies, Ecuador, and NW Brazil.

SLATE-THROATED REDSTART

Myioborus miniatus Pl. 42(7)

Candelita Pechinegra

DESCRIPTION. 4¾″ (12cm); 10g. Bill broad at base; rictal bristles very long; long tail with conspicuous white corners; blackish throat contrasting with deep yellow breast and belly. **Adults:** center of crown dark rufous; rest of upperparts, sides of head, throat, and chest blackish-slate; tail black, the outer 3 rectrices tipped with white (outermost mostly white); breast and belly orange-yellow; especially in south, chest tinged with cinnamon-orange; vent and crissum white. Bill black; legs blackish. **Young:** head and upperparts dark sooty-gray, smudged with brownish; no chestnut in crown; face and throat dark sooty-brown; breast and belly dull buff with dusky fringes; less white in tail.

HABITS. Frequents upper understory to middle canopy of highland forest, coming lower at edges and gaps and in adjacent second growth and hedgerows along pastures; hops and flits through foliage or over trunks and along branches, often opening and closing its tail, which may help to flush insects that are then captured in intricate aerial pursuits, or less often gleaned from bark or foliage; eats protein corpuscles of *Cecropia*; in pairs throughout year; often accompanies mixed flocks.

VOICE. Call a sharp but not loud *pik* or *chip* or a drier *chet*; song a varied series of whistled notes, some slurred up or down, sometimes accelerating toward end: a simple *chee chee chee chee chee* to an accelerating *tew tsee tsew tsew-tseu-tsu-tsu-tseweep* or a more complex *tsi-chew tsi-chew tsi-chew chwee chwee chew chwee chew*, etc.; tone thin and dry to fairly mellow.

NEST. A bulky, roofed structure with a side entrance, built on an ample foundation of broad dead leaves, usually in a niche in a vertical bank, or a little hollow amid herbage on a steep slope, or amid epiphytes on a fallen log. Eggs 2, more often 3, white, with speckles and blotches ranging from bright brown to chocolate. April–May.

STATUS. Common resident throughout mountains, from ca. 2500ft (750m) on Caribbean slope and 3600ft (1100m) on Pacific side, to ca. 7000ft (2150m) except in drier areas such as Valle Central.

RANGE. N Mexico to N Bolivia and Guyana.

COLLARED REDSTART

Myioborus torquatus Pl. 42(6)

Candelita Collareja (Amigo de Hombre)

DESCRIPTION. 5″ (12.5cm); 11g. Resembles Slate-throated Redstart but easily distinguished by its yellow face and throat and dark breast-band. **Adults:** forecrown black, rest of crown rufous-chestnut, bordered by black; rest of upperparts blackish-slate; tail black, the outer 2 rectrices mostly white; forehead, face, and underparts bright yellow (paler on crissum), with blackish band across breast. Bill and feet black; narrow bare eyelid blackish, giving a big-eyed look. **Immatures:** like adults but retain some brown-edged wing-coverts. **Young:** upperparts dark slate, washed with brown; entire head, throat, and breast dull slate, shading to pale dull yellow on belly; no yellow on face or rufous on crown.

HABITS. Frequents mossy highland forests and adjacent bushy ravines, pastures, and second growth; forages like Slate-throated Redstart, often pirouetting with fanned tail, sallying in pursuit of startled insects; often joins mixed flocks, sometimes sallying for prey flushed by other birds; occasionally follows cattle in pastures and in more remote areas may follow humans, evidently because they also flush prey; paired throughout year.

VOICE. Call a sharp *pit* or *tzip*, more penetrating than that of Slate-throated; song longer, more mellow and varied, including warbles and trills as well as slurred whistles.

NEST. A roofed structure with a round entrance in the side, of dry bamboo leaves and fine vegetable fibers or brown scales of tree ferns, in a niche in a vertical bank, or depression in a grassy slope, or beneath a fallen log. Eggs 2–3, white, sprinkled all over with pale brown. March–May.

STATUS. Common resident of high mountains from Cordillera de Tilarán S; mainly above 5250ft (1600m) in Cordillera de Tilarán and from ca. 6000ft (1850m) up to timberline in Cordillera Central and Cordillera de Talamanca; especially late in rainy season, a few may descend to ca. 5000ft (1500m).

RANGE. Costa Rica and W Panama.

THREE-STRIPED WARBLER

Basileuterus tristriatus Pl. 40(20)

Reinita Cabecilistada

DESCRIPTION. 5″ (13cm); 12g. Facial pattern more contrasting than that of Golden-crowned Warbler, with curving black stripe behind and below eye (lacking in Worm-eating Warbler). **Adults:** above dull olive-green; head dull buff with broad black stripes on side of crown and through eye; broad black bar down auriculars joins eye-stripe with narrow, indistinct malar streak; feathers of central crown-stripe orange basally; chin buffy-white; below pale dull yellow, palest on throat; breast, sides, and flanks washed with olive. Bill pale horn with blackish base of culmen; legs reddish-brown. **Young:** above duller, duskier, with indistinct head-stripes; trace of pale wing-bars; below with strong brownish tinge.

HABITS. Frequents understory and adjacent edges of primary and tall secondary highland forest, in territorial groups of 5–8 birds that travel noisily in their foraging rounds, often attended by 1–2 individuals of Spotted Barbtails, Plain Antvireos, foliage-gleaners, etc.; forages actively, rummaging and gleaning in foliage, making fluttering sallies in pursuit of startled or dislodged insects.

VOICE. Foraging birds incessantly repeat high, sharp, staccato chips and twitters, lighter and less scratchy than calls of Golden-crowned Warblers. Song a rather siskinlike, rapid medley of short trills, twitters, warbles, and buzzy notes.

NEST. Undescribed (?).

STATUS. Common resident of wet middle elevations between 3300ft and 7200ft (1000–2200m) locally, chiefly on Caribbean slope from Cordillera de Tilarán S to Panama; to Pacific slope in upper Cordillera de Tilarán and around N end of Cordillera de Talamanca to Dota region; locally along Cordillera de Talamanca (especially in upper Coto Brus valley).

RANGE. Costa Rica to E Peru and N Bolivia.

GOLDEN-CROWNED WARBLER

Basileuterus culicivorus Pl. 40(18)

Reinita Coronidorada

DESCRIPTION. 4¾″ (12cm); 10.5g. Conspicuously striped crown, grayish upperparts, yellow underparts and orangish legs facilitate separation from Three-striped Warbler. **Adults:** feathers of center of crown with cinnamon-orange or yellow bases, olive tips; broad black stripe on each side of crown; rest of upperparts olive-gray; superciliaries olive-green to dull yellowish; a broken yellow eye-ring; indistinct grayish eye-stripe and cheeks; underparts yellow, washed with olive on throat, breast, sides, and flanks. Upper mandible dusky, lower horn-color with pale tomia; legs pale orange. **Young:** crown dull sooty-brown; upperparts dull gray with brown fringes; wing-coverts dusky with narrow buffy or brownish tips; lores and face mottled with yellowish; sides of head brownish-olive; throat dull buffy-olive, passing to buffy-brown on chest and sides, and dull buff below; crissum yellowish.

HABITS. Prefers undergrowth of mountain forests and adjacent second growth and edges; in pairs or small groups of 3–5 wandering through territory with antwrens, antvireos, redstarts, and other small birds, actively gleaning in foliage, poking into leafy tangles, occasionally making darting sallies after startled insects; takes berries; often cocks tail; continually flicks wings.

VOICE. Foraging groups keep up an almost continuous dry ticking or sputtering of staccato *chut* notes; song a simple series of 3–5 whistled notes that slide rapidly upward, the last note loudest and strongly slurred: *chee wee wee weet?*

NEST. A roughly globular structure with a side entrance, of black rootlets, fibers, strips of palm leaves, stems of foliose liverworts, densely lined with fine brownish fibers, hidden beneath fallen leaves on the forest floor. Eggs 3, white, heavily wreathed with dark brown blotches. April (1 Costa Rican record).

STATUS. Resident in mountains throughout, at 1000–5000ft (300–1500m) on Caribbean slope and 3000–7000ft (900–2150m) on Pacific.

RANGE. NE Mexico (casually S Texas) to N Argentina.

BLACK-CHEEKED WARBLER

Basileuterus melanogenys Pl. 40(17)

Reinita Carinegra

DESCRIPTION. 5¼″ (13.5cm); 13g. Large, with conspicuous white superciliaries above black cheeks; tail long; legs long and strong. **Adults:** crown chestnut-rufous, bordered narrowly by black; sides of head black; rest of upperparts dull olive; throat dull white, mottled with dusky on chin; chest pale grayish-olive; sides and flanks buffy-olive, shading to yellowish-white on belly and crissum. Culmen blackish; rest of bill and legs flesh-color to pale horn. **Young:** pileum and sides of head sooty-brown; superciliaries dull, tinged with olive; above browner, with 2 narrow buffy wing-bars; below dull yellowish, washed with brownish on chest, sides, and flanks.

HABITS. Frequents bamboo-choked ravines and understory of oak forest in high mountains, entering páramo but usually not clearings or pastures; in pairs or groups of 3–4 that travel over large territories, usually alone but sometimes with other small birds; actively rummages and gleans foliage, especially axillary tufts of bamboos, often hanging acrobatically; flits up to snatch prey from under leaves; eats spiders, moths, caterpillars, beetles, crickets, and ants; and also takes berries.

VOICE. Call a high, thin *tsit*, often in a rapid series when alarmed or varied to a sputter followed by a thin whistle: *t-t-t-t-tew*; a high *pit-tew*; song a slight, lisping, sputtering jumble of notes, *tsi tsi wee tsi tsi wu tsi wee*, shorter and weaker than that of Rufous-capped Warbler.

NEST. A bulky, roofed structure with a side entrance, of bamboo and other leaves mixed with pieces of fern, slender roots and rhizomes, thickly lined with shredded vegetable fibers and brown treefern scales, on a mossy bank or in a niche in the steep side of a ravine. Eggs 2, glossy white, speckled with cinnamon-rufous. April–June.

STATUS. Resident in upper parts of Cordillera Central and Cordillera de Talamanca, including Dota region, from 5300ft (1600m) up to timberline and somewhat beyond in páramo.

RANGE. Costa Rica and W Panama.

RUFOUS-CAPPED WARBLER

Basileuterus rufifrons Pl. 40(16)

Reinita Cabecicastaña

DESCRIPTION. 5″ (12.5cm); 11.5g. Larger than most migrant warblers, with different shape: thicker bill, heavier legs and body, shorter wings, and longer, narrower tail; no other yellow-breasted warbler has head mostly rufous. **Adults:** pileum and auriculars chestnut-rufous; lores and orbital area black; superciliaries, spot below eye and chin white; sides and back of neck grayish; rest of upperparts olive-green; below bright yellow, washed with olive on sides and flanks. Bill black; legs brownish-horn. **Young:** upperparts, including pileum and sides of head, dull brown with sooty fringes; indistinct buffy-olive superciliaries; 2 cinnamon-buff wing-bars; throat and chest olive-buff, tinged with yellow, shading to pale yellowish on breast and belly.

HABITS. In dry NW occupies interior and edges of deciduous forest; elsewhere prefers light woods, coffee plantations, tall second growth, brushy ravines, and hedgerows; forages at a slower pace than migrant warblers, spending longer intervals scanning actively before hopping or flitting to pluck small insects, caterpillars, or spiders from foliage; takes some berries; often cocks and wags tail, flicks wings; in pairs year-round.

Voice. Call an emphatic, rather metallic *chink* or *cheek*, often in chattering series when excited; song a bright, jerky, rambling medley, variable but usually beginning with 1–2 chirping notes and ending with a clear, whistled note: *chip, chirp, chipa-chupity chipity cha-cheweet*, delivered almost continuously from high perches at dawn and at intervals through day during much of year.

Nest. Oven-shaped, with a side entrance, of various vegetable materials, thickly lined with shredded bast fibers, well hidden amid litter on level or sloping ground, beside or between rocks or fallen logs, sometimes in a niche in a high bank. Eggs 2–3, white, finely speckled with cinnamon. April–July.

Status. Common resident throughout lowlands of N Pacific slope, S to about Tivives and Orotina; to 4000ft (1200m) or more on slopes of N cordilleras and locally to 6600ft (2000m) in Valle Central and Dota region; E in central highlands to beyond Paraíso; at 2000–5000ft (600–1500m) in Terraba–Coto Brus region.

Range. NW Mexico (casual in S Arizona and Texas) to NW Venezuela.

Note. Birds from SE Guatemala and C El Salvador S are often considered a separate species, *B. delattrii*, Chestnut-capped Warbler, from the northern, white-bellied race, *rufifrons*, but complete intergradation occurs in El Salvador.

BUFF-RUMPED WARBLER

Phaeothlypis fulvicauda Pl. 40(23)

Reinita Guardaribera

Description. 5″ (13cm); 14.5g. Pale rump and base of tail contrast with dark upperparts; tail rather broad; legs long; bill heavy. **Adults:** pileum dark sooty-olive; back and shoulders dark olive; wings and distal half of tail blackish; upper and lower tail-coverts and basal half of tail pale yellowish-buff; superciliaries dull buff; eye-stripe dusky; face dark olive with fine, buffy streaking; underparts dull whitish, clouded or indistinctly spotted with grayish to buffy-olive on breast and sides; flanks dull olive to buffy-brown. Bill black; legs dusky to yellowish-horn. **Young:** above browner than adult, including face, with sooty-black fringes and smudges; wingbars faint olive-buff; throat grayish-brown; breast and sides dull buff; belly whitish; buff of tail-coverts and tail paler than in adults.

Habits. Forages along narrow, shady watercourses or banks of wider rivers, flitting between exposed rocks and roots, hopping along logs and bare shore, picking up small invertebrates from water's edge and sallying up to catch flying insects; constantly flags spread tail from side to side, pumping it up and down; in wet weather, when streams flood, may forage on roadsides, woodland trails, lawns, or pastures; rarely walks for a few steps instead of hopping. Pairs defend linear territories throughout the year.

Voice. Call a hard, sharp *chut, chut-it*, or *chut-t-t*; song of male commences with a soft introductory warble, then a rapid level or slightly descending series of ca. 15 clear whistles that increase in volume to a ringing crescendo, which female occasionally answers with a softer, shorter song of sweet warbles and slurred notes.

Nest. A bulky domed or oven-shaped structure of diverse vegetable materials, lined with fine shredded fibers and fragments of leaf skeletons, with the wide doorway facing outward; in sloping bank, often beside a stream or path. Eggs 2, glossy white, spotted and blotched with brown or rufous, chiefly on the thicker end. April–August.

Status. Common resident throughout Caribbean and S Pacific lowlands, N to Carara; locally to 3500ft (1100m) on Caribbean slope and 5000ft (1500m) on Pacific.

Range. Honduras to NW Peru and W Amazonian Brazil.

Note. Often included in genus *Basileuterus*; by a few authors merged with the River Warbler, *B. rivularis*, of E South America.

ZELEDONIA

Zeledonia coronata Pl. 32(13)

Zeledonia

Description. 4½″ (11.5cm); 21g. Plump, dark, short-tailed, long-legged, terrestrial. **Adults:** crown bright cinnamon-orange, bordered by black stripes; rest of upperparts dark olive; face and underparts dark slate-gray, tinged with olive on flanks and crissum. Bill black; legs brownish. **Young:** orange of crown largely obscured by sooty-olive feather tips; below more brownish with sooty fringes, giving a smudgy or scaly effect.

Habits. Prefers bamboo-choked thickets in or beside cool, wet highland forests, along streams or ravines, in dense second-growth or páramo vegetation; hops about on ground or in low, dense herbage, gleaning larvae and pupae of lepidopterans and beetles, other small insects and spiders; flicks wings nervously; flies seldom and weakly; pairs for breeding, possibly year-round.

Voice. Call a high, thin, colorless *pseee* or *psss* with rising inflection, higher and thinner than similar calls of Black-billed Nightingale-Thrush or Timberline Wren. Song a variable short phrase of 3–5 high, piercing whistles: *see-see-suu seeep*, or *see-suu-seep*, often repeated; male and female sing back and forth.

Nest. A hollow ball of tightly packed mosses with side entrance, well concealed in niche in vertical bank overgrown with mosses and other plants. Eggs 2, whitish, heavily speckled with pale brown. April–June (Hunt).

Status. Resident on all higher mountains N at least to Volcán Miravalles in Cordillera de Guanacaste, from mountaintops down to 5000ft (1500m) in N cordilleras, 5500ft (1700m) in Cordillera Central, and 6000ft (1850m) in Cordillera de Talamanca (center of abundance ca. 8200ft (2500m); locally common but rarely seen; best detected by calls.

Range. Costa Rica and Panama.

Note. Often called Wrenthrush; long separated in its own family, the Zeledoniidae, and thought to be related to wrens and/or thrushes, but recent biochemical and anatomical studies indicate that *Zeledonia* is an aberrant warbler probably most closely related to *Basileuterus*.

FAMILY **Icteridae:** American Orioles and Blackbirds

The American orioles, oropendolas, caciques, grackles, cowbirds, meadowlarks, and their allies, collectively called icterids, form a varied family of some 90 species, confined to the continents and islands of the Western Hemisphere. Northern species tend to be migratory, but most members of the family are permanent residents in the tropics, from warm coastlands to high, frosty mountains. In size 6–22″ (15–56cm), they have sharp-pointed, conical bills with the culmen extending far up the forehead and sometimes expanded into a casque; the legs and feet are usually stout and strong. Most species are black, wholly or in combination with red, yellow, or orange; a few (especially females) are brown or gray, or streaked. The sexes differ in color in most migratory icterids but are often alike in even the most brilliantly colored tropical species; males and females may differ considerably in size. Icterids inhabit open areas, thickets, light woodland, marshes, and the canopy of dense forests, but rarely the understory; they eat insects and other small animals, fruits, seeds, and nectar; some are quite omnivorous. Their notes are as varied as their appearances and habits; many are superb songsters. Many nest in colonies, weaving long, pensile pouches that rival the work of African weaver-birds. Others build cups that are often reinforced with a layer of mud or cow dung, or hide open or roofed nests amid meadow grasses. Nests are built by females, exceptionally with a male's help. Eggs, 2–3 in many tropical species, often 4–6 or more at high latitudes, are white, blue-green, or blue, immaculate or more often spotted or scrawled with rufous, brown, or black. As far as is known, only females incubate for periods of 11–14 days. Hatched with closed eyes and sparse natal down, rarely naked, nestlings are fed by both parents of monogamous species and only rarely by males of polygamous, colonial species. Nestling periods range from 9 or 10 days in the Brown-headed Cowbird to a month or more in the large oropendolas. The 6 species of cowbirds, brood parasites that build no nest, are an anomaly in a family that includes the most skillful weavers in the Western world.

CHESTNUT-HEADED OROPENDOLA

Psarocolius wagleri Pl. 44(9)

Oropéndola Cabecicastaña

Description. ♂ 14″ (35cm), 225g; ♀ 11″ (28cm), 125g. Large, dark, with conspicuous yellow tail; pale eyes; long,pale bill with swollen casque on forehead; sparse crest of a few narrow, elongated feathers; tips of male's outer 5 primaries greatly attenuated. **Adults:** head, neck, rump, lower belly, and tail-coverts dark chestnut; back, wings, and central rectrices black; rest of underparts black in male, dark chestnut-brown in female; rest of tail yellow, but outer rectrices edged with black. Iris pale blue; bill greenish-ivory to pale yellow; legs black. **Immatures:** like adults but black areas duller and browner, chestnut duller and duskier. Probably 2 years to full adult dress. **Young:** head, neck, and underparts dark sooty-brown, blacker on back and wings, shading to dull chestnut on sides and flanks and cinnamon-brown on crissum; a tuft of pale yellow on each side of forehead; bill brownish.

Habits. Prefers forested regions but nests at forest edge or in isolated trees and clearings; forages mostly high in canopy, hopping and running along branches with speed and agility, probing moss and epiphytes, sometimes hanging acrobatically to scrutinize branch undersides or crevices; eats much fruit as well as insects, small frogs and lizards, and nectar of large flowers like those of balsa; troops about in straggling, noisy flocks of few to many birds that fly at treetop level; females usually outnumber males at colonies.

VOICE. Common call a deep, rough, resonant *kok* or *chek*; various short, liquid-sounding notes: *poik, ploop*, etc.; a loud, machine-gunlike rattle; a nasal, whining *heaaanh*, alarm or warning call an explosive *cack-cack!* Song a loud, liquid gurgle preceded by a harsh crashing or rustling noise. Many of the calls are deeper and more resonant than corresponding ones of Montezuma Oropendola; song less elaborate.

NEST. A brownish, pyriform pouch 1 yard (1m) long, with opening at the top, skillfully woven of fibers and slender vines or Spanish moss (*Tillandsia*), attached to a terminal twig. A dozen to 50 or more of these nests hang close together in the crown of a tall tree standing apart from others, in a colony conspicuous from afar, attended by birds coming and going all day long. Eggs 2, pale blue, marked with brownish-black, most heavily on the thicker end. January–June.

STATUS. Resident on Caribbean slope from lowlands to 4000ft (1200m), in breeding season locally to 5500ft (1700m); most numerous in foothills; local in Coto Brus and Golfo Dulce regions of S Pacific slope, occasional in lower Valle del General.

RANGE. S Mexico to NW Ecuador.

NOTE. Formerly classified in the monotypic genus *Zarhynchus*.

MONTEZUMA OROPENDOLA

Psarocolius montezuma Pl. 44(8)

Oropéndola de Moctezuma

DESCRIPTION. ♂ 20″ (50cm), 520g; ♀ 15″ (38cm), 230g. Very large; told from Chestnut-headed Oropendola by pale facial skin, dark eyes, and orange-tipped bill; outer primaries of male emarginated on distal half. **Adults:** head, neck, and chest black; rest of body deep chestnut-maroon (males) or chestnut (females), below with some black smudging, mainly on belly and thighs; remiges chestnut with blackish tips; rectrices yellow except central pair black. Skin of cheeks pale blue; that covering base of lower mandible purplish-pink; that around base of casque orange; base of bill black, tip orange; legs black. **Immatures:** like adults but head duller black, underparts dark, dull chestnut with faint, indistinct, black shaft-streaks. **Young:** head dark sooty-gray to blackish; median underparts sooty-gray, shading to dull chestnut on sides and flanks; above duller chestnut; bill dusky, indistinctly tipped with pale orangish.

HABITS. Usually nests in dense colonies in isolated trees in clearings or semi-open but often enters forest canopy to forage; flies slowly at treetop height, in straggling flocks of few to many birds, with measured wingbeats that, in males, produce a deep thrumming; searches for a variety of invertebrate and small vertebrate prey in foliage, epiphytes, bark crevices, and undersides of branches and fronds, gleaning, poking, and prying; takes much fruit and arillate seeds, including ripe cultivated bananas and *Cecropia* spikes; drinks nectar from flowers of banana, balsa, and *Norantea*; shy and wary while foraging but confident at nest trees.

VOICE. Calls include a low, guttural, throaty *chuck* or *cluck*; a yelping *eeuh*; nasal, whining notes; a loud, harsh call, like the ripping of strong linen, given by male; alarm call a stentorian *cack*. An agonized scream expresses concern for offspring. Song of male a far-carrying, liquid gurgle, melodious at a distance but with screeching, metallic overtones at close range, often preceded by a loud rustling or crashing sound, also produced vocally, as he bows deeply on a high perch, raising his wings and tail.

NEST. A tightly woven, massive, pyriform pouch 2–6ft (0.6–1.8m) long, attached to high terminal twig in dense colony. Eggs 2, white or buffy, lie on a bed of small leaves or leaf fragments at bottom of the pouch. January–August.

STATUS. Common resident in Caribbean lowlands, locally to 2600ft (800m) in most areas, but through Reventazón drainage and across Valle Central, to ca. 5000ft (1500m) on adjacent slopes; scarce and local in dry NW lowlands, mainly near bases of cordilleras.

RANGE. S Mexico to C Panama.

NOTE. Formerly placed in the genus *Gymnostinops*.

SCARLET-RUMPED CACIQUE

Cacicus uropygialis Pl. 44(11)

Cacique Lomiescarlata (Plío)

DESCRIPTION. ♂ 9″ (23cm), 68g; ♀ 8″ (20cm), 53g. Sharp-pointed pale bill; pale eyes; scarlet rump often hidden beneath wings; male Scarlet-rumped Tanager smaller, stockier, with shorter, thicker bill and obvious red rump. **Adults:** rump flame-scarlet; rest of plumage black, glossier in male. Iris pale blue; bill ivory-white to yellowish, silvery toward tip; legs black. **Young:** head dull sooty-black; body plumage dull brownish-black, below smudged with blackish; rump cinnamon-orange; iris dark.

Habits. Groups of 4–10 troop noisily through upper levels of forest, neighboring semi-open, clearings with scattered trees, and second growth, searching and rummaging in foliage, especially of palms, prying open rolled leaves, examining crevices, spiderwebs, and undersides of branches, often hanging acrobatically, occasionally descending to ground, taking insects, especially caterpillars, spiders, and occasional small vertebrates; holds large insect beneath a foot to dismember it. This species often associates with flocks of oropendolas, fruitcrows, or Black-faced Grosbeaks, the attendant commotion attracting followers like nunbirds, trogons, and flycatchers that sally for insects startled from concealment; also takes berries, arillate seeds, and nectar; roosts gregariously in trees.

Voice. Calls include loud, ringing, urgent-sounding whistles, *pleeeo* or *preeer*, often in series, *preeo-preeeer-peeo-peeo-peo*, etc. Courting males utter shrill notes, often with a whining quality, and a short, gurgling, wheezy *pu-zeek-eew*, suggesting louder songs of oropendolas. Females sing in higher, weaker voices, often in their nests.

Nest. A pyriform pouch, skillfully woven of fibrous materials, 15–25″ (38–64cm) long, the bottom padded with seed down, hanging conspicuously from outer branch of tree 11–100ft (3.5–30m) up, sometimes lower over water; often close to occupied wasps' nest; not in colonies, but current nest often very close to one used previously. Eggs 2, long-oval, white with a few pale brown and blackish spots and scrawls, mostly near large end. February–June. The young may be fed by flock members other than their parents.

Status. Common resident of Caribbean lowlands, except rare in drier regions S of Lago de Nicaragua; in S Pacific lowlands, N to Carara; occasionally, perhaps seasonally, to 3600ft (1100m).

Range. NE Honduras to W Ecuador and NE Peru.

Note. Birds of Central America and South America W of Andes are often separated as *C. microrhynchus*, Small-billed, or Flame-rumped, Cacique, from the larger *C. uropygialis* E of Andes.

YELLOW-BILLED CACIQUE

Amblycercus holosericeus Pl. 44(12)

Cacique Picoplata (Pico de Plata)

Description. ♂ 9″ (23cm), 70g; ♀ 8″ (20.5cm), 60g. Long conical bill flattened and chisel-like at tip; heavy-set, with rather broad, rounded tail; no other black bird in Costa Rica has bill and eyes yellow. **Adults:** plumage entirely black. Iris orange-yellow; bill pale yellow, tinged with gray at base, and with silvery sheen toward tip; legs gray. **Young:** plumage dull blackish, with dull black fringes giving a smudgy aspect; iris brown.

Habits. Frequents low impenetrable thickets in young second growth, canebrakes, streamsides, and forest edge; at high elevations in bamboo-choked forest understory and ravines; ransacks tangles of dead leaves; pokes bill into rolled leaves and clasping monocot leaf bases, pries or slits them open by gaping; also splits open slender stems by gaping after pecking, woodpeckerlike, into them; peers into crannies and hollows of all sorts in vegetation; eats crickets, roaches, beetles, and other invertebrates; takes some berries; paired all year.

Voice. Wrenlike; two distinct song types, sung as duet by mated pair: a short, rich-toned phrase of slurred whistles, *chewee-chewoo* or *cheer-chew-chew*, repeated 3–4 times; and a longer downslurred whistle followed by a prolonged rattling churr: *cheeeeeeeuu-chrrrrr-rrrr*. Scold a heavy, nasal *ramp-ramp-ramp* or a chattering, heavy *kwuk-kwuk-kwk*; other harsh or clear whistled notes.

Nest. A well-made, thick-walled cup, the outer layer of slender dry vines and fibrous leaf-sheaths of banana plants; middle layer of cane leaves or strips of broader leaves; lining of fine dry vines; 3–10ft (1–3m) up in thickets or weed-choked stands of sugarcane. Eggs 2, pale blue, wreathed with heavy and fine black scrawls and spots. February–June.

Status. Uncommon to very common resident at all elevations on Caribbean slope, Península de Nicoya, central highlands, and mainland Pacific slope from Golfo de Nicoya S, up to 10,000ft (3000m) or more in high mountains; rare or absent only in N Guanacaste.

Range. C Mexico to NW Peru and N Bolivia.

Note. Also called Chisel-billed or Prevost's Cacique; sometimes included in the genus *Cacicus*, but its bill, habitat, vocalizations, behavior, and nest are quite distinct.

GIANT COWBIRD

Scaphidura oryzivora Pl. 44(10)

Vaquero Grande

Description. ♂ 13″ (33cm), 200g; ♀ 11″ (28cm), 140g. Large, black; wings long, with longest primaries emarginated on outer web;

tail rather long, nearly truncate; bill thick, with flattened, convex culmen; often appears small-headed; conspicuous red iris. **Adult ♂:** head and body, including large, erectile neck-ruff, glossy purplish-black; wings and tail black with bluish gloss. ♀: smaller, black with faint bluish gloss; rudimentary neck-ruff. Bill and legs black. **Young:** duller, more sooty-black, with glossy purplish-black fringes. Bill ivory-yellow; iris yellow or brown; neck-ruff rudimentary (males) to absent (females).

Habits. Frequents open country with scattered trees, semi-open, gaps and clearings in forest; flies with a few quick, shallow flaps and a glide with wingtips bent upward or folds wings and swoops briefly after several strokes; male's wingbeats often produce a deep whir; often in flocks, foraging much on ground, shoving or turning stones and other small objects by pushing forward with bill; catches insects stirred up by grazing cattle and alights on their backs to pick off ticks; eats seeds and sometimes fruits. Courting male, standing in front of female on ground, stretches up, arches neck until bill is pressed down onto foreneck, erects neck-ruff into an iridescent cape, then bobs up and down.

Voice. Male, especially in courtship, utters an unattractive, spluttering, ascending screech: *fwrreeeeee?* Female flies from irate oropendolas voicing harsh, nasal whistles.

Nest. Parasitic, depositing its white eggs in the long pouches of oropendolas and caciques. Females spend long periods on conspicuous perches observing oropendola colonies, awaiting an opportunity to approach and lay. Usually promptly chased from nest tree, they sometimes succeed in entering a nest, especially when oropendolas all leave colony in a "panic reaction."

Status. Uncommon to locally common resident throughout Caribbean lowlands, locally in Valle Central, and in Golfo Dulce region, from sea level to 5500ft (1700m).

Range. C Mexico to W Ecuador, NE Argentina, and NE Brazil.

Note. Sometimes called Rice Grackle; formerly placed in genus *Psomocolax*.

BRONZED COWBIRD

Molothrus aeneus Pls. 44(15), 52(7)

Vaquero Ojirrojo (Pius)

Description. ♂ 8″ (20cm), 68g; ♀ 7¼″ (18.5cm), 56g. Rather chunky and short-tailed with thick, conical bill. **Adult ♂:** body and head, including erectile neck-ruff, black with strong greenish-bronze gloss; wings and tail black with bluish gloss. ♀: above duller black, below browner; rudimentary neck-ruff. Iris red; bill and legs black. **Young:** dull blackish-brown, blackest on crown; below with paler grayish fringes, giving faint scaly pattern; iris brown.

Habits. Frequents open country, especially agricultural areas; often seen along roadsides, in towns and urban parks; highly gregarious, flying and foraging in compact small groups or large flocks; forages on ground, gathering fallen grains and weed seeds, or insects stirred up by grazing cattle, from whose backs it sometimes plucks ticks; lifts, pushes aside, or turns over stones by shoving with bill, eating what it finds beneath. Courting male, with all plumage fluffed out, bobs up and down in front of female, or hovers on rapidly beating wings ca. 3ft (1m) above her, fluttering slowly straight down before her.

Voice. Males make a variety of long-drawn-out, high, thin whistles that will remind northern birders of calls of European Starlings; both sexes utter a thin, short, whistled *tswee*.

Nest. Parasitic, depositing pale bluish-green, unspotted eggs into nests of a variety of small birds; in Costa Rica favorite hosts include Yellow-throated Brush-Finch, Prevost's Ground-Sparrow, *Arremonops* sparrows. March– or April–July.

Status. Common resident of deforested parts of Caribbean slope, Valle Central and S Pacific slope; uncommon and local in most of Guanacaste, General-Térraba region, and in extensively forested areas; has spread with deforestation in last 30 years; sea level to 6000ft (1850m).

Range. SW USA to C Panama.

Note. Also called Red-eyed Cowbird; sometimes separated in genus *Tangavius*.

MELODIOUS BLACKBIRD

Dives dives Pl. 52(6)

Tordo Cantor

Description. ♂ 10″ (25.5cm), 108g; ♀ 9″ (23cm), 95g. The only all-black icterid in Costa Rica; bill conical, longer than those of cowbirds; tail of medium length, rounded, rather broad. **Adult ♂:** plumage entirely black, fairly strongly glossed with blue above and below. ♀: similar but duller, more brownish-black. **Immatures:** resemble female but plumage duller, more sooty, without gloss.

Habits. Frequents borders of fields and pastures, scrub, as well as light woodland and forest edge, especially along streams; walks over ground with head-bobbing gait, picking up insects, sometimes turning over leaves or stones with bill by gaping to expose prey; also gleans insects, especially caterpillars, from foliage, clinging rather awkwardly to slender twigs, sometimes gaping to open rolled leaves; takes nectar, especially of balsa, and may attack ripening ears of maize; usually in pairs that may accompany flocks of other blackbirds or anis; continually flicks partly spread tail up and down.

VOICE. Calls include a sharp, metallic *puit! puit! puit!*, a more ringing *pweet! pweet!* in excitement; in flight, a clear *tink tink tink.* A variety of loud, clear, mellow, usually downslurred whistles, each often preceded by one or more sharp notes: *ch'-pyeeeeer*; *ch-peeeer-peeeeeur*; *whit whit whit wheeer.* Male and female may call back and forth, the former rapidly repeating a loud clear *k'keeer*, the latter a slightly burry, softer *kipwaaherr.*

NEST. In Guatemala, a deep woven cup built by both sexes, the foundation of coarse fibers; 10–20ft (3–6m) up in dense foliage of dooryard tree. Eggs 3, light blue with scattered large and small black dots, especially near large end.

STATUS. Known in Costa Rica from a single bird photographed by S. G. Howell as it accompanied a flock of Grooved-billed Anis in trees beside a pasture near Sarmiento, ca. 5km N Puente Lagartos in extreme NW Provincia de Puntarenas. To be expected elsewhere in NW lowlands, because the species has been expanding its range S on the Pacific slope of Nicaragua in recent years (Martínez).

RANGE. SE Mexico to SW Nicaragua and NW Costa Rica.

NOTE. Sometimes *D. warszewicsi*, Scrub Blackbird of Ecuador and Peru is considered conspecific with *D. dives.*

GREAT-TAILED GRACKLE

Quiscalus mexicanus Pl. 44(16)

Clarinero o Zanate Grande (Sanate, Zanate)

DESCRIPTION. ♂ 17″ (43cm), 230g; ♀ 13″ (33cm), 125g. Long, stout bill; large legs and feet; long tail, in male folded into a deep V. **Adult ♂:** head and body glossy purplish-black; wings and tail black with greenish-blue gloss. ♀: above dull dark brown; wings and tail blackish and slightly glossy; superciliaries dull grayish-buff; eye-stripe dusky; below grayish-brown, paler and tinged with buff on throat, darker and more sooty-brown on flanks. Iris pale yellow; bill and legs black. **Young:** resemble adult female (though difference in size is evident between sexes at fledging) but above more sooty-brown with brighter fringes; below brighter, buffier; throat and center of belly pale grayish; underparts indistinctly streaked with dusky; iris dark gray or brown.

HABITS. From its original habitat of open marshes, coastal mudflats, mangroves, and estuaries has spread and occupied nonforest habitats including agricultural areas, savannas, towns, and suburbs, often roosting and nesting in village shade trees or urban parks; highly gregarious; eats berries and larger fruits, newly planted and ripening grain, larvae extracted from ground, ticks plucked from cattle, various intertidal invertebrates, lizards, small fishes, eggs and nestlings of other birds, carrion, scraps, and offal; forages mostly on ground, sometimes with cowbirds; turns over stones with bill, especially along shore; males in aggressive or courtship display pose with bill pointing straight upward.

VOICE. Common contact calls include a low, guttural *chuck*, and, from females, a rapid clicking *thick thick thick*; also chattering and rattling calls; males have an exceedingly diverse repertoire of sharp, loud notes, including a prolonged, squeaky, rising whistle, a stentorian, buglelike tooting, metallic whistles suggesting a little tin horn, and various guttural or creaking notes.

NEST. A bulky bowl with a foundation of miscellaneous plant matter, outer wall of coarse fibers, middle layer of mud or cow dung, and lining of fine rootlets or other fibers; singly or in loose colonies in mangroves, or in crowded colonies in palms or large trees inland, mostly 17–65ft (5–20m) up. Eggs 2–3, pale to bright blue, with dots, blotches or intricate scrawls of brown or black. January–July; 2 broods in Guatemala.

STATUS. Resident. Known only along Pacific coast early in this century, has been spreading inland; reached Caribbean lowlands in early 1960s, Valle Central by 1970; now widespread from both coasts (still rare in extensively forested regions) to at least 5000ft (1500m), and widely regarded as a pest.

RANGE. SW USA to NW Peru and NW Venezuela.

NOTE. Formerly placed in genus *Cassidix.*

NICARAGUAN GRACKLE

Quiscalus nicaraguensis Pl. 44(17)

Clarinero o Zanate de Laguna (Totí, Garrapatero)

DESCRIPTION. ♂ 11½″ (29cm), 94g; ♀ 9½″ (24cm), 58g. A small version of Great-tailed Grackle; males have even more strikingly "keeled" tail; female much paler below. **Adult ♂:** plumage entirely black, strongly glossed with purplish. ♀: crown and face dusky olive-brown, shading to dark sooty-brown on back and rump; wings and tail blackish; superciliaries, throat, and breast grayish-buff, brightest and buffiest on chest; belly, flanks, and crissum dark sooty-brown. Iris pale yellow; bill and legs black.

HABITS. Prefers open, marshy areas, lake shores, riverbanks, wet pastures and scrub; travels and forages in small groups to large flocks, walking about and picking up insects, seeds, etc., from ground or water's edge, often turning over small stones, leaves and debris; sometimes feeds beside or upon cows, in company of cowbirds (*Molothrus*) or at periphery of large flocks of Great-tailed Grackles.

VOICE. Usual notes a nasal *jep*, a sharp, dry *chik*; also a variety of whistled notes higher in pitch, weaker, less varied than those of Great-tailed Grackle; song of male composed of accelerating whistles with rising inflection: *kleeee klookleekleekleekleeee*?

NEST. Coarsely but strongly made cup of strips of grass and sedge leaves and rootlets; in colonies, in trees overlooking marshes, or well hidden in grass tussocks in marsh vegetation. Eggs 2–3, blue with spots, blotches, and scrawls of dark brown and black, mostly at large end. March–September.

STATUS. Resident, with local seasonal movements, in Río Frío region, especially at Lago Caño Negro, and in swampy areas S of Lago de Nicaragua (e.g., Las Camelias, San José de Upala).

RANGE. SW Nicaragua and Costa Rica.

ORCHARD ORIOLE

Icterus spurius Pl. 44(6)

Bolsero Castaño (Cacique Ahumado)

DESCRIPTION. 6¼″ (16cm); 20g. Small, slender; male unmistakable; females and young nondescript, greenish, with notably slender bills, dull white wing-bars. **Adult ♂:** head, neck, chest, back, wings, and tail black with brownish edgings that mostly wear away by spring; 1 buffy-white wing-bar; shoulders, rump, tail-coverts, and rest of underparts chestnut. ♀: above olive-green, brightest on forecrown and rump; wings dusky with 2 buffy-white wing-bars; below greenish-yellow, tinged with ochraceous on breast and with olive on sides and flanks. Bill blackish, except base of lower mandible silvery; legs bluish-gray. **Immatures:** resemble adult female but above duller and duskier; wing-bars duller, more grayish; below paler, without ochraceous tinge; males usually show some black on throat, which increases until by spring entire throat and face are usually black.

HABITS. Frequents open or semi-open areas including scrubby second growth, hedgerows, savanna, roadside groves, and gardens, usually in groups or small flocks; sometimes congregates in large numbers to sip nectar of flowering trees (*Erythrina*, *Gliricidia*, *Eugenia*), vines (*Combretum*), or banana plants; takes much fruit (notably mistletoes) as well as insects gleaned from foliage; roosts gregariously, sometimes with Northern (Baltimore) Orioles.

VOICE. Usual call a heavy *chek* or *cluck*, irregularly repeated; also a heavy, jerky chatter, deeper and less regular than that of Northern Oriole. For a short period in fall and spring, males sing a varied, hurried medley of mellow whistles, chips, and chatters.

STATUS. Common locally during migration in fall (late July or early August through October) and spring (March–April) from lowlands to ca. 5000ft (1500m) on both slopes; uncommon to rather rare winter resident, principally in lowlands of N Pacific slope and Río Frío region, locally at low or middle elevations elsewhere.

RANGE. Breeds from S Canada to C Mexico; winters from NW and C Mexico to N Colombia and NW Venezuela.

BLACK-COWLED ORIOLE

Icterus dominicensis Pl. 44(5)

Bolsero Capuchinegro (Cacique Amarillo)

DESCRIPTION. 7½″ (19cm); 32g. Medium-sized, black and yellow oriole with no white in wings or tail; slender, slightly decurved bill. **Adults:** mostly black; shoulder, rump, upper tail-coverts, wing-linings, lower breast, and crissum bright yellow; a touch of chestnut between black and yellow of breast. Bill black except base of lower mandible silvery; legs plumbeous. **Immatures:** forehead, face, and throat dull black; wings and tail blackish; rest of upperparts dull yellow-olive, feathers of back with more or less concealed blackish bases; below dull yellow, washed with olive laterally. Suggests young male Orchard Oriole but larger, with dark tail, no wing-bars; Yellow-tailed Oriole has yellow in tail in all plumages, is much larger.

HABITS. Frequents forest edge along streams and rivers, banana plantations, open and semi-open country with scattered trees and palms, tall second growth; gleans, probes, and pries amid foliage for insects and spiders, often hanging head downward to pluck prey from undersides of palm fronds; eats berries and *Cecropia* spikes; takes nectar from flowers of banana, epiphytes (e.g., *Columnea*), and trees like *Inga*; roosts gregariously amid tall grass. Juveniles sometimes travel in flocks.

Voice. Call a scratchy *weet* with rising inflection, sometimes varied to a nasal *deep*; when disturbed, a series of sharp *cherp* and *chep* notes; song a series of clear, mellow, slurred whistles and metallic notes in a seemingly random order, of no great carrying power, *wheet chewp chep wheet wheew wheew chep wheet chewp chewp. . . .*

Nest. A short pouch or deep cup, woven of fine, pale-colored fibers, suspended beneath the midrib, or a fold near the edge, of a leaf of banana, palm, or other large-leaved plant by strands laced through perforations that the oriole makes in the leaf, 5–35ft (1.5–11m) up; sometimes beneath the eaves of a house, hung from nails or service wires. Eggs unknown (?). March–July.

Status. Fairly common resident throughout Caribbean lowlands, to 2300ft (700m) locally, occasionally wandering to 4000ft (1200m).

Range. S Mexico to W Panama; Bahamas and Greater Antilles.

Note. Central American populations often considered a separate species, *I. prosthemelas*.

YELLOW-TAILED ORIOLE

Icterus mesomelas Pl. 44(3)

Bolsero Coliamarillo (Chiltote, Chiltotel)

Description. 9″ (23cm); 70g. Large, long-tailed, thick-billed oriole, mostly yellow, with conspicuously yellow-sided tail. **Adults:** lores, orbital area, cheeks, throat, center of chest, back, most of wings, and center of tail black; rest of head, neck underparts, rump, upper tail-coverts, and all but bases of 3 outer rectrices bright yellow; shoulder and part of wing-coverts yellow, forming a stripe separating black of wings and back. Bill black, except base of mandible silvery; legs bluish-gray. **Immatures:** head, neck, and underparts bright greenish-yellow, sometimes with ochraceous tinge; chest sometimes smudged with dusky; back, central rectrices, and wings dusky, with broad olive-green edgings, which may form 1 or 2 wing-bars; rump olive-green; outer 3 rectrices pale yellow with olive edgings. Young Black-cowled Oriole smaller, with all-dark tail and black face, throat, and chest (thus more like adult Yellow-tailed).

Habits. Prefers dense, vine-tangled thickets, canebrakes, crowded stands of *Heliconia* or Manila hemp, and other dense growth on low, swampy terrain; forages through banana groves, roadside trees, and dooryard shrubbery near water; usually in pairs that poke, pry, and glean in dense foliage for insects and spiders; occasionally takes fruit; flies rather seldom and for short distances between thickets, with pumping tail, and sometimes making a crashing noise, apparently with wings.

Voice. Call a mellow *chup chup chup*; also sharp, emphatic, metallic notes uttered singly or in a regular sequence; when disturbed, mechanically repeats a loud, harsh *kik-cher kik-cher kik-cher*. Singing male repeats over and over a phrase of 3–6 rich, mellow notes, then reiterates a wholly different phrase in the same manner; his mate may answer with a shorter, less varied song of repeated 2- or 3-note phrases, sometimes singing from the nest; the total effect somewhat wrenlike.

Nest. A deep cup attached by its rim, 7ft (2m) up in a thorny shrub beside a brook in scrubby pasture; woven of clean, bright vegetable fibers, so thin that the eggs are visible through the bottom. Eggs 3, white, tinged with blue, heavily marked on the thicker end with pale chocolate and black. April (only 1 nest, in Guatemala; no Costa Rican nest reported, but recently fledged juveniles in May and July).

Status. Resident in Caribbean lowlands. Formerly common and widespread but now greatly reduced in most areas by persecution for cage-bird trade (it is highly esteemed as a singer); in most areas from sea level to 1000ft (300m); an isolated colony at 3300ft (1000m) at Lago Dabagri in foothills of Cordillera de Talamanca in SE.

Range. S Mexico to W Peru and NW Venezuela.

SPOTTED-BREASTED ORIOLE

Icterus pectoralis Pl. 44(1)

Bolsero Pechimanchado (Chiltote)

Description. 8¼″ (21cm); 50g. Black spotting on breast of adults diagnostic; back solid black; wings with white on remiges, not coverts. **Adults:** lores, orbital area, throat, and center of chest black; sides of chest with light to very heavy black spotting; back and most of tail and wings black, except inner secondaries and bases of primaries broadly edged with white, outer rectrices edged with gray at tips;

rest of head, neck, underparts, shoulders, rump, and upper tail-coverts yellow-orange to brilliant orange (both intensity of orange and amount of black spotting may increase with age). Bill black except basal half of lower mandible silvery; legs bluish-gray. **Immatures:** head, neck, and underparts pale yellow to dull orange-yellow, washed with olive on crown, with little or no black on face, throat, or breast; back olive; rump ochraceous-olive; wings dusky with narrow dull white edgings; tail blackish with yellowish shafts. As with other resident *Icterus*, this juvenile plumage is largely replaced on the body, but not usually the wings and tail, by the time the birds are a year old and capable of breeding.

Habits. Frequents open woodland, shade trees around ranches and villages, edges of gallery forest and dry scrub; gleans insects in foliage, often prying open rolled leaves; often visits flowers of trees like *Caesalpinia, Gliricidia*, or *Erythrina* for nectar; often in small flocks when not nesting.

Voice. Calls include loud, nasal notes, singly or in a jerky series; song of male a prolonged, caroling liquid series of rich, clear, slow whistles, often quite beautiful; female's song simpler, with thinner tone.

Nest. A pyriform pouch, up to 18″ (46cm) long, with entrance at top, woven of fibers and thin roots of epiphytes, the bottom thickly padded with fine fibers, suspended in the terminal fork of a branch 20–60ft (6–18m) up. Eggs white, scrawled all over with black and pale lilac (in Guatemala; no Costa Rican nest examined). May–July.

Status. Uncommon and local resident of N Pacific lowlands and foothills, principally around Golfo de Nicoya N from Puntarenas, Tempisque basin, and locally along other rivers, up to ca. 1650ft (500m) on approaches to Cordillera de Guanacaste. Introduced and apparently established on Cocos I.; small numbers found in all major river valleys.

Range. C Mexico to Costa Rica; Florida and Cocos I. (introduced).

NORTHERN (BALTIMORE) ORIOLE

Icterus g. galbula Pl. 44(7)

Bolsero Norteño (Cacicón, Cacique Veranero)

Description. 7″ (18cm); 34g. Male the only orange oriole in Costa Rica with a black head; females and young above browner than other orange-breasted orioles, which either lack wing-bars or have streaked backs. **Adult ♂:** head, neck, upper breast, back, and most of wings black; 1 broad white wing-bar, remiges broadly edged with white; rest of underparts, shoulders, rump, upper tail-coverts, and corners of tail orange. ♀: above olive, tinged with orange; feathers of crown and back with black bases giving a more or less spotted appearance; wings blackish with 2 white wing-bars, the anterior one tinged with orange; breast and crissum orange, fading to whitish on chin and pale orange to grayish on belly; head and throat often irregularly mottled or blotched with black. Upper mandible black, lower silvery; legs bluish-gray. **Immatures:** resemble adult female but no black on head, throat, or bases of back feathers; above dull olive, "frosted" with gray; wings duller; chin and throat yellowish; breast more ochraceous; belly more grayish. Females duller, less orange than males, especially on rump and crown. In late winter and spring young males acquire much orange and black, but remain duller and patchier than adult.

Habits. Perches and flies at treetop level in cacao and coffee plantations, semi-open, forest canopy, and savanna groves, but freely descends to shrub level to feed along edges and in second growth; usually in groups of 2–5, rarely 15 birds that traverse definite home ranges; sometimes forms large communal roosts; fond of nectar, especially of trees like *Erythrina*, *Inga*, and *Calliandra*, vines (e.g., *Combretum*), and epiphytes (e.g., *Norantea*); also of bananas and other fruits (sometimes a pest of tomato or citrus crops); gleans and probes branches and foliage for insects and spiders, sometimes opening rolled leaves by gaping.

Voice. Most frequently heard are sharp, metallic, slurred whistles and squeaks and a loud, dry chatter, often prolonged. Occasionally over winter, frequently toward spring migration, males sing 1 to several rich, melodious whistles with no well-defined pattern.

Status. Common to abundant migrant and winter resident (early September–early May) from sea level to ca. 5000ft (1500m) on both slopes, and in smaller numbers as high as 6600ft (2000m); most abundant on Pacific slope, especially in NW lowlands and Valle Central.

Range. Breeds SE Canada and E USA; winters from C Mexico to N South America, in small numbers in SE USA and Greater Antilles.

Note. The western form of this complex, Bullock's Oriole, *I. g. bullockii* (breeds SW Canada to W Mexico, winters mainly to N Central America), has been seen several times on

Pacific slope, mainly in Guanacaste. Adult males differ from Baltimores in having head mostly orange; only crown, eye-stripe, and throat black; large white wing-patches. Females and young resemble young Baltimores but are paler below, with belly extensively white.

STREAKED-BACKED ORIOLE

Icterus pustulatus Pl. 44(2)

Bolsero Dorsilistado (Chorcha)

DESCRIPTION. 7½″ (19cm); 45g. Streaked back diagnostic in all plumages; differs further from Spotted-breasted Oriole in having more white on wing, no spots on breast. **Adults:** lores, orbital area, center of throat and chest, heavy streaking on back, and most of wings and tail black; rest of body plumage bright orange-yellow, shading to orange on head, this color averaging more intense in males; shoulders orange-yellow, middle wing-coverts mostly white, greater coverts and remiges edged with white; outer 3 rectrices tipped with grayish-white. Bill black, except base of lower mandible bluish-gray; legs bluish-gray. **Young:** above olive-green, with heavy but indistinct black streaking on back; tail dusky-olive; wings dusky with narrow, dull white wing-bars and edges of remiges; below dull greenish-yellow, palest on belly. Within a few months of fledging, immature birds have acquired body plumage like that of adult but duller, with little or no orange; retain juvenile flight feathers through next breeding season.

HABITS. Frequents edges and canopy of deciduous and gallery forest, savannas and scrub with scattered trees, *Parkinsonia* swamps, and roadside and village groves; in pairs for much of year and in family groups following breeding; gleans beetles, caterpillars, and other insects from foliage, holding large prey beneath a foot to dismember it; pokes into rotting wood for beetle grubs; takes more fruit and arillate seeds than other orioles; visits flowering trees for nectar and to catch wasps and bees.

VOICE. Calls include a bubbling rattle like that of a meadowlark, a dry chatter higher-pitched and faster than that of Northern Oriole; song of rather halting, vireolike phrases of rich, throaty, whistled notes, *chew-chewy, chip-cheer, chew . . . chirp . . . weet-chewy.*

NEST. A pyriform pouch 10–20″ (25–50cm) long, usually of finely woven, black fungal rhizomorphs with some grass and pieces of vine incorporated but sometimes nearly entirely of pale-colored fibers, 10–50ft (3–15m) up at tip of drooping branch, often of a bull's-horn acacia inhabited by stinging ants. Eggs 3–4, white or pale bluish, with blotches and scrawls of dark brown, mostly near large end. May–June.

STATUS. Fairly common resident of dry N Pacific lowlands, S to about Cañas and the head of Golfo de Nicoya, to 1500ft (450m) on approaches to Cordillera de Guanacaste.

RANGE. NW Mexico to Costa Rica.

RED-WINGED BLACKBIRD

Agelaius phoeniceus Pl. 44(14)

Tordo Sargento (Sargento)

DESCRIPTION. ♂ 8¾″ (22cm), 64g; ♀ 7″ (18cm), 36g. Black male with red shoulders unmistakable; heavy streaking below of females and young distinguishes them from all open-country birds of similar size. **Adult ♂:** lesser wing-coverts scarlet, middle coverts rich buff; rest of plumage glossy black. Bill and feet black. ♀: above blackish-brown, streaked with buff and chestnut-brown; the wing-coverts edged with brown and tipped with buff; shoulders suffused with dark brick-red; superciliaries buffy; eye-stripe dusky; cheeks brown, speckled with buff; throat pale buff, often tinged with pinkish; breast buff, belly whitish, both heavily streaked with blackish brown; lower belly, flanks, and crissum dark sooty-brown. Upper mandible black, lower dusky-horn. **Immature ♂:** dull black, with broad, brownish fringes; shoulders duller, mottled with black. **Young:** resemble adult female but below paler, less buffy, with dark streaking blacker, narrower, less regular; belly more grayish; darker above with less distinct streaking; no red on shoulder.

HABITS. Breeds in seasonal marshes and herbaceous vegetation bordering ponds, rivers, and drainage canals, where males defend territories within which 1 to several females nest; forages amid aquatic vegetation for insects and spiders, often gaping to force apart leaf sheaths and expose prey; seeks insects and seeds on open ground in pastures and agricultural fields; in large flocks outside the breeding season; often a pest of newly sown grain; forms communal roosts of thousands in cattail beds or trees during dry season.

VOICE. Calls include a dry *chek*, a scratchy *pik* with wooden resonance, and a nasal *neeah*; song of male in Guanacaste *ko-leeurlee* or *kurshleeuee*, the last note a wavering, buzzy gurgle; song at Caño Negro a single gurgling note and 2–3 clear whistles, *zhww-koleeyu*. When chased or courted by males, females give a burry *ch-ch-chree yar-yar*.

Nest. A strongly woven cup of grass stems and leaves, sprigs of vines, and rootlets, lined with finer grasses, 1–5ft (0.3–1.5m) above water, in fork of open shrub or supported by several vertical stems of cattails or other herbaceous vegetation. Eggs 2–3, bluish-white with blackish blotches and scrawls mostly at thick end. April–August, sometimes as late as November, at least in Río Frío region.

Status. Locally common to abundant resident of Tempisque and lower Bebedero basins, appearing sporadically elsewhere in lowland Guanacaste, and in Río Frío region of Caribbean slope.

Range. Alaska, C and SE Canada to Costa Rica; northern populations migratory, wintering S as far as C Mexico.

YELLOW-HEADED BLACKBIRD

Xanthocephalus xanthocephalus Pl. 44(4)

Tordo Cabecidorado

Description. ♂ 9½″ (24cm), 80g; ♀ 8″ (20cm), 50g. Unmistakable: mostly brownish or black, with face and chest, or entire head, yellow. **Adult ♂:** head, nape, throat, and breast bright yellow; orbital area, rest of body, wings, and tail black; primary coverts white; crown and neck clouded with dusky in winter. ♀: crown, back, wings, tail, and posterior underparts dull brownish-black; face and chest ochre-yellow; throat sometimes whitish; auriculars dusky; upper belly streaked with white. Bill and legs black. **Immatures:** like adult female, but male more blackish; crown, nape, and cheeks marked with blackish; yellow areas of female duller and less extensive.

Habits. In freshwater marshes and flooded rice fields, usually in flocks of up to 20, often associated with much larger flocks of Red-winged Blackbirds; forages on ground or in vegetation at water's edge, taking insects and seeds.

Voice. A hoarse, rattling *kruuk* or *gruk*.

Status. Casual winter visitor to NW lowlands (Palo Verde, Pelón de la Bajura) chiefly in January and February.

Range. Breeds from SW Canada to SW USA; winters regularly from SW USA to S Mexico, casually to Costa Rica (accidental in Panama).

RED-BREASTED BLACKBIRD

Sturnella militaris Pl. 44(13)

Tordo Pechirrojo

Description. 6½″ (16.5cm); ♂ 48g, ♀ 41g. Chunky, short-tailed; male unmistakable; red shoulders conspicuous in flight; red tinge on breast of female diagnostic. **Adult ♂:** throat and breast bright red; head and rest of body plumage black, with buffy feather-edgings that disappear with wear; rectrices and remiges barred with grayish-brown. Bill black; feet dusky. ♀: above blackish-brown, streaked with buff; superciliaries and median crown-stripe buffy; postocular stripe dusky; grayish-brown and blackish barring on remiges, rump, and tail; below dark buff, streaked with blackish posteriorly and on sides; breast suffused with pinkish-red. Upper mandible dusky, lower mandible and legs horn-color. **Immature ♂:** like adult but browner, with heavy buffy fringes, trace of buffy crown-stripe. ♀: like adult but paler, little or no red on breast.

Habits. In wet savannas and pastures, rice fields, and low, wet scrub; forages on or near ground for beetles, grasshoppers, caterpillars, grass and weed seeds; takes fruit from low shrubs; outside breeding season wanders erratically in flocks, sometimes associated with bobolinks or meadowlarks.

Voice. Calls include a metallic *plik*! and a dry, sputtering *jrrrt*, like meadowlark but softer; song a staccato note, followed by 2 buzzes on different pitches, *kip-bzzzzz-baaaaa*, in flight or while perched on fence post, shrub or log.

Nest. A deep, well-made cup of grass stems and leaves, lined with finer material, on or near ground in dense grass; sometimes in loose colonies. Eggs 2–3, creamy- or bluish-white with rusty blotches and scrawls. May–August.

Status. Locally common breeding resident (often not in same area in successive years) in Golfo Dulce region, N at least to Sierpe-Palmar area. Erratically wandering postbreeding flocks have reached as far NW as lower Valle del General (Volcán); first observed in Costa Rica in 1974, still extending range N.

Range. Costa Rica to N Argentina.

Note. Often separated in genus *Leistes*.

EASTERN MEADOWLARK

Sturnella magna Pl. 50(16)

Zacatero Común (Zacatera, Carmelo)

Description. 8″ (20cm); 85g. Chunky, with short, conspicuously white-edged tail, long, sharp-pointed bill, and long, stout legs and feet. **Adults:** head buffy-white with broad blackish stripes on each side of pileum and behind eye; supraloral stripe yellow; upperparts blackish-brown, streaked with buffy-brown; remiges and central rectrices barred with dusky and buffy-brown, outer 3 rectrices white; below bright yellow, the upper breast crossed by a broad U- or V-shaped black band; sides, flanks, and crissum buffy-white, streaked with blackish. Culmen blackish, shading to pale grayish on tomia and lower mandible; legs dusky-flesh. **Young:** pale stripes on head rich buff; upperparts rich brown, spotted and barred with black and edged with bright buff; below very pale yellow, brighter and washed with buff on breast; breast-band narrow, largely obscured by buffy fringes.

Habits. Walks over ground in grassy or weedy meadows and savannas, cultivated fields, and marshes, gathering insects, their larvae, and other small invertebrates, supplemented with seeds and fruits; in flight, alternates rapid wingstrokes and short glides with spread tail; when alarmed, flashes white outer tail feathers by rapidly spreading and closing the tail; in pairs or family groups most of year.

Voice. Calls include a bubbling chatter; a harsh, nasal, rolling *dzeerrt*; a chattering *chick-chechecheer* in excitement or alarm; a shorter *chik*. Song of 2–6 rich, clear whistles with flutelike transitions, the last notes usually descending in pitch: *peeealoo, pleeuu, peeeu, peealur*, or *cheero, cheerio*, etc., delivered from fence post, top of tree or shrub or in flight while male puffs out his breast, spreads his tail, and glides down to a perch.

Nest. A roofed structure with a wide side entrance, of interlaced grasses, lined with finer materials, often horsehair, usually in a slight depression in the ground amid dense concealing herbage. Eggs (in North America; not seen in Costa Rica) white or pinkish, profusely speckled and blotched all over with shades of brown. One Costa Rican nest held 3 nestlings in May; fledglings seen between early April and September.

Status. Common resident countrywide where agriculture and cattle raising are well established, from lowlands to 8200ft (2500m); colonized much of Caribbean slope in last 30 years, S Pacific sector in last 20 years, and still spreading with destruction of forests.

Range. SE Canada to Amazonian Brazil; migratory in north, sedentary in tropics.

BOBOLINK

Dolichonyx oryzivorus Pl. 50(18)

Tordo Arrocero

Description. 6″ (15cm); 25g. Slimmer, with more conical bill than other open-country icterids; female Red-winged Blackbird lacks crown-stripes, has streaked breast. **Adult ♂** Winter: head boldly striped with black and buff; an indistinct russet nuchal collar; back streaked with black, buff and buffy-brown; rump buffy-olive streaked with black; wings and tail brownish-black with buffy margins; trace of a pale buffy wing-bar; rectrices sharp-pointed; below rich buff, shading to white on chin and center of belly; breast obscurely mottled with black; flanks streaked with black. Breeding: head and underparts mostly black; nape buff; back streaked with black and buff; rump and scapulars white. ♀ Winter: like winter male but breast immaculate. Breeding: similar but above blacker, below more heavily streaked. Bill blackish, except basal part of lower mandible and tomia of upper, grayish to horn-color; legs pale brownish. **Immatures:** like adult female but below much more yellowish; belly and chin pale yellow.

Habits. In grassy clearings, fields, scrub, rice fields, either singly or in flocks, sometimes associated with Red-breasted or Red-winged blackbirds; feeds mainly on seeds of weeds and grasses, including rice.

Voice. A sharp, clear *chit* or *plit*, less metallic than note of Red-breasted Blackbird; also a rather mellow *peech* or *peek*.

Status. Rare fall migrant (mid-September–October), principally along Caribbean coast, sporadically inland, to middle elevations (e.g., near Cartago), mostly as single birds. Small flocks occasionally appear in rice-growing areas of Pacific slope, especially in SW (Laurel, Coto). No spring records for mainland, but has strayed to Cocos I. in April and July.

Range. Breeds in S Canada and N USA; winters mainly from Peru and C Brazil to N Argentina.

DICKCISSEL

Spiza americana Pl. 50(19)

Sabanero Arrocero (Pius)

Description. 6″ (15cm); 28g. Thick-billed, streaked-backed, sparrowlike, but always at least a trace of yellow on breast and superciliaries, rufous on shoulders. **Adult ♂:** head mostly grayish, the crown tinged with olive and streaked with black; superciliaries yellow, often becoming white behind eye; rest of upperparts pale brown, the back streaked with black; tail and wings dusky with buffy edges, except lesser and middle coverts rufous-chestnut, with pale fringes in fall; suborbital spot, chin, and sides of throat white; a black bib on lower throat, largely veiled by pale grayish fringes that wear away by spring; center of breast bright yellow, shading to buffy-white on belly and crissum; sides grayish; flanks brownish. ♀: above like male but more heavily streaked with black; head more brownish; shoulders paler rufous, with dusky feather bases exposed; yellow superciliaries duller, narrower; a narrow black malar streak; throat white, with a few black spots posteriorly but no bib; rest of underparts like male but yellow less extensive; sides, flanks, and sometimes breast streaked finely with black. Culmen blackish, shading to grayish or horn-color on rest of bill; legs brown. **Immatures:** like adult female but pileum browner and more spotted with black; breast and indistinct superciliaries tinged with buff; scapulars and tertials edged sharply with buff; streaking below more dusky; males have more rufous in shoulders and streaking below; females have less streaking below and may have little or no yellow on breast or rufous on shoulder.

Habits. Highly gregarious, in small groups to large flocks, especially in rice-growing areas; favors open, grassy areas, notably fond of seeds of guinea grass (*Panicum maximum*) as well as rice; picks seeds off seedhead or from ground. Compact flocks rise high, wheel swiftly and in unison like shorebirds, often alighting in tops of trees or bushes to rest after feeding.

Voice. Call an unmistakable, low-pitched, rather harsh and grunting *drrt* or *djrrt*. Increasingly from February onward, wintering males sing, often several together, an animated chirping that rises and falls in intensity, punctuated by frequent, detached *hips*.

Status. Common migrant in fall (early September–late October) and spring (early April to mid-May) on both slopes, from lowlands to 5000ft (1500m) in central highlands; irregularly rare to extraordinarily abundant winter resident in rice-growing areas of Pacific lowlands, sporadically elsewhere.

Range. Breeds in S-C Canada, C and E USA; winters from C Mexico to N South America.

Note. Recent anatomical studies show that this species, long considered a somewhat aberrant member of the family Emberizidae (or Fringillidae), is best placed in the Icteridae.

FAMILY **Thraupidae:** Tanagers and Honeycreepers

With the exception of about half a dozen that migrate well into the temperate zones of North or South America to breed, the 230 species of tanagers and honeycreepers are confined to the tropics and subtropics of the Western Hemisphere, including the Antilles. Tanagers are 3–12″ (7.5–30cm) long, with usually short, moderately thick bills, often with a tooth on the cutting edge, for eating fruits and arillate seeds that are the mainstay of most; they vary their diet with insects. Honeycreepers, long classified in a separate family with the Bananaquit and flowerpiercers, are currently considered to be tanagers with thinner, sharper bills better able to reach the nectar of flowers, although they eat mostly fruits and insects. Tanagers live at sea level to the upper limits of flowering plants, most abundantly in humid forests and clearings but also in arid regions. Most are arboreal, but a few rummage for seeds and insects on the ground. No other family contributes so much color to tropical American bird life, and the largest genus, *Tangara*, seems to exhaust the color patterns possible on sparrow-sized birds. Exceptionally, tanagers are black and white, plain black, brown, gray, or olive. The sexes may have similar plumage, even in many of the most colorful species. Although most tanagers are poorly endowed vocally, a few sing delightfully. Nearly all are monogamous, and many live in pairs throughout the year. The nests of most species are thick or thin cups, built in trees or shrubs, rarely in cavities or on the ground, usually by the female, often assisted by her mate. In the euphonias, both sexes build a roofed nest with a side entrance, usually tucked into a cranny. The eggs, 2 or 3 in most tropical species, sometimes 4 or 5 in euphonias, range from white to bright blue and are usually marked with brown, lilac, or black. As far as is known, only the female incubates for a period of 12–18 days. The nestlings, hatched with sparse down and red mouth-linings, are fed by both parents and remain in the nest for 11–24 days; the tiny euphonias and chlorophonias have the longest recorded incubation and nestling periods.

GOLDEN-BROWED CHLOROPHONIA

Chlorophonia callophrys Pl. 45(10)

Clorofonia Cejidorada (Rualdo)

Description. 5″ (13cm); 25g. Stout, with thick bill and short tail; bright green with blue on crown, yellow on belly. **Adult ♂:** forehead and broad superciliaries golden-yellow; crown and nape pale violet-blue; a line of turquoise from sides of neck around hindneck; rest of upperparts, sides of head and neck, throat, chest, and flanks bright green; rest of underparts bright yellow, separated from green of chest by a thin black line. Bill black, except base of lower mandible gray; legs gray. ♀: blue of crown duller and less extensive; lacks yellow superciliaries; green of upperparts duller and darker; green of underparts duller, tinged with olive, and shading to yellow on center of belly and crissum. Base of lower mandible dull yellowish to horn-color. **Immature ♂:** similar to adult female but crown brighter blue; forehead and superciliaries green with golden gloss; green of face and throat brighter. Passes first breeding season in this plumage. **Young:** resemble adult female but above duller, especially on crown which is at most tinged with blue; below with more yellowish-olive.

HABITS. Frequents canopy and edges of wet, epiphyte-burdened highland forests and adjacent semi-open and pastures with scattered trees, sometimes coming low in second growth; through much of year, wanders about in flocks of up to 12 that break up into pairs for breeding; eats mainly fruits of mistletoes (*Gaiadendron, Psittacanthus*), berries of Ericaceae, Araceae, and other epiphytes, figs; leans to peer beneath twigs for insects and spiders.

VOICE. Calls include a resonant, melancholy, melodious whistled *koow* or *keeeu*; contact note of foraging flocks a nasal chattering of short notes, *neck, jip*, or *jup*, with a rising inflection; a sharp, short, two-toned *kook*, the whole performance sometimes sounding like a gate lazily swinging on rusty hinges; song a jumble of short, clear, melodious but disconnected whistles.

NEST. A mossy globe- or oven-shaped structure with a side entrance, tucked amid mosses and bromeliads or other epiphytes on trees in clearing of mountain forest, 35–100ft (11–30m) up. At least 3 eggs, undescribed (?). February–June; 2 broods. Feeds young by regurgitation.

STATUS. Common resident of mountains the length of both slopes, from timberline down locally to 3000ft (900m) on Caribbean side and to 5000ft (1500m) on Pacific side; may descend much lower, especially on Caribbean slope, during wet season when fruit is scarce at high elevations.

RANGE. Costa Rica and W Panama.

NOTE. Often merged with *C. occipitalis*, Blue-crowned Chlorophonia, of N Central America.

BLUE-HOODED EUPHONIA

Euphonia elegantissima Pl. 45(9)

Eufonia Capuchiceleste (Caciquita)

DESCRIPTION. 4¼" (11cm); 15g. A small-billed euphonia easily distinguished from all others of the genus by its sky-blue hood. **Adult ♂:** forehead rufous, bordered by a line of black; crown, hindneck, and line down each side of neck pale blue; rest of head, neck, upperparts, and throat, glossy blue-black; remaining underparts tawny-orange, darkest on upper breast; wing-linings white. ♀: forehead dull rufous; blue on head and neck like male; sides of head and rest of upperparts bright olive-green; throat bright cinnamon-buff; breast, sides, and flanks yellowish-olive, shading to greenish-yellow on belly and crissum. Base of lower mandible silvery, rest of bill black; legs grayish. **Young:** forehead dull yellowish-olive; crown and hindneck gray, mixed with blue on hindcrown; above duller and grayer than adult female; remiges of male edged with blue-black basally; below grayish-olive, tinged with yellow on throat and belly.

HABITS. Wanders through upper levels of highland forest in small flocks, coming lower at edges and in semi-open and second growth; eats soft berries, principally of mistletoes, from which it removes the skin before swallowing pulp and seeds; sometimes at least 50 birds gather from different directions to sleep together in dense tufts of foliage, dispersing again at dawn.

VOICE. Calls varied, including a reedy chirp, a high, thin *tsee* or *cheezee*, a sharp, sibilant *schew*, a lower and more mellow *chewp* that, alternated with the *tsee*, sounds like 2 different birds. Song a prolonged, rambling sequence of gurgling trills, sharp whistles, chips, and tinkling notes in no fixed order, delivered from high treetop.

NEST. A globular structure with a side entrance, finely woven of slender rootlets and dry grass stems, lined with finer grasses, in a niche in a vertical bank 7ft (2m) up, concealed by grasses (only Costa Rican record). Eggs 3, white with reddish-brown spots, heaviest near thick end. April. Elsewhere, nests hidden amid green moss, or gray Spanish moss (*Tillandsia*), other epiphytes, or in crotch of tree, 16–60ft (5–18m) up. March–May.

STATUS. Resident; uncommon to rare during breeding season on Pacific slope of central highlands and Cordillera de Talamanca, 4300–6600ft (1300–2000m), sporadically lower; outside breeding season, largely moves to Caribbean slope; may wander as low as 2500ft (750m) in humid areas; numbers undoubtedly reduced by intense persecution to secure cage birds.

RANGE. NW Mexico to W Panama.

Note. Sometimes considered conspecific with *E. musica* of West Indies. The genus *Euphonia* was formerly called *Tanagra*.

TAWNY-CAPPED EUPHONIA

Euphonia anneae Pl. 45(6)

Eufonia Gorricanela (Barranquilla)

Description. 4¼″ (11cm); 16g. A fairly large, chunky euphonia with orangish or rufous in crown of adults. **Adult ♂:** pileum bright cinnamon-orange; rest of head, neck, throat, and upperparts glossy blue-black, tinged with violet on nape; crissum white; rest of underparts bright yellow, with orangish tinge on chest. ♀: forehead and forecrown cinnamon-rufous, spotted with blackish, shading to dark gray on nape; rest of upperparts dark olive-green with steely gloss on back; wings and tail blackish with olive edgings; chin yellowish-olive, breast and upper belly dull gray, shading to bright olive-green on sides and flanks, and dull buffy on center of belly. Bill black with base of lower mandible pale gray; legs gray. **Young:** above dull, dark olive-green, washed with dark gray on crown and nape; face, breast, and sides dull olive-green, shading to dull grayish-olive on throat and grayish-buff on belly; remiges of male edged basally with blue-black. Males attain adult body plumage before their first breeding season.

Habits. Frequents canopy and edge of dense, wet, epiphyte-festooned mountain forests, coming lower in adjacent second growth and clearings with scattered trees, or at treefall gaps; in pairs or family groups that often accompany mixed flocks of tanagers, honeycreepers, and other small birds; eats a wider variety of fruits than most euphonias, including mistletoes, melastomes, figs, Ericaceae, and especially *Anthurium*, often mashing them in bill and swallowing pulp and small seeds, discarding husks.

Voice. Calls include a scratchy, downslurred *tsewp*, often doubled; a whining, nasal vireo-like *teer*, possibly an alarm or scolding note; a metallic rolling *chrrrt* like that of Olive-backed Euphonia, and a *chrrrt-tew tew*, suggesting call of Black-faced Grosbeak; song a series of piercing and slurred whistles, with occasionally a little buzzy trill: *see see-seep tsew see-seep tsew tsew see chrrrt see-see tsew. . . .*

Nest. A globular structure of mosses and fibers, in tuft of moss on a dangling vine or in a niche amid epiphytes on a slender branch, 25–35ft (8–11m) up at forest edge. Eggs undescribed (?). March–June.

Status. Common resident of foothills the length of Caribbean slope, at 2000–5600ft (600–1700m) in N, and 3000–6600ft (900–2000m) in SE; outside breeding season may wander down to ca. 1000ft (300m).

Range. Costa Rica to extreme NW Colombia.

WHITE-VENTED EUPHONIA

Euphonia minuta Pl. 45(7)

Eufonia Menuda (Agüío, Canaria)

Description. 3¾″ (9.5cm); 10g. A very small, slender-billed euphonia distinguished by white lower belly and crissum in both sexes. **Adult ♂:** forehead and most of underparts bright yellow; rest of head, neck, throat, and upperparts glossy blue-black, tinged with violet on head and nape; wing-linings and most of inner webs of outer 3 rectrices white. ♀: above dull olive-green; crown and nape tinged with gray and forehead with yellowish; a faint, pale superciliary and dark eye-stripe; cheeks olive-green; throat and crissum pale gray; chin, breast, sides, and flanks olive-yellow; center of belly white. Bill black except base of lower mandible pale gray; legs bluish-gray. **Young:** above duller, somewhat browner than adult female; below more uniform dull olive-yellow, including throat. Male has dark bases of feathers of forehead and crown, producing obscure spotting; remiges edged with blue-black basally; crissum white; acquires nearly complete adult plumage before first breeding season.

Habits. Wanders widely through upper levels of humid forests, descending lower along edges and at gaps, in clearings with scattered trees, semi-open, and crowns of small trees in second growth; eats mainly berries, especially of mistletoes; hunts small insects on high, thin dead twigs, along which it advances by hopping, by sidling along for short distances and by about-facing, often bending down to examine the undersides, and incessantly wagging half-spread tail from side to side; in pairs or small groups.

Voice. Calls include a thin, sharp, warbler-like chip, constantly repeated by both sexes as they attend a nest; while foraging, *chip-worthy*, *chip-wertily*, or *chip-chewee*. Song a rambling medley of staccato chips, chirps, and warbles, lacking the metallic rolling notes so prominent in repertoires of Spotted-crowned and Olive-backed euphonias.

Nest. A globe with a side entrance, of green moss, small filmy ferns, and orchid roots, lined with fine, pale-colored plant fibers, 9–60ft (3–18m) up amid mosses and liverworts or larger epiphytes. Eggs 3–5, white, heavily spotted and blotched with brown. February–July; 2 broods.

Status. Uncommon resident of lowlands and lower foothills, to 3500ft (1100m), on Caribbean slope and S Pacific slope; sometimes wanders to 5000ft (1500m) in early wet season.

Range. S Mexico to W Ecuador, N Bolivia, and E Brazil.

SCRUB EUPHONIA

Euphonia affinis Pl. 45(4)

Eufonia Gargantinegra (Agüío, Monjita Fina)

Description. 3¾″ (9.5cm); 10.5g. Very small; male is only dark-throated euphonia in dry northwest; adult differs from White-vented in having crissum and lower belly yellow. **Adult ♂:** forehead, forecrown, and underparts posterior to throat bright yellow; rest of head, neck, throat, and upperparts glossy blue-black, tinged with violet on head and throat; wing-linings and most of inner webs of outer 2 rectrices white. ♀: forehead yellow-olive; crown, nape, and auriculars gray; rest of upperparts olive-green, tinged with gray on back; throat, cheeks, breast, sides, and flanks olive-yellow, darkest and dullest on breast, shading to yellow on lower belly and crissum. Bill black except base of lower mandible pale gray; legs blackish. **Young:** like adult female but above more olive, without gray on head; remiges of young male edged basally with blue-black. Males acquire adult body plumage when a few months old.

Habits. In pairs or small groups of up to 5, wanders through canopy of deciduous and gallery forest, coming lower at edges and in dry scrub and ranging into isolated trees in pastures and savannas; eats mistletoe berries and such other soft fruits as figs and *Muntingia*; examines undersides of thin twigs for insects, sometimes wagging half-spread tail from side to side in manner of White-vented Euphonia; in early morning often perches conspicuously on exposed twigs in treetops.

Voice. Call 2–3 thin, high whistles, often followed by 3more on slightly lower pitch: *pwee pwee, pwee-pwee-pwee*, somewhat thinner and more rapid than similar call of Yellow-crowned Euphonia; song a medley of twitters, chirps, and warbles, with *pwee* notes interspersed; a rapid, flycatcherlike *ts-tsairoo*, the *air* long and buzzy (Slud).

Nest. Oven-shaped, with side entrance, of grass stems, leaf petioles, bits of leaves, strips of bark, fine rootlets, with fungal fibers in lining; 12–25ft (3.5–8m) up on horizontal branch, in fork of sapling, or beneath large dead leaf of *Cecropia* wedged in fork. Eggs 2–3, white to buffy-white, with reddish-brown spots or paler blotches forming a wreath or cap. November–April.

Status. Common resident of dry NW lowlands, S at least to Puntarenas; in small numbers to 3600ft (1100m) on slopes of N cordilleras and lower part of Valle Central.

Range. N Mexico to Costa Rica.

YELLOW-CROWNED EUPHONIA

Euphonia luteicapilla Pl. 45(1)

Eufonia Coroniamarilla (Monjita, Agüío)

Description. 3¾″ (9.5cm); 12.5g. Bill slender; male distinguished by combination of dark throat and wholly yellow pileum; female unpatterned. **Adult ♂:** forehead, crown, and underparts posterior to throat bright yellow; rest of head, throat, and upperparts glossy blue-black; wing-linings white; no white in tail. ♀: above olive-green, tinged with yellow on forehead; below olive-yellow, shading to yellow on center of belly and crissum. Bill black with base of lower mandible and rami of upper silvery; legs gray. **Young:** resemble adult female; male below somewhat deeper yellow. Male, when a few months old, acquires some adult feathers on throat, face and forecrown but not on body; he wears this plumage through the first breeding season.

Habits. Frequents forest edge, light woodland, semi-open, and open country with scattered trees and shrubs; wanders widely in pairs or small groups of 3 or 4, occasionally alone; eats soft fruits, principally mistletoe berries; examines small bare twigs for insects; usually perches high, flies at treetop level.

Voice. Most frequent calls high, clear whistles, in 2s and 3s: *beee-beee-beee*, or *beebee, peepeepee*, the second group lower in pitch; staccato *tsit* notes in flight or as contact calls among members of foraging group; song a short, wiry *chut-tsip ts'tsew tsiuweer* or similar dry, sputtering phrases.

Nest. A small globe with a side entrance, of mosses, fine rootlets, and grass stems, lined with finer grasses and animal hair or varied soft materials, in a snug cranny in a fence post, stump, or snag, amid orchids or other epiphytes or between clustered twigs, 3–100ft (1–30m), usually 6–9ft (2–3m) up. Eggs 2–4, usually 3, white, heavily mottled with brown. February–May; 2 broods.

Status. Common resident of lowlands and foothills throughout Caribbean slope, and locally N to Carara on S Pacific slope; sea level to 4000ft (1200m). Reduced in many areas by persecution for cage-bird trade.

RANGE. E Nicaragua to E Panama.

THICK-BILLED EUPHONIA

Euphonia laniirostris Pl. 45(8)

Eufonia Piquigruesa (Agüío)

DESCRIPTION. 4¼″ (11cm); 15g. Bill thick; adults differ from Yellow-throated Euphonia in lacking white on belly, crown-patch of male larger. **Adult ♂:** forehead, crown, and underparts bright yellow, tinged with orange on breast; rest of upperparts, sides of head, and neck glossy blue-black; wing-linings and most of inner webs of outer 2 rectrices white. ♀: above olive-green; below olive-yellow, paler on throat, and shading to pale yellow on lower belly. Base of lower mandible pale gray, rest of bill black; legs gray. **Young:** very similar to adult female; male has brighter yellow belly; female's belly washed with olive. Male acquires adult plumage on throat, face, and forecrown, but no more than a few scattered feathers elsewhere, when a few months old, and wears this plumage through first breeding season.

HABITS. In small groups, frequents forest borders, open woods, semi-open, clearings with scattered trees, second growth, and gardens; may accompany flocks of tanagers, honeycreepers, and other small birds; eats berries of mistletoes and occasionally of melastomes or figs; reported to mimic alarm calls of other species when its nest is threatened, inducing them to mob the predator (Morton).

VOICE. Calls include a sweet, chirpy *chweet* or *tseeyoop*; a buzzy *bee-wit*; a heavier, burry *burrweer*; and a clear *peet* or *peet peet* suggestive of Yellow-crowned Euphonia but shorter and softer. Song a rambling, rather sweet medley of chirps, warbles, and staccato notes; often includes imitations of notes of Mistletoe Tyrannulet, Variable Seedeater, and other small birds.

NEST. A globular structure with a side entrance inclined downward and protected by a visor, of grasses, leaves, mosses, pieces of plastic, and thread, lined with finer grasses and leaves, 8ft (2.5m) up in citrus tree (only Costa Rican record). Eggs 4, white, thickly streaked and speckled with brown, mainly on large end. March–September (in Panama and probably Costa Rica).

STATUS. Locally common resident in Coto Brus district and Golfo Dulce lowlands adjacent to Panama border, rare northward; scattered old reports N to C Costa Rica on both slopes may refer to Yellow-throated Euphonia.

RANGE. Costa Rica and N Venezuela to W Peru, N Bolivia, and E Brazil.

YELLOW-THROATED EUPHONIA

Euphonia hirundinacea Pl. 45(5)

Eufonia Gorgiamarilla (Agüío, Caciquita)

DESCRIPTION. 4¼″ (11cm); 15g. Bill thick; from Thick-billed Euphonia, distinguished by white on belly and male with less yellow on head. **Adult ♂:** forehead and most of underparts, including throat, bright yellow; rest of head, neck, and upperparts glossy blue-black; center of belly, wing-linings and inner webs of outer 2 rectrices white. ♀: above dull olive-green; throat yellowish; breast, sides, flanks, and crissum yellowish-olive; belly extensively grayish-white. Bill black, except base of lower mandible pale gray; legs gray. **Young:** resemble adult female but below mostly grayish-white. Male undergoes a partial body molt when a few months old, resulting in a mixture of adult and juvenile plumage over entire head and body, which persists through first breeding season.

HABITS. In dry areas prefers gallery forest and moist evergreen woodland; in wetter areas inhabits forest edge, semi-open, plantations, second growth, and clearings with scattered trees and shrubs; travels in pairs and small groups, eating many berries, chiefly of mistletoes, and occasional small insects gleaned from bark or twigs.

VOICE. Calls include a rolling *cheek-chereeg*, a loud, scratchy *chee-cheet*, or a single, metallic *weet*; female gives thin, dry notes, sometimes high-pitched and almost trilled. Male sings short whistles and brief warbles in choppy phrases of 2–4 notes, clear and bell-like or with plaintive intonation.

NEST. Globular, with a side entrance, in a cranny in decaying fence post or stump, axil of a palm frond, niche in a vertical bank, clump of epiphytes, or Royal Flycatcher's pensile nest, 3–50ft (1–15m) up. Eggs 3–5, white, heavily blotched with brown. March–June.

STATUS. Resident, local and uncommon in dry NW and Valle Central, becoming common to abundant on Pacific approaches to N cordilleras, and on Península de Nicoya; very common from Río Frío region W on Caribbean slope; rare on S Pacific slope; sea level to 4600ft (1400m).

RANGE. NE Mexico to W Panama.

NOTE. Sometimes called Bonaparte's Euphonia; was long known as *Euphonia* (or *Tanagra*) *lauta*.

OLIVE-BACKED EUPHONIA

Euphonia gouldi Pl. 45(3)

Eufonia Olivácea (Agüío)

Description. 3¾″ (9.5cm); 12g. Distinctive green euphonia with rufous belly; forehead yellow in male, rufous in female. **Adult ♂:** forehead yellow, obscurely spotted with greenish-black; rest of upperparts glossy olive-green to steely-green; face, throat, and chest olive-green; sides and flanks olive-green mixed with yellow; lower breast, belly, and crissum chestnut-rufous. ♀: forehead chestnut-rufous; rest of upperparts olive-green, with some steely-green gloss; below yellowish-olive, darker laterally, with center of lower breast and belly rufous. Bill black except base of lower mandible silvery; legs gray. **Young:** above dark olive-green, tinged with sooty on crown and nape; fringes of forehead feathers yellow in male, rufous in female; below dull olive-green, tinged with yellowish on throat and belly, and with rufous on crissum. Full adult body plumage is attained when a few months of age.

Habits. Frequents canopy and middle levels of wet forest, coming lower at gaps and edges, in semi-open, tall second growth, and clearings with scattered trees and shrubs; in pairs or small groups, often accompanying mixed flocks of other small birds, including tanagers, honeycreepers, greenlets, and warblers; eats a wide variety of fruits, including berries of *Anthurium*, melastomes, *Trema, Urera*, mistletoes, and *Cecropia* spikes; mashes soft fruits in bill, discarding husk and sometimes seeds, swallowing pulp.

Voice. Usual call a metallic, rolling *chrrr-chrrr* or *chur-chur-chur*; song a rapid, jerky, rambling medley of rolling *chrrr*s, staccato notes, short, clear whistles, and more nasal, slurred whistles, often long continued.

Nest. A globular structure with a side entrance, of thin, wiry roots and rhizomes mixed with green moss, fine palm, and bast fibers, lined with fine grasses or other fibers, 7–50ft (2–15m) up, in clump of moss or epiphytes. Eggs 3, dull white with brown spots and blotches, mostly in a wreath around large end. February–July.

Status. Common resident throughout Caribbean lowlands, to 2500ft (750m) in foothills, sometimes 3300ft (1000m) in SE; reaches Pacific side of Cordillera de Guanacaste via low passes.

Range. S Mexico to W Panama.

Note. Occasionally called Gould's Euphonia.

SPOTTED-CROWNED EUPHONIA

Euphonia imitans Pl. 45(2)

Eufonia Vientrirrojiza (Agüío Barranquillo)

Description. 4″ (10cm); 14g. Male very like Yellow-crowned Euphonia but bill thicker, smaller yellow crown-patch obscurely spotted with blue; no other female euphonia in its range has rufous forehead and belly. **Adult ♂:** forehead and forecrown yellow, with extensive blue-black feather bases; rest of head, neck, throat, and upperparts glossy blue-black; remaining underparts bright yellow. ♀: forehead rufous; rest of upperparts olive-green with bluish-green gloss, except on wings and tail; below yellowish-olive, paler on throat; center of belly and crissum cinnamon-rufous. Bill black except base of lower mandible silvery; legs gray. **Young:** resemble adult female but above duller, with little or no gloss; lack rufous on forehead and belly; acquire adult plumage when a few months old.

Habits. Wanders through middle and upper levels of humid forest and into adjacent shady clearings and tall second growth, singly or in pairs, sometimes with mixed-species flocks; eats a variety of berries, discarding skin; pecks into larger fruits like guavas; takes nectar; visits feeders for bananas; searches mossy branches and tangles of dead leaves for insects; sleeps singly in nooks amid mosses or larger epiphytes.

Voice. Calls include a rolling, metallic *chrrrt*, often given in groups of 2–4, sometimes preceded by 1–2 downslurred whistled *tewp*; a sibilant, "sizzling" *yerrlirr*; scolds with a harsh, rolling *jurry-jurry*. Song a medley of clear whistles, dry or wheezy notes, short warbles: *chip a cher weet; chip tuck tuck; wee churee-cha* . . ., often long continued.

Nest. A ball of mosses and fibrous materials with a side entrance, embedded among mosses or larger epiphytes, often in a dangling tangle of epiphytic roots or a detached, swinging length of an epiphyte-covered branch, 6–25ft (2–8m) up, in forest or nearby clearing. Eggs 2–3, white or pinkish, speckled with reddish-brown or chocolate. March–June; 2 broods.

Status. Resident on S Pacific slope from lowlands and valleys to 4500ft (1400m), common N to the hills above Parrita, in smaller numbers N to Carara.

Range. Costa Rica and W Panama.

Note. Also called Tawny-bellied Euphonia, for the more distinctive female.

EMERALD TANAGER

Tangara florida Pl. 46(5)

Tangara Orejinegra

Description. 4¾″ (12cm); 19g. Pale green with streaked back, black in wings and tail, square black ear-patch. **Adult ♂:** mostly bright, glossy, pale green, brightest on face, shading to pale yellow on center of belly and crissum; black around base of bill and on auriculars; back streaked with black; hindcrown and rump golden-yellow; wings and tail black, with broad, pale green edgings except primaries and their coverts wholly black. ♀: similar but overall duller, darker green; crown and rump green with golden gloss. Bill black; legs bluish-gray. **Immature ♂:** closely resembles adult female. **Young:** much duller; above green areas with olive tinge; crown with indistinct blackish spotting; back with narrow, indistinct, dusky streaks; rump and underparts olive-green, tinged with gray on throat and yellow on belly.

Habits. Frequents canopy and edge of foothill forest and adjacent second growth, in pairs or small groups that nearly always accompany mixed flocks of tanagers, honeycreepers, warblers, etc.; hops rapidly along branches, switching from side to side to peer beneath them, reaching down to glean insects and spiders; takes many small fruits including melastomes, *Phytolacca, Schefflera*, and *Trema*.

Voice. Weak, sibilant, chipping or "tsipping" notes, sometimes accelerated to a rapid twitter; a possible song is an evenly spaced series of loud *cheet* notes.

Nest. A cup of strips of *Heliconia* leaf, rootlets, fungal and other fibers, decorated with bits of moss and fern, beautifully concealed in crevice among mosses and bromeliads in fork of branch, from 6ft (1.8m) up in tree in clearing to 30ft (9m) or more in forest. Eggs undescribed (?). March–May.

Status. Fairly common resident of Caribbean foothills, from Cordillera de Guanacaste (N at least to Volcán Miravalles) S to Panama, between 1150 and 3600ft (350–1100m); in SE, mostly above 2000ft (600m). Some local movement to lower elevations outside breeding season.

Range. Costa Rica to W Ecuador.

SPECKLED TANAGER

Tangara guttata Pl. 46(8)

Tangara Moteada (Zebra)

Description. 5″ (13cm); 20g. No other tanager appears densely speckled with black above and below. **Adult ♂:** feathers of upperparts black, edged with bright grass-green, giving effect of black spotting; rump and flanks solid green; wings and tail black with blue-green edgings; black lores and eye-stripe on yellow face; feathers of throat, breast, and sides black with pale blue-green to whitish edgings, passing to white on belly; crissum yellow-green, spotted with black. ♀: similar but black markings less extensive on lower breast; above green edgings broader and less clean-cut. Upper mandible black, lower silvery with black at tip; legs bluish-gray. **Young:** above duller green, black markings reduced to dusky smudging; lores dusky; face and forehead tinged with yellow; below dull whitish, mottled with dusky on throat and breast; sides and flanks washed with olive-yellow; crissum yellowish.

Habits. Travels through upper levels of humid forests and neighboring second growth and clearings with scattered trees and shrubs in pairs or small parties of 3–6, often in mixed flocks of frugivorous birds; eats berries, arillate seeds, and other fruits, including *Cecropia* spikes; searches foliage and bends over to examine undersides of horizontal boughs for insects and spiders; rarely sallies for flying insects.

Voice. Staccato, thin chips and chatters, thinner and sharper than notes of Golden-hooded Tanager; a series of short, clear, bell-like notes, accelerating almost to a trill.

Nest. A neat, compact cup of slender rachises of small compound leaves and fragments of leaf blades or strips of banana leaves lined with finer materials; lacks green moss; 10–25ft (3–8m) up amid foliage. Eggs 2, white, heavily mottled with brown. April–June. At 1 nest 3 adults fed the young.

Status. Uncommon to fairly common resident of foothills at 1300–3300ft (400–1000m) on Caribbean slope from Cordillera de Tilarán S, and on S Pacific slope 1000–4600ft (300–1400m); somewhat higher near Panama border on both slopes.

Range. Costa Rica and Trinidad to SE Colombia and N Brazil.

Note. *T. chrysophrys* was formerly used for this species.

SILVER-THROATED TANAGER

Tangara icterocephala Pl. 46(6)

Tangara Dorada (Juanita)

Description. 5″ (13cm); 21g. Mostly yellow, with whitish throat, black malar stripe and streaks on back. **Adult ♂:** upperparts mostly glossy bright yellow; back heavily streaked with black; wings and tail black with pale green edgings; black stripe from bill to sides of neck; throat glossy greenish-white, this color continuing as a narrow, partly concealed collar around hindneck; rest of underparts bright yellow, washed with greenish on flanks. ♀: similar but duller yellow, tinged with greenish; crown often faintly mottled with blackish. Bill black; legs gray. **Young:** much duller and greener; above dull green, tinged with olive; head and back indistinctly streaked with sooty-blackish; edges of wing and tail feathers dull green; malar stripe dusky; throat dull grayish; breast olive-green; posterior underparts dull yellow, tinged with olive; distinguished from Emerald Tanager by dark malar stripe and lack of dark ear-patch.

Habits. Frequents middle and upper levels of humid forests, edges and adjacent semi-open, tall second growth, and shady clearings; travels in pairs or, more often, small groups, often with other tanagers, honeycreepers, warblers, and other small birds, sometimes alone; eats berries, especially of melastomes, figs, and arillate seeds; searches foliage for insects; hops along mossy branches, swiveling to face opposite direction with each hop and bending down to examine underside of branch; creeps along diagonal branches, stretching to peer at underside; rarely hovers or sallies.

Voice. Usual call a characteristic sharp, buzzy *bzeet* or *zzeep*; short, weak, often harsh insectlike notes; not known to sing.

Nest. A compact cup of mosses and fibers amid foliage of trees or shrubs, often on mossy branch, 3–42ft (1–13m) up. Eggs 2, dull white or grayish, heavily mottled with brown, especially toward large end. February–September; 2 broods.

Status. Common to abundant resident at 2000–5500ft (600–1700m) in humid mountain forests, mainly on Caribbean slope but reaching Pacific slope in N cordilleras, and along Cordillera de Talamanca, Dota region, upper General and Coto Brus valleys, and coastal ranges, including hills of Península de Osa; descends to sea level on both slopes in latter part of wet season during heaviest rains.

Range. Costa Rica to W Ecuador.

GOLDEN-HOODED TANAGER

Tangara larvata Pl. 46(13)

Tangara Capuchidorada (Siete Colores, Mariposa)

Description. 5″ (13cm); 19g. Black mask on golden head, mostly black body, and white belly unmistakable. **Adult ♂:** chin, lores, and orbital area black; forehead and cheeks violet-blue, separated by a narrow line of turquoise from tawny-gold of rest of head, neck, and throat; breast and rest of upperparts black except shoulders and rump pale turquoise; wings and tail narrowly edged with pale golden-green; sides blue; flanks more greenish; center of belly white; crissum tinged with greenish. ♀: very similar but gold of head with greenish tinge; crown often speckled with black; below with white more extensive. Bill black; legs dusky. **Young:** head dull grayish-green with golden gloss; lores and orbital area duskier; back dull grayish-olive; rump more yellow-green; wings and tail dusky blackish; throat yellowish-green, mottled with dusky; breast and sides dull grayish-buff to olive, shading to dull yellowish on belly.

Habits. Frequents upper levels of heavy forest, clearings with scattered trees, semi-open, tall second growth, and shady gardens; in pairs, sometimes family groups, occasionally several adults together, sometimes traveling with mixed flocks of other small birds; eats many berries, arillate seeds, *Cecropia* and *Piper* spikes, also searches foliage and mossy boughs for insects, sometimes even catching them in the air.

Voice. Calls include high-pitched, sharp *tsit* or *chik* notes with a distinctive, scratchy, metallic tone; many such notes accelerate into a sharp twitter or chatter; a rapid series of dry *ticks* sometimes repeated for several seconds or up to many minutes at daybreak, forming a tuneless dawn song.

Nest. A compact cup of small dead leaves, papery bark, with finer materials in lining, often covered with green moss, in a mossy crotch, amid foliage, in a bunch of green bananas or rarely in a cavity in a tree, 5–90ft (1.5–28m) up. Eggs 2, dull white, thickly sprinkled all over with brown or chocolate, most heavily on larger end. March–September; often 2 broods, young from the first helping to feed nestlings of the second. Some nests are attended by 3 or 4 adults.

Status. Common resident of lowlands and foothills of Caribbean slope and S Pacific slope (N to about Carara), from sea level to ca. 5000ft (1500m), W to between Paraíso and Cartago in central highlands.

Range. S Mexico to W Ecuador.

Note. Also called Golden-masked Tanager (although the mask is the only part of the head *not* golden); sometimes lumped with the blue-hooded *T. nigrocincta* of E South America and called Masked Tanager.

PLAIN-COLORED TANAGER

Tangara inornata Pl. 46(11)

Tangara Cenicienta

Description. 4¾″ (12cm); 19g. Dull gray with contrasting blackish wings that suggest the larger, greener Palm Tanager. **Adults:** above slate-gray, darker and duskier on lores; cheeks, throat, breast, and sides paler gray, tinged with buff, shading to pinkish-buff on belly and crissum; shoulder violet-blue, usually concealed; rest of wings and tail dusky blackish. Bill black; legs blue-gray. **Young:** similar but upperparts tinged with brown; no blue on shoulder; wing-coverts broadly edged with paler gray.

Habits. Travels through treetops at forest edge, especially along streams and in adjacent semi-open, tall second growth, and trees in pastures; eats many berries, including those of melastomes, mistletoes, *Trema* and *Hasseltia*, also spikes of *Cecropia*; hops along twigs, turning body to peer beneath first one side, then the other, in search of small insects and spiders; may hang far below perch to reach an insect or fruit; in parties of 3–6, less often in pairs, sometimes accompanied by other small birds.

Voice. Calls include high, thin, colorless notes: *tsit, tseep, sip*, etc., sometimes accelerated into a dry twitter; song consists of high, thin whistles and short staccato notes and sputters in no particular order: *tseee tsp tsp tseee tsp tseee tseee tsp tsp tsp*, sometimes long continued, especially at dawn.

Nest. A neat cup of pale-colored fibers covered with green moss, bound together with spiderweb and cocoons, ca. 30ft (9m) up amid foliage of fruit trees in clearing. Eggs 2, whitish, speckled. March–August (in Panama; no Costa Rican record).

Status. Fairly common resident of Caribbean lowlands, locally to 1300ft (400m), from Panama N to upper Sarapiquí watershed (La Selva).

Range. Costa Rica to N Colombia.

RUFOUS-WINGED TANAGER

Tangara lavinia Pl. 46(10)

Tangara Alirrufa

Description. 5¼″ (13.5cm); 24g. Male distinguished from Bay-headed Tanager by rufous wings, green throat and breast; female told from honeycreepers by pale, rather heavy bill. **Adult ♂:** top and sides of head bright chestnut-rufous; center of belly blue; rest of body plumage bright grass-green, with strong golden-yellow gloss on hindneck and upper back; wing-coverts and edges of remiges chestnut-rufous; thighs rufous. ♀: no chestnut on head, but face, crown, and upper back with golden-bronze gloss; body plumage duller green, above darker and below paler; belly paler blue; chestnut-rufous on wings replaced with bronzy-olive, tinged with rufous. Bill horn-color with dusky culmen; legs gray. **Young:** resemble adult female but duller; crown obscurely mottled with dusky; face tinged with olive; belly mixed with dull whitish; crissum pale yellowish; head of male brighter, tinged with ochraceous; wings more reddish-bronze.

Habits. Frequents canopy of forest and second growth, descending to shrub level along thickety edges; travels about in pairs or small groups, often associating with other tanagers, honeycreepers, greenlets, etc.; gleans branches for insects; takes many kinds of berries, as do other *Tangara*.

Voice. High, thin tsittering notes in flight, higher, drier, more piercing than notes of Golden-hooded Tanager. Calls include a sharp *tseeup* or *tseep* and a piercing *zeek*. Song *seetsir-tsirtsir-tsirtsir* (Slud).

Nest. A cup of mosses, mixed with fine rootlets and grass stems, lined with fine, black fungal fibers, 6ft (1.8m) up in citrus tree in pasture at forest edge. Eggs 3, white, spotted and speckled with shades of brown, especially near thick end (only 1 record). May (Marín).

Status. Fairly common resident in a narrow belt at ca. 800–2500ft (250–750m), at base of mountains the length of Caribbean slope, descending to near sea level in SE in wet season.

Range. E Guatemala to W Ecuador.

BAY-HEADED TANAGER

Tangara gyrola Pl. 46(9)

Tangara Cabecicastaña

Description. 5¼″ (13.5cm); 23g. Above green, below blue, with chestnut head in both sexes. **Adult ♂:** top and sides of head, around to chin, chestnut, separated by a line of golden-yellow from bright grass-green of back, wings, and tail; shoulder golden-yellow; rump and underparts bright cerulean-blue, shading to green on lower belly, flanks, and crissum; thighs russet. ♀: similar but head paler chestnut-rufous, mixed with dusky on pileum; back duller green; less blue on rump; below paler blue, green more extensive posteriorly; less yellow on hindneck and shoulder. Bill blackish, shading to grayish-horn on base of lower mandible; legs grayish-flesh. **Young:** feathers of upperparts and wings dusky, with grass-green tips and edges, giving a mottled effect; face tinged with bronzy, sometimes with rufous at base of bill; feathers of underparts grayish at base, blue at tips, giving mottled or smudgy effect.

Habits. Frequents upper and middle levels of humid forest and nearby clearings with scattered trees, semi-open and tall second growth; in pairs or family groups throughout year, often accompanying mixed flocks of tanagers, honeycreepers, and other small birds; eats many berries, especially those of melastomes and the epiphytic shrub *Lycianthes synanthera*, figs, *Coussapoa* fruits, *Cecropia* catkins, and arillate seeds; hops along horizontal branches covered with mosses or lichens, bending down first on one side, then the other, to search underside for small insects.

Voice. Calls include a sibilant *tsip* and a wiry, squeaky *tzeeik* or *weeit*; song a series of 4–6 thin, wiry notes descending in pitch, with a peculiar whining twang.

Nest. A cup covered with green moss mixed with wiry rootlets and rhizomes of small epiphytic ferns; middle layer of small leaves and grass blades; lining of finer materials; 7–25ft (2–8m) up amid screening foliage. Eggs 2, dull white, spotted with brown or chocolate, chiefly in a wreath around thicker end. February–September; sometimes 3 broods.

Status. Common resident of wet foothills of Caribbean slope at 2000–5000ft (600–1500m), from Cordillera de Tilarán S; common to abundant on S Pacific slope (N to Carara), from lowlands to 5000ft (1500m). In latter part of rainy season much of population moves downslope, reaching 330ft (100m) or less on Caribbean side.

Range. Costa Rica to W Ecuador, N Bolivia, and E Brazil.

SPANGLED-CHEEKED TANAGER

Tangara dowii Pl. 46(12)

Tangara Vientricastaña

Description. 5″ (13cm); 20g. Head and upperparts mostly black, belly cinnamon, with pale scaling and spotting on chest and head. **Adults:** head, neck, breast, back, wings, and tail black; chest scaled with pale bluish-green (more extensively in male); sides of head and hindneck spotted with pale golden-green; a patch of golden-brown on hindcrown; feathers of wings and tail edged with dark blue; shoulder blue; rump pale green, passing to blue on upper tail-coverts; posterior underparts cinnamon. Bill black except base of lower mandible pale gray; legs gray. **Young:** resemble adults but black of head and body duller, more sooty; pale scaling on chest and spots on head broader, duller, and less distinct; no crown-patch; rump dull green; less blue on shoulder; edges of wing feathers dull greenish-blue.

Habits. Frequents upper levels of epiphyte-laden, mossy mountain forests, coming lower along edges and in nearby second growth, pastures with scattered trees; in pairs or small groups year-round, often accompanying flocks of bush-tanagers or other small frugivorous birds; hops actively about searching foliage for small insects and spiders, sometimes stretching down to examine undersides of twigs and mossy branches; eats many berries, including those of melastomes, epiphytes of many sorts, especially Ericaceae, and shrubs like *Fuchsia*.

Voice. A high, thin, penetrating *tsip* or *seek*, often in twittering series; song not recorded.

Nest. A bulky cup of green liverworts and mosses mixed with dry pieces of inflorescences, lined with strips and shreds of dry bromeliad leaves, sometimes decorated with bits of fern, 11–33ft (3.5–10m) up amid epiphytes on stump or branch or in crotch of tree, often at or near forest edge. Eggs 2, undescribed (?). April–August.

Status. Common resident in mountains at 4000–9000ft (1200–2750m), in smaller numbers down to 2600ft (800m) and up to 10,000ft (3000m), from Cordillera de Tilarán S to Panama.

Range. Costa Rica and W Panama.

GREEN HONEYCREEPER

Chlorophanes spiza Pl. 46(7)

Mielero Verde (Rey de Trepadores)

Description. 5″ (13cm); 19g. Green; male with black half-hood; bill thick at base, sharply pointed, conspicuously bicolored. **Adult ♂:** top and sides of head and chin black; rest of body bright, glossy bluish-green, with somewhat darker, more bluish tertials, wing-coverts, and edgings on wings and tail. Iris brick-red; culmen black, rest of bill bright yellow; legs grayish. ♀: above bright yellowish-green, glossy on back, duller and tinged with olive on head, wings, and tail; breast and sides bright yellowish-green; throat, belly, and crissum paler and duller. Culmen black, rest of bill dull yellow, shading to horn-color at base of lower mandible. **Immatures:** like adult female but paler and duller below, belly more grayish; iris reddish-brown. Young males acquire full adult plumage in the months preceding their first breeding season.

Habits. Frequents canopy of dense forest but often descends low along edges, in clearings with scattered trees and shrubbery, semi-open, second growth, and gardens; usually in pairs that often join birds of other species at fruiting trees; often travels through canopy with tanagers, warblers, and other small birds; eats fruits and arillate seeds; often pugnacious at fruiting trees; probes flowers for nectar while perching; varies diet with insects caught in air or gleaned from foliage; attracted to fruit at feeders.

Voice. Calls include a sharp, warblerlike *chip*; a high, piercing *pseet*; a low, dry rattle when excited or perturbed. Males sometimes repeat a sharp, dry *tsup tsup tsup* up to 10 times in series, which may serve as the song.

Nest. A slight, shallow cup, almost wholly of small, dry dicotyledonous leaves mixed with rachises of compound leaves, tendrils and fungal filaments, 5–40ft (1.5–12m) up in tree or shrub in shady clearing near forest. Eggs 2, white, finely speckled with reddish-brown. April–July.

Status. Resident on Caribbean slope from sea level to ca. 3300ft (1000m); uncommon in lowlands, common at 1000–2600ft (300–800m); common in lowlands and foothills of S Pacific slope, to ca. 4000ft (1200m).

Range. S Mexico to NW Peru, C Bolivia, and SE Brazil.

RED-LEGGED HONEYCREEPER

Cyanerpes cyaneus Pl. 46(2)

Mielero Patirrojo (Picudo, Tucuso, Trepador)

Description. 4½″ (11.5cm); 13.5g. Bill long, slender, decurved; legs reddish. **Adult ♂ Breeding**: crown glossy, pale turquoise; lores, orbital area, back, most of wings, and tail velvety black; rest of head and body, including scapulars, bluish-violet; wing-linings, including inner webs of most remiges, bright yellow. Bill black; legs carmine. Nonbreeding "eclipse": like adult female but retaining black wings and tail. ♀: above olive-green, sometimes tinged with gray or, faintly, with bluish; wings and tail dusky with greenish edgings; narrow superciliaries yellowish; whitish eye-ring interrupted by dusky eye-stripe; sides of head, chest, and sides pale dull green with paler, often whitish shaft-streaks; throat indistinctly streaked with greenish and dull yellowish; flanks dull green; center of belly pale yellow to whitish. Legs dull rufous to dull brick-red. **Immatures:** like adult female, below averaging less heavily streaked. Adult males wear "eclipse" plumage between about June and November; young males molt into adult-type plumage between December and February, may start wing molt by October; thus between June and February one can encounter "patchy" males.

Habits. Spends most of year in small flocks that wander widely through crowns of trees in forest and semi-open, coming low in second growth, along edges and in gardens, often acompanying other small birds; breeds in isolated pairs; eats many arillate seeds, notably those of *Clusia*, berries, and other fruits; while perching, extracts nectar from flowers, especially of *Inga, Calliandra*, and other mimosoid legumes, also *Erythrina, Genipa*; gleans small insects from foliage while hopping and creeping along fine twigs, or catches them in the air; probes rolled leaves and crevices in bark.

Voice. Calls include a high, thin, piercing *tseet* and a thin, weak, nasal *chaa*, rather gnatcatcherlike; scolds with a thin, scratchy *wee-peew*; rarely heard dawn song a slight *tsip tsip chaa, tsip tsip chaa*, repeated for many minutes.

Nest. A small, thin, hemispheric cup composed of fine fibers, rootlets, and grass inflorescences, 10–45ft (3–14m) up amid foliage of slender branches. Eggs 2, white, speckled with pale brown. February–June.

Status. Resident; common in dry N lowlands and mountains, uncommon E to about Grecia in Valle Central; uncommmon to fairly common S on Pacific slope to Panama, sea level to 4000ft (1200m); on Caribbean side abundant in extreme NW to Río Frío region but rare or sporadic elsewhere, up to 3300ft (1000m); has declined in some areas as human population has increased (e.g., Valle del General); abundant on Isla del Caño, off coast of Península de Osa.

Range. N Mexico to W Ecuador, N Bolivia, and E Brazil; Cuba (where possibly introduced).

Note. Also called Blue Honeycreeper.

SHINING HONEYCREEPER

Cyanerpes lucidus Pl. 46(1)

Mielero Luciente (Picudo, Patiamarillo, Trepador)

Description. 4″ (10cm); 11g. Smaller than Red-legged Honeycreeper, with yellow legs, short tail, no yellow in wings; male throat black; female breast heavily streaked. **Adult ♂:** lores, orbital area, throat, wings, and tail black; rest of head and body violet-blue, palest on forehead. Bill black; legs bright yellow. ♀: head dull greenish-blue, streaked narrowly with buff on forehead and sides of head, shading to bright violet-blue on malar area; rest of upperparts, sides, and flanks green; throat buffy; breast violet-blue with broad buffy streaking; belly and crissum pale yellowish. Legs greenish-yellow. **Immatures:** like adult female but blue of breast less extensive, head more greenish; males molt to adult plumage November–January.

Habits. Frequents upper levels of forest and neighboring semi-open, coming lower along edges and in second growth; alone, in pairs or family groups, but never forms flocks as does Red-legged; often with flocks of warblers, greenlets, gnatcatchers, etc.; eats many berries and arillate seeds, nectar from flowers that it probes while perching, and small insects and spiders that it gleans from thin, leafless, dangling vines or exposed twigs, to which it clings in all attitudes, often hanging head downward, yellow legs prominent; like Red-legged, attracted to fruit at feeders.

Voice. Calls include high, sharp, staccato chittering notes and a little chatter, also a very high, thin *seee seee seee seeuu*; males sing a high, thin, *pit pit pit pit-pit pit-peet pit pit-peet pit pit . . .* from a treetop, for minutes on end; parents concerned for nestlings repeat a low, hard, metallic *click* or *tick*.

Nest. A shallow cup attached by rim to thin horizontal twigs, much like a manakin's nest, of slender dark strands, so thin that eggs are visible through the bottom, 20–30ft (6–9m) up in clearings beside forest. Eggs 2, undescribed (?). April–September (fledglings in May, July, October).

Status. Common to locally abundant resident of lowlands and foothills throughout Caribbean slope and on S Pacific slope, N sparingly to about Carara; sea level to 4000ft (1200m).

Range. S Mexico to extreme NW Colombia.

Note. Sometimes considered conspecific with Purple Honeycreeper (*C. caeruleus*) of South America, but the two are sympatric in Darién and NW Colombia.

BLUE DACNIS

Dacnis cayana Pl. 46(3)

Mielero Azulejo

Description. 4½″ (11.5cm); 13.5g. Bill rather thick and pinkish at base and sharply pointed; male paler blue than *Cyanerpes*; female green with blue head. **Adult ♂:** mostly bright blue; throat dusky-blackish; lores, orbital area, back, tail, and most of wings black; scapulars, wing-coverts, and remiges edged with blue. ♀: top and sides of head greenish-blue; lores dusky; throat gray; rest of body, including edges of wing-coverts, remiges, and rectrices, grass-green, darker and duller on back; shoulders tinged with blue. Iris brick-red; bill blackish with basal half of lower mandible pinkish; legs flesh-color. **Immatures:** like adult female; males acquire adult plumage before first breeding season, but retain some green-edged remiges and green on belly.

Habits. Frequents sunlit upper levels of forest, descending lower in adjacent clearings with scattered trees and bushes, second growth and edges; in pairs or small family-sized groups, often accompanying flocks of tanagers, warblers, other honeycreepers; searches for insects in foliage like a warbler, sometimes hanging and reaching acrobatically; eats arillate seeds and fruits, including bananas at feeders; probes flowers for nectar while perching.

Voice. A high, thin *tsip* or *tseep*, sometimes repeated; an unhoneycreeperlike *skwirk-irk* (Slud); has not been heard to sing; a generally quiet bird.

Nest. A deep cup suspended between 2 leafy twigs, of fine bast fibers and soft seed down compactly matted with secondary rachises of small compound leaves, 18–25ft (5.5–8m) up in isolated tree. Eggs 2, white with dark markings. May–October (fledglings in November). At 1 nest 3 adults fed the young.

Status. Resident; uncommon throughout Caribbean lowlands, locally up to 3000ft (900m); common in lowlands and foothills of S Pacific slope, N to about Carara, locally to 4000ft (1200m).

Range. NE Honduras to W Ecuador, NE Argentina, and S Brazil.

SCARLET-THIGHED DACNIS

Dacnis venusta Pl. 46(4)

Mielero Celeste y Negro

Description. 4¾″ (12cm); 16g. Bill rather slender, sharp-pointed; blue half-hood. **Adult ♂:** forehead, lores, orbital area, wings, and tail black; throat, underparts, and sides of back greenish-black, except thighs red; top and sides of head and neck, scapulars, center of back, and rump turquoise-blue. ♀: above dusky bluish-green, brightest on sides of neck and malar area, dullest on sides of back, shading to turquoise on rump; lores, wings, and tail dusky; throat gray; breast, sides, and flanks grayish-olive, the breast somewhat tinged with blue, shading to buffy-yellow on belly and crissum; thighs cinnamon-orange. Iris brick-red; bill black; legs gray. **Immatures:** like adult female but duller and grayer, especially on head. Young males acquire some adult plumage on the head and body from September onward but remain much duller than older birds through first breeding season.

Habits. Frequents canopy and edges of humid forest and clearings with scattered fruiting trees, often far from forest; usually travels in pairs or small groups, often with mixed flocks of tanagers, other honeycreepers, and warblers; after breeding may wander widely in small flocks; eats berries and arillate seeds, including those of *Clusia* and *Zanthoxylum*; gleans insects in foliage.

Voice. A chirping *woit* or *wheit* with upward inflection, or a scratchy *bzik*; song 2–3 scratchy, insectlike trills ending in high, squeaky notes: *bzzeek bzzeek bzzeek*; a female defending her nest utters a low, nasal cry.

Nest. A slight, shallow bowl or hammock, stretched between 2 slender twigs, of coarse, wiry materials including rootlets, tendrils, and slender rachises of ferns, the thin fabric completely covered below with green pieces of fern, 40–50ft (12–15m) up, amid concealing foliage (2 records). Eggs undescribed (?). March–May.

Status. Resident; breeds in foothills at 1650–4000ft (500–1200m) on Caribbean slope and 3000–5000ft (900–1500m) on S Pacific slope; following breeding often moves to lowlands in search of fruiting trees.

Range. Costa Rica to NW Ecuador.

BLUE-AND-GOLD TANAGER

Buthraupis arcaei Pl. 45(11)

Tangara de Costillas Negras

Description. 6″ (15cm); 46g. Suggests an overgrown euphonia with red eye, heavy bill, and dark sides. **Adults:** upperparts dark, dull blue, brightest on crown, becoming blackish on lores and throat, and dull blue-black on sides, flanks, and thighs; breast and belly bright yellow, upper breast tinged with orange. Bill black; iris brick-red; legs dark gray. **Young:** similar but duller; iris brown.

Habits. Frequents middle and upper canopy in forest, descending to shrub level at edges and gaps; usually in pairs, or groups of 3–4 after breeding, generally associating with mixed flocks including tanagers and warblers; alternates periods of sluggish perching with sudden flights and fast, heavy hops along branches; peers and probes deliberately into tufts of moss and epiphytes on branches for roaches, spiders, and other invertebrates; perches to pluck berries, often reaching far downward; plucks flowers of epiphytes (e.g., *Satyria*) to squeeze out nectar.

Voice. Call a high, piercing *zeek* or *keep* and a squeaky *spit* or *chit*; song a halting medley of high-pitched, short, sibilant notes and thin, squeaky or wheezy whistles: *tsip tseee tsup tseeeur weeuwee tsip*. . . .

Nest. A bulky cup of plant fibers and mosses, hidden in mass of epiphytes, usually well up in tree at forest edge. Eggs undescribed (?). March–June.

Status. Locally common resident along lower Caribbean slope of Cordillera Central and Cordillera de Tilarán, N in smaller numbers to Cordillera de Guanacaste (Volcán Miravalles), at 1300–4000ft (400–1200m); center of abundance 2300–3300ft (700–1000m). Not yet recorded from Cordillera de Talamanca.

Range. Costa Rica to C Panama.

Note. Often placed in genus *Bangsia*.

BLUE-GRAY TANAGER

Thraupis episcopus Pl. 46(15)

Tangara Azuleja (Viuda)

Description. 6″ (15cm); 32g. One of the commonest, best-known, and unmistakable birds of Costa Rica. **Adults:** head, throat, and underparts pale gray, tinged with greenish-blue, becoming darker and more bluish on flanks; back darker blue-gray; rump and tail brighter blue; wings bright sky-blue, darker on primary coverts and shoulders, tinged with greenish-gray on tertials. Male brighter, female duller and grayer overall. Upper mandible black, lower blue-gray; legs blackish. **Young:** much duller and grayer, with at most a slight bluish tinge on head and underparts; a narrow, broken white eye-ring; back duller and more greenish; rump grayish-blue; wings and tail dingy greenish-blue, tinged with olive on wing-coverts and little or no blue on shoulder.

Habits. In all types of open country with trees and shrubs, suburban gardens, town parks and plazas, second growth; edges of humid forest, even ranging into the canopy; in pairs year-round, sometimes in loose, straggling flocks following breeding; eats many kinds of berries and arillate seeds, sometimes congregating in large numbers at fig trees in particular; is readily attracted to feeders; takes nectar from flowers of balsa and *Erythrina*; searches for insects and spiders amid foliage and by bending down to examine undersides of horizontal boughs, railings, electric wires, etc.; eats tender green leaves.

Voice. A variety of slurred, squeaky whistles; song a rather animated but unstructured sing-song of squeaky or wiry whistles, some slurred up or down but all on about the same pitch.

Nest. A neat cup of soft materials bound together with spiderweb, middle layer of strips of leaves or grass blades, lining of fine fibers, often with much green moss on outside; ground level to 100ft (0–30m), usually in a tree or shrub at no great height, occasionally in an open shed or in a bunch of green bananas; sometimes steals nest of a smaller tanager and incubates builder's eggs along with its own. Eggs 2, pale blue-gray, heavily mottled with brown or chocolate. March–July; 2 broods. A nest with 4 eggs was attended by both females of a bigamous male.

Status. Common to abundant resident through most of country from lowlands to 6600ft (2000m), occasionally to 7500ft (2300m), except in dry NW, where uncommon and local in humid areas around houses.

Range. C Mexico to NW Bolivia and Amazonian Brazil.

PALM TANAGER

Thraupis palmarum Pl. 45(19)

Tangara Palmera

Description. 6¼″ (16cm); 38g. Dull olive with contrasting blackish remiges and rectrices. **Adults:** face and forehead olive-green, shading to glossy grayish-olive over rest of head and body; above somewhat darker, the feathers with indistinct, dull violaceous tips, broader and brighter in males; wing-coverts and bases of remiges pale dull olive; rest of remiges and rectrices blackish. Bill black except base of lower mandible silvery; legs gray. **Young:** duller and duskier overall, without silky gloss or violaceous tinge.

Habits. Prefers open country with scattered trees, especially palms, including light woods, plantations of oil, coconut or pejibaye palms, pastures, parks, and gardens; in pairs year-round; eats fruits, especially figs and *Cecropia* spikes; searches foliage, including that of palms, for insects and spiders, often hanging from edge of frond or large leaf to scan underside; occasionally sallies for flying insects; does not associate with mixed flocks in Costa Rica.

Voice. Calls include various sharp, high-pitched, short whistled notes, shriller than those of Blue-gray Tanager but less thin and wiry, and sometimes with metallic tone; song resembles that of Blue-gray but with a more hurried delivery, broken into more or less distinct phrases: *su'suri su'suri su'suri sree sree sree su'suri susuri*, etc. (Eisenmann), sometimes with interpolated metallic sputters.

Nest. A neat cup of grasses, strips of dry leaves, palm fibers, and sometimes mosses, usually well concealed in crown of palm or other tree with large, clustered leaves, amid epiphytes, in cavity in tree, or crevice in a building, 20–100ft (6–30m) up. Eggs 2, pale bluish, heavily spotted with brown and pale lilac. March–June.

Status. Fairly common resident countrywide, except rare and local in dry NW, from lowlands to 5000ft (1500m), rarely higher; especially abundant along Caribbean coast in coconut groves; nearly everywhere less numerous than Blue-gray Tanager.

Range. E Honduras to W Ecuador, N Bolivia, and S Brazil.

Note. Between late 1985 and early 1987, at least 2 Yellow-winged Tanagers (*Thraupis abbas*) were seen repeatedly on the grounds of CATIE, the agricultural institute at Turrialba (Sánchez and Campos). The species is generally similar to the Palm Tanager but with a bright purplish blue head and chest, a black back scaled with blue, and a bright yellow patch at the base of the primaries. Given the distance to the species' normal range (Mexico to E-C Nicaragua) and the number of visitors to the institute from other Central American countries, the birds might have been escapees rather than strays; apparently they did not become established.

SCARLET-RUMPED TANAGER

Ramphocelus passerinii Pl. 47(4)

Tangara Lomiescarlata (Sargento, Rabadilla Tinta, Terciopelo, Sangre de Toro)

Description. 6¼″ (16cm); 31g. Thick bill with inflated silvery rami of lower mandible; 2 races on coastal slopes of Costa Rica that differ mainly in color of the females. **Adult ♂:** velvety black, except lower back, rump, and upper tail-coverts brilliant scarlet (slightly more orange in Pacific race). Iris chestnut to dark red; bill silvery with black tomia and tip; legs blackish. ♀: (Caribbean race) head brownish-gray; upperparts ochraceous-olive, paler and brighter on rump; wings and tail dusky; throat grayish-buff; rest of underparts ochraceous-olive, brightest on breast; bill often duller gray. Pacific race similar but underparts more ochraceous, with a broad band of dark orange across breast; rump and upper tail-coverts ochraceous-orange (breeding females vary considerably in color of breast and rump); back with ochraceous tinge. **Young:** like adult female (Pacific race) with breast less orange; immature males often with scattered black and red feathers, wings and tail mixed with black and olive, gradually acquiring adult plumage over their first 6 months or so.

Habits. Prefers lighter second growth, lush thickets, bushy pastures, gardens and dooryard shrubbery, semi-open, and forest edge, sometimes entering humid forest for a short distance; travels through home range in straggling flocks that are often joined by saltators, other tanagers, warblers, and other small birds; eats many fruits, including berries, *Cecropia* and *Piper* spikes; hunts insects and spiders amid foliage, sometimes descending to ground, occasionally sallying for flying insects; male often perches conspicuously with rump feathers fluffed, scarlet visible from afar; sleeps amid dense shrubbery or ferns in compact groups of up to a dozen.

Voice. Calls include a sharp *ac* or *wac*, a dry, scratchy *chuck*, a dry *pzzt*, and a sharp *whip* when alarmed; a sharp descending chatter in interactions; males of Pacific race sing simple phrases of 3 or 4 clear, pleasant notes, often tirelessly repeated; Caribbean males sing shorter phrases, much less frequently or persistently.

Nest. A scanty to fairly bulky cup, mainly of dry strips of leaves bound together with cordlike or fibrous materials, lined with rootlets, tendrils, fungal hyphae; often with a spray or 2 of a small living fern on the outside; 1–20ft (0.3–6m) up amid foliage of shrub or tree. Eggs 2, very rarely 3, pale blue or gray, rarely whitish, blotched, dotted and scrawled with black, brown, and pale lilac. March–August; occasionally 2 broods.

Status. Abundant resident of Caribbean slope and S Pacific slope N to near Puriscal and Carara, from lowlands to 4000ft (1200m), rarely to 5500ft (1700m); locally N in Pacific lowlands to beyond Puntarenas, and in humid pockets and passes on Pacific side of N cordilleras.

Range. S Mexico to W Panama.

CRIMSON-COLLARED TANAGER

Phlogothraupis sanguinolenta Pl. 47(3)

Tangara Capuchirroja (Sangre de Toro)

Description. 7¼″ (18.5cm); 40g. Heavy, conical bill; rather long tail; deep red or rufous hood surrounds black face and throat. **Adults:** forehead, face, and throat black; crown, back, and sides of neck and breast dark scarlet-red to crimson, as are lower rump and upper and lower tail-coverts; rest of plumage black; females average duller than males but not always distinguishable. Iris red; bill silvery-white, often with thin black line along tomia; legs black. **Young:** red areas of adult are dull rufous to brick-red, with dusky feather bases giving mottled effect; black areas dull brownish-black; bill duller, tomia horn-color.

Habits. Frequents thickety forest edge, tall second growth, semi-open, and rank hedgerows, generally staying higher than Scarlet-rumped Tanager; in pairs year-round, briefly in small groups after breeding; occasionally accompanies mixed flocks of other tanagers, saltators, etc.; forages for insects in thick foliage, eats many fruits; males sing from elevated perches, are much more territorial than Scarlet-rumped.

Voice. Call a thin, querulous *tsewee* or *weet weee* with rising inflection; song a leisurely sequence of rising and falling whistled notes separated by pauses, *tueee-teew, chu-cheweee chew, teweee*.

Nest. A solidly built cup with outer layer of moss, rootlets, fine stems, tendrils, and sometimes a spray of small living fern, middle layer of sheets of papery bark or strips of *Heliconia* or banana leaf, with variable lining of fine stems, 6–20ft (2–6m) up in tree at forest edge or in clearing. Eggs 2, pale blue with large and small black spots. April–June.

Status. Uncommon to locally common resident of Caribbean lowlands, to 3500ft (1100m) in foothills, extending to Pacific side of N cordilleras through low passes.

Range. S Mexico to C Panama.

Note. Sometimes placed in *Ramphocelus*; we keep it separate on the basis of differences from all species of that genus in bill shape, plumage sequence, voice, social system, and eggs.

SUMMER TANAGER

Piranga rubra Pl. 47(5)

Tangara Veranera (Cardenal Veranero)

Description. 6½″ (16.5cm); 30g. Medium size; unpatterned red or yellow-brown; thick pale bill. **Adult ♂:** below rosy-red; above darker and duskier rose-red; remiges dusky with rose-red edgings. **♀:** above olive; below yellow-olive, with an overall ochraceous or orange tinge, brightest on crissum. Upper mandible horn-color to yellowish-horn, lower pale yellowish; legs grayish. **Immature ♂:** like adult female but below averaging brighter yellow, often tinged with orange and often with scattered red feathers, especially about head, breast, and back. ♀: averages duller than adult female, above greener and below more buffy; feathers of wings edged with brownish or grayish.

Habits. Frequents canopy and edges of evergreen and deciduous forest, light woodland, semi-open, tall second growth, shady gardens, scattered trees in clearings and pastures; solitary and territorial in winter, though often joins mixed flocks of other tanagers, warblers, etc.; eats fruits of various sorts; tears open nests of wasps and stingless bees to extract larvae, pupae, sometimes adults; catches many insects in flight, especially stingless bees; visits feeders for bananas.

Voice. Call a sharp, distinctive *KITtituck* or *KI-ti-ti-tuk*; song a slightly burry, robinlike warble, occasionally heard in fall, usually in subdued form, as recent arrivals are establishing territories.

Status. Common migrant and winter resident between mid-September and mid to late April, in lowlands of both slopes and upward in decreasing abundance to 8200ft (2500m), sometimes higher in dry season.

Range. Breeds from S USA to N Mexico; winters from C Mexico to W Ecuador, N Bolivia, and Amazonian Brazil.

HEPATIC TANAGER

Piranga flava Pl. 47(6)

Tangara Bermeja (Cardenal)

Description. 7″ (18cm); 40g. Heavy blackish bill with prominently toothed tomia; lores and often cheeks smudged with dusky. **Adult ♂:** entirely dusky-red to dark brick-red, paler and brighter on throat and lower belly. ♀: above olive-green; below yellowish-olive, washed with olive-green on breast, sides, and flanks, and brightest yellow on belly and crissum. Upper mandible blackish, lower gray with dusky tip; legs dark gray. **Immature ♂:** like adult female but with varying amounts of brick-red tinge or spotting about head and throat; underparts brighter, more ochraceous.

Habits. Frequents canopy and edge in mountainous terrain, usually in treetops but sometimes descending lower in adjacent tall second growth and clearings with scattered tall trees; hops deliberately in foliage, scanning up and down, sometimes bending over to peer beneath perch in search of insects and spiders; occasionally sallies for flying insects; mashes flowers in bill; eats many berries and arillate seeds; usually in pairs, alone or accompanying mixed flocks including other tanagers, honeycreepers, warblers; continually flicks tail through a short arc.

Voice. Call a quick series of 3–4 sharp notes, rising in pitch, and all more or less equally accented, *chu-chit-it* or *chudidit*; song a series of warbled phrases of 4–8 notes, with a rich, slightly burry or squeaky tone, higher in pitch than song of Flame-colored Tanager; typical phrase: *chew-chewee-cheer-cheeur-chuee*.

Nest. A rather shallow bowl of rootlets, grass stems, and fibers, in niche among bromeliads and other epiphytes on high branch or dead vertical trunk, 20–40ft (6–12m) up. Eggs 2–3, white or greenish, spotted with brown (in North America; eggs not seen in Costa Rica). March–May.

Status. Common resident the length of Caribbean slope at 2000–5000ft (600–1500m) in N and C mountains, and 3000–5500ft (900–1700m) in SE; fairly common on Pacific slope of N cordilleras above 3500ft (1100m) and along slopes of Dota region and Cordillera de Talamanca, mostly at 3300–6000ft (1000–1850m); also breeds in higher hills of Península de Osa; largely descends to lower elevations outside of breeding season.

Range. SW USA to N Argentina.

SCARLET TANAGER

Piranga olivacea Pl. 47(8)

Tangara Escarlata

Description. 6¼″ (16cm); 28g. Females and immatures usually greener than Summer Tanagers, with whitish (not yellow) wing-linings; Western Tanagers have definite wing-bars. **Adult** ♂ Winter: above olive-green; below bright greenish-yellow, washed with olive on breast, sides and flanks; wings and tail black. Breeding: head and body plumage scarlet; wings and tail black. ♀ Winter: like male but wings and tail dusky-olive with olive-green edgings; usually duller and greener on throat and chest. Breeding: similar but often with orange tinge over head, throat, and breast. Bill yellowish-horn, often grayish toward base; legs bluish-gray. **Immature ♂:** like adult female but wings more dusky, usually some or all of wing-coverts black. ♀: like adult but duller, browner overall, especially on wings; rectrices narrower, more pointed.

Habits. Frequents canopy and edge of forest, taller second growth and semi-open, sometimes lower in bushy areas; gleans insects and spiders from foliage, sometimes sallying to pluck them from leaves or twigs; eats fruits of many sorts; usually in small, loose groups or singly in fall, sometimes in larger flocks in spring.

VOICE. Call a low *chibirr* or *chip-churr*, rarely heard in migration.

STATUS. Uncommon fall migrant (late September–early November), mainly in Caribbean lowlands and in central highlands; sporadically common spring migrant (late March–early May) from lowlands of both slopes up to ca. 5000ft (1500m); no winter records.

RANGE. Breeds from S Canada to SE USA; winters from Panama (rarely) and Colombia to E Peru and NW Bolivia.

WHITE-WINGED TANAGER

Piranga leucoptera Pl. 47(7)

Tangara Aliblanca (Cardenalito)

DESCRIPTION. 5¼″ (13.5cm); 15g. Resembles Flame-colored Tanager but much smaller, with broader wing-bars and unstreaked back; male deeper red with black mask. **Adult ♂:** lores, orbital area, cheeks, and chin black; rest of head and neck, center of back, rump, and entire underparts bright red, paler posteriorly; sides of back, wings and tail black, with 2 broad white wing-bars. ♀: above olive-green, tinged with yellow on head; lores and orbital area dusky-olive; tail and wings dusky with olive edgings, white wing-bars; below rich yellow, tinged with orange on throat and breast. Upper mandible black, lower pale gray with black tip; legs blackish. **Young:** like adult female but below duller and paler; back tinged with grayish; males have dusky mask. When a few months old, male acquires some or all red plumage on head and body (although often paler than adult), but retains juvenile wings and tail.

HABITS. In mountain forests, frequents canopy and edges, and trees in neighboring shady clearings; actively gleans spiders, beetles, roaches, crickets from foliage and crevices in bark; extracts larvae and pupae of wasps from their cells; takes fruit; singly or in pairs, accompanies mixed flocks of other tanagers, warblers, honeycreepers and other small birds.

VOICE. Call a squeaky *chick-it* or *chick-up*; repeats soft *chick, chick* in flight; song *chut-weet, chut-weet weet*, the *chut* low and hoarse, the *weet* thin and piercing or weak and lisping.

NEST. A small cup, 45ft (14m) up on mossy branch of tree in mountain pasture, built by female; not accessible for examination. Eggs unknown (?). April–May.

STATUS. Uncommon resident of upper parts of N cordilleras, Caribbean slope of Cordillera Central, and both slopes of Cordillera de Talamanca, including Dota region and S coastal ranges, at 3500–6000ft (1100–1850m); most numerous in upper Coto Brus valley and adjacent slopes.

RANGE. C Mexico to W Ecuador, NW Bolivia, and NW Brazil.

WESTERN TANAGER

Piranga ludoviciana Pl. 47(2)

Tangara Carirroja

DESCRIPTION. 6½″ (16.5cm); 28g. Differs from other migrant tanagers in having distinct wing-bars, and from Flame-colored Tanager in its unstreaked back. **Adult ♂** WINTER: above olive-green, tinged with orange on face; feathers of back black with broad olive-green tips; rump brighter, yellowish-olive; wings and tail black, with 2 pale yellow wing-bars; below greenish-yellow tinged with olive laterally; face often tinged with orange. BREEDING: back black; nape, rump, and underparts bright yellow; rest of head orange-red. ♀: above olive-green, washed with grayish on back; wings and tail dusky; wing-bars whitish; below duller and greener than male. **Immatures:** resemble adult female; males blacker on back and more yellowish on head and rump; females duller, with wings more brownish.

HABITS. Frequents canopy of deciduous forest, crowns of small trees in scrub and second growth, semi-open, and hedgerows; solitary and possibly territorial in winter, in small loose groups in migration; forages for insects in crowns of trees, often flying out to catch them; eats many fruits.

VOICE. Call a sharp, ascending set of 2–5 notes, *chirdididick cherdit* or *pit-er-ick*, sometimes followed by a lower *chert-it*; song, occasionally heard in spring, a prolonged robinlike warble, slightly hoarse and burry in tone.

STATUS. Winter resident, uncommon to fairly common in dry NW, rare in central highlands and on S Pacific slope; accidental in Caribbean lowlands; arrives late October and departs in April.

RANGE. Breeds in W North America, S to N Baja California; winters from C Mexico to Costa Rica, casually to W Panama.

FLAME-COLORED TANAGER

Piranga bidentata Pl. 47(9)

Tangara Dorsirrayada (Cardenal)

DESCRIPTION. 7″ (18cm); 38g. No other tanager combines streaked back and wing-bars; heavy bill usually has 2 teeth on each tomium. **Adult ♂:** head, neck, and underparts orange-red, paler posteriorly; lores and cheeks mixed with dusky; a blackish patch on auriculars; back and rump duller, tinged with olive, back streaked with black; tail and wings blackish with reddish edgings, 2 pinkish-white wing-bars; outer 3 rectrices tipped with grayish-white. ♀: head yellowish-olive, mixed with dusky on lores, cheeks, and auriculars; pileum narrowly streaked with blackish; rest of upperparts olive-green, heavily streaked with blackish on back; tail and wings dusky with olive edgings, white wing-bars; below greenish-yellow, paler posteriorly. Upper mandible black, lower gray; legs gray. **Immature ♂:** (through first breeding season) duller orange than adult, more or less mixed with olive-green; retains dull juvenile wings and tail. **Young:** resemble adult female but duller and paler, especially below.

HABITS. Frequents treetops in mountain forests and neighboring shady pastures, coffee plantations, gardens, and cypress trees along roadsides and between pastures; forages for insects in foliage; eats many fruits, especially figs and berries of melastomes and Ericaceae; often mashes fruit in bill, discards husk; sometimes descends to shrubs or ground for insects or berries; usually in pairs that sometimes accompany mixed flocks.

VOICE. Call a sharp, full-voiced *per-DECK* or *chi-DICK*. Song, delivered from high treetops, a long series of phrases of 3–6 slurred, warbling notes; tone rich but slightly hoarse or burry: *chewee-very-vire, chewee-very-vire-very, cheery-cheweea*, etc.

NEST. A loosely constructed cup of stiff rootlets, slender twigs, and vine tendrils, lined with fine grass stems and inflorescences; 16–35ft (5–11m) up in densely-foliaged tree, often cypress. Eggs 2–3, pale blue, speckled with reddish-brown and dark lavender, mostly at large end. April–June.

STATUS. Resident in Cordillera Central, Cordillera de Talamanca and Dota region; on Caribbean slope uncommon at ca. 6000–9300ft (1850–2850m); more common and widespread on Pacific slope, from timberline down to 4000ft (1200m), rarely to 3000ft (900m).

RANGE. NW Mexico to W Panama.

NOTE. Also called Streaked-backed Tanager.

OLIVE TANAGER

Chlorothraupis carmioli Pl. 45(16)

Tangara Aceitunada o de Carmiol

DESCRIPTION. 6¾″ (17cm); 38g. Large, noisy, green, without contrasting markings. **Adults:** above olive-green; below paler, brightest on throat, which is sometimes strongly tinged with yellowish. Bill black, legs plumbeous. **Young:** iris dull gray rather than dark brown.

HABITS. Travels through understory and middle levels of wet forest and adjacent shady second growth in noisy flocks of 5–20 that seem to follow regular routes and perhaps to defend territories; pairs evidently drop out when nesting. Foraging birds hop heavily in vegetation, rummaging and peering for insects, sometimes dashing rapidly along twig or hanging below to reach beetles, roaches, crickets, katydids, or fruit; take many berries, especially melastomes, often entire flock congregating briefly in fruiting tree. May bite pieces from large, soft fruits (e.g., *Coussarea*) or hold large insect with foot, pull it apart with bill. Flocks often travel with Tawny-crested Tanagers, are joined by ovenbirds, woodcreepers, and other birds.

VOICE. Calls of moving flock include a high, piercing, slightly buzzy *seeet* or *seee-seeee* and staccato *tik* or *tip* in bursts when about to fly or in flight; when alarmed or agitated, a raucous, nasal, scratchy *nyaaah* or *cheeyah*, often repeated; song a rapid stream of clear, grating, or wheezy short notes, in groups of 3–8, often prolonged (Isler).

NEST. A neat cup of mosses and long vegetable fibers with some leaves in the base, lined with fine rootlets and black fungal rhizomorphs, 5–8ft (1.5–2.5m) up in fork of understory shrub or tree. Eggs 2, white, with fine specks and large blotches of dark and pale brown, gray, and reddish-brown, mostly at large end. March–May.

STATUS. Common resident of wet lowlands and foothills the length of Caribbean slope, sea level to 3300ft (1000m), occasionally to 4600ft (1400m); barely reaches Pacific slope via passes in Cordillera de Guanacaste.

Range. E Nicaragua to E Peru and NW Bolivia.

Note. Also called Carmiol's Tanager. Sometimes considered conspecific with *C. olivacea*, Lemon-browed Tanager, of E Panama and South America.

RED-CROWNED ANT-TANAGER

Habia rubica Pl. 47(10)

Tangara Hormiguera Coronirroja

Description. 7″ (18cm); 38g. Distinguished from other ant-tanagers by the bright, erectile crown-patch, which forms a conspicuous bifid crest in male. **Adult ♂:** usually concealed crown-patch pinkish-scarlet, bordered by reddish-black; rest of upperparts, including wings and tail, dusky-red; underparts paler dull red, brightest on throat, which may be streaked minutely with brighter red; sides and flanks darker; belly and crissum paler and washed with grayish. ♀: upperparts olive-green; center of crown golden-yellow; throat and chest paler olive, the throat minutely streaked or suffused with yellowish; posterior underparts duller and paler olive. Bill blackish; legs brownish-gray. **Immatures:** like adult female, except males have varying amounts of red; crown-patch more orangish, less extensive than adult. **Young:** above brownish-olive, crown feathers with obscure yellowish bases; below dull buff, smudged or mottled with brownish-olive on sides and chest.

Habits. In understory of humid and subhumid evergreen forest and adjacent tall, shady second growth; in pairs or family groups, often travels with Tawny-crowned Greenlets, Buff-throated Foliage-gleaners, and Slaty Antwrens; restless, active, in constant motion, rummages and gleans in foliage and branches for insects, especially beetles; eats many fruits; despite its name only infrequently forages with army ants.

Voice. Progress of flock through forest marked by soft chips and peeps and dry, explosive, staccato chatters; common call a rasping *chek-chek*. Chiefly at dawn, males repeat for many minutes a phrase of 2–4 loud, mellow, clear, whistled notes: *peter peter peter* . . . or *intervene, intervene, intervene*. . . . At other times they warble softly and sweetly.

Nest. A shallow cup of rootlets and fibers, so thinly bound with spiderweb that eggs are visible through bottom, in sapling or slender bush 5½ft to, rarely, 18ft (1.6–5.5m) up in forest. Eggs 2, less often 3, dull white or pale blue, speckled and blotched with cinnamon and chocolate. February–June.

Status. Resident, mainly in foothills and interior valleys, the length of Pacific slope; common in hills of Península de Nicoya and along base of N cordilleras and W edge of Valle Central but rare and local in dry lowlands; common in remaining forests of General–Térraba–Coto Brus regions but absent from Golfo Dulce lowlands; to 4000ft (1200m) or slightly higher in Coto Brus region.

Range. C Mexico to N Bolivia, NE Argentina, and SE Brazil.

RED-THROATED ANT-TANAGER

Habia fuscicauda Pl. 47(12)

Tangara Hormiguera Gorgirroja

Description. 7½″ (19cm); 40g. Fairly large; bright crown-patch usually concealed, without black border; bright, contrasting throat in adults. **Adult ♂:** center of crown bright red; rest of upperparts dark dusky-red; below slightly paler, dusky-red, shading to bright scarlet on throat; chin and lores dusky. ♀: above dark brownish-olive, with small, dull yellow crown-patch; face dusky-olive; throat olive-yellow; breast, sides, and flanks olive, belly paler grayish-olive. Bill blackish; legs dusky to gray. **Immature ♂:** like adult female but above often browner; face sometimes tinged with rufous, throat with dull orange. **Young:** above dark rich brown, below paler brown; no crown patch; throat dull ochraceous.

Habits. Prefers thickety growth along forest edge, including banks of streams and rivers, overgrown plantations and dense, shady second growth, usually not penetrating interior of forest; in pairs or noisy groups of 4–8 that search for insects and fruits in foliage with incessant nervous activity; sometimes attends army ant raids, especially in lowlands of extreme NW Caribbean slope, where antbirds are rare. If a juvenile is menaced by a predator, the entire flock may give spread-wing-and-tail displays, posturing and fluttering from perch to perch.

Voice. Calls include dry, raucous, wrenlike scolding notes that are longer, more upwardly inflected than scolds of its neighbor, the Bay Wren: *raaah*; staccato, nasal *pip, pik* and *chut* notes that are often interspersed with harsh, dry chatters as the flock travels; song a throaty, mellow, whistled *cherry-quick cherry-quick cherry-quick cherry-chewer* or similar phrase, over and over at dawn in breeding season.

Nest. A bulky, loosely made cup of fine plant fibers on a foundation of matted leaves, held together with vines and rootlets and often decorated with a living sprig of fern; lining of fine, black fungal rhizomorphs; in a fork of shrub or small tree, 4–10ft (1.2–3m) up. Eggs 2–3, white. April–June.

Status. Common resident of lowlands and foothills throughout Caribbean slope, from sea level to 2000ft (600m), locally higher.

Range. SE Mexico to E Panama.

Note. Also called Dusky-tailed Ant-Tanager.

BLACK-CHEEKED ANT-TANAGER

Habia atrimaxillaris Pl. 47(11)

Tangara Hormiguera Carinegra

Description. 7″ (18cm); 38g. Head mostly blackish with contrasting salmon throat; no other ant-tanager in its restricted range. **Adult ♂:** partly concealed central crown-patch bright salmon-orange; front and sides of crown black with slaty feather tips; rest of upperparts dark gray, tinged with dull reddish; wings and tail blackish, edged with slate-gray; chin and sides of head black; throat bright salmon, shading to dusky-red on breast and dull pinkish-gray posteriorly; darker and duskier laterally. ♀: similar but crown-patch small and dull; rest of head dull blackish; throat slightly duller salmon. Iris dark red; bill black; legs dark horn. **Young:** above duller, browner, no crown-patch; sides of head and chin dark sooty-brown; throat duller, more grayish; rest of underparts with brownish tinge and faint, sooty shaft-streaks; flanks and crissum dusky.

Habits. Restricted to understory of dense lowland forest, barely entering adjacent tall second growth; travels in pairs and small groups, often accompanying mixed flocks of antwrens, foliage-gleaners, greenlets, and other small birds; gleans small insects from leaves and twigs; rummages in dead leaves collected in crowns of small palms, and investigates rolled dead leaves; occasionally forages at army ant swarms; eats berries of melastomes, Solanaceae, etc., squeezing out and swallowing pulp and discarding husks.

Voice. Traveling groups keep up a continual flow of low, nasal *chit* and *chet* notes, alternating with raucous, dry, scolding notes like those of Red-throated Ant-Tanager but shorter and more clipped, sounding like *rrak* or *rruk*. Song, heard only at dawn in presumed breeding season, a clear, mellow, whistled *tu-see, tu-see, tu-see*, or *tu-seeur tu-swee tu-seeur*.

Nest. Unknown (?). Bird with nesting material in March; fledglings in May.

Status. Endemic to Golfo Dulce lowlands of Costa Rica, increasingly scarce as its forest habitat is reduced; still fairly common in lowlands of Península de Osa and around Golfo Dulce to Golfito where forest still remains, but within a few years the entire population may be confined to Parque Nacional Corcovado.

Range. Costa Rica.

WHITE-THROATED SHRIKE-TANAGER

Lanio leucothorax Pl. 47(1)

Tangara Piquiganchuda

Description. 8″ (20cm); 40g. Slender, long-tailed, with prominently hooked bill with tooth on cutting edge. **Adult ♂:** head, nape, most of scapulars and wings, upper tail-coverts, tail, and thighs black; a partly concealed white patch on scapulars and shoulder; throat white, shading to buffy on chest; rest of body golden-yellow (Caribbean race) or with rump and crissum also black (Pacific race). ♀: pileum dark olive-green; face brownish; rest of upperparts bright russet-brown; throat and chest buffy-brown, tinged with gray except on center of throat and shading through olive to golden-yellow on belly; sides, flanks, and crissum tinged with russet. Bill black; legs blackish. **Young:** like female but more rufescent; pileum russet with dusky scaling; rest of upperparts dull chestnut-rufous, the wing-coverts and remiges edged with brighter rufous; throat ochraceous-buff, shading to bright ochraceous on chest and sides and buffy-yellow on belly; flanks and crissum pale rufous; lower mandible grayish.

Habits. Frequents middle and upper levels of tall wet forest, usually in hilly terrain; in pairs or small family groups of 3 or 4, accompanies mixed flocks of tanagers, antbirds, ovenbirds, and flycatchers; perches erectly, scanning alertly all around, then sallies abruptly to snatch insects, often those fleeing from or dislodged by other flock members, from foliage, bark, or midair, carries them to perch, against which it often beats larger prey before swallowing; noisy, excitable, quick to sound alarm at approach of predator.

Voice. Calls include short, sharp, sometimes scratchy chips and chatters and loud, ringing, slightly burry, whistled notes: *cheew, cheer-churr-churr*, etc.; song a low, leisurely medley followed by rather high, plaintive whistles, repeated rapidly or deliberately in different tones.

Nest. A cup in a low bush near a forest stream. Eggs 2, white, with bold chocolate and grayish-brown spots, more numerous toward larger end. May (fledglings seen in July).

Status. Uncommon to fairly common resident in foothills and adjacent lowlands the length of Caribbean slope at 330–3000ft (100–900m) but not below 1000ft (300m) in SE; in lowlands and foothills of S Pacific slope, N to Carara, up to 2500ft (750m).

Range. E Honduras to W Panama.

Note. Sometimes considered conspecific with Black-throated Shrike-Tanager of Mexico to Honduras, in which case the enlarged species is called Great Shrike-Tanager, *L. aurantius*.

WHITE-LINED TANAGER

Tachyphonus rufus Pl. 46(16)

Tangara Forriblanca (Fraile)

Description. 6¾″ (17cm); 32g. Black male flashes white in flight; female differs from other rufous birds by tanager bill, habitat, voice; association of black and rufous birds in scrub is distinctive. **Adult ♂:** glossy black, except for small, usually concealed patch of white on scapulars and white wing-linings. ♀: above rufous-brown, duller and duskier on pileum, brighter rufous on tail coverts, tail, and wing-coverts; below paler and more tawny, brightest on breast and crissum, palest on belly. Upper mandible black, lower pale gray with black tip; legs dusky. **Young:** like adult female but duller; below extensive grayish feather-bases give obscurely mottled appearance. Young males acquire full adult plumage before first breeding season.

Habits. Frequents scrubby, dense young second growth, open woods; sometimes found around rural houses; forages low, eating many fruits, usually pressing out and swallowing the pulp and small seeds, discarding the husks; gleans insects in foliage or captures them on short aerial sallies, or pounces from a low perch onto insects on the ground; in pairs throughout year, not generally associating with mixed flocks.

Voice. Calls include a short, rather finchlike *kchip* or *check*, a weak, thin *seep*; song a long series of irregularly spaced, melodious, chirping notes inflected upward or downward: *chip cheep cherp cherp cheet chep cherp chip chip. . . .*

Nest. A bulky cup of leaves, vine tendrils and rootlets, lined with thin fibers, 4–20ft (1.2–6m) up in shrub, tree, or banana plant. Eggs 2, gray, spotted and scrawled with black (in Panama; not seen in Costa Rica). April–July.

Status. Uncommon to fairly common resident of most of Caribbean slope, though not yet recorded W of Río Frío; in decreasing numbers up Reventazón drainage to around Cartago; sporadic farther W in Valle Central; lowlands to 4600ft (1400m); invaded Golfo Dulce lowlands in last 15–20 years; continues to expand its range on both slopes with deforestation.

Range. Costa Rica and Trinidad to NW Ecuador, N Bolivia, NE Argentina, and SE Brazil.

WHITE-SHOULDERED TANAGER

Tachyphonus luctuosus Pl. 46(18)

Tangara Caponiblanca

Description. 5½″ (14cm); 16g. Conspicuously white-shouldered male unmistakable; female differs from most other similar species in whitish throat and lack of white on belly or gray on head. **Adult ♂:** glossy black, with lesser and middle wing-coverts and wing-linings white; in Pacific race, a usually concealed crown-patch of buffy-orange. ♀: above olive-green, tinged with gray on crown; throat whitish (more buffy in Pacific race); below fairly bright yellow, paler posteriorly, and tinged with olive on sides and flanks. Bill black, except base of lower mandible silvery; legs gray. **Young:** like adult female but above somewhat browner; wing-coverts with paler tips and edges; below entirely olive-yellow, including throat. Males acquire full adult plumage before first breeding season.

Habits. Frequents middle and upper levels of wet forest, taller second-growth woods, nearby shady clearings, and semi-open; usually in pairs, sometimes larger groups, typically accompanying mixed flocks of other tanagers, warblers, greenlets, honeycreepers, etc.; searches actively for insects amid foliage much like a warbler but usually more slowly and deliberately; eats many small fruits; rarely descends to ground.

Voice. Calls include a piercing, thin *tsst* or a weaker *tsip*; a slightly nasal *nit*, often repeated rapidly; a dry *chirr*; song a squeaky *tseek chur-chur-chur*, often repeated many times.

Nest. A fairly deep cup of dry grass lined with fine fibers, in low vegetation within a few feet (m) of ground. Eggs 3, pale cream to rich buff, marked with reddish- or blackish-brown. April–May (in Trinidad; no Costa Rican record).

Status. Uncommon to fairly common resident throughout Caribbean slope and common on S Pacific slope, N in decreasing numbers to Carara; sea level to 2500ft (750m).

Range. E Honduras and Trinidad to W Ecuador, N Bolivia, and E Brazil.

TAWNY-CRESTED TANAGER

Tachyphonus delattrii Pl. 46(17)

Tangara Coronidorada

Description. 5¾″ (14.5cm); 19g. Black male has conspicuous, erectile, rounded crest; dark brown female could be mistaken for an antbird but has typical tanager form and bill. **Adult ♂:** dull black except for elongated, tawny-orange feathers on most of crown. **♀:** dark brown to olive-brown above, slightly paler below, more blackish on wings, tail-coverts, and tail. Bill black, adult male with silvery base to lower mandible; legs blackish. **Young:** like adult female but above and below averaging duller, more sooty-brown. Males acquire full adult plumage before their first breeding season.

Habits. Travels rapidly through lower and middle levels of dense, wet forest in noisy flocks of up to 20, often with bands of Olive Tanagers, and attended by a few woodcreepers, furnariids, and other birds, often including a pair of White-throated Shrike-Tanagers; sometimes visits neighboring tall second growth; searches actively in foliage, sometimes hanging acrobatically, to glean insects, including ants, crickets, and beetles, and spiders; eats many fruits, especially *Ossaea* and other melastomes.

Voice. Calls include a metallic *tchit* or *zick*, a higher *tsip* or *tchewp*, and a distinctive high, thin, sibilant or slightly scratchy *zeeet* or *pseeet*; song not reported.

Nest. A bulky, deep cup, one wall higher and forming a partial roof, mostly of green moss with lining of fine rootlets and fungal rhizomorphs, 5ft (1.5m) up in dense shrub beside forest trail; eggs 2, pale green with a wreath of fine black blotches and scrawls. Pairs separate from flock to nest. April (1 record). Probable breeding season March–June, from gonad data and presence of juveniles.

Status. Common to abundant resident of wet lowlands and foothills the length of Caribbean slope, except rare in drier area S of Lago de Nicaragua; sea level to 3300ft (1000m), or to 4000ft (1200m) in SE.

Range. E Honduras to W Ecuador.

SULPHUR-RUMPED TANAGER

Heterospingus rubrifrons Pl. 45(15)

Tangara Lomiazufrada

Description. 6¼″ (16cm); 30g. Bill rather long and slender, slightly hooked at tip; mostly gray, with white breast tufts and bright yellow rump. **Adults:** above blackish-slate; wings and tail blackish; upper and lower tail-coverts olive-green (this color duller, sometimes lacking, in females); below slate-gray, tinged with brownish on belly in female; tuft at each side of breast overlapping wings, and wing-linings, white. Bill black; legs blue-gray. **Young:** similar to adults but duller, above more sooty, below tinged with olive; wing-coverts edged with dull olive; smaller yellow patch on rump.

Habits. Frequents middle and upper levels of wet forest and adjacent tall second growth and semi-open, coming lower at edges and clearings; in pairs and small groups, often with mixed flocks or bands of Black-faced Grosbeaks or Tawny-crested Tanagers; actively gleans small insects from outer foliage of trees; eats many small fruits, including *Trema*, mistletoes, melastomes, and *Cecropia* spikes.

Voice. Call a sharp, buzzy *dzeet*; a higher, thinner *tseet* in flight; a short, rapid chatter; song not recorded.

Nest. Unknown (?); evidently located on high, epiphyte-covered branches (Ridgely).

Status. Uncommon resident of lowlands and foothills of S Caribbean slope, from Río Reventazón drainage S and from sea level to 2300ft (700m).

Range. Costa Rica to E Panama.

Note. Sometimes considered conspecific with *H. xanthopygius*, Scarlet-browed Tanager, of extreme E Panama and N South America.

GRAY-HEADED TANAGER

Eucometis penicillata Pl. 45(17)

Tangara Cabecigrís

Description. 6½″ (16cm); 31g. Gray head diagnostic; appears long-tailed and small-headed, with a short, fluffy crest. **Adults:** head and neck slate-gray, throat paler gray; rest of upperparts bright olive-green; below rich yellow, tinged with ochraceous on breast and sides, and with olive on flanks; breast and sides sometimes show faint olive streaking (especially Pacific race). Iris reddish-brown; bill blackish; legs flesh-color. **Young:** head and throat olive-green like upperparts; breast tawny-orange with heavy, smudgy olive streaking; rest of underparts paler yellow than adults; wing-coverts edged with ochraceous.

Habits. In understory of humid forests and taller second growth, pairs or family groups of 3–5 regularly accompany army ants, sometimes following them into adjacent clearings, catching insects fleeing from ants; also travels alone, gleaning insects from foliage and taking many small fruits; shy and nervous, constantly twitching wings and tail, retreating from ant swarm at observer's approach before most antbirds do; occasionally follows scratching chickens or wood-quail coveys instead of ants.

Voice. Calls include a squeaky, high, thin *tseeet*, often repeated 2–3 times; a high-pitched, thin *chip* or *pit*, sometimes repeated incessantly; song a high, thin, rather sibilant medley, *whichis whichis whicheery whichis whichu', tsee tseep SEEUr tsp-tsp tseeur tsp-tsp seeur ts-suur*, etc.

Nest. A slight cup through which light passes, often in a spiny palm in the forest or a shrub in a neighboring clearing, 2–10ft (0.6–3m) up. Eggs 2, rarely 3, pale blue-gray, heavily mottled with dark brown. March–July; 2 broods.

Status. Resident; uncommon in dry NW lowlands where limited to moist sites, becoming very common in wetter parts of Península de Nicoya and lower slopes of Cordillera de Guanacaste; on extreme NW Caribbean slope, E to Río Frío; common throughout lowlands and foothills of S Pacific slope; sea level to 4000ft (1200m).

Range. S Mexico to N Bolivia and E Brazil.

DUSKY-FACED TANAGER

Mitrospingus cassinii Pl. 45(18)

Tangara Carinegruzca

Description. 7¼″ (18.5cm); 42g. The rather long and sharp-pointed bill, pale eyes and dark coloration suggest an icterid. **Adults:** forehead, lores, orbital area, and cheeks black, the feathers stiff and erect, shading to dark gray on throat; crown and hindneck bright olive-green; rest of upperparts blackish, sometimes tinged with olive on back; rest of underparts olive-green, brighter and tinged with yellow on breast, darker and duskier on flanks; crissum dull rufous. Iris pale gray to greenish-white; upper mandible blackish shading to horn-color on tomia, lower pale gray; legs dusky. **Young:** crown dark sooty-olive; upperparts more brownish; face blackish; throat mottled with blackish; breast olive-brown smudged with dusky, shading to dull ochraceous-buff smudged with sooty-brown on belly; crissum dull cinnamon.

Habits. In noisy parties of 4–10, moves restlessly through tangled, thickety vegetation along streams and edges of forest, around marshy openings, and through dense, shady second growth and overgrown cacao plantations; foraging birds hop along branches, peck at twigs, rummage in foliage, probe curled dead leaves for beetles, grasshoppers, katydids, caterpillars, other insects, and spiders; takes berries of melastomes, Rubiaceae, and Solanaceae, seeds and arils; constantly flicks wings and tail nervously; usually shy.

Voice. Calls include a harsh, rolling, hard *chrrrt*; a high, thin *zeet* and a sputtering chatter; song, repeated persistently at dawn, a sharp, high-pitched, very emphatic *zeep zeep zeep zeep-it zeepity zeep.*

Nest. A rather flimsy to bulky cup composed of long, brown pistillate inflorescences of *Myriocarpa*, rootlets and sometimes strands of climbing fern, which also decorate the outside, with lining of black fungal rhizomorphs, slung basketlike between 2 to several upright stems, 5–10ft (1.5–3m) up beside forest stream. Eggs 1–2, long oval, white, spotted with brown and reddish-brown, very heavily at large end. March–April. At 1 nest, 2 nestlings were attended by 7 adults.

Status. Locally common resident of Caribbean lowlands, except not reported N or W of Lago Caño Negro, to 2000ft (600m) in foothills.

Range. Costa Rica to W Ecuador.

ROSY THRUSH-TANAGER

Rhodinocichla rosea Pl. 47(13)

Tangara Pechirrosada (Queo)

Description. 7½″ (19cm); 50g. Large, long-tailed, with long, rather slender bill (for a tanager); legs and feet stout. **Adult ♂:** above slate-black, tinged with brown on crown; sides of head, sides, flanks, thighs, and crissum sooty-black; wing-coverts edged with gray; long superciliaries rose-red in front of eye, shading to white behind; throat and central underparts rose-red. ♀: pattern similar, but dark areas duller and browner, rose-red replaced by cinnamon-rufous, darkest on breast. Iris gray-brown; upper mandible black, lower gray; legs dark brown. **Young:** like female but superciliaries dull white, extending to behind eye; below duller with trace of dusky streaking on breast; belly smudged with whitish; wing-coverts edged with ochraceous.

Habits. Forages on ground beneath second-growth thickets and canebrakes, tossing aside dead leaves with bill to expose beetles, bugs, other insects, spiders, small frogs and lizards, and seeds, in dry weather the noise of rustling leaves often revealing its presence; in pairs throughout year, sometimes accompanied by young; shy and retiring.

Voice. Call a mellow, rather querulous *kweeo, kyew*, or *querup*, tirelessly repeated; mates answer each other with liquid notes that sound like "gold" and "silver." Song a varied repertoire of short, 4- to 6-note phrases, each repeated several times in rapid succession, the sweet, rich tone reminiscent of an oriole or certain wrens (e.g., Black-bellied); pairs join in duets, the female's voice being slightly weaker, the total effect wrenlike. Typical phrases: *cheer, cheer, cheerily cheer; cheeo-cheera cheerily cho*.

Nest. A shallow, well-made bowl of rachises of an acacialike shrub, on a foundation of coarse sticks, sometimes lined with black fungal rhizomorphs, ca. 3ft (1m) up in tangled thicket. Eggs 2, white, with blackish spots and scrawls. January–September.

Status. Rare and local resident in General–Térraba–Coto Brus valleys of S Pacific at 800–3000ft (250–900m); reported from Coto district of Golfo Dulce lowlands.

Range. Discontinuously from W Mexico to C Colombia and N Venezuela.

Note. Also called Rose-breasted Thrush-Tanager.

BLACK-AND-YELLOW TANAGER

Chrysothlypis chrysomelas Pl. 46(14)

Tangara Negro y Dorado

Description. 4¾" (12cm); 12.5g. Small, dainty, warblerlike, with contrasting black bill. **Adult ♂:** head, rump, and entire underparts bright yellow, tinged with orange on head; eye-ring, back, wings, and tail velvety black (appears big-eyed because of dark eye-ring); wing-linings white. ♀: above olive-green, tinged with yellow on crown and face; below yellow, tinged with olive on breast and grayish on sides and flanks; belly whitish; crissum brighter yellow; usually a tuft of whitish at edge of breast, overlapping wings. Legs olive-gray. **Immature ♂:** like adult female but below somewhat brighter yellow, especially on belly; wing-coverts with paler edgings.

Habits. Frequents canopy of wet, mossy forest and adjacent tall second growth, descending at times to shrub level at edges and gaps; in groups of 3–6 that usually accompany the large, diverse canopy flocks of small frugivorous and insectivorous birds characteristic of lower middle elevations; actively gleans insects and fruits in the outer foliage of trees and tall shrubs; takes *Cecropia* protein corpuscles; occasionally makes short, fluttering sallies for flying insects.

Voice. Usual call a scratchy *tsew* or *tseerw*, higher, faster, more sibilant than note of Silver-throated Tanager, singly or followed by several shorter notes: *tsew tsu-tsu-tsu*.

Nest. A neat cup of fine rootlets, lined with black fungal rhizomorphs with a middle layer of small dead leaves and an outer layer of green moss; near tip of mossy branch of *Inga* tree, 25ft (8m) up at forest edge (only 1 record). Eggs unknown (?). June.

Status. Locally common resident of foothills and lower mountain slopes on Caribbean side of Cordillera Central and Cordillera de Talamanca, N in small numbers at least to Cerro Santa María in Cordillera de Guanacaste, mainly at 2000–4000ft (600–1200m), locally down to 1300ft (400m) (e.g., Carrillo) in N and C Costa Rica; rare below 3300ft (1000m) in SE.

Range. Costa Rica to E Panama.

COMMON BUSH-TANAGER

Chlorospingus ophthalmicus Pl. 45(13)

Tangara de Monte Ojeruda

Description. 5¼" (13.5cm); 20g. Greenish, with center of belly whitish; head dark with conspicuous postocular spot. **Adults:** top and sides of head dusky-brown, shading to buffy-brown on malar area; white spot behind and partly surrounding eye; rest of upperparts uniform greenish-olive; throat pale grayish, more or less mottled with dusky; chest, sides, flanks, and crissum yellowish-olive; center of belly grayish-white. Iris reddish-brown; bill black; legs gray. **Young:** above more brownish, shading to sooty-brown on crown; postocular spot tinged with olive, face with dusky; olive areas of underparts darker, with faint dusky streaking; throat and belly pale dull yellowish with darker smudging.

Habits. Ranges through cool, wet, epiphyte-laden highland forests from lower understory to middle of canopy and out into neighboring shady or bushy clearings, in flocks of 6–12, that are often accompanied by other tanagers, antvireos, antwrens, ovenbirds, woodcreepers, and other small birds; gleans amid foliage, rummages and probes in mosses and epiphytes on branches for insects and spiders; eats many small fruits, especially those of melastomes and Ericaceae, usually mashing them in the bill, swallowing pulp and small seeds and discarding husks; plucks flowers and presses out nectar in same manner; sleeps in loose groups in dense foliage.

VOICE. Noisy, incessantly uttering thin, sharp, penetrating notes like *tsip, tseep* and a sharp *tseek* that rushes into a rapid, descending twitter; a rattling *tsrrrrr* and a staccato *cut* in interactions. At dawn male repeats a piercing *seeek* or *tseer* for minutes on end from a high perch, sometimes inserting a twitter.

NEST. A bulky cup of grass blades, broader leaves, rootlets, herbaceous stems, lined with rootlets, animal hairs, fungal rhizomorphs, moss setae, and other fine materials, usually with much green moss on outside, on a grassy bank, sloping ground, or to 50ft (15m) up in a tree, where hidden amid thick moss or large epiphytes. Eggs 2, dull white, spotted with cinnamon and brown, often forming a wreath. April–June; 2 broods.

STATUS. Abundant resident of wet highlands with at least remnant strips of forest along streams, from 1300ft (400m) on Caribbean slope and 3500ft (1100m) on Pacific, up to 7500ft (2300m), rarely higher.

RANGE. C Mexico to Bolivia and NW Argentina.

NOTE. Also called Brown-headed Bush-Tanager.

SOOTY-CAPPED BUSH-TANAGER

Chlorospingus pileatus Pl. 45(14)

Tangara de Monte Cejiblanca

DESCRIPTION. 5¼″ (13.5cm); 20g. Resembles Common Bush-Tanager but head blacker, with broad broken white superciliary; at higher elevations. **Adults:** top and sides of head sooty-black; long white superciliary breaks above eye; rest of upperparts greenish-olive; throat pale gray with an indistinct dusky malar streak, and speckled with dusky; chest, sides, flanks, and crissum olive-yellow; center of belly pale gray. Iris brick-red; bill black; legs blackish. **Young:** like adults but head dull brownish-black; superciliaries duller and tinged with olive; above somewhat duller and more brownish; throat and belly dull yellowish; entire underparts indistinctly streaked with dusky; iris brown.

HABITS. Travels in straggling flocks of 5–20 through mossy mountain forest and tall second growth, ranging from lower understory to middle of canopy and out into brushy clearings and páramo; searches along mossy branches, hops up through dense shrubbery, pokes into moss, lichens, and bromeliads, sometimes hanging acrobatically, seeking small insects and spiders; eats many berries, especially *Miconia, Fuchsia*, blackberries, and Ericaceae; often accompanied by Ruddy Treerunners, warblers, and other small birds.

VOICE. Calls include a high, double *zit* or *zeet*; a variety of high, scratchy notes like *tsip, tseep*; a short chatter and a twitter. At dawn and sometimes more briefly at other times, males sing a scratchy, rapid *see-chur see-chur see-chur see see see-chur. . . .*

NEST. A soft, bulky cup of mosses, liverworts, and beard lichen lined with fine grasses or fine, black rootlets; in niche or atop a vertical bank, under a tuft of moss in a dense, low shrub, or in a mass of green mosses or epiphytes on a branch to 35ft (11m) up. Eggs 1–2, white, speckled and blotched with pinkish-brown, mostly at large end. February–July. Often 3 or more adults attend nestlings.

STATUS. Abundant resident of higher parts of Cordillera Central and Cordillera de Talamanca, from ca. 6500ft (2000m) to above timberline; fairly common in Cordillera de Tilarán, above 5200ft (1600m).

RANGE. Costa Rica and W Panama.

NOTE. Some individuals on Irazú-Turrialba massif are duller, more grayish throughout, without any yellow tinge below. Often formerly classified as a distinct species, *C. zeledoni*, they are now known to be a color phase of *pileatus*.

ASHY-THROATED BUSH-TANAGER

Chlorospingus canigularis Pl. 45(12)

Tangara de Monte Gargantigrís

DESCRIPTION. 5″ (13cm); 18g. Somewhat smaller and slimmer than other bush-tanagers, without white markings on head; best field marks are contrast between bright yellowish-olive chest-band and sides and grayish-white lower breast and belly, and dark malar stripe. **Adults:** side of head grayish-olive, sometimes with an inconspicuous paler postocular streak; rest of upperparts olive-green, darkest on crown; throat pale gray with a fairly distinct, dark gray malar stripe. Iris dark red to chestnut; bill blackish; legs dark gray. **Young:** below extensively dull yellowish, chest-band not distinct; breast and sides with faint darker streaks.

HABITS. Roams through forest canopy, often descending to upper understory along edges, at gaps, or in second growth beside forest; usually in pairs, or family-sized groups following breeding; often accompanies mixed flocks of tanagers (e.g. *Tangara, Tachyphonus, Chrysothlypis*), honeycreepers, and other small birds; actively gleans and searches for insects in foliage and along branches like other bush-tanagers, also takes berries (melastomes, Rubiaceae, *Hedyosmum*, etc.).

VOICE. Call a high, thin, sharp, penetrating *zeeZIT* or *dzee DZIT*, sometimes *zee zee ZIT*; possibly the song is a sibilant, slightly ascending *tse tse tse tsee*.

NEST. Undescribed (?). Birds with nesting material in May; fledglings in July.

STATUS. Local, uncommon to rare resident of lower Caribbean slope of Cordillera Central, from N foothills of Volcán Poás to Turrialba area, at 1300–4000ft (400–1200m), at higher elevations mainly in deep river canyons; not reported from Cordillera de Talamanca, or from Panama.

RANGE. Costa Rica; NW Venezuela to N Peru.

NOTE. The isolated Costa Rican race, *C. c. olivaceiceps*, may best be considered a species (or semispecies) separate from the gray-headed Andean forms, which are also evidently more social (Hilty and Brown 1986).

FAMILY **Emberizidae:** New World Sparrows, Finches, Grosbeaks, and Allies

The limits of this large family have been variously drawn by different authors. The latest AOU Check-List expands the family to encompass the entire assemblage of "9-primaried" oscines, including warblers, icterids, tanagers and the bananaquit as well as the grosbeaks, finches, and sparrows—a huge, unwieldy, and heterogeneous assemblage in terms of behavior, ecology, and external anatomy. We prefer to restrict the family to 2 groups: the grosbeaks and cardinals (Cardinalinae) and the finches, seedeaters, and sparrows (Emberizinae), which, with the addition of the flowerpiercers, give a total of about 320 species.

Although well represented in the Old World, principally by the bunting genus *Emberiza*, the emberizids are far more diverse in the New World, where they spread from the Arctic islands to Tierra del Fuego and from tropical lowlands to above treeline on cold mountains. In length they are 3.5–10″ (9–25cm); nearly all except the flowerpiercers have thick, more or less conical bills. Their colors are varied; some of the Cardinalinae are brilliantly colored, with pronounced sexual differences; in others, like the saltators, the sexes are alike and lack bright colors. Among the Emberizinae, some finches, flowerpiercers, and the gregarious seedeaters and grassquits have males that are black, gray, or blue, often combined with white, whereas females are some shade of brown. The permanently paired brush-finches, ground-sparrows, and their allies are often strikingly patterned, with the sexes alike or nearly so, as are the sparrows, which among resident species are in pairs or family groups year-round.

Costa Rican members of this family tend to forage low, often obscurely on the ground beneath thickets, gathering seeds, insects, and fruits, for which some of them ascend into shrubs and trees. Some rarely rise to high, open perches to sing; many of the more gifted songsters, including grosbeaks, saltators, seedeaters, and seed-finches, regularly perform from higher, more exposed perches. Unassisted, females build flimsy cups, more ample bowls, or bulky roofed structures with a side entrance, commonly at no great height in bushes and trees and sometimes on the ground. Rarely, as in the Yellow-faced Grassquit, the male takes the lead in nest construction. The 2 eggs, seldom more than 3 in Costa Rica, are, with rare exceptions, incubated by the female alone and hatch in 12–14, rarely 15, days. Both sexes feed the young, by regurgitation in seedeaters and grassquits, directly from the bill in other species. Hatched with sparse down or less often naked, the young remain in the nest for 10–15 days and may leave in plumage quite different from that of their parents.

BLACK-HEADED SALTATOR

Saltator atriceps Pl. 48(1)

Saltator Cabecinegro

DESCRIPTION. 9½″ (24cm); 85g. Resembles Buff-throated Saltator but larger, with blacker head, paler upperparts, and white throat. **Adults:** pileum black; sides of head, nape, and most of underparts slate-gray; entire upperparts, wings, and tail yellowish-olive; lores more dusky; narrow superciliaries and throat white, the throat with broad black border extending from chin to side of breast but incomplete medially; flanks brownish-gray; crissum dull rufous. Bill black; legs dark brown. **Young:** like adults but black of head and sides of throat more sooty, less sharply defined; throat with dusky mottling; breast scaled or smudged with dusky; superciliaries dull and indistinct; upperparts darker.

HABITS. Prefers low, thickety second growth, brushy pastures, plantations, gardens, and forest edge, often near water; travels in small, noisy, family-sized groups, sometimes with other small birds; eats many small fruits and seeds; gleans insects in foliage; sometimes takes flowers and tender buds; usually shy.

Voice. Calls sharp, loud, raucous, and parrotlike: a single descending *deeuh* or *deeer* or several such notes ending in a harsh, explosive chatter. Bowing deeply while turning from side to side, a saltator utters a series of such loud, scratchy notes ending with a prolonged, upwardly inflected whistle: *cher ch-ch cher jur jur weeeee*; mates often duet. A high, thin dawn song *tseety, tseety-tseety-tseety-tseety* (Wetmore).

Nest. A very bulky, shallow cup of grass blades and stems, strips of banana leaves, and the like, on a loose foundation of twigs or vines, lined with finer grass or vines, 5–10ft (1.5–3m) up in thicket. Eggs 2, pale blue, with a wreath of heavy black scrawls on larger end. April–July.

Status. Fairly common resident throughout Caribbean lowlands, except absent from Río Frío region W; extends up Reventazón drainage regularly to just E of Cartago (Paraíso, Las Cońcavas); occasionally strays W to San José; sea level to 4300ft (1330m).

Range. C Mexico to E Panama.

BUFF-THROATED SALTATOR

Saltator maximus Pl. 48(2)

Saltator Gorgianteado (Sinsonte Verde)

Description. 8″ (20cm); 50g. Large; heavy, convex bill; buffy throat broadly bordered with black. **Adults:** head mostly dark slate, tinged with olive on hindcrown; narrow superciliaries white; rest of upperparts olive-green; sides of throat and broad band across lower foreneck black, enclosing white chin and buffy center of throat; rest of underparts gray, tinged with buff shading to brownish on flanks and tawny-buff on crissum. Bill and legs blackish. **Young:** pileum dull olive-green; superciliaries tinged with olive; throat and chest blackish, mottled with whitish on chin and throat; rest of underparts tinged with buffy or brownish and scaled or smudged with dusky. Bill pale grayish to horn-color, with dusky base of culmen. Young acquire adult plumage when 3–4months of age.

Habits. Frequents second-growth thickets, brushy pastures, shady plantations, semi-open, gardens, and dense vegetation at edge of forest, which it enters a short way to forage and rarely to nest; throughout year in pairs, which often accompany groups of Scarlet-rumped Tanagers or other small frugivorous birds; eats insects, a wide variety of fruits, nectar that it presses from flowers, sometimes soft flowers and buds; usually shy and wary.

Voice. Call a high, thin, sibilant *seeent* or *tseeeet*; song a prolonged, richly varied flow of soft, lilting notes. Mates answer each other with melodious, descending, rather abrupt warbled phrases: *cheery cheery*, answered by *cheer to you* or *cheer-cheerdle-chew*, often combined with call notes.

Nest. A well-made, bulky cup with foundation of long pieces of dry vines and weed stalks, middle layer of grass blades, strips of banana leaves, papery bark, and other flat materials, and lining of thin vines, tendrils, rootlets, or fibers; in thicket, hedge, plantation, dooryard shrubbery, fruit tree, usually 10″–7ft (0.25–2m) up, rarely as high as 30ft (9m) amid dense foliage. Eggs 2, pale blue, wreathed with scrawls and dots of black. March–August; 2 broods.

Status. Resident; rare and local in dry NW, very uncommon in Valle Central, otherwise common to abundant countrywide from sea level to 4000ft (1200m), occasionally to 5000ft (1500m).

Range. SE Mexico to W Ecuador, N Bolivia, and SE Brazil.

GRAYISH SALTATOR

Saltator coerulescens Pl. 48(3)

Saltator Grisáceo (Sensontle, Sinsonte)

Description. 8″ (20cm); 52g. Differs from other saltators in darker, more uniformly grayish coloration. **Adults:** above dark slate-gray, tinged with brown posteriorly; superciliary and spot on lower eyelid white; throat white, framed by black malar stripes; breast dull slate, tinged with buff on sides and flanks; center of belly and crissum pale cinnamon. Bill black; legs dark brownish-gray. **Young:** above dull olive; superciliary and suborbital spot yellow; chin and throat yellowish, bordered with dusky-olive; breast dull ochraceous-olive, shading to ochraceous-buff on crissum, dusky-olive on flanks. Postjuvenile molt often incomplete, first-year birds often have considerable olive when breeding.

Habits. Frequents second growth and scrub, coffee plantations and semi-open, shady lawns and gardens, hedgerows; in pairs throughout year; eats many fruits, flowers, tender young leaves, buds, young shoots of vines, and slow-moving insects; usually forages low, but often sings from tops of small trees; often rather shy.

Voice. Call a sharp *tsit* or *tseet*, higher than note of Rose-breasted Grosbeak, shorter and sharper than that of Buff-throated Saltator; song a jerky series of mellow whistles, warbles, and short, sputtering notes, usually ending in an upwardly or downwardly inflected whistle: *speaking-of-wheeet, pee-pitpit, pit-cheeew, chick a chichreee*, or *pee pipichewpi*, etc.; sometimes male and female duet with shorter phrases. Voice more raucous in NW Caribbean lowlands, where Black-headed Saltator is absent.

Nest. A rather bulky cup with a foundation of twigs, middle layer of small leaves or broad strips of larger ones, thickly lined with fine fibers, 6–13ft (1.8–4m) up in dense shrubbery or a leafy tree, usually well hidden. Eggs 2, occasionally 3, bright blue, with a wreath of black scrawls and blotches around thick end. April–July.

Status. Fairly common resident of Valle Central and adjacent slopes up to ca. 6000ft (1850m), E to below Turrialba at 1500ft (450m) and W to lowlands around Golfo de Nicoya; abundant in Río Frío region and lowlands to W, and evidently expanding its range S on Caribbean slope; resident in Sarapiquí region since ca. 1975.

Range. NW Mexico to Costa Rica and N Colombia, and Trinidad to E Peru, N Argentina, and E Brazil.

Note. Sometimes the Central American populations are considered a separate species, *S. grandis*, Middle American Saltator.

STREAKED SALTATOR

Saltator albicollis Pl. 48(4)

Saltator Listado

Description. 7¼″ (18.5cm); 40g. Only the much browner female Rose-breasted Grosbeak has such heavy streaking below, but is also streaked above and has a pale bill. **Adults:** above olive-green, shading to gray on upper tail-coverts and tail; broken eye-ring and supraloral stripe whitish; lores dusky; below whitish, heavily streaked with dusky-olive on sides of throat and neck, breast, sides, flanks, and upper belly; chest tinged with yellowish, belly with pale yellow. Iris gray-brown; bill black, often with tip, tomia, and rictus yellow; legs olive-gray. **Young:** like adults but above tinged with brownish; broken eye-ring yellow; loral stripe faint or absent; throat duller, tinged with grayish-buff, malar area with gray; streaking below narrower.

Habits. A retiring inhabitant of low second-growth thickets, abandoned pastures, semi-open, occasionally shady gardens and forest edge, but avoids interior of woodland; in pairs throughout year; eats ants, beetles, other slow-moving insects, a wide variety of fruits, flowers, and tender buds.

Voice. Calls include a loud *spit* or *speet*, sharper and shorter that of Buff-throated Saltator; a scratchy *cheu*; a squeaky *whit-chaaaar*, the second note a dry rattle; song a varied phrase of loud, sweet-toned whistles, often with a little sputtering roll at the end: *chu chu chu ch'ree rerrr-chup, prrr-chow chow cheeer purrrit* or a simpler *cheer che here to greet you*; in flight, constantly calls *qua qua qua*.

Nest. Resembles that of Buff-throated Saltator but less bulky; 2–20ft (0.6–6m), mostly 4–8ft (1.2–2.4m) up in thicket or vine-draped tree. Eggs 2, blue, with a wreath of black spots and scrawls. March–June.

Status. Resident on S Pacific slope; common in General–Térraba–Coto Brus region; uncommon but increasing with deforestation in Golfo Dulce area; in recent years scattered records from Caribbean slope, from Panama N to Sarapiquí; possibly expanding its range W and N on this slope as well (N. G. Smith).

Range. Costa Rica and Lesser Antilles to Peru and Venezuela.

BLACK-FACED GROSBEAK

Caryothraustes poliogaster Pl. 48(5)

Picogrueso Carinegro

Description. 6½″ (16.5cm); 36g. Bill heavy, hooked at tip, with convex culmen; black face contrasts with mustard-yellow head and breast. **Adults:** lores, orbital area, cheeks, and throat black; foreneck, sides of neck to forehead yellow-ochre, shading to yellowish-olive on breast, hindneck, and upper back; wings and tail olive-green; rump, entire belly and upper and lower tail-coverts gray, palest on belly. Bill black except base of lower mandible silvery; legs bluish-gray. **Young:** like adults but duller: face more sooty, less sharply defined; yellow areas washed with olive.

Habits. Troops through middle and upper levels of dense, wet forest and adjacent clearings, semi-open, and tall second growth in large, noisy, straggling flocks often containing 20 or more birds, regularly accompanied by tanagers, honeycreepers, and other birds; searches for beetles, caterpillars, and other insects in foliage with slow, deliberate movements, often hanging head downward to reach them, sometimes darting out to catch them in flight; eats many fruits and arillate seeds, takes nectar from open flowers or bracts of epiphytes (e.g., *Norantea*).

Voice. Call a sharp, rolling to buzzy note followed by several sharp, downwardly inflected whistles: *prrrrt chew-chew-chew*; a staccato, sometimes metallic *chick* or *chip*, often repeated, and a soft *tweet*; song of 3–6 mellow-toned but sharp whistles: *cher che weet, cher che weet; chew tee chu-chuweet; weet-cher cher-weet-cher*, etc.

NEST. A shallow bowl loosely made of slender, creeping epiphytic rhizomes, pieces of dead dicotyledonous leaves in middle layer, and long, thin, green leaves of a bromeliad in lining, 9–20ft (3–6m) up in a small tree or spiny palm in shady clearing. Eggs 3, dull white, mottled and spotted with bright brown. April–June. Three or 4 adults attend a nest.

STATUS. Common resident throughout Caribbean lowlands and foothills, except drier areas just S of Lago de Nicaragua; sea level to 3000ft (900m).

RANGE. SE Mexico to C Panama.

NOTE. Sometimes merged with *C. canadensis* of South America.

SLATE-COLORED GROSBEAK

Pitylus grossus Pl. 48(6)

Picogrueso Piquirrojo

DESCRIPTION. 7½" (19cm); 50g. Thick, reddish, almost parrotlike bill with convex culmen, hooked tip, and prominent tooth on cutting edge, contrasts with white throat, dark head and body. **Adult ♂:** center of chin and throat white; face, sides of neck, and chest black; rest of plumage dark bluish-slate, below paler. Bill orange-red; legs dark gray to blackish. ♀: above similar to male but averaging duller, less bluish; little or no black on face, cheeks, neck, or chest; below much duller, more brownish-gray, sometimes a trace of pale brownish streaking on chest below white throat. Bill orange. **Young:** uniform dull sooty-gray, lacking black or white on face and throat; bill grayish-horn.

HABITS. Prefers upper and middle levels of tall, humid forest, forest edge, and neighboring shady plantations and semi-open, where often found much lower; usually in pairs, sometimes alone, infrequently joining mixed flocks; eats various fruits, not only small berries but also larger fruits an inch (2–3cm) or more long, which it rests on a horizontal branch without using feet to hold them, while it extracts seeds from the flesh and removes embryos from seed coats; gleans insects from foliage.

VOICE. Calls include a sharp, metallic *pink* or *spit*, higher than note of Blue-black Grosbeak; a whining, nasal, jaylike, slurred *nyaaah* or *caaw*; song consists of melodious but monotonously repeated phrases of 3–7 rich whistled notes and warbles such as *ch'woo-ch'weeeu wheeeuu, ch'woo chuweeet* or *George o George*.

NEST. Unknown (?). Fledglings May–June.

STATUS. Uncommon to fairly common resident of lowlands and foothills the length of Caribbean slope, from sea level to 3500ft (1100m) and to 4000ft (1200m) in SE; rare in drier lowlands S of Lago de Nicaragua.

RANGE. NE Honduras to W Ecuador, NW Bolivia, and E Brazil.

BLACK-THIGHED GROSBEAK

Pheucticus tibialis Pl. 48(7)

Picogrueso Vientriamarillo

DESCRIPTION. 8" (20cm); 70g. Large, with very heavy bill; mostly yellowish with black wings, tail, and back; prominent white wing-patch. **Adult ♂:** lores and orbital area blackish; rest of head, neck, and breast glossy yellow-ochre, clouded with dusky; posterior underparts bright yellow, crissum partly white; back feathers black, edged with yellowish-olive; scapulars, tail, and wings black except bases of primaries white; rump yellow, sometimes with black streaking anteriorly; thighs black. ♀: similar but head and breast usually paler, less glossy; olive scaling on back more extensive, extending to scapulars; usually less white in wing. Upper mandible black, lower pale gray with blackish tip; legs gray. **Immatures:** like adults but head and breast duller, more olive; crown with fine, dark shaft-streaks; breast often mottled with dusky; rump more streaked.

HABITS. Frequents canopy and edges of mountain forests, semi-open, pastures with scattered tall trees, and old second growth; usually forages high in trees, near tips of branches for insects, gleaning, sallying to snatch them from foliage or catch them in the air; eats many fruits and seeds, for which sometimes descends to shrub level at openings; solitary or in pairs.

VOICE. Call a high, sharp *pik* or *spit* much like that of Rose-breasted Grosbeak but higher and thinner; song high-pitched, rich-toned, clear, with a scratchy undertone audible at close range, consisting of a leisurely series of phrases that include whistles, warbles, and slurred notes and often end with a repeated note that may accelerate into a canarylike trill; successive phrases often quite different. One of Costa Rica's best songsters.

NEST. A bulky, thick foundation of twigs and mosses supporting a frail cup of rootlets, in a small tree or vine tangle on a stub in a small clearing beside forest, 10–30ft (3–9m) up.

Eggs 2, pale blue, with dots and blotches of dull chestnut-brown and purplish-lilac (Carriker). March–May.

Status. Resident in highlands, chiefly from ca. 8500ft (2600m), rarely higher, down to 5000ft (1500m) on Pacific slope, and locally to 3300ft (1000m) on Caribbean slope, N to Cordillera de Guanacaste (Volcán Miravalles).

Range. Costa Rica and W Panama.

Note. Sometimes classified as a race of the Yellow Grosbeak, *P. chrysopeplus*, of Mexico and Guatemala.

ROSE-BREASTED GROSBEAK

Pheucticus ludovicianus Pl. 48(8)

Picogrueso Pechirrosado (Calandria)

Description. 7″ (18cm); 45g. Medium-sized, with thick, pale bill; in fall, streaked above and below, with white bars or patches on wings. **Adult** ♂ Winter: upperparts black, feathers of head, neck, and back with broad, brownish edgings; feathers of rump white, tipped with black and brown; middle and inner greater wing-coverts, tertials, and inner webs of most rectrices broadly tipped with white; bases of primaries white; superciliaries, malar area, and most of underparts white; throat and breast rose-red, mixed with brown; breast, sides, and flanks sparsely spotted or streaked with black. Breeding: head, throat, and back black; breast and center of upper belly rose-red; rump and rest of underparts white. ♀: central crown-stripe buff; rest of upperparts brown, darker on wings and tail; crown and back heavily streaked with black; 2 buffy-white wing-bars; superciliaries and underparts white; sides of throat, breast, sides, and flanks tinged with buff and streaked with blackish; wing-linings yellow. Bill pale horn, shading to dusky on culmen; legs grayish. **Immature** ♂: like adult female but with wing-linings rose-pink; streaking of throat and breast more irregular; breast often tinged with pink; scapulars and wing-coverts with varying amounts of black. By first breeding season resembles adult but with brown edges on feathers of head and back, retaining juvenile wings.

Habits. In winter frequents open woods, semi-open, forest edge, scattered trees in clearings, gardens, and plantations, in flocks of 3–6, rarely up to 20; usually remains high but sometimes descends to ground, avoiding interior of closed forest; eats berries, palm fruits, arillate and other seeds, including winged seeds of jacaranda and explosive pods of the introduced "china" (*Impatiens*); takes insects and nectar.

Voice. Call a sharp, metallic *pink* or *pick*. Shortly before spring departure, a male may rarely utter a mellow, warbled song.

Status. Uncommon to locally common winter resident and passage migrant, virtually countrywide from lowlands to 5000ft (1500m); mainly mid-October to mid-April, occasionally by early September or as late as early May. Sometimes kept as a cage bird.

Range. Breeds in S Canada, E and C USA; winters from C Mexico to Venezuela and Peru.

BLACK-HEADED GROSBEAK

Pheucticus melanocephalus Pl. 48(9)

Picogrueso Cabecinegro

Description. 7½″ (19cm); 47g. Size and shape of Rose-breasted Grosbeak but much more buffy to rufescent. **Adult** ♂: head black; body mostly bright cinnamon-buff to cinnamon-rufous, the back streaked with black; wings and tail black, boldly marked with white; belly and wing-linings yellow; in winter head scaled with buffy. ♀: similar to female Rose-breasted Grosbeak but buffier overall, especially below, with fine, sparse black streaking on sides and sometimes on chest.

Habits. Generally similar to those of Rose-breasted Grosbeak, with which lone individuals may associate.

Voice. Resembles that of Rose-breasted Grosbeak.

Status. Accidental. Sight reports of a female associating with Rose-breasted Grosbeaks in December 1977 in San Isidro del General (MacKay); and several birds of both sexes, perhaps a stray migrating flock, in October 1978 in Parque Nacional Santa Rosa (Chaves). Most likely to appear in Guanacaste.

Range. Breeds from SW Canada to W Mexico; normally winters to S Mexico, accidental farther south.

Note. Hybridizes on Great Plains with Rose-breasted Grosbeak; the 2 are sometimes considered conspecific.

BLUE GROSBEAK

Guiraca caerulea Pl. 48(11)

Picogrueso Azul

Description. 6½″ (16.5cm); 31g. Cinnamon, buffy, or rufous wing bars distinguish males from other blue birds; females and immatures resemble female Indigo Bunting but much larger. **Adult** ♂: mostly bright violet-blue (partly obscured by buffy fringes in fresh plumage, especially in fall migrants); lores black; feathers of crissum tipped with whitish; wings and tail black with blue edgings; anterior wing-bar chestnut-rufous, posterior bar cinnamon-rufous. Upper mandible black, lower silvery. ♀: above olive-brown, faintly (residents) to strongly (migrants) streaked with dusky on back; wings and tail dusky with buffy edgings and 2 cinnamon-buff wing-bars; below buffy-brown, darkest across chest; flanks streaked faintly with dusky. Upper mandible dusky-horn, lower paler; legs dusky. **Immature** ♂**:** like adult female but rump and tail tinged with blue; anterior wing-bar more rufous; feathers below with blue-gray bases, producing a smudgy effect. **Young:** like adult female but brighter, above more tawny, with faint paler scaling; 2 buffy wing-bars; below dull buff, tinged with cinnamon on breast.

Habits. Frequents agricultural areas with scattered trees and hedgerows, second growth, open woodland, edges of deciduous forest, and dry, thorny scrub; residents usually in pairs, migrants solitary or in small groups; often erects crown feathers, appearing to have a short, shaggy crest and flat forehead; eats seeds and small insects gleaned from low vegetation or ground; makes short sallies for flying insects; visits roadsides or streams to pick up grit; migrants, in particular, incessantly flick and partly spread tail.

Voice. Calls include a metallic *tchip* or *zik*; a high, rolling *preet* or *tcheep*, in alarm or excitement; song a clear, sweet, subdued warble.

Nest. A neat cup composed externally of fine, short twigs bound together with spiderweb, middle layer of strips of fibrous bark, thick lining of fine rootlets, 4ft (1.2m) up in a shrub (3 nestlings in July, in Guatemala; no Costa Rican record). Eggs (in USA) 3 or 4, bluish-white (in Costa Rica, females gathering nesting material seen in May, June, and July).

Status. Uncommon to fairly common resident in lowlands of dry NW, to 3600ft (1100m) in foothills of N cordilleras and to W side of San José (Pavas) in Valle Central; uncommon fall migrant, principally along Caribbean coast, rare inland (October–November); very uncommon winter resident in NW and Caribbean lowlands; departs by mid-April.

Range. Breeds from S and C USA to Costa Rica; northern birds winter to C Panama, casually to South America.

Note. Sometimes included in genus *Passerina*.

BLUE-BLACK GROSBEAK

Cyanocompsa cyanoides Pl. 48(10)

Picogrueso Negro Azulado

Description. 6½″ (16.5cm); 32g. Dark, unpatterned; very heavy, dark, conical bill; larger than other similar species; female like female Pink-billed Seed-Finch but slightly larger, found in denser habitat. **Adult** ♂**:** blue-black, blacker on wings, tail, and lores, with brighter blue forehead and short, indistinct superciliaries. ♀: above rich, dark brown; below slightly paler, darker on breast; wings and tail dusky. Bill black, with base of lower mandible silvery; legs black. **Young:** resemble adult female but above with faint, sooty smudging; throat paler, tinged with cinnamon and often mottled with dusky; belly tinged with gray; below often with dusky shaft-streaks. Young males retain at least some of this plumage for a full year.

Habits. Frequents understory of humid forest, dark ravines, tall second growth, shady semi-open; enters lower second growth and grainfields to forage; eats a variety of fruits and seeds, including those of *Heliconia* and ripening or dry maize, crushing hard seeds in its powerful bill; takes beetles and roaches, gleaned from low vegetation; usually in pairs, sometimes alone; may accompany mixed flocks in forest.

Voice. Calls include a sharp, hard *spek* or *chik*, and low, nasal notes; song consists of ca. 6 successively lower, clear, mellow whistles, usually ending with a sharp little warble: *seee seee sewee suwee sweet seee, temporarily*. Females sing almost as well as males, often answering their songs while sitting in the nest. The sweet, prolonged flight song is rarely heard.

Nest. An ample cup, frail or strong, of rootlets, tendrils, vines, lined with black fungal rhizomorphs, often permeable to light, 16″–16ft (0.4–5m) up, rarely higher, in a spiny palm or tree fern in forest, orange tree or maize plant in clearings; may refurbish an old nest of its own or some other bird. Eggs 2, white or blue-tinged, with cinnamon, rufous, and pale lilac speckles and blotches. March–September, chiefly July–August; 2 or 3 broods.

Status. Common resident of Caribbean and S Pacific slopes; locally common, mostly in gallery forest or wet ravines, over most of Península de Nicoya and along Golfo de Nicoya to lower Tempisque valley; absent in N Guanacaste and Valle Central; sea level to 4000ft (1200m).

RANGE. S Mexico to W Ecuador, N Bolivia, and Amazonian Brazil.

NOTE. Sometimes included in the genus *Passerina*.

INDIGO BUNTING

Passerina cyanea Pls. 48(12), 50(6)

Azulillo Norteño (Indris)

DESCRIPTION. 4¾″ (12cm); 13.5g. Fairly small, slimmer than most seedeaters, with overall buffy tone; males usually show some blue; females below much less streaked than female Blue-black Grassquit. **Adult ♂** WINTER: above tawny-brown; below buffy-brown, shading to white on belly; especially on rump, shoulders, and underparts, more or less mixed with blue; wings and tail blackish, edged with greenish-blue, mixed with brown on wing-coverts and secondaries. BREEDING: deep blue, above more greenish, tinged with violet on head and breast; lores black. Bill grayish-horn with silvery rami; legs dusky. ♀: above tawny-brown; tail and wings dusky with brown to blue-gray edgings; shoulder tinged with blue; 2 indistinct buffy wing-bars; below buffy, darkest on breast and sides, where faintly streaked with dusky, shading to white on belly. Bill dark horn, more yellowish at base. **Immature ♂:** like adult female but stronger blue tinge to rump, tail, and shoulders; below duller, whiter; feathers below with dull blue-gray bases, producing effect of obscure streaking or mottling over most of underparts. ♀: like adult but little or no blue tinge; below whiter, streaked more extensively with dusky; back faintly streaked. Most young birds molt to an adult winter plumage soon after arrival on wintering grounds.

HABITS. Frequents weedy fields, savanna edges, recently harvested croplands, low second growth, and semi-open, sometimes singly or in small groups but more often in flocks of 20 or more that in winter move about following the seeding of grasses, notably guinea (*Panicum maximum*); feeds on ground or plucks seeds directly from grass inflorescences; takes small insects and berries.

VOICE. Calls include a short buzzy *bzeet* and a sharp *tsick*; not known to sing in Costa Rica.

STATUS. Widespread, sporadically common migrant and winter resident on both slopes between early to mid-October and late April; most abundant in drier regions of Pacific slope and Valle Central, though as dry season advances many move to wetter areas; lowlands to 5000ft (1500m).

RANGE. Breeds in SE Canada, E and C USA; winters from C Mexico to NW Colombia and Greater Antilles.

PAINTED BUNTING

Passerina ciris Pl. 48(13)

Azulillo Sietecolores (Siete colores)

DESCRIPTION. 5″ (13cm); 15g. Brilliant adult male unmistakable; female above more greenish, below more yellowish than other seedeaters or grassquits. **Adult ♂:** top and sides of head and neck, to sides of breast, violet-blue; back yellow-green; rump and upper tail-coverts pinkish-red; shoulder violet; middle coverts purplish; greater coverts and tertials greenish; rectrices and remiges dusky-purplish; rest of underparts bright red; eye-ring rose-red. ♀: above bright olive-green to dull grass-green, often tinged with bluish on sides of crown or nape; below greenish-yellow, washed with green on breast and tinged with olive on sides and flanks; belly sometimes strongly tinged with buff. Upper mandible blackish, lower grayish; legs dusky. **Immatures:** like adult female but duller, above more brownish; males below often brighter yellow; females duller and buffier.

HABITS. Forages low amid dense cover of brushy second growth, overgrown pastures, high grass, or riverside stands of wild cane; often alone, or in small groups, as when visiting water holes in dry areas; by March a male may accompany a female as though already mated; eats seeds, some fruits and insects.

VOICE. Call a sharp *chirp* or *chip*; not known to sing in Costa Rica.

STATUS. Uncommon and local winter resident on Pacific slope, chiefly in Tempisque basin and around Golfo de Nicoya and in Térraba region; late October–late March; sea level to 4500ft (1350m) in Valle Central.

RANGE. Breeds in S USA and N Mexico; winters from N Mexico and Florida to W Panama and West Indies.

YELLOW-FACED GRASSQUIT

Tiaris olivacea Pl. 49(6)

Semillerito Cariamarillo (Gallito)

DESCRIPTION. 4″ (10cm); 10g. Small, greenish; bill sharper and more conical than in *Sporophila* seedeaters; striking face pattern in male is faintly suggested in female. **Adult ♂:** superciliaries, spot on lower eyelid, and throat bright yellow; forehead, rest of face, foreneck, breast, and upper belly dull black, shading posteriorly to dull olive-green above and grayish-olive below. Bill black; legs grayish. ♀: above dull olive-green, below paler, more grayish on belly; narrow superciliaries, spot on lower eyelid, and chin dull yellow; breast more or less smudged with black. Upper mandible dusky, lower dark horn. **Young:** like adult female but duller and grayer; male begins acquiring adult plumage during first year.

Habits. Frequents weedy fields, pastures, roadsides, unkempt lawns with seeding grasses; usually in loose flocks, sometimes singly or in pairs; often breeds in loose colonies; plucks seeds from inflorescences or gathers them from open ground; when grass seeds are scarce, gathers berries from shrubs and trees, or, warblerlike, searches foliage for insects; often associates with other seedeaters in mixed flocks.

Voice. Song a series of high, thin, rapid trills, varying in pitch, speed and intonation, often prolonged, weak but sweet-toned; sometimes delivered by several males perching together in a shrub. Courting male rapidly vibrates his wings as he sings in front of female perched a few inches (cms) away.

Nest. An ovoid or roughly globular structure with a narrow side entrance, of grasses and weed stems thickly compacted, lined with pieces of grass inflorescences or shredded bast fibers, 6–12″ (15–30cm) above ground, rarely as high as 5ft (1.5m), often on a grassy or weedy bank. Eggs 2–3, rarely 4, dull white, speckled with brown. Chiefly May–August, sparingly until January.

Status. Common to abundant resident in open areas of Caribbean slope and Pacific slope except dry NW lowlands, but including E Valle Central and approaches and passes in N cordilleras; sea level locally to 6600ft (2000m).

Range. C Mexico to Colombia and NW Venezuela; Greater Antilles.

SLATE-COLORED SEEDEATER

Sporophila schistacea Pl. 49(5)

Espiguero Pizarroso

Description. 4¼″ (11cm); 13cm. Rather large, heavy-billed, with at least a tinge of gray in plumage and yellow in bill. **Adult ♂:** upperparts, head, breast, and sides slate-gray; belly, patch at base of primaries, and sides of throat white. Bill yellow-orange; legs dark gray. ♀: above olive-brown, below paler, shading to grayish-white or buffy-white on belly; throat tinged with gray. Bill dark horn with yellow base and gape. **Immature ♂:** like adult female but grayer overall; may have a distinct second-year plumage nearly as gray as adult with white in primaries but lacking white on throat; bill yellow-orange with dusky culmen.

Habits. Irruptive, possibly nomadic, appearing in small flocks that settle to breed where large crop of grass seeds is maturing, often disappearing afterward, seldom breeding in same place in 2 consecutive years; prefers forest edges or thinned woodland where bamboo is fruiting, and new pastures, especially on swampy ground where scattered trees remain; forages for grass seeds, including bamboo, either on ground or on inflorescences. Males sing from perches 15–33ft (5–10m) up or on slow fluttery flights between perches.

Voice. Calls include a thin *tsew* or *tseet* and a sibilant, metallic *zheet* with rising inflection. Song high, thin, ventriloquial, the first notes very penetrating; starts with 1–3 loud notes, ends with 3–10 similar notes in measured cadence: *tlee lee lee see-see-see-see-see* or *tilee zizizet tsi-tsi-tsi-tsi-tsi*.

Nest. A cup of leaves, lined with rootlets, 16ft (5m) up in bamboo vine (in Panama; no Costa Rican nest seen). Eggs undescribed (?). June–September (as suggested by nest-building, fledglings, and gonad data).

Status. Nomadic resident of Golfo Dulce district, probably also Térraba region (has definitely bred there in past; no recent records), and of Caribbean lowlands, principally near coast, from Panama N at least to Parismina; has bred at 3300ft (1000m) above Carrillo in Cordillera Central (July 1980, when bamboo fruited); evidently expanding its range rapidly N, having invaded Caribbean slope of Costa Rica since about 1975.

Range. Locally in S Mexico (?) and N Honduras; Costa Rica to NW Ecuador, N Bolivia, NE Brazil.

WHITE-COLLARED SEEDEATER

Sporophila torqueola Pl. 49(2)

Espiguero Collarejo (Setillero)

Description. 4″ (10cm); 9.5g. Small, with very thick, convex bill; in all plumages distinguished from other Costa Rican seedeaters by broad wing-bars. **Adult ♂:** head and upperparts mostly black; rump buffy-brown to buffy-white; wing-bars, bases of inner primaries, and edges of tertials, throat, spot on lower eyelid, and nuchal collar white; chest and thighs black, rest of underparts buffy-white. Bill and legs black. ♀: above olive-brown, palest on rump; wings and tail dusky with olive-brown edgings; 2 broad buffy-white wing-bars, tertials edged with pale brownish; below rich buff to buffy-white, darkest on chest. Upper mandible blackish, lower dark horn. **Young:** like adult female but below sometimes duller and darker. Males acquire some black feathers on head and chest by first breeding season and attain full adult plumage at 2 years.

Habits. Frequents open areas with seeding grasses, including pastures, roadsides, weedy fields, scrub, agricultural areas, and marshes covered with tall, coarse grass or *Polygonum*, another favorite seed source; usually feeds in flocks, sometimes with other seedeating birds; sometimes a pest on ripening sorghum and rice; eats insects and spiders and occasionally berries.

Voice. Calls include a variety of soft *chirp* and *cheep* notes and a harsher, more strident *preep* in alarm or distress; a more nasal *chea*. Song an often pleasant and melodious, variable series of clear, rich, whistled notes, sometimes incorporating canarylike trills and often ending with dry, chaffy or buzzy trill; 1 version: *swee-swee-swee-swee, tee'oo-tee'oo-tee'oo, tewtewtewtew* (Eisenmann).

Nest. A slight cup of fine rootlets, fibers, pieces of grass inflorescences, bound with spiderweb, often lined with horsehairs, so thin that light passes through it, 2–10ft (0.6–3m) up in coarse herbs, shrub, or tree. Eggs 2, occasionally 3, pale blue or blue-gray, thickly mottled with brown, chocolate and sometimes heavy black blotches. April–December.

Status. Resident almost countrywide from sea level to ca. 5000ft (1500m), occasionally higher. Expanded range greatly in last 25 years on Caribbean and S Pacific slopes with deforestation but still rare in Valle del General though common in more coastal areas; uncommon and local in much of Valle Central due to persecution for cage-bird trade, its song being much esteemed; some local dialects, particularly popular among *pajareros*, are becoming increasingly rare.

Range. S Texas to W Panama.

VARIABLE SEEDEATER

Sporophila aurita Pl. 49(3)

Espiguero Variable

Description. 4″ (10.5cm); 11g. On Caribbean slope males are best distinguished from other black seedeaters by convex *Sporophila* bill; female duller and browner than other seedeaters. On Pacific slope male differs from male White-collared in lacking white throat, complete collar, and wing-bars; female unpatterned except for whitish belly. **Adult ♂:** Caribbean race entirely black except for white bases of inner primaries, wing-linings, and midline of belly. Pacific race above black, rump white; wings as in Caribbean race; broad white patches on each side of neck narrowing or disappearing on foreneck; otherwise head, neck, throat, chest, and thighs black; rest of underparts white, indistinctly barred with blackish on sides and flanks. Bill black; legs blackish. ♀: Caribbean race above dark olive-brown, below paler; wing-linings white. Pacific race above olive, below paler, shading to dull yellowish-white on belly and crissum; wing-linings white. **Young:** like adult female of their race. Young males may not acquire full adult plumage in their first year and breed in more or less variegated plumage.

Habits. In small groups, pairs or singly in low thickets, gardens, roadsides, and at forest edge, but often flocking with other seedeaters, grassquits, and seed-finches in large stands of tall seeding grasses, weedy fields, or pastures; in addition to grass seeds, eats berries and seeds of trees and shrubs, also some insects; traverses belts of forest when moving between clearings; males often sing from well up in tree.

Voice. Calls include mellow, slurred chirps, a harsher *chur*, and dry chatters in interactions. Song a prolonged, varied warble, from a breezy start accelerating into a rushing jumble of notes, which may taper off in chaffy trills; often twitters are incorporated, goldfinchlike. Birds on Pacific slope tend to sing more elaborate songs.

Nest. A slight, thin-walled cup of wiry rootlets and fibers lined with a few fungal filaments or horsehairs, attached by spiderweb to a forked twig of tree or shrub, 30″–20ft (0.8–6m), rarely 30ft (9m) up. Eggs 2, rarely 3, pale gray or blue-gray, speckled and blotched with brown and often spotted or scrawled with black. May–August; 2 broods; a minor breeding period November–January.

Status. Abundant resident throughout Caribbean lowlands and foothills; on Pacific slope common from Panama N to about Carara, and uncommon to rare N to around Cañas and head of Golfo de Nicoya; not known from Valle Central. Sea level to 5000ft (1500m), occasionally higher on Caribbean slope.

Range. S Mexico to NW Peru.

Note. Sometimes lumped with *S. americana* of eastern South America. The all-black Caribbean race was once considered a distinct species, *S. corvina*, Black Seedeater, but the 2 races intergrade in C Panama.

YELLOW-BELLIED SEEDEATER
Sporophila nigricollis Pl. 49(4)
Espiguero Vientriamarillo

Description. 3¾″ (9.5cm); 9g. Contrastingly marked adult male unmistakable; females and young resemble those of Variable Seedeater but paler and below more extensively yellow. At close range, note distinctly smaller bill. **Adult ♂:** forehead, face, throat, sides of neck, and breast black, shading to dark grayish-olive over rest of upperparts; rest of underparts pale yellow to almost white, obscurely smudged or streaked with dusky on flanks; wing-linings white. Bill silver-gray; legs dark gray. ♀: above brownish-olive, shading to paler olive on throat and chest; rest of underparts pale yellow to buffy-yellow, tinged with olive on sides and flanks. Bill dusky-horn, paler at base of lower mandible. **Immatures:** like adult female, but male usually has some black blotching on face and breast, acquiring adult plumage after first breeding season. **Young:** like adult female but more brownish, below buffy; wing-coverts tipped with buff.

Habits. Frequents open, grassy places, including abandoned pastures, young second growth, roadsides, sometimes dooryard shrubbery; often in milpas when maize is in tassle; males often sing from trees; when not breeding found in groups of 20 or more, often with other seedeaters in flocks that appear to move nomadically following seeding of grasses, seeking wetter areas as dry season progresses.

Voice. Calls include dry *chips* and *chirps*, less mellow than those of Variable Seedeater; song of clear, slurred or short whistles, often ending with a dry trill or buzzy note; like that of White-collared Seedeater but shorter and with a somewhat burry or lisping tone: *seee seeuu se seeuu seeoo tsr-r-r-r*.

Nest. A deep cup of grasses, rootlets, or palm fibers so thin that light passes through it, 5–10ft (1.5–3m) up amid foliage of a shrub or tree. Eggs 2, occasionally 3, greenish or buffy, heavily mottled with brown or purplish. July–September (in Panama; no Costa Rican nest, though singing males and fledglings August–October).

Status. Permanent or seasonal resident; nomadic, particularly in dry season; movements poorly understood; recorded from General-Térraba valleys and Golfo Dulce lowlands of S Pacific slope; in recent years small numbers apparently have bred (fledglings) near Cartago (El Radio) and E of San José (Lourdes), in Valle Central.

Range. Costa Rica and Lesser Antilles to NE Argentina and E Brazil.

RUDDY-BREASTED SEEDEATER
Sporophila minuta Pl. 49(1)
Espiguero Menudo

Description. 3½″ (9cm); 8g. Red-breasted adult male unmistakable; female and young very buffy, lack the distinct wing-bars of bigger, larger-billed White-collared Seedeater. **Adult ♂:** upperparts mostly dull olive-gray, darkest on wings and tail; wing-linings and bases of inner primaries white; rump and entire underparts cinnamon-rufous. Bill blackish, sometimes yellowish at base; legs brownish. ♀: above olive-brown, paler and brighter on rump; tail and wings dusky, edged with brown, the wing-coverts and tertials edged with buff; below buff, washed with brown or cinnamon on chest. Bill dusky-horn, base of lower mandible horn-color. **Immatures:** like adult female; males more heavily washed with cinnamon-brown on throat and chest.

Habits. Frequents open, grassy savannas, marshes, bushy scrub, sometimes gardens and dooryards; usually forages in groups or flocks, often with other seedeaters; flocks tend to be larger and wander more outside breeding season; feeds mainly on grass seeds, supplemented with small insects and berries.

Voice. Calls include slurred chirps and whistles, surprisingly piercing for so small a bird; song a medley of clear, slurred whistles, often in 2s, less elaborate than that of White-collared and lacking trills: *weet weet chercher teeaweet, weet weet cher-weet weet cher cher teeawee*, etc.

Nest. A small cup of grass stems and rootlets, in tussock of grass or dense, low shrub, generally below 3ft (1m) but occasionally to 8ft (2.4m) in small tree. Eggs 2–3, white or pale buff, spotted with brown, most heavily at large end (in Panama; eggs not seen in Costa Rica, where two nests held 2 and 3 nestlings in July). June–August, possibly also November–December (male in breeding condition).

Status. Locally common resident in savannas of Térraba and Río Frío regions; in recent years becoming widely established in Coto region of Golfo Dulce lowlands, following deforestation; probably still expanding its range.

Range. C Mexico and Trinidad to NW Ecuador, N Argentina, and Uruguay.

BLUE SEEDEATER
Amaurospiza concolor Pl. 49(11)

Semillero Azulado

Description. 4½″ (11.5cm); 13g. Males usually appear dull blackish, their blue tints apparent only in good light; females brighter, more rufescent than other seedeaters, without streaks; bill heavy but conical, not convex as in *Sporophila*; habitat distinctive. **Adult ♂:** entirely dull blue-black; no white in plumage. Bill black, lower mandible grayish at base; legs gray. ♀: upperparts bright brown to cinnamon-brown; below uniform cinnamon to dull tawny-chestnut. Bill often more bicolored, with base of lower mandible paler gray. **Young:** like adult female but brighter, below more rufescent, with fairly conspicuous dull cinnamon wing-bars.

Habits. Prefers ravines, openings, and edges in primary or secondary forest where bamboo grows profusely; singly, in pairs, or family groups, hops and flits actively in foliage of bamboo, nibbling at or pulling out tender growing shoots at tips of stems and eating the bases; takes insects and seeds; sometimes accompanies groups of Black-cheeked Warblers or Yellow-thighed Finches; usually forages 10–40ft (3–12m) up.

Voice. Call a distinctive, sharp *hu-tseet* or *chu-tsee*; song a pleasant warble, often ending in a little sputter, similar in quality to that of Slate-throated Redstart, *ch'swee ch'swee ch'switto ch'seeu ptr-r-r*, or a simpler, somewhat plaintive *chi chi wee chee chee*.

Nest. A cup of coarse grasses, lined with finer grasses and other plant fibers, 10ft (3m) up in fork of tall shrub. Eggs 2, unmarked pale blue (in Mexico, of the race *relicta*, by some considered a distinct species; nest and eggs of *concolor* unknown (?), though fledglings seen in early September in Costa Rica).

Status. Uncommon to rare and very local in highlands; recorded from all cordilleras, mainly on Pacific slope but also on Caribbean slope of Cordillera de Tilarán and Cordillera Central; most regular on Pacific face of Cordillera Central at 5500–6600ft (1700–2000m), elsewhere 3300–7200ft (1000–2200m).

Range. C Mexico to NW Ecuador.

PINK-BILLED SEED-FINCH

Oryzoborus nuttingi Pl. 49(13)

Semillero Piquirrosado

Description. 5¾″ (14.5cm); 24g. Sparrow-sized, with massive bill; larger than Thick-billed Seed-Finch, with darker female, pale-billed male. **Adult ♂:** entirely black, except inner webs of primaries grayish-white basally. Bill flesh-horn, often becoming bright salmon-pink during breeding; legs blackish. ♀: above dark brown; below dark rufous to chestnut, palest on throat. Bill blackish, shading to flesh-horn at base. **Immatures:** like adult female but male averages darker, below less rufescent. **Young:** like adult female but below more buffy, with 2 buffy-brown wing-bars.

Habits. Prefers wet grassy areas, usually near open water or marsh, and flanked by taller vegetation; in pairs during breeding; outside breeding season in small, wandering groups which may join flocks of other seedeating finches; eats mostly grass seeds. Males sing in flight or from trees or shrubs overlooking grassy breeding sites.

Voice. Call a short, sharp *chek* or *chip*; song a caroling series of slurred, whistled notes, many doubled or tripled, slower, deeper, richer in tone than song of Thick-billed Seed-Finch.

Nest. Flimsy cup of grass stems and rootlets, lined with fungal rhizomorphs and other fine fibers, in grass, sedge, or small shrub low over ground or water. Eggs 2, greenish-white, heavily spotted with shades of brown, in wreath at thick end. February–July.

Status. Uncommon and local resident throughout Caribbean lowlands up to 1600ft (500m), wandering to 3000ft (900m). Range has evidently expanded S from Nicaragua in last 30 years.

Range. E Nicaragua to NW Panama.

Note. Often called Nicaraguan Seed-Finch. Sometimes considered a race of the South American *O. maximiliani*, Great-billed Seed-Finch, but distinct color differences suggest relationship at superspecies level.

THICK-BILLED SEED-FINCH

Oryzoborus funereus Pl. 49(10)

Semillero Picogrueso

Description. 4½″ (11.5cm); 13.5g. Bill much heavier and more conical than those of smaller *Sporophila* seedeaters; female richer brown than other small seedeaters; female Pink-billed Seed-Finch considerably larger. **Adult ♂:** deep black except wing-linings and bases of inner primaries white (usually only visible in flight); sometimes a few white feathers in midline of belly. Bill black; legs dark gray. ♀: above rich, dark brown; wings and tail blackish, edged with brown; below dark cinnamon-brown to chestnut-brown, darker and duller on throat, breast and flanks; wing-linings white. Bill blackish, shading to dark horn on tomia. **Immatures:** like adult female, the male below averaging deeper chestnut, not acquiring full adult plumage until second year; occasionally breeds in immature plumage.

Habits. Frequents grassy and weedy fields with scattered shrubs and trees, second growth with tall grass, thickets, scrubby marshes, and forest edge; less gregarious and more arboreal than *Sporophila* seedeaters, usually alone or in pairs that sometimes join flocks of seedeaters and grassquits; eats mostly grass seeds, some berries and insects; often wags tail from side to side nervously; sings high in tree.

Voice. Call an abrupt, slurred *tchew*; also a staccato *dit* or *dik*; song in some areas a medley of slurred whistles and short warbles; in others more complex, incorporating trills and twitters to give a jumbled, rich-toned, varied performance lasting commonly 4–8sec, up to 20sec; somewhat resembles song of Variable Seedeater but sweeter and longer. Brown young males sing as well as do black adults.

Nest. A slight cup of fine rootlets, pieces of grass fibers or inflorescences, sometimes lined with horsehair, so thin that light passes through it; 30″–8ft (0.8–2.5m) up in shrub, small tree, or vine tangle. Eggs 2, dull white, heavily mottled with brown and pale lilac. April–September.

Status. Common resident of lowlands and foothills of Caribbean and S Pacific slopes, to 3500ft (1100m), rarely higher; birds of upper part of this range may descend to lower elevations after breeding, return when wet season is well established.

Range. SE Mexico to W Ecuador.

Note. Sometimes lumped with the Chestnut-bellied Seed-Finch, *O. angolensis*, with which it apparently hybridizes in Colombia. The genus *Oryzoborus* is sometimes merged in *Sporophila*.

BLUE-BLACK GRASSQUIT

Volatinia jacarina Pl. 49(7)

Semillerito Negro Azulado (Brea, Pius)

Description. 4″ (10cm); 9.5g. Very small, with conical, sharp-pointed bill; male more glossy and female more heavily streaked below than other small seedeaters. **Adult ♂:** entirely glossy blue-black, except for patch of white on each side of breast, exposed in flight; immediately following molt, blue may be partly obscured by brown fringes that soon wear away; there may even be a more or less "eclipse" (femalelike) plumage, worn briefly during dry season: more data needed. Upper mandible black, lower gray with black tip; legs dusky. ♀: above dull brown; wings dusky with buffy-brown edgings, tail dusky; below buffy-white, washed with grayish on throat and rich buff on breast; flanks tinged with brown; throat, breast, and sides streaked with dusky. Upper mandible blackish, lower horn-color with dusky tip. **Immature ♂:** body plumage blue-black with very heavy brown and rich buffy fringes that give a femalelike appearance but darker and mottled; wings and tail mostly blue-black. These fringes vary in extent, below may be whitish; plumage sequence not well understood. **Young:** like female but overall duller and dingier; above more sooty, below browner, with heavier blackish streaking; wing-coverts fringed with buff.

Habits. Frequents grassy or weedy fields, low scrub, bushy areas with grassy openings and roadsides; usually in loose flocks, breeding in pairs in dense local populations; lurks and forages low in vegetation, unseen until flushed; food includes mainly grass seeds, supplemented with insects, berries, and *Cecropia* protein corpuscles; often alights on paved roads to pick up grit and small seeds.

Voice. Calls include a soft, dry ticking, a high *peet*; song of male a buzzy *b'zeeer* or *zizheeer* or a simpler, plaintive *t-s-e-u*, from low perch or as he flits vertically upward for 1–3ft (30–90cm) and flutters back down to the same perch; often many song–jumps are made every few seconds, for half an hour or more.

Nest. A tiny, thin-walled cup, strongly made of fibrous rootlets, straws, and tendrils, lined with finer rootlets, attached by spiderweb to grasses or bush, 3″ to, rarely, 9ft (0.75–2.7m) up. Eggs 2 or 3, white or bluish, heavily speckled with shades of brown, especially on the thicker end. June–October.

Status. Common resident countrywide, from both coasts locally to 5500ft (1700m) in central highlands.

Range. NW Mexico and S Lesser Antilles to N Chile and N Argentina.

Note. Considered by some to be related to, even congeneric with, the finches of the Galápagos (*Geospiza*). Although we might agree that there is a relationship, we keep *Volatinia* as a separate genus.

COCOS FINCH

Pinaroloxias inornata Pl. 49(8)

Pinzón de la Isla del Coco

Description. 4¾″ (12cm); 12.5g. Chunky, short-tailed, with fluffy plumage, strong legs and feet, sharp-pointed, slightly decurved bill. **Adult ♂:** entire plumage black. ♀: above blackish-brown, variably and indistinctly scaled and streaked with olive-brown; forehead, lores, and superciliaries sometimes extensively buffy; wing-bars ochraceous-buff; below buff, paler on belly, more or less heavily and indistinctly streaked with black; crissum buff. Bill and legs black. **Young:** like adult female but bill yellow. Occasionally a male with black plumage and yellow bill is seen.

HABITS. Occupies every available habitat on Cocos I., moving about busily in pairs or small flocks, incessantly, methodically foraging with an astonishing variety of techniques: gleans from leaves, twigs, branches, and ground; probes bark crevices and rolled leaves; extracts leaf-mining larvae from damaged leaves; turns small stones; pierces and probes flowers; visits extrafloral nectaries; plucks fruits. Evidently most individuals specialize in 1 or a few of these methods, different birds foraging differently (Werner and Sherry). Often seemingly oblivious to presence of people more than 2–3 yards (m) away.

VOICE. Notes include a burry, rough *djrr*, often doubled; a whistled *tyew*, sharpened in aggressive encounters to a grating *jeew* or *jeeak*; song of male prolonged and buzzy, rising at end, often preceded by a high, piercing, metallic note: *tseek ka-jeeuuuurrreek*.

NEST. A messy globe of grass stems, twigs, and dead leaves, with a side entrance, at tip of branch, usually 6–30ft (2–9m) up. Eggs 2, white, spotted with brown. January–May but some nesting throughout year (Werner and Sherry).

STATUS. Resident throughout Cocos I.; abundant in *Hibiscus* thickets along coast, sparser in wet highland forests.

RANGE. Endemic to Cocos I.

GRASSLAND YELLOW-FINCH

Sicalis luteola Pl. 50(3)

Chirigüe Sabanero

DESCRIPTION. 4½″ (11.5cm); 11g. Small, streaked above and yellow below; bill short, thick, convex, almost *Sporophila*-like. **Adult ♂:** pileum olive-green, streaked with dusky; back brownish-olive, streaked with brownish-black; rump and upper tail-coverts yellowish-olive; wings and tail dusky, edged with grayish-buff; cheeks olive; superciliaries and entire underparts bright yellow. Iris reddish-brown, upper mandible blackish, lower pale grayish; legs blackish. ♀: similar but duller, with little or no green on head; rump olive-green; superciliaries narrow and dull; below paler yellow, strongly tinged with buffy-brown on sides of throat, breast, sides, and flanks. **Young:** like adult female but above more buffy, especially on edgings of wing-coverts; below paler; chest and sides streaked with dark brown.

HABITS. In open, scrubby grasslands and burned pastures; gregarious, in flocks of up to 50, also singly or in pairs; forages on ground, hopping and running, picking up grass seeds; may flutter up and alight on grass stem, bend it down to pluck off seeds; flight undulating.

VOICE. Individuals make a guttural bicker or buzzy or chirping twitter, which sounds insectlike when many individuals in a flock are heard (Slud); song a thin, buzzy trill, delivered from exposed perch or in flight (Ridgely).

NEST. A compact cup of dry grasses, lined with finer grasses or rootlets, on the ground or up to 1ft (30cm). Eggs 3–5, white or tinted with blue or green, thickly spotted with brown [in South America; no Central American record; probable breeding noted in June in Panama (Wetmore)].

STATUS. Only Costa Rican occurrence of apparently wandering or migrating flocks in late May 1955 on W approach to Volcán Miravalles, Guanacaste (Slud). Closest probable breeding areas are Coclé, W Panama, and pine savannas of NE Nicaragua.

RANGE. C Mexico to S Chile and S Argentina; distribution discontinuous in Central America.

NOTE. Another terrestrial open-country bird possible along the southern Panama border as deforestation continues is the Yellowish Pipit (*Anthus lutescens*, Bisbita Amarillenta), a small (5″, 12.5cm), slender bird with slim bill, blackish-brown upperparts streaked and edged with buff; pale buffy to yellowish underparts with dusky streaking on chest; a narrow whitish eye-ring; and white outer rectrices that flash in flight. Walks over ground in savannas and pastures; occurs on Pacific slope of Panama W to W Chiriquí, and widely in South America, S to N Chile and N Argentina.

SLATY FINCH

Haplospiza rustica Pl. 49(14)

Fringilo Plomizo

DESCRIPTION. 5″ (12.5cm); 17g. Resembles Peg-billed Finch but with conical bill and darker, more uniform coloration. **Adult ♂:** slate-gray, darker and with bluish tinge on head and back. ♀: above dark grayish-brown;

below paler; breast tinged with olive; throat paler, more buffy; faint, sparse dusky streaking on breast and sides. Upper mandible blackish, lower horn-color to gray with paler tomia; legs pale brownish. **Immature ♂:** body mostly slate-gray, but above with brown feather-tips; below with some brownish wash; remiges edged with brown.

Habits. Frequents clearings, pastures, and second growth in forested country; usually in small flocks that gather grass seeds and grit from ground or strip seeds from grass inflorescences with bill; flight low, direct, with little gliding; males sing from shrubs or low trees along edge.

Voice. Call a colorless, high, thin *seep* or *sssp*, which becomes louder and harsher in interactions; song a very high, thin, sibilant medley of staccato notes and longer whistles or trills: *tsit-tsit-tsit psssss ts'-pseee*.

Nest. Unknown (?).

Status. Rare or very local resident (though with irregular, perhaps nomadic, seasonal movements) of all major mountain ranges, except not recorded from Cordillera de Guanacaste. Above 5000ft (1500m) in Cordillera de Tilarán and from 6000ft (1850m) up to timberline in Cordilleras Central and de Talamanca.

Range. S Mexico to NW Bolivia.

Note. Sometimes segregated in genus *Spodiornis*.

PEG-BILLED FINCH

Acanthidops bairdii Pl. 49(12)

Fringilo Piquiagudo

Description. 5¼″ (13.5cm); 16g. Larger, slimmer than *Diglossa*; tail often looks notched; female with distinct wing-bars. Note rather long, slender, slightly upturned bill. **Adult ♂:** slate-gray, paler on belly. ♀: olive-brown, paler on throat and belly; head, upper back, sometimes throat and chest strongly washed with gray; below sometimes with faint, dusky streaking; pale superciliaries; wing-bars bright cinnamon-buff. Upper mandible black, lower yellowish-horn to bright yellow; legs flesh-color to yellowish-horn. **Immature ♂:** like adult male but above tinged with brown; retains juvenile remiges; 1 buffy wing-bar. **Young:** like adult female but below paler; wing-bars paler buff.

Habits. Chiefly at edges and breaks in highland forest, scrubby or bamboo-choked second growth, and shady pastures; singly, in pairs or small groups, either alone or with flocks of bush-tanagers, warblers, etc.; probes small flowers for nectar, actively gleans foliage for insects or catches them in the air; takes bamboo seeds from inflorescences or ground; hops on ground picking up grass seeds; pierces berries with bill, then carries them to a horizontal perch against which it presses out the juice, then discards the skins. Flight fast, low, direct; hops actively along branches.

Voice. A rolling, dry *pzeep*; a fast descending twitter similar to that of *Diglossa*; song of several high, squeaky whistles followed by a lower, buzzy note: *chee-schee-schee-schee-paah*.

Nest. Undescribed (?); probable breeding season March–May or June.

Status. Uncommon to sporadically common locally; resident, possibly nomadic, in highlands from Cordillera de Tilarán S to Panama, from 5000ft (1500m) on Cordillera de Tilarán and 6900ft (2100m) on Cordilleras Central and de Talamanca up to timberline; in rainy season descends occasionally to 4000ft (1200m), as at San Pedro.

Range. Costa Rica and extreme W Panama.

SLATY FLOWERPIERCER

Diglossa plumbea Pl. 49(9)

Pinchaflor Plomizo

Description. 4″ (10cm); 9g. Warbler-sized, plain-colored; distinctive upturned bill with upper mandible hooked, lower awl-tipped. **Adult ♂:** above dark gray to bluish-gray, wings and tail blackish with dark gray edgings; throat and breast lead-gray, shading to paler slate-gray posteriorly. ♀: above brownish-olive, the greater coverts and tertials edged with dull olive-buff; throat and breast paler, shading to dull olive-buff on belly; flanks washed with dull olive; crissum tinged with cinnamon. Bill black, except base of lower mandible flesh-color to dusky-flesh; legs brownish. **Young:** like adult female but usually 2 tawny wing-bars; below dull buffy-yellowish, washed with olive on throat and breast, with faint dusky smudging or streaking on breast. Males acquire adult plumage on body and wing-coverts during first year but retain juvenile remiges until after first breeding season.

Habits. Frequents canopy and edges of mountain forest but more numerous at sunny openings, clearings and treeless summits densely overgrown with flowering shrubs; singly or in pairs, visits flowers of epiphytes and shrubs, hooking tip of upper mandible over base of tubular corolla, which it perforates with lower mandible, and with brush-tipped tongue extracts nectar through the hole; gleans tiny insects from foliage or catches them in flight; eludes attacks of hummingbirds defending flowers by retreating to dense foliage.

VOICE. Call a high, thin, weak *tsip* or a piercing *tseep*; several such notes run together in a sharp twitter, especially in interactions. Song a varied high, thin, rapidly delivered medley of slurred whistles, short warbles, chips, and weak trills: *see-chew see-chew see-chew seer seer surrtseep, tsee tseew tsewery tseer tsewery sewy tsink-tsink-tsink*; often has a tinkling quality.

NEST. A bulky, substantial cup of mosses, shreds of decaying leaves, coarse vegetable fibers, lined with finer fibers, 15″–13ft (0.4–4m) up in spiny palm, dense shrub, or grass tussock in mountain pasture. Eggs 2, pale blue, finely speckled with brown. August–December, sometimes through March.

STATUS. Common resident of higher elevations countrywide, upward from 4000ft (1200m) in Cordillera de Guanacaste, 5000ft (1500m) in Cordillera de Tilarán, and 6300ft (1900m) in Cordilleras Central and de Talamanca to well above timberline in páramo.

RANGE. Costa Rica to W Panama.

NOTE. Formerly included in the honeycreeper family, Coerebidae; by some authors considered conspecific with *D. baritula* of N Middle America.

LARGE-FOOTED FINCH

Pezopetes capitalis Pl. 48(20)

Saltón Patigrande

DESCRIPTION. 8″ (20cm); 56g. Large, chunky, with rather slender bill and very large, stout legs and feet; mostly olive with gray and black head. **Adults:** forecrown, sides of crown and nape, lores, and upper throat black; rest of head, throat, and neck dark slate-gray; rest of upperparts dark olive; wings and tail black with olive edgings; below bright olive, washed with brown on flanks and crissum. Iris reddish; bill black; legs dark brown. **Young:** gray on head replaced by dark olive, smudged or streaked with blackish; upperparts scaled with sooty-black; underparts buffy-olive, heavily streaked with blackish; flanks washed with brown.

HABITS. Hops over ground in bamboo-choked ravines, forest understory, second growth, and brushy pastures in high mountains, exposing seeds and insects in ground litter with powerful double-scratches that may send leaves flying a foot (30cm) or more behind; plucks berries from low shrubs or small trees. Male sings from ground or low perch. In pairs throughout year.

VOICE. Calls include a high, thin, soft *psee* or *seet*; peeping and clucking notes between members of pair; a greeting duet when pair comes together, consisting of several high, thin, scratchy notes followed by a loud series of hard, thin, almost buzzy trills, successively lower in pitch and faster, cascading down to a short, warbled final note. Song of male a medley of short, rich, throaty whistles and warbles, often incorporating rattles, trills, chatters and imitations of other birds, most unfinchlike.

NEST. A bulky, loosely made cup of mosses, bamboo leaves, and herbaceous stems, lined with fine grasses and often built upon a thick foundation of bamboo leaves, 1–7ft (0.3–2m) up in dense shrub or bamboo. Eggs 1–2, dull white to bluish-white, with pale brown blotches and dark brown spots and specks, mostly toward large end. March–June.

STATUS. Common in upper parts of Cordillera Central and Cordillera de Talamanca from 7000ft (2150m) up to shrubby páramo, ca. 11,000ft (3350m).

RANGE. Costa Rica and W Panama.

YELLOW-THIGHED FINCH

Pselliophorus tibialis Pl. 48(22)

Saltón de Muslos Amarillos

DESCRIPTION. 7¼″ (18.5cm); 31g. Dark, slender, long-tailed, with puffy yellow thighs; bill rather long and slender; *Atlapetes*-like. **Adults:** pileum and hindneck black; rest of upperparts slate-black; throat blackish, shading to blackish-slate over rest of underparts; breast and belly tinged with olive-green. Iris chestnut; bill black; legs dusky. Some birds have a small spot of yellow at bend of wing. **Young:** similar but above duller, more sooty-black; below strongly tinged with brownish; thighs pale olive-brown.

HABITS. Travels in pairs or family groups through humid highland forests, bushy openings, second growth, bamboo-choked ravines, and nearby shady pastures, often accompanying mixed flocks of bush-tanagers, warblers, and other small birds; forages at all levels from high in trees to ground, rummaging and gleaning for insects and spiders; eats many berries; plucks tubular corollas to press out nectar.

VOICE. Call a heavy, metallic *tchuk* or *tchek*, sometimes several of these notes accelerate into a hard chatter. Mates greet each other with a prolonged rush of tinkling notes, often sounding like a gurgling twitter. Song of male a short, breezy phrase in a high, dry tone: *cheedle tweep cher tweeip, pity me sweet sweet*, or *tee tidi-dee dee wink wink*, repeated over and over at dawn in nesting season and sporadically at other times.

Nest. A bulky cup of straws, grass blades, dry bamboo leaves, 30″–15ft (0.7–4.6m) up in coarse grasses, bamboo, or densely foliaged tree in woods or clearing. Eggs 2, white or bluish, speckled and blotched with brown and lilac. March–May.

Status. Common resident of upper Cordillera Central and Cordillera de Talamanca, from timberline down to 5500ft (1700m), occasionally lower; outside breeding season some may descend to 4000ft (1200m) on Caribbean slope; fairly common above ca. 5000ft (1500m) on Cordillera de Tilarán.

Range. Costa Rica to W Panama.

YELLOW-THROATED BRUSH-FINCH

Atlapetes gutturalis Pl. 48(15)

Saltón Gargantiamarilla (Comepuntas, Purisco)

Description. 6¾″ (17cm); 33g. Fairly slender, long-tailed; conspicuous white stripe on crown; yellow throat inconspicuous in poor light. **Adults:** top and sides of head black except white median stripe from midcrown to hindneck; rest of upperparts slate-black; wing-coverts sometimes tipped with slate; throat bright yellow; breast and sides grayish-white, shading to white on belly and grayish-brown on flanks and crissum. Iris chestnut; bill black; legs brown. **Young:** head sooty-black, crown stripe indistinct and tinged with brown; upperparts dull blackish-brown; wing-coverts edged with brown; throat paler, more buffy; breast, sides and flanks buffy-brown, shading to pale, dull buff on belly; below extensively and finely streaked with dusky.

Habits. Frequents tall second growth and scrub, brushy clearings, thickety forest edge, weedy coffee plantations, hedgerows, garden and dooryard shrubbery, moving about in pairs or family groups; boisterous and bounding, constantly posturing erectly, then ducking its head and wagging its long, mobile tail; forages on ground, moving in springy hops, poking through litter with bill for insects, spiders, and seeds but rarely scratching; gleans in foliage, plucks many berries; in some areas may accompany Chestnut-capped Brush-Finch; flies flutteringly with pumping tail.

Voice. Call a repeated, piercing, high, thin *tsit* or *tseep*, a staccato *tsit*ting; song a high-pitched, thin, deliberate series of sibilant notes: *tsee tseea tseea tseea tseer* or *o see me, o see, I'm weary, pity moo*.

Nest. A bulky cup of straws, grass blades, and weed stems, lined with finer fibers, 1–7ft (0.3–2m) up in tangle of vegetation, pasture grasses, coffee shrub, or dense hedge. Eggs 2–3, white or pale blue, immaculate or with a few brown specks. March–June.

Status. Common resident the length of the country at ca. 3000–7200ft (900–2200m), mainly on Pacific slope but following deforested areas well onto Caribbean slope, especially in central highlands; very abundant in upper parts of Valle Central.

Range. Extreme S Mexico to Colombia.

Note. Sometimes merged with the White-naped Brush-Finch, *A. albinucha*, of S Mexico.

CHESTNUT-CAPPED BRUSH-FINCH

Atlapetes brunneinucha Pl. 48(16)

Saltón Cabecicastaño

Description. 7¼″ (18.5cm); 42g. Robust, largely terrestrial, with rather long, slender bill, white-spotted face, and white throat that is often conspicuously puffed out in excitement or irritation. **Adults:** forehead and sides of head black, with white spot in center and one on each side of forehead; crown and nape chestnut-rufous, bordered on each side by stripe of yellow-ochre; rest of upperparts dark olive-green, darker on wings and tail; throat and center of breast and belly white, a black band across chest; sides of breast and sides slaty; flanks and crissum olive-green. Bill black; legs brown to blackish. **Young:** pileum sooty-brown; face sooty-black; upperparts dull brownish-olive to sooty-brown; throat and breast sooty-olive, shading to dull yellow-olive posteriorly, heavily streaked with dusky; base of lower mandible yellow.

Habits. In pairs or family groups in understory of wet highland forest or lower secondary forest and shady ravines; hops over ground, throwing or pushing aside litter with bill, to expose beetles, roaches, crickets, moths, spiders, small centipedes, and other invertebrates; rises to low shrubs to pluck berries and glean from foliage; when surprised may fly to a low perch and posture from side to side with throat puffed, peering at intruder; often keeps company with wood-wrens, *Basileuterus* warblers, Yellow-throated Brush-Finches, etc.

Voice. Calls include a weak, high-pitched *chink* or *pink* and a sibilant, piercing, very high-pitched *psssst;* song a jumble of high-pitched, thin, slurred whistles and staccato notes, like that of Yellow-throated Brush-Finch but more patterned, prolonged and forceful.

Nest. A bulky cup of coarse herbaceous stems, dead leaves and twigs lined with rootlets or horsehairs, 2–8ft (0.6–2.5m) up in a shrub or sapling in dense forest or brushy ravine. Eggs 2, glossy, immaculate white or pale blue. April–August.

Status. Common resident of wet middle elevations on both slopes, from ca. 8200ft (2500m), rarely higher, down to 4000ft (1200m) on Pacific and 3000ft (900m) on Caribbean.

Range. C Mexico to SE Peru.

BLACK-HEADED BRUSH-FINCH

Atlapetes atricapillus Pl. 48(14)

Saltón Cabecinegro

Description. 7½″ (19cm); 45g. Large, robust; white throat and breast contrast with black cheeks and gray sides; Black-striped Sparrow has less contrast between grayish cheeks and sides, grayer throat and breast. **Adults:** crown broadly striped with slate-gray and black; sides of head black; rest of upperparts olive-green, darker on wings and tail; bend of wings yellow; below mostly white; sides slate-gray; flanks and crissum olive-green. Bill black; iris chestnut; legs blackish-brown. **Young:** upperparts, including head, greenish-olive; cheeks and crown-stripes sooty-black and less distinct than in adults; below gray, shading to dull buff on belly.

Habits. Hops over ground in deep shade of humid second growth thickets and woods, sometimes venturing a short way into primary forest or shady plantations; flicks aside fallen leaves with bill to expose insects, spiders and other invertebrates; rises into shrubs and low trees to pluck berries or rummage in tangled foliage; puffs throat in excitement; paired throughout year.

Voice. Call a thin, piercing *tsst* or *tseet*, or a shorter *tsit*, several of which may accelerate into a sharp twitter; song, delivered from low perch in thicket, chiefly at dawn, a rapidly flowing medley of high, thin, piercing whistles, twitters and trills, more varied than that of Chestnut-capped Brush-Finch but similar in tone; females sing in a weaker voice.

Nest. A bulky cup, chiefly of many dry leaves, on a foundation of coarse, dry weed stems or vines, lined with fine rootlets, fungal filaments, or slender rachises of compound leaves, 4–20ft (1.2–6m) up in a bush or tangle of vines. Eggs 2, immaculate white or faintly blue-tinged. April–August.

Status. Resident in foothills, at ca. 1000–4000ft (300–1200m), on S Pacific slope, N to above Carara; fairly common in General–Térraba–Coto Brus region; rare and local elsewhere.

Range. Costa Rica to N Colombia.

Note. Often merged with several South American forms into *A. torquatus* and called Striped or Stripe-headed Brush-Finch.

SOOTY-FACED FINCH

Lysurus crassirostris Pl. 48(21)

Pinzón Barranquero

Description. 6¼″ (16cm); 33g. Medium-sized, dark, chunky, with contrasting yellow belly and indistinct white malar stripe. **Adults:** pileum rufous-chestnut; rest of upperparts, breast, sides, flanks, and crissum dark olive-green; face and throat blackish-slate; bases of feathers of malar area and sometimes throat white; center of belly yellow. Upper mandible black, lower silvery; legs dusky. **Young:** above rich olive-brown, slightly paler on pileum; face and throat sooty-olive; indistinct paler olive malar stripe; breast, sides and flanks brownish-olive; center of belly dull olive-yellow, smudged or scaled with blackish.

Habits. Prefers dense, dank undergrowth in deep ravines and stream bottoms in or beside wet forest in hilly terrain; in pairs or family groups; hops heavily over ground, continually flicking wings and twitching tail, picking up insects, spiders, fallen fruits, seeds, and grit; gleans and plucks berries in dense understory; sometimes follows army ants for insects, or briefly joins mixed flocks of small understory birds.

Voice. Usual call a high, thin, sibilant, emphatic *pseee-PSEET* or *pssss-PSSSST* with the second syllable higher; sometimes only 1 or up to 4 notes; song a high, thin, whistled *see see seeya sue seeya sue sisi see*, sometimes prolonged, especially at dawn.

Nest. Unknown (?). Fledglings in July and August.

Status. Locally common resident of Caribbean slope from Cordillera de Tilarán S to Panama, ca. 2000–5000ft (600–1500m), sometimes to 6000ft (1850m), especially in SE; on S Pacific slope, reported from Dota region.

Range. Costa Rica to E Panama.
Note. Sometimes lumped with *L. castaneiceps* of South America.

ORANGE-BILLED SPARROW

Arremon aurantiirostris Pl. 48(17)
Pinzón Piquinaranja

Description. 6″ (15.5cm); 35g. Bright orange bill and white throat present sharp contrast with dark upperparts; chunky and terrestrial. **Adults:** chin and most of head black; superciliaries white; median crown-stripe and sides of neck gray; rest of upperparts olive-green, darker on wings and tail; shoulder yellow; below mostly white; a broad black band across breast; sides gray; flanks and crissum brownish-gray to olive-green. Legs flesh-color to dark brown. **Immatures:** like adults but bill black, changing to orange in first year. **Young:** upperparts and breast sooty-olive; throat and superciliaries slightly paler; belly dingy buffy-olive, darker and browner on flanks and crissum; bill black with yellow base and rictus.

Habits. Frequents dark understory of humid forests and tall second growth woods; in pairs and family groups; hops over ground, sometimes scratching, in search of insects, spiders, seeds, and fallen berries; occasionally rises to low branch to glean an insect or pluck a berry; rarely attends army ants for insects; sings from low perch in thicket, log, or ground; seldom ascends more than 10ft (3m).

Voice. Call a sharp, sibilant *tchip* or *tsuk*; a short sputter in excitement. Song on Caribbean slope alternates staccato notes and high, thin whistles: *ts'seeew ts'seeew ts' seeew seer* or *t'seeeer t'seeeew-tsup*; on Pacific slope a tinkling series of high-pitched, thin, metallic or squeaky notes, alternately rising and falling. Females sing in weaker voices, sometimes in nest.

Nest. A very bulky covered structure with a side entrance, of dead leaves and other coarse materials, usually including many pieces of green living ferns or *Selaginella*, with more strewn in front of the doorway, well lined with rootlets and fibers; on the ground, often on a low bank, gentle slope, or mound. Eggs 2, glossy white, sparsely speckled with shades of brown and sometimes black, mostly at large end. April–August.

Status. Common resident throughout Caribbean lowlands, to 2600ft (800m) locally in foothills; on S Pacific slope common in lowlands and foothills, to 4000ft (1200m) locally, N to Carara; local in gallery forests N to Puntarenas; reaches Pacific side of N cordilleras through low passes.

Range. S Mexico to NW Ecuador.

OLIVE SPARROW

Arremonops rufivirgatus Pl. 50(15)
Pinzón Aceitunado

Description. 5¼″ (13.5cm); 25g. Only sparrow in interior of dry forest; combination of striped head and plain back diagnostic. **Adults:** head grayish-buff, brightest on cheeks, broadly striped with blackish-brown; rest of upperparts olive-green, tinged with gray on back; throat pale, dull buff; breast, sides, and flanks grayish-buff; belly dull white; crissum pale buff. Upper mandible dusky, lower pale horn with dusky tip; legs pale flesh. **Young:** above browner, indistinctly streaked with dusky; trace of buffy wing-bars; head stripes indistinct; breast and sides streaked with dark brown.

Habits. Frequents dense shrubbery in understory of deciduous woods, evergreen gallery forest, and adjacent tall second growth; usually singly or in pairs, hopping on ground or through dense tangles, gleaning insects, small spiders, and seeds in leaf litter, pushing aside leaves with bill but rarely scratching; gleans from low vegetation; occasionally eats fruits; often cocks or flicks tail. Males sing from upper understory, sometimes from lower canopy.

Voice. Call a thin *tsip*, often rapidly repeated; song an accelerating series of thin, sibilant *tsew* notes that usually drops in pitch toward end, sometimes includes an opening motif of sharp, clear whistles or staccato notes.

Nest. A globular structure with side entrance, of rootlets and grasses, roofed with woven grasses, on ground or log, amid dense undergrowth in dry or gallery forest. Eggs 2–3, white. April–July.

Status. Common resident of Pacific lowlands and foothills, from N Guanacaste S to Río Tárcoles and across Valle Central at least as far as Grecia and Santa Ana, chiefly in wooded river bottoms; from sea level to ca. 2600ft (800m) on approaches to Cordillera de Guanacaste, and to 3000ft (900m) on Península de Nicoya and Valle Central.

Range. NW Mexico and S Texas to Costa Rica.

Note. Sometimes the W Mexican and Costa Rican birds are classified as *A. superciliosus*, Pacific Sparrow, a species distinct from *A. rufivirgatus*.

BLACK-STRIPED SPARROW
Arremonops conirostris Pl. 50(14)
Pinzón Cabecilistado

Description. 6½″ (16.5cm); 37.5g. No other open-country finch of humid areas has boldly striped head and plain back; no overlap with smaller Olive Sparrow. **Adults:** head gray, a broad black stripe on each side of crown and a narrower one through eye; rest of upperparts olive-green; bend of wing yellow; throat and belly dull white, shading to pale gray on sides of neck, breast, and sides; flanks and crissum olive. Iris reddish-brown; bill black; legs grayish-flesh to dusky. **Young:** head yellowish-olive, striped with dark brown and narrowly streaked with dusky; rest of upperparts brownish-olive streaked with dusky; below yellowish-olive washed with brown on breast and flanks and streaked with dark brown; bill with yellow at base. Adult plumage attained when 4–5 months old.

Habits. Frequents low thickets, weedy fields, young second growth, shady plantations and gardens, woodland edges in humid regions; in pairs or family groups at all seasons, never flocking; hops over ground seeking small invertebrates and seeds; gleans insects from low foliage and regularly ascends into shrubs and low trees for berries; often sings from well up in dense, small trees or shrubs; shy and retiring.

Voice. Calls include a metallic to nasal *churk, tsook* or *churip*. Mates greet each other with series of whining notes, inflected upward or downward and uttered in unison, preceded by clear, thin whistles or slurred notes: *seee-seee-seee churry-churry-churry-chur* or *wot-woit-whoit-shaip whaip whaip*. Song a striking performance lasting 10sec or more, a long series of alternating high and low throaty notes that gradually speeds up, suddenly breaking into an accelerating trill that ends abruptly, often with a sharp note.

Nest. A bulky domed structure of coarse vegetable materials, with a wide side entrance, in dense vegetation, usually 6″–3ft (15–90cm) up, rarely on ground or 6ft (1.8m) above it. Eggs 2, rarely 3, immaculate white. February–July; 2 broods.

Status. Common resident of lowlands and foothills of entire Caribbean slope and S Pacific slope, N to Carara in coastal lowlands; reaches Pacific slope along N cordilleras through low passes but absent from dry NW; lowlands to 5000ft (1500m) locally, rarely higher; may be seasonal breeding resident at higher elevations.

Range. N Honduras to W Ecuador, N Brazil, and Venezuela.

PREVOST'S GROUND-SPARROW
Melozone biarcuatum Pl. 48(18)
Pinzón Cafetalero (Mercenario, Rey de Comemaiz)

Description. 6″ (15cm); 28g. A chunky, unstreaked sparrow with a striking facial pattern. **Adults:** forehead and cheeks black; lores, eye-ring, and postocular spot white; rest of head chestnut-rufous; sides of neck slate; rest of upperparts olive-brown, tinged with tawny on edges of remiges; below mostly white, with black malar streak and large breast-spot; lower belly and flanks buffy-brown to grayish-brown; crissum buff. Iris reddish-brown; bill black; legs horn-color. **Young:** above brighter brown, streaked with blackish; rufous of head and facial pattern dull and indistinct, obscured by sooty streaking and mottling; below yellowish-white with indistinct dusky streaking and breast-spot.

Habits. Frequents undergrowth and thickets of scrubby to tall second growth, patchy woodland, coffee plantations, and hedgerows; forages mainly on ground, hopping about gleaning, sometimes double-scratching in litter for seeds and small insects; shy and retiring, but ventures out of thick cover to forage in early morning; paired year-round, but male sings only in wet season from perch in dense vegetation.

Voice. Calls include weak, high, staccato *tsit*ting and chitters; a greeting duet of sharp, whistled notes: *pseee psee psee*; a sharp, thin *pseeer* when excited or aggressive. Song consists of a high, thin, staccato sputter, a buzzy note, and/or a slow trill followed by 3–4 clear, forceful whistles: *Pst-t-t-t peee-peer-peer, bzeew whee-whee-whee*, or *bzeew tow-hewhewhewhe peer-peer-peer-peer*, less loud and explosive than song of White-eared Ground-Sparrow; often heard at dawn in breeding season.

Nest. A sturdy cup of grass, rootlets, stems, and dead leaves, lined with fine grasses, 1–6ft (0.25–2m) up in large grass tussock or dense bush. Eggs 2–3, white, heavily spotted and blotched with reddish-brown to dull red, especially on large end. May–September.

Status. Common resident in Valle Central, roughly from San Ramón and Atenas E across continental divide to Turrialba in Reventazón drainage, but rare E of Cartago, at 2000–5200ft (600–1600m).

Range. S Mexico to W Honduras; Costa Rica.

Note. Also called White-faced Ground-Sparrow. The Costa Rican population, separated from those of N Central America by a wide gap, is quite different in appearance and probably represents a distinct species (or semispecies), *M. cabanisi*.

WHITE-EARED GROUND-SPARROW

Melozone leucotis Pl. 48(19)

Pinzón Orejiblanco (Cuatro Ojos)

Description. 7″ (17.5cm); 43g. Large, chunky, big-footed, with black foreparts, striking white and yellow face pattern. **Adults:** forehead and sides of head mostly black; crown slaty, shading to olive-green on hindneck; lores, broken eye-ring, spot on auriculars white; narrow postocular streak and broad bar down sides of neck bright yellow; rest of upperparts olive-brown, washed with gray on back; wings edged with olive-green; throat black, narrowly separated from large black breast patch by a narrow band of gray and rufous; belly white; sides gray; flanks and crissum buffy-brown. Iris reddish-brown; bill black; legs horn-color. **Young:** black of head and breast duller and less extensive; facial markings indistinct, suffused with olive; back scaled with sooty; belly dull yellowish, streaked and scaled with dusky.

Habits. Prefers dark thickets in moist ravines, forest edge and patchy woodland, dense second growth; usually stays on or near ground (though males may sing from upper understory); hops bouncily, continually flicking tail up slightly, dropping it slowly; forages by gleaning insects, spiders, other arthropods, and seeds from leaf litter, often exposing food by powerful double-scratching. Flight heavy, fluttery, not prolonged. Paired year-round.

Voice. Call a high, piercing *tsip* or *tzip*; characteristic greeting duet a buzzy, sibilant, accelerating and descending *pseeeer-zeeeer-zeeer-zeeer-zur*; song of explosive short phrases of staccato notes and loud, often rather hoarse or penetrating whistles: *spit-CHUR see-see-see, PSEET-seecha seecha seecha*, etc.

Nest. A massive bowl of narrow dead leaves and stems, stout petioles, thin twigs, lined with finer or more tightly interlaced materials, including twigs of *Casuarina*; well hidden among banana plants, orchids, etc., on ground or 28″ (71cm) or more up. Eggs 2, white, streaked and spotted with pale brown or cinnamon. April–July (Winnett-Murray, in part).

Status. Locally uncommon to fairly common resident of Pacific slopes of Cordillera de Tilarán and Cordillera Central, including upper Valle Central, reaching Caribbean slope locally in Cartago area and upper Reventazón valley, Bajo La Hondura (to Carrillo, at least formerly), and above San Ramón; ca. 3300ft (1000m) to 5000ft (1500m) on Cordillera de Tilarán and 6500ft (2000m) on Pacific slope above Valle Central, locally down to ca. 1600ft (500m) on Caribbean side.

Range. S Mexico to Costa Rica.

SAVANNAH SPARROW

Passerculus sandwichensis Not Illustrated

Sabanero Zanjero

Description. 5¼″ (13.5cm); 20g. Heavily streaked, with short, notched tail. Crown blackish with pale median stripe; rest of upperparts grayish brown, streaked with black and buffy-white; wings and tail dusky, edged with pale brown; pale superciliaries becoming yellowish above lores; cheeks brownish, set off by dusky stripes below and through eye; dark malar streak; below white, breast and sides heavily streaked with blackish; outer rectrices often noticeably pale. Bill flesh-color with dusky culmen; legs pale flesh. Lincoln's Sparrow much buffier on breast, with finer streaking and gray head; young Rufous-collared Sparrow much browner, no yellowish on superciliaries; Grasshopper Sparrow lacks streaks below.

Habits. Feeds on ground in open, grassy or brushy sites, taking mostly small seeds of grasses; often difficult to flush; flies a short distance and drops back into grass but may respond to squeaking.

Voice. Call a high, thin *tsit* or "ticking."

Status. Accidental. One sighting (13 April 1963) on Cocos I. by Slud. An occasional individual may overshoot its normal wintering range and reach our area; most likely to appear in open, brushy pastures or savannas. Specimen confirmation desirable.

Range. Breeds from Alaska and N Canada to SW Guatemala; normally winters to N Honduras; accidental on Cocos I.

GRASSHOPPER SPARROW

Ammodramus savannarum Pl. 50(8)

Sabanero Colicorto

Description. 4½″ (11.5cm); 17.5g. A rather small, buffy, flat-headed, short-tailed sparrow of open grassy areas. **Adults:** above blackish-brown, spotted with gray and scaled and streaked with buff; rufous flecking on nape, scapulars, and rump; a median buffy crown-stripe; lores and bend of wing yellow; face, throat, breast and sides bright buff; belly immaculate whitish. **Young:** above browner, more scaled with buff; below paler, streaked with dark brown on breast. Wintering northern birds similar but gray and rufous areas more extensive on upperparts. See Botteri's and Lincoln's sparrows.

Habits. Prefers dry, open savanna grasslands, often with scattered volcanic boulders; in Valle Central frequents wet, brushy fields and pastures; mostly territorial; walks on ground, nervously jerking tail and flicking wings, or runs mouselike with head low; when flushed, flies flutteringly and soon dives into cover; eats mainly grass seeds and some small insects. Male sings from boulder, small shrub or grass stem.

Voice. Calls include high, thin, tinkling, whistled notes; when alarmed, a piercing *tseee* or *tweeet*. Song high, thin, and insect-like, with several wiry, buzzy trills on different pitches, interspersed with staccato notes: *tsk-tsk sweeeeezeeee tik-tik zweeeee tik tiptip sweeeee*, etc.

Nest. A cup of grasses, lined with rootlets and hairs, on ground in or under a clump of grass or low herbs. Eggs 4–5, white, speckled with reddish-brown around thick end (in North America; no Costa Rican record). Probably breeds April or May–July (song and gonad data).

Status. Fairly common resident in NE Guanacaste, in grasslands at base of Cordillera de Guanacaste, at 330–2600ft (100–800m), occasionally higher; a winter resident in small numbers in Valle Central, especially near Cartago and (formerly?) above San José.

Range. Breeds from S Canada to NW Mexico, locally from C Mexico to W Ecuador, and in West Indies. Breeding birds of North America winter from S USA to Costa Rica.

STRIPED-HEADED SPARROW

Aimophila ruficauda Pl. 50(7)

Sabanero Cabecilistado (Albarda Nueva, Ratoncillo)

Description. 7″ (18cm); 35g. Large sparrow with conspicuously striped head, stout bill, and long, cinnamon tail. **Adults:** top and sides of head black, with a broad white stripe over center of crown and above each eye; a narrow gray nuchal collar streaked with dusky; back pale brown, streaked with black; rump pale buffy-brown; tail dull cinnamon; shoulders rufous; wing-coverts edged with buffy and remiges with grayish-buff; below mostly white; breast gray, more or less mixed with white; flanks and crissum buffy. Upper mandible black, lower mandible and legs pale flesh. **Young:** head pattern of buffy and blackish stripes obscured by dusky and brown streaking; above paler, more buffy and less heavily streaked than adult; tail pale rufous; throat and belly duller white, throat speckled with dusky; breast dull buff, streaked with dark brown.

Habits. Frequents brushy savanna, second growth and edges of deciduous woods; in family groups or small, straggling flocks of 3–7; forages mainly on open ground, moving in springy hops, picking up grass seeds and occasional small insects and spiders; flies low, flutteringly, with long tail held low; sings high in tree or atop a tall cactus; sleeps in compact groups on branch well up in small tree.

Voice. A group keeps up a continual sputtering, mouselike jumble of high, squeaky notes, some sibilant, others nasal, that rises to a crescendo as group members gather in a low bush when alarmed. Male sings a long series of dry, thin notes, sometimes accelerated to a rapid, chaffy trill; mates greet each other with a chattering duet.

Nest. A rather deep cup of coarse grasses and fine twigs, lined with finer grasses and horsehair, 1–4ft (0.3–1.2m) up in dense, often spiny shrub, usually amid open grassy area. Eggs 2–3, pale blue, unmarked. July–September or October. Cooperative groups attend a nest.

Status. Common resident of dry N Pacific lowlands, to 2600ft (800m) on slopes of Cordillera de Guanacaste and hills on Península de Nicoya; S to between Puntarenas and Orotina.

Range. N-C Mexico to Costa Rica.

RUSTY SPARROW

Aimophila rufescens Pl. 50(4)

Sabanero Rojizo

Description. 7¼″ (18.5cm); 36g. Large, with rufous cap, conspicuous malar stripe. **Adults:** pileum chestnut-rufous with indistinct dull buffy or grayish median stripe and bordered with black; back brown, heavily streaked with black; rump brown; wings and tail dusky with dull rufous edgings; superciliaries pale gray, whitish anteriorly; lores and postocular streak dusky; dull whitish eye-ring; cheeks olive-gray; sides of neck grayish; throat white with black malar stripes; breast, sides, and flanks pale grayish-brown; belly buffy-white; crissum buffy. Upper mandible black, lower gray; legs brown. **Young:** pileum duller, streaked with black, with broader pale central stripe; back more narrowly streaked; below tinged with yellow, breast and sides heavily streaked with dusky.

HABITS. Only in windswept grassland with scattered small trees and patches of dense shrubbery on lower Pacific slopes of Guanacaste volcanoes; in pairs or small family groups after nesting; forages mostly by hopping over ground, picking up seeds and small invertebrates; deliberately gleans insects in low foliage; flies flutteringly, with pumping tail.

VOICE. A dry, rattling *churr* in alarm or excitement; song an explosive, throaty *chirup chirup chirup see-bore*, with variations. Mates greet each other with a series of squealing notes, often combined with *churr*s.

NEST. A cup of grasses, on ground or up to 8ft (2.4m) in a bush. Eggs 2–3, bluish-white, unmarked. May–July (in Mexico; no Costa Rican record).

STATUS. Uncommon and local resident at ca. 2000–3000ft (600–900m) along Pacific side of Cordillera de Guanacaste.

RANGE. NW Mexico to Costa Rica.

BOTTERI'S SPARROW

Aimophila botterii Pl. 50(5)

Sabanero Pechianteado

DESCRIPTION. 5½″ (14cm); 20g. Nondescript; above grayer and more streaked, below duller than Grasshopper Sparrow, with short, narrow, blackish malar streak. **Adults:** above slate-gray, with heavy, blotchy black and chestnut streaking; wings and tail dusky, with grayish-buff to brownish edgings; postocular stripe dark chestnut mixed with black; cheeks mottled with black; below pale, dull grayish-buff, paler on throat, shading to whitish on belly. Upper mandible dark brown, lower brownish-horn; legs pale brownish. **Young:** above blackish-brown, broadly scaled with buff; below streaked with blackish, lightly on throat, heavily on breast and sides.

HABITS. Frequents open, barren, boulder-strewn grasslands; forages on ground, hopping unobtrusively under grass tussocks, running across openings, taking seeds and small insects; in pairs; often cocks tail; flies flutteringly, jerkily, low over ground; usually shy and wary.

VOICE. Call a sharp *tsit* or *swip*, strong and rasping or weak and thin; song several pairs of short double notes, followed by a trill and sometimes 1–2 final whistled notes; or a prolonged, unstructured medley of chipping and chirping notes and phrases (Wolf).

NEST. A compact, deep, rather frail cup of dry leaves and stems of grasses, unlined or lined with finer grasses, on ground or low in clump of grass. Eggs 3–4, white. May–September (in Texas and Mexico; no Costa Rican record).

STATUS. Rare and local resident of lower Pacific slope of Cordillera de Guanacaste at ca. 1300–3600ft (400–1100m), S to Volcán Miravalles.

RANGE. SW USA to Costa Rica.

VOLCANO JUNCO

Junco vulcani Pl. 50(11)

Junco Paramero

DESCRIPTION. 6.25in (16cm); 28g. Fairly large, chunky sparrow with conspicuous pale eye and bill, dull gray underparts. **Adults:** upperparts dull brown, streaked with black, lightly on crown and heavily on back; wing-coverts edged with tawny-brown, each greater covert with a small white spot at tip; rectrices and remiges edged with grayish-brown; lores and below eye blackish; cheeks brown, mottled with black; superciliaries slate-gray; sides of neck and underparts dull gray, with brownish tinge strongest on flanks; belly tinged with buff. Iris bright yellow; legs and bill flesh-color, culmen sometimes tinged with dusky. **Young:** above brighter brown with heavier black streaking; below more buffy, shading to white on belly; throat and breast with short dusky streaks.

HABITS. In open, grassy or brushy areas of páramo, stunted scrub produced by volcanic eruptions, second growth, overgrown pastures, and roadsides at very high elevations; usually in pairs, sometimes alone or in family group; forages on ground, hopping and running, picking up seeds, fallen berries, small insects, and spiders; flies heavily and flutteringly, rarely far.

VOICE. Among its varied calls are a high, thin, *tsee* or *tsee-tsee*; a scolding, clear, or scratchy *whew* or *jeew*; a repeated *tchup* or *tchip* when excited, and a fast, rhythmical *cher-we cher-we cher-we*. . . . Song consists of short, choppy phrases of squeaky, burbling, or buzzy notes: *k'cheew chu k'wee, chip-chip ts'chew tsi'weet*, etc.

NEST. A well-made cup of grass stems lined with animal hair and thistle floss, sometimes decorated on outside with moss, on ground under rock, log or bush, or in a niche in a mossy bank. Eggs 2, pale blue, with a few brown or lilac dots near large end. March–July.

Status. Fairly common resident above timberline on Irazú-Turrialba massif of Cordillera Central and on all major peaks of Cordillera de Talamanca, mostly above 10,000ft (3000m); an isolated population at ca. 6900ft (2100m) in Sabana Dúrika, an extensive open bog on Caribbean slope of this range. On Cerro de la Muerte, forest clearing has allowed it to expand its range downward, locally to 8500ft (2600m).

Range. Costa Rica and W Panama.

CHIPPING SPARROW

Spizella passerina Pl. 50(9)

Chimbito Común

Description. 5″ (12.5cm); 13g. Smaller and slimmer than Rufous-collared Sparrow, with rufous crown, no crest, no black on chest. **Adults:** forehead black; crown rufous, streaked with black posteriorly and with narrow gray central streak; black stripe through eye; superciliaries white; short, faint, dark malar streak; breast and sides of head pale gray, shading to white on throat and belly; cheeks and flanks washed with brownish; back buffy-brown, streaked with black; rump gray; wing and tail feathers dusky, edged with brown to rufous; narrow wing-bars white. Bill horn-color, shading to blackish on culmen; legs pale flesh.

Habits. Generally prefers open, grassy woodlands and savannas; the single bird recorded in Costa Rica was accompanying a foraging flock of Rufous-collared Sparrows in a brushy overgrown field, feeding on grass seeds.

Voice. A weak *tsip*, softer than note of Rufous-collared Sparrow.

Status. Accidental; one record for E Valle Central, San Pedro at 4100ft (1250m), 9–13 September 1977; probably a northern bird that strayed during fall migration as the nearest breeding population in N Nicaragua is apparently nonmigratory.

Range. Breeds from Alaska and N Canada to N Nicaragua; northern populations winter S to S Mexico; accidental in Costa Rica. Central American populations sedentary.

RUFOUS-COLLARED SPARROW

Zonotrichia capensis Pl. 50(13)

Chingolo (Comemaiz, Pirrís)

Description. 5¼″ (13.5cm); 20g. Jaunty, medium-sized with a short crest, black bow tie on chest, broad rufous nuchal collar. **Adults:** head pale gray with a broad, black stripe on each side of crown, narrower stripes through eye and below cheeks; broken white eye-ring; back and sides of neck and sides of breast rufous; back and edgings of wings and tail buffy-brown, back heavily streaked with black; rump olive-brown; 2 narrow whitish wing-bars; throat white; black patch on chest narrow at center, broad at sides; sides and flanks brownish; rest of underparts dull white. Upper mandible dusky-horn, lower horn-color with dusky tip; legs pale flesh. **Young:** head dull buff with dark brown stripes; median crown-stripe streaked with dusky; no rufous collar; throat spotted with dusky; breast and sides buffy, streaked narrowly with dusky.

Habits. Frequents urban, suburban, and country lawns, dooryards, parks, gardens, agricultural fields, and scrubby second growth; breeding pairs hold small territories throughout year; nonterritorial birds live furtively in the same areas or form flocks that wander locally; forages chiefly on ground, hopping about, gathering grass and weed seeds, fallen grain, and insects; sometimes makes short, fluttering sallies for low-flying insects; males sing from elevated perches.

Voice. Calls include a sharp, warblerlike *chip*, a short sputter; in alarm a high, thin *tsit*ting. Song of 1 to several long, clear, slurred whistles and a trill: *seee-seeur te-e-e-e-e, seeeeur tse-e-e-e-e-e*; in some areas the trill is reduced to a few notes or is absent, the song then being 3–5 or 6 clear, somewhat plaintive whistles: *weeer heer too, too, too*, the last notes in a falling voice.

Nest. A neat and compact to bulky cup of grasses, rootlets, etc., lined with finer grasses, sometimes animal hair, in a niche in a wall or bank, on ground amid matted vegetation or 1–6ft, rarely 16ft (0.3–5m) up in a dense shrub or small tree, sometimes in open shed. Eggs 2–3, pale greenish-blue, everywhere densely blotched and speckled with shades of brown. Virtually throughout year; in some rural areas possibly with peaks February–April and June–August.

Status. Abundant resident throughout highlands at 2000–10,000ft (600–3000m), locally down to 1300ft (400m) on Caribbean slope; occasionally appears in lowlands (e.g., Puerto Limón).

Range. Extreme SE Mexico to Tierra del Fuego; Hispaniola.

LINCOLN'S SPARROW

Melospiza lincolnii Pl. 50(12)

Sabanero de Lincoln

Description. 5¼″ (13.5cm); 16g. Slimmer

than Rufous-collared Sparrow, without crest. Above grayish-brown, streaked with black; bright brown crown-stripe and wings; pale wing-bars; sides of head and neck gray; narrow whitish eye-ring; cheek outlined in dusky; breast, sides and flanks buffy, streaked finely with blackish, the streaks sometimes coalescing into a central spot; throat and belly white. Upper mandible dusky, lower brownish; legs brownish-flesh. Young Rufous-collared Sparrow browner (no gray on head), crested; below more broadly streaked with brown. Grasshopper Sparrow lacks gray on head and streaks on breast; has yellow lores.

Habits. Prefers brushy fields, either dry or marshy; seen singly, sometimes with Rufous-collared Sparrows; forages on ground under grass and brush for seeds and small insects.

Voice. Call a low *tsup* or *zut*.

Status. Casual to rare fall migrant and winter resident (mid-November–late February) in highlands (Monteverde in Cordillera de Tilarán; vicinity of Cartago and San Pedro in upper Valle Central); to be expected in fall along Caribbean coast, as a bird was banded and released in late October in W Bocas del Toro, Panama.

Range. Breeds from Alaska and N Canada to W and NE USA; winters regularly S to Honduras, casually to C Panama.

WEDGE-TAILED GRASS-FINCH

Emberizoides herbicola Pl. 50(10)

Sabanero Coludo (Chicharrón)

Description. 7″ (18cm); 22g. Sparrowlike, with bicolored bill and long, graduated tail, the rectrices with greatly attenuated tips somewhat like those of a spinetail (*Synallaxis*). **Adults:** above buffy-brown, broadly streaked with brownish-black; wing-coverts and primaries edged with bright olive-green, secondaries and rectrices with dull buff; bend of wing yellow; broken white eye-ring; superciliaries whitish; cheeks brownish; throat and belly white; breast, sides, flanks, and crissum buffy, sides and flanks sparsely streaked with dusky. Upper mandible mostly black, tomia and lower mandible bright yellow; legs pale brown to flesh-color. **Young:** above more richly colored, streaked with cinnamon-buff and blackish; superciliaries and underparts largely pale yellow; breast and sides heavily streaked with brownish-black.

Habits. Frequents grassy savannas with scattered trees and shrubs, especially where grass is tall and not heavily grazed or burned; in pairs and family groups; forages on ground for insects and seeds; when alarmed, flushes abruptly, flies low over vegetation with tail pumping, and dives into concealment after a short distance; when excited perches erectly, switches tail about like a gnatcatcher, flicks wings; rises to low bush or swaying grass stem to sing.

Voice. Calls include a sharp, dry, metallic *spit* or *plit*, repeated over and over when agitated, and a soft, low *chip* or *chrip*. Song a short, decelerating "chippy" trill of 4–6 notes: *chidididideer*.

Nest. A cup of dried grasses and a few rootlets, from ground level to ca. 1ft (30cm) up amid grasses. Eggs 2, white, with a few blackish spots on thicker end. May–September (gonad data and presence of juveniles).

Status. Uncommon and very local resident in Térraba region of S Pacific slope at 500–1500ft (150–450m).

Range. Costa Rica to Bolivia and NE Argentina

FAMILY **Fringillidae:** Siskins, Canaries, Finches, and Allies

With the removal of the 9-primaried Emberizidae, long considered to belong to this family, the Fringillidae now comprises about 125 species, all with 10 primaries, to which should probably be added the 30 or so species, living or extinct, of Hawaiian honeycreepers. Represented on all continents except Australia, this family is much more diverse in the Old World than in the New, where, however, it is found from the high Arctic to Tierra del Fuego and in the West Indies. Only the genus *Carduelis* is widespread in the New World tropics, including Costa Rica. Fringillids are ca. 3.5–10″ (9–25cm) in size and are diversely attired, with males often much more colorful than females. Primarily eaters of seeds, they have bills that are short and conical, in some species thick and powerful for cracking hard seed coats, in crossbills with twisted, overlapping mandibles for extracting seeds from pine cones. Many species are excellent songsters, and nearly all are gregarious for most of the year. Nests of fringillids are usually well-made compact cups, of various soft materials, situated in a shrub or tree, rarely on the ground. Tropical species lay 2–4 eggs, northern species up to 7, that may be white, greenish, blue, olive, or brownish, unmarked or spotted or streaked with brown or rufous. The female, who commonly builds the nest without help, usually incubates alone and often is fed so liberally by her mate that she is able to sit continuously for long intervals. Incubation lasts 10–14 days. Fed by regurgitation by both parents, the young of most finches remain in the nest for 10–15 days, but crossbills sometimes stay as long as 24 days. Unlike nearly all passerine birds, parent finches, careful at first, frequently neglect nest sanitation as their nestlings grow older. A heavy white deposit on the rim of a recently abandoned, cup-shaped nest is an almost certain sign that it belonged to a member of this family.

LESSER GOLDFINCH

Carduelis psaltria Pl. 50(2)

Jilguero Menor (Mozotillo de Charral)

DESCRIPTION. 4″ (10cm); 10g. Small, with rather thick, conical, sharp-pointed bill, white patches in wings and tail. **Adult ♂:** above glossy blue-black, including sides of head and neck; bases of inner primaries, and all but tips of inner webs of outer 2 rectrices, white; tertials spotted with yellowish-white at tips; entire underparts bright yellow. Bill yellowish, shading to black on culmen; legs dusky. ♀: above dull olive-green, brighter on head; wings and tail dusky, the white areas smaller than in male; below bright yellow, washed with olive on throat, chest, and flanks. Bill and legs more brownish than in male. **Immature ♂:** like adult but retains juvenile primaries, and upperparts mixed with olive through first breeding season. **Young:** like adult female but above more brownish, little white in primaries or rectrices; wing-bars buffy-white; tertials tipped with grayish-white; throat, breast, sides and flanks yellowish-buff; belly and crissum pale yellow.

HABITS. Prefers open country in highlands, including scrubby second growth, weedy fields, plantations, and clearings with scattered trees; usually in small flocks of 10 or fewer that fly undulatingly, high over the ground, alighting in treetops where several males often sing together and descending to forage for grass seeds on or near the ground, when often shy and wary; in coffee plantations often feeds on orange flowers of *Grevillea* shade trees.

VOICE. Calls include a high, thin, 2-toned, plaintive *pseee*; a clear, descending *tseer*; an ascending, slightly burry *chirreee*; a high, thin *peeyooo*. Song a lively, varied, prolonged outpouring of clear notes, twitters, and trills, rising and falling in pitch.

NEST. A compact, thick-walled cup of grasses, decaying herbaceous stems, rootlets, and fibers, lined with finer materials, 10–20ft (3–6m) up in tree or shrub beside road or in scrubby field. Eggs 3 or 4, white, unmarked. August–January.

STATUS. Uncommon, nomadic resident in upper parts of Valle Central and Reventazón drainage, especially on slopes of Volcán Irazú and Volcán Turrialba, also along Pacific slope of Cordillera de Talamanca, including Valle del General and San Vito area, at 2800–7000ft (850–2150m), occasionally to 9000ft (2750m). Relentless persecution for cage-bird trade has greatly reduced its numbers in many areas.

RANGE. NW USA to Venezuela and NW Peru.

NOTE. Also known as Dark-backed Goldfinch; formerly placed in genus *Spinus*.

YELLOW-BELLIED SISKIN

Carduelis xanthogastra Pl. 50(1)

Jilguero Vientriamarillo (Mozotillo de Montaña)

DESCRIPTION. 4″ (10.5cm); 12g. Darker and with slimmer bill than Lesser Goldfinch; large yellow wing-patches, conspicuous in flight. **Adult ♂:** entire head, throat, and chest glossy black; bases of primaries and all but central rectrices and rest of underparts bright yellow except flanks largely dull black. Bill blackish; legs dusky. ♀: above dark, dull olive-green; wings and tail blackish, with yellow areas smaller; throat and chest olive-green, tinged with yellow, becoming olive-yellow posteriorly; flanks dull olive. Bill dusky-horn. **Young:** like adult female but above with dusky or sooty fringes, much less yellow in wing and almost none in tail; below faintly smudged or streaked with dull brown.

HABITS. Frequents canopy of mountain oak forests and adjacent clearings, sometimes second growth or pastures some distance from forest; usually seen singly, in pairs, or small groups but in remote districts still occurs in large flocks that settle in treetops, the males singing socially when not breeding; feeds on small insects, flowers of oaks, possibly berries high in trees, and seeds of grasses and composites in clearings.

VOICE. Calls include a high, piercing *pyee* and a harsher, scratchy *bziee*; song a rapid outpouring of twittering, chattering notes, a note or phrase often repeated several times in rapid succession, less musical and varied than that of Lesser Goldfinch.

NEST. A shallow, thick-walled cup of rootlets, fungal filaments, fibrous bark, and gray beard lichen (*Usnea*), covered on the outside with green moss and well lined with fine black fungal strands, 8–12ft (2.4–3.7m) up amid dense foliage of small tree in clearing. Eggs 2–3, white, tinged with green, unmarked or speckled. April–May (2 records).

Status. Resident in higher parts of Cordillera Central and Cordillera de Talamanca, wandering widely and seldom staying long in any one locality except when nesting; mostly 6600–10,000ft (2000–3000m), sometimes down to 6000ft (1850m). Intense persecution for cage-bird trade has made this species very uncommon to rare in Cordillera Central and northernmost Cordillera de Talamanca, but it is still common in some remote parts of this range.

Range. Costa Rica to SW Ecuador and C Bolivia.

Note. Formerly classified in the genus *Spinus*.

FAMILY **Passeridae:** Old World Sparrows

The 39 species in this family were formerly classified with the weavers in the Ploceidae. With the exception of 2 species introduced into the Americas, all are confined to the Old World, where they are best represented in Africa. Most are 4–8″ (10–20cm) in length and have short, rather thick bills. Lacking brilliant colors, they are mainly brown, gray, black, and white, in some species adorned with chestnut or yellow. The sexes look alike or different; the more ornate males of some species change plumage colors seasonally, as do the bills of males of many other species. Members of this family prefer open country, including savannas, scrub, deserts, high mountains, farms, and, notably in the case of the House Sparrow, cities and suburbs. They forage chiefly on the ground, gathering seeds and insects. They also eat buds and fruits and, in urban areas, bread and other scraps thrown out by people. A number have voices more melodious than that of the noisy, chirping, chattering House Sparrow. Old World sparrows tend to be gregarious or to live in closely integrated, cooperatively breeding groups. Their nests, made by both sexes, range from untidy accumulations of straws and feathers to roofed nests with side entrances and enormous communal ''apartment houses,'' the largest of all birds' nests. Their eggs, in sets of 2–7, are white, creamy, pinkish, greenish, or bluish, usually marked with darker colors. Incubated by both adults or by the female alone, they hatch in 12–14 days. The young, hatched with sparse down or naked, are commonly fed by both parents, in a number of species aided by nonbreeding helpers. They remain in the nest for 15–24 days.

HOUSE SPARROW

Passer domesticus Pl. 50(17)

Gorrión Común

Description. 5½″ (14cm); 26g. Larger bill, flatter head than Rufous-collared Sparrow; shorter legs. **Adult ♂:** pileum dull gray; broad postocular stripe and nape chestnut; lores and below eye black; back cinnamon-brown, streaked with black; rump grayish-brown; shoulders chestnut; middle wing-coverts broadly tipped with white, greater coverts tawny-brown; remiges edged with dull buff; tail dusky; cheeks white; chin, throat, and chest black; rest of underparts dull grayish. Bill black (all year in Costa Rica); legs pale brown. ♀: pileum grayish-brown; superciliaries pale grayish-buff; rest of upperparts and wings much grayer than male, without chestnut or cinnamon tones; wing-coverts edged with tawny-buff; rest of head, breast, and sides grayish-brown; belly dull whitish. Bill dark horn. **Immature ♂:** like adult female but crown darker, superciliaries brighter, with faint grayish bib. **Young:** like adult female but paler, below more buffy.

Habits. In towns and cities, especially around parks, markets, grain silos, and railway stations; gregarious, often breeding and traveling in groups; forages on ground, often in noisy, quarrelsome crowds, taking seeds, grain, bread and other scraps, and insects; sometimes sallies for flying insects, especially in early morning.

Voice. A loud, monotonous, often incessant *cheep* or *chirp*; alarm note a harsher *chirp*; male has a song of shorter, more rapid *chirps*.

Nest. A messy, hollow ball with a side entrance, of straw, often with bits of string, plastic, paper, other trash, and a few feathers in lining; in niche or crevice in building or mass of bromeliads, or among dead palm leaves. Eggs usually 3 (more at higher latitudes), dull white, speckled with brown. Mainly March–August but perhaps some nesting throughout year.

Status. Since invading Costa Rica from N in about 1974, it has steadily extended its range and is now established and still increasing in most cities and large towns, especially in drier regions. It often seems to invade new areas in flocks that establish compact colonies; as these increase, more and more birds nest as solitary pairs as well.

Range. Widespread in Old World; introduced in New World, resident in much of North, Central and South America and still spreading.

Birding in Costa Rica

A major reason why fieldwork in Costa Rica is so pleasant is the inherent friendliness and educational level of Costa Ricans. Many *ticos* have had sufficient contact with birders to view their mania with sympathy (if not complete understanding) and if addressed courteously will do their best to be helpful. Especially in the countryside, where traditional attitudes prevail, old-fashioned courtesy is all-important. Even if your Spanish is limited or nonexistent, sign language can work wonders as long as you remain smiling and friendly. As Costa Ricans themselves admit wryly, "everything in this country moves by sugar or fiscal stamps" (a reference to the government bureaucracy, which is every bit as exasperating to Costa Ricans as it will be to you, should you have to deal with it).

The list of birding areas in the next chapter is by no means exhaustive; one can find interesting birds in many a remnant patch of forest or lighter woodland, even in areas long denuded of extensive forest, such as the Valle Central. Many such areas are on private land, and before opening a gate or jumping a fence, you should always try to obtain permission to enter from the owner or foreman (*mandador*) of the property in question. If properly informed, people are nearly always cooperative and may even indicate the best birding areas, but they have every reason to be less than kind to trespassers. Some national parks (Volcán Poás, Santa Rosa, Cahuita, and Mañuel Antonio) are equipped to handle visitors at all times, but to visit other parks and reserves, you should first check with the Servicio de Parques Nacionales office (across from the Calderón Guardia Hospital in San José). Palo Verde and Tapantí wildlife refuges can also be visited without prior notice, but to see Caño Negro and Curú you need permission from the Departamento de Vida Silvestre, whose office is in the Dirección Forestal building on Avenida 2 in San José.

Most birders respect the property of others, but a few lamentable exceptions prompt us to offer the following suggestions. Do not damage fences, especially on private land, and close behind you any gate you must open to pass through; cut no trees and do not tramp across cropland or gardens. In national parks or wildlife refuges, resist the temptation to pick flowers or to leave the trails in pursuit of birds. Do not station yourself beside active nests to see or photograph the birds; you might disturb them and even cause them to abandon their nests. Finally, remember that mist nets may be used only in conjunction with legitimate research or banding projects, with permission from the Departamento de Vida Silvestre. It is of course strictly forbidden to capture wild birds for pets (except in the case of licensed birdcatchers, who can take only limited numbers of certain species at certain seasons). It is illegal to buy, sell, or export wild birds with few exceptions. Fortunately most birders are interested only in observing or photographing birds and are thoroughly willing to cooperate with local efforts for conservation.

You will find birding in Costa Rica more enjoyable if you are properly equipped. Steep, muddy roads and trails are the rule in most parts of this wet, mountainous country; good footwear is therefore important. Comfortable boots, high enough for ankle support and with a good tread for traction, are superior to tennis shoes once you get off paved roads and beaches. We wear calf-high rubber boots that, in addition to traction, provide some protection against poisonous snakes and chiggers and keep our feet dry. Insect repellent is important, not because the insects themselves are so unbearable most of the time, but because it is difficult to hold your binoculars steady while slapping a mosquito . . . and the noise and movement as you slap may be enough to frighten those shy understory antbirds and tinamous you're trying so hard to see! A liberal application of repellent around your socks and waistband can save you much (but usually not all!) agony from chiggers and ticks.

Clothing is also important: We recommend work shirts and blue jeans or slacks with lots of pockets; these should allow air circulation but be of a fabric heavy enough to deter most biting insects. T-shirts may be slightly cooler in the lowlands but in many areas are an open invitation to dinner for many a mosquito and blackfly. Avoid colorful or strongly patterned clothes, especially in the forest, where even a white shirt is conspicuous. A touch of red or yellow, such as a bandanna or a feather in your hat, may excite a hummingbird's curiosity, but most birds react with horror to aloha shirts! Except in Guanacaste or the Valle Central in the dry season, some sort of rain gear is a must, preferably a lightweight poncho that takes up little room in your pack or pocket.

We can think of no other tropical country as healthy for fieldwork as Costa Rica. Typhoid, malaria, and yellow fever are at most extremely rare—although you might consider antimalarial tablets if you plan to spend much time in the wet lowlands near the Nicaraguan border. A tetanus shot before you come and a tube of antibiotic ointment will guard against problems from ordinary cuts and bruises. Other insect-borne diseases exist but are rare enough so that judicious use of insect repellent will reduce your risk.

Poisonous snakes are potentially a more serious hazard. Although you are most unlikely to see a venomous snake in the course of a short visit, a few simple precautions are sensible. Good footwear is one; another is to be careful where you step—particularly in second growth or off trails, never put your feet where you cannot see the ground. On woodland trails, it is prudent to look down as you advance and stop before you look up and around. Someone in your party should have a snakebite kit with antivenin (available in Costa Rica) and should know how to use it. A hospital should be the first recourse for emergency treatment if you are less than 3hr away from one.

A less severe but more frequent hazard is the sting of a scorpion and stinging insects, especially ants. Here, too, the best advice is to watch where you put your feet—and hands. Especially in the northwest, when dressing in the morning, check to be sure that a scorpion hasn't gotten into your clothes or boots before you! If you are sensitive to stings, be sure to carry antihistamine just in case. A growing hazard in dry areas is Africanized (''killer'') bees, which react en masse to any loud noise or other disturbance around their nests (these are usually in holes in the ground or trees or in caves or crevices). A nylon-mesh head net that fits in your pocket is a useful precaution (to be worn with a full-brimmed hat). If you are attacked by a swarm, run away (standing still will not cause the bees to stop attacking as it does with other honeybees), and get under water if possible. A person allergic to bee stings should carry an epinephrine kit; antihistimine treatment should be the first recourse followed by emergency treatment at a hospital if indicated.

Probably the most frequent cause of discomfort to visitors in Costa Rica (or any other tropical country) is upset stomach and diarrhea, or *turista*, usually brought on by careless use of water. The water in Costa Rican cities and most small towns is usually free of dangerous organisms, but for a short visit you can minimize your risk by drinking only processed liquids, without ice, and using water purification tablets in your canteen. If you are planning a long stay, drink the water and put up with a possible few days of discomfort. In the field, water in forest springs and streams is nearly always potable;

avoid drinking stagnant water anywhere or water from streams that flow through pastures, agricultural areas, or inhabited districts in general: hepatitis or amoebic dysentery could result. Costa Rican fruits and vegetables are generally excellent but should be washed and/or peeled before use.

For those not accustomed to observing birds in tropical vegetation, a few tips on birding are in order. Study your guide beforehand; ideally you should be thoroughly familiar with the plates before setting out. Once in the field, carry a pocket notebook in which to make memory-jogging notes and sketches. Do not attempt to look up each bird as you see it or you may miss many members of a mixed flock. Rather, at your next rest stop, car, or field station, check the most interesting or confusing birds in your guide. Brief notes on voice, habitat, or behavior may help to clinch many identifications and can recall the bird to your mind's eye long afterward.

We recommend 8×–10× binoculars for picking out birds in the canopy; close focusing is a must for seeing understory birds. A telescope mounted on a tripod is useful in open marshes, salinas or mudflats and sometimes in savannas but is generally dead weight in the forest. For photographing birds, you will want at least a 200mm telephoto, and fast film is often essential in forest understory and toward dawn or dusk. Beware 2× or 4× converters: they reduce your light levels too much to be useful in many habitats. Finally, for night walks, have a headlamp that throws a sharp beam to help you spot owls and nightjars in vegetation.

A major part of birding in tropical forest is learning and using the voices of the birds. In many groups, such as flycatchers and woodcreepers, birds very similar in appearance have distinct voices. Often your first clue to a bird's presence in dense vegetation will be its voice. Many understory species communicate and defend their territories mainly by voice, and even a poor imitation of its song can bring many an antbird into close range. Recording the song and playing it back can be effective, particularly with wrens, but the trick can also be overdone. Playing back five songs to the bird's one, or at too high a volume, will probably frighten your quarry away—so take your cue from the bird's normal manner of singing. Also, do not continue playbacks for more than a few minutes, to avoid disrupting the bird's territorial behavior.

''Spishing'' or ''squeaking'' to attract birds is most effective in dry or open habitats but usually frightens birds away in wet forest understory. It is effective during or after the breeding season, say March through August, for most species. In areas where pygmy-owls occur, imitations or playbacks of their calls often attract many small birds to mob the owl, particularly during or just after the nesting season. Pygmy-owl calls combined with spishing, sounding like a mobbing episode already under way, can be even more effective in attracting small birds, but do not overdo it.

Spotting birds in dense vegetation can be tricky. Scan for movement with your eyes, which have a much broader field and depth of focus than your binoculars. The moment you detect a movement, zero in on it with your binoculars; if you do not see the bird, lower the binoculars and continue scanning by eye. Scanning with binoculars is likely to be productive only in open habitats or when you have already pinned down a bird's approximate location in dense vegetation by voice. Proficiency in the use of your binoculars can be acquired only by practice—preferably before you step into the field. Many further hints on observing birds appear in *The Audubon Society Handbook for Birders* (Kress 1981).

Some Costa Rican Birding Localities

In this section we describe a number of good birding localities, including principal habitats, and we give directions for travel by car, by public transport, and on foot. An asterisk indicates places where roads leading to a site require a four-wheel-drive vehicle. Although the list is not exhaustive, it is a representative selection of habitats in different sections of the country: most Costa Rican birds can be seen in one or more of these localities.

The first 18 localities, indicated on Map 3, are for people who have only a brief time for birding from a base in San José. All the localities mentioned can be reached by bus and on foot; some involve fairly strenuous hikes. Other localities in the rest of the country are indicated by number on Map 2 (pp. 4–5).

1. Universidad de Costa Rica Campus. Lawns, gardens, streams, second-growth woods. Take San Pedro bus from center of San José, get off at San Pedro church, walk 3 blocks N. Best before 7:00 A.M. or on Sundays. Also good areas of second growth N and E of campus.

2. Los Chigüites Watershed Protection Area, Tres Ríos. Small patch of old second-growth forest, tree plantations, adjoining coffee plantations—good for hummingbirds April–July when understory *Heliconia* flowers are in bloom. Take Tres Ríos bus to end of line; from NE corner of central park in Tres Ríos, walk 1mi (1.5km) uphill on road to Dulce Nombre (or take Dulce Nombre bus); look for large blue gate on L, walk in 100m to SNAA cement-topped water tanks; woods beyond.

3. La Carpintera. Tall hill to R as one goes from San José to Cartago, just beyond Tres Ríos. Take Cartago bus, get off where old road from Tres Ríos joins new highway; follow signs to Campamento Scout (Boy Scout camp) on highest part of hill (ask for permission at scout headquarters in San José); walk of ca. 1hr from highway (steep). Nice old second-growth forest, scrub, meadows.

4. Ochomogo. Wet fields, scrub, patches of second-growth forest, the best on R as you approach continental divide going from San José to Cartago, opposite Kativo paint factory. Take Cartago bus, and walk across fields to woods.

5. Coris. Interesting wet pastures and marsh with volcanic hot springs. Take San Isidro bypass around Cartago, pass gas station on R and take first road to R. Follow signs for pig farm (Granja Porcina Americana), heading W toward back side of La Carpintera. About 3mi (5km) from highway, note small dairy farm on L; turn in and ask permission to enter. Go ca. 300yd (m)* to end of farm road, walk across very muddy (!!) pastures diagonally to R to marsh, ca. ½mi (0.8km); also ponds behind pig farm.

6. Las Cóncavas–Jardín Lankester. From Cartago, go E toward Paraíso (or take Paraíso bus). At Campo Ayala, ca. 3mi (5km) beyond Cartago, look for colorful sign for Jardín Lankester on R; the gardens are ca. ½mi (0.8km) in, ca. 200yd (m) after taking first R turn at power station. Beautiful gardens, tall second growth. Hacienda Las Cóncavas, below the gardens (best reached by going straight at first intersection, then bearing R) has 2 interesting ponds, also pastures. Ask permission to enter.

7. Reserva de Vida Silvestre Tapantí. From Paraíso, follow signs to Orosi (or take Orosi–Río Macho bus); drive straight through Orosi and on through coffee plantations and pastures for ca. 4mi (7km) to bridge over Río Grande de Orosi, passing Río Macho

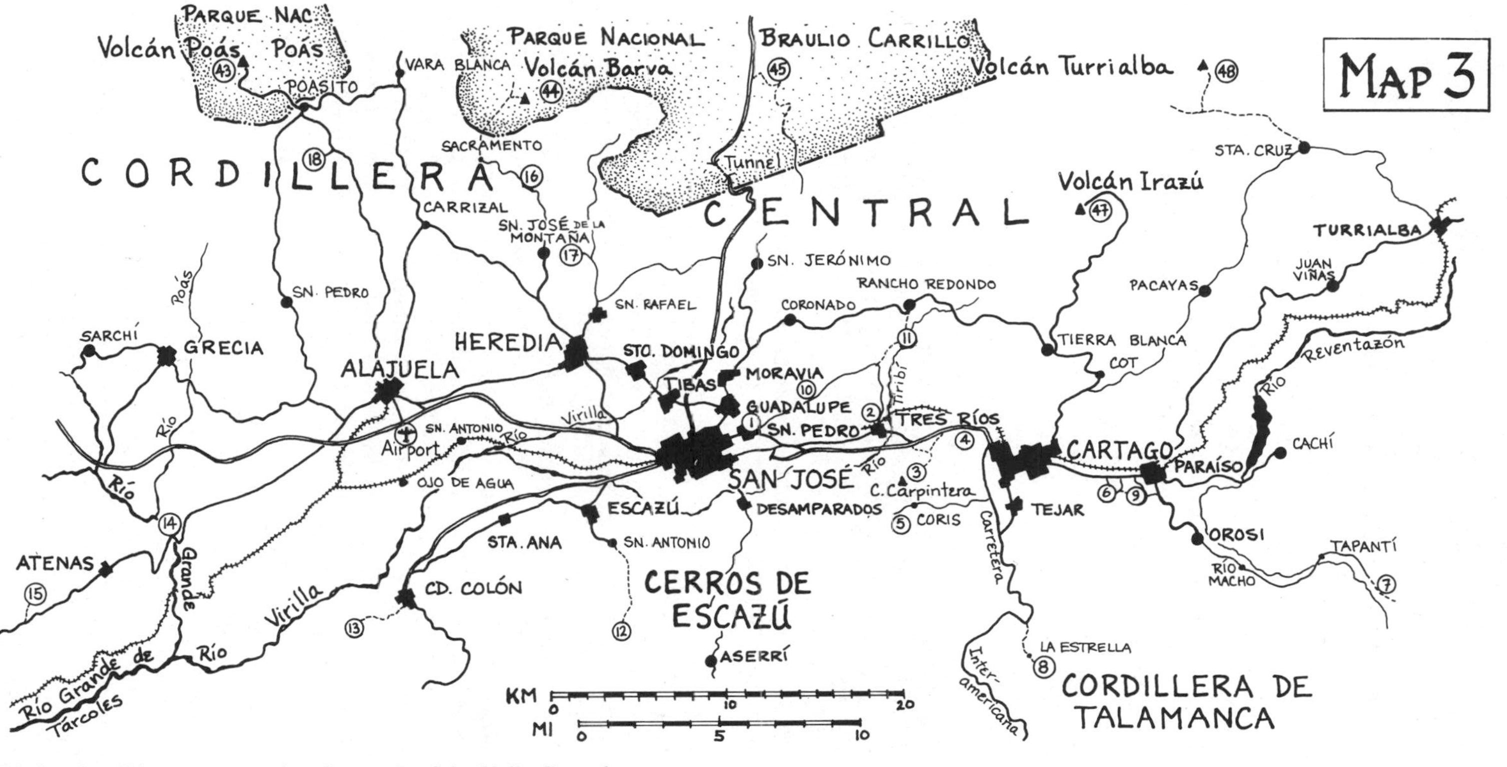

Birding localities, towns, and major roads of the Valle Central.

electric plant on R (bus to here). Bridge is good for river birds; check alder woods on R; continue on 1¼mi (2km) beyond bridge to refuge entrance, pay fee, and enter; road continues through steeply sloping, forested terrain; several good trails, spectacular scenery, often good birding near road.

8. La Estrella. Take San Isidro bypass around Cartago; ca. 3mi (5km) after road starts to climb up into the Talamancas, turn L and wind steeply down dirt road* to river and up other side to good tract of second-growth forest.

9. Laguna Doña Ana. Pretty park with pond, lawns, second-growth scrub; reached either by turning R at far end of squatter settlement in aborted housing development just past Campo Ayala on road to Paraíso, continuing beyond development and bearing L; or by turning L at first intersection on road in to Jardín Lankester;* or by taking road to Orosi out of Paraíso, after 1¼mi (2km) turn R, following signs to Parque La Expresión.

10. Parque del Este. Take San Pedro–San Ramón bus from center of San José, or take Sabanilla bus to end of line (statue of Christ at intersection). Walk up road ⅝mi (1km). Lawns, gardens, second-growth scrub, woodland. Best on weekdays (fewer people).

11. Los Lotes–Río Tiribí. Take San Pedro–San Ramón bus to end of line (Pulpería Bella Vista). Continue up road; where paved road turns R (ca. ⅓mi, 0.5km), continue straight on dirt road* for 1¼mi (2km), bearing R at intersections; go through gate marked Finca Los Lotes, and continue on to tiny María del Rosario hydroelectric plant. Walk up river gorge trail; especially good for swifts, some highland birds.

12. Pico Blanco, Cerros de Escazú. Take San Antonio de Escazú bus from San José and get off at Guardia Rural station; ca. 3mi (5km), hike straight up steep dirt road* to ridge top. After passing crest, go through first gate on L, continue up through pastures and patches of forest. Spectacular views; a good all-day outing for those who like to hike.

13. El Rodeo. Best forest in lower Valle Central, a mixture of middle-elevation and dry-forest birds. Take bus to Ciudad Colón (ca. 1hr); from Ciudad Colón follow signs to Universidad para La Paz (Peace University) to W; road passes through patches of forest, with best tracts opposite old hacienda and beyond the university buildings, across pastures; ca. 3mi (5km) in all.

14. La Garita. Take bus to Atenas from Alajuela or San José, or drive on highway to Puntarenas and take first exit after airport and Alajuela turnoff, turning L toward Atenas. After following a ridge for ca. 5–6mi (8–10km), road drops steeply to canyon of Río Grande. At gate to electric plant, obtain permission to enter: good birding down toward river in wooded canyon, along pasture edges.

15. Atenas Watershed Reserve. Drive through Atenas on road to Orotina and Jacó (or take Jacó bus from San José); ca. 5mi (8km) past Atenas, note fair-sized patch of woods on R, shortly before reaching hilltop and starting down coastal side. Look for paths into woods; mostly middle-elevation birds, some moist lowland species.

16. San José de la Montaña. Very pretty mountain town, reached by bus from Heredia; best birding areas are beyond and farther up mountain (steep hills). Best to take Paso Llano bus to end of line and walk up through pastures before stream to R (ask permission) to attractive patch of forest. Road to R continues up to Sacramento and Volcán Barva, for those wanting a stiff hike.

17. Río de la Hoja. Pretty park with old cypresses, natural forest; from Heredia take bus to San Rafael and Monte La Cruz. About 2mi (3km) N of San Rafael, get off at La Casita restaurant; park is ca. 2mi (3km) walk up road to L. Best on weekdays (fewer people), as is the following locality.

18. Laguna Fraijanes. Pretty pond, pastureland, gardens, second-growth woodland.

If you are driving, take turnoff to Alajuela at overpass just beyond (not before) airport and continue straight N through Alajuela and up for 9mi (15km). Turnoff to park is on L, well marked. Poasito bus from Alajuela stops at entrance.

Note: Those with a vehicle can find interesting patches of remnant streamside forest along almost any of the roads climbing out of the Valle Central to the volcanoes of the Cordillera Central (e.g., the roads to Rancho Redondo, Las Nubes, Carrizal, and Vara Blanca and the new highway to Guápiles).

19. Parque Nacional Santa Rosa. 21mi (35km) N of Liberia on Carretera Interamericana (Highway), take Peñas Blancas bus from San José or Liberia, get off at park entrance; 4mi (7km) in to camping area and park headquarters through savanna, dry deciduous forest and some evergreen forest. Fine forests, mangroves, beach, salina on road* to Playa Naranjo (9mi, 15km).

20. Parque Nacional Rincón de la Vieja. Beautiful moist forests, interesting scrub-savanna, hot springs; trail up volcano for those who want a fine 2–3-day hike. 1½hr E of Liberia by rough road (especially rough* in wet season).

21. Refugio de Fauna Silvestre Rafael Lucas Rodriguez Caballero (= Palo Verde). 15mi (25km) W of Bagaces (follow signs) on rough road (especially rough* and often impassable in wet season); fine dry forest with patches of evergreen forest, savanna, extensive freshwater marshes.

22. Parque Nacional Palo Verde. Habitats similar to those in wildlife refuge (see above), but most marshes shallower, dry out earlier in dry season; forests less extensive; ca. 10mi (16km) SW of Cañas (follow signs); road* may be impassable in wet season.

23. Finca la Pacífica. 3mi (5km) N of Cañas on Carretera Interamericana (Pan-American Highway). Good dry and gallery forest, savannas, restaurant, accommodations. Ask permission.

24. Reserva de Bosque Nuboso Monteverde. Private biological reserve run by Tropical Science Center, San José. Beautiful cloud and wet montane forests, moist forests lower down. Accommodations in and around community of Monteverde (several hotels); or at field station of Reserve by prior arrangement; 25mi (40km) up rough road from Puente Lagartos on Carretera Interamericana; take bus from Puntarenas to nearby community of Santa Elena. For the ambitious hiker, the Peñas Blancas valley offers many interesting Caribbean birds.

25. Colorado. Town on Golfo de Nicoya, with several salinas nearby; best are Salina Conchal, ⅓mi (0.5km) W of town, and Salina Bonilla, 1mi (1.5km) N; both excellent for shorebirds at high tide.

26. Chomes. Town on Golfo de Nicoya, 6mi (9km) south of Carretera Interamericana; often excellent for gulls, terns, shorebirds, herons, etc., near shrimp-culture ponds S of town (Criadero de Camarones de Chomes; ask permission).

27. Puntarenas. Good mangroves across estuary from city; can sometimes hire boats to take you into Golfo de Nicoya; best areas usually around Isla Guayabos (flat-topped island just across from Puntarenas) and waters between Puntarenas and Caldera to S. Coming into town, check Laguna Chacarita on R just past cemetery; in spite of trashy aspect, sometimes interesting.

28. Reserva Biológica Cabo Blanco. Very isolated; a long, rough trip,* but beautiful moist forest, beach, huge offshore rocks where Brown Boobies nest (ask permission of Servicio de Parques Nacionales in San José). If you make this trip, allow 2–3 days; interesting areas en route include Curú wildlife refuge with fine moist forest and beach, also Bahía Ballena, with its mangroves, beach, and sometimes seabirds; there is also some good forest at Montezuma as well as accommodations.

29. Sámara. One of a number of resort communities on outer Península de Nicoya; good moist forest in river bottoms, some mangroves, fine gallery forest along Río Buena Vista a few mi (km) N; beautiful beach, good area for seabirds if you can hire a small boat to go offshore.

30. Río Cañas. Very interesting freshwater marshes and river pools; about 4mi (7km) N of Santa Cruz, take L (toward NW) beside huge rice-processing plant, go 1¼mi (2km) on dirt road* (may be impassible mud in wet season) to lagoons.

31. Cerro Vista al Mar. High mountain some 10mi (17km) S of Santa Cruz; past Arado, turn R at river, climb very steep, long ridge; at and beyond radio towers are patches of moist-wet forest with breeding Three-wattled Bellbirds and other species.

32. Tamarindo–Playa Grande. Resort community; beaches and rocky points good for shorebirds and seabirds, especially in migration; extensive mangrove swamp (can hire boats in Tamarindo); some freshwater lagoons on adjacent hacienda to S.

33. Playas del Coco, Playa Hermosa. Beach communities with some patchy forest; hire a boat for seabirding—an interesting all-day trip to Islas Santa Catalina, where Bridled Terns breed.

34. Estero Madrigal. Beautiful river pools, gallery forest, marsh along Río Lajas, reached via Hacienda Solimar. On road to Tempisque ferry, turn R 1¼mi (2km) after passing turnoff to Colorado on L; go 5mi (8km) to hacienda, ask permission and if possible hire guide; drive to river in dry season,* often impassible in wet (but walk takes ca. 1hr).

35. Bijagua and Vicinity. Pretty valley on Caribbean side of Cordillera de Guanacaste. Those with vehicle* and enthusiasm for a challenging hike can take road to Zapote, first L after crossing river; at second river head up pastures to L into forest. Small patch of forest beyond valley on road to Upala.

36. Los Chiles. Can be reached by air or by new highway, along which a few patches of forest still remain, also some good marshes; wet pastures, marshes, and gallery forest S of Los Chiles along Río Frío. Hire a boat to go up Río Frío to Caño Negro; often good for waterbirds.

37. Lago Caño Negro. Reached in wet season by boat from Los Chiles or San Rafael (Guatuso); in dry season by road* from Upala; large seasonal lake and swamp and gallery forests, marshes, savannas. Check at wildlife refuge headquarters at small town of Caño Negro. Lake mostly disappears in dry season.

38. Tortuguero. Beautiful swamp forest, freshwater sloughs, beach, river mouth, second growth; forested hills just N of river mouth and inside national park, good trails around park headquarters. Accessible by boat through inland waterway from Limón or by charter plane. Can be exciting during migrations.

39. Barra del Colorado. Habitats similar to Tortuguero but estuary better for waterbirds; accessible via new highway, along which some patches of forest remain, also scattered marshes and swamps.

40. Estación Biológica la Selva. Biological research station run by Organization for Tropical Studies (OTS). Superb forest and trail system. Accommodations limited; check with OTS office in Moravia for reservations; accommodations may be available in nearby Puerto Viejo. One can hike from La Selva into lower part (Zona Protectora) of Parque Nacional Braulio Carrillo.

41. Rara Avis–Finca Plástico. Field station presently being developed in spectacular foothill forest; access difficult except in dry season, soon to be improved; 1–2hr by vehicle* or 3hr hike up from Las Horquetas; make arrangements with A. Bien in San José. Excellent for birds of Caribbean foothills.

42. Virgen del Socorro. Excellent spot for lower middle-elevation birds of Caribbean

slope, including many tanagers; from San José, take main paved road through Vara Blanca toward Sarapiquí lowlands (La Paz waterfall worth a short stop). Take first R after passing El Angel marmalade factory (ca. 2mi, 3km). Rough unpaved road* goes down into forested gorge of Río Sarapiquí; good trail along river, also good birding along road.

43. Parque Nacional Volcán Poás. Highland forest, second growth, spectacular crater, good visitor center; reached by good roads via Alajuela or Heredia.

44. Volcán Barva. In Parque Nacional Braulio Carrillo; go through San José de la Montana from Heredia, continue to Paso Llano, and take R to Sacramento* (park guard station). From here hike ca. 3mi (5km) up into beautiful highland forests and take trail into crater lake.

45. Bajo La Hondura. Scenic valley in upper watershed of Río La Hondura, at end of road* from San José through Moravia and San Jerónimo; both disturbed and relatively intact forest, especially down toward Río La Hondura in Parque Nacional Braulio Carrillo.

46. Carrillo. In Parque Nacional Braulio Carrillo; take new highway to Guápiles; trail upstream just beyond tunnel often productive; along highway, in spectacular scenery, often good birding, especially around La Montura (ca. 5mi, 8km, below tunnel) and at old Carrillo ranger station, where interesting trails take off down forested ridges. Excellent for lower middle-elevation species.

47. Parque Nacional Volcán Irazú. Reached by road from Cartago; birding best in scrub and forest below crater or around Prusia (turnoff to L after Tierra Blanca), where main habitats are planted woods, pastures, natural forest along streams.

48. Volcán Turrialba. Reached by jeep road* up from Turrialba or from Cartago via Pacayas; from Hacienda Central a hike of 1–2hr up through pastures, forest, and high scrub around crater; a good day trip (but leave early).

49. CATIE-Turrialba. Agricultural university 3mi (5km) E of town of Turrialba; best areas for birds include lily pond near entrance, forest trail behind main building, and campus itself early in morning. For the adventurous and amphibious, swampy pastures to R beyond entrance offer possible herons and rails.

50. Madreselva. Take Carretera Interamericana S from Cartago; at km66 a dirt road goes off to R down into beautiful oak forest; ca. 200m farther up highway, another road goes L up into lower, stunted forest; both offer fine birding (if weather cooperates).

51. Cerro de la Muerte. Good area for high mountain birds; well-preserved páramo at Cerro Asunción (km88–89 of Carretera Interamericana); oak forest behind La Georgina (km95) and over ridge in front. If you continue down toward San Isidro for ca. 1mi (1.7km), a road turns off through a pole gate on L just past pulpería, leads steeply down to thinned oak forest, often excellent birding. Trail continues N through superb forest of Cerro Abarca.

52. Parque Nacional Manuel Antonio. Lovely beaches, rocky shores, some impressive forest; reached by fair road from Quepos (can take bus or drive); if you are driving, turn off at Parrita and check out mouth of Río Parrita, often good for sea- and shorebirds; fine mangroves to S along Estero Palo Seco.

53. Reserva Biológica Carara. Northernmost moist-wet forest in Pacific lowlands, limit of distribution for many wet-forest birds; take highway to coast through Atenas (shorter coastal road from Puntarenas-Caldera should be completed by 1989). Reserve headquarters are beside highway, 1mi (1.5km) after Río Tárcoles bridge. The mouth of Río Tarcoles is interesting (turn off to R just after reserve). If you can, hire a boat there; the mangroves of Estero Guacalillos-Pigres are excellent for birds.

54. Tivives Mangrove Reserve. Beautiful mangroves reached by rough dirt road over

hill from Caldera, S of Puntarenas; excellent for mangrove birds; mouth of river good for shorebirds; best to enter mangroves from above (turn L after coming down hill, and go in by old well. Mud).

55. Limón-Moín. Only extensive mangroves on Caribbean coast at Moín; Limón harbor sometimes good for birds, also city parks in town; can hire boat to visit Isla Uvita, which has small booby colony on nearby islet.

56. Matina. Small coastal town reached by inland waterway or by dirt road along beach. River mouth and estuary often excellent for waterbirds, as is river mouth at Parismina, farther N along inland waterway (artificial canals, natural sloughs and lagoons—good birding in early morning).

57. Río Vizcaya. River mouth ca. 6mi (10km) SE of Limón often excellent for observing migration; turn L before bridge, explore scrub around former resort; often many shorebirds, etc., at river mouth.

58. Parque Nacional Cahuita. Fine swamp forest, beaches; coral reefs. Best birding on Puerto Vargas side, beyond turnoff for town of Cahuita; camping possible.

59. Manzanillo. At end of coast road. Good birding along forest trail to Punta Mona, in old cacao plantations along road; during migrations, scrub behind town and also around Puerto Viejo can be exciting; an interesting swamp behind Manzanillo (hire guide in town). Nearest hotel: Puerto Viejo. A new road from Punta Uva back into broken forest toward Sixaola also interesting,* in dry weather.

60. Bribrí. Follow paved road to R at Hone Creek. About halfway to Bribrí, cross Río Sandbox; an interesting walk upstream; at Bribrí walk up old railroad tracks to R, look for trails into forest and old cacao plantation up ridge to R. Old cacao and swamps on road to Sixaola can be productive; forested ridge to L.

61. Suretka. A number of roads* (dry weather) through forest made by oil company during drilling operations in Campo Diablo area; ask permission from RECOPE in San José or Limón.

62. Reserva Biológica Hitoy-Cerere. Turn R at bridge over Río La Estrella, take road to Vesta, follow signs to reserve; old second growth, forest on hills overlooking Río Cerere, little marsh at reserve headquarters.

63. Parque Nacional Chirripó. For those interested in hiking and mountaineering. Drive to town of San Gerardo out of San Isidro; from there a stiff 1-day hike (or take it easy in 2 days) up to shelter in páramo; glacier lakes, páramo, fine mountain scenery, some fine forest (and much burned-over forest) en route. Can hire horses in San Gerardo; trail well marked.

64. Rincón de Osa. Take new road from Piedras Blancas around N end of Golfo Dulce; some nice forest but being cut in many places. At Rincón is a good patch of mangroves; take turn to airstrip, beyond which is nice forest up valley of Quebrada Aguabuena. Second growth along airstrip can also be interesting. Before Piedras Blancas turnoff, a representative patch of forest on L (Esquinas forest) as you go S.

65. Golfito. Steeply sloping forests beyond town and old United Fruit Company buildings behind airstrip; interesting second growth and patches of forest along road in from Carretera Interamericana (but being cut); scruffy mangroves near town; can hire boat here to explore Golfo Dulce (sometimes good for seabirds, especially toward mouth of Río Coto, which also has good mangroves).

66. Parque Nacional Corcovado. Beautiful forests, though mostly disturbed near park center at Sirena; best areas at N end of park around Llorona and San Pedrillo (hike up beach from Sirena) and beyond Pavo or at Los Patos, up Río Rincón from Golfo Dulce. Access by air to Sirena or hike in from any of several points; ask permission at Servicio

de Parques Nacionales office in San José. Another option is the private Marenco Biological Station located just north of park; make arrangements in San José.

67. Buenos Aires–Paso Real Savannas. Best savannas around Buenos Aires destroyed to make pineapple plantation, but good savannas remain toward Paso Real and up dirt road* to right of highway (only in dry weather); small pond a short way from highway on road to Buenos Aires often productive.

68. Jardín Botánico Catherine y Robert Wilson (Las Cruces). Beautiful botanical garden and patch of forest ca. 2¼mi (4km) from San Vito on road to Ciudad Neily; best to cross on ferry at Paso Real and reach San Vito on road through Sabanilla. Good field station and cabins, also hotel in San Vito. Check with OTS office in Moravia.

69. Las Tablas. Majestic moist to wet mountain forest;* inquire at Servicio de Parques Nacionales in San José. From San Vito drive to Panama border at Sabalito and turn L on road to Las Mellizas (CAFROSA). From here take road to Cotoncito and Las Tablas; can camp at Cotoncito; roads and trails through forest to Coto Brus and Las Tablas.

70. Las Alturas. Coffee *finca* (plantation) SE of Las Tablas. Take road through Rionegro* (may be difficult in wet season) from Sabalito. At Las Alturas a trail up from Bella Vista quarry leads into fine forest, as do road and trail to Río Cedros and trail up Río Cotón. Ask permission at farm office in San José.

Note. The adventurous (and physically fit) birder may wish to arrange to hike into parts of Parque Nacional la Amistad in the Cordillera de Talamanca; check with Servicio de Parques Nacionales in San José regarding arrangements. You can hike across the Cordillera de Talamanca in a week or so with a guide.

Natural history tourism is on the upswing in Costa Rica, and a number of new field stations, hotels, and nature preserves are being planned or constructed as this guide goes to press. Also, road conditions, bus routes, and access to other parks and reserves change from year to year. The best way to keep abreast of such developments is to buy the latest edition of Beatrice Blake and Ann Becher's *The New Key to Costa Rica*, available in many bookstores in the United States and in most bookstores in the Valle Central of Costa Rica.

Bibliography

Our objective is to list principal sources of published information and to provide the interested reader with access to the literature on plumages, distribution, natural history, breeding biology, and conservation of Costa Rican birds. We mention works focusing on other Neotropical areas where these provide information relevant to, or not available for, birds in Costa Rica per se. However, we do not list references to species breeding in North America except where they refer to migration or wintering in the neotropics; we assume that users of this guide will already be familiar with North American bird guides. We include only books and articles that discuss major groups or broad segments of the avifauna rather than studies of particular species; references to the latter can usually be found in literature citations of the works listed here.

American Ornithologists' Union. 1983. *Check-list of North American Birds*. 6th ed. Lawrence, Kans.: Allen Press.

Arnold, K. A. 1966. Notes on Costa Rican birds. *Wilson Bulletin* 78:316–317.

Blake, E. R. 1958. Birds of the Volcán de Chiriquí, Panama. *Fieldiana: Zoology* 36:499–577.

Blake, E. R. 1977. *Manual of Neotropical Birds*, vol. 1. Chicago: University of Chicago Press.

Bond, J. 1971. *Birds of the West Indies*, 2d ed. Boston: Houghton Mifflin.

Brown, L., and D. Amadon. 1968. *Eagles, Hawks, and Falcons of the World*. London: Country Life Press.

Buckley, P. A., M. S. Foster, E. S. Morton, R. S. Ridgely, and F. G. Buckley (eds.). 1985. *Neotropical Ornithology*. Ornithological Monographs no. 36.

Buskirk, W. H. 1973. Four new migrants for Costa Rica. *Condor* 75:363–364.

Calvert, A. S., and P. P. Calvert. 1917. *A Year of Costa Rican Natural History*. New York: Macmillan.

Carriker, M. A. 1910. *An Annotated List of the Birds of Costa Rica, including Cocos Island*. Annals of the Carnegie Museum of Natural History 6:314–915.

Chapman, F. M. 1929. *My Tropical Air Castle*. New York: D. Appleton.

Cherrie, G. K. 1916. A contribution to the ornithology of the Orinoco region. *Bulletin of the Museum of the Brooklyn Institute of Arts and Sciences* 2:133a–374.

Davis, L. I. 1972. *A Field Guide to the Birds of Mexico and Central America*. Austin: University of Texas Press.

Delacour, J., and D. Amadon. 1973. *Curassows and Related Birds*. New York: American Museum of Natural History.

Dickerman, R. W. 1971. Further notes on Costa Rican birds. *Condor* 73:252–255.

Dickey, D. R., and A. J. van Rossem. 1938. *The Birds of El Salvador*. Zoological Series, vol. 23, Field Museum of Natural History Publication no. 406.

Dunning, J. B., Jr. 1984. *Body Weights of 686 Species of North American Birds*. Western Bird Banding Association Monograph no. 1.

Eisenmann, E. 1952. Annotated list of birds of Barro Colorado Island, Panama Canal Zone. *Smithsonian Miscellaneous Collection* 117(5):1–62.

Feinsinger, P. 1977. Notes on the hummingbirds of Monteverde, Cordillera de Tilarán, Costa Rica. *Wilson Bulletin* 89:159–164.

ffrench, R. 1973. *A Guide to the Birds of Trinidad and Tobago*. Wynnewood, Pa.: Livingston.

Foster, M. S., and N. K. Johnson. 1974. Notes on the birds of Costa Rica. *Wilson Bulletin* 86:58–63.

Hancock, J., and J. Kushlan. 1984. *The Herons Handbook*. London: Croom Helm.

Harrison, P. 1983. *Seabirds: An Identification Guide*. Boston: Houghton Mifflin.

Haverschmidt, F. 1968. *The Birds of Surinam*. Edinburgh, Scotland: Oliver and Boyd.

Hayman, P., J. Marchant, and T. Prater. 1986. *Shorebirds: An Identification Guide to the Waders of the World*. London: Croom Helm.

Hilty, S. L., and W. L. Brown. 1986. *A Guide to the Birds of Colombia*. Princeton, N.J.: Princeton University Press.

Howell, T. R. 1957. Birds of a second-growth rainforest of Nicaragua. *Condor* 59:73–111.

Huber, W. 1932. Birds collected in northeastern Nicaragua in 1922. *Proceedings of the Academy of Natural Sciences of Philadelphia* 84:205–249.

Isler, M. L., and P. R. Isler. 1987. *The Tanagers: Natural History, Distribution, and Identification*. Washington, D.C.: Smithsonian Institution Press.

Janzen, D. H. (ed.). 1983. *Costa Rican Natural History*. Chicago: University of Chicago Press.

Jehl, J. R., Jr. 1974. The near-shore avifauna of the Middle American west coast. *Auk* 91:681–699.

Keast, A., and E. S. Morton (eds.). 1980. *Migrant Birds in the Neotropics: Ecology, Behavior, Distribution, and Conservation*. Washington, D.C.: Smithsonian Institution Press.

Kiff, L. F. 1975. Notes on southwestern Costa Rican birds. *Condor* 77:101–103.

Kress, S. W. 1981. *The Audubon Society Handbook for Birders*. New York: Charles Scribner's Sons.

Land, H. C. 1970. *Birds of Guatemala*. Wynnewood, Pa.: Livingston.

Leigh, E. G., Jr., A. S. Rand, and D. M. Windsor (eds.). 1982. *The Ecology of a Tropical Forest: Seasonal Rhythms and Long-term Changes*. Washington, D.C.: Smithsonian Institution Press.

Leopold, A. S. 1950. *Wildlife of Mexico: The Game Birds and Mammals*. Berkeley: University of California Press.

Lewis, T. J., and F. G. Stiles. 1980. Locational checklist of the birds of Costa Rica. San José: Costa Rica Expeditions.

Méndez, E. 1979. *Las Aves de Caza de Panama*. Panama: Published by the author.

Monroe, B. L., Jr. 1968. *A Distributional Survey of the Birds of Honduras*. Ornithological Monographs no. 7.

Murphy, R. C. 1936. *Oceanic Birds of South America*. 2 vols. New York: Macmillan and the American Museum of Natural History.

Orians, G. H., and D. R. Paulson. 1969. Notes on Costa Rican birds. *Condor* 71:426–431.

Peterson, R. T., and E. L. Chalif. 1973. *A Field Guide to Mexican Birds*. Boston: Houghton Mifflin.

Phillips, A. R., M. A. Howe, and W. E. Lanyon. 1966. Identification of the flycatchers of eastern North America, with special emphasis on the genus *Empidonax*. *Bird-Banding* 37:153–171.

Prater, A. J., J. H. Marchant, and J. Vuorinen. 1977. *Guide to the Identification and Aging of Holarctic Waders*. Tring, Herts.: British Trust for Ornithology.

Ridgely, R. S. 1981. *A Guide to the Birds of Panama*, 2d ed. Princeton, N.J.: Princeton University Press.

Ridgway, R. 1901–1950. *The Birds of North and Middle America*, Pts. 1–11. *Bulletin of the U.S. National Museum of Natural History*, vol. 50 (pts. 9–11 continued by H. Friedmann).

Rowley, J. S. 1966. Breeding birds of the Sierra Madre del Sur, Oaxaca, Mexico. *Proceedings of the Western Foundation for Vertebrate Zoology* 1:107–204.

Rowley, J. S. 1984. Breeding records of land birds in Oaxaca, Mexico. *Proceedings of the Western Foundation for Vertebrate Zoology* 2:73–224.

Russell, S. M. 1964. *A Distributional Survey of the Birds of British Honduras*. Ornithological Monographs no. 1.

Skutch, A. F. 1950. The nesting seasons of Central American birds in relation to climate and food supply. *Ibis* 92:185–222.

Skutch, A. F. 1954. *Life Histories of Central American Birds. Pacific Coast Avifauna* no. 31.

Skutch, A. F. 1960. *Life Histories of Central American Birds II. Pacific Coast Avifauna* no. 34.

Skutch, A. F. 1966. A breeding bird census and nesting success in Central America. *Ibis* 108:1–16.

Skutch, A. F. 1967. *Life Histories of Central American Highland Birds*. Publications of the Nuttall Ornithological Club no. 7.

Skutch, A. F. 1969. *Life Histories of Central American Birds III. Pacific Coast Avifauna* no. 35.

Skutch, A. F. 1971. *A Naturalist in Costa Rica*. Gainesville: University of Florida Press.

Skutch, A. F. 1972. *Studies of Tropical American Birds*. Publications of the Nuttall Ornithological Club no. 10.

Skutch, A. F. 1976. *Parent Birds and Their Young*. Austin: University of Texas Press.

Skutch, A. F. 1981. *New Studies of Tropical American Birds*. Publications of the Nuttall Ornithological Club no. 19.

Skutch, A. F. 1983. *Birds of Tropical America*. Austin: University of Texas Press.

Slud, P. 1960. The birds of Finca "La Selva," a tropical wet forest locality. *Bulletin of the American Museum of Natural History* 121:49–148.

Slud, P. 1964. *The Birds of Costa Rica: Distribution and Ecology*. Bulletin of the American Museum of Natural History, vol. 128.

Slud, P. 1967. The birds of Cocos Island, Costa Rica. *Bulletin of the American Museum of Natural History* 134:261–296.

Slud, P. 1980. The birds of Hacienda Palo Verde, Guanacaste, Costa Rica. *Smithsonian Contributions to Zoology* 292:1–92.

Smithe, F. B. 1966. *The Birds of Tikal*. Garden City, N.Y.: Natural History Press.

Snow, D. W. [1976] 1985. *The Web of Adaptation*. Reprint, with foreword by Alexander F. Skutch. Ithaca, N.Y.: Cornell University Press.

Snow, D. W. 1982. *The Cotingas: Bellbirds, Umbrellabirds, and Their Allies*. Ithaca, N.Y.: Cornell University Press.

Stiles, F. G. 1980. The annual cycle in a tropical wet forest hummingbird community. *Ibis* 122:322–343.

Stiles, F. G. 1983. Status and conservation of seabirds in Costa Rica. *In* J. P. Croxall, P. G. H. Evans, and R. W. Schreiber (eds.), *Status and Conservation of the World's Seabirds*, pp. 223–229. Technical Publication no. 2. Cambridge: International Council for Bird Preservation.

Stiles, F. G. 1985. Conservation of forest birds in Costa Rica: Problems and perspectives. *In* A. W. Diamond and T. S. Lovejoy (eds.), *Conservation of Tropical Forest Birds*, pp. 141–168. Technical Publication no. 4. Cambridge: International Council for Bird Preservation.

Stiles, F. G. 1988. Notes on the status and distribution of certain birds in Costa Rica. *Condor* 90: 931–933.

Stiles, F. G., and S. M. Smith. 1977. New information on Costa Rican waterbirds. *Condor* 79:90–97.

Stiles, F. G., and S. M. Smith. 1980. Notes on bird distribution in Costa Rica. *Brenesia* 17:137–156.

Thurber, W. A., J. F. Serrano, A. Sermeño, and M. Benítez. 1987. Status of common and previously unreported birds of El Salvador. Proceedings of the Western Foundation of Vertebrate Zoology 3:109–293.

Wetmore, A. 1965–1984. *The Birds of the Republic of Panama, Parts 1–4*. Smithsonian Miscellaneous Collection, vol. 150 (pt. 4 completed by R. F. Pasquier and S. L. Olson).

Wheelwright, N. T., W. A. Haber, K. G. Murray, and C. Guindon. 1984. Tropical fruit-eating birds and their food plants: a survey of a Costa Rican lower montane forest. *Biotropica* 16:173–192.

Wolf, L. L. 1976. Birds of the Cerro de la Muerte area, Cordillera de Talamanca, Costa Rica. *American Museum of Natural History Novitates* no. 2606.

Index

The preferred English names of species, i.e., those used to head the species accounts, and Latin names of taxa higher than genus are in **BOLDFACE CAPITAL LETTERS.** Alternative English names, English names of species mentioned only in the Note sections of the species accounts, and alternative Latin names of taxa higher than genus are in CAPITAL AND SMALL CAPITAL LETTERS. Latin names used to head the species accounts are in ***boldface italic letters.*** Alternative Latin names and Latin names of species mentioned only in the Note sections are in *italic letters*. Recommended Spanish names are in CAPITAL LETTERS. Local Spanish names are in Capital and Lowercase Letters.

Acanthidops bairdii, 454, Pl. 49(12)
Accipiter bicolor, 103, Pl. 16(8)
Accipiter chionogaster, 102
Accipiter cooperii, 102, Pl. 16(4)
Accipiter erythronemius, 102
Accipiter striatus, 102, Pl. 16(3)
Accipiter superciliosus, 102, Pl. 16(2)
ACCIPITRIDAE, 98
Acrochordopus. See ***Phyllomyias***
Actitis macularia, 143, Pl. 11(8)
Aegolius acadicus, 196
Aegolius ridgwayi, 196, Pl. 20(13)
Agamia agami, 87, Pl. 5(1)
Agelaius phoeniceus, 413, Pl. 44(14)
AGUILA ARPIA, 110
AGUILA CRESTADA, 109
AGUILA PESCADORA, 97
AGUILA SOLITARIA, 105
AGUILILLO BLANCO Y NEGRO, 110
AGUILILLO NEGRO, 111
AGUILILLO PENACHUDO, 111
AGUILUCHO NORTEÑO, 101
Agüío, 418–421
Agüío Barranquillo, 421
AGUJA CANELA, 140
AGUJA LOMIBLANCA, 140
AGUJETA COMUN, 145
AGUJETA SILBONA O PIQUILARGA, 145
Aimophila botterii, 462, Pl. 50(5)
Aimophila rufescens, 461, Pl. 50(4)
Aimophila ruficauda, 461, Pl. 50(7)
Ajaia ajaja, 90, Pl. 4(7)
ALA DE SABLE VIOLACEO, 213
Albarda Nueva, 461
Alcaraván, 135
ALCARAVAN AMERICANO, 135
ALCATRAZ, 76
Alcedinidae, 237
Alzacolita, 143
Amaurolimnas concolor, 125, Pl. 6(8)
Amaurospiza concolor, 450, Pl. 49(11)
AMAZILIA CANELA, 222
AMAZILIA COLIAZUL, 221
AMAZILIA CORONA DE BERILO, 219
AMAZILIA CULIAZUL, 221
AMAZILIA GORRIAZUL, 221
AMAZILIA MANGLERA, 220
AMAZILIA PECHIAZUL, 220
AMAZILIA PECHIBLANCA, 219
AMAZILIA RABIRRUFA, 222
AMAZILIA VIENTRIBLANCA, 222
Amazilia alfaroana, 221
Amazilia amabilis, 220, Pl. 24(14)
Amazilia boucardi, 220, Pl. 24(2)
Amazilia candida, 219, Pl. 24(1)
Amazilia cyanifrons, 221
Amazilia cyanura, 221, Pl. 24(16)
Amazilia decora, 219, Pl. 24(18)
Amazilia edward, 222, Pl. 24(3)
Amazilia rutila, 222, Pl. 24(11)
Amazilia saucerrottei, 221, Pl. 24(15)
Amazilia sophiae, 221
Amazilia tzacatl, 222, Pl. 24(10)
Amazona albifrons, 182, Pl. 19(6)
Amazona auropalliata, 182, Pl. 19(5)
Amazona autumnalis, 182, Pl. 19(4)
Amazona farinosa, 183, Pl. 19(3)
Amazona ochrocephala, 183
Amblycercus holosericeus, 407, Pl. 44(12)
Amigo de Hombre, 402
Ammodramus savannarum, 460, Pl. 50(8)
Anabacerthia striaticollis, 271
Anabacerthia variegaticeps, 271, Pl. 30(2)
AÑAPERO COLICORTO, 199
AÑAPERO MENOR, 200
AÑAPERO ZUMBON, 199
Anas acuta, 93, Pl. 8(1)
Anas americana, 92, Pl. 8(6)
Anas carolinensis, 92
Anas clypeata, 94, Pl. 8(2)
Anas crecca, 92, Pl. 8(4)
Anas cyanoptera, 93, Pl. 8(3)
Anas discors, 93, Pl. 8(5)
Anas platyrhynchos, 92
ANATIDAE, 90
ANDARRIOS MACULADO, 143
ANDARRIOS SOLITARIO, 142
ANHINGA, 78, Pl. 4(3)
Anhinga anhinga, 78, Pl. 4(3)
ANHINGIDAE, 78

ANINGA, 78
ANI, GROOVE-BILLED, 186, Pl. 21(9)
ANI, SMOOTH-BILLED, 186, Pl. 21(8)
Anous alba, 164
Anous minutus, 163, Pl. 2(7)
Anous stolidus, 163, Pl. 2(8)
Anous tenuirostris, 164
ANSERIFORMES, 90
ANTBIRD, BARE-CROWNED, 283, Pl. 31(13)
ANTBIRD, BICOLORED, 284, Pl. 31(9)
ANTBIRD, CHESTNUT-BACKED, 283, Pl. 31(11)
ANTBIRD, DULL-MANTLED, 284, Pl. 31(14)
ANTBIRD, DUSKY, 282, Pl. 31(8)
ANTBIRD, IMMACULATE, 284, Pl. 31(10)
ANTBIRD, OCELLATED, 285, Pl. 31(7)
ANTBIRD, SPOTTED, 285, Pl. 31(12)
Antbird, Tyrannine, 282
Antbird, Wing-banded, 286, Pl. 30(21)
Anthracothorax prevostii, 214, Pl. 23(13)
Anthracothorax veraguensis, 215
Anthus lutescens, 453
ANTIFACITO CORONIGRIS, 399
ANTIFACITO CORONIOLIVO, 398
ANTIFACITO NORTEÑO, 397
ANTIFACITO SUREÑO, 398
ANTPITTA, BLACK-CROWNED, 287, Pl. 30(19)
ANTPITTA, FULVOUS-BELLIED, 289, Pl. 30(14)
ANTPITTA, OCHRE-BREASTED, 289, Pl. 30(13)
ANTPITTA, SCALED, 288, Pl. 30(20)
ANTPITTA, SPECTACLED, 288, Pl. 30(18)
Antpitta, Streak-chested, 289
ANTSHRIKE, BARRED, 276, Pl. 31(3)
ANTSHRIKE, BLACK-HOODED, 277, Pl. 31(6)
ANTSHRIKE, FASCIATED, 276, Pl. 31(1)
ANTSHRIKE, GREAT, 276, Pl. 31(2)
ANTSHRIKE, RUSSET, 278, Pl. 31(4)
ANTSHRIKE, SLATY, 277, Pl. 31(5)
ANT-TANAGER, BLACK-CHEEKED, 435, Pl. 47(11)
Ant-Tanager, Dusky-tailed, 435
ANT-TANAGER, RED-CROWNED, 434, Pl. 47(10)
ANT-TANAGER, RED-THROATED, 434, Pl. 47(12)
ANTTHRUSH, BLACK-FACED, 286, Pl. 30(15)
ANTTHRUSH, BLACK-HEADED, 286, Pl. 30(17)
Antthrush, Ocellated, 286
ANTTHRUSH, RUFOUS-BREASTED, 287, Pl. 30(16)
ANTVIREO, PLAIN, 278, Pl. 32(6)
ANTVIREO, SPOTTED-CROWNED, 279, Pl. 32(8)
ANTVIREO, STREAKED-CROWNED, 279, Pl. 32(7)
ANTWREN, CHECKER-THROATED, 280, Pl. 32(5)
ANTWREN, DOTTED-WINGED, 281, Pl. 32(1)
Antwren, Fulvous-bellied, 280
ANTWREN, RUFOUS-RUMPED, 281, Pl. 32(10)
ANTWREN, SLATY, 281, Pl. 32(3)
Antwren, Velvety, 282
ANTWREN, WHITE-FLANKED, 280, Pl. 32(2)
Aphanotriccus capitalis, 324, Pl. 36(10)
Aphriza virgata, 145, Pl. 10(7)
APODIDAE, 203
APODIFORMES, 203
Ara ambigua, 177, Pl. 19(2)
Ara macao, 177, Pl. 19(1)
ARACARI, COLLARED, 248, Pl. 27(15)
ARACARI, FIERY-BILLED, 249, Pl. 27(16)
ARAMIDAE, 123
Aramides axillaris, 125, Pl. 6(14)
Aramides cajanea, 125, Pl. 6(13)
Aramus guarauna, 124, Pl. 5(5)
Aratinga astec, 178
Aratinga canicularis, 178, Pl. 19(13)
Aratinga finschi, 177, Pl. 19(10)
Aratinga nana, 178, Pl. 19(11)
Archilochus colubris, 230, Pl. 25(9)
Ardea alba, 86
Ardea herodias, 86, Pl. 5(6)
ARDEIDAE, 80
Ardeola. See ***Bubulcus***
Arenaria interpres, 144, Pl. 10(6)
ARQUITECTO GÜITIO, 267
ARQUITECTO PLOMIZO, 267
Arremendado, 368
Arremon aurantiirostris, 458, Pl. 48(17)
Arremonops conirostris, 459, Pl. 50(14)
Arremonops rufivirgatus, 458, Pl. 50(15)
Arremonops superciliosus, 458
Arrocero, 415
Asio clamator, 195, Pl. 20(2)
Asio flammeus, 196
Asturina nitida, 107
Athene cunicularia, 194, Pl. 20(14)
ATILA LOMIAMARILLA, 310
Atlapetes albinucha, 456
Atlapetes atricapillus, 457, Pl. 48(14)
Atlapetes brunneinucha, 456, Pl. 48(16)
Atlapetes gutturalis, 456, Pl. 48(15)
Atlapetes torquatus, 457
Atticora. See ***Notiochelidon***
ATTILA, BRIGHT-RUMPED, 310, Pl. 35(6)
Attila, Polymorphic, 310
Attila spadiceus, 310, Pl. 35(6)
Aulacorhynchus caeruleogularis, 248
Aulacorhynchus prasinus, 248, Pl. 27(17)
Automolus, Buff-throated, 272
Automolus, Ruddy, 273
Automolus ochrolaemus, 272, Pl. 30(3)
Automolus rubiginosus, 272, Pl. 30(5)
Avefría, 136

AVETORILLO PANTANERO, 81
AVETORO NEOTROPICAL, 80
AVETORO NORTEÑO, 80
AVOCET, AMERICAN, 135, Pl. 9(11)
AVOCETA AMERICANA, 135
Aythya affinis, 94, Pl. 7(6)
Aythya collaris, 94, Pl. 7(7)
Aythya marila, 95, Pl. 7(5)
AZULILLO NORTEÑO, 447
AZULILLO SIETECOLORES, 447

Bailarín, 303
BALDPATE, 92
BANANAQUIT, 302, Pl. 40(24)
Bangsia. See ***Buthraupis***
BARBET, PRONG-BILLED, 247, Pl. 28(1)
BARBET, RED-HEADED, 246, Pl. 28(2)
BARBTAIL, SPOTTED, 268, Pl. 29(5)
BARBTHROAT, BAND-TAILED, 210, Pl. 23(5)
BARBUDO CABECIRROJO, 246
BARBUDO COCORA, 247
BARN-OWL, COMMON, 189, Pl. 20(9)
Barranquilla, 418
Bartramia longicauda, 141, Pl. 11(17)
Baryphthengus martii, 241, Pl. 27(7)
Baryphthengus ruficapillus, 242
Basileuterus culicivorus, 402, Pl. 40(18)
Basileuterus delattrii, 404
Basileuterus fulvicauda, 404
Basileuterus melanogenys, 403, Pl. 40(17)
Basileuterus rivularis, 404
Basileuterus rufifrons, 403, Pl. 40(16)
Basileuterus tristriatus, 402, Pl. 40(20)
BATARA BARRETEADO, 276
BATARA CAFE, 278
BATARA GRANDE, 276
BATARA LINEADO, 276
BATARA NEGRUZCO, 277
BATARA PLOMIZO, 277
BATARITO CABECIGRIS, 278
BATARITO CABECIPUNTEADO, 279
BATARITO PECHIRRAYADO, 279
BEARDLESS-TYRANNULET, NORTHERN, 336, Pl. 37(19)
BEARDLESS-TYRANNULET, SOUTHERN, 336, Pl. 37(20)
Becacina, 141, 142
BECACINA COMUN, 144
Becada, 144
BECARD, BARRED, 291, Pl. 33(14)
BECARD, BLACK-AND-WHITE, 292, Pl. 33(15)
BECARD, CINNAMON, 291, Pl. 33(11)
BECARD, ROSE-THROATED, 293, Pl. 33(12)
BECARD, WHITE-WINGED, 292, Pl. 33(13)
BELLBIRD, THREE-WATTLED, 298, Pl. 34(12)
BENTBILL, 331
BENTBILL, NORTHERN, 331, Pl. 37(6)
BIENTEVEO GRANDE, 313
BITTERN, AMERICAN, 80, Pl. 5(12)
BITTERN, LEAST, 81, Pl. 6(3)
BITTERN, PINNATED, 80, Pl. 5(11)
BLACKBIRD, MELODIOUS, 408, Pl. 52(6)
BLACKBIRD, RED-BREASTED, 414, Pl. 44(13)
BLACKBIRD, RED-WINGED, 413, Pl. 44(14)
BLACKBIRD, SCRUB, 414
BLACKBIRD, YELLOW-HEADED, 414, Pl. 44(4)
BLACK-HAWK, COMMON, 104, Pl. 13(6)
BLACK-HAWK, GREAT, 105, Pl. 13(7)
BLACK-HAWK, MANGROVE, 104
Bobillo, 333
Bobo Chiso, 185
BOBOLINK, 415, Pl. 50(18)
BOBWHITE, CRESTED, 121, Pl. 12(14)
BOBWHITE, SPOTTED-BELLIED, 120, Pl. 12(13)
Bolborhynchus lineola, 179, Pl. 19(16)
BOLSERO CAPUCHINEGRO, 410
BOLSERO CASTAÑO, 410
BOLSERO COLIAMARILLO, 411
BOLSERO DORSILISTADO, 413
BOLSERO NORTEÑO, 412
BOLSERO PECHIMANCHADO, 411
Bombycilla cedrorum, 373, Pl. 39(14)
BOMBYCILLIDAE, 372
BOOBY, BLUE-FACED, 76
BOOBY, BLUE-FOOTED, 76, Pl. 1(4)
BOOBY, BROWN, 76, Pl. 1(1)
BOOBY, MASKED, 76, Pl. 1(3)
BOOBY, RED-FOOTED, 77, Pl. 1(2)
BOOBY, WHITE, 76
Botaurus lentiginosus, 80, Pl. 5(12)
Botaurus pinnatus, 80, Pl. 5(11)
Brea, 452
BRILLANTE FRENTIVERDE, 227
BRILLIANT, GREEN-CROWNED, 227, Pl. 23(15)
Brotogeris jugularis, 180, Pl. 19(14)
BROWN-JAY, WHITE-TIPPED, 348
Bruja, 197
BRUSH-FINCH, BLACK-HEADED, 457, Pl. 48(14)
BRUSH-FINCH, CHESTNUT-CAPPED, 456, Pl. 48(16)
BRUSH-FINCH, STRIPED, 457
BRUSH-FINCH, STRIPE-HEADED, 457
BRUSH-FINCH, WHITE-NAPED, 456
BRUSH-FINCH, YELLOW-THROATED, 456, Pl. 48(15)
Bubo virginianus, 192, Pl. 20(1)
Bubulcus ibis, 83, Pl. 5(13)
Bucco macrorhynchos, 244, Pl. 28(3)
Bucco tectus, 244, Pl. 28(7)
BUCCONIDAE, 244
Buchón, 75
BUCO BARBON, 245
BUCO COLLAREJO, 244

BUCO PINTO, 244
Buho, 189, 192–196
BUHO DE ANTEOJOS, 192
BUHO GRANDE, 192
BUHO LISTADO, 195
BUHO PENACHUDO, 192
BUNTING, INDIGO, 447, Pls. 48(12), 50(6)
BUNTING, PAINTED, 447, Pl. 48(13)
BURHINIDAE, 135
Burhinus bistriatus, 135, Pl. 9(10)
Busarellus nigricollis, 105, Pl. 17(5)
BUSH-TANAGER, ASHY-THROATED, 440, Pl. 45(12)
Bush-Tanager, Brown-headed, 440
BUSH-TANAGER, COMMON, 439, Pl. 45(13)
BUSH-TANAGER, SOOTY-CAPPED, 440, Pl. 45(14)
Buteo albicaudatus, 108, Pls. 14(7), 17(6)
Buteo albonotatus, 109, Pls. 13(1), 14(2)
Buteo brachyurus, 107, Pls. 14(1), 16(11)
Buteo jamaicensis, 109, Pl. 17(8)
Buteo magnirostris, 107, Pl. 16(12)
Buteo nitidus, 106, Pl. 16(14)
Buteo plagiatus, 107
Buteo platypterus, 107, Pl. 16(13)
Buteo swainsoni, 108, Pls. 14(3), 17(7)
Buteogallus anthracinus, 104, Pl. 13(6)
Buteogallus meridionalis, 106, Pl. 17(4)
Buteogallus subtilis, 104
Buteogallus urubitinga, 105, Pl. 13(7)
Buthraupis arcaei, 428, Pl. 45(11)
Butorides striatus, 84, Pl. 6(1,2)
Butorides virescens, 84

CABEZON ALIBLANCO, 292
CABEZON CANELO, 291
CABEZON CEJIBLANCO, 292
CABEZON ONDEADO, 291
CABEZON PLOMIZO, 293
Cacao, 112
Cacicón, 412
Cacicus holosericeus, 407
Cacicus microrhynchus, 407
Cacicus uropygialis, 406, Pl. 44(11)
Cacique, Chisel-billed, 407
Cacique, Flame-rumped, 407
Cacique, Prevost's, 407
CACIQUE, SCARLET-RUMPED, 406, Pl. 44(11)
Cacique, Small-billed, 407
CACIQUE, YELLOW-BILLED, 407, Pl. 44(12)
Cacique Ahumado, 410
Cacique Amarillo, 410
CACIQUE LOMIESCARLATA, 406
CACIQUE PICOPLATA, 407
Cacique Veranero, 412
Caciquita, 417
Cairina moschata, 92, Pl. 8(10)
Calamón Morada, 129
Calandria, 298, 445
Calidris alba, 146, Pl. 11(11)
Calidris alpina, 149, Pl. 11(10)
Calidris bairdii, 148, Pl. 11(15)
Calidris canutus, 146, Pl. 11(19)
Calidris ferruginea, 149, Pl. 11(9)
Calidris fuscicollis, 148, Pl. 11(16)
Calidris himantopus, 149, Pl. 11(6)
Calidris mauri, 146, Pl. 11(12)
Calidris melanotos, 148, Pl. 11(7)
Calidris minutilla, 147, Pl. 11(14)
Calidris pusilla,147, Pl. 11(13)
Calliphlox bryantae, 229, Pl. 25(2)
Calocitta colliei, 347
Calocitta formosa, 347, Pl. 39(18)
Camaleón, 102, 103, 115
CAMPANERO TRICARUNCULADO, 298
Campephilus guatemalensis, 258, Pl. 27(14)
Camptostoma imberbe, 336, Pl. 37(19)
Camptostoma obsoletum, 336, Pl. 37(20)
Campylopterus cuvierii, 213
Campylopterus hemileucurus, 213, Pl. 23(9)
Campylorhamphus pusillus, 266, Pl. 29(14)
Campylorhynchus rufinucha, 352, Pl. 38(2)
Campylorhynchus zonatus, 351, Pl. 38(3)
Canaria, 387
CANDELITA COLLAREJA, 402
CANDELITA NORTEÑA, 401
CANDELITA PECHINEGRA, 401
Capella. See ***Gallinago***
Capito. See ***Eubucco***
CAPITONIDAE, 246
CAPRIMULGIDAE, 198
CAPRIMULGIFORMES, 196
Caprimulgus carolinensis, 201, Pl. 21(16)
Caprimulgus cayennensis, 203, Pl. 21(20)
Caprimulgus rufus, 202, Pl. 21(15)
Caprimulgus saturatus, 202, Pl. 21(19)
Caprimulgus vociferus, 202, Pl. 21(17)
Capsiempis flaveola, 333, Pl. 37(11)
CAPULINERO COLILARGO, 373
CAPULINERO NEGRO Y AMARILLO, 374
Cara de Gato, 189
CARACARA, CRESTED, 112, Pl. 14(12)
CARACARA, RED-THROATED, 112, Pl. 14(11)
CARACARA, YELLOW-HEADED, 113, Pl. 15(9)
CARACARA AVISPERA, 112
CARACARA CABECIGUALDO, 113
CARACARA CARGAHUESOS, 112
CARAO, 124
Cardenal, 431
Cardenal Veranero, 430
Cardenalito, 432
Carduelis psaltria, 465, Pl. 50(2)
Carduelis xanthogastra, 465, Pl. 50(1)
Cargahuesos, 112
Carmelo, 414
CARPENTERITO OLIVACEO, 251
Carpintero, 251–258

CARPINTERO ALIRRUFO, 256
CARPINTERO BEBEDOR, 254
CARPINTERO CANELO, 257
CARPINTERO CARETO, 252
CARPINTERO CARINEGRO, 252
CARPINTERO CASTAÑO, 257
Carpintero Chiricano, 258
CARPINTERO DE HOFFMANN, 253
CARPINTERO LINEADO, 257
CARPINTERO LOMIRROJO, 255
CARPINTERO NUQUIDORADO, 252
CARPINTERO NUQUIRROJO, 253
CARPINTERO PARDO, 255
CARPINTERO PICOPLATA, 258
CARPINTERO SERRANERO, 254
CARPINTERO VELLOSO, 254
CARPINTERO VERDE DORADO, 256
Carpodectes antoniae, 297, Pl. 34(3)
Carpodectes nitidus, 296, Pl. 34(4)
Carraco, 92
Caryothraustes canadensis, 444
Caryothraustes poliogaster, 443, Pl. 48(5)
Casmerodius albus, 86, Pl. 5(14)
Cassidix. See ***Quiscalus***
CASTLEBUILDER, PALE-BREASTED, 267
CASTLEBUILDER, SLATY, 268
Catano, 178
CATBIRD, GRAY, 362, Pl. 39(12)
Catharacta antarctica, 153
Catharacta chilensis, 153
Catharacta maccormicki, 153, Pl. 3(12)
Catharacta skua, 153
Cathartes aura, 96, Pl. 13(3)
Cathartes burrovianus, 96, Pl. 13(2)
CATHARTIDAE, 95
Catharus aurantiirostris, 369, Pl. 38(26)
Catharus frantzii, 369, Pl. 38(27)
Catharus fuscater, 368, Pl. 38(23)
Catharus fuscescens, 367, Pl. 39(3)
Catharus gracilirostris, 370, Pl. 38(24)
Catharus griseiceps, 369
Catharus mexicanus, 368, Pl. 38(25)
Catharus minimus, 367, Pl. 39(1)
Catharus ustulatus, 366, Pl. 39(2)
Catoptrophorus semipalmatus, 143, Pl. 9(6)
Celeus castaneus, 257, Pl. 28(8)
Celeus loricatus, 257, Pl. 28(9)
Centurus. See ***Melanerpes***
Cephalopterus glabricollis, 297, Pl. 34(13)
Cephalopterus ornatus, 298
CERCETA ALIAZUL, 93
CERCETA ALIVERDE, 92
CERCETA CASTAÑA, 93
Cercomacra tyrannina, 282, Pl. 31(8)
CERNICALO AMERICANO, 115
Ceryle alcyon, 238, Pl. 27(2)
Ceryle torquata, 237, Pl. 27(1)
Chachalaca, 118
CHACHALACA, GRAY-HEADED, 118, Pl. 12(1)
CHACHALACA, PLAIN, 118, Pl. 12(2)
CHACHALACA CABECIGRIS, 118
CHACHALACA OLIVACEA, 118
Chaetura cinereiventris, 207, Pl. 22(10)
Chaetura martinica, 207
Chaetura pelagica, 206, Pl. 22(8)
Chaetura richmondi, 207
Chaetura rutila, 206
Chaetura spinicauda, 207, Pl. 22(11)
Chaetura vauxi, 207, Pl. 22(9)
Chalybura melanorrhoa, 225
Chalybura urochrysia, 225, Pl. 23(12)
Chamaepetes unicolor, 119, Pl. 12(5)
Chamaethlypis poliocephala, 399
CHARADRIIDAE, 136
CHARADRIIFORMES, 131
Charadrius alexandrinus, 138, Pl. 10(4)
Charadrius collaris, 139, Pl. 10(3)
Charadrius hiaticula, 137
Charadrius melodus, 137
Charadrius nivosus, 139
Charadrius semipalmatus, 137, Pl. 10(5)
Charadrius vociferus, 138, Pl. 10(1)
Charadrius wilsonia, 138, Pl. 10(2)
CHARRAN BLANCO, 164
CHARRAN CHICO, 161
CHARRAN COMUN, 159
CHARRAN DE FORSTER, 160
CHARRAN EMBRIDADO, 160
CHARRAN MENUDO, 161
CHARRAN PIQUINEGRO, 159
CHARRAN SOMBRIO, 161
CHARRANCITO NEGRO, 158
CHAT, GROUND, 399
CHAT, YELLOW-BREASTED, 399, Pl. 42(9)
Cherelá, 140
Chicharrón, 464
Chico Piojo, 352
Chiltote, 411
Chiltotel, 411
CHIMBITO COMUN, 463
Chinchirigüí, 352
CHINGOLO, 463
CHIRIGÜE SABANERO, 453
Chirincoco, 125
Chiroxiphia lanceolata, 301, Pl. 33(5)
Chiroxiphia linearis, 301, Pl. 33(4)
Chirrascuá, 120, 122
CHISPITA GORGINARANJA, 230
CHISPITA VOLCANERA, 231
Chispitas, 230, 231
Chlidonias niger, 158, Pl. 2(6)
Chloroceryle aenea, 239, Pl. 27(6)
Chloroceryle amazona, 238, Pl. 27(4)
Chloroceryle americana, 238, Pl. 27(5)
Chloroceryle inda, 239, Pl. 27(3)
Chlorophanes spiza, 425, Pl. 46(7)
CHLOROPHONIA, BLUE-CROWNED, 417
CHLOROPHONIA, GOLDEN-BROWED, 416, Pl. 45(10)

Chlorophonia callophrys, 416, Pl. 45(10)
Chlorophonia occipitalis, 417
Chlorospingus canigularis, 440, Pl. 45(12)
Chlorospingus canigularis olivaceiceps, 441
Chlorospingus ophthalmicus, 439, Pl. 45(13)
Chlorospingus pileatus, 440, Pl. 45(14)
Chlorospingus zeledoni, 440
Chlorostilbon assimilis, 217
Chlorostilbon canivetii, 217, Pls. 24(13), 25(10)
Chlorostilbon mellisugus, 217
Chlorothraupis carmioli, 433, Pl. 45(16)
Chlorothraupis olivacea, 434
Chocuaca, 82
Chocuaco, 83, 84
Chondrohierax uncinatus, 98, Pls. 14(9), 16(9)
Chorcha, 413
Chordeiles acutipennis, 200, Pl. 21(12)
Chordeiles minor, 199, Pl. 21(13)
CHORLITEJO COLLAREJO, 139
CHORLITEJO DE DOS COLLARES, 138
CHORLITEJO PATINEGRO, 138
CHORLITEJO PIQUIGRUESO, 138
CHORLITEJO SEMIPALMADO, 137
CHORLITEJO TILDIO, 138
Chorlito, 137–139
CHORLITO DE ROMPIENTES, 145
CHORLITO DORADO MENOR, 136
CHORLITO GRIS, 136
Chorlito Gritón, 138
CHOTACABRAS COLIBLANCO, 203
CHOTACABRAS DE PASO, 201
CHOTACABRAS GRITON O RUIDOSO, 202
CHOTACABRAS OCELADO, 201
CHOTACABRAS ROJIZO, 202
CHOTACABRAS SOMBRIO, 202
Chrysothlypis chrysomelas, 439, Pl. 46(14)
CHUCK-WILL'S-WIDOW, 201, Pl. 21(16)
Chucuyo, 200
Ciccaba huhula, 195
Ciccaba nigrolineata, 195, Pl. 20(7)
Ciccaba virgata, 194, Pl. 20(6)
CICONIIDAE, 87
CICONIIFORMES, 80
CIGÜEÑON, 87
CIGÜEÑUELA CUELLINEGRO, 134
CINCLIDAE, 350
Cinclus mexicanus, 350, Pl. 39(11)
Circus cyaneus, 101, Pl. 15(7)
Cirujano, 132
Cistothorus platensis, 351, Pl. 38(1)
Cistothorus stellaris, 351
Claravis mondetoura, 172, Pl. 18(10)
Claravis pretiosa, 171, Pl. 18(6)
CLARINERO DE LAGUNA, 409
CLARINERO GRANDE, 409
CLOROFONIA CEJIDORADA, 416
Coccyzus americanus, 184, Pl. 21(2)
Coccyzus erythropthalmus, 184, Pl. 21(1)
Coccyzus ferrugineus, 185, Pl. 21(3)
Coccyzus melacoryphus, 185
Coccyzus minor, 185, Pl. 21(4)
COCHLEARIIDAE, 83
Cochlearius cochlearius, 83, Pl. 5(2)
Coco, 89
Coco Negro, 88
Cocora, 247
Codorniz, 120–123
CODORNIZ CARIRROJA, 121
CODORNIZ CARIRRUFA, 123
CODORNIZ CRESTADA, 121
CODORNIZ MOTEADA, 122
CODORNIZ PECHICASTAÑA, 122
CODORNIZ PECHINEGRA, 122
CODORNIZ PINTADA, 122
CODORNIZ VIENTRIMANCHADA, 120
Coereba flaveola, 382, Pl. 40(24)
COEREBIDAE, 381
COLAESPINA CARIRROJA, 268
Coliblanca, 172
COLIBRI CABECIAZUL, 215
COLIBRI COLIDORADO, 219
COLIBRI COLIRRAYADO, 223
COLIBRI GARGANTA DE FUEGO, 218
COLIBRI GARGANTA DE RUBI, 230
COLIBRI GARGANTA DE ZAFIRO, 218
COLIBRI MAGNIFICO, 227
COLIBRI MONTAÑES COLIGRIS, 226
COLIBRI MONTAÑES GORGIMORADO, 227
COLIBRI MONTAÑES VIENTRIBLANCO, 225
Colibrí Mosca, 230, 231
COLIBRI OREJIVIOLACEO PARDO, 214
COLIBRI OREJIVIOLACEO VERDE, 214
COLIBRI PATIRROJO, 225
COLIBRI PECHIESCAMADO, 212
COLIBRI PECHINEGRO, 223
COLIBRI PICOPUNZON, 228
COLIBRI PIQUILARGO, 229
COLIBRI POCHOTERO, 228
Colibri delphinae, 214, Pl. 23(6)
Colibri thalassinus, 214, Pl. 23(7)
COLICERDA VERDE, 216
Colinus cristatus, 121, Pl. 12(14)
Colinus leucopogn, 120, Pl. 12(13)
Colonia colonus, 305, Pl. 36(6)
Columba albilinea, 167
Columba cayennensis, 167, Pl. 18(3)
Columba fasciata, 167, Pl. 18(1)
Columba flavirostris, 168, Pl. 18(2)
Columba leucocephala, 166, Pl. 51(1)
Columba livia, 166
Columba nigrirostris, 168, Pl. 18(5)
Columba speciosa, 166, Pl. 18(4)
Columba subvinacea, 168
COLUMBIDAE, 165
COLUMBIFORMES, 165
Columbigallina. See ***Columbina***
Columbina inca, 171, Pl. 18(11)
Columbina minuta, 170, Pl. 18(9)
Columbina passerina, 170, Pl. 18(8)
Columbina talpacoti, 170, Pl. 18(7)

COMBATIENTE, 150
Come-cacao, 112
Comemaiz, 463
Comepuntas, 456
Conchita, 170
Conopias. See ***Coryphotriccus***
Contopus borealis, 317, Pl. 36(4)
Contopus cinereus, 318, Pl. 36(9)
Contopus fumigatus, 319
Contopus lugubris, 319, Pl. 36(2)
Contopus ochraceus, 319, Pl. 36(3)
Contopus sordidulus, 317, Pl. 36(7)
Contopus virens, 318, Pl. 36(8)
COOT, AMERICAN, 130, Pl. 7(1)
COPETE DE NIEVE, 225
COPETON COLIPARDO, 314
COPETON CRESTIOSCURO, 316
COPETON CRESTIPARDO, 315
COPETON GARGANTICENIZA, 316
COPETON DE NUTTING, 315
COPETON VIAJERO, 316
Copetoncillo, 333
COQUETA CRESTIBLANCA, 215
COQUETA CRESTINEGRA, 216
COQUETA CRESTIRROJIZA, 216
Coquette, Adorable, 216
COQUETTE, BLACK-CRESTED, 216, Pl. 25(6)
COQUETTE, RUFOUS-CRESTED, 216, Pl. 25(4)
COQUETTE, WHITE-CRESTED, 215, Pl. 25(5)
CORACIIFORMES, 237
Coragyps atratus, 96, Pl. 13(4)
Corapipo altera, 302
Corapipo leucorrhoa, 302, Pl. 33(9)
CORCOVADO, 121
CORMORAN NEOTROPICAL, 77
Cormorant, Neotropical, 77
CORMORANT, OLIVACEOUS, 77, Pl. 4(4)
Correa, 124
CORRELIMOS DE BAIRD, 148
CORRELIMOS GRANDE, 146
CORRELIMOS LOMIBLANCO, 148
CORRELIMOS MENUDO (MENUDILLO), 147
CORRELIMOS OCCIDENTAL, 146
CORRELIMOS PATILARGO, 149
CORRELIMOS PECHINEGRO, 149
CORRELIMOS PECHIRRAYADO O PECTORAL, 148
CORRELIMOS SEMIPALMADO, 147
CORRELIMOS VAGAMUNDO, 143
CORVIDAE, 347
Coryphotriccus albovittatus, 309, Pl. 35(16)
Cotinga, Antonia's, 297
COTINGA, LOVELY, 295, Pl. 34(6)
COTINGA, SNOWY, 296, Pl. 34(4)
COTINGA, TURQUOISE, 296, Pl. 34(5)
COTINGA, YELLOW-BILLED, 297, Pl. 34(3)
COTINGA LINDA, 295
COTINGA NIVOSA, 296
COTINGA PIQUIAMARILLO, 297
COTINGA TURQUESA, 296
Cotinga amabilis, 295, Pl. 34(6)
Cotinga ridgwayi, 296, Pl. 34(5)
COTINGIDAE, 294
Cotorra, 177
Coturnicops. See ***Micropygia***
COWBIRD, BRONZED, 408, Pls. 44(15), 52(7)
COWBIRD, GIANT, 407, Pl. 44(10)
Cowbird, Red-eyed, 408
CRACIDAE, 117
Crake, Colombian, 128
CRAKE, GRAY-BREASTED, 127, Pl. 6(6)
CRAKE, OCELLATED, 128, Pl. 6(7)
CRAKE, PAINTED-BILLED, 128, Pl. 51(12)
CRAKE, RUDDY, 128, Pl. 6(11)
CRAKE, UNIFORM, 125, Pl. 6(8)
CRAKE, WHITE-THROATED, 127, Pl. 6(9)
CRAKE, YELLOW-BREASTED, 126, Pl. 6(10)
Cranioleuca erythrops, 268, Pl. 30(12)
Crax rubra, 119, Pl. 12(3)
Creagrus furcatus, 158
Cristofué, 313
Crotophaga ani, 186, Pl. 21(8)
Crotophaga sulcirostris, 186, Pl. 21(9)
Crypturellus boucardi, 67, Pl. 12(8)
Crypturellus cinnamomeus, 66, Pl. 12(18)
Crypturellus columbianus, 67
Crypturellus idoneus, 67
Crypturellus kerriae, 67
Crypturellus noctivagus, 67
Crypturellus soui, 66, Pl. 12(17)
Cuaca, 83
Cuatro Ojos, 194, 460
Cucarachero, 357
CUCKOO, BLACK-BILLED, 184, Pl. 21(1)
CUCKOO, COCOS, 185, Pl. 21(3)
CUCKOO, MANGROVE, 185, Pl. 21(4)
CUCKOO, PHEASANT, 188, Pl. 21(6)
CUCKOO, SQUIRREL, 185, Pl. 21(7)
CUCKOO, STRIPED, 187, Pl. 21(5)
CUCKOO, YELLOW-BILLED, 184, Pl. 21(2)
CUCLILLO DE ANTIFAZ, 185
CUCLILLO DE LA ISLA DEL COCO, 185
CUCLILLO FAISAN, 188
CUCLILLO LISTADO, 187
CUCLILLO OREJINEGRO, 185
CUCLILLO PIQUIGUALDO, 184
CUCLILLO PIQUINEGRO, 184
CUCLILLO SABANERO, 187
CUCO ARDILLA, 185
CUCO HORMIGUERO, 188
CUCULIDAE, 183
CUCULIFORMES, 183
Cuitiento, 370
CURASSOW, GREAT, 119, Pl. 12(3)
CURLEW, LONG-BILLED, 141, Pl. 9(13)
Curré, 248
Curré Negro, 250

Cusingo, 249
Cuyeo, 300
Cyanerpes caeruleus, 427
Cyanerpes cyaneus, 426, Pl. 46(2)
Cyanerpes lucidus, 426, Pl. 46(1)
Cyanocompsa cyanoides, 446, Pl. 48(10)
Cyanocorax affinis, 348, Pl. 39(20)
Cyanocorax morio, 348, Pl. 39(19)
Cyanolyca argentigula, 349, Pl. 39(16)
Cyanolyca cucullata, 349, Pl. 39(17)
Cyclarhis gujanensis, 375, Pl. 40(2)
Cymbilaimus lineatus, 276, Pl. 31(1)
Cyphorhinus aradus, 361
Cyphorhinus phaeocephalus, 361, Pl. 38(22)
Cypseloides cherriei, 205, Pl. 22(5)
Cypseloides cryptus, 205, Pl. 22(4)
Cypseloides niger, 204, Pl. 22(2)
Cypseloides rutilus, 205, Pl. 22(3)
DACNIS, BLUE, 427, Pl. 46(3)
DACNIS, SCARLET-THIGHED, 427, Pl. 46(4)
Dacnis cayana, 427, Pl. 46(3)
Dacnis venusta, 427, Pl. 46(4)
Damophila julie, 218
Daptrius americanus, 112, Pl. 14(11)
Deconychura longicauda, 260, Pl. 29(10)
Deconychura typica, 261
Dendrocincla anabatina, 260, Pl. 29(12)
Dendrocincla fuliginosa, 259, Pl. 29(11)
Dendrocincla homochroa, 260, Pl. 29(13)
Dendrocolaptes certhia, 262, Pl. 29(19)
Dendrocolaptes picumnus, 263, Pl. 29(15)
DENDROCOLAPTIDAE, 259
Dendrocopos. See ***Picoides***
Dendrocygna autumnalis, 90, Pl. 8(7)
Dendrocygna bicolor, 91, Pl. 8(9)
Dendrocygna viduata, 91, Pl. 8(8)
Dendroica auduboni, 390
Dendroica caerulescens, 389, Pl. 43(1)
Dendroica castanea, 393, Pl. 41(9)
Dendroica cerulea, 391, Pl. 43(2)
Dendroica coronata, 389, Pl. 43(11)
Dendroica discolor, 394, Pl. 43(5)
Dendroica dominica, 392, Pl. 41(12)
Dendroica fusca, 392, Pl. 41(8)
Dendroica magnolia, 388, Pl. 43(10)
Dendroica occidentalis, 391, Pl. 43(7)
Dendroica palmarum, 394, Pl. 43(3)
Dendroica pensylvanica, 392, Pl. 43(4)
Dendroica petechia, 387, Pl. 42(2)
Dendroica petechia aureola, 388
Dendroica petechia erithachorides, 388, Pl. 42(3)
Dendroica pinus, 394, Pl. 41(14)
Dendroica striata, 393, Pl. 41(11)
Dendroica tigrina, 389, Pl. 43(6)
Dendroica townsendi, 390, Pl. 43(8)
Dendroica virens, 390, Pl. 43(9)
Dendrortyx leucophrys, 120, Pl. 12(16)
Deslenguado, 112
Dichromanassa rufescens, 86
DICKCISSEL, 415, Pl. 50(19)
Diglossa baritula, 455
Diglossa plumbea, 454, Pl. 49(9)
Dios-te-dé, 250
DIPPER, AMERICAN, 350, Pl. 39(11)
Discosura conversii, 216, Pl. 25(1)
Dives dives, 408, Pl. 52(6)
Dives warszewiczi, 409
Dolichonyx oryzivorus, 415, Pl. 50(18)
Doryfera ludoviciae, 212, Pl. 23(11)
Dos Golpes, 258
Dos-tontos-son, 168
DOVE, GRAY-CHESTED, 173, Pl. 18(16)
DOVE, GRAY-FRONTED, 173, Pl. 18(15)
Dove, Gray-headed, 173
DOVE, INCA, 171, Pl. 18(11)
DOVE, MOURNING, 169, Pl. 18(13)
DOVE, ROCK, 166
Dove, Rufous-naped, 174
Dove, White-fronted, 172
DOVE, WHITE-TIPPED, 172, Pl. 18(14)
DOVE, WHITE-WINGED, 169, Pl. 18(12)
Dowitcher, Common, 145
DOWITCHER, LONG-BILLED, 145, Pl. 9(4)
DOWITCHER, SHORT-BILLED, 145, Pls. 9(3), 11(5)
Dromococcyx phasianellus, 188, Pl. 21(6)
Dryocopus lineatus, 257, Pl. 27(13)
DUCK, MASKED, 95, Pl. 7(8)
DUCK, MUSCOVY, 92, Pl. 8(10)
DUCK, RING-NECKED, 94, Pl. 7(7)
Duck, Ruddy, 95
Dumetella carolinensis, 362, Pl. 39(12)
DUNLIN, 149, Pl. 11(10)
Dysithamnus mentalis, 278, Pl. 32(6)
Dysithamnus puncticeps, 279, Pl. 32(8)
Dysithamnus striaticeps, 279, Pl. 32(7)

EAGLE, CRESTED, 109, 17(10)
EAGLE, HARPY, 110, Pl. 17(9)
EAGLE, SOLITARY, 105, Pl. 13(8)
Egret, American, 86
EGRET, CATTLE, 83, Pl. 5(13)
EGRET, GREAT, 86, Pl. 5(14)
EGRET, REDDISH, 85, Pl. 5(7)
EGRET, SNOWY, 86, Pl. 5(10)
Egretta alba, 86
Egretta caerulea, 84, Pl. 5(9)
Egretta rufescens, 85, Pl. 5(7)
Egretta thula, 86, Pl. 5(10)
Egretta tricolon, 85, Pl. 5(8)
ELAENIA, GREENISH, 335, Pl. 37(21)
ELAENIA, LESSER, 334, Pl. 37(25)
ELAENIA, MOUNTAIN, 334, Pl. 37(24)
ELAENIA, YELLOW-BELLIED, 333, Pl. 37(26)
Elaenia chiriquensis, 334, Pl. 37(25)
Elaenia flavogaster, 333, Pl. 37(26)
Elaenia frantzii, 334, Pl. 37(24)
ELAINIA COPETONA, 333
ELAINIA MONTAÑERA, 334

ELAINIA SABANERA, 334
ELAINIA VERDOSA, 335
ELANIO CARACOLERO, 99
ELANIO COLIBLANCO, 99
ELANIO COLINEGRO, 101
ELANIO PLOMIZO, 100
ELANIO TIJERETA, 99
***Elanoides forficatus*,** 99, Pl. 15(2)
***Elanus caeruleus*,** 99, Pl. 15(1)
Elanus leucurus, 99
***Electron carinatum*,** 241, Pl. 27(10)
***Electron platyrhynchum*,** 240, Pl. 27(9)
***Elvira chionura*,** 224, Pls. 24(19), 25(16)
***Elvira cupreiceps*,** 224, Pls. 24(20), 25(17)
EMBERIZIDAE, 441
***Emberizoides herbicola*,** 464, Pl. 50(10)
EMERALD, COPPERY-HEADED, 224, Pls. 24(20), 25(17)
EMERALD, FORK-TAILED, 217, Pls. 24(13), 25(10)
EMERALD, GARDEN, 217
EMERALD, WHITE-BELLIED, 219, Pl. 24(1)
EMERALD, WHITE-TAILED, 224, Pls. 24(19), 25(16)
***Empidonax albigularis*,** 322, Pl. 36(13)
***Empidonax alnorum*,** 321, Pl. 36(14)
***Empidonax atriceps*,** 323, Pl. 36(16)
***Empidonax flavescens*,** 322, Pl. 36(19)
***Empidonax flaviventris*,** 320, Pl. 36(20)
***Empidonax minimus*,** 322, Pl. 36(15)
***Empidonax traillii*,** 321, Pl. 36(14)
***Empidonax virescens*,** 320, Pl. 36(12)
Erator inquisitor, 294
ERMITAÑO BARBUDO, 210
ERMITAÑO BRONCEADO, 210
ERMITAÑO COLILARGO, 211
ERMITAÑO ENANO, 211
ERMITAÑO VERDE, 211
Escarchado, 365
Escarchero, 365
ESMERALDA COLIBLANCA, 224
ESMERALDA DE CORONILLA COBRIZA, 224
ESMERALDA RABIHORCADA, 217
ESMEREJON, 115
ESPATULA ROSADA, 90
ESPATULILLA CABECIGRIS, 330
ESPATULILLA CABECINEGRA, 329
ESPATULILLA COMUN, 330
ESPIGUERO COLLAREJO, 448
ESPIGUERO MENUDO, 450
ESPIGUERO PIZARROSO, 448
ESPIGUERO VARIABLE, 449
ESPIGUERO VIENTRIAMARILLO, 450
ESTRELLITA GORGIMORADA, 229
Estucurú, 190, 191
***Eubucco bourcierii*,** 246, Pl. 28(2)
***Eucometis penicillata*,** 437, Pl. 45(17)
***Eudocimus albus*,** 89, Pl. 4(8)
EUFONIA CAPUCHICELESTE, 417
EUFONIA CORONIAMARILLA, 419
EUFONIA GARGANTINEGRA, 419
EUFONIA GORGIAMARILLA, 420
EUFONIA GORRICANELA, 418
EUFONIA MENUDA, 418
EUFONIA OLIVACEA, 421
EUFONIA PIQUIGRUESA, 420
EUFONIA VIENTRIRROJIZA, 421
***Eugenes fulgens*,** 227, Pl. 23(16)
Eugenes spectabilis, 228
***Eumomota superciliosa*,** 241, Pl. 27(11)
***Eupherusa eximia*,** 223, Pls. 24(17), 25(14)
***Eupherusa nigriventris*,** 223, Pls. 24(21), 25(12)
EUPHONIA, BLUE-HOODED, 417, Pl. 45(9)
EUPHONIA, BONAPARTE'S, 420
EUPHONIA, GOULD'S, 421
EUPHONIA, OLIVE-BACKED, 421, Pl. 45(3)
EUPHONIA, SCRUB, 419, Pl. 45(4)
EUPHONIA, SPOTTED-CHROWNED, 421, Pl. 45(2)
EUPHONIA, TAWNY-BELLIED, 422
EUPHONIA, TAWNY-CAPPED, 418, Pl. 45(6)
EUPHONIA, THICK-BILLED, 420, Pl. 45(8)
EUPHONIA, WHITE-VENTED, 418, Pl. 45(7)
EUPHONIA, YELLOW-CROWNED, 419, Pl. 45(1)
EUPHONIA, YELLOW-THROATED, 420, Pl. 45(5)
***Euphonia affinis*,** 419, Pl. 45(4)
***Euphonia anneae*,** 418, Pl. 45(6)
***Euphonia elegantissima*,** 417, Pl. 45(9)
***Euphonia gouldi*,** 421, Pl. 45(3)
***Euphonia hirundinacea*,** 420, Pl. 45(5)
***Euphonia imitans*,** 421, Pl. 45(2)
***Euphonia laniirostris*,** 420, Pl. 45(8)
Euphonia lauta, 420
***Euphonia luteicapilla*,** 419, Pl. 45(1)
***Euphonia minuta*,** 418, Pl. 45(7)
Euphonia musica, 417
***Eurypyga helias*,** 131, Pl 6(16)
EURYPYGIDAE, 131
***Eutoxeres aquila*,** 209, Pl. 23(8)

FAIRY, BLACK-EARED, 228
FAIRY, PURPLE-CROWNED, 228, Pl. 23(14)
FALAROPO DE WILSON, 152
FALAROPO PICOFINO, 151
FALAROPO ROJO, 151
FALAROPO TRICOLOR, 152
***Falco columbarius*,** 115, Pl. 15(15)
***Falco deiroleucus*,** 116, Pl. 15(11)
***Falco femoralis*,** 116, Pl. 15(10)
***Falco peregrinus*,** 117, Pl. 15(5)
***Falco rufigularis*,** 116, Pl. 15(14)
***Falco sparverius*,** 115, Pl. 15(13)
FALCON, APLOMADO, 116, Pl. 15(10)
FALCON, BAT, 116, Pl. 15(14)
FALCON, LAUGHING, 113, Pl. 15(8)
FALCON, ORANGE-BREASTED, 116, Pl. 15(11)
FALCON, PEREGRINE, 117, Pl. 15(5)

FALCONIDAE, 112
FALCONIFORMES, 95
Felix, 248
FINCH, COCOS, 452, Pl. 49(8)
FINCH, LARGE-FOOTED, 455, Pl. 48(20)
FINCH, PEG-BILLED, 454, Pl. 49(12)
FINCH, SLATY, 453, Pl. 49(14)
FINCH, SOOTY-FACED, 457, Pl. 48(21)
FINCH, YELLOW-THIGHED, 455, Pl. 48(22)
Finfoot, American, 131
FLATBILL, EYE-RINGED, 329, Pl. 37(23)
Flatbill, Yellow-olive, 328
Florida caerulea, 84
Florisuga mellivora, 213, Pl. 23(17)
FLOWERPIERCER, SLATY, 454, Pl. 49(9)
FLYCATCHER, ACADIAN, 320, Pl. 36(12)
FLYCATCHER, ALDER, 321, Pl. 36(14)
FLYCATCHER, ASH-THROATED, 316, Pl. 35(19)
FLYCATCHER, BLACK-CAPPED, 323, Pl. 36(16)
FLYCATCHER, BLACK-TAILED, 325, Pl. 36(22)
FLYCATCHER, BOAT-BILLED, 309, Pl. 35(12)
FLYCATCHER, BRAN-COLORED, 326, Pl. 36(17)
FLYCATCHER, BROWN-CRESTED, 315, Pl. 35(18)
FLYCATCHER, COCOS, 324, Pl. 36(18)
FLYCATCHER, DUSKY-CAPPED, 316, Pl. 35(21)
FLYCATCHER, FORK-TAILED, 306, Pl. 35(4)
FLYCATCHER, GOLDEN-BELLIED, 312, Pl. 35(9)
Flycatcher, Golden-crowned, 312
FLYCATCHER, GRAY-CAPPED, 312, Pl. 35(15)
FLYCATCHER, GREAT CRESTED, 316, Pl. 35(17)
FLYCATCHER, LEAST, 322, Pl. 36(14)
FLYCATCHER, NUTTING'S, 315, Pl. 35(22)
FLYCATCHER, OCHRE-BELLIED, 340, Pl. 36(25)
FLYCATCHER, OLIVE-SIDED, 317, Pl. 36(4)
FLYCATCHER, OLIVE-STRIPED, 340, Pl. 36(24)
Flycatcher, Pale-throated, 315
FLYCATCHER, PANAMA, 314, Pl. 35(20)
FLYCATCHER, PIRATIC, 308, Pl. 35(8)
FLYCATCHER, ROYAL, 326, Pl. 35(23)
FLYCATCHER, RUDDY-TAILED, 324, Pl. 36(23)
FLYCATCHER, SCISSOR-TAILED, 306, Pl. 35(5)
FLYCATCHER, SCRUB, 336, Pl. 37(22)
FLYCATCHER, SEPIA-CAPPED, 340, Pl. 36(27)
Flycatcher, Short-billed, 337
Flycatcher, Short-crested, 315
FLYCATCHER, SLATY-CAPPED, 339, Pl. 36(26)
FLYCATHER, SOCIAL, 313, Pl. 35(14)
FLYCATCHER, STREAKED, 311, Pl. 35(11)
FLYCATCHER, SULPHUR BELLIED, 310, Pl. 35(10)
FLYCATCHER, SULPHUR-RUMPED, 325, Pl. 36(21)
FLYCATCHER, TAWNY-CHESTED, 324, Pl. 36(10)
Flycatcher, Traill's, 321
FLYCATCHER, TUFTED, 323, Pl. 36(11)
Flycatcher, Vermilion-crowned, 313
FLYCATCHER, WHITE-RINGED, 306, Pl. 35(16)
FLYCATCHER, WHITE-THROATED, 322, Pl. 36(13)
Flycatcher, Wied's Crested, 315
FLYCATCHER, WILLOW, 321, Pl. 36(14)
FLYCATCHER, YELLOW-BELLIED, 320, Pl. 36(20)
FLYCATCHER, YELLOWISH, 322, Pl. 36(19)
FLYCATCHER, YELLOW-MARGINED, 329, Pl. 37(17)
FLYCATCHER, YELLOW-OLIVE, 328, Pl. 37(16)
FOCHA AMERICANA, 130
FOLIAGE-GLEANER, BUFF-FRONTED, 271, Pl. 30(6)
FOLIAGE-GLEANER, BUFF-THROATED, 272, Pl. 30(3)
FOLIAGE-GLEANER, LINEATED, 270, Pl. 30(8)
FOLIAGE-GLEANER, RUDDY, 272, Pl. 30(5)
Foliage-gleaner, Scaly-Throated, 271
FOLIAGE-GLEANER, SPECTACLED, 271, Pl. 30(2)
FOLIAGE-GLEANER, STRIPED, 270, Pl. 30(4)
FOREST-FALCON, BARRED, 114, Pl. 16(5)
FOREST-FALCON, COLLARED, 114, Pls. 14(10), 16(7)
FOREST-FALCON, SLATY-BACKED, 114, Pl. 16(6)
FORMICARIIDAE, 275
Formicarius analis, 286, Pl. 30(15)
Formicarius nigricapillus, 286, Pl. 30(17)
Formicarius rufipectus, 287, Pl. 30(16)
Fregata magnificens, 79, Pl. 1(6)
Fregata minor, 79, Pl. 1(5)
FREGATIDAE, 78
Freidora, 127
FRIGATEBIRD, GREAT, 79, Pl. 1(5)
FRIGATEBIRD, MAGNIFICENT, 79, Pl. 1(6)
FRINGILLIDAE, 464
FRINGILO PIQUIAGUDO, 454
FRINGILO PLOMIZO, 453
FRUITCROW, PURPLE-THROATED, 297, Pl. 34(11)

Fulica americana, 130, Pl. 7(1)
FUMAREL NEGRO, 158
FURNARIIDAE, 266

Galán Sin Ventura, 88
Galbula melanogenia, 243
Galbula ruficauda, 243, Pl. 26(12)
GALBULIDAE, 243
Gallardo, 67
GALLARETA FRENTIRROJA, 129
GALLARETA MORADA, 129
GALLIFORMES, 117
Gallina de Agua, 129
Gallina de Monte, 65
Gallinago gallinago, 144, Pl. 9(5)
Gallinazo, 96
Gallinita, 122
Gallinita de Monte, 122
Gallinula chloropus, 129, Pl. 7(2)
GALLINULE, COMMON, 129, Pl. 7(2)
GALLINULE, PURPLE, 129, Pl. 6(15)
Gallito, 447
Gallito de Agua, 132
GALLITO HORMIGUERO CABECINEGRO, 286
GALLITO HORMIGUERO CARINEGRO, 286
GALLITO HORMIGUERO PECHICASTANO, 287
Gampsonyx swainsonii, 115, Pl. 15(12)
Gansa, 141
GARCETA AZUL, 84
GARCETA GRANDE, 86
GARCETA NIVOSA, 86
GARCETA ROJIZA, 85
GARCETA TRICOLOR, 85
GARCILLA BUEYERA, 83
GARCILLA VERDE, 84
Garrapatero, 186
GARRAPATERO PIQUIESTRIADO, 186
GARRAPATERO PIQUILISO, 186
GARRAPATERO TIJO, 186
GARUMA, 157
Garza Ceniza, 86
Garza del Ganado, 83
GARZA DEL SOL, 131
Garza Morena, 90
GARZA PECHICASTANA, 87
Garza Real, 86
Garza Rosada, 90
GARZA-TIGRE CUELLINUDA, 82
GARZA-TIGRE DE RIO, 81
GARZA-TIGRE DE SELVA, 81
Garzón, 87
GARZON AZULADO, 86
GAVILAN ALICASTAÑO, 106
GAVILAN ALUDO, 107
Gavilán Bailarín, 99
GAVILAN BICOLOR, 103
GAVILAN BLANCO, 104
GAVILAN CABECIGRIS, 98
GAVILAN CANGREJERO, 104
Gavilán Caracolero, 99
GAVILAN CHAPULINERO, 107
GAVILAN COLIBLANCO, 108
GAVILAN COLICORTO, 107
GAVILAN COLIFAJEADO, 109
GAVILAN COLIRROJO, 109
GAVILAN DE CIENEGA, 105
GAVILAN DE COOPER, 102
GAVILAN DE SWAINSON, 108
GAVILAN DORSIPLOMIZO, 103
GAVILAN ENANO, 102
GAVILAN GORGIRRAYADO, 100
GAVILAN GRIS, 106
GAVILAN NEGRO MAYOR, 105
GAVILAN PAJARERO, 102
GAVILAN PECHINEGRO, 105
Gavilán Pescador, 97, 105
GAVILAN PIQUIGANCHUDO, 98
Gavilán Pollero, 106, 107
GAVILAN RANERO, 101
Gavilán Sabanero, 108
Gavilán Silbero, 105
Gavilán Tijerilla, 99
Gavilán Valdivia, 109
GAVIOTA ARGENTEA, 155
GAVIOTA DE BONAPARTE, 154
GAVIOTA DE FRANKLIN, 156
GAVIOTA DE SABINE, 158
GAVIOTA PIQUIANILLADA, 155
GAVIOTA REIDORA, 156
Gelochelidon nilotica, 159
Geospiza. See ***Volatinia***
Geothlypis aequinoctialis, 398, Pl. 42(12)
Geothlypis chiriquensis, 398
Geothlypis poliocephala, 399, Pl. 42(13)
Geothlypis semiflava, 398, Pl. 42(11)
Geothlypis trichas, 397, Pl. 42(10)
Geotrygon albifacies, 175
Geotrygon chiriquensis, 174, Pl. 18(18)
Geotrygon costaricensis, 174, Pl. 18(19)
Geotrygon lawrencii, 174, Pl. 18(21)
Geotrygon linearis, 175
Geotrygon montana, 176, Pl. 18(17)
Geotrygon veraguensis, 175, Pl. 18(22)
Geotrygon violacea, 175, Pl. 18(20)
Geranospiza caerulescens, 101, Pl. 14(4)
Glaucidium brasilianum, 194, Pl. 20(17)
Galucidium gnoma, 194
Glaucidium jardinii, 193, Pl. 20(18)
Glaucidium minutissimum, 193, Pl. 20(16)
Glaucis aenea, 210, Pl. 23(4)
Glaucis hirsuta, 210
Glyphorhynchus spirurus, 261, Pl. 29(6)
GNATCATCHER, TROPICAL, 371, Pl. 41(2)
GNATCATCHER, WHITE-LORED, 370, Pl. 41(1)
GNATWREN, HALF-COLLARED, 372
GNATWREN, LONG-BILLED, 371, Pl. 32(15)

Gnatwren, Straight-billed, 371
GNATWREN, TAWNY-FACED, 372, Pl. 32(14)
GODWIT, HUDSONIAN, 140, Pl. 9(7)
GODWIT, MARBLED, 140, Pl. 9(8)
Golden-Plover, American, 136
Golden-Plover, Asiatic, 136
GOLDEN-PLOVER, LESSER, 136, Pl. 9(2)
GOLDENTAIL, BLUE-THROATED, 219, Pl. 24(9)
Goldfinch, Dark-backed, 465
GOLDFINCH, LESSER, 465, Pl. 50(2)
GOLONDRINA ALIRRASPOSA NORTEÑA, 344
GOLONDRINA ALIRRASPOSA SUREÑA, 344
GOLONDRINA AZUL Y BLANCO, 345
GOLONDRINA BICOLOR, 346
GOLONDRINA LOMIBLANCA, 346
GOLONDRINA RIBEREÑA, 345
GOLONDRINA RISQUERA, 343
GOLONDRINA TIJERETA, 343
GOLONDRINA VERDE VIOLACEA, 347
Golondrón, 206, 342
Gongolona, 65, 66
GORRION COMUN, 466
Gorrión de Montaña, 243
GRACKLE, GREAT-TAILED, 409, Pl. 44(16)
GRACKLE, NICARAGUAN, 409, Pl. 44(17)
Grackle, Rice, 408
Grallaria guatimalensis, 288, Pl. 30(20)
Grallaria perspicillata, 289
Grallaricula flavirostris, 289, Pl. 30(13)
Gran Curré Negro, 250
Granadera, 119
GRASS-FINCH, WEDGE-TAILED, 464, Pl. 50(10)
GRASSQUIT, BLUE-BLACK, 452, Pl. 49(7)
GRASSQUIT, YELLOW-FACED, 447, Pl. 49(6)
GREBE, EARED, 68, Pl. 51(2)
GREBE, LEAST, 68, Pls. 7(4), 51(3)
GREBE, PIED-BILLED, 67, Pl. 7(3)
Greenlet, Gray-headed, 381
GREENLET, LESSER, 381, Pl. 40(7)
GREENLET, SCRUB, 380, Pl. 40(8)
GREENLET, TAWNY-CROWNED, 380, Pl. 32(4)
GROSBEAK, BLACK-FACED, 443, Pl. 48(5)
GROSBEAK, BLACK-HEADED, 445, Pl. 48(9)
GROSBEAK, BLACK-THIGHED, 444, Pl. 48(7)
GROSBEAK, BLUE, 445, Pl. 48(11)
GROSBEAK, BLUE-BLACK, 446, Pl. 48(10)
GROSBEAK, ROSE-BREASTED, 445, Pl. 48(8)
GROSBEAK, SLATE-COLORED, 444, Pl. 48(6)
Grosbeak, Yellow, 445
GROUND-CUCKOO, LESSER, 187, Pl. 21(11)
GROUND-CUCKOO, RUFOUS-VENTED, 188, Pl. 21(10)
GROUND-DOVE, BLUE, 171, Pl. 18(6)
GROUND-DOVE, COMMON, 170, Pl. 18(8)
GROUND-DOVE, MAROON-CHESTED, 172, Pl. 18(10)
GROUND-DOVE, PLAIN-BREASTED, 170, Pl. 18(9)
GROUND-DOVE, RUDDY, 170, Pl. 18(7)
Ground-Dove, Scaly-breasted, 170
GROUND-SPARROW, PREVOST'S, 459, Pl. 48(18)
GROUND-SPARROW, WHITE-EARED, 460, Pl. 48(19)
Ground-Sparrow, White-faced, 460
GRUIFORMES, 123
GUACAMAYO ROJO, 177
GUACAMAYO VERDE MAYOR, 177
GUACHARO, 197
GUACO, 113
Guairón, 87
GUAN, BLACK, 119, Pl. 12(5)
GUAN, CRESTED, 118, Pl. 12(4)
Guiraca caerulea, 445, Pl. 48(11)
Gull, Black-headed, 157
GULL, BONAPARTE'S, 157, Pl. 3(6)
GULL, FRANKLIN'S, 156, Pl. 3(5)
GULL, GRAY, 157, Pl. 51(4)
Gull, Heermann's, 158, Pl. 51(5)
GULL, HERRING, 155, Pl. 3(10)
GULL, LAUGHING, 156, Pls. 3(8), 51(6)
GULL, RING-BILLED, 155, Pl. 3(9)
GULL, SABINE'S, 158, Pl. 3(7)
Gull, Swallow-tailed, 158
Gurrión, 209–231
Gygis alba, 164, Pl. 2(10)
Gymnocichla nudiceps, 283, Pl. 31(13)
Gymnopithys bicolor, 283
Gymnopithys leucaspis, 284, Pl. 31(9)
Gymnostinops montezuma, 406

Habia atrimaxillaris, 435, Pl. 47(11)
Habia fuscicauda, 434, Pl. 47(12)
Habia rubica, 434, Pl. 47(10)
HAEMATOPODIDAE, 133
Haematopus ostralegus, 134
Haematopus palliatus, 133, Pl. 9(9)
HALCON APLOMADO, 116
HALCON CUELLIBLANCO, 116
HALCON DE MONTE BARRETEADO, 114
HALCON DE MONTE COLLAREJO, 114
HALCON DE MONTE DORSIGRIS, 114
HALCON PECHIRRUFO, 116
HALCON PEREGRINO, 117
Halocyptena microsoma, 72
Haplospiza rustica, 453, Pl. 49(14)
Harpagus bidentatus, 100, Pl. 16(1)
Harpia harpyja, 110, Pl. 17(9)
Harpyhaliaetus solitarius, 105, Pl. 13(8)
HARRIER, NORTHERN, 101, Pl. 15(7)
Hawk, Barred, 103
HAWK, BAY-WINGED, 106, Pl. 14(5)
HAWK, BICOLORED, 103, Pl. 16(8)

HAWK, BLACK-CHESTED, 103, Pl. 17(1)
HAWK, BLACK-COLLARED, 105, Pl. 17(5)
HAWK, BROAD-WINGED, 107, Pl. 16(13)
HAWK, COOPER'S, 102, Pl. 16(4)
HAWK, CRANE, 101, Pl. 14(4)
HAWK, GRAY, 106, Pl. 16(14)
HAWK, HARRIS'S, 106
HAWK, MARSH, 102
HAWK, PRINCE, 103
HAWK, RED-TAILED, 109, Pl. 17(8)
HAWK, ROADSIDE, 107, Pl. 16(12)
HAWK, RUFOUS-THIGHED, 102
HAWK, SAVANNA, 106, Pl. 17(4)
HAWK, SEMIPLUMBEOUS, 103, Pl. 16(10)
HAWK, SHARP-SHINNED, 102, Pl. 16(3)
HAWK, SHORT-TAILED, 107, Pls. 14(1), 16(11)
HAWK, SWAINSON'S, 108, Pls. 14(3), 17(7)
HAWK, TINY, 102, Pl. 16(2)
HAWK, WHITE, 104, Pl. 17(2)
HAWK, WHITE-BREASTED, 102
HAWK, WHITE-TAILED, 108, Pls. 14(7), 17(6)
HAWK, ZONE-TAILED, 109, Pls. 13(1), 14(2)
HAWK-EAGLE, BLACK, 111, Pls. 13(9), 17(12)
HAWK-EAGLE, BLACK-AND-WHITE, 110, Pl. 17(13)
HAWK-EAGLE, ORNATE, 111, Pl. 17(11)
Heliodoxa jacula, 227, Pl. 23(15)
Heliomaster constantii, 228, Pl. 23(19)
Heliomaster longirostris, 229, Pl. 23(18)
Heliornis fulica, 130, Pl. 7(9)
HELIORNITHIDAE, 130
Heliothryx aurita, 228
Heliothryx barroti, 228, Pl. 23(14)
Helmitheros vermivorus, 384, Pl. 40(21)
Henicorhina leucophrys, 359, Pl. 38(14)
Henicorhina leucosticta, 359, Pl. 38(15)
HERMIT, BRONZY, 210, Pl. 23(4)
HERMIT, GREEN, 211, Pl. 23(3)
HERMIT, GUY'S, 211
HERMIT, LITTLE, 211, Pl. 23(1)
HERMIT, LONG-TAILED, 211, Pl. 23(2)
HERMIT, RUFOUS-BREASTED, 210
HERON, AGAMI, 87
HERON, BOAT-BILLED, 83, Pl. 5(2)
HERON, CHESTNUT-BELLIED, 87, Pl. 5(1)
HERON, GREAT BLUE, 86, Pl. 5(6)
HERON, GREEN-BACKED, 84, Pl. 6(1,2)
HERON, LITTLE BLUE, 84, Pl. 5(9)
HERON, TRICOLORED, 85, Pl. 5(8)
Herpetotheres cachinnans, 113, Pl. 15(8)
Heterocnus mexicanum, 82
Heteroscelus incanus, 143, Pl. 10(8)
Heterospingus rubrifrons, 437, Pl. 45(15)
Heterospingus xanthopygius, 437
Heterospizias meridionalis, 106
Himantopus himantopus, 135
Himantopus mexicanus, 134, Pl. 9(12)
HIRUNDINIDAE, 341
Hirundo fulva, 343, Pl. 52(2)
Hirundo pyrrhonota, 343, Pls. 22(13), 52(1)
Hirundo rustica, 343, Pl. 22(12)
HOJARRASQUERO GORGIANTEADO, 272
HOJARRASQUERO ROJIZO, 272
Hombrecillo, 302
HONEYCREEPER, BLUE, 426
HONEYCREEPER, GREEN, 425, Pl. 46(7)
HONEYCREEPER, PURPLE, 427
HONEYCREEPER, RED-LEGGED, 426, Pl. 46(2)
HONEYCREEPER, SHINING, 426, Pl. 46(1)
Horera, 187
HORMIGUERITO ALIPUNTEADO, 281
HORMIGUERITO CAFE, 280
HORMIGUERITO FLANQUIBLANCO, 280
HORMIGUERITO LOMIRRUFO, 281
HORMIGUERITO PIZARROSO, 281
HORMIGUERO ALIMACULADO, 284
HORMIGUERO BICOLOR, 284
HORMIGUERO CALVO, 283
HORMIGUERO DORSICASTAÑO, 283
HORMIGUERO INMACULADO, 284
HORMIGUERO MOTEADO, 285
HORMIGUERO NEGRUZCO, 282
HORMIGUERO OCELADO, 285
HOUSE-WREN, NORTHERN, 358
HOUSE-WREN, SOUTHERN, 358
Hú de Leon, 194
Huevo Frito, 127
HUMMINGBIRD, ALFARO'S, 221
HUMMINGBIRD, BERYL-CROWNED, 219, Pl. 24(18)
HUMMINGBIRD, BLACK-BELLIED, 223, Pls. 24(21), 25(12)
HUMMINGBIRD, BLUE-CHESTED, 220, Pl. 24(14)
HUMMINGBIRD, BLUE-TAILED, 221, Pl. 24(16)
HUMMINGBIRD, BLUE-VENTED, 221
HUMMINGBIRD, CERISE-THROATED, 231
HUMMINGBIRD, CHARMING, 220
HUMMINGBIRD, CINNAMON, 222, Pl. 24(11)
HUMMINGBIRD, FIERY-THROATED, 218, Pl. 24(12)
HUMMINGBIRD, INDIGO-CAPPED, 221
HUMMINGBIRD, MAGNIFICENT, 227, Pl. 23(16)
HUMMINGBIRD, MANGROVE, 220, Pl. 24(2)
HUMMINGBIRD, RIVOLI'S, 228
HUMMINGBIRD, RUBY-THROATED, 230, Pl. 25(9)
HUMMINGBIRD, RUFOUS-TAILED, 222, Pl. 24(10)
HUMMINGBIRD, SAPPHIRE-THROATED, 218, Pls. 24(8), 25(13)
HUMMINGBIRD, SCALY-BREASTED, 212, Pl. 23(10)

HUMMINGBIRD, SCINTILLANT, 230, Pl. 25(7)
HUMMINGBIRD, SNOWY-BELLIED, 222, Pl. 24(3)
HUMMINGBIRD, SNOWY-BREASTED, 222
HUMMINGBIRD, STEELY-VENTED, 221, Pl. 24(15)
HUMMINGBIRD, STRIPED-TAILED, 223, Pls. 24(17), 25(14)
HUMMINGBIRD, VIOLET-BELLIED, 218
HUMMINGBIRD, VIOLET-HEADED, 215, Pl. 25(11)
HUMMINGBIRD, VOLCANO, 231, Pl. 25(3)
Hydranassa tricolor, 85
HYDROBATIDAE, 71
Hydroprogne caspia, 159
Hylocharis eliciae, 219, Pl. 24(9)
Hylocichla fuscescens, 367
Hylocichla minima, 367
Hylocichla mustelina, 366, Pl. 39(4)
Hylocichla ustulata, 366
Hyloctistes subulatus, 270, Pl. 30(4)
Hylomanes momotula, 240, Pl. 27(12)
Hylopezus fulviventris, 289, Pl. 30(14)
Hylopezus perspicillatus, 288, Pl. 30(18)
Hylophilus decurtatus, 381, Pl. 40(7)
Hylophilus flavipes, 380, Pl. 40(8)
Hylophilus minor, 381
Hylophilus ochraceiceps, 380, Pl. 32(4)
Hylophylax naevioidem, 285, Pl. 31(12)

IBIS, GLOSSY, 89, Pl. 4(9)
IBIS, GREEN, 88, Pl. 4(10)
IBIS, WHITE, 89, Pl. 4(8)
IBIS, WHITE-FACED, 89
IBIS BLANCO, 89
IBIS CARIBLANCO, 89
IBIS MORITO, 89
IBIS VERDE, 88
Icteria virens, 399, Pl. 42(9)
ICTERIDAE, 405
Icterus dominicensis, 410, Pl. 44(5)
Icterus galbula bullockii, 412
Icterus g. galbula, 412, Pl. 44(7)
Icterus mesomelas, 411, Pl. 44(3)
Icterus pectoralis, 411, Pl. 44(1)
Icterus prosthemelas, 411
Icterus pustulatus, 413, Pl. 44(2)
Icterus spurius, 410, Pl. 44(6)
Ictinia mississippiensis, 101, Pl. 15(3)
Ictinia plumbea, 100, Pls. 14(8), 15(4)
Indris, 447
Inglesito, 369
Iridoprocne. See ***Tachycineta***
Ixobrychus exilis, 81, Pl. 6(3)

JABIRU, 88
JABIRU, 88, Pl. 4(5)
Jabiru mycteria, 88, Pl. 4(5)
JACAMAR, BLACK-CHINNED, 243
JACAMAR, GREAT, 243, Pl. 26(11)
JACAMAR, RUFOUS-TAILED, 243, Pl. 26(12)
JACAMAR GRANDE, 243
JACAMAR RABIRRUFO, 243
Jacamerops aurea, 243, Pl. 26(11)
JACANA, NORTHERN, 132, Pl. 6(18)
JACANA, WATTLED, 132, Pl. 6(17)
JACANA CENTROAMERICANA, 132
Jacana jacana, 132, Pl. 6(17)
Jacana spinosa, 132, Pl. 6(18)
JACANA SUREÑA, 132
JACANIDAE, 132
JACOBIN, WHITE-NECKED, 213, Pl. 23(17)
JACOBINO NUQUIBLANCO, 213
JAEGER, LONG-TAILED, 154, Pl. 3(15)
JAEGER, PARASITIC, 154, Pl. 3(14)
JAEGER, POMARINE, 153, Pl. 3(13)
JAY, AZURE-HOODED, 349, Pl. 39(17)
JAY, BLACK-CHESTED, 348, Pl. 39(20)
JAY, BROWN, 348, Pl. 39(19)
JAY, SILVERY-THROATED, 349, Pl. 39(16)
Jilguerillo, 368
Jilguero, 365
JILGUERO MENOR, 465
JILGUERO VIENTRIAMARILLO, 465
Juanita, 422
Julío, 246
JUNCO, VOLCANO, 462, Pl. 50(11)
JUNCO PARAMERO, 462
Junco vulcani, 462, Pl. 50(11)

KESTREL, AMERICAN, 115, Pl. 15(13)
KILLDEER, 138, Pl. 10(1)
KINGBIRD, EASTERN, 307, Pl. 35(3)
KINGBIRD, GRAY, 308, Pl. 35(7)
KINGBIRD, TROPICAL, 307, Pl. 35(1)
KINGBIRD, WESTERN, 308, Pl. 35(2)
KINGFISHER, AMAZON, 238, Pl. 27(4)
KINGFISHER, AMERICAN PYGMY, 239, Pl. 27(6)
KINGFISHER, BELTED, 238, Pl. 27(2)
KINGFISHER, GREEN, 238, Pl. 27(5)
KINGFISHER, GREEN-AND-RUFOUS, 239, Pl. 27(3)
KINGFISHER, PYGMY, 239
KINGFISHER, RINGED, 237, Pl. 27(1)
KISKADEE, GREAT, 313, Pl. 35(13)
KITE, AMERICAN SWALLOW-TAILED, 99, Pl. 15(2)
KITE, BLACK-SHOULDERED, 99, Pl. 15(1)
KITE, DOUBLE-TOOTHED, 100, Pl. 16(1)
KITE, GRAY-HEADED, 98, Pl. 17(3)
KITE, HOOK-BILLED, 98, Pls. 14(9), 16(9)
KITE, MISSISSIPPI, 101, Pl. 15(3)
KITE, PEARL, 115, Pl. 15(12)
KITE, PLUMBEOUS, 100, Pls. 14(8), 15(4)
KITE, SNAIL, 99, Pls. 14(6), 15(6)
KITE, SWALLOW-TAILED, 99
KITE, WHITE-TAILED, 99
Klais guimeti, 215, Pl. 25(11)

Klis-klis, 115
KNOT, RED, 146, Pl. 11(19)

Lampornis castaneoventris, 226
Lampornis (castaneoventris) calolaema, 226, Pl. 24(7)
Lampornis (castaneoventris) cinereicauda, 227, Pl. 24(6)
Lampornis hemileucus, 225, Pl. 24(5)
Lampornis sybillae, 226
Lampornis viridipallens, 226
LANCEBILL, GREEN-FRONTED, 212, Pl. 23(11)
Lanio aurantius, 436
Lanio leucothorax, 435, Pl. 47(1)
Laniocera rufescens, 295, Pl. 34(8)
Lapa Colorada, 177
Lapa Roja, 177
Lapa Verde, 177
LARIDAE, 154
Larus argentatus, 155, Pl. 3(10)
Larus atricilla, 156, Pls. 3(8), 51(6)
Larus delawarensis, 155, Pl. 3(9)
Larus heermanni, 157, Pl. 51(5)
Larus modestus, 157, Pl. 51(4)
Larus philadelphia, 157, Pl. 3(6)
Larus pipixcan, 156, Pl. 3(5)
Larus ridibundus, 157
Laterallus albigularis, 127, Pl. 6(9)
Laterallus exilis, 127, Pl. 6(6)
Laterallus jamaicensis, 126, Pl. 6(5)
Laterallus melanophaius, 127
Laterallus ruber, 128, Pl. 6(11)
LEAFSCRAPER. See **LEAFTOSSER**
LEAFTOSSER, GRAY-THROATED, 273, Pl. 30(10)
LEAFTOSSER, SCALY-THROATED, 274, Pl. 30(9)
LEAFTOSSER, TAWNY-THROATED, 273, Pl. 30(11)
LECHUCITA NEOTROPICAL, 190
LECHUCITA PARDA, 196
LECHUCITA SABANERA, 190
LECHUCITA SERRANERA, 191
LECHUCITA VERMICULADA, 191
LECHUZA BLANCO Y NEGRO, 195
LECHUZA CAFE, 194
LECHUZA CAMPESTRE, 196
Lechuza de Campanario, 189
LECHUZA LLANERA, 194
LECHUZA RATONERA, 189
LECHUZA TERRESTRE, 194
Legatus leucophaius, 308, Pl. 35(8)
Leistes militaris, 414
Leona, 197
Lepidocolaptes affinis, 265, Pl. 29(9)
Lepidocolaptes souleyetii, 265, Pl. 29(8)
Lepidopyga coeruleogularis, 218, Pls. 24(8), 25(13)
Leptodon cayanensis, 98, Pl. 17(3)
Leptopogon amaurocephalus, 340, Pl. 36(27)
Leptopogon superciliaris, 339, Pl. 36(26)
Leptotila cassinii, 173, Pl. 18(16)
Leptotila plumbeiceps, 173
Leptotila rufaxilla, 173, Pl. 18(15)
Leptotila rufinucha, 174
Leptotila verreauxi, 172, Pl. 18(14)
Leucophoyx thula, 86
Leucopternis albicollis, 104, Pl. 17(2)
Leucopternis princeps, 103, Pl. 17(1)
Leucopternis semiplumbea, 103, Pl. 16(10)
Limnodromus griseus, 145, Pls. 9(3), 11(5)
Limnodromus scolopaceus, 145, Pl. 9(4)
Limosa fedoa, 140, Pl. 9(8)
Limosa haemastica, 140, Pl. 9(7)
LIMPKIN, 124, Pl. 5(5)
Lipaugus unirufus, 295, Pl. 34(10)
Lobipes lobatus, 152
Loomelania melania, 74
Lophornis adorabilis, 215, Pl. 25(5)
Lophornis delattrei, 216, Pl. 25(4)
Lophornis helenae, 216, Pl. 25(6)
Lophostrix cristata, 192, Pl. 20(3)
Lophotriccus pileatus, 331, Pl. 37(4)
Lora, 180–183
LORO CABECIAZUL, 181
LORO CABECIPARDO, 180
LORO CORONIBLANCO, 181
LORO DE NUCA AMARILLA, 182
LORO FRENTIBLANCO, 182
LORO FRENTIRROJO, 182
LORO VERDE, 183
Lurocalis semitorquatus, 199, Pl. 21(14)
Lysurus castaneiceps, 458
Lysurus crassirostris, 457, Pl. 48(21)

MACAW, GREAT GREEN, 177, Pl. 19(2)
MACAW, SCARLET, 177, Pl. 19(1)
Macuá, 208
MAGPIE JAY, 348
MAGPIE-JAY, BLACK-THROATED, 348
MAGPIE-JAY, WHITE-THROATED, 347, Pl. 39(18)
Mahafierro, 194
Malacoptila panamensis, 245, Pl. 28(6)
MALLARD, 92
Manacus aurantiacus, 302, Pl. 33(2)
Manacus candei, 303, Pl. 33(1)
Manacus vitellinus, 303
MANAKIN, BLUE-CROWNED, 300, Pl. 33(7)
MANAKIN, BROWN, 304
MANAKIN, GOLDEN-COLLARED, 303
MANAKIN, GRAY-HEADED, 303, Pl. 33(3)
MANAKIN, LANCE-TAILED, 301, Pl. 33(5)
MANAKIN, LONG-TAILED, 301, Pl. 33(4)
MANAKIN, ORANGE-COLLARED, 302, Pl. 33(2)
MANAKIN, RED-CAPPED, 299, Pl. 33(6)
MANAKIN, THRUSHLIKE, 304, Pl. 33(10)
MANAKIN, VELVETY, 300

Manakin, White-bibbed, 302
MANAKIN, WHITE-COLLARED, 303, Pl. 33(1)
MANAKIN, WHITE-CROWNED, 300, Pl. 33(8)
MANAKIN, WHITE-RUFFED, 302, Pl. 33(9)
Manakin, Yellow-thighed, 300
MANGO, GREEN-BREASTED, 214, Pl. 23(13)
MANGUITO PECHIVERDE, 214
Mareca americana, 92
Margarornis rubiginosus, 269, Pl. 29(4)
Mariposa, 423
Marsh-Wren, Short-billed, 351
MARTIN, BROWN-CHESTED, 342, Pl. 22(14)
MARTIN, GRAY-BREASTED, 342, Pl. 22(15)
MARTIN, PURPLE, 342, Pl. 22(16)
Martin, Southern, 342
MARTIN DE RIOS, 342
MARTIN PECHIGRIS, 342
Martín Peña, 81–84
Martín Pescador, 237–239
MARTIN PESCADOR AMAZONICO, 238
MARTIN PESCADOR COLLAREJO, 237
MARTIN PESCADOR ENANO, 239
MARTIN PESCADOR NORTEÑO, 238
MARTIN PESCADOR VERDE, 238
MARTIN PESCADOR VIENTRIRRUFO, 239
MARTIN PURPUREA, 342
MARTINETE CABECIPINTO, 83
MARTINETE CORONINEGRO, 82
MEADOWLARK, EASTERN, 414, Pl. 50(16)
Megaceryle torquatus, 237
Megarhynchus pitangua, 309, Pl. 35(12)
Melanerpes aurifrons, 253
Melanerpes chrysauchen, 252, Pl. 28(18)
Melanerpes formicivorus, 252, Pl. 28(15)
Melanerpes hoffmannii, 253, Pl. 28(16)
Melanerpes pucherani, 252, Pl. 28(14)
Melanerpes rubricapillus, 253, Pl. 28(17)
Melospiza lincolnii, 463, Pl. 50(12)
Melozone biarcuatum, 459, Pl. 48(18)
Melozone cabanisi, 460
Melozone leucotis, 460, Pl. 48(19)
Menea Cola, 395
Mercenario, 459
MERLIN, 115, Pl. 15(15)
Mesembrinibis cayennensis, 88, Pl. 4(10)
Micrastur mirandollei, 114, Pl. 16(6)
Micrastur ruficollis, 114, Pl. 16(5)
Micrastur semitorquatus, 114, Pls. 14(10), 16(7)
Microbates cinereiventris, 372, Pl. 32(14)
Microcerculus luscinia, 361, Pl. 38(20)
Microcerculus marginatus, 361
Microcerculus philomela, 360, Pl. 38(21)
Microchera albocoronata, 225, Pl. 25(8)
Micromonacha lanceolata, 245, Pl. 28(5)
Micropalama himantopus, 150
Micropygia schomburgkii, 128, Pl. 6(7)
Microrhopias quixensis, 281, Pl. 32(1)
Microtriccus. See ***Ornithion***
MIELERO AZULEJO, 427
MIELERO CELESTE Y NEGRO, 427
MIELERO LUCIENTE, 426
MIELERO PATIRROJO, 426
MIELERO VERDE, 425
Milvago chimachima, 113, Pl. 15(9)
MIMIDAE, 362
Mimus gilvus, 362
Mionectes oleagineus, 340, Pl. 36(25)
Mionectes olivaceus, 340, Pl. 36(24)
Mirasol, 80
MIRLO ACUATICO PLOMIZO, 350
MIRLO GORGIBLANCO, 363
MIRLO MONTAÑERO, 365
MIRLO NEGRUZCO, 365
MIRLO PARDO, 363
MIRLO VIENTRIBLANCO, 364
Mitrephanes phaeocercus, 323, Pl. 36(11)
Mitrospingus cassinii, 438, Pl. 45(18)
Mniotilta varia, 383, Pl. 41(13)
MOCHUELO COMUN, 194
MOCHUELO ENANO, 193
MOCHUELO MONTAÑERO, 193
Mockingbird, Tropical, 362
Molothrus aeneus, 408, Pls. 44(15), 52(7)
MOMOTIDAE, 239
MOMOTO CANELO MAYOR, 241
MOMOTO CEJICELESTE, 241
MOMOTO COMUN, 242
MOMOTO ENANO, 240
MOMOTO PICO QUILLA, 241
MOMOTO PIQUIANCHO, 240
Momotus momota, 242, Pl. 27(8)
Monasa morphoeus, 245, Pl. 28(4)
MONJA FRENTIBLANCA, 245
Monjita, 76, 419
Monjita Fina, 419
MONJITO RAYADO, 245
MONKLET, LANCEOLATED, 245, Pl. 28(5)
Moorhen, Common, 129
Morococcyx erythropygius, 187, Pl. 21(11)
Morphnus guianensis, 109, Pl. 17(10)
MOSQUERITO ACEITUNADO, 340
MOSQUERITO AMARILLENTO, 322
MOSQUERITO AMARILLO, 333
MOSQUERITO CABECINEGRO, 323
MOSQUERITO CABECIPARDO, 340
MOSQUERITO CEJIBLANCO, 339
MOSQUERITO CEJIGRIS, 337
MOSQUERITO CEJIRRUFO, 332
MOSQUERITO CHEBEC, 322
MOSQUERITO CHILLON, 336
MOSQUERITO COLICORTO, 332
MOSQUERITO COLINEGRO, 325
MOSQUERITO COLIRRUFO, 324
MOSQUERITO CORONIAMARILLO, 337
MOSQUERITO DE CHARRAL, 321
MOSQUERITO DE LA ISLA DEL COCO, 324
MOSQUERITO DE TRAILL, 321
MOSQUERITO DE YELMO, 331
MOSQUERITO FRENTIBLANCO, 338

MOSQUERITO GARGANTIBLANCO, 322
MOSQUERITO GORRICAFE, 338
MOSQUERITO GUARDARRIOS, 330
MOSQUERITO LOMIAMARILLO, 325
MOSQUERITO MOÑUDO, 323
MOSQUERITO OJIMANCHADO, 340
MOSQUERITO OREJINEGRO, 339
MOSQUERITO PECHILEONADO, 324
MOSQUERITO PECHIRRAYADO, 326
MOSQUERITO SILBADOR, 336
MOSQUERITO VERDOSO, 320
MOSQUERITO VIENTRIAMARILLO, 320
MOSQUERO CABECIANILLADO, 309
MOSQUERO CABECIGRIS, 312
MOSQUERO CEJIBLANCO, 313
MOSQUERO COLUDO, 305
MOSQUERO DE AGUA, 305
MOSQUERO GORGIGRIS, 336
MOSQUERO LISTADO, 311
MOSQUERO PIRATA, 308
MOSQUERO REAL, 326
MOSQUERO VIENTRIAZUFRADO, 310
MOSQUERO VIENTRIDORADO, 312
MOSQUERON PICUDO, 309
MOTMOT, BLUE-CROWNED, 242, Pl. 27(8)
Motmot, Blue-Diademed, 242
MOTMOT, BROAD-BILLED, 240, Pl. 27(9)
MOTMOT, KEEL-BILLED, 241, Pl. 27(10)
MOTMOT, RUFOUS, 241, Pl. 27(7)
Motmot, Rufous-capped, 241
MOTMOT, TODY, 240, Pl. 27(12)
MOTMOT, TURQUOISE-BROWED,, 241, Pl. 27(11)
MOUNTAIN-GEM, GRAY-TAILED, 227, Pl. 24(6)
MOUNTAIN-GEM, PURPLE-THROATED, 226, Pl. 24(7)
Mountain-gem, Variable, 226
MOUNTAIN-GEM, WHITE-BELLIED, 225, Pl. 24(5)
Mourner, Brown, 304
MOURNER, RUFOUS, 314, Pl. 34(9)
MOURNER, SPECKLED, 295, Pl. 34(8)
Mourner, Thrushlike, 304
Mozotillo de Charral, 465
Mozotillo de Montaña, 465
Mulita, 132
Muscivora forficata, 306
Muscivora tyrannus, 307
Myadestes melanops, 365, Pl. 39(13)
Myadestes ralloides, 366
Mycteria americana, 87, Pl. 4(6)
Myiarchus cinerascens, 316, Pl. 35(19)
Myiarchus crinitus, 316, Pl. 35(17)
Myiarchus ferox, 315
Myiarchus nuttingi, 315, Pl. 35(22)
Myiarchus panamensis, 314, Pl. 35(20)
Myiarchus tuberculifer, 315, Pl. 35(21)
Myiarchus tyrannulus, 315, Pl. 35(18)
Myiobius atricaudus, 325, Pl. 36(22)
Myiobius barbatus, 325
Myiobius sulphureipygius, 325, Pl. 36(21)
Myioborus miniatus, 401, Pl. 42(7)
Myioborus torquatus, 402, Pl. 42(6)
Myiochanes richardsonii, 318
Myiochanes virens, 318
Myiodynastes chrysocephalus, 312
Myiodynastes hemichrysus, 312, Pl. 35(9)
Myiodynastes luteiventris, 310, Pl. 35(10)
Myiodynastes maculatus, 311, Pl. 35(11)
Myiopagis viridicata, 335, Pl. 37(21)
Myiophobus fasciatus, 326, Pl. 36(17)
Myiornis atricapillus, 332, Pl. 37(5)
Myiornis ecaudatus, 332
Myiozetetes granadensis, 312, Pl. 35(15)
Myiozetetes similis, 313, Pl. 35(14)
Myiozetetes texensis, 313
Myrmeciza exsul, 283, Pl. 31(11)
Myrmeciza immaculata, 284, Pl. 31(10)
Myrmeciza laemosticta, 284, Pl. 31(14)
Myrmornis torquata, 286, Pl. 30(21)
Myrmotherula axillaris, 280, Pl. 32(2)
Myrmotherula fulviventris, 280, Pl. 32(5)
Myrmotherula schisticolor, 281, Pl. 32(3)

Neocrex columbianus, 128
Neocrex erythrops, 128, Pl. 51(12)
Neomorphus geoffroyi, 188, Pl. 21(10)
Neophoecetes niger, 205
Nesotriccus ridgwayi, 324, Pl. 36(18)
NICTIBIO COMUN, 198
NICTIBIO GRANDE, 197
NIGHTHAWK, COMMON, 199, Pl. 21(13)
NIGHTHAWK, LESSER, 200, Pl. 21(12)
NIGHTHAWK, SHORT-TAILED, 199, Pl. 21(14)
NIGHTINGALE-THRUSH, BLACK-BILLED, 370, Pl. 38(24)
NIGHTINGALE-THRUSH, BLACK-HEADED, 368, Pl. 38(25)
NIGHTINGALE-THRUSH, ORANGE-BILLED, 369, Pl. 38(26)
NIGHTINGALE-THRUSH, RUDDY-CAPPED, 369, Pl. 38(27)
NIGHTINGALE-THRUSH, SLATY-BACKED, 368, Pl. 38(23)
NIGHTJAR, DUSKY, 202, Pl. 21(19)
NIGHTJAR, RUFOUS, 202, Pl. 21(15)
NIGHTJAR, WHITE-TAILED, 203, Pl. 21(20)
NIGHT-HERON, BLACK-CROWNED, 82, Pl. 5(4)
NIGHT-HERON, YELLOW-CROWNED, 83, Pl. 5(3)
NINFA VIOLETA Y VERDE, 217
NODDY, BLACK, 163, Pl. 2(7)
NODDY, BROWN, 163, Pl. 2(8)
Noddy, Lesser, 164
Noneca, 96
Notharchus. See **Bucco**
Nothocercus bonapartei, 65, Pl. 12(7)

Notiochelidon cyanoleuca, 345, Pl. 22(20)
Numenius americanus, 140, Pl. 9(13)
Numenius phaeopus, 140, Pl. 9(14)
NUNBIRD, WHITE-FRONTED, 245, Pl. 28(4)
Nuttallornis borealis, 317
Nyctanassa violacea, 83, Pl. 5(3)
NYCTIBIIDAE, 197
Nyctibius grandis, 197, Pl. 20(5)
Nyctibius griseus, 198, Pls. 20(4), 51(7)
Nyctibius jamaicensis, 198
Nycticorax nycticorax, 82, Pl. 5(4)
Nycticorax violaceus, 83
Nyctidromus albicollis, 200, Pls. 21(18), 51(10)
Nyctiphrynus ocellatus, 201, Pl. 51(9)

Oceanites gracilis, 72
Oceanites oceanicus, 71, Pl. 2(15)
Oceanodroma castro, 72, Pl. 2(14)
Oceanodroma homochroa, 74
Oceanodroma leucorhoa, 73, Pl. 2(16)
Oceanodroma markhami, 73
Oceanodroma melania, 73, Pl. 2(17)
Oceanodroma microsoma, 72, Pl. 2(13)
Oceanodroma tethys, 72, Pl. 2(12)
Odontophorus erythrops, 122, Pl. 12(12)
Odontophorus gujanensis, 121, Pl. 12(9)
Odontophorus guttatus, 122, Pl. 12(11)
Odontophorus leucolaemus, 122, Pl. 12(10)
OILBIRD, 197, Pl. 51(8)
Oncostoma cinereigulare, 331, Pl. 37(6)
Oncostoma olivaceum, 331
Onychorhynchus coronatus, 326, Pl. 35(23)
Onychorhynchus mexicanus, 327
Oporornis agilis, 397, Pl. 52(4)
Oporornis formosus, 396, Pl. 42(15)
Oporornis philadelphia, 396, Pls. 42(14), 52(5)
Oporornis tolmiei, 397, Pl. 42(16)
Oreopyra. See ***Lampornis***
ORIOLE, BLACK-COWLED, 410, Pl. 44(5)
ORIOLE, BULLOCK'S, 412
ORIOLE, NORTHERN (BALTIMORE), 412, Pl. 44(7)
ORIOLE, ORCHARD, 410, Pl. 44(6)
ORIOLE, SPOTTED-BREASTED, 411, Pl. 44(1)
ORIOLE, STREAKED-BACKED, 413, Pl. 44(2)
ORIOLE, YELLOW-TAILED, 411, Pl. 44(3)
Ornithion brunneicapillum, 338, Pl. 37(13)
Ornithion semiflavum, 339, Pl. 37(14)
Oropéndola, 405, 406
OROPENDOLA, CHESTNUT-HEADED, 405, Pl. 44(9)
OROPENDOLA, MONTEZUMA, 406, Pl. 44(8)
OROPENDOLA CABECICASTAÑA, 405
OROPENDOLA DE MOCTEZUMA, 406
Oropopo, 192
Ortalis cinereiceps, 118, Pl. 12(1)
Ortalis leucogastra, 118
Ortalis vetula, 118, Pl. 12(2)
Oryzoborus angolensis, 452
Oryzoborus funereus, 451, Pl. 49(10)
Oryzoborus maximiliani, 451
Oryzoborus nuttingi, 451, Pl. 49(13)
OSPREY, 97, Pl. 17(14)
OSTRERO AMERICANO, 133
Otus asio, 191
Otus choliba, 190, Pl. 20(15)
Otus clarkii, 191, Pl. 20(10)
Otus cooperi, 190, Pl. 20(11)
Otus guatemalae, 191, Pl. 20(12)
Otus vermiculatus, 191
OVENBIRD, 395, Pl. 43(12)
OWL, BLACK-AND-WHITE, 195, Pl. 20(7)
OWL, BURROWING, 194, Pl. 20(14)
OWL, CRESTED, 192, Pl. 20(3)
OWL, GREAT HORNED, 192, Pl. 20(1)
OWL, MOTTLED, 194, Pl. 20(6)
OWL, SHORT-EARED, 196
OWL, SPECTACLED, 192, Pl. 20(8)
OWL, STRIPED, 195, Pl. 20(2)
OWL, UNSPOTTED SAW-WHET, 196, Pl. 20(13)
Oxyruncus cristatus, 299, Pl. 34(7)
Oxyura dominica, 95, Pl. 7(8)
Oxyura jamaicensis, 95
OYSTERCATCHER, AMERICAN, 133, Pl. 9(9)

Pachyramphus aglaiae, 293, Pl. 33(12)
Pachyramphus albogriseus, 292, Pl. 33(15)
Pachyramphus cinnamomeus, 291, Pl. 33(11)
Pachyramphus polychopterus, 292, Pl. 33(13)
Pachyramphus versicolor, 291, Pl. 33(14)
PAGALO COLILARGO, 154
PAGALO PARASITO, 154
PAGALO POMARINO, 153
PAGAZA ELEGANTE, 163
PAGAZA MAYOR, 159
PAGAZA PIQUIRROJO, 159
PAGAZA PUNTIAMARILLA, 162
PAGAZA REAL, 162
PAIÑO DANZARIN, 72
PAIÑO DE LEACH, 73
PAIÑO DE MARKHAM, 73
PAIÑO DE WILSON, 71
PAIÑO MENUDO, 72
PAIÑO NEGRO, 73
PAIÑO PECHIALBO, 72
PAIÑO RABIFAJEADO, 72
Pájaro Bobo, 242
Pájaro Campana, 298
Pájaro Chancho, 293
Pájaro Danta, 297
Pájaro Estaca, 198
Pájaro Palo, 198
Pájaro Vaco, 81, 82
PAJARO-GATO GRIS, 362
PAJARO-SOMBRILLA CUELLINUDO, 297
Paloma Ala Blanca, 169
PALOMA ALIBLANCA, 169
PALOMA COLIBLANCA, 172

PALOMA COLLAREJA, 167
PALOMA COLORADA, 167
PALOMA CORONIBLANCA, 166
PALOMA CORONIGRIS, 173
Paloma de Castilla, 166
PALOMA DOMESTICA, 166
PALOMA ESCAMOSA, 166
Paloma Morada, 168
PALOMA PECHIGRIS, 173
PALOMA PIQUICORTA, 168
PALOMA PIQUIRROJA, 168
PALOMA RABUDA, 169
PALOMA ROJIZA, 168
PALOMA-PERDIZ BIGOTIBLANCA, 175
PALOMA-PERDIZ COSTARRIQUEÑA, 174
PALOMA-PERDIZ PECHICANELA, 174
PALOMA-PERDIZ ROJIZA, 176
PALOMA-PERDIZ SOMBRIA, 174
PALOMA-PERDIZ VIOLACEA, 175
Palomita, 170
Palomita Colorada, 170
Palomita del Espíritu Santo, 164
Pandion haliaetus, 97, Pl. 17(14)
PANDIONIDAE, 97
Panterpe insignis, 218, Pl. 24(12)
Panyptila cayennensis, 208, Pl. 22(7)
Panyptila sanctihieronymi, 208, Pl. 22(6)
Paphosia. See ***Lophornis***
Parabuteo unicinctus, 106, Pl.14(5)
PARAKEET, BARRED, 179, Pl. 19(16)
PARAKEET, BROWN-SHOULDERED, 180
PARAKEET, CRIMSON-FRONTED, 177, Pl. 19(10)
PARAKEET, OLIVE-THROATED, 178, Pl. 19(11)
PARAKEET, ORANGE-CHINNED, 180, Pl. 19(14)
PARAKEET, ORANGE-FRONTED, 178, Pl. 19(13)
PARAKEET, SULFUR-WINGED, 179, Pl. 19(12)
PARDELA BLANCA COMUN, 70
PARDELA COLICORTA, 71
PARDELA COLICUÑA, 70
PARDELA DE AUDUBON, 71
PARDELA SOMBRIA, 70
Pardirallus maculatus, 124, Pl. 6(4)
PARROT, BLUE-HEADED, 181, Pl. 19(8)
PARROT, BROWN-HOODED, 180, Pl. 19(9)
PARROT, MEALY, 183, Pl. 19(3)
PARROT, RED-LORED, 182, Pl. 19(4)
PARROT, WHITE-CROWNED, 181, Pl. 19(7)
PARROT, WHITE-FRONTED, 182, Pl. 19(6)
PARROT, YELLOW-NAPED, 182, Pl. 19(5)
PARROTLET, RED-FRONTED, 180, Pl. 19(15)
PARROTLET, RED-WINGED, 180
PARULA, NORTHERN, 386, Pl. 41(4)
PARULA, TROPICAL, 387, Pl. 41(3)
Parula americana, 386, Pl. 41(4)
Parula gutturalis, 386, Pl. 41(7)
PARULA NORTEÑA, 386
Parula pitiayumi, 387, Pl. 41(3)
PARULA TROPICAL, 387
PARULIDAE, 382
Passer domesticus, 466, Pl. 50(17)
Passerculus sandwichensis, 460
PASSERIDAE, 466
PASSERIFORMES, 258
Passerina caerulea, 446
Passerina ciris, 447, Pl. 48(13)
Passerina cyanea, 447, Pls. 48(12), 50(6)
Passerina cyanoides, 447
Patiamarillo, 426
PATIAMARILLO MAYOR, 141
PATIAMARILLO MENOR, 142
Patillo, 68
PATO AGUJA, 78
PATO CABECIVERDE, 92
PATO CALVO, 92
Pato Canadiense, 93
PATO CANTIL, 130
Pato Chancho, 77
PATO CUCHARA, 94
Pato Chuchara, 90
Pato de Agua, 67, 68, 77
PATO ENMASCARADO, 95
Pato Perulero, 92
PATO RABUDO, 93
PATO REAL, 92
Pato Rosado, 90
Patudo, 146–149
PAURAQUE, COMMON, 200, Pls. 21(18), 51(10)
Pava, 118
PAVA CRESTADA, 118
PAVA NEGRA, 119
Pavita, 118
Pavón, 119
PAVON GRANDE, 119
Pecho Amarillo, 307–313, 333
Pelagodroma marina, 72, Pl. 2(11)
PELECANIDAE, 74
PELECANIFORMES, 74
Pelecanus erythrorhynchos, 75, Pl. 4(2)
Pelecanus occidentalis, 75, Pl. 4(1)
PELICAN, AMERICAN WHITE, 75, Pl. 4(2)
PELICAN, BROWN, 75, Pl. 4(1)
Pelícano, 75
PELICANO BLANCO AMERICANO, 75
PELICANO PARDO, 75
Penelope purpurascens, 118, Pl. 12(4)
PEPPERSHRIKE, RUFOUS-BROWED, 375, Pl. 40(2)
Perdiz, 65, 66
PERDIZ MONTAÑERA, 120
Perico, 177–180
PERICO ALIAZUFRADO, 179
PERICO AZTECO, 178
PERICO FRENTINARANJA, 178
PERICO FRENTIRROJO, 177

Periquito, 178–180
PERIQUITO ALIRROJO, 180
PERIQUITO BARBINARANJA, 180
PERIQUITO LISTADO, 179
Perissotriccus. See ***Myiornis***
PERLITA CABECINEGRA, 370
PERLITA TROPICAL, 371
Perrito de Agua, 130
PETREL, BLACK, 70
PETREL, BLACK-CAPPED, 69, Pl. 1(15)
PETREL, DARK-RUMPED, 69, Pl. 1(11)
PETREL, PARKINSON'S, 70, Pl. 1(12)
PETREL DE PARKINSON, 70
PETREL GORRINEGRO, 69
PETREL LOMIOSCURO, 69
Petrochelidon pyrrhonota, 343
PEWEE, DARK, 319, Pl. 36(2)
PEWEE, OCHRACEOUS, 319, Pl. 36(3)
PEWEE, TROPICAL, 318, Pl. 36(9)
Pezopetes capitalis, 455, Pl. 48(20)
Phaenostictus mcleannani, 285, Pl. 31(7)
Phaeochroa cuvierii, 212, Pl. 23(10)
Phaeoprogne tapera, 342, Pl. 22(14)
Phaeothlypis fulvicauda, 404, Pl. 40(23)
Phaethon aethereus, 74, Pl. 1(7)
Phaethon lepturus, 74
PHAETHONTIDAE, 74
Phaethornis guy, 211, Pl. 23(3)
Phaethornis longuemareus, 211, Pl. 23(1)
Phaethornis superciliosus, 211, Pl. 23(2)
Phainoptila melanoxantha, 374, Pl. 39(10)
PHALACROCORACIDAE, 77
Phalacrocorax olivaceus, 77, Pl. 4(4)
PHALAROPE, RED, 151, Pl. 10(11)
PHALAROPE, RED-NECKED, 151, Pl. 10(10)
PHALAROPE, WILSON'S, 152, Pl. 10(9)
PHALAROPODIDAE, 151
Phalaropus fulicarius, 151, Pl. 10(11)
Phalaropus lobatus, 151, Pl. 10(10)
Phalaropus tricolor, 152
Pharomachrus mocinno, 232, Pl. 26(1)
PHASIANIDAE, 120
Pheucticus chrysopeplus, 445
Pheucticus ludovicianus, 445, Pl. 48(8)
Pheucticus melanocephalus, 445, Pl. 48(9)
Pheucticus tibialis, 444, Pl. 48(7)
Philodice. See ***Calliphlox***
Philomachus pugnax, 150, Pl. 11(1)
Philydor rufus, 271, Pl. 30(6)
Philydor subalaris, 270
Philydor subulatus, 270
Philydor variegaticeps, 271
Phlogothraupis sanguinolenta, 430, Pl. 47(3)
PHOEBE, BLACK, 305, Pl. 36(5)
Phyllomyias burmeisteri, 338
Phyllomyias leucogonys, 338
Phyllomyias zeledoni, 338, Pl. 37(18)
Phylloscartes flaveola, 333
Phylloscartes superciliaris, 332, Pl. 37(12)
Piapia, 348
Piapia Azul, 347
Piapia de Montaña, 349
Piaya cayana, 185, Pl. 21(7)
PIBI BOREAL, 317
PIBI OCCIDENTAL, 317
PIBI OCRACEO, 319
PIBI ORIENTAL, 318
PIBI SOMBRIO, 319
PIBI TROPICAL, 318
Piche, 90
Piche Careto, 91
PICIDAE, 251
PICIFORMES, 242
PICO DE HOZ, 209
PICO DE LANZA FRENTIVERDE, 212
Pico de Plata, 407
PICOAGUDO, 299
PICOGRUESO AZUL, 445
PICOGRUESO CABECINEGRO, 445
PICOGRUESO CARINEGRO, 443
PICOGRUESO NEGRO AZULADO, 446
PICOGRUESO PECHIRROSADO, 445
PICOGRUESO PIQUIRROJO, 444
PICOGRUESO VIENTRIAMARILLO, 444
PICO-CHUCHARA, 83
Picoides villosus, 254, Pl. 28(19)
Picudo, 426
PICULET, OLIVACEOUS, 251, Pl. 29(1)
Piculus leucolaemus, 256
Piculus rubiginosus, 257, Pl. 28(11)
Piculus simplex, 256, Pl. 28(10)
Picumnus olivaceus, 251, Pl. 29(1)
PIGEON, BAND-TAILED, 167, Pl. 18(1)
PIGEON, PALE-VENTED, 167, Pl. 18(3)
PIGEON, RED-BILLED, 168, Pl. 18(2)
PIGEON, RUDDY, 168
PIGEON, SCALED, 166, Pl. 18(4)
PIGEON, SHORT-BILLED, 168, Pl. 18(5)
PIGEON, WHITE-CROWNED, 166, Pl. 51(1)
PIGEON, WHITE-NAPED, 167
PIGÜILO, 143
PIHA, RUFOUS, 295, Pl. 34(10)
Pijije, 141, 142
PIJIJE CANELO, 91
PIJIJE CARIBLANCO, 90
PIJIJE COMUN, 90
Pinaroloxias inornata, 452, Pl. 49(8)
Pinchaflor, 454
PINCHAFLOR PLOMIZO, 454
PINTAIL, NORTHERN, 93, Pl. 8(1)
PINZON ACEITUNADO, 458
PINZON BARRANQUERO, 457
PINZON CABECILISTADO, 459
PINZON CAFETALERO, 459
PINZON OREJIBLANCO, 460
PINZON PIQUINARANJA, 458
Pionopsitta haematotis, 180, Pl. 19(9)
Pionus menstruus, 181, Pl. 19(8)
Pionus senilis, 181, Pl. 19(7)
PIPIT, YELLOWISH, 453

Pipra coronata, 300, Pl. 33(7)
Pipra mentalis, 299, Pl. 33(6)
Pipra pipra, 300, Pl. 33(8)
PIPRIDAE, 299
Piprites griseiceps, 303, Pl. 33(3)
PIPROMORPHA, OLEAGINOUS, 341
Pipromorpha oleaginea, 341
PIQUERO BLANCO, 76
PIQUERO MORENO, 76
PIQUERO PATIAZUL, 76
PIQUERO PATIRROJO, 77
PIQUICHATO CORONIRRUFO, 328
PIQUICHATO GARGANTIBLANCO, 327
PIQUICHATO NORTEÑO, 327
PIQUIPLANO ALIAMARILLO, 329
PIQUIPLANO AZUFRADO, 328
PIQUIPLANO DE ANTEOJOS, 329
PIQUITORCIDO NORTEÑO, 331
Piranga bidentata, 433, Pl. 47(9)
Piranga flava, 431, Pl. 47(6)
Piranga leucoptera, 432, Pl. 47(7)
Piranga ludoviciana, 432, Pl. 47(2)
Piranga olivacea, 431, Pl. 47(8)
Piranga rubra, 430, Pl. 47(5)
Piririza, 143
Pirrís, 463
Pitangus sulphuratus, 313, Pl. 35(13)
Pitorreal, 373
Pittasoma michleri, 287, Pl. 30(19)
Pitylus grossus, 44, Pl. 48(6)
Pius, 408, 452
PLAÑIDERA MOTEADA, 295
PLAÑIDERA ROJIZA, 314
Platalea. See ***Ajaia***
Platypsaris aglaiae, 293
Platyrinchus cancrominus, 327, Pl. 37(2)
Platyrinchus coronatus, 328, Pl. 37(3)
Platyrinchus mystaceus, 327, Pl. 37(1)
PLAYERO ARENERO, 146
Plegadis chihi, 89
Plegadis falcinellus, 89, Pl. 4(9)
Plío, 406
PLOVER, BLACK-BELLIED, 136, Pl. 9(1)
PLOVER, COLLARED, 139, Pl. 10(3)
PLOVER, GREY, 136
PLOVER, PIPING, 137
PLOVER, SEMIPALMATED, 137, Pl. 10(5)
PLOVER, SNOWY, 138, Pl. 10(4)
PLOVER, THICK-BILLED, 138
PLOVER, UPLAND, 141
PLOVER, WILSON'S, 138, Pl. 10(2)
PLUMELETEER, BRONZE-TAILED, 225
PLUMELETEER, RED-FOOTED, 225, Pl. 23(12)
Pluvialis dominica, 136, Pl. 9(2)
Pluvialis fulva, 136
Pluvialis squatarola, 136, Pl. 9(1)
Podiceps andinus, 68
Podiceps caspicus, 68
Podiceps dominicus, 68
Podiceps nigricollis, 68, Pl. 51(2)
PODICIPEDIDAE, 67
PODICIPEDIFORMES, 67
Podilymbus podiceps, 67, Pl. 7(3)
Polioptila albiloris, 370, Pl. 41(1)
Polioptila plumbea, 371, Pl. 41(2)
POLLUELA COLORADA, 128
POLLUELA GARGANTIBLANCA, 127
POLLUELA NEGRA, 126
POLLUELA NORTEÑA, 126
POLLUELA OCELADA, 128
POLLUELA PECHIAMARILLA, 126
POLLUELA PECHIGRIS, 127
POLLUELA PIQUIRROJA, 128
POLLUELA SORA, 126
Polyborus plancus, 112, Pl. 14(12)
Pomponé, 125
Pone-pone, 125
POORWILL, OCELLATED, 201, Pl. 51(9)
Popelairia. See ***Discosura***
Porphyrula martinica, 129, Pl. 6(15)
PORRON COLLAREJO, 94
PORRON MAYOR, 95
PORRON MENOR, 94
Porzana carolina, 126, Pls. 6(12), 51(11)
Porzana flaviventer, 126, Pl. 6(10)
POTOO, COMMON, 198, Pls. 20(4), 51(7)
POTOO, GREAT, 197, Pl. 20(5)
POTOO, JAMAICAN, 198
POTOO, NORTHERN, 198
PRADERITO PECHIANTEADO, 150
PRADERO, 141
Premnoplex brunnescens, 268, Pl. 29(5)
Procellaria parkinsoni, 70, Pl. 1(12)
PROCELLARIIDAE, 69
PROCELLARIIFORMES, 68
Procnias tricarunculata, 298, Pl. 34(12)
Progne chalybea, 342, Pl. 22(15)
Progne elegans, 342
Progne subis, 342, Pl. 22(16)
Progne tapera, 343
Protonotaria citrea, 383, Pl. 42(1)
Psarocolius montezuma, 406, Pl. 44(8)
Psarocolius wagleri, 405, Pl. 44(9)
Pselliophorus tibialis, 455, Pl. 48(22)
Pseudocolaptes boissonneautii, 270
Pseudocolaptes lawrencii, 269, Pl. 30(1)
Psilorhinus mexicanus, 348
Psilorhinus morio, 348
PSITTACIDAE, 176
PSITTACIFORMES, 176
Psomocolax. See ***Scaphidura***
Pterdoroma externa, 69
Pterodroma hasitata, 69, Pl. 1(15)
Pterodroma phaeopygia, 69, Pl. 1(11)
Pteroglossus frantzii, 249, Pl. 27(16)
Pteroglossus torquatus, 248, Pl. 27(15)
PTILOGONATIDAE, 373
Ptilogonys caudatus, 373, Pl. 39(15)
Pucuyo, 200

PUFFBIRD, PIED, 244, Pl. 28(7)
PUFFBIRD, WHITE-NECKED, 244, Pl. 28(3)
PUFFBIRD, WHITE-WHISKERED, 245, Pl. 28(6)
Puffinus auricularis, 70
Puffinus carneipes, 70
Puffinus creatopus, 70, Pl. 1(8)
Puffinus gravis, 69
Puffinus griseus, 70, Pl. 1(9)
Puffinus lherminieri, 71, Pl. 1(14)
Puffinus opisthomelas, 70
Puffinus pacificus, 70, Pl. 1(10)
Puffinus tenuirostris, 71, Pl. 1(13)
Pulsatrix perspicillata, 192, Pl. 20(8)
Puncus, 80
Purisco, 456
PYGMY-OWL, ANDEAN, 193, Pl. 20(18)
PYGMY-OWL, FERRUGINOUS, 194, Pl. 20(17)
PYGMY-OWL, LEAST, 193, Pl. 20(16)
PYGMY-TYRANT, BLACK-CAPPED, 332, Pl. 37(5)
PYGMY-TYRANT, SCALE-CRESTED, 331, Pl. 37(4)
Pygmy-Tyrant, Short-tailed, 332
Pygochelidon. See ***Notiochelidon***
Pyrrhura hoffmanni, 179, Pl. 19(12)

QUAIL, TAWNY-FACED, 123, Pl. 12(15)
QUAIL-DOVE, BUFF-FRONTED, 174, Pl. 18(19)
QUAIL-DOVE, CHIRIQUI, 174, Pl. 18(18)
Quail-Dove, Lined, 175
QUAIL-DOVE, OLIVE-BACKED, 175, Pl. 18(22)
QUAIL-DOVE, PURPLISH-BACKED, 174, Pl. 18(21)
Quail-Dove, Rufous-breasted, 175
QUAIL-DOVE, RUDDY, 176, Pl. 18(17)
QUAIL-DOVE, VIOLACEOUS, 175, Pl. 18(20)
Quail-Dove, White-faced, 175
Quebrantahuesos, 112
Queo, 438
Querque, 112
QUERULA GORGIMORADA, 297
Querula purpurata, 297, Pl. 34(11)
QUETZAL, 232
Quetzal, 232
QUETZAL, RESPLENDENT, 232, Pl. 26(1)
Quetzal Macho, 235
Quioro, 250
Quiscalus mexicanus, 409, Pl. 44(16)
Quiscalus nicaraguensis, 409, Pl. 44(17)

Rabadilla Tinta, 429
RABIHORCADO GRANDE, 79
RABIHORCADO MAGNO, 79
RABIJUNCO PIQUIRROJO, 74
RAIL, BLACK, 126, Pl. 6(5)
RAIL, SPOTTED, 124, Pl. 6(4)
RALLIDAE, 124
RAMPHASTIDAE, 247
Ramphastos ambiguus, 251
Ramphastos sulfuratus, 250, Pl. 27(18)
Ramphastos swainsonii, 250, Pl. 27(19)
Ramphocaenus melanurus, 371, Pl. 32(15)
Ramphocaenus rufiventris, 372
Ramphocelus passerinii, 429, Pl. 47(4)
Ramphocelus sanguinolentus, 430
RASCON CAFE, 125
RASCON CUELLIGRIS, 125
RASCON CUELLIRRUFO, 125
RASCON MOTEADO, 124
Ratoncillo, 461
Raya Roja, 401
RAYADOR NEGRO, 165
Recurvirostra americana, 135, Pl. 9(11)
RECURVIROSTRIDAE, 134
REDSTART, AMERICAN, 401, Pl. 41(10)
REDSTART, COLLARED, 402, Pl. 42(6)
REDSTART, SLATE-THROATED, 401, Pl. 42(7)
REINITA ACUATICA NORTEÑA, 395
REINITA ACUATICA PIQUIGRANDE, 395
REINITA ALIAZUL, 384
REINITA ALIDORADA, 384
REINITA AMARILLA, 387
REINITA AZUL Y NEGRO, 389
REINITA CABECICASTANA, 403
REINITA CABECIDORADA, 383
REINITA CABECIGUALDA, 391
REINITA CABECILISTADA, 402
REINITA CACHETIGRIS, 385
REINITA CACHETINEGRA, 396
REINITA CARIAMARILLA, 390
REINITA CARINEGRA, 403
REINITA CASTAÑA, 393
REINITA CERULEA, 391
REINITA COLIFAJEADA, 388
REINITA CORONICASTAÑA, 394
REINITA CORONIDORADA, 402
REINITA DE COSTILLAS CASTAÑAS, 392
REINITA DE MANGLAR, 388
REINITA DE PINOS, 394
REINITA DE TOWNSEND, 390
REINITA DE TUPIDERO, 397
REINITA ENCAPUCHADA, 399
REINITA ENLUTADA, 396
REINITA GALANA, 394
REINITA GARGANTA DE FUEGO, 386
REINITA GORGIAMARILLA, 392
REINITA GORGINARANJA, 392
REINITA GORRINEGRA, 400
REINITA GRANDE, 399
REINITA GUARDARIBERA, 404
REINITA GUSANERA, 384
REINITA HORNERA, 395
REINITA LOMIAMARILLA, 389
REINITA MIELERA, 382
REINITA OJIANILLADA, 397

REINITA OLIVADA, 385
REINITA PECHIRRAYADA, 400
REINITA RAYADA, 393
REINITA TIGRINA, 389
REINITA TREPADORA, 383
REINITA VERDILLA, 385
Relinchero, 264
Rey de Comemaiz, 459
Rey de Trepadores, 426
Rey de Zopilotes, 97
Rey Gallinazo, 97
RHINOCRYPTIDAE, 283
Rhinoptynx clamator, 196
Rhodinochichla rosea, 438, Pl. 47(13)
Rhynchocyclus brevirostris, 329, Pl. 37(23)
Rhynchortyx cinctus, 123, Pl. 12(15)
Rhytipterna holerythra, 314, Pl. 34(9)
Rin-ran, 298
Riparia riparia, 345, Pl. 22(17)
ROBIN, CLAY-COLORED, 363, Pl. 39(8)
ROBIN, MOUNTAIN, 365, Pl. 39(6)
ROBIN, PALE-VENTED, 364, Pl. 39(5)
ROBIN, SOOTY, 365, Pl. 39(7)
Robin, White-necked, 363
ROBIN, WHITE-THROATED, 363, Pl. 39(9)
Rostrhamus sociabilis, 99, Pls. 14(6), 15(6)
Rualdo, 416
RUFF, 150, Pl. 11(1)
RYNCHOPIDAE, 164
Rynchops niger, 165, Pl. 3(11)

SABANERO ARROCERO, 415
SABANERO CABECILISTADO, 461
SABANERO COLICORTO, 460
SABANERO COLUDO, 460
SABANERO DE LINCOLN, 463
SABANERO PECHIANTEADO, 462
SABANERO ROJIZO, 461
SABANERO ZANJERO, 460
SABREWING, VIOLET, 213, Pl. 23(9)
Salpinctes obsoletus, 360, Pl. 38(4)
Salta Pinuela, 352
SALTARIN CABECIGRIS, 303
SALTARIN CABECIRROJO, 299
SALTARIN COLUDO, 301
SALTARIN CORONIBLANCO, 300
SALTARIN CORONICELESTE, 300
SALTARIN CUELLIANARANJADO, 302
SALTARIN CUELLIBLANCO, 303
SALTARIN GORGIBLANCO, 302
SALTARIN TOLEDO, 301
SALTATOR, BLACK-HEADED, 441, Pl. 48(1)
SALTATOR, BUFF-THROATED, 442, Pl. 48(2)
SALTATOR, GRAYISH, 442, Pl. 48(3)
Saltator, Middle American, 443
SALTATOR, STREAKED, 443, Pl. 48(4)
Saltator albicollis, 443, Pl. 48(4)
Saltator atriceps, 441, Pl. 48(1)
SALTATOR CABECINEGRO, 441
Saltator coerulescens, 442, Pl. 48(3)
SALTATOR GORGIANTEADO, 442
Saltator grandis, 443
SALTATOR GRISACEO, 442
SALTATOR LISTADO, 443
Saltator maximus, 442, Pl. 48(2)
SALTEADOR POLAR, 153
SALTON CABECICASTAÑO, 456
SALTON CABECINEGRO, 457
SALTON DE MUSLOS AMARILLOS, 455
SALTON GARGANTIAMARILLA, 456
SALTON PATIGRANDE, 455
San Juan, 171
Sanate, 409
SANDERLING, 146, Pl. 11(11)
SANDPIPER, BAIRD'S, 148, Pl. 11(15)
SANDPIPER, BUFF-BREASTED, 150, Pl. 11(18)
SANDPIPER, CURLEW, 149, Pl. 11(9)
SANDPIPER, LEAST, 147, Pl. 11(14)
SANDPIPER, PECTORAL, 148, Pl. 11(7)
SANDPIPER, SEMIPALMATED, 147, Pl. 11(13)
SANDPIPER, SOLITARY, 142, Pl. 11(4)
SANDPIPER, SPOTTED, 143, Pl. 11(8)
SANDPIPER, STILT, 149, Pl. 11(6)
SANDPIPER, UPLAND, 141, Pl. 11(17)
SANDPIPER, WESTERN, 146, Pl. 11(12)
SANDPIPER, WHITE-RUMPED, 148, Pl. 11(16)
Sangre de Toro, 429, 430
Santa Marta, 382
SAPSUCKER, YELLOW-BELLIED, 254, Pl. 28(20)
Sarcoramphus papa, 97, Pl. 13(5)
Sargento, 413, 429
Sayornis nigricans, 305, Pl. 36(5)
Scaphidura oryzivora, 407, Pl. 44(10)
Scardafella inca, 171
SCAUP, GREATER, 95, Pl. 7(5)
SCAUP, LESSER, 94, Pl. 7(6)
Schiffornis turdinus, 304, Pl. 33(10)
Schiffornis veraepacis, 304
Sclerurus albigularis, 273, Pl. 30(10)
Sclerurus guatemalensis, 274, Pl. 30(9)
Sclerurus mexicanus, 273, Pl. 30(11)
SCOLOPACIDAE, 139
SCREECH-OWL, BARE-SHANKED, 191, Pl. 20(10)
Screech-Owl, Middle American, 191
SCREECH-OWL, PACIFIC, 190, Pl. 20(11)
SCREECH-OWL, TROPICAL, 190, Pl. 20(15)
SCREECH-OWL, VERMICULATED, 191, Pl. 20(12)
Scytalopus argentifrons, 290, Pl. 32(12)
SCYTHEBILL, BROWN-BILLED, 266, Pl. 29(14)
Seedeater, Black, 449
SEEDEATER, BLUE, 450, Pl. 49(11)
SEEDEATER, RUDDY-BREASTED, 450, Pl. 49(1)

SEEDEATER, SLATE-COLORED, 448, Pl. 49(5)
SEEDEATER, VARIABLE, 449, Pl. 49(3)
SEEDEATER, WHITE-COLLARED, 448, Pl. 49(2)
SEEDEATER, YELLOW-BELLIED, 450, Pl. 49(4)
Seed-Finch, Chestnut-bellied, 452
Seed-Finch, Great-billed, 451
Seed-Finch, Nicaraguan, 451
SEED-FINCH, PINK-BILLED, 451, Pl. 49(13)
SEED-FINCH, THICK-BILLED, 451, Pl. 49(10)
Seiurus aurocapillus, 395, Pl. 43(12)
Seiurus motacilla, 395, Pl. 43(13)
Seiurus noveboracensis, 395, Pl. 43(14)
Selasphorus flammula, 231, Pl. 25(3)
Selasphorus scintilla, 230, Pl. 25(7)
Selasphorus simoni, 231
Selenidera spectabilis, 249, Pl. 27(20)
SEMILLERITO CARIAMARILLO, 447
SEMILLERITO NEGRO AZULADO, 452
SEMILLERO AZULADO, 450
SEMILLERO PICOGRUESO, 451
SEMILLERO PIQUIRROSADO, 451
Semnornis frantzii, 247, Pl. 28(1)
Sensontle, 442
Serpophaga cinerea, 333, Pl. 36(1)
Setillero, 448–450
Setophaga ruticilla, 401, Pl. 41(10)
SHARPBILL, 299, Pl. 34(7)
SHEARWATER, AUDUBON'S, 71, Pl. 1(14)
Shearwater, Black-vented, 70
Shearwater, Flesh-footed, 70
SHEARWATER, PINK-FOOTED, 70, Pl. 1(8)
SHEARWATER, SHORT-TAILED, 71, Pl. 1(13)
SHEARWATER, SOOTY, 70, Pl. 1(9)
Shearwater, Townsend's, 70
SHEARWATER, WEDGE-TAILED, 70, Pl. 1(10)
SHOVELER, NORTHERN, 94, Pl. 8(2)
Shrike-Tanager, Black-throated, 436
Shrike-Tanager, Great, 436
SHRIKE-TANAGER, WHITE-THROATED, 435, Pl. 47(1)
SHRIKE-VIREO, GREEN, 375, Pl. 40(1)
Sicalis luteola, 453, Pl. 50(3)
SICKLEBILL, WHITE-TIPPED, 209, Pl. 23(8)
Siete Colores, 423, 447
SILKY-FLYCATCHER, BLACK-AND-YELLOW, 374, Pl. 39(10)
SILKY-FLYCATCHER, LONG-TAILED, 373, Pl. 39(15)
Sinsonte, 442
Sinsonte Verde, 442
SISKIN, YELLOW-BELLIED, 465, Pl. 50(1)
Sittasomus griseicapillus, 261, Pl. 29(7)
SKIMMER, BLACK, 165, Pl. 3(11)
Skua, Brown, 153
Skua, Chilean, 153
Skua, Great (Northern), 153
SKUA, SOUTH POLAR, 153, Pl. 3(12)
Smaragdolanius pulchellus, 376
SNIPE, COMMON, 144, Pl. 9(5)
SNOWCAP, 225, Pl. 25(8)
Softwing, White-whiskered, 245
Soldadito, 134
Solitaire, Andean, 366
SOLITAIRE, BLACK-FACED, 365, Pl. 39(13)
SOLITARIO CARINEGRO, 365
SORA, 126, Pls. 6(12), 51(11)
Sorococa, 190, 191
SOTERILLO CARICAFE, 372
SOTERILLO PICUDO, 371
Soterrey, 357
SOTERREY CANORO, 361
SOTERREY CARIMOTEADO, 357
SOTERREY CASTAÑO, 354
SOTERREY CHINCHIRIGUI, 352
SOTERREY CUCARACHERO, 357
SOTERREY DE COSTILLAS BARRETEADAS, 355
SOTERREY DEL BAMBU, 358
SOTERREY GORGINEGRO, 356
SOTERREY MATRAQUERO, 351
SOTERREY NUQUIRRUFO, 352
SOTERREY OCROSO, 358
SOTERREY PECHIBARRETEADO, 354
SOTERREY PECHIBLANCO, 359
SOTERREY PECHIGRIS, 359
SOTERREY PECHIMOTEADO, 356
SOTERREY PECHIRRAYADO, 354
SOTERREY ROQUERO, 360
SOTERREY RUFO Y BLANCO, 353
SOTERREY RUISEÑOR, 360
SOTERREY SABANERO, 351
SOTERREY SILBADOR, 361
SOTERREY VIENTRINEGRO, 356
SPADEBILL, GOLDEN-CROWNED, 328, Pl. 37(3)
SPADEBILL, STUB-TAILED, 327, Pl. 37(2)
SPADEBILL, WHITE-THROATED, 327, Pl. 37(1)
SPARROW, BLACK-STRIPED, 459, Pl. 50(14)
SPARROW, BOTTERI'S, 462, Pl. 50(5)
SPARROW, CHIPPING, 463, Pl. 50(9)
SPARROW, GRASSHOPPER, 460, Pl. 50(8)
SPARROW, HOUSE, 466, Pl. 50(17)
SPARROW, LINCOLN'S, 463, Pl. 50(12)
SPARROW, OLIVE, 458, Pl. 50(15)
SPARROW, ORANGE-BILLED, 458, Pl. 48(17)
Sparrow, Pacific, 459
SPARROW, RUFOUS-COLLARED, 463, Pl. 50(13)
SPARROW, RUSTY, 461, Pl. 50(4)
SPARROW, SAVANNAH, 460
SPARROW, STRIPED-HEADED, 461, Pl. 50(7)
Spatula clypeata, 94
Speotyto. See ***Athene***
Sphyrapicus varius, 254, Pl. 28(20)
SPINETAIL, PALE-BREASTED, 267, Pl. 32(11)
SPINETAIL, RED-FACED, 268, Pl. 30(12)

SPINETAIL, SLATY, 267, Pl. 32(9)
Spinus. See ***Carduelis***
Spiza americana, 415, Pl. 50(19)
Spizaetus ornatus, 111, Pl. 17(11)
Spizaetus tyrannus, 111, Pls. 13(9), 17(12)
Spizastur melanoleucus, 110, Pl. 17(13)
Spizella passerina, 463, Pl. 50(9)
Spodiornis. See ***Haplospiza***
SPOONBILL, ROSEATE, 90, Pl. 4(7)
Sporophila americana, 449
Sporophila aurita, 449, Pl. 49(3)
Sporophila corvina, 449
Sporophila minuta, 450, Pl. 49(1)
Sporophila nigricollis, 450, Pl. 49(4)
Sporophila schistacea, 448, Pl. 49(5)
Sporophila torqueola, 448, Pl. 49(2)
STARTHROAT, LONG-BILLED, 229, Pl. 23(18)
STARTHROAT, PLAIN-CAPPED, 228, Pl. 23(19)
Steatornis caripensis, 197, Pl. 51(8)
STEATORNITHIDAE, 197
Steganopus tricolor, 152, Pl. 10(9)
Stelgidopteryx ruficollis, 344, Pl. 22(19)
Stelgidopteryx serripennis, 344, Pl. 22(18)
STERCORARIIDAE, 152
Stercorarius longicaudus, 154, Pl. 3(15)
Stercorarius parasiticus, 154, Pl. 3(14)
Stercorarius pomarinus, 153, Pl. 3(13)
Sterna albifrons, 162
Sterna anaethetus, 160, Pl. 2(2)
Sterna antillarum, 161, Pl. 2(4)
Sterna caspia, 159, Pl. 3(1)
Sterna dougallii, 160
Sterna elegans, 163, Pl. 3(3)
Sterna eurygnatha, 162
Sterna forsteri, 160, Pl. 2(9)
Sterna fuscata, 161, Pl. 2(3)
Sterna hirundo, 159, Pl. 2(5)
Sterna maxima, 162, Pl. 3(2)
Sterna nilotica, 159, Pl. 2(1)
Sterna paradisaea, 160
Sterna sandvicensis, 162, Pl. 3(4)
STILT, BLACK-NECKED, 134, Pl. 9(12)
STILT, COMMON, 135
STORK, WOOD, 87, Pl. 4(6)
STORM-PETREL, ASHY, 74
STORM-PETREL, BAND-RUMPED, 72, Pl. 2(14)
STORM-PETREL, BLACK, 73, Pl. 2(17)
STORM-PETREL, HARCOURT'S, 73
STORM-PETREL, LEACH'S, 73, Pl. 2(16)
STORM-PETREL, LEAST, 72, Pl. 2(13)
STORM-PETREL, MADEIRA, 73
STORM-PETREL, MARKHAM'S, 73
STORM-PETREL, WEDGE-RUMPED, 72, Pl. 2(12)
STORM-PETREL, WHITE-FACED, 72, Pl. 2(11)
STORM-PETREL, WHITE-VENTED, 72
STORM-PETREL, WILSON'S, 71, Pl. 2(15)
Streptoprocne zonaris, 206, Pl. 22(1)
STRIGIDAE, 189
STRIGIFORMES, 188
Sturnella magna, 414, Pl. 50(16)
Sturnella militaris, 414, Pl. 44(13)
SUBEPALO MOTEADO, 268
SUBEPALO ROJIZO, 269
Sublegatus arenarum, 336
Sublegatus modestus, 336, Pl. 37(22)
Sula dactylatra, 76, Pl. 1(3)
Sula leucogaster, 76, Pl. 1(1)
Sula nebouxii, 76, Pl. 1(4)
Sula sula, 77, Pl. 1(2)
SULIDAE, 75
SUNBITTERN, 131, Pl. 6(16)
SUNGREBE, 130, Pl. 7(9)
SURFBIRD, 145, Pl. 10(7)
SWALLOW, BANK, 345, Pl. 22(17)
SWALLOW, BARN, 343, Pl. 22(12)
SWALLOW, BLUE-AND-WHITE, 345, Pl. 22(20)
SWALLOW, CAVE, 343, Pl. 52(2)
SWALLOW, CLIFF, 343, Pls. 22(13), 52(1)
SWALLOW MANGROVE, 346, Pl. 22(23)
SWALLOW, NORTHERN ROUGH-WINGED, 344, Pl. 22(18)
SWALLOW, ROUGH-WINGED, 345
SWALLOW, SOUTHERN ROUGH-WINGED, 344, Pl. 22(19)
SWALLOW, TREE, 346, Pl. 22(21)
SWALLOW, VIOLET-GREEN, 347, Pl. 22(22)
SWIFT, BAND-RUMPED, 207, Pl. 22(11)
SWIFT, BLACK, 204, Pl. 22(2)
SWIFT, CHESTNUT-COLLARED, 206, Pl. 22(3)
SWIFT, CHIMNEY, 206, Pl. 22(8)
SWIFT, DUSKY-BACKED, 207
SWIFT, GRAY-RUMPED, 207, Pl. 22(10)
SWIFT, GREAT SWALLOW-TAILED, 208, Pl. 22(6)
SWIFT, LESSER SWALLOW-TAILED, 208, Pl. 22(7)
SWIFT, SPOT-FRONTED, 205, Pl. 22(5)
SWIFT, VAUX'S, 207, Pl. 22(9)
SWIFT, WHITE-CHINNED, 205, Pl. 22(4)
SWIFT, WHITE-COLLARED, 206, Pl. 22(1)
SYLVIIDAE, 370
Synallaxis albescens, 267, Pl. 32(11)
Synallaxis brachyura, 267, Pl. 32(9)
Syndactyla subalaris, 270, Pl. 30(8)

Tachybaptus dominicus, 68, Pls. 7(4), 51(3)
Tachycineta albilinea, 346, Pl. 22(23)
Tachycineta bicolor, 346, Pl. 22(21)
Tachycineta thalassina, 347, Pl. 22(22)
Tachyphonus delattrii, 437, Pl. 46(17)
Tachyphonus luctuosus, 436, Pl. 46(18)
Tachyphonus rufus, 436, Pl. 46(16)
TANAGER, BAY-HEADED, 424, Pl. 46(9)
TANAGER, BLACK-AND-YELLOW, 439, Pl. 46(14)

TANAGER, BLUE-AND-GOLD, 428, Pl. 45(11)
TANAGER, BLUE-GRAY, 428, Pl. 46(15)
Tanager, Carmiol's, 434
TANAGER, CRIMSON-COLLARED, 430, Pl. 47(3)
TANAGER, DUSKY-FACED, 438, Pl. 45(18)
TANAGER, EMERALD, 422, Pl. 46(5)
TANAGER, FLAME-COLORED, 433, Pl. 47(9)
TANAGER, GOLDEN-HOODED, 423, Pl. 46(13)
Tanager, Golden-masked, 423
TANAGER, GRAY-HEADED, 437, Pl. 45(17)
TANAGER, HEPATIC, 431, Pl. 47(6)
Tanager, Lemon-Browed, 434
Tanager, Masked, 423
TANAGER, OLIVE, 433, Pl. 45(16)
TANAGER, PALM, 429, Pl. 45(19)
TANAGER, PLAIN-COLORED, 424, Pl. 46(11)
TANAGER, RUFOUS-WINGED, 424, Pl. 46(10)
TANAGER, SCARLET, 431, Pl. 47(8)
Tanager, Scarlet-browed, 437
TANAGER, SCARLET-RUMPED, 429, Pl. 47(4)
TANAGER, SILVER-THROATED, 422, Pl. 46(6)
TANAGER, SPANGLED-CHEEKED, 425, Pl. 46(12)
TANAGER, SPECKLED, 422, Pl. 46(8)
Tanager, Streaked-backed, 433
TANAGER, SULPHUR-RUMPED, 437, Pl. 45(15)
TANAGER, SUMMER, 430, Pl. 47(5)
TANAGER, TAWNY-CRESTED, 437, Pl. 46(17)
TANAGER, WESTERN, 432, Pl. 47(2)
TANAGER, WHITE-LINED, 436, Pl. 46(16)
TANAGER, WHITE-SHOULDERED, 436, Pl. 46(18)
TANAGER, WHITE-WINGED, 432, Pl. 47(7)
Tanager, Yellow-winged, 429
Tanagra. See ***Euphonia***
TANGARA ACEITUNADA, 433
TANGARA ALIBLANCA, 432
TANGARA ALIRRUFA, 424
TANGARA AZULEJA, 428
TANGARA BERMEJA, 431
TANGARA CABECICASTAÑA, 424
TANGARA CABECIGRIS, 437
TANGARA CAPONIBLANCA, 436
TANGARA CAPUCHIDORADA, 423
TANGARA CAPUCHIRROJA, 430
TANGARA CARINEGRUZCA, 438
TANGARA CARIRROJA, 432
TANGARA CENICIENTA, 424
TANGARA CORONIDORADA, 437
TANGARA DE CARMIOL, 433
TANGARA DE COSTILLAS NEGRAS, 428
TANGARA DE MONTE CEJIBLANCA, 440
TANGARA DE MONTE GARGANTIGRIS, 440
TANGARA DE MONTE OJERUDA, 439
TANGARA DORADA, 422
TANGARA DORSIRRAYADA, 433
TANGARA ESCARLATA, 431
TANGARA FORRIBLANCA, 436
TANGARA HORMIGUERA CARINEGRA, 435
TANGARA HORMIGUERA CORONIRROJA, 434
TANGARA HORMIGUERA GORGIRROJA, 435
TANGARA LOMIAZUFRADA, 437
TANGARA LOMIESCARLATA, 429
TANGARA MOTEADA, 422
TANGARA NEGRO Y DORADO, 439
TANGARA OREJINEGRA, 422
TANGARA PALMERA, 429
TANGARA PECHIRROSADA, 438
TANGARA PIQUIGANCHUDA, 435
TANGARA VERANERA, 430
TANGARA VIENTRICASTAÑA, 425
Tangara chrysophrys, 422
Tangara dowii, 425, Pl. 46(12)
Tangara florida, 422, Pl. 46(5)
Tangara guttata, 422, Pl. 46(8)
Tangara gyrola, 424, Pl. 46(9)
Tangara icterocephala, 422, Pl. 46(6)
Tangara inornata, 424, Pl. 46(11)
Tangara larvata, 423, Pl. 46(13)
Tangara lavinia, 424, Pl. 46(10)
Tangara nigrocincta, 423
Tangavius. See ***Molothrus***
TAPACAMINOS COMUN, 200
TAPACULO, SILVERY-FRONTED, 290, Pl. 32(12)
TAPACULO FRENTIPLATEADO, 290
Tapera naevia, 187, Pl. 21(5)
Taraba major, 276, Pl. 31(2)
TATTLER, WANDERING, 143, Pl. 10(8)
TEAL, BLUE-WINGED, 93, Pl. 8(5)
TEAL, CINNAMON, 93, Pl. 8(3)
TEAL, GREEN-WINGED, 92, Pl. 8(4)
Telegrafista, 251
Terciopelo, 429
Terenotriccus erythrurus, 324, Pl. 36(23)
Terenura callinota, 281, Pl. 32(10)
Tern, Arctic, 160
TERN, BLACK, 158, Pl. 2(6)
TERN, BRIDLED, 160, Pl. 2(2)
TERN, CASPIAN, 159, Pl. 3(1)
TERN, COMMON, 159, Pl. 2(5)
TERN, ELEGANT, 163, Pl. 3(3)
Tern, Fairy, 164
TERN, FORSTER'S, 160, Pl. 2(9)
TERN, GULL-BILLED, 159, Pl. 2(1)
TERN, LEAST, 161, Pl. 2(4)
Tern, Roseate, 160
TERN, ROYAL, 162, Pl. 3(2)
TERN, SANDWICH, 162, Pl. 3(4)
TERN, SOOTY, 161, Pl. 2(3)
TERN, WHITE, 164, Pl. 2(10)
Thalasseus elegans, 163
Thalasseus maximus, 162
Thalasseus sandvicensis, 162

Thalurania colombica, 217, Pls. 24(4), 25(15)
Thalurania furcata, 218
Thamnistes anabatinus, 278, Pl. 31(4)
Thamnophilus bridgesi, 277, Pl. 31(6)
Thamnophilus doliatus, 276, Pl. 31(3)
Thamnophilus punctatus, 277, Pl. 31(5)
THICK-KNEE, DOUBLE-STRIPED, 135, Pl. 9(10)
THORNTAIL, GREEN, 216, Pl. 25(1)
THRAUPIDAE, 416
Thraupis abbas, 429
Thraupis episcopus, 428, Pl. 46(15)
Thraupis palmarum, 429, Pl. 45(19)
Threnetes ruckeri, 210, Pl. 23(5)
THRESKIORNITHIDAE, 88
Thripadectes rufobrunneus, 271, Pl. 30(7)
Thrush, Garden, 364
THRUSH, GRAY-CHEEKED, 367, Pl. 39(1)
Thrush, Gray's, 364
Thrush, Olive-backed, 367
Thrush, Pale-vented, 364
Thrush, Russet-backed, 367
THRUSH, SWAINSON'S, 366, Pl. 39(2)
Thrush, White-necked, 363
Thrush, White-throated, 363
THRUSH, WOOD, 366, Pl. 39(4)
Thrush-Tanager, Rose-breasted, 439
THRUSH-TANAGER, ROSY, 438, Pl. 47(13)
Thryorchilus browni, 358, Pl. 38(16)
Thryothorus atrogularis, 356, Pl. 38(13)
Thryothorus fasciatoventris, 356, Pl. 38(11)
Thryothorus maculipectus, 356, Pl. 38(7)
Thryothorus modestus, 352, Pl. 38(17)
Thryothorus nigricapillus, 354, Pl. 38(12)
Thryothorus pleurostictus, 355, Pl. 38(6)
Thryothorus rufalbus, 353, Pl. 38(9)
Thryothorus rutilus, 357, Pl. 38(8)
Thryothorus sclateri, 357
Thryothorus semibadius, 354, Pl. 38(10)
Thryothorus spadix, 356
Thryothorus thoracicus, 354, Pl. 38(5)
Thryothorus zeledoni, 353
Tiaris olivacea, 447, Pl. 49(6)
TIGER-HERON, BARE-THROATED, 82, Pl. 5(16)
TIGER-HERON, FASCIATED, 81, Pl. 5(15)
TIGER-HERON, RUFESCENT, 81, Pl. 5(17)
Tigrisoma fasciatum, 81, Pl. 5(15)
Tigrisoma lineatum, 81, Pl. 5(17)
Tigrisoma mexicanum, 82, Pl. 5(16)
Tigrisoma salmoni, 81
Tigüiza, 142
Tijereta del Mar, 79
TIJERETA ROSADA, 306
TIJERETA SABANERA, 306
Tijerilla, 79, 99, 306
Tijo, 186
Tildío, 138
Timbre, 374
TINAMIDAE, 65
TINAMIFORMES, 65
TINAMOU, GREAT, 65, Pl. 12(6)
TINAMOU, HIGHLAND, 65, Pl. 12(7)
TINAMOU, LITTLE, 66, Pl. 12(17)
TINAMOU, SLATY-BREASTED, 67, Pl. 12(8)
TINAMOU, THICKETT, 66, Pl. 12(18)
TINAMU CANELO, 66
TINAMU CHICO, 66
TINAMU GRANDE, 65
TINAMU PIZARROSO, 67
TINAMU SERRANO, 65
Tinamus major, 65, Pl. 12(6)
Tinco, 186
TIÑOSA COMUN, 163
TIÑOSA NEGRA, 163
TIRAHOJAS BARBIESCAMADO, 274
TIRAHOJAS GARGANTIGRIS, 273
TIRAHOJAS PECHIRRUFO, 273
TIRANO GRIS, 308
TIRANO NORTENO, 307
TIRANO OCCIDENTAL, 308
TIRANO TROPICAL, 307
Tirirí, 330
Tití, 248
TITYRA, BLACK-CROWNED, 294, Pl. 34(2)
TITYRA, MASKED, 293, Pl. 34(1)
TITYRA CARIRROJA, 293
TITYRA CORONINEGRA, 294
Tityra inquisitor, 294, Pl. 34(2)
Tityra semifasciata, 293, Pl. 34(1)
TITYRIDAE, 290
Toboba, 130
Todirostrum cinereum, 330, Pl. 37(7)
Todirostrum nigriceps, 329, Pl. 37(8)
Todirostrum sylvia, 330, Pl. 37(9)
Tody-Flycatcher, Black-fronted, 330
TODY-FLYCATCHER, BLACK-HEADED, 329, Pl. 37(8)
TODY-FLYCATCHER, COMMON, 330, Pl. 37(7)
TODY-FLYCATCHER, SLATE-HEADED, 330, Pl. 37(9)
Toledo, 301
Tolmomyias assimilis, 329, Pl. 37(17)
Tolmomyias sulphurescens, 328, Pl. 37(16)
Tontillo, 333
TORDO ARROCERO, 415
TORDO CABECIDORADO, 414
TORDO CANTOR, 408
Tordo de Agua, 395
TORDO PECHIRROJO, 414
TORDO SARGENTO, 413
TORDO-SALTARIN, 304
TORERO, 158
TOROROI DORSIESCAMADO, 288
TOROROI PECHICANELO, 289
TOROROI PECHIESCAMOSO, 287
TOROROI PECHILISTADO, 288
TOROROI PIQUIGUALDO, 289
Tortolita, 170–172

TORTOLITA AZULADA, 171
TORTOLITA COLILARGA, 171
Tortolita Colorada, 170
TORTOLITA COMUN, 170
TORTOLITA MENUDA, 170
TORTOLITA ROJIZA, 170
TORTOLITA SERRANERA, 172
Totí, 409
TOUCAN, CHESTNUT-MANDIBLED, 250, Pl. 27(19)
TOUCAN, KEEL-BILLED, 250, Pl. 27(18)
TOUCAN, RAINBOW-BILLED, 250
TOUCAN, SWAINSON'S, 251
TOUCANET, BLUE-THROATED, 248
TOUCANET, EMERALD, 248, Pl. 27(17)
TOUCANET, YELLOW-EARED, 249, Pl. 27(20)
Touit costaricensis, 180, Pl. 19(15)
Touit dilectissima, 180
TREEHUNTER, STREAKED-BREASTED, 271, Pl. 30(7)
TREERUNNER, RUDDY, 269, Pl. 29(4)
TREE-DUCK, BLACK-BELLIED, 90
TREE-DUCK, FULVOUS, 91
TREE-DUCK, WHITE-FACED, 90
Trepador, 426
TREPADOR ALIRRUBIO, 260
TREPADOR BARRETEADO, 262
TREPADOR CABECIPUNTEADO, 265
TREPADOR CABECIRRAYADO, 265
TREPADOR DELGADO, 260
TREPADOR GIGANTE, 262
TREPADOR GORGIANTEADO, 263
TREPADOR MANCHADO, 264
TREPADOR PARDO, 259
TREPADOR PICO DE HOZ, 266
TREPADOR PINTO, 264
TREPADOR PIQUICLARO, 264
TREPADOR ROJIZO, 260
TREPADOR VIENTRIBARRETEADO, 263
TREPADORCITO ACEITUNADO, 261
TREPADORCITO PICO DE CUÑA, 261
TREPAMUSGO CACHETON, 269
TREPAMUSGO CUELLIROJIZO, 271
TREPAMUSGO DE ANTEOJOS, 271
TREPAMUSGO LINEADO, 270
TREPAMUSGO RAYADO, 270
TREPAMUSGO ROJIZO, 271
Tres Pesos, 187
Tringa flavipes, 142, Pl. 11(3)
Tringa melanoleuca, 141, Pl. 11(2)
Tringa solitaria, 142, Pl. 11(4)
Tripsurus. See ***Melanerpes***
TROCHILIDAE, 208
Troglodytes aedon, 357, Pl. 38(18)
Troglodytes browni, 359
Troglodytes musculus, 358
Troglodytes ochraceus, 358, Pl. 38(19)
Troglodytes rufociliatus, 358
Troglodytes solstitialis, 358
TROGLODYTIDAE, 350
TROGON, BAIRD'S, 233, Pl. 26(6)
TROGON, BAR-TAILED, 235
TROGON, BLACK-HEADED, 234, Pl. 26(10)
TROGON, BLACK-THROATED, 236, Pl. 26(9)
TROGON, CITREOLINE, 234
TROGON, COLLARED, 235, Pl. 26(5)
TROGON, COPPERY-TAILED, 235
TROGON, ELEGANT, 234, Pl. 26(7)
TROGON, GARTERED, 237
TROGON, LATTICE-TAILED, 233, Pl. 26(3)
TROGON, ORANGE-BELLIED, 235, Pl. 26(4)
TROGON, SLATY-TAILED, 232, Pl. 26(2)
TROGON, VIOLACEOUS, 236, Pl. 26(8)
TROGON, WHITE-TAILED, 234
TROGON CABECINEGRO, 234
TROGON CABECIVERDE, 236
TROGON COLIPLOMIZO, 232
TROGON COLLAREJO, 235
TROGON ELEGANTE, 234
TROGON OJIBLANCO, 233
TROGON VIENTRIANARANJADO, 235
TROGON VIENTRIBERMEJO, 233
TROGON VIOLACEO, 236
Trogon aurantiiventris, 235, Pl. 26(4)
Trogon bairdii, 233, Pl. 26(6)
Trogon citreolus, 234
Trogon clathratus, 233, Pl. 26(3)
Trogon collaris, 235, Pl. 26(5)
Trogon elegans, 234, Pl. 26(7)
Trogon massena, 232, Pl. 26(2)
Trogon melanocephalus, 234, Pl. 26(10)
Trogon rufus, 236, Pl. 26(9)
Trogon violaceus, 236, Pl. 26(8)
Trogon viridis, 234
TROGONIDAE, 231
TROGONIFORMES, 231
TROPICBIRD, RED-BILLED, 74, Pl. 1(7)
TROPICBIRD, WHITE-TAILED, 74
Tryngites subruficollis, 150, Pl. 11(18)
TUCAN DE SWAINSON, 250
TUCAN PICO IRIS, 250
TUCANCILLO COLLAREJO, 248
TUCANCILLO OREJIAMARILLO, 249
TUCANCILLO PIQUIANARANJADO, 249
TUCANCILLO VERDE, 248
Tucuso, 426
TUFTEDCHEEK, BUFFY, 269, Pl. 30(1)
TURDIDAE, 362
Turdus albicollis, 363
Turdus assimilis, 363, Pl. 39(9)
Turdus fumigatus, 364
Turdus grayi, 363, Pl. 39(8)
Turdus nigrescens, 365, Pl. 39(7)
Turdus obsoletus, 364, Pl. 39(5)
Turdus plebejus, 365, Pl. 39(6)
Turillo, 137–139
TURNSTONE, RUDDY, 144, Pl. 10(6)
TYRANNIDAE, 304
Tyranniscus. See ***Zimmerius***
TYRANNULET, BROWN-CAPPED, 338, Pl. 37(13)

TYRANNULET, MISTLETOE, 337, Pl. 37(10)
Tyrannulet, Paltry, 337
Tyrannulet, Rough-legged, 338
TYRANNULET, RUFOUS-BROWED, 332, Pl. 37(12)
TYRANNULET, TORRENT, 333, Pl. 36(1)
Tyrannulet, White-Fronted, 338
TYRANNULET, YELLOW, 333, Pl. 37(11)
TYRANNULET, YELLOW-BELLIED, 339, Pl. 37(14)
TYRANNULET, YELLOW-CROWNED, 337, Pl. 37(15)
TYRANNULET, ZELEDON'S, 338, Pl. 37(18)
Tyrannulus elatus, 337, Pl. 37(15)
Tyrannus dominicensis, 308, Pl. 35(7)
Tyrannus forficatus, 306, Pl. 35(5)
Tyrannus melancholicus, 307, Pl. 35(1)
Tyrannus savana, 306, Pl. 35(4)
Tyrannus tyrannus, 307, Pl. 35(3)
Tyrannus verticalis, 308, Pl. 35(2)
TYRANT, LONG-TAILED, 305, Pl. 36(6)
Tyto alba, 189, Pl. 20(9)
TYTONIDAE, 189

UMBRELLABIRD, BARE-NECKED, 297, Pl. 34(13)
Urraca, 347
URRACA COPETONA, 347
URRACA DE TOCA CELESTE, 349
URRACA GORGIPLATEADA, 349
URRACA PARDA, 348
URRACA PECHINEGRA, 348
Urubitornis. See ***Harpyhaliaetus***

VAQUERO GRANDE, 407
VAQUERO OJIRROJO, 408
VEERY, 367, Pl. 39(3)
VENCEJO COMUN O GRISACEO, 207
VENCEJO CUELLICASTAÑO, 205
VENCEJO DE CHERRIE, 205
VENCEJO DE PASO, 206
VENCEJO DE RABADILLA CLARA, 207
VENCEJO LOMIGRIS, 207
VENCEJO NEGRO, 204
VENCEJO SOMBRIO, 205
VENCEJO TIJERETA MAYOR, 208
VENCEJO TIJERETA MENOR, 208
VENCEJON COLLAREJO, 206
Veniliornis fumigatus, 255, Pl. 28(13)
Veniliornis kirkii, 255, Pl. 28(12)
VERDILLO LEONADO, 380
VERDILLO MATORRALERO, 380
VERDILLO MENUDO, 381
Vermivora celata, 385, Pl. 40(19)
Vermivora chrysoptera, 384, Pl. 41(5)
Vermivora peregrina, 385, Pl. 40(22)
Vermivora pinus, 384, Pl. 41(6)
Vermivora ruficapilla, 385, Pl. 52(3)
Veterano, 88
VIOLET-EAR, BROWN, 214, Pl. 23(6)
VIOLET-EAR, GREEN, 214, Pl. 23(7)
VIREO, BLACK-WHISKERED, 379, Pl. 40(4)
VIREO, BROWN-CAPPED, 379, Pl. 40(14)
VIREO, MANGROVE, 376, Pl. 40(11)
VIREO, PHILADELPHIA, 379, Pl. 40(15)
VIREO, RED-EYED, 378, Pl. 40(3)
VIREO, SOLITARY, 377, Pl. 40(9)
VIREO, WARBLING, 379, Pl. 40(13)
VIREO, WHITE-EYED, 376, Pl. 40(12)
VIREO, YELLOW-GREEN, 378, Pl. 40(5)
VIREO, YELLOW-THROATED, 377, Pl. 40(6)
VIREO, YELLOW-WINGED, 376, Pl. 40(10)
VIREO ALIAMARILLO, 376
VIREO AMARILLENTO, 379
VIREO BIGOTUDO, 379
VIREO CABECIGRIS, 378
VIREO CANORO, 379
VIREO DE MANGLAR, 376
VIREO MONTAÑERO, 379
VIREO OJIBLANCO, 376
VIREO OJIRROJO, 378
VIREO PECHIAMARILLO, 377
VIREO SOLITARIO, 377
Vireo altiloquus, 379, Pl. 40(14)
Vireo carmioli, 376, Pl. 40(10)
Vireo flavifrons, 377, Pl. 40(6)
Vireo flavoviridis, 378, Pl. 40(5)
Vireo gilvus, 379, Pl. 40(13)
Vireo griseus, 376, Pl. 40(12)
Vireo leucophrys, 379, Pl. 40(14)
Vireo olivaceus, 378, Pl. 40(3)
Vireo pallens, 376, Pl. 40(11)
Vireo philadelphicus, 379, Pl. 40(15)
Vireo solitarius, 377, Pl. 40(9)
Vireolanius pulchellus, 375, Pl. 40(1)
VIREON CEJIRRUFO, 375
VIREON ESMERALDINO, 375
VIREONIDAE, 374
Viuda, 428
Viuda Amarilla, 234
Viuda Roja, 232
Volatinia jacarina, 452, Pl. 49(7)
VUELVEPIEDRAS ROJIZO, 144
VULTURE, BLACK, 96, Pl. 13(4)
VULTURE, KING, 97, Pl. 13(5)
VULTURE, LESSER YELLOW-HEADED, 96, Pl. 13(2)
VULTURE, TURKEY, 96, Pl. 13(3)

Warbler, Audubon's, 390, Pl. 43(11)
WARBLER, BAY-BREASTED, 393, Pl. 41(9)
WARBLER, BLACK-AND-WHITE, 383, Pl. 41(13)
WARBLER, BLACKBURNIAN, 392, Pl. 41(8)
WARBLER, BLACK-CHEEKED, 403, Pl. 40(17)
WARBLER, BLACKPOLL, 393, Pl. 41(11)
WARBLER, BLACK-THROATED BLUE, 389, Pl. 43(1)
WARBLER, BLACK-THROATED GREEN, 390, Pl. 43(9)
WARBLER, BLUE-WINGED, 384, Pl. 41(6)

Warbler, Brewster's, 384
WARBLER, BUFF-RUMPED, 404, Pl. 40(23)
WARBLER, CANADA, 400, Pl. 42(8)
WARBLER, CAPE MAY, 389, Pl. 43(6)
WARBLER, CERULEAN, 391, Pl. 43(2)
Warbler, Chestnut-capped, 404
WARBLER, CHESTNUT-SIDED, 392, Pl. 43(4)
WARBLER, CONNECTICUT, 397, Pl. 52(4)
WARBLER, FLAME-THROATED, 386, Pl. 41(7)
WARBLER, GOLDEN-CROWNED, 402, Pl. 40(18)
WARBLER, GOLDEN-WINGED, 384, Pl. 41(5)
WARBLER, HERMIT, 391, Pl. 43(7)
WARBLER, HOODED, 399, Pl. 42(5)
WARBLER, KENTUCKY, 396, Pl. 42(15)
WARBLER, MACGILLIVRAY'S, 397, Pl. 42(16)
WARBLER, MAGNOLIA, 388, Pl. 43(10)
WARBLER, MANGROVE, 388, Pl. 42(3)
WARBLER, MOURNING, 396, Pls. 42(14), 52(5)
WARBLER, NASHVILLE, 385, Pl. 52(3)
Warbler, Olive-backed, 387
WARBLER, ORANGE-CROWNED, 385, Pl. 40(19)
WARBLER, PALM, 394, Pl. 43(3)
WARBLER, PINE, 394, Pl. 41(14)
WARBLER, PRAIRIE, 394, Pl. 43(5)
WARBLER, PROTHONOTARY, 383, Pl. 42(1)
Warbler, River, 404
WARBLER, RUFOUS-CAPPED, 403, Pl. 40(16)
WARBLER, TENNESSEE, 385, Pl. 40(22)
WARBLER, THREE-STRIPED, 402, Pl. 40(20)
WARBLER, TOWNSEND'S, 390, Pl. 43(8)
WARBLER, WILSON'S, 400, Pl. 42(4)
WARBLER, WORM-EATING, 384, Pl. 40(21)
WARBLER, YELLOW, 387, Pl. 42(2)
WARBLER, YELLOW-RUMPED (MYRTLE), 389, Pl. 43(11)
WARBLER, YELLOW-THROATED, 392, Pl. 41(12)
WATERTHRUSH, LOUISIANA, 395, Pl. 43(13)
WATERTHRUSH, NORTHERN, 395, Pl. 43(14)
WAXWING, CEDAR, 373, Pl. 39(14)
WHIMBREL, 140, Pl. 9(14)
WHIP-POOR-WILL, 202, Pl. 21(17)
WHISTLING-DUCK, BLACK-BELLIED, 90, Pl. 8(7)
WHISTLING-DUCK, FULVOUS, 91, Pl. 8(9)
WHISTLING-DUCK, WHITE-FACED, 91, Pl. 8(8)
WIGEON, AMERICAN, 92, Pl. 8(6)
WILLET, 143, Pl. 9(6)
Wilsonia canadensis, 400, Pl. 42(8)
Wilsonia citrina, 399, Pl. 42(5)
Wilsonia pusilla, 400, Pl. 42(4)
WOODCREEPER, BARRED, 262, Pl. 29(19)
WOODCREEPER, BLACK-BANDED, 263, Pl. 29(15)
WOODCREEPER, BLACK-STRIPED, 264, Pl. 29(16)
WOODCREEPER, BUFF-THROATED, 263, Pl. 29(17)
Woodcreeper, Cherrie's, 261
WOODCREEPER, IVORY-BILLED, 264, Pl. 29(21)
WOODCREEPER, LONG-TAILED, 260, Pl. 29(10)
WOODCREEPER, OLIVACEOUS, 261, Pl. 29(7)
WOODCREEPER, PLAIN-BROWN, 259, Pl. 29(11)
WOODCREEPER, RUDDY, 260, Pl. 29(13)
WOODCREEPER, SPOTTED, 264, Pl. 29(20)
WOODCREEPER, SPOTTED-CROWNED, 265, Pl. 29(9)
WOODCREEPER, STREAKED-HEADED, 265, Pl. 29(8)
WOODCREEPER, STRONG-BILLED, 262, Pl. 29(18)
WOODCREEPER, TAWNY-WINGED, 260, Pl. 29(12)
WOODCREEPER, WEDGE-BILLED, 261, Pl. 29(6)
Woodhaunter, Striped, 270
WOODNYMPH, CROWNED, 217, Pl. 24 (4), 25 (15)
Wood-Owl, Mottled, 195
WOOD-PARTRIDGE, BUFFY-CROWNED, 120, Pl. 12(16)
Wood-Partridge, Buff-fronted, 120
WOODPECKER, ACORN, 252, Pl. 28(15)
WOODPECKER, BLACK-CHEEKED, 252, Pl. 28(14)
WOODPECKER, CHESTNUT-COLORED, 257, Pl. 28(8)
WOODPECKER, CINNAMON, 257, Pl. 28(9)
Woodpecker, Flint-billed, 258
Woodpecker, golden-fronted, 253
WOODPECKER, GOLDEN-NAPED, 252, Pl. 28(18)
WOODPECKER, GOLDEN-OLIVE, 256, Pl. 28(11)
WOODPECKER, HAIRY, 254, Pl. 28(19)
WOODPECKER, HOFFMANN'S, 253, Pl. 28(16)
WOODPECKER, LINEATED, 257, Pl. 27(13)
WOODPECKER, PALE-BILLED, 258, Pl. 27(14)
WOODPECKER, RED-CROWNED, 253, Pl. 28(17)
WOODPECKER, RED-RUMPED, 255, Pl. 28(12)
WOODPECKER, RUFOUS-WINGED, 256, Pl. 28(10)
WOODPECKER, SMOKY-BROWN, 255, Pl. 28(13)
WOOD-PEWEE, EASTERN, 318, Pl. 36(8)
WOOD-PEWEE, WESTERN, 317, Pl. 36(7)

WOOD-QUAIL, BLACK-BREASTED, 122, Pl. 12(10)
WOOD-QUAIL, MARBLED, 121, Pl. 12(9)
WOOD-QUAIL, RUFOUS-FRONTED, 122, Pl. 12(12)
WOOD-QUAIL, SPOTTED, 122, Pl. 12(11)
Wood-Quail, White-Throated, 123
WOOD-RAIL, GRAY-NECKED, 125, Pl. 6(13)
WOOD-RAIL, RUFOUS-NECKED, 125, Pl. 6(14)
WOODSTAR, MAGENTA-THROATED, 229, Pl. 25(2)
WOOD-WREN, GRAY-BREASTED, 359, Pl. 38(14)
Wood-Wren, Lowland, 359
WOOD-WREN, WHITE-BREASTED, 359, Pl. 38(15)
WREN, BANDED, 355, Pl. 38(6)
WREN, BANDED-BACKED, 351, Pl. 38(3)
WREN, BAY, 354, Pl. 38(12)
WREN, BLACK-BELLIED, 356, Pl. 38(11)
WREN, BLACK-THROATED, 356, Pl. 38(13)
Wren, Canebrake, 352
WREN, HOUSE, 357, Pl. 38(18)
Wren, Mountain, 358
Wren, Musician, 362
WREN, NIGHTINGALE, 360, Pl. 38(21)
WREN, OCHRACEOUS, 358, Pl. 38(19)
WREN, PLAIN, 352, Pl. 38(17)
WREN, ROCK, 360, Pl. 38(4)
WREN, RIVERSIDE, 354, Pl. 38(10)
WREN, RUFOUS-AND-WHITE, 353, Pl. 38(9)
WREN, RUFOUS-BREASTED, 357, Pl. 38(8)
WREN, RUFOUS-NAPED, 352, Pl. 38(2)
Wren, Scaly-breasted, 361
WREN, SEDGE, 351, Pl. 38(1)
WREN, SONG, 361, Pl. 38(22)
Wren, Sooty-headed, 356
Wren, Speckled, 357
WREN, SPOTTED-BREASTED, 356, Pl. 38(7)
WREN, STRIPED-BREASTED, 354, Pl. 38(5)
WREN, TIMBERLINE, 358, Pl. 38(16)
WREN, WHISTLING, 361, Pl. 38(20)
Wrenthrush, 405

Xanthocephalus xanthocephalus, 414, Pl. 44(4)
Xema sabini, 158, Pl. 3(7)
XENOPS, PLAIN, 274, Pl. 29(3)
XENOPS, STREAKED, 275, Pl. 29(2)
XENOPS COMUN, 274
Xenops minutus, 274, Pl. 29(3)
XENOPS RAYADO, 275
Xenops rutilans, 275, Pl. 29(2)
Xiphocolaptes promeropirhynchus, 262, Pl. 29(18)
Xiphorhynchus erythropygius, 264, Pl. 29(20)
Xiphorhynchus flavigaster, 264, Pl. 29(21)
Xiphorhynchus guttatus, 263, Pl. 29(17)
Xiphorhynchus lachrymosus, 264, Pl. 29(16)

YELLOW-FINCH, GRASSLAND, 453, Pl. 50(3)
YELLOWLEGS, GREATER, 141, Pl. 11(2)
YELLOWLEGS, LESSER, 142, Pl. 11(3)
Yellowthroat, Chiriqui, 398
YELLOWTHROAT, COMMON, 397, Pl. 42(10)
YELLOWTHROAT, GRAY-CROWNED, 399, Pl. 42(13)
YELLOWTHROAT, MASKED, 398, Pl. 42(12)
YELLOWTHROAT, OLIVE-CROWNED, 398, Pl. 42(11)
Yerre, 66
Yigüirro, 363
Yigüirro Collarejo, 363
Yigüirro de Montaña, 364, 365
Yuré, 172

Zacatera, 414
ZACATERO COMUN, 414
ZAMBULLIDOR ENANO, 68
ZAMBULLIDOR MEDIANO, 68
ZAMBULLIDOR PIQUIPINTO, 67
Zanate, 409
ZANATE DE LAGUNA, 409
ZANATE GRANDE, 409
Zapoyol, 178
Zapoyolito, 180
ZARAPITO PIQUILARGO, 141
ZARAPITO TRINADOR, 140
Zarceta, 92–93, 140–142, 150
Zarhynchus wagleri, 406
Zebra, 422
ZELEDONIA, 404
ZELEDONIA, 404, Pl. 32(13)
Zeledonia coronata, 404, Pl. 32(13)
Zeledoniidae, 405
Zenaida asiatica, 169, Pl. 18(12)
Zenaida macroura, 169, Pl. 18(13)
Zenaidura macroura, 169
Zimmerius vilissimus, 337, Pl. 37(10)
Zonchiche, 96
Zoncho, 97
Zonotrichia capensis, 463, Pl. 50(13)
ZOPILOTE CABECIGUALDO, 96
ZOPILOTE CABECIRROJO, 96
Zopilote de Mar, 79
ZOPILOTE NEGRO, 97
ZOPILOTE REY, 97
Zopilotillo, 186
ZORZAL CABECINEGRO, 368
ZORZAL CARIGRIS, 367
ZORZAL DE SWAINSON, 366
ZORZAL DEL BOSQUE, 366
ZORZAL DORSIRROJIZO, 367
ZORZAL GORRIROJIZO, 369
ZORZAL PIQUIANARANJADO, 369
ZORZAL PIQUINEGRO, 370
ZORZAL SOMBRIO, 368
Zoterré, 357